FOR INSTRUCTORS

Wiley**PLUS** is built around the activities you perform in your class each day. With Wiley**PLUS** you can:

Prepare & Present

Create outstanding class presentations using a wealth of resources such as PowerPoint™ slides, image galleries, interactive simulations, and more. You can even add materials you have created yourself.

Create Assignments

Automate the assigning and grading of homework or quizzes by using the provided question banks, or by writing your own.

Track Student Progress

Keep track of your students' progress and analyze individual and overall class results.

Now Available with WebCT and Blackboard!

"It has been a great help, and I believe it has helped me to achieve a better grade."

Michael Morris,
Columbia Basin College

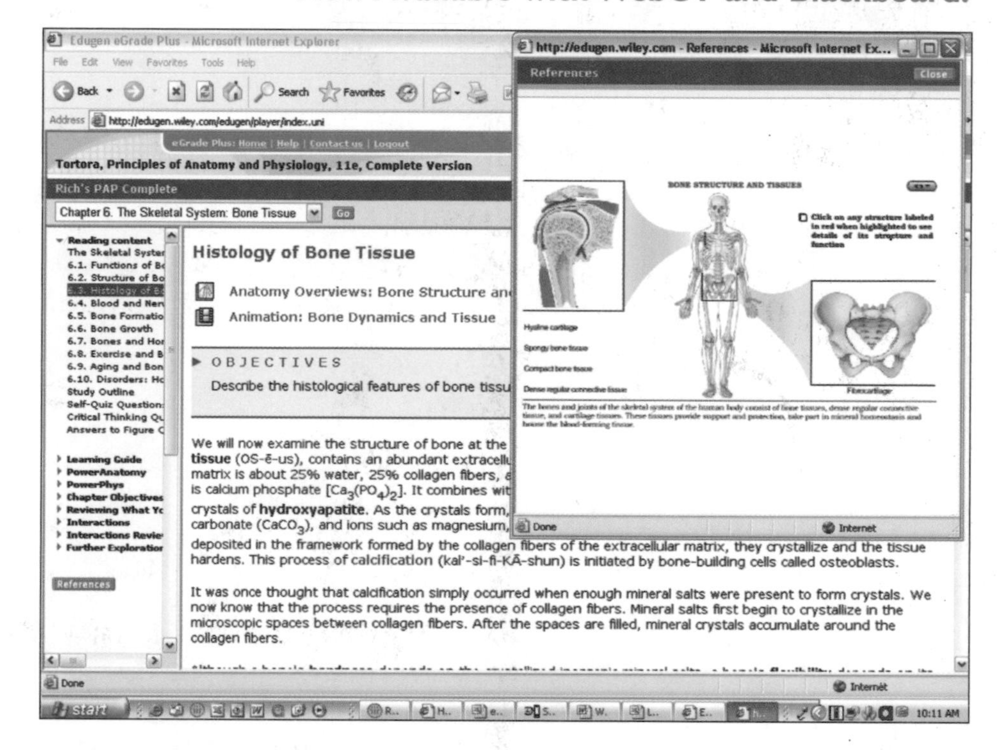

FOR STUDENTS

You have the potential to make a difference!

WileyPLUS is a powerful online system packed with features to help you make the most of your potential and get the best grade you can!

With Wiley**PLUS** you get:

- A complete online version of your text and other study resources.
- Problem-solving help, instant grading, and feedback on your homework and quizzes.
- The ability to track your progress and grades throughout the term.

For more information on what *WileyPLUS* can do to help you and your students reach their potential, please visit www.wiley.com/college/*wileyplus*.

76% of students surveyed said it made them better prepared for tests. *

*Based on a survey of 972 student users of *WileyPLUS*

Seventh Edition

Introduction to the Human Body

the essentials of anatomy and physiology

GERARD J. TORTORA
Bergen Community College

BRYAN DERRICKSON
Valencia Community College

WILEY

JOHN WILEY & SONS, INC.

Executive Editor	Bonnie Roesch
Executive Marketing Manager	Clay Stone
Developmental Editor	Karen Trost
Production Manager	Kelly Tavares
Senior Production Editor	Kelly Tavares
Text and Cover Designer	Karin Gerdes Kincheloe
Art Coordinator	Edward T. Starr
Photo Editor	Felicia Ruocco and Hilary Newman
Cover Photo	©2005 Lois Greenfield

Illustration and photo credits follow the Glossary.

This book was typeset by Progressive Information Technologies and printed and bound by Von Hoffmann Press, Inc. The cover was also printed by Von Hoffman Press, Inc.

The paper in this book was manufactured by a mill whose forest management programs include sustained yield harvesting of its timberlands. Sustained yield harvesting principles ensure that the number of trees cut each year does not exceed the amount of new growth.

This title published by John Wiley & Sons, Inc.

ISBN-13978-0-471-69123-5
ISBN-100-471-69123-2
Printed in the United States of America
10 9 8 7 6 5 4 3 2 1

ABOUT THE AUTHORS

Gerard J. Tortora is Professor of Biology and former Coordinator at Bergen Community College in Paramus, New Jersey, where he teaches human anatomy and physiology as well as microbiology. He received his bachelor's degree in biology from Fairleigh Dickinson University and his master's degree in science education from Montclair State College. He is a member of many professional organizations, such as the Human Anatomy and Physiology Society (HAPS), the American Society of Microbiology (ASM), American Association for the Advancement of Science (AAAS), National Education Association (NEA), and the Metropolitan Association of College and University Biologists (MACUB).

Above all, Jerry is devoted to his students and their aspirations. In recognition of this commitment, Jerry was the recipient of MACUB's 1992 President's Memorial Award. In 1996, he received a National Institute for Staff and Organizational Development (NISOD) excellence award from the University of Texas and was selected to represent Bergen Community College in a campaign to increase awareness of the contributions of community colleges to higher education.

Jerry is the author of several best-selling science textbooks and laboratory manuals, a calling that often requires an additional 40 hours per week beyond his teaching responsibilities. Nevertheless, he still makes time for four or five weekly aerobic workouts that include biking and running. He also enjoys attending college basketball and professional hockey games and performances at the Metropolitan Opera House.

To my son, Kenneth S. Tortora, and his wife Dawn, who have welcomed into their hearts and home seven children whom they have adopted and given unconditional love, guidance, and acceptance. Ken is truly my hero and Dawn is his strength. With all my love, Dad.

Bryan Derrickson is Professor of Biology at Valencia Community College in Orlando, Florida, where he teaches human anatomy and physiology as well as general biology and human sexuality. He received his bachelor's degree in biology from Morehouse college and his Ph.D. in Cell Biology from Duke University. Bryan's study at Duke was in the Physiology Division within the Department of Cell Biology, so while his degree is in Cell Biology his training focused on physiology. At Valencia, he frequently serves on faculty hiring committees. He has served as a member of the Faculty Senate, which is the governing body of the college and as a member of the Faculty Academy committee (now called the Teaching and Learning Academy), which sets the standards for the acquisition of tenure by faculty members. Nationally, he is a member of the Human Anatomy and Physiology Society (HAPS) and the National Association of Biology Teachers (NABT).

Bryan has always wanted to teach. Inspired by several biology professors while in college, he decided to pursue physiology with an eye to teaching at the college level. He is completely dedicated to the success of his students. He particularly enjoys the challenges of his diverse student population, in terms of their age, ethnicity, and academic ability, and finds being able to teach all of them, despite their differences, a rewarding experience. His students continually recognize Bryan's efforts and care by nominating him for a campus award known as the "Valencia Professor Who Makes Valencia A Better Place to Start." Bryan has received this award three times in the past six years.

PREFACE

Introduction to the Human Body: The Essentials of Anatomy and Physiology, Seventh Edition, is designed for courses in human anatomy and physiology or in human biology. It assumes no previous study of the human body. The successful approach of the previous editions—to provide students with a basic understanding of the structure and functions of the human body with an emphasis on homeostasis—has been retained. In the development of the seventh edition, we focused on improving all of the acknowledged strengths of the text.

ORGANIZATION AND CONTENT IMPROVEMENTS

Like previous editions of *Introduction to the Human Body*, the seventh edition of the book is divided into 24 chapters. It approaches the study of the human body system by system, beginning with the integumentary system in Chapter 5. In response to feedback from professors and students facing the challenges of teaching and learning the complexities of the material in a single semester, this new edition of *Introduction to the Human Body* has been streamlined without sacrificing completeness of coverage. Every chapter in the seventh edition incorporates improvements to both the text and the art, many suggested by reviewers, educators, and students.

Some of the significant changes in selected chapters include the following:

- **Chapter 1** Figures 1.1, 1.4, 1.6, and 1.8 have been redrawn and a Medical Terminology and Conditions list has been added.
- **Chapter 2** Information on essential fatty acids in health and disease has been added.
- **Chapter 3** Figures 3.8, 3.10, and 3.11 have been redrawn.
- **Chapter 4** The art in Tables 4.1 and 4.2 has been redrawn.
- **Chapter 5** Information on cosmetic anti-aging treatments has been added to the section on aging.
- **Chapter 6** The figures illustrating the bones of the skeletal system have been revised to include new bone art. A new illustration of intramembranous ossification has been added.
- **Chapter 7** This chapter has been reorganized, so that the types of synovial joints (planar, hinge, pivot, condyloid, saddle, and ball-and-socket) are now described after the types of movement at synovial joints (gliding, flexion, extension, etc.). In addition, most of the figures have been redrawn for better clarity and depth.
- **Chapter 8** Figures 8.2, 8.4, 8.6, 8.7, and 8.11 have been redrawn and a new figure of ATP production for muscle contraction has been added.
- **Chapter 9** Table 9.1 was revised to include new art on neuroglia and Figures 9.6 and 9.7 were redrawn.
- **Chapter 10** Figures 10.6 and 10.15 have been redrawn, Tables 10.1 and 10.2 have been revised, and a new figure of the limbic system has been added.

- **Chapter 11** Figure 11.1 has been redrawn to include several examples of autonomic motor neuron pathways.
- **Chapter 12** The art in Tables 12.2 and 12.3 has been redrawn.
- **Chapter 13** Figure 13.3 was revised in order to clarify the mechanism of action of water-soluble hormones.
- **Chapter 14** The section on blood groups and blood types has been revised.
- **Chapter 15** Figure 15.8 was redrawn and information on coronary artery disease in the common disorders section has been updated.
- **Chapter 16** Figure 16.3 has been redrawn and information on hypertension in the common disorders section has been updated.
- **Chapter 17** Figures 17.7, 17.8, 17.9, and 17.10 have been redrawn
- **Chapter 18** Figure 18.9 has been revised to include the values of the lung volumes and lung capacities for women.
- **Chapter 19** New to this chapter is the addition of a section on phases of digestion that describes the cephalic, gastric, and intestinal phases of digestion and the major gastrointestinal hormones.
- **Chapter 20** Figure 20.1 has been redrawn and updated to include the new My Pyramid introduced by the United States Department of Agriculture (USDA).
- **Chapter 21** Figure 21.4 has been redrawn.
- **Chapter 22** Figure 22.4 has been revised.
- **Chapter 23** Figures 23.2, 23.3, 23.4, and 23.8 have been redrawn and a Focus on Homeostasis: The Reproductive Systems section has been added at the end of the chapter.
- **Chapter 24** The clinical application on stem cells has been updated.

On the following pages, the many features of this text are described and illustrated. These include the "Did You Know" chapter introductions, "Looking Back to Move Ahead" concept review, Chapter Objectives, Checkpoint Questions, and Clinical Applications. Highly praised features such as Focus on Wellness and Focus on Homeostasis are also illustrated and explained. The revised illustration program is presented along with the effective and supportive pedagogy that surrounds every illustration. An overview of the many ancillaries that support *Introduction to the Human Body*, including *Wiley PLUS*, is provided as well.

FEATURES DESIGNED FOR STUDENT SUCCESS

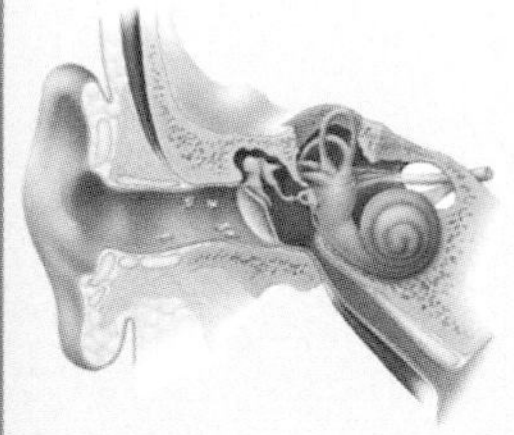

Consider whatwould happen if youcould not feel the pain of a hot pot handleor an inflamed appendix, or if you couldnot see an oncoming car, hear a baby'scry, smell smoke, taste your favoritedessert, or maintain your balance on a flight of stairs. In short, if you could not "sense" your environment and make the necessary homeostatic adjustments, you could not survive very well on your own.

did you know? *Some things improve with age, but hearing is not one of them. Damage to the hair cells that convert sound waves into nerve impulses accumulates over a lifetime, and by the time hearing loss is discovered, irreversible damage has already occurred. Exposure to excessive noise is the most common cause of hair cell damage. Damage increases with both the intensity and duration of exposure. The hair cells appear to be less traumatized by short periods of loud noise, such as a fire alarm going off, than by chronic exposure to moderately loud noise, such as the noise of vacuum cleaners, power tools, engines, and loud music.*

Focus on Wellness, page 289

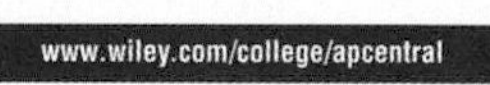

looking back to move ahead . . .

- Sensory Nerve Endings and Sensory Receptors in the Skin (page 100)
- Somatic Sensory Pathways (page 259)

284

looking back to move ahead . . .

- Sensory Nerve Endings and Sensory Receptors in the Skin (page 100)
- Somatic Sensory Pathways (page 259)

DID YOU KNOW?

Each chapter of the 7th edition is introduced with a short introduction linking the chapter content to come with some relevant, everyday scenario. Carefully crafted by Barbara Brehm, the author of the very popular Focus on Wellness essays, these are designed to pique the students' curiosity about the contents of the upcoming chapter by making connections to issues concerning their own health. In addition, these chapter-opening teasers are linked to web-based activities that allow students to explore further the connection of the chapter content to this particular wellness issue.

LOOKING BACK TO MOVE AHEAD

Beginning with Chapter 2, chapter-opening pages also include a listing of concepts that were previously covered, but that will be particularly relevant to understanding the chapter at hand. Complete with page numbers for easy reference, these background basics allow the student to make the connections between important concepts developed in earlier chapters that are so crucial to success in the course.

STRUCTURE AND ORGANIZATION OF THE HEART

OBJECTIVES • Describe the location of the heart and the structure and functions of the pericardium.
- **Describe the layers of the heart wall and the chambers of the heart.**
- **Identify the major blood vessels that enter and exit the heart.**
- **Describe the structure and functions of the valves of the heart.**

CHAPTER OBJECTIVES

Feedback from users indicates that many students do better when the material in each chapter is developed so that they can focus on fewer learning objectives at a time. To address this need, chapter objectives are interspersed throughout each chapter. Each major heading begins with a short, focused list of the important concepts to learn for that section.

■ **CHECKPOINT**

1. Identify the location of the heart.
2. Describe the various layers of the pericardium and the heart wall.
3. How do atria and ventricles differ in structure and function?
4. Which blood vessels that enter and exit the heart carry oxygenated blood? Which carry deoxygenated blood?
5. In correct sequence, which heart chambers, heart valves, and blood vessels would a drop of blood encounter from the time it flows out of the right atrium until it reaches the aorta?

CHECKPOINTS

Complementing the format for objectives, **Checkpoint Questions** appear at strategic intervals within chapters to give students the chance to test their understanding of what they have just learned. Answers are found easily within the chapter content immediately preceding the checkpoint; where checkpoints appear within exhibits, the answers can be found within the same exhibit.

Carbon monoxide (CO) is a colorless and odorless gas found in tobacco smoke and in exhaust fumes from automobiles, gas furnaces, and space heaters. CO binds to the heme group of hemoglobin, just as O_2 does, except that CO binds over 200 times more strongly. At a concentration as small as 0.1%, CO combines with half the available hemoglobin molecules and reduces the oxygen-carrying capacity of the blood by 50%. Elevated blood levels of CO cause **carbon monoxide poisoning**, which can cause the lips and oral mucosa to appear bright, cherry-red (the color of hemoglobin with carbon monoxide bound to it). Administering pure oxygen, which speeds up the separation of carbon monoxide from hemoglobin, may rescue the person.

CLINICAL APPLICATIONS

A perennial favorite among students, the intriguing **clinical applications** in every chapter explore the clinical, professional, or everyday relevance of a particular anatomical structure or its related function. A colored screen highlights selected examples discussing health issues and the treatment of disease.

FOCUS ON WELLNESS

A popular feature of the last four editions has been the wellness essays, written by Barbara Brehm Curtis of Smith College. These essays increase students' appreciation of the relevance to the maintenance of good health to the concepts and details of anatomy and physiology presented in the text. The wellness philosophy supports the notion that lifestyle choices that individuals make throughout the years have an important influence on their mental and physical wellbeing.

In addition, each essay includes a **"Think It Over"** concept application exercise. We believe that the information contained in these Focus on Wellness essays is timely and interesting; we hope students and instructors continue to feel this way, too.

Adaptive Immunity 431

FOCUS ON WELLNESS

Lifestyle, Immune Function, and Resistance to Disease

If you want to observe the relationship between lifestyle and immune function, visit a college campus. As the semester progresses and the workload accumulates, an increasing number of students can be found in the waiting rooms of student health services.

Is Stress the Culprit?

Stress has been implicated as hazardous to immune function. Researchers in the field of *psychoneuroimmunology (PNI)* have found many communication pathways that link the nervous, endocrine, and immune systems. Chronic stress affects the immune system in several ways. For example, cortisol, a hormone secreted by the adrenal cortex in association with the stress response, inhibits immune system activity, perhaps one of its energy conservation effects. PNI research supports what many people have observed since the beginning of time: Your thoughts, feelings, moods, and beliefs influence your level of health and the course of disease. Especially toxic to the immune system are feelings of helplessness, hopelessness, fear, and social isolation.

People resistant to the negative health effects of stress are more likely to experience a sense of control over the future, a commitment to their work, expectations of generally positive outcomes for themselves, and feelings of social support. To increase your stress resistance, cultivate an optimistic outlook, get involved in your work, and build good relationships with others.

Or Is Lifestyle at Fault?

When work and stress pile up, health habits can change. Many people smoke or consume more alcohol when stressed, two habits detrimental to optimal immune function. Under stress, people are less likely to eat well or exercise regularly, two habits that enhance immunity.

Adequate sleep and relaxation are especially important for a healthy immune system. But when there aren't enough hours in the day, you may be tempted to steal some from the night. While skipping sleep may give you a few more hours of productive time in the short run, in the long run you end up even farther behind, especially if getting sick keeps you out of commission for several days, blurs your concentration, and blocks your creativity.

Even if you make time to get eight hours of sleep, stress can cause insomnia. If you find yourself tossing and turning at night, it's time to improve your stress management and relaxation skills! Be sure to unwind from the day before going to bed.

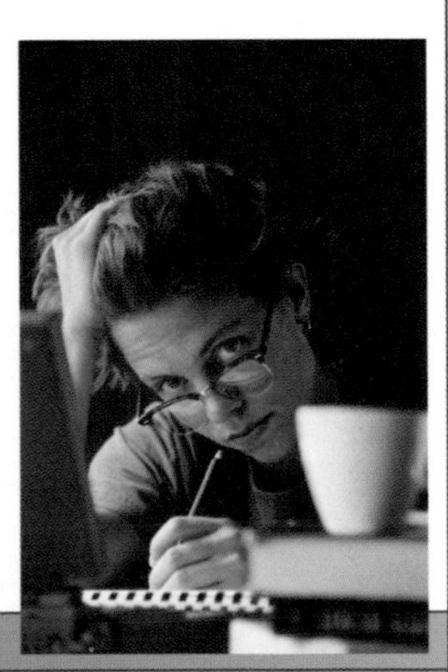

► THINK IT OVER . . .

► *Have you ever observed a connection between stress and illness in your own life?*

FOCUS ON HOMEOSTASIS

Eleven **Focus on Homeostasis** pages, one each for the integumentary, skeletal, muscular, nervous, endocrine, cardiovascular, lymphatic, respiratory, digestive, urinary, and reproductive systems are included at the end of the respective system chapters. Incorporating both graphic and narrative elements, these pages explain, clearly and succinctly, how the system under consideration contributes to the homeostasis of each of the other body systems. Use of this feature will enhance student understanding of the links among body systems and how interactions among systems contribute to the homeostasis of the body as a whole.

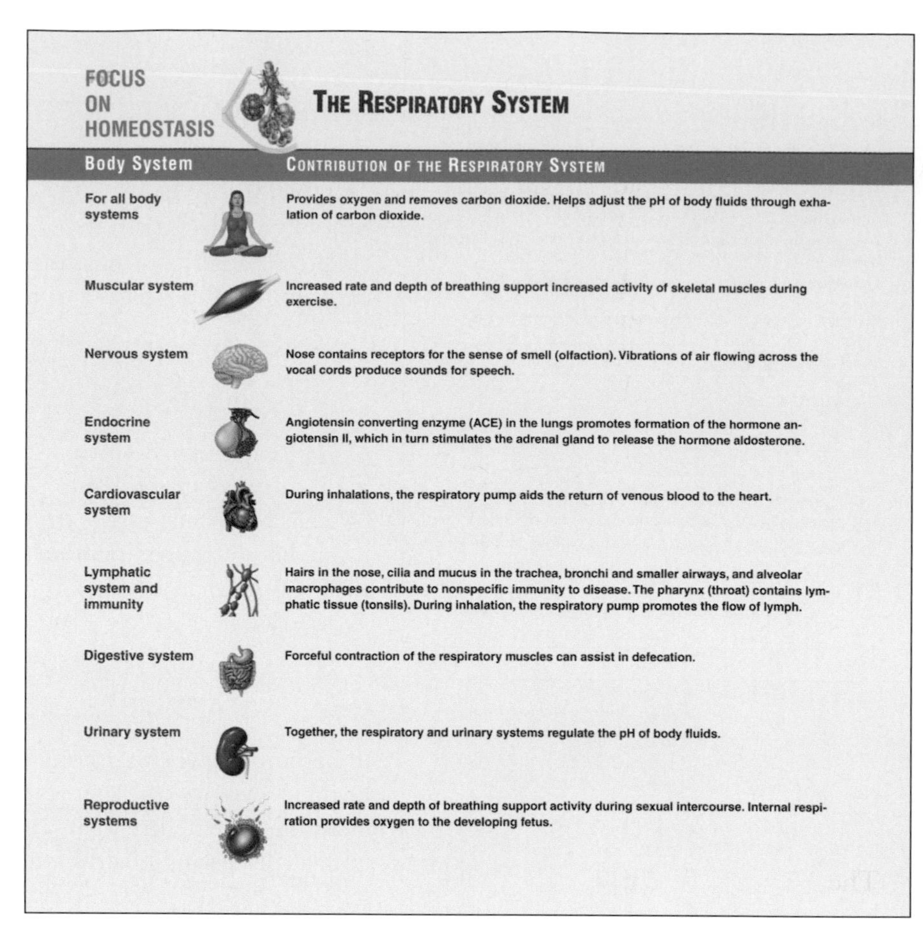

FOCUS ON HOMEOSTASIS — THE RESPIRATORY SYSTEM

Body System	Contribution of the Respiratory System
For all body systems	Provides oxygen and removes carbon dioxide. Helps adjust the pH of body fluids through exhalation of carbon dioxide.
Muscular system	Increased rate and depth of breathing support increased activity of skeletal muscles during exercise.
Nervous system	Nose contains receptors for the sense of smell (olfaction). Vibrations of air flowing across the vocal cords produce sounds for speech.
Endocrine system	Angiotensin converting enzyme (ACE) in the lungs promotes formation of the hormone angiotensin II, which in turn stimulates the adrenal gland to release the hormone aldosterone.
Cardiovascular system	During inhalations, the respiratory pump aids the return of venous blood to the heart.
Lymphatic system and immunity	Hairs in the nose, cilia and mucus in the trachea, bronchi and smaller airways, and alveolar macrophages contribute to nonspecific immunity to disease. The pharynx (throat) contains lymphatic tissue (tonsils). During inhalation, the respiratory pump promotes the flow of lymph.
Digestive system	Forceful contraction of the respiratory muscles can assist in defecation.
Urinary system	Together, the respiratory and urinary systems regulate the pH of body fluids.
Reproductive systems	Increased rate and depth of breathing support activity during sexual intercourse. Internal respiration provides oxygen to the developing fetus.

EXHIBITS

The Exhibits are self-contained features designed to give students the extra help that they need to learn the numerous structures that constitute certain body systems—most notably skeletal muscles, articulations, blood vessels, and nerves. Each Exhibit consists of an overview, a tabular summary of relevant anatomy, and appropriate illustrations. Many also include a relevant clinical correlation. Students will find this clear and concise presentation to be the ideal study vehicle for learning anatomically complex body systems.

210 Chapter 8 The Muscular System — Principal Skeletal Muscles 211

Exhibit 8.10 Muscles That Move the Vertebral Column (Backbone) *(Figure 8.22)*

OBJECTIVE • Describe the origin, insertion, and action of the muscles that move the vertebral column.

Overview: The ***erector spinae muscles*** form the largest muscular mass of the back, forming a prominent bulge on either side of the vertebral column (Figure 8.22). It consists of three groups of overlapping muscles: ***iliocostalis group*** (il′-ē-ō-kos-TĀ-lis), ***longissimus group*** (lon′-JI-si-mus), and ***spinalis group*** (spi-NĀ-lis). Other muscles that move the vertebral column include the ***sternocleidomastoid, quadratus lumborum, rectus abdominis*** (see Exhibit 8.4), ***psoas major*** (see Exhibit 8.11), and ***iliacus*** (see Exhibit 8.11).

Full flexion at the waist, as in touching your toes, overstretches the erector spinae muscles and muscles that are overstretched cannot contract effectively. Straightening up from such a position is therefore initiated by the hamstring muscles on the back of the thigh and the gluteus maximus muscles of the buttocks. The erector spinae muscles join in as the degree of flexion decreases. **Improperly lifting a heavy weight**, however, can strain the erector spinae muscles. The result can be painful muscle spasms, tearing of tendons and ligaments of the lower back, and rupturing of intervertebral discs. The lumbar muscles are adapted for maintaining posture, not for lifting. This is why it is important to kneel and use the powerful extensor muscles of the thighs and buttocks while lifting a heavy load.

Relating muscles to movements: Arrange the muscles in this exhibit according to the following actions on the vertebral column: (1) flexion and (2) extension.

■ CHECKPOINT
Which groups of muscles make up the erector spinae?

Muscle	Origin	Insertion	Action
Erector spinae (e-REK-tor SPI-nē; *erector* = raise; *spinae* = of the spine) (iliocostalis group, longissimus group, and spinalis group)	All ribs plus cervical, thoracic, and lumbar vertebrae.	Occipital bone, temporal bone, ribs, and vertebrae.	Extends head; extends and laterally flexes vertebral column.
Sternocleidomastoid (ster′-nō-klī′-dō-MAS-toid; *sternum* = breastbone; *cleido-* = clavicle; *mastoid* = mastoid process of temporal bone) (See Figure 8.13b.)	Sternum and clavicle.	Temporal bone.	Contractions of both muscles flex cervical part of the vertebral column and flex the head; contraction of one muscle rotates head toward side opposite contracting muscle.
Quadratus lumborum (kwod-RĀ-tus lum-BOR-um; *quadratus* = four-sided; *lumbo* = lumbar region). (See Figure 8.17b.)	Ilium.	Twelfth rib and upper four lumbar vertebrae.	Contractions of both muscles extend lumbar part of the vertebral column; contraction of one muscle flexes lumbar part of vertebral column.

Figure 8.22 Major muscles that move the vertebral column (back bone).

The erector spinae muscles extend the vertebral column.

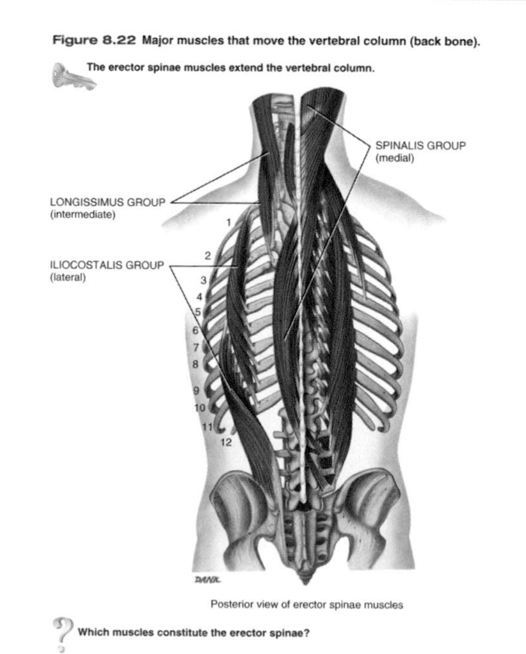

Posterior view of erector spinae muscles

Which muscles constitute the erector spinae?

AGING

Anatomy and physiology is not static. As the body ages, its structure and related functions change in subtle and not so subtle ways. Many students will go on to careers in health related fields, in which the average age of the client population is steadily advancing. For this reason, discussions of this professionally relevant topic are included in chapters 1, 3, 4, 7, 8, 13, 16, 17, 18, 19, 22, and 23.

AGING AND THE DIGESTIVE SYSTEM

OBJECTIVE • Describe the effects of aging on the digestive system.

Changes in the digestive system associated with aging include decreased secretory mechanisms, decreased motility of the digestive organs, loss of strength and tone of the muscular tissue and its supporting structures, changes in sensory feedback regarding enzyme and hormone release, and diminished response to pain and internal sensations. In the upper portion of the GI tract, common changes include reduced sensitivity to mouth irritations and sores, loss of taste, periodontal disease, difficulty in swallowing, hiatal hernia, gastritis, and peptic ulcer disease. Changes that may appear in the small intestine include duodenal ulcers, maldigestion, and malabsorption. Other pathologies that increase in incidence with age are appendicitis, gallbladder problems, jaundice, cirrhosis of the liver, and acute pancreatitis. Changes in the large intestine such as constipation, hemorrhoids, and diverticular disease may also occur. The incidence of cancer of the colon or rectum increases with age.

■ **CHECKPOINT**

25. List several changes in the upper and lower portions of the GI tract associated with aging.

• • •

COMMON DISORDERS AND MEDICAL TERMINOLOGY AND CONDITIONS

The Common Disorders and Medical Terminology and Conditions sections have been completely updated and revised for this edition. Marked with distinctive icons, these sections are located at the ends of appropriate chapters. The problems considered in the Common Disorders sections are selected to provide a review of normal body processes and demonstrate the importance of the study of anatomy and physiology to careers in health-related fields. The glossaries of selected medical terms and conditions are designed to build vocabulary and enhance understanding. Many of the terms include pronunciation guides and word roots.

COMMON DISORDERS

Asthma

Asthma (AZ-ma = panting) is a disorder characterized by chronic airway inflammation, airway hypersensitivity to a variety of stimuli, and airway obstruction. The airway obstruction may be due to smooth muscle spasms in the walls of smaller bronchi and bronchioles, swelling of the mucosa of the airways, increased mucus secretion, or damage to the epithelium of the airway. Asthma is at least partially reversible, either spontaneously or with treatment. It affects 3–5% of the U.S. population and is becoming increasingly common in children.

Asthmatics typically react to low concentrations of stimuli that do not normally cause symptoms in people without asthma. Sometimes the trigger is an allergen such as pollen, dust mites, molds, or a ticular food. Other common triggers include emotional upset, as- n, sulfiting agents (used in wine and beer and to keep greens fresh salad bars), exercise, and breathing cold air or cigarette smoke. nptoms include difficult breathing, coughing, wheezing, chest tness, tachycardia, fatigue, moist skin, and anxiety.

ronic Obstructive Pulmonary Disease

ronic obstructive pulmonary disease (COPD) is a respiratory dis- er characterized by chronic obstruction of airflow. The principal es of COPD are emphysema and chronic bronchitis. In most es, COPD is preventable because its most common cause is ciga- e smoking or breathing secondhand smoke. Other causes include pollution, pulmonary infection, occupational exposure to dusts gases, and genetic factors.

physema

physema (em′-fi-SĒ-ma = blown up or full of air) is a disorder char- rized by destruction of the walls of the alveoli, which produces ab- mally large air spaces that remain filled with air during exhalation. h less surface area for gas exchange, O_2 diffusion across the respira- membrane is reduced. Blood O_2 level is somewhat lowered, and mild exercise that raises the O_2 requirements of the cells leaves the ent breathless. As increasing numbers of alveolar walls are damaged, g elastic recoil decreases due to loss of elastic fibers, and an increas- amount of air becomes trapped in the lungs at the end of exhalation. r several years, added respiratory exertion increases the size of the st cage, resulting in a "barrel chest." Emphysema is a common pre- sor to the development of lung cancer.

ronic Bronchitis

onic bronchitis is a disorder characterized by excessive secretion of nchial mucus accompanied by a cough. Inhaled irritants lead to onic inflammation with an increase in the size and number of mucous nds and goblet cells in the airway epithelium. The thickened and ex- ive mucus produced narrows the airway and impairs the action of cilia. us, inhaled pathogens become embedded in airway secretions and tiply rapidly. Besides a cough, symptoms of chronic bronchitis are rtness of breath, wheezing, cyanosis, and pulmonary hypertension.

Lung Cancer

In the United States, ***lung cancer*** is the leading cause of cancer death in both males and females. At the time of diagnosis, lung cancer is usually well advanced. Most people with lung cancer die within a year of the diagnosis, and the overall survival rate is only 10–15%. About 85% of lung cancer cases are due to smoking, and the disease is 10 to 30 times more common in smokers than nonsmokers. Exposure to secondhand smoke also causes lung cancer and heart disease. Other causes of lung cancer are ionizing radiation, such as x-rays, and inhaled irritants, such as asbestos and radon gas.

Symptoms of lung cancer may include a chronic cough, spitting blood from the respiratory tract, wheezing, shortness of breath, chest pain, hoarseness, difficulty swallowing, weight loss, anorexia, fatigue, bone pain, confusion, problems with balance, headache, anemia, low blood platelet count, and jaundice.

Pneumonia

Pneumonia or ***pneumonitis*** is an acute infection or inflammation of the alveoli. It is the most common infectious cause of death in the United States, where an estimated 4 million cases occur annually. When certain microbes enter the lungs of susceptible individuals, they release damaging toxins, stimulating inflammation and immune responses that have damaging side effects. The toxins and immune response damage alveoli and bronchial mucous membranes; inflammation and edema cause the alveoli to fill with debris and fluid, interfering with ventilation and gas exchange. The most common cause is the bacterium *Streptococcus pneumoniae*, but other bacteria, viruses, or fungi may also cause pneumonia.

Tuberculosis

The bacterium *Mycobacterium tuberculosis* produces an infectious, communicable disease called ***tuberculosis (TB)*** that most often affects the lungs and the pleurae but may involve other parts of the body. Once the bacteria are inside the lungs, they multiply and cause inflammation, which stimulates neutrophils and macrophages to migrate to the area and engulf the bacteria to prevent their spread. If the immune system is not impaired, the bacteria may remain dormant for life. Impaired immunity may enable the bacteria to escape into blood and lymph to infect other organs. In many people, symptoms—fatigue, weight loss, lethargy, anorexia, a low-grade fever, night sweats, cough, dyspnea, chest pain, and spitting blood (hemoptysis)—do not develop until the disease is advanced.

Coryza and Influenza

Hundreds of viruses, especially the *rhinoviruses* (*rhin-* = nose), can cause ***coryza*** (ko-RĪ-za) or the ***common cold***. Typical symptoms include sneezing, excessive nasal secretion, dry cough, and congestion. The uncomplicated common cold is not usually accompanied by a fever. Complications include sinusitis, asthma, bronchitis, ear infections, and laryngitis.

Influenza (flu) is also caused by a virus. Its symptoms include chills, fever (usually higher than 101°F, or 38°C), headache, and muscular aches. Coldlike symptoms appear as the fever subsides.

Pulmonary Edema

Pulmonary edema is an abnormal accumulation of interstitial fluid in the interstitial spaces and alveoli of the lungs. The edema may arise from increased pulmonary capillary permeability (pulmonary origin) or increased pulmonary capillary pressure due to congestive heart failure (cardiac origin). The most common symptom is painful or labored breathing. Other symptoms include wheezing, rapid breathing rate, restlessness, a feeling of suffocation, cyanosis, paleness, and excessive perspiration.

MEDICAL TERMINOLOGY AND CONDITIONS

Abdominal thrust maneuver First-aid procedure to clear the airways of obstructing objects. It is performed by applying a quick upward thrust between the navel and lower ribs that causes sudden elevation of the diaphragm and forceful, rapid expulsion of air from the lungs, forcing air out of the trachea to eject the obstructing object. Also used to expel water from the lungs of near-drowning victims before resuscitation is begun. Also known as the ***Heimlich maneuver*** (HĪM-lik ma-NOO-ver).

Asphyxia (as-FIK-sē-a; *sphyxia* = pulse) Oxygen starvation due to low atmospheric oxygen or interference with ventilation, external respiration, or internal respiration.

Aspiration (as′-pi-RĀ-shun) Inhalation into the bronchial tree of a substance other than air, for instance, water, food, or a foreign body.

Bronchoscopy The visual examination of the bronchi through a ***bronchoscope***, an illuminated, tubular instrument that is passed through the mouth (or nose), larynx, and trachea into the bronchi.

Cystic fibrosis (CF) An inherited disease of secretory epithelia that affects the airways, liver, pancreas, small intestine, and sweat glands. Clogging and infection of the airways leads to difficulty in breathing and eventual destruction of lung tissue.

Dyspnea (DISP-nē-a; *dys-* = painful, difficult) Painful or labored breathing.

Epistaxis (ep′-i-STAK-sis) Loss of blood from the nose due to trauma, infection, allergy, malignant growths, or bleeding disorders. It can be arrested by cautery with silver nitrate, electrocautery, or firm packing. Also called ***nosebleed.***

Hypoxia (hī-POK-sē-a; *hypo-* = below or under) A deficiency of O_2 at the tissue level that may be caused by a low P_{O_2} in arterial blood, as from high altitudes; too little functioning hemoglobin in the blood, as in anemia; inability of the blood to carry O_2 to tissues fast enough to sustain their needs, as in heart failure; or inability of tissues to use O_2 properly, as in cyanide poisoning.

Mechanical ventilation The use of an automatically cycling device (ventilator or respirator) to assist breathing. A plastic tube is inserted into the nose or mouth and the tube is attached to a device that forces air into the lungs. Exhalation occurs passively due to the elastic recoil of the lungs.

Pleurisy Inflammation of the pleural membranes, which causes friction during breathing that can be quite painful when the swollen membranes rub against each other. Also known as ***pleuritis***.

Rales (RĀLS) Sounds sometimes heard in the lungs that resemble bubbling or rattling. Different types are due to the presence of an abnormal type or amount of fluid or mucus within the bronchi or alveoli, or to bronchoconstriction that causes turbulent airflow.

Respiratory distress syndrome (RDS) A breathing disorder of premature newborns in which the alveoli do not remain open due to a lack of surfactant. Surfactant reduces surface tension and is necessary to prevent the collapse of alveoli during exhalation.

Respiratory failure A condition in which the respiratory system either cannot supply enough O_2 to maintain metabolism or cannot eliminate enough CO_2 to prevent respiratory acidosis (a higher-than-normal H^+ level in interstitial fluid).

Rhinitis (rī-NĪ-tis; *rhin-* = nose) Chronic or acute inflammation of the mucous membrane of the nose.

Sudden infant death syndrome (SIDS) Death of infants between the ages of 1 week and 12 months thought to be due to hypoxia that occurs while sleeping in a prone position (on the stomach) and rebreathing exhaled air trapped in a depression of the mattress. It is now recommended that normal newborns be placed on their backs for sleeping (remember: "back to sleep").

Tachypnea (tak′-ip-NĒ-a; *tachy-* = rapid) Rapid breathing rate.

Wheeze (HWĒZ) A whistling, squeaking, or musical high-pitched sound during breathing resulting from a partially obstructed airway.

THE ILLUSTRATION PROGRAM

Beautiful artwork, carefully chosen photographs and photomicrographs, and unique pedagogical enhancements all combine to make the visual appeal and usefulness of the illustration program in *Introduction to the Human Body* distinctive. The seventh edition has been carefully reviewed, revised, and updated to uphold the standard of excellence that instructors and students alike have come to expect.

COLOR CODING

Colors are used in a consistent and meaningful manner throughout the text to emphasize structural and functional relations. For example, sensory structures, sensory neurons, and sensory regions of the brain are shades of blue, whereas motor structures are red. Membrane phospholipids are gray and aqua, the cytosol is sand, and extracellular fluid is blue. Illustrations of our distinctive negative and positive feedback loops also use color cues to aid the students in recognizing and understanding these concepts.

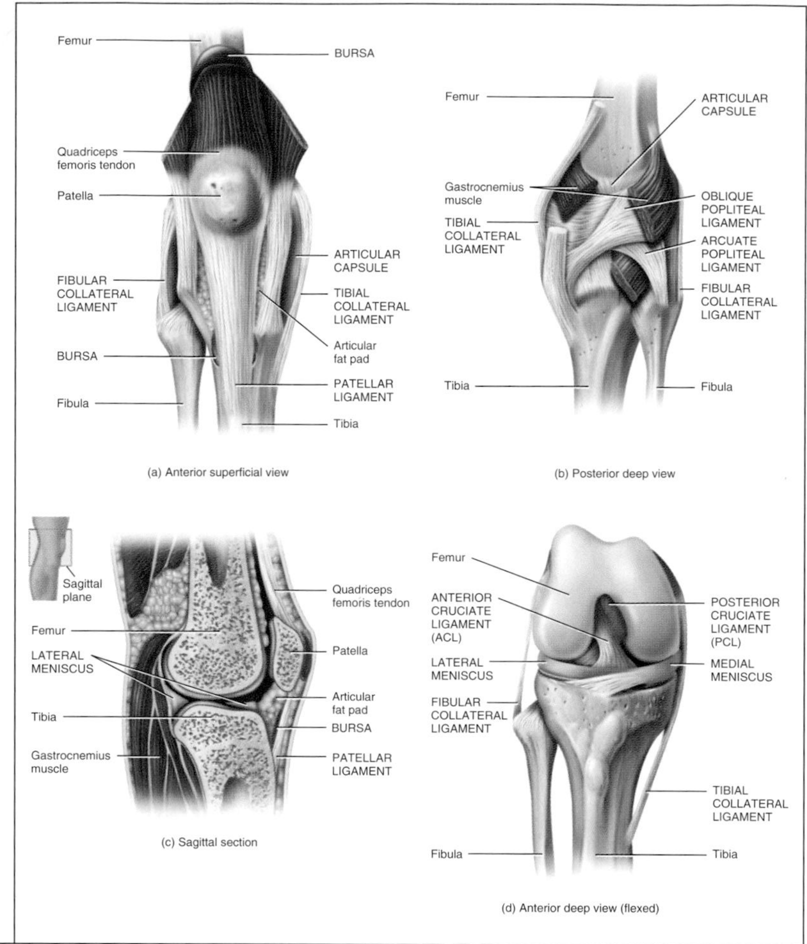

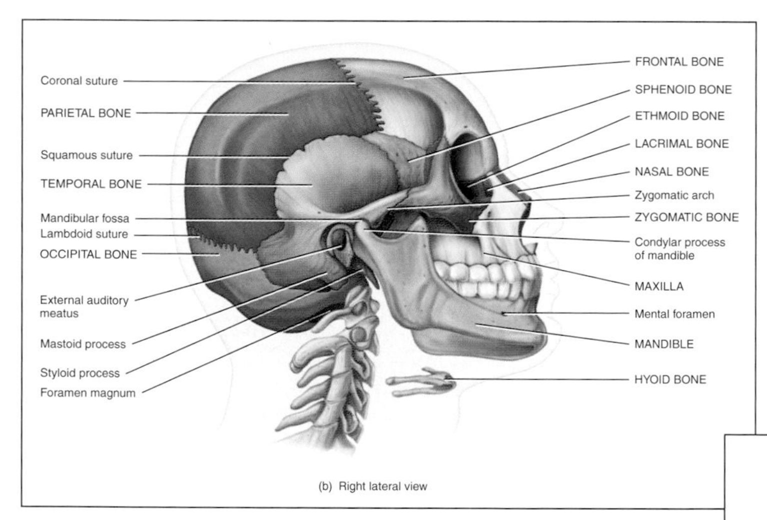

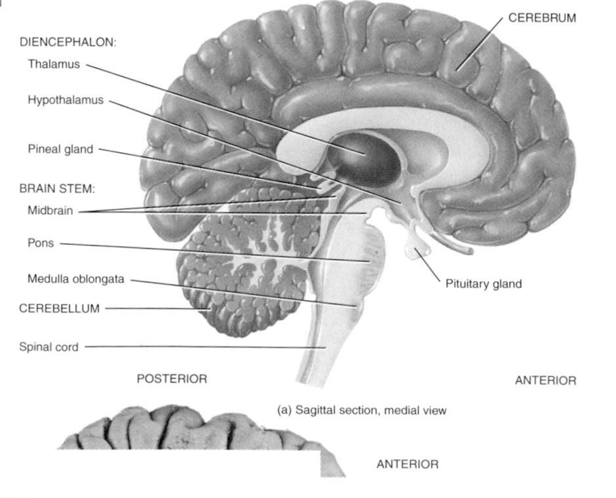

NEW ARTWORK

Exciting new three-dimensional paintings grace the pages of many chapters. Many other line drawings have been newly rendered for this edition, and nearly every figure has been revised or improved in some way. For example, virtually all of the art in the chapter on skeletal system (Chapter 6) has been redrawn.

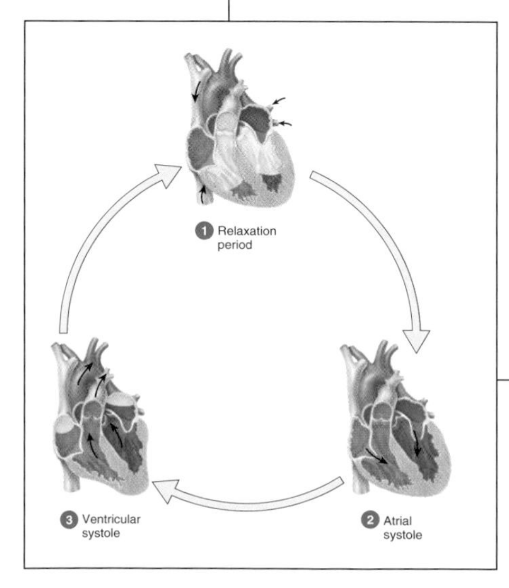

FEEDBACK LOOP ILLUSTRATIONS

As in past editions, this popular series of illustrations captures and clarifies the body's dynamic counterbalancing act in maintaining homeostasis. The feedback loops visually accentuate the roles that receptors, control centers, and effectors play in modifying a controlled physiological condition.

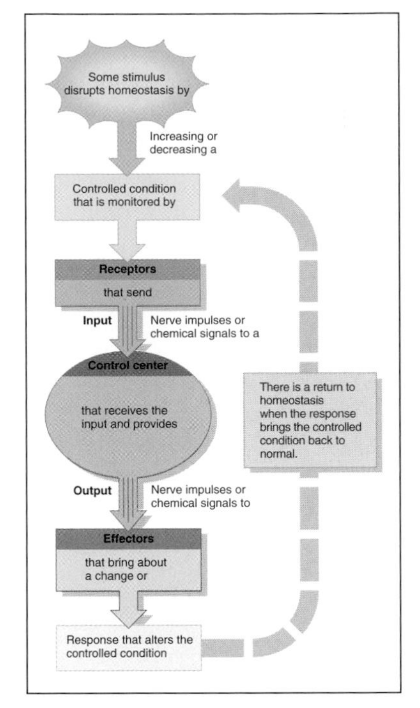

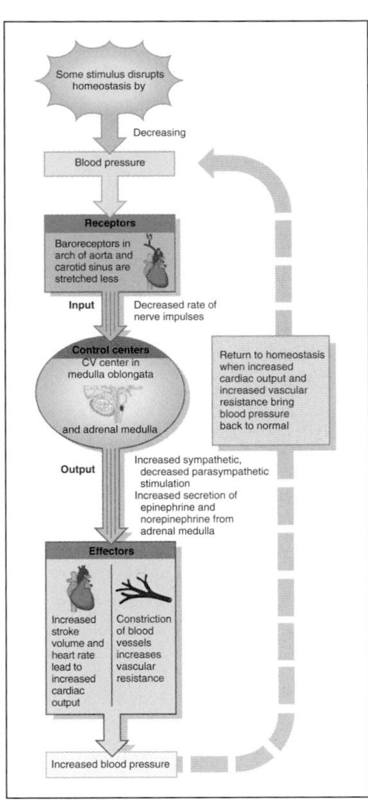

FUNCTIONS OVERVIEW

The Functions Overview is a feature that juxtaposes the anatomical components and a brief functional overview for each body system. These function "boxes" accompany the first figure of chapters dealing with body systems. They help students to integrate visually the structure and function of a body system and make the connection between the interactions of various systems.

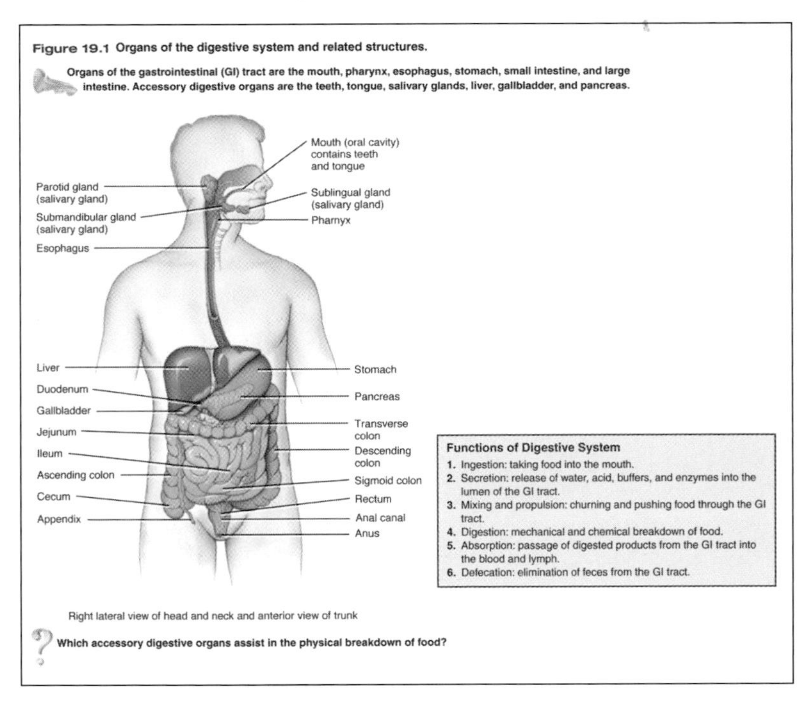

ORIENTATION DIAGRAMS

Students sometimes need help figuring out the perspective of structural illustrations—descriptions alone do not always suffice. An orientation diagram depicting and explaining the perspective of the view represented in the figure accompanies most anatomy and histology illustrations. The orientation diagrams are of three general types: (1) planes indicating where certain sections are made when a part of the body is cut; (2) diagrams containing a directional arrow and the word "View" to indicate the direction from which the body part is viewed, e.g. superior, inferior, posterior, anterior; (3) diagrams with arrows to direct attention to enlarged and detailed parts of illustrations.

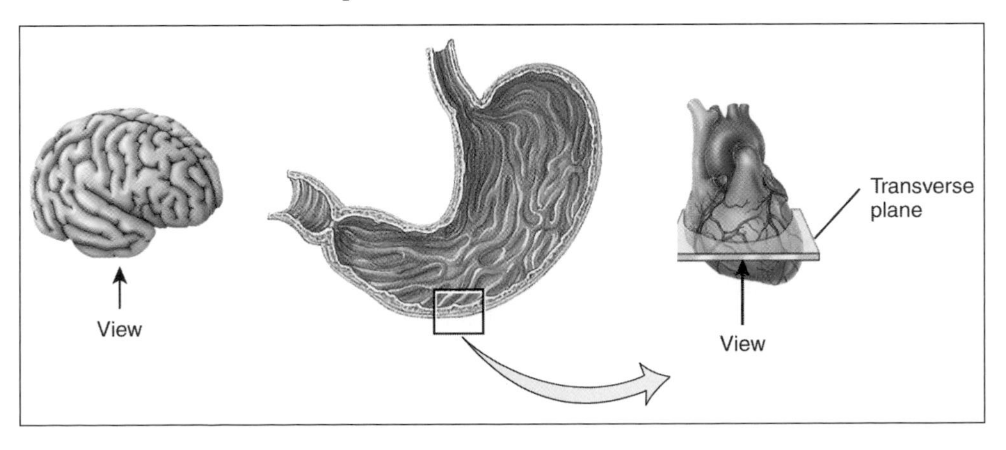

CORRELATION OF SEQUENTIAL PROCESSES

Correlation of sequential processes in text and art is achieved through the use of numbered lists in the narrative that correspond both visually and numerically to numbered segments in the accompanying figure. This approach is used extensively throughout the book to lend clarity to the flow of complex processes.

Events at a Synapse

Although the presynaptic and postsynaptic neurons are in close proximity at a synapse, their plasma membranes do not touch. They are separated by the ***synaptic cleft***, a tiny space filled with interstitial fluid. Because nerve impulses cannot conduct across the synaptic cleft, an alternate, indirect form of communication occurs across this space. A typical synapse operates as follows (Figure 9.7):

1. A nerve impulse arrives at a synaptic end bulb of a presynaptic axon.
2. The depolarizing phase of the nerve impulse opens *voltage-gated Ca^{2+} channels*, which are present in the membrane of synaptic end bulbs. Because calcium ions are more concentrated in th into the synaptic end bulb t
3. An increase in the concen synaptic end bulb triggers synaptic vesicles, which relea mitter molecules into the syn
4. The neurotransmitter molec tic cleft and bind to ***neuro*** postsynaptic neuron's plasma
5. Binding of neurotransmitter which allows certain ions to fl
6. As ions flow through the o across the membrane chang

Figure 9.7 Synaptic transmission at a chemical synapse. Exocytosis of synaptic vesicles from a presynaptic neuron releases neurotransmitter molecules, which bind to receptors in the plasma membrane of the postsynaptic neuron.

At a chemical synapse, a presynaptic electrical signal (nerve impulse) is converted into a chemical signal (neurotransmitter release). The chemical signal is then converted back into an electrical signal (depolarization or hyperpolarization) in the postsynaptic cell.

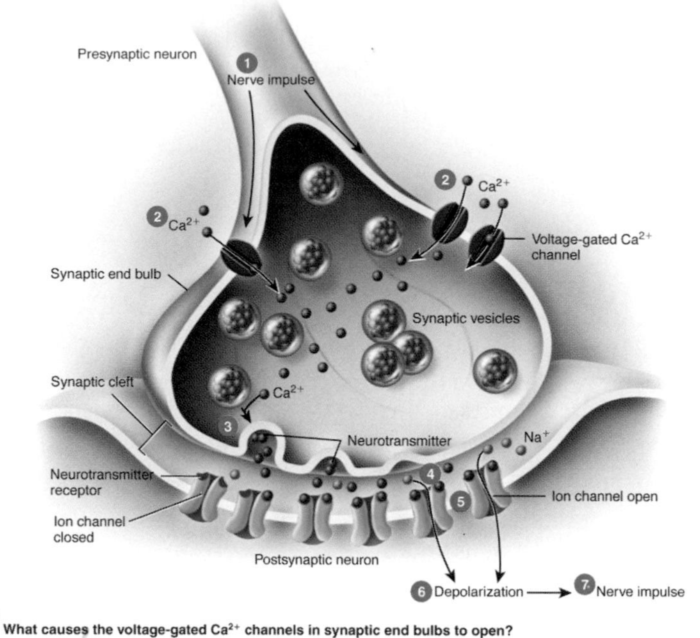

What causes the voltage-gated Ca^{2+} channels in synaptic end bulbs to open?

KEY CONCEPT STATEMENTS

Included above every figure and denoted by the "key" icon, this feature summarizes an idea that is discussed in the text and demonstrated in a figure. A pedagogical feature unique to our text, these help students keep focused on the relevance of the figure to their understanding of specific content.

FIGURE QUESTIONS

This highly applauded feature asks readers to synthesize verbal and visual information, think critically, or draw conclusions about what they see in a figure. Each Figure Question appears below its illustration and is highlighted by a distinctive Question Mark icon. Answers are located at the end of each chapter.

Figure 11.1 Comparison of somatic and autonomic motor neuron pathways to their effector tissues.

Stimulation by the autonomic motor neurons can either excite or inhibit smooth muscle, cardiac muscle, and glands. Stimulation by somatic motor neurons always causes contraction of skeletal muscle.

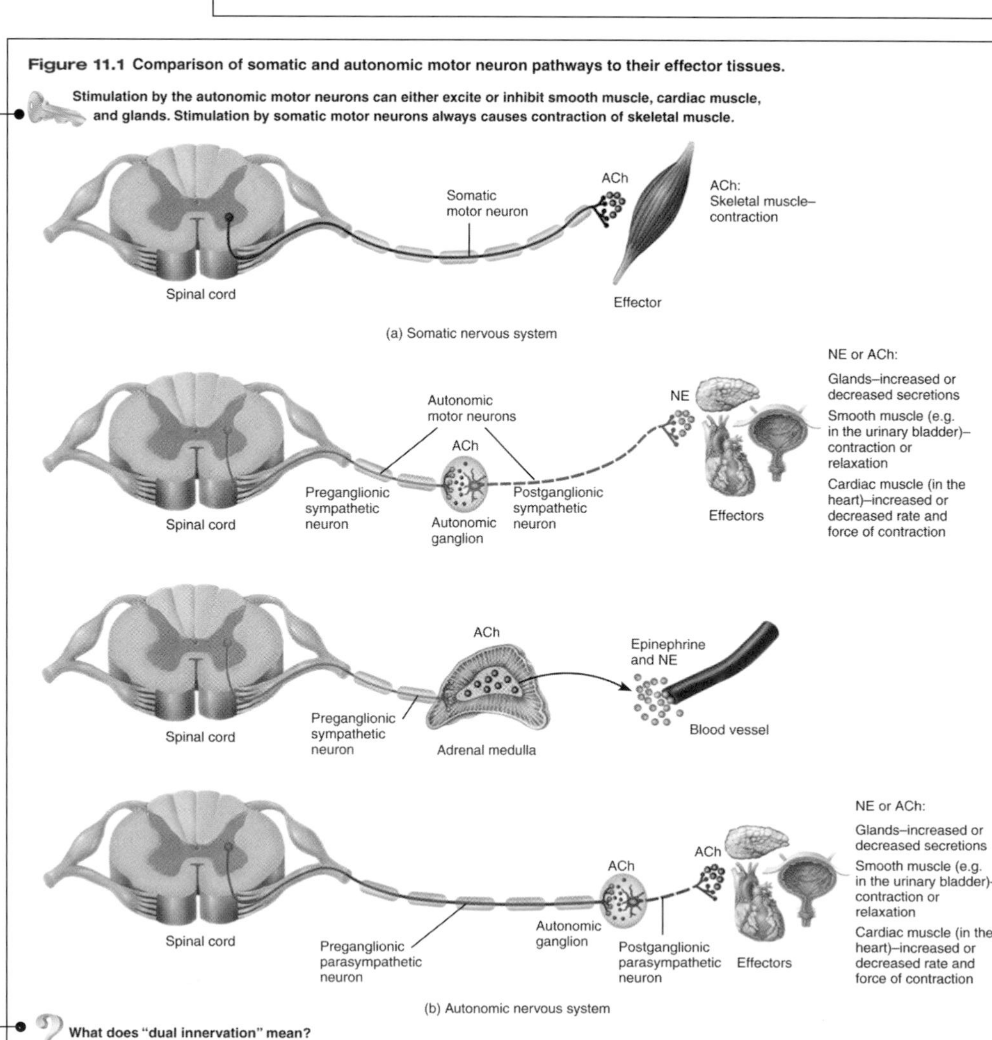

What does "dual innervation" mean?

LEARNING AIDS

In response to users of the previous edition of *Introduction to the Human Body*, we have retained the learning aids that students and instructors find most useful and have tried to improve them wherever possible. Several learning aids already discussed are the chapter objectives and the checkpoint questions. All learning aids—such as critical thinking questions and end of chapter quizzes—have been revised to reflect the enhancements to the text and art.

CROSS REFERENCES

This new edition features cross-references that guide the reader to specific pages and figures. Most will help students relate new concepts to previously learned material. However, we acknowledge the really ambitious student by including some cross-references to material that has yet to be considered.

ANATOMICAL TERMS

The anatomical terms included in the seventh edition have been updated completely to correspond to the *Terminology Anatomica*, the latest word on the use of international anatomical terminology.

PHONETIC PRONUNCIATIONS AND WORD ROOTS

We have carefully revised all phonetic pronunciations, which appear in parentheses after many anatomical and physiological terms. The pronunciations are given both in the text, where the term is introduced, and in the Glossary at the end of the book. Many new pronunciations have been added to this edition. Word roots, derivations designed to provide an understanding of the meaning of new terms, appear in parentheses when a term is introduced.

STUDY OUTLINE

The Study Outline at the end of each chapter summarizes major topics and includes specific page references so that students can easily find full text discussions of the topics they need to review.

SELF-QUIZZES

Self-Quizzes at the end of every chapter include fill-in-the-blank, multiple choice, and matching questions. These quizzes are meant not only for students to test their ability to memorize the facts presented in each chapter, but also to sharpen their critical thinking skills by applying the concepts and processes that are part of the way the human body is structured and how it functions. Answers to the Self-Quiz items are presented in an appendix at the end of the book.

CRITICAL THINKING APPLICATIONS

Critical Thinking Applications, included at the end of each chapter, are essay-style problems that encourage students to think about and apply the concepts they have studied. Although there is no single right answer to this type of question, suggested answers are available in an appendix at the end of the text so the students can check to see if they are on the right track.

GLOSSARY

A full glossary of terms with phonetic pronunciations appears at the end of the book. It includes over 1700 entries, providing students with a terrific reference.

INSIDE COVER MATERIALS

The endpapers at the back of this text provide students with useful information about prefixes, suffixes, word roots, and combining forms for the terminology used in the study of anatomy and physiology. At the front of the text, you will find descriptions of the many supplemental materials available to support both teaching and study.

A COMPLETE TEACHING AND LEARNING PACKAGE

The following ancillaries are available to accompany the seventh edition of *Introduction to the Human Body*. Each of our supplementary products is specifically created with one goal in mind: to help teachers teach and students learn. *Please contact your Wiley sales representative for additional information about any of the following supplementary products.*

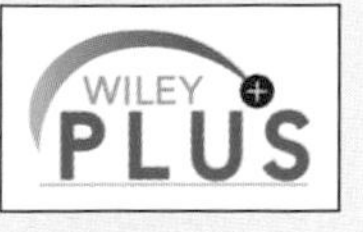

Wiley PLUS Helping Teachers Teach and Students Learn

www.wiley.com/college/tortora

This title is available with *Wiley PLUS*, a powerful online tool that provides instructors and students with an integrated suite of teaching and learning resources in one easy to use Website. *Wiley PLUS* is organized around the essential activities you and your students perform in class.

FOR INSTRUCTORS

Prepare & Present: Create class presentations using a wealth of Wiley-provided resources – such as an online version of the textbook, PowerPoint slides, animations, overviews, and interactive case studies from *Interactions*, and images from the Wiley A&P Visual Library – making your preparation time more efficient. You may easily adapt, customize, and add to this content to meet the needs of your course.

Create Assignments: Automate the assigning and grading of homework or quizzes by using Wiley-provided question banks, or by writing your own. Student results will be automatically graded and recorded in your gradebook. *Wiley PLUS* can link the pre-lecture quizzes and test bank questions to the relevant section of the online text, providing students with context-sensitive help.

Track Student Progress: Keep track of your students' progress via and instructor's gradebook, which allow you to analyze individual and overall class results to determine their progress and level of understanding.

Administer Your Course: *Wiley PLUS* can easily be integrated with another course management system, gradebook, or other resources you are using in your class, providing you with the flexibility to build your course, your way.

FOR STUDENTS

Wiley PLUS provides immediate feedback on student assignments and a wealth of support materials. This powerful study tool will help your students develop their conceptual understanding of the class material and increase their ability to answer questions.

A **"Study and Practice"** area links directly to text content, allowing students to review the text while they study and answer. Resources include all of the *Interactions* content, inclusive of animations, interactive exercises and concept maps, and animated case studies. Also included are practice quizzes, anatomy drill and practice, a flash card tool, pronunciation dictionary, web explorations, pre-lecture quizzes, and other resources for study.

An **"Assignment"** area keeps all the work you want your students to complete in one location, making it easy for them to stay "on task". Students will have access to a variety of interactive self-assessment tools, as well as other resources for building their confidence and understanding. In addition, all of the pre-lecture quizzes contain a link to the relevant section of the multimedia book, providing students with context-sensitive help that allows them to conquer problem-solving obstacles as they arise.

A **Personal Gradebook** for each student will allow students to view their results from past assignments at any time.

Please view our online demo at **www.wiley.com/college/wileyplus.** Here you will find additional information about the features and benefits of *Wiley PLUS*, how to request a "test drive" of *Wiley PLUS* for this title, and how to adopt it for class use.

MATERIALS AVAILABLE FOR INSTRUCTORS

Instructor's Companion Website This dedicated website provides on-line resources including a chapter-by-chapter synopsis, suggested lecture outlines, learning objectives, and teaching tips. Each chapter also includes a description of what is new to this edition. All of these resources are fully downloadable. The electronic format provides you with the opportunity to customize lecture outlines or activities for delivery either in print or electronically to your students. Also available are three sets of **PowerPoint Slides** to ease in lecture presentation – Illustrated Lecture Slides, slides of all illustrations in the text by chapter, and slides of all text tables.

Wiley's Visual Library for Anatomy & Physiology 3.0 (0-471-736791) This all-new cross-platform DVD includes all of the illustrations from the textbook in labeled, unlabeled, and leader lines only format. In addition, illustrations and photographs not included in the text, but which could easily be added to enhance lecture or lab, are included. Search for images by chapter, or by using key words. The Visual Library can also be accessed online at the Instructor's Companion Website, or through *Wiley PLUS*.

Full-color Overhead Transparencies (0-471-76103-6) A set of full-color overheads includes all of the figures in the text, including histology micrographs. All transparencies have been color-enhanced and carefully reviewed to maximize the labels for clear projection in the classroom.

Test Bank A complete testbank is available on the Instructor's Companion Website as printable word document. A variety of formats—multiple choice, short answer, matching and essay—are provided to accommodate different testing preferences. It is also available as a cross-platform CD-ROM **(0-471-77069-8)** or integrated into *Wiley PLUS*, users can easily view, edit and add questions. Users can create questions in six different formats, import graphics and create graphs.

Personal Response System A full set of questions to use with Personal Response Systems are available. For more information, see your Wiley representative, or go to **www.wiley.com/college/prs.**

Faculty Resource Network Wiley's support structure to help instructors implement the dynamic new media that supports this text into their classrooms, laboratories, or online courses. Consult with your Wiley representative for details about this program or visit **www.wherefacultyconnect.com**.

MATERIALS AVAILABLE FOR STUDENTS

Dedicated Book Companion Website A dynamic website rich with many activities for review and exploration includes: Chapter Overview and Objectives, Self-Quizzes for each chapter, Anatomy Drag and Drops, Cadaver Practicals, Pronunciation Dictionary with Flash Card Option and Terminology Quiz, Insights and Explorations – Web-based Activities, Crossword Puzzles, Disorder Search linked to Chapter Content, Weblinks linked to Chapter Content, Medical Tests and Procedures linked to Chapter Content, Essays on Wellness. In addition there are sections on study tips, determining your learning style, and correlations of what assets will work best with a specific learning style.

Illustrated Notebook (0-471-76100-4)—Students can organize their note taking and improve their understanding of anatomical structures and physiological processes by using this handy illustrated notebook. Following the illustration sequence in the textbook, each left-handed page displays an unlabeled, black-and-white copy of every text figure. Students can fill in the labels during lecture or lab at the instructor's directions and take additional notes on the blank right-hand pages.

Student Learning Guide (0-471-76105-2)—by Kathleen Schmidt Prezbindowski, College of Mt. St. Joseph. This study guide is designed to appeal to a broad range of learning styles. It includes multiple activities for each chapter, including *Framework*—visual maps of the chapter content; *Wordbytes*—to help master vocabulary; *Checkpoints*—a series of study activities such as questions to answers, illustrations to label, tables to complete; *Critical Thinking Questions*; *Mastery Test*.

FOR THE LABORATORY

Essentials of Anatomy and Physiology Laboratory Manual (0-471-46516-X) by Connie Allen and Valerie Harper, Edison Community College. This laboratory manual presents material covered in the 1-semester undergraduate anatomy and physiology lab course in brief, clear, and concise way. The manual is very hands-on and contains activities and experiments that enhance students' ability to both visualize anatomical structures and understand key physiological topics.

Cat Dissection Manual 2e (0-471-70141-6) by Connie Allen, Edison Community College, and Valerie Harper. This manual includes photographs and illustrations of the cat along with guidelines for dissection. All photographs of the cat dissection, provided by Dennis Strete of McClennan CC, are new to this edition. It is available independently as well as bundled with the main manual depending upon your adoption needs.

Fetal Pig Dissection Manual 2e (0-471-701386) by Connie Allen, Edison Community College, and Valerie Harper. This manual includes photographs and illustrations of the fetal pig along with guidelines for dissection. All photographs of the fetal pig dissection, provided by Dennis Strete of McClennan CC, are new to this edition. It is available independently as well as bundled with the main manual depending upon your adoption needs.

A Photographic Atlas of the Human Body with Selected Cat, Sheep, and Cow Dissections, 2nd edition by Gerard Tortora **(0-471-42064-6)** This four-colored atlas is designed to support both study and laboratory experiences. Organized by body systems, the clearly labeled photographs provide a stunning visual reference to gross anatomy. Histological micrographs are also included.

New! PowerPhys (0-471-66289-5) by Allen, Harper, Lancraft, and Ivlev. – 10 Self Contained Lab Modules for exploring physiological principles. Each module contains objectives with illustrated and animated review material, pre-lab quizzes, pre-lab reporting, data collection and analysis, and a full lab report with discussion and application questions. Experiments contain randomly generated data, allowing users to experiment multiple times, but still arrive at the same conclusions. Available as a stand-alone product, PowerPhys is also bundled with every new copy of the Allen and Harper Laboratory Manual and integrated into eGrade Plus. View a demo at
http://www.wiley.com/college/powerphysdemo/.

PowerAnatomy (0-471-44558-4) by Allen, Harper and Baxley - Developed in conjunctions with Primal Pictures, U.K., this is an on-line human anatomy laboratory manual, combining beautiful 3-D images of the human body alongside text, exercises and review questions focused on the undergraduate anatomy and anatomy & physiology student. Users can rotate the images, click on linked terms to see structures, and then answer self-assessing questions to test their knowledge. Included is a free 6-month subscription to Primal's acclaimed Anatomy.tv website. To view a demo of this product, go to
www.wiley.com/college/apcentral.

ACKNOWLEDGEMENTS

We wish to especially thank several people for their helpful contributions to this edition. Barbara Brehm Curtis of Smith College enhanced her work on the Focus on Wellness Essays, by creating the new "Did You Know" scenarios for each chapter opener. She also created the coordinating web activities. Once again, Kathy Prezbindowski has provided students with an excellent review in the accompanying Learning Guide. We are sincerely grateful to Robert Amitrano for development of the Instructor's Resources, Charles Wert for the Website Study Quizzes, Richard Connett for the PowerPoint lecture slides, LuAnne Clark for crafting a useful testbank, James Crowder for researching and summarizing appropriate WebLinks, and Caryl Tickner for the PRS (Personal Response System) and *Wiley PLUS* pre-test questions. We are grateful to all for their fine work in making this an even better edition for students to use.

Bonnie Roesch, Executive Editor continues to provide the guidance, creativity, and professionalism, which have impacted so profoundly on all of our books. There is no way that we can adequately express our gratitude to Bonnie. Clay Stone, Senior Marketing Manager, has been a vital component of all of our publications with Wiley. Clay coordinates a network of individuals who present the salient features of our books to professors. The feedback that Clay provides us from professors and students is invaluable. Karen Trost, Developmental Editor, provided insightful comments and suggestions during the revision cycle, helping to keep the focus on the needs of both professors and students. Kelly Tavares, Production Manager, has once again distinguished herself as a "super" coordinator for all aspects of the production process. Her commitment to and passion for her job is so obvious and appreciated. Karin Gerdes Kincheloe, Senior Designer, is responsible for the cover and text design. Throughout, the pages are visually attractive, pedagogically effective, and student-friendly. Edward Starr, Art Coordinator, has provided us with a superior artistic vision and tremendous organziational skills. Hillary Newman, Photo Editor, provided us with all of the photos we requested and did it with efficiency, accuracy, and professionalism. We also want to acknowledge Mary O'Sullivan, Project Editor, and Alicia Romano, Editorial Assistant, for their outstanding efforts in coordinating the development of supplements for this text.

Finally, we would also like to thank all those who have corresponded with us to offer feedback on the usefulness of this text. Your input helps so much in revising. We particularly want to express our gratitude to the following reviewers who took the time to read and evaluate the draft manuscripts prior to the production of this edition.

Michele Barr, California State University- Fullerton
Christy Carmack, Davidson County Community College
Richard Connett, Monroe Community College
Brent Graves, Northern Michigan University
Susan Hovey, Community College Southern Nevada
David Quadagno, Florida State University
Kent C. Robbins, California State University – Northridge
April Rottman, Rock Valley College
Bradley A. Sarchet, Manatee Community College
Marilyn Shopper, Johnson County Community College
Janice Toyoshima, Evergreen Valley College

We would like to invite all readers and users of the book to continue the tradition of sending comments and suggestions to us so that we can include them in the next edition.

Gerard J. Tortora
Department of Science and Technology, S229
Bergen Community College
400 Paramus Road
Paramus, NJ 07652

Bryan Derrickson
Department of Science
Valencia Community College
PO Box 3028
Orlando, FL 32802
bderrickson@valenciacc.edu

NOTE TO THE STUDENT

Your book has a variety of special features that will make your study of anatomy and physiology a more rewarding experience. These have been developed based on feedback from students—like you—who have used previous editions of the text. Below are some hints for using some of these helpful aids. A review of the preface will give you insight, both visually and in narrative, to all of the text's distinctive features.

Our experience in the classroom has taught us that students appreciate a hint at the beginning of the chapter about what to expect from its content.

chapter 12 **SOMATIC SENSES AND SPECIAL SENSES**

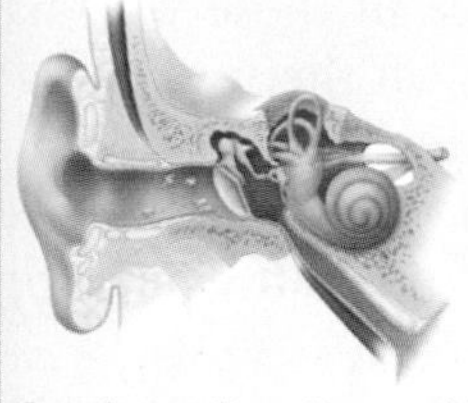

Consider whatwould happen if youcould not feel the pain of a hot pot handleor an inflamed appendix, or if you couldnot see an oncoming car, hear a baby'scry, smell smoke, taste your favoritedessert, or maintain your balance on a flight of stairs. In short, if you could not "sense" your environment and make the necessary homeostatic adjustments, you could not survive very well on your own.

did you know? ***Some things improve with age, but hearing is not one of them. Damage to the hair cells that convert sound waves into nerve impulses accumulates over a lifetime, and by the time hearing loss is discovered, irreversible damage has already occurred. Exposure to excessive noise is the most common cause of hair cell damage. Damage increases with both the intensity and duration of exposure. The hair cells appear to be less traumatized by short periods of loud noise, such as a fire alarm going off, than by chronic exposure to moderately loud noise, such as the noise of vacuum cleaners, power tools, engines, and loud music.***

Focus on Wellness, page 289

www.wiley.com/college/apcentral

looking back to move ahead . . .

- Sensory Nerve Endings and Sensory Receptors in the Skin (page 100)
- Somatic Sensory Pathways (page 259)

284

The **Did You Know?** sections on the opening page of each chapter will help you by connecting the content of the upcoming chapter to issues relevant to your everyday experiences. A look back at what you have already learned will also help to prepare you for what is to come.

Looking Back to Move Ahead is a listing of concepts, complete with page numbers, which you will need to know to be successful in learning the material in the current chapter.

As you begin each narrative section of the chapter, be sure to take note of the **objectives** at the beginning of the section to help you focus on what is important as you read it.

OVERVIEW OF SENSATIONS

OBJECTIVE • Define a sensation and describe the conditions needed for a sensation to occur.

Most of us are aware of sensory input to the central nervous system (CNS) from structures associated with smell, taste, vision, hearing, and balance. These five senses are known as the *special senses*. Th... and include both so...

At the end of the section, take time to try and answer the **Checkpoint** questions placed there. If you can, then you are ready to move on to the next section. If you experience difficulty answering the questions, you may want to re-read the section before continuing.

■ CHECKPOINT

8. How do olfactory receptors and gustatory receptor cells differ in structure and function?
9. Compare the olfactory and gustatory pathways.

Studying the **Figures** (illustrations that include artwork and photographs) in this book is as important as reading the text. To get the most out of the visual parts of this book, use the tools we have added to the figures to help you understand the concepts being presented.

Start by reading the **caption**, which explains what the figure is about. Next, study the **key concept statement**, which reveals a basic idea portrayed in the figure. Added to many figures you will also find an **orientation diagram** to help you understand the perspective from which you are viewing a particular piece of anatomical art. Finally, at the bottom of each figure you will find a **Figure Question**. If you try to answer these questions as you go along, they will serve as additional self-checks to help you understand the material. Often it will be possible to answer a question by examining the figure itself. Other questions will encourage you to integrate the knowledge you've gained by carefully reading the text associated with the figure. Still other questions may prompt you to think critically about the topic at hand or predict a consequence in advance of its description in the text. You will find the answers to figure questions at the end of the chapter in which they appear.

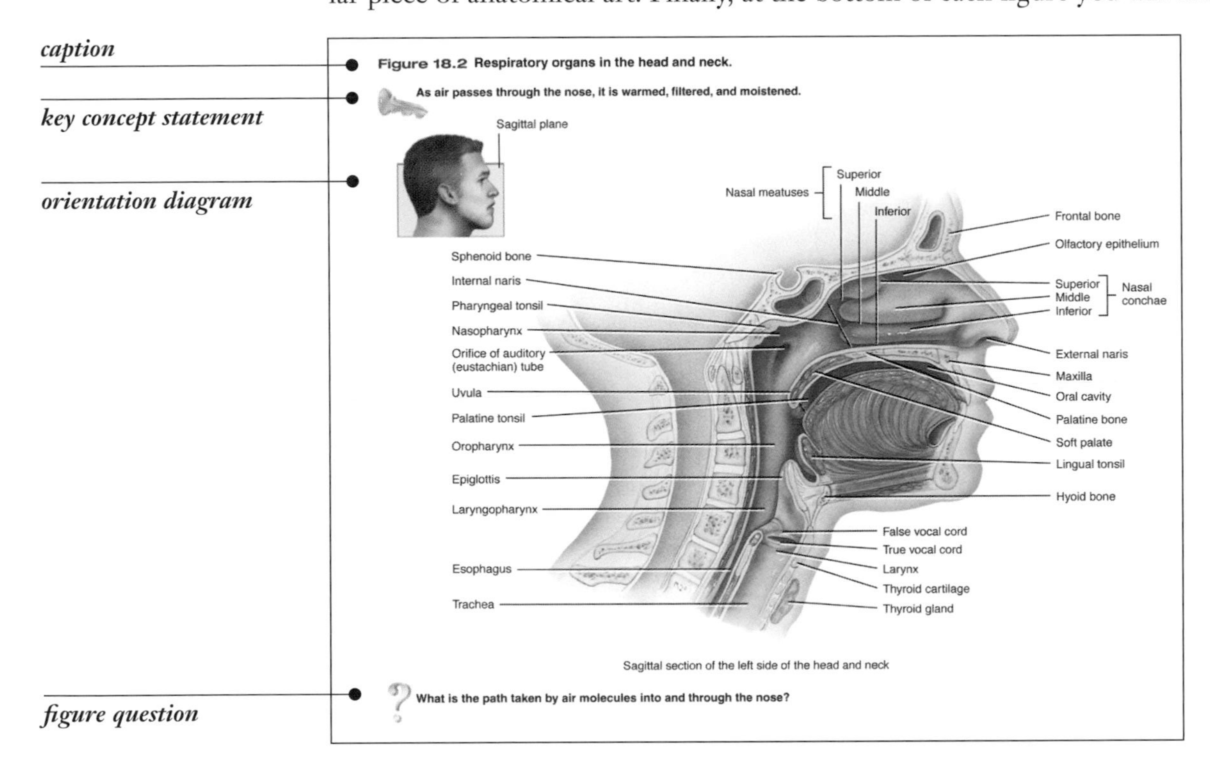

At the end of each chapter are other resources that you will find useful. The **Study Outline** is a concise summary of important topics discussed in the chapter. Page numbers are listed next to key concepts so you can easily refer to specific passages in the text for clarification or amplification. The **Self-Quiz** is a quick objective test designed to help you evaluate your understanding of the chapter contents. Once you have completed the self-quiz, you can grade yourself by checking your answers against the key at the back of the book.

Critical Thinking Applications are word problems that allow you to apply the concepts you have studied in the chapter to specific situations. Because these types of questions are intended to spark creative thinking, they have no one right answer; however, you can check if you are on the right track by comparing your answers against the suggested answers that appear at the end of the book.

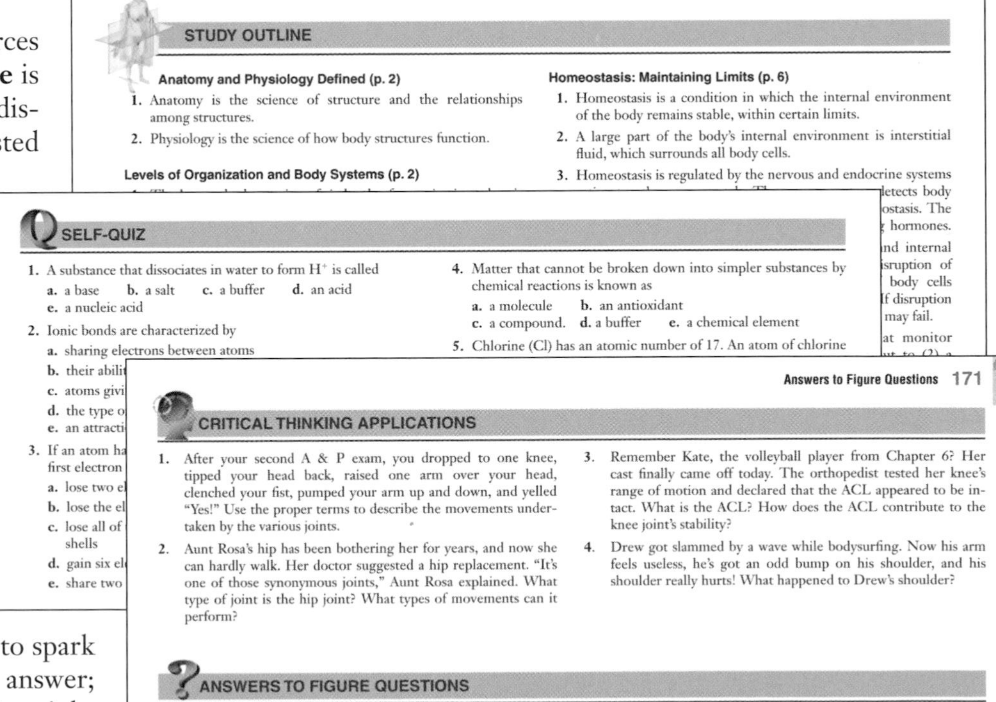
STUDY OUTLINE

Anatomy and Physiology Defined (p. 2)

1. Anatomy is the science of structure and the relationships among structures.
2. Physiology is the science of how body structures function.

Levels of Organization and Body Systems (p. 2)

Homeostasis: Maintaining Limits (p. 6)

1. Homeostasis is a condition in which the internal environment of the body remains stable, within certain limits.
2. A large part of the body's internal environment is interstitial fluid, which surrounds all body cells.
3. Homeostasis is regulated by the nervous and endocrine systems

SELF-QUIZ

1. A substance that dissociates in water to form H^+ is called
a. a base b. a salt c. a buffer d. an acid
e. a nucleic acid
2. Ionic bonds are characterized by
a. sharing electrons between atoms
b. their abili
c. atoms givi
d. the type o
e. an attracti
3. If an atom ha first electron
a. lose two e
b. lose the el
c. lose all of shells
d. gain six el
e. share two
4. Matter that cannot be broken down into simpler substances by chemical reactions is known as
a. a molecule b. an antioxidant
c. a compound. d. a buffer e. a chemical element
5. Chlorine (Cl) has an atomic number of 17. An atom of chlorine

Answers to Figure Questions 171

CRITICAL THINKING APPLICATIONS

1. After your second A & P exam, you dropped to one knee, tipped your head back, raised one arm over your head, clenched your fist, pumped your arm up and down, and yelled "Yes!" Use the proper terms to describe the movements undertaken by the various joints.
2. Aunt Rosa's hip has been bothering her for years, and now she can hardly walk. Her doctor suggested a hip replacement. "It's one of those synonymous joints," Aunt Rosa explained. What type of joint is the hip joint? What types of movements can it perform?
3. Remember Kate, the volleyball player from Chapter 6? Her cast finally came off today. The orthopedist tested her knee's range of motion and declared that the ACL appeared to be intact. What is the ACL? How does the ACL contribute to the knee joint's stability?
4. Drew got slammed by a wave while bodysurfing. Now his arm feels useless, he's got an odd bump on his shoulder, and his shoulder really hurts! What happened to Drew's shoulder?

ANSWERS TO FIGURE QUESTIONS

7.1 Sutures are synarthroses because they are immovable; syndesmoses are classified as amphiarthroses because they are slightly movable.
7.2 Hyaline cartilage holds a synchondrosis together, and fibro-
7.6 Circumduction can occur at the shoulder joint and at the hip joint.
7.7 The anterior surface of a bone or limb rotates toward the midline in medial rotation, and away from the midline in lat-

Throughout the text we have included **Pronunciation Guides** and, sometimes, **Word Roots**, for many terms that may be new to you. These appear in parentheses immediately following the new words, and the pronunciations are repeated in the glossary at the back of the book. Look at the words carefully and say them out loud several times. Learning to pronounce a new word will help you remember it and make it a useful part of your medical vocabulary. Take a few minutes now to read the following pronunciation key, so it will be familiar as you encounter new words. The key is repeated at the beginning of the Glossary, page G-1.

PRONUNCIATION KEY

1. The most strongly accented syllable appears in capital letters, for example, bilateral (bī-LAT-er-al) and diagnosis (dī-ag-NŌ-sis).
2. If there is a secondary accent, it is noted by a prime (′), for example, constitution (kon′-sti-TOO-shun) and physiology (fiz′-ē-OL-ō-jē). Any additional secondary accents are also noted by a prime, for example, decarboxylation (dē-kar-bok′-si-LĀ-shun).
3. Vowels marked by a line above the letter are pronounced with the long sound, as in the following common words:
   ā as in *māke* *ō* as in *pōle*
   ē as in *bē* *ū* as in *cute*
   ī as in *īvy*
4. Vowels not marked by a line above the letter are pronounced with the short sound, as in the following words:
   a as in *above* or *at* *o* as in *not*
   e as in *bet* *u* as in *bud*
   i as in *sip*
5. Other vowel sounds are indicated as follows:
   oy as in *oil* *oo* as in *root*
6. Consonant sounds are pronounced as in the following words:
   b as in *bat* *m* as in *mother*
   ch as in *chair* *n* as in *no*
   d as in *dog* *p* as in *pick*
   f as in *father* *r* as in *rib*
   g as in *get* *s* as in *so*
   h as in *hat* *t* as in *tea*
   j as in *jump* *v* as in *very*
   k as in *can* *w* as in *welcome*
   ks as in *tax* *z* as in *zero*
   kw as in *quit* *zh* as in *lesion*
   l as in *let*

BRIEF CONTENTS

chapters

1 Organization of the Human Body 1
2 Introductory Chemistry 22
3 Cells 44
4 Tissues 72
5 The Integumentary System 97
6 The Skeletal System 113
7 Joints 156
8 The Muscular System 172
9 Nervous Tissue 225
10 Central Nervous System, Spinal Nerves, and Cranial Nerves 242
11 Autonomic Nervous System 271
12 Somatic Senses and Special Senses 284
13 The Endocrine System 315
14 The Cardiovascular System: Blood 345
15 The Cardiovascular System: Heart 364
16 The Cardiovascular System: Blood Vessels and Circulation 385
17 The Lymphatic System and Immunity 420
18 The Respiratory System 445
19 The Digestive System 472
20 Nutrition and Metabolism 503
21 The Urinary System 523
22 Fluid, Electrolyte, and Acid-Base Balance 550
23 The Reproductive Systems 556
24 Development and Inheritance 586

Answers to Self-Quizzes and Critical Thinking Applications A-1
Glossary G-1
Credits C-1
Index I-1

CONTENTS

chapter 1

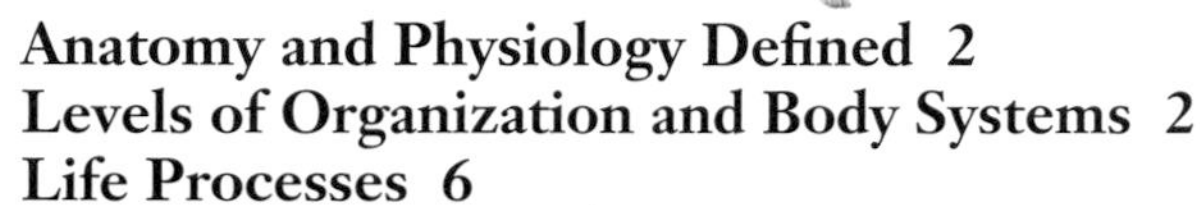

ORGANIZATION OF THE HUMAN BODY 1

Anatomy and Physiology Defined 2
Levels of Organization and Body Systems 2
Life Processes 6
Homeostasis: Maintaining Limits 6
- **Control of Homeostasis: Feedback Systems 7**
  - Negative Feedback Systems 8
  - Positive Feedback Systems 8
- **Homeostasis and Disease 8**

Aging and Homeostasis 9
Anatomical Terms 9
- **Names of Body Regions 9**

Focus on Wellness:
Good Health—Homeostasis Is the Basis 10

- **Directional Terms 10**
- **Planes and Sections 14**

Body Cavities 15
- **Abdominopelvic Regions and Quadrants 17**
- **Medical Terminology and Conditions 18**
  - *Study Outline 18 Self-Quiz 19*
  - *Critical Thinking Applications 21*
  - *Answers to Figure Questions 21*

chapter 2

INTRODUCTORY CHEMISTRY 22

Introduction to Chemistry 23
- **Chemical Elements and Atoms 23**
- **Ions, Molecules, and Compounds 25**
- **Chemical Bonds 25**
  - Ionic Bonds 26
  - Covalent Bonds 26
  - Hydrogen Bonds 28

- **Chemical Reactions 28**
  - Forms of Energy and Chemical Reactions 28
  - Synthesis Reactions 28
  - Decomposition Reactions 28
  - Exchange Reactions 29
  - Reversible Reactions 29

Chemical Compounds and Life Processes 29
- **Inorganic Compounds 29**
  - **Water 29**
  - **Inorganic Acids, Bases, and Salts 30**
  - **Acid-Base Balance: the Concept of pH 30**
  - **Maintaining pH: Buffer Systems 30**
- **Organic Compounds 31**
  - Carbohydrates 31
  - Lipids 32
  - Proteins 35

Focus on Wellness:
Herbal Supplements—They're Natural but Are They Safe? 36

  - **Enzymes 37**
  - **Nucleic Acids: Deoxyribonucleic Acid (DNA) and Ribonucleic Acid (RNA) 38**
  - **Adenosine Triphosphate 38**
  - *Study Outline 40 Self-Quiz 41*
  - *Critical Thinking Applications 43*
  - *Answers to Figure Questions 43*

chapter 3

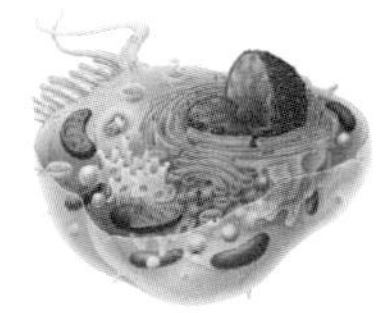

CELLS 44

A Generalized View of the Cell 45
The Plasma Membrane 46
Transport across the Plasma Membrane 47
- **Passive Processes 47**
  - Diffusion: The Principle 47
  - Osmosis 49
- **Active Processes 50**
  - Active Transport 50
  - Transport in Vesicles 51

Cytoplasm 52
Cytosol 52
Organelles 52
The Cytoskeleton 53
Centrosome 54
Cilia and Flagella 54
Ribosomes 54
Endoplasmic Reticulum 55
Golgi Complex 56
Lysosomes 56
Peroxisomes 56
Proteasomes 57
Mitochondria 57
Nucleus 57
Gene Action: Protein Synthesis 58
Transcription 58
Translation 60
Somatic Cell Division 62
Interphase 62
Mitotic Phase 62
Nuclear Division: Mitosis 62
Focus on Wellness:
Phytochemicals—Protecting Cellular Function 64
Cytoplasmic Division: Cytokinesis 64
Aging and Cells 65
Common Disorders 66
Medical Terminology and Conditions 67
Study Outline 67 *Self-Quiz* 69
Critical Thinking Applications 71
Answers to Figure Questions 71

chapter 4
TISSUES 72
Types of Tissues 73
Epithelial Tissue 73
General Features of Epithelial Tissue 73
Covering and Lining Epithelium 73
Simple Epithelium 75
Pseudostratified Columnar Epithelium 75
Stratified Epithelium 75
Glandular Epithelium 82
Connective Tissue 82
General Features of Connective Tissue 82
Connective Tissue Cells 83
Connective Tissue Extracellular Matrix 83
Ground Substance 83
Fibers 84
Classification of Connective Tissue 84
Loose Connective Tissue 84
Dense Connective Tissue 86
Cartilage 89
Bone Tissue 89
Liquid Connective Tissue 90
Muscular Tissue 90
Nervous Tissue 90
Membranes 90
Mucous Membranes 90
Focus on Wellness:
Excess Adiposity—Too Much of a Good Thing 91
Serous Membranes 91
Synovial Membranes 91
Tissue Repair: Restoring Homeostasis 92
Aging and Tissues 92
Common Disorders 93
Medical Terminology and Conditions 93
Study Outline 93 *Self-Quiz* 95
Critical Thinking Applications 96
Answer to Figure Questions 96

chapter 5
THE INTEGUMENTARY SYSTEM 97
Skin 98
Structure of Skin 98
Epidermis 99
Dermis 100
Skin Color 101
Accessory Structures of the Skin 101
Hair 101
Glands 103
Sebaceous Glands 103
Sudoriferous Glands 103
Ceruminous Glands 104
Nails 104
Functions of the Skin 104

Focus on Wellness:
Skin Care for Active Lifestyles 105
Aging and the Integumentary System 106
Common Disorders 107
Medical Terminology and Conditions 108
Focus on Homeostasis: The Integumentary System 109
Study Outline 110 *Self-Quiz* 110
Critical Thinking Applications 112
Answers to Figure Questions 112

chapter 6
THE SKELETAL SYSTEM 113
Functions of Bone and the Skeletal System 114
Types of Bones 114
Structure of Bone 114
Macroscopic Structure of Bone 114
Microscopic Structure of Bone 116

Compact Bone Tissue 116
Spongy Bone Tissue 118
Bone Formation 118
Initial Bone Formation in an Embryo and Fetus 118
Intramembranous Ossification 118
Endochondral Ossification 120
Bone Growth in Length and Thickness 121
Growth in Length 121
Growth in Thickness 121
Bone Remodeling 121
Fractures 122
Factors Affecting Bone Growth 122
Bone's Role in Calcium Homeostasis 122
Exercise and Bone Tissue 123
Divisions of the Skeletal System 124
Skull and Hyoid Bone 125
Cranial Bones 127
Facial Bones 129
Unique Features of the Skull 131
Sutures 131
Paranasal Sinuses 131
Fontanels 132
Hyoid Bone 132
Vertebral Column 133
Regions of the Vertebral Column 133
Normal Curves of the Vertebral Column 134
Vertebrae 134
Thorax 137
Sternum 137
Ribs 138
Pectoral (Shoulder) Girdle 139
Clavicle 139
Scapula 139
Upper Limb 140
Humerus 140
Ulna and Radius 140
Carpals, Metacarpals, and Phalanges 141
Pelvic (Hip) Girdle 143
Lower Limb 144
Femur 144
Patella 144
Tibia and Fibula 145
Tarsals, Metatarsals, and Phalanges 145
Comparison of Female and Male Skeletons 146
Aging and the Skeletal System 147
Focus on Wellness: Steps to Healthy Feet 148
Focus on Homeostasis: The Skeletal System 149
Common Disorders 150
Medical Terminology and Conditions 151
Study Outline 152 Self-Quiz 153
Critical Thinking Applications 155
Answers to Figure Questions 155

chapter 7
JOINTS 156
Joints 157
Fibrous Joints 157
Cartilaginous Joints 158
Synovial Joints 158
Structure of Synovial Joints 158
Types of Movements at Synovial Joints 160
Gliding 160
Angular Movements 160
Rotation 161
Special Movements 162
Types of Synovial Joints 164
Details of a Synovial Joint: The Knee Joint 165
Aging and Joints 165
Focus on Wellness: Joint Care: Prevent Repetitive Motion Injury 167
Common Disorders 168
Medical Terminology and Conditions 169
Study Outline 169 Self-Quiz 170
Critical Thinking Applications 171
Answers to Figure Questions 171

chapter 8
THE MUSCULAR SYSTEM 172
Overview of Muscular Tissue 173
Types of Muscular Tissue 173
Functions of Muscular Tissue 173
Skeletal Muscle Tissue 173
Connective Tissue Components 173
Nerve and Blood Supply 175
Histology 175
Contraction and Relaxation of Skeletal Muscle 177
Sliding-Filament Mechanism 177
Neuromuscular Junction 177
Physiology of Contraction 179
Relaxation 180
Muscle Tone 180
Metabolism of Skeletal Muscle Tissue 180
Energy for Contraction 180
Muscle Fatigue 183
Oxygen Consumption after Exercise 183
Control of Muscle Tension 183
Twitch Contraction 184
Frequency of Stimulation 184
Motor Unit Recruitment 184
Types of Skeletal Muscle Fibers 184
Isometric and Isotonic Contractions 184
Exercise and Skeletal Muscle Tissue 186
Cardiac Muscle Tissue 186

Smooth Muscle Tissue 186
Aging and Muscular Tissue 188
How Skeletal Muscles Produce Movement 188
Origin and Insertion 188
Group Actions 189
Principal Skeletal Muscles 189
Focus on Wellness: Effective Stretching Increases Muscle Flexibility 190
Focus on Homeostasis: The Muscular System 218
Common Disorders 219
Medical Terminology and Conditions 220
Study Outline 220 *Self-Quiz* 222
Critical Thinking Applications 224
Answers to Figure Questions 224

chapter 9
NERVOUS TISSUE 225
Overview of the Nervous System 226
Structures of the Nervous System 226
Functions of the Nervous System 226
Organization of the Nervous System 227
Histology of Nervous Tissue 228
Neurons 228
Myelination 228
Gray and White Matter 228
Neuroglia 230
Action Potentials 230
Ion Channels 230
Resting Membrane Potential 232
Generation of Action Potentials 232
Conduction of Nerve Impulses 233
Synaptic Transmission 234
Events at a Synapse 235
Focus on Wellness: Neurotransmitters—Why Food Affects Mood 236
Neurotransmitters 237
Common Disorders 237
Medical Terminology and Conditions 238
Study Outline 238 *Self-Quiz* 239
Critical Thinking Applications 241
Answers to Figure Questions 241

chapter 10
CENTRAL NERVOUS SYSTEM, SPINAL NERVES, AND CRANIAL NERVES 242
Spinal Cord 243
Protection and Coverings: Vertebral Canal and Meninges 243
Gross Anatomy of the Spinal Cord 243
Internal Structure of the Spinal Cord 245
Spinal Nerves 246
Spinal Nerve Composition and Coverings 246
Distribution of Spinal Nerves 246
Plexuses 246
Spinal Cord Functions 247
Brain 248
Major Parts and Protective Coverings 248
Brain Blood Supply and the Blood-Brain Barrier 250
Cerebrospinal Fluid 250
Brain Stem 250
Medulla Oblongata 250
Pons 253
Midbrain 253
Reticular Formation 253
Diencephalon 254
Thalamus 254
Hypothalamus 254
Pineal Gland 255
Cerebellum 255
Cerebrum 255
Focus on Wellness: Coffee Nerves—the Health Risks of Caffeine 257
Limbic System 257
Functional Areas of the Cerebral Cortex 258
Somatic Sensory and Somatic Motor Pathways 259
Hemispheric Lateralization 261
Memory 262
Electroencephalogram (EEG) 262
Cranial Nerves 263
Aging and the Nervous System 265
Common Disorders 265
Medical Terminology and Conditions 267
Study Outline 267 *Self-Quiz* 269
Critical Thinking Applications 270
Answers to Figure Questions 270

chapter 11
AUTONOMIC NERVOUS SYSTEM 271
Comparison of Somatic and Autonomic Nervous Systems 272
Structure of the Autonomic Nervous System 273
Organization of the Sympathetic Division 274
Organization of the Parasympathetic Division 275
Functions of the Autonomic Nervous System 276
ANS Neurotransmitters 276
Activities of the ANS 276
Sympathetic Activities 276
Focus on Wellness: Mind-Body Exercise—an Antidote to Stress 278
Parasympathetic Activities 278
Common Disorders 280

Study Outline 280 Self-Quiz 281
Critical Thinking Applications 283
Answers to Figure Questions 283

chapter 12

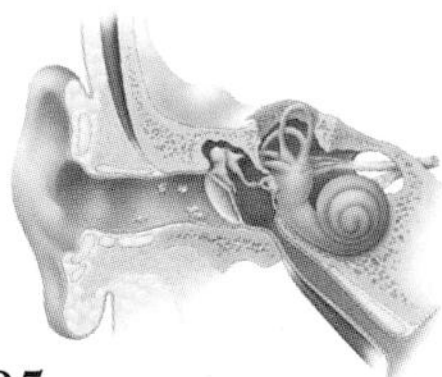

SOMATIC SENSES AND SPECIAL SENSES 284

Overview of Sensations 285
Definition of Sensation 285
Characteristics of Sensations 285
Types of Sensory Receptors 285
Somatic Senses 286
Tactile Sensations 287
Touch 287
Pressure and Vibration 287
Itch and Tickle 287
Thermal Sensations 287
Pain Sensations 287
Focus on Wellness:
Pain Management—Sensation Modulation 289
Proprioceptive Sensations 289
Special Senses 290
Olfaction: Sense of Smell 290
Structure of the Olfactory Epithelium 290
Stimulation of Olfactory Receptors 290
The Olfactory Pathway 290
Gustation: Sense of Taste 291
Structure of Taste Buds 292
Stimulation of Gustatory Receptors 293
The Gustatory Pathway 293
Vision 293
Accessory Structures of the Eye 293
Layers of the Eyeball 294
Fibrous Tunic 294
Vascular Tunic 294
Retina 296
Interior of the Eyeball 297
Image Formation and Binocular Vision 297
Refraction of Light Rays 297
Accommodation 299
Constriction of the Pupil 300
Convergence 300
Stimulation of Photoreceptors 300
The Visual Pathway 300
Hearing and Equilibrium 301
Structure of the Ear 301
Outer Ear 301
Middle Ear 301
Internal Ear 302
Physiology of Hearing 304
Auditory Pathway 305
Physiology of Equilibrium 305
Static Equilibrium 305
Dynamic Equilibrium 305
Equilibrium Pathways 306
Focus on Homeostasis: The Nervous System 309
Common Disorders 310
Medical Terminology and Conditions 310
Study Outline 311 Self-Quiz 312
Critical Thinking Applications 314
Answers to Figure Questions 314

chapter 13

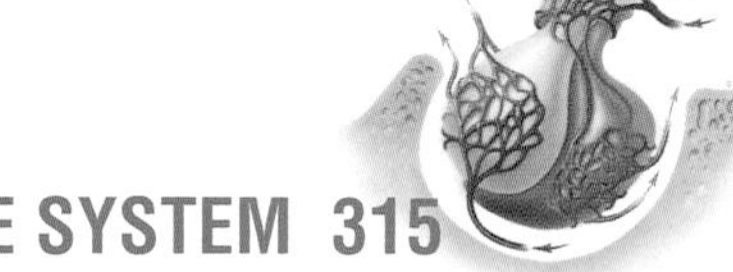

THE ENDOCRINE SYSTEM 315

Introduction 316
Hormone Action 317
Target Cells and Hormone Receptors 317
Chemistry of Hormones 317
Mechanisms of Hormone Action 317
Action of Lipid-soluble Hormones 317
Action of Water-soluble Hormones 318
Control of Hormone Secretions 318
Hypothalamus and Pituitary Gland 319
Anterior Pituitary Hormones 320
Human Growth Hormone and Insulinlike Growth Factors 320
Thyroid-stimulating Hormone 320
Follicle-Stimulating Hormone and Luteinizing Hormone 320
Prolactin 320
Adrenocorticotropic Hormone 320
Melanocyte-stimulating Hormone 321
Posterior Pituitary Hormones 321
Oxytocin 322
Antidiuretic Hormone 322
Thyroid Gland 323
Actions of Thyroid Hormones 323
Control of Thyroid Hormone Secretion 325
Calcitonin 325
Parathyroid Glands 325
Pancreatic Islets 327
Actions of Glucagon and Insulin 327
Adrenal Glands 329
Adrenal Cortex Hormones 329
Mineralocorticoids 331
Glucocorticoids 331
Focus on Wellness:
Insulin Resistance—a Metabolic Medley 332
Androgens 332
Adrenal Medulla Hormones 333
Ovaries and Testes 333
Pineal Gland 333
Other Hormones 334
Hormones from Other Endocrine Cells 334

Prostaglandins and Leukotrienes 334
The Stress Response 335
Aging and the Endocrine System 336
Focus on Homeostasis: The Endocrine System 337
Common Disorders 338
Medical Terminology and Conditions 340
Study Outline 340 *Self-Quiz* 342
Critical Thinking Applications 343
Answers to Figure Questions 344

chapter 14

THE CARDIOVASCULAR SYSTEM: BLOOD 345

Functions of Blood 346
Components of Whole Blood 346
Blood Plasma 346
Formed Elements 346
Formation of Blood Cells 346
Red Blood Cells 349
White Blood Cells 353
Platelets 353
Hemostasis 354
Vascular Spasm 354
Platelet Plug Formation 354
Blood Clotting 354
Clot Retraction and Blood Vessel Repair 356
Hemostatic Control Mechanisms 357
Clotting in Blood Vessels 357
Blood Groups and Blood Types 357
ABO Blood Group 357
Rh Blood Group 358
Transfusions 358
Focus on Wellness: Lifestyle and Blood Circulation—Let It Flow 359
Common Disorders 360
Medical Terminology and Conditions 361
Study Outline 361 *Self-Quiz* 362
Critical Thinking Applications 363
Answers to Figure Questions 363

chapter 15

THE CARDIOVASCULAR SYSTEM: HEART 364

Structure and Organization of the Heart 365
Location and Coverings of the Heart 365
Heart Wall 366
Chambers of the Heart 367
Great Vessels of the Heart 369
Valves of the Heart 369
Blood Flow and Blood Supply of the Heart 371
Blood Flow Through the Heart 371
Blood Supply of the Heart 372
Conduction System of the Heart 372
Electrocardiogram 374
The Cardiac Cycle 374
Heart Sounds 375
Cardiac Output 375
Regulation of Stroke Volume 375
Focus on Wellness: Sudden Cardiac Death During Exercise—What's the Risk? 376
Regulation of Heart Rate 377
Autonomic Regulation of Heart Rate 377
Chemical Regulation of Heart Rate 378
Other Factors in Heart Rate Regulation 378
Exercise and the Heart 378
Common Disorders 379
Medical Terminology and Conditions 381
Study Outline 381 *Self-Quiz* 382
Critical Thinking Applications 384
Answers to Figure Questions 384

chapter 16

THE CARDIOVASCULAR SYSTEM: BLOOD VESSELS AND CIRCULATION 385

Blood Vessel Structure and Function 386
Arteries and Arterioles 386
Capillaries 386
Structure of Capillaries 387
Capillary Exchange 387
Venules and Veins 389
Structure of Venules and Veins 389
Venous Return 390
Blood Flow Through Blood Vessels 390
Blood Pressure 390
Resistance 391
Regulation of Blood Pressure and Blood Flow 391
Role of the Cardiovascular Center 391
Hormonal Regulation of Blood Pressure and Blood Flow 393
Checking Circulation 394
Pulse 394
Measurement of Blood Pressure 394
Circulatory Routes 394
Systemic Circulation 394
Focus on Wellness: Arterial Health—Undoing the Damage of Atherosclerosis 410
Pulmonary Circulation 410
Hepatic Portal Circulation 410
Fetal Circulation 412
Aging and the Cardiovascular System 412
Focus on Homeostasis: The Cardiovascular System 414

Common Disorders 415
Medical Terminology and Conditions 415
Study Outline 416 Self-Quiz 417
Critical Thinking Applications 419
Answers to Figure Questions 419

chapter 17
THE LYMPHATIC SYSTEM AND IMMUNITY 420

Lymphatic System 421
Lymphatic Vessels and Lymph Circulation 421
Lymphatic Organs and Tissues 424
Thymus 424
Lymph Nodes 424
Spleen 425
Lymphatic Nodules 425
Innate Immunity 425
First Line of Defense: Skin and Mucous Membranes 425
Second Line of Defense: Internal Defenses 426
Internal Antimicrobial Proteins 426
Phagocytes and Natural Killer Cells 426
Inflammation 426
Fever 427
Adaptive Immunity 428
Maturation of T Cells and B Cells 429
Types of Adaptive Immune Responses 429
Antigens and Antibodies 429
Processing and Presenting Antigens 430
Focus on Wellness: Lifestyle, Immune Function, and Resistance to Disease 431
T Cells and Cell-mediated Immunity 432
B Cells and Antibody-mediated Immunity 435
Immunological Memory 436
Primary and Secondary Response 436
Naturally Acquired and Artificially Acquired Immunity 437
Aging and the Immune System 437
Focus on Homeostasis: The Lymphatic and Immune System 438
Common Disorders 439
Medical Terminology and Conditions 441
Study Outline 441 Self-Quiz 443
Critical Thinking Applications 444
Answers to Figure Questions 444

chapter 18
THE RESPIRATORY SYSTEM 445

Organs of the Respiratory System 446
Nose 447
Pharynx 448
Larynx 448
The Structures of Voice Production 448
Trachea 450
Bronchi and Bronchioles 451
Lungs 451
Alveoli 451
Pulmonary Ventilation 453
Muscles of Inhalation and Exhalation 453
Pressure Changes During Ventilation 455
Lung Volumes and Capacities 456
Breathing Patterns and Modified Respiratory Movements 457
Exchange of Oxygen and Carbon Dioxide 457
External Respiration: Pulmonary Gas Exchange 458
Internal Respiration: Systemic Gas Exchange 459
Transport of Respiratory Gases 459
Oxygen Transport 459
Carbon Dioxide Transport 461
Control of Respiration 461
Respiratory Center 462
Regulation of the Respiratory Center 462
Cortical Influences on Respiration 462

Focus on Wellness: Smoking—a Breathtaking Experience 463
Chemoreceptor Regulation of Respiration 463
Other Influences on Respiration 464
Exercise and the Respiratory System 465
Aging and the Respiratory System 465
Focus on Homeostasis: The Respiratory System 466
Common Disorders 467
Medical Terminology and Conditions 468
Study Outline 468 Self-Quiz 469
Critical Thinking Applications 471
Answers to Figure Questions 471

chapter 19
THE DIGESTIVE SYSTEM 472

Overview of the Digestive System 473
Layers of the GI Tract and the Omentum 474
Mouth 476
Tongue 477
Salivary Glands 477
Teeth 477
Digestion in the Mouth 478
Pharynx and Esophagus 478
Stomach 480
Structure of the Stomach 481
Digestion and Absorption in the Stomach 482
Pancreas 483
Structure of the Pancreas 483
Pancreatic Juice 484
Liver and Gallbladder 484
Structure of the Liver and Gallbladder 484
Bile 484

Functions of the Liver 485

Small Intestine 486

Structure of the Small Intestine 486
Intestinal Juice 488
Mechanical Digestion in the Small Intestine 488
Chemical Digestion in the Small Intestine 488
Absorption in the Small Intestine 488

Absorption of Monosaccharides 488
Absorption of Amino Acids 489
Absorption of Ions and Water 489
Absorption of Lipids and Bile Salts 489
Absorption of Vitamins 491

Large Intestine 491

Structure of the Large Intestine 491
Digestion and Absorption in the Large Intestine 493
The Defecation Reflex 493

Phases of Digestion 493
Cephalic Phase 493

Focus on Wellness:
Emotional Eating—Consumed by Food 494

Gastric Phase 494

Intestinal Phase 494

Aging and the Digestive System 495
Focus on Homeostasis: The Digestive System 496
Common Disorders 497
Medical Terminology and Conditions 498

Study Outline 498 Self-Quiz 500
Critical Thinking Applications 501
Answers to Figure Questions 502

chapter 20

NUTRITION AND METABOLISM 503

Nutrients 504

Guidelines for Healthy Eating 504
Minerals 504
Vitamins 505

Metabolism 507

Carbohydrate Metabolism 510

Glucose Catabolism 510
Glucose Anabolism 511

Lipid Metabolism 512

Lipid Catabolism 512
Lipid Anabolism 512
Lipid Transport in Blood 512

Protein Metabolism 514

Protein Catabolism 514
Protein Anabolism 514

Metabolism and Body Heat 515

Measuring Heat 515
Body Temperature Homeostasis 515

Body Heat Production 515
Body Heat Loss 516

Regulation of Body Temperature 516

Focus on Wellness:
Exercise Training—Metabolic Workout 518

Common Disorders 519
Medical Terminology and Conditions 519

Study Outline 519 Self-Quiz 521
Critical Thinking Applications 522
Answers to Figure Questions 522

chapter 21

THE URINARY SYSTEM 523

Overview of the Urinary System 524
Structure of the Kidneys 525

External Anatomy of the Kidneys 525
Internal Anatomy of the Kidneys 526
Renal Blood Supply 526
Nephrons 526

Functions of the Nephron 528

Glomerular Filtration 530

Net Filtration Pressure 530
Glomerular Filtration Rate 531

Tubular Reabsorption and Secretion 531

Hormonal Regulation of Nephron Functions 533

Components of Urine 533

Focus on Wellness:
Infection Prevention for Recurrent UTIs 535

Transportation, Storage, and Elimination of Urine 535

Ureters 535
Urinary Bladder 536
Urethra 536
Micturition 537

Aging and the Urinary System 537

Focus on Homeostasis: The Urinary System 538
Common Disorders 539
Medical Terminology and Conditions 539

Study Outline 540 Self-Quiz 541
Critical Thinking Applications 542
Answers to Figure Questions 542

chapter 22

FLUID, ELECTROLYTE, AND ACID-BASE BALANCE 543

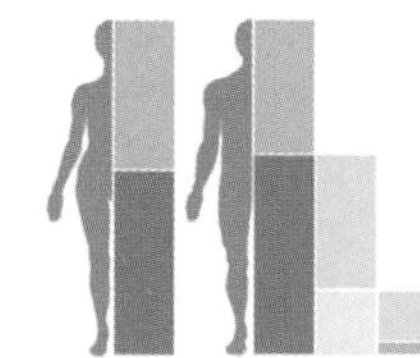
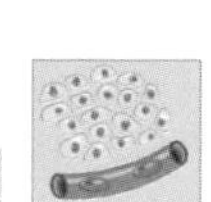

Fluid Compartments and Fluid Balance 544

Sources of Body Water Gain and Loss 545
Regulation of Body Water Gain 545
Regulation of Water and Solute Loss 545
Movement of Water Between Fluid Compartments 547

Electrolytes in Body Fluids 548
Acid-Base Balance 550
The Actions of Buffer Systems 550
Protein Buffer System 550
Carbonic Acid-Bicarbonate Buffer System 550
Phosphate Buffer System 550

Focus on Wellness: Prolonged Physical Activity—a Challenge to Fluid and Electrolyte Balance 551
Exhalation of Carbon Dioxide 551
Kidney Excretion of H^+ 552
Acid-Base Imbalances 552
Aging and Fluid, Electrolyte, and Acid-Base Balance 553
Study Outline 553 Self-Quiz 554
Critical Thinking Applications 555
Answers to Figure Questions 555

chapter 23
THE REPRODUCTIVE SYSTEMS 556
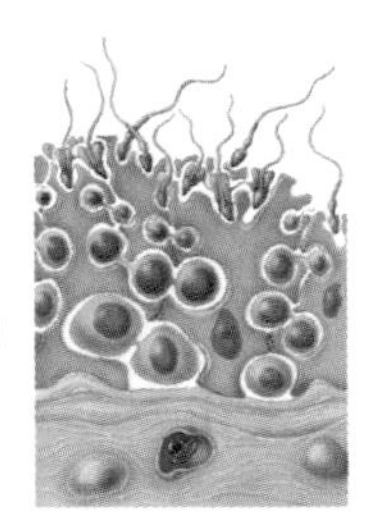
Male Reproductive System 557
Scrotum 557
Testes 558
Spermatogenesis 558
Sperm 560
Hormonal Control of the Testes 561
Ducts 562
Epididymis 562
Ductus Deferens 562
Ejaculatory Ducts 562
Urethra 562
Accessory Sex Glands 562
Semen 563
Penis 563
Female Reproductive System 564
Ovaries 564
Oogenesis 565
Uterine Tubes 566
Uterus 566
Vagina 568
Perineum and Vulva 568
Mammary Glands 569
The Female Reproductive Cycle 570
Hormonal Regulation of the Female Reproductive Cycle 570
Phases of the Female Reproductive Cycle 572
Menstrual Phase 572
Preovulatory Phase 572
Ovulation 572
Postovulatory Phase 572
Birth Control Methods and Abortion 574
Surgical Sterilization 574
Hormonal Methods 574
Intrauterine Devices 575
Spermicides 575
Barrier Methods 575
Periodic Abstinence 575
Focus on Wellness: The Female Athlete Triad—Disordered Eating, Amenorrhea, and Premature Osteoporosis 576
Abortion 577
Aging and the Reproductive Systems 577
Focus on Homeostasis: The Reproductive Systems 578
Common Disorders 579
Medical Terminology and Conditions 581
Study Outline 581 Self-Quiz 583
Critical Thinking Applications 584
Answers to Figure Questions 585

chapter 24
DEVELOPMENT AND INHERITANCE 586
Embryonic Period 587
First Week of Development 587
Fertilization 587
Early Embryonic Development 588
Second Week of Development 590
Third Week of Development 592
Gastrulation 592
Development of the Allantois, Chorionic Villi, and Placenta 592
Fourth Through Eighth Weeks of Development 594
Fetal Period 596
Maternal Changes During Pregnancy 600
Hormones of Pregnancy 600
Changes during Pregnancy 600
Exercise and Pregnancy 601
Labor and Delivery 601
Lactation 602

Focus on Wellness: Breast Milk—Mother Nature's Approach to Infection Prevention 603
Inheritance 604
Genotype and Phenotype 604
Autosomes and Sex Chromosomes 605
Common Disorders 607
Medical Terminology and Conditions 607
Study Outline 608 Self-Quiz 610
Critical Thinking Applications 611
Answers to Figure Questions 611

Answers to Self-Quizzes and Critical Thinking Applications A-1
Glossary G-1
Credits C-1
Index I-1

Introduction to the Human Body

ORGANIZATION OF THE HUMAN BODY

chapter 1

did you know?

The body's ability to maintain homeostasis gives it tremendous healing power and a remarkable resistance to abuse. The physiological processes responsible for maintaining homeostasis are in large part also responsible for your good health. For most people, lifelong good health is not something that just happens. Two of the many factors in this balance called health are the environment and your own behavior. Your body's homeostasis is affected by the air you breathe, the food you eat, and even by the thoughts you think. The way you live your life can either support or interfere with your body's ability to maintain homeostasis and good health.

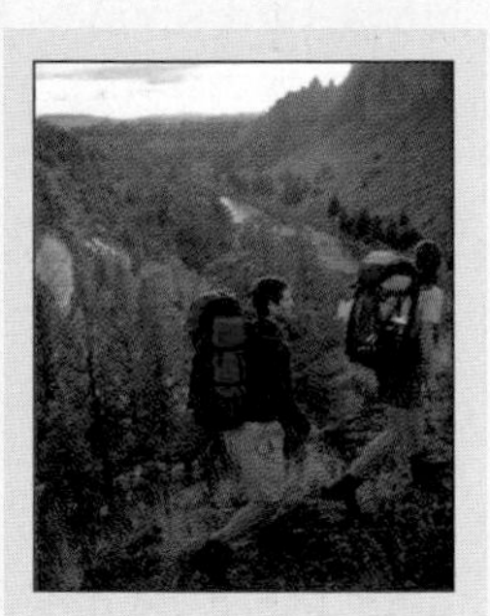

Focus on Wellness, page 10

www.wiley.com/college/apcentral

You are beginning a fascinating exploration of the human body in which you'll learn how it is organized and how it functions. First you will be introduced to the scientific disciplines of anatomy and physiology; we'll consider the levels of organization that characterize living things and the properties that all living things share. Then, we will examine how the body is constantly regulating its internal environment. This ceaseless process, called homeostasis, is a major theme in every chapter of this book. We will also discuss how the various individual systems that compose the human body cooperate with one another to maintain the health of the body as a whole. Finally, we will establish a basic vocabulary that allows us to speak about the body in a way that is understood by scientists and health-care professionals alike.

ANATOMY AND PHYSIOLOGY DEFINED

OBJECTIVE • **Define anatomy and physiology.**

The sciences of anatomy and physiology are the foundation for understanding the structures and functions of the human body. ***Anatomy*** (a-NAT-ō-mē; *ana-* = up; *-tomy* = process of cutting) is the science of *structure* and the relationships among structures. ***Physiology*** (fiz′-ē-OL-ō-jē; *physio-* = nature, *-logy* = study of) is the science of body *functions*, that is, how the body parts work. Because function can never be separated completely from structure, we can understand the human body best by studying anatomy and physiology together. We will look at how each structure of the body is designed to carry out a particular function and how the structure of a part often determines the functions it can perform. The bones of the skull, for example, are tightly joined to form a rigid case that protects the brain. The bones of the fingers, by contrast, are more loosely joined, which enables them to perform a variety of movements, such as turning the pages of this book.

■ **CHECKPOINT**

1. What is the basic difference between anatomy and physiology?
2. Give your own example of how the structure of a part of the body is related to its function.

LEVELS OF ORGANIZATION AND BODY SYSTEMS

OBJECTIVES • **Describe the structural organization of the human body.**

• **Define the body systems and explain how they relate to one another.**

The structures of the human body are organized on several levels, similar to the way letters of the alphabet, words, sentences, and paragraphs make up language. Listed here, from smallest to largest, are the six levels of organization of the human body: chemical, cellular, tissue, organ, system, and organismal (Figure 1.1).

1. The ***chemical level***, which can be compared to letters of the alphabet, includes ***atoms***, the smallest units of matter that participate in chemical reactions, and ***molecules***, two or more atoms joined together. Certain atoms, such as carbon (C), hydrogen (H), oxygen (O), nitrogen (N), calcium (Ca), and others, are essential for maintaining life. Familiar examples of molecules found in the body are DNA (deoxyribonucleic acid), the genetic material passed on from one generation to another; hemoglobin, which carries oxygen in the blood; glucose, commonly known as blood sugar; and vitamins, which are needed for a variety of chemical processes. Chapters 2 and 20 focus on the chemical level of organization.
2. Molecules combine to form structures at the next level of organization—the ***cellular level. Cells*** are the basic structural and functional units of an organism. Just as words are the smallest elements of language, cells are the smallest living units in the human body. Among the many types of cells in your body are muscle cells, nerve cells, and blood cells. Figure 1.1 shows a smooth muscle cell, one of three different kinds of muscle cells in your body. As you will see in Chapter 3, cells contain specialized structures called *organelles*, such as the nucleus, mitochondria, and lysosomes, that perform specific functions.
3. The ***tissue level*** is the next level of structural organization. ***Tissues*** are groups of cells and the materials surrounding them that work together to perform a particular function, similar to the way words are put together to form sentences. The four basic types of tissue in your body are *epithelial tissue*, *connective tissue*, *muscular tissue*, and *nervous tissue*. The similarities and differences among the different types of tissues are the focus of Chapter 4. Note in Figure 1.1 that smooth muscle tissue consists of tightly packed smooth muscle cells.
4. At the ***organ level***, different kinds of tissues join together to form body structures. Similar to the relationship between sentences and paragraphs, ***organs*** usually have a recognizable shape, are composed of two or more different types of tissues, and have specific functions. Examples of organs are the stomach, heart, liver, lungs, and brain. Figure 1.1 shows several tissues that make up the stomach. The *serous membrane* is a layer around the outside of the stomach that protects it and reduces friction when the stomach moves and rubs against other organs. Underneath the serous membrane are the *smooth muscle tissue layers*, which contract to churn and mix food and push it on to the next digestive organ, the small intestine. The innermost lining of the stomach is an *epithelial tissue layer*, which contributes fluid and chemicals that aid digestion.
5. The next level of structural organization in the body is the ***system level***. A ***system*** (or chapter in our analogy) consists of related organs (paragraphs) that have a common function. The example shown in Figure 1.1 is the digestive system, which breaks down and absorbs molecules in food. In the chapters that follow, we will explore the anatomy and physiology of each of the body systems. Table 1.1 on pages 4–5 introduces the

Figure 1.1 Levels of structural organization in the human body.

The levels of structural organization are the chemical, cellular, tissue, organ, system, and organismal.

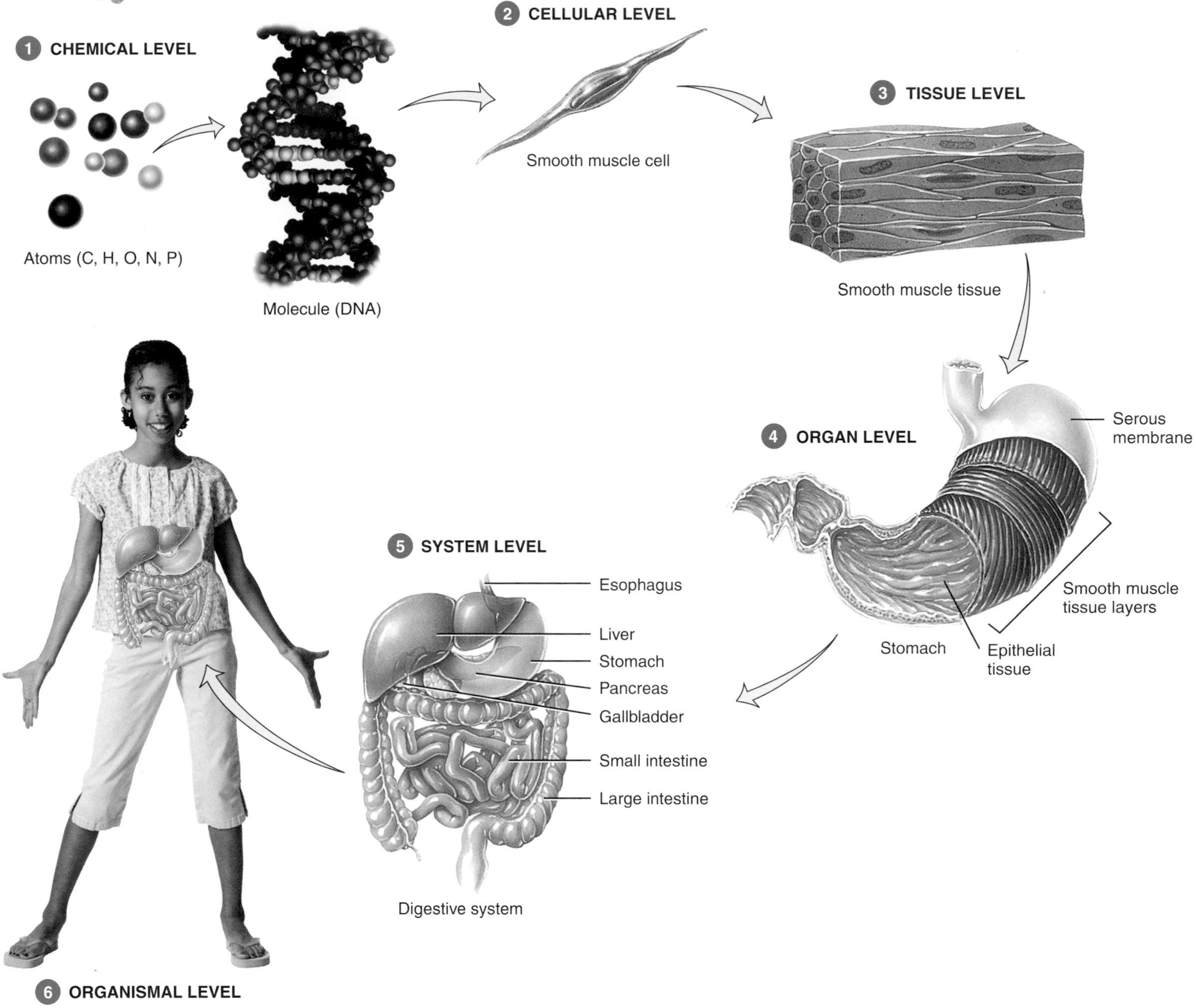

1 CHEMICAL LEVEL

Atoms (C, H, O, N, P)

Molecule (DNA)

2 CELLULAR LEVEL

Smooth muscle cell

3 TISSUE LEVEL

Smooth muscle tissue

4 ORGAN LEVEL

Serous membrane

Smooth muscle tissue layers

Stomach

Epithelial tissue

5 SYSTEM LEVEL

Esophagus

Liver

Stomach

Pancreas

Gallbladder

Small intestine

Large intestine

Digestive system

6 ORGANISMAL LEVEL

Which level of structural organization usually has a recognizable shape and is composed of two or more different types of tissues that have a specific function?

components and functions of these systems. As you study the body systems, you will discover how they work together to maintain health, protect you from disease, and allow for reproduction of the species.

6 The ***organismal level*** is the largest level of organization. All the systems of the body combine to make up an ***organism***, that is, one human being. An organism can be compared to a book in our analogy.

Table 1.1 Components and Functions of the Eleven Principal Systems of the Human Body

1. Integumentary System (Chapter 5)

Components: Skin and structures derived from it, such as hair, nails, and sweat and oil glands.

Functions: Helps regulate body temperature; protects the body; eliminates some wastes; helps make vitamin D; detects sensations such as touch, pressure, pain, warmth, and cold.

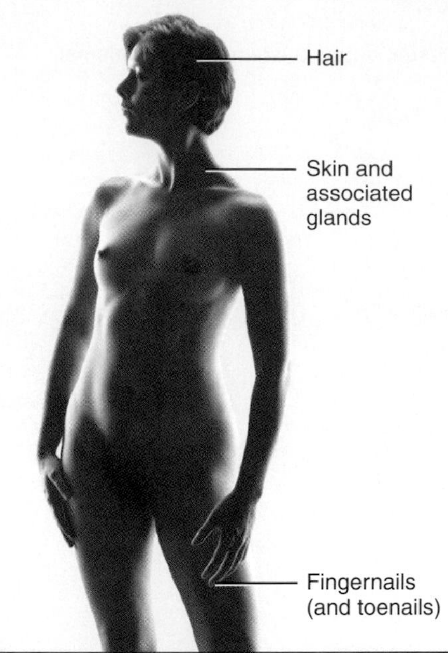

2. Skeletal System (Chapters 6 and 7)

Components: All the bones and joints of the body and their associated cartilages.

Functions: Supports and protects the body, provides a specific area for muscle attachment, assists with body movements, stores cells that produce blood cells, and stores minerals and lipids (fats).

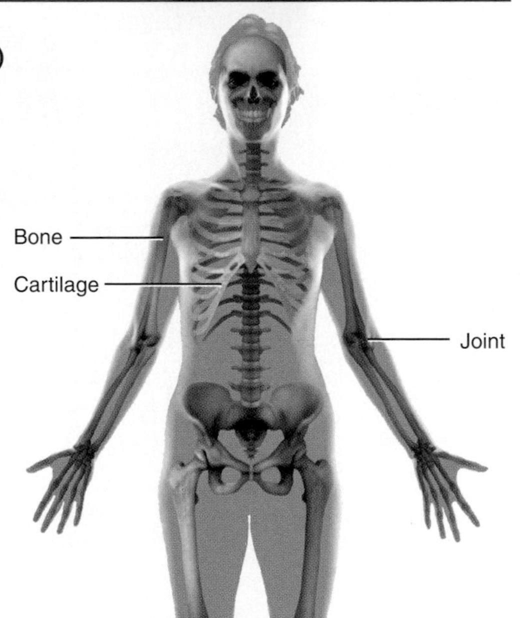

3. Muscular System (Chapter 8)

Components: Specifically refers to skeletal muscle tissue, which is muscle usually attached to bones (other muscle tissues include smooth and cardiac).

Functions: Participates in bringing about body movements, maintains posture, and produces heat.

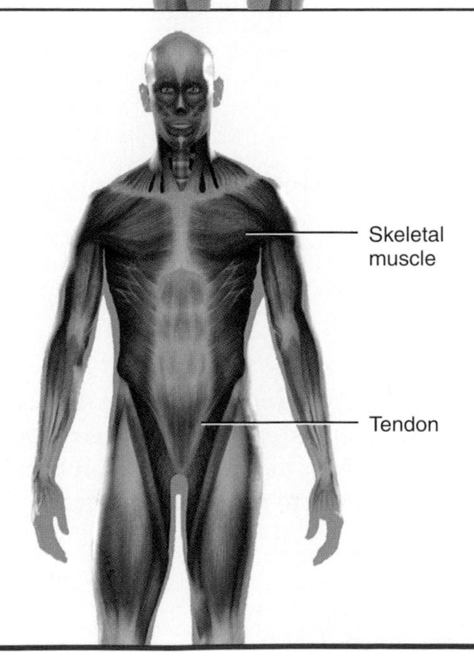

4. Nervous System (Chapters 9–12)

Components: Brain, spinal cord, nerves, and sense organs such as the eyes and ears.

Functions: Regulates body activities through nerve impulses by detecting changes in the environment, interpreting the changes, and responding to the changes by bringing about muscular contractions or glandular secretions.

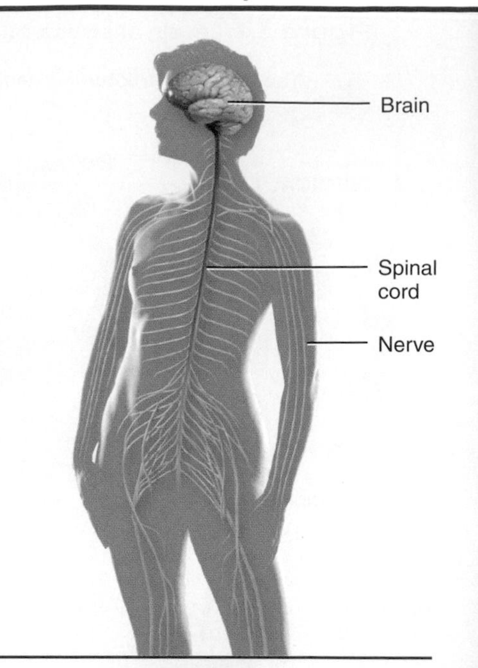

5. Endocrine System (Chapter 13)

Components: All glands and tissues that produce chemical regulators of body functions, called hormones.

Functions: Regulates body activities through hormones transported by the blood to various target organs.

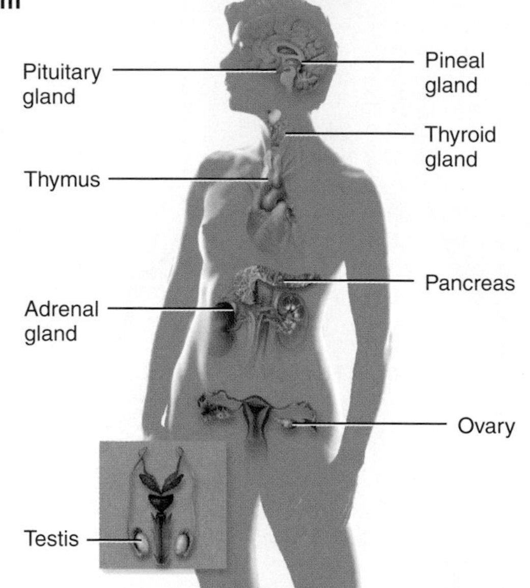

6. Cardiovascular System (Chapters 14–16)

Components: Blood, heart, and blood vessels.

Functions: Heart pumps blood through blood vessels; blood carries oxygen and nutrients to cells and carbon dioxide and wastes away from cells, and helps regulate acidity, temperature, and water content of body fluids; blood components help defend against disease and mend damaged blood vessels.

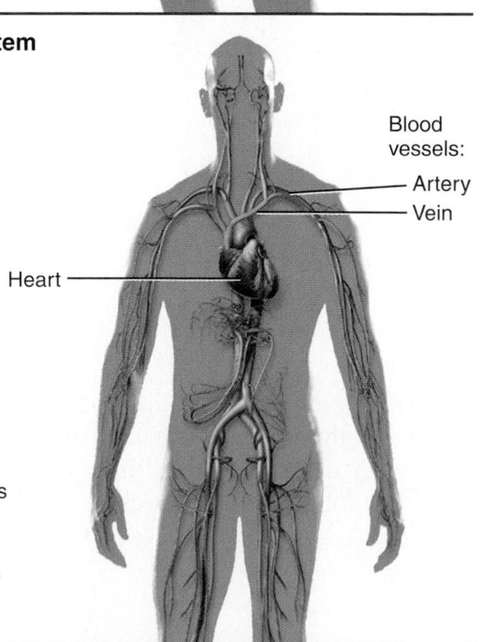

7. Lymphatic System and Immunity (Chapter 17)

Components: Lymphatic fluid and vessels; spleen, thymus, lymph nodes, and tonsils; cells that carry out immune responses (B cells, T cells, and others).

Functions: Returns proteins and fluid to blood; carries lipids from gastrointestinal tract to blood; contains sites of maturation and proliferation of B cells and T cells that protect against disease-causing microbes.

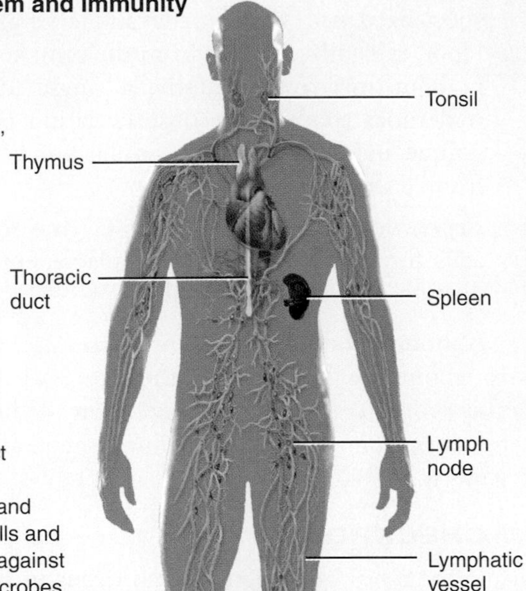

9. Digestive System (Chapter 19)

Components: Organs of gastrointestinal tract, including the mouth, pharynx (throat), esophagus, stomach, small and large intestines, rectum, and anus; also includes accessory digestive organs that assist in digestive processes, such as the salivary glands, liver, gallbladder, and pancreas.

Functions: Achieves physical and chemical breakdown of food; absorbs nutrients; eliminates solid wastes.

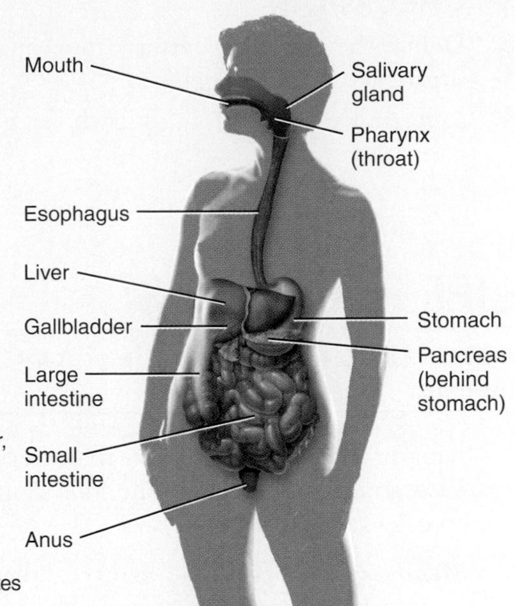

8. Respiratory System (Chapter 18)

Components: Lungs and air passageways such as the pharynx (throat), larynx (voice box), trachea (windpipe), and bronchial tubes leading into and out of them.

Functions: Transfers oxygen from inhaled air to blood and carbon dioxide from blood to exhaled air; helps regulate acidity of body fluids; air flowing out of lungs through vocal cords produces sounds.

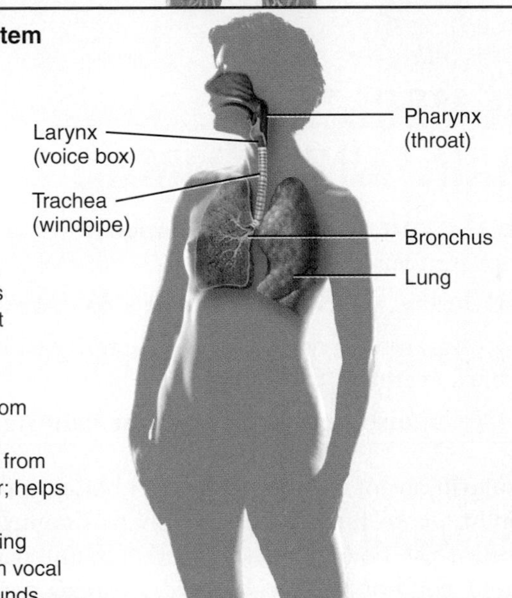

10. Urinary System (Chapter 21)

Components: Kidneys, ureters, urinary bladder, and urethra.

Functions: Produces, stores, and eliminates urine; eliminates wastes and regulates volume and chemical composition of blood; helps regulate acidity of body fluids; maintains body's mineral balance; helps regulate red blood cell production.

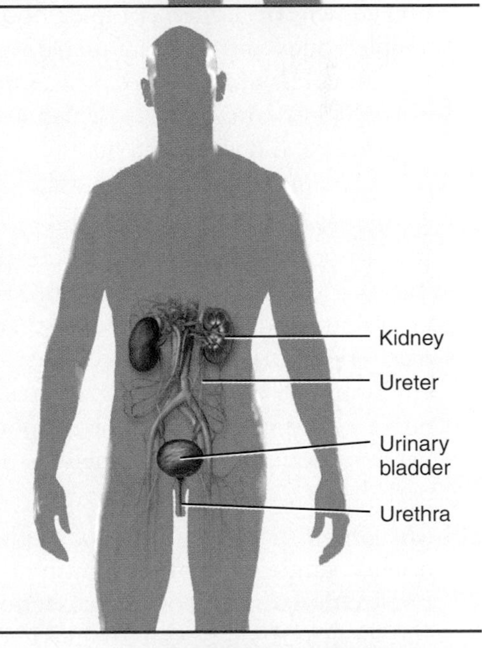

11. Reproductive Systems (Chapter 23)

Components: Gonads (testes or ovaries) and associated organs: uterine tubes, uterus, and vagina in females, and epididymis, ductus (vas) deferens, and penis in males. Also, mammary glands in females.

Functions: Gonads produce gametes (sperm or oocytes) that unite to form a new organism and release hormones that regulate reproduction and other body processes; associated organs transport and store gametes. Mammary glands produce milk.

Mammary gland
Uterine (fallopian) tube
Ovary
Uterus
Vagina

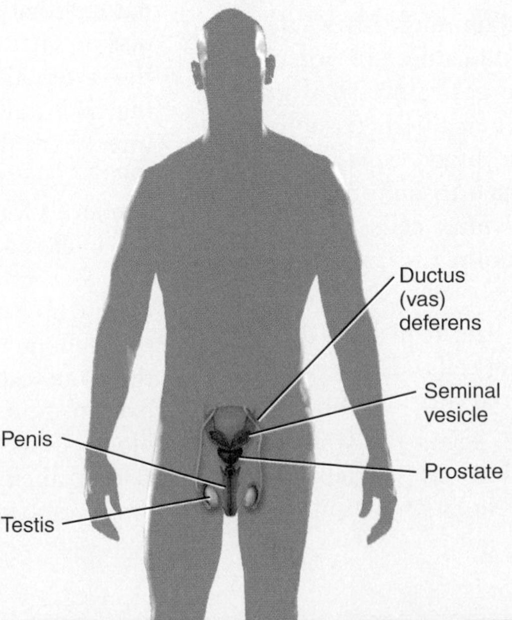

■ **CHECKPOINT**

3. Define the following terms: atom, molecule, cell, tissue, organ, system, and organism.
4. Referring to Table 1.1, which body systems help eliminate wastes?

LIFE PROCESSES

OBJECTIVE • Define the important life processes of humans.

All living organisms have certain characteristics that set them apart from nonliving things. The following are six important life processes of humans:

1. ***Metabolism*** (me-TAB-ō-lizm) is the sum of all the chemical processes that occur in the body. It includes the breakdown of large, complex molecules into smaller, simpler ones and the building up of complex molecules from smaller, simpler ones. For example, proteins in food are split into amino acids, which are the building blocks of proteins. These amino acids can then be used to build new proteins that make up muscles and bones.
2. ***Responsiveness*** is the body's ability to detect and respond to changes in its internal (inside the body) or external (outside the body) environment. Different cells in the body detect different sorts of changes and respond in characteristic ways. Nerve cells respond to changes in the environment by generating electrical signals, known as nerve impulses. Muscle cells respond to nerve impulses by contracting, which generates force to move body parts.
3. ***Movement*** includes motion of the whole body, individual organs, single cells, and even tiny organelles inside cells. For example, the coordinated action of several muscles and bones enables you to move your body from one place to another by walking or running. After you eat a meal that contains fats, your gallbladder (an organ) contracts and squirts bile into the gastrointestinal tract to help in the digestion of fats. When a body tissue is damaged or infected, certain white blood cells move from the blood into the affected tissue to help clean up and repair the area. And inside individual cells, various parts move from one position to another to carry out their functions.
4. ***Growth*** is an increase in body size. It may be due to an increase in (1) the size of existing cells, (2) the number of cells, or (3) the amount of material surrounding cells.
5. ***Differentiation*** (dif′-er-en-shē-Ā-shun) is the process whereby unspecialized cells become specialized cells. Specialized cells differ in structure and function from the unspecialized cells that gave rise to them. For example, specialized red blood cells and several types of white blood cells differentiate from the same unspecialized cells in bone marrow. Similarly, a single fertilized egg cell undergoes tremendous differentiation to develop into a unique individual who is similar to, yet quite different from, either of the parents.
6. ***Reproduction*** refers to either (1) the formation of new cells for growth, repair, or replacement or (2) the production of a new individual.

Although not all of these processes are occurring in cells throughout the body all of the time, when they cease to occur properly cell death may occur. When cell death is extensive and leads to organ failure, the result is death of the organism.

■ **CHECKPOINT**

5. What types of movement can occur in the human body?

HOMEOSTASIS: MAINTAINING LIMITS

OBJECTIVES • Define homeostasis and explain its importance.

- **Describe the components of a feedback system.**
- **Compare the operation of negative and positive feedback systems.**
- **Distinguish between symptoms and signs of a disease.**

The trillions of cells of the human body need relatively stable conditions to function effectively and contribute to the survival of the body as a whole. The maintenance of relatively stable conditions is called ***homeostasis*** (hō′-mē-ō-STĀ-sis; *homeo-* = sameness; *-stasis* = standing still). Homeostasis ensures that the body's internal environment remains steady despite changes inside and outside the body. A large part of the internal environment consists of the fluid surrounding body cells, called *interstitial fluid.* Homeostasis keeps the interstitial fluid at a proper temperature of 37° Celsius (98.6° Fahrenheit) and maintains adequate nutrient and oxygen levels for body cells to flourish.

Each body system contributes to homeostasis in some way. For instance, in the cardiovascular system, alternating contraction and relaxation of the heart propels blood throughout the body's blood vessels. As blood flows through the blood capillaries, the smallest blood vessels, nutrients and oxygen move into interstitial fluid and wastes move into the blood. Cells, in turn, remove nutrients and oxygen from and release their wastes into interstitial fluid. Homeostasis is *dynamic*; that is, it can change over a narrow range that is compatible with maintaining cellular life processes. For example, the level of glucose in the blood is maintained within a narrow range. It normally does not fall too low between meals or rise too high even after eating a high-glucose meal. The brain needs a steady supply of glucose to keep functioning—a low blood glucose level may lead to unconsciousness or even death. A prolonged high blood glucose level, by contrast, can damage blood vessels and cause excessive loss of water in the urine.

Control of Homeostasis: Feedback Systems

Fortunately, every body structure, from cells to systems, has one or more homeostatic devices that work to keep the internal environment within normal limits. The homeostatic mechanisms of the body are mainly under the control of two systems, the nervous system and the endocrine system. The nervous system detects changes from the balanced state and sends messages in the form of *nerve impulses* to organs that can counteract the change. For example, when body temperature rises, nerve impulses cause sweat glands to release more sweat, which cools the body as it evaporates. The endocrine system corrects changes by secreting molecules called *hormones* into the blood. Hormones affect specific body cells where they cause responses that restore homeostasis. For example, the hormone insulin reduces blood glucose level when it is too high. Nerve impulses typically cause rapid corrections; whereas hormones usually work more slowly.

Homeostasis is maintained by means of many feedback systems. A ***feedback system*** or *feedback loop* is a cycle of events in which a condition in the body is continually monitored, evaluated, changed, remonitored, reevaluated, and so on. Each monitored condition, such as body temperature, blood pressure, or blood glucose level, is termed a *controlled condition.* Any disruption that causes a change in a controlled condition is called a *stimulus.* Some stimuli come from the external environment, such as intense heat or lack of oxygen. Others originate in the internal environment, such as a blood glucose level that is too low. Homeostatic imbalances may also occur due to psychological stresses in our social environment—the demands of work and school, for example. In most cases, the disruption of homeostasis is mild and temporary, and the responses of body cells quickly restore balance in the internal environment. In other cases, the disruption of homeostasis may be intense and prolonged, as in poisoning, overexposure to temperature extremes, severe infection, or death of a loved one.

Three basic components make up a feedback system: a receptor, a control center, and an effector (Figure 1.2).

Figure 1.2 Parts of a feedback system. The dashed return arrow symbolizes negative feedback.

The three basic elements of a feedback system are the receptor, control center, and effector.

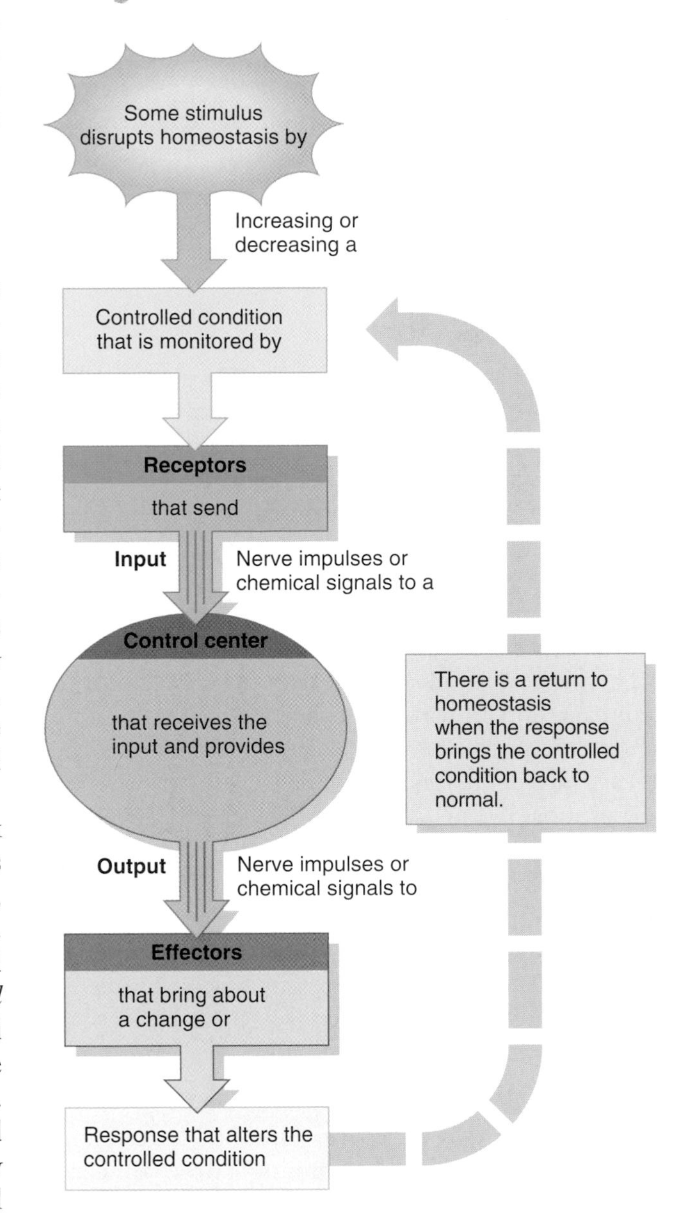

What is the basic difference between negative and positive feedback systems?

1. A ***receptor*** is a body structure that monitors changes in a controlled condition and sends information called the *input* to a control center. Input is in the form of nerve impulses or chemical signals. Nerve endings in the skin that sense temperature are one of the hundreds of different kinds of receptors in the body.
2. A ***control center*** in the body, for example, the brain, sets the range of values within which a controlled condition should be maintained, evaluates the input it receives from receptors, and generates output commands when they are needed. *Output* is information, in the form of nerve

impulses or chemical signals, that is relayed from the control center to an effector.

3. An ***effector*** is a body structure that receives output from the control center and produces a *response* that changes the controlled condition. Nearly every organ or tissue in the body can behave as an effector. For example, when your body temperature drops sharply, your brain (control center) sends nerve impulses to your skeletal muscles (effectors) that cause you to shiver, which generates heat and raises your temperature.

Feedback systems can be classified as either negative feedback systems or positive feedback systems.

Negative Feedback Systems

A ***negative feedback system*** *reverses* a change in a controlled condition. Consider one negative feedback system that helps regulate blood pressure. *Blood pressure (BP)* is the force exerted by blood as it presses against the walls of blood vessels. When the heart beats faster or harder, BP increases. If a stimulus causes blood pressure (controlled condition) to rise, the following sequence of events occurs (Figure 1.3). The higher pressure is detected by *baroreceptors*, pressure-sensitive nerve cells located in the walls of certain blood vessels (the receptors). The baroreceptors send nerve impulses (input) to the brain (control center), which interprets the impulses and responds by sending nerve impulses (output) to the heart (the effector). Heart rate decreases, which causes blood pressure to decrease (response). This sequence of events returns the controlled condition—blood pressure—to normal, and homeostasis is restored. This is a negative feedback system because the activity of the effector produces a result, a drop in blood pressure, that reverses the effect of the stimulus. Negative feedback systems tend to regulate conditions in the body that are held fairly stable over long periods of time, such as blood pressure, blood glucose level, and body temperature.

Positive Feedback Systems

A ***positive feedback system*** *strengthens* a change in a controlled condition. Normal positive feedback systems tend to reinforce conditions that don't happen very often, such as childbirth, ovulation, and blood clotting. Because a positive feedback system continually reinforces a change in a controlled condition, it must be shut off by some event outside the system. If the action of a positive feedback system isn't stopped, it can "run away" and produce life-threatening changes in the body.

Homeostasis and Disease

As long as all of the body's controlled conditions remain within certain narrow limits, body cells function efficiently, homeostasis is maintained, and the body stays healthy.

Figure 1.3 Homeostasis of blood pressure by a negative feedback system. Note that the response is fed back into the system, and the system continues to lower blood pressure until there is a return to normal blood pressure (homeostasis).

If the response reverses a change in a controlled condition, a system is operating by negative feedback.

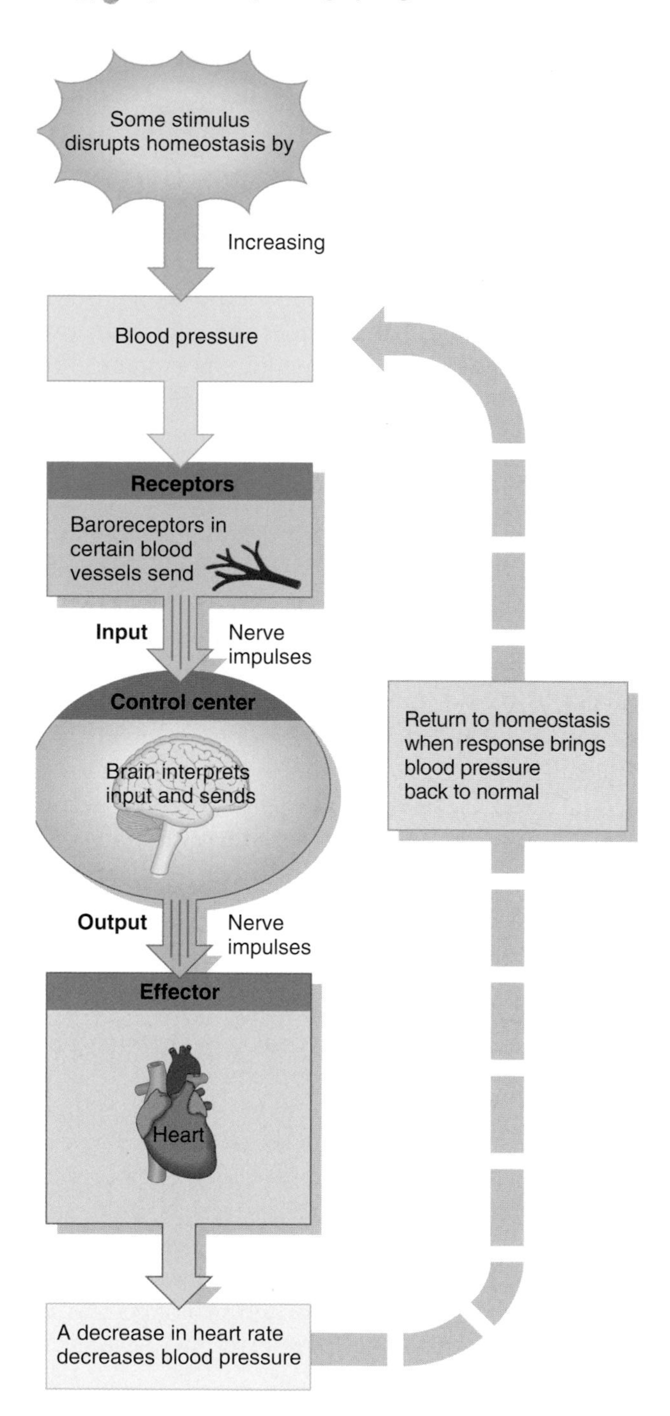

What would happen to the heart rate if some stimulus caused blood pressure to decrease? Would this occur by positive or negative feedback?

Should one or more components of the body lose their ability to contribute to homeostasis, however, the normal balance among all of the body's processes may be disturbed. If the homeostatic imbalance is moderate, a disorder or disease may occur; if it is severe, death may result.

A ***disorder*** is any abnormality of structure and/or function. ***Disease*** is a more specific term for an illness characterized by a recognizable set of symptoms and signs. ***Symptoms*** are *subjective* changes in body functions that are not apparent to an observer, for example, headache or nausea. ***Signs*** are *objective* changes that a clinician can observe and measure, such as bleeding, swelling, vomiting, diarrhea, fever, a rash, or paralysis. Specific diseases alter body structure and function in characteristic ways, usually producing a recognizable set of symptoms and signs.

Diagnosis (dī′-ag-NŌ-sis; *dia-* = through; *-gnosis* = knowledge) is the identification of a disease or disorder based on a scientific evaluation of the patient's symptoms and signs, medical history, physical examination, and sometimes data from laboratory tests. Taking a *medical history* consists of collecting information about events that might be related to a patient's illness, including the chief complaint, history of present illness, past medical problems, family medical problems, and social history. A *physical examination* is an orderly evaluation of the body and its functions. This process includes *inspection* (looking at or into the body with various instruments), *palpation* (feeling body surfaces with the hands), auscultation (listening to body sounds, often using a stethoscope), *percussion* (tapping on body surfaces and listening to the resulting echo), and measuring vital signs (temperature, pulse, respiratory rate, and blood pressure). Some common laboratory tests include analyses of blood and urine.

■ **CHECKPOINT**

6. What types of disturbances can act as stimuli that initiate a feedback system?
7. How are negative and positive feedback systems similar? How are they different?
8. Contrast and give examples of symptoms and signs of a disease.

AGING AND HOMEOSTASIS

OBJECTIVE • **Describe some of the effects of aging.**

As you will see later, ***aging*** is a normal process characterized by a progressive decline in the body's ability to restore homeostasis. Aging produces observable changes in structure and function and increases vulnerability to stress and disease. The changes associated with aging are apparent in all body systems. Examples include wrinkled skin, gray hair, loss of bone mass, decreased muscle mass and strength, diminished reflexes, decreased production of some hormones, increased incidence of heart disease, increased susceptibility to infections and cancer, decreased lung capacity, less efficient functioning of the digestive system, decreased kidney function, menopause, and enlarged prostate. These and other effects of aging will be discussed in detail in later chapters.

■ **CHECKPOINT**

9. What are some of the signs of aging?

ANATOMICAL TERMS

OBJECTIVES • **Describe the anatomical position.**

- **Identify the major regions of the body and relate the common names to the corresponding anatomical terms for various parts of the body.**
- **Define the directional terms and the anatomical planes and sections used to locate parts of the human body.**

The language of anatomy and physiology is very precise. When describing where the wrist is located, is it correct to say "the wrist is above the fingers"? This description is true if your arms are at your sides. But if you hold your hands up above your head, your fingers would be above your wrists. To prevent this kind of confusion, scientists and health-care professionals refer to one standard anatomical position and use a special vocabulary for relating body parts to one another.

In the study of anatomy, descriptions of any part of the human body assume that the body is in a specific stance called the ***anatomical position***. In the anatomical position, the subject stands erect facing the observer, with the head level and the eyes facing forward. The feet are flat on the floor and directed forward, and the arms are at the sides with the palms turned forward (Figure 1.4 on page 11).

Names of Body Regions

The human body is divided into several major regions that can be identified externally. These are the head, neck, trunk, upper limbs, and lower limbs (Figure 1.4). The ***head*** consists of the skull and face. The *skull* is the part of the head that encloses and protects the brain, and the *face* is the front portion of the head that includes the eyes, nose, mouth, forehead, cheeks, and chin. The ***neck*** supports the head and attaches it to the trunk. The ***trunk*** consists of the chest, abdomen, and pelvis. Each ***upper limb*** is attached to the trunk and consists

Focus on Wellness

Good Health—Homeostasis Is the Basis

You've seen *homeostasis* defined as a condition in which the body's internal environment remains relatively stable. What does this mean to you in your everyday life?

Homeostasis: The Power to Heal

The body's ability to maintain homeostasis gives it tremendous healing power and a remarkable resistance to abuse. The physiological processes responsible for maintaining homeostasis are in large part also responsible for your good health.

For most people, lifelong good health is not something that just happens. Two of the many factors in this balance called health are the environment and your own behavior. Also important is your genetic makeup. Your body's homeostasis is affected by the air you breathe, the food you eat, and even the thoughts you think. The way you live your life can either support or interfere with your body's ability to maintain homeostasis and recover from the inevitable stresses life throws your way.

Let's consider the common cold. You support your natural healing processes when you take care of yourself. Plenty of rest, fluids, and chicken soup allow the immune system to do its job. The cold runs its course, and you are soon back on your feet. If, instead of taking care of yourself, you continue to smoke two packs of cigarettes a day, skip meals, and pull several all nighters studying for an anatomy and physiology exam, you interfere with the immune system's ability to fend off attacking microbes and bring the body back to homeostasis and good health. Other infections take advantage of your weakened state, and pretty soon the cold has "turned into" bronchitis or pneumonia.

Homeostasis and Disease Prevention

Many diseases are the result of years of poor health behavior that interferes with the body's natural drive to maintain homeostasis. An obvious example is smoking-related illness. Smoking tobacco exposes sensitive lung tissue to a multitude of chemicals that cause cancer and damage the lung's ability to repair itself. Because diseases such as emphysema and lung cancer are difficult to treat and very rarely cured, it is much wiser to quit smoking—or never start—than to hope a doctor can fix you once you are diagnosed with a lung disease. Developing a lifestyle that works with, rather than against, your body's homeostatic processes helps you maximize your personal potential for optimal health and well-being.

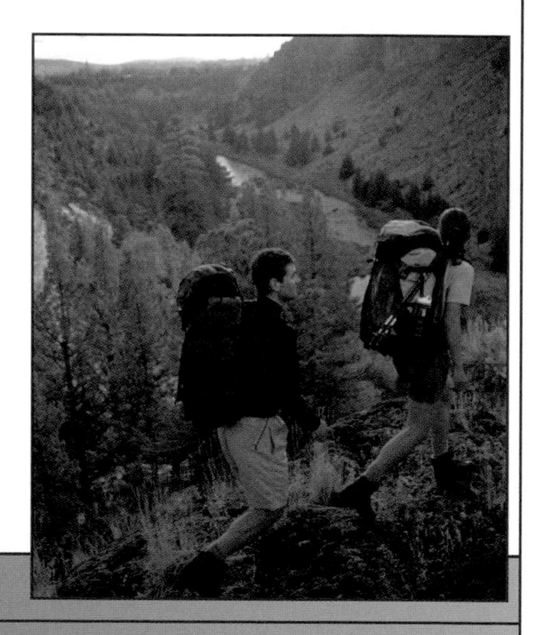

► **Think It Over . . .**

► ***What health habits have you developed over the past several years to prevent disease or enhance your body's ability to maintain health and homeostasis?***

of the shoulder, armpit, arm (portion of the limb from the shoulder to the elbow), forearm (portion of the limb from the elbow to the wrist), wrist, and hand. Each ***lower limb*** is also attached to the trunk and consists of the buttock, thigh (portion of the limb from the hip to the knee), leg (portion of the limb from the knee to the ankle), ankle, and foot. The *groin* is the area on the front surface of the body, marked by a crease on each side, where the trunk attaches to the thighs.

In Figure 1.4, the corresponding anatomical adjective for each part of the body appears in parentheses next to the common name. For example, if you receive a tetanus shot in your *buttock*, it is a *gluteal* injection. The descriptive form of a body part is based on a Greek or Latin word or "root" for the same part or area. The Latin word for armpit is *axilla* (ak-SIL-a), for example, and thus one of the nerves passing within the armpit is named the axillary nerve. You will learn more about the word roots of anatomical and physiological terms as you read this book.

Directional Terms

To locate various body structures, anatomists use specific ***directional terms***, words that describe the position of one body part relative to another. Several directional terms can be grouped in pairs that have opposite meanings, for example, anterior (front) and posterior (back). Study Exhibit 1.1 on page 12 and Figure 1.5 on page 13 to determine, among other things, whether your stomach is superior to your lungs.

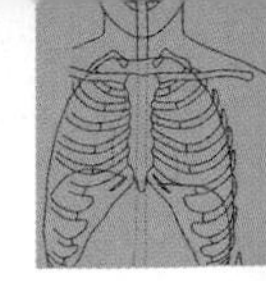

Figure 1.4 The anatomical position. The common names and corresponding anatomical terms (in parentheses) indicate specific body regions. For example, the head is the cephalic region.

In the anatomical position, the subject stands erect facing the observer, with the head level and the eyes facing forward. The feet are flat on the floor and directed forward, and the arms are at the sides with the palms facing forward.

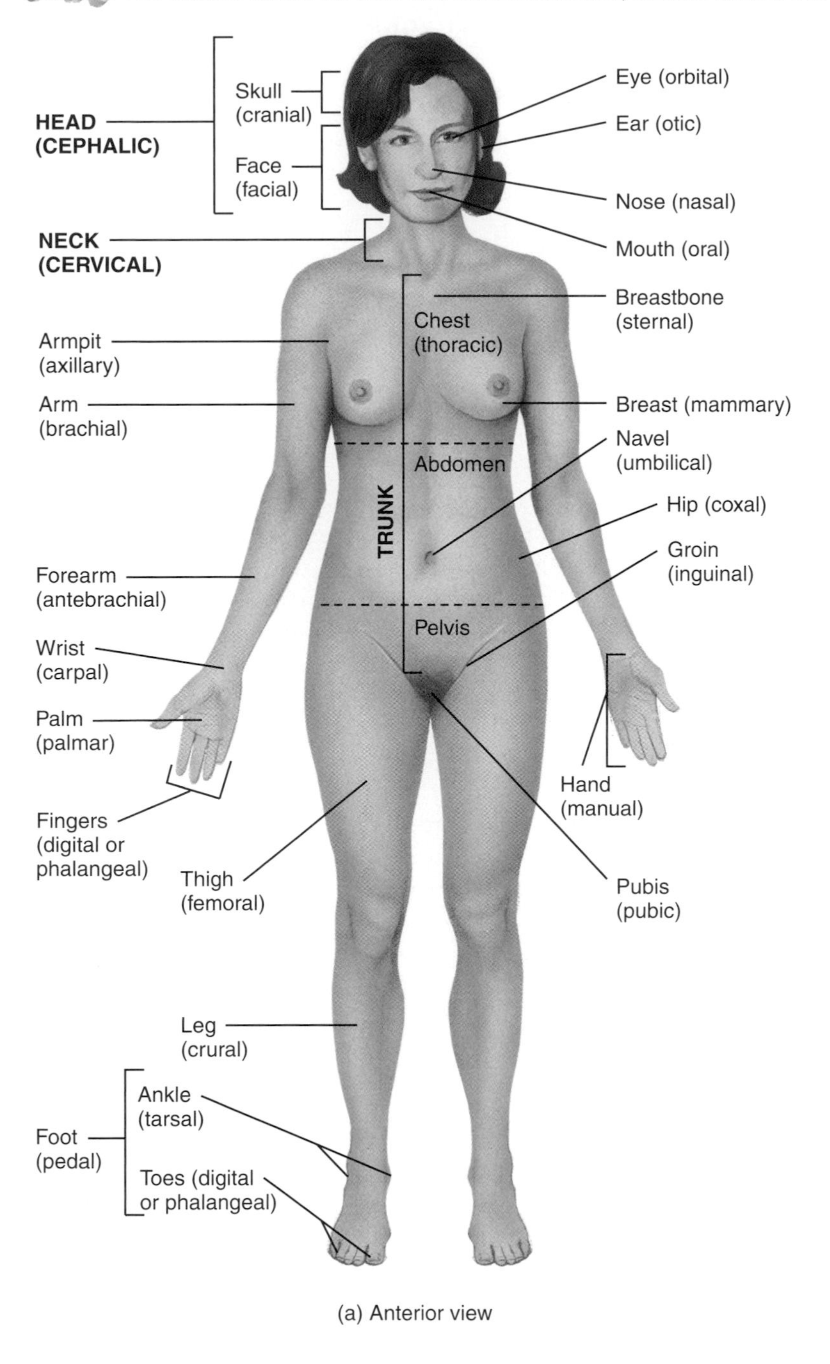

(a) Anterior view

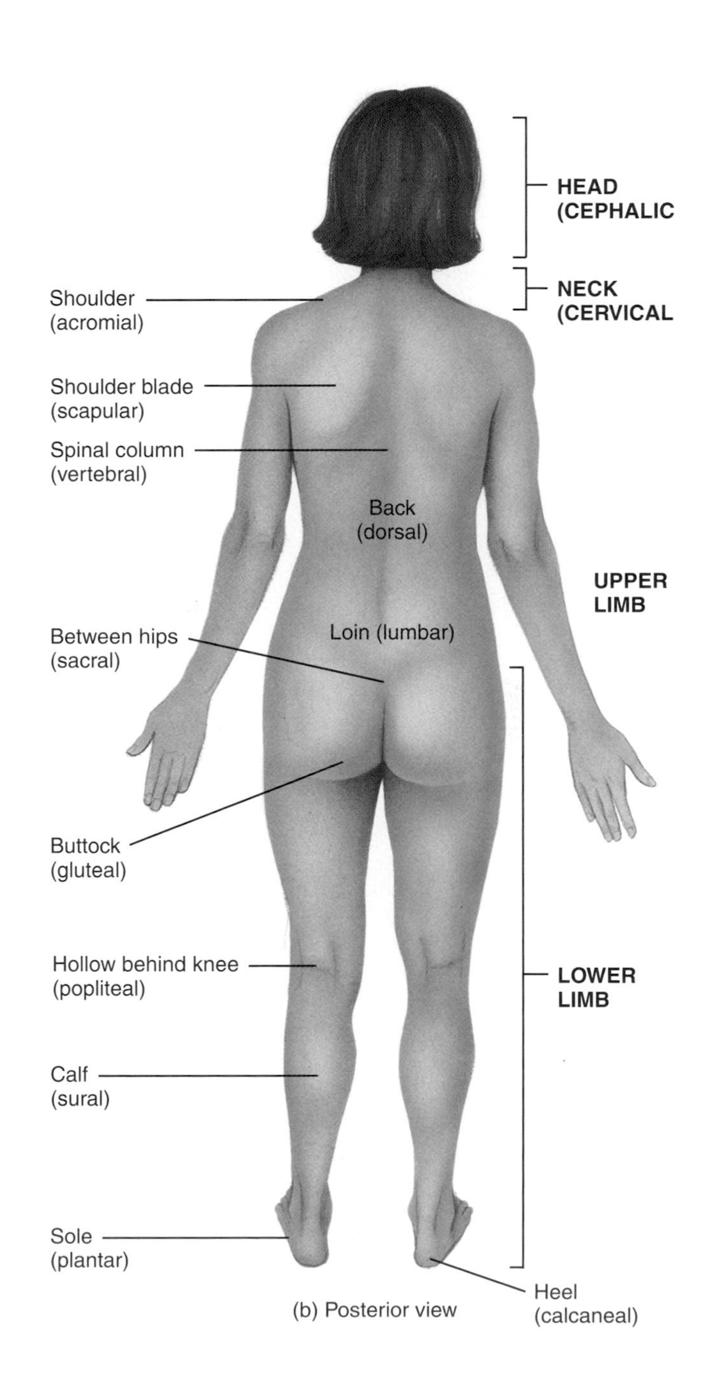

(b) Posterior view

Where is a plantar wart located?

Exhibit 1.1 Directional Terms *(Figure 1.5)*

OBJECTIVE • Define each directional term used to describe the human body.

Most of the directional terms used to describe the human body can be grouped into pairs that have opposite meanings. For example, ***superior*** means toward the upper part of the body, and ***inferior*** means toward the lower part of the body. It is important to understand that directional terms have *relative* meanings; they only make sense when used to describe the position of one structure relative to another. For example, your knee is superior to your ankle, even though both are located in the inferior half of the body. Study the directional terms and the example of how each is used. As you read each example, refer to Figure 1.5 to see the location of the structures mentioned.

■ CHECKPOINT

Which directional terms can be used to specify the relationships between (1) the elbow and the shoulder, (2) the left and right shoulders, (3) the sternum and the humerus, and (4) the heart and the diaphragm?

Directional Term	Definition	Example of Use
Superior (soo′-PEER-ē-or) (**cephalic** or **cranial**)	Toward the head, or the upper part of a structure.	The heart is superior to the liver.
Inferior (in′-FEER-ē-or) (**caudal**)	Away from the head, or the lower part of a structure.	The stomach is inferior to the lungs.
Anterior (an-TEER-ē-or) (**ventral**)	Nearer to or at the front of the body.	The sternum (breastbone) is anterior to heart.
Posterior (pos-TEER-ē-or) (**dorsal**)	Nearer to or at the back of the body.	The esophagus (food tube) is posterior to the trachea (windpipe).
Medial (MĒ-dē-al)	Nearer to the midline† or midsagittal plane.	The ulna is medial to the radius.
Lateral (LAT-er-al)	Farther from the midline or midsagittal plane.	The lungs are lateral to the heart.
Proximal (PROK-si-mal)	Nearer to the attachment of a limb to the trunk; nearer to the point of origin or the beginning.	The humerus is proximal to the radius.
Distal (DIS-tal)	Farther from the attachment of a limb to the trunk; farther from the point of origin or the beginning.	The phalanges are distal to the carpals.
Superficial (soo′-per-FISH-al)	Toward or on the surface of the body.	The ribs are superficial to the lungs.
Deep (DĒP)	Away from the surface of the body.	The ribs are deep to the skin of the chest and back.

†The midline is an imaginary vertical line that divides the body into equal right and left sides.

Figure 1.5 Directional terms.

Directional terms precisely locate various parts of the body in relation to one another.

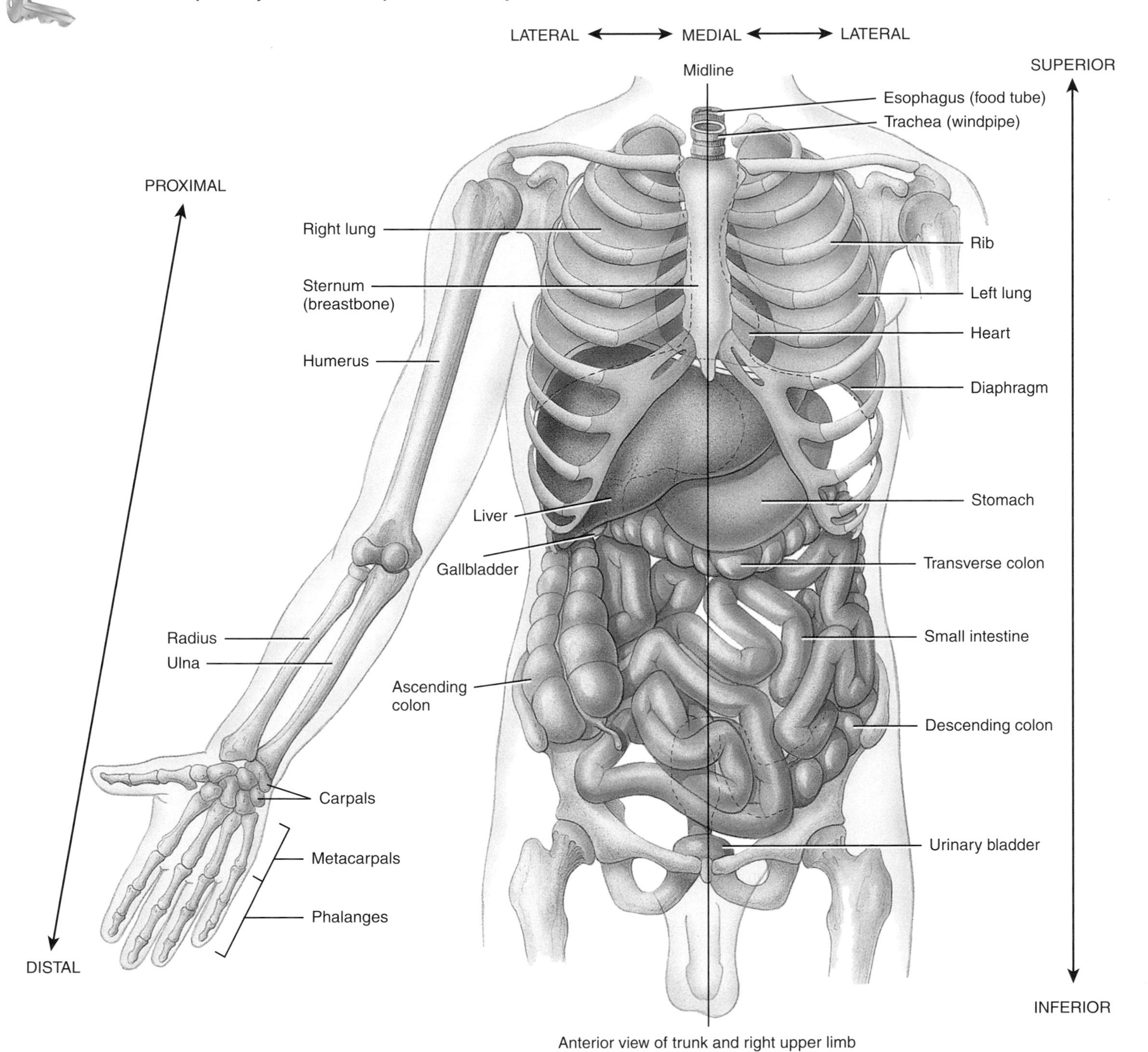

Anterior view of trunk and right upper limb

Is the radius proximal to the humerus? Is the esophagus anterior to the trachea? Are the ribs superficial to the lungs? Is the urinary bladder medial to the ascending colon? Is the sternum lateral to the descending colon?

Planes and Sections

You will also study parts of the body in four major ***planes***, that is, imaginary flat surfaces that pass through the body parts (Figure 1.6): sagittal, frontal, transverse, and oblique. A ***sagittal plane*** (SAJ-i-tal; *sagitt-* = arrow) is a vertical plane that divides the body or an organ into right and left sides. More specifically, when such a plane passes through the midline of the body or organ and divides it into *equal* right and left sides, it is called a ***midsagittal plane***. If the sagittal plane does not pass through the midline but instead divides the body or an organ into *unequal* right and left sides, it is called a ***parasagittal plane*** (*para-* = near). A ***frontal plane*** or *coronal plane* divides the body or an organ into anterior (front) and posterior (back) portions. A ***transverse plane*** divides the body or an organ into superior (upper) and inferior (lower) portions. A transverse plane may also be termed a *cross-sectional* or *horizontal plane*. Sagittal, frontal, and transverse planes are all at right angles to one another. An ***oblique plane***, by contrast, passes through the body or an organ at an angle between the transverse plane and a sagittal plane or between the transverse plane and the frontal plane.

When you study a body region, you will often view it in ***section***, meaning that you look at only one flat surface of the three-dimensional structure. It is important to know the plane of the section so you can understand the anatomical relationship of one part to another. Figure 1.7 indicates how

Figure 1.6 Planes through the human body.

Frontal, transverse, sagittal, and oblique planes divide the body in specific ways.

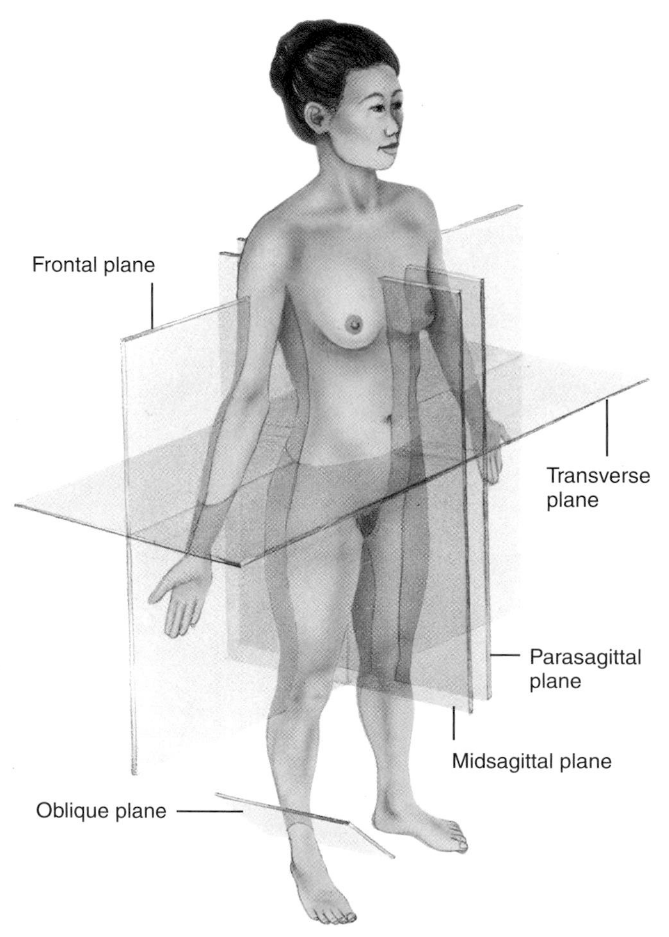

Which plane divides the heart into anterior and posterior portions?

Figure 1.7 Planes and sections through different parts of the brain. The diagrams (left) show the planes, and the photographs (right) show the resulting sections. (**Note:** The "view" arrows in the diagrams indicate the direction from which each section is viewed. This aid is used throughout the book to indicate viewing perspective.)

Planes divide the body in various ways to produce sections.

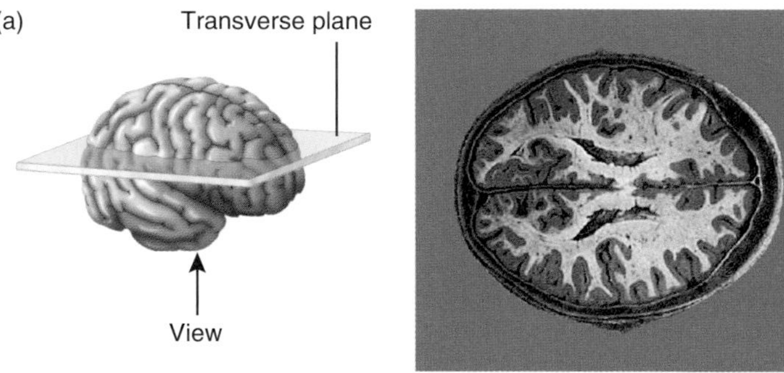

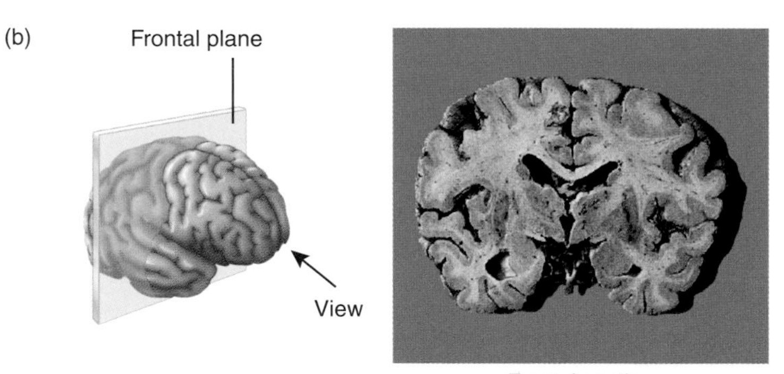

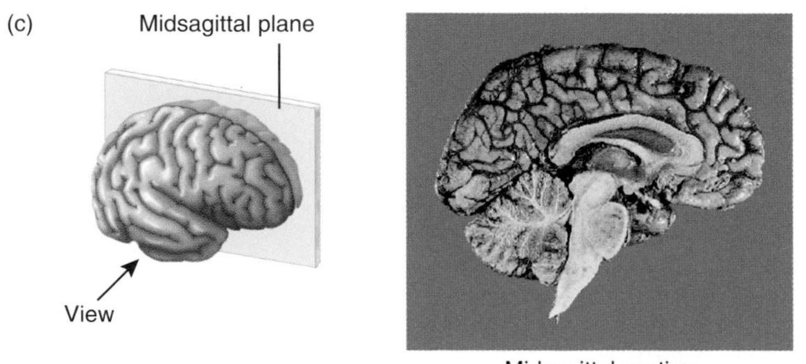

Which plane divides the brain into equal right and left sides?

three different sections—a *transverse (cross) section*, a *frontal section*, and a *midsagittal section*—provide different views of the brain.

■ **CHECKPOINT**

10. Describe the anatomical position and explain why it is used.

11. Locate each region on your own body, and then identify it by its common name and the corresponding anatomical descriptive form.

12. For each directional term listed in Exhibit 1.1 on page 12, provide your own example.

13. What are the various planes that may be passed through the body? Explain how each divides the body.

BODY CAVITIES

OBJECTIVES • **Describe the principal body cavities and the organs they contain.**

• **Explain why the abdominopelvic cavity is divided into regions and quadrants.**

Spaces within the body that contain, protect, separate, and support internal organs are called ***body cavities***. Here we discuss several of the larger body cavities (Figure 1.8).

The ***cranial cavity*** is formed by the cranial (skull) bones and contains the brain. The ***vertebral (spinal) cavity*** is formed by the bones of the vertebral column (backbone) and contains the spinal cord.

The major body cavities of the trunk are the thoracic and abdominopelvic cavities. The ***thoracic cavity*** (thor-AS-ik;

Figure 1.8 Body cavities. The dashed lines indicate the border between the abdominal and pelvic cavities.

The major body cavities of the trunk are the thoracic and abdominopelvic cavities.

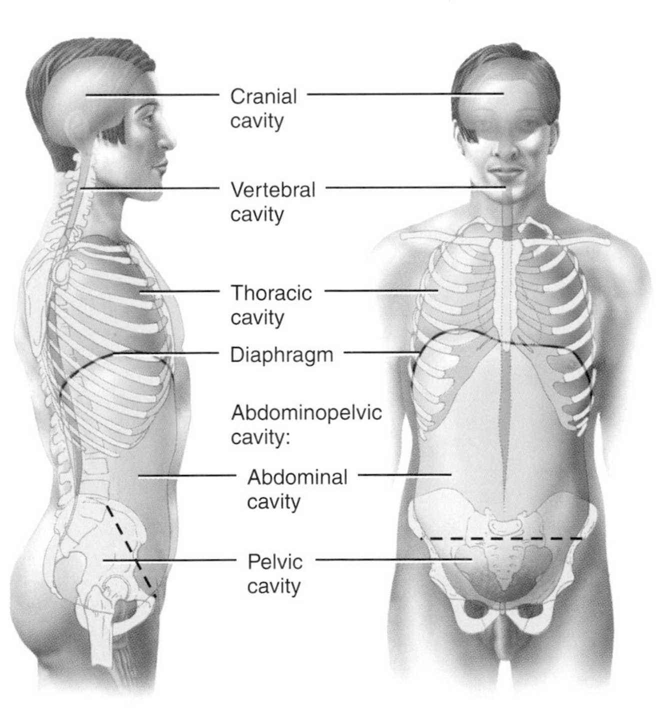

CAVITY	COMMENTS
Cranial cavity	Formed by cranial bones and contains brain.
Vertebral cavity	Formed by vertebral column and contains spinal cord and the beginnings of spinal nerves.
Thoracic cavity*	Chest cavity; contains pleural and pericardial cavities and mediastinum.
Pleural cavity	Each surrounds a lung; the serous membrane of the pleural cavities is the pleura.
Pericardial cavity	Surrounds the heart; the serous membrane of the pericardial cavity is the pericardium.
Mediastinum	Central portion of thoracic cavity between the lungs; extends from sternum to vertebral column and from neck to diaphragm; contains heart, thymus, esophagus, trachea, and several large blood vessels.
Abdominopelvic cavity	Subdivided into abdominal and pelvic cavities.
Abdominal cavity	Contains stomach, spleen, liver, gallbladder, small intestine, and most of large intestine; the serous membrane of the abdominal cavity is the peritoneum.
Pelvic cavity	Contains urinary bladder, portions of large intestine, and internal organs of reproduction.

* See Figure 1.9 for details of the thoracic cavity.

? In which cavities are the following organs located: urinary bladder, stomach, heart, small intestine, lungs, internal female reproductive organs, thymus, spleen, liver? Use the following symbols for your response: T = thoracic cavity, A = abdominal cavity, or P = pelvic cavity.

Figure 1.9 The thoracic cavity. The dashed lines indicate the borders of the mediastinum. Notice that the pericardial cavity surrounds the heart, and that the pleural cavities surround the lungs.

The mediastinum is medial to the lungs; it extends from the sternum to the vertebral column and from the neck to the diaphragm.

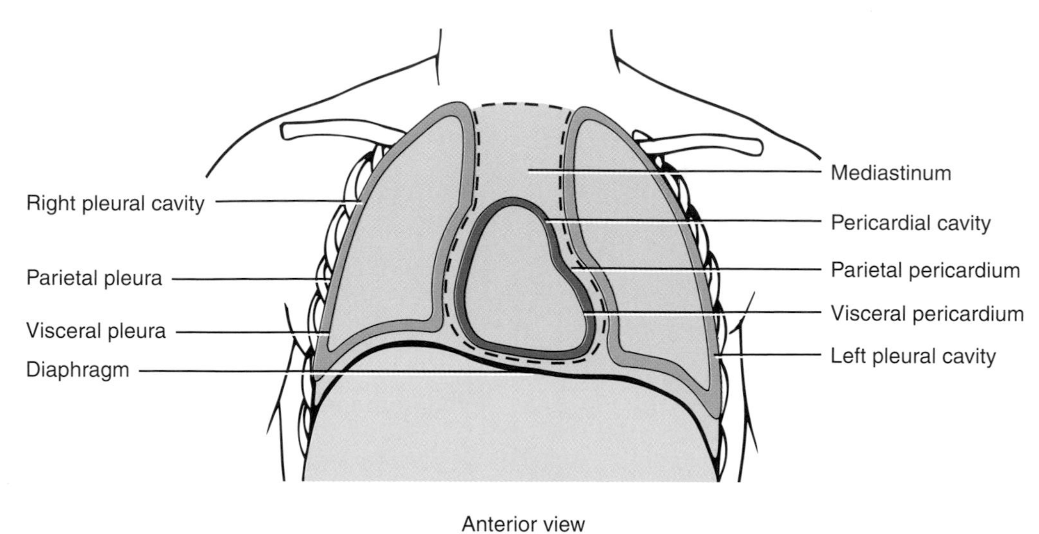

Anterior view

Which of the following structures are contained in the mediastinum: right lung, heart, esophagus, spinal cord, aorta, left pleural cavity?

Figure 1.10 The nine regions of the abdominopelvic cavity. The internal reproductive organs in the pelvic cavity are shown in Figures 23.1 on page 557 and 23.6 on page 564.

The nine-region designation is used for anatomical studies.

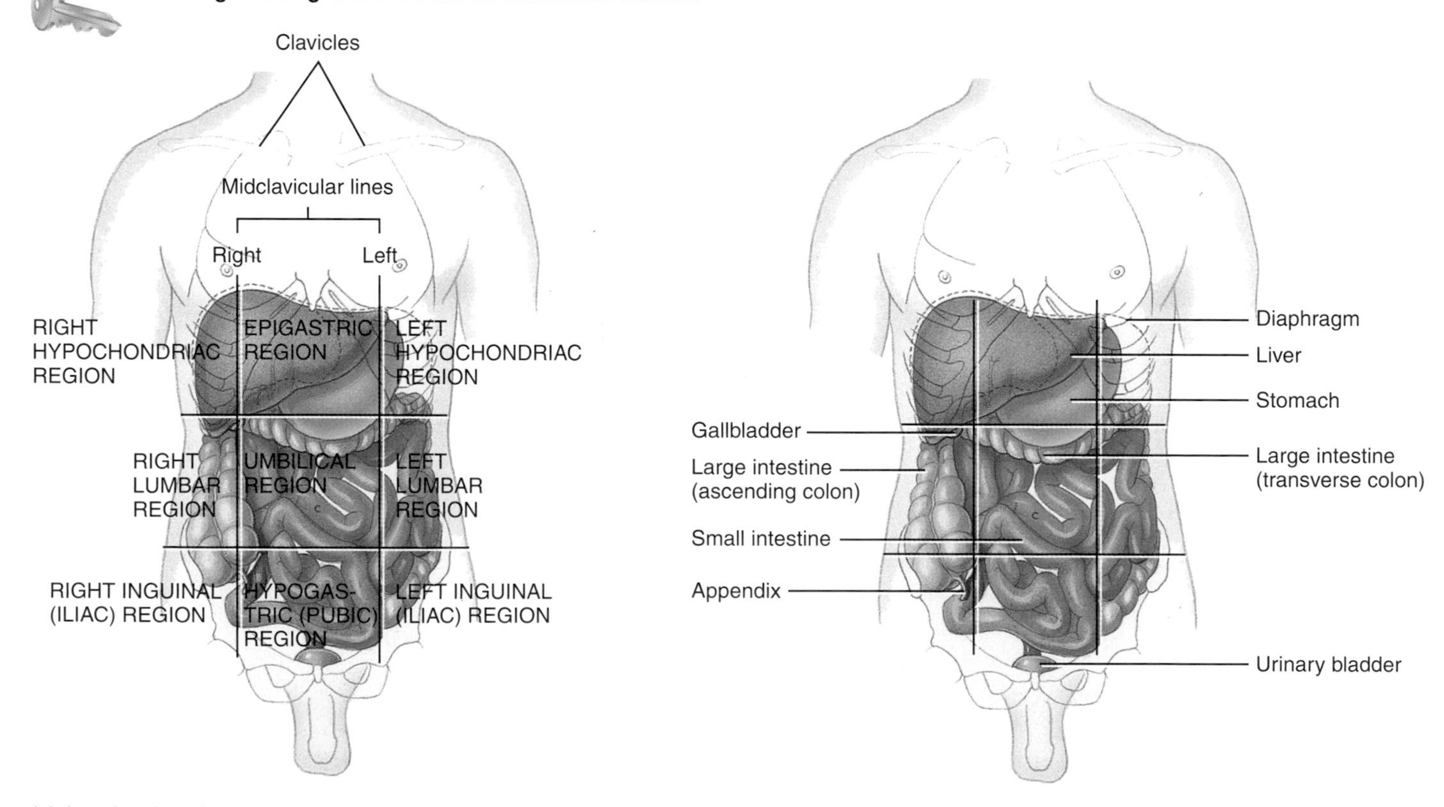

(a) Anterior view showing location of abdominopelvic regions

(b) Anterior superficial view of organs in abdominopelvic regions

In which abdominopelvic region is each of the following found: most of the liver, ascending colon, urinary bladder, appendix?

thorac- = chest) is the chest cavity. Within the thoracic cavity are three smaller cavities: the ***pericardial cavity*** (per′-i-KAR-dē-al; *peri-* = around; *-cardial* = heart), a fluid-filled space that surrounds the heart, and two ***pleural cavities*** (PLOOR-al; *pleur-* = rib or side), each of which surrounds one lung and contains a small amount of fluid (Figure 1.9 on page 16). The central portion of the thoracic cavity is called the ***mediastinum*** (mē′-dē-a-STĪ-num; *media-* = middle; *-stinum* = partition). It is between the lungs, extending from the sternum (breastbone) to the vertebral column (backbone), and from the neck to the diaphragm (Figure 1.9). The mediastinum contains all thoracic organs except the lungs themselves. Among the structures in the mediastinum are the heart, esophagus, trachea, and several large blood vessels. The ***diaphragm*** (DĪ-a-fram = partition or wall) is a dome-shaped muscle that powers breathing and separates the thoracic cavity from the abdominopelvic cavity.

The ***abdominopelvic cavity*** (ab-dom′-i-no-PEL-vic) extends from the diaphragm to the groin. As the name suggests, the abdominopelvic cavity is divided into two portions, although no wall separates them (see Figure 1.8). The upper portion, the ***abdominal cavity*** (*abdomin-* = belly) contains the stomach, spleen, liver, gallbladder, small intestine, and most of the large intestine. The lower portion, the ***pelvic cavity*** (*pelv-* = basin) contains the urinary bladder, portions of the large intestine, and internal organs of the reproductive system. The pelvic cavity is located below the dashed line in Figure 1.8. Organs inside the thoracic and abdominopelvic cavities are called ***viscera*** (VIS-e-ra).

Abdominopelvic Regions and Quadrants

To describe the location of the many abdominal and pelvic organs more precisely, the abdominopelvic cavity may be divided into smaller compartments. In one method, two horizontal and two vertical lines, like a tic-tac-toe grid, partition the cavity into nine ***abdominopelvic regions*** (Figure 1.10 on page 16). The names of the nine abdominopelvic regions are the *right hypochondriac* (hī′-pō-KON-drē-ak), *epigastric* (ep-i-GAS-trik), *left hypochondriac, right lumbar, umbilical* (um-BIL-i-kal), *left lumbar, right inguinal (iliac)* (IL-ē-ak), *hypogastric* (hī′-pō-GAS-trik), and *left inguinal (iliac).* In another method, one horizontal and one vertical line passing through the *umbilicus* (um-BIL-i-kus or um-bi-LĪ-kus; *umbilic-* = navel) or belly button divide the abdominopelvic cavity into ***quadrants*** (KWOD-rantz; *quad-* = one-fourth) (Figure 1.11). The names of the abdominopelvic quadrants are the *right upper quadrant (RUQ), left upper quadrant (LUQ), right lower quadrant (RLQ),* and *left lower quadrant (LLQ).*

Figure 1.11 Quadrants of the abdominopelvic cavity. The two lines cross at right angles at the umbilicus (navel).

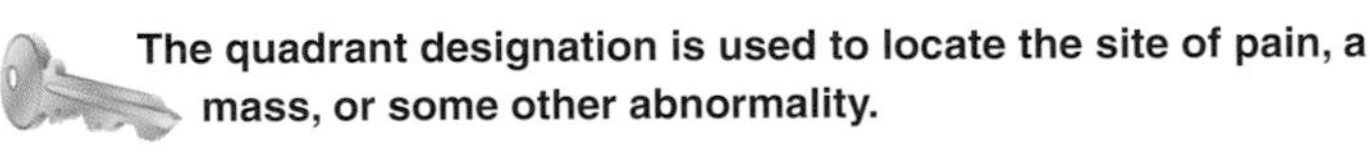

The quadrant designation is used to locate the site of pain, a mass, or some other abnormality.

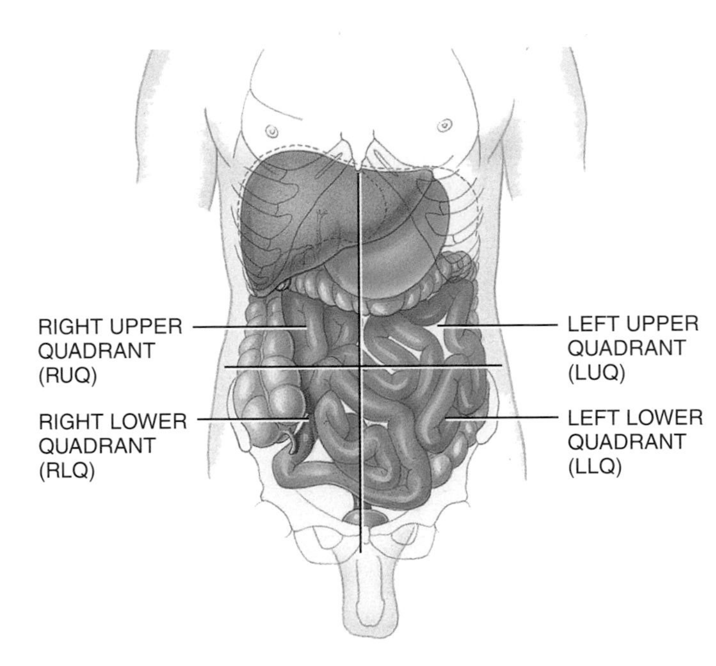

Anterior view showing location of abdominopelvic quadrants

In which abdominopelvic quadrant would the pain from appendicitis (inflammation of the appendix) be felt?

The nine-region division is more widely used for anatomical studies, and quadrants are more commonly used by clinicians to describe the site of an abdominopelvic pain, mass, or other abnormality.

■ CHECKPOINT

14. What landmarks separate the various body cavities from one another?

15. Locate the nine abdominopelvic regions and the four abdominopelvic quadrants on yourself, and list some of the organs found in each.

• • •

We will next examine the chemical level of organization in Chapter 2. You will learn about the various groups of chemicals in your body, how they function, and how they contribute to the homeostasis of your body.

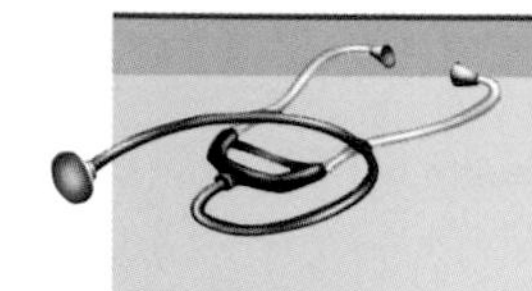

MEDICAL TERMINOLOGY AND CONDITIONS

Most chapters in this text are followed by a glossary of key medical terms that include both normal and pathological conditions. You should familiarize yourself with these terms because they will play an essential role in your medical vocabulary.

Some of these conditions, as well as ones discussed in the text, are referred to as local or systemic. A *local disease* is one that affects one part or a limited area of the body. A *systemic disease* affects the entire body or several parts.

Epidemiology (ep′-i-dē-mē-OL-ō-jē; *epi-* = upon; *-demi* = people) The science that deals with why, when, and where diseases occur and how they are transmitted within a defined human population.

Geriatrics (jer′-ē-AT-riks; *ger-* = old; *-iatrics* = medicine) The science that deals with the medical problems and care of elderly persons.

Pathology (pa-THOL-ō-jē; *patho-* = disease) The science that deals with the nature, causes, and development of abnormal conditions and the structural and functional changes that diseases produce.

Pharmacology (far-ma-KOL-ō-jē; *pharmac-* = drug) The science that deals with the effects and use of drugs in the treatment of disease.

STUDY OUTLINE

Anatomy and Physiology Defined (p. 2)

1. Anatomy is the science of structure and the relationships among structures.
2. Physiology is the science of how body structures function.

Levels of Organization and Body Systems (p. 2)

1. The human body consists of six levels of organization: chemical, cellular, tissue, organ, system, and organismal.
2. Cells are the basic structural and functional units of an organism and the smallest living units in the human body.
3. Tissues consist of groups of cells and the materials surrounding them that work together to perform a particular function.
4. Organs usually have recognizable shapes, are composed of two or more different types of tissues, and have specific functions.
5. Systems consist of related organs that have a common function.
6. Table 1.1 on pages 4–5 introduces the eleven systems of the human body: integumentary, skeletal, muscular, nervous, endocrine, cardiovascular, lymphatic, respiratory, digestive, urinary, and reproductive.
7. The human organism is a collection of structurally and functionally integrated systems.
8. Body systems work together to maintain health, protect against disease, and allow for reproduction of the species.

Life Processes (p. 6)

1. All living organisms have certain characteristics that set them apart from nonliving things.
2. Among the life processes in humans are metabolism, responsiveness, movement, growth, differentiation, and reproduction.

Homeostasis: Maintaining Limits (p. 6)

1. Homeostasis is a condition in which the internal environment of the body remains stable, within certain limits.
2. A large part of the body's internal environment is interstitial fluid, which surrounds all body cells.
3. Homeostasis is regulated by the nervous and endocrine systems acting together or separately. The nervous system detects body changes and sends nerve impulses to maintain homeostasis. The endocrine system regulates homeostasis by secreting hormones.
4. Disruptions of homeostasis come from external and internal stimuli and from psychological stresses. When disruption of homeostasis is mild and temporary, responses of body cells quickly restore balance in the internal environment. If disruption is extreme, the body's attempts to restore homeostasis may fail.
5. A feedback system consists of (1) receptors that monitor changes in a controlled condition and send input to (2) a control center that sets the value at which a controlled condition should be maintained, evaluates the input it receives, and generates output commands when they are needed, and (3) effectors that receive output from the control center and produce a response (effect) that alters the controlled condition.
6. If a response reverses a change in a controlled condition, the system is called a negative feedback system. If a response strengthens a change in a controlled condition, the system is referred to as a positive feedback system.
7. One example of negative feedback is the system that regulates blood pressure. If a stimulus causes blood pressure (controlled condition) to rise, baroreceptors (pressure-sensitive nerve cells, the receptors) in blood vessels send impulses (input) to the brain (control center). The brain sends impulses (output) to the heart

(effector). As a result, heart rate decreases (response), and blood pressure drops back to normal (restoration of homeostasis).

8. Disruptions of homeostasis—homeostatic imbalances—can lead to disorders, disease, and even death.
9. A disorder is any abnormality of structure and/or function. Disease is a more specific term for an illness with a definite set of signs and symptoms.
10. Symptoms are subjective changes in body functions that are not apparent to an observer, whereas signs are objective changes that can be observed and measured.
11. Diagnosis of disease involves identification of symptoms and signs, a medical history, physical examination, and sometimes laboratory tests.

Aging and Homeostasis (p. 9)

1. Aging produces observable changes in structure and function and increases vulnerability to stress and disease.
2. Changes associated with aging occur in all body systems.

Anatomical Terms (p. 9)

1. Descriptions of any region of the body assume the body is in the anatomical position, in which the subject stands erect facing the observer, with the head level and the eyes facing forward, the feet flat on the floor and directed forward, and the arms at the sides, with the palms turned forward.
2. The human body is divided into several major regions: the head, neck, trunk, upper limbs, and lower limbs.
3. Within body regions, specific body parts have common names and corresponding anatomical descriptive forms (adjectives). Examples are chest (thoracic), nose (nasal), and wrist (carpal).
4. Directional terms indicate the relationship of one part of the body to another. Exhibit 1.1 on page 12 summarizes commonly used directional terms.
5. Planes are imaginary flat surfaces that divide the body or organs into two parts. A midsagittal plane divides the body or an organ into equal right and left sides. A parasagittal plane divides the body or an organ into unequal right and left sides. A frontal plane divides the body or an organ into anterior and posterior portions. A transverse plane divides the body or an organ into superior and inferior portions. An oblique plane passes through the body or an organ at an angle between a transverse plane and a sagittal plane, or between a transverse plane and a frontal plane.
6. Sections result from cuts through body structures. They are named according to the plane on which the cut is made: transverse, frontal, or sagittal.

Body Cavities (p. 15)

1. Spaces in the body that contain, protect, separate, and support internal organs are called body cavities.
2. The cranial cavity contains the brain, and the vertebral cavity contains the spinal cord.
3. The thoracic cavity is subdivided into three smaller cavities: a pericardial cavity, which contains the heart, and two pleural cavities, which each contain a lung.
4. The central portion of the thoracic cavity is the mediastinum. It is located between the lungs and extends from the sternum to the vertebral column and from the neck to the diaphragm. It contains all thoracic organs except the lungs.
5. The abdominopelvic cavity is separated from the thoracic cavity by the diaphragm and is divided into a superior abdominal cavity and an inferior pelvic cavity.
6. Organs in the thoracic and abdominopelvic cavities are called viscera.
7. Viscera of the abdominal cavity include the stomach, spleen, liver, gallbladder, small intestine, and most of the large intestine.
8. Viscera of the pelvic cavity include the urinary bladder, portions of the large intestine, and internal organs of the reproductive system.
9. To describe the location of organs easily, the abdominopelvic cavity may be divided into nine abdominopelvic regions by two horizontal and two vertical lines.
10. The names of the nine abdominopelvic regions are right hypochondriac, epigastric, left hypochondriac, right lumbar, umbilical, left lumbar, right inguinal, hypogastric, and left inguinal.
11. The abdominopelvic cavity may also be divided into quadrants by passing one horizontal and one vertical line through the umbilicus (navel).
12. The names of the abdominopelvic quadrants are right upper quadrant (RUQ), left upper quadrant (LUQ), right lower quadrant (RLQ), and left lower quadrant (LLQ).

SELF-QUIZ

1. To properly reconnect the disconnected bones of a human skeleton, you would need to have a good understanding of
   a. physiology
   b. homeostasis
   c. chemistry
   d. anatomy
   e. feedback systems
2. Which of the following best illustrates the idea of increasing levels of organizational complexity?
   a. chemical → tissue → cellular → organ → organismal → system
   b. chemical → cellular → tissue → organ → system → organismal
   c. cellular → chemical → tissue → organismal → organ → system
   d. chemical → cellular → tissue → system → organ → organismal
   e. tissue → cellular → chemical → organ → system → organismal

3. Match the following:

____ a. transports oxygen, nutrients, and carbon dioxide		A. urinary system
____ b. breaks down and absorbsfood		B. digestive system
____ c. functions in body movement, posture, and heat production		C. endocrine system
____ d. regulates body activities through hormones		D. integumentary system
____ e. supports and protects the body		E. muscular system
____ f. eliminates wastes and regulates the chemical composition and volume of blood		F. skeletal system
____ g. protects the body, detects sensations, and helps regulate body temperature		G. cardiovascular system

4. Fill in the missing blanks in the following table.

System	Major Organs	Functions
a	b	Regulates body activities by nerve impulses
c	Lymph vessels, spleen, thymus, tonsils, lymph nodes	d
e	f	Supplies oxygen to cells, eliminates carbon dioxide, regulates acid–base balance
Reproductive	g	h

5. Homeostasis is
   a. the sum of all of the chemical processes in the body
   b. the sign of a disorder or disease
   c. the combination of growth, repair, and energy release that is basic to life
   d. the tendency to maintain constant, favorable internal body conditions
   e. caused by stress

6. Which of the following is NOT true concerning the life processes?
   a. The pupils of your eyes becoming smaller when exposed to strong light is an example of differentiation.
   b. The ability to walk to your car following class is a result of the life process called movement.
   c. The repair of injured skin would involve the life process of reproduction.
   d. Digesting and absorbing food is an example of metabolism.
   e. Sweating on a hot summer day involves responsiveness.

7. In a negative feedback system,
   a. the controlled condition is never disrupted
   b. there tends to be a "runaway" body response
   c. the change in the controlled condition is reversed
   d. the body part that responds to the output is known as the receptor
   e. the response results in a reinforcement of the original stimulus

8. The part of a feedback system that receives the input and generates the output command is the
   a. effector b. receptor c. feedback loop
   d. response e. control center

9. Match the following:

a. observable, measurable change	A. systemic
b. abnormality of function	B. symptom
c. affects the entire body	C. sign
d. subjective changes that aren't easily observed	D. disorder

10. An itch in your axillary region would cause you to scratch
    a. your armpit b. the front of your elbow
    c. your neck d. the top of your head e. your calf

11. If you were facing a person who is in the correct anatomical position, you could observe the
    a. crural region b. lumbar region c. gluteal region
    d. popliteal region e. scapular region

12. Where would you look for the femoral artery?
    a. wrist b. forearm c. face
    d. thigh e. shoulder

13. The right ear is ________ to the right nostril.
    a. intermediate b. inferior c. lateral
    d. distal e. medial

14. Your chin is ________ in relation to your lips.
    a. lateral b. superior c. deep
    d. posterior e. inferior

15. Your skull is ________ in relation to your brain.
    a. intermediate b. superior c. deep
    d. superficial e. proximal

16. A magician is about to separate his assistant's body into superior and inferior portions. The plane through which he will pass his magic wand is the
    a. midsagittal b. frontal c. transverse
    d. parasagittal e. oblique

17. Which statement is NOT true of body cavities?
    a. The diaphragm separates the thoracic and abdominopelvic cavities.
    b. The organs in the cranial and vertebral cavities are called viscera.
    c. The urinary bladder is in the pelvic cavity.
    d. The abdominal cavity is below the thoracic cavity.
    e. The pelvic cavity terminates below the groin.

18. If Jamie is having her appendix removed, the surgeon would prepare which area for surgery?
    a. right upper quadrant
    b. right lower quadrant
    c. left upper quadrant
    d. left lower quadrant
    e. left hypochondriac region

19. To find the urinary bladder, you would look in the
    a. hypochondriac region
    b. umbilical region
    c. epigastric region
    d. iliac region
    e. hypogastric region

20. Match the following:

____	a. contains the urinary bladder and reproductive organs	A.	cranial cavity
____	b. contains the brain	B.	abdominal cavity
____	c. contains the heart	C.	vertebral cavity
____	d. region between the lungs, from the breastbone to the backbone	D.	pelvic cavity
____	e. separates the thoracic and abdominal cavities	E.	pleural cavity
____	f. contains a lung	F.	mediastinum
____	g. contains the spinal cord	G.	diaphragm
____	h. contains the stomach and liver	H.	pericardial cavity

CRITICAL THINKING APPLICATIONS

1. Taylor was going for the playground record for the longest upside-down hang from the monkey bars. She didn't make it and may have broken her arm. The emergency room technician would like an x-ray film of Taylor's arm in the anatomical position. Use the proper anatomical terms to describe the position of Taylor's arm in the x-ray film.

2. Imagine that a manned space flight lands on Mars. The astronaut life specialist observes lumpy shapes that may be life forms. What are some characteristics of living organisms that may help the astronaut determine if these are life forms or mud balls?

3. Guy was trying to impress Jenna with a tale about his last rugby match. "The coach said I suffered a caudal injury to the dorsal sural in my groin." Jenna responded, "I think either you or your coach suffered a cephalic injury." Why wasn't Jenna impressed by Guy's athletic prowess?

4. There's a special fun-house mirror that hides half your body and doubles the image of your other side. In the mirror, you can do amazing feats such as lifting both legs off the ground. Along what plane is the mirror dividing your body? A different mirror in the next room shows your reflection with two heads, four arms and no legs. Along what plane is this mirror dividing your body?

ANSWERS TO FIGURE QUESTIONS

1.1 Organs have a recognizable shape and consist of two or more different types of tissues that have a specific function.

1.2 The basic difference between negative and positive feedback systems is that in negative feedback systems, the response reverses a change in a controlled condition, and in positive feedback systems, the response strengthens the change in a controlled condition.

1.3 If a stimulus caused blood pressure to decrease, the heart rate would increase due to the operation of this negative feedback system.

1.4 A plantar wart is found on the sole.

1.5 No, the radius is distal to the humerus; No, the esophagus is posterior to the trachea; Yes, the ribs are superficial to the lungs; Yes, the urinary bladder is medial to the ascending colon; No, the sternum is medial to the descending colon.

1.6 The frontal plane divides the heart into anterior and posterior portions.

1.7 The midsagittal plane divides the brain into equal right and left sides.

1.8 Urinary bladder = P, stomach = A, heart = T, small intestine = A, lungs = T, internal female reproductive organs = P, thymus = T, spleen = A, liver = A.

1.9 Some structures in the mediastinum are the heart, esophagus, and aorta.

1.10 The liver is mostly in the epigastric region; the ascending colon is in the right lumbar region; the urinary bladder is in the hypogastric region; the appendix is in the right inguinal region.

1.11 The pain associated with appendicitis would be felt in the right lower quadrant (RLQ).

chapter 2 INTRODUCTORY CHEMISTRY

did you know?

How often have you heard people talk about dietary fat? You have probably heard many debates regarding the health benefits of monounsaturated and polyunsaturated versus saturated fats (fatty acids). The chemistry of fatty acids is responsible for the physiological roles they play. A great deal of research has focused on the behavior of the various kinds of fatty acids in the body. At one time many scientists thought that total fat intake should be kept low to prevent heart disease. Now scientists believe that certain types of unsaturated fatty acids, such as those found in fish oils, may actually reduce heart disease risk.

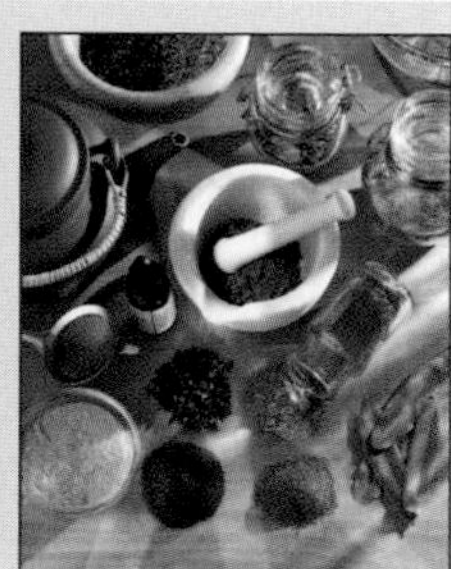

Focus on Wellness, page 36

www.wiley.com/college/apcentral

Many common substances we eat and drink—water, sugar, table salt, proteins, starches, fats—play vital roles in keeping us alive. In this chapter, you will learn how these substances function in your body. Because your body is composed of chemicals and all body activities are chemical in nature, it is important to become familiar with the language and basic ideas of chemistry to understand human anatomy and physiology.

looking back to move ahead . . .

- Levels of Organization and Body Systems (p. 2)

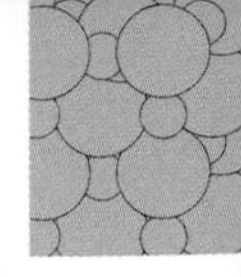

INTRODUCTION TO CHEMISTRY

OBJECTIVES • Define a chemical element, atom, ion, molecule, and compound.

- **Explain how chemical bonds form.**
- **Describe what happens in a chemical reaction and explain why it is important to the human body.**

Chemistry (KEM-is-trē) is the science of the structure and interactions of ***matter***, which is anything that occupies space and has mass. ***Mass*** is the amount of matter in any living organism or nonliving thing.

Chemical Elements and Atoms

All forms of matter are made up of a limited number of building blocks called ***chemical elements***, substances that cannot be broken down into a simpler form by ordinary chemical means. At present, scientists recognize 112 different elements. Each element is designated by a ***chemical symbol***, one or two letters of the element's name in English, Latin, or another language. Examples are H for hydrogen, C for carbon, O for oxygen, N for nitrogen, K for potassium, Na for sodium, Fe for iron, and Ca for calcium.

Twenty-six different elements normally are present in your body. Just four elements, called the *major elements*, constitute about 96% of the body's mass: oxygen, carbon, hydrogen, and nitrogen. Eight others, the *lesser elements*, contribute 3.8% of the body's mass: calcium (Ca), phosphorus (P), potassium (K), sulfur (S), sodium (Na), chlorine (Cl), magnesium (Mg), and iron (Fe). An additional 14 elements—the *trace elements*—are present in tiny amounts. Together, they account for the remaining 0.2% of the body's mass. Several trace elements have important functions in the body. For example, iodine (I) is needed to make thyroid hormones. The functions of some trace elements are unknown. Table 2.1 lists the main chemical elements of the human body.

Each element is made up of ***atoms***, the smallest units of matter that retain the properties and characteristics of the element. A sample of the element carbon, such as pure coal, contains only carbon atoms, and a tank of helium gas contains only helium atoms.

Table 2.1 Main Chemical Elements In the Body

Chemical Element (Symbol)	% of Total Body Mass	Significance
MAJOR ELEMENTS		
Oxygen (O)	65.0	Part of water and many organic (carbon-containing) molecules; used to generate ATP, a molecule used by cells to temporarily store chemical energy.
Carbon (C)	18.5	Forms backbone chains and rings of all organic molecules: carbohydrates, lipids (fats), proteins, and nucleic acids (DNA and RNA).
Hydrogen (H)	9.5	Constituent of water and most organic molecules; ionized form (H^+) makes body fluids more acidic.
Nitrogen (N)	3.2	Component of all proteins and nucleic acids.
LESSER ELEMENTS		
Calcium (Ca)	1.5	Contributes to hardness of bones and teeth; ionized form (Ca^{2+}) needed for blood clotting, release of hormones, contraction of muscle, and many other processes.
Phosphorus (P)	1.0	Component of nucleic acids and ATP; required for normal bone and tooth structure.
Potassium (K)	0.35	Ionized form (K^+) is the most plentiful cation (positively charged particle) in intracellular fluid; needed to generate action potentials.
Sulfur (S)	0.25	Component of some vitamins and many proteins.
Sodium (Na)	0.2	Ionized form (Na^+) is the most plentiful cation in extracellular fluid; essential for maintaining water balance; needed to generate action potentials.
Chlorine (Cl)	0.2	Ionized form (Cl^-) is the most plentiful anion (negatively charged particle) in extracellular fluid; essential for maintaining water balance.
Magnesium (Mg)	0.1	Ionized form (Mg^{2+}) needed for action of many enzymes, molecules that increase the rate of chemical reactions in organisms.
Iron (Fe)	0.005	Ionized forms (Fe^{2+} and Fe^{3+}) are part of hemoglobin (oxygen-carrying protein in red blood cells) and some enzymes (proteins that catalyze chemical reactions in living cells).
TRACE ELEMENTS		
	0.2	Aluminum (Al), Boron (B), Chromium (Cr), Cobalt (Co), Copper (Cu), Fluorine (F), Iodine (I), Manganese (Mn), Molybdenum (Mo), Selenium (Se), Silicon (Si), Tin (Sn), Vanadium (V), and Zinc (Zn).

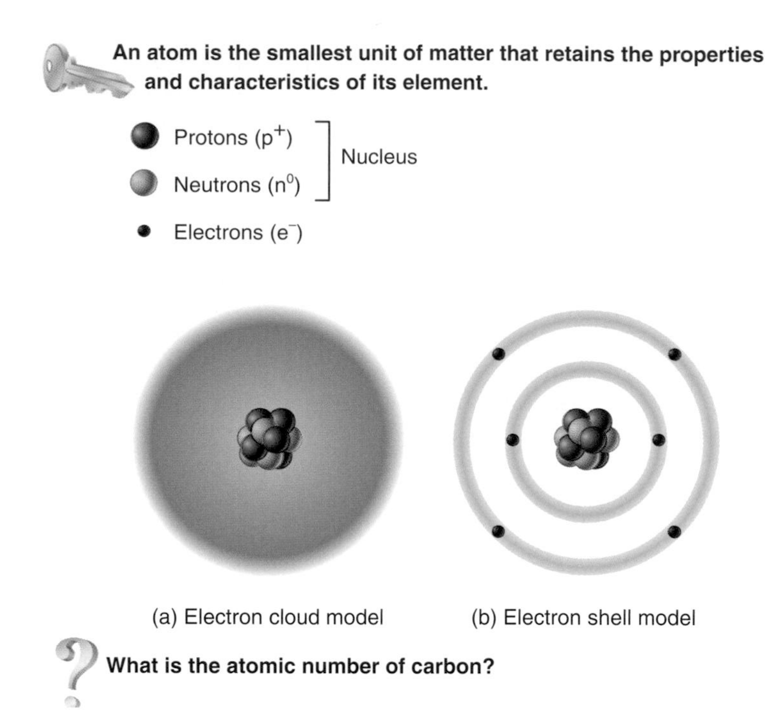

Figure 2.1 Two representations of the structure of an atom. Electrons move about the nucleus, which contains neutrons and protons. (a) In the electron cloud model of an atom, the shading represents the chance of finding an electron in regions outside the nucleus. (b) In the electron shell model, filled circles represent individual electrons, which are grouped into concentric circles according to the shells they occupy. Both models depict a carbon atom, with six protons, six neutrons, and six electrons.

An atom consists of two basic parts: a nucleus and one or more electrons (Figure 2.1). The centrally located ***nucleus*** contains positively charged ***protons*** (p^+) and uncharged (neutral) ***neutrons*** (n^0). Because each proton has one positive charge, the nucleus is positively charged. The ***electrons*** (e^-) are tiny, negatively charged particles that move about in a large space surrounding the nucleus. They do not follow a fixed path or orbit but instead form a negatively charged "cloud" that surrounds the nucleus (Figure 2.1a). The number of electrons in an atom equals the number of protons. Because each electron carries one negative charge, the negatively charged electrons and the positively charged protons balance each other. As a result, each atom is electrically neutral, meaning its total charge is zero.

The number of protons in the nucleus of an atom is called the atom's ***atomic number***. The atoms of each different kind of element have a different number of protons in the nucleus: A hydrogen atom has 1 proton, a carbon atom has 6 protons, a sodium atom has 11 protons, a chlorine atom has 17 protons, and so on (Figure 2.2). Thus, each type of atom, or element, has a different atomic number. The total number

Figure 2.2 Atomic structures of several atoms that have important roles in the human body.

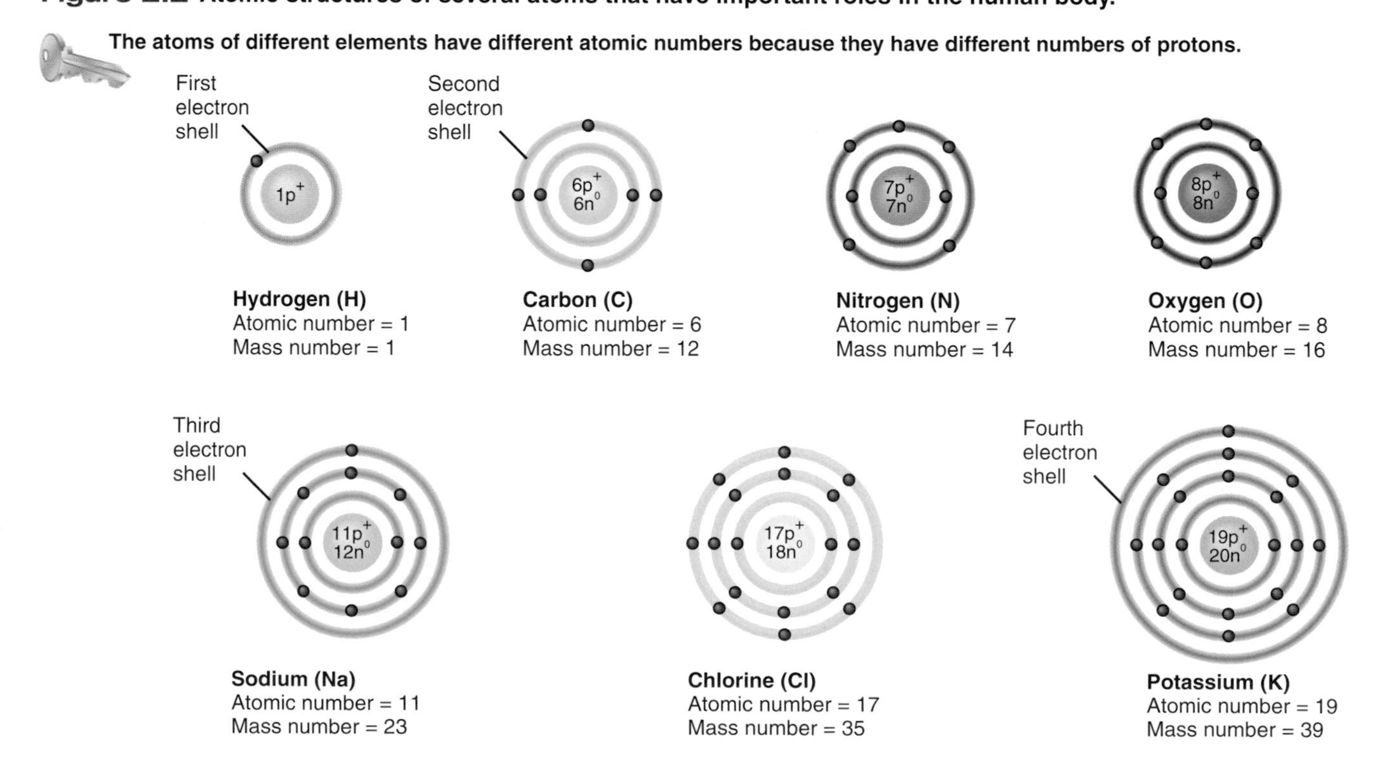

Which four of these elements are most abundant in living organisms?

of protons plus neutrons in an atom is its ***mass number***. For instance, an atom of sodium, with 11 protons and 12 neutrons in its nucleus, has a mass number of 23.

Even though their exact positions cannot be predicted, specific groups of electrons are most likely to move about within certain regions around the nucleus. These regions are called ***electron shells***, which are depicted as circles in Figures 2.1b and 2.2 even though some of their shapes are not spherical. The electron shell nearest the nucleus—the first electron shell—can hold a maximum of 2 electrons. The second electron shell can hold a maximum of 8 electrons, and the third can hold up to 18 electrons. Higher electron shells (there are as many as seven) can contain many more electrons. The electron shells are filled with electrons in a specific order, beginning with the first shell.

Ions, Molecules, and Compounds

The atoms of each element have a characteristic way of losing, gaining, or sharing their electrons when interacting with other atoms. If an atom either *gives up* or *gains* electrons, it becomes an ***ion*** (Ī-on), an atom that has a positive or negative charge due to unequal numbers of protons and electrons. An ion of an atom is symbolized by writing its chemical symbol followed by the number of its positive (+) or negative (−) charges. For example, Ca^{2+} stands for a calcium ion that has two positive charges because it has given up two electrons. Refer to Table 2.1 on page 23 for the important functions of several ions in the body.

In contrast, when two or more atoms *share* electrons, the resulting combination of atoms is called a ***molecule*** (MOL-e-kūl). A *molecular formula* indicates the number and type of atoms that make up a molecule. A molecule may consist of two or more atoms of the same element, such as an oxygen molecule or a hydrogen molecule, or two or more atoms of different elements, such as a water molecule (Figure 2.3). The molecular formula for a molecule of oxygen is O_2. The subscript 2 indicates there are two atoms of oxygen in the oxygen molecule. In the water molecule, H_2O, one atom of oxygen shares electrons with two atoms of hydrogen. Notice that two hydrogen molecules can combine with one oxygen molecule to form two water molecules (Figure 2.3).

Figure 2.3 Molecules.

A molecule may consist of two or more atoms of the same element or two or more atoms of different elements.

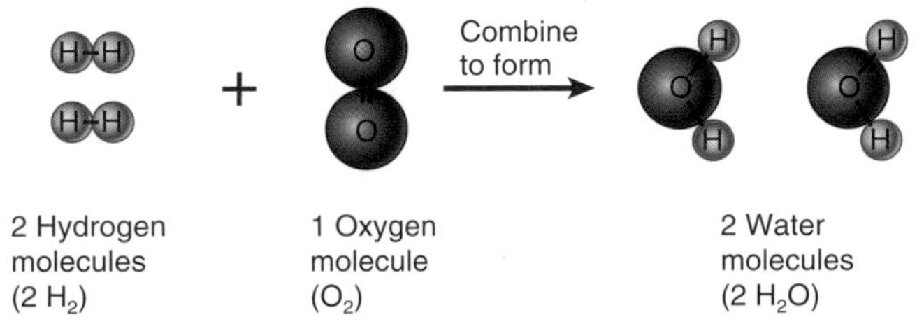

Which of the molecules shown here is a compound?

A ***compound*** is a substance containing atoms of two or more different elements. Most of the atoms in your body are joined into compounds, for example, water (H_2O). A molecule of oxygen (O_2) is *not* a compound because it consists of atoms of only one element.

A ***free radical*** is an electrically charged ion or molecule that has an unpaired electron in its outermost shell. (Most of an atom's electrons associate in pairs.) A common example of a free radical is *superoxide*, which is formed by the addition of an electron to an oxygen molecule. Having an unpaired electron makes a free radical unstable and destructive to nearby molecules. Free radicals break apart important body molecules by either giving up their unpaired electron to or taking on an electron from another molecule.

In our bodies, several processes can generate free radicals. They may result from exposure to ultraviolet radiation in sunlight or to x-rays. Some reactions that occur during normal metabolic processes produce free radicals. Moreover, certain harmful substances, such as carbon tetrachloride (a solvent used in dry cleaning), give rise to free radicals when they participate in metabolic reactions in the body. Among the many disorders and diseases linked to oxygen-derived free radicals are cancer, the buildup of fatty materials in blood vessels (atherosclerosis), Alzheimer disease, emphysema, diabetes mellitus, cataracts, macular degeneration, rheumatoid arthritis, and deterioration associated with aging. Consuming more **antioxidants**—substances that inactivate oxygen-derived free radicals—is thought to slow the pace of damage caused by free radicals. Important dietary antioxidants include selenium, zinc, beta-carotene, and vitamins C and E.

Chemical Bonds

The forces that bind the atoms of molecules and compounds together, resisting their separation, are ***chemical bonds***. The chance that an atom will form a chemical bond with another atom depends on the number of electrons in its outermost shell, also called the ***valence shell***. An atom with an outer shell holding eight electrons is *chemically stable*, which means it is unlikely to form chemical bonds with other atoms. Neon, for example, has eight electrons in its outer shell, and for this reason it rarely forms bonds with other atoms.

The atoms of most biologically important elements do not have eight electrons in their outer shells. Given the right conditions, two or more such atoms can interact or bond in ways that produce a chemically stable arrangement of eight electrons in the outer shell of each atom (*octet rule*). Three general types of chemical bonds are ionic bonds, covalent bonds, and hydrogen bonds.

Ionic Bonds

Positively charged ions and negatively charged ions are attracted to one another. This force of attraction between ions of opposite charges is called an ***ionic bond***. Consider sodium and chlorine atoms to see how an ionic bond forms (Figure 2.4). Sodium has one outer shell electron (Figure 2.4a). If a sodium atom *loses* this electron, it is left with the eight electrons in its second shell. However, the total number of protons (11) now exceeds the number of electrons (10). As a result, the sodium atom becomes a ***cation*** (KAT-ī-on), a positively charged ion. A sodium ion has a charge of 1+ and is written Na^+. On the other hand, chlorine has seven outer shell electrons (Figure 2.4b), too many to lose. But if chlorine *accepts* one electron from a neighboring atom, it will have eight electrons in its third electron shell. When this happens, the total number of electrons (18) exceeds the number of protons (17), and the chlorine atom becomes an ***anion*** (AN-ī-on), a negatively charged ion. The ionic form of chlorine is called a chloride ion. It has a charge of 1− and is written Cl^-. When an atom of sodium donates its sole outer shell electron to an atom of chlorine, the resulting positive and negative charges attract each other to form an ionic bond (Figure 2.4c). The resulting ionic compound is sodium chloride, written NaCl.

In the body, ionic bonds are found mainly in teeth and bones, where they give great strength to the tissue. Most other ions in the body are dissolved in body fluids. An ionic compound that breaks apart into cations and anions when dissolved is called an ***electrolyte*** (e-LEK-trō-līt) because the solution can conduct an electric current. As you will see in later chapters, electrolytes have many important functions. For example, they are critical for controlling water movement within the body, maintaining acid–base balance, and producing nerve impulses.

Figure 2.4 Ions and ionic bond formation. (a) A sodium atom can attain the stability of eight electrons in its outermost shell by losing its one valence electron; it then becomes a sodium ion, Na^+. (b) A chlorine atom can attain the stability of eight electrons in its outermost shell by accepting one electron; it then becomes a chloride ion, Cl^-. (c) An ionic bond holds Na^+ and Cl^- together in the ionic compound sodium chloride, NaCl. The electron that is donated or accepted is colored red.

An ionic bond is the force of attraction that holds together oppositely charged ions.

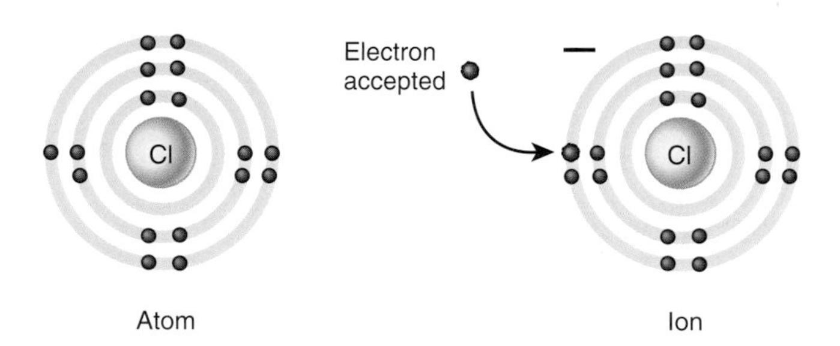

(b) Chlorine: 7 valence electrons

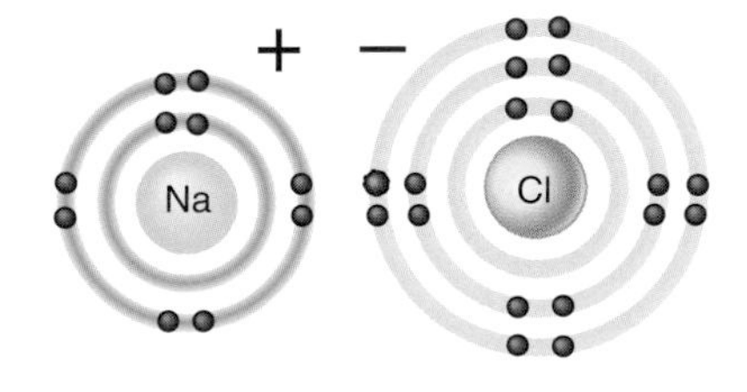

(c) Ionic bond in sodium chloride (NaCl)

Will the element potassium (K) be more likely to form an anion or a cation? Why? (Hint: Look back to Figure 2.2 for the atomic structure of K.)

Covalent Bonds

When a ***covalent bond*** forms, neither of the combining atoms loses or gains electrons. Instead, the atoms form a molecule by *sharing* one, two, or three pairs of their outer shell electrons. The greater the number of electron pairs shared between two atoms, the stronger the covalent bond. Covalent bonds are the most common chemical bonds in the body, and the compounds that result from them form most of the body's structures. Unlike ionic bonds, most covalent bonds do not break apart when the molecule is dissolved in water.

It is easiest to understand the nature of covalent bonds by considering those that form between atoms of the same element (Figure 2.5). A *single covalent bond* results when two atoms share one electron pair. For example, a molecule of hydrogen forms when two hydrogen atoms share their single valence electrons (Figure 2.5a), which allows both atoms to have a full valence shell. (Recall that the first electron shell holds only two electrons.) A *double covalent bond* (Figure 2.5b) or a *triple covalent bond* (Figure 2.5c) results when two atoms share two or three pairs of electrons. Notice the *structural formulas* for covalently bonded molecules in Figure 2.5. The number of lines between the chemical symbols for two atoms indicates whether the bond is a single (−), double (=), or triple (≡) covalent bond.

The same principles of covalent bonding that apply to atoms of the same element also apply to covalent bonds between atoms of different elements. Methane (CH_4), a gas,

Figure 2.5 Covalent bond formation. The red electrons are shared equally in (a)-(d) and unequally in (e). To the right are simpler ways to represent these molecules. In a structural formula, each covalent bond is denoted by a straight line between the chemical symbols for two atoms. In a molecular formula, the number of atoms in each molecule is noted by subscripts.

In a covalent bond, two atoms share one, two, or three pairs of electrons in the outer shell.

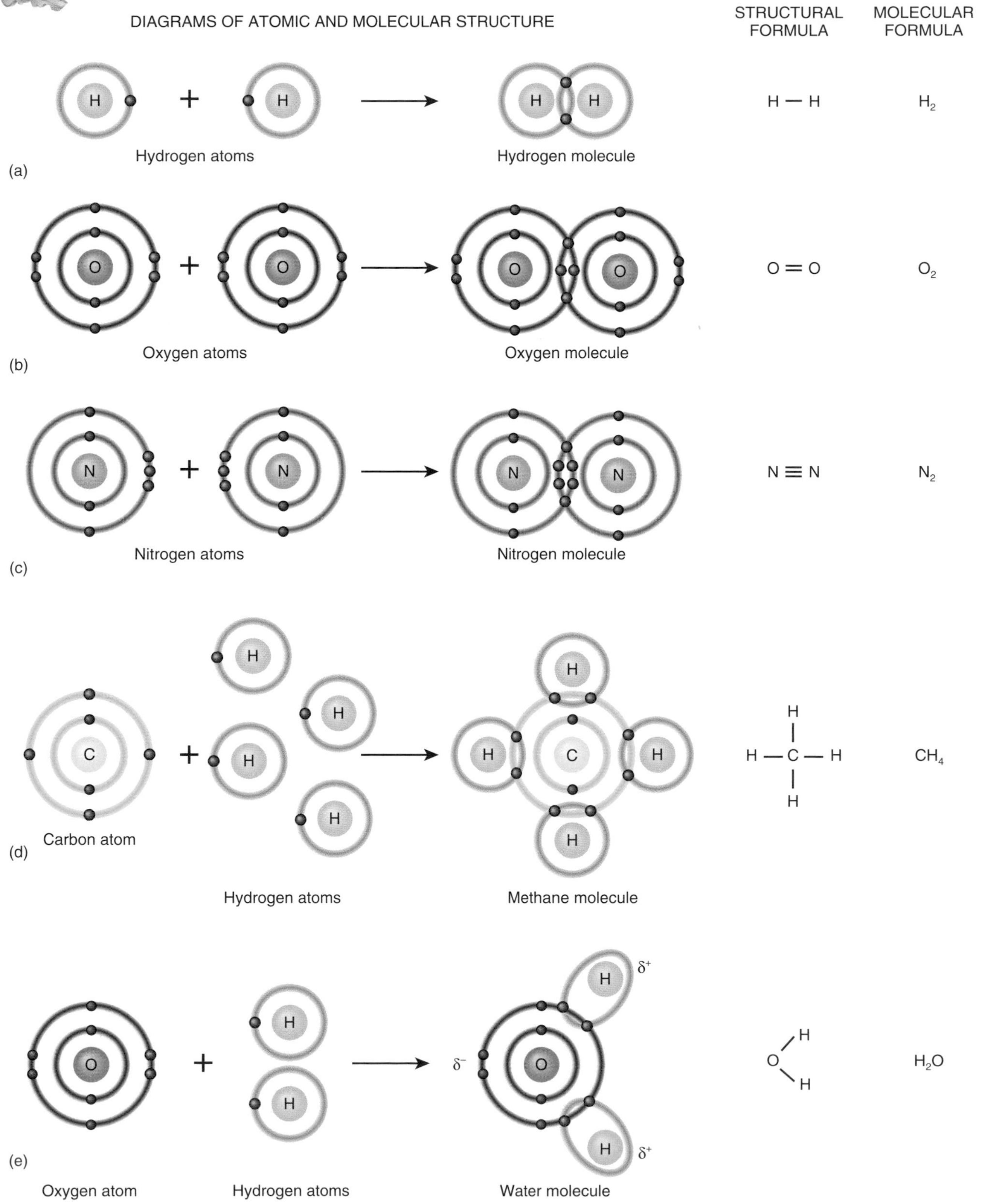

What is the main difference between an ionic bond and a covalent bond?

contains four separate single covalent bonds; each hydrogen atom shares one pair of electrons with the carbon atom (Figure 2.5d).

In some covalent bonds, atoms share the electrons equally—one atom does not attract the shared electrons more strongly than the other atom. This is called a *nonpolar covalent bond*. The bonds between two identical atoms always are nonpolar covalent bonds (Figure 2.5a–c). Another example of a nonpolar covalent bond is the single covalent bond that forms between carbon and each atom of hydrogen in a methane molecule (Figure 2.5d).

In a *polar covalent bond*, the sharing of electrons between atoms is unequal—one atom attracts the shared electrons more strongly than the other. The partial charges are indicated by a lowercase Greek delta (δ) with a minus or plus sign. For example, when polar covalent bonds form, the resulting molecule has a partial negative charge, written δ^-, near the atom that attracts electrons more strongly. At least one other atom in the molecule then will have a partial positive charge, written δ^+. A very important example of a polar covalent bond in living systems is the bond between oxygen and hydrogen in a molecule of water (Figure 2.5e).

Hydrogen Bonds

The polar covalent bonds that form between hydrogen atoms and other atoms can give rise to a third type of chemical bond, a hydrogen bond. A ***hydrogen bond*** forms when a hydrogen atom with a partial positive charge (δ^+) attracts the partial negative charge (δ^-) of neighboring electronegative atoms, most often oxygen or nitrogen. Thus, hydrogen bonds result from attraction of oppositely charged parts of molecules rather than from sharing of electrons as in covalent bonds. Hydrogen bonds are weak when compared to ionic and covalent bonds. Thus, they cannot bind atoms into molecules. However, hydrogen bonds do establish important links between molecules or between different parts of a large molecule, such as deoxyribonucleic acid (DNA). See Figure 2.15.

Chemical Reactions

A ***chemical reaction*** occurs when new bonds form and/or old bonds break between atoms. Through chemical reactions, body structures are built and body functions are carried out, processes that involve transfers of energy.

Forms of Energy and Chemical Reactions

Energy (*en-* = in; *-ergy* = work) is the capacity to do work. The two main forms of energy are ***potential energy***, energy stored by matter due to its *position*, and ***kinetic energy***, the energy of matter *in motion*. For example, the energy stored in a battery or in a person poised to jump down some steps is potential energy. When the battery is used to run a clock or the person jumps, potential energy is converted into kinetic energy. ***Chemical energy*** is a form of potential energy that is stored in the bonds of molecules. In your body, chemical energy in the foods you eat is eventually converted into various forms of kinetic energy, such as mechanical energy, used to walk and talk, and heat energy, used to maintain body temperature. In chemical reactions, breaking old bonds requires an input of energy and forming new bonds releases energy. Because most chemical reactions involve both breaking old bonds and forming new bonds, the *overall reaction* may either release energy or require energy.

Synthesis Reactions

When two or more atoms, ions, or molecules combine to form new and larger molecules, the process is a ***synthesis reaction***. The word *synthesis* means "to put together." Synthesis reactions can be expressed as follows:

$$A + B \xrightarrow{\text{Combine to form}} AB$$

Atom, ion, or molecule A; Atom, ion, or molecule B; New molecule AB

An example of a synthesis reaction is the synthesis of water from hydrogen and oxygen molecules (see Figure 2.3):

$$2\,H_2 + O_2 \xrightarrow{\text{Combine to form}} 2\,H_2O$$

Two hydrogen molecules; One oxygen molecule; Two water molecules

All the synthesis reactions that occur in your body are collectively referred to as ***anabolism*** (a-NAB-ō-lizm). Combining simple molecules like amino acids (discussed shortly) to form large molecules such as proteins is an example of anabolism.

Decomposition Reactions

In a ***decomposition reaction***, a molecule is split apart. The word *decompose* means to break down into smaller parts. Large molecules are split into smaller molecules, ions, or atoms. A decomposition reaction occurs in this way:

$$AB \xrightarrow{\text{Breaks down into}} A + B$$

Molecule AB; Atom, ion, or molecule A; Atom, ion, or molecule B

For example, under the proper conditions, a methane molecule can decompose into one carbon atom and two hydrogen molecules:

$$CH_4 \xrightarrow{\text{Breaks down into}} C + 2\,H_2$$

One methane molecule; One carbon atom; Two hydrogen molecules

The decomposition reactions that occur in your body are collectively referred to as ***catabolism*** (ka-TAB-ō-lizm). The breakdown of large starch molecules into many small glucose molecules during digestion is an example of catabolism.

In general, energy-releasing reactions occur as nutrients, such as glucose, are broken down via decomposition reac-

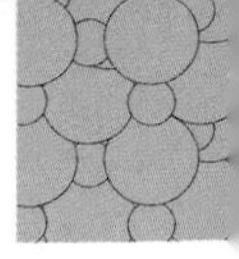

tions. Some of the energy released is temporarily stored in a special molecule called ***adenosine triphosphate (ATP)***, which will be discussed more fully later in this chapter. The energy transferred to the ATP molecules is then used to drive the energy-requiring synthesis reactions that lead to the building of body structures such as muscles and bones.

Exchange Reactions

Many reactions in the body are ***exchange reactions***; they consist of both synthesis and decomposition reactions. One type of exchange reaction works like this:

$$AB + CD \longrightarrow AD + BC$$

The bonds between A and B and between C and D break (decomposition), and new bonds then form (synthesis) between A and D and between B and C. An example of an exchange reaction is:

$$\underset{\text{Hydrochloric acid}}{HCl} + \underset{\text{Sodium bicarbonate}}{NaHCO_3} \longrightarrow \underset{\text{Carbonic acid}}{H_2CO_3} + \underset{\text{Sodium chloride}}{NaCl}$$

Notice that the ions in both compounds have "switched partners": The hydrogen ion (H^+) from HCl has combined with the bicarbonate ion (HCO_3^-) from $NaHCO_3$, and the sodium ion (Na^+) from $NaHCO_3$ has combined with the chloride ion (Cl^-) from HCl.

Reversible Reactions

Some chemical reactions proceed in only one direction, as previously indicated by the single arrows. Other chemical reactions may be reversible. ***Reversible reactions*** can go in either direction under different conditions and are indicated by two half arrows pointing in opposite directions:

$$AB \underset{\text{Combine to form}}{\overset{\text{Breaks down into}}{\rightleftharpoons}} A + B$$

Some reactions are reversible only under special conditions:

$$AB \underset{\text{Heat}}{\overset{\text{Water}}{\rightleftharpoons}} A + B$$

Whatever is written above or below the arrows indicates the condition needed for the reaction to occur. In these reactions, AB breaks down into A and B only when water is added, and A and B react to produce AB only when heat is applied.

■ CHECKPOINT

1. Compare the meanings of atomic number, mass number, ion, and molecule.
2. What is the significance of the valence (outer) electron shell of an atom?
3. Distinguish among ionic, covalent, and hydrogen bonds.
4. Explain the difference between anabolism and catabolism. Which involves synthesis reactions?

CHEMICAL COMPOUNDS AND LIFE PROCESSES

OBJECTIVES • **Discuss the functions of water and inorganic acids, bases, and salts.**

- **Define pH and explain how the body attempts to keep pH within the limits of homeostasis.**
- **Discuss the functions of carbohydrates, lipids, and proteins.**
- **Explain the importance of deoxyribonucleic acid (DNA), ribonucleic acid (RNA), and adenosine triphosphate (ATP).**

Chemicals in the body can be divided into two main classes of compounds: inorganic and organic. ***Inorganic compounds*** usually lack carbon, are structurally simple, and are held together by ionic or covalent bonds. They include water, many salts, acids, and bases. Two inorganic compounds that contain carbon are carbon dioxide (CO_2) and bicarbonate ion (HCO_3^-). ***Organic compounds***, by contrast, always contain carbon, usually contain hydrogen, and always have covalent bonds. Examples include carbohydrates, lipids, proteins, nucleic acids, and adenosine triphosphate (ATP). Organic compounds are discussed in detail in Chapters 19 and 20. Large organic molecules called *macromolecules* are formed by covalent bonding of many identical or similar building-block subunits termed *monomers*.

Inorganic Compounds

Water

Water is the most important and most abundant inorganic compound in all living systems, making up 55% to 60% of body mass in lean adults. With few exceptions, most of the volume of cells and body fluids is water. Several of its properties explain why water is such a vital compound for life.

1. **Water is an excellent solvent.** A ***solvent*** is a liquid or gas in which some other material, called a ***solute***, has been dissolved. The combination of solvent plus solute is called a ***solution***. Water is the solvent that carries nutrients, oxygen, and wastes throughout the body. The versatility of water as a solvent is due to its polar covalent bonds and its "bent" shape (see Figure 2.5e), which allow each water molecule to interact with several neighboring ions or molecules. Solutes that are charged or contain polar covalent bonds are ***hydrophilic*** (*hydro-* = water; *-philic* = loving), which means they dissolve easily in water. Common examples of hydrophilic solutes are sugar and salt. Molecules that contain mainly nonpolar covalent bonds, by contrast, are ***hydrophobic*** (*-phobic* = fearing). They are not very water soluble. Examples of hydrophobic compounds include animal fats and vegetable oils.

2. **Water participates in chemical reactions**. Because water can dissolve so many different substances, it is an ideal medium for chemical reactions. Water also is an active participant in some decomposition and synthesis reactions. During digestion, for example, decomposition reactions break down large nutrient molecules into smaller molecules by the addition of water molecules. This type of reaction is called ***hydrolysis*** (hī-DROL-i-sis; *-lysis* = to loosen or break apart) (see Figure 2.8). Hydrolysis reactions enable dietary nutrients to be absorbed into the body.
3. **Water absorbs and releases heat very slowly**. In comparison to most other substances, water can absorb or release a relatively large amount of heat with only a slight change in its own temperature. The large amount of water in the body thus moderates the effect of changes in the environmental temperature, thereby helping maintain the homeostasis of body temperature.
4. **Water requires a large amount of heat to change from a liquid to a gas**. When the water in sweat evaporates from the skin surface, it takes with it large quantities of heat and provides an excellent cooling mechanism.
5. **Water serves as a lubricant**. Water is a major part of saliva, mucus, and other lubricating fluids. Lubrication is especially necessary in the thoracic and abdominal cavities, where internal organs touch and slide over one another. It is also needed at joints, where bones, ligaments, and tendons rub against one another.

Inorganic Acids, Bases, and Salts

Many inorganic compounds can be classified as acids, bases, or salts. An ***acid*** is a substance that breaks apart or *dissociates* (dis-SŌ-sē-āts´) into one or more *hydrogen ions* (H^+) when it dissolves in water (Figure 2.6a). A ***base***, by contrast, usually dissociates into one or more *hydroxide ions* (OH^-) when it dissolves in water (Figure 2.6b). A ***salt***, when dissolved in water, dissociates into cations and anions, neither of which is H^+ or OH^- (Figure 2.6c).

Acids and bases react with one another to form salts. For example, the reaction of hydrochloric acid (HCl) and potassium hydroxide (KOH), a base, produces the salt potassium chloride (KCl), along with water (H_2O). This exchange reaction can be written as follows:

$$\underset{\text{Acid}}{HCl} + \underset{\text{Base}}{KOH} \longrightarrow \underset{\text{Salt}}{KCl} + \underset{\text{Water}}{H_2O}$$

Acid–Base Balance: The Concept of pH

To ensure homeostasis, body fluids must contain almost balanced quantities of acids and bases. The more hydrogen ions (H^+) dissolved in a solution, the more acidic the solution; conversely, the more hydroxide ions (OH^-), the more basic (alkaline) the solution. The chemical reactions that take place in the body are very sensitive to even small changes in the acidity or alkalinity of the body fluids in which they occur. Any departure from the narrow limits of normal H^+ and OH^- concentrations greatly disrupts body functions.

Figure 2.6 Acids, bases, and salts. (a) When placed in water, hydrochloric acid (HCl) ionizes into H^+ and Cl^-. (b) When the base potassium hydroxide (KOH) is placed in water, it ionizes into OH^- and K^+. (c) When the salt potassium chloride (KCl) is placed in water, it ionizes into positive and negative ions (K^+ and Cl^-), neither of which is H^+ or OH^-.

Ionization is the separation of inorganic acids, bases, and salts into ions in a solution.

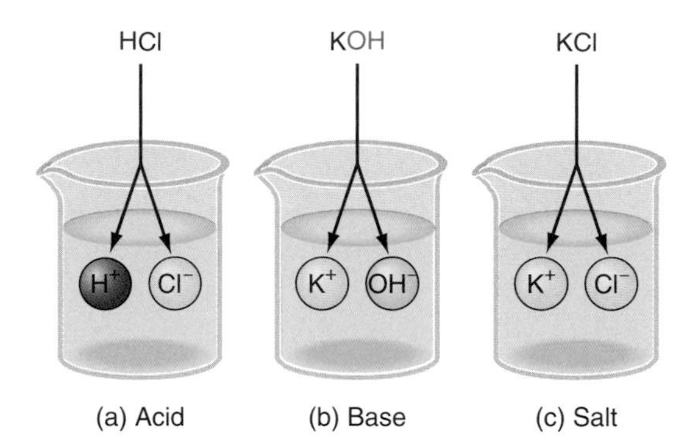

The compound $CaCO_3$ (calcium carbonate) dissociates into a calcium ion (Ca^{2+}) and a carbonate ion (CO_3^{2-}). Is it an acid, a base, or a salt? What about H_2SO_4, which dissociates into two H^+ and one SO_4^{2-}?

A solution's acidity or alkalinity is expressed on the ***pH scale***, which extends from 0 to 14 (Figure 2.7). This scale is based on the number of hydrogen ions in a solution. The midpoint of the pH scale is 7, where the numbers of H^+ and OH^- are equal. A solution with a pH of 7, such as pure water, is neutral—neither acidic nor alkaline. A solution that has more H^+ than OH^- is ***acidic*** and has a pH below 7. A solution that has more OH^- than H^+ is ***basic (alkaline)*** and has a pH above 7. A change of one whole number on the pH scale represents a *10-fold* change in the number of H^+. At a pH of 6, there are 10 times more H^+ than at a pH of 7. Put another way, a pH of 6 is 10 times more acidic than a pH of 7, and a pH of 9 is 100 times more alkaline than a pH of 7.

Maintaining pH: Buffer Systems

Although the pH of various body fluids may differ, the normal limits for each are quite narrow. Table 2.2 shows the pH values for certain body fluids compared with those of common household substances. Homeostatic mechanisms maintain the pH of blood between 7.35 and 7.45, so that it is slightly more basic than pure water. Even though strong acids and bases may be taken into the body or be formed by body cells, the pH of fluids inside and outside cells remains almost constant. One important reason is the presence of ***buffer systems***, in which chemical compounds called *buffers* convert strong acids or bases into weak acids or bases. (More will be said about buffers in Chapter 22.)

Figure 2.7 The pH scale. A pH below 7 indicates an acidic solution, or more H^+ than OH^-. The lower the numerical value of the pH, the more acidic the solution because the H^+ concentration becomes progressively greater. A pH above 7 indicates a basic (alkaline) solution; that is, there are more OH^- than H^+. The higher the pH, the more basic the solution.

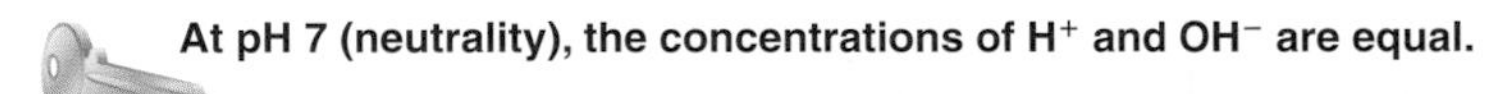

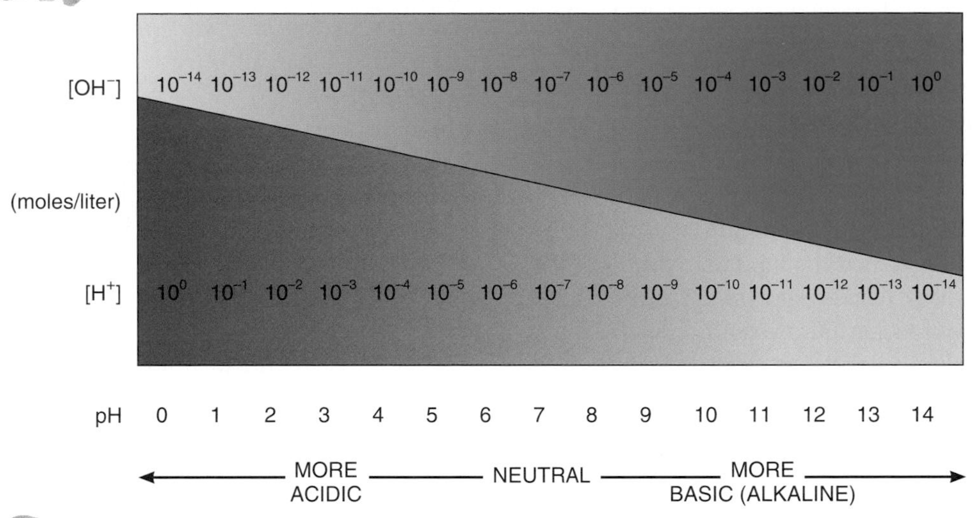

Which pH is more acidic, 6.82 or 6.91? Which pH is closer to neutral, 8.41 or 5.59?

Organic Compounds

Carbohydrates

Carbohydrates include sugars, glycogen, starches, and cellulose. The elements present in carbohydrates are carbon, hydrogen, and oxygen. The ratio of hydrogen to oxygen atoms is usually 2:1, as in water (H_2O), and the number of carbon and oxygen atoms is the same or nearly the same. For example, the molecular formula for the small carbohydrate glucose is $C_6H_{12}O_6$. Carbohydrates are divided into three major groups based on their size: monosaccharides, disaccharides, and polysaccharides. Monosaccharides and disaccharides are termed ***simple sugars***, and polysaccharides are also known as ***complex carbohydrates***.

1. ***Monosaccharides*** (mon′-ō-SAK-a-rīds; *mono-* = one; *sacchar-* = sugar) are the building blocks of carbohydrates. In your body, the principal function of the monosaccharide glucose is to serve as a source of chemical energy for generating the ATP that fuels metabolic reactions. Ribose and deoxyribose are monosaccharides used to make ribonucleic acid (RNA) and deoxyribonucleic acid (DNA), which are described on pages 38-39.
2. ***Disaccharides*** (dī-SAK-a-rīds; *di-* = two) are simple sugars that consist of two monosaccharides joined by a covalent bond. When two monosaccharides (smaller molecules) combine to form a disaccharide (a larger molecule), a molecule of water is formed and removed. Such a reaction is called ***dehydration synthesis***

Table 2.2 pH Values of Selected Substances

Substance*	pH Value
Gastric juice (digestive juice of the stomach)	1.2–3.0
Lemon juice	2.3
Grapefruit juice, vinegar, wine	3.0
Carbonated soft drink	3.0–3.5
Orange juice	3.5
Vaginal fluid	3.5–4.5
Tomato juice	4.2
Coffee	5.0
Urine	4.6–8.0
Saliva	6.35–6.85
Cow's milk	6.8
Distilled (pure) water	7.0
Blood	7.35–7.45
Semen (fluid containing sperm)	7.20–7.60
Cerebrospinal fluid (fluid associated with the nervous system)	7.4
Pancreatic juice (digestive juice of the pancreas)	7.1–8.2
Bile (liver secretion that aids fat digestion)	7.6–8.6
Milk of magnesia	10.5
Lye	14.0

*Substances in the human body are highlighted in gold.

(*de-* = from, down, or out; *hydra-* = water). Such reactions occur during synthesis of large molecules. For example, the monosaccharides glucose and fructose combine to form the disaccharide sucrose (table sugar) as shown in Figure 2.8. Disaccharides can be split into monosaccharides by adding a molecule of water, a hydrolysis reaction. Sucrose, for example, may be hydrolyzed into its components of glucose and fructose by the addition of water (Figure 2.8). Other disaccharides include maltose (glucose + glucose), or malt sugar, and lactose (glucose + galactose), the sugar in milk.

3. ***Polysaccharides*** (pol′-ē-SAK-a-rīds; *poly-* = many) are large, complex carbohydrates that contain tens or hundreds of monosaccharides joined through dehydration synthesis reactions. Like disaccharides, polysaccharides can be broken down into monosaccharides through hydrolysis reactions. The main polysaccharide in the human body is *glycogen*, which is made entirely of glucose units joined together in branching chains (Figure 2.9). Glycogen is stored in cells of the liver and in skeletal muscles. If energy demands of the body are high, glycogen is broken down into glucose; when energy demands are low, glucose is built back up into glycogen. *Starches* are also made of glucose units and are polysaccharides made mostly by plants. We digest starches to glucose as another energy source. *Cellulose* is a polysaccharide found in plant cell walls. Although humans cannot digest cellulose, it does provide bulk (roughage or fiber) that helps move feces through the large intestine. Unlike simple sugars, polysaccharides usually are not soluble in water and do not taste sweet.

Lipids

Like carbohydrates, ***lipids*** (*lip-* = fat) contain carbon, hydrogen, and oxygen. Unlike carbohydrates, they do not have a 2:1 ratio of hydrogen to oxygen. The proportion of oxygen atoms in lipids is usually smaller than in carbohydrates, so there are fewer polar covalent bonds. As a result, most lipids are hydrophobic; that is, they are insoluble in water (see page 29).

The diverse lipid family includes triglycerides (fats and oils), phospholipids (lipids that contain phosphorus), steroids, fatty acids, and fat-soluble vitamins (vitamins A, D, E, and K).

The most plentiful lipids in your body and in your diet are the ***triglycerides*** (trī-GLI-cer-īdes; *tri-* = three). At room temperature, triglycerides may be either solids (fats) or liquids (oils). They are the body's most highly concentrated form of chemical energy, storing more than twice as much chemical energy per gram as carbohydrates or proteins. Our capacity to store triglycerides in fat tissue, called adipose tissue, for all practical purposes, is unlimited. Excess dietary carbohydrates, proteins, fats, and oils all have the same fate: They are deposited in adipose tissue as triglycerides.

Figure 2.8 Dehydration synthesis and hydrolysis of a molecule of sucrose. In the dehydration synthesis reaction (read from left to right), two smaller molecules, glucose and fructose, are joined to form a larger molecule of sucrose. Note the loss of a water molecule. In the hydrolysis reaction (read from right to left), the larger sucrose molecule is broken down into two smaller molecules, glucose and fructose. Here, a molecule of water is added to sucrose for the reaction to occur.

Monosaccharides are the building blocks of carbohydrates.

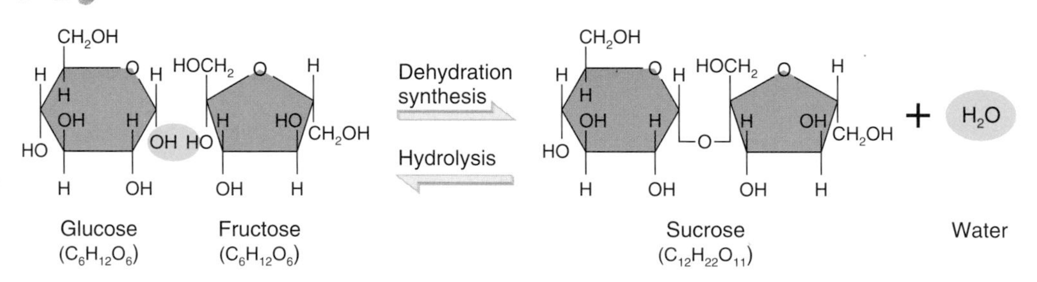

(a) Dehydration synthesis and hydrolysis of sucrose

All atoms written out

Standard shorthand

(b) Alternate chemical structures of organic molecules (shown here is glucose)

How many carbons are there in fructose? in sucrose?

Figure 2.9 Part of a glycogen molecule, the main polysaccharide in the human body.

Glycogen is made up of glucose units and is the storage form of carbohydrate in the human body.

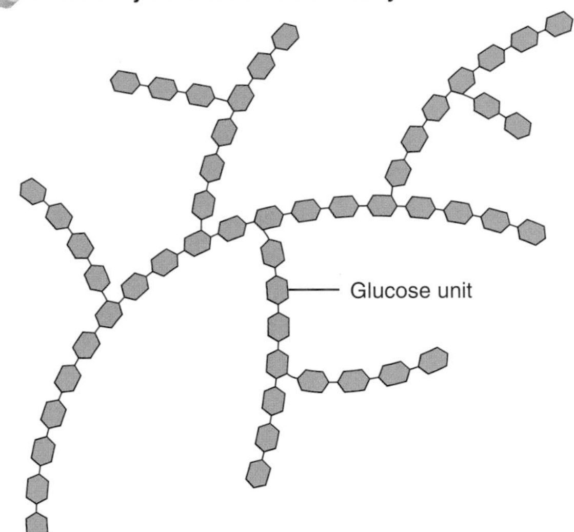

Which body cells store glycogen?

A triglyceride consists of two types of building blocks: a single glycerol molecule and three fatty acid molecules. A three-carbon ***glycerol*** molecule forms the backbone of a triglyceride (Figure 2.10). Three ***fatty acids*** are attached, by dehydration synthesis reactions, one to each carbon of the glycerol backbone. The fatty acid chains of a triglyceride may be saturated, monounsaturated, or polyunsaturated. ***Saturated fats*** contain only *single covalent bonds* between fatty acid carbon atoms. Because they do not contain any double bonds between fatty acid carbon atoms, each carbon atom is *saturated with hydrogen atoms* (see palmitic acid and stearic acid in Figure 2.10). Triglycerides with mainly saturated fatty acids are solid at room temperature and occur mostly in meats (especially red meats) and nonskim dairy products (whole milk, cheese, and butter). They also occur in a few tropical plants, such as cocoa, palm, and coconut. Diets that contain large amounts of saturated fats are associated with disorders such as heart disease and colorectal cancer. ***Monounsaturated fats*** (*mono-* = one) contain fatty acids with *one double covalent bond* between two fatty acid carbon atoms and thus are not completely saturated with hydrogen atoms (see oleic acid in Figure 2.10). Olive oil, peanut oil, canola oil, most nuts, and avocados are rich in triglycerides with monounsaturated fatty acids. Monosaturated fats are thought to decrease the risk of heart disease. ***Polyunsaturated fats*** (*poly-* = many) contain *more than one double covalent bond* between fatty acid carbon atoms. Corn oil, safflower oil, sunflower oil, soybean oil, and fatty fish (salmon, tuna, and mackerel) contain a high percentage of polyunsaturated fatty acids. Polysaturated fats are also believed to decrease the risk of heart disease. However, when products such as margarine and vegetable shortening are made from polyunsaturated fats, compounds called *trans* fatty acids are produced. Trans fatty acids, like saturated fats, increase the risk of cardiovascular disease.

A group of fatty acids called **essential fatty acids (EFAs)** are essential to human health. However, they cannot be made by the human body and must be obtained from foods or

Figure 2.10 Triglycerides consist of three fatty acids attached to a glycerol backbone. The fatty acids vary in length and the number and location of double bonds between carbon atoms (C═C). Shown here is a triglyceride molecule that contains two saturated fatty acids and one monounsaturated fatty acid.

A triglyceride consists of two types of building blocks: a single glycerol molecule and three fatty acid molecules.

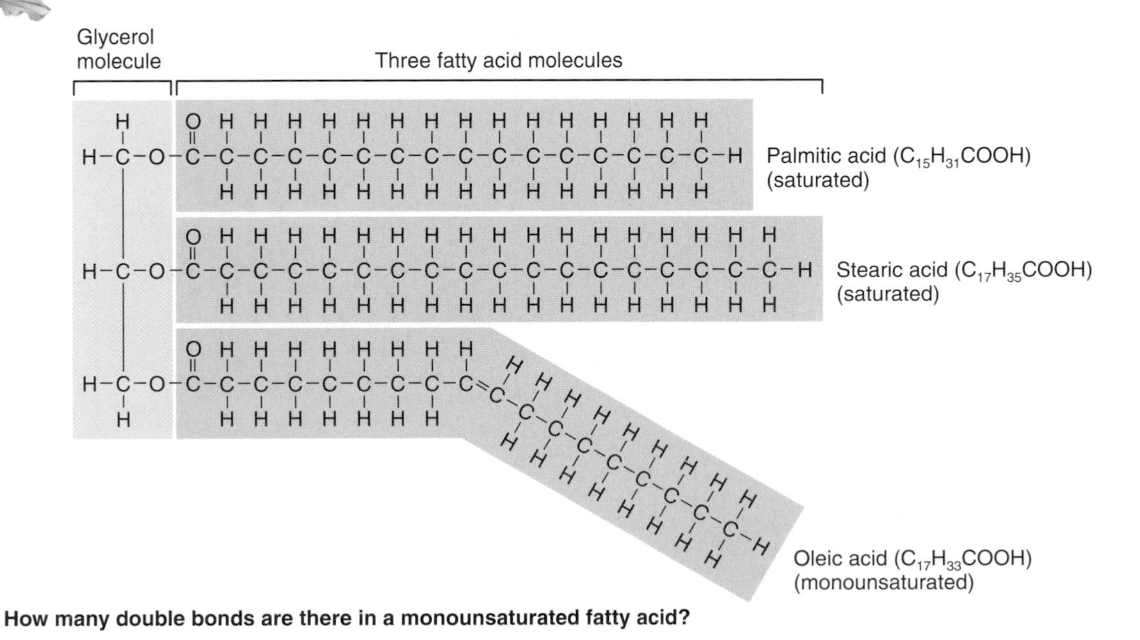

How many double bonds are there in a monounsaturated fatty acid?

supplements. Among the more important EFAs are *omega-3 fatty acids*, *omega-6 fatty acids*, and cis-*fatty acids*.

Omega-3 and omega-6 fatty acids are polyunsaturated fatty acids that may have a protective effect against heart disease and stroke by lowering total cholesterol, raising HDL (high-density lipoproteins or "good cholesterol") and lowering LDL (low-density lipoproteins or "bad cholesterol"). In addition, they decrease bone loss; reduce symptoms of arthritis due to inflammation; promote wound healing; improve certain skin disorders (psoriasis, eczema, and acne); and improve mental functions. Primary sources of omega-3 fatty acids include flaxseed, fatty fish, oils that have large amounts of polyunsaturated fats, fish oils, and walnuts. Primary sources of omega-6 fatty acids include most processed foods (cereals, breads, white rice), eggs, baked goods, oils with large amounts of polyunsaturated fats, and meats (especially organ meats, such as liver).

Cis-fatty acids are nutritionally beneficial monosaturated fatty acids that are used by the body to produce hormone-like regulators and cell membranes. However, when *cis*-fatty acids are heated, pressurized, and combined with a catalyst (usually nickel) in a process called *hydrogenation*, they are changed to unhealthy *trans* fatty acids. Hydrogenation is used by manufacturers to make vegetable oils solid at room temperature and less likely to turn rancid. Hydrogenated or *trans* fatty acids are common in commercially baked goods (crackers, cakes, and cookies), salty snack foods, some margarines, and fried foods (donuts and french fries). If a product label contains the words hydrogenated or partially hydrogenated, then the product contains *trans* fatty acids. Among the adverse effects of *trans* fatty acids are an increase in total cholesterol, a decrease in HDL, an increase in LDL, and an increase in triglycerides. These effects, which can increase the risk of heart disease and other cardiovascular diseases, are similar to those caused by saturated fats.

Like triglycerides, ***phospholipids*** have a glycerol backbone and two fatty acids attached to the first two carbons (Figure 2.11a). Attached to the third carbon is a phosphate

Figure 2.11 Phospholipids. (a) In the synthesis of phospholipids, two fatty acids attach to the first two carbons of the glycerol backbone. A phosphate group links a small charged group to the third carbon in glycerol. In (b), the circle represents the polar head region, and the two wavy lines represent the two nonpolar tails.

Phospholipids are the main lipids in cell membranes.

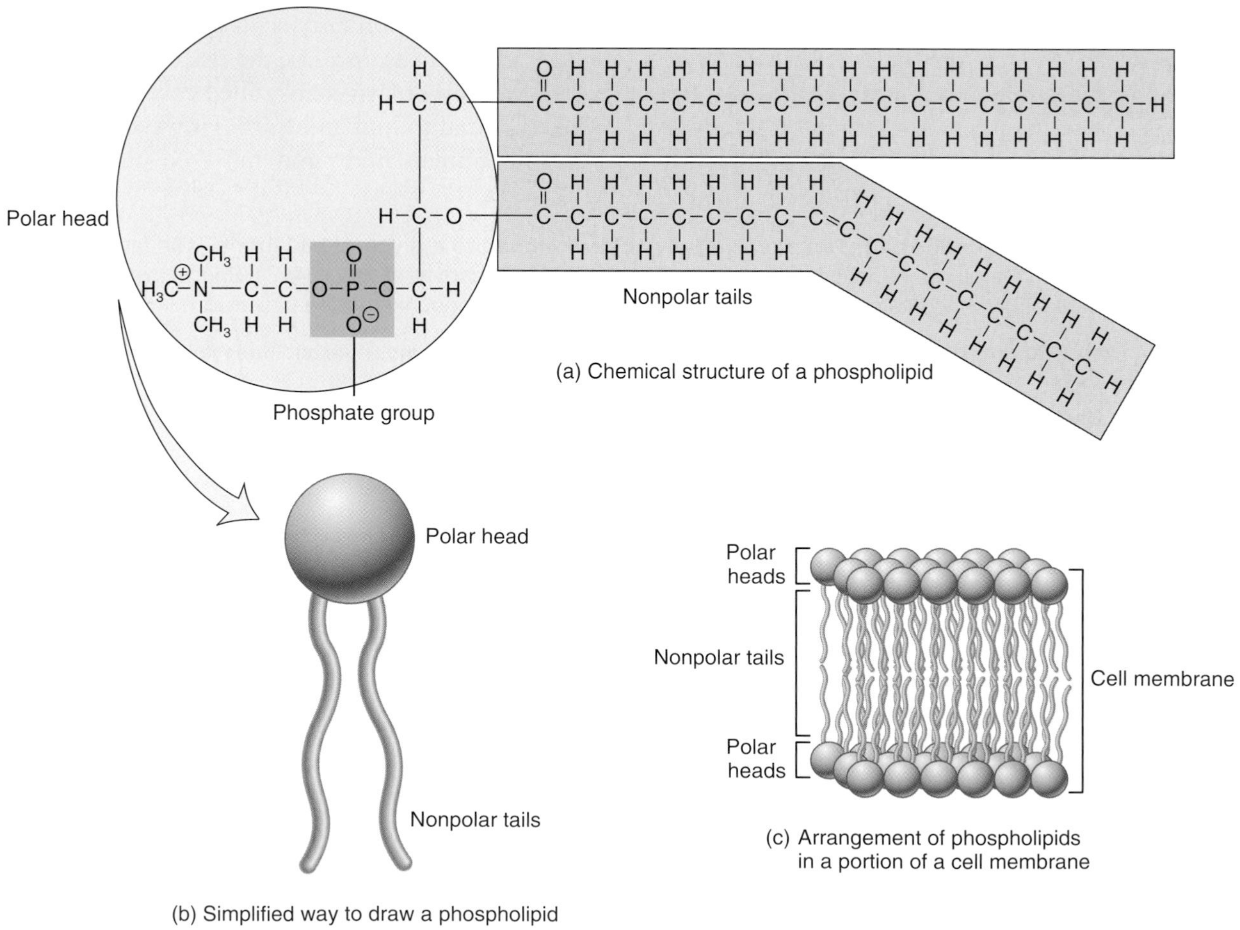

(a) Chemical structure of a phospholipid

(b) Simplified way to draw a phospholipid

(c) Arrangement of phospholipids in a portion of a cell membrane

How does a phospholipid differ from a triglyceride?

Figure 2.12 Steroids. All steroids have four rings of carbon atoms.

Cholesterol is the starting material for synthesis of other steroids in the body.

4 rings

(a) Cholesterol

(b) Estradiol (an estrogen or female sex hormone)

Which dietary lipids are thought to contribute to atherosclerosis?

group (PO_4^{3-}) that links a small charged group to the glycerol backbone. Whereas the nonpolar fatty acids form the hydrophobic "tails" of a phospholipid, the polar phosphate group and charged group form the hydrophilic "head" (Figure 2.11b). Phospholipids line up tails-to-tails in a double row to make up much of the membrane that surrounds each cell (Figure 2.11c).

The structure of ***steroids***, with their four rings of carbon atoms, differs considerably from that of the triglycerides and phospholipids. Cholesterol (Figure 2.12a), which is needed for membrane structure, is the steroid from which other steroids may be synthesized by body cells. For example, cells in the ovaries of females synthesize estradiol (Figure 2.12b), which is one of the estrogens or female sex hormones. Estrogens regulate sexual functions. Other steroids include testosterone (the main male sex hormone), which also regulates sexual functions; cortisol, which is necessary for maintaining normal blood sugar levels; bile salts, which are needed for lipid digestion and absorption; and vitamin D, which is related to bone growth.

Proteins

Proteins are large molecules that contain carbon, hydrogen, oxygen, and nitrogen; some proteins also contain sulfur. Much more complex in structure than carbohydrates or lipids, proteins have many roles in the body and are largely responsible for the structure of body cells. For example, proteins termed enzymes speed up particular chemical reactions, other proteins are responsible for contraction of muscles, proteins called antibodies help defend the body against invading microbes, and some hormones are proteins.

Amino acids (a-MĒ-nō) are the building blocks of proteins. All amino acids have an *amino group* ($—NH_2$) at one end and a *carboxyl group* (—COOH) at the other end. Each of the 20 different amino acids has a different *side chain* (R group) (Figure 2.13a). The covalent bonds that join amino acids together to form more complex molecules are called *peptide bonds* (Figure 2.13b).

Figure 2.13 Amino acids. (a) In keeping with their name, amino acids have an amino group (shaded blue) and a carboxyl (acid) group (shaded red). The side chain (R group) is shaded gold and is different in each type of amino acid. (b) When two amino acids are chemically united by dehydration synthesis (read from left to right), the resulting covalent bond between them is called a peptide bond. The peptide bond is formed at the point where water is lost. Here, the amino acids glycine and alanine are joined to form the dipeptide glycylalanine. Breaking a peptide bond occurs by hydrolysis (read from right to left).

Amino acids are the building blocks of proteins.

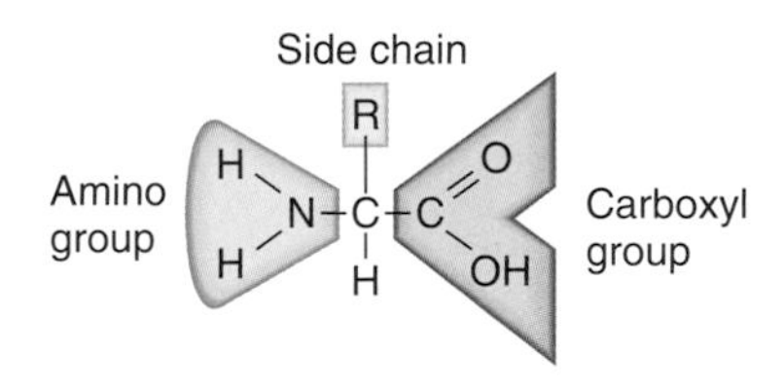

(a) Structure of an amino acid

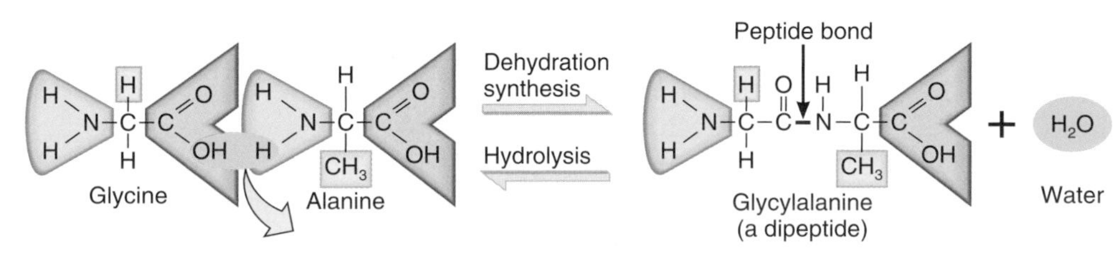

(b) Protein formation

How many peptide bonds would there be in a tripeptide?

Focus on Wellness

Herbal Supplements—They're Natural but Are They Safe?

Sales of herbal supplements are booming. Preparations of ginseng and echinacea stand next to bottles of vitamin C and aspirin in medicine cabinets across North America. But beware: Although some herbal supplements are helpful for specific problems, others are a waste of money, and many can be harmful to your health.

Does Natural Mean Safe?

Herbal supplements are preparations made from the leaves, flowers, bark, berries, or roots of plants. Herbal preparations have been used throughout the ages in cultures around the world to relieve pain, heal wounds, chase away evil spirits, and even to kill. Many of the active ingredients in drugs we use today were originally isolated from herbs. For example, the heart drug digitalis comes from the foxglove plant.

Anyone who knows some chemistry can understand why "natural" does not necessarily mean "safe." Natural chemicals are still chemicals. They participate in chemical reactions in your body. They have chemical effects in the same way that manufactured drugs do.

Herbal products can't be effective and harmless at the same time because anything that has a physiological effect can be harmful at some dose. All drugs become toxic if you take too much of them.

Handle with Care

If you want to use herbal supplements, you must also use your head. Because regulation of these supplements is currently fairly loose in most countries, you can't believe everything the manufacturer says on the label or in advertising literature. If a product sounds too good to be true, beware!

Health-care professionals are especially concerned about the lack of data on the long-term safety of many herbal products. Scientists are just beginning to investigate the use of herbs, and our understanding of these remedies is still in its infancy.

Talk to Your Doctor

If you decide to try herbal supplements for an ailment, talk to your health-care provider to be sure you are not overlooking beneficial medical treatments. If you are taking any medications, ask your pharmacist whether you should be concerned about possible interactions between the supplement and your drugs. For example, it is dangerous to take ginkgo biloba and aspirin together, because both have potent blood-thinning effects that can lead to dangerous bleeding.

Women who are pregnant, intending to become pregnant, or nursing a baby should avoid supplements in the same way that they avoid drugs.

► **Think It Over . . .**

► ***Your Aunt Mary tells you she is taking an herbal weight-loss supplement. "It's natural, so it's safe," she says. In fact, it's not working as well as it was two weeks ago, so she is now taking double the recommended dose. What would you say to her?***

The union of two or more amino acids produces a ***peptide*** (PEP-tīd). When two amino acids combine, the molecule is called a ***dipeptide*** (Figure 2.13b). Adding another amino acid to a dipeptide produces a ***tripeptide***. A ***polypeptide*** contains a large number of amino acids. Proteins are polypeptides that contain as few as 50 or as many as 2000 amino acids. Because each variation in the number and sequence of amino acids produces a different protein, a great variety of proteins is possible. The situation is similar to using an alphabet of 20 letters to form words. Each letter would be equivalent to an amino acid, and each word would be a different protein.

> An alteration in the sequence of amino acids can have serious consequences. For example, a single substitution of an amino acid in hemoglobin, a blood protein, can result in a deformed molecule that produces **sickle-cell disease** (page 360).

A protein may consist of only one polypeptide or several intertwined polypeptides. A given type of protein has a unique three-dimensional shape because of the ways that each individual polypeptide twists and folds as associated polypeptides come together. If a protein encounters a hostile environment in which temperature, pH, or ion concentration is significantly altered, it may unravel and lose its characteristic shape. This process is called ***denaturation*** (dē-nā′-chur-Ā-shun). Denatured proteins are no longer functional. A common example of denaturation is seen in frying an egg. In a raw egg, the egg-white protein (albumin) is soluble and the egg white appears as a clear, viscous fluid. When heat is applied to the egg, however, the albumin denatures; it changes shape, becomes insoluble, and turns white.

Enzymes

As we have seen, chemical reactions occur when chemical bonds are made or broken as atoms, ions, or molecules collide with one another. At normal body temperature, such collisions occur too infrequently to maintain life. ***Enzymes*** (EN-zīms) are the living cell's solution to this problem, because they speed up chemical reactions by increasing the frequency of collisions and by properly orienting the colliding molecules. Substances such as enzymes that can speed up chemical reactions without themselves being altered are called ***catalysts*** (KAT-a-lists). In living cells, most enzymes are proteins. The names of enzymes usually end in ***-ase***. All enzymes can be grouped according to the types of chemical reactions they catalyze. For example, *oxidases* add oxygen, *kinases* add phosphate, *dehydrogenases* remove hydrogen, *anhydrases* remove water, *ATPases* split ATP, *proteases* break down proteins, and *lipases* break down lipids.

Enzymes catalyze selected reactions with great efficiency and with many built-in controls. Three important properties of enzymes are their specificity, efficiency, and control.

1. **Specificity.** Enzymes are highly specific. Each particular enzyme catalyzes a particular chemical reaction that involves specific ***substrates***, the molecules on which the enzyme acts, and that gives rise to specific ***products***, the molecules produced by the reaction. In some cases, the enzyme fits the substrate like a key fits in a lock. In other cases, the enzyme changes its shape to fit snugly around the substrate once the substrate and enzyme come together. Each of the more than 1000 known enzymes in your body has a characteristic three-dimensional shape with a specific surface configuration that allows it to fit specific substrates.
2. **Efficiency**. Under optimal conditions, enzymes can catalyze reactions at rates that are millions to billions of times more rapid than those of similar reactions occurring without enzymes. A single enzyme molecule can convert substrate molecules to product molecules at rates as high as 600,000 per second.
3. **Control**. Enzymes are subject to a variety of cellular controls. Their rate of synthesis and their concentration at any given time are under the control of a cell's genes. Substances within the cell may either enhance or inhibit activity of a given enzyme. Many enzymes exist in both active and inactive forms within the cell. The rate at which the inactive form becomes active or vice versa is determined by the chemical environment inside the cell. Many enzymes require a nonprotein substance, known as a ***cofactor*** or ***coenzyme***, to operate properly. Ions of iron, zinc, magnesium, or calcium are cofactors; niacin or riboflavin, derivatives of B vitamins, act as coenzymes.

Figure 2.14 illustrates the actions of an enzyme.

1. The substrates attach to the ***active site*** of the enzyme molecule, the specific part of the enzyme that catalyzes the reaction, forming a temporary compound called the ***enzyme–substrate complex***. In this reaction, the substrates are the disaccharide sucrose and a molecule of water.
2. The substrate molecules are transformed by the rearrangement of existing atoms, the breakdown of the substrate molecule, or the combination of several substrate molecules into products of the reaction. Here the products are two monosaccharides: glucose and fructose.
3. After the reaction is completed and the reaction products move away from the enzyme, the unchanged enzyme is free to attach to another substrate molecule.

Figure 2.14 How an enzyme works.

An enzyme speeds up a chemical reaction without being altered or consumed.

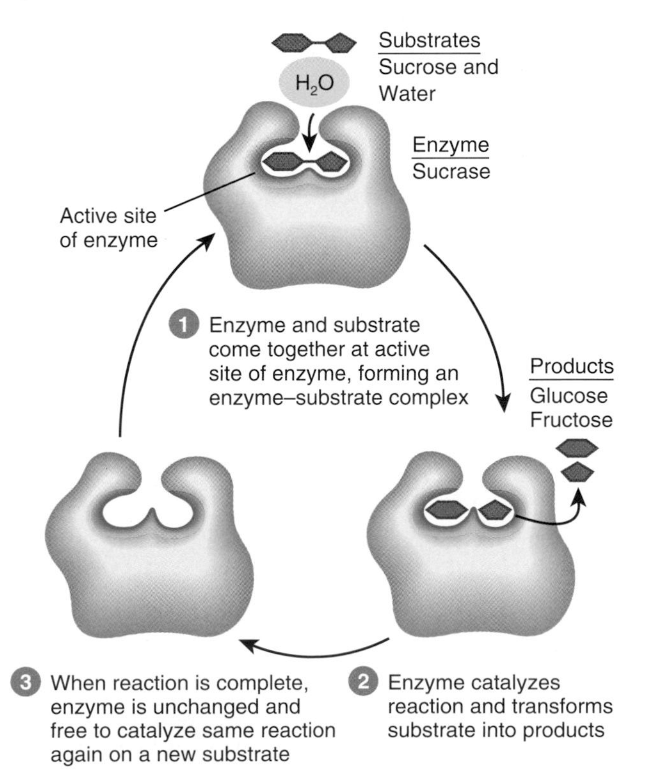

What part of an enzyme combines with its substrate?

Enzyme deficiencies may lead to certain disorders. For example, some people do not produce enough lactase, an enzyme that breaks down the disaccharide lactose into the monosaccharides glucose and galactose. This deficiency causes a condition called **lactose intolerance**, in which undigested lactose retains fluid in the feces, and bacterial fermentation of lactose results in the production of gases. Symptoms of lactose intolerance include diarrhea, gas, bloating, and abdominal cramps after consumption of milk and other dairy products. The severity of symptoms varies from relatively minor to sufficiently serious to require medical attention. Persons with lactose intolerance can take dietary enzyme supplements to aid in the digestion of lactose.

Nucleic Acids: Deoxyribonucleic Acid (DNA) and Ribonucleic Acid (RNA)

Nucleic acids (noo-KLĒ-ic), so named because they were first discovered in the nuclei of cells, are huge organic molecules that contain carbon, hydrogen, oxygen, nitrogen, and phosphorus. The two kinds of nucleic acids are ***deoxyribonucleic acid (DNA)*** (dē-ok′-sē-rī′-bō-noo-KLĒ-ik) and ***ribonucleic acid (RNA)***.

A nucleic acid molecule is composed of repeating building blocks called ***nucleotides***. Each nucleotide of DNA consists of three parts (Figure 2.15a):

- One of four different ***nitrogenous bases***, ring-shaped molecules that contain atoms of C, H, O, and N.
- A five-carbon monosaccharide called *deoxyribose*.
- A *phosphate group* (PO_4^{3-}).

In DNA, the four bases are adenine (A), thymine (T), cytosine (C), and guanine (G). Figure 2.15b shows the following structural characteristics of the DNA molecule:

1. The molecule consists of two strands, with crossbars. The strands twist about each other in the form of a ***double helix*** so that the shape resembles a twisted rope ladder.
2. The uprights (strands) of the DNA ladder consist of alternating phosphate groups and the deoxyribose portions of the nucleotides.
3. The rungs of the ladder contain paired nitrogenous bases, which are held together by hydrogen bonds. Adenine always pairs with thymine, and cytosine always pairs with guanine.

About 1000 rungs of DNA comprise a ***gene***, a portion of a DNA strand that performs a specific function, for example, providing instructions to synthesize the hormone insulin. Humans have about 30,000 genes. Genes determine which traits we inherit, and they control all the activities that take place in our cells throughout a lifetime. Any change that occurs in the sequence of nitrogenous bases of a gene is called a *mutation*. Some mutations can result in the death of a cell, cause cancer, or produce genetic defects in future generations.

RNA, the second kind of nucleic acid, is copied from DNA but differs from DNA in several respects. DNA is double-stranded, RNA is single-stranded. The sugar in the RNA nucleotide is ribose, and RNA contains the nitrogenous base uracil (U) rather than thymine. Cells contain three different kinds of RNA: messenger RNA, ribosomal RNA, and transfer RNA. Each has a specific role to perform in carrying out the instructions encoded in DNA, as will be described in Chapter 3.

Adenosine Triphosphate

Adenosine triphosphate (a-DEN-ō-sēn trī-FOS-fāt) or ***ATP*** is the "energy currency" of living organisms. As you learned earlier in the chapter, ATP transfers energy from energy-releasing reactions to energy-requiring reactions that maintain cellular activities. Among these cellular activities are contraction of muscles, movement of chromosomes during cell division, movement of structures within cells, transport of substances across cell membranes, and synthesis of larger molecules from smaller ones.

Structurally, ATP consists of three phosphate groups attached to adenosine, which is composed of adenine and ribose (Figure 2.16). The energy-transferring reaction occurs via hydrolysis: Removal of the last phosphate group (PO_4^{3-}), symbolized by Ⓟ in the following discussion, by addition of a water molecule liberates energy and leaves a molecule called ***adenosine diphosphate (ADP)***. The enzyme that catalyzes the hydrolysis of ATP is called *ATPase*. This reaction may be represented as follows:

$$\underset{\text{Adenosine triphosphate}}{\text{ATP}} + \underset{\text{Water}}{H_2O} \xrightarrow{\textit{ATPase}} \underset{\text{Phosphate group}}{\text{Ⓟ}} + \underset{\text{Energy}}{\text{E}}$$

The energy released by the breakdown of ATP into ADP is constantly being used by the cell. As the supply of ATP at any given time is limited, a mechanism exists to replenish it: The enzyme *ATP synthase* promotes the addition of a phosphate group to ADP. The reaction may be represented as follows:

$$\underset{\text{Adenosine diphosphate}}{\text{ADP}} + \underset{\text{Phosphate group}}{\text{Ⓟ}} + \underset{\text{Energy}}{\text{E}} \xrightarrow{\textit{ATP synthase}} \underset{\text{Adenosine triphosphate}}{\text{ATP}} + \underset{\text{Water}}{H_2O}$$

As you can see from this reaction, energy is required to produce ATP. The energy needed to attach a phosphate group to ADP is supplied mainly by the breakdown of glucose in a process called cellular respiration, which you will learn more about in Chapter 20.

Figure 2.15 DNA molecule. (a) A nucleotide consists of a nitrogenous base, a five-carbon sugar, and a phosphate group. (b) The paired nitrogenous bases project toward the center of the double helix. The structure is stabilized by hydrogen bonds (dotted lines) between each base pair. There are two hydrogen bonds between adenine and thymine and three between cytosine and guanine.

Nucleotides are the building blocks of nucleic acids.

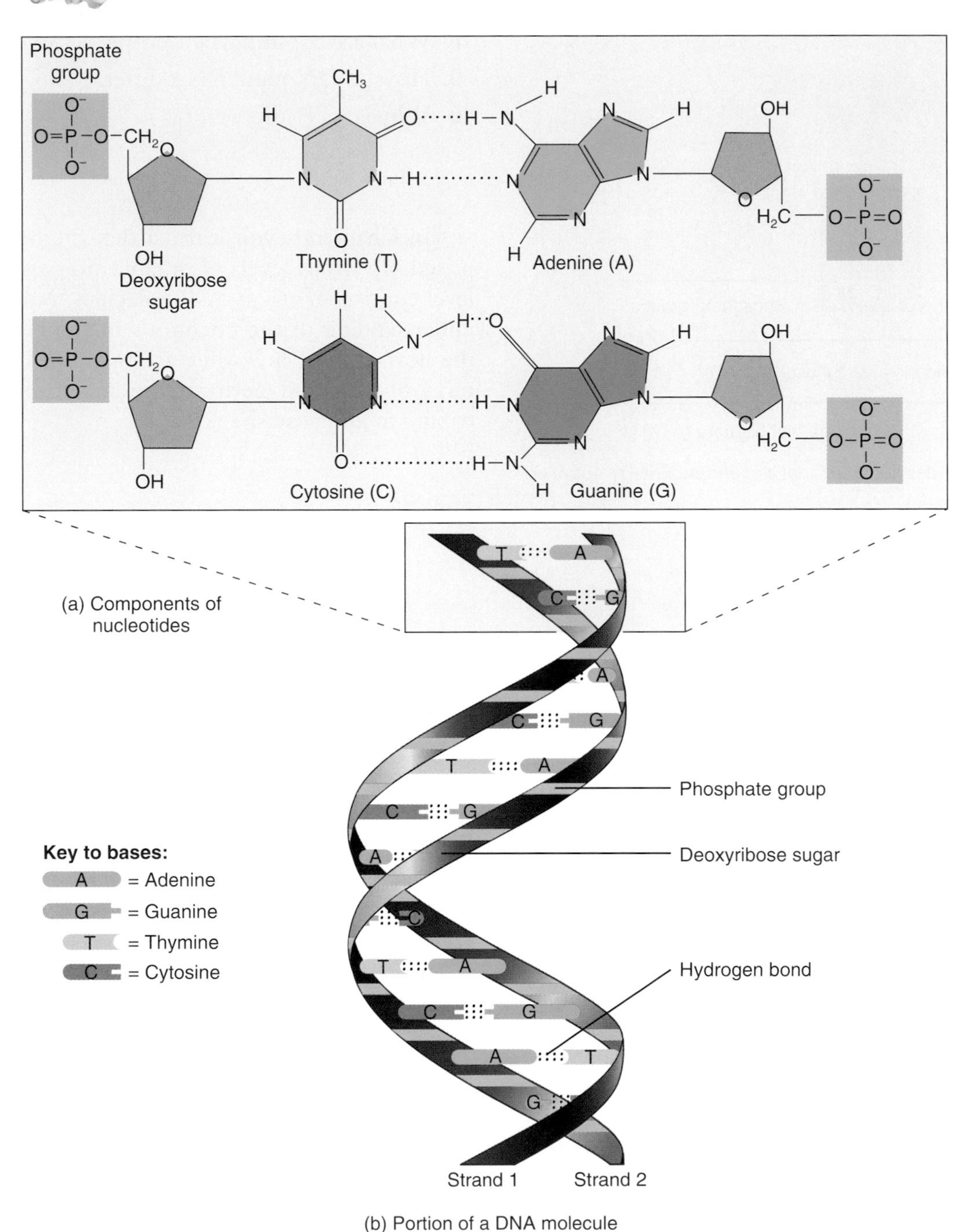

(b) Portion of a DNA molecule

Which nitrogenous base is not present in RNA? Which nitrogenous base is not present in DNA?

Figure 2.16 Structure of ATP and ADP. The two phosphate bonds that can be used to transfer energy are indicated in red. Most often energy transfer involves hydrolysis of the terminal phosphate bond of ATP.

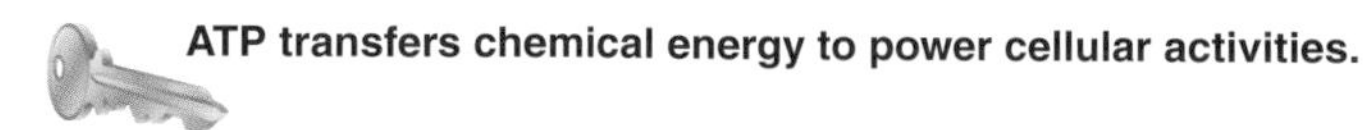

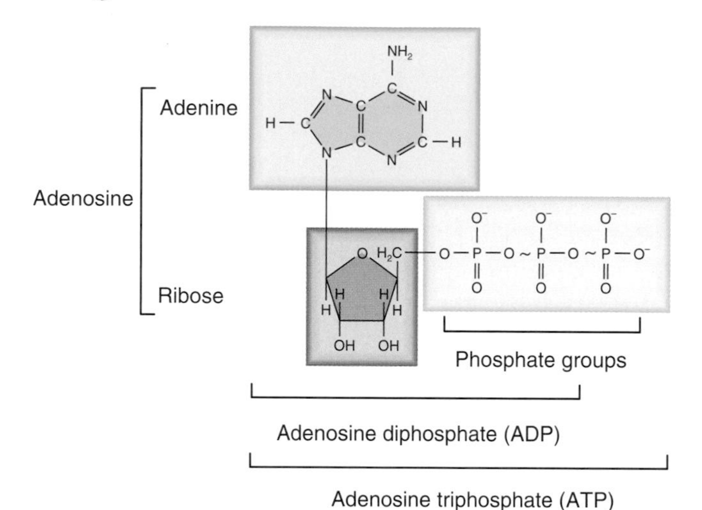

What are some cellular activities that depend on energy supplied by ATP?

■ **CHECKPOINT**

5. How do inorganic compounds differ from organic compounds?
6. What functions does water perform in the body?
7. Distinguish among saturated, monounsaturated, and polyunsaturated fats.
8. What are the important properties of enzymes?
9. How do DNA and RNA differ?
10. Why is ATP important?

• • •

In Chapter 1, you learned that the human body is comprised of various levels of organization and that the chemical level consists of atoms and molecules. Now that you have an understanding of the chemicals in the body, you will see in the next chapter how they are organized to form the structures of cells and perform the activities of cells that contribute to homeostasis.

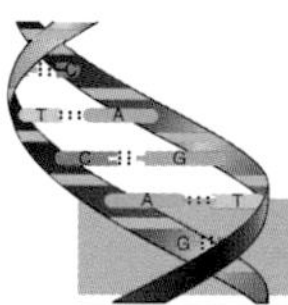

STUDY OUTLINE

Introduction to Chemistry (p. 23)

1. Chemistry is the science of the structure and interactions of matter, which is anything that occupies space and has mass. Matter is made up of chemical elements.
2. The elements oxygen (O), carbon (C), hydrogen (H), and nitrogen (N) make up 96% of the body's mass.
3. Each element is made up of units called atoms, which consist of a nucleus that contains protons and neutrons, and electrons that move about the nucleus in electron shells. The number of electrons is equal to the number of protons in an atom.
4. The atomic number, the number of protons, distinguishes the atoms of one element from those of another element.
5. The combined total of protons and neutrons in an atom is its mass number.
6. An atom that *gives up* or *gains* electrons becomes an ion—an atom that has a positive or negative charge due to having unequal numbers of protons and electrons.
7. A molecule is a substance that consists of two or more chemically combined atoms. The molecular formula indicates the number and type of atoms that make up a molecule.
8. A compound is a substance that can be broken down into two or more different elements by ordinary chemical means.
9. A free radical is a destructive, electrically charged ion or molecule that has an unpaired electron in its outermost shell.
10. Chemical bonds hold the atoms of a molecule together.
11. Electrons in the valence (outermost) shell are the parts of an atom that participate in chemical reactions.
12. When outer shell electrons are transferred from one atom to another, the transfer forms ions, whose unlike charges attract each other and form ionic bonds. Positively charged ions are called cations; negatively charged ions are called anions.
13. In a covalent bond, pairs of outer shell electrons are shared between two atoms.
14. Hydrogen bonds are weak bonds between hydrogen and certain other atoms within large complex molecules such as proteins and nucleic acids. They add strength and stability and help determine the molecule's three-dimensional shape.
15. Energy is the capacity to do work. Potential energy is energy stored by matter due to its position. Kinetic energy is the energy of matter in motion. Chemical energy is a form of potential energy stored in the bonds of molecules.
16. In chemical reactions, breaking old bonds requires energy and forming new bonds releases energy.

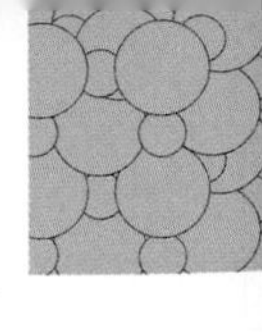

17. In a synthesis (anabolic) reaction, two or more atoms, ions, or molecules combine to form a new and larger molecule. In a decomposition (catabolic) reaction, a molecule is split apart into smaller molecules, ions, or atoms.
18. When nutrients, such as glucose, are broken down via decomposition reactions, some of the energy released is temporarily stored in adenosine triphosphate (ATP) and then later used to drive energy-requiring synthesis reactions that build body structures, such as muscles and bones.
19. Exchange reactions are combination synthesis and decomposition reactions. Reversible reactions can proceed in both directions under different conditions.

Chemical Compounds and Life Processes (p. 29)

1. Inorganic compounds usually are structurally simple and lack carbon. Organic substances always contain carbon, usually contain hydrogen, and always have covalent bonds.
2. Water is the most abundant substance in the body. It is an excellent solvent, participates in chemical reactions, absorbs and releases heat slowly, requires a large amount of heat to change from a liquid to a gas, and serves as a lubricant.
3. Inorganic acids, bases, and salts dissociate into ions in water. An acid ionizes into hydrogen ions (H^-); a base usually ionizes into hydroxide ions (OH^+). A salt ionizes into neither H^+ nor OH^- ions.
4. The pH of body fluids must remain fairly constant for the body to maintain homeostasis. On the pH scale, 7 represents neutrality. Values below 7 indicate acidic solutions, and values above 7 indicate alkaline solutions.
5. Buffer systems help maintain pH by converting strong acids or bases into weak acids or bases.
6. Carbohydrates include sugars, glycogen, and starches. They may be monosaccharides, disaccharides, or polysaccharides. Carbohydrates provide most of the chemical energy needed to generate ATP. Carbohydrates, and other large, organic molecules, are synthesized via dehydration synthesis reactions, in which a molecule of water is lost. In the reverse process, called hydrolysis, large molecules are broken down into smaller ones upon the addition of water.
7. Lipids are a diverse group of compounds that include triglycerides (fats and oils), phospholipids, and steroids. Triglycerides protect, insulate, provide energy, and are stored in adipose tissue. Phospholipids are important membrane components. Steroids are synthesized from cholesterol.
8. Proteins are constructed from amino acids. They give structure to the body, regulate processes, provide protection, help muscles to contract, transport substances, and serve as enzymes.
9. Enzymes are molecules, usually proteins, that speed up chemical reactions and are subject to a variety of cellular controls.
10. Deoxyribonucleic acid (DNA) and ribonucleic acid (RNA) are nucleic acids consisting of nitrogenous bases, five-carbon sugars, and phosphate groups. DNA is a double helix and is the primary chemical in genes. RNA differs in structure and chemical composition from DNA; its main function is to carry out the instructions encoded in DNA.
11. Adenosine triphosphate (ATP) is the principal energy-transferring molecule in living systems. When it transfers energy, ATP is decomposed by hydrolysis to adenosine diphosphate (ADP) and Ⓟ. ATP is synthesized from ADP and Ⓟ using primarily the energy supplied by the breakdown of glucose.

SELF-QUIZ

1. A substance that dissociates in water to form H^+ is called
   a. a base b. a salt c. a buffer d. an acid
   e. a nucleic acid
2. Ionic bonds are characterized by
   a. sharing electrons between atoms
   b. their ability to form strong, stable bonds
   c. atoms giving away and taking electrons
   d. the type of bonding formed in most organic compounds
   e. an attraction between water molecules
3. If an atom has two electrons in its second electron shell and its first electron shell is filled, it will tend to
   a. lose two electrons from its second electron shell
   b. lose the electrons from its first electron shell
   c. lose all of the electrons from its first and second electron shells
   d. gain six electrons in its second electron shell
   e. share two electrons in its second electron shell
4. Matter that cannot be broken down into simpler substances by chemical reactions is known as
   a. a molecule b. an antioxidant
   c. a compound. d. a buffer e. a chemical element
5. Chlorine (Cl) has an atomic number of 17. An atom of chlorine may become a chloride ion (Cl^-) by
   a. losing one electron b. losing one neutron
   c. gaining one proton d. gaining one electron
   e. gaining two electrons
6. Which of the following is NOT true?
   a. A substance that separates in water to form some cation other than H^+ and some anion other than OH^- is known as a salt.
   b. A solution that has a pH of 9.4 is acidic.
   c. A solution with a pH of 5 is 100 times more acidic than distilled water, which has a pH of 7.
   d. Buffers help to make the body's pH more stable.
   e. Amino acids are linked by peptide bonds.

7. Which of the following organic compounds are NOT paired with their correct subunits (building blocks)?
   a. glycogen, glucose
   b. proteins, monosaccharides
   c. DNA, nucleotides
   d. lipids, glycerol and fatty acids
   e. ATP, ADP and P

8. The type of reaction by which a disaccharide is formed from two monosaccharides is known as a
   a. decomposition reaction
   b. hydrolysis reaction
   c. dehydration synthesis reaction
   d. reversible reaction
   e. dissociation reaction

9. Which of the following contains the genetic code in human cells?
   a. DNA b. enzymes c. RNA d. glucose
   e. ATP

10. What is the principal energy-transferring molecule in the body?
    a. ADP b. RNA c. DNA d. ATP e. NAD

11. Which of the following statements about water is NOT true?
    a. It is involved in many chemical reactions in the body.
    b. It is an important solvent in the human body.
    c. It helps lubricate a variety of structures in the body.
    d. It can absorb a large amount of heat without changing its temperature.
    e. It requires very little heat to change from a liquid to a gas.

12. The difference in H^+ concentration between solutions with a pH of 3 and a pH of 5 is that the solution with the pH of 3 has ____ H^+.
    a. 2 times more
    b. 5 times more
    c. 10 times more
    d. 100 times more
    e. 200 times less

13. Which of the following is NOT a true statement about enzyme activity?
    a. Enzymes form a temporary complex with their substrates
    b. Enzymes are not permanently altered by the chemical reactions they catalyze.
    c. All proteins are enzymes.
    d. Enzymes are considered to be organic catalysts.
    e. Enzymes are subject to cellular control.

14. For each item in the following list, place an R if it applies to RNA or a D if it refers to DNA; use R and D if it applies to both RNA and DNA.
    ____ a. composed of nucleotides
    ____ b. forms a double helix
    ____ c. contains thymine
    ____ d. contains the sugar ribose
    ____ e. contains the nitrogenous base uracil
    ____ f. is the hereditary material of cells
    ____ g. contains the sugar deoxyribose
    ____ h. is single-stranded
    ____ i. contains adenine
    ____ j. contains phosphate groups

15. An organic compound that consists of C, H, and O and that may be broken down into glycerol and fatty acids is a
    a. triglyceride b. nucleic acid c. monosaccharide
    d. carbohydrate e. protein

16. Why is it important to consume foods that contain antioxidants?
    a. They provide an energy source for the body.
    b. They help inactivate damaging free radicals.
    c. They make up the body's genes.
    d. They act as buffers to help maintain the blood's pH.
    e. They are important solvents in the body.

17. If an enzyme is exposed to an extremely high temperature, it will
    a. divide b. release energy c. become an electrolyte
    d. form hydrogen bonds e. denature

18. In what form are lipids stored in the adipose (fat) tissue of the body?
    a. triglycerides b. glycogen c. cholesterol
    d. polypeptides e. disaccharides

19. Approximately 96% of your body's mass is composed of which of the following elements? Place an X beside each correct answer.

____ calcium	____ iron	____ nitrogen
____ phosphorus	____ sodium	____ chlorine
____ carbon	____ oxygen	____ sulfur
____ hydrogen	____ potassium	____ magnesium

20. Match the following:

__ a. inorganic compound	A. glycogen
__ b. monosaccharide	B. enzyme
__ c. polysaccharide	C. glucose
__ d. component of triglycerides	D. water
__ e. lipase	E. glycerol

CRITICAL THINKING APPLICATIONS

1. While having a tea party, your three-year-old cousin Sabrina added milk, lemon juice, and lots of sugar to her tea. The tea now has strange white lumps floating in it. What caused the milk to curdle?
2. Joy is very proud of her healthy diet. "I drink only pure spring water and eat organic foods. I have a chemical-free body." Sonia replied, "Ever hear of H_2O?" Explain the error in Joy's reasoning.
3. Albert, Jr., was trying out the new Super Genius Home Chemistry Kit that he got for his birthday. He decided to check the pH of his secret formula: lemon juice and diet cola. The pH was 2.5. Next he added tomato juice. Now he has a really disgusting mixture with a pH of 3.5. "Wow! That's twice as strong!" Does Albert, Jr., have the makings of a "Super Genius"? Explain.
4. During chemistry lab, Maria places sucrose (table sugar) in a glass beaker, adds water, and stirs. As the table sugar disappears, she loudly proclaims that she has chemically broken down the sucrose into fructose and glucose. Is Maria's chemical analysis correct?

ANSWERS TO FIGURE QUESTIONS

2.1 The atomic number of carbon is 6.

2.2 The four most plentiful elements in living organisms are oxygen, carbon, hydrogen, and nitrogen.

2.3 Water is a compound because it contains atoms of both hydrogen and oxygen.

2.4 K is an electron donor; when it ionizes, it becomes a cation, K^+, because losing one electron from the fourth electron shell leaves eight electrons in the third shell.

2.5 An ionic bond involves the *loss* and *gain* of electrons; a covalent bond involves the *sharing* of pairs of electrons.

2.6 $CaCO_3$ is a salt, and H_2SO_4 is an acid.

2.7 A pH of 6.82 is more acidic than a pH of 6.91. Both pH = 8.41 and pH = 5.59 are 1.41 pH units from neutral (pH = 7).

2.8 There are 6 carbons in fructose, 12 in sucrose.

2.9 Glycogen is stored in liver and skeletal muscle cells.

2.10 A monounsaturated fatty acid has one double bond.

2.11 A triglyceride has three fatty acid molecules attached to a glycerol backbone, and a phospholipid has two fatty acid tails and a phosphate group attached to a glycerol backbone.

2.12 The dietary lipids thought to contribute to atherosclerosis are cholesterol and saturated fats.

2.13 A tripeptide would have two peptide bonds, each linking two amino acids.

2.14 The enzyme's active site combines with the substrate.

2.15 Thymine is present in DNA but not in RNA, and uracil is present in RNA but not in DNA.

2.16 A few cellular activities that depend on energy supplied by ATP are muscular contractions, movement of chromosomes, transport of substances across cell membranes, and synthesis reactions.

chapter 3 CELLS

did you know?

Why is it so important to eat a variety of fruits and vegetables? Because your parents wouldn't let you have dessert unless you did? Another good reason to eat plenty of fruits and vegetables is that these foods contain important compounds, known as phytochemicals (literally, "plant chemicals"), which help to keep cells healthy. Some phytochemicals block chemicals that can cause damage to cells. Others enhance your body's production of enzymes that render potentially cancer-causing substances harmless. Collectively, the actions of phytochemicals promote healthy cellular function, and prevent the types of cellular damage associated with cancer, aging, and heart disease.

Focus on Wellness, page 64

About 200 different types of cells compose your body. Each ***cell*** is a living structural and functional unit that is enclosed by a membrane. All cells arise from existing cells by the process of ***cell division***, in which one cell divides into two new cells. In your body, different types of cells fulfill unique roles that support homeostasis and contribute to the many functional capabilities of the human organism. ***Cell biology*** is the study of cellular structure and function. As you study the various parts of a cell and their relationships to each other, you will learn that cell structure and function are intimately related.

looking back to move ahead . . .

- **Levels of Organization and Body Systems (page 2)**
- **Free Radicals (page 25)**
- **Carbohydrates (page 31)**
- **Lipids (page 32)**
- **Proteins (page 35)**
- **Nucleic Acids: Deoxyribonucleic Acid (DNA) and Ribonucleic Acid (RNA) (page 38)**

A GENERALIZED VIEW OF THE CELL

OBJECTIVE • **Name and describe the three main parts of a cell.**

Figure 3.1 is a generalized view of a cell that shows the main cellular components. Though some body cells lack some cellular structures shown in this diagram, many body cells include most of these components. For ease of study, we can divide a cell into 3 main parts: the plasma membrane, cytoplasm, and nucleus.

- The ***plasma membrane*** forms a cell's outer surface, separating the cell's internal environment (inside the cell) from its external environment (outside the cell). It regulates the flow of materials into and out of a cell to maintain the appropriate environment for normal cellular activities. The plasma membrane also plays a key role in communication among cells and between cells and their external environment.
- The ***cytoplasm*** (SĪ-tō-plazm; *-plasm* = formed or molded) consists of all the cellular contents between the plasma membrane and the nucleus. This compartment can be divided into two components: cytosol and organelles. ***Cytosol*** (SĪ-tō-sol) is the fluid portion of cytoplasm that consists mostly of water plus dissolved solutes and suspended particles. Within the cytosol are several different types of ***organelles*** (or-ga-NELZ = little organs), each of which has a characteristic structure and specific functions.
- The ***nucleus*** (NOO-klē-us = nut kernel) is the largest organelle of a cell. The nucleus acts as the control center for a cell because it contains the genes, which control cellular structure and most cellular activities.

CHECKPOINT

1. What are the general functions of the three main parts of a cell?

Figure 3.1 Generalized view of a body cell.

The cell is the basic, living, structural and functional unit of the body.

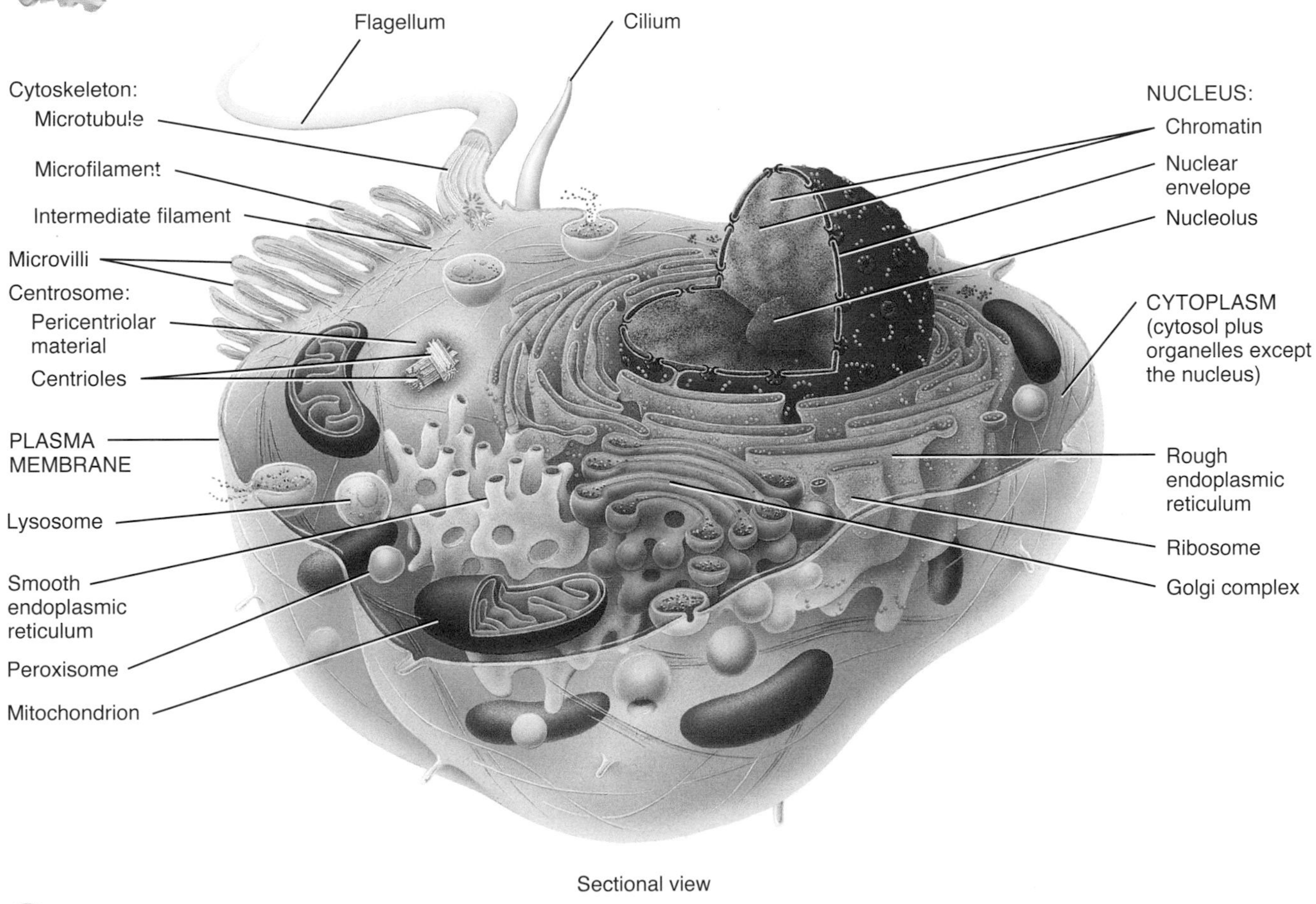

What are the three principal parts of a cell?

THE PLASMA MEMBRANE

OBJECTIVE • Describe the structure and functions of the plasma membrane.

The ***plasma membrane*** is a flexible yet sturdy barrier that consists mostly of phospholipids (lipids that contain phosphorus) and proteins. Virtually all membrane proteins are ***glycoproteins***, proteins with attached carbohydrates. Other molecules present in lesser amounts in the plasma membrane are cholesterol and glycolipids (lipids with attached carbohydrates). The basic framework of the plasma membrane is the ***lipid bilayer***, two back-to-back layers made up of three types of lipid molecules: phospholipids, cholesterol, and glycolipids (Figure 3.2). The proteins in a membrane are of two types—integral and peripheral (Figure 3.2). ***Integral proteins*** extend into or through the lipid bilayer among the fatty acid tails. ***Peripheral proteins*** are loosely attached to the exterior or interior surface of the membrane. Although many of the proteins can float laterally in the lipid bilayer, each individual protein has a specific orientation with respect to the "inside" and "outside" faces of the membrane.

The plasma membrane allows some substances to move into and out of the cell but restricts the passage of other substances. This property of membranes is called ***selective permeability*** (per′-mē-a-BIL-i-tē). The lipid bilayer part of the membrane is permeable to water and to nonpolar (lipid-soluble) molecules, such as fatty acids, fat-soluble vitamins, steroids, oxygen, and carbon dioxide. The lipid bilayer is *not* permeable to ions and large, uncharged polar molecules such as glucose and amino acids. These small and medium-sized water-soluble materials may cross the membrane with the assistance of integral proteins. Some integral proteins form ***ion channels*** through which specific substances can move into and out of cells (Figure 3.2). Other membrane proteins act as ***transporters***, which change shape as they move a substance from one side of the membrane to the other. Large molecules such as proteins are unable to pass through the plasma membrane except by transport within vesicles (discussed later in this chapter).

Most functions of the plasma membrane depend on the types of proteins that are present. Integral proteins called ***receptors*** recognize and bind a specific molecule that governs some cellular function, for example, a hormone such as insulin. Some integral and peripheral proteins act as ***enzymes***, speeding up specific chemical reactions. Membrane glycoproteins and glycolipids often are ***cell identity markers***. They enable a cell to recognize other cells of its own kind during tissue formation, or to recognize and respond to potentially dangerous foreign cells.

Figure 3.2 Chemistry and structure of the plasma membrane.

The plasma membrane consists mostly of phospholipids, arranged in a bilayer, and proteins, most of which are glycoproteins.

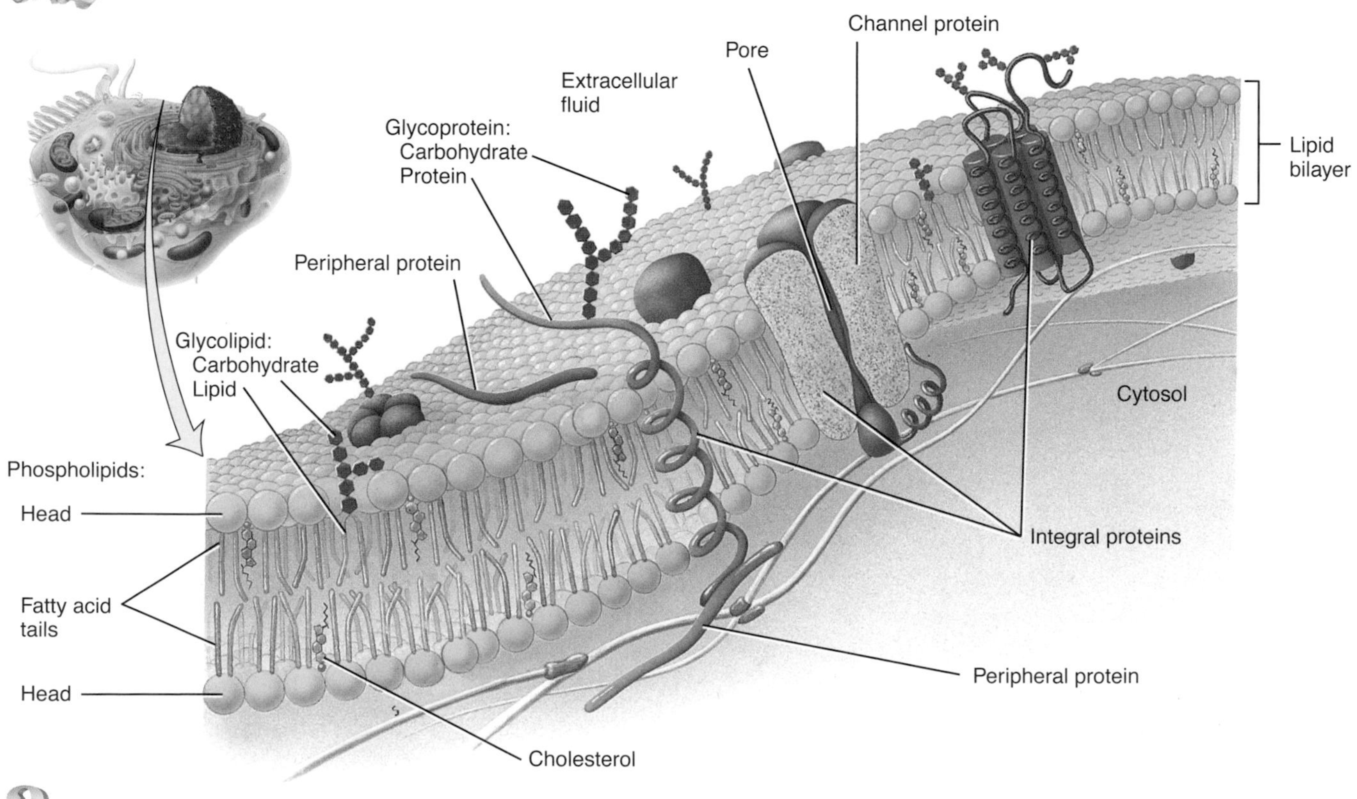

Name several functions carried out by membrane proteins.

■ **CHECKPOINT**

2. What molecules make up the plasma membrane and what are their functions?
3. What is meant by selective permeability?

TRANSPORT ACROSS THE PLASMA MEMBRANE

OBJECTIVE • Describe the processes that transport substances across the plasma membrane.

Movement of materials across its plasma membrane is essential to the life of a cell. Certain substances must move into the cell to support metabolic reactions. Other materials must be moved out because they have been produced by the cell for export or are cellular waste products. Before discussing how materials move into and out of a cell, we need to understand what exactly is being moved as well as the form it needs to take to make its journey.

About two-thirds of the fluid in your body is contained inside body cells and is called ***intracellular fluid*** or ***ICF*** (*intra-* = within). ICF is actually the cytosol of a cell. Fluid outside body cells is called ***extracellular fluid*** or ***ECF*** (*extra-* = outside). The ECF in the microscopic spaces between the cells of tissues is ***interstitial fluid*** (in′-ter-STISH-al; *inter-* = between). The ECF in blood vessels is called ***plasma***, and that in lymphatic vessels is called ***lymph***.

Materials dissolved in body fluids include gases, nutrients, ions, and other substances needed to maintain life. Any material dissolved in a fluid is called a ***solute***, and the fluid in which it is dissolved is the ***solvent***. Body fluids are dilute solutions in which a variety of solutes are dissolved in a very familiar solvent, water. The amount of a solute in a solution is its ***concentration***. A ***concentration gradient*** is a difference in concentration between two different areas, for example, the ICF and ECF. Solutes moving from a high-concentration area (where there are more of them) to a low-concentration area (where there are fewer of them) are said to move *down* or *with* the concentration gradient. Solutes moving from a low-concentration area to a high-concentration area are said to move *up* or *against* the concentration gradient.

Substances move across cellular membranes by passive processes and active processes. ***Passive processes***, in which a substance moves down its concentration gradient through the membrane, using only its own energy of motion (kinetic energy), include simple diffusion and osmosis. In ***active processes***, cellular energy, usually in the form of ATP, is used to "push" the substance through the membrane "uphill" against its concentration gradient. An example is active transport. Another way that some substances may enter and leave cells is an active process in which tiny membrane sacs referred to as ***vesicles*** are used (see Figure 3.10).

Passive Processes

Diffusion: The Principle

Diffusion (di-FŪ-zhun; *diffus-* = spreading) is a passive process in which a substance moves from one place to another due to the substance's kinetic energy. If a particular substance is present in high concentration in one area and in low concentration in another area, more particles of the substance diffuse from the region of high concentration to the region of low concentration than diffuse in the opposite direction. The diffusion of more molecules in one direction than the other is called *net* diffusion. Substances undergoing net diffusion move from a high to a low concentration, or *down their concentration gradient*. After some time, ***equilibrium*** (ē′-kwi-LIB-rē-um) is reached: The substance becomes evenly distributed throughout the solution and the concentration gradient disappears.

Placing a crystal of dye in a water-filled container provides an example of diffusion (Figure 3.3). At the beginning, the color is most intense just next to the crystal because the crystal is dissolving and the dye concentration is greatest there. At increasing distances, the color is lighter and lighter because the dye concentration is lower and lower. The dye molecules undergo net diffusion, down their concentration gradient, until they are evenly mixed in the water. At equilibrium the solution has a uniform color. In the example of dye diffusion, no membrane was involved. Substances may also diffuse across a membrane, if the membrane is permeable to them.

Figure 3.3 Principle of diffusion. A crystal of dye placed in a cylinder of water dissolves (beginning), and there is net diffusion from the region of higher dye concentration to regions of lower dye concentration (intermediate). At equilibrium, dye concentration is uniform throughout the solution.

At equilibrium, net diffusion stops but random movements continue.

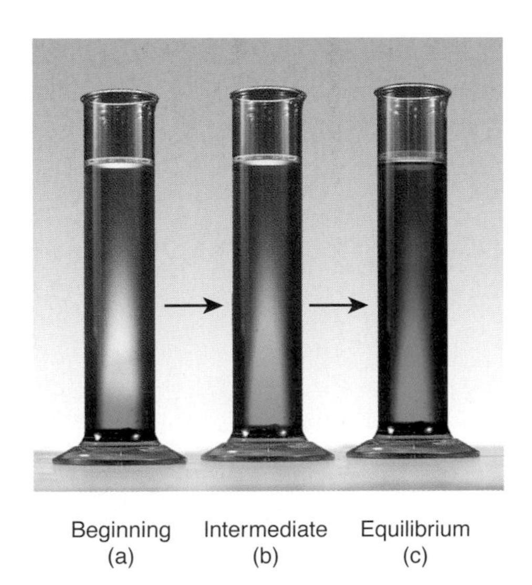

How does simple diffusion differ from facilitated diffusion?

Now that you have a basic understanding of the nature of diffusion, we will consider two types of diffusion: simple diffusion and facilitated diffusion.

SIMPLE DIFFUSION In ***simple diffusion***, substances diffuse across a membrane in one of two ways: lipid-soluble substances diffuse through the lipid bilayer, and ions diffuse through pores of ion channels formed by integral proteins (Figure 3.4). Lipid-soluble substances that move across membranes by simple diffusion through the lipid bilayer include oxygen, carbon dioxide, and nitrogen gases; fatty acids; steroids; and fat-soluble vitamins (A, D, E, and K). Polar molecules such as water and urea also move through the lipid bilayer. Simple diffusion through the lipid bilayer is important in the exchange of oxygen and carbon dioxide between blood and body cells and between blood and air within the lungs during breathing. It also is the transport method for absorption of lipid-soluble nutrients and release of some wastes from body cells.

Most membrane channels are *ion channels*, which allow a specific type of ion to move across the membrane by simple diffusion through the channel's pore. In typical plasma membranes, the most common ion channels are selective for K^+ (potassium ions) or Cl^- (chloride ions); fewer channels are available for Na^+ (sodium ions) or Ca^{2+} (calcium ions). Many ion channels are gated; that is, a portion of the channel protein acts as a "gate," moving in one direction to open the pore and in another direction to close it (Figure 3.5). When the gates are open, ions diffuse into or out of cells, down their concentration gradient. Gated channels are important for the production of electrical signals by body cells.

Figure 3.5 Diffusion of potassium ions (K^+) through a gated K^+ channel. A gated channel is one in which a portion of the channel protein acts as a gate to open or close the channel's pore to the passage of ions.

Channels are integral membrane proteins that allow specific small, inorganic ions to pass across the membrane by simple diffusion.

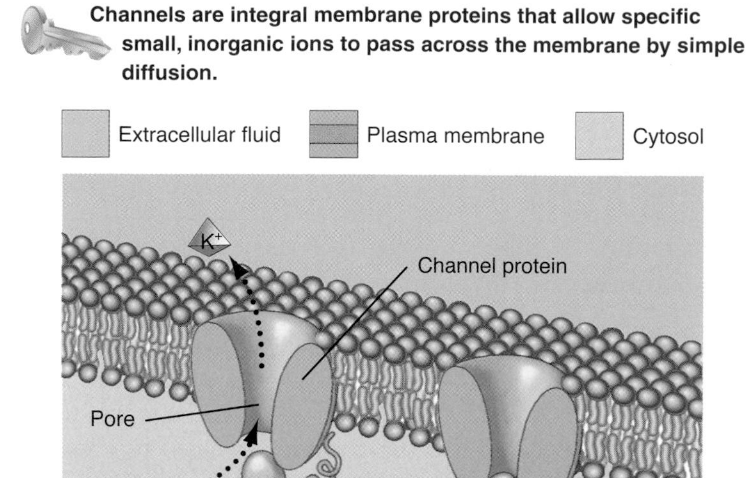

Details of the K^+ channel

Is the concentration of K^+ in body cells higher in the cytosol or in the extracellular fluid?

Figure 3.4 Simple diffusion. Lipid-soluble molecules may diffuse through the lipid bilayer, and ions may diffuse through pores of ion channels in integral proteins. Plasma membranes have channels, formed by integral proteins, that are selective for potassium ions (K^+), sodium ions (Na^+), calcium ions (Ca^{2+}), and chloride ions (Cl^-).

In simple diffusion there is a net (greater) movement of substances from a region of their higher concentration to a region of their lower concentration.

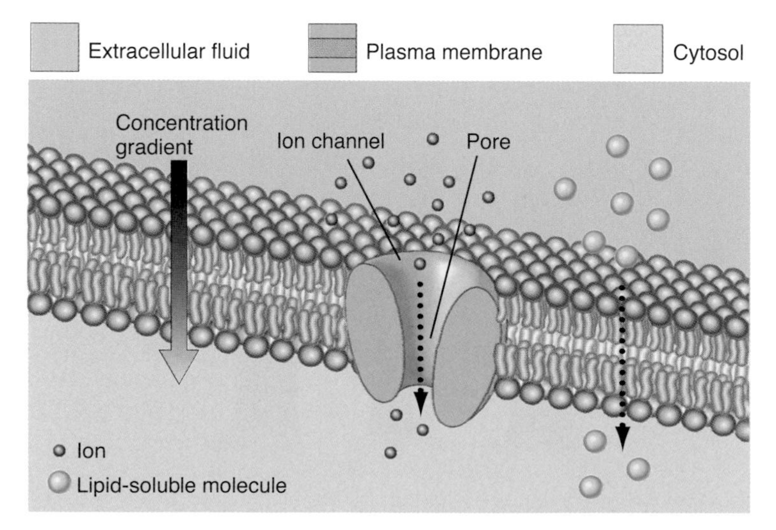

What are some examples of substances that diffuse through the lipid bilayer?

FACILITATED DIFFUSION Some substances that cannot diffuse through the lipid bilayer or through ion channels do cross the plasma membrane by a passive process called ***facilitated diffusion***. In this process, an integral membrane protein assists a specific substance across the membrane. The substance binds to a specific *transporter* on one side of the membrane and is released on the other side after the transporter undergoes a change in shape. As is true for simple diffusion, facilitated diffusion moves a substance down a concentration gradient—from a region of higher concentration to a region of lower concentration—and does not require cellular energy in the form of ATP.

Substances that move across plasma membranes by facilitated diffusion include glucose, fructose, galactose, and some vitamins. Glucose enters many body cells by facilitated diffusion as follows (Figure 3.6):

1. Glucose binds to a glucose transporter protein on the outside surface of the membrane.
2. As the transporter undergoes a change in shape, glucose passes through the membrane.
3. The transporter releases glucose on the other side of the membrane.

Figure 3.6 Facilitated diffusion of glucose across a plasma membrane. The transporter protein binds to glucose in the extracellular fluid and releases it into the cytosol.

Facilitated diffusion across a membrane requires a transporter protein but does not use ATP.

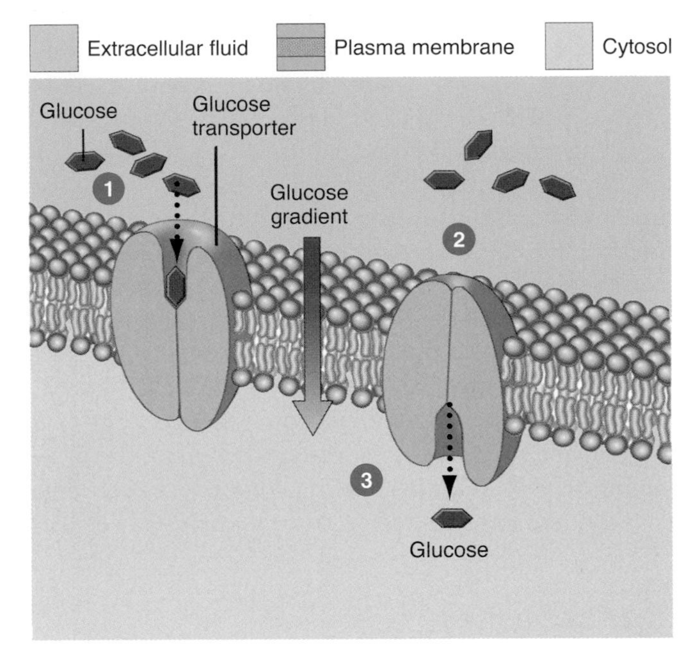

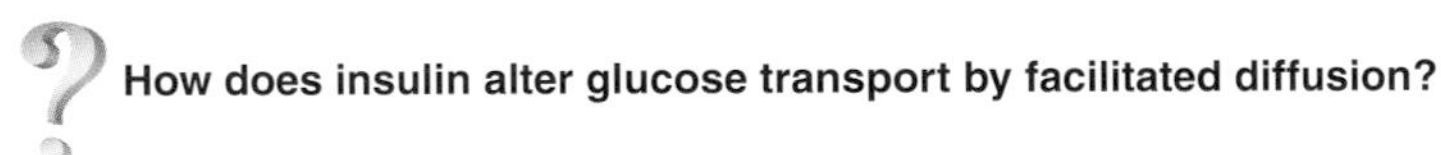
How does insulin alter glucose transport by facilitated diffusion?

The selective permeability of the plasma membrane is often regulated to achieve homeostasis. For example, the hormone insulin promotes the insertion of glucose transporters into the plasma membranes of certain cells. Thus, the effect of insulin is to increase entry of glucose into body cells by means of facilitated diffusion.

Osmosis

Osmosis (oz-MŌ-sis) is a passive process in which there is a net movement of water through a selectively permeable membrane. Water moves by osmosis from an area of *higher water concentration* to an area of *lower water concentration* (or from an area of *lower solute concentration* to an area of *higher solute concentration*). Water molecules pass through plasma membranes in two places: through the lipid bilayer and through integral membrane proteins that function as water channels.

The device in Figure 3.7 demonstrates osmosis. A sac made of cellophane, a selectively permeable membrane that permits water but not sucrose (sugar) molecules to pass, is filled with a solution that is 20% sucrose and 80% water. The upper part of the cellophane sac is wrapped tightly about a stopper through which a glass tube is fitted. The sac is then placed into a beaker containing pure (100%) water (Figure 3.7a). Notice that the cellophane now separates two fluids having different water concentrations. As a result, water begins to move by osmosis from the region where its concentration is higher (100% water in the beaker) through the cellophane to where its concentration is lower (80% water inside the sac). Because the cellophane is not permeable to sucrose, however, all the sucrose molecules remain inside the sac. As water moves into the sac, the volume of the sucrose solution increases and the fluid rises into the glass tube (Figure 3.7b). As the fluid rises in the tube, its water pressure forces some water molecules from the sac back into the beaker. At equilibrium, just as many water molecules are moving into the beaker due to the water pressure as are moving into the sac due to osmosis.

A solution containing solute particles that cannot pass through a membrane exerts a pressure on the membrane, called ***osmotic pressure***. The osmotic pressure of a solution depends on the concentration of its solute particles—the higher the solute concentration, the higher the solution's osmotic pressure. Because the osmotic pressure of cytosol and

Figure 3.7 Principle of osmosis. (a) At the start of the experiment, a cellophane sac—a selectively permeable membrane that permits water but not sucrose molecules to pass—containing a 20% sucrose solution is immersed in a beaker of pure (100%) water. Osmosis begins (arrows) as water moves down its concentration gradient into the sac. (b) As the volume of the sucrose solution increases, the solution moves up the glass tubing. The added fluid in the tube exerts a pressure that drives some water molecules back into the beaker. At equilibrium, osmosis has stopped because the number of water molecules entering and the number leaving the cellophane sac are equal.

Osmosis is the net movement of water molecules through a selectively permeable membrane.

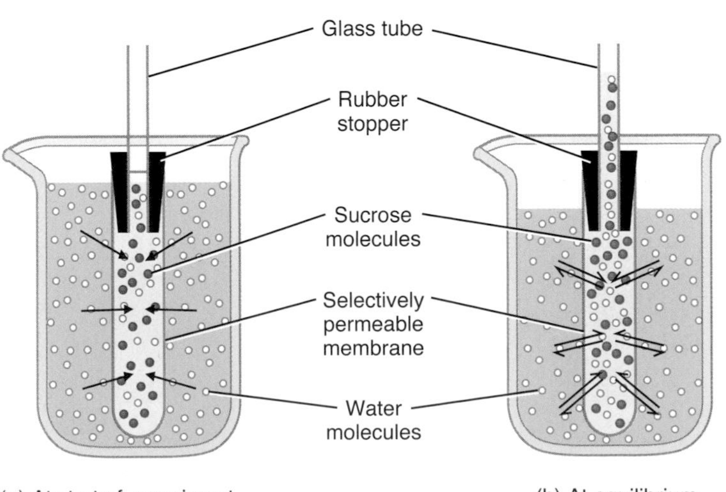

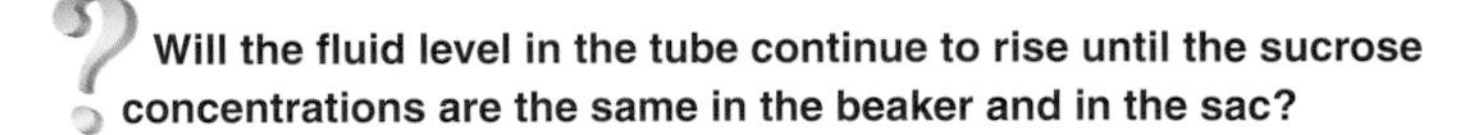
Will the fluid level in the tube continue to rise until the sucrose concentrations are the same in the beaker and in the sac?

interstitial fluid is the same, cell volume remains constant. Cells neither shrink due to water loss by osmosis nor swell due to water gain by osmosis.

Any solution in which cells maintain their normal shape and volume is called an ***isotonic solution*** (*iso-* = same; *tonic-* = tension) (Figure 3.8). This is a solution in which the concentrations of solutes are the *same* on both sides. For example, a 0.9% NaCl (sodium chloride, or table salt) solution, called a *normal saline solution*, is isotonic for red blood cells. When red blood cells are bathed in 0.9% NaCl, water molecules enter and exit the cells at the same rate, allowing the red blood cells to maintain their normal shape and volume.

If red blood cells are placed in a ***hypotonic solution*** (*hypo-* = less than), a solution that has a *lower* concentration of solutes (higher concentration of water) than the cytosol inside the red blood cells (Figure 3.8), water molecules enter the cells by osmosis faster than they leave. This situation causes the red blood cells to swell and eventually to burst. Rupture of red blood cells is called ***hemolysis*** (hē-MOL-i-sis). A ***hypertonic solution*** (*hyper-* = greater than) has a *higher* concentration of solutes (lower concentration of water) than does the cytosol inside red blood cells (Figure 3.8). When cells are placed in a hypertonic solution, water molecules move out of the cells by osmosis faster than they enter, causing the cells to shrink. Such shrinkage of red blood cells is called ***crenation*** (kre-NĀ-shun).

Figure 3.8 Principle of osmosis applied to red blood cells (RBCs). The arrows indicate the direction and degree of water movement into and out of the cells. One example of an isotonic solution for RBCs is 0.9% NaCl.

An isotonic solution is one in which cells maintain their normal shape and volume.

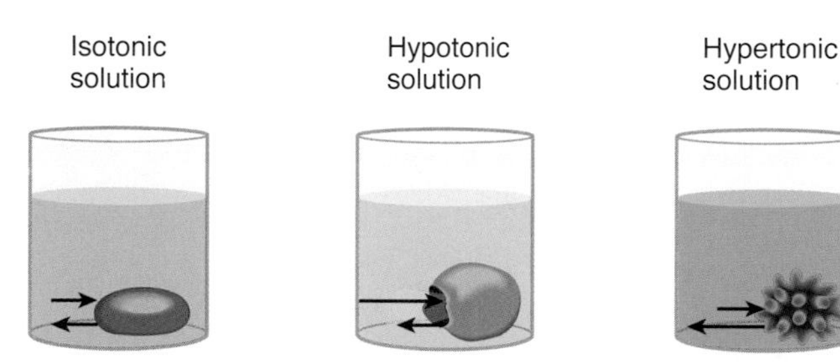

(a) Illustrations showing direction of water movement

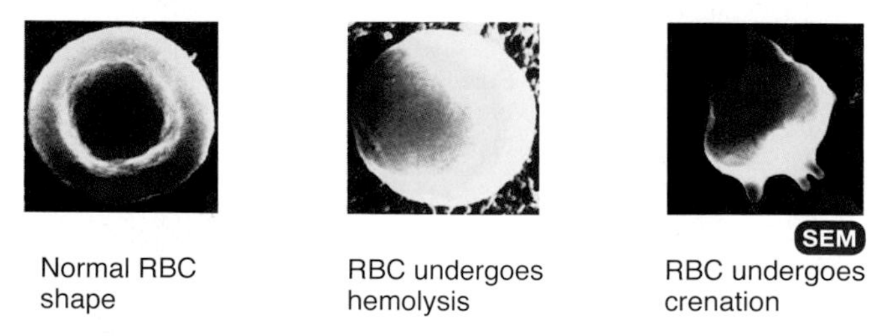

(b) Scanning electron micrographs (all 800x)

Will a 2% solution of NaCl cause hemolysis or crenation of RBCs?

RBCs and other body cells may be damaged or destroyed if exposed to hypertonic or hypotonic solutions. For this reason, most **intravenous (IV) solutions**, liquids infused into the blood of a vein, are isotonic. Examples are isotonic saline (0.9% NaCl) and D5W, which stands for dextrose 5% in water. Sometimes infusion of a hypertonic solution is useful to treat patients who have *cerebral edema*, excess interstitial fluid in the brain. Infusion of such a solution relieves fluid overload by causing osmosis of water from interstitial fluid into the blood. The kidneys then excrete the excess water from the blood into the urine. Hypotonic solutions, given either orally or through an IV, can be used to treat people who are dehydrated. The water in the hypotonic solution moves from the blood into interstitial fluid and then into body cells to rehydrate them. Water and most sports drinks that you consume to "rehydrate" after a workout are hypotonic relative to your body cells.

Active Processes

Active Transport

Active transport is an active process in which cellular energy is used to transport substances across the membrane against a concentration gradient (from an area of low to an area of high concentration).

Energy derived from splitting ATP changes the shape of a transporter protein, called a ***pump***, which moves a substance across a cellular membrane against its concentration gradient. A typical body cell expends about 40% of its ATP on active transport. Drugs that turn off ATP production, such as the poison cyanide, are lethal because they shut down active transport in cells throughout the body. Substances transported across the plasma membrane by active transport are mainly ions, primarily Na^+, K^+, H^+, Ca^{2+}, I^-, and Cl^-.

The most important active transport pump expels sodium ions (Na^+) from cells and brings in potassium ions (K^+). The pump protein also acts as an enzyme to split ATP. Because of the ions it moves, this pump is called the ***sodium-potassium (Na^+/K^+) pump***. All cells have thousands of sodium-potassium pumps in their plasma membranes. These pumps maintain a low concentration of sodium ions in the cytosol by pumping Na^+ into the extracellular fluid against the Na^+ concentration gradient. At the same time, the pump moves potassium ions into cells against the K^+ concentration gradient. Because K^+ and Na^+ slowly leak back across the plasma membrane down their gradients, the sodium-potassium pumps must operate continually to maintain a low concentration of Na^+ and a high concentration of K^+ in the cytosol. These differing concentrations are crucial for osmotic

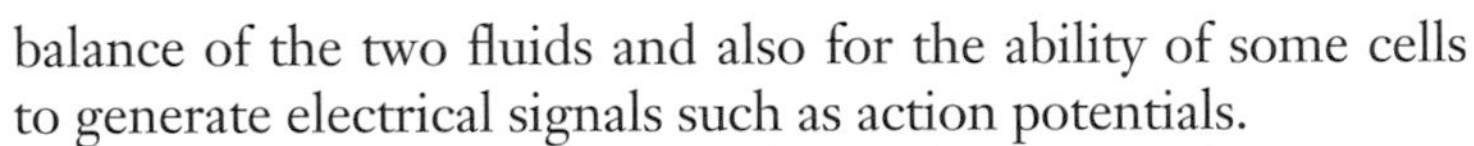

balance of the two fluids and also for the ability of some cells to generate electrical signals such as action potentials.

Figure 3.9 shows how the sodium-potassium pump operates.

1. Three sodium ions (Na^+) in the cytosol bind to the pump protein.
2. Na^+ binding triggers the splitting of ATP into ADP plus a phosphate group (Ⓟ), which also becomes attached to the pump protein. This chemical reaction changes the shape of the pump protein, expelling the three Na^+ into the extracellular fluid. The changed shape of the pump protein then favors binding of two potassium ions (K^+) in the extracellular fluid to the pump protein.
3. The binding of K^+ causes the pump protein to release the phosphate group, which causes the pump protein to return to its original shape.
4. As the pump protein returns to its original shape, it releases the two K^+ into the cytosol. At this point, the pump is ready again to bind Na^+, and the cycle repeats.

Transport in Vesicles

A ***vesicle*** (VES-i-kul) is a small round sac formed by budding off from an existing membrane. Vesicles transport substances from one structure to another within cells, take in substances from extracellular fluid, and release substances into extracellular fluid. Movement of vesicles requires energy supplied by ATP and is therefore an active process. The two main types of transport in vesicles between a cell and the extracellular fluid that surrounds it are (1) ***endocytosis*** (*endo-* = within), in which materials move *into* a cell in a vesicle formed from the plasma membrane, and (2) ***exocytosis*** (*exo-* = out), in which materials move *out of* a cell by the fusion of a vesicle formed inside a cell with the plasma membrane.

ENDOCYTOSIS Substances brought into the cell by endocytosis are surrounded by a piece of the plasma membrane, which buds off inside the cell to form a vesicle containing the ingested substances. The two types of endocytosis we will consider are phagocytosis and bulk-phase endocytosis.

1. **Phagocytosis.** In ***phagocytosis*** (fag′-ō-sī-TŌ-sis; *phago-* = to eat), large solid particles, such as whole bacteria or viruses or aged or dead cells, are taken in by the cell (Figure 3.10). Phagocytosis begins as the particle binds to a plasma membrane receptor, causing the cell to extend projections of its plasma membrane and cytoplasm, called ***pseudopods*** (SOO-dō-pods; *pseudo-* = false; *-pods* = feet). Two or more pseudopods surround the particle, and portions of their membranes fuse to form a vesicle that enters the cytoplasm. The vesicle fuses with one or more lysosomes, and lysosomal enzymes break down the ingested material. In most cases, any undigested materials remain indefinitely in a vesicle called a *residual body.*

   Phagocytosis occurs only in ***phagocytes***, cells that are specialized to engulf and destroy bacteria and other foreign substances. Phagocytes include certain types of white blood cells and macrophages, which are present in most body tissues. The process of phagocytosis is a vital defense mechanism that helps protect the body from disease.

Figure 3.9 Operation of the sodium-potassium pump. Sodium ions (Na^+) are expelled from the cell, and potassium ions (K^+) are imported into the cell. The pump does not work unless Na^+ and ATP are present in the cytosol and K^+ is present in the extracellular fluid.

The sodium-potassium pump maintains a low intracellular concentration of Na^+.

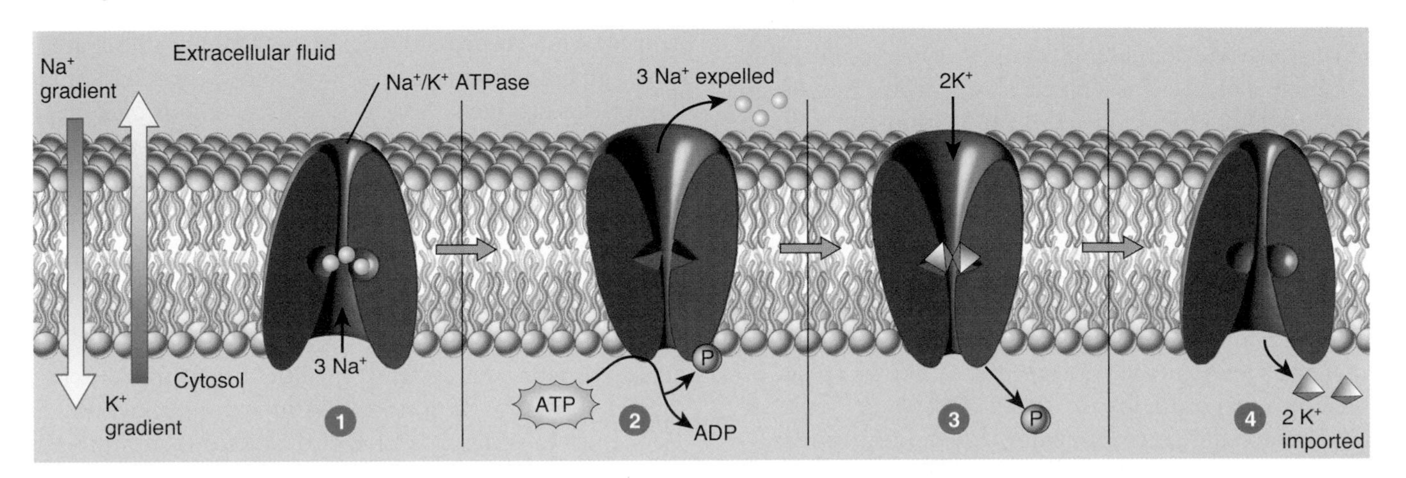

What is the role of ATP in the operation of this pump?

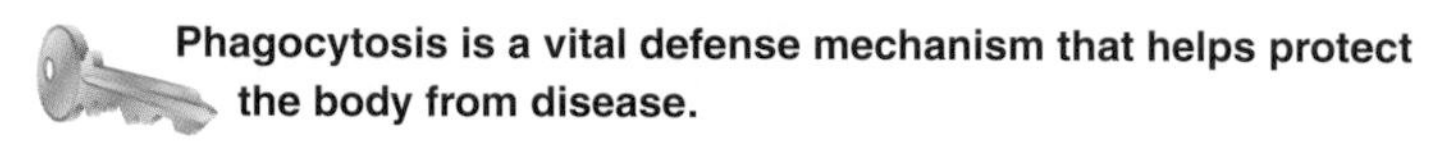

Figure 3.10 Phagocytosis.

Phagocytosis is a vital defense mechanism that helps protect the body from disease.

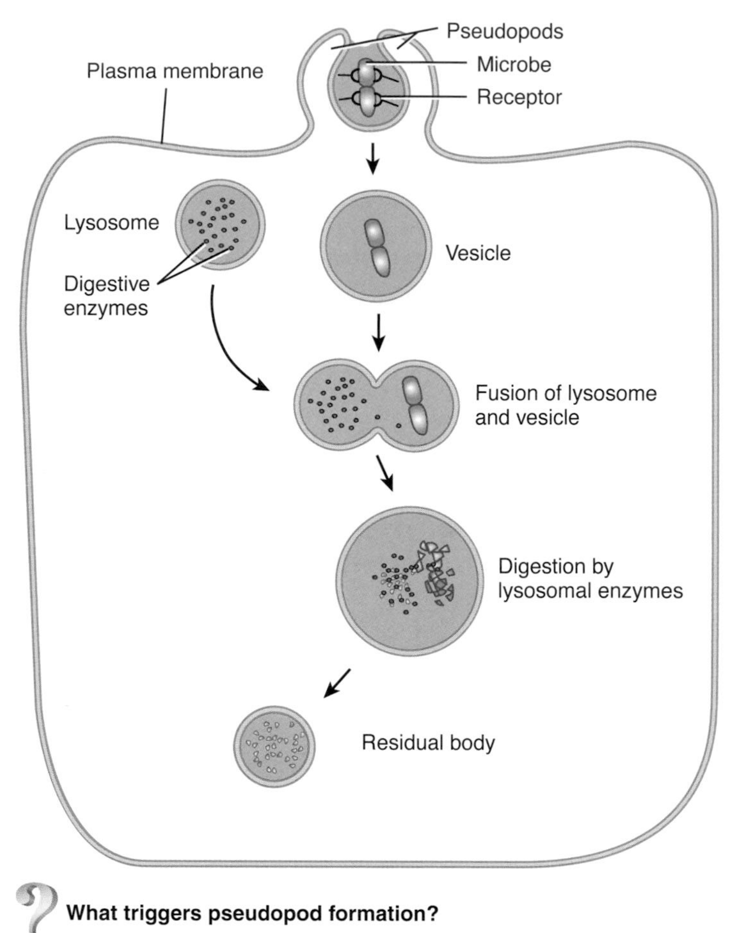

What triggers pseudopod formation?

2. **Bulk-phase Endocytosis**. In ***bulk-phase endocytosis (pinocytosis)***, cells take up tiny droplets of extracellular fluid. The process occurs in most body cells and takes in any and all solutes dissolved in the extracellular fluid. During bulk-phase endocytosis the plasma membrane folds inward and forms a vesicle containing a droplet of extracellular fluid. The vesicle detaches or "pinches off" from the plasma membrane and enters the cytosol. Within the cell, the vesicle fuses with a lysosome, where enzymes degrade the engulfed solutes. The resulting smaller molecules, such as amino acids and fatty acids, leave the lysosome to be used elsewhere in the cell.

EXOCYTOSIS In contrast with endocytosis, which brings materials into a cell, exocytosis results in *secretion*, the liberation of materials from a cell. All cells carry out exocytosis, but it is especially important in two types of cells: (1) secretory cells that liberate digestive enzymes, hormones, mucus, or other secretions; (2) nerve cells that release substances called *neurotransmitters* via exocytosis (see Figure 9.7 on page 235). During exocytosis, membrane-enclosed vesicles called *secretory vesicles* form inside the cell, fuse with the plasma membrane, and release their contents into the extracellular fluid.

Segments of the plasma membrane lost through endocytosis are recovered or recycled by exocytosis. The balance between endocytosis and exocytosis keeps the surface area of a cell's plasma membrane relatively constant.

Table 3.1 summarizes the processes by which materials move into and out of cells.

■ **CHECKPOINT**

4. What is the key difference between passive and active processes?
5. How does diffusion through membrane channels compare to facilitated diffusion?
6. In what ways are endocytosis and exocytosis similar and different?

CYTOPLASM

OBJECTIVE • Describe the structure and functions of cytoplasm, cytosol, and organelles.

Cytoplasm consists of all of the cellular contents between the plasma membrane and the nucleus and includes both cytosol and organelles.

Cytosol

The ***cytosol*** (*intracellular fluid*) is the fluid portion of the cytoplasm that surrounds organelles and accounts for about 55% of the total cell volume. Although cytosol varies in composition and consistency from one part of a cell to another, typically it is 75% to 90% water plus various dissolved solutes and suspended particles. Among these are various ions, glucose, amino acids, fatty acids, proteins, lipids, ATP, and waste products. Some cells also contain *lipid droplets* that contain triglycerides and *glycogen granules*, clusters of glycogen molecules. The cytosol is the site of many of the chemical reactions that maintain cell structures and allow cellular growth.

Organelles

Organelles are specialized structures inside cells that have characteristic shapes and specific functions. Each type of organelle is a functional compartment where specific processes take place, and each has its own unique set of enzymes.

Table 3.1 Transport of Materials Into and Out of Cells

Transport Process	Description	Substances Transported
Passive Processes	Movement of substances down a concentration gradient until equilibrium is reached; do not require cellular energy in the form of ATP.	
Diffusion		
Simple diffusion		
Diffusion through the lipid bilayer	Passive movement of a substance through the lipid bilayer of the plasma membrane.	Lipid-soluble molecules: oxygen, carbon dioxide, and nitrogen gases; fatty acids, steroids, and fat-soluble vitamins (A, D, E, K). Polar molecules: water and urea.
Diffusion through membrane channels	Passive movement of a substance down its gradient through channels that span a lipid bilayer; some channels are gated.	Mainly ions: K^+, Cl^-, Na^+, and Ca^{2+}. Water.
Facilitated diffusion	Passive movement of a substance down its concentration gradient aided by membrane proteins known as transporters.	Glucose, fructose, galactose, and some vitamins.
Osmosis	Movement of water molecules across a selectively permeable membrane from an area of higher water concentration to an area of lower water concentration.	Water.
Active Processes	Movement of substances against a concentration gradient; requires cellular energy in the form of ATP.	
Active Transport	Transport in which cell expends energy to move a substance across the membrane against its concentration gradient aided by membrane proteins that act as pumps; these integral membrane proteins use energy supplied by ATP.	Na^+, K^+, Ca^{2+}, H^+, I^-, Cl^-, and other ions.
Transport In Vesicles	Movement of substances into or out of a cell in vesicles that bud from the plasma membrane; requires energy supplied by ATP.	
Endocytosis	Movement of substances into a cell in vesicles.	
Phagocytosis	"Cell eating"; movement of a solid particle into a cell after pseudopods engulf it.	Bacteria, viruses, and aged or dead cells.
Bulk-phase endocytosis	"Cell drinking"; movement of extracellular fluid into a cell by infolding of plasma membrane.	Solutes in extracellular fluid.
Exocytosis	Movement of substances out of a cell in secretory vesicles that fuse with the plasma membrane and release their contents into the extracellular fluid.	Neurotransmitters, hormones, and digestive enzymes.

The Cytoskeleton

Extending throughout the cytosol, the ***cytoskeleton*** is a network of three different types of protein filaments: microfilaments, intermediate filaments, and microtubules (Figure 3.11).

The thinnest elements of the cytoskeleton are the ***microfilaments***, which are concentrated at the periphery of a cell and contribute to the cell's strength and shape (Figure 3.11a). Microfilaments have two general functions: providing mechanical support and helping generate movements. They also anchor the cytoskeleton to integral proteins in the plasma membrane and provide support for microscopic, fingerlike projections of the plasma membrane called ***microvilli*** (*micro-* = small; *-villi* = tufts of hair; singular is *microvillus*). Because they greatly increase the surface area of the cell, microvilli are abundant on cells involved in absorption, such as the cells that line the small intestine. Some microfilaments extend beyond the plasma membrane and help cells attach to one another or to extracellular materials.

With respect to movement, microfilaments are involved in muscle contraction, cell division, and cell locomotion. Microfilament-assisted movements include the migration of embryonic cells during development, the invasion of tissues by white blood cells to fight infection, and the migration of skin cells during wound healing.

As their name suggests, ***intermediate filaments*** are thicker than microfilaments but thinner than microtubules (Figure 3.11b). They are found in parts of cells subject to tension (such as stretching), help hold organelles such as the nucleus in place, and help attach cells to one another.

The largest of the cytoskeletal components, ***microtubules*** are long, hollow tubes (Figure 3.11c). Microtubules help determine cell shape and function in both the movement of organelles, such as secretory vesicles, within a cell and the migration of chromosomes during cell division. They also are responsible for movements of cilia and flagella.

Figure 3.11 Cytoskeleton.

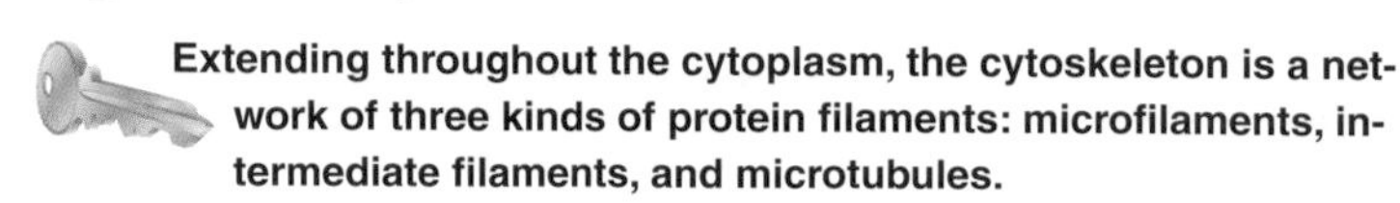

Extending throughout the cytoplasm, the cytoskeleton is a network of three kinds of protein filaments: microfilaments, intermediate filaments, and microtubules.

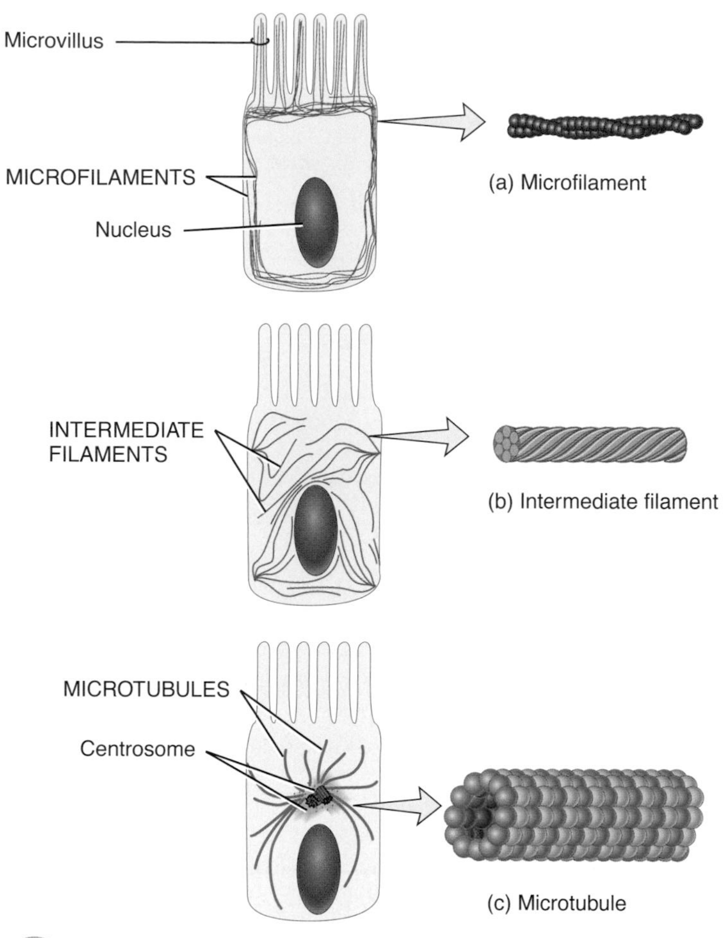

Which cytoskeletal components help form the structure of centrioles, cilia, and flagella?

Centrosome

The ***centrosome***, located near the nucleus, has two components—a pair of centrioles and pericentriolar material (Figure 3.12). The two *centrioles* are cylindrical structures, each of which is composed of nine clusters of three microtubules (a triplet) arranged in a circular pattern. Surrounding the centrioles is the *pericentriolar material* (per′-ē-sen′-trē-Ō-lar), containing hundreds of ring-shaped proteins called *tubulins*. The tubulins are the organizing centers for growth of the mitotic spindle, which plays a critical role in cell division, and for microtubule formation in nondividing cells.

Cilia and Flagella

Microtubules are the main structural and functional components of cilia and flagella, both of which are motile projections of the cell surface. ***Cilia*** (SIL-ē-a; singular is *cilium* = eyelash) are numerous, short, hairlike projections that extend from the surface of the cell (see Figure 3.1). In the human body, cilia propel fluids across the surfaces of cells that are firmly anchored in place. The coordinated movement of many cilia on the surface of a cell causes a steady movement of fluid along the cell's surface. Many cells of the respiratory tract, for example, have hundreds of cilia that help sweep foreign particles trapped in mucus away from the lungs. Their movement is paralyzed by nicotine in cigarette smoke. For this reason, smokers cough often to remove foreign particles from their airways. Cells that line the uterine (fallopian) tubes also have cilia that sweep oocytes (egg cells) toward the uterus.

Flagella (fla-JEL-a; singular is *flagellum* = whip) are similar in structure to cilia but are much longer (see Figure 3.1). Flagella usually move an entire cell. The only example of a flagellum in the human body is a sperm cell's tail, which propels the sperm toward its possible union with an oocyte.

Ribosomes

Ribosomes (RĪ-bō-sōms; *-somes* = bodies) are the sites of protein synthesis. Ribosomes are named for their high content of *ribo*nucleic acid (RNA). Besides ribosomal RNA (rRNA), these tiny organelles contain ribosomal proteins. Structurally, a ribosome consists of two subunits, large and small, one about half the size of the other (Figure 3.13). The large and

Figure 3.12 Centrosome.

The pericentriolar material of a centrosome organizes the mitotic spindle during cell division.

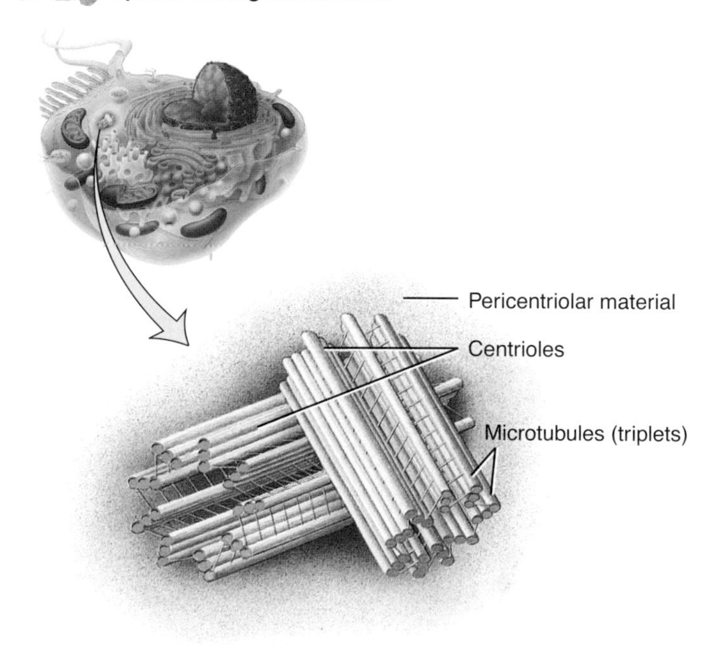

What are the components of the centrosome?

Figure 3.13 Ribosomes.

Ribosomes, the sites of protein synthesis, consist of a large subunit and a small subunit.

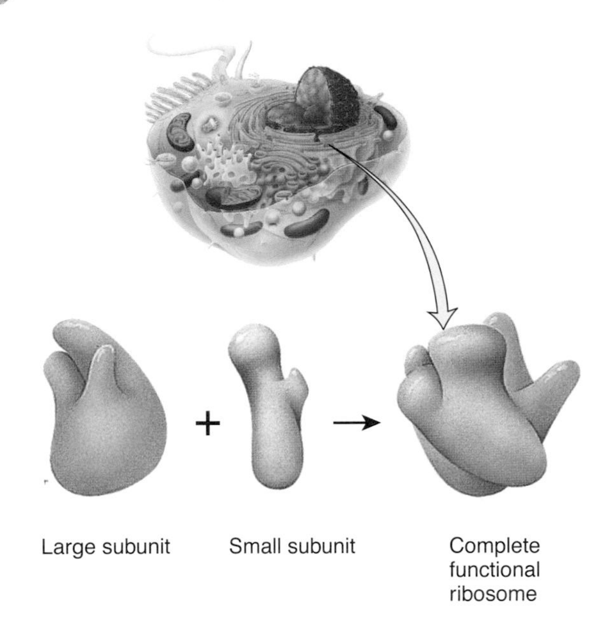

Where are ribosomal subunits synthesized and assembled?

small subunits are made in the nucleolus of the nucleus. Later, they exit the nucleus and are assembled in the cytoplasm, where they form a functional ribosome.

Some ribosomes are attached to the outer surface of the nuclear membrane and to an extensively folded membrane called the endoplasmic reticulum. These ribosomes synthesize proteins destined for specific organelles, for insertion in the plasma membrane, or for export from the cell. Other ribosomes are "free" or unattached to other cytoplasmic structures. Free ribosomes synthesize proteins used in the cytosol. Ribosomes are also located within mitochondria, where they synthesize mitochondrial proteins.

Endoplasmic Reticulum

The ***endoplasmic reticulum*** (en′-dō-PLAS-mik re-TIK-ū-lum; *-plasmic* = cytoplasm; *reticulum* = network) or ***ER*** is a network of folded membranes (Figure 3.14). The ER extends throughout the cytoplasm and is so extensive that it constitutes more than half of the membranous surfaces within the cytoplasm of most cells.

Cells contain two distinct forms of ER that differ in structure and function. ***Rough ER*** extends from the nuclear envelope (membrane around the nucleus) and appears "rough" because its outer surface is studded with ribosomes. Proteins synthesized by ribosomes attached to rough ER enter the spaces within the ER for processing and sorting. These molecules (glycoproteins and phospholipids) may be incorporated into organelle membranes or the plasma membrane. Thus, rough ER is a factory for synthesizing secretory proteins and membrane molecules.

Smooth ER extends from the rough ER to form a network of membranous tubules (Figure 3.14). As you may already have guessed, smooth ER appears "smooth" because it lacks ribosomes. Smooth ER is where fatty acids and steroids such as estrogens and testosterone are synthesized. In liver cells, enzymes of the smooth ER also help release glucose into the bloodstream and inactivate or detoxify a variety of drugs and potentially harmful substances, including alcohol, pesticides, and carcinogens (cancer-causing agents).

One of the functions of smooth ER, as noted earlier, is to detoxify certain drugs. Individuals who repeatedly take such drugs, such as the sedative phenobarbital, develop changes in the smooth ER in their liver cells. Prolonged administration of phenobarbital results in increased tolerance to the drug; the same dose no longer produces the same degree of sedation. With repeated exposure to the drug, the amount of smooth ER and its enzymes increases to protect the cell from its toxic effects. As the amount of smooth ER increases, higher and higher dosages of the drug are needed to achieve the original effect.

Figure 3.14 Endoplasmic reticulum (ER).

The ER is a network of folded membranes that extend throughout the cytoplasm and connect to the nuclear envelope.

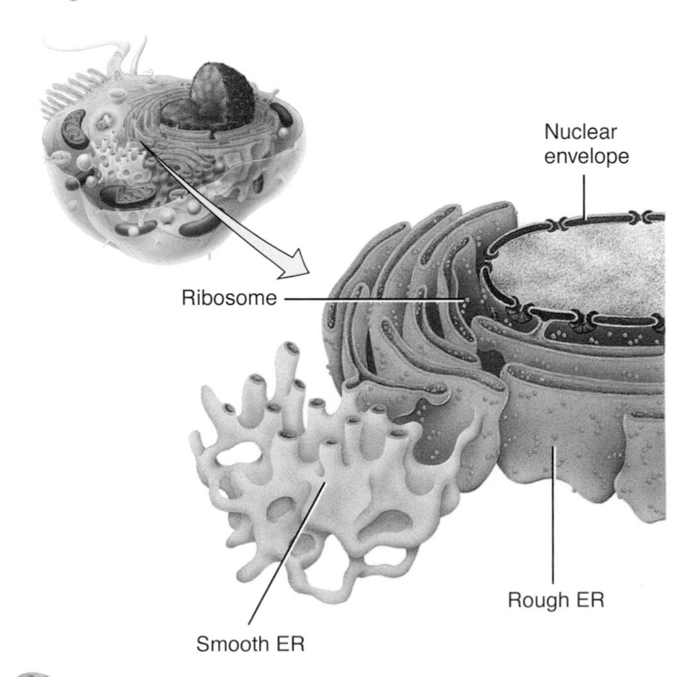

How do rough ER and smooth ER differ structurally and functionally?

Golgi Complex

After proteins are synthesized on a ribosome attached to rough ER, they usually are transported to another region of the cell. The first step in the transport pathway is through an organelle called the ***Golgi complex*** (GOL-jē). It consists of 3 to 20 ***cisterns*** (SIS-terns = cavities), flattened membranous sacs with bulging edges, piled on each other like a stack of pita bread (Figure 3.15). Most cells have several Golgi complexes. The Golgi complex is more extensive in cells that secrete proteins.

The main function of the Golgi complex is to modify and package proteins. Proteins synthesized by ribosomes on rough ER enter the Golgi complex and are modified to form glycoproteins and lipoproteins. Then, they are sorted and packaged. Some of the processed proteins are discharged from the cell by exocytosis. Certain cells of the pancreas release the hormone insulin this way. Other processed proteins become part of the plasma membrane as existing parts of the membrane are lost. Still other processed proteins become incorporated into organelles called lysosomes.

Lysosomes

Lysosomes (LĪ-sō-sōms; *lyso-* = dissolving; *-somes* = bodies) are membrane-enclosed vesicles (see Figure 3.1) that may contain as many as 60 different digestive enzymes; these enzymes can break down a wide variety of molecules once the lysosome fuses with vesicles formed during endocytosis. The lysosomal membrane allows the final products of digestion, such as monosaccharides, fatty acids, and amino acids, to be transported into the cytosol.

Figure 3.15 Golgi complex.

Most proteins synthesized by ribosomes attached to rough ER pass through the Golgi complex for processing.

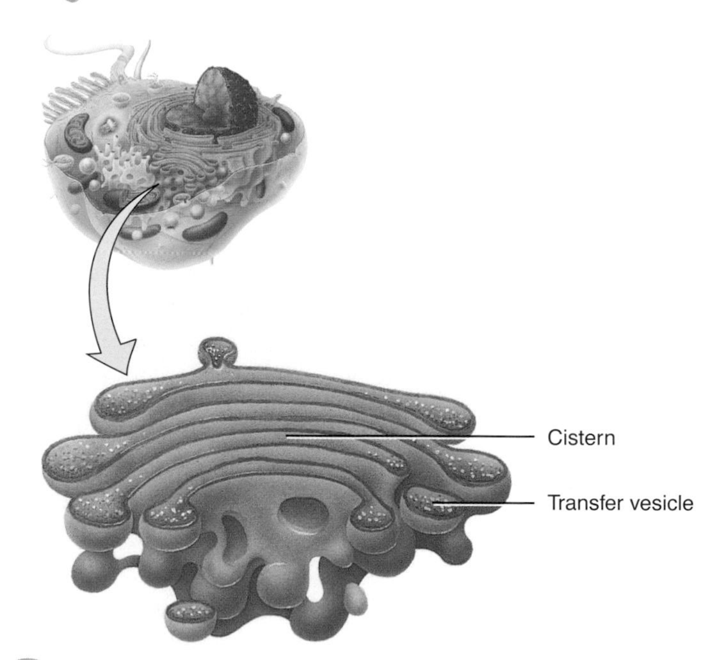

What types of body cells are likely to have extensive Golgi complexes?

Lysosomal enzymes also help recycle worn-out structures. A lysosome can engulf another organelle, digest it, and return the digested components to the cytosol for reuse. In this way, old organelles are continually replaced. The process by which worn-out organelles are digested is called ***autophagy*** (aw-TOF-a-jē; *auto-* = self; *-phagy* = eating). During autophagy, the organelle to be digested is enclosed by a membrane derived from the ER to create a vesicle that then fuses with a lysosome. In this way, a human liver cell, for example, recycles about half its contents every week. Lysosomal enzymes may also destroy the entire cell, a process known as ***autolysis*** (aw-TOL-i-sis). Autolysis occurs in some pathological conditions and also is responsible for the tissue deterioration that occurs just after death.

Some disorders are caused by faulty or absent lysosomal enzymes. For instance, **Tay-Sachs disease**, which most often affects children of Ashkenazi (eastern European Jewish) descent, is an inherited condition characterized by the absence of a single lysosomal enzyme. This enzyme normally breaks down a membrane glycolipid called ganglioside G_{M2} that is especially prevalent in nerve cells. As the excess ganglioside G_{M2} accumulates, because it is not broken down, the nerve cells function less efficiently. Children with Tay-Sachs disease typically experience seizures and muscle rigidity. They gradually become blind, demented, and uncoordinated and usually die before the age of 5. Tests can now reveal whether an adult is a carrier of the defective gene.

Peroxisomes

Another group of organelles similar in structure to lysosomes, but smaller, are called ***peroxisomes*** (per-OK-si-sōms; *peroxi-* = peroxide; see Figure 3.1). Peroxisomes contain several *oxidases*, which are enzymes that can oxidize (remove hydrogen atoms from) various organic substances. For example, amino acids and fatty acids are oxidized in peroxisomes as part of normal metabolism. In addition, enzymes in peroxisomes oxidize toxic substances. Thus, peroxisomes are very abundant in the liver, where detoxification of alcohol and other damaging substances takes place. A byproduct of the oxidation reactions is hydrogen peroxide (H_2O_2), a potentially toxic compound. However, peroxisomes also contain an enzyme called *catalase* that decomposes the H_2O_2. Because the generation and degradation of H_2O_2 occurs within the same organelle, peroxisomes protect other parts of the cell from the toxic effects of H_2O_2.

Proteasomes

Although lysosomes degrade proteins delivered to them in vesicles, proteins in the cytosol also require disposal at certain times in the life of a cell. Continuous destruction of unneeded, damaged, or faulty proteins is the function of tiny barrel-shaped structures called ***proteasomes*** (PRŌ-tē-a-sōmes = protein bodies). A typical body cell contains many thousands of proteasomes, in both the cytosol and the nucleus. Proteasomes were so named because they contain myriad *proteases*, enzymes that cut proteins into small peptides. Once the enzymes of a proteasome have chopped up a protein into smaller chunks, other enzymes then break down the peptides into amino acids, which can be recycled into new proteins.

Mitochondria

Because they are the site of most ATP production, the "powerhouses" of a cell are its ***mitochondria*** (mī-tō-KON-drē-a; *mito-* = thread; *-chondria* = granules; singular is *mitochondrion*). A cell may have as few as one hundred or as many as several thousand mitochondria, depending on how active the cell is. For example, active cells such as those found in muscles, the liver, and kidneys use ATP at a high rate and have large numbers of mitochondria. A mitochondrion consists of two membranes, each of which is similar in structure to the plasma membrane (Figure 3.16). The *outer mitochondrial membrane* is smooth, but the *inner mitochondrial membrane* is arranged in a series of folds called ***cristae*** (KRIS-tē; singular is *crista* = ridge). The large central fluid-filled cavity of a mitochondrion, enclosed by the inner membrane and cristae, is the ***matrix***. The elaborate folds of the cristae provide an enormous surface area for a series of chemical reactions that provide most of a cell's ATP. Enzymes that catalyze these reactions are located in the matrix and on the cristae. Mitochondria also contain a small number of genes and a few ribosomes, enabling them to synthesize some proteins.

Figure 3.16 Mitochondrion.

Within mitochondria, chemical reactions generate most of a cell's ATP.

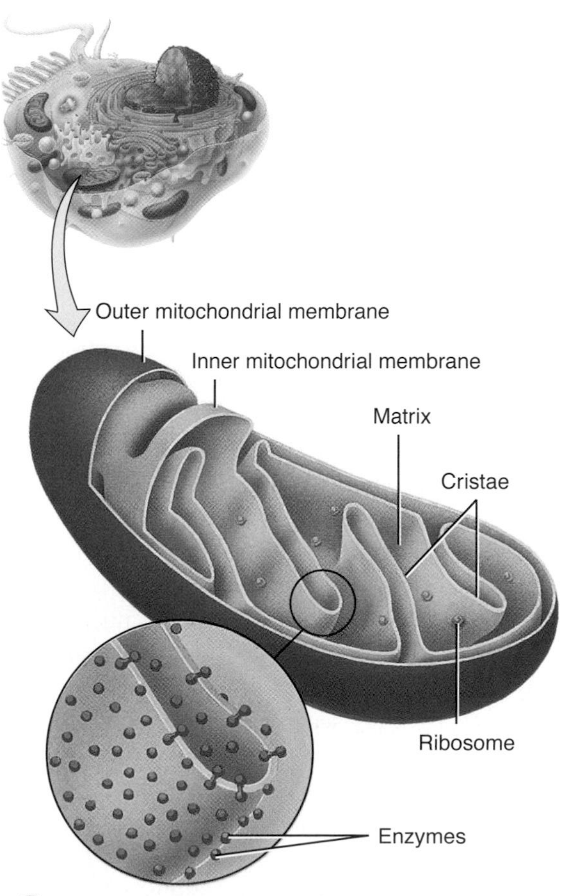

How do the cristae of a mitochondrion contribute to its ATP-producing function?

■ **CHECKPOINT**

7. What does cytoplasm have that cytosol does not?
8. What is an organelle?
9. Describe the structure and function of ribosomes, the Golgi complex, and mitochondria.

NUCLEUS

OBJECTIVE • **Describe the structure and functions of the nucleus.**

The ***nucleus*** is a spherical or oval structure that usually is the most prominent feature of a cell (Figure 3.17). Most body cells have a single nucleus, although some, such as mature red blood cells, have none. In contrast, skeletal muscle cells and a few other types of cells have several nuclei. A double membrane called the ***nuclear envelope*** separates the nucleus from the cytoplasm. Both layers of the nuclear envelope are lipid bilayers similar to the plasma membrane. The outer membrane of the nuclear envelope is continuous with the rough endoplasmic reticulum and resembles it in structure. Many openings called ***nuclear pores*** pierce the nuclear envelope. Nuclear pores control the movement of substances between the nucleus and the cytoplasm.

Inside the nucleus are one or more spherical bodies called ***nucleoli*** (noo′-KLĒ-ō-lī; singular is *nucleolus*). These clusters of protein, DNA, and RNA are the sites of assembly of ribosomes, which exit the nucleus through nuclear pores and participate in protein synthesis in the cytoplasm. Cells that synthesize large amounts of protein, such as muscle and liver cells, have prominent nucleoli.

Figure 3.17 Nucleus.

The nucleus contains most of a cell's genes, which are located on chromosomes.

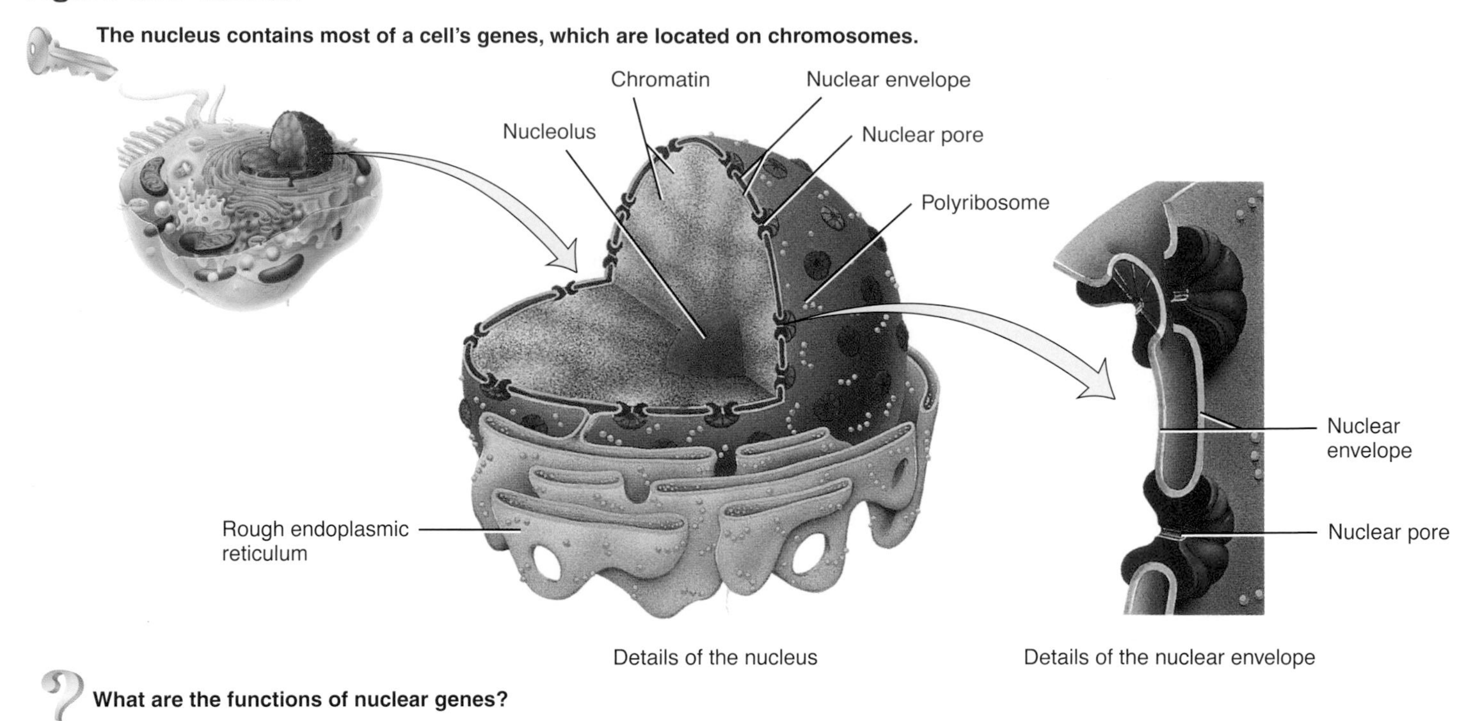

What are the functions of nuclear genes?

Also within the nucleus are most of the cell's hereditary units, called ***genes***, which control cellular structure and direct most cellular activities. The nuclear genes are arranged along ***chromosomes*** (*chromo-* = colored) (see Figure 3.21). Human somatic (body) cells have 46 chromosomes, 23 inherited from each parent. In a cell that is not dividing, the 46 chromosomes appear as a diffuse, granular mass, which is called ***chromatin*** (Figure 3.17). The total genetic information carried in a cell or organism is called its ***genome***.

In the last decade of the twentieth century, the genomes of humans, mice, fruit flies, and more than 50 microbes were sequenced. As a result, research in the field of ***genomics***, the study of the relationships between the genome and the biological functions of an organism, has flourished. The Human Genome Project began in 1990 as an effort to sequence all of the nearly 3.2 billion nucleotides of our genome and was completed in April 2003. Scientists now know that the total number of genes in the human genome is about 30,000. Information regarding the human genome and how it is affected by the environment seeks to identify and discover the functions of the specific genes that play a role in genetic diseases. Genomic medicine also aims to design new drugs and to provide screening tests to enable physicians to provide more effective counseling and treatment for disorders with significant genetic components such as hypertension (high blood pressure), obesity, diabetes, and cancer.

The main parts of a cell and their functions are summarized in Table 3.2.

■ **CHECKPOINT**

10. Why is the nucleus so important in the life of a cell?

GENE ACTION: PROTEIN SYNTHESIS

OBJECTIVE • Outline the sequence of events involved in protein synthesis.

Although cells synthesize many chemicals to maintain homeostasis, much of the cellular machinery is devoted to protein production. Cells constantly synthesize large numbers of diverse proteins. The proteins, in turn, determine the physical and chemical characteristics of cells and, on a larger scale, of organisms.

The DNA contained in genes provides the instructions for making proteins. To synthesize a protein, the information contained in a specific region of DNA is first *transcribed* (copied) to produce a specific molecule of RNA (ribonucleic acid). The RNA then attaches to a ribosome, where the information contained in the RNA is *translated* into a corresponding specific sequence of amino acids to form a new protein molecule (Figure 3.18 on page 60).

Information is stored in DNA in four types of nucleotides, the repeating units of nucleic acids (see Figure 2.15 on page 39). Each sequence of three DNA nucleotides is transcribed as a complementary (corresponding) sequence of three RNA nucleotides. Such a sequence of three successive DNA nucleotides is called a ***base triplet***. Each DNA base triplet is transcribed as a complementary sequence of three successive RNA nucleotides. The three successive RNA nucleotides are called a ***codon***. When translated, a given codon specifies a particular amino acid.

Transcription

During ***transcription***, which occurs in the nucleus, the genetic information in DNA base triplets is copied into a com-

Table 3.2 Cell Parts and Their Functions

Part	Structure	Functions
Plasma Membrane	Composed of a lipid bilayer consisting of phospholipids, cholesterol, and glycolipids with various proteins inserted; surrounds cytoplasm.	Protects cellular contents; makes contact with other cells; contains channels, transporters, receptors, enzymes, and cell-identity markers; mediates the entry and exit of substances.
Cytoplasm	Cellular contents between the plasma membrane and nucleus, including cytosol and organelles.	Site of all intracellular activities except those occurring in the nucleus.
Cytosol	Composed of water, solutes, suspended particles, lipid droplets, and glycogen granules.	Medium in which many of the cell's chemical reactions occur.
Organelles	Specialized cellular structures with characteristic shapes and specific functions.	Each organelle has one or more specific functions.
Cytoskeleton	Network composed of three protein filaments: microfilaments, intermediate filaments, and microtubules.	Maintains shape and general organization of cellular contents; responsible for cell movements.
Centrosome	Paired centrioles plus pericentriolar material.	Pericentriolar material is organizing center for microtubules and mitotic spindle.
Cilia and flagella	Motile cell surface projections with inner core of microtubules.	Cilia move fluids over a cell's surface; a flagellum moves an entire cell.
Ribosome	Composed of two subunits containing ribosomal RNA and proteins; may be free in cytosol or attached to rough ER.	Protein synthesis.
Endoplasmic reticulum (ER)	Membranous network of folded membranes. Rough ER is studded with ribosomes and is attached to the nuclear membrane; smooth ER lacks ribosomes.	Rough ER is the site of synthesis of glycoproteins and phospholipids; smooth ER is the site of fatty acid and steroid synthesis. Smooth ER also releases glucose into the bloodstream, inactivates or detoxifies drugs and potentially harmful substances, and stores calcium ions for muscle contraction.
Golgi complex	A stack of 3–20 flattened membranous sacs called cisterns.	Accepts proteins from rough ER; forms glycoproteins and lipoproteins; stores, packages, and exports proteins.
Lysosome	Vesicle formed from Golgi complex; contains digestive enzymes.	Fuses with and digests contents of vesicles; digests worn-out organelles (autophagy), entire cells (autolysis), and extracellular materials.
Peroxisome	Vesicle containing oxidative enzymes.	Detoxifies harmful substances.
Proteasome	Tiny structure that contains proteases, enzymes that cut proteins.	Degrades unneeded, damaged, or faulty proteins by cutting them into small peptides.
Mitochondrion	Consists of an outer and inner membranes, cristae, and matrix.	Site of reactions that produce most of a cell's ATP.
Nucleus	Consists of nuclear envelope with pores, nucleoli, and chromatin (or chromosomes).	Contains genes, which control cellular structure and direct most cellular activities.

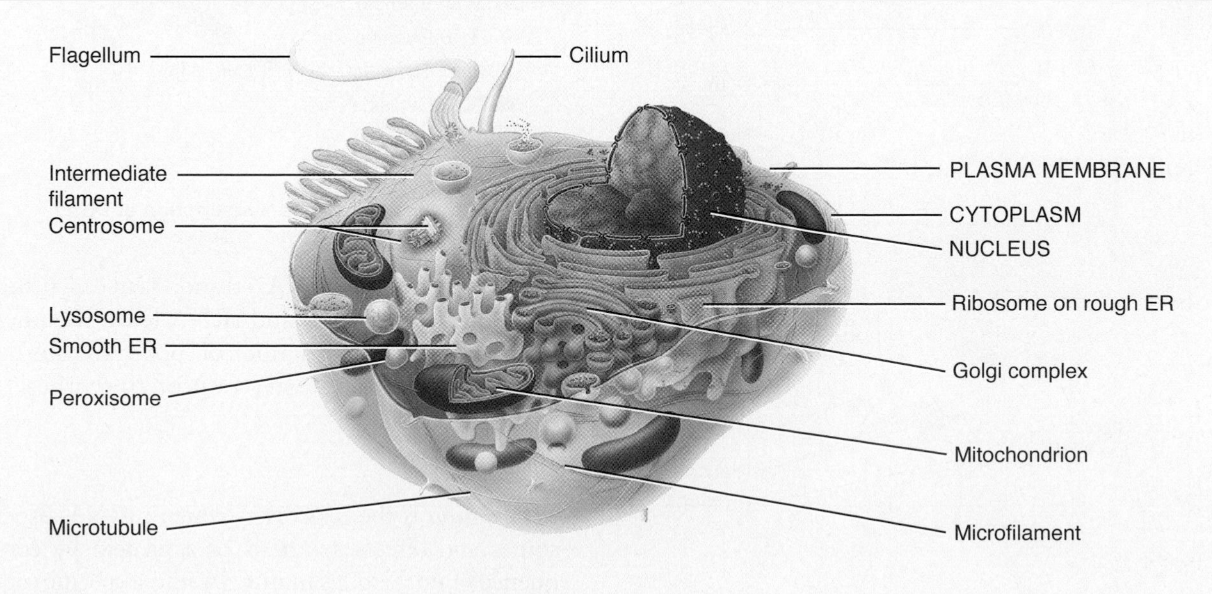

plementary sequence of codons in a strand of RNA. Transcription of DNA is catalyzed by the enzyme *RNA polymerase*, which must be instructed where to start the transcription process and where to end it. The segment of DNA where RNA polymerase attaches to it is a special sequence of nucleotides called a ***promoter***, located near the beginning of a gene (Figure 3.19a). Three kinds of RNA are made from DNA:

- ***Messenger RNA (mRNA)*** directs synthesis of a protein.
- ***Ribosomal RNA (rRNA)*** joins with ribosomal proteins to make ribosomes.
- ***Transfer RNA (tRNA)*** binds to an amino acid and holds it in place on a ribosome until it is incorporated into a protein during translation. Each of the more than 20 different types of tRNA binds to only one of the 20 different amino acids.

During transcription, nucleotides pair in a complementary manner: The nitrogenous base cytosine (C) in DNA dictates the complementary nitrogenous base guanine (G) in the new RNA strand, a G in DNA dictates a C in RNA, a thymine (T) in DNA dictates an adenine (A) in RNA, and an A in DNA dictates a uracil (U) in RNA. As an example, if a segment of DNA had the base sequence ATGCAT, the newly transcribed RNA strand would have the complementary base sequence UACGUA.

Transcription of DNA ends at another special nucleotide sequence on DNA called a ***terminator***, which specifies the end of the gene (Figure 3.19a). Upon reaching the terminator, RNA polymerase detaches from the transcribed RNA

Figure 3.18 Overview of transcription and translation.

Transcription occurs in the nucleus; translation takes place in the cytoplasm.

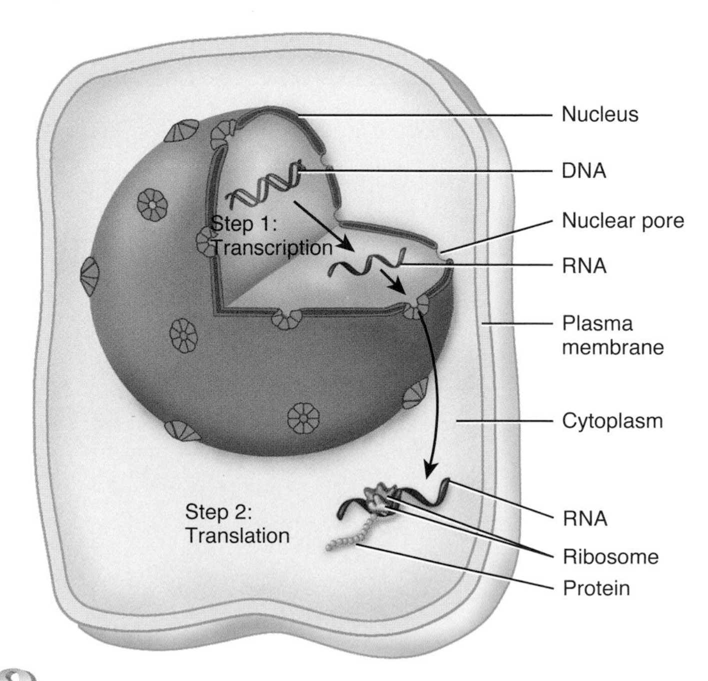

Why are proteins important in the life of a cell?

Figure 3.19 Transcription.

During transcription, the genetic information in DNA is copied to RNA.

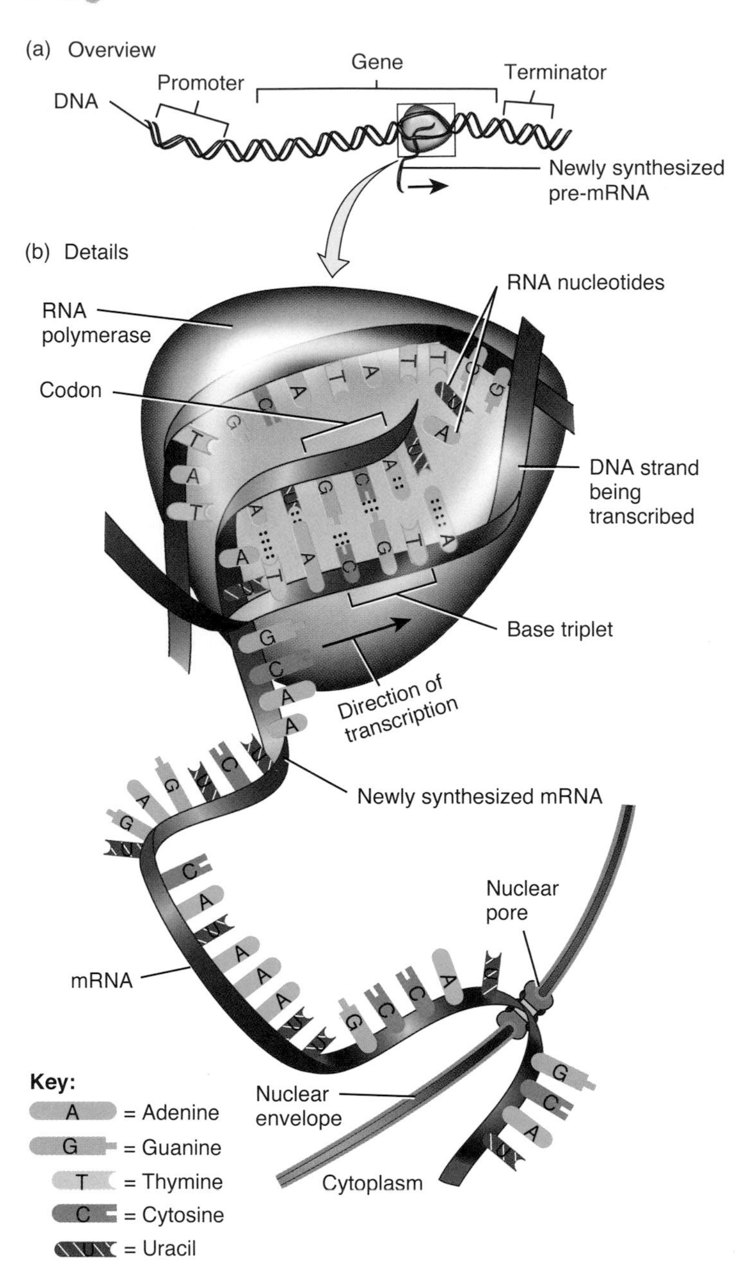

What enzyme catalyzes transcription of DNA?

molecule and the DNA strand. Once synthesized, mRNA, rRNA (in ribosomes), and tRNA leave the nucleus of the cell by passing through a nuclear pore. In the cytoplasm, they participate in the next step in protein synthesis, translation.

Translation

Translation is the process in which mRNA associates with ribosomes and directs synthesis of a protein by converting the sequence of nucleotides in mRNA into a specific sequence of amino acids. Translation occurs in the following way (Figure 3.20):

Figure 3.20 **Protein elongation and termination of protein synthesis during translation.**

During protein synthesis the ribosomal subunits join, but they separate when the process is complete.

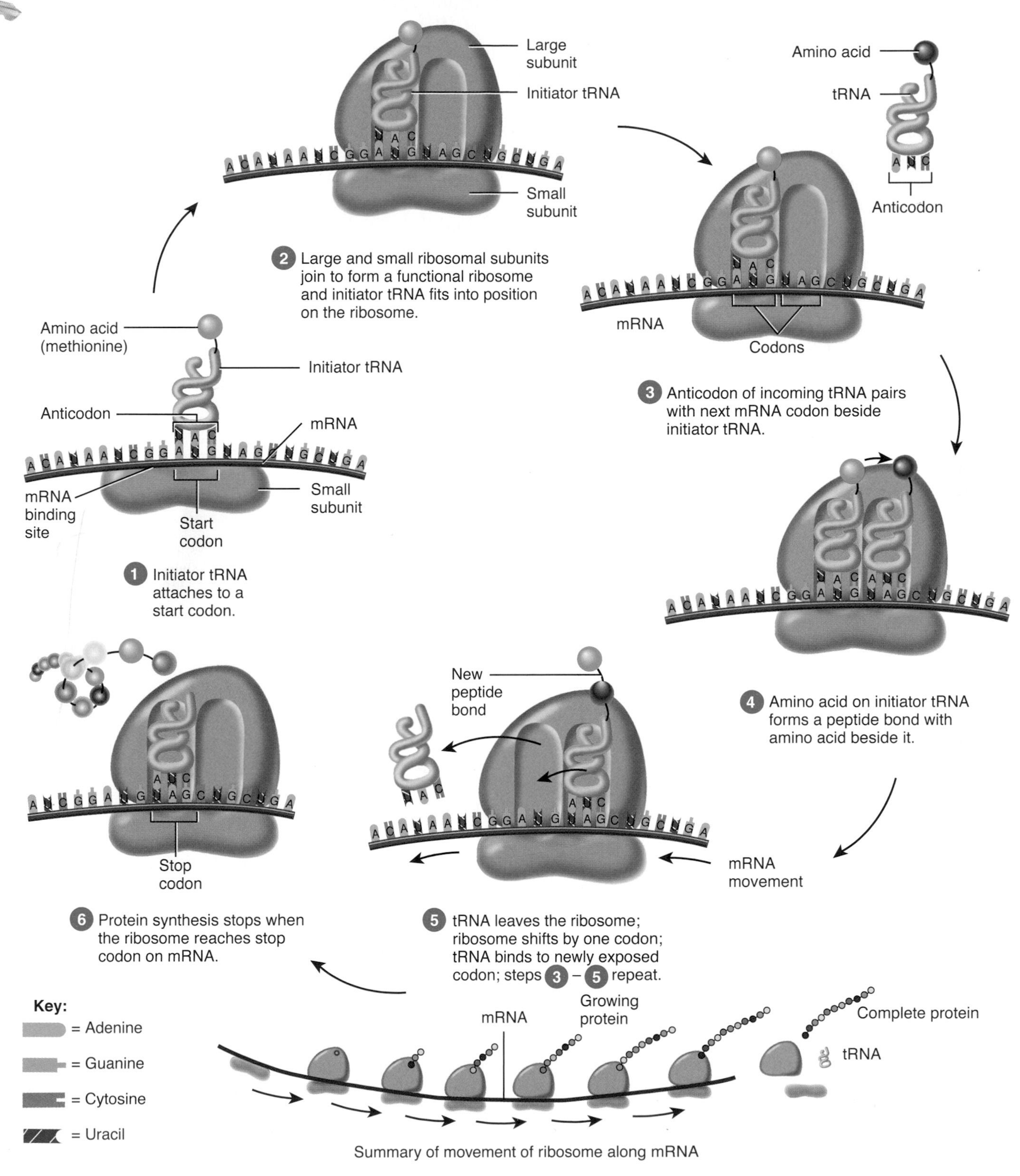

What is the function of a stop codon?

1. An mRNA molecule binds to the small ribosomal subunit, and a special tRNA, called *initiator tRNA*, binds to the start codon (AUG) on mRNA, where translation begins.
2. The large ribosomal subunit attaches to the small subunit, creating a functional ribosome. The initiator tRNA fits into position on the ribosome. One end of a tRNA carries a specific amino acid, and the opposite end consists of a triplet of nucleotides called an ***anticodon***. By pairing between complementary nitrogenous bases, the tRNA anticodon attaches to the mRNA codon. For example, if the mRNA codon is AUG, then a tRNA with the anticodon UAC would attach to it.
3. The anticodon of another tRNA with its amino acid attaches to the complementary mRNA codon next to the initiator tRNA.
4. A peptide bond is formed between the amino acids carried by the initiator tRNA and the tRNA next to it.
5. After the peptide bond forms, the tRNA detaches from the ribosome, and the ribosome shifts the mRNA strand by one codon. As the tRNA bearing the newly forming protein shifts, another tRNA with its amino acid binds to a newly exposed codon. Steps 3 through 5 repeat again and again as the protein lengthens.
6. Protein synthesis ends when the ribosome reaches a stop codon, at which time the completed protein detaches from the final tRNA. When the tRNA vacates the ribosome, the ribosome splits into its large and small subunits.

Protein synthesis progresses at a rate of about 15 amino acids per second. As the ribosome moves along the mRNA and before it completes synthesis of the whole protein, another ribosome may attach behind it and begin translation of the same mRNA strand. In this way, several ribosomes may be attached to the same mRNA, an assembly called a ***polyribosome***. The simultaneous movement of several ribosomes along the same mRNA strand permits a large amount of protein to be produced from each mRNA.

■ CHECKPOINT

11. Define protein synthesis.

12. Distinguish between transcription and translation.

SOMATIC CELL DIVISION

OBJECTIVE • Discuss the stages, events, and significance of somatic cell division.

As body cells become damaged, diseased, or worn out, they are replaced by ***cell division***, the process whereby cells reproduce themselves. The two types of cell division are reproductive cell division and somatic cell division. ***Reproductive cell division*** or ***meiosis*** is the process that produces gametes—sperm and oocytes—the cells needed to form the next generation of sexually reproducing organisms. This is described in Chapter 23; here we will focus on somatic cell division.

All body cells, except the gametes, are called ***somatic*** (*soma* = body) ***cells***. In ***somatic cell division***, a cell divides into two identical cells. An important part of somatic cell division is replication (duplication) of the DNA sequences that make up genes and chromosomes so that the same genetic material can be passed on to the newly formed cells. After somatic cell division, each newly formed cell has the same number of chromosomes as the original cell. Somatic cell division replaces dead or injured cells and adds new ones for tissue growth. For example, skin cells are continually replaced by somatic cell divisions.

The ***cell cycle*** is the name for the sequence of changes that a cell undergoes from the time it forms until it duplicates its contents and divides into two cells. In somatic cells, the cell cycle consists of two major periods: interphase, when a cell is not dividing, and the mitotic (M) phase, when a cell is dividing.

Interphase

During ***interphase*** the cell replicates its DNA. It also manufactures additional organelles and cytosolic components in anticipation of cell division. Interphase is a state of high metabolic activity, and during this time the cell does most of its growing.

A microscopic view of a cell during interphase shows a clearly defined nuclear envelope, a nucleolus, and a tangled mass of chromatin (Figure 3.21a). Once a cell completes its replication of DNA and other activities of interphase, the mitotic phase begins.

Mitotic Phase

The ***mitotic phase*** (mī-TOT-ik) of the cell cycle consists of *mitosis*, division of the nucleus, followed by *cytokinesis*, division of the cytoplasm into two cells. The events that take place during mitosis and cytokinesis are plainly visible under a microscope because chromatin condenses into chromosomes.

Nuclear Division: Mitosis

During ***mitosis*** (mī-TŌ-sis; *mitos* = thread), the duplicated chromosomes become exactly segregated, one set into each of two separate nuclei. For convenience, biologists divide the process into four stages: prophase, metaphase, anaphase, and telophase. However, mitosis is a continuous process, with one stage merging imperceptibly into the next.

PROPHASE During early prophase, the chromatin fibers condense and shorten into chromosomes that are visible under the light microscope (Figure 3.21b). The condensation process may prevent entangling of the long DNA strands as they move during mitosis. Recall that DNA replication took place during interphase. Thus, each prophase chromosome consists of a pair of identical, double-stranded ***chromatids***. A

Figure 3.21 Cell division: mitosis and cytokinesis. Begin the sequence at (a) at the top of the figure and read clockwise until you complete the process.

In somatic cell division, a single cell divides to produce two identical cells.

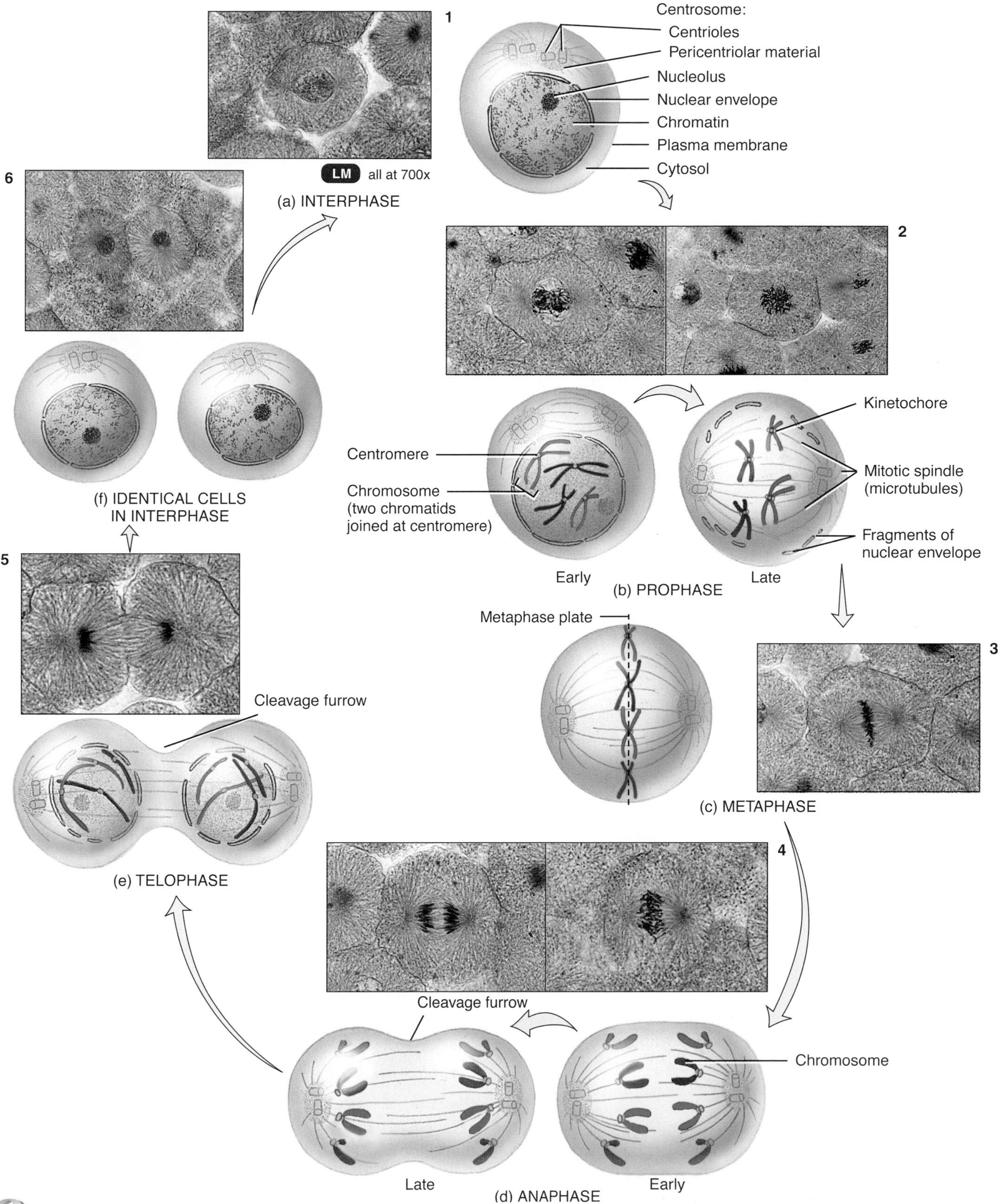

When does cytokinesis begin?

FOCUS ON WELLNESS

Phytochemicals—Protecting Cellular Function

Many studies over the years have shown that people who consume plenty of plant foods, including vegetables, beans, fruits, and grains, have a lower risk of cancer and heart disease than their meat-and-potato-eating peers. Scientists are just beginning to uncover the biochemical explanations for these associations. Their investigations have led to the discovery of compounds in plants that appear to promote healthy cellular function, and to prevent the types of cellular damage associated with cancer, aging, and heart disease. Collectively, these compounds are called *phytochemicals*, literally "plant chemicals."

A Radical Notion?

Phytochemicals appear to protect cells and interrupt cancerous tumor growth in a number of interesting ways. Some phytochemicals block chemicals that can cause oxidative damage to cells. You will learn about the process of oxidation in Chapter 20 when you read about metabolism. Oxidative damage commonly occurs in cells when byproducts of metabolism, known as oxygen free radicals, "steal" electrons from other molecules. This electron theft causes chain reactions of electron transfers that can damage cell membranes, the membranes of cellular organelles, and even the cells' genetic material.

Some phytochemicals act as antioxidants, donating electrons to free radical molecules, thus protecting cellular structures. Antioxidants include polyphenols, which are found in green tea, and lycopenes, which are found in tomato products.

Disabling the Opponent

Many substances entering the body are potentially carcinogenic, depending upon their interaction with certain enzymes in the liver. Some phytochemicals, such as the allyl sulfides in garlic and onions, enhance the production of enzymes that may render potentially carcinogenic substances harmless. The sulforaphane in broccoli, cauliflower, and other cruciferous vegetables performs a similar function.

Promoting Health

Some phytochemicals protect against cancer by blocking the action of substances called promoters. Promoters encourage the aggressive cellular division of cells that have undergone cancer-causing genetic changes. For example, estrogens are hormones that promote the division of cancerous cells in the breast. Isoflavonoids, found in soy products, weaken the action of estrogens in breast tissue.

Variety is the key to consuming more phytochemicals. Try to eat two to four servings of fruit and three to five servings of vegetables each day.

► THINK IT OVER . . .

► ***What are some dietary changes you could make that would increase your intake of helpful phytochemicals?***

constricted region of the chromosome, called a ***centromere***, holds the chromatid pair together.

Later in prophase, the pericentriolar material of the two centrosomes starts to form the ***mitotic spindle***, a football-shaped assembly of microtubules (Figure 3.21b). Lengthening of the microtubules between centrosomes pushes the centrosomes to opposite poles (ends) of the cell. Finally, the spindle extends from pole to pole. Then the nucleolus and nuclear envelope break down.

METAPHASE During metaphase, the centromeres of the chromatid pairs are aligned along the microtubules of the mitotic spindle at the exact center of the mitotic spindle (Figure 3.21c). This midpoint region is called the ***metaphase plate***.

ANAPHASE During anaphase the centromeres split, separating the two members of each chromatid pair, which move to opposite poles of the cell (Figure 3.21d). Once separated, the chromatids are called chromosomes. As the chromosomes are pulled by the microtubules of the mitotic spindle during anaphase, they appear V-shaped because the centromeres lead the way and seem to drag the trailing arms of the chromosomes toward the pole.

TELOPHASE The final stage of mitosis, telophase, begins after chromosomal movement stops (Figure 3.21e). The identical sets of chromosomes, now at opposite poles of the cell, uncoil and revert to the threadlike chromatin form. A new nuclear envelope forms around each chromatin mass, nucleoli appear, and eventually the mitotic spindle breaks up.

Cytoplasmic Division: Cytokinesis

Division of a cell's cytoplasm and organelles is called ***cytokinesis*** (sī′-tō-ki-NĒ-sis; *-kinesis* = motion). This process usually begins late in anaphase with formation of a ***cleavage furrow***, a slight indentation of the plasma membrane, that extends around the center of the cell (Figure 3.21d,e). Microfilaments in the cleavage furrow pull the plasma membrane progressively inward, constricting the center of the cell like a belt around a waist, and ultimately pinching it in two. After cytokinesis there are two new and separate cells, each with equal portions of cytoplasm and organelles and identical sets of chromosomes. When cytokinesis is complete, interphase begins (Figure 3.21f).

One of the distinguishing features of cancer cells is uncontrolled division. The mass of cells resulting from this division is called a neoplasm or tumor. One of the ways to treat cancer is by *chemotherapy*, the use of anticancer drugs. Some of these drugs stop cell division by inhibiting the formation of the mitotic spindle. Unfortunately, these types of anticancer drugs also kill all types of rapidly dividing cells in the body, causing side effects such as nausea, diarrhea, hair loss, fatigue, and decreased resistance to disease.

■ **CHECKPOINT**

13. Distinguish between somatic and reproductive cell division. Why is each important?

14. What are the major events of each stage of the mitotic phase?

AGING AND CELLS

OBJECTIVE • **Describe the cellular changes that occur with aging.**

Aging is a normal process accompanied by a progressive alteration of the body's homeostatic adaptive responses. It produces observable changes in structure and function and increases vulnerability to environmental stress and disease. The specialized branch of medicine that deals with the medical problems and care of elderly persons is ***geriatrics*** (jer′-ē-AT-riks; *ger-* = old age; *-iatrics* = medicine). ***Gerontology*** (jer′-on-TOL-ō-jē) is the scientific study of the process and problems associated with aging.

Although many millions of new cells normally are produced each minute, several kinds of cells in the body—skeletal muscle cells and nerve cells—do not divide. Experiments have shown that many other cell types have only a limited capability to divide. Normal cells grown outside the body divide only a certain number of times and then stop. These observations suggest that cessation of mitosis is a normal, genetically programmed event. According to this view, "aging genes" are part of the genetic blueprint at birth. These genes have an important function in normal cells, but their activities slow over time. They bring about aging by slowing down or halting processes vital to life.

Another aspect of aging involves ***telomeres*** (TĒ-lō-merz), specific DNA sequences found only at the tips of each chromosome. These pieces of DNA protect the tips of chromosomes from erosion and from sticking to one another. However, in most normal body cells each cycle of cell division shortens the telomeres. Eventually, after many cycles of cell division, the telomeres can be completely gone, and even some of the functional chromosomal material may be lost. These observations suggest that erosion of DNA from the tips of our chromosomes contributes greatly to the aging and death of cells.

Glucose, the most abundant sugar in the body, plays a role in the aging process. It is haphazardly added to proteins inside and outside cells, forming irreversible cross-links between adjacent protein molecules. With advancing age, more cross-links form, which contributes to the stiffening and loss of elasticity that occur in aging tissues.

Free radicals produce oxidative damage in lipids, proteins, or nucleic acids. Some effects are wrinkled skin, stiff joints, and hardened arteries. Naturally occurring enzymes in peroxisomes and in the cytosol normally dispose of free radicals. Certain dietary substances, such as vitamin E, vitamin C, beta carotene, and selenium, are antioxidants that inhibit free radical formation.

Some theories of aging explain the process at the cellular level, while others concentrate on regulatory mechanisms operating within the entire organism. For example, the immune system may start to attack the body's own cells. This *autoimmune response* might be caused by changes in certain plasma membrane glycoproteins and glycolipids (cell-identity markers) that cause antibodies to attach to and mark the cell for destruction. As changes in the proteins on the plasma membrane of cells increase, the autoimmune response intensifies, producing the well-known signs of aging.

Progeria (prō-JER-ē-a) is a disease characterized by normal development in the first year of life followed by rapid aging. It is caused by a genetic defect in which telomeres are considerably shorter than normal. Symptoms include dry and wrinkled skin, total baldness, and birdlike facial features. Death usually occurs around age 13.

Werner syndrome is a rare, inherited disease that causes a rapid acceleration of aging, usually while the person is only in his or her twenties. It is characterized by wrinkling of the skin, graying of the hair and baldness, cataracts, muscular atrophy, and a tendency to develop diabetes mellitus, cancer, and cardiovascular disease. Most afflicted individuals die before age 50. Recently, the gene that causes Werner syndrome has been identified. Researchers hope to use the information to gain insight into the mechanisms of aging, as well as to help those suffering from the disorder.

■ **CHECKPOINT**

15. Briefly outline the cellular changes involved in aging.

• • •

Next, in Chapter 4, we will explore how cells associate to form the tissues and organs that we will discuss later in the text.

Cancer

Cancer is a group of diseases characterized by uncontrolled or abnormal cell proliferation. When cells in a part of the body divide without control, the excess tissue that develops is called a ***tumor*** or ***neoplasm*** (NĒ-ō-plazm; *neo-* = new). The study of tumors is called ***oncology*** (on-KOL-ō-jē; *onco-* = swelling or mass). Tumors may be cancerous and often fatal, or they may be harmless. A cancerous neoplasm is called a ***malignant tumor*** or ***malignancy***. One property of most malignant tumors is their ability to undergo ***metastasis*** (me-TAS-ta-sis), the spread of cancerous cells to other parts of the body. A ***benign tumor*** is a neoplasm that does not metastasize. An example is a wart. Most benign tumors may be surgically removed if this interferes with normal body function or they become disfiguring. Some can be inoperable and perhaps fatal.

Growth and Spread of Cancer

Cells of malignant tumors duplicate rapidly and continuously. As malignant cells invade surrounding tissues, they often trigger ***angiogenesis***, the growth of new networks of blood vessels. As the cancer grows, it begins to compete with normal tissues for space and nutrients. Eventually, the normal tissue decreases in size and dies. Some malignant cells may detach from the initial (primary) tumor and invade a body cavity or enter the blood or lymph, then circulate to and invade other body tissues, establishing secondary tumors. The pain associated with cancer develops when the tumor presses on nerves or blocks a passageway in an organ so that secretions build up pressure.

Causes of Cancer

Several factors may trigger a normal cell to lose control and become cancerous. One cause is environmental agents: substances in the air we breathe, the water we drink, and the food we eat. A chemical agent or radiation that produces cancer is called a ***carcinogen*** (car-SIN-ō-jen). Carcinogens induce ***mutations***, permanent changes in the DNA base sequence of a gene. The World Health Organization estimates that carcinogens are associated with 60–90% of all human cancers. Examples of carcinogens are hydrocarbons found in cigarette tar, radon gas from the earth, and ultraviolet (UV) radiation in sunlight.

Intensive research efforts are now directed toward studying cancer-causing genes, or ***oncogenes*** (ON-kō-jēnz). When inappropriately activated, these genes have the ability to transform a normal cell into a cancerous cell. Most oncogenes derive from normal genes called ***proto-oncogenes*** that regulate growth and development. The proto-oncogene undergoes some change that either causes it to be expressed inappropriately or make its products in excessive amounts or at the wrong time. Some oncogenes cause excessive production of growth factors, chemicals that stimulate cell growth. Others may trigger changes in a cell-surface receptor, causing it to send signals as though it were being activated by a growth factor. As a result, the growth pattern of the cell becomes abnormal.

Some cancers have a viral origin. Viruses are tiny packages of nucleic acids, either RNA or DNA, that can reproduce only while inside the cells they infect. Some viruses, termed ***oncogenic viruses***, cause cancer by stimulating abnormal proliferation of cells. For instance, the *human papillomavirus (HPV)* causes virtually all cervical cancers in women.

Recent studies suggest that certain cancers may be linked to a cell having abnormal numbers of chromosomes. As a result, the cell could potentially have extra copies of oncogenes or too few copies of tumor-suppressor genes, which in either case could lead to uncontrolled cell proliferation. There is also some evidence suggesting that cancer may be caused by normal stem cells that develop into cancerous stem cells capable of forming malignant tumors.

Carcinogenesis: A Multistep Process

Carcinogenesis (kar′-si-nō-JEN-e-sis), the process by which cancer develops, is a multistep process in which as many as 10 distinct mutations may have to accumulate in a cell before it becomes cancerous. In colon cancer, the tumor begins as an area of increased cell proliferation that results from one mutation. This growth then progresses to abnormal, but noncancerous, growths called adenomas. After several more mutations, a carcinoma develops. The fact that so many mutations are needed for a cancer to develop indicates that cell growth is normally controlled with many sets of checks and balances.

Treatment of Cancer

Many cancers are removed surgically. However, when cancer is widely distributed throughout the body or exists in organs such as the brain whose functioning would be greatly harmed by surgery, chemotherapy and radiation therapy may be used instead. Sometimes surgery, chemotherapy, and radiation therapy are used in combination. Chemotherapy involves administering drugs that cause the death of cancerous cells. Radiation therapy breaks chromosomes, thus blocking cell division. Because cancerous cells divide rapidly, they are more vulnerable to the destructive effects of chemotherapy and radiation therapy than are normal cells. Unfortunately for the patients, hair follicle cells, red bone marrow cells, and cells lining the gastrointestinal tract also are rapidly dividing. Hence, the side effects of chemotherapy and radiation therapy include hair loss due to death of hair follicle cells, vomiting and nausea due to death of cells lining the stomach and intestines, and susceptibility to infection due to slowed production of white blood cells in red bone marrow.

MEDICAL TERMINOLOGY AND CONDITIONS

Anaplasia (an′-a-PLĀ-zē-a; *an-* = not; *-plasia* = to shape) The loss of tissue differentiation and function that is characteristic of most malignancies.

Apoptosis (ap′-op-TŌ-sis; a falling off, like dead leaves from a tree) An orderly, genetically programmed cell death in which "cell-suicide" genes become activated. Enzymes produced by these genes disrupt the cytoskeleton and nucleus; the cell shrinks and pulls away from neighboring cells; the DNA within the nucleus fragments; and the cytoplasm shrinks, although the plasma membrane remains intact. Phagocytes in the vicinity then ingest the dying cell. Apoptosis removes unneeded cells during development before birth and continues after birth both to regulate the number of cells in a tissue and to eliminate potentially dangerous cells such as cancer cells.

Atrophy (AT-rō-fē; *a-* = without; *-trophy* = nourishment) A decrease in the size of cells with subsequent decrease in the size of the affected tissue or organ; wasting away.

Biopsy (BĪ-op-sē; *bio-* = life; *-opsy* = viewing) The removal and microscopic examination of tissue from the living body for diagnosis.

Dysplasia (dis-PLĀ-zē-a; *dys-* = abnormal) Alteration in the size, shape, and organization of cells due to chronic irritation or inflammation; may progress to a neoplasm (tumor formation, usually malignant) or revert to normal if the irritation is removed.

Hyperplasia (hī′-per-PLĀ-zē-a; *hyper-* = over) Increase in the number of cells of a tissue due to an increase in the frequency of cell division.

Hypertrophy (hī-PER-trō-fē) Increase in the size of cells in a tissue without cell division.

Metaplasia (met′-a-PLĀ-zē-a; *meta-* = change) The transformation of one type of cell into another.

Necrosis (ne-KRŌ-sis = death) A pathological type of cell death, resulting from tissue injury, in which many adjacent cells swell, burst, and spill their cytoplasm into the interstitial fluid; the cellular debris usually stimulates an inflammatory response, which does not occur in apoptosis.

Progeny (PROJ-e-nē; *pro-* = forward; *-geny* = production) Offspring or descendants.

Proteomics (prō′-tē-Ō-miks; *proteo-* = protein) The study of the proteome (all of an organism's proteins) in order to identify all the proteins produced; it involves determining how the proteins interact and ascertaining the three-dimensional structure of proteins so that drugs can be designed to alter protein activity to help in the treatment and diagnosis of disease.

Tumor marker A substance introduced into circulation by tumor cells that indicates the presence of a tumor, as well as the specific type. Tumor markers may be used to screen, diagnose, make a prognosis, evaluate a response to treatment, and monitor for recurrence of cancer.

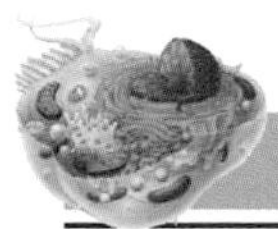

STUDY OUTLINE

Introduction (p. 44)

1. A cell is the basic, living, structural and functional unit of the body.
2. Cell biology is the study of cell structure and function.

A Generalized View of the Cell (p. 45)

1. Figure 3.1 on page 45 shows a generalized view of a cell that is a composite of many different cells in the body.
2. The principal parts of a cell are the plasma membrane; the cytoplasm, which consists of cytosol and organelles; and the nucleus.

The Plasma Membrane (p. 46)

1. The plasma membrane surrounds and contains the cytoplasm of a cell; it is composed of proteins and lipids.
2. The lipid bilayer consists of two back-to-back layers of phospholipids, cholesterol, and glycolipids.
3. Integral proteins extend into or through the lipid bilayer; peripheral proteins associate with the inner or outer surface of the membrane.
4. The membrane's selective permeability permits some substances to pass across it more easily than others. The lipid bilayer is permeable to water and to most lipid-soluble molecules. Small- and medium-sized water-soluble materials may cross the membrane with the assistance of integral proteins.
5. Membrane proteins have several functions. Channels and transporters are integral proteins that help specific solutes across the membrane; receptors serve as cellular recognition sites; some membrane proteins are enzymes; and others are cell identity markers.

Transport Across the Plasma Membrane (p. 47)

1. Fluid inside body cells is called intracellular fluid (ICF); fluid outside body cells is extracellular fluid (ECF). The ECF in the microscopic spaces between the cells of tissues is interstitial

fluid. The ECF in blood vessels is plasma, and that in lymphatic vessels is lymph.

2. Any material dissolved in a fluid is called a solute, and the fluid that dissolves materials is the solvent. Body fluids are dilute solutions in which a variety of solutes are dissolved in the solvent water.
3. The selective permeability of the plasma membrane supports the existence of concentration gradients, differences in the concentration of chemicals between one side of the membrane and the other.
4. Materials move through cell membranes by passive processes or by active processes. In passive processes, a substance moves down its concentration gradient across the membrane. In active transport, cellular energy is used to drive the substance "uphill" against its concentration gradient.
5. In transport in vesicles, tiny vesicles either detach from the plasma membrane while bringing materials into the cell or merge with the plasma membrane to release materials from the cell.
6. Diffusion is the movement of substances due to their kinetic energy. In net diffusion, substances move from an area of higher concentration to an area of lower concentration until equilibrium is reached. At equilibrium the concentration is the same throughout the solution.
7. In simple diffusion, substances move through the lipid bilayer or through channels in integral proteins. Ion channels selective for K^+, Cl^-, Na^+, and Ca^{2+} allow these ions to diffuse across the plasma membrane by simple diffusion. In facilitated diffusion, substances cross the membrane with the assistance of transporters, which bind to a specific substance on one side of the membrane and release it on the other side after the transporter undergoes a change in shape.
8. Osmosis is the movement of water molecules through a selectively permeable membrane from an area of higher to an area of lower water concentration.
9. In an isotonic solution, red blood cells maintain their normal shape; in a hypotonic solution, they gain water and undergo hemolysis; in a hypertonic solution, they lose water and undergo crenation.
10. With the expenditure of cellular energy, usually in the form of ATP, solutes can cross the membrane against their concentration gradient by means of active transport. Actively transported solutes include several ions such as Na^+, K^+, H^+, Ca^{2+}, I^-, and Cl^-; amino acids; and monosaccharides.
11. The most important active transport pump is the sodium-potassium pump, which expels Na^+ from cells and brings K^+ in.
12. Transport in vesicles includes both endocytosis (phagocytosis and bulk-phase endocytosis) and exocytosis.
13. Phagocytosis is the ingestion of solid particles. It is an important process used by some white blood cells to destroy bacteria that enter the body. Bulk-phase endocytosis is the ingestion of extracellular fluid.
14. Exocytosis involves movement of secretory or waste products out of a cell by fusion of vesicles with the plasma membrane.

Cytoplasm (p. 52)

1. Cytoplasm includes all the cellular contents between the plasma membrane and nucleus; it consists of cytosol and organelles.
2. The fluid portion of cytoplasm is cytosol, composed mostly of water, plus ions, glucose, amino acids, fatty acids, proteins, lipids, ATP, and waste products; the cytosol is the site of many chemical reactions required for a cell's existence.
3. Organelles are specialized cellular structures with characteristic shapes and specific functions.
4. The cytoskeleton is a network of several kinds of protein filaments that extend throughout the cytoplasm; they provide a structural framework for the cell and generate movements. Components of the cytoskeleton include microfilaments, intermediate filaments, and microtubules.
5. The centrosome consists of two centrioles and pericentriolar material. The centrosome serves as a center for organizing microtubules in interphase cells and the mitotic spindle during cell division.
6. Cilia and flagella are motile projections of the cell surface. Cilia move fluid along the cell surface; a flagellum moves an entire cell.
7. Ribosomes, composed of ribosomal RNA and ribosomal proteins, consist of two subunits and are the sites of protein synthesis.
8. Endoplasmic reticulum (ER) is a network of membranes that extends from the nuclear envelope throughout the cytoplasm.
9. Rough ER is studded with ribosomes. Proteins synthesized on the ribosomes enter the ER for processing and sorting. The ER is also where glycoproteins and phospholipids form.
10. Smooth ER lacks ribosomes. It is the site where fatty acids and steroids are synthesized. Smooth ER also participates in releasing glucose from the liver into the bloodstream, inactivating or detoxifying drugs and other potentially harmful substances, and releasing calcium ions that trigger contraction in muscle cells.
11. The Golgi complex consists of flattened sacs called cisterns that receive proteins synthesized in the rough ER. Within the Golgi cisterns the proteins are modified, sorted, and packaged into vesicles for transport to different destinations. Some processed proteins leave the cell in secretory vesicles, some are incorporated into the plasma membrane, and some enter lysosomes.
12. Lysosomes are membrane-enclosed vesicles that contain digestive enzymes. They function in digestion of worn-out organelles (autophagy) and even in digestion of their own cell (autolysis).
13. Peroxisomes are similar to lysosomes but smaller. They oxidize various organic substances such as amino acids, fatty acids, and toxic substances and, in the process, produce hydrogen peroxide. The hydrogen peroxide is degraded by an enzyme in peroxisomes called catalase.
14. Proteasomes contain proteases that continually degrade unneeded, damaged, or faulty proteins.

15. Mitochondria consist of a smooth outer membrane, an inner membrane containing cristae, and a fluid-filled cavity called the matrix. They are called "powerhouses" of the cell because they produce most of a cell's ATP.

Nucleus (p. 57)

1. The nucleus consists of a double nuclear envelope; nuclear pores, which control the movement of substances between the nucleus and cytoplasm; nucleoli, which produce ribosomes; and genes arranged on chromosomes.
2. Most body cells have a single nucleus; some (red blood cells) have none, and others (skeletal muscle cells) have several.
3. Genes control cellular structure and most cellular functions.

Gene Action: Protein Synthesis (p. 58)

1. Most of the cellular machinery is devoted to protein synthesis.
2. Cells make proteins by transcribing and translating the genetic information encoded in the sequence of four types of nitrogenous bases in DNA.
3. In transcription, genetic information encoded in the DNA base sequence is copied into a complementary sequence of bases in a strand of messenger RNA (mRNA). Transcription begins on DNA in a region called a promoter.
4. Translation is the process in which mRNA associates with ribosomes and directs synthesis of a protein, converting the nucleotide sequence in mRNA into a specific sequence of amino acids.
5. In translation, mRNA binds to a ribosome, specific amino acids attach to tRNA, and anticodons of tRNA bind to codons of mRNA, bringing specific amino acids into position on a growing protein.
6. Translation begins at the start codon and terminates at the stop codon.

Somatic Cell Division (p. 62)

1. Cell division is the process by which cells reproduce themselves.
2. Cell division that results in an increase in the number of body cells is called somatic cell division; it involves a nuclear division called mitosis plus division of cytoplasm, called cytokinesis.
3. Cell division that results in the production of sperm and oocytes is called reproductive cell division.
4. The cell cycle is an orderly sequence of events in which a cell duplicates its contents and divides in two. It consists of interphase and a mitotic phase.
5. Before the mitotic phase, the DNA molecules, or chromosomes, replicate themselves so that identical chromosomes can be passed on to the next generation of cells.
6. A cell that is between divisions and is carrying on every life process except division is said to be in interphase.
7. Mitosis is the replication and distribution of two sets of chromosomes into separate and equal nuclei; it consists of prophase, metaphase, anaphase, and telophase.
8. Cytokinesis usually begins late in anaphase and ends in telophase.
9. A cleavage furrow forms and progresses inward, cutting through the cell to form two separate identical cells, each with equal portions of cytoplasm, organelles, and chromosomes.

Aging and Cells (p. 65)

1. Aging is a normal process accompanied by progressive alteration of the body's homeostatic adaptive responses.
2. Many theories of aging have been proposed, including genetically programmed cessation of cell division, shortening of telomeres, addition of glucose to proteins, buildup of free radicals, and an intensified autoimmune response.

SELF-QUIZ

1. If the extracellular fluid contains a greater concentration of solutes than the cytosol of the cell, the extracellular fluid is said to be
   a. isotonic b. hypertonic c. hypotonic
   d. cytotonic e. epitonic

2. The proteins found in the plasma membrane
   a. are primarily glycoproteins
   b. allow the passage of many substances into the cell
   c. allow cells to recognize other cells
   d. help anchor cells to each other
   e. have all of the above functions

3. To enter many body cells, glucose must bind to a specific membrane transport protein, which assists glucose to cross the membrane without using ATP. This type of movement is known as
   a. facilitated diffusion b. simple diffusion
   c. vesicular transport d. osmosis e. active transport

4. A red blood cell placed in a hypotonic solution undergoes
   a. hemolysis b. crenation c. equilibrium
   d. a decrease in osmotic pressure e. shrinkage

5. Which of the following normally pass through the plasma membrane only by transport in vesicles?
   a. water molecules b. sodium ions c. proteins
   d. oxygen molecules e. hydrogen ions

. Which of the following statements concerning diffusion is NOT true?
  a. Diffusion speeds up as body temperature rises.
  b. A small surface area slows down the rate of diffusion.
  c. A low-weight particle diffuses faster than a high-weight particle.
  d. It moves materials from an area of low concentration to an area of high concentration by kinetic energy.
  e. Diffusion over a greater distance takes longer than diffusion over a short distance.

7. Which of the following processes requires ATP?
  a. diffusion b. active transport c. osmosis
  d. facilitated diffusion e. net diffusion

8. Nicotine in cigarette smoke interferes with the ability of cells to rid the breathing passageways of debris. Which organelles are "paralyzed" by nicotine?
  a. flagella b. ribosomes c. microfilaments
  d. cilia e. lysosomes

9. Many proteins found in the plasma membrane are formed by the ______ and packaged by the ______.
  a. ribosomes, Golgi complex
  b. smooth endoplasmic reticulum, Golgi complex
  c. Golgi complex, lysosomes
  d. mitochondria, Golgi complex
  e. nucleus, smooth endoplasmic reticulum

10. Match the following:

____ a. cellular movement	A. centrosome
____ b. selective permeability	B. cytoskeleton
____ c. protein synthesis	C. Golgi complex
____ d. lipid synthesis, detoxification	D. lysosomes
____ e. packages proteins and lipids	E. mitochondria
____ f. ATP production	F. plasma membrane
____ g. digest bacteria and worn-out organelles	G. ribosomes
____ h. forms mitotic spindle	H. smooth ER

11. If the smooth endoplasmic reticulum were destroyed, a cell would not be able to
  a. form lysosomes b. synthesize certain proteins
  c. generate energy d. phagocytize bacteria
  e. synthesize fatty acids and steroids

12. Water moves into and out of red blood cells through the process of
  a. endocytosis b. phagocytosis c. osmosis
  d. active transport e. facilitated diffusion

13. A cell undergoing mitosis goes through the following stages in which sequence?
  a. interphase, metaphase, prophase, cytokinesis
  b. interphase, prophase, cytokinesis, telophase
  c. anaphase, metaphase, prophase, telophase
  d. anaphase, metaphase, prophase, cytokinesis
  e. prophase, metaphase, anaphase, telophase

14. Transcription involves
  a. transferring information from the mRNA to tRNA
  b. codon binding with anticodons c. joining amino acids by peptide bonds d. copying information contained in the DNA to mRNA e. synthesizing the protein on the ribosome

15. If a DNA strand has a nitrogenous base sequence TACGA, then the sequence of bases on the corresponding mRNA would be
  a. ATGCT b. AUGCU c. GUACU d. CTGAT
  e. AUCUG

16. Place the following events of protein synthesis in the proper order.
  1. DNA uncoils and mRNA is transcribed. 2. tRNA with an attached amino acid pairs with mRNA. 3. mRNA passes from the nucleus into the cytoplasm and attaches to a ribosome. 4. Protein is formed. 5. Two amino acids are linked by a peptide bond.
  a. 1, 2, 3, 4, 5 b. 1, 3, 2, 5, 4 c. 1, 2, 3, 5, 4
  d. 1, 5, 3, 2, 4 e. 2, 1, 3, 4, 5

17. Match the following descriptions with the phases shown.
  ____ a. nuclear envelope (membrane) and nucleoli reappear
  ____ b. centromeres of the chromatid pairs line up in the center of the mitotic spindle
  ____ c. DNA duplicates
  ____ d. cleavage furrow splits cell into two identical cells
  ____ e. chromosomes move toward opposite poles of cell
  ____ f. chromatids are attached at centromeres; mitotic spindle forms

  A. prophase
  B. cytokinesis
  C. telophase
  D. anaphase
  E. metaphase
  F. interphase

18. In which phase is a cell highly active and growing?
  a. anaphase b. prophase c. metaphase
  d. telophase e. interphase

19. If a virus were to enter a cell and destroy its ribosomes, how would the cell be affected?
  a. It would be unable to undergo mitosis.
  b. It could no longer produce ATP.
  c. Movement of the cell would cease.
  d. It would undergo autophagy.
  e. It would be unable to synthesize proteins.

20. Which of the following statements concerning cancer is NOT true?
  a. A benign tumor is noncancerous.
  b. When a cancerous growth presses on nerves, it can cause pain.
  c. Angiogenesis is the spread of cancerous cells to other parts of the body.
  d. Ultraviolet radiation and radon gas are carcinogens.
  e. Cancer is uncontrolled mitosis in abnormal cells.

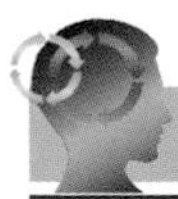

CRITICAL THINKING APPLICATIONS

1. One of your bones' functions is to store minerals, especially calcium. The bone tissue must be dissolved to release the calcium for use by the body's systems. Which organelle would be involved in breaking down bone tissue?
2. In your dream, you're floating on a raft in the middle of the ocean. The sun's hot, you're very thirsty, and you're surrounded by water. You want to take a long, cool drink of seawater, but something you learned in A&P (you knew that was coming) stops you from drinking and saves your life! Why shouldn't you drink seawater?
3. Mucin is a glycoprotein present in saliva. When mixed with water, mucin becomes the slippery substance known as mucus. Trace the route taken by mucin through the cells of the salivary glands, starting with the organelle where it is synthesized and ending with its release from the cells.
4. Jethro loves his french fries super-sized with extra salt. He dropped A&P last semester but remembers something about a pump. "Hey, I'm a big guy. I need the extra salt to power my sodium pump." How well does Jethro remember his A&P?

ANSWERS TO FIGURE QUESTIONS

3.1 The three main parts of a cell are the plasma membrane, cytoplasm, and nucleus.

3.2 Some integral proteins function as channels or transporters to move substances across membranes. Other integral proteins function as receptors. Membrane glycolipids and glycoproteins are involved in cellular recognition.

3.3 In simple diffusion, substances cross a membrane through the lipid bilayer and pores of ion channels; in facilitated diffusion, transporters are involved.

3.4 Oxygen, carbon dioxide, fatty acids, fat-soluble vitamins, and steroids can cross the plasma membrane by simple diffusion through the lipid bilayer.

3.5 The concentration of K^+ is higher in the cytosol of body cells than in extracellular fluids.

3.6 Insulin promotes insertion of glucose transporters in the plasma membrane, which increases cellular glucose uptake by facilitated diffusion.

3.7 No, the water concentrations can never be the same because the beaker always contains pure (100%) water and the sac contains a solution that is less than 100% water.

3.8 A 2% solution of NaCl will cause crenation of RBCs because it is hypertonic.

3.9 ATP adds a phosphate group to the pump protein, which changes the pump's three-dimensional shape.

3.10 The trigger that causes pseudopod extension is binding of a particle to a membrane receptor.

3.11 Clusters of microtubules form the structure of centrioles, cilia, and flagella.

3.12 The components of the centrosome are two centrioles and the pericentriolar material.

3.13 Large and small ribosomal subunits are synthesized in a nucleolus in the nucleus and then join together in the cytoplasm.

3.14 Rough ER has attached ribosomes where proteins that will be exported from the cell are synthesized; smooth ER lacks ribosomes and is associated with lipid synthesis and other metabolic reactions.

3.15 Cells that secrete proteins into extracellular fluid have extensive Golgi complexes.

3.16 Mitochondrial cristae provide a large surface area for chemical reactions and contain enzymes needed for ATP production.

3.17 Nuclear genes control cellular structure and direct most cellular activities.

3.18 Proteins determine the physical and chemical characteristics of cells.

3.19 RNA polymerase catalyzes transcription of DNA.

3.20 When a ribosome encounters a stop codon in mRNA, the completed protein detaches from the final tRNA.

3.21 Cytokinesis usually begins late in anaphase.

chapter 4 TISSUES

did you know?

People who have never taken biology courses have been known to remark, "I don't want to lift weights, because if I build muscular tissue, I am afraid it will turn to fat when I stop lifting." Of course, muscular tissue is specialized for its contractile function, and does not "turn into" fat tissue, which is specialized for energy storage. As people age, skeletal muscles do atrophy, or shrink, over the years. Adipose tissue tends to grow. While some of these changes in body composition appear to be an inevitable part of the aging process, scientists believe much of the loss of muscular tissue can be prevented with exercise training.

Focus on Wellness, page 91

www.wiley.com/college/apcentral

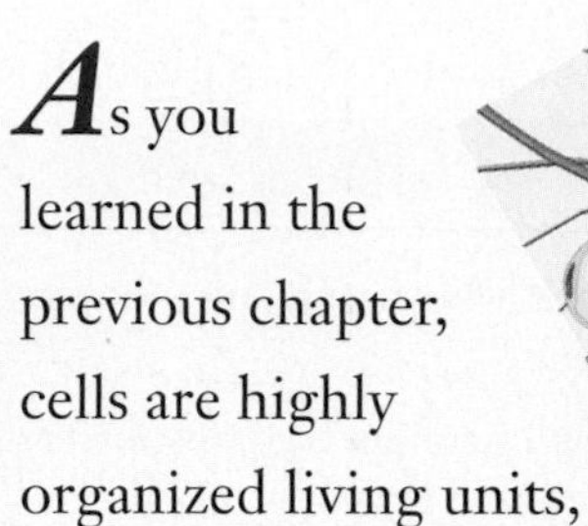

As you learned in the previous chapter, cells are highly organized living units, but they typically do not function alone. Instead, cells work together in groups called tissues. A ***tissue*** is a group of similar cells, usually with a common embryonic origin, that function together to carry out specialized activities. ***Histology*** (hiss-TOL-ō-jē; *hist-* = tissue; *-logy* = study of) is the science that deals with the study of tissues. A ***pathologist*** (pa-THOL-ō-gist; *patho-* = disease) is a physician who specializes in laboratory studies of cells and tissues to help other physicians make accurate diagnoses. One of the principal functions of a pathologist is to examine tissues for any changes that might indicate disease.

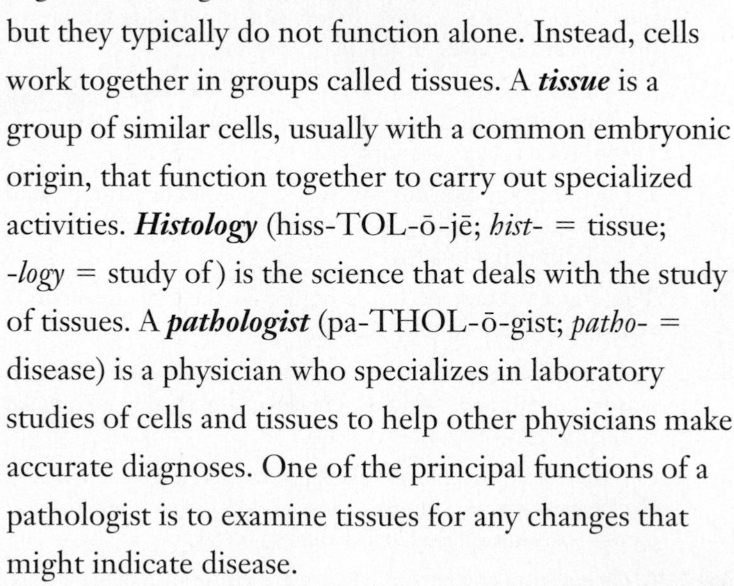

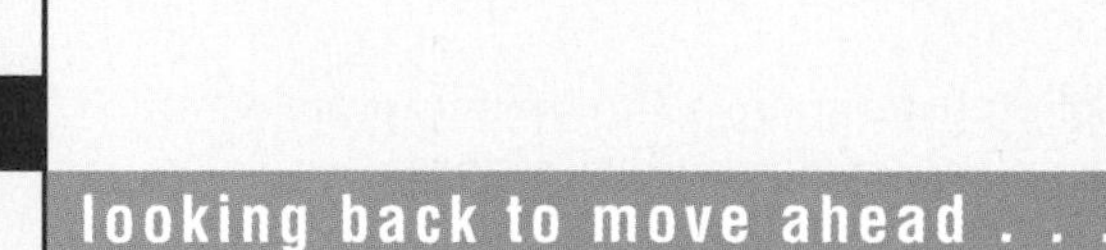

looking back to move ahead . . .

- **Levels of Organization and Body Systems (page 2)**
- **A Generalized View of the Cell (page 45)**
- **Phagocytosis (page 51)**
- **Cytosol (page 52)**
- **Organelles (page 52)**
- **Cilia (page 54)**

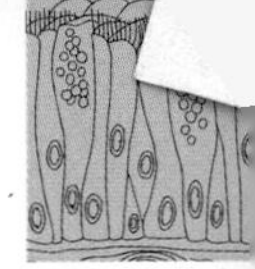

TYPES OF TISSUES

OBJECTIVE • Name four basic types of tissue that make up the human body and state the characteristics of each.

Body tissues are classified into four basic types based on their structure and functions:

1. ***Epithelial tissue*** (ep′-i-THĒ-lē-al) covers body surfaces; lines body cavities, hollow organs, and ducts (tubes); and forms glands.
2. ***Connective tissue*** protects and supports the body and its organs, binds organs together, stores energy reserves as fat, and provides immunity.
3. ***Muscular tissue*** generates the physical force needed to make body structures move.
4. ***Nervous tissue*** detects changes inside and outside the body and initiates and transmits nerve impulses (action potentials) that coordinate body activities to help maintain homeostasis.

Epithelial tissue and most types of connective tissue are discussed in detail in this chapter. The structure and functions of bone tissue and blood (connective tissues), muscular tissue, and nervous tissue are examined in detail in later chapters.

Most epithelial cells and some muscle and nerve cells are tightly joined into functional units by points of contact between their plasma membranes called ***cell junctions***. Some cell junctions fuse cells together so tightly that they prevent substances from passing between the cells. This fusion is very important for tissues that line the stomach, intestines, and urinary bladder because it prevents the contents of these organs from leaking out. Other cell junctions hold cells together so that they don't separate while performing their functions. Still other cell junctions form channels that allow ions and molecules to pass between cells. This permits cells in a tissue to communicate with each other and it also enables nerve or muscle impulses to spread rapidly among cells.

■ **CHECKPOINT**

1. Define a tissue. What are the four basic types of body tissues?
2. Why are cell junctions important?

EPITHELIAL TISSUE

OBJECTIVES • Discuss the general features of epithelial tissue.

• Describe the structure, location, and function of the various types of epithelial tissue.

Epithelial tissue, or more simply ***epithelium*** (plural is *epithelia*), may be divided into two types: (1) *covering and lining epithelium* and (2) *glandular epithelium*. As its name suggests, covering and lining epithelium forms the outer covering of the skin and the outer covering of some internal organs. It also lines body cavities; blood vessels; ducts; and the interiors of the respiratory, digestive, urinary, and reproductive systems. It makes up, along with nervous tissue, the parts of the sense organs for hearing, vision, and touch. Glandular epithelium makes up the secreting portion of glands, such as sweat glands.

General Features of Epithelial Tissue

As you will see shortly, there are many different types of epithelia, each with characteristic structure and functions. However, all of the different types of epithelial tissue also have features in common. General features of epithelial tissue include the following:

1. Epithelium consists largely or entirely of closely packed cells with little extracellular material between them, and the cells are arranged in continuous sheets, in either single or multiple layers.
2. Epithelial cells have an *apical* (free) *surface*, which is exposed to a body cavity, lining of an internal organ, or the exterior of the body; *lateral surfaces*, which face adjacent cells on either side; and a *basal surface*, which is attached to a basement membrane. In discussing epithelia with multiple layers, the term *apical layer* refers to the most superficial layer of cells, whereas the term *basal layer* refers to the deepest layer of cells. The ***basement membrane*** is a thin extracellular structure composed mostly of protein fibers. It is located between the epithelium and the underlying connective tissue layer and helps bind and support the epithelium.
3. Epithelia are ***avascular*** (*a-* = without; *vascular* = blood vessels); that is, they lack blood vessels. The vessels that supply nutrients to and remove wastes from epithelia are located in adjacent connective tissues. The exchange of materials between epithelium and connective tissue occurs by diffusion.
4. Epithelia have a nerve supply.
5. Because epithelium is subject to a certain amount of wear and tear and injury, it has a high capacity for renewal by cell division.

Covering and Lining Epithelium

Covering and lining epithelium, which covers or lines various parts of the body, is classified according to the arrangement of cells into layers and the shape of the cells (Figure 4.1):

1. ***Arrangement of cells in layers.*** The cells of covering and lining epithelia are arranged in one or more layers depending on the functions the epithelium performs:

Figure 4.1 Cell shapes and arrangement of layers for covering and lining epithelium.

Cell shapes and arrangement of layers are the bases for classifying covering and lining epithelium.

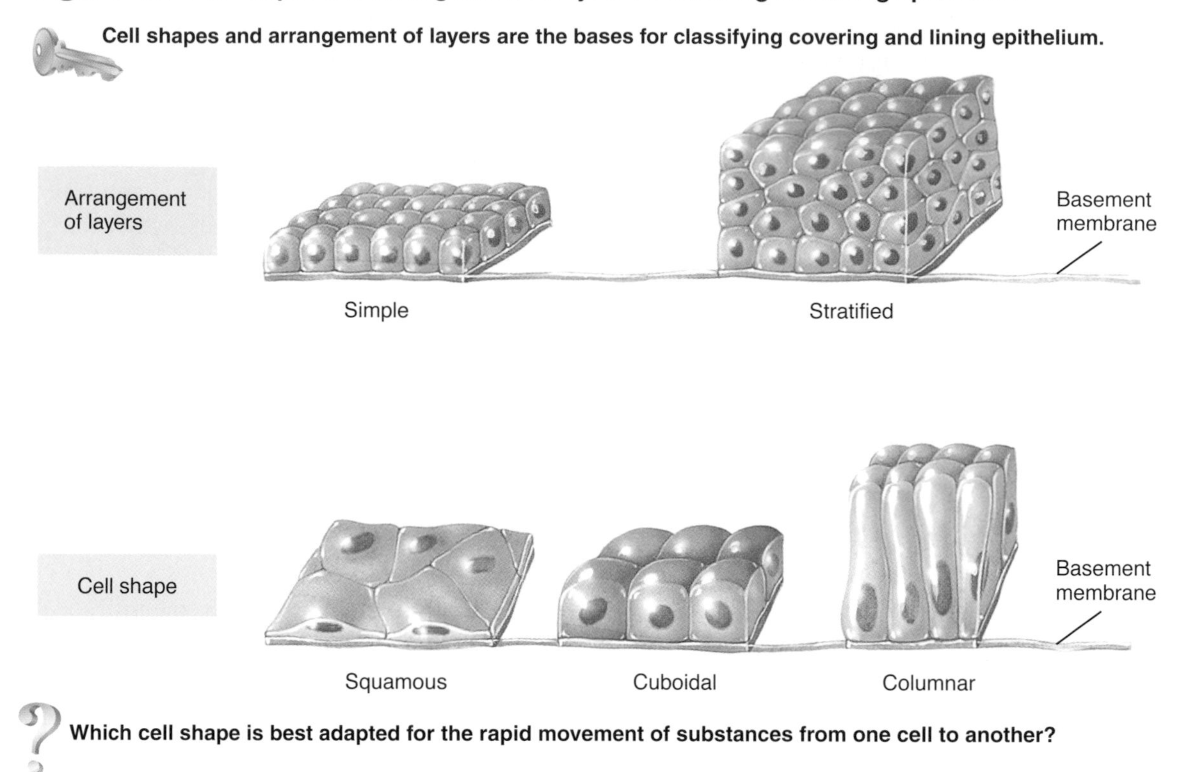

Which cell shape is best adapted for the rapid movement of substances from one cell to another?

a. *Simple epithelium* is a single layer of cells that functions in diffusion, osmosis, filtration, secretion, and absorption. ***Secretion*** is the production and release of substances such as mucus, sweat, or enzymes. ***Absorption*** is the intake of fluids or other substances such as digested food from the intestinal tract.

b. *Pseudostratified epithelium* (*pseudo-* = false) appears to have multiple layers of cells because the cell nuclei lie at different levels and not all cells reach the apical surface. Cells that do extend to the apical surface may contain cilia; others (goblet cells) secrete mucus. Pseudostratified epithelium is actually a simple epithelium because all its cells rest on the basement membrane.

c. *Stratified epithelium* (*stratum* = layer) consists of two or more layers of cells that protect underlying tissues in locations where there is considerable wear and tear.

2. ***Cell shapes.***

a. *Squamous* cells (SKWĀ-mus = flat) are arranged like floor tiles and are thin, which allows for the rapid passage of substances.

b. *Cuboidal* cells are as tall as they are wide and are shaped like cubes or hexagons. They may have microvilli at their apical surface and function in either secretion or absorption.

c. *Columnar* cells are much taller than they are wide, like columns, and protect underlying tissues. Their apical surfaces may have cilia or microvilli, and they often are specialized for secretion and absorption.

d. *Transitional* cells change shape, from flat to cuboidal, as organs such as the urinary bladder stretch (distend) to a larger size and then collapse to a smaller size.

Combining the two characteristics (arrangements of layers and cell shapes), the types of covering and lining epithelia are as follows:

I. Simple epithelium

- **A.** Simple squamous epithelium
- **B.** Simple cuboidal epithelium
- **C.** Simple columnar epithelium (nonciliated and ciliated)
- **D.** Pseudostratified columnar epithelium (nonciliated and ciliated)

II. Stratified epithelium

- **A.** Stratified squamous epithelium (keratinized and nonkeratinized)*
- **B.** Stratified cuboidal epithelium*
- **C.** Stratified columnar epithelium*
- **D.** Transitional epithelium

*This classification is based on the shape of the cells in the *apical* layer.

Each of these covering and lining epithelia is described in the following sections and illustrated in Table 4.1. The illustration of each type consists of a photomicrograph, a corresponding diagram, and an inset that identifies a major location of the tissue in the body. Descriptions, locations, and functions of the tissues accompany each illustration.

Simple Epithelium

SIMPLE SQUAMOUS EPITHELIUM This tissue consists of a single layer of flat cells that resembles a tiled floor when viewed from its apical surface (Table 4.1A). The nucleus of each cell is a flattened oval or sphere and is centrally located. Simple squamous epithelium is found in parts of the body where filtration (kidneys) or diffusion (lungs) are priority processes. It is not found in body areas that are subjected to wear and tear.

The simple squamous epithelium that lines the heart, blood vessels, and lymphatic vessels is known as ***endothelium*** (*endo-* = within; *-thelium* = covering); the type that forms the epithelial layer of serous membranes, such as the peritoneum, is called ***mesothelium*** (*meso-* = middle).

SIMPLE CUBOIDAL EPITHELIUM The cuboidal shape of the cells in this tissue (Table 4.1B) is obvious only when the tissue is sectioned and viewed from the side. Cell nuclei are usually round and centrally located. Simple cuboidal epithelium is found in organs such as the thyroid gland and kidneys and performs the functions of secretion and absorption.

SIMPLE COLUMNAR EPITHELIUM When viewed from the side, the cells of simple columnar epithelium appear like columns with oval nuclei near the base of the cells. Simple columnar epithelium exists in two forms: nonciliated simple columnar epithelium and ciliated simple columnar epithelium.

Nonciliated simple columnar epithelium contains two types of cells–columnar epithelial cells with microvilli at their apical surface, and goblet cells (Table 4.1C). ***Microvilli***, microscopic fingerlike projections, increase the surface area of the plasma membrane (see Figure 3.1 on page 45), thus increasing the rate of absorption by the cell. ***Goblet cells*** are modified columnar cells that secrete mucus, a slightly sticky fluid, at their apical surfaces. Before it is released, mucus accumulates in the upper portion of the cell, causing that area to bulge out. The whole cell then resembles a goblet or wine glass. Secreted mucus serves as a lubricant for the linings of the digestive, respiratory, reproductive, and most of the urinary tracts. Mucus also helps to trap dust entering the respiratory tract, and it prevents destruction of the stomach lining by acid secreted by the stomach.

Ciliated simple columnar epithelium (Table 4.1D) contains cells with cilia at their apical surface. In a few parts of the upper respiratory tract, ciliated columnar cells are interspersed with goblet cells. Mucus secreted by the goblet cells forms a film over the respiratory surface that traps inhaled foreign particles. The cilia wave in unison and move the mucus and any trapped foreign particles toward the throat, where it can be coughed up and swallowed or spit out. Cilia also help to move oocytes expelled by the ovaries through the uterine tubes into the uterus.

Pseudostratified Columnar Epithelium

As noted earlier, pseudostratified columnar epithelium appears to have several layers because the nuclei of the cells are at various depths (Table 4.1E). Even though all the cells are attached to the basement membrane in a single layer, some cells do not extend to the apical surface. When viewed from the side, these features give the false impression of a multilayered tissue—thus the name *pseudo*stratified epithelium (*pseudo-* = false). In *pseudostratified ciliated columnar epithelium*, the cells that extend to the surface either secrete mucus (goblet cells) or bear cilia. The secreted mucus traps foreign particles and the cilia sweep away mucus for eventual elimination from the body. *Pseudostratified nonciliated columnar epithelium* contains cells without cilia and lacks goblet cells.

Stratified Epithelium

Stratified epithelium contains two or more layers of cells used for protection of underlying tissues in areas where there is considerable wear and tear. Some cells of stratified epithelia also produce secretions. The name of the specific kind of stratified epithelium depends on the shape of the cells in the apical layer.

STRATIFIED SQUAMOUS EPITHELIUM Cells in the apical layer of this type of epithelium are flat; those in the deep layers vary in shape from cuboidal to columnar (Table 4.1F). The basal (deepest) cells continually undergo cell division. As new cells grow, the cells of the basal layer are pushed upward toward the surface. As they move farther from the deeper layers and from their blood supply in the underlying connective tissue, they become dehydrated, shrunken, and harder. At the apical layer, the cells lose their cell junctions and are sloughed off, but they are replaced as new cells continually emerge from the basal cells.

Stratified squamous epithelium exists in both keratinized and nonkeratinized forms. In *keratinized stratified squamous epithelium*, a tough layer of keratin is deposited in the apical layer and several layers deep to it. *Keratin* is a tough protein that helps protect the skin and underlying tissues from microbes, heat, and chemicals. *Nonkeratinized stratified squamous epithelium*, which is found, for example, lining the mouth, does not contain keratin in the apical layer and several layers deep to it and remains moist. Stratified squamous epithelium forms the first line of defense against microbes.

Table 4.1 Epithelial Tissues

Covering and Lining Epithelium

Simple Epithelium

A. Simple squamous epithelium

Description: Single layer of flat cells; centrally located nucleus.

Location: Lines heart, blood vessels, lymphatic vessels, air sacs of lungs, glomerular (Bowman's) capsule of kidneys, and inner surface of the tympanic membrance (eardrum); forms epithelial layer of serous membranes, such as the peritoneum.

Function: Filtration, diffusion, osmosis, and secretion in serous membranes.

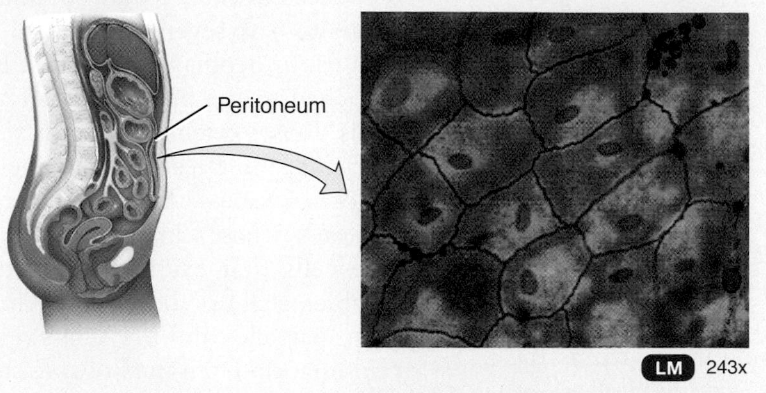

Surface view of simple squamous epithelium of mesothelial lining of peritoneum

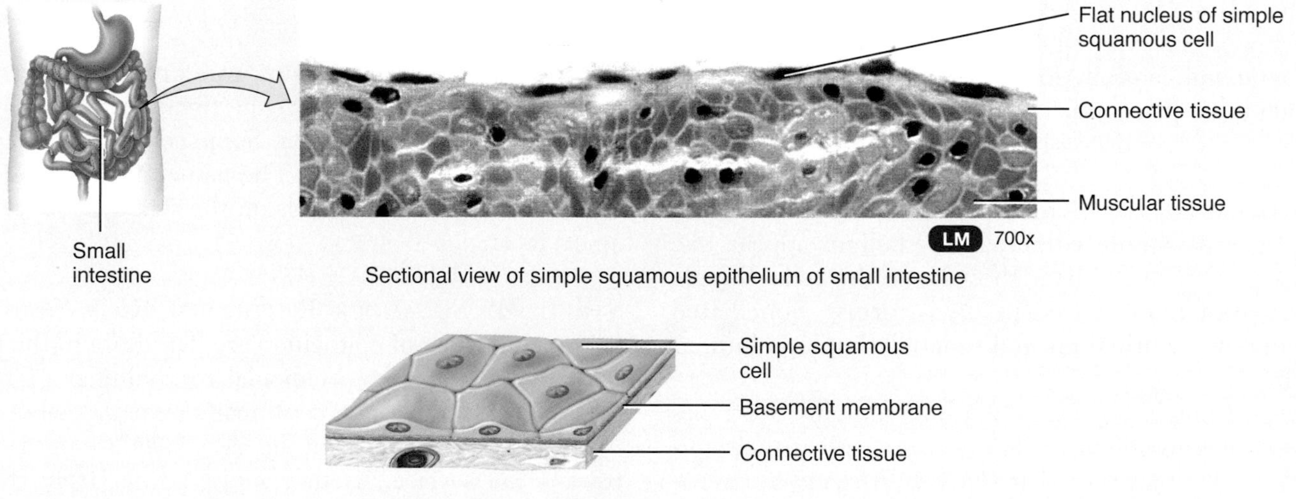

Sectional view of simple squamous epithelium of small intestine

Simple squamous epithelium

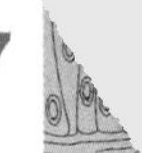

Covering and Lining Epithelium

B. Simple cuboidal epithelium

Description: Single layer of cube-shaped cells; centrally located nucleus.

Location: Covers surface of ovary, lines anterior surface of capsule of the lens of the eye, forms the pigmented epithelium at the posterior surface of the eye, lines kidney tubules and smaller ducts of many glands, and makes up the secreting portion of some glands such as the thyroid gland.

Function: Secretion and absorption.

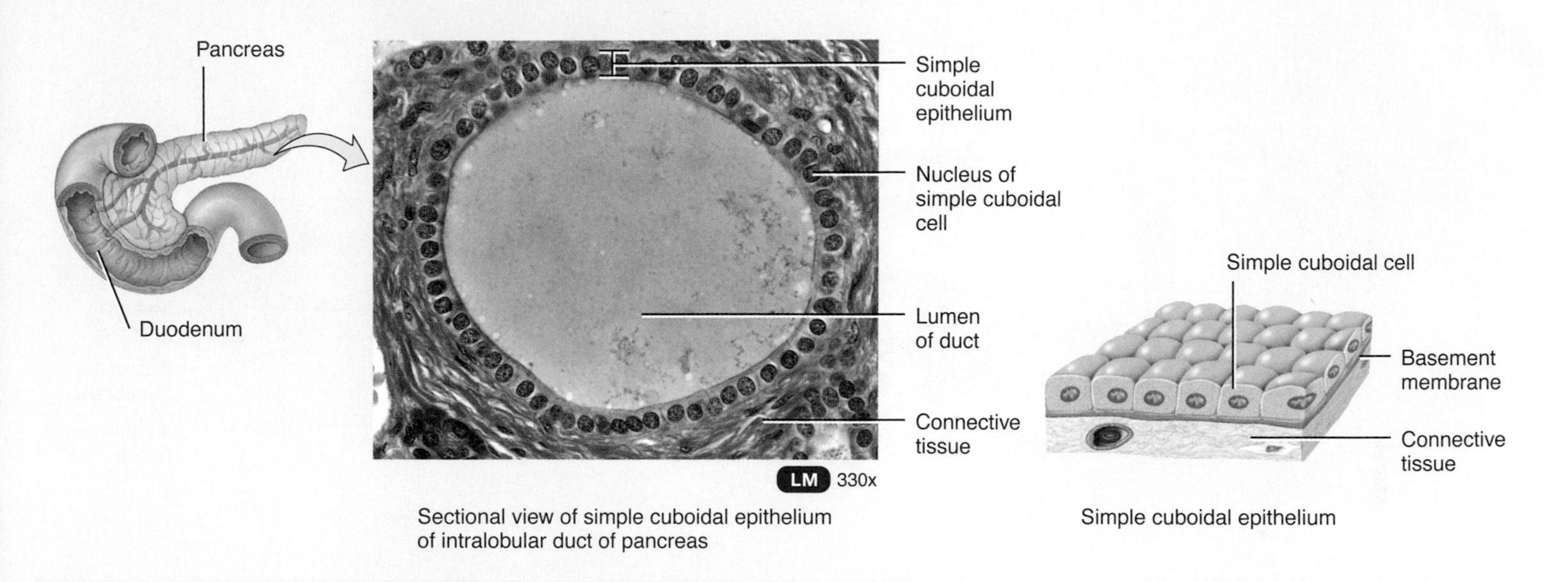

Sectional view of simple cuboidal epithelium of intralobular duct of pancreas

Simple cuboidal epithelium

C. Nonciliated simple columnar epithelium

Description: Single layer of nonciliated column-like cells with nuclei near bases of cells; contains goblet cells and cells with microvilli in some locations.

Location: Lines most of the gastrointestinal tract (from the stomach to the anus), ducts of many glands, and gallbladder.

Function: Secretion and absorption.

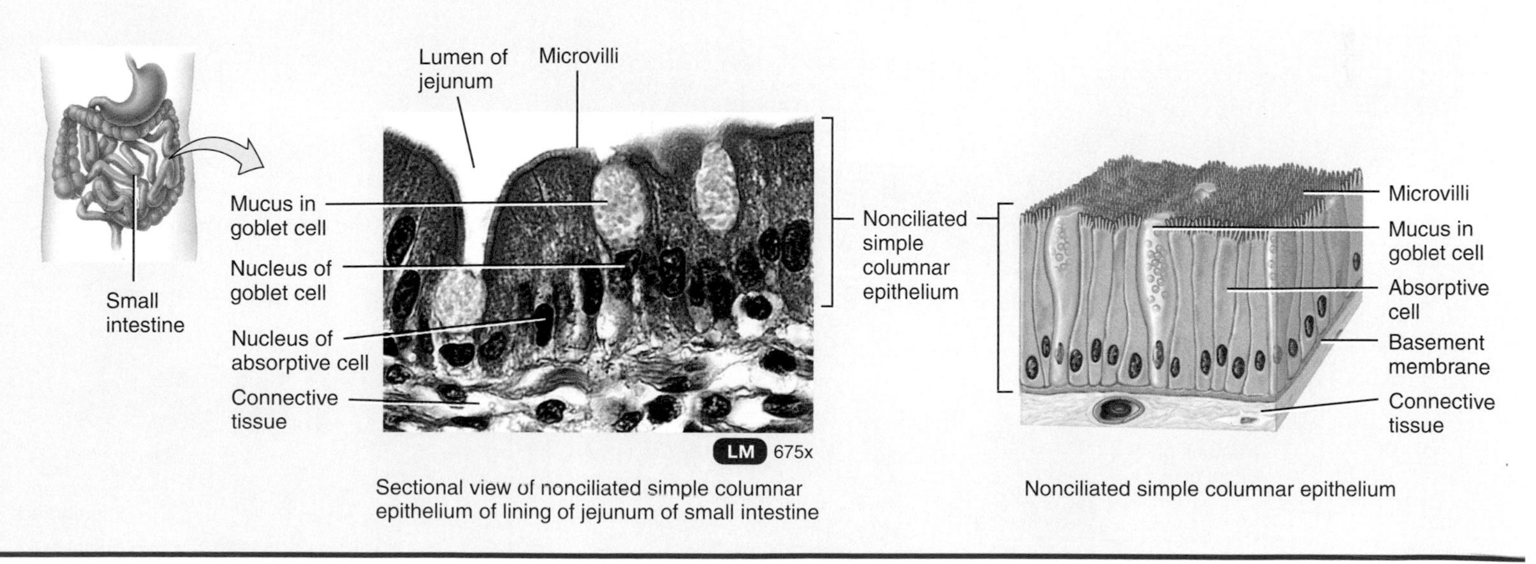

Sectional view of nonciliated simple columnar epithelium of lining of jejunum of small intestine

Nonciliated simple columnar epithelium

(continues)

Table 4.1 Epithelial Tissues (continued)

Covering and Lining Epithelium

D. Ciliated simple columnar epithelium

Description: Single layer of ciliated column-like cells with nuclei near bases; contains goblet cells in some locations.

Location: Lines a few portions of upper respiratory tract, uterine (fallopian) tubes, uterus, some paranasal sinuses, and central canal of spinal cord.

Function: Moves mucus and other substances by ciliary action.

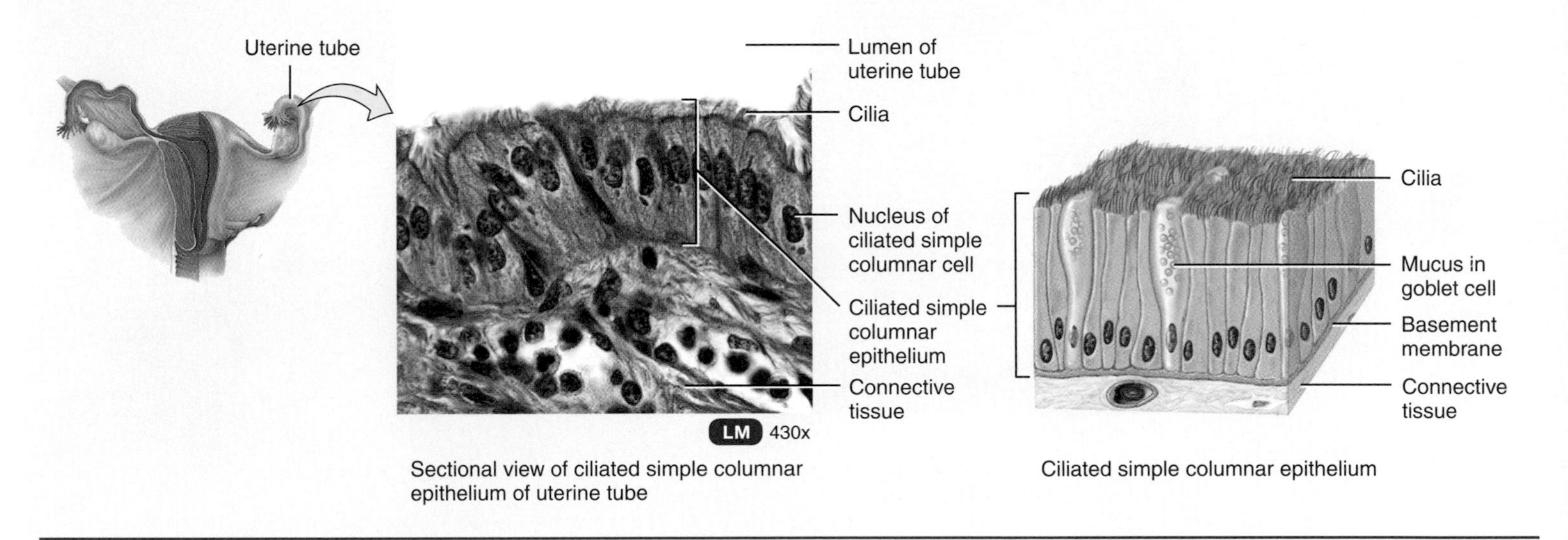

Sectional view of ciliated simple columnar epithelium of uterine tube

Ciliated simple columnar epithelium

E. Pseudostratified columnar epithelium

Description: Not a true stratified tissue; nuclei of cells are at different levels; all cells are attached to basement membrane, but not all reach the apical surface.

Location: Pseudostratified ciliated columnar epithelium lines the airways of most of upper respiratory tract; pseudostratified nonciliated columnar epithelium lines larger ducts of many glands, epididymis, and part of male urethra.

Function: Secretion and movement of mucus by ciliary action.

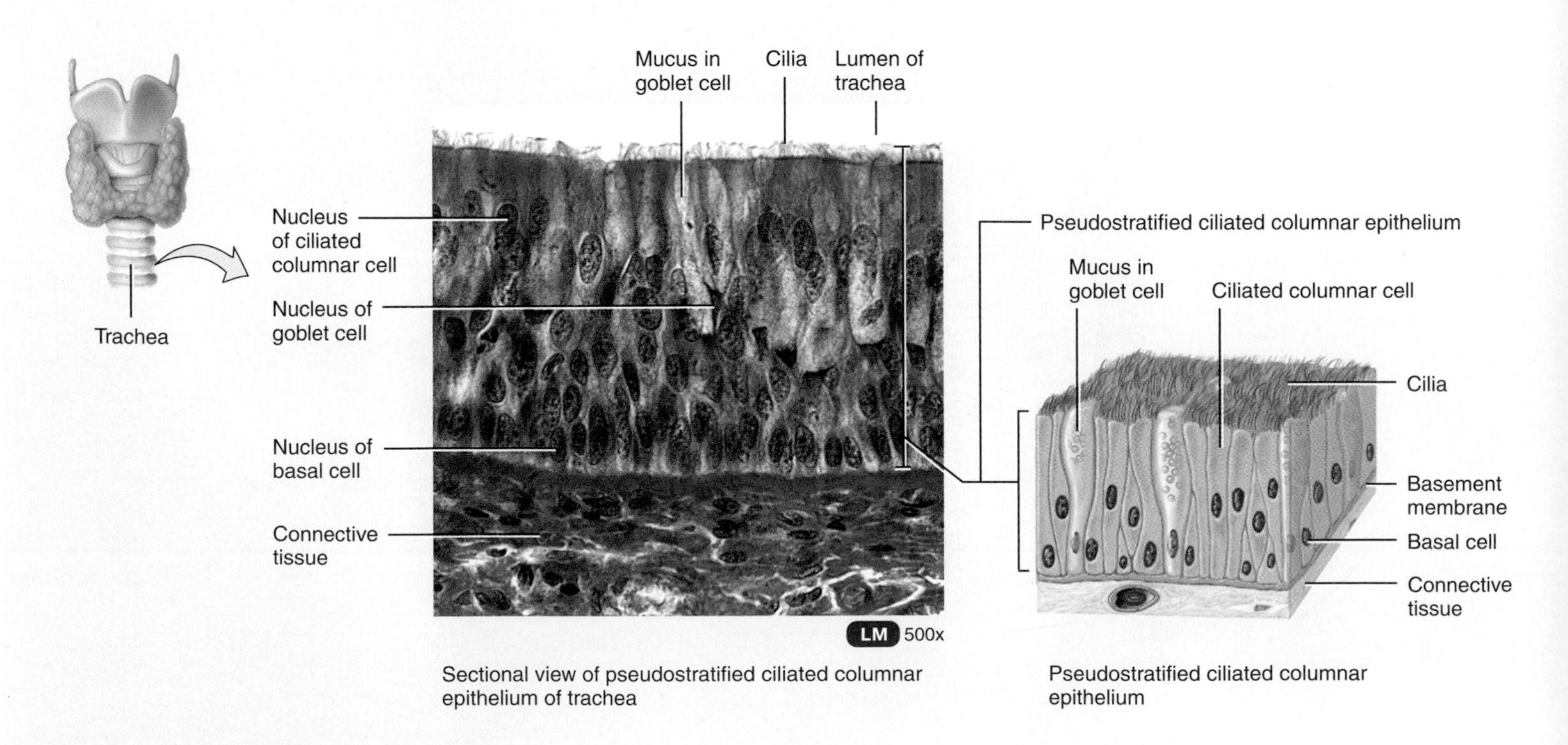

Sectional view of pseudostratified ciliated columnar epithelium of trachea

Pseudostratified ciliated columnar epithelium

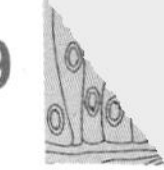

Covering and Lining Epithelium

Stratified Epithelium

F. Stratified squamous epithelium

Description: Several layers of cells; cuboidal to columnar shape in deep layers; squamous cells form the apical layer and several layers deep to it; cells from the basal layer replace surface cells as they are lost.

Location: Keratinized variety forms superficial layer of skin; nonkeratinized variety lines wet surfaces, such as lining of the mouth, esophagus, part of epiglottis, part of pharynx, and vagina, and covers the tongue.

Function: Protection.

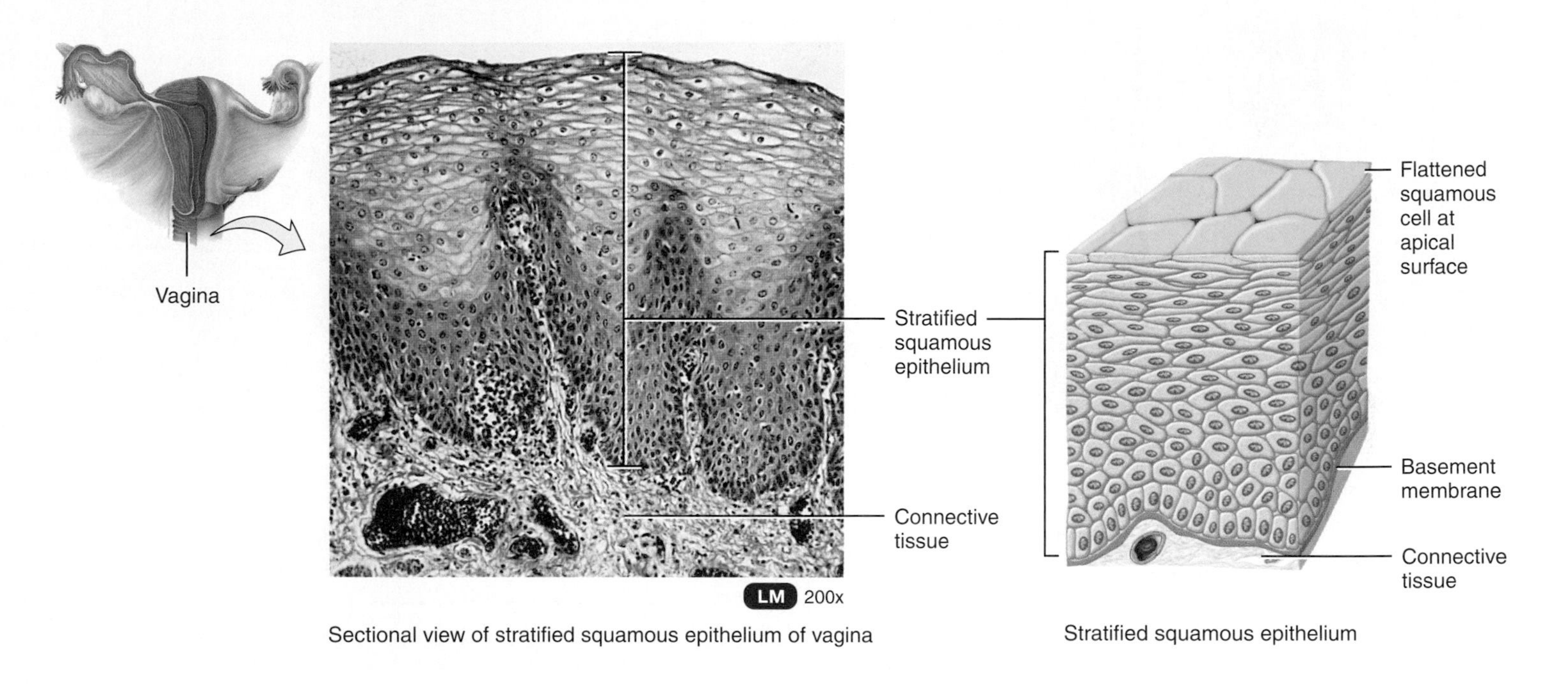

Sectional view of stratified squamous epithelium of vagina

Stratified squamous epithelium

G. Stratified cuboidal epithelium

Description: Two or more layers of cells in which cells in the apical layer are cube-shaped.

Location: Ducts of adult sweat glands and esophageal glands and part of male urethra.

Function: Protection and limited secretion and absorption.

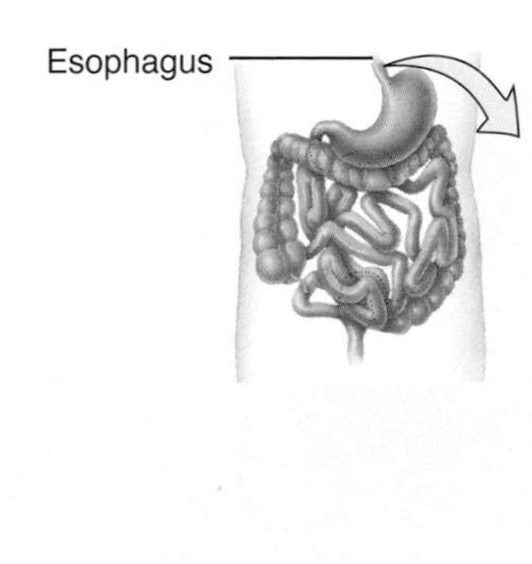

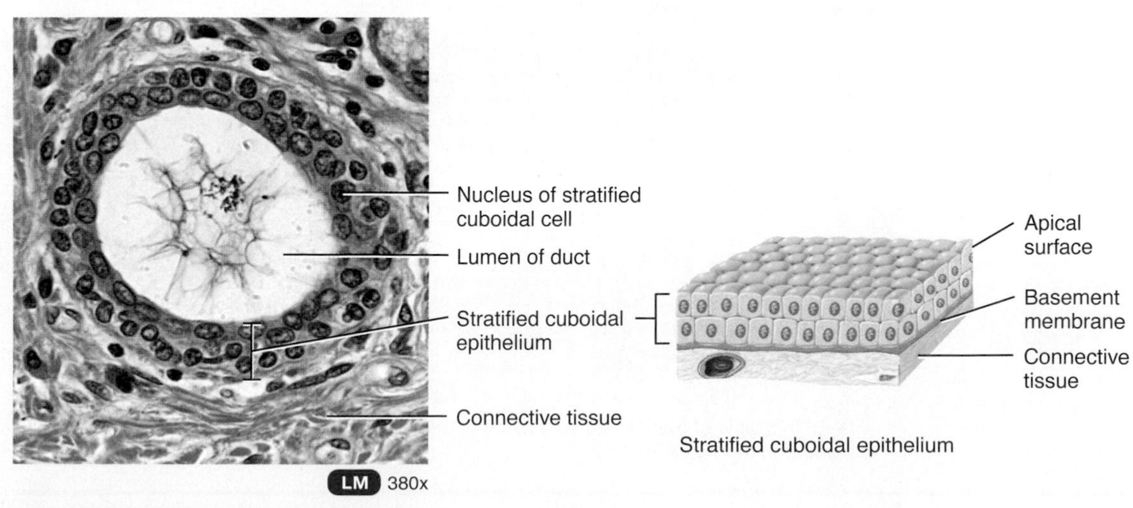

Sectional view of stratified cuboidal epithelium of the duct of an esophageal gland

Stratified cuboidal epithelium

(continues)

Table 4.1 Epithelial Tissues (continued)

Covering and Lining Epithelium

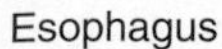

H. Stratified columnar epithelium

Description: Several layers of irregularly shaped cells; only the apical layer has columnar cells.

Location: Lines part of urethra, large excretory ducts of some glands such as esophageal glands, small areas in anal mucous membrane, and a part of the conjunctiva of the eye.

Function: Protection and secretion.

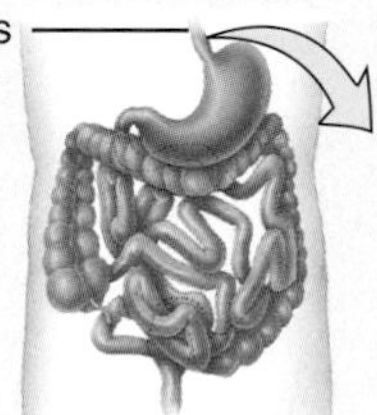

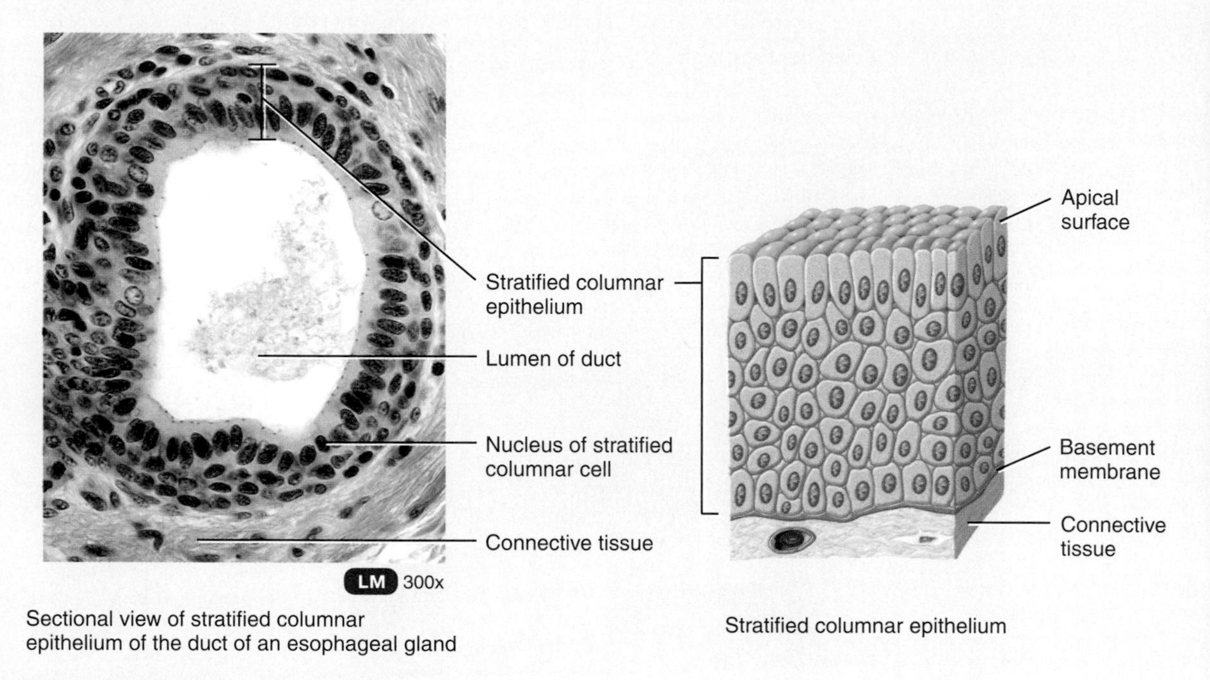

Sectional view of stratified columnar epithelium of the duct of an esophageal gland

Stratified columnar epithelium

I. Transitional epithelium

Description: Appearance is variable (transitional); shape of cells in apical layer ranges from squamous (when stretched) to cuboidal (when relaxed).

Location: Lines urinary bladder and portions of ureters and urethra.

Function: Permits distention.

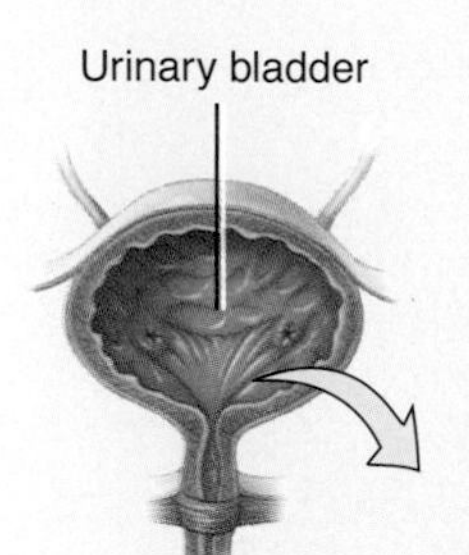

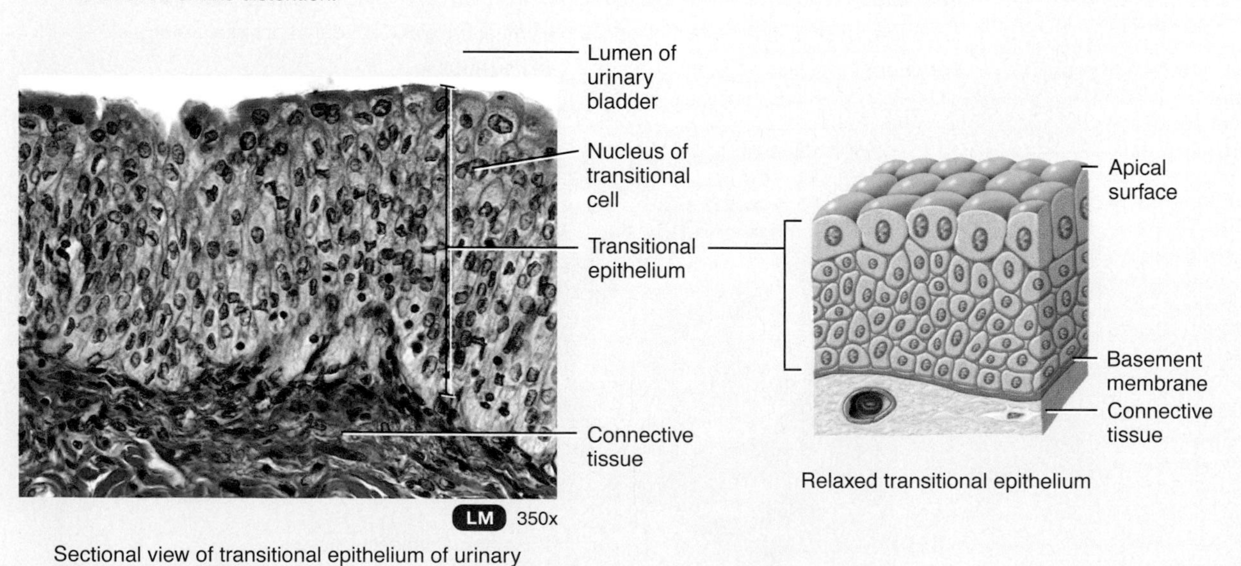

Sectional view of transitional epithelium of urinary bladder in relaxed state

Relaxed transitional epithelium

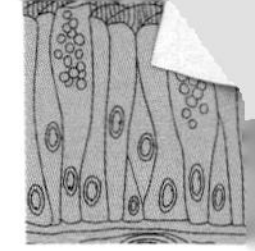

Covering and Lining Epithelium

Glandular Epithelium

J. Endocrine glands

Description: Secretory products (hormones) diffuse into blood after passing through interstitial fluid.

Location: Examples include pituitary gland at base of brain, pineal gland in brain, thyroid and parathyroid glands near larynx (voice box), adrenal glands superior to kidneys, pancreas near stomach, ovaries in pelvic cavity, testes in scrotum, and thymus in thoracic cavity.

Function: Produce hormones that regulate various body activities.

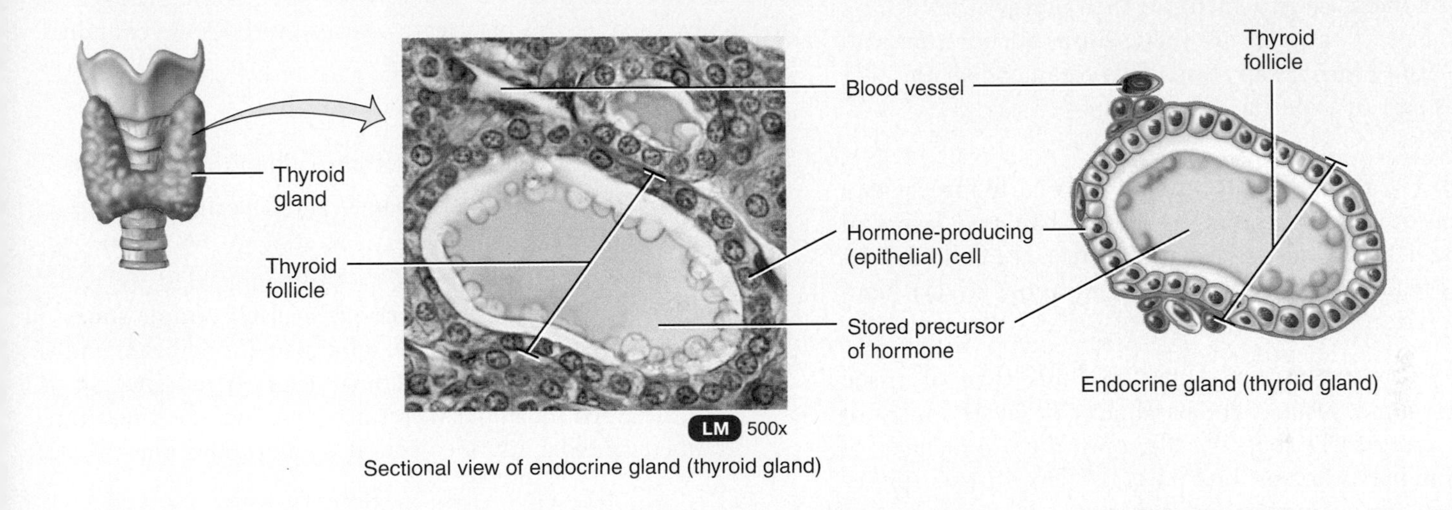

Sectional view of endocrine gland (thyroid gland)

K. Exocrine glands

Description: Secretory products released into ducts.

Location: Sweat, oil, and earwax glands of the skin; digestive glands such as salivary glands, which secrete into mouth cavity, and pancreas, which secretes into the small intestine.

Function: Produce substances such as sweat, oil, earwax, saliva, or digestive enzymes.

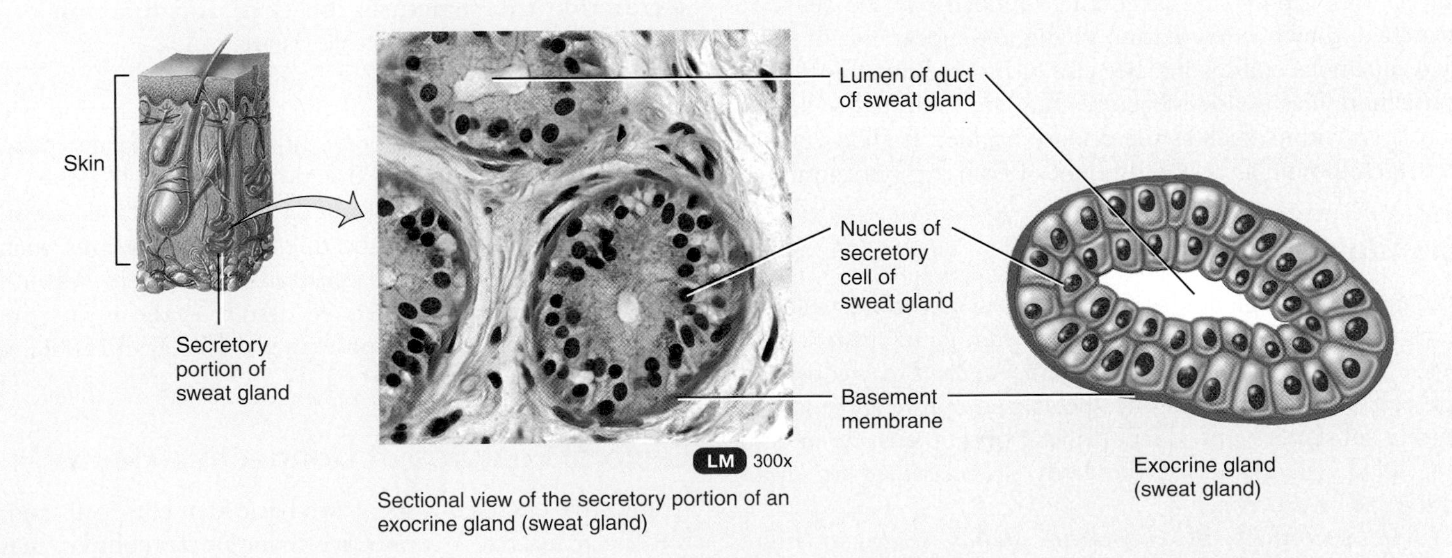

Sectional view of the secretory portion of an exocrine gland (sweat gland)

A **Papanicolaou test** (pa-pa-NI-kō-lō), also called a **Pap test** or **Pap smear**, involves collection and microscopic examination of epithelial cells that have been scraped off the apical layer of a tissue. A very common type of Pap test involves examining the cells from the nonkeratinized stratified squamous epithelium of the vagina and cervix (inferior portion) of the uterus. This type of Pap test is performed mainly to detect early changes in the cells of the female reproductive system that may indicate cancer or a precancerous condition. In performing a Pap smear, a physician collects cells, which are then smeared on a microscope slide. The slides are then sent to a laboratory for analysis. An annual Pap test is recommended for all women as part of a routine pelvic exam.

STRATIFIED CUBOIDAL EPITHELIUM This fairly rare type of epithelium sometimes consists of more than two layers of cells (Table 4.1G). Cells in the apical layer are cuboidal. Its function is mainly protective; in some locations it also functions in secretion and absorption.

STRATIFIED COLUMNAR EPITHELIUM This type of tissue also is uncommon. Usually the basal layer or layers consist of shortened, irregularly shaped cells; only the apical layer of cells is columnar in form (Table 4.1H). This type of epithelium functions in protection and secretion.

TRANSITIONAL EPITHELIUM This type of stratified epithelium is variable in appearance, depending on whether the organ it lines is unstretched or stretched. In its unstretched state (Table 4.1I), transitional epithelium looks similar to stratified cuboidal epithelium, except that the cells in the apical layer tend to be large and rounded. As the cells are stretched, they become flatter, giving the appearance of stratified squamous epithelium. Because of its elasticity, transitional epithelium lines hollow structures that are subjected to expansion from within, such as the urinary bladder. It allows organs to stretch to hold a variable amount of fluid without rupturing.

Glandular Epithelium

The function of glandular epithelium is secretion, which is accomplished by glandular cells that often lie in clusters deep to the covering and lining epithelium. A ***gland*** may consist of one cell or a group of highly specialized epithelial cells that secrete substances into ducts (tubes), onto a surface, or into the blood. All glands of the body are classified as either endocrine or exocrine.

The secretions of ***endocrine glands*** (*endo-* = within; *-crine* = secretion) (Table 4.1J) enter the interstitial fluid and then diffuse into the bloodstream without flowing through a duct. These secretions, called *hormones*, regulate many metabolic and physiological activities to maintain homeostasis. The pituitary, thyroid, and adrenal glands are examples of endocrine glands. Endocrine glands will be described in detail in Chapter 13.

Exocrine glands (*exo-* = outside; Table 4.1K) secrete their products into ducts that empty at the surface of covering and lining epithelium such as the skin surface or the lumen (interior space) of a hollow organ. The secretions of exocrine glands include mucus, perspiration, oil, earwax, milk, saliva, and digestive enzymes. Examples of exocrine glands are sweat glands, which produce perspiration to help lower body temperature, and salivary glands, which secrete mucus and digestive enzymes. As you will see later, some glands of the body, such as the pancreas, ovaries, and testes, contain both endocrine and exocrine tissue.

■ **CHECKPOINT**

3. What characteristics are common to all epithelial tissues?
4. Describe the various cell shapes and layering arrangements of epithelium.
5. Explain how the structure of the following kinds of epithelium is related to the functions of each: simple squamous, simple cuboidal, simple columnar (nonciliated and ciliated), pseudostratified columnar (nonciliated and ciliated), stratified squamous (keratinized and nonkeratinized), stratified cuboidal, stratified columnar, and transitional.

CONNECTIVE TISSUE

OBJECTIVES

- **Discuss the general features of connective tissue.**
- **Describe the structure, location, and function of the various types of connective tissue.**

Connective tissue is one of the most abundant and widely distributed tissues in the body. In its various forms, connective tissue has a variety of functions. It binds together, supports, and strengthens other body tissues; protects and insulates internal organs; compartmentalizes structures such as skeletal muscles; is the major transport system within the body (blood, a fluid connective tissue); is the major site of stored energy reserves (adipose, or fat tissue); and is the main site of immune responses.

General Features of Connective Tissue

Connective tissue consists of two basic elements: cells and extracellular matrix. A connective tissue's **extracellular matrix** is the material between its widely spaced cells. The extracellular matrix consists of protein fibers and ground substance, the material between the cells and the fibers. The extracellular matrix is usually secreted by the connective tissue cells and determines the tissue's qualities. For instance, in cartilage,

Figure 4.2 Representative cells and fibers present in connective tissues.

Fibroblasts are usually the most numerous connective tissue cells.

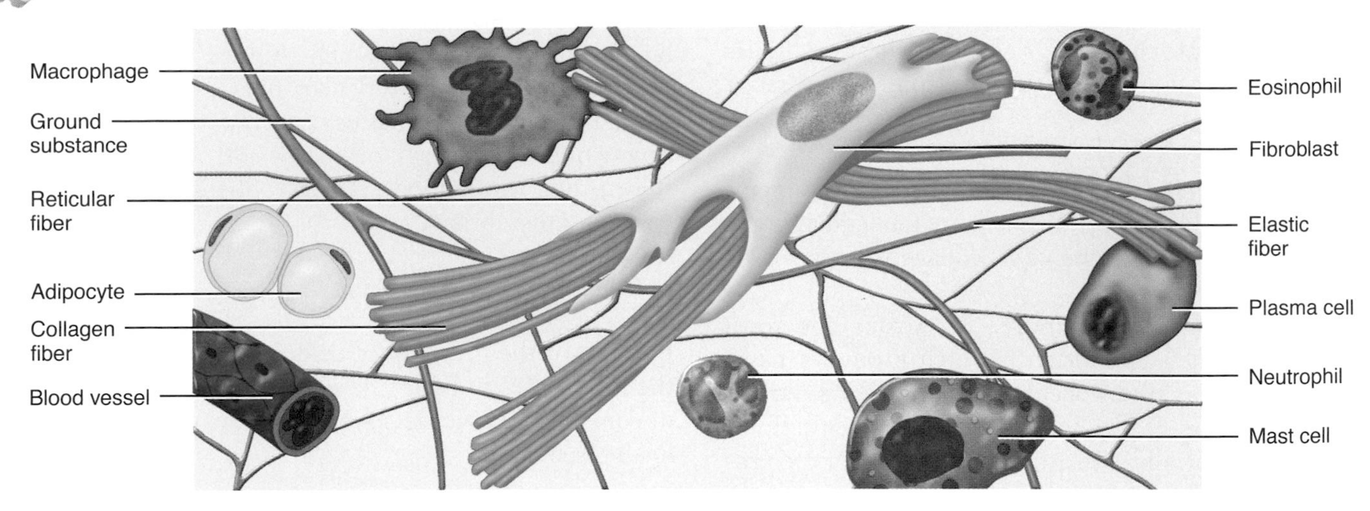

What is the function of fibroblasts?

the extracellular matrix is firm but pliable. The extracellular matrix of bone, by contrast, is hard and not pliable.

In contrast to epithelia, connective tissues do not usually occur on body surfaces. Also, unlike epithelia, connective tissues usually are highly vascular; that is, they have a rich blood supply. Exceptions include cartilage, which is avascular, and tendons, with a scanty blood supply. Except for cartilage, connective tissues, like epithelia, are supplied with nerves.

Connective Tissue Cells

The types of connective tissue cells vary according to the type of tissue and include the following (Figure 4.2):

1. ***Fibroblasts*** (FĪ-brō-blasts; *fibro-* = fibers) are large, flat cells with branching processes. They are present in several connective tissues, and usually are the most numerous. Fibroblasts migrate through the connective tissue, secreting the fibers and ground substance of the extracellular matrix.
2. ***Macrophages*** (MAK-rō-fā-jez; *macro-* = large; *-phages* = eaters) develop from monocytes, a type of white blood cell. Macrophages have an irregular shape with short branching projections and are capable of engulfing bacteria and cellular debris by phagocytosis.
3. ***Plasma cells*** are small cells that develop from a type of white blood cell called a B lymphocyte. Plasma cells secrete antibodies, proteins that attack or neutralize foreign substances in the body. Thus, plasma cells are an important part of the body's immune response.
4. ***Mast cells*** are abundant alongside the blood vessels that supply connective tissue. They produce histamine, a chemical that dilates small blood vessels as part of the inflammatory response, the body's reaction to injury or infection. Mast cells can also kill bacteria.
5. ***Adipocytes***, also called fat cells or adipose cells, are connective tissue cells that store triglycerides (fats). They are found below the skin and around organs such as the heart and kidneys.

Connective Tissue Extracellular Matrix

Each type of connective tissue has unique properties, based on the specific extracellular materials between the cells. The extracellular matrix consists of a fluid, gel, or solid ground substance plus protein fibers.

Ground Substance

Ground substance, the component of a connective tissue between the cells and fibers, supports cells, binds them together, and provides a medium through which substances are exchanged between the blood and cells. The ground substance plays an active role in how tissues develop, migrate, proliferate, and change shape, and in how they carry out their metabolic functions.

Ground substance contains water and an assortment of large organic molecules, many of which are complex combinations of polysaccharides and proteins. For example, ***hyaluronic acid*** (hī′-a-loo-RON-ik) is a viscous, slippery substance that binds cells together, lubricates joints, and helps maintain the shape of the eyeballs. It also appears to play a role in helping phagocytes migrate through connective tissue during development and wound repair. White blood cells, sperm cells, and some bacteria produce *hyaluronidase*, an enzyme

...at breaks apart hyaluronic acid and causes the ground substance of connective tissue to become watery. The ability to produce hyaluronidase enables white blood cells to move through connective tissues to reach sites of infection and sperm cells to penetrate the ovum during fertilization. It also accounts for how bacteria spread through connective tissues.

Fibers

Fibers in the extracellular matrix strengthen and support connective tissues. Three types of fibers are embedded in the extracellular matrix between the cells: collagen fibers, elastic fibers, and reticular fibers.

Collagen fibers (*colla* = glue) are very strong and resist pulling forces, but they are not stiff, which promotes tissue flexibility. These fibers often occur in bundles lying parallel to one another (Figure 4.2). The bundle arrangement affords great strength. Chemically, collagen fibers consist of the protein *collagen.* This is the most abundant protein in your body, representing about 25% of total protein. Collagen fibers are found in most types of connective tissues, especially bone, cartilage, tendons, and ligaments.

> Despite their strength, ligaments may be stressed beyond their normal capacity. This results in **sprain**, a stretched or torn ligament. The ankle joint is most frequently sprained. Because of their poor blood supply, the healing of even partially torn ligaments is a very slow process; completely torn ligaments require surgical repair.

Elastic fibers, which are smaller in diameter than collagen fibers, branch and join together to form a network within a tissue. An elastic fiber consists of molecules of a protein called *elastin* surrounded by a glycoprotein named *fibrillin,* which is essential to the stability of an elastic fiber. Elastic fibers are strong but can be stretched up to one-and-a-half times their relaxed length without breaking. Equally important, elastic fibers have the ability to return to their original shape after being stretched, a property called *elasticity.* Elastic fibers are plentiful in skin, blood vessel walls, and lung tissue.

> **Marfan syndrome** (MAR-fan) is an inherited disorder caused by a defective fibrillin gene. The result is abnormal development of elastic fibers. Tissues rich in elastic fibers are malformed or weakened. Structures affected most seriously are the covering layer of bones (periosteum), the ligament that suspends the lens of the eye, and the walls of the large arteries. People with Marfan syndrome tend to be tall and have disproportionately long arms, legs, fingers, and toes. A common symptom is blurred vision caused by displacement of the lens of the eye. The most life-threatening complication of Marfan syndrome is weakening of the aorta (the main artery that emerges from the heart), which can suddenly burst.

Reticular fibers (*reticul-* = net), consisting of *collagen* and a coating of glycoprotein, provide support in the walls of blood vessels and form branching networks around fat cells, nerve fibers, and skeletal and smooth muscle cells. Produced by fibroblasts, they are much thinner than collagen fibers. Like collagen fibers, reticular fibers provide support and strength and also form the ***stroma*** (= bed or covering) or supporting framework of many soft organs, such as the spleen and lymph nodes. These fibers also help form the basement membrane.

Classification of Connective Tissues

Because of the diversity of cells and extracellular matrix and the differences in their relative proportions, the classification of connective tissues is not always clear-cut. We offer the following scheme:

I. Loose connective tissue
  A. Areolar connective tissue
  B. Adipose tissue
  C. Reticular connective tissue

II. Dense connective tissue
  D. Dense regular connective tissue
  E. Dense irregular connective tissue
  F. Elastic connective tissue

III. Cartilage
  G. Hyaline cartilage
  H. Fibrocartilage
  I. Elastic cartilage

IV. Bone tissue

V. Liquid connective tissue (blood tissue and lymph)

Loose Connective Tissue

The fibers in ***loose connective tissue*** are loosely intertwined among the many cells. The types of loose connective tissue are areolar connective tissue, adipose tissue, and reticular connective tissue.

AREOLAR CONNECTIVE TISSUE One of the most widely distributed connective tissues in the body is ***areolar connective tissue*** (a-RĒ-ō-lar; *areol-* = a small space). It contains several kinds of cells, including fibroblasts, macrophages, plasma cells, mast cells, adipocytes, and a few white blood cells (Table 4.2A). All three types of fibers—collagen, elastic, and reticular—are arranged randomly throughout the tissue. Combined with adipose tissue, areolar connective tissue forms the *subcutaneous layer,* the layer of tissue that attaches the skin to underlying tissues and organs.

ADIPOSE TISSUE Adipose tissue is a loose connective tissue in which the cells, called ***adipocytes*** (*adipo-* = fat), are specialized for storage of triglycerides (fats) (Table 4.2B). Because the cell fills up with a single, large triglyceride droplet, the cytoplasm and nucleus are pushed to the periphery of the cell. Adipose tissue is found wherever areolar connective tissue is located. Adipose tissue is a good insulator and can therefore reduce heat loss through the skin. It is a major energy reserve and generally supports and protects various organs. As the amount of adipose tissue increases with weight gain, new blood vessels form. Thus, an obese person has many more blood vessels than does a lean person, a situation that can cause high blood pressure, since the heart has to work harder.

A surgical procedure, called **liposuction** (*lip-* = fat) or **suction lipectomy** (*-ectomy* = to cut out), involves suctioning out small amounts of adipose tissue from various areas of the body. The technique can be used as a body-contouring procedure in regions such as the thighs, buttocks, arms, breasts, and abdomen. Postsurgical complications that may develop include fat emboli (clots), infection, fluid depletion, injury to internal structures, and severe postoperative pain.

Table 4.2 Connective Tissues

Loose Connective Tissue

A. Areolar connective tissue

Description: Consists of fibers (collagen, elastic, and reticular) and several kinds of cells (fibroblasts, macrophages, plasma cells, adipocytes, and mast cells) embedded in a semifluid ground substance.

Location: Subcutaneous layer deep to skin; superficial region of dermis of skin; lamina propria of mucous membranes; and around blood vessels, nerves, and body organs.

Function: Strength, elasticity, and support.

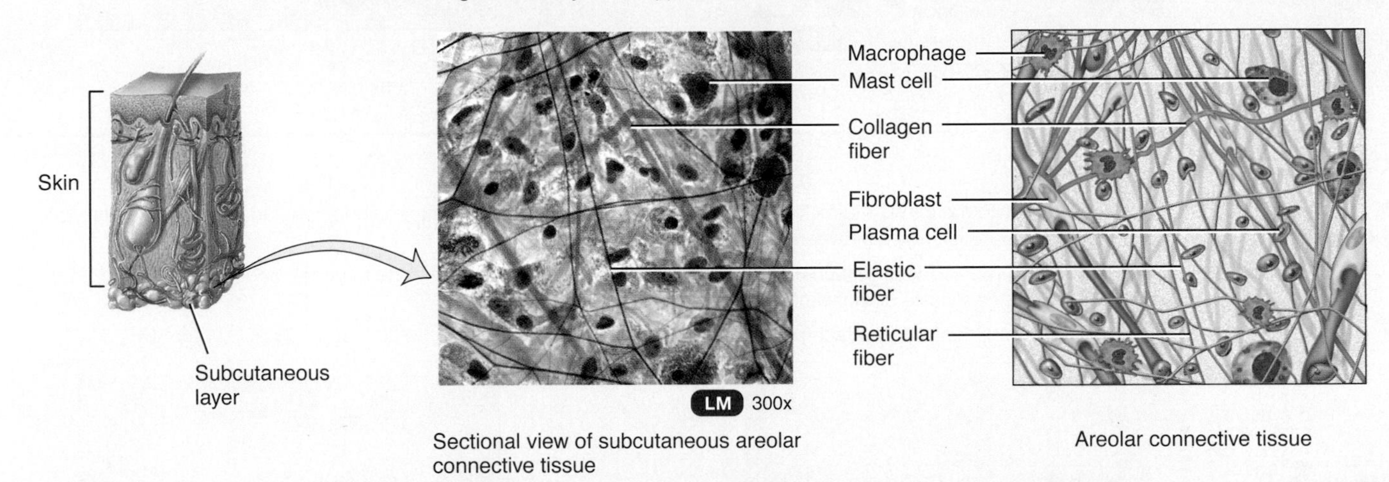

Sectional view of subcutaneous areolar connective tissue

Areolar connective tissue

B. Adipose tissue

Description: Consists of adipocytes, cells specialized to store triglycerides (fats) as a large centrally located droplet; nucleus and cytoplasm are peripherally located.

Location: Subcutaneous layer deep to skin, around heart and kidneys, yellow bone marrow, and padding around joints and behind eyeball in eye socket.

Function: Reduces heat loss through skin, serves as an energy reserve, supports, and protects.

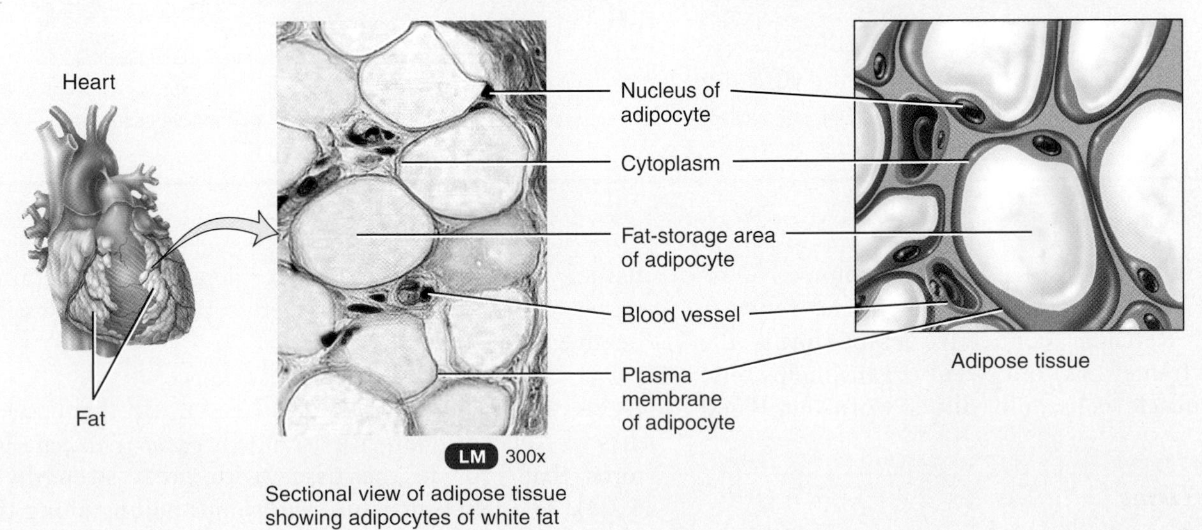

Sectional view of adipose tissue showing adipocytes of white fat

Adipose tissue

(continues)

Table 4.2 Connective Tissues (continued)

Loose Connective Tissue

C. Reticular connective tissue

Description: A network of interlacing reticular fibers and reticular cells.

Location: Stroma (supporting framework) of liver, spleen, lymph nodes; red bone marrow, which gives rise to blood cells; reticular lamina of the basement membrane; and around blood vessels and muscles.

Function: Forms stroma of organs; binds together smooth muscle tissue cells; filters and removes worn-out blood cells in the spleen and microbes in lymph nodes.

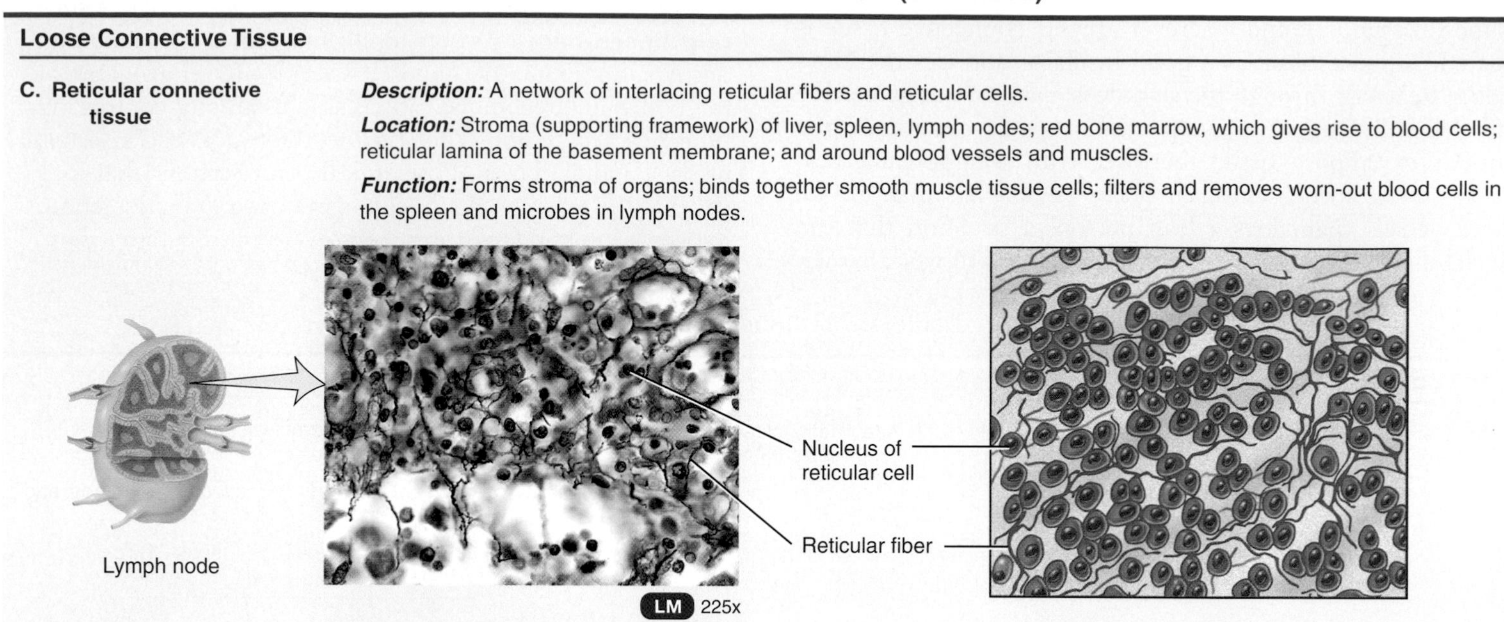

Sectional view of reticular connective tissue of a lymph node

Reticular connective tissue

Dense Connective Tissue

D. Dense regular connective tissue

Description: Extracellular matrix looks shiny white; consists mainly of collagen fibers regularly arranged in bundles; fibroblasts present in rows between bundles.

Location: Forms tendons (attach muscle to bone), most ligaments (attach bone to bone), and aponeuroses (sheetlike tendons that attach muscle to muscle or muscle to bone).

Function: Provides strong attachment between various structures.

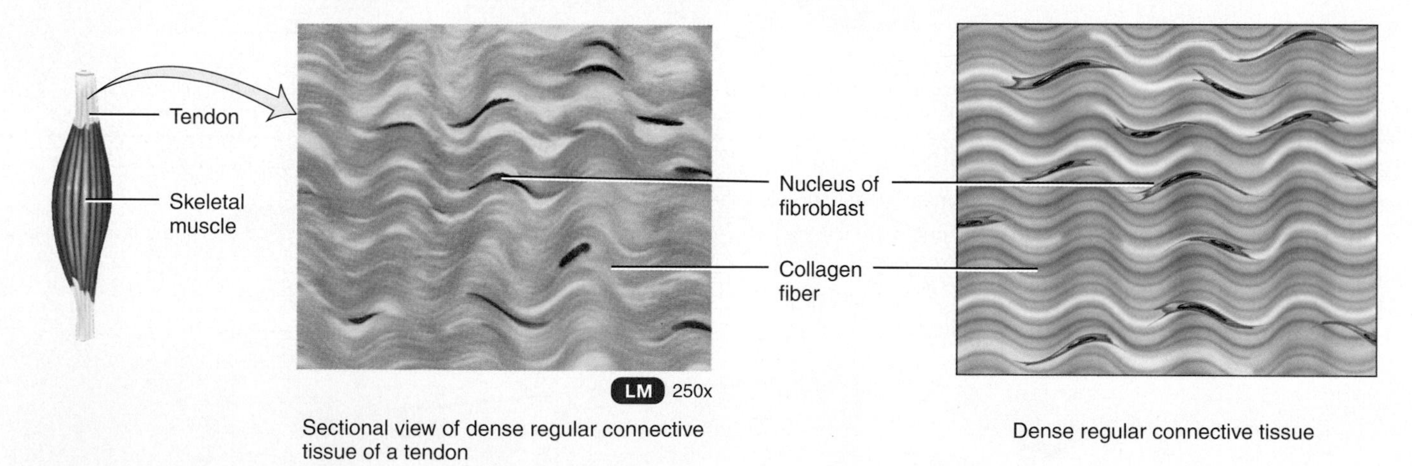

Sectional view of dense regular connective tissue of a tendon

Dense regular connective tissue

RECTICULAR CONNECTIVE TISSUE Reticular connective tissue consists of fine interlacing reticular fibers and reticular cells (Table 4.2C). Reticular connective tissue forms the stroma (supporting framework) of certain organs, helps bind together smooth muscle cells, and filters worn-out blood cells and bacteria.

Dense Connective Tissue

Dense connective tissue contains more numerous, thicker, and denser fibers but fewer cells than loose connective tissue. There are three types: dense regular connective tissue, dense irregular connective tissue, and elastic connective tissue.

DENSE REGULAR CONNECTIVE TISSUE In this tissue, bundles of collagen fibers are arranged *regularly* in parallel patterns that provide the tissue with great strength (Table 4.2D). The tissue structure withstands pulling along the axis of the fibers. Fibroblasts, which produce the fibers and ground substance, appear in rows between the fibers. The

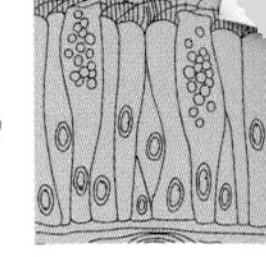

Loose Connective Tissue

E. Dense irregular connective tissue

Description: Consists predominantly of collagen fibers randomly arranged and a few fibroblasts.

Location: Fasciae (tissue beneath skin and around muscles and other organs), deeper region of dermis of skin, periosteum of bone, perichondrium of cartilage, joint capsules, membrane capsules around various organs (kidneys, liver, testes, lymph nodes), pericardium of the heart, and heart valves.

Function: Provides strength.

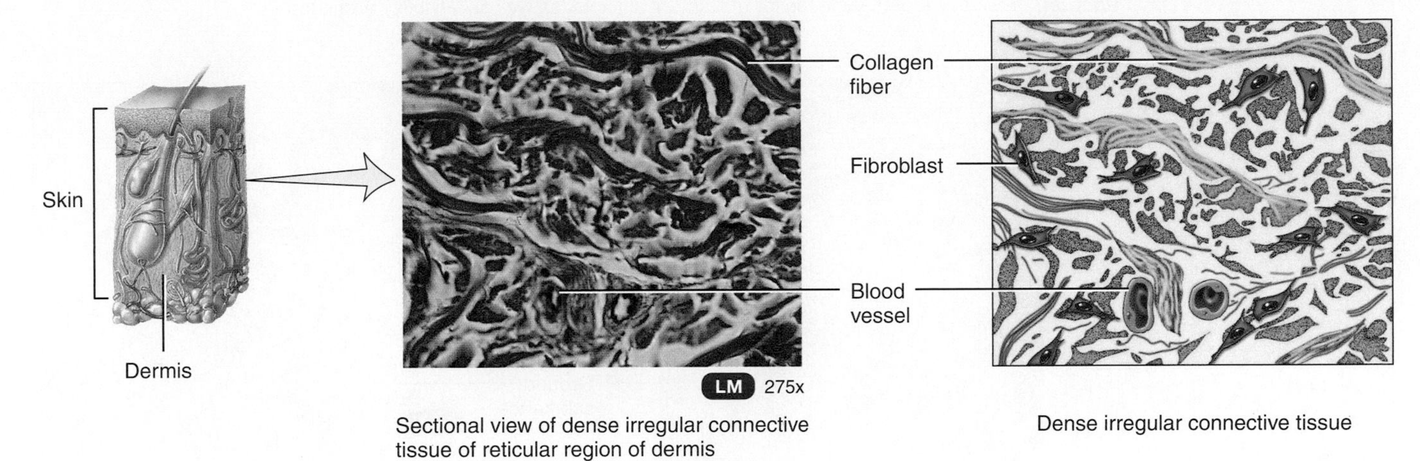

Sectional view of dense irregular connective tissue of reticular region of dermis

Dense irregular connective tissue

F. Elastic connective tissue

Description: Consists predominantly of freely branching elastic fibers; fibroblasts are present in spaces between fibers.

Location: Lung tissue, walls of elastic arteries, trachea, bronchial tubes, true vocal cords, suspensory ligament of penis, and ligaments between vertebrae.

Function: Allows stretching of various organs.

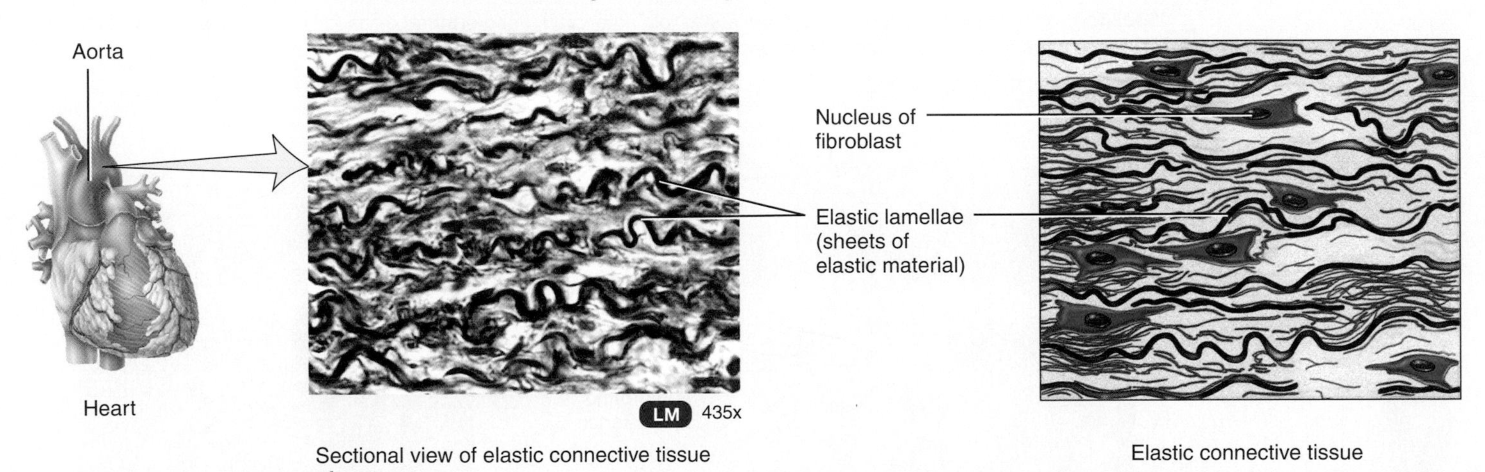

Sectional view of elastic connective tissue of aorta

Elastic connective tissue

(continues)

tissue is silvery white and tough, yet somewhat pliable. Examples are tendons and most ligaments.

Dense Irregular Connective Tissue This tissue contains collagen fibers that are packed more closely together than in loose connective tissue and are usually *irregularly* arranged (Table 4.2E). It is found in parts of the body where pulling forces are exerted in various directions. The tissue usually occurs in sheets, such as in the dermis of the skin, which underlies the epidermis. Heart valves, the perichondrium (the membrane surrounding cartilage), and the periosteum (the membrane surrounding bone) are considered dense irregular connective tissues, despite a fairly orderly arrangement of their collagen fibers.

Elastic Connective Tissue Branching elastic fibers predominate in elastic connective tissue (Table 4.2F), giving the unstained tissue a yellowish color. Fibroblasts are present in the spaces between the fibers. Elastic connective tissue is quite strong and can recoil to its original shape after

Table 4.2 Connective Tissues (continued)

Cartilage

G. Hyaline cartilage

Description: Consists of a bluish-white, shiny ground substance with fine collagen fibers and many chondrocytes; most abundant type of cartilage.

Location: Ends of long bones, anterior ends of ribs, nose, parts of larynx, trachea, bronchi, bronchial tubes, and embryonic and fetal skeleton.

Function: Provides smooth surfaces for movement at joints, as well as flexibility and support.

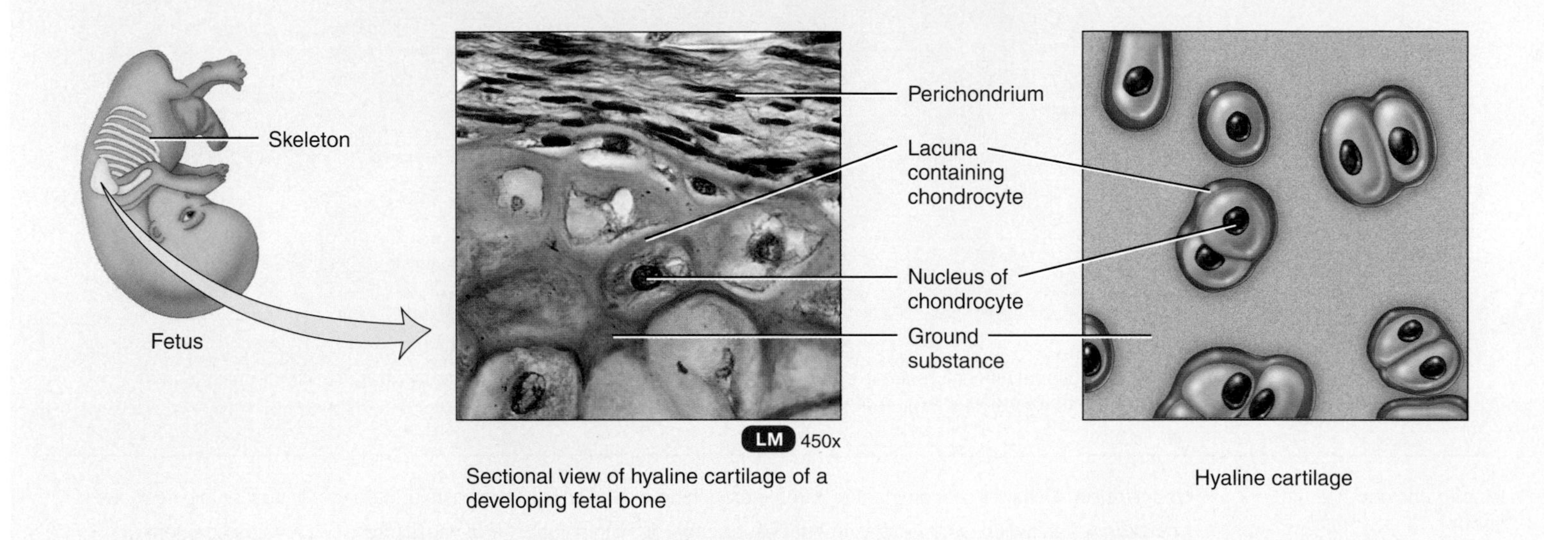

Sectional view of hyaline cartilage of a developing fetal bone

Hyaline cartilage

H. Fibrocartilage

Description: Consists of chondrocytes scattered among bundles of collagen fibers within the extracellular matrix.

Location: Pubic symphysis (point where hip bones join anteriorly), intervertebral discs (discs between vertebrae), menisci (cartilage pads) of knee, and portions of tendons that insert into cartilage.

Function: Support and fusion.

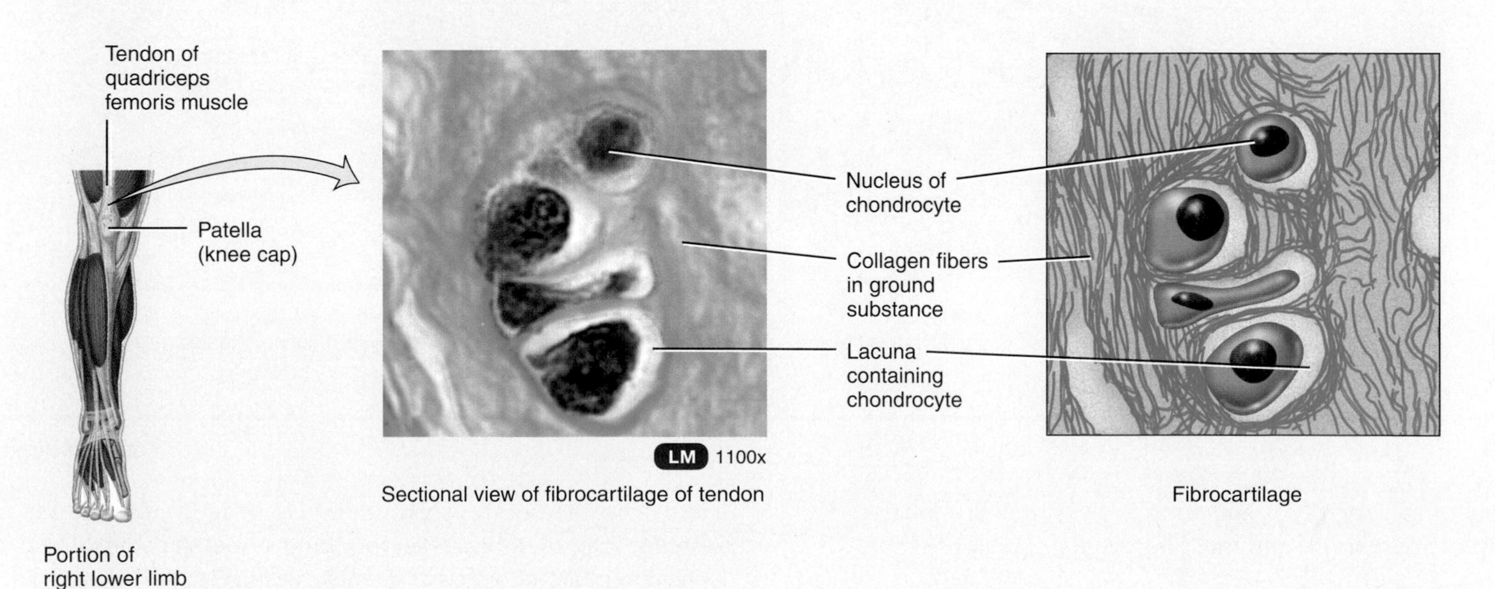

Portion of right lower limb

Sectional view of fibrocartilage of tendon

Fibrocartilage

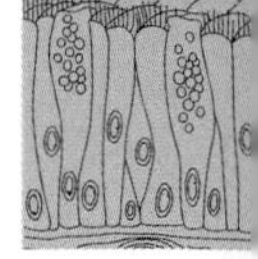

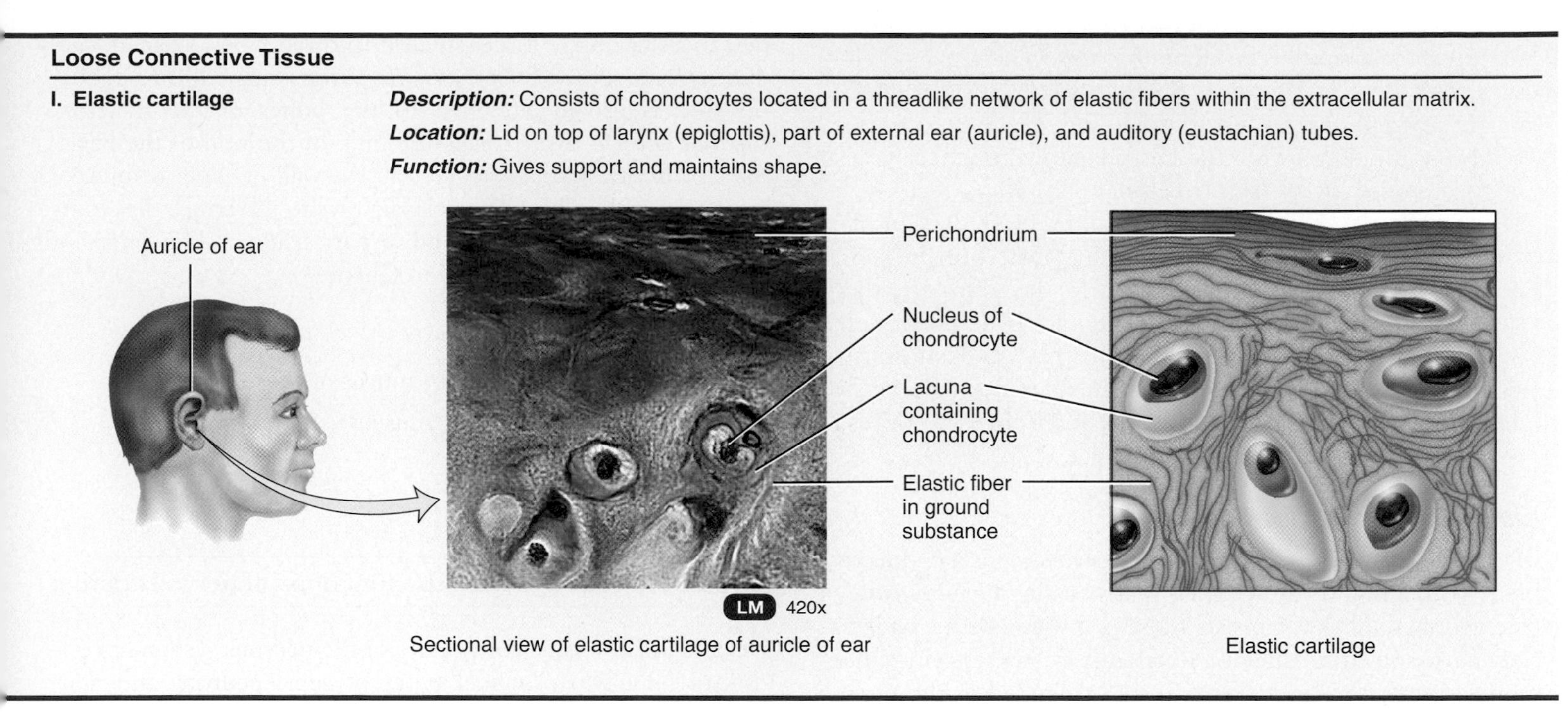

Sectional view of elastic cartilage of auricle of ear

Elastic cartilage

being stretched. Elasticity is important to the normal functioning of lung tissue, which recoils as you exhale, and elastic arteries, whose recoil between heart beats helps maintain blood flow.

Cartilage

Cartilage consists of a dense network of collagen fibers and elastic fibers firmly embedded in chondroitin sulfate, a rubbery component of the ground substance. Cartilage can endure considerably more stress than loose and dense connective tissues. While the strength of cartilage is due to its collagen fibers, its *resilience* (ability to assume its original shape after deformation) is due to chondroitin sulfate.

The cells of mature cartilage, called ***chondrocytes*** (KON-drō-sīts; *chondro-* = cartilage), occur singly or in groups within spaces called ***lacunae*** (la-KOO-nē = little lakes; singular is *lacuna*) in the extracellular matrix. The surface of most cartilage is surrounded by a membrane of dense irregular connective tissue called the ***perichondrium*** (per′-i-KON-drē-um; *peri-* = around). Unlike other connective tissues, cartilage has no blood vessels or nerves, except in the perichondrium. Since cartilage has no blood supply, it heals poorly following an injury. The three types of cartilage are hyaline cartilage, fibrocartilage, and elastic cartilage.

HYALINE CARTILAGE This type of cartilage contains a resilient gel as its ground substance and appears in the body as a bluish-white, shiny substance. The fine collagen fibers are not visible with ordinary staining techniques, and prominent chondrocytes are found in lacunae (Table 4.2G). Most hyaline cartilage is surrounded by a perichondrium. The exceptions are the articular cartilage in joints and the epiphyseal plates, the regions where bones lengthen as a person grows. Hyaline cartilage is the most abundant cartilage in the body. It affords flexibility and support and, at joints, reduces friction and absorbs shock. Hyaline cartilage is the weakest of the three types of cartilage.

FIBROCARTILAGE Chondrocytes are scattered among clearly visible bundles of collagen fibers within the extracellular matrix of this type of cartilage (Table 4.2H). Fibrocartilage lacks a perichondrium. This tissue combines strength and rigidity and is the strongest of the three types of cartilage. One location of fibrocartilage is in the discs between vertebrae (backbones).

ELASTIC CARTILAGE In elastic cartilage, chondrocytes are located within a threadlike network of elastic fibers within the extracellular matrix (Table 4.2I). A perichondrium is present. Elastic cartilage provides strength and elasticity and maintains the shape of certain structures, such as the external ear.

Bone Tissue

Bones are organs composed of several different connective tissues, including ***bone*** or *osseous tissue* (OS-ē-us). Bone tissue has several functions. It supports soft tissues, protects delicate structures, and works with skeletal muscles to generate movement. Bone stores calcium and phosphorus; stores red bone marrow, which produces blood cells; and houses yellow bone marrow, a storage site for triglycerides. The details of bone tissue are presented in Chapter 6.

The technology of **tissue engineering** has allowed scientists to grow new tissues in the laboratory to replace damaged tissues in the body. Tissue engineers have already developed laboratory-grown versions of skin and cartilage. In the procedure, scaffolding beds of biodegradable synthetic materials or collagen are used as substrates that permit body cells such as skin cells or cartilage cells to be cultured. As the cells divide and assemble, the scaffolding degrades, and the new, permanent tissue is then implanted in the patient. Other structures being developed by tissue engineers include bones, tendons, heart valves, bone marrow, and intestines. Work is also underway to develop insulin-producing cells for diabetics, dopamine-producing cells for Parkinson disease patients, and even entire livers and kidneys.

Liquid Connective Tissue

BLOOD TISSUE ***Blood tissue*** (or simply *blood*) is a connective tissue with a liquid extracellular matrix called ***blood plasma***, a pale yellow fluid that consists mostly of water with a wide variety of dissolved substances: nutrients, wastes, enzymes, hormones, respiratory gases, and ions. Suspended in the plasma are red blood cells, white blood cells, and platelets. ***Red blood cells*** transport oxygen to body cells and remove carbon dioxide from them. ***White blood cells*** are involved in phagocytosis, immunity, and allergic reactions. ***Platelets*** participate in blood clotting. The details of blood are considered in Chapter 14.

LYMPH ***Lymph*** is a fluid that flows in lymphatic vessels. It is a connective tissue that consists of several types of cells in a clear extracellular matrix similar to blood plasma but with much less protein. The details of lymph are considered in Chapter 17.

■ **CHECKPOINT**

6. What are the features of the cells, ground substance, and fibers that make up connective tissue?
7. How are the structures of the following connective tissues related to their functions: areolar connective tissue, adipose tissue, reticular connective tissue, dense regular connective tissue, dense irregular connective tissue, elastic connective tissue, hyaline cartilage, fibrocartilage, elastic cartilage, bone tissue, blood tissue, and lymph?

MUSCULAR TISSUE

OBJECTIVES • **Describe the functions of muscular tissue.**

• **Contrast the locations of the three types of muscular tissue.**

Muscular tissue consists of elongated cells called *muscle fibers* that are highly specialized to generate force. As a result of this characteristic, muscular tissue produces motion, maintains posture, and generates heat. It also offers protection. Based on its location and certain structural and functional characteristics, muscular tissue is classified into three types: skeletal, cardiac, and smooth. ***Skeletal muscle tissue*** is named for its location—it is usually attached to the bones of the skeleton. ***Cardiac muscle tissue*** forms the bulk of the wall of the heart. ***Smooth muscle tissue*** is located in the walls of hollow internal structures such as blood vessels, airways to the lungs, the stomach, intestines, gallbladder, and urinary bladder. The details of muscular tissue are presented in Chapter 8.

■ **CHECKPOINT**

8. What are the functions of muscular tissue?
9. Name the three types of muscular tissue.

NERVOUS TISSUE

OBJECTIVE • **Describe the functions of nervous tissue.**

Despite the awesome complexity of the nervous system, it consists of only two principal types of cells: neurons and neuroglia. ***Neurons*** (*neur-* = nerve, nerve tissue, nervous system) or *nerve cells*, are sensitive to various stimuli. They convert stimuli into nerve impulses (action potentials) and conduct these impulses to other neurons, to muscle fibers, or to glands. ***Neuroglia*** (noo-RŌG-lē-a; *-glia* = glue) do not generate or conduct nerve impulses, but they do have many other important supportive functions. The detailed structure and function of neurons and neuroglia are considered in Chapter 9.

■ **CHECKPOINT**

10. How do neurons differ from neuroglia?

MEMBRANES

OBJECTIVES • **Define a membrane.**

• **Describe the classification of membranes.**

Membranes are flat sheets of pliable tissue that cover or line a part of the body. The combination of an epithelial layer and an underlying connective tissue layer constitutes an ***epithelial membrane***. The principal epithelial membranes of the body are mucous membranes, serous membranes, and the cutaneous membrane, or skin. (Skin is discussed in detail in Chapter 5 and will not be discussed here.) Another kind of membrane, a synovial membrane, lines joints and contains connective tissue but no epithelium.

Mucous Membranes

A ***mucous membrane*** or ***mucosa*** (mū-KŌ-sa) lines a body cavity that opens directly to the exterior. Mucous membranes line the entire digestive, respiratory, and reproductive systems and much of the urinary system. The epithelial layer of a

FOCUS ON WELLNESS

Excess Adiposity—Too Much of a Good Thing

Adipose tissue contains adipocytes, cells specialized for the function of energy storage. Adequate energy storage has been vital to the survival of our species through the ages. But what happens when we have too much of a good thing?

Excess Adiposity: Too Much Fat

Adiposity becomes too much of a good thing when it leads to health problems, which can include hypertension (high blood pressure), poor blood sugar regulation (including type 2 diabetes), heart disease, certain cancers, gallstones, arthritis, and lower backaches. In general, the risk of health problems associated with excess adipose tissue increases in a dose-dependent fashion: The greater the excess weight, the greater the risk. People with a great deal of excess fat have a much higher risk for health problems than people who are only slightly too fat. But the health effects of excess fat depend on several important factors besides quantity of adipose tissue.

Location of Adipose Tissue

People who carry extra fat on the torso are at greater risk for hypertension, type 2 diabetes, and artery disease than people whose extra fat resides in the hips and thighs. Especially risky is excess fat stored around the viscera (such as the abdominal organs). Adipocytes in this area appear to be more "metabolically active" than those under the skin. Visceral fat affects blood sugar and blood fat regulation, which in turn can lead to the health problems mentioned above.

Family Medical History and Age

Excess body fat is especially risky for people who have already developed fat-related health problems, or who have a family history of these disorders. On the other hand, people over 70 years old may benefit from a little extra adipose tissue. Many health professionals recommend an extra 10 or 15 pounds for people over 70 to help them resist wasting if they should become ill.

Beware the Deadly Sins: Gluttony and Sloth

People with a moderate amount of excess adipose tissue who eat a healthy diet and exercise regularly have health risks similar to those of their leaner peers. This observation suggests that some of the health risks seen in overweight people may be caused by poor health habits (such as too much food, too much alcohol, or too little exercise) rather than by the presence of excess adipose tissue.

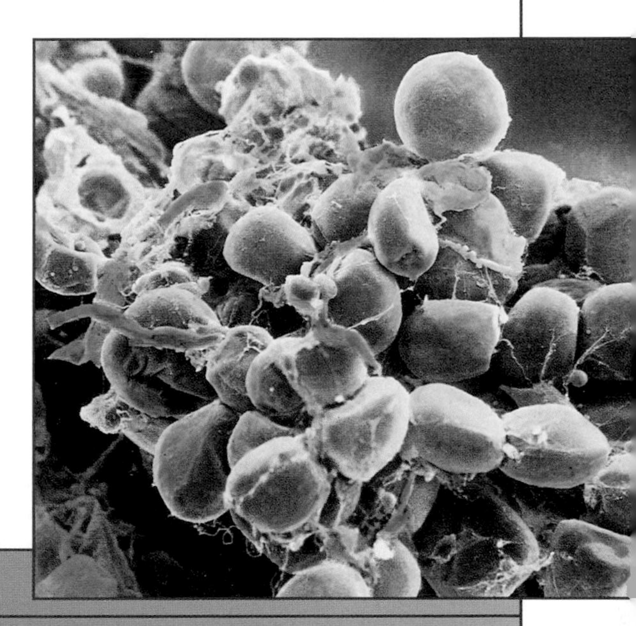

► THINK IT OVER . . .

► ***Why do you think body weight alone is not always a good measure of adiposity, or a good predictor of health risks associated with excess adipose tissue?***

mucous membrane secretes mucus, which prevents the cavities from drying out. It also traps particles in the respiratory passageways, lubricates and absorbs food as it moves through the gastrointestinal tract, and secretes digestive enzymes. The connective tissue layer helps bind the epithelium to the underlying structures. It also provides the epithelium with oxygen and nutrients and removes wastes via its blood vessels.

Serous Membranes

A ***serous membrane*** (*serous* = watery) lines a body cavity that does not open directly to the exterior, and it also covers the organs that lie within the cavity. Serous membranes consist of two parts: a parietal layer and a visceral layer. The ***parietal layer*** (pa-RĪ-e-tal; *pariet-* = wall) is the part attached to the cavity wall, and the ***visceral layer*** (*viscer-* = body organ) is the part that covers and attaches to the organs inside these cavities. Each layer consists of areolar connective tissue covered by *mesothelium.* Mesothelium is a simple squamous epithelium. It secretes ***serous fluid***, a watery lubricating fluid that allows organs to glide easily over one another or to slide against the walls of cavities.

The serous membrane lining the thoracic cavity and covering the lungs is the ***pleura***. The serous membrane lining the heart cavity and covering the heart is the ***pericardium***. The serous membrane lining the abdominal cavity and covering the abdominal organs is the ***peritoneum***.

Synovial Membranes

Synovial membranes (sin-Ō-vē-al) line the cavities of some joints. They are composed of areolar connective tissue and adipose tissue with collagen fibers; they do not have an epithelial layer. Synovial membranes contain cells (synoviocytes) which secrete ***synovial fluid***. This fluid lubricates the ends of bones as they move at joints, nourishes the cartilage covering the bones, and removes microbes and debris from the joint cavity.

■ **CHECKPOINT**

11. Define the following kinds of membranes: mucous, serous, cutaneous, and synovial.
12. Where is each type of membrane located in the body? What are their functions?

TISSUE REPAIR: RESTORING HOMEOSTASIS

OBJECTIVE • Describe the role of tissue repair in restoring homeostasis.

Tissue repair is the process that replaces worn-out, damaged, or dead cells. New cells originate by cell division from the ***stroma***, the supporting connective tissue, or from the ***parenchyma***, cells that constitute the functioning part of the tissue or organ. In adults, each of the four basic tissue types (epithelial, connective, muscular, and nervous) has a different capacity for replenishing parenchymal cells lost by damage, disease, or other processes.

Epithelial cells, which endure considerable wear and tear (and even injury) in some locations, have a continuous capacity for renewal. In some cases, immature, undifferentiated cells called ***stem cells*** divide to replace lost or damaged cells. For example, stem cells reside in protected locations in the epithelia of the skin and gastrointestinal tract to replenish cells sloughed from the apical layer.

Some connective tissues also have a continuous capacity for renewal. One example is bone, which has an ample blood supply. Other connective tissues such as cartilage can replenish cells less readily in part because of a poor blood supply.

Muscular tissue has a relatively poor capacity for renewal of lost cells. Cardiac muscle fibers can be produced from stem cells under special conditions (see page 367). Skeletal muscle tissue does not divide rapidly enough to replace extensively damaged muscle fibers. Smooth muscle fibers can proliferate to some extent, but they do so much more slowly than the cells of epithelial or connective tissues.

Nervous tissue has the poorest capacity for renewal. Although experiments have revealed the presence of some stem cells in the brain, they normally do not undergo mitosis to replace damaged neurons.

If parenchymal cells accomplish the repair, ***tissue regeneration*** is possible, and a near-perfect reconstruction of the injured tissue may occur. However, if fibroblasts of the stroma are active in the repair, the replacement tissue will be a new connective tissue. The fibroblasts synthesize collagen and other extracellular matrix materials that aggregate to form scar tissue, a process known as ***fibrosis***. Because scar tissue is not specialized to perform the functions of the parenchymal tissue, the original function of the tissue or organ is impaired.

> Scar tissue can form **adhesions**, abnormal joining of tissues. Adhesions commonly form in the abdomen around a site of previous inflammation such as an inflamed appendix, and they can develop after surgery. Although adhesions do not always cause problems, they can decrease tissue flexibility, cause obstruction (such as in the intestine), and make a subsequent operation more difficult.

■ **CHECKPOINT**

13. How are stromal and parenchymal repair of a tissue different?

AGING AND TISSUES

OBJECTIVE • Describe the effects of aging on tissues.

Generally, tissues heal faster and leave less obvious scars in the young than in the aged. In fact, surgery performed on fetuses leaves no scars. The younger body is generally in a better nutritional state, its tissues have a better blood supply, and its cells have a higher metabolic rate. Thus, cells can synthesize needed materials and divide more quickly. The extracellular components of tissues also change with age. Glucose, the most abundant sugar in the body, plays a role in the aging process. Glucose is haphazardly added to proteins inside and outside cells, forming irreversible cross-links between adjacent protein molecules. With advancing age, more cross-links form, which contributes to the stiffening and loss of elasticity that occur in aging tissues. Collagen fibers, responsible for the strength of tendons, increase in number and change in quality with aging. Elastin, another extracellular component, is responsible for the elasticity of blood vessels and skin. It thickens, fragments, and acquires a greater affinity for calcium with age—changes that may also be associated with the development of atherosclerosis, the deposition of fatty materials in arterial walls.

■ **CHECKPOINT**

14. What common changes occur in epithelial and connective tissues with aging?

• • •

Now that you have an understanding of tissues, we will look at the organization of tissues into organs and organs into systems. In the next chapter we will consider how the skin and other organs function as components of the integumentary system.

COMMON DISORDERS

Sjögren's Syndrome

Sjögren's syndrome (SHŌ-grenz) is a common autoimmune disease that causes inflammation and destruction of exocrine glands, especially the lacrimal (tear) glands and salivary glands. Signs of Sjögren's syndrome include dryness of the eyes, mouth, nose, ears, skin, and vagina, and salivary gland enlargement. Systemic effects include fatigue, arthritis, difficulty in swallowing, pancreatitis (inflammation of the pancreas), pleuritis (inflammation of the pleurae of the lungs), and muscle and joint pain. The disorder affects females more than males by a ratio of 9 to 1. About 20% of older adults experience some signs of Sjögren's syndroms. Treatment is supportive, including using artifical tears to moisten the eyes, sipping fluids, chewing sugarless gum, using a saliva substitute to moisten the mouth, and using moisturing creams for the skin.

Systemic Lupus Erythematosus

Systemic lupus erythematosus (er-i-thē-ma-TŌ-sus), ***SLE***, or simply *lupus*, is a chronic inflammatory disease of connective tissue occurring mostly in nonwhite women during their childbearing years. It is an autoimmune disease that can cause tissue damage in every body system. The disease, which can range from a mild condition in most patients to a rapidly fatal disease, is marked by periods of exacerbation and remission. Although the cause of SLE is unknown, genetic, environmental, and hormonal factors all have been implicated. The genetic component is suggested by studies of twins and family history. Environmental factors include viruses, bacteria, chemicals, drugs, exposure to excessive sunlight, and emotional stress. Sex hormones, such as estrogens, may also trigger SLE.

Signs and symptoms of SLE include painful joints, low-grade fever, fatigue, mouth ulcers, weight loss, enlarged lymph nodes and spleen, sensitivity to sunlight, rapid loss of large amounts of scalp hair, and anorexia. A distinguishing feature of lupus is an eruption across the bridge of the nose and cheeks called a "butterfly rash." Other skin lesions may occur, including blistering and ulceration. The erosive nature of some SLE skin lesions was thought to resemble the damage inflicted by the bite of a wolf—thus, the name *lupus* (= wolf). The most serious complications of the disease involve inflammation of the kidneys, liver, spleen, lungs, heart, brain, and gastrointestinal tract. Because there is no cure for SLE, treatment is supportive, including anti-inflammatory drugs, such as aspirin, and immunosuppressive drugs.

MEDICAL TERMINOLOGY AND CONDITIONS

Tissue rejection An immune response of the body directed at foreign proteins in a transplanted tissue or organ; immunosuppressive drugs, such as cyclosporine, have largely overcome tissue rejection in heart-, kidney-, and liver-transplant patients.

Tissue transplantation The replacement of a diseased or injured tissue or organ; the most successful transplants involve use of a person's own tissues or those from an identical twin.

Xenotransplantation (zen′-ō-trans′-plan-TĀ-shun; *xeno-* = strange, foreign) The replacement of a diseased or injured tissue or organ with cells or tissues from an animal. Only a few cases of successful xenotransplantation exist to date.

STUDY OUTLINE

Types of Tissues (p. 73)

1. A tissue is a group of similar cells that usually has a similar embryological origin and is specialized for a particular function.
2. The various tissues of the body are classified into four basic types: epithelial, connective, muscular, and nervous.

Epithelial Tissue (p. 73)

1. The general types of epithelia include covering and lining epithelium and glandular epithelium.
2. Some general characteristics of epithelium: It consists mostly of cells with little extracellular material, is arranged in sheets, is attached to connective tissue by a basement membrane, is avascular (no blood vessels), has a nerve supply, and can replace itself.
3. Epithelial layers can be simple (one layer) or stratified (several layers). The cell shapes may be squamous (flat), cuboidal (cubelike), columnar (rectangular), or transitional (variable).
4. Simple squamous epithelium consists of a single layer of flat cells (Table 4.1A). It is found in parts of the body where filtration or diffusion are priority processes. One type, endothelium, lines the heart and blood vessels. Another type, mesothelium, forms the serous membranes that line the thoracic and abdominal cavities and cover the organs within them.

5. Simple cuboidal epithelium consists of a single layer of cube-shaped cells that function in secretion and absorption (Table 4.1B). It is found covering the ovaries, in the kidneys and eyes, and lining some glandular ducts.
6. Nonciliated simple columnar epithelium consists of a single layer of nonciliated rectangular cells (Table 4.1C). It lines most of the gastrointestinal tract. Specialized cells containing microvilli perform absorption. Goblet cells secrete mucus.
7. Ciliated simple columnar epithelium consists of a single layer of ciliated rectangular cells (Table 4.1D). It is found in a few portions of the upper respiratory tract, where it moves foreign particles trapped in mucus out of the respiratory tract.
8. Pseudostratified columnar epithelium has only one layer but gives the appearance of many (Table 4.1E).
9. Stratified squamous epithelium consists of several layers of cells; cells in the apical layer and several layers deep to it are flat (Table 4.1F). It is protective. A nonkeratinized variety lines the mouth; a keratinized variety forms the epidermis, the most superficial layer of the skin.
10. Stratified cuboidal epithelium consists of several layers of cells; cells in the apical layer are cube-shaped (Table 4.1G). It is found in adult sweat glands and a portion of the male urethra.
11. Stratified columnar epithelium consists of several layers of cells; cells in the apical layer are column-shaped (Table 4.1H). It is found in a portion of the male urethra and large excretory ducts of some glands.
12. Transitional epithelium consists of several layers of cells whose appearance varies with the degree of stretching (Table 4.1I). It lines the urinary bladder.
13. A gland is a single cell or a group of epithelial cells adapted for secretion.
14. Endocrine glands secrete hormones into interstitial fluid and then the blood (Table 4.1J).
15. Exocrine glands (mucous, sweat, oil, and digestive glands) secrete into ducts or directly onto a free surface (Table 4.1K).

Connective Tissue (p. 82)

1. Connective tissue is one of the most abundant body tissues.
2. Connective tissue consists of cells and an extracellular matrix of ground substance and fibers; it has abundant matrix with relatively few cells. It does not usually occur on free surfaces, has a nerve supply (except for cartilage), and is highly vascular (except for cartilage, tendons, and ligaments).
3. Cells in connective tissue include fibroblasts (secrete matrix), macrophages (perform phagocytosis), mast cells (produce histamine), and adipocytes (store fat).
4. The ground substance and fibers make up the extracellular matrix.
5. The ground substance supports and binds cells together, provides a medium for the exchange of materials, and is active in influencing cell functions.
6. The fibers in the extracellular matrix provide strength and support and are of three types: (a) collagen fibers (composed of collagen) are found in large amounts in bone, tendons, and ligaments; (b) elastic fibers (composed of elastin, fibrillin, and other glycoproteins) are found in skin, blood vessel walls, and lungs; and (c) reticular fibers (composed of collagen and glycoprotein) are found around fat cells, nerve fibers, and skeletal and smooth muscle cells.
7. Connective tissue is subdivided into loose connective tissue, dense connective tissue, cartilage, bone tissue, and liquid connective tissue (blood tissue and lymph).
8. Loose connective tissue includes areolar connective tissue, adipose tissue, and reticular connective tissue.
9. Areolar connective tissue consists of the three types of fibers, several cells, and a semifluid ground substance (Table 4.2A). It is found in the subcutaneous layer; in mucous membranes; and around blood vessels, nerves, and body organs.
10. Adipose tissue consists of adipocytes, which store triglycerides (Table 4.2B). It is found in the subcutaneous layer, around organs, and in the yellow bone marrow.
11. Reticular connective tissue consists of reticular fibers and reticular cells and is found in the liver, spleen, and lymph nodes (Table 4.2C).
12. Dense connective tissue includes dense regular connective tissue, dense irregular connective tissue, and elastic connective tissue.
13. Dense regular connective tissue consists of parallel bundles of collagen fibers and fibroblasts (Table 4.2D). It forms tendons, most ligaments, and aponeuroses.
14. Dense irregular connective tissue consists of usually randomly arranged collagen fibers and a few fibroblasts (Table 4.2E). It is found in fasciae, the dermis of skin, and membrane capsules around organs.
15. Elastic connective tissue consists of branching elastic fibers and fibroblasts (Table 4.2F). It is found in the walls of large arteries, lungs, trachea, and bronchial tubes.
16. Cartilage contains chondrocytes and has a rubbery matrix (chondroitin sulfate) containing collagen and elastic fibers.
17. Hyaline cartilage is found in the embryonic skeleton, at the ends of bones, in the nose, and in respiratory structures (Table 4.2G). It is flexible, allows movement, and provides support.
18. Fibrocartilage is found in the pubic symphysis, intervertebral discs, and menisci (cartilage pads) of the knee joint (Table 4.2H).
19. Elastic cartilage maintains the shape of organs such as the epiglottis of the larynx, auditory (eustachian) tubes, and external ear (Table 4.2I).
20. Bone or osseous tissue supports, protects, helps provide movement, stores minerals, and houses blood-forming tissue.
21. Blood tissue is liquid connective tissue that consists of blood plasma and formed elements—red blood cells, white blood cells, and platelets. Its cells transport oxygen and carbon dioxide, carry on phagocytosis, participate in allergic reactions, provide immunity, and bring about blood clotting. Lymph, the extracellular fluid that flows in lymphatic vessels, is also a liquid connective tissue. It is a clear fluid similar to blood plasma but with less protein.

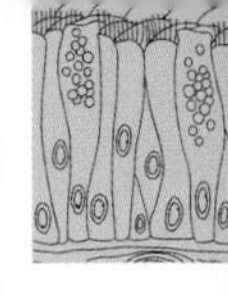

Muscular Tissue (p. 90)

1. Muscular tissue consists of cells (called muscle fibers) that are specialized for contraction. It provides motion, maintenance of posture, heat production, and protection.
2. Skeletal muscle tissue is attached to bones, cardiac muscle tissue forms most of the heart wall, and smooth muscle tissue is found in the walls of hollow internal structures (blood vessels and viscera).

Nervous Tissue (p. 90)

1. The nervous system is composed of neurons (nerve cells) and neuroglia (protective and supporting cells).
2. Neurons are sensitive to stimuli, convert stimuli into nerve impulses, and conduct nerve impulses.

Membranes (p. 90)

1. An epithelial membrane consists of an epithelial layer overlying a connective tissue layer. Examples are mucous, serous, and cutaneous membranes.
2. Mucous membranes line cavities that open to the exterior, such as the gastrointestinal tract.
3. Serous membranes line closed cavities (pleura, pericardium, peritoneum) and cover the organs in the cavities. These membranes consist of parietal and visceral layers.
4. Synovial membranes line joint cavities, bursae, and tendon sheaths and consist of areolar connective tissue instead of epithelium.

Tissue Repair: Restoring Homeostasis (p. 92)

1. Tissue repair is the replacement of worn-out, damaged, or dead cells by healthy ones.
2. Stem cells may divide to replace lost or damaged cells.

Aging and Tissues (p. 92)

1. Tissues heal faster and leave less obvious scars in the young than in the aged; surgery performed on fetuses leaves no scars.
2. The extracellular components of tissues, such as collagen and elastic fibers, also change with age.

SELF-QUIZ

1. Epithelial tissue functions in
   a. conducting nerve impulses **b.** storing fat
   c. covering and lining the body and its parts
   d. movement **e.** storing minerals
2. Epithelial tissue is classified according to
   a. its location
   b. its function
   c. the composition of the extracellular matrix
   d. the shape and arrangement of its cells
   e. whether it is under voluntary or involuntary control
3. Mucous membranes are
   a. composed of three layers
   b. found in body cavities that open to the body's exterior
   c. located at the ends of bones
   d. found lining the thoracic cavity
   e. capable of producing synovial fluid
4. Which of the following is NOT a type of connective tissue?
   a. blood **b.** adipose **c.** reticular
   d. simple cuboidal **e.** cartilage
5. Which of the following is true concerning connective tissue?
   a. Except for cartilage, connective tissue has a rich blood supply.
   b. Connective tissue is classified according to cell shape and arrangement.
   c. The cells of connective tissue are generally closely joined.
   d. Loose connective tissue consists of many fibers arranged in a regular pattern.
   e. The fibers in connective tissue are composed of lipids.
6. Match the following tissue types with their descriptions.

____ **a.** fat storage	**A.** simple cuboidal epithelium
____ **b.** waterproofs the skin	**B.** simple squamous epithelium
____ **c.** forms the stroma (framework) of many organs	**C.** adipose
____ **d.** composes the intervertebral discs	**D.** fibrocartilage
____ **e.** stores red bone marrow, protects, supports	**E.** reticular connective
____ **f.** found in the walls of hollow organs	**F.** smooth muscle
____ **g.** found in lungs, involved in diffusion	**G.** keratinized stratified squamous epithelium
____ **h.** found in kidney tubules, involved in absorption	**H.** bone

7. If you were going to design a hollow organ that needed to expand and have stretchability, which of the following epithelial and connective tissues might you use?
   a. transitional epithelium and elastic connective tissue
   b. stratified columnar epithelium and adipose tissue
   c. simple columnar epithelium and dense regular connective tissue
   d. simple squamous epithelium and hyaline cartilage
   e. transitional epithelium and reticular connective tissue

8. Which of the following statements is NOT true concerning epithelial tissue?
   a. The cells of epithelial tissue are closely packed.
   b. The basal layer of cells rests on a basement membrane.
   c. Epithelial tissue has a nerve supply.
   d. Epithelial tissue undergoes rapid rates of cell division.
   e. Epithelial tissue is well supplied with blood vessels.

9. Where would you find smooth muscle tissue?
   a. the heart b. attached to bones c. in joints d. in the discs between the vertebrae e. in the walls of hollow organs

10. A connective tissue with a liquid extracellular matrix is
    a. elastic cartilage b. blood c. areolar d. reticular e. osseous

11. The interior of your nose is lined with
    a. a mucous membrane
    b. smooth muscle tissue
    c. a synovial membrane
    d. keratinized stratified squamous epithelium
    e. a serous membrane

12. The four main types of tissue are
    a. epithelial, embryonic, blood, nervous
    b. blood, connective, muscular, nervous
    c. connective, epithelial, muscular, nervous
    d. stratified, muscular, striated, nervous
    e. epithelial, connective, muscular, membranous

13. Which of the following materials would NOT be found in the extracellular matrix of connective tissue?
    a. collagen fibers b. elastic fibers c. keratin d. reticular fibers e. hyaluronic acid

14. Which connective tissue cells secrete antibodies?
    a. mast cells b. adipocytes c. macrophages d. plasma cells e. chondrocytes

15. Modified columnar epithelial cells that secrete mucus are ________ cells.
    a. ciliated b. keratinized c. mast d. fibroblast e. goblet

16. Which tissue forms the bulk of the heart wall?
    a. skeletal muscle b. nervous c. bone d. cardiac muscle e. smooth muscle

17. Stratified squamous epithelium functions in
    a. protection and secretion b. contraction c. absorption d. stretching e. transport

18. What tissue type is found in tendons?
    a. dense irregular connective tissue b. elastic connective tissue c. dense regular connective tissue d. pseudostratified epithelium e. areolar tissue

19. In what tissue type would you find stores of calcium and phosphorus?
    a. bone b. hyaline cartilage c. fibrocartilage d. dense irregular connective tissue e. elastic cartilage

20. Which of the following statements is true concerning glandular tissue?
    a. Endocrine glands are composed of connective tissue; exocrine glands are composed of modified epithelium.
    b. Endocrine gland secretions diffuse into the bloodstream; exocrine gland secretions enter ducts.
    c. A sweat gland is an example of an endocrine gland.
    d. Endocrine glands contain ducts; exocrine glands do not.
    e. Exocrine glands produce substances known as hormones.

CRITICAL THINKING APPLICATIONS

1. Your young nephew can't wait to get his eyebrow pierced like his big brother. In the meantime, he's walking around with sewing needles stuck through his fingertips. There is no visible bleeding. What type of tissue has he pierced? (Be specific.) How do you know?

2. Collagen is the new "miracle" cosmetic. It's advertised to give you shiny hair and glowing skin, and can be injected to reduce wrinkles. What is collagen? If you wanted to launch your own line of cosmetics, what tissue or structure would supply you with abundant collagen?

3. Your lab partner Samir put a tissue slide labeled uterine tube under the microscope. He focused the slide and exclaimed, "Look! It's all hairy." Explain to Samir what the "hair" really is.

4. You've gone out to eat at your favorite fast-food joint, Goodbody's Fried Chicken Emporium. A health-food zealot grabs your chicken leg and declares, "This is all fat!" Using your knowledge of tissues, defend your dinner choice.

ANSWERS TO FIGURE QUESTIONS

4.1 Substances would move most rapidly through squamous cells because they are so thin.

4.2 Fibroblasts secrete the fibers and ground substance of the extracellular matrix.

THE INTEGUMENTARY SYSTEM

did you know? ***Protecting skin from the sun's damaging ultraviolet (UV) rays helps prevent premature aging and cancers of the skin. Avoiding direct sunlight is the best strategy, but when not practical, a sunscreen should be used on exposed skin. These do not shield the skin completely, but they do reduce the damaging effects of the ultraviolet rays. Evidence suggests that the skin can repair some damage when sunscreens are applied consistently. But researchers warn that sunscreens can provide a false sense of security. Because they prevent burning, sunscreens may lull us into thinking the sun is not hurting us, while damage is occurring.***

Focus on Wellness, page 105

www.wiley.com/college/apcentral

Of all the body's organs, none is more easily inspected or more exposed to infection, disease, and injury than the skin. Because of its visibility, skin reflects our emotions and some aspects of normal physiology, as evidenced by frowning, blushing, and sweating. Changes in skin color or condition may indicate homeostatic imbalances in the body. For example, a skin rash such as occurs in chickenpox reveals a systemic infection, but a yellowing of the skin is an indication of jaundice, usually due to disease of the liver, an internal organ. Other disorders may be limited to the skin, such as warts, age spots, or pimples. The skin's location makes it vulnerable to damage from trauma, sunlight, microbes, or pollutants in the environment. Major damage to the skin, as occurs in third-degree burns, can be life threatening due to the loss of the protective skin functions.

Many interrelated factors may affect both the appearance and health of the skin, including nutrition, hygiene, circulation, age, immunity, genetic traits, psychological state, and drugs. So important is the skin to body image that people spend much time and money to restore it to a more youthful appearance. ***Dermatology*** (der´-ma-TOL-ō-jē; *dermato-* = skin; *-logy* = study of) is the branch of medicine that specializes in diagnosing and treating skin disorders.

looking back to move ahead . . .

- **Types of Tissues (page 73)**
- **General Features of Epithelial Tissue (page 73)**
- **Stratified Squamous Epithelium (page 75)**
- **General Features of Connective Tissue (page 82)**
- **Areolar Connective Tissue (page 84)**
- **Dense Irregular Connective Tissue (page 87)**

SKIN

OBJECTIVES • **Describe the structure and functions of the skin.**

• **Explain the basis for different skin colors.**

Recall from Chapter 1 that a system consists of a group of organs working together to perform specific activities. The ***integumentary system*** (in-teg-ū-MEN-tar-ē; *inte-* = whole; *-gument* = body covering) is composed of organs such as the skin and hairs, and other structures such as nails. The ***skin*** or ***cutaneous membrane*** covers the external surface of the body. It is the largest organ of the body in surface area and weight. In adults, the skin covers an area of about 2 square meters (22 square feet) and weighs 4.5–5kg (10–11 lb), about 16% of total body weight.

Structure of Skin

Structurally, the skin consists of two main parts (Figure 5.1). The superficial, thinner portion, which is composed of *ep-*

Figure 5.1 Components of the integumentary system. The skin consists of a thin, superficial epidermis and a deep, thicker dermis. Deep to the skin is the subcutaneous layer, which attaches the dermis to underlying organs and tissues.

The integumentary system includes the skin and its accessory structures—hair, nails, and glands—along with associated muscles and nerves.

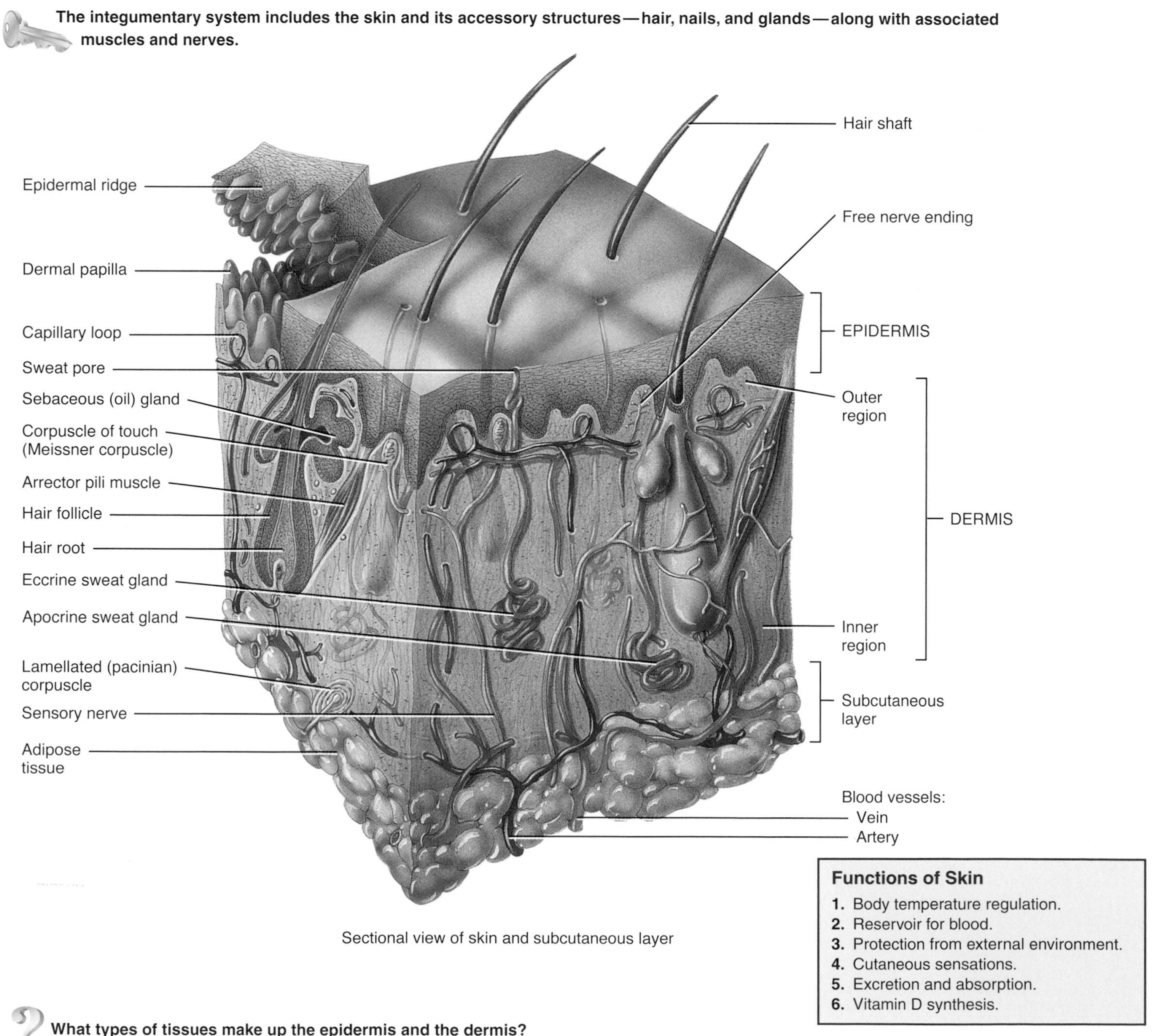

Sectional view of skin and subcutaneous layer

Functions of Skin

1. Body temperature regulation.
2. Reservoir for blood.
3. Protection from external environment.
4. Cutaneous sensations.
5. Excretion and absorption.
6. Vitamin D synthesis.

What types of tissues make up the epidermis and the dermis?

ithelial tissue, is the ***epidermis*** (ep′-i-DERM-is; *epi-* = above). The deeper, thicker *connective tissue* portion is the ***dermis.***

Deep to the dermis, but not part of the skin, is the ***subcutaneous (subQ) layer.*** Also called the ***hypodermis*** (*hypo-* = below), this layer consists of areolar and adipose tissues. Fibers that extend from the dermis anchor the skin to the subcutaneous layer, which, in turn, attaches to underlying tissues and organs. The subcutaneous layer serves as a storage depot for fat and contains large blood vessels that supply the skin. This region (and sometimes the dermis) also contains nerve endings called ***lamellated (pacinian) corpuscles*** (pa-SIN-ē-an) that are sensitive to pressure (Figure 5.1).

Epidermis

The ***epidermis*** is composed of keratinized stratified squamous epithelium. It contains four principal types of cells: keratinocytes, melanocytes, Langerhans cells, and Merkel cells (see Figure 5.2). About 90% of epidermal cells are ***keratinocytes*** (ker-a-TIN-ō-sīts; *keratino-* = hornlike; *-cytes* = cells), which are arranged in four or five layers and produce the protein ***keratin.*** Recall from Chapter 4 that keratin is a tough, fibrous protein that helps protect the skin and underlying tissues from heat, microbes, and chemicals. Keratinocytes also produce lamellar granules, which release a water-repellent sealant.

About 8% of the epidermal cells are ***melanocytes*** (MEL-a-nō-sīts; *melano-* = black), which produce the pigment melanin. Their long, slender projections extend between the keratinocytes and transfer melanin granules to them. ***Melanin*** is a yellow-red or brown-black pigment that contributes to skin color and absorbs damaging ultraviolet (UV) light. Although keratinocytes gain some protection from melanin granules, melanocytes themselves are particularly susceptible to damage by UV light.

Langerhans cells (LANG-er-hans) participate in immune responses mounted against microbes that invade the skin. Langerhans cells, macrophages, and B cells help other cells of the immune system recognize an antigen (foreign microbe or substance) so that it can be destroyed (Chapter 17). Langerhans cells are easily damaged by UV light.

Figure 5.2 Layers of the epidermis.

The epidermis consists of keratinized stratified squamous epithelium.

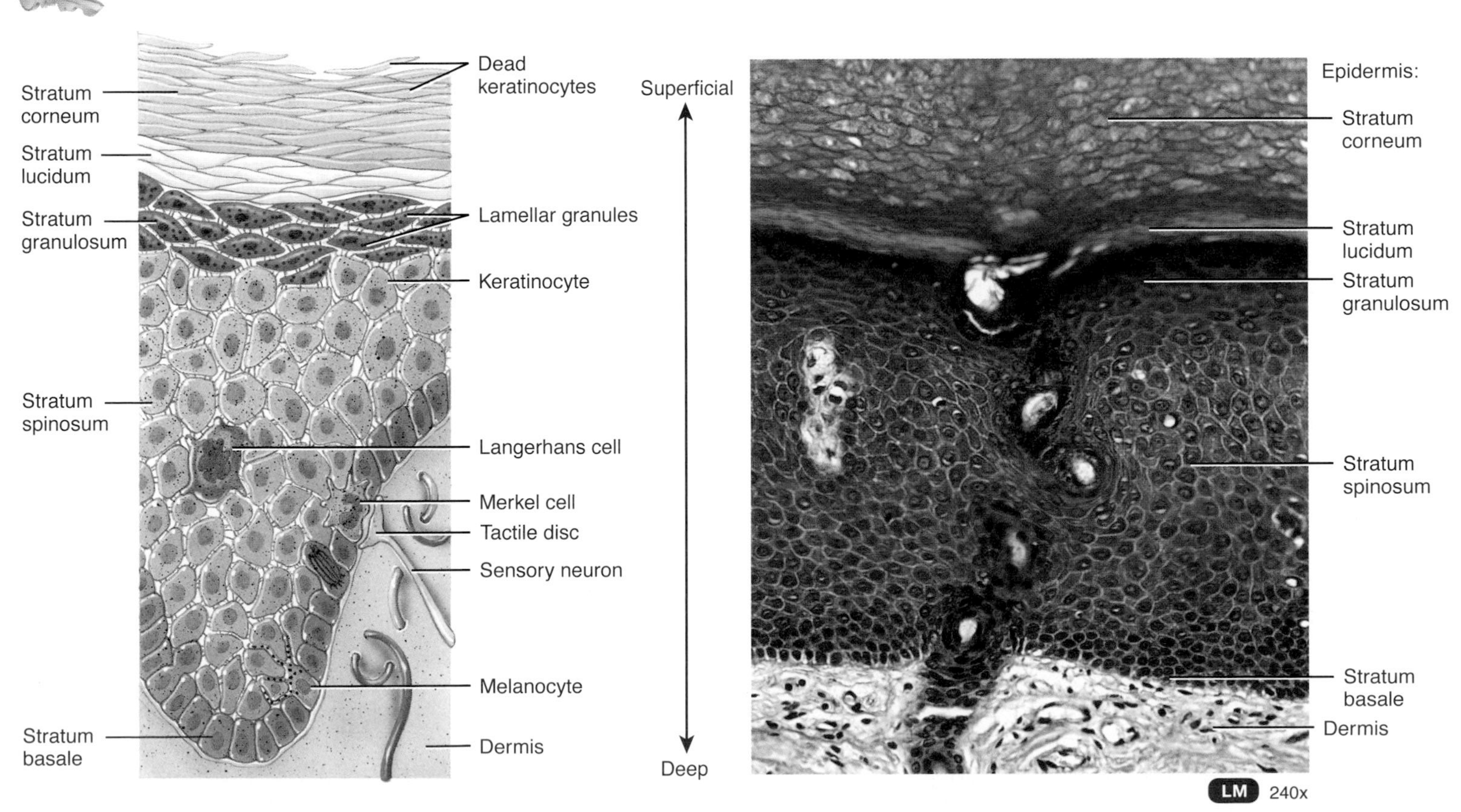

(a) Four principal cell types in epidermis

(b) Photomicrograph of a portion of the skin

Which epidermal layer includes stem cells that continually undergo cell division?

Merkel cells contact the flattened process of a sensory neuron (nerve cell), a structure called a ***tactile (Merkel) disc.*** Merkel cells and tactile discs detect different aspects of touch sensations.

Several distinct layers of keratinocytes in various stages of development form the epidermis (Figure 5.2). In most regions of the body the epidermis has four strata or layers—stratum basale, stratum spinosum, stratum granulosum, and a thin stratum corneum. This is called *thin skin.* Where exposure to friction is greatest, such as in the fingertips, palms, and soles, the epidermis has five layers—stratum basale, stratum spinosum, stratum granulosum, stratum lucidum, and a thick stratum corneum. This is referred to as *thick skin.*

The deepest layer of the epidermis is the ***stratum basale*** (ba-SA-lē; *basal-* = base), composed of a single row of cuboidal or columnar keratinocytes. Some cells in this layer are *stem cells* that undergo cell division to continually produce new keratinocytes.

New skin cannot regenerate if an injury destroys the stratum basale and its stem cells. Skin wounds of this magnitude require skin grafts in order to heal. A **skin graft** is the transfer of a patch of healthy skin taken from a donor site to cover a wound. To avoid tissue rejection, the transplanted skin is usually taken from the same individual (*autograft*) or an identical twin (*isograft*). If skin damage is so extensive that an autograft would cause harm, a self-donation procedure called *autologous skin transplantation* (aw-TOL-ō-gus) may be used. In this procedure, performed most often for severely burned patients, small amounts of an individual's epidermis are removed, and the keratinocytes are cultured in the laboratory to produce thin sheets of skin. The new skin is transplanted back to the patient so that it covers the burn wound and generates a permanent skin. Also available as skin grafts are products grown in the laboratory from the foreskins of circumcised infants, and synthetic materials, which are products of tissue engineering.

Superficial to the stratum basale is the ***stratum spinosum*** (spi-NŌ-sum; *spinos-* = thornlike), where 8 to 10 layers of many-sided keratinocytes fit closely together. Cells in the more superficial portions of this layer become somewhat flattened.

At about the middle of the epidermis, the ***stratum granulosum*** (gran-ū-LŌ-sum; *granulos-* = little grains) consists of three to five layers of flattened keratinocytes that are undergoing apoptosis, genetically programmed cell death in which the nucleus fragments before the cells die. The nuclei and other organelles of these cells begin to degenerate. A distinctive feature of cells in this layer is the presence of keratin. Also present in the keratinocytes are membrane-enclosed ***lamellar granules***, which release a lipid-rich secretion that acts as a water-repellent sealant, retarding loss of body fluids and entry of foreign materials.

The ***stratum lucidum*** (LOO-si-dum; *lucid-* = clear) is present only in the thick skin of the fingertips, palms, and soles. It consists of three to five layers of flattened clear, dead keratinocytes that contain large amounts of keratin.

The ***stratum corneum*** (COR-nē-um; *corne-* = horn or horny) consists of 25 to 30 layers of flattened dead keratinocytes. These cells are continuously shed and replaced by cells from the deeper strata. The interior of the cells contains mostly keratin. Its multiple layers of dead cells help to protect deeper layers from injury and microbial invasion. Constant exposure of skin to friction stimulates the formation of a *callus*, an abnormal thickening of the stratum corneum.

Newly formed cells in the stratum basale are slowly pushed to the surface. As the cells move from one epidermal layer to the next, they accumulate more and more keratin, a process called ***keratinization*** (ker′-a-tin-i-ZĀ-shun). Eventually the keratinized cells slough off and are replaced by underlying cells that, in turn, become keratinized. The whole process by which cells form in the stratum basale, rise to the surface, become keratinized, and slough off, takes about four weeks in an average epidermis of 0.1 mm (0.004 in.) thickness. An excessive amount of keratinized cells shed from the skin of the scalp is called *dandruff.*

Dermis

The second, deeper part of the skin, the ***dermis***, is composed mainly of connective tissue containing collagen and elastic fibers. The superficial part of the dermis makes up about one-fifth of the thickness of the total layer (see Figure 5.1). It consists of areolar connective tissue containing fine elastic fibers. Its surface area is greatly increased by small, fingerlike projections called ***dermal papillae*** (pa-PIL-ē = nipples). These nipple-shaped structures indent the epidermis. Some contain *capillary loops* (blood capillaries). Other dermal papillae also contain tactile receptors called *corpuscles of touch* or *Meissner corpuscles*, nerve endings that are sensitive to touch. Also present in the dermal papillae are *free nerve endings* that are associated with sensations of warmth, coolness, pain, tickling, and itching.

The deeper part of the dermis, which is attached to the subcutaneous layer, consists of dense irregular connective tissue containing bundles of collagen and some coarse elastic fibers. Adipose cells, hair follicles, nerves, oil glands, and sweat glands are found between the fibers.

The combination of collagen and elastic fibers in the deeper part of the dermis provides the skin with strength, *extensibility* (ability to stretch), and *elasticity* (ability to return to original shape after stretching). The extensibility of skin can readily be seen in pregnancy and obesity. Extreme stretching, however, may produce small tears in the dermis, causing *striae* (STRĪ-ē = streaks), or stretch marks, that are visible as red or silvery white streaks on the skin surface.

Skin Color

Melanin, hemoglobin, and carotene are three pigments that impart a wide variety of colors to skin. The amount of ***melanin*** causes the skin's color to vary from pale yellow to red to tan to black. Melanocytes are most plentiful in the epidermis of the penis, nipples of the breasts, the area just around the nipples (areolae), face, and limbs. They are also present in mucous membranes. Because the *number* of melanocytes is about the same in all people, differences in skin color are due mainly to the *amount of pigment* the melanocytes produce and transfer to keratinocytes. In some people, melanin accumulates in patches called ***freckles***. As a person grows older, ***age (liver) spots*** may develop. These flat blemishes look like freckles and range in color from light brown to black. Like freckles, age spots are accumulations of melanin. A round, flat, or raised area that represents a benign localized overgrowth of melanocytes and usually develops in childhood or adolescence is called a ***nevus*** (NĒ-vus), or a ***mole***.

Exposure to UV light stimulates melanin production. Both the amount and darkness of melanin increase, which gives the skin a tanned appearance and further protects the body against UV radiation. Thus, within limits, melanin serves a protective function. Nevertheless, repeatedly exposing the skin to UV light causes skin cancer. A tan is lost when the melanin-containing keratinocytes are shed from the stratum corneum. ***Albinism*** (AL-bin-izm; *albin-* = white) is the inherited inability of an individual to produce melanin. Most ***albinos*** (al-BĪ-nōs), people affected by albinism, do not have melanin in their hair, eyes, and skin.

Dark-skinned individuals have large amounts of melanin in the epidermis. Consequently, the epidermis has a dark pigmentation and skin color ranges from yellow to red to tan to black. Light-skinned individuals have little melanin in the epidermis. Thus, the epidermis appears translucent and skin color ranges from pink to red depending on the amount and oxygen content of the blood moving through capillaries in the dermis. The red color is due to ***hemoglobin***, the oxygen-carrying pigment in red blood cells.

Carotene (KAR-ō-tēn; *carot* = carrot) is a yellow-orange pigment that gives egg yolk and carrots their color. This precursor of vitamin A, which is used to synthesize pigments needed for vision, accumulates in the stratum corneum and fatty areas of the dermis and subcutaneous layer in response to excessive dietary intake. In fact, so much carotene may be deposited in the skin after eating large amounts of carotene-rich foods that the skin color actually turns orange, which is especially apparent in light-skinned individuals.

The color of skin and mucous membranes can provide clues for diagnosing certain conditions. When blood is not picking up an adequate amount of oxygen in the lungs, as in someone who has stopped breathing, the mucous membranes, nail beds, and skin appear bluish or **cyanotic** (sī-a-NOT-ik; *cyan-* = blue). **Jaundice** (JON-dis; *jaund-* = yellow) is due to a buildup of the yellow pigment bilirubin in the skin. This condition gives a yellowish appearance to the skin and the whites of the eyes, and usually indicates liver disease. **Erythema** (er-i-THĒ-ma; *eryth-* = red), redness of the skin, is caused by engorgement of capillaries in the dermis with blood due to skin injury, exposure to heat, infection, inflammation, or allergic reactions. **Pallor** (PAL-or), or paleness of the skin, may occur in conditions such as shock and anemia. All skin color changes are observed most readily in people with lighter-colored skin and may be more difficult to discern in people with darker skin. However, examination of the nail beds and gums can provide some information about circulation in individuals with darker skin.

■ **CHECKPOINT**

1. What structures are included in the integumentary system?
2. What are the main differences between the epidermis and dermis of the skin?
3. What are the three pigments found in the skin, and how do they contribute to skin color?

ACCESSORY STRUCTURES OF THE SKIN

OBJECTIVE • Describe the structure and functions of hair, skin glands, and nails.

Accessory structures of the skin that develop from the epidermis of an embryo—hair, glands, and nails—perform vital functions. Hair and nails protect the body. Sweat glands help regulate body temperature.

Hair

Hairs, or ***pili*** (PI-lē), are present on most skin surfaces except the palms, palmar surfaces of the fingers, soles, and plantar surfaces of the toes. In adults, hair usually is most heavily distributed across the scalp, over the brows of the eyes, and around the external genitalia. Genetic and hormonal influences largely determine the thickness and pattern of distribution of hair. Hair on the head guards the scalp from injury and the sun's rays; eyebrows and eyelashes protect the eyes from foreign particles; and hair in the nostrils protects against inhaling insects and foreign particles.

Each hair is a thread of fused, dead, keratinized cells that consists of a shaft and a root (Figure 5.3). The ***shaft*** is the superficial portion, most of which projects above the surface of the skin. The ***root*** is the portion below the surface that penetrates into the dermis and sometimes into the subcutaneous layer. Surrounding the root is the ***hair follicle***, which is composed of two layers of epidermal cells: *external* and *internal root sheaths* surrounded by a *connective tissue sheath*. Surrounding each hair follicle are nerve endings, called *hair root plexuses*, that are sensitive to touch. If a hair shaft is moved, its hair root plexus responds.

The base of each follicle is enlarged into an onion-shaped structure, the *bulb*. In the bulb is a nipple-shaped indentation, the *papilla of the hair*, that contains many blood vessels and provides nourishment for the growing hair. The bulb also contains a region of cells called the *matrix*, which produces new hairs by cell division when older hairs are shed.

Chemotherapy is the treatment of disease, usually cancer, by means of chemical substances or drugs. Chemotherapeutic agents interrupt the life cycle of rapidly dividing cancer cells. Unfortunately, the drugs also affect other rapidly dividing cells in the body, such as the matrix cells of a hair. It is for this reason that individuals undergoing chemotherapy experience hair loss. Since about 15% of the matrix cells of scalp hairs are in the resting stage, these cells are not affected by chemotherapy. Once chemotherapy is stopped, the matrix cells replace lost hair follicles and hair growth resumes.

Figure 5.3 Hair.

Hairs are growths of epidermis composed of dead, keratinized cells.

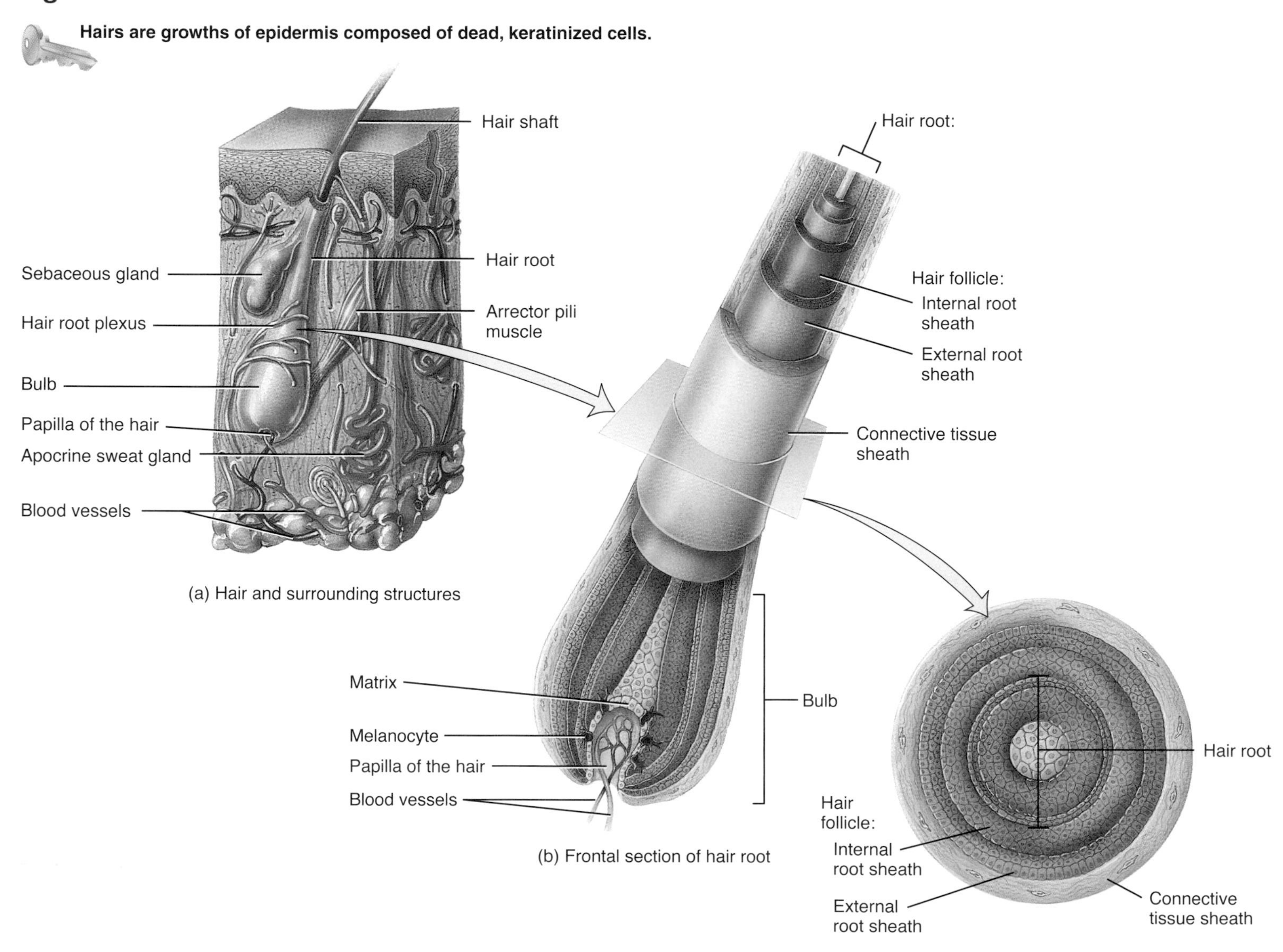

Which part of a hair produces a new hair by cell division?

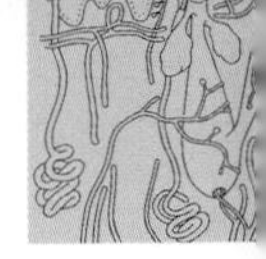

Sebaceous (oil) glands (discussed shortly) and a bundle of smooth muscle cells are also associated with hairs. The smooth muscle is called ***arrector pili*** (a-REK-tor PI-lē; *arrect* = to raise). It extends from the upper dermis to the side of the hair follicle. In its normal position, hair emerges at an angle to the surface of the skin. Under stress, such as cold or fright, nerve endings stimulate the arrector pili muscles to contract, which pulls the hair shafts perpendicular to the skin surface. This action causes "goose bumps" because the skin around the shaft forms slight elevations.

The color of hair is due to melanin. It is synthesized by melanocytes in the matrix of the bulb and passes into cells of the root and shaft. Dark-colored hair contains mostly true melanin. Blond and red hair contain variants of melanin in which there is iron and more sulfur. Gray hair occurs with a decline in the synthesis of melanin. White hair results from accumulation of air bubbles in the hair shaft.

At puberty, when the testes begin secreting significant quantities of androgens (masculinizing sex hormones), males develop the typical male pattern of hair growth, including a beard and a hairy chest. In females at puberty, the ovaries and the adrenal glands produce small quantities of androgens, which promote hair growth in the axillae and pubic region. Occasionally, a tumor of the adrenal glands, testes, or ovaries produces an excessive amount of androgens. The result in females or prepubertal males is ***hirsutism*** (HER-soo-tizm; *hirsut-* = shaggy), a condition of excessive body hair.

Surprisingly, androgens also must be present for occurrence of the most common form of baldness, ***androgenic alopecia*** or ***male-pattern baldness***. In genetically predisposed adults, androgens inhibit hair growth. On men, hair loss is most obvious at the temples and crown. Women are more likely to have thinning of hair on top of the head. The first drug approved for enhancing scalp hair growth was minoxidil (Rogaine®). It causes vasodilation (widening of blood vessels), thus increasing circulation. In about a third of the people who try it, minoxidil improves hair growth, causing scalp follicles to enlarge and lengthening the growth cycle. For many, however, the hair growth is meager. Minoxidil does not help people who already are bald.

Glands

Recall from Chapter 4 that glands are single or groups of epithelial cells that secrete a substance. The glands associated with the skin include sebaceous, sudoriferous, and ceruminous glands.

Sebaceous Glands

Sebaceous glands (se-BĀ-shus; *sebace-* = greasy) or ***oil glands***, with few exceptions, are connected to hair follicles (Figure 5.3a). The secreting portions of the glands lie in the dermis and open into the hair follicles or directly onto a skin surface. There are no sebaceous glands in the palms and soles.

Sebaceous glands secrete an oily substance called ***sebum*** (SĒ-bum). Sebum keeps hair from drying out, prevents excessive evaporation of water from the skin, keeps the skin soft, and inhibits the growth of certain bacteria.

When sebaceous glands of the face become enlarged because of accumulated sebum, ***blackheads*** develop. Because sebum is nutritive to certain bacteria, ***pimples*** or ***boils*** often result. The color of blackheads is due to melanin and oxidized oil, not dirt. Sebaceous gland activity increases during adolescence.

Sudoriferous Glands

There are three to four million ***sudoriferous glands*** (soo′-dor-IF-er-us; *sudori-* = sweat; *-ferous* = bearing), or ***sweat glands***, divided into two main types: eccrine and apocrine.

Eccrine sweat glands (*eccrine* = secreting outwardly) are much more common than apocrine sweat glands. They are distributed throughout the skin of most parts of the body, except for the margins of the lips, nail beds of the fingers and toes, glans penis, glans clitoris, labia minora, and eardrums. Eccrine sweat glands are most numerous in the skin of the forehead, palms, and soles; their density can be as high as 450 per square centimeter (3000 per square inch) in the palms. The secretory portion of eccrine sweat glands is located mostly in the deep dermis (sometimes in the upper subcutaneous layer). The excretory duct projects through the dermis and epidermis and ends as a pore at the surface of the epidermis (see Figure 5.1). The sweat produced by eccrine sweat glands (about 600 mL per day) consists of water, ions (mostly Na^+ and Cl^-), urea, uric acid, ammonia, amino acids, glucose, and lactic acid. The main function of eccrine sweat glands is to help regulate body temperature through evaporation. As sweat evaporates, large quantities of heat energy leave the body surface.

Apocrine sweat glands are found mainly in the skin of the axilla (armpit), groin, areolae (pigmented areas around the nipples) of the breasts, and bearded regions of the face in adult males. The secretory portion of these sweat glands is located mostly in the subcutaneous layer, and the excretory duct opens into hair follicles (see Figure 5.1). Their secretory product is slightly viscous compared to eccrine secretions and contains the same components as eccrine sweat plus lipids and proteins. Eccrine sweat glands start to function soon after birth, but apocrine sweat glands do not begin to function until puberty. Apocrine sweat glands are stimulated during emotional stress and sexual excitement; these secretions are commonly known as a "cold sweat."

Ceruminous Glands

Ceruminous glands (se-ROO-mi-nus; *cer-* = wax) are present in the external auditory canal, the outer ear canal. The combined secretion of the ceruminous and sebaceous glands is called ***cerumen*** or earwax. Cerumen and the hairs in the external auditory meatus provide a sticky barrier against foreign bodies.

Nails

Nails are plates of tightly packed, hard, dead, keratinized cells of the epidermis. Each nail (Figure 5.4) consists of a nail body, a free edge, and a nail root. The ***nail body*** is the portion of the nail that is visible; the ***free edge*** is the part of the body that extends past the end of the finger or toe; the ***nail root*** is the portion that is not visible. Most of the nail body is pink because of the underlying blood capillaries. The whitish semilunar area near the nail root is called the ***lunula*** (LOO-nyū-la = little moon). It appears whitish because the vascular tissue underneath does not show through due to the thickened stratum basale in the area. Nail growth occurs by the transformation of superficial cells of the ***nail matrix*** into nail cells. The average growth of fingernails is about 1 mm (0.04 inch) per week. The ***cuticle*** consists of stratum corneum.

Functionally, nails help us grasp and manipulate small objects, provide protection to the ends of the fingers and toes, and allow us to scratch various parts of the body.

■ CHECKPOINT

4. Describe the structure of a hair. What causes "goose bumps"?
5. Contrast the locations and functions of sebaceous (oil) glands and sudoriferous (sweat) glands.
6. Describe the parts of a nail.

FUNCTIONS OF THE SKIN

OBJECTIVE • Describe how the skin contributes to the regulation of body temperature, protection, sensation, excretion and absorption, and synthesis of vitamin D.

Following are the major functions of the skin:

1. **Body temperature regulation.** The skin contributes to the homeostatic regulation of body temperature by liberating sweat at its surface and by adjusting the flow of blood in the dermis (discussed in detail in Chapter 20).
2. **Protection.** Keratin in the skin protects underlying tissues from microbes, abrasion, heat, and chemicals, and the tightly interlocked keratinocytes resist invasion by microbes. Lipids released by lamellar granules inhibit evaporation of water from the skin surface, thus protecting the body from dehydration. Oily sebum prevents hairs from drying out and contains bactericidal chemicals that

Figure 5.4 Nails. Shown is a fingernail.

Nail cells arise by transformation of superficial cells of the nail matrix into nail cells.

(a) Dorsal view

(b) Sagittal section showing internal detail

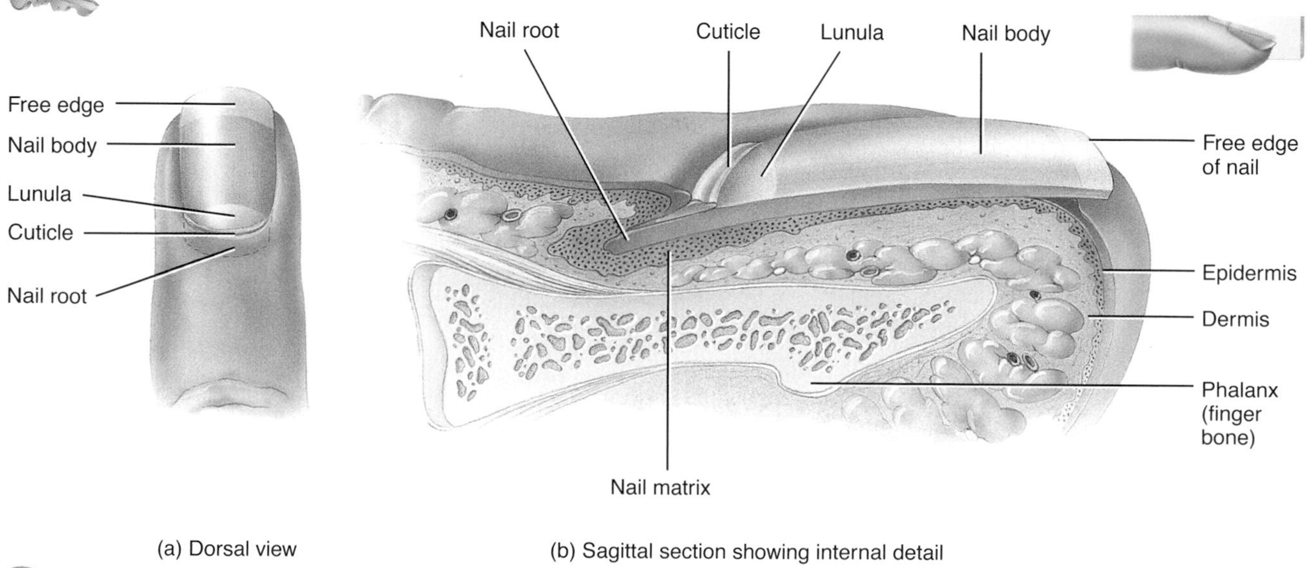

Why are nails so hard?

Focus on Wellness

Skin Care for Active Lifestyles

Physical activity is good for your skin. During exercise, the body shunts blood to the skin to help release excess heat produced by the contracting muscles. This increased blood flow provides the skin with nutrients and gets rid of wastes.

Fun in the Sun

From your skin's point of view, the main problem with exercise is that it often occurs outdoors, where sun exposure over the years can lead to wrinkles, age spots, and cancers of the skin. To prevent these, do what you can to minimize sun exposure. The most effective skin protection is some form of sun block. Tightly woven clothing (hold it up to a light and see how much shines through) helps keep the sun's rays from reaching the skin, and wide-brimmed hats provide some protection. Zinc oxide blocks the sun and is good for noses and lips when long-term exposure is unavoidable.

When a sun block is not practical, a sunscreen should be used. These do not shield the skin completely, but they do reduce the damaging effects of the ultraviolet rays. Evidence suggests that the skin can repair some damage when sunscreens are consistently applied. But researchers warn that sunscreens can provide a false sense of security. Because they prevent burning, sunscreens may lull us into thinking the sun is not hurting us, while damage may still be occurring.

Barriers to Skin Protection

Chemists have yet to invent a sunscreen that is fun to wear. Many exercisers can't take the grease, especially, as an avid bicyclist put it, "as it mingles with sweat and dead bugs." Advice for heavy sweaters is to exercise in the early or late part of the day, take as shady a route as possible, wear a hat and protective clothing, and use as much sunscreen as you can tolerate.

Swimmers should note that "waterproof" sunscreen stays on for only about 30 minutes in the water and should be reapplied at that time.

Dry Skin Care

Although not life threatening, dry skin can be very uncomfortable. Frequent showers and water exposure can strip the skin of its natural protective oils. The only solution is frequent moisturizing. Use of a good moisturizing cream immediately after drying off will counteract the drying effect of a "wash-and-wear" lifestyle.

► **Think It Over . . .**

► ***Imagine you have a friend who is training for a marathon and must exercise outdoors for an hour or more on most days. He has fair skin and a family history of skin cancer. What advice would you give him for minimizing sun exposure while continuing his training?***

kill surface bacteria. The acidic pH of perspiration retards the growth of some microbes. Melanin provides some protection against the damaging effects of UV light. Hair and nails also have protective functions.

3. **Cutaneous sensations.** ***Cutaneous sensations*** are those that arise in the skin. These include tactile sensations—touch, pressure, vibration, and tickling—as well as thermal sensations such as warmth and coolness. Another cutaneous sensation, pain, usually is an indication of impending or actual tissue damage. Chapter 12 provides more details on the topic of cutaneous sensations.
4. **Excretion and absorption.** The skin normally has a small role in *excretion*, the elimination of substances from the body, and *absorption*, the passage of materials from the external environment into body cells.

Most drugs are either absorbed into the body through the digestive system or injected into subcutaneous tissue or muscle. An alternative route, **transdermal (transcutaneous) drug administration**, enables a drug contained within an adhesive skin patch to pass across the epidermis and into the blood vessels of the dermis. The drug is released continuously at a controlled rate over one to several days. A growing number of drugs are available for transdermal administration, including nitroglycerin, for prevention of angina pectoris (chest pain associated with heart disease); scopolamine, for motion sickness; estradiol, used for estrogen-replacement therapy during menopause; ethinyl estradiol and norelgestromin in contraceptive patches; nicotine, used to help people stop smoking; and fentanyl, used to relieve severe pain in cancer patients.

5. **Synthesis of vitamin D.** Exposure of the skin to ultraviolet radiation activates vitamin D. Ultimately vitamin D is converted to its active form, a hormone called calcitriol, that aids in the absorption of calcium and phosphorus from the gastrointestinal tract into the blood. People who avoid sun exposure and individuals who live in colder, northern climates may experience vitamin D deficiency if it is not included in their diet or as supplements.

■ **CHECKPOINT**

7. In what two ways does the skin help regulate body temperature?
8. In what ways does the skin serve as a protective barrier?
9. What sensations arise from stimulation of neurons in the skin?

AGING AND THE INTEGUMENTARY SYSTEM

OBJECTIVE • Describe the effects of aging on the integumentary system.

Most infants and children encounter relatively few problems with the skin as it ages. With the arrival of adolescence, however, some teens develop acne. The pronounced effects of skin aging do not become noticeable until people reach their late forties. Most of the age-related changes occur in the dermis. Collagen fibers in the dermis begin to decrease in number, stiffen, break apart, and disorganize into a shapeless, matted tangle. Elastic fibers lose some of their elasticity, thicken into clumps, and fray, an effect that is greatly accelerated in the skin of smokers. Fibroblasts, which produce both collagen and elastic fibers, decrease in number. As a result, the skin forms the characteristic crevices and furrows known as *wrinkles*.

With further aging, Langerhans cells dwindle and macrophages become less-efficient phagocytes, thus decreasing the skin's immune responsiveness. Moreover, decreased size of sebaceous glands leads to dry and broken skin that is more susceptible to infection. Production of sweat diminishes, which probably contributes to the increased incidence of heat stroke in the elderly. There is a decrease in the number of functioning melanocytes, resulting in gray hair and atypical skin pigmentation. An increase in the size of some melanocytes produces pigmented blotching (age spots). Walls of blood vessels in the dermis become thicker and less permeable, and subcutaneous fat is lost. Aged skin (especially the dermis) is thinner than young skin, and the migration of cells from the basal layer to the epidermal surface slows considerably. With the onset of old age, skin heals poorly and becomes more susceptible to pathological conditions such as skin cancer, itching, and pressure ulcers.

Growth of nails and hair begins to slow during the second and third decades of life. The nails also may become more brittle with age, often due to dehydration or repeated use of cuticle remover or nail polish.

Several cosmetic anti-aging treatments are available to diminish the effects of aging or sun-damaged skin, including ***topical products*** that bleach the skin to tone down blotches and blemishes (hydroquinone) or decrease fine wrinkles and roughness (retinoic acid); ***microdermabrasion*** (mī-krō-DER-ma-brā-zhun; *mikros-* = small, *derm* = skin; *-abrasio* = to wear away), the use of tiny crystals under pressure to remove and vacuum the skin's surface cells to improve skin texture and reduce blemishes; ***chemical peel***, the application of a mild acid (such as glycolic acid) to the skin to remove surface cells to improve skin texture and reduce blemishes; ***laser resurfacing***, the use of a laser to clear up blood vessels near the skin surface, even out blotches and blemishes, and decrease fine wrinkles; ***dermal fillers***, injections of collagen from cows, hyaluronic acid, or calcium hydroxylapatite that plumps up the skin to smooth out wrinkles and fill in furrows, such as those around the nose and mouth and between the eyebrows; ***fat transplantation***, in which fat from one part of the body is injected into another location such as around the eyes; ***botulinum toxin*** or ***Botox®***, a diluted version of the toxin that causes food poisoning, which is injected into the skin to paralyze muscles that cause the skin to wrinkle; ***radio frequency nonsurgical facelift***, the use of radio frequency emissions to tighten skin of the jowls, neck, and sagging eyebrows and eyelids; and ***facelift***, ***browlift***, or ***necklift***, invasive surgery in which loose skin and fat are removed surgically and the underlying connective tissue and muscle are tightened.

■ **CHECKPOINT**

10. Which portion of the skin is involved in most age-related changes? Give several examples.

• • •

To appreciate the many ways that skin contributes to homeostasis of other body systems, examine the Focus on Homeostasis: The Integumentary System on page 109. This focus box is the first of ten, found at the end of selected chapters, that explain how the body system under consideration contributes to homeostasis of all the other body systems. The Focus on Homeostasis feature will help you understand how the individual body systems interact to contribute to the homeostasis of the entire body. Next, in Chapter 6, we will explore how bone tissue is formed and how bones are assembled into the skeletal system, which protects many of our internal organs.

COMMON DISORDERS

Skin Cancer

Excessive exposure to the sun causes virtually all of the one million cases of ***skin cancer*** diagnosed annually in the United States. There are three common forms of skin cancer. ***Basal cell carcinomas*** account for about 78% of all skin cancers. The tumors arise from cells in the stratum basale of the epidermis and rarely metastasize. ***Squamous cell carcinomas,*** which account for about 20% of all skin cancers, arise from squamous cells of the epidermis, and they have a variable tendency to metastasize. Most arise from preexisting lesions of damaged tissue on sun-exposed skin. Basal and squamous cell carcinomas are together known as *nonmelanoma skin cancer.* They are 50% more common in males than in females.

Malignant melanomas arise from melanocytes and account for about 2% of all skin cancers. They are the most prevalent life-threatening cancer in young women. The estimated lifetime risk of developing melanoma is now 1 in 75, double the risk only 15 years ago. In part, this increase is due to depletion of the ozone layer, which absorbs some UV light high in the atmosphere. But the main reason for the increase is that more people are spending more time in the sun and in tanning beds. Malignant melanomas metastasize rapidly and can kill a person within months of diagnosis.

The key to successful treatment of malignant melanoma is early detection. The early warning signs of malignant melanoma are identified by the acronym ABCD (Figure 5.5). *A* is for *asymmetry;* malignant melanomas tend to lack symmetry. *B* is for *border;* malignant melanomas have irregular—notched, indented, scalloped, or indistinct—borders. *C* is for *color;* malignant melanomas have uneven coloration and may contain several colors. *D* is for *diameter;* ordinary moles typically are smaller than 6 mm (0.25 in.), about the size of a pencil eraser. Once a malignant melanoma has the characteristics of A, B, and C, it is usually larger than 6 mm.

Among the risk factors for skin cancer are the following:

1. ***Skin type.*** Individuals with light-colored skin who never tan but always burn are at high risk.
2. ***Sun exposure.*** People who live in areas with many days of sunlight per year and at high altitudes (where ultraviolet light is more intense) have a higher risk of developing skin cancer. Likewise, people who engage in outdoor occupations and those who have suffered three or more severe sunburns have a higher risk.
3. ***Family history.*** Skin cancer rates are higher in some families than in others.
4. ***Age.*** Older people are more prone to skin cancer owing to longer total exposure to sunlight.
5. ***Immunological status.*** Individuals who are immunosuppressed have a higher incidence of skin cancer.

Figure 5.5 Comparison of a normal nevus (mole) and a malignant melanoma.

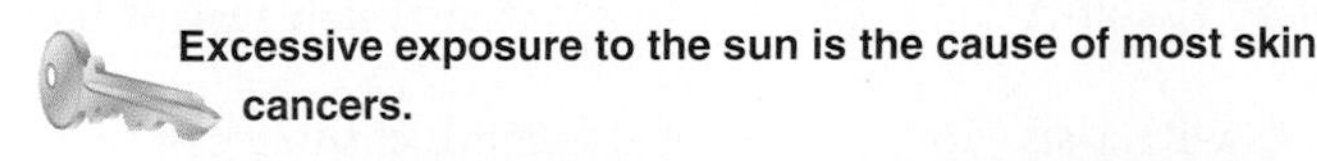

Excessive exposure to the sun is the cause of most skin cancers.

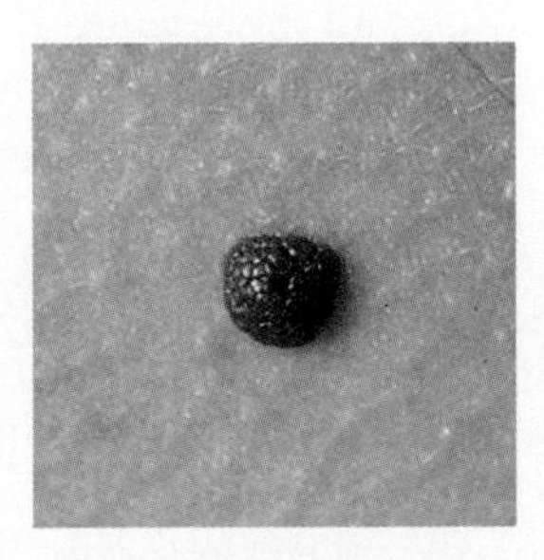

(a) Normal nevus (mole)

(b) Malignant melanoma

Which type of skin cancer is the most common type?

Sun Damage

Although basking in the warmth of the sun may feel good, it is not a healthy practice. There are two forms of ultraviolet radiation that affect the health of the skin. Longer-wavelength ultraviolet A (UVA) rays make up nearly 95% of the ultraviolet radiation that reaches the earth. UVA rays are not absorbed by the ozone layer. They penetrate the furthest into the skin, where they are absorbed by melanocytes and thus are involved in sun tanning. UVA rays also depress the immune system. Shorter-wavelength ultraviolet B (UVB) rays are partially absorbed by the ozone layer and do not penetrate the skin as deeply as UVA rays. UVB rays cause sunburn and are responsible for most of the tissue damage (production of oxygen free radicals that disrupt collagen and elastic fibers) that results in wrinkling and aging of the skin and cataract formation. Both UVA and UVB rays are thought to cause skin cancer. Long-term overexposure to sunlight results in dilated blood vessels, age spots, freckles, and changes in skin texture.

Exposure to ultraviolet radiation (either natural sunlight or the artificial light of a tanning booth) may also produce ***photosensitivity,*** a heightened reaction of the skin after consumption of certain medications or contact with certain substances. Photosensitivity is characterized by redness, itching, blistering, peeling, hives, and even shock. Among the medications or substances that may cause a photosensitivity reaction are certain antibiotics (tetracycline), nonsteroidal anti-inflammatory drugs (ibuprofen or naproxen), certain herbal supplements (St. John's Wort), some birth control pills, some high blood pressure medications, some antihistamines, and certain artificial sweeteners, perfumes, aftershaves, lotions, detergents, and medicated cosmetics.

Burns

A ***burn*** is tissue damage caused by excessive heat, electricity, radioactivity, or corrosive chemicals that denature the proteins in the skin cells. Burns destroy some of the skin's important contributions to homeostasis—protection against microbial invasion and desiccation, and regulation of body temperature.

Burns are graded according to their severity. A *first-degree burn* involves only the epidermis. It is characterized by mild pain and redness but no blisters. Skin functions remain intact. A *second-degree burn* destroys a portion of the epidermis and part of the dermis. Some skin functions are lost. In a second-degree burn there is redness, blister formation, edema, and pain, and scarring may result. First- and second-degree burns are collectively referred to as *partial-thickness burns.*

A *third-degree burn* or *full-thickness burn* destroys the epidermis, the underlying dermis, and subcutaneous layer. Most skin functions are lost. Skin grafting may be required to promote healing and to minimize scarring.

The seriousness of a burn is determined by its depth and extent of area involved, as well as the person's age and general health. According to the American Burn Association's classification of burn injury, a major burn includes third-degree burns over 10% of body surface area; or second-degree burns over 25% of body surface area; or any third-degree burns on the face, hands, feet, or *perineum* (per-i-NĒ-um, which includes the anal and urogenital regions). When the burn area exceeds 70%, more than half the victims die.

Pressure Ulcers

Pressure ulcers, also known as *decubitus ulcers* (dē-KŪ-bi-tus) or *bedsores*, are a shedding of epithelium caused by a constant deficiency of blood flow to tissues. Typically the affected tissue overlies a bony projection that has been subjected to prolonged pressure against an object such as a bed, cast, or splint. If the pressure is relieved in a few hours, redness occurs but no lasting tissue damage results. Prolonged pressure causes tissue ulceration. Small breaks in the epidermis become infected, and the sensitive subcutaneous layer and deeper tissues are damaged. Eventually, the tissue dies. Pressure ulcers occur most often in bedridden patients. With proper care, pressure ulcers are preventable.

Acne

Acne is an inflammation of sebaceous glands that usually begins at puberty, when the sebaceous glands grow in size and increase their production of sebum. Androgens from the testes, ovaries, and adrenal glands play the greatest role in stimulating sebaceous glands. Acne occurs predominantly in sebaceous follicles that have been colonized by bacteria, some of which thrive in the lipid-rich sebum. The infection may cause a cyst (sac of connective tissue cells) to form, which can destroy and displace epidermal cells. This condition, called ***cystic acne***, can permanently scar the epidermis.

MEDICAL TERMINOLOGY AND CONDITIONS

Abrasion (a-BRĀ-shun; *ab-* = away; *-raison* = scraped) A portion of the epidermis that has been scraped away.

Athlete's (ATH-lēts) ***foot*** A superficial fungus infection of the skin of the foot.

Blister A collection of serous fluid within the epidermis or between the epidermis and dermis, due to short-term but severe friction.

Cold sore A lesion, usually in the oral mucous membrane, caused by type 1 herpes simplex virus (HSV) transmitted by oral or respiratory routes. The virus remains dormant until triggered by factors such as ultraviolet light, hormonal changes, and emotional stress. Also called a ***fever blister***.

Contact dermatitis (der′-ma-TĪ-tis; *dermat-* = skin; *-itis* = inflammation) Inflammation of the skin characterized by redness, itching, and swelling and caused by exposure of the skin to chemicals that bring about an allergic reaction, such as poison ivy toxin.

Corn (KORN) A painful thickening of the stratum corneum of the epidermis found principally over toe joints and between the toes, often caused by friction or pressure. Corns may be hard or soft, depending on their location. Hard corns are usually found over toe joints, and soft corns are usually found between the fourth and fifth toes.

Hemangioma (hē-man′-jē-Ō-ma; *hem-* = blood; *-angi* = blood vessel; *-oma* = tumor) Localized tumor of the skin and subcutaneous layer that results from an abnormal increase in blood vessels. One type is a ***port-wine stain***, a flat pink, red, or purple lesion present at birth, usually at the nape of the neck.

Hives (HĪVZ) Skin condition marked by reddened elevated patches that are often itchy. Most commonly caused by infections, physical trauma, medications, emotional stress, food additives, and certain food allergies.

Impetigo (im′-pe-TĪ-gō) Superficial skin infection caused by *Staphylococcus* bacteria; most common in children.

Intradermal (in-tra-DER-mal; *intra-* = within) Within the skin. Also called ***intracutaneous***.

Keratosis (ker′-a-TŌ-sis; *kera-* = horn) Formation of a hardened growth of epidermal tissue, such as a ***solar keratosis***, a premalignant lesion of the sun-exposed skin of the face and hands.

Laceration (las-er-Ā-shun; *lacer-* = torn) An irregular tear of the skin.

Psoriasis (sō-RĪ-a-sis, *psora* = itch) A common, chronic skin disorder in which keratinocytes divide and move more quickly than normal from the stratum basale to the stratum corneum and form flaky scales, most often on the knees, elbows, and scalp.

Pruritus (proo-RĪ-tus; *pruri-* = to itch) Itching, one of the most common dermatological disorders. It may be caused by skin disorders (infections), systemic disorders (cancer, kidney failure), psychogenic factors (emotional stress), or allergic reactions.

Topical Refers to a medication applied to the skin surface rather than ingested or injected.

Wart Mass produced by uncontrolled growth of epithelial skin cells, caused by a papilloma virus. Most warts are noncancerous.

FOCUS ON HOMEOSTASIS

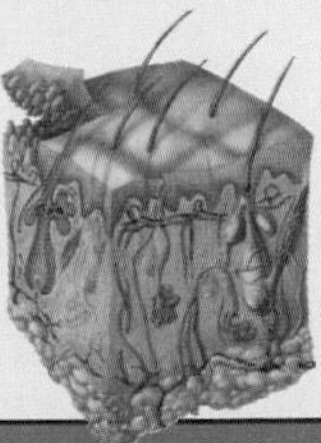

THE INTEGUMENTARY SYSTEM

Body System		Contribution of the Integumentary System
For all body systems		The skin and hair provide barriers that protect all internal organs from damaging agents in the external environment; sweat glands and skin blood vessels help regulate body temperature, needed for proper functioning of other body systems.
Skeletal system	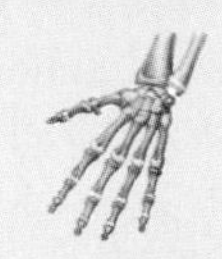	The skin helps activate vitamin D, needed for proper absorption of dietary calcium and phosphorus to build and maintain bones.
Muscular system		Through activation of vitamin D, the skin helps provide calcium ions, needed for muscle contraction; the skin also rids the body of heat produced by muscular activity.
Nervous system	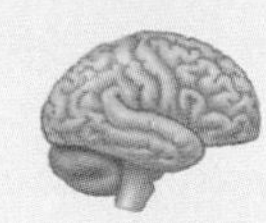	Nerve endings in the skin and subcutaneous tissue provide input to the brain for touch, pressure, thermal, and pain sensations.
Endocrine system	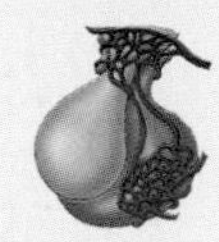	Keratinocytes in the skin help activate vitamin D, initiating its conversion to calcitriol, a hormone that aids absorption of dietary calcium and phosphorus.
Cardiovascular system	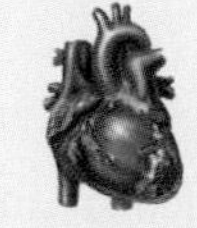	Local chemical changes in the dermis cause widening and narrowing of skin blood vessels, which help adjust blood flow to the skin.
Lymphatic system and immunity	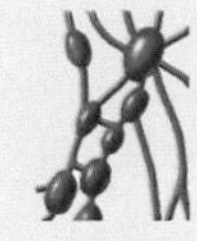	The skin is the "first line of defense" in immunity, providing mechanical barriers and chemical secretions that discourage penetration and growth of microbes; Langerhans cells in the epidermis participate in immune responses by recognizing foreign antigens for destruction by immune cells.
Respiratory system	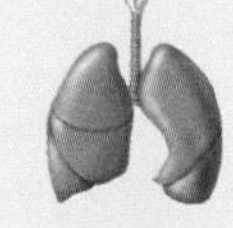	Hairs in the nose filter dust particles from inhaled air; stimulation of pain nerve endings in the skin may alter breathing rate.
Digestive system	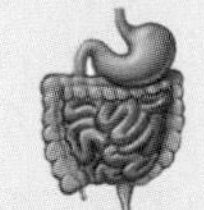	The skin helps activate vitamin D to become the hormone calcitriol, which promotes absorption of dietary calcium and phosphorus in the small intestine.
Urinary system	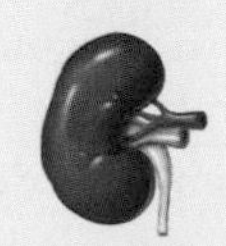	Kidney cells receive partially activated vitamin D hormone from the skin and convert it to calcitriol; some waste products are excreted from the body in sweat, contributing to excretion by the urinary system.
Reproductive systems	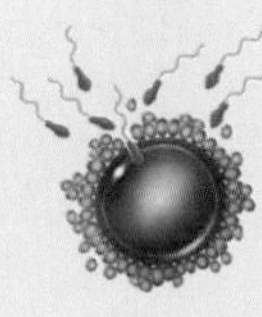	Nerve endings in the skin and subcutaneous tissue respond to erotic stimuli, thereby contributing to sexual pleasure; suckling of a baby stimulates nerve endings in the skin, leading to milk ejection; mammary glands (modified sweat glands) produce milk; the skin stretches during pregnancy as the fetus enlarges.

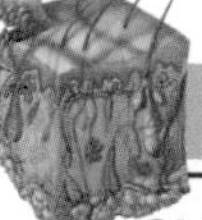

STUDY OUTLINE

Skin (p. 98)

1. The skin and hairs and other structures such as nails constitute the integumentary system.
2. The principal parts of the skin are the outer epidermis and inner dermis. The dermis overlies and attaches to the subcutaneous layer.
3. Epidermal cells include keratinocytes, melanocytes, Langerhans cells, and Merkel cells. The epidermal layers, from deepest to most superficial, are the stratum basale, stratum spinosum, stratum granulosum, stratum lucidum, and stratum corneum. The stratum basale undergoes continuous cell division and produces all other layers.
4. The dermis consists of two regions. The superficial region is areolar connective tissue containing blood vessels, nerves, hair follicles, dermal papillae, and corpuscles of touch (Meissner corpuscles). The deeper region is dense, irregularly arranged connective tissue containing adipose tissue, hair follicles, nerves, sebaceous (oil) glands, and ducts of sudoriferous (sweat) glands.
5. Skin color is due to the pigments melanin, carotene, and hemoglobin.

Accessory Structures of the Skin (p. 101)

1. Accessory structures of the skin develop from the epidermis of an embryo.
2. They include hair, skin glands (sebaceous, sudoriferous, and ceruminous), and nails.
3. Hairs are threads of fused, dead keratinized cells that function in protection.
4. Hairs consist of a shaft above the surface, a root that penetrates the dermis and subcutaneous layer, and a hair follicle.
5. Associated with hairs are bundles of smooth muscle called arrector pili and sebaceous (oil) glands.
6. Sebaceous (oil) glands are usually connected to hair follicles; they are absent in the palms and soles. Sebaceous glands produce sebum, which moistens hairs and waterproofs the skin.
7. There are two types of sudoriferous (sweat) glands: eccrine and apocrine. Eccrine sweat glands have an extensive distribution; their ducts terminate at pores at the surface of the epidermis, and their main function is to help regulate body temperature. Apocrine sweat glands are limited in distribution, and their ducts open into hair follicles. They begin functioning at puberty and are stimulated during emotional stress and sexual excitement.
8. Ceruminous glands are modified sudoriferous glands that secrete cerumen. They are found in the external auditory canal.
9. Nails are hard, dead, keratinized epidermal cells covering the terminal portions of the fingers and toes.
10. The principal parts of a nail are the body, free edge, root, lunula, cuticle, and matrix. Cell division of the matrix cells produces new nails.

Functions of the Skin (p. 104)

1. Skin functions include body temperature regulation, protection, sensation, excretion and absorption, and synthesis of vitamin D.
2. The skin participates in body temperature regulation by liberating sweat at its surface and by adjusting the flow of blood in the dermis.
3. The skin provides physical, chemical, and biological barriers that help protect the body.
4. Cutaneous sensations include tactile sensations, thermal sensations, and pain.

Aging and the Integumentary System (p. 106)

1. Most effects of aging occur when an individual reaches the late forties.
2. Among the effects of aging are wrinkling, loss of subcutaneous fat, atrophy of sebaceous glands, and decrease in the number of melanocytes and Langerhans cells.

SELF-QUIZ

1. Hair follicles
   a. consist of dead cells
   b. extend above the surface of the skin
   c. can increase in number as you age
   d. contain cells undergoing mitosis
   e. are another name for arrector pili muscles
2. Skin coloration
   a. is due to melanin found in the subcutaneous layer
   b. in European-Americans is due mainly to carotene
   c. is related to apocrine glands
   d. is stimulated by exposure to the sun
   e. is produced by Merkel cells
3. In which portion of the skin will you find dermal papillae?
   a. superficial region of the dermis **b.** epidermis
   c. hypodermis **d.** stratum spinosum
   e. deeper region of the dermis
4. If you pricked your fingertip with a needle, the first layer of epidermis that it would penetrate is the
   a. stratum basale **b.** stratum spinosum
   c. stratum granulosum **d.** stratum lucidum
   e. stratum corneum
5. A person with albinism has a defect in the production of
   a. carotene **b.** keratin **c.** collagen **d.** cerumen
   e. melanin

6. The red or pink tones seen in some skin are due to
   a. hemoglobin in the blood moving through capillaries in the dermis
   b. the presence of carotene
   c. the lack of oxygen
   d. a buildup of bilirubin in the blood
   e. an increased production of melanin

7. When you have your hair cut, scissors are cutting through the hair
   a. follicle   b. root   c. shaft   d. papilla   e. bulb

8. Which of the following is NOT true concerning eccrine sweat glands?
   a. They are most numerous on the palms and the soles.
   b. They help regulate body temperature.
   c. They produce a viscous secretion.
   d. They function throughout life.
   e. They terminate at pores on the skin's surface.

9. Which tissue is the main type found in the inner region of the dermis?
   a. dense irregular connective   b. stratified squamous epithelium   c. smooth muscle   d. nervous
   e. cartilage

10. Which of the following is NOT a function of skin?
    a. calcium production   b. vitamin D synthesis
    c. protection   d. immunity   e. temperature regulation

11. Which of the following is NOT true concerning hair?
    a. Hair is mainly composed of keratin.
    b. Hirsutism is another name for male-pattern baldness.
    c. Hair color is due to melanin.
    d. Sebaceous glands are associated with hair.
    e. Contraction of the arrector pili muscles makes hair stand erect.

12. Sebaceous glands
    a. secrete an oily substance
    b. are located on the palms and soles
    c. are responsible for breaking out in a "cold sweat"
    d. are involved in body temperature regulation
    e. are found in the external auditory meatus

13. As keratinocytes in the stratum basale are pushed toward the skin's surface, they
    a. begin to divide more rapidly   b. become more elastic
    c. begin to die   d. lose their melanin
    e. begin to assume a columnar shape

14. To produce vitamin D, the skin cells need to be exposed to
    a. calcium and phosphorus   b. ultraviolet light
    c. heat   d. pressure   e. keratin

15. To prevent an unwanted hair from growing back, you must destroy which structure?
    a. shaft   b. root sheath   c. lunula   d. hair matrix
    e. arrector pili

16. Aging can result in
    a. an increase in collagen and elastic fibers in the skin
    b. a steady increase in the activity of sudoriferous glands
    c. a greater immune response from Langerhans cells
    d. more efficient activity by macrophages
    e. a decline in the activity of sebaceous glands

17. The portion of the nail that is responsible for nail growth is the
    a. cuticle   b. nail matrix   c. lunula   d. nail body
    e. nail root

18. Match the following:

____	a. Langerhans cell	A. earwax
____	b. Merkel cell	B. silvery white streaks
____	c. keratin	C. yellow-orange precursor of vitamin A
____	d. melanin	D. function in immune responses
____	e. lamellated (pacinian) corpuscle	E. protective protein of skin, hair
____	f. cerumen	F. touch receptor found in epidermis
____	g. carotene	G. yellow to black pigment
____	h. striae	H. nerve endings sensitive to pressure
____	i. corpuscle of touch (Meissner corpuscle)	I. touch receptor found in dermal papillae

19. Which of the following is NOT an accessory structure of the skin?
    a. dermal papillae   b. sudoriferous glands
    c. sebaceous glands   d. ceruminous glands   e. nails

20. What is the response by effectors when the body temperature is elevated?
    a. Blood vessels in the dermis constrict.
    b. Sweat glands increase production of sweat.
    c. Skeletal muscles begin to contract involuntarily.
    d. The body's metabolic rate increases.
    e. The ceruminous glands increase production.

CRITICAL THINKING APPLICATIONS

1. Three-year-old Michael was having his first haircut. As the barber started to snip his hair, Michael cried, "Stop! You're killing it!" He then pulled his own hair, yelling, "Ouch! See! It's alive!" Is Michael right about his hair?
2. Michael's twin sister Michelle scraped her knee at the playground. She told her mother that she wanted "new skin that doesn't leak." Her mother promised that new skin would soon appear under the bandage. How does new skin grow?
3. Andrew is training for the Megaman triathlon. After hours in running shoes and damp locker rooms, his feet are a mess! He has calluses, warts, and athlete's foot. What are the causes of his misery?
4. Fifteen-year-old Jeremy has a bad case of "blackheads." According to his Aunt Frieda, Jeremy's skin problems are from too much late-night TV, frozen pizza, and cheddar popcorn. Explain the real cause of blackheads to Aunt Frieda.

ANSWERS TO FIGURE QUESTIONS

5.1 The epidermis is made up of epithelial tissue, and the dermis is composed of connective tissue.

5.2 The stratum basale is the layer of the epidermis that contains stem cells that continually undergo cell division.

5.3 The matrix produces a new hair by cell division.

5.4 Nails are hard because they are composed of tightly packed, hard, dead keratinized epidermal cells.

5.5 Basal cell carcinoma is the most common type of skin cancer.

THE SKELETAL SYSTEM

chapter 6

did you know?

People are encouraged to limit food intake and exercise regularly to prevent obesity and stay healthy. But people who go to extremes can compromise the health of their bones. Premature osteoporosis can occur in young women who experience prolonged menstrual irregularity, which is often caused by extreme dieting and/or excessive exercise. These women have low levels of the estrogens, hormones which help to keep bones strong. Extreme dieting behavior may mean a minimal calcium intake, which then limits the body's bone-building ability. People who build strong bones during adolescence and young adulthood reduce their likelihood of developing osteoporosis later in life.

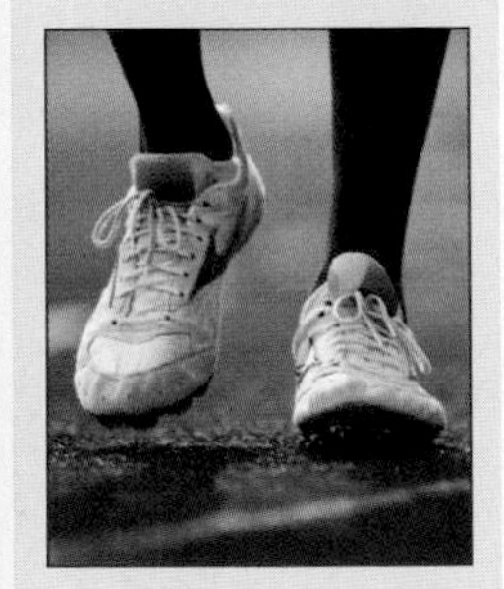

Focus on Wellness, page 148

www.wiley.com/college/apcentral

Despite its simple appearance, bone is a complex and dynamic living tissue that is remodeled continuously—new bone is built while old bone is broken down. Each individual bone is an organ composed of several different tissues working together: bone, cartilage, dense connective tissues, epithelium, blood-forming tissue, adipose tissue, and nervous tissue. The entire framework of bones and their cartilages constitute the ***skeletal system***. The study of bone structure and the treatment of bone disorders is termed ***osteology*** (os′-tē-OL-ō-jē; *osteo-* = bone; *-logy* = study of).

looking back to move ahead . . .

- **Connective Tissue Extracellular Matrix (page 83)**
- **Cartilage (page 89)**
- **Bone Tissue (page 89)**
- **Collagen Fibers (page 84)**
- **Dense Irregular Connective Tissue (page 87)**

FUNCTIONS OF BONE AND THE SKELETAL SYSTEM

OBJECTIVE • Discuss the functions of bone and the skeletal system.

Bone tissue and the skeletal system perform several basic functions:

1. **Support.** The skeleton provides a structural framework for the body by supporting soft tissues and providing points of attachment for most skeletal muscles.
2. **Protection.** The skeleton protects many internal organs from injury. For example, cranial bones protect the brain, vertebrae (backbones) protect the spinal cord, and the rib cage protects the heart and lungs.
3. **Assisting in movement.** Because most skeletal muscles attach to bones, when muscles contract, they pull on bones. Together bones and muscles produce movement. This function is discussed in detail in Chapter 8.
4. **Mineral homeostasis.** Bone tissue stores several minerals, especially calcium and phosphorus. On demand, bone releases minerals into the blood to maintain critical mineral balances (homeostasis) and to distribute the minerals to other parts of the body.
5. **Production of blood cells.** Within certain bones a connective tissue called ***red bone marrow*** produces red blood cells, white blood cells, and platelets, a process called ***hemopoiesis*** (hēm-ō-poy-Ē-sis; *hemo-* = blood; *poiesis* = making). Red bone marrow consists of developing blood cells, adipocytes, fibroblasts, and macrophages. It is present in developing bones of the fetus and in some adult bones, such as the pelvis, ribs, sternum (breastbone), vertebrae (backbones), skull, and ends of the arm bones and thigh bones.
6. **Triglyceride storage.** ***Yellow bone marrow*** consists mainly of adipose cells, which store triglycerides. The stored triglycerides are a potential chemical energy reserve. Yellow bone marrow also contains a few blood cells. In the newborn, all bone marrow is red and is involved in hemopoiesis. With increasing age, much of the bone marrow changes from red to yellow.

■ **CHECKPOINT**

1. What kinds of tissues make up the skeletal system?
2. How do red and yellow bone marrow differ in composition, location, and function?

TYPES OF BONES

OBJECTIVE • Classify bones on the basis of their shape and location.

Almost all the bones of the body may be classified into four main types based on their shape: long, short, flat, or irregular. ***Long bones*** have greater length than width and consist of a shaft and a variable number of ends. They are usually somewhat curved for strength. Long bones include those in the thigh (femur), leg (tibia and fibula), arm (humerus), forearm (ulna and radius), and fingers and toes (phalanges).

Short bones are somewhat cube-shaped and nearly equal in length and width. Examples of short bones include most wrist and ankle bones.

Flat bones are generally thin, afford considerable protection, and provide extensive surfaces for muscle attachment. Bones classified as flat bones include the cranial bones, which protect the brain; the sternum (breastbone) and ribs, which protect organs in the thorax; and the scapulae (shoulder blades).

Irregular bones have complex shapes and cannot be grouped into any of the previous categories. Such bones include the vertebrae and some facial bones.

■ **CHECKPOINT**

3. Give several examples of long, short, flat, and irregular bones.

STRUCTURE OF BONE

OBJECTIVES • Describe the parts of a long bone. • Describe the histological features of bone tissue.

We will now explore the structure of bone at both the macroscopic and microscopic levels.

Macroscopic Structure of Bone

The structure of a bone may be analyzed by considering the parts of a long bone, for instance, the humerus (the arm bone), as shown in Figure 6.1. A typical long bone consists of the following parts:

1. The ***diaphysis*** (dī-AF-i-sis = growing between) is the bone's shaft or body—the long, cylindrical, main portion of the bone.
2. The ***epiphyses*** (e-PIF-i-sēz = growing over; singular is *epiphysis*) are the distal and proximal ends of the bone.

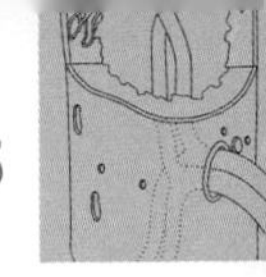

Figure 6.1 Parts of a long bone: epiphysis, metaphysis, and diaphysis. The spongy bone of the epiphysis and metaphysis contains red bone marrow, and the medullary cavity of the diaphysis contains yellow bone marrow in an adult.

A long bone is covered by articular cartilage at its proximal and distal epiphyses and by periosteum around the remainder of the bone.

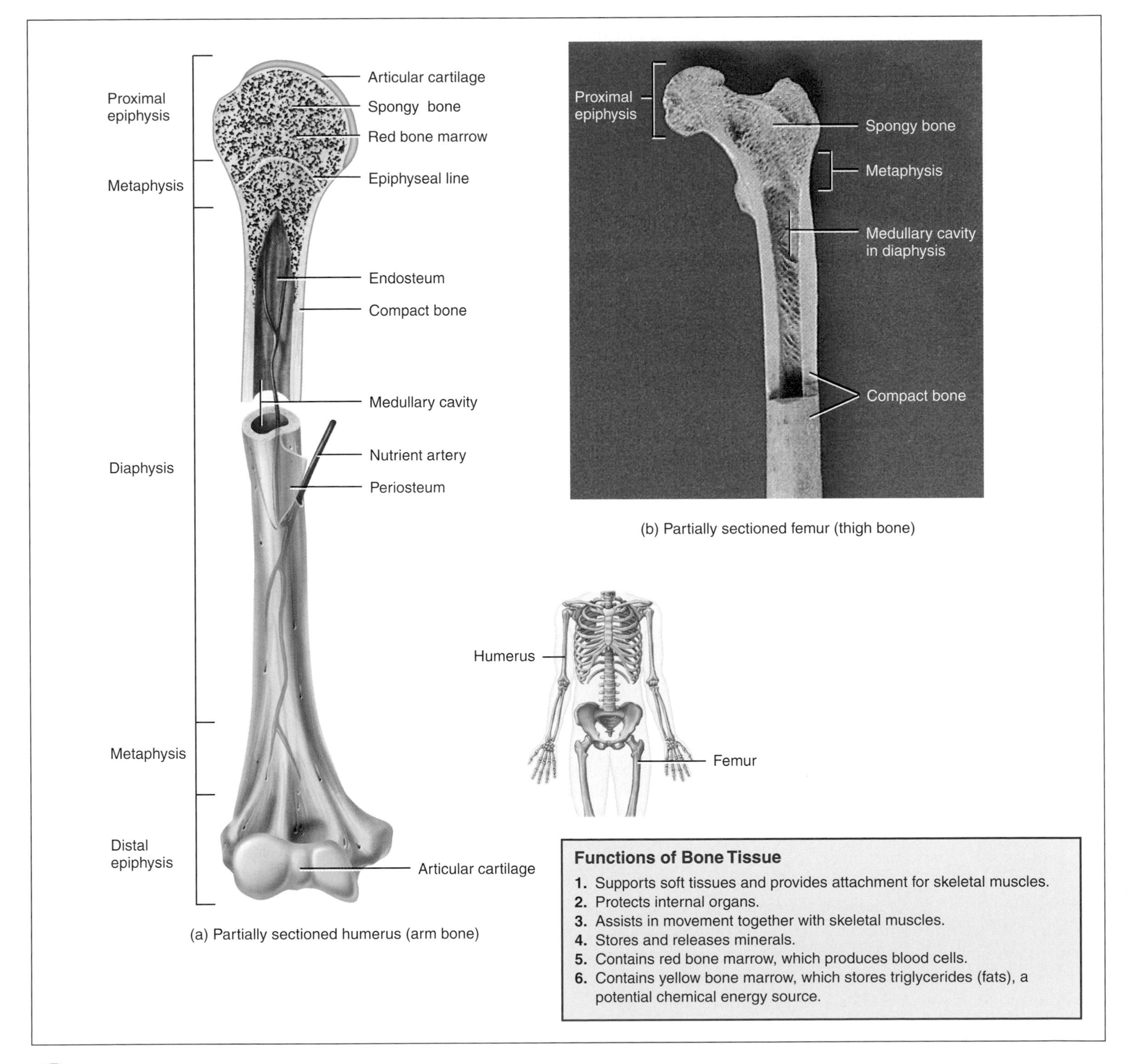

(a) Partially sectioned humerus (arm bone)

(b) Partially sectioned femur (thigh bone)

Functions of Bone Tissue

1. Supports soft tissues and provides attachment for skeletal muscles.
2. Protects internal organs.
3. Assists in movement together with skeletal muscles.
4. Stores and releases minerals.
5. Contains red bone marrow, which produces blood cells.
6. Contains yellow bone marrow, which stores triglycerides (fats), a potential chemical energy source.

Which part of a bone reduces friction at joints? Produces blood cells? Lines the medullary cavity?

3. The ***metaphyses*** (me-TAF-i-sēz; *meta-* = between; singular is *metaphysis*) are the regions in a *mature bone* where the diaphysis joins the epiphyses. In a *growing bone*, each metaphysis contains an *epiphyseal plate* (ep′-i-FIZ-ē-al), a layer of hyaline cartilage that allows the diaphysis of the bone to grow in length (described later in the chapter). When bone growth in length stops, the cartilage in the epiphyseal plate is replaced by bone and the resulting bony structure is known as the *epiphyseal line*.
4. The ***articular cartilage*** is a thin layer of hyaline cartilage covering the part of the epiphysis where the bone forms an articulation (joint) with another bone. Articular cartilage reduces friction and absorbs shock at freely movable joints. Because articular cartilage lacks a perichondrium, repair of damage is limited.
5. The ***periosteum*** (per′-ē-OS-tē-um; *peri-* = around) is a tough sheath of dense irregular connective tissue that surrounds the bone surface wherever it is not covered by articular cartilage. The periosteum contains bone-forming cells that enable bone to grow in diameter or thickness, but not in length. It also protects the bone, assists in fracture repair, helps nourish bone tissue, and serves as an attachment point for ligaments and tendons.
6. The ***medullary cavity*** (MED-ū-lar′-ē; *medulla-* = marrow, pith) or *marrow cavity* is a hollow, cylindrical space within the diaphysis that contains fatty yellow bone marrow in adults.
7. The ***endosteum*** (end-OS-tē-um; *endo-* = within) is a thin membrane that lines the medullary cavity. It contains a single layer of bone-forming cells.

Microscopic Structure of Bone

Like other connective tissues, ***bone***, or ***osseous tissue*** (OS-ē-us) contains abundant extracellular matrix that surrounds widely separated cells. The extracellular matrix is about 25% water, 25% collagen fibers, and 50% crystallized mineral salts. As these mineral salts are deposited in the framework formed by the collagen fibers of the extracellular matrix, they crystallize and the tissue hardens. This process of ***calcification*** is initiated by osteoblasts, the bone-building cells.

Although a bone's *hardness* depends on the crystallized inorganic mineral salts, a bone's *flexibility* depends on its collagen fibers. Like reinforcing metal rods in concrete, collagen fibers and other organic molecules provide *tensile strength*, which is resistance to being stretched or torn apart.

Four major types of cells are present in bone tissue: osteogenic cells, osteoblasts, osteocytes, and osteoclasts (Figure 6.2a).

1. ***Osteogenic cells*** (os-tē-ō-JEN-ik; *-genic* = producing) are unspecialized stem cells derived from mesenchyme, the tissue from which all connective tissues are formed. They are the only bone cells to undergo cell division; the resulting cells develop into osteoblasts. Osteogenic cells are found along the inner portion of the periosteum, in the endosteum, and in the canals within bone that contain blood vessels.
2. ***Osteoblasts*** (OS-tē-ō-blasts′; *-blasts* = buds or sprouts) are bone-building cells. They synthesize and secrete collagen fibers and other organic components needed to build the extracellular matrix of bone tissue. As osteoblasts surround themselves with matrix, they become trapped in their secretions and become osteocytes. (Note: *Blasts* in bone or any other connective tissue secrete matrix.)
3. ***Osteocytes*** (OS-tē-ō-sīts′; *-cytes* = cells), mature bone cells, are the main cells in bone tissue and maintain its daily metabolism, such as the exchange of nutrients and wastes with the blood. Like osteoblasts, osteocytes do not undergo cell division. (Note: *Cytes* in bone or any other tissue maintain the tissue.)
4. ***Osteoclasts*** (OS-tē-ō-clasts′; *-clast* = break) are huge cells derived from the fusion of as many as 50 monocytes (a type of white blood cell) and are concentrated in the endosteum. They release powerful lysosomal enzymes and acids that digest the protein and mineral components of the bone extracellular matrix. This breakdown of bone extracellular matrix, termed *resorption*, is part of the normal development, growth, maintenance, and repair of bone. (Note: *Clasts* in bone break down extracellular matrix.)

Bone is not completely solid but has many small spaces between its cells and extracellular matrix components. Some spaces are channels for blood vessels that supply bone cells with nutrients. Other spaces are storage areas for red bone marrow. Depending on the size and distribution of the spaces, the regions of a bone may be categorized as compact or spongy (see Figure 6.1). Overall, about 80% of the skeleton is compact bone and 20% is spongy bone.

Compact Bone Tissue

Compact bone tissue contains few spaces and is arranged in repeating units called ***osteons*** or ***haversian systems*** (Figure 6.2c). Each osteon consists of a central (haversian) canal with its concentrically arranged lamellae, lacunae, osteocytes, and canaliculi. A ***central*** or ***haversian*** (ha-VER-shun) ***canal*** is a channel that contains blood vessels, nerves, and lymphatic vessels. The central canals run longitudinally through the bone. Around the canals are ***concentric lamellae*** (la-MEL-ē)—rings of hard, calcified extracellular matrix. Between the lamellae are small spaces called ***lacunae*** (la-KOO-nē = little lakes; singular is *lacuna*), which contain osteocytes. Radiating in all directions from the lacunae are tiny ***canaliculi*** (kan′-a-LIK-ū-lī = small channels), which are filled with extracellular fluid. Inside the canaliculi are slender fingerlike processes of osteocytes (see inset at right of Figure 6.2c). The canaliculi connect lacunae with one another and with the central canals. Thus, an intricate, miniature canal system throughout the bone provides many routes for nutrients and oxygen to reach the osteocytes and for wastes to diffuse away. This is very important because diffusion through the lamellae is extremely slow.

Figure 6.2 **Histology of bone.**

Osteocytes lie in lacunae arranged in concentric circles around a central (haversian) canal in compact bone, and in lacunae arranged irregularly in the trabeculae of spongy bone.

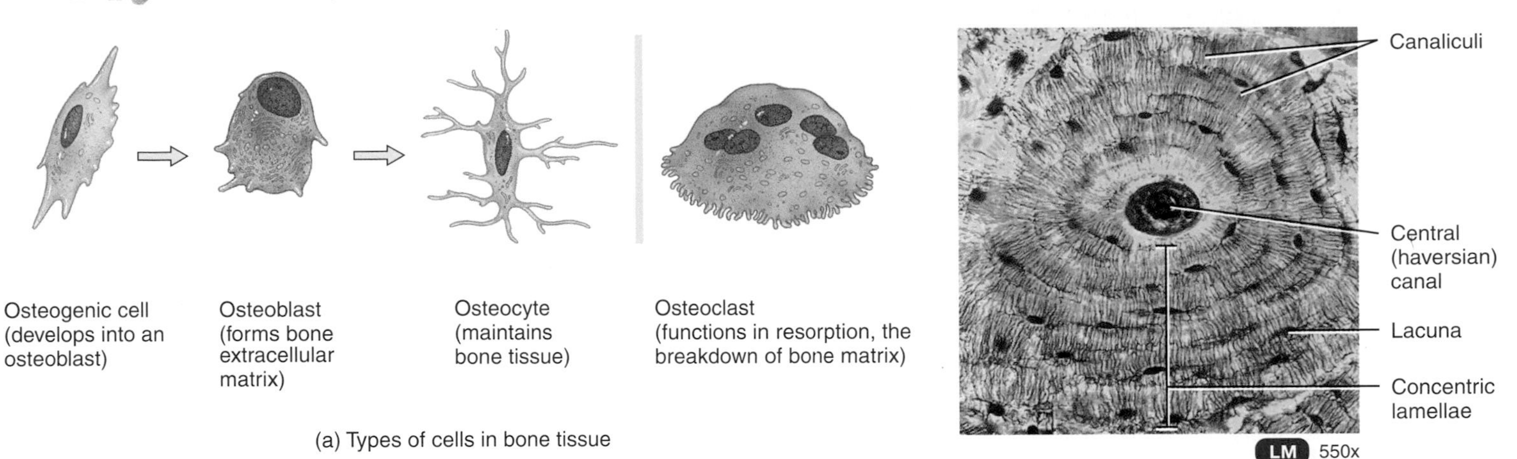

(a) Types of cells in bone tissue

(b) Sectional view of an osteon (haversian system)

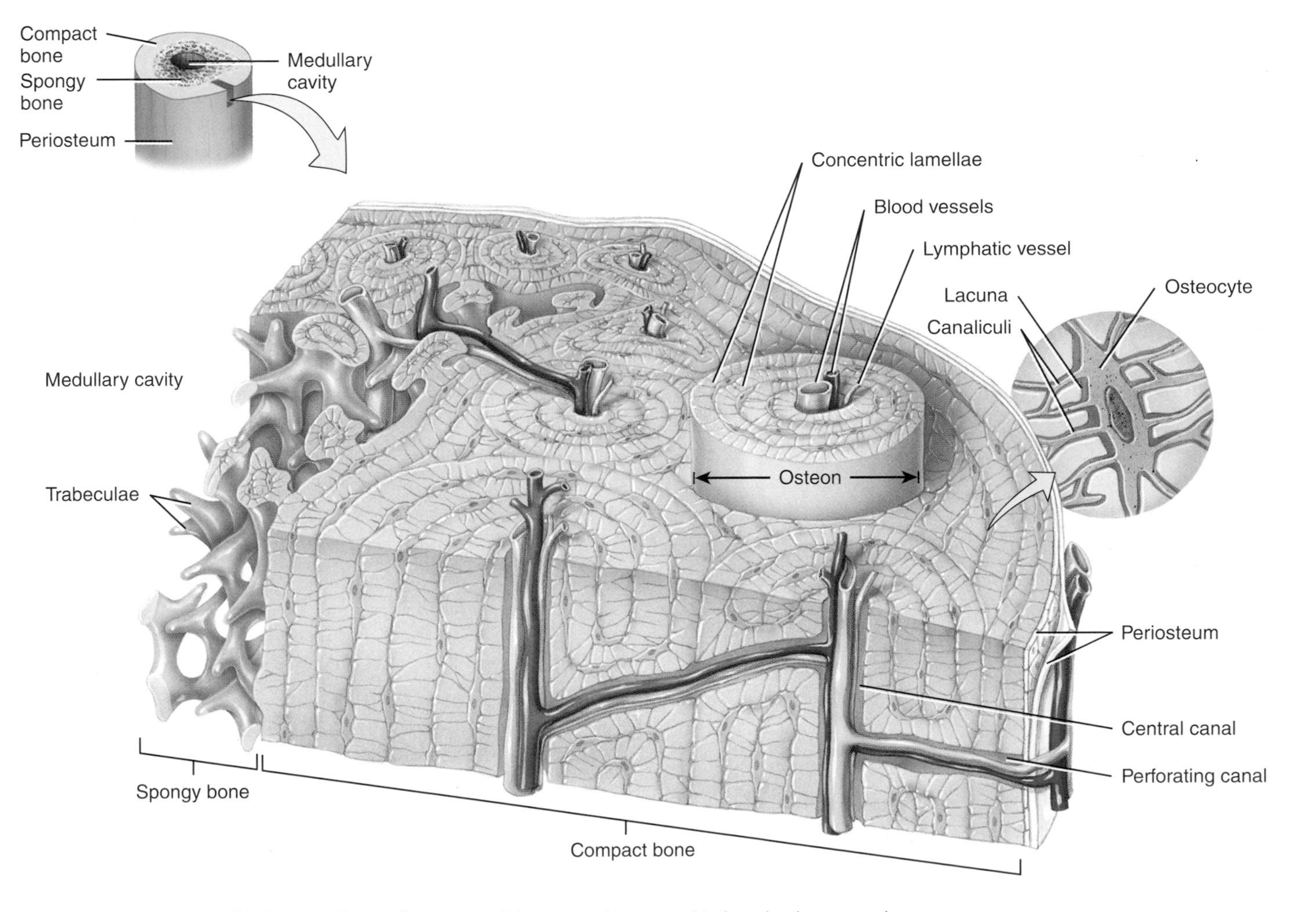

(c) Osteons (haversian systems) in compact bone and trabeculae in spongy bone

As people age, some central (haversian) canals may become blocked. What effect would this have on the osteocytes?

Blood vessels, lymphatic vessels, and nerves from the periosteum penetrate the compact bone through transverse ***perforating (volkmann's) canals***. The vessels and nerves of the perforating canals connect with those of the medullary cavity, periosteum, and central (haversian) canals.

Compact bone tissue contains few spaces. It is found beneath the periosteum of all bones and makes up the bulk of the diaphyses of long bones. Compact bone tissue provides protection and support and resists the stresses produced by weight and movement.

Spongy Bone Tissue

In contrast to compact bone tissue, ***spongy bone tissue*** does not contain osteons. As shown in Figure 6.2c it consists of units called ***trabeculae*** (tra-BEK-ū-lē = little beams; singular is *trabecula*), an irregular latticework of thin columns of bone. The macroscopic spaces between the trabeculae of some bones are filled with red bone marrow. Within each trabecula are osteocytes that lie in lacunae. Radiating from the lacunae are canaliculi.

Spongy bone tissue makes up most of the bone tissue of short, flat, and irregularly shaped bones. It also forms most of the epiphyses of long bones and a narrow rim around the medullary cavity of the diaphysis of long bones.

Spongy bone tissue is different from compact bone tissue in two respects. First, spongy bone tissue is light, which reduces the overall weight of a bone so that it moves more readily when pulled by a skeletal muscle. Second, the trabeculae of spongy bone tissue support and protect the red bone marrow. The spongy bone tissue in the hip bones, ribs, breastbone, backbones, and the ends of long bones is the only site where red bone marrow is found and, thus, the site of blood cell production in adults.

In a **bone scan**, a small amount of a radioactive tracer compound that is readily absorbed by bone is injected intravenously. The degree of uptake of the tracer is related to the amount of blood flow to the bone. Normal bone tissue is identified by a consistent gray color throughout because of its uniform uptake of the radioactive tracer. Darker or lighter areas may indicate bone abnormalities. Darker areas called "hot spots" are areas of increased metabolism that absorb more of the radioactive tracer due to increased blood flow. Hot spots may indicate bone cancer, abnormal healing of fractures, or abnormal bone growth. Lighter areas called "cold spots" are areas of decreased metabolism that absorb less of the radioactive tracer due to decreased blood flow. Cold spots may indicate problems such as degenerative bone disease, decalcified bone, fractures, bone infections, Paget's disease, or rheumatoid arthritis. A bone scan detects abnormalities 3 to 6 months sooner than standard x-ray procedures and exposes the patient to less radiation. A bone scan is the standard test for bone density screening, particularly for osteoporosis in females.

CHECKPOINT

4. Diagram the parts of a long bone, and list the functions of each part.
5. What are the four types of cells in bone tissue?
6. How are spongy and compact bone tissue different in terms of their microscopic appearance, location, and function?

BONE FORMATION

OBJECTIVES • **Explain the importance of bone formation during different phases of a person's lifetime.**
- **Describe the factors that affect bone growth during a person's lifetime.**

The process by which bone forms is called ***ossification*** (os′-i-fi-KĀ-shun; *ossi-* = bone; *-fication* = making). Bone formation occurs in four principal situations: (1) the initial formation of bones in an embryo and fetus, (2) the growth of bones during infancy, childhood, and adolescence until their adult sizes are reached, (3) the remodeling of bone (replacement of old bone tissue by new bone tissue throughout life); and (4) the repair of fractures (breaks in bones) throughout life.

Initial Bone Formation in an Embryo and Fetus

We will first consider the initial formation of bone in an embryo and fetus. The embryonic "skeleton" is at first composed of mesenchyme shaped like bones and are the sites where ossification occurs. These "bones" provide the template for subsequent ossification, which begins during the sixth week of embryonic development and follows one of two patterns.

The two methods of bone formation, which both involve the replacement of a preexisting connective tissue with bone, do not lead to differences in the structure of mature bones, but are simply different methods of bone development. In the first type of ossification, called ***intramembranous ossification*** (in′-tra-MEM-bra-nus; *intra-* = within; *membram-* = membrane), bone forms directly within mesenchyme arranged in sheetlike layers that resemble membranes. In the second type, ***endochondral ossification*** (en′-dō-KON-dral; *endo-* = within; *-chondral* = cartilage), bone forms within hyaline cartilage that develops from mesenchyme.

Intramembranous Ossification

Intramembranous ossification is the simpler of the two methods of bone formation. The flat bones of the skull and mandible (lower jawbone) are formed in this way. Also, the "soft spots" that help the fetal skull pass through the birth canal later harden as they undergo intramembranous ossification, which occurs as follows (Figure 6.3):

1. ***Development of the ossification center.*** At the site where bone will develop, called the ***ossification center***, cells in mesenchyme cluster together and differentiate, first into osteogenic cells and then into osteoblasts. Osteoblasts secrete the organic extracellular matrix of bone.
2. ***Calcification.*** Next, the secretion of extracellular matrix stops and the cells, now called osteocytes, lie in lacunae and extend their narrow cytoplasmic processes into canaliculi that radiate in all directions. Within a few days, calcium and other mineral salts are deposited and the extracellular matrix hardens or calcifies (calcification).
3. ***Formation of trabeculae.*** As the bone extracellular matrix forms, it develops into trabeculae that fuse with one another to form spongy bone. Blood vessels grow into the spaces between the trabeculae. Connective tissue that is associated with the blood vessels in the trabeculae differentiates into red bone marrow.

Figure 6.3 Intramembranous ossification. Illustrations 1 and 2 show a smaller field of vision at higher magnification than illustrations 3 and 4.

Intramembranous ossification involves the formation of bone within mesenchyme arranged in sheetlike layers that resemble membranes.

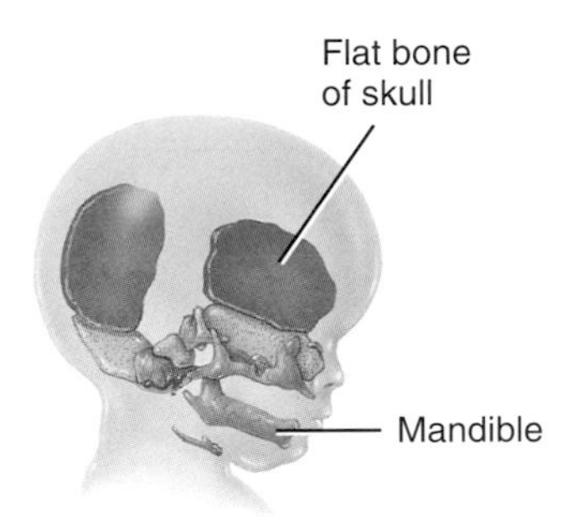

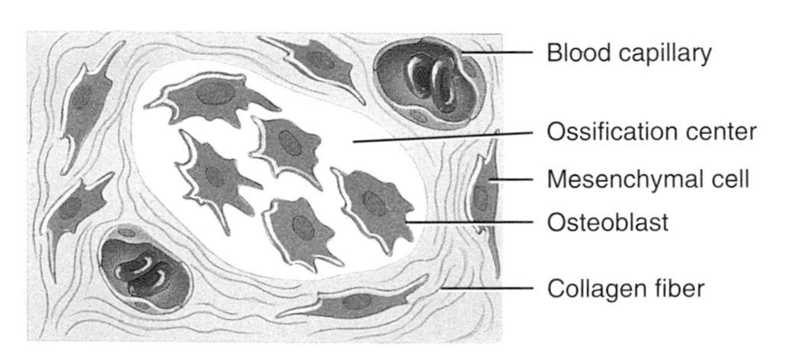

1 Development of ossification center

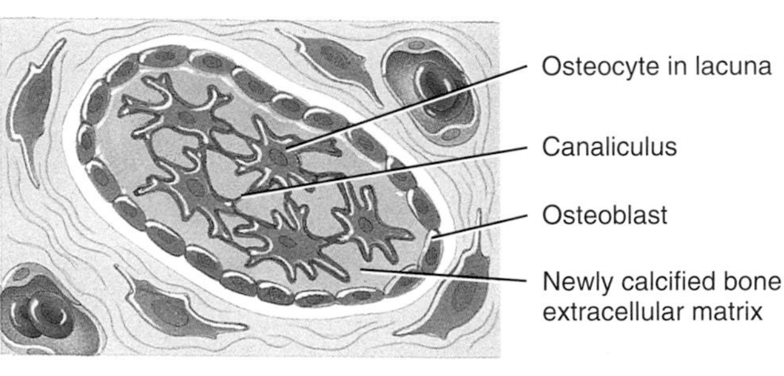

2 Calcification

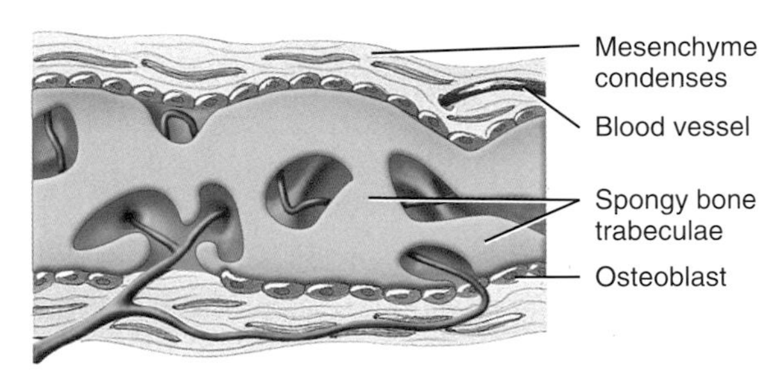

3 Formation of trabeculae

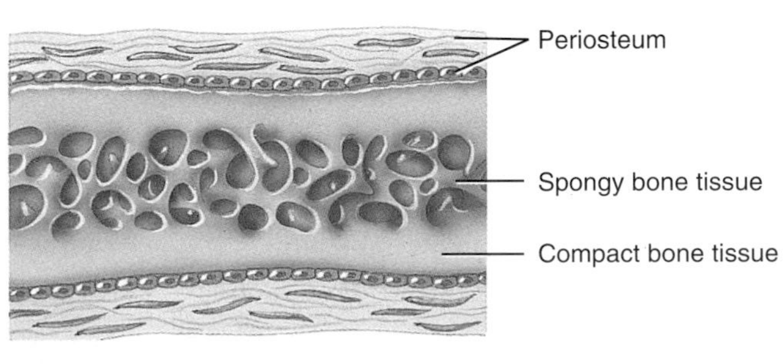

4 Development of the periosteum

Which bones of the body develop by intramembranous ossification?

4. ***Development of the periosteum.*** At the periphery of the bone, the mesenchyme condenses and develops into the periosteum. Eventually, a thin layer of compact bone replaces the surface layers of the spongy bone, but spongy bone remains in the center.

Endochondral Ossification

The replacement of cartilage by bone is called ***endochondral ossification.*** Most bones of the body are formed in this way, but as shown in Figure 6.4, this type of ossification is best observed in a long bone:

Figure 6.4 Endochondral ossification of the tibia (shin bone).

During endochondral ossification, bone gradually replaces a cartilage model.

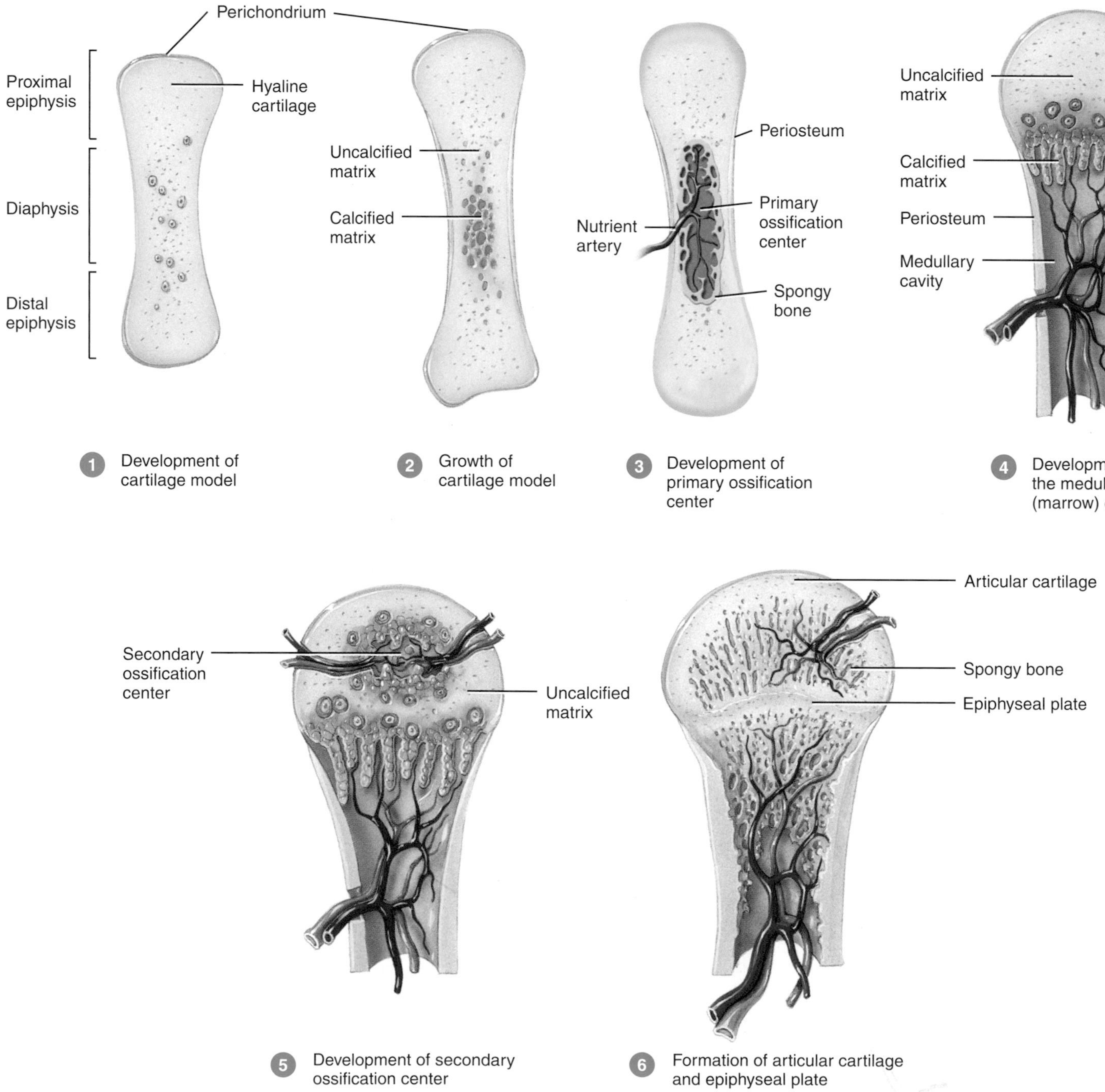

1. Development of cartilage model
2. Growth of cartilage model
3. Development of primary ossification center
4. Development of the medullary (marrow) cavity
5. Development of secondary ossification center
6. Formation of articular cartilage and epiphyseal plate

Which structure signals that bone growth in length has stopped?

1. **Development of the cartilage model.** At the site where the bone is going to form, cells in mesenchyme crowd together in the shape of the future bone and then develop into chondroblasts. The chondroblasts secrete cartilage extracellular matrix, producing a ***cartilage model*** consisting of hyaline cartilage. A membrane called the ***perichondrium*** (per′-i-KON-drē-um) develops around the cartilage model.
2. **Growth of the cartilage model.** Once chondroblasts become deeply buried in cartilage extracellular matrix, they are called chondrocytes. As the cartilage model continues to grow, chondrocytes in its mid-region increase in size and the surrounding extracellular matrix begins to calcify. Other chondrocytes within the calcifying cartilage die because nutrients can no longer diffuse quickly enough through the extracellular matrix. As chondrocytes die, lacunae form and eventually merge into small cavities.
3. **Development of the primary ossification center.** Primary ossification proceeds *inward* from the external surface of the bone. A nutrient artery penetrates the perichondrium and the calcifying cartilage model in the mid-region of the cartilage model, stimulating osteogenic cells in the perichondrium to differentiate into osteoblasts. Once the perichondrium starts to form bone, it is known as the ***periosteum***. Near the middle of the model, blood vessels grow into the disintegrating calcified cartilage and induce growth of a ***primary ossification center***, a region where bone tissue will replace most of the cartilage. Osteoblasts then begin to deposit bone extracellular matrix over the remnants of calcified cartilage, forming spongy bone trabeculae.
4. As the primary ossification center grows toward the ends of the bone, osteoclasts break down some of the newly formed spongy bone trabeculae. This activity leaves a cavity, the medullary (marrow) cavity, in the diaphysis (shaft). The medullary cavity then fills with red bone marrow. Most of the wall of the diaphysis is replaced by compact bone.
5. **Development of the secondary ossification centers.** When blood vessels enter the epiphyses, ***secondary ossification centers*** develop, usually around the time of birth. Bone formation is similar to that in primary ossification centers except that spongy bone remains in the interior of the epiphyses (no medullary cavities are formed there). Secondary ossification proceeds *outward* from the center of the epiphysis toward the outer surface of the bone.
6. **Formation of articular cartilage and the epiphyseal plate.** The hyaline cartilage that covers the epiphyses becomes the ***articular cartilage***. Prior to adulthood, hyaline cartilage remains between the diaphysis and epiphysis as the ***epiphyseal plate***, which is responsible for the lengthwise growth of long bones.

Bone Growth in Length and Thickness

During infancy, childood, and adolescence, long bones grow in length and thickness.

Growth in Length

Bone growth in length is related to the activity of the epiphyseal plate. Within the epiphyseal plate is a group of young chondrocytes that are constantly dividing. As a bone grows in length, new chondrocytes are formed on the epiphyseal side of the plate, while old chondrocytes on the diaphyseal side of the plate are replaced by bone. In this way the thickness of the epiphyseal plate remains relatively constant, but the bone on the diaphyseal side increases in length. When adolescence comes to an end, the formation of new cells and extracellular matrix decreases and eventually stops between ages 18 and 25. At this point, bone replaces all the cartilage, leaving a bony structure called the ***epiphyseal line***. With the appearance of the epiphyseal line, bone growth in length stops. If a bone fracture damages the epiphyseal plate, the fractured bone may be shorter than normal once adult stature is reached. This is because damage to cartilage, which is avascular, accelerates closure of the epiphyseal plate, thus inhibiting lengthwise growth of the bone.

Growth in Thickness

As long bones lengthen, they also grow in thickness (width). At the bone surface, cells in the perichondrium differentiate into osteoblasts, which secrete bone extracellular matrix. Then the osteoblasts develop into osteocytes, lamellae are added to the surface of the bone, and new osteons of compact bone tissue are formed. At the same time, osteoclasts in the endosteum destroy the bone tissue lining the medullary cavity. Bone destruction on the inside of the bone by osteoclasts occurs at a slower rate than bone formation on the outside of the bone. Thus, the medullary cavity enlarges as the bone increases in thickness.

Bone Remodeling

Like skin, bone forms before birth but continually renews itself thereafter. ***Bone remodeling*** is the ongoing replacement of old bone tissue by new bone tissue. It involves ***bone resorption***, the removal of minerals and collagen fibers from bone by osteoclasts, and ***bone deposition***, the addition of minerals and collagen fibers to bone by osteoblasts. Thus, bone resorption results in the destruction of bone extracellular matrix, while bone deposition results in the formation of bone extracellular matrix. Remodeling takes place at different rates in different regions of the body. Even after bones have reached their adult shapes and sizes, old bone is continually destroyed and new bone is formed in its place. Remodeling also removes injured bone, replacing it with new bone tissue. Remodeling may be triggered by factors such as exercise, sedentary lifestyle, and changes in diet.

Orthodontics (or-thō-DON-tiks) is the branch of dentistry concerned with the prevention and correction of poorly aligned teeth. The movement of teeth by braces places a stress on the bone that forms the sockets that anchor the teeth. In response to this artificial stress, osteoclasts and osteblasts remodel the sockets so that the teeth align properly.

A delicate balance exists between the actions of osteoclasts and osteoblasts. Should too much new tissue be formed, the bones become abnormally thick and heavy. If too much mineral material is deposited in the bone, the surplus may form thick bumps, called *spurs*, on the bone that interfere with movement at joints. Excessive loss of calcium or tissue weakens the bones, and they may break, as occurs in osteoporosis, or they may become too flexible, as in rickets and osteomalacia. (For more on these disorders, see the Disorders section at the end of the chapter.) Abnormal acceleration of the remodeling process results in a condition called Paget's disease, in which the newly formed bone, especially that of the pelvis, limbs, lower vertebrae, and skull, becomes hard and brittle and fractures easily.

Fractures

A ***fracture*** is any break in a bone. Types of fractures include the following:

1. **Partial:** an incomplete break across the bone, such as a crack.
2. **Complete:** a complete break across the bone; that is, the bone is broken into two or more pieces.
3. **Closed (simple):** the fractured bone does not break through the skin.
4. **Open (compound):** the broken ends of the bone protrude through the skin.

Repair of a fracture involves several steps. First, phagocytes begin to remove any dead bone tissue. Then, chondroblasts form fibrocartilage at the fracture site and this bridges the broken ends of the bone. Next, the fibrocartilage is converted to spongy bone tissue by osteoblasts. Finally, bone remodeling occurs, in which dead portions of bone are absorbed by osteoclasts and spongy bone is converted to compact bone.

Although bone has a generous blood supply, healing sometimes takes months. The calcium and phosphorus needed to strengthen and harden new bone are deposited only gradually, and bone cells generally grow and reproduce slowly. The temporary disruption in their blood supply also helps explain the slowness of healing of severely fractured bones.

Factors Affecting Bone Growth

Bone growth in the young, bone remodeling in the adult, and the repair of fractured bone depend on several factors. These include (1) adequate minerals, most importantly calcium, phosphorus, and magnesium; (2) vitamins A, C, and D; (3) several hormones; and (4) weight-bearing exercise (exercise that places stress on bones). Before puberty, the main hormones that stimulate bone growth are human growth hormone (hGH), which is produced by the anterior lobe of the pituitary gland, and insulinlike growth factors (IGFs), which are produced locally by bone and also by the liver in response to hGH stimulation. Oversecretion of hGH produces gigantism, in which a person becomes much taller and heavier than normal, and undersecretion of hGH produces dwarfism (short stature). Thyroid hormones, from the thyroid gland, and insulin, from the pancreas, also stimulate normal bone growth. At puberty, estrogens (sex hormones produced by the ovaries) and androgens (sex hormones produced by the testes in males and the adrenal glands in both sexes) start to be released in larger quantities. These hormones are responsible for the sudden growth spurt that occurs during the teenage years. Estrogens also promote changes in the skeleton that are typical of females, for example, widening of the pelvis.

Bone's Role in Calcium Homeostasis

Bone is the major reservoir of calcium, storing 99% of the total amount of calcium present in the body. Calcium (Ca^{2+}) becomes available to other tissues when bone is broken down during remodeling. However, even small changes in blood calcium levels can be deadly—the heart may stop (cardiac arrest) if the level is too high or breathing may cease (respiratory arrest) if the level is too low. In addition, most functions of nerve cells depend on just the right level of Ca^{2+}, many enzymes require Ca^{2+} as a cofactor, and blood clotting requires Ca^{2+}. The role of bone in calcium homeostasis is to "buffer" the blood calcium level, releasing Ca^{2+} to the blood when the blood calcium level falls (using osteoclasts) and depositing Ca^{2+} back in bone when the blood level rises (using osteoblasts).

The most important hormone that regulates Ca^{2+} exchange between bone and blood is ***parathyroid hormone (PTH)***, secreted by the parathyroid glands (see Figure 13.10 on page 327). PTH secretion operates via a negative feedback system (Figure 6.5). If some stimulus causes blood Ca^{2+} level to decrease, parathyroid gland cells (receptors) detect this change and increase their production of a molecule known as cyclic adenosine monophosphate (cyclic AMP). The gene for PTH within the nucleus of a parathyroid gland cell, which acts as the control center, detects the increased production of cyclic AMP (the input). As a result, PTH synthesis speeds up, and more PTH (the output) is released into

Figure 6.5 Negative feedback system for the regulation of blood calcium (Ca^{2+}) level.

Release of calcium from bone extracellular matrix and retention of calcium by the kidneys are the two main ways that blood calcium level can be increased.

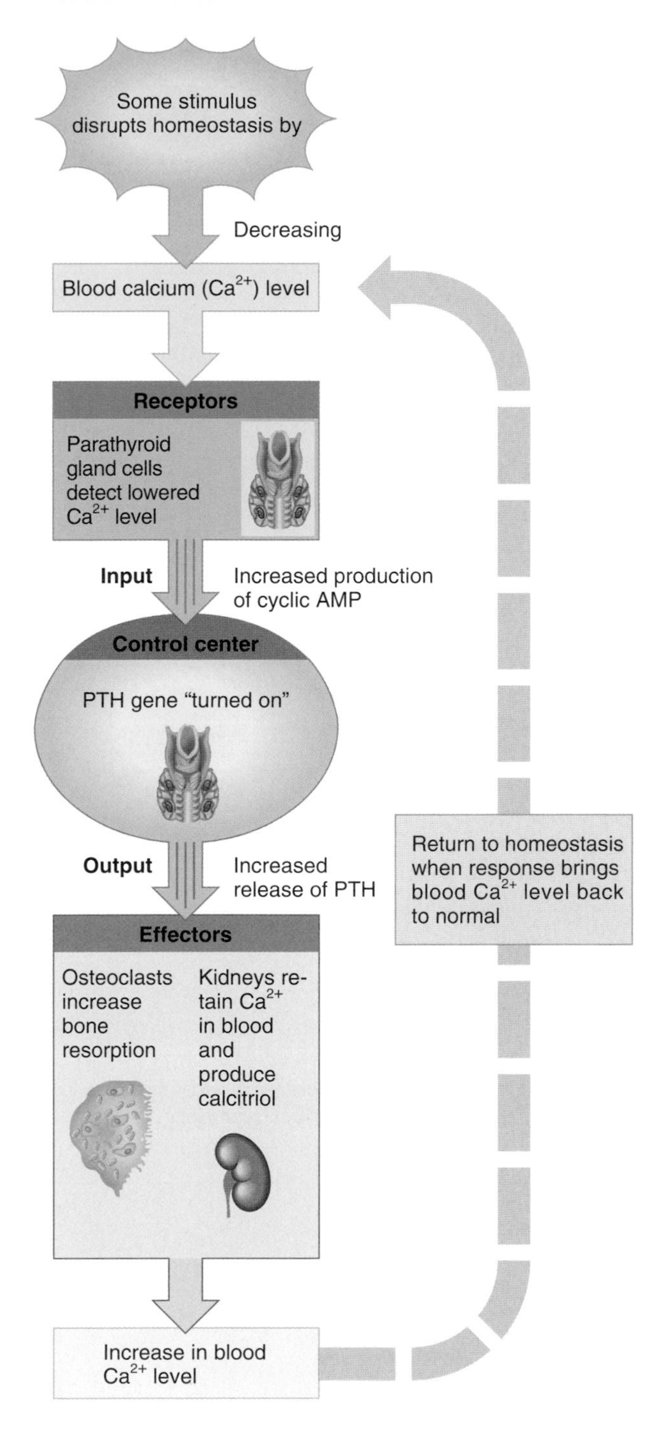

What body functions depend on proper levels of Ca^{2+}?

the blood. The presence of higher levels of PTH increases the number and activity of osteoclasts (effectors), which step up the pace of bone resorption. The resulting release of Ca^{2+} from bone into blood returns the blood Ca^{2+} level to normal.

PTH also decreases loss of Ca^{2+} in the urine, so more is retained in the blood, and it stimulates formation of calcitriol, a hormone that promotes absorption of calcium from the gastrointestinal tract. Both of these effects also help elevate the blood Ca^{2+} level.

As you will learn in Chapter 13, another hormone involved in calcium homeostasis is *calcitonin (CT)*. This hormone is produced by the thyroid gland and decreases blood Ca^{2+} level by inhibiting the action of osteoclasts, thus decreasing bone resorption.

■ **CHECKPOINT**

7. Distinguish between intramembranous and endochondral ossification.
8. Explain how bones grow in length and thickness.
9. What is bone remodeling? Why is it important?
10. Define a fracture and explain how fracture repair occurs.
11. What factors affect bone growth?
12. What are some of the important functions of calcium in the body?

EXERCISE AND BONE TISSUE

OBJECTIVE • **Describe how exercise and mechanical stress affect bone tissue.**

Within limits, bone tissue has the ability to alter its strength in response to mechanical stress. When placed under stress, bone tissue becomes stronger through increased deposition of mineral salts and production of collagen fibers. Without mechanical stress, bone does not remodel normally because resorption outpaces bone formation. The absence of mechanical stress weakens bone through decreased numbers of collagen fibers and ***demineralization***, loss of bone minerals.

The main mechanical stresses on bone are those that result from the pull of skeletal muscles and the pull of gravity. If a person is bedridden or has a fractured bone in a cast, the strength of the unstressed bones diminishes. Astronauts subjected to the weightlessness of space also lose bone mass. In both cases, the bone loss can be dramatic, as much as 1% per week. Bones of athletes, which are repetitively and highly stressed, become notably thicker than those of nonathletes. Weight-bearing activities, such as walking or moderate weightlifting, help build and retain bone mass. Adolescents and young adults should engage in regular weight-bearing exercise prior to the closure of the epiphyseal plates to help build total mass before its inevitable reduction with aging.

However, the benefits of exercise do not end in young adulthood. Even elderly people can strengthen their bones by engaging in weight-bearing exercise.

Table 6.1 summarizes the factors that influence bone metabolism: growth, remodeling, and repair of fractured bones.

■ **CHECKPOINT**

13. What types of mechanical stress may be used to strengthen bone tissue?

DIVISIONS OF THE SKELETAL SYSTEM

OBJECTIVE • Group the bones of the body into axial and appendicular divisions.

Because the skeletal system forms the framework of the body, a familiarity with the names, shapes, and positions of individual bones will help you locate other organs. For example, the radial artery, the site where the pulse is usually taken, is named for its closeness to the radius, the lateral bone of the forearm. The ulnar nerve is named for its closeness to the ulna, the medial bone of the forearm. The frontal lobe of the brain lies deep to the frontal (forehead) bone. The tibialis anterior muscle lies along the anterior surface of the tibia (shin bone).

The adult human skeleton consists of 206 bones grouped in two principal divisions: 80 in the ***axial skeleton*** and 126 in the ***appendicular skeleton*** (Table 6.2 and Figure 6.6). The axial skeleton consists of the bones that lie around the longitudinal *axis* of the human body, an imaginary line that runs through the body's center of gravity from the head to the space between the feet: the bones of the skull, auditory ossicles (ear bones), hyoid bone, ribs, sternum, and vertebrae. The appendicular skeleton contains the bones of the upper and lower limbs plus the bone groups called *girdles* that connect the limbs to the axial skeleton. The skeletons of infants and children have more than 206 bones because some of their bones, such as the hip bones and vertebrae, fuse later in life.

■ **CHECKPOINT**

14. How are the limbs connected to the axial skeleton?

Table 6.1 Summary of Factors That Influence Bone Metabolism

Factor	Comment
Minerals	
Calcium and phosphorus	Make bone extracellular matrix hard.
Magnesium	Needed for normal activity of osteoblasts.
Vitamins	
Vitamin A	Needed for the activity of osteoblasts during remodeling of bone; deficiency stunts bone growth; toxic in high doses.
Vitamin C	Helps maintain bone extracellular matrix; deficiency leads to decreased collagen production, which slows down bone growth and delays repair of broken bones.
Vitamin D	Active form (calcitriol) is formed in the skin and kidneys; helps build bone by increasing absorption of calcium from small intestine into blood; deficiency causes faulty calcification and slows down bone growth; may reduce the risk of osteoporosis but is toxic if taken in high doses.
Hormones	
Human growth hormone (hGH)	Secreted by the anterior lobe of the pituitary gland; promotes general growth of all body tissues, including bone, mainly by stimulating production of insulinlike growth factors.
Insulinlike growth factors (IGFs)	Secreted by the liver, bones, and other tissues upon stimulation by human growth hormone; stimulate the uptake of amino acids and synthesis of proteins; promote tissue repair and bone growth.
Insulin	Secreted by the pancreas; promotes normal bone growth.
Thyroid hormones (thyroxine and triiodothyronine)	Secreted by thyroid gland; promote normal bone growth.
Parathyroid hormone (PTH)	Secreted by the parathyroid glands; promotes bone resorption by osteoclasts; enhances recovery of Ca^{2+} from urine; promotes formation of the active form of vitamin D (calcitriol).
Calcitonin (CT)	Secreted by the thyroid gland; inhibits bone resorption by osteoclasts.
Exercise	Weight-bearing activities help build thicker, stronger bones and retard the loss of bone mass that occurs as people age.

Figure 6.6 Divisions of the skeletal system. The axial skeleton is indicated in green. (Note the position of the hyoid bone in Figure 6.7b.)

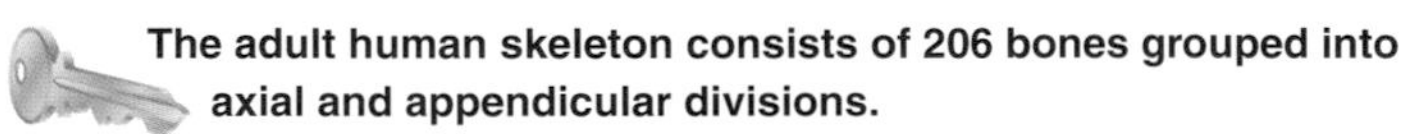

The adult human skeleton consists of 206 bones grouped into axial and appendicular divisions.

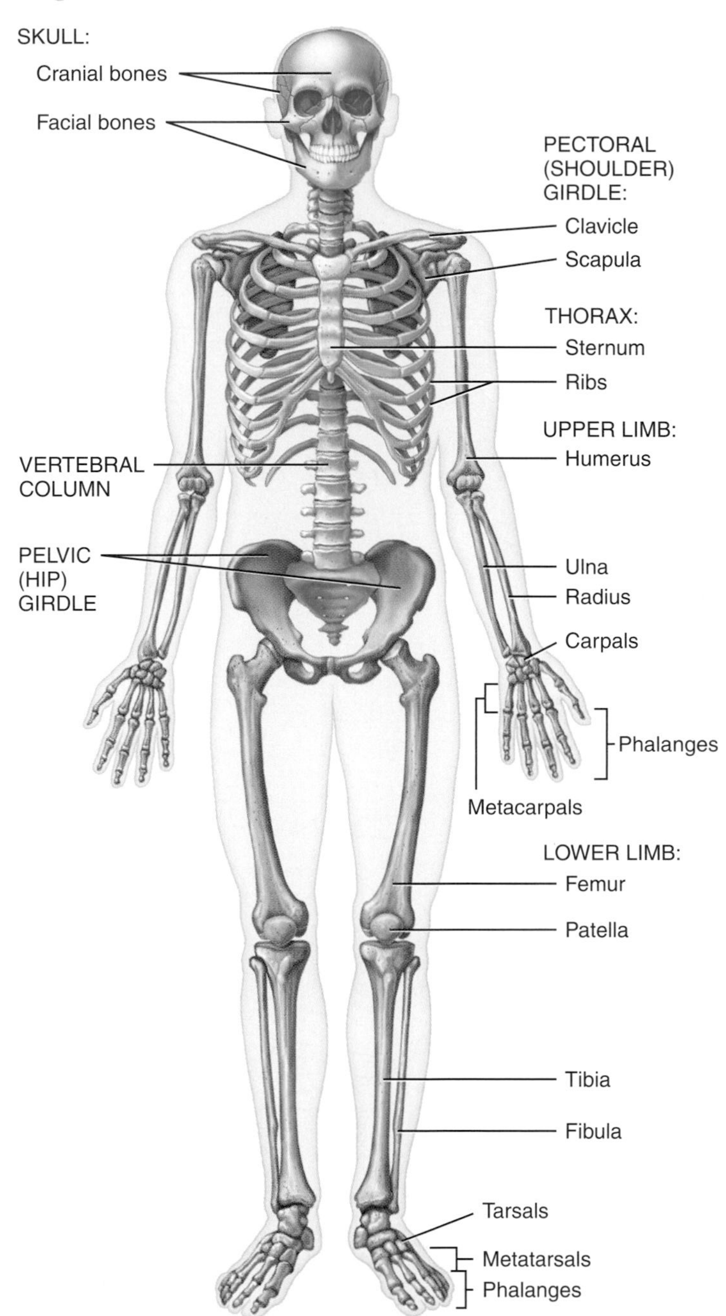

Anterior view

Identify each of the following bones as part of the axial skeleton or the appendicular skeleton: skull, clavicle, vertebral column, shoulder girdle, humerus, pelvic girdle, and femur.

Table 6.2 The Bones of the Adult Skeletal System

Division of the Skeleton	Structure	Number of Bones
Axial Skeleton	**Skull**	
	Cranium	8
	Face	14
	Hyoid	1
	Auditory ossicles	6
	Vertebral column	26
	Thorax	
	Sternum	1
	Ribs	24
		Subtotal = 80
Appendicular Skeleton	**Pectoral (shoulder) girdles**	
	Clavicle	2
	Scapula	2
	Upper limbs	
	Humerus	2
	Ulna	2
	Radius	2
	Carpals	16
	Metacarpals	10
	Phalanges	28
	Pelvic (hip) girdle	
	Hip or pelvic bone	2
	Lower limbs	
	Femur	2
	Patella	2
	Fibula	2
	Tibia	2
	Tarsals	14
	Metatarsals	10
	Phalanges	28
		Subtotal = 126
		Total = 206

SKULL AND HYOID BONE

OBJECTIVE • Name the cranial and facial bones and indicate their locations and major structural features.

The ***skull***, which contains 22 bones, rests on top of the vertebral column. It includes two sets of bones: cranial bones and facial bones. The eight ***cranial bones*** form the cranial cavity that encloses and protects the brain. They are the frontal bone, two parietal bones, two temporal bones, occipital bone,

Figure 6.7 Skull. Although the hyoid bone is not part of the skull, it is included in (b) and (c) for reference.

The skull consists of two sets of bones: Eight cranial bones form the cranial cavity and fourteen facial bones form the face.

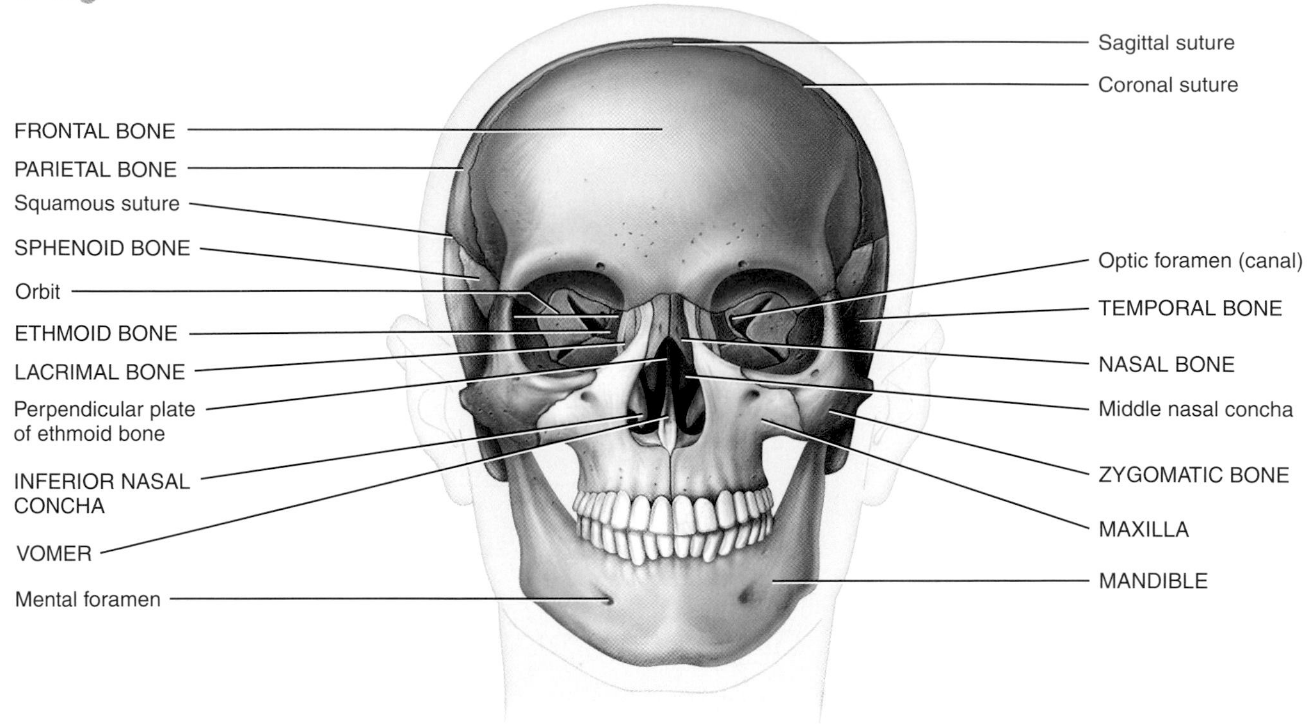

(a) Anterior view

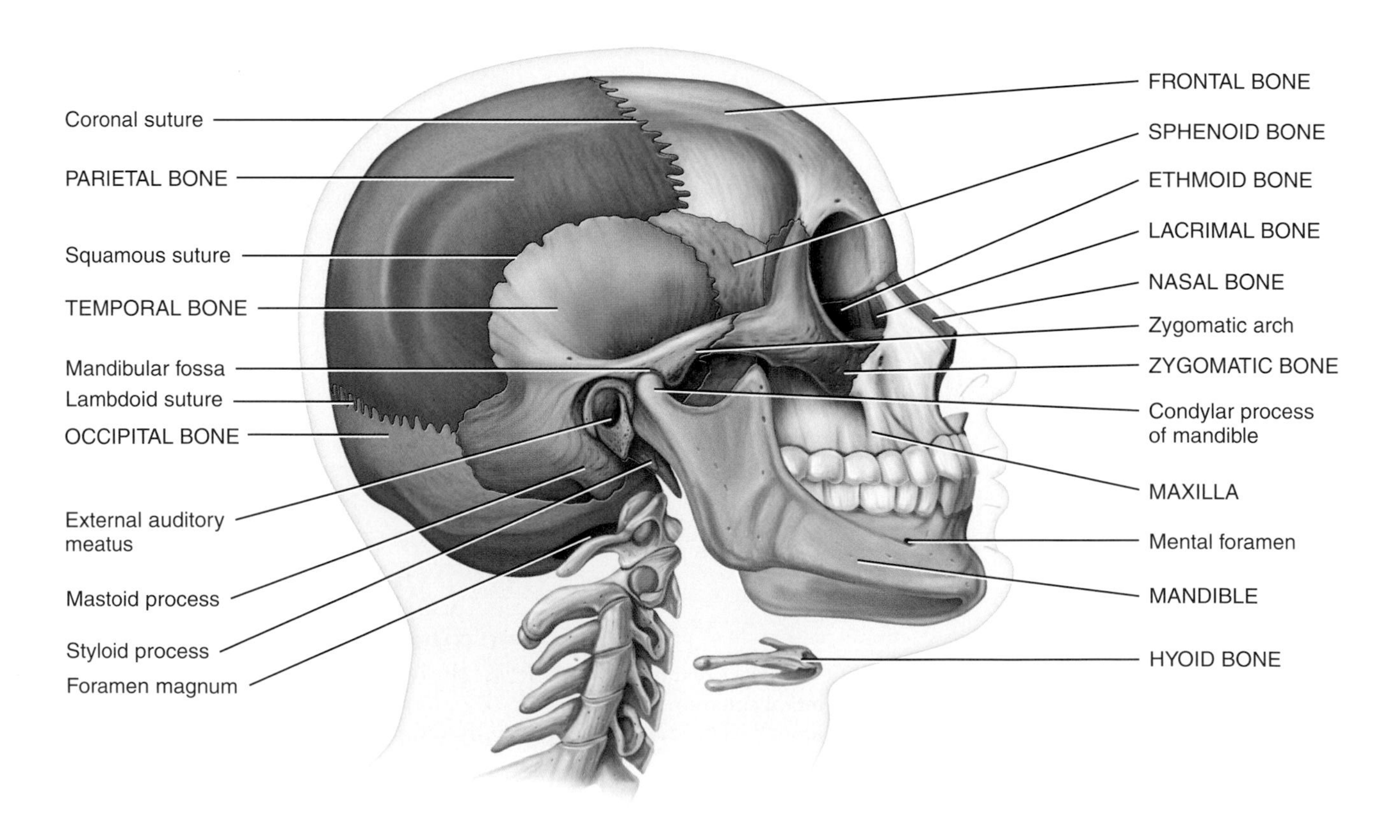

(b) Right lateral view

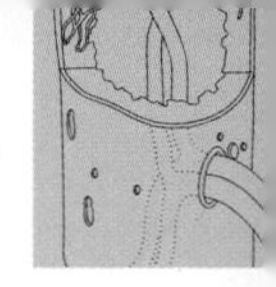

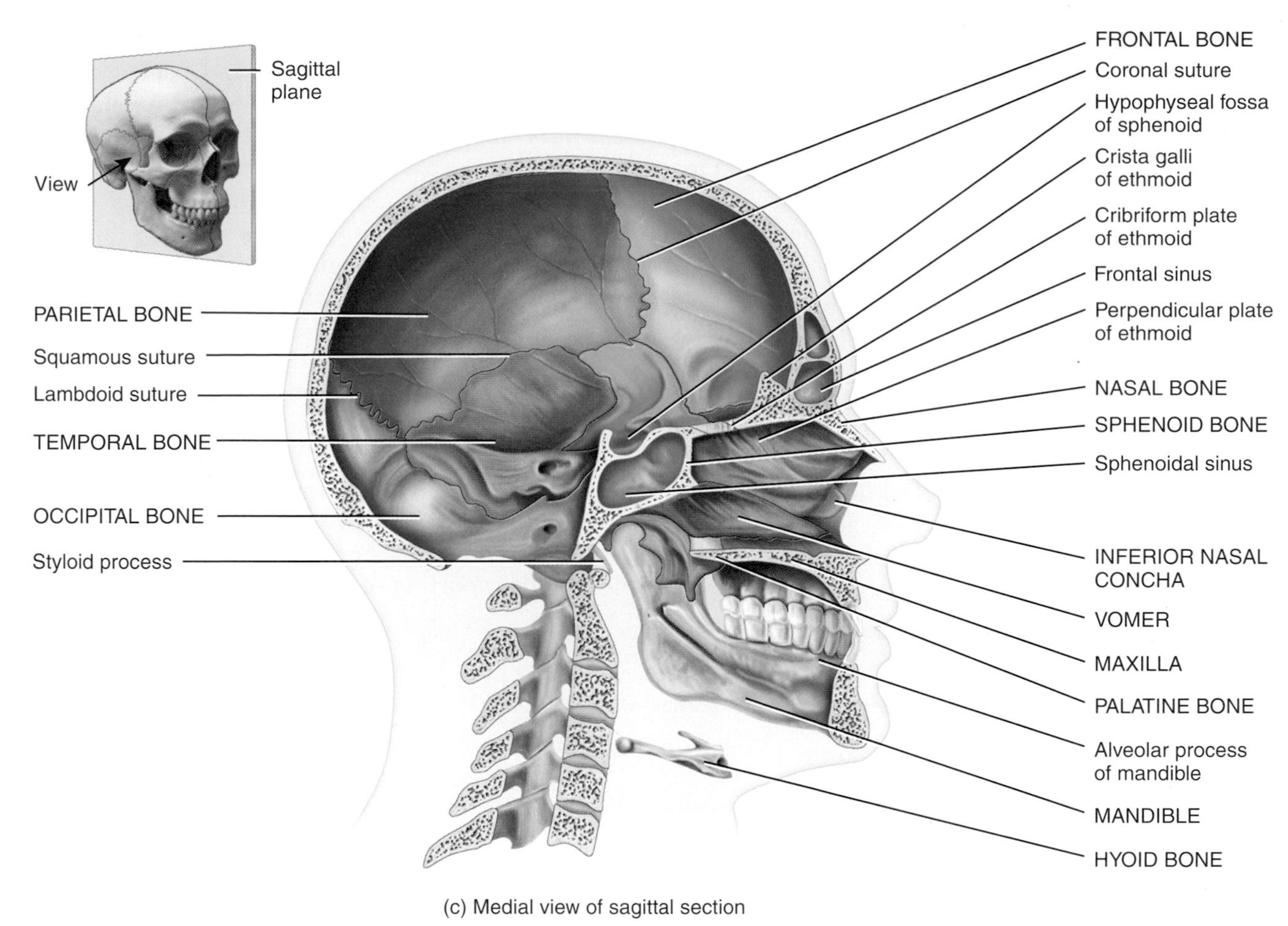

(c) Medial view of sagittal section

What are the names of the cranial bones?

sphenoid bone, and ethmoid bone. Fourteen ***facial bones*** form the face: two nasal bones, two maxillae, two zygomatic bones, the mandible, two lacrimal bones, two palatine bones, two inferior nasal conchae, and the vomer. Figure 6.7 shows these bones from three different viewing directions to permit you to view all of the bones from their best perspective.

The cranial bones have functions besides protection of the brain. Their inner surfaces attach to membranes (meninges) that stabilize the positions of the brain, blood vessels, and nerves. Their outer surfaces provide large areas of attachment for muscles that move various parts of the head. Besides forming the framework of the face, the facial bones protect and provide support for the entrances to the digestive and respiratory systems. The facial bones also provide attachment for some muscles that are involved in producing various facial expressions. Together, the cranial and facial bones protect and support the delicate special sense organs for vision, taste, smell, hearing, and equilibrium (balance).

Cranial Bones

The ***frontal bone*** forms the forehead (the anterior part of the cranium), the roofs of the *orbits* (eye sockets; Figure 6.7a), and most of the anterior (front) part of the cranial floor. The *frontal sinuses* lie deep within the frontal bone (Figure 6.7c). These mucous membrane-lined cavities act as sound chambers that give the voice resonance. Other functions of the sinuses are given on page 131.

The two ***parietal bones*** (pa-RĪ-e-tal; *pariet-* = wall) form most of the sides and roof of the cranial cavity (Figure 6.7).

The two ***temporal bones*** (*tempor-* = temples) form the inferior (lower) sides of the cranium and part of the cranial floor. In the lateral view of the skull (Figure 6.7b), note that the temporal and zygomatic bones join to form the *zygomatic arch*. The *mandibular fossa* (depression) forms a joint with a projection on the mandible (lower jawbone) called the condylar process to form the *temporomandibular joint (TMJ)*. The mandibular fossa can be seen in Figure 6.8. The *external auditory meatus* is the canal in the temporal bone that leads to the middle ear. The *mastoid process* (*mastoid* = breast-shaped; see Figure 6.7b) is a rounded projection of the temporal bone posterior to (behind) the external auditory meatus. It serves as a point of attachment for several neck muscles. The *styloid process* (*styl-* = stake or pole; see Figure 6.7b) is a slender projection that points downward from the undersurface of the temporal bone and serves as a point of attachment for muscles and ligaments of the tongue and neck. The *carotid foramen* (Figure 6.8) is a hole through which the carotid artery passes.

The ***occipital bone*** (ok-SIP-i-tal; *occipit-* = back of head) forms the posterior part and most of the base of the cranium (Figures 6.7b, c and 6.8). The *foramen magnum* (*magnum* = large), the largest foramen in the skull, passes through the occipital bone (Figures 6.7b and 6.8). Within this foramen are the medulla oblongata of the brain, connecting to the spinal cord, and the vertebral and spinal arteries. The *occipital condyles* are oval processes, one on either side of the foramen magnum (Figure 6.8), that articulate (connect) with the first cervical vertebra.

The ***sphenoid bone*** (SFĒ-noyd = wedge-shaped) lies at the middle part of the base of the skull (Figures 6.7, 6.8, and 6.9). This bone is called the keystone of the cranial floor because it articulates with all the other cranial bones, holding them together. The shape of the sphenoid bone resembles a bat with outstretched wings. The cubelike central portion of the sphenoid bone contains the *sphenoidal sinuses*, which drain into the nasal cavity (see Figures 6.7a and 6.11). On the superior surface of the sphenoid is a depression called the ***hypophyseal fossa*** (hī-pō-FIZ-ē-al), which contains the pituitary gland. Two nerves pass through foramina in the sphenoid bone: the mandibular nerve through the *foramen ovale* and the optic nerve through the *optic foramen (canal)*.

The ***ethmoid bone*** (ETH-moid = sievelike) is spongelike in appearance and is located in the anterior part of the cranial floor between the orbits (Figure 6.10 on page 130). It forms part of the anterior portion of the cranial floor, the medial wall of the orbits, the superior portions of the nasal septum, a partition that divides the nasal cavity into right and left sides, and most of the side walls of the nasal cavity. The ethmoid bone contains 3 to 18 air spaces, or "cells," that give this bone a sievelike appearance. The ethmoidal cells together form the *ethmoidal sinuses* (see Figure 6.11). The *perpendicular plate* forms the upper portion of the nasal septum. The *cribri-*

Figure 6.8 Skull.

The occipital bone forms most of the posterior and inferior portion of the cranium.

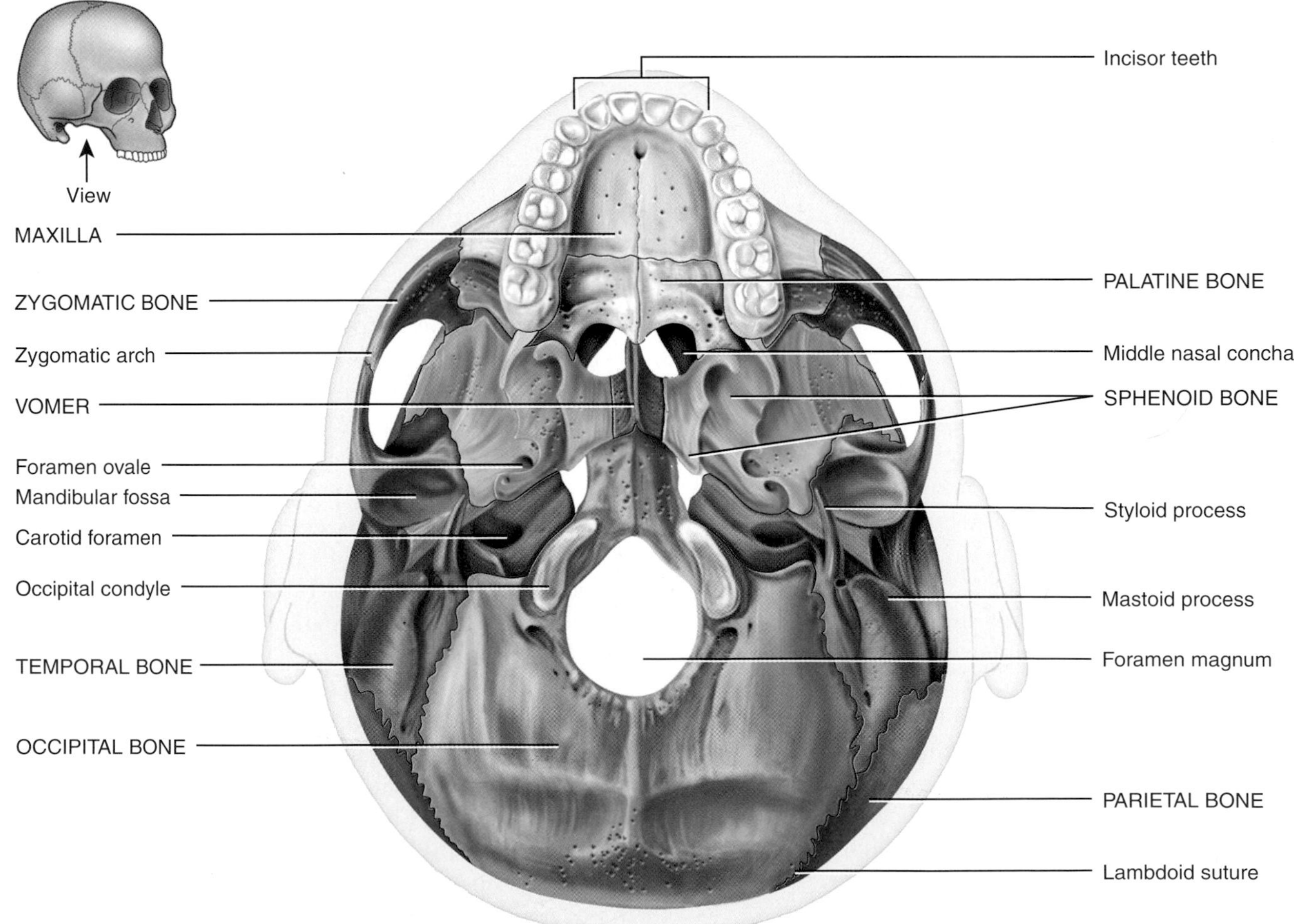

Inferior view, mandible removed

What is the largest foramen in the skull?

Figure 6.9 **Sphenoid bone.**

The sphenoid bone is called the keystone of the cranial floor because it articulates with all other cranial bones, holding them together.

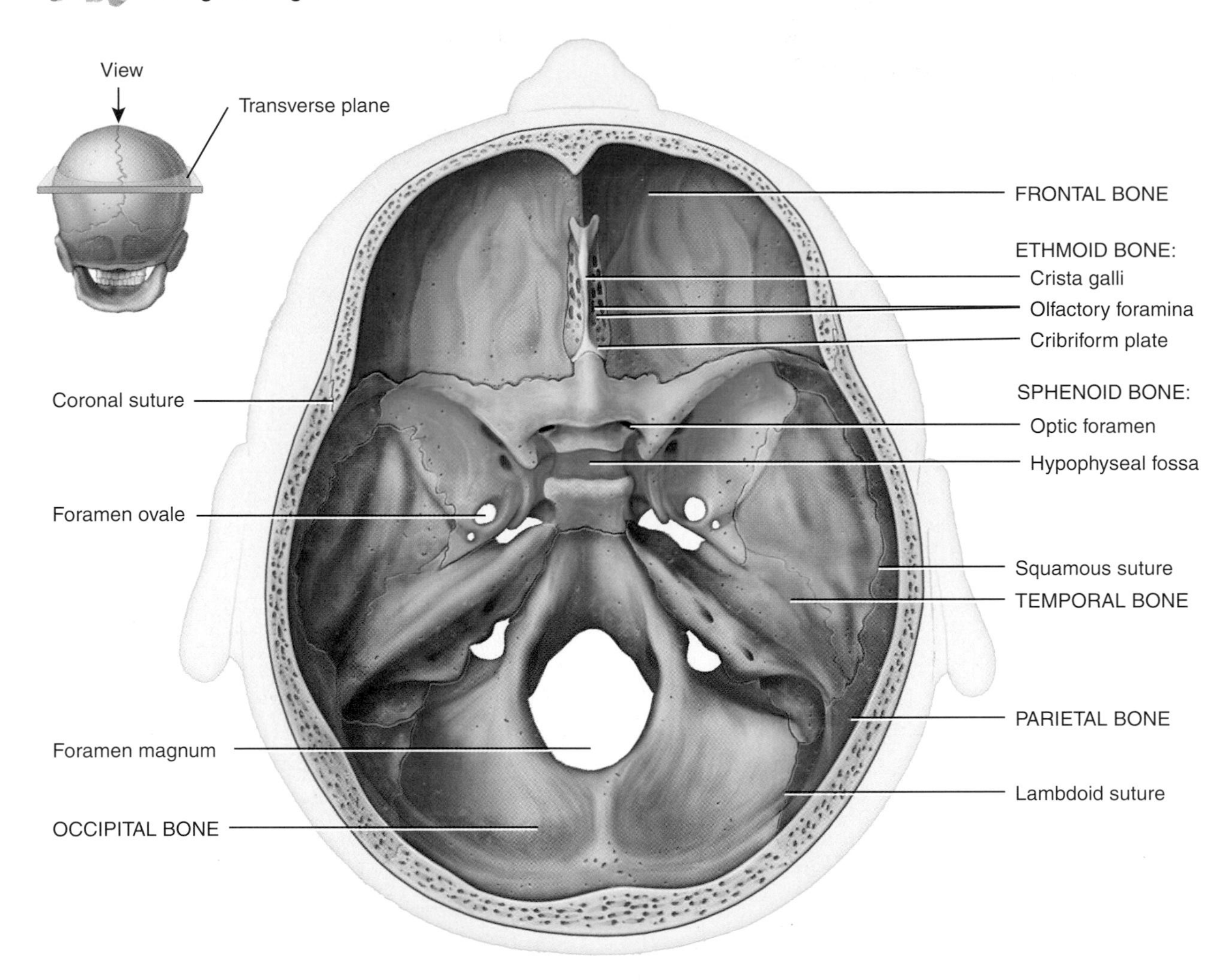

Viewed from above in floor of cranium

Starting at the crista galli of the ethmoid bone and going in a clockwise direction, what are the names of the bones that articulate with the sphenoid bone?

form plate (KRIB-ri-form) forms the roof of the nasal cavity (Figure 6.10). It contains the *olfactory foramina* (*olfact-* = to smell), holes through which fibers of the olfactory nerves pass (see Figure 6.9). Projecting upward from the cribriform plate is a triangular process called the *crista galli* (= cock's comb), which serves as a point of attachment for the membranes (meninges) that cover the brain (Figure 6.10).

Also part of the ethmoid bone are two thin, scroll-shaped bones on either side of the nasal septum. These are called the *superior nasal concha* (KONG-ka; *conch-* = shell) and the *middle nasal concha*. The plural is *conchae* (KONG-kē). The conchae cause turbulence in inhaled air, which results in many inhaled particles striking and becoming trapped in the mucus that lines the nasal passageways. This turbulence thus cleanses the inhaled air before it passes into the rest of the respiratory tract.

Facial Bones

The shape of the face changes dramatically during the first two years after birth. The brain and cranial bones expand, the teeth form and erupt (emerge), and the paranasal sinuses increase in size. Growth of the face ceases at about 16 years of age.

The paired ***nasal bones*** form part of the bridge of the nose (see Figure 6.7a). The rest of the supporting tissue of the nose consists of cartilage.

The paired ***maxillae*** (mak-SIL-ē = jawbones; singular is *maxilla*) unite to form the upper jawbone and articulate with every bone of the face except the mandible (lower jawbone) (see Figure 6.7). Each maxilla contains a *maxillary sinus* that empties into the nasal cavity (see Figure 6.11). The *alveolar*

Figure 6.10 Ethmoid bone.

The ethmoid bone is the major supporting structure of the nasal cavity.

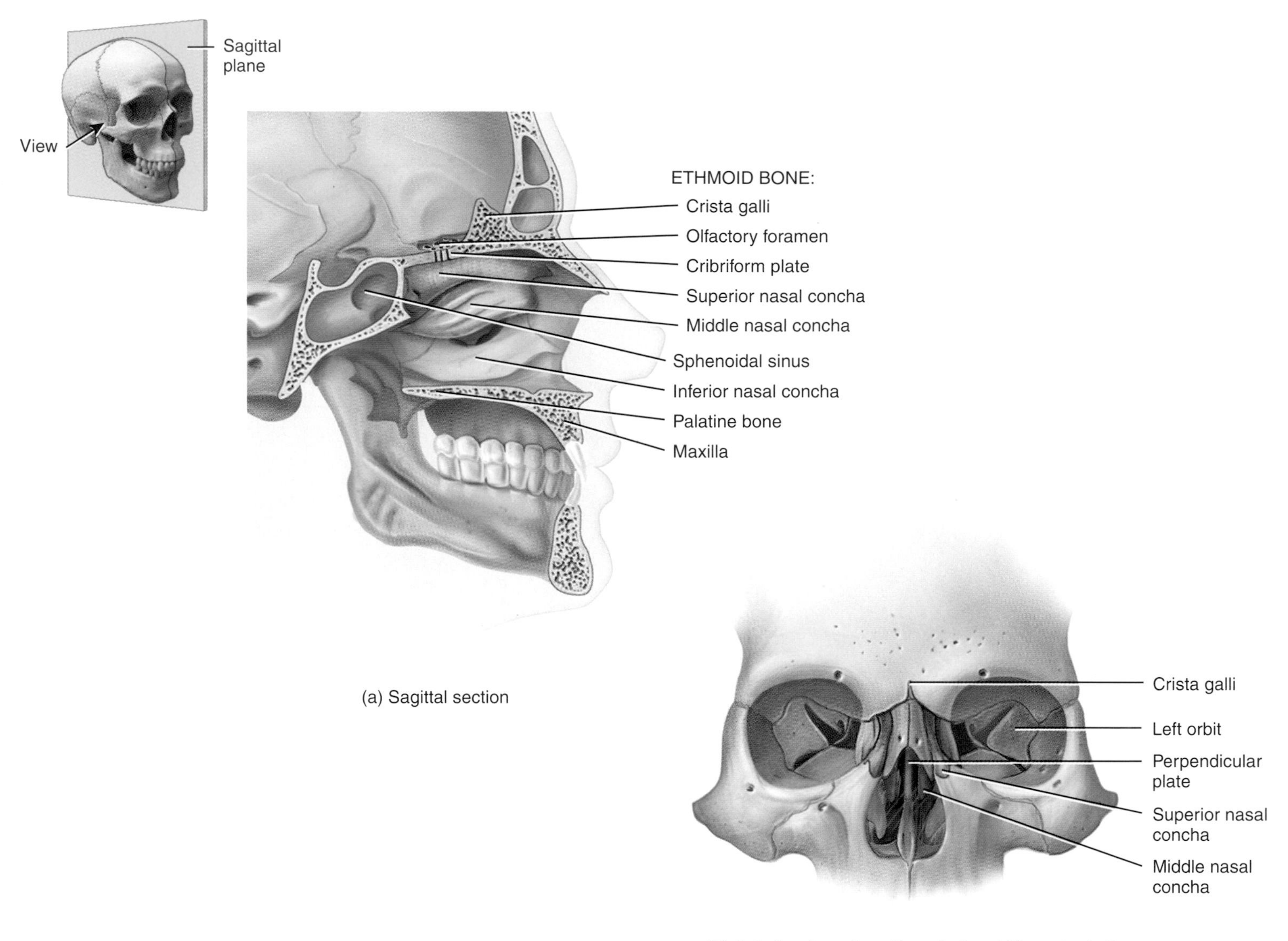

(a) Sagittal section

(b) Anterior view of position of ethmoid bone in skull

What part of the ethmoid bone forms the top part of the nasal septum?

process (al-VĒ-ō-lar; *alveol-* = small cavity) of the maxilla is an arch that contains the *alveoli* (sockets) for the maxillary (upper) teeth. The maxilla forms the anterior three-quarters of the hard palate.

Usually the left and right maxillary bones unite during weeks 10 to 12 of embryonic development. Failure to do so can result in one type of **cleft palate.** The condition may also involve incomplete fusion of the palatine bones (see Figure 6.8). Another form of this condition, called **cleft lip,** involves a split in the upper lip. Cleft lip and cleft palate often occur together. Depending on the extent and position of the cleft, speech and swallowing may be affected. Facial and oral surgeons recommend closure of cleft lip during the first few weeks following birth, and surgical results are excellent. Repair of cleft palate typically is completed between 12 and 18 months of age, ideally before the child begins to talk. Because the palate is important for pronouncing consonants, speech therapy may be required, and orthodontic therapy may be needed to align the teeth. Again, results are usually excellent. Supplementation with folic acid (one of the B vitamins) during pregnancy decreases the incidence of cleft palate and cleft lip.

The two L-shaped ***palatine bones*** (PAL-a-tīn; *palat-* = roof of mouth) are fused and form the posterior portion of the hard palate, part of the floor and lateral wall of the nasal cavity, and a small portion of the floors of the orbits (see Figure 6.8). In cleft palate, the palatine bones may also be incompletely fused.

The ***mandible*** (*mand-* = to chew), or lower jawbone, is the largest, strongest facial bone (see Figure 6.7b). It is the only movable skull bone. Recall from our discussion of the temporal bone that the mandible has a *condylar process* (KON-di-lar). This process articulates with the mandibular fossa of the temporal bone to form the temporomandibular joint. The mandible, like the maxilla, has an *alveolar process* containing the *alveoli* (sockets) for the mandibular (lower) teeth (see Figure 6.7c). The *mental foramen* (*ment-* = chin) is a hole in the mandible that can be used by dentists to reach the mental nerve when injecting anesthetics (see Figure 6.7a).

One problem associated with the temporomandibular joint (TMJ) is **temporomandibular joint (TMJ) syndrome.** It is characterized by dull pain around the ear, tenderness of the jaw muscles, a clicking or popping noise when opening or closing the mouth, limited or abnormal opening of the mouth, headache, tooth sensitivity, and abnormal wearing of the teeth. TMJ syndrome can be caused by improperly aligned teeth, grinding or clenching the teeth, trauma to the head and neck, or arthritis. Treatments include applying moist heat or ice, eating a soft diet, taking pain relievers such as aspirin, muscle retraining, adjusting or reshaping the teeth, orthodontic treatment, or surgery.

The two ***zygomatic bones*** (*zygo-* = like a yoke), commonly called cheekbones, form the prominences of the cheeks and part of the lateral wall and floor of each orbit (see Figure 6.7a, b). They articulate with the frontal, maxilla, sphenoid, and temporal bones.

The paired ***lacrimal bones*** (LAK-ri-mal; *lacrim-* = teardrop), the smallest bones of the face, are thin and roughly resemble a fingernail in size and shape. The lacrimal bones can be seen in the anterior and lateral views of the skull in parts a and b of Figure 6.7.

The two ***inferior nasal conchae*** are scroll-like bones that project into the nasal cavity below the superior and middle nasal conchae of the ethmoid bone (see Figures 6.7a, c and 6.10). They serve the same function as the other nasal conchae: the filtration of air before it passes into the lungs.

The ***vomer*** (VŌ-mer = plowshare) is a roughly triangular bone on the floor of the nasal cavity that articulates inferiorly with both the maxillae and palatine bones along the midline of the skull. The vomer, clearly seen in the anterior view of the skull in Figure 6.7a and the inferior view in Figure 6.8, is one of the components of the *nasal septum*, a partition that divides the nasal cavity into right and left sides. The nasal septum is formed by the vomer, septal cartilage, and the perpendicular plate of the ethmoid bone (see Figure 6.7a and c). The anterior border of the vomer articulates with the septal cartilage (hyaline cartilage) to form the anterior portion of the septum. The upper border of the vomer articulates with the perpendicular plate of the ethmoid bone to form the remainder of the nasal septum.

A **deviated nasal septum** is one that does not run along the midline of the nasal cavity. It deviates (bends) to one side. Septal deviations may occur due to a developmental abnormality or trauma. If the deviation is severe, it may block the nasal passageway entirely. Even a partial blockage may lead to infection. If inflammation occurs, it may cause nasal congestion, blockage of the paranasal sinus openings, chronic sinusitis, headache, and nosebleeds. The condition usually can be corrected, or at least improved, surgically.

Unique Features of the Skull

Now that you are familiar with the names of the skull bones, we will take a closer look at three unique features of the skull: sutures, paranasal sinuses, and fontanels.

Sutures

A ***suture*** (SOO-chur = seam) is an immovable joint in an adult that is found only between skull bones. Sutures hold skull bones together. Of the many sutures that are found in the skull, we will identify only four prominent ones (see Figure 6.7):

1. The ***coronal suture*** (kō-RŌ-nal; *coron-* = crown) unites the frontal bone and two parietal bones.
2. The ***sagittal suture*** (SAJ-i-tal; *sagitt-* = arrow) unites the two parietal bones.
3. The ***lambdoid suture*** (LAM-doyd; so named because its shape resembles the Greek letter lambda, Λ) unites the parietal bones to the occipital bone.
4. The ***squamous sutures*** (SKWĀ-mus; *squam-* = flat) unite the parietal bones to the temporal bones.

Paranasal Sinuses

Paired cavities, the ***paranasal sinuses*** (*para-* = beside), are located in certain skull bones near the nasal cavity (Figure 6.11). The paranasal sinuses are lined with mucous membranes that are continuous with the lining of the nasal cavity. Skull bones containing paranasal sinuses include the frontal bone (*frontal sinus*), sphenoid bone (*sphenoid sinus*), ethmoid bone (*ethmoidal sinuses*), and maxillae (*maxillary sinuses*). Besides producing mucus, the paranasal sinuses serve as resonating chambers, producing the unique sounds of each of our speaking and singing voices, and lighten the weight of the skull.

Secretions produced by the mucous membranes of the paranasal sinuses drain into the nasal cavity. An inflammation of the membranes due to an allergic reaction or infection is called **sinusitis**. If the membranes swell enough to block drainage into the nasal cavity, fluid pressure builds up in the paranasal sinuses, resulting in a sinus headache.

Figure 6.11 **Paranasal sinuses.**

Paranasal sinuses are mucous membrane-lined spaces in the frontal, sphenoid, ethmoid, and maxillary bones that connect to the nasal cavity.

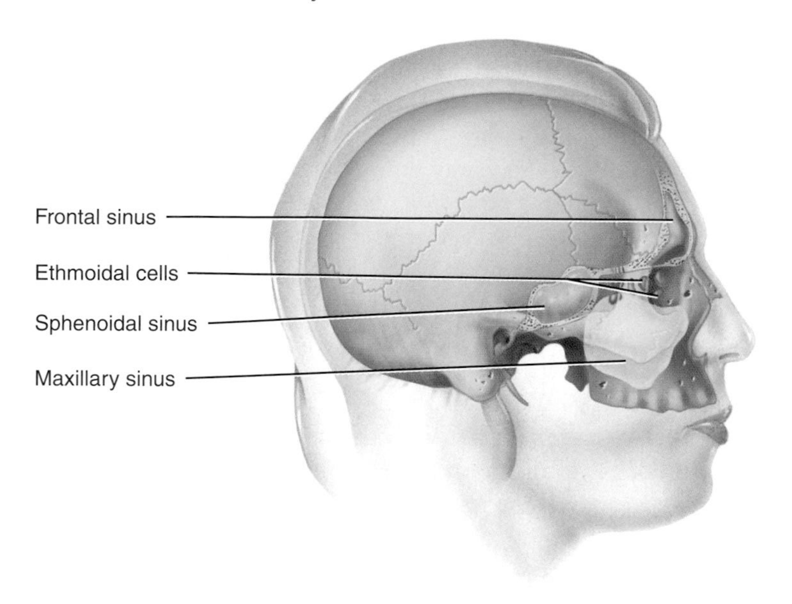

Right lateral view

What are two main functions of the paranasal sinuses?

Fontanels

Recall that the skeleton of a newly formed embryo consists of cartilage or mesenchyme arranged like membranes shaped like bones. Gradually, ossification occurs—bone replaces the cartilage or mesenchyme. Mesenchyme-filled spaces called ***fontanels*** (fonta-NELZ = little fountains) or "soft spots" are found between cranial bones at birth. They include the anterior fontanel, the posterior fontanel, the anterolateral fontanels, and the posterolateral fontanels. These areas of unossified mesenchyme will eventually be replaced with bone by intramembranous ossification and become sutures. Functionally, the fontanels enable the fetal skull to be compressed as it passes through the birth canal and permit rapid growth of the brain during infancy. The form and location of several fontanels are shown and described in Table 6.3.

Hyoid Bone

The single ***hyoid bone*** (HĪ-oyd = U-shaped) is a unique component of the axial skeleton because it does not articulate with or attach to any other bone. Rather, it is suspended from the styloid processes of the temporal bones by ligaments and muscles. The hyoid bone is located in the neck between the mandible and larynx (see Figure 6.7b). It supports the tongue and provides attachment sites for some tongue muscles and for muscles of the neck and pharynx. The hyoid bone, as well

Table 6.3 Fontanels

Fontanel	Location	Description
Anterior	Between the two parietal bones and the frontal bone.	Roughly diamond-shaped, the largest of the fontanels; usually closes 18–24 months after birth.
Posterior	Between the two parietal bones and the occipital bone. Anterior Posterior	Diamond-shaped, considerably smaller than the anterior fontanel; generally closes about 2 months after birth.
Anterolateral	One on each side of the skull between the frontal, parietal, temporal, and sphenoid bones.	Small and irregular in shape; normally close about 3 months after birth.
Posterolateral	One on each side of the skull between the parietal, occipital, and temporal bones. Anterolateral Posterolateral	Irregularly shaped; begin to close 1 or 2 months after birth, but closure is generally not complete until 12 months.

as the cartilage of the larynx and trachea, is often fractured during strangulation. As a result, they are carefully examined in an autopsy when strangulation is suspected.

■ **CHECKPOINT**

15. Describe the general features of the skull.

16. Define the following: suture, foramen, nasal septum, paranasal sinus, and fontanel.

VERTEBRAL COLUMN

OBJECTIVE • **Identify the regions and normal curves of the vertebral column and describe its structural and functional features.**

The ***vertebral column***, also called the *spine*, *spinal column*, or *backbone*, is composed of a series of bones called ***vertebrae*** (VER-te-brē; singular is *vertebra*). The vertebral column functions as a strong, flexible rod that can rotate and move forward, backward, and sideways. It encloses and protects the spinal cord, supports the head, and serves as a point of attachment for the ribs, pelvic girdle, and the muscles of the back.

Regions of the Vertebral Column

The total number of vertebrae during early development is 33. Then, several vertebrae in the sacral and coccygeal regions fuse. As a result, the adult vertebral column typically contains 26 vertebrae (Figure 6.12). These are distributed as follows:

- **7 *cervical vertebrae*** (*cervic-* = neck) in the neck region
- **12 *thoracic vertebrae*** (*thorax* = chest) posterior to the thoracic cavity
- **5 *lumbar vertebrae*** (*lumb-* = loin) supporting the lower back
- **1 *sacrum*** (SĀ-krum = sacred bone) consisting of five fused *sacral vertebrae*
- **1 *coccyx*** (KOK-siks = cuckoo, because the shape resembles the bill of a cuckoo bird) usually consisting of four fused *coccygeal vertebrae* (kok-SIJ-ē-al)

Figure 6.12 Vertebral column.

The adult vertebral column typically contains 26 vertebrae.

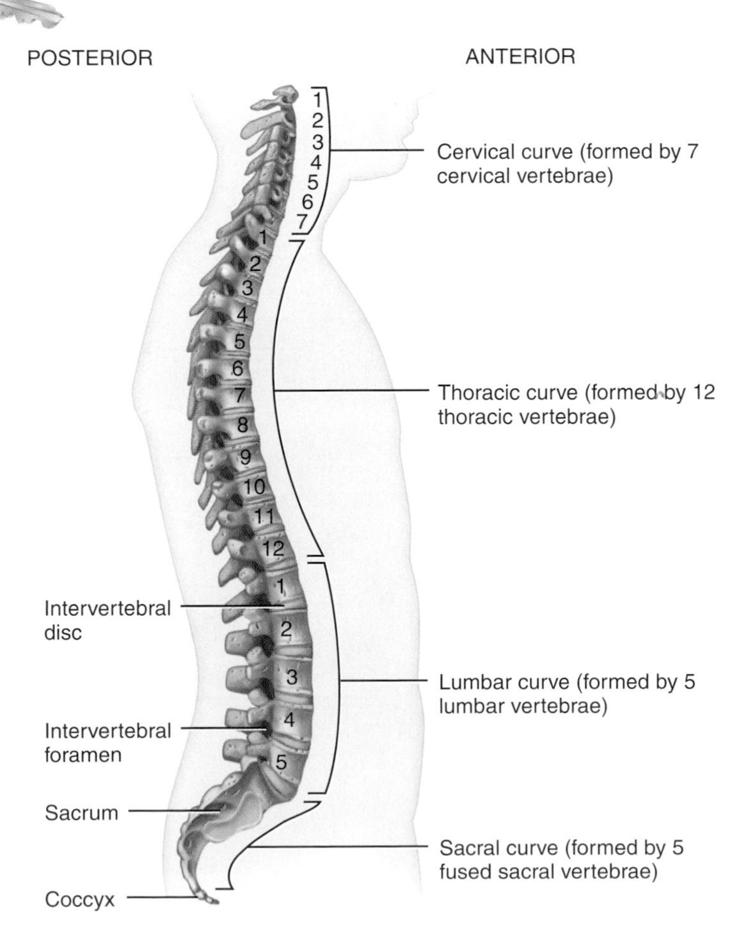

(a) Right lateral view showing four normal curves

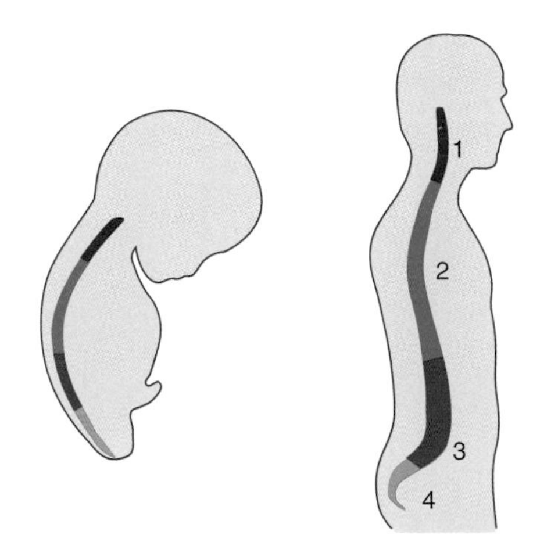

(b) Fetal and adult curves

Which curves are concave (relative to the front of the body)?

Functions of the Vertebral Column

1. Permits movement.
2. Encloses and protects the spinal cord.
3. Serves as a point of attachment for the ribs and muscles of the back.

The cervical, thoracic, and lumbar vertebrae are movable, but the sacrum and coccyx are immovable. Between adjacent vertebrae from the second cervical vertebra to the sacrum are ***intervertebral discs*** (*inter-* = between). Each disc has an outer ring of fibrocartilage and a soft, pulpy, highly elastic interior. The discs form strong joints, permit various movements of the vertebral column, and absorb vertical shock.

Normal Curves of the Vertebral Column

When viewed from the side, the vertebral column shows four slight bends called ***normal curves*** (Figure 6.12). Relative to the front of the body, the ***cervical*** and ***lumbar curves*** are convex (bulging out), and the ***thoracic*** and ***sacral curves*** are concave (cupping in). The curves of the vertebral column increase its strength, help maintain balance in the upright position, absorb shocks during walking and running, and help protect the vertebrae from breaks.

In the fetus, there is a single concave curve (Figure 6.12b). At about the third month after birth, when an infant begins to hold its head erect, the cervical curve develops. Later, when the child sits up, stands, and walks, the lumbar curve develops.

Vertebrae

Vertebrae in different regions of the spinal column vary in size, shape, and detail, but they are similar enough that we can discuss the structure and functions of a typical vertebra (Figure 6.13).

1. The ***body***, the thick, disc-shaped front portion, is the weight-bearing part of a vertebra.
2. The ***vertebral arch*** extends backwards from the body of the vertebra. It is formed by two short, thick processes, the *pedicles* (PED-i-kuls = little feet), which project backward from the body to unite with the laminae. The *laminae* (LAM-i-nē = thin layers) are the flat parts of the arch and end in a single sharp, slender projection called a *spinous process*. The hole between the vertebral arch and body contains the spinal cord and is known as the *vertebral foramen*. Together, the vertebral foramina of all vertebrae form the *vertebral cavity*. When the vertebrae are stacked on top of one another, there is an opening between adjoining vertebrae on both sides of the column. Each opening, called an *intervertebral foramen*, permits the passage of a single spinal nerve.

Figure 6.13 Structure of a typical vertebra, as illustrated by a thoracic vertebra. (Note the facets for the ribs, which vertebrae other than the thoracic vertebrae do not have.) In (b), only one spinal nerve has been included, and it has been extended beyond the intervertebral foramen for clarity.

A vertebra consists of a body, a vertebral arch, and several processes.

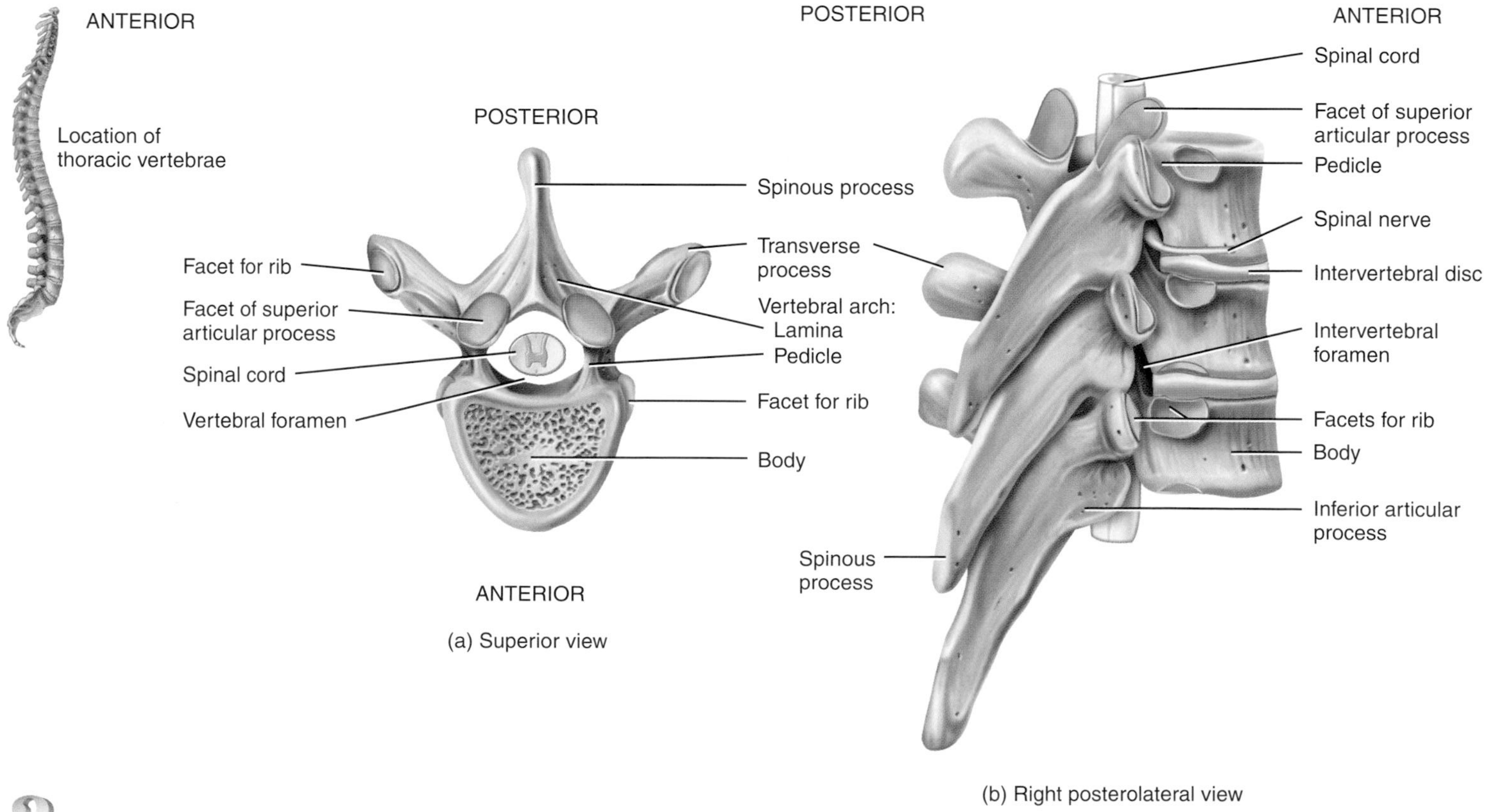

(a) Superior view

(b) Right posterolateral view

What are the functions of the vertebral and intervertebral foramina?

3. Seven ***processes*** arise from the vertebral arch. At the point where a lamina and pedicle join, a *transverse process* extends laterally on each side. A single *spinous process (spine)* projects from the junction of the laminae. These three processes serve as points of attachment for muscles. The remaining four processes form joints with other vertebrae above or below. The two *superior articular processes* of a vertebra articulate with the vertebra immediately above them. The two *inferior articular processes* of a vertebra articulate with the vertebra immediately below them. The smooth articulating surfaces of the articular processes are called *facets* (= little faces), which are covered with hyaline cartilage.

Vertebrae in each region are numbered in sequence from top to bottom. The seven ***cervical vertebrae*** are termed C1 through C7 (Figure 6.14). The spinous processes of the

Figure 6.14 Cervical vertebrae.

The cervical vertebrae are found in the neck region.

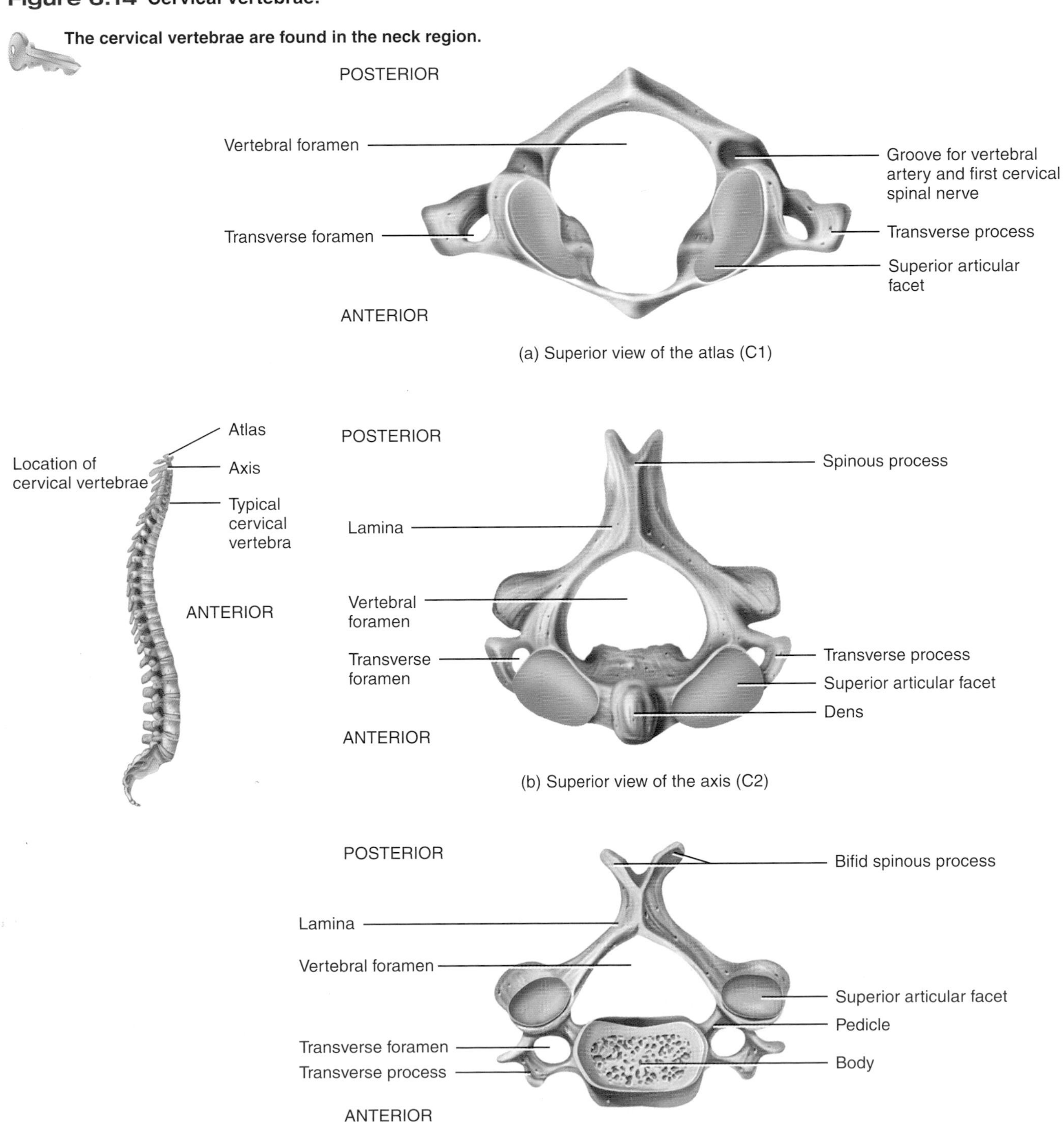

(a) Superior view of the atlas (C1)

(b) Superior view of the axis (C2)

(c) Superior view of a typical cervical vertebra

Which bones permit the movement of the head to signify "no"?

second through sixth cervical vertebrae are often *bifid*, or split into two parts (Figure 6.14b, c). All cervical vertebrae have three foramina: one vertebral foramen and two transverse foramina. Each cervical transverse process contains a *transverse foramen* through which blood vessels and nerves pass.

The first two cervical vertebrae differ considerably from the others. The first cervical vertebra (C1), the ***atlas***, supports the head and is named for the mythological Atlas who supported the world on his shoulders. The atlas lacks a body and a spinous process. The upper surface contains *superior articular facets* that articulate with the occipital bone of the skull. This articulation permits you to nod your head to indicate "yes." The inferior surface contains *inferior articular facets* that articulate with the second cervical vertebra.

The second cervical vertebra (C2), the ***axis***, does have a body and a spinous process. A tooth-shaped process called the *dens* (= tooth) projects up through the vertebral foramen of the atlas. The dens is a pivot on which the atlas and head move, as in side-to-side movement of the head to signify "no."

The third through sixth cervical vertebrae (C3 through C6), represented by the vertebra in Figure 6.14c, correspond to the structural pattern of the typical cervical vertebra described previously. The seventh cervical vertebra (C7), called the *vertebra prominens*, is somewhat different. It is marked by a single, large spinous process that can be seen and felt at the base of the neck.

Thoracic vertebrae (T1 through T12) are considerably larger and stronger than cervical vertebrae. Distinguishing features of the thoracic vertebrae are their facets for articulating with the ribs (see Figure 6.13). Movements of the thoracic region are limited by the attachment of the ribs to the sternum.

The ***lumbar vertebrae*** (L1 through L5) are the largest and strongest in the column (Figure 6.15). Their various projections are short and thick, and the spinous processes are well adapted for the attachment of the large back muscles.

The ***sacrum*** is a triangular bone formed by the fusion of five sacral vertebrae, indicated in Figure 6.16 as S1 through S5. The fusion of the sacral vertebrae begins between ages 16 and 18 years and is usually completed by age 30. The sacrum serves as a strong foundation for the pelvic girdle. It is positioned at the back of the pelvic cavity medial to the two hip bones.

The anterior and posterior sides of the sacrum contain four pairs of *sacral foramina*. Nerves and blood vessels pass through the foramina. The *sacral canal* is a continuation of the vertebral cavity. The lower entrance is called the *sacral hiatus* (hī-Ā-tus = opening). The anterior top border of the sacrum has a projection, called the *sacral promontory* (PROM-on-tō′-rē), which is used as a landmark for measuring the pelvis prior to childbirth.

> Anesthetic agents that act on the sacral and coccygeal nerves are sometimes injected through the sacral hiatus, a procedure called **caudal anesthesia** or **epidural block.**

Figure 6.15 Lumbar vertebrae.

Lumbar vertebrae are found in the lower back.

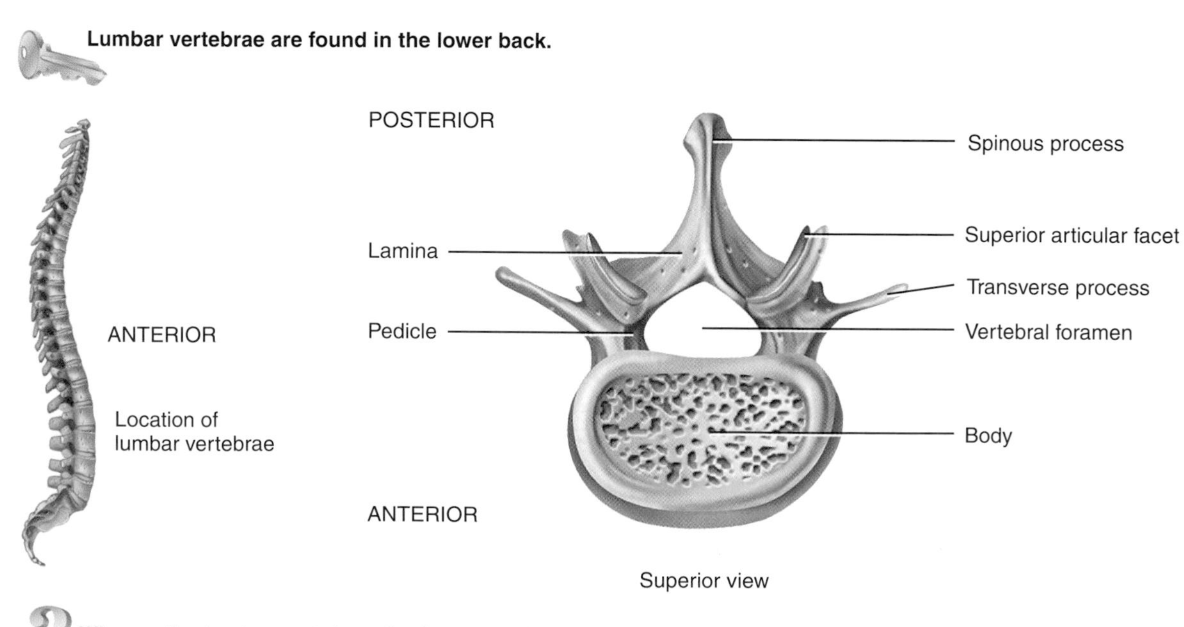

Why are the lumbar vertebrae the largest and strongest in the vertebral column?

> The procedure is used most often to relieve pain during labor and to provide anesthesia to the perineal area. Anesthetic agents may also be injected through the posterior sacral foramina.

The *coccyx*, like the sacrum, is triangular in shape and is formed by the fusion of the four coccygeal vertebrae. These are indicated in Figure 6.16 as Co1 through Co4. The top of the coccyx articulates with the sacrum.

■ **CHECKPOINT**

17. What are the functions of the vertebral column?

18. What are the main distinguishing characteristics of the bones of the various regions of the vertebral column?

THORAX

OBJECTIVE • Identify the bones of the thorax and their principal markings.

The term ***thorax*** refers to the entire chest. The skeletal portion of the thorax, the ***thoracic cage***, is a bony cage formed by the sternum, costal cartilages, ribs, and the bodies of the thoracic vertebrae (Figure 6.17). The thoracic cage encloses and protects the organs in the thoracic cavity and upper abdominal cavity. It also provides support for the bones of the shoulder girdle and upper limbs.

Sternum

The ***sternum***, or breastbone, is a flat, narrow bone located in the center of the anterior thoracic wall and consists of three parts that usually fuse by age 25 (Figure 6.17). The upper part is the *manubrium* (ma-NOO-brē-um = handle-like); the middle and largest part is the *body*; and the lowest, smallest part is the *xiphoid process* (ZĪ-foyd = sword-shaped).

The manubrium articulates with the clavicles and the first and second ribs. The body of the sternum articulates directly or indirectly with the second through tenth ribs. The xiphoid process consists of hyaline cartilage during infancy and childhood and does not ossify completely until about age 40. It has no ribs attached to it but provides attachment for some abdominal muscles. If the hands of a rescuer are incorrectly positioned during cardiopulmonary resuscitation (CPR), there is danger of fracturing the xiphoid process and driving it into internal organs.

Figure 6.16 Sacrum and coccyx.

The sacrum is formed by the union of five sacral vertebrae, and the coccyx is formed by the union of usually four coccygeal vertebrae.

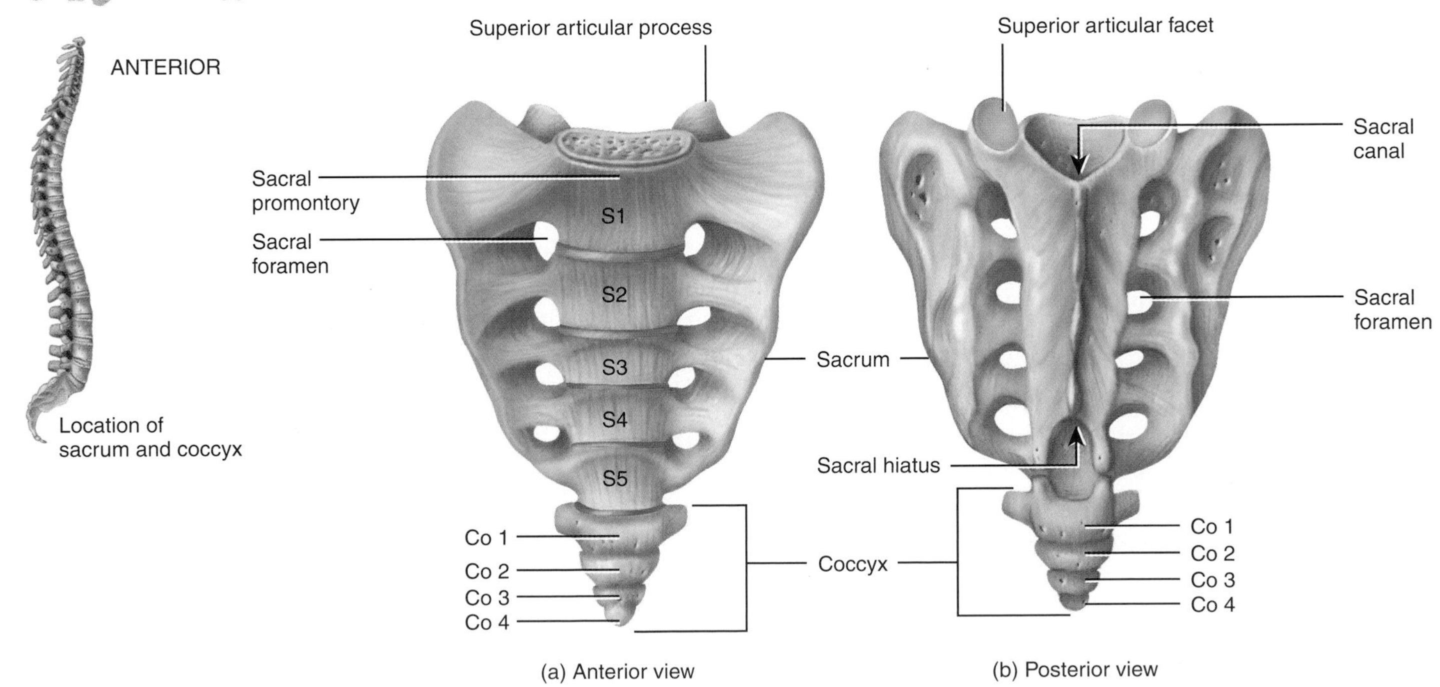

What is the function of the sacral foramina?

Figure 6.17 Skeleton of the thorax.

The bones of the thorax enclose and protect organs in the thoracic cavity and upper abdominal cavity.

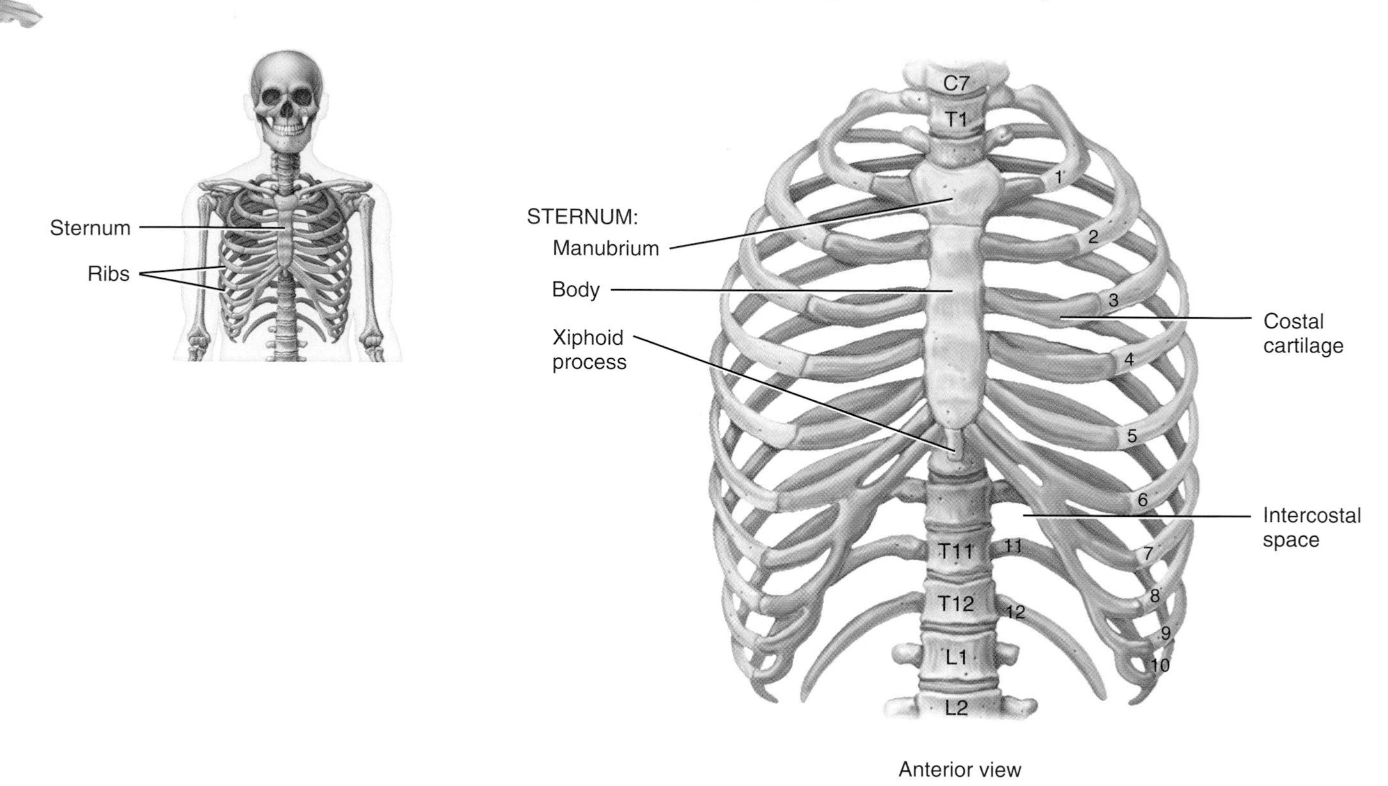

Which ribs are true ribs? False ribs? Floating ribs?

Ribs

Twelve pairs of ***ribs*** make up the sides of the thoracic cavity (Figure 6.17). The ribs increase in length from the first through seventh ribs, then decrease in length to the twelfth rib. Each rib articulates posteriorly with its corresponding thoracic vertebra.

The first through seventh pairs of ribs have a direct anterior attachment to the sternum by a strip of hyaline cartilage called *costal cartilage* (*cost-* = rib). These ribs are called *true ribs*. The remaining five pairs of ribs are termed *false ribs* because their costal cartilages either attach indirectly to the sternum or do not attach to the sternum at all. The cartilages of the eighth, ninth, and tenth pairs of ribs attach to each other and then to the cartilages of the seventh pair of ribs. The eleventh and twelfth false ribs are also known as *floating ribs* because the costal cartilage at their anterior ends does not attach to the sternum at all. Floating ribs attach only posteriorly to the thoracic vertebrae. Spaces between ribs, called *intercostal spaces*, are occupied by intercostal muscles, blood vessels, and nerves.

Rib fractures are the most common chest injuries, and they usually result from direct blows, most often from impact with a steering wheel, falls, and crushing injuries to the chest. In some cases, fractured ribs may puncture the heart, great vessels of the heart, lungs, trachea, bronchi, esophagus, spleen, liver, and kidneys. Rib fractures are usually quite painful. Rib fractures are no longer bound with bandages because of the pneumonia that would result from lack of proper lung ventilation.

■ **CHECKPOINT**

19. What are the functions of the bones of the thorax?

20. What are the parts of the sternum?

PECTORAL (SHOULDER) GIRDLE

OBJECTIVE • Identify the bones of the pectoral (shoulder) girdle and their principal markings.

The ***pectoral girdles*** (PEK-tō-ral) or ***shoulder girdles*** attach the bones of the upper limbs to the axial skeleton (Figure 6.18). The right and left pectoral girdles each consist of two bones: a clavicle and a scapula. The clavicle, the anterior component, articulates with the sternum, and the scapula, the posterior component, articulates with the clavicle and the humerus. The pectoral girdles do not articulate with the vertebral column. The joints of the shoulder girdles are freely movable and thus allow movements in many directions.

Clavicle

Each ***clavicle*** (KLAV-i-kul = key) or collarbone is a long, slender S-shaped bone that is positioned horizontally above the first rib. The medial end of the clavicle articulates with the sternum, and the lateral end articulates with the acromion of the scapula (Figure 6.18). Because of its position, the clavicle transmits mechanical force from the upper limb to the trunk. If the force transmitted to the clavicle is excessive, as when you fall on your outstretched arm, a *fractured clavicle* may result.

Scapula

Each ***scapula*** (SCAP-yū-la), or ***shoulder blade***, is a large, flat, triangular bone situated in the posterior part of the thorax (Figure 6.18). A sharp ridge, the *spine*, runs diagonally across the posterior surface of the flattened, triangular *body* of the scapula. The lateral end of the spine, the *acromion* (a-KRŌ-mē-on; *acrom-* = topmost), is easily felt as the high point of the shoulder and is the site of articulation with the clavicle. Inferior to the acromion is a depression called the *glenoid cavity*. This cavity articulates with the head of the humerus (arm bone) to form the shoulder joint. Also present on the scapula is a projection called the *coracoid process* (KOR-a-koyd = like a crow's beak) to which muscles attach.

■ **CHECKPOINT**

21. What bones make up the pectoral girdle? What is the function of the pectoral girdle?

Figure 6.18 Right pectoral (shoulder) girdle.

The pectoral girdle attaches the bones of the upper limb to the axial skeleton.

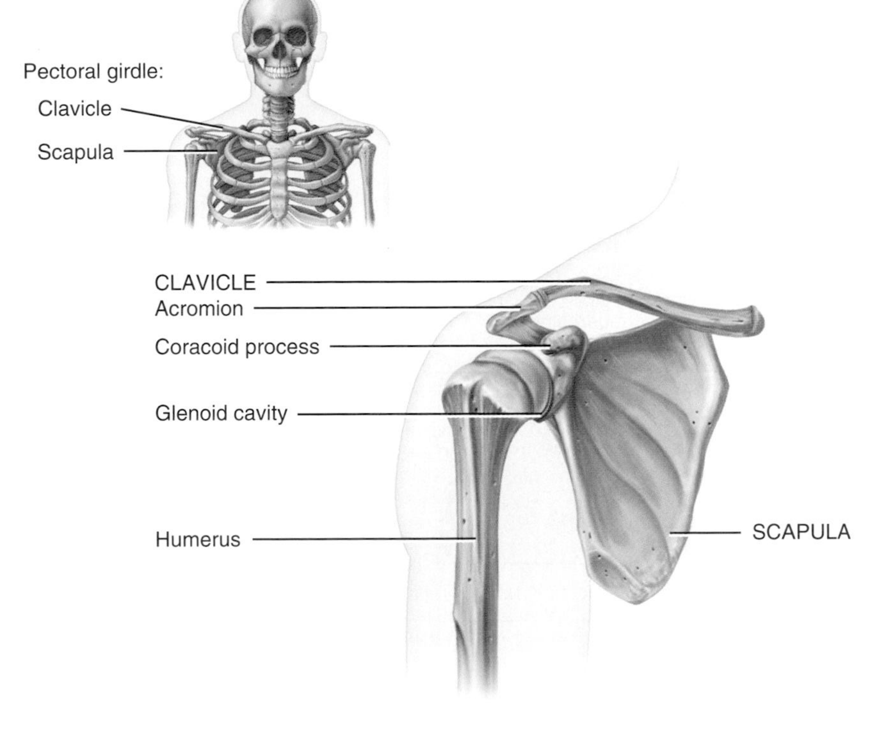

(a) Anterior view

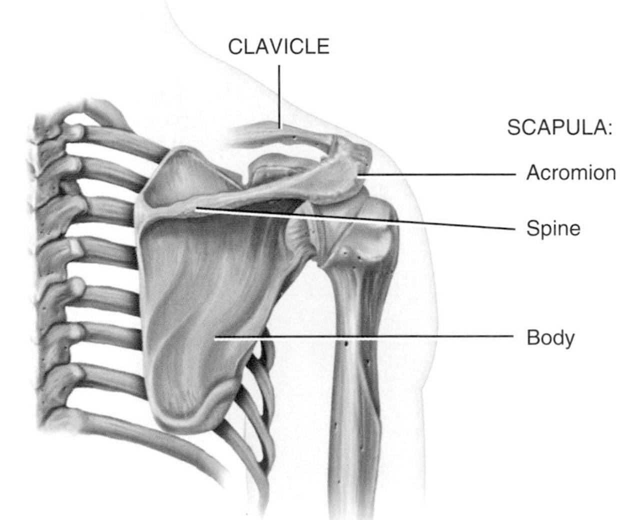

(b) Posterior view

Which bones make up a pectoral girdle?

UPPER LIMB

OBJECTIVE • Identify the bones of the upper limb and their principal markings.

Each ***upper limb*** consists of 30 bones. Each upper limb includes a humerus in the arm; ulna and radius in the forearm; and 8 carpals (wrist bones), 5 metacarpals (palm bones), and 14 phalanges (finger bones) in the hand (see Figure 6.6).

Humerus

The ***humerus*** (HŪ-mer-us), or arm bone, is the longest and largest bone of the upper limb (Figure 6.19). At the shoulder it articulates with the scapula, and at the elbow it articulates with both the ulna and radius. The proximal end of the humerus consists of a *head* that articulates with the glenoid cavity of the scapula. It also has an *anatomical neck*, the former site of the epiphyseal plate, which is a groove just distal to the head. The *body* of the humerus contains a roughened, V-shaped area called the *deltoid tuberosity* where the deltoid muscle attaches. At the distal end of the humerus, the *capitulum* (ka-PIT-ū-lum = small head), is a rounded knob that articulates with the head of the radius. The *radial fossa* is a depression that receives the head of the radius when the forearm is flexed (bent). The *trochlea* (TRŌK-lē-a) is a spool-shaped surface that articulates with the ulna. The *coronoid fossa* (KOR-ō-noyd = crown-shaped) is a depression that receives part of the ulna when the forearm is flexed. The *olecranon fossa* (ō-LEK-ra-non) is a depression on the back of the bone that receives the olecranon of the ulna when the forearm is extended (straightened).

Ulna and Radius

The ***ulna*** is on the medial aspect (little-finger side) of the forearm and is longer than the radius (Figure 6.20). At the proximal end of the ulna is the *olecranon*, which forms the prominence of the elbow. The *coronoid process*, together with the olecranon, receives the trochlea of the humerus. The trochlea of the humerus also fits into the *trochlear notch*, a large curved area between the olecranon and the coronoid process. The *radial notch* is a depression for the head of the radius. A *styloid process* is at the distal end of the ulna.

The ***radius*** is located on the lateral aspect (thumb side) of the forearm. The proximal end of the radius has a disc-

Figure 6.19 Right humerus in relation to the scapula, ulna, and radius.

The humerus is the longest and largest bone of the upper limb.

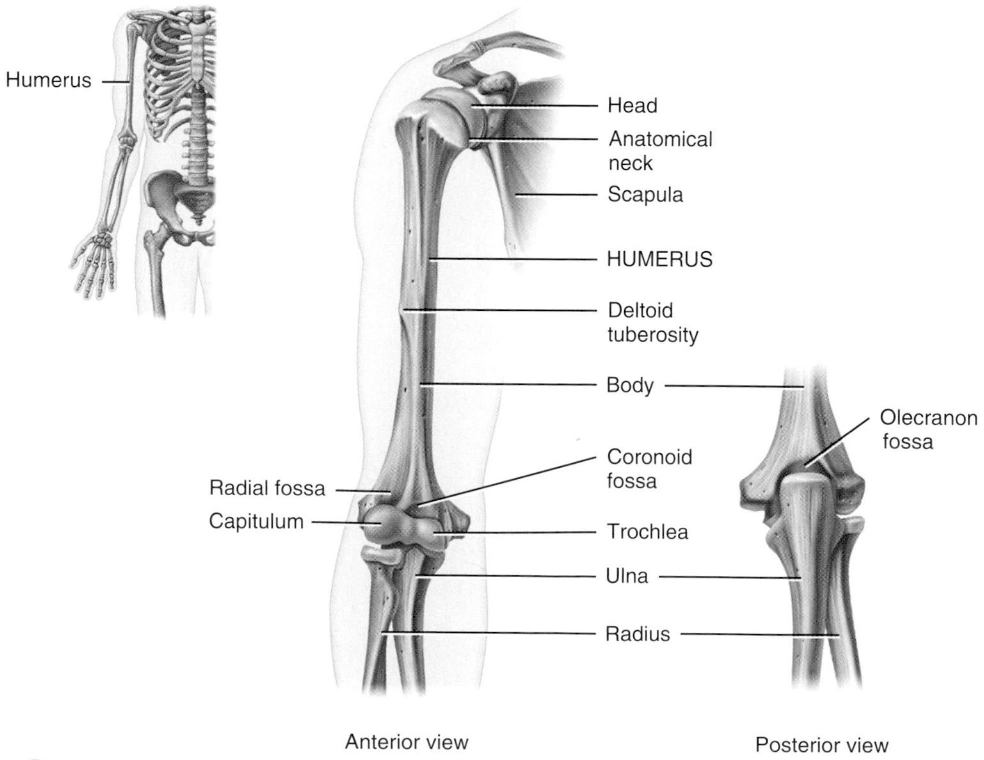

With which part of the scapula does the humerus articulate?

Figure 6.20 Right ulna and radius in relation to the humerus and carpals.

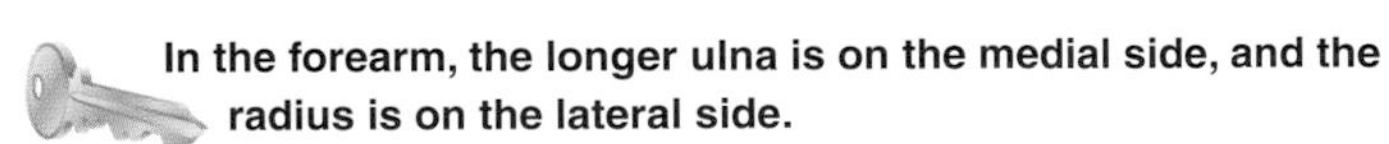

In the forearm, the longer ulna is on the medial side, and the radius is on the lateral side.

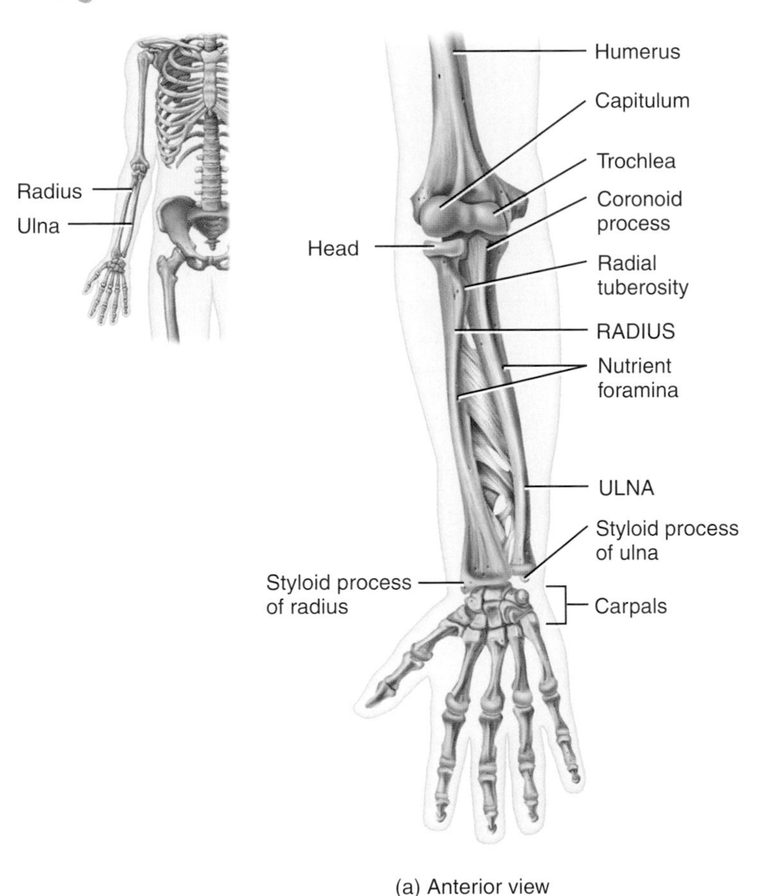

(a) Anterior view

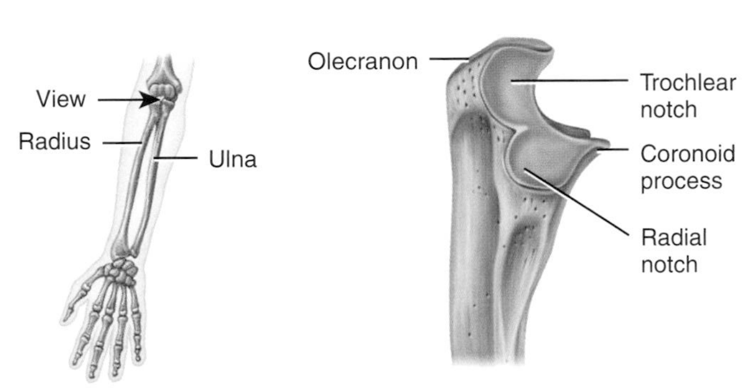

(b) Lateral view of proximal end of ulna

What part of the ulna is called the elbow?

shaped *head* that articulates with the capitulum of the humerus and radial notch of the ulna. It has a raised, roughened area called the *radial tuberosity* that provides a point of attachment for the biceps brachii muscle. The distal end of the radius articulates with three carpal bones of the wrist. Also at the distal end is a *styloid process*. Fracture of the distal end of the radius is the most common fracture in adults older than 50 years.

Carpals, Metacarpals, and Phalanges

The ***carpus (wrist)*** of the hand contains eight small bones, the ***carpals***, held together by ligaments (Figure 6.21). The carpals are arranged in two transverse rows, with four bones in each row, and they are named for their shapes. In the anatomical position, the carpals in the top row, from the

Figure 6.21 Right wrist and hand in relation to the ulna and radius.

The skeleton of the hand consists of the carpals, metacarpals, and phalanges.

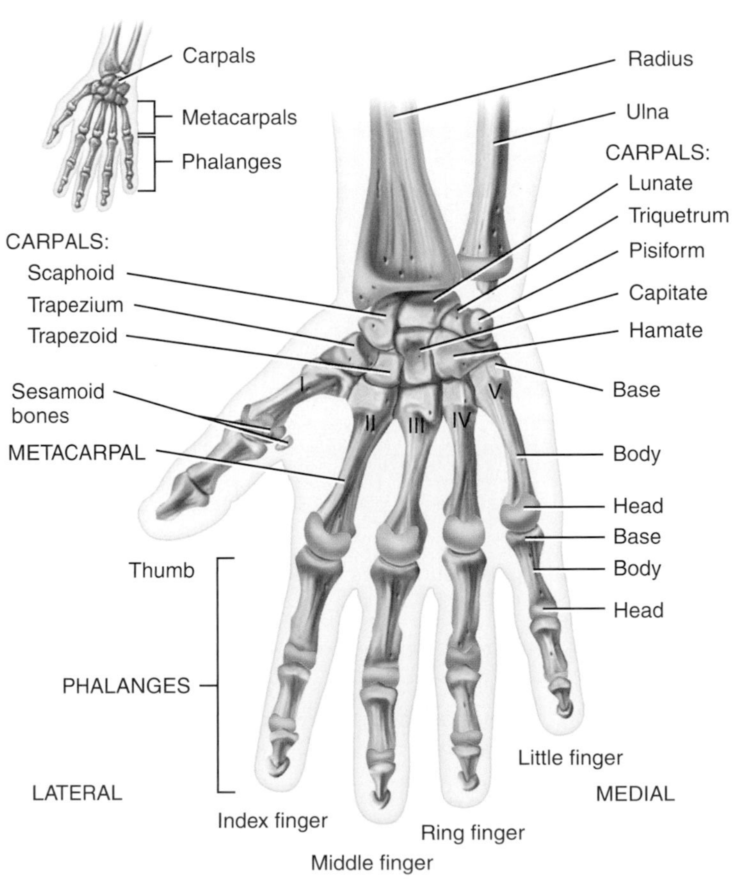

Anterior view

What part of which bones are commonly called the knuckles?

Figure 6.22 **Female pelvic (hip) girdle.**

The hip bones are united in front at the pubic symphysis and in back at the sacrum.

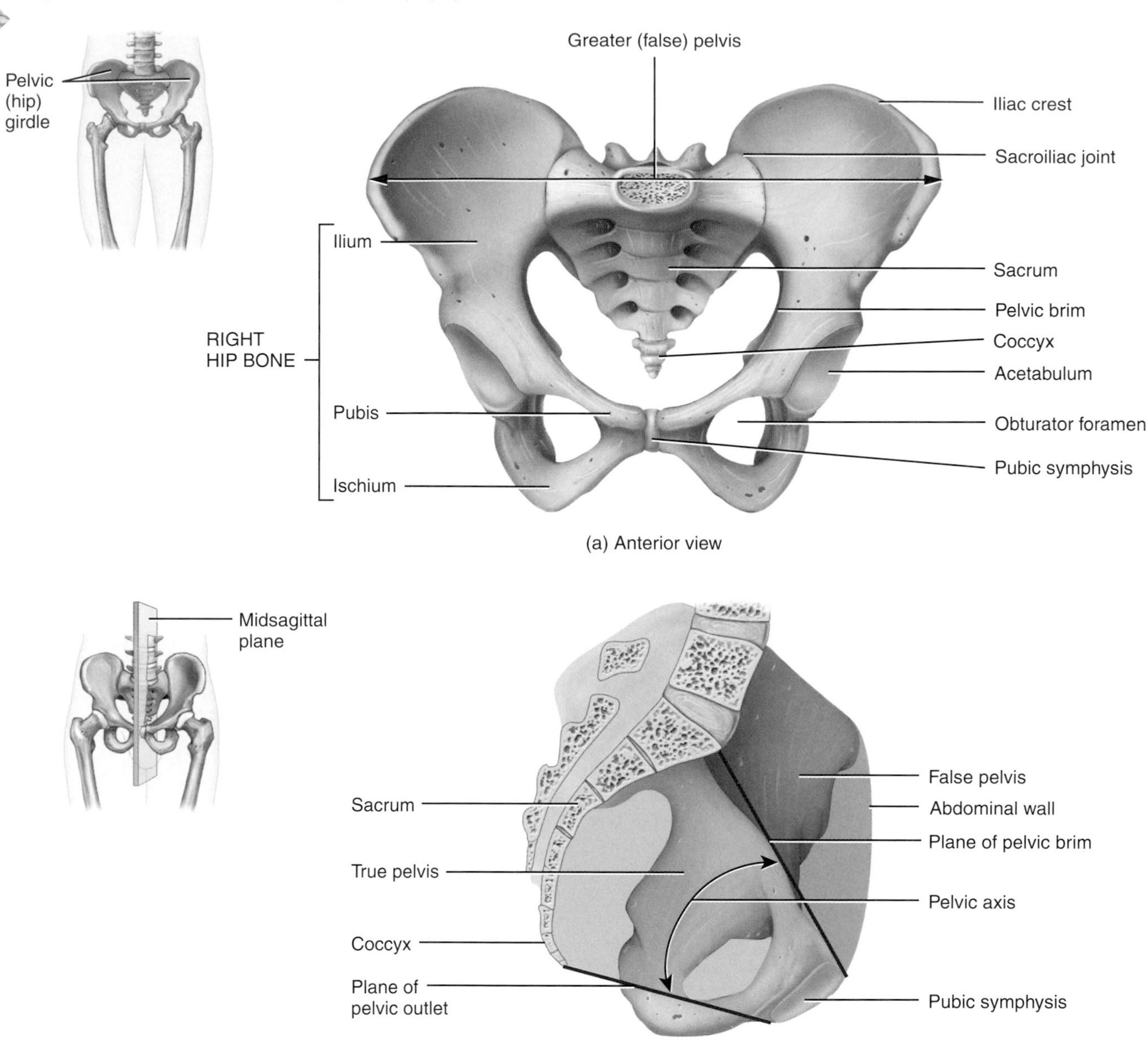

(a) Anterior view

(b) Midsagittal section indicating locations of true and false pelves

What part of the pelvis surrounds the pelvic organs in the pelvic cavity?

lateral to medial position, are the ***scaphoid*** (SKAF-oid = boatlike), ***lunate*** (LOO-nāt = moon-shaped), ***triquetrum*** (trī-KWĒ-trum = three cornered), and ***pisiform*** (PĪ-si-form = pea-shaped). In about 70% of carpal fractures, only the scaphoid is broken because of the force transmitted through it to the radius. The carpals in the bottom row, from the lateral to medial position, are the ***trapezium*** (tra-PĒ-zē-um = four-sided figure with no two sides parallel), ***trapezoid*** (TRAP-e-zoid = four-sided figure with two sides parallel), ***capitate*** (KAP-i-tāt = head-shaped; the largest carpal bone, whose rounded projection, the head, articulates with the lunate), and ***hamate*** (HAM-āt = hooked; named for a large hook-shaped projection on its anterior surface). Together, the concavity formed by the pisiform and hamate (on the ulnar side) and the scaphoid and trapezium (on the radial side) constitute a space called the ***carpal tunnel***. Through it pass the long flexor tendons of the digits and thumb and the median nerve.

Narrowing of the carpal tunnel gives rise to a condition called **carpal tunnel syndrome,** in which the median nerve is compressed. The nerve compression causes pain, numbness, tingling, and muscle weakness in the hand.

The ***metacarpus (palm)*** of the hand contains five bones called ***metacarpals*** (*meta-* = after or beyond). Each metacarpal bone consists of a proximal *base*, an intermediate *body*, and a distal *head*. The metacarpal bones are numbered I through V (or 1 to 5), starting with the lateral bone in the thumb. The heads of the metacarpals are commonly called the "knuckles" and are readily visible in a clenched fist.

The ***phalanges*** (fa-LAN-jēz = battle lines) are the bones of the fingers. They number 14 in each hand. Like the metacarpals, the phalanges are numbered I through V (or 1 to 5), beginning with the thumb. A single bone of a finger or toe is termed a *phalanx* (FĀ-lanks). Like the metacarpals, each phalanx consists of a proximal *base*, an intermediate *body*, and a distal *head*. There are two phalanges (proximal and distal) in the thumb and three phalanges (proximal, middle, and distal) in each of the other four digits. In order from the thumb, these other four digits are commonly referred to as the index finger, middle finger, ring finger, and little finger (Figure 6.21).

■ **CHECKPOINT**

22. What bones form the upper limb, from proximal to distal?

PELVIC (HIP) GIRDLE

OBJECTIVE • Identify the bones of the pelvic (hip) girdle and their principal markings.

The ***pelvic (hip) girdle*** consists of the two ***hip bones***, also called ***coxal bones*** (Figure 6.22 on page 142). The pelvic girdle provides a strong, stable support for the vertebral column, protects the pelvic viscera, and attaches the lower limbs to the axial skeleton. The hip bones are united to each other in front at a joint called the ***pubic symphysis*** (PŪ-bik SIM-fi-sis); posteriorly they unite with the sacrum at the sacroiliac joint.

Together with the sacrum and coccyx, the two hip bones of the pelvic girdle form a basinlike structure called the ***pelvis*** (plural is *pelvises* or *pelves*). In turn, the bony pelvis is divided into upper and lower portions by a boundary called the ***pelvic brim*** (Figure 6.22). The part of the pelvis above the pelvic brim is called the ***false (greater) pelvis***. The false pelvis is actually part of the abdomen and does not contain any pelvic organs, except for the urinary bladder, when it is full, and the uterus during pregnancy. The part of the pelvis below the pelvic brim is called the ***true (lesser) pelvis***. The true pelvis surrounds the pelvic cavity (see Figure 1.8 on page 15). The upper opening of the true pelvis is called the ***pelvic inlet***, and the lower opening of the true pelvis is called the ***pelvic outlet***. The ***pelvic axis*** is an imaginary curved line passing through the true pelvis; it joins the central points of the planes of the pelvic inlet and outlet. During childbirth, the pelvic axis is the course taken by the baby's head as it descends through the pelvis.

Pelvimetry is the measurement of the size of the inlet and outlet of the birth canal, which may be done by ultrasonography or physical examination. Measurement of the pelvic outlet in pregnant females is important because it must become large enough for the fetus to pass through at birth.

Each of the two hip bones of a newborn is composed of three parts: the ilium, the pubis, and the ischium (Figure 6.23). The ***ilium*** (= flank) is the largest of the three subdivisions of the hip bone. Its upper border is the *iliac crest*. On

Figure 6.23 Right hip bone. The lines of fusion of the ilium, ischium, and pubis are not always visible in an adult hip bone.

The two hip bones form the pelvic girdle, which attaches the lower limbs to the axial skeleton and supports the vertebral column and viscera.

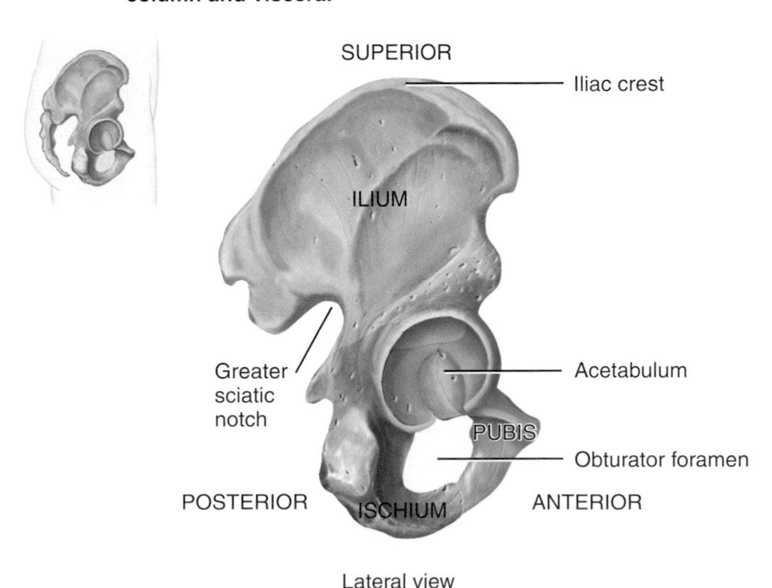

Which bone fits into the socket formed by the acetabulum?

the lower surface is the *greater sciatic notch* (sī-AT-ik) through which the sciatic nerve, the longest nerve in the body, passes. The ***ischium*** (IS-kē-um = lip) is the lower, posterior part of the hipbone. The ***pubis*** (PŪ-bis = pubic hair) is the lower, anterior part of the hipbone. By age 23 years, the three separate bones have fused into one. The deep fossa (depression) where the three bones meet is the *acetabulum* (as-e-TAB-ū-lum = vinegar cup). It is the socket for the head of the femur. The ischium joins with the pubis, and together they surround the *obturator foramen* (OB-too-rā-ter), the largest foramen in the skeleton.

■ **CHECKPOINT**

23. What bones make up the pelvic girdle? What is the function of the pelvic girdle?

LOWER LIMB

OBJECTIVE • List the skeletal components of the lower limb and their principal markings.

Each ***lower limb*** is composed of 30 bones: the femur in the thigh; the patella (kneecap); the tibia and fibula in the leg (the part of the lower limb between the knee and the ankle); and 7 tarsals (ankle bones), 5 metatarsals, and 14 phalanges (toes) in the foot (see Figure 6.6).

Femur

The ***femur*** (thigh bone) is the longest, heaviest, and strongest bone in the body (Figure 6.24). Its proximal end articulates with the hip bone, and its distal end articulates with the tibia and patella. The body of the femur bends medially, and as a result, the knee joints are brought nearer to the midline of the body. The bend is greater in females because the female pelvis is broader.

The *head* of the femur articulates with the acetabulum of the hip bone to form the *hip joint*. The *neck* of the femur is a constricted region below the head. A fairly common fracture in the elderly occurs at the neck of the femur, which becomes so weak that it fails to support the weight of the body. Although it is actually the femur that is fractured, this condition is commonly known as a broken hip. The *greater trochanter* (trō-KAN-ter) is a projection felt and seen in front of the hollow on the side of the hip. It is where some of the thigh and buttock muscles attach and serves as a landmark for intramuscular injections in the thigh.

The distal end of the femur expands into the *medial condyle* and *lateral condyle*, projections which articulate with the tibia. The *patellar surface* is located on the anterior surface of the femur between the condyles.

Figure 6.24 Right femur in relation to the hip bone, patella, tibia, and fibula.

The head of the femur articulates with the acetabulum of the hip bone to form the hip joint.

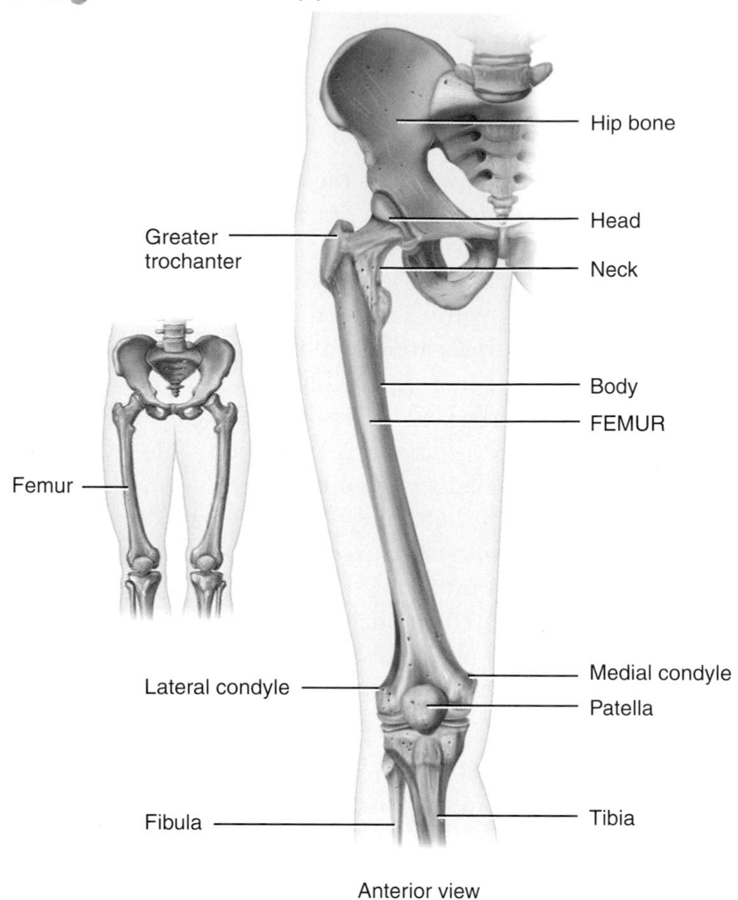

With which bones does the distal end of the femur articulate?

Patella

The ***patella*** (= little dish), or kneecap, is a small, triangular bone in front of the joint between the femur and tibia, commonly known as the knee joint (Figure 6.24). The patella develops in the tendon of the quadriceps femoris muscle. Its functions are to increase the leverage of the tendon, maintain the position of the tendon when the knee is flexed, and protect the knee joint. During normal flexion and extension of the knee, the patella tracks (glides) up and down in the groove between the two femoral condyles.

In "runner's knee," or **patellofemoral stress syndrome,** normal tracking does not occur. Instead, the patella tracks laterally, and the increased pressure of abnormal tracking causes the associated pain. A common cause of runner's knee is constantly walking, running, or jogging on the same side of the road. Because roads are high in the middle and slope down on the sides, the slope stresses the knee that is closer to the center of the road.

Tibia and Fibula

The ***tibia***, or shin bone, is the larger, medial, weight-bearing bone of the leg (Figure 6.25). The tibia articulates at its proximal end with the femur and fibula, and at its distal end with the fibula and talus of the ankle. The proximal end of the tibia expands into a *lateral condyle* and a *medial condyle*, projections which articulate with the condyles of the femur to form the *knee joint*. The *tibial tuberosity* is on the anterior surface below the condyles and is a point of attachment for the patellar ligament. The medial surface of the distal end of the tibia forms the *medial malleolus* (ma-LĒ-ō-lus = little hammer), which articulates with the talus of the ankle and forms the prominence that can be felt on the medial surface of your ankle.

Shin splints is the name given to soreness or pain along the tibia. Probably caused by inflammation of the periosteum brought about by repeated tugging of the attached muscles and tendons, it is often the result of walking or running up and down hills.

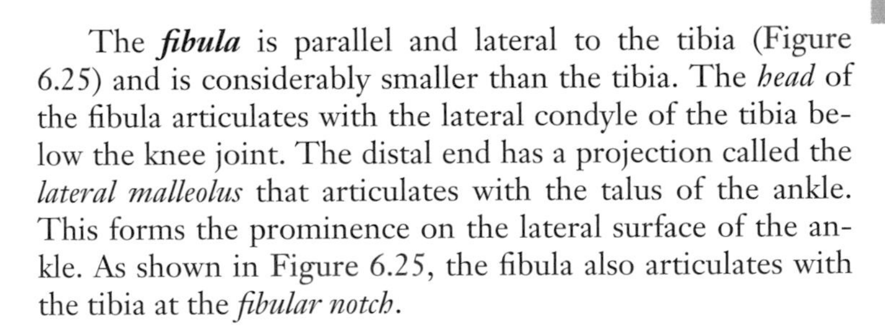

The ***fibula*** is parallel and lateral to the tibia (Figure 6.25) and is considerably smaller than the tibia. The *head* of the fibula articulates with the lateral condyle of the tibia below the knee joint. The distal end has a projection called the *lateral malleolus* that articulates with the talus of the ankle. This forms the prominence on the lateral surface of the ankle. As shown in Figure 6.25, the fibula also articulates with the tibia at the *fibular notch*.

Tarsals, Metatarsals, and Phalanges

The ***tarsus (ankle)*** of the foot contains seven bones, the ***tarsals***, held together by ligaments (Figure 6.26). Of these, the ***talus*** (TĀ-lus = ankle bone) and ***calcaneus*** (kal-KĀ-nē-us = heel bone) are located on the posterior part of the foot. The anterior part of the ankle contains the ***cuboid*** (KŪ-boyd), ***navicular*** (na-VIK-ū-lar), and three ***cuneiform bones*** (KŪ-nē-i-form) called the *first*, *second*, and *third cuneiforms*. The talus is the only bone of the foot that articulates with the fibula and tibia. It articulates medially with the medial malleolus of the tibia and laterally with the lateral malleolus of the fibula. During walking, the talus initially bears the entire weight of the body. About half the weight is then trans-

Figure 6.25 Right tibia and fibula in relation to the femur, patella, and talus.

The tibia articulates with the femur and fibula proximally and with the fibula and talus distally, while the fibula articulates proximally with the tibia below the knee joint and distally with the talus.

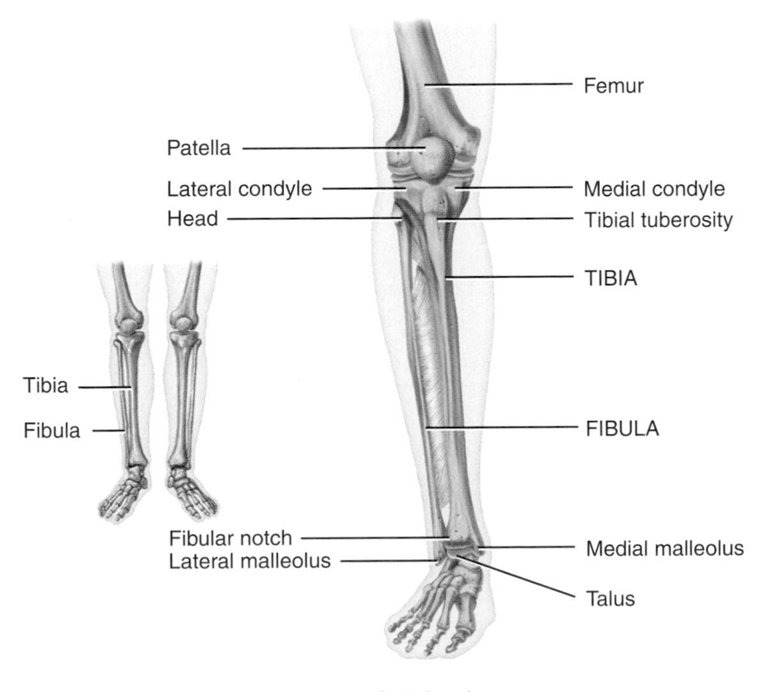

Which leg bone bears the weight of the body?

Figure 6.26 Right foot.

The skeleton of the foot consists of the tarsals, metatarsals, and phalanges.

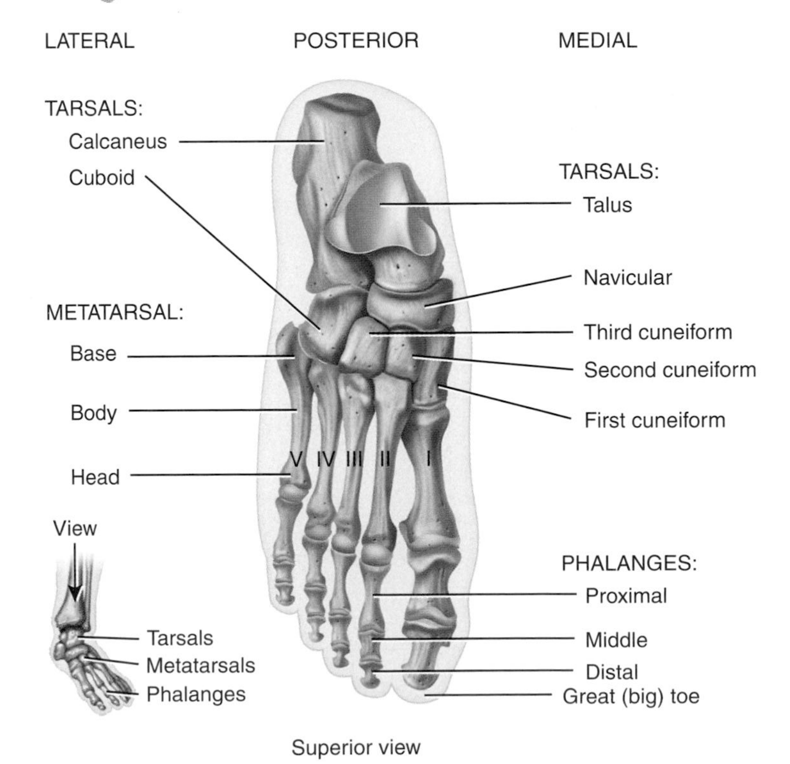

Which tarsal bone articulates with the tibia and fibula?

mitted to the calcaneus. The remainder is transmitted to the other tarsal bones. The calcaneus is the largest and strongest of the tarsals.

Five bones called ***metatarsals*** and numbered I to V (or 1 to 5) from the medial to lateral position form the skeleton of the *metatarsus*. Like the metacarpals of the palm, each metatarsal consists of a proximal *base*, an intermediate *body*, and a distal *head*. The first metatarsal, which is connected to the big toe, is thicker than the others because it bears more weight.

The ***phalanges*** of the foot resemble those of the hand both in number and arrangement. Each also consists of a proximal *base*, an intermediate *body*, and a distal *head*. The great or big toe (*hallux*) has two large, heavy phalanges—proximal and distal. The other four toes each have three phalanges—proximal, middle, and distal.

The bones of the foot are arranged in two ***arches*** (Figure 6.27). These arches enable the foot to support the weight of the body, provide an ideal distribution of body weight over the hard and soft tissues of the foot, and provide leverage while walking. The arches are not rigid—they yield as weight is applied and spring back when the weight is lifted, thus helping to absorb shocks. The *longitudinal arch* extends from the front to the back of the foot and has two parts, medial and lateral. The *transverse arch* is formed by the navicular, three cuneiforms, and the bases of the five metatarsals.

The bones composing the arches are held in position by ligaments and tendons. If these ligaments and tendons are weakened by excess weight, postural abnormalities, or genetic predisposition, the height of the medial longitudinal arch may decrease or "fall." The result is a condition called **flatfoot.**

■ **CHECKPOINT**

24. What bones form the lower limb, from proximal to distal?

25. What are the functions of the arches of the foot?

COMPARISON OF FEMALE AND MALE SKELETONS

OBJECTIVE • Identify the principal structural differences between female and male skeletons.

The bones of a male are generally larger and heavier than those of a female. The articular ends are thicker in relation to the shafts. In addition, because certain muscles of the male are larger than those of the female, the points of muscle attachment—tuberosities, lines, and ridges—are larger in the male skeleton.

Many significant structural differences between the skeletons of females and males are related to pregnancy and childbirth. Because the female's pelvis is wider and shallower than the male's, there is more space in the true pelvis of the female, especially in the pelvic inlet and pelvic outlet, which accommodate the passage of the infant's head at birth. Several of the significant differences between the female and male pelves are shown in Table 6.4.

■ **CHECKPOINT**

26. Explain the major structural differences between female and male skeletons related to pregnancy and childbirth.

Figure 6.27 Arches of the right foot.

Arches help the foot support and distribute the weight of the body and provide leverage during walking.

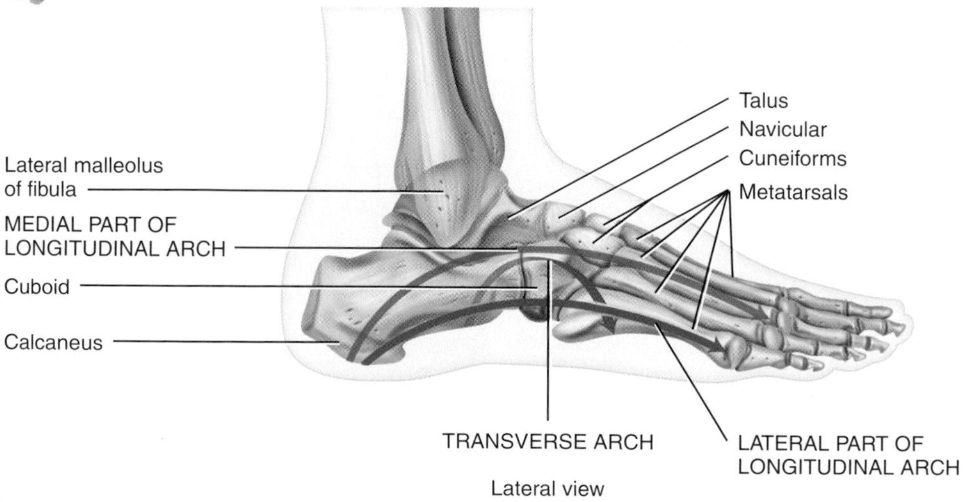

Lateral malleolus of fibula
MEDIAL PART OF LONGITUDINAL ARCH
Cuboid
Calcaneus
Talus
Navicular
Cuneiforms
Metatarsals
TRANSVERSE ARCH
LATERAL PART OF LONGITUDINAL ARCH

Lateral view

What structural aspect of the arches allows them to absorb shocks?

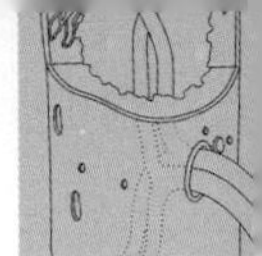

AGING AND THE SKELETAL SYSTEM

OBJECTIVE • Describe the effects of aging on the skeletal system.

From birth through adolescence, more bone is produced than is lost during bone remodeling. In young adults, the rates of bone production and loss are about the same. As the levels of sex steroids diminish during middle age, especially in women after menopause, a decrease in bone mass occurs because bone destruction outpaces bone formation. Because women's bones generally are smaller than men's bones to begin with, loss of bone mass in old age typically causes greater problems in women. These factors contribute to a higher incidence of osteoporosis in women.

Aging has two main effects on the skeletal system: Bones become more brittle and lose mass. Bone brittleness results from a decrease in the rate of protein synthesis and in the production of human growth hormone, which diminishes the production of the collagen fibers that give bone its strength and flexibility. As a result, inorganic minerals gradually constitute a greater proportion of the bone extracellular matrix. Loss of bone mass results from demineralization and usually begins after age 30 in females, accelerates greatly around age 45 as levels of estrogens decrease, and continues until as much as 30% of the calcium in bones is lost by age 70. Once bone loss begins in females, about 8% of bone mass is lost every 10 years. In males, calcium loss from bone typically does not begin until after age 60, and about 3% of bone mass is lost every 10 years. The loss of calcium from bones is one of the problems in osteoporosis (described on page 150). Loss of bone mass also leads to bone deformity, pain, stiffness, some loss of height, and loss of teeth.

■ **CHECKPOINT**

27. How does aging affect the brittleness of bone and the loss of bone mass?

• • •

To appreciate the many ways that the skeletal system contributes to homeostasis of other body systems, examine Focus on Homeostasis: The Skeletal System on page 149. Next, in Chapter 7, we will see how joints both hold the skeleton together and permit it to participate in movements.

Table 6.4 Comparison of the Pelvis in Females and Males

Point of Comparison	Female	Male
General structure	Light and thin.	Heavy and thick.
False (greater) pelvis	Shallow.	Deep.
Pelvic inlet	Larger and more oval.	Smaller and heart-shaped.
Acetabulum	Small and faces anteriorly.	Large and faces laterally.
Obturator foramen	Oval.	Round.
Pubic arch	Greater than 90° angle.	Less than 90° angle.

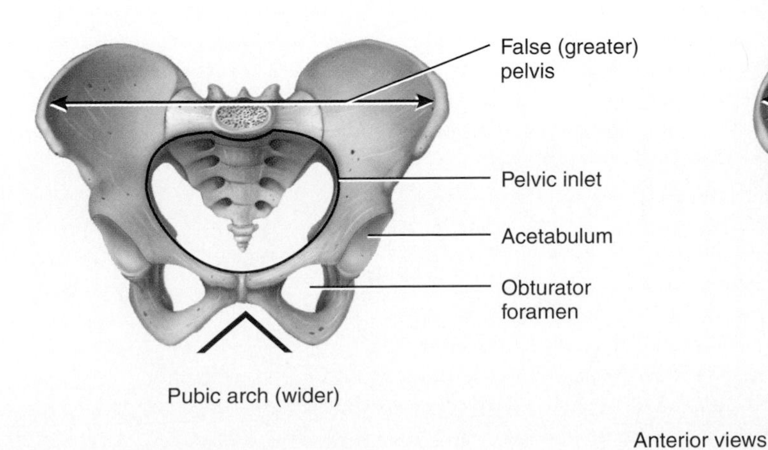

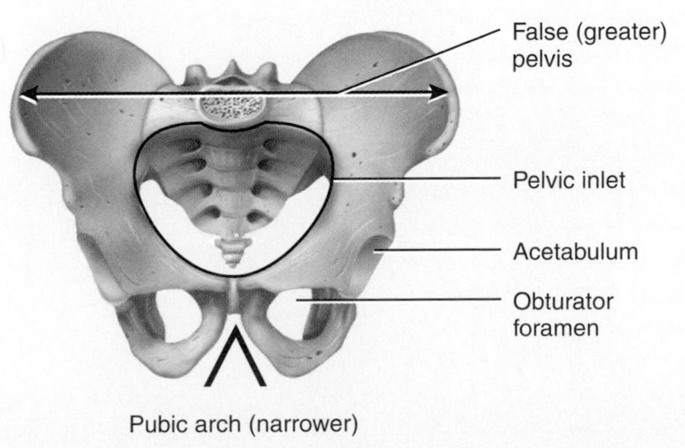

Anterior views

Focus on Wellness

Steps to Healthy Feet

We take the structure and function of our feet for granted—until they start to hurt. And even then we often continue to mistreat them, cramming them into shoes that are too tight, and then walking on concrete sidewalks and taking long shopping expeditions. No wonder foot problems are such a common complaint! Fortunately, most foot problems are preventable by understanding the foot's structure and function and then using good footwear to support them in their work.

These Feet Were Made for Walking

Each time you take a step, your heel strikes the ground first. Then you roll through the arches, over the ball of your foot, and onto your toes. Your arches flatten slightly as they absorb the weight of your body. One foot continues to bear your weight until the heel of the other foot touches the ground. As you walk, your big toe maintains your balance while the other toes give your foot some resiliency. The two outer metatarsals move to accommodate uneven surfaces, while the inner three stay rigid for support.

The most common cause of foot problems is ill-fitting shoes, which stress the structure and interfere with the function of the foot. The high heel is a case in point, which explains why 80% of those suffering from foot problems are women. Although many people think high heels look good and are fun to wear, they should not be used for walking because they make the body's weight fall onto the forefoot. Thus, the arches of the foot are not allowed to absorb the force of the body's weight. This unnatural stress can injure soft-tissue structures, joints, and bones.

Good Shoes for Happy Feet

Choosing shoes that are "good" to your feet can prevent many foot problems, an especially important consideration if you are doing any amount of walking. A good shoe has a sole that is strong and flexible and provides a good gripping surface. Cushioned insoles help protect feet from hard surfaces. Arch supports help distribute weight over a broader area, just like the arches in your foot.

Many people spend a great deal of time researching which brand of shoes to buy but do not spend adequate time evaluating whether or not the shoes suit their feet. A high-quality shoe is only worth buying if it fits! Shop for shoes in the late afternoon when your feet are at their largest. One foot is often bigger than the other; always buy for the bigger foot. The shoes you try on should feel comfortable immediately—don't plan on shoes stretching with wear. The heel should fit snugly, and the instep should not gape open. The toe box should be wide enough to wiggle all your toes.

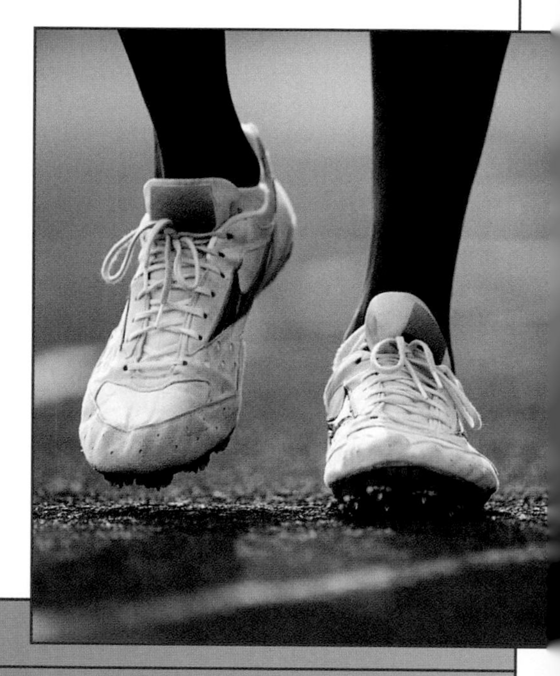

► **Think It Over . . .**

► ***Why do you think excess body weight is associated with an increased risk of foot problems?***

FOCUS ON HOMEOSTASIS

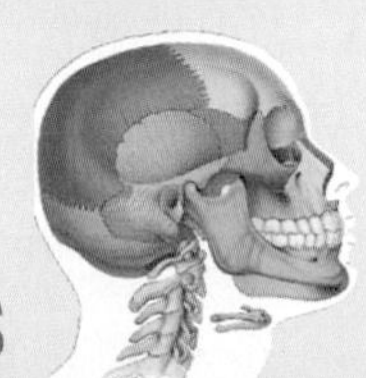

THE SKELETAL SYSTEM

Body System		Contribution Of The Skeletal System
For all body systems		Bones provide support and protection for internal organs; bones store and release calcium, which is needed for proper functioning of most body tissues.
Integumentary system	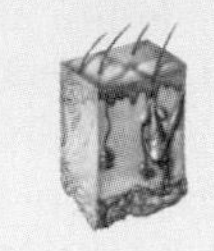	Bones provide strong support for overlying muscles and skin; joints provide flexibility while skin compensates for the change in joint angle.
Muscular system	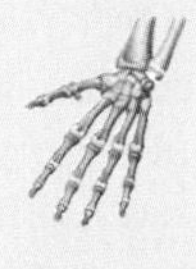	Bones provide attachment points for skeletal muscles and leverage for the muscles to bring about body movements; contraction of skeletal muscle requires calcium ions.
Nervous system	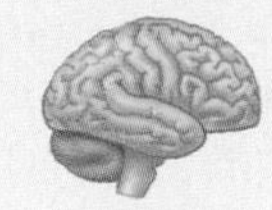	The skull and vertebrae protect the brain and spinal cord; a normal blood level of calcium is needed for normal functioning of neurons and neuroglia.
Endocrine system	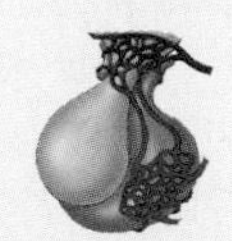	Bones store and release calcium, needed for normal actions of many hormones.
Cardiovascular system	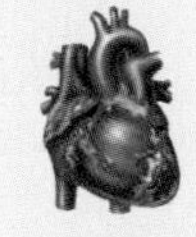	Red bone marrow carries out hemopoiesis (blood cell formation); rhythmic beating of the heart requires calcium ions.
Lymphatic system and immunity		Red bone marrow produces white blood cells involved in immune responses.
Respiratory system	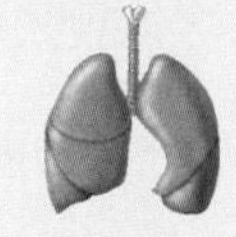	The axial skeleton of the thorax protects the lungs; rib movements assist breathing; some muscles used for breathing attach to bones by means of tendons.
Digestive system	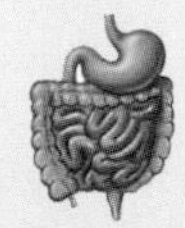	Teeth masticate (chew) food; the rib cage protects the esophagus, stomach, and liver; the pelvis protects portions of the intestines.
Urinary system	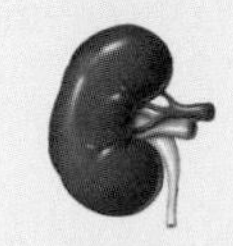	Ribs partially protect the kidneys, and the pelvis protects the urinary bladder and urethra.
Reproductive systems	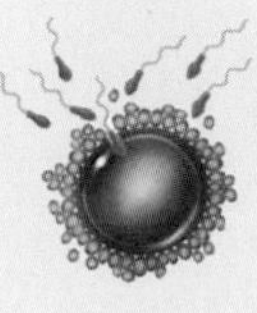	The pelvis protects the ovaries, uterine (fallopian) tubes, and uterus in females and part of the ductus (vas) deferens and accessory glands in males; bones are an important source of calcium needed for milk synthesis during lactation.

COMMON DISORDERS

Osteoporosis

Osteoporosis (os′-tē-ō-pō-RŌ-sis; *por-* = passageway; *-osis* = condition) is literally a condition of porous bones (Figure 6.28). The basic problem is that bone destruction outpaces bone formation. In large part this is due to depletion of calcium from the body—more calcium is lost in urine, feces, and sweat than is absorbed from the diet. Bone mass becomes so depleted that bones fracture, often spontaneously, under the mechanical stresses of everyday living. For example, a hip fracture might result from simply sitting down too quickly. In the United States, osteoporosis causes more than a million fractures a year, mainly in the hip, wrist, and vertebrae. Osteoporosis afflicts the entire skeletal system. In addition to fractures, osteoporosis causes shrinkage of vertebrae, height loss, hunched backs, and bone pain.

Thirty million people in the United States suffer from osteoporosis. The disorder primarily affects middle-aged and elderly people, 80% of them women. Older women suffer from osteoporosis more often than men for two reasons: Women's bones are less massive than men's bones, and production of estrogens in women declines dramatically at menopause; production of the main androgen, testosterone, in older men wanes gradually and only slightly. Estrogens and testosterone stimulate osteoblast activity and synthesis of bone extracellular matrix. Besides gender, risk factors for developing osteoporosis include a family history of the disease, European or Asian ancestry, thin or small body build, an inactive lifestyle, cigarette smoking, a diet low in calcium and vitamin D, more than two alcoholic drinks a day, and the use of certain medications.

In postmenopausal women, treatment of osteoporosis may include estrogen replacement therapy (ERT; low doses of estrogens) or hormone replacement therapy (HRT; a combination of estrogens and progesterone, another sex steroid). Although such treatments help combat osteoporosis, they increase a woman's risk of breast cancer. The drug Raloxifene® (Evista) mimics the beneficial effects of estrogens on bone without increasing the risk of breast cancer. Another drug that may be used is the nonhormone drug Alendronate (Fosamax®), which blocks resorption of bone by osteoclasts.

Perhaps more important than treatment is prevention. Adequate calcium intake and weight-bearing exercise in her early years may be more beneficial to a woman than drugs and calcium supplements when she is older.

Figure 6.28 Comparison of spongy bone tissue from (a) a normal young adult and (b) a person with osteoporosis. Notice the weakened trabeculae in (b). Compact bone tissue is similarly affected by osteoporosis.

In osteoporosis, bone resorption outpaces bone formation, so bone mass decreases.

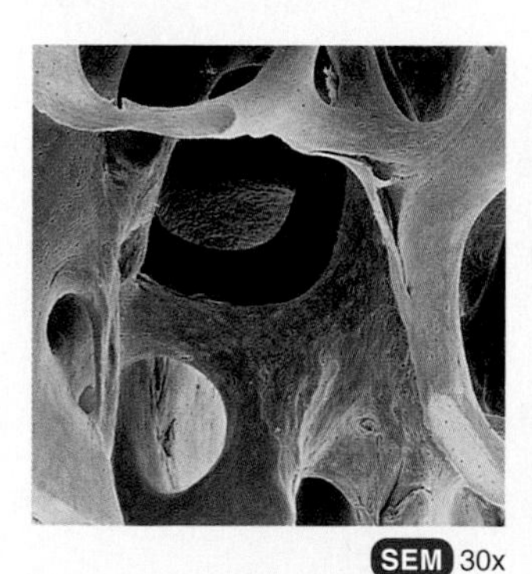

(a) Normal bone

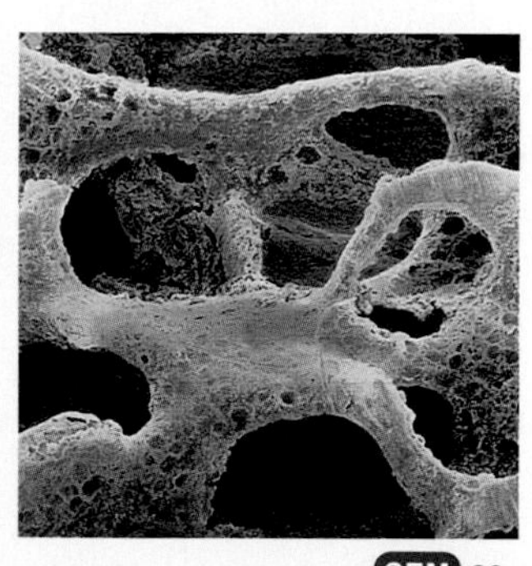

(b) Osteoporotic bone

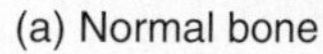

? If you wanted to develop a drug to lessen the effects of osteoporosis, would you look for a chemical that inhibits the activity of osteoblasts or that of osteoclasts?

Rickets and Osteomalacia

Rickets and ***osteomalacia*** (os′-tē-ō-ma-LĀ-she-ah; *-malacia* = softness) are disorders in which bone calcification fails. The bones become soft or rubbery and are easily deformed. Rickets affects the growing bones of children, but osteomalacia affects the bones of adults.

Herniated Disc

If the ligaments of the intervertebral discs become injured or weakened, the resulting pressure may be great enough to rupture the surrounding fibrocartilage. When this occurs, the material inside may herniate (protrude). This condition is called a ***herniated (slipped) disc***. It occurs most often in the lumbar region because that part of the vertebral column bears much of the weight of the body and is the region of the most bending.

Spina Bifida

Spina bifida (SPĪ-na BIF-i-da) is a congenital defect of the vertebral column in which laminae fail to unite at the midline. In serious cases, protrusion of the membranes (meninges) around the spinal cord or the spinal cord itself may produce partial or complete paralysis, partial or complete loss of urinary bladder control, and the absence of reflexes. Because an increased risk of spina bifida is associated with a low level of folic acid (one of the B vitamins) early in pregnancy, all women who might become pregnant are encouraged to take folic acid supplements.

Hip Fracture

Although any region of the hip girdle may fracture, the term ***hip fracture*** most commonly applies to a break in the bones associated with the hip joint—the head, neck, or trochanteric regions of the femur, or the bones that form the acetabulum. In the United States, 300,000 to 500,000 people sustain hip fractures each year. The incidence of hip fractures is increasing, in part due to longer life spans. Decreases in bone mass due to osteoporosis and an increased tendency to fall predispose elderly people to hip fractures.

Hip fractures often require surgical treatment, the goal of which is to repair and stabilize the fracture, increase mobility, and decrease pain. Sometimes the repair is accomplished by using surgical pins, screws, nails, and plates to secure the head of the femur. In severe hip fractures, the femoral head or the acetabulum of the hip bone may be replaced by prostheses (artificial devices). The procedure of replacing either the femoral head or the acetabulum is *hemiarthroplasty* (hem-ē-AR-thrō-plas-tē; *hemi-* = one half; *-arthro-* = joint; *-plasty* = molding). Replacement of both the femoral head and acetabulum is *total hip arthroplasty*. The acetabular prosthesis is made of plastic, and the femoral prosthesis is metal; both are designed to withstand a high degree of stress. The prostheses are attached to healthy portions of bone with acrylic cement and screws.

MEDICAL TERMINOLOGY AND CONDITIONS

Bunion (BUN-yun) A deformity of the great toe that typically is caused by wearing tightly fitting shoes. The condition produces inflammation of bursae (fluid-filled sacs at the joint), bone spurs, and calluses.

Clawfoot A condition in which the medial part of the longitudinal arch is abnormally elevated. It is often caused by muscle deformities, such as may result from diabetes.

Kyphosis (kī-FŌ-sis; *kypho-* = bent; *-osis* = condition) An exaggeration of the thoracic curve of the vertebral column. In the elderly, degeneration of the intervertebral discs leads to kyphosis; it may also be caused by osteoporosis, rickets, and poor posture.

Lordosis (lor-DŌ-sis; *lord-* = bent backward) An exaggeration of the lumbar curve of the vertebral column, also called *hollow back*. It may result from increased weight of the abdomen as in pregnancy or extreme obesity, poor posture, rickets, or tuberculosis of the spine.

Osteoarthritis (os′-tē-ō-ar-THRĪ-tis; *arthr* = joint) The degeneration of articular cartilage such that the bony ends touch; the resulting friction of bone against bone worsens the condition. Usually associated with the elderly.

Osteogenic sarcoma (os′-tē-Ō-JEN-ik sar-KŌ-ma; *sarcoma* = connective tissue tumor) Bone cancer that primarily affects osteoblasts and occurs most often in teenagers during their growth spurt; the most common sites are the metaphyses of the thigh bone (femur), shin bone (tibia), and arm bone (humerus). Metastases occur most often in lungs; treatment consists of multidrug chemotherapy and removal of the malignant growth, or amputation of the limb.

Osteomyelitis (os′-tē-ō-mī-e-LĪ-tis) An infection of bone characterized by high fever, sweating, chills, pain, nausea, pus formation, edema, and warmth over the affected bone and rigid overlying muscles. Bacteria, usually *Staphylococcus aureus*, often cause it. The bacteria may reach the bone from outside the body (through open fractures, penetrating wounds, or orthopedic surgical procedures); from other sites of infection in the body (abscessed teeth, burn infections, urinary tract infections, or upper respiratory infections) via the blood; and from adjacent soft tissue infections (as occurs in diabetes mellitus).

Osteopenia (os′-tē-ō-PĒ-nē-a; *penia* = poverty) Reduced bone mass due to a decrease in the rate of bone synthesis to a level insufficient to compensate for normal bone resorption; any decrease in bone mass below normal. An example is osteoporosis.

Scoliosis (skō′-lē-O-sis; *scolio-* = crooked) A sideways bending of the vertebral column, usually in the thoracic region. It may result from congenitally (present at birth) malformed vertebrae, chronic sciatica, paralysis of muscles on one side of the vertebral column, poor posture, or one leg being shorter than the other.

Whiplash injury Injury to the neck region due to severe hyperextension (backward tilting) of the head followed by severe hyperflexion (forward tilting) of the head, usually associated with a rear-end automobile collision. Symptoms are related to stretching and tearing of ligaments and muscles, vertebral fractures, and herniated vertebral discs.

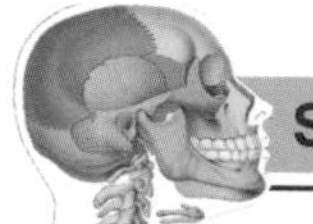

STUDY OUTLINE

Functions of Bone and the Skeletal System (p. 114)

1. The skeletal system consists of all bones attached at joints and cartilage between joints.
2. The functions of the skeletal system include support, protection, movement, mineral homeostasis, housing blood-forming tissue, and storage of energy.

Types of Bones (p. 114)

1. On the basis of shape, bones are classified as long, short, flat, or irregular.

Structure of Bone (p. 114)

1. Parts of a long bone include the diaphysis (shaft), epiphyses (ends), metaphysis, articular cartilage, periosteum, medullary (marrow) cavity, and endosteum.
2. The diaphysis is covered by periosteum.
3. Bone tissue consists of widely separated cells surrounded by large amounts of extracellular matrix (intercellular substance). The four principal types of cells are osteogenic cells, osteoblasts, osteocytes, and osteoclasts. The extracellular matrix contains collagen fibers (organic) and mineral salts that consist mainly of calcium phosphate (inorganic).
4. Compact (dense) bone tissue consists of osteons (haversian systems) with little space between them. Compact bone composes most of the bone tissue of the diaphysis. Functionally, compact bone protects, supports, and resists stress.
5. Spongy bone tissue consists of trabeculae surrounding many red bone marrow–filled spaces. It forms most of the structure of short, flat, and irregular bones and the epiphyses of long bones. Functionally, spongy bone stores red bone marrow and provides some support.

Bone Formation (p. 118)

1. Bone forms by a process called ossification.
2. Bone formation in an embryo or fetus occurs by intramembranous and endochondral ossification, which involve the replacement of preexisting connective tissue with bone.
3. Intramembranous ossification occurs within mesenchyme arranged in sheetlike layers that resemble membranes.
4. Endochondral ossification occurs within a hyaline cartilage derived from mesenchyme. The primary ossification center of a long bone is in the diaphysis. Cartilage degenerates, leaving cavities that merge to form the medullary (marrow) cavity. Osteoblasts lay down bone. Next, ossification occurs in the epiphyses, where bone replaces cartilage, except for articular cartilage and the epiphyseal plate.
5. Because of the activity of the epiphyseal plate, the diaphysis of a bone increases in length.
6. Bone grows in diameter as a result of the addition of new bone tissue around the outer surface of the bone.
7. Old bone is constantly destroyed by osteoclasts, while new bone is constructed by osteoblasts. This process is called remodeling.
8. A fracture is any break in a bone. Fracture repair involves remodeling.
9. Normal growth depends on minerals (calcium, phosphorus, magnesium), vitamins (A, C, D), and hormones (human growth hormone, insulinlike growth factors, insulin, thyroid hormones, sex hormones, and parathyroid hormone).
10. Bones store and release calcium and phosphate, controlled mainly by parathyroid hormone (PTH). PTH raises blood calcium level.
11. Calcitonin (CT) lowers blood calcium level.

Exercise and Bone Tissue (p. 123)

1. Mechanical stress increases bone strength by increasing deposition of mineral salts and production of collagen fibers.
2. Removal of mechanical stress weakens bone through demineralization and collagen fiber reduction.

Divisions of the Skeletal System (p. 124)

1. The axial skeleton consists of bones arranged along the longitudinal axis of the body. The parts of the axial skeleton are the skull, hyoid bone, auditory ossicles, vertebral column, sternum, and ribs.
2. The appendicular skeleton consists of the bones of the girdles and the upper and lower limbs. The parts of the appendicular skeleton are the pectoral (shoulder) girdles, bones of the upper limbs, pelvic (hip) girdle, and bones of the lower limbs.

Skull and Hyoid Bone (p. 125)

1. The skull consists of cranial bones and facial bones.
2. The eight cranial bones include the frontal (1), parietal (2), temporal (2), occipital (1), sphenoid (1), and ethmoid (1).
3. The 14 facial bones are the nasal (2), maxillae (2), zygomatic (2), mandible (1), lacrimal (2), palatine (2), inferior nasal conchae (2), and vomer (1).
4. The hyoid bone, a U-shaped bone that does not articulate with any other bone, supports the tongue and provides attachment for some of its muscles as well as some neck muscles.
5. Sutures are immovable joints between bones of the skull. Examples are the coronal, sagittal, lambdoid, and squamous sutures.
6. Paranasal sinuses are cavities in bones of the skull that communicate with the nasal cavity. They are lined by mucous membranes. Cranial bones containing paranasal sinuses are the frontal, sphenoid, ethmoid, and maxillae.
7. Fontanels are mesenchyme-filled spaces between the cranial bones of fetuses and infants. The major fontanels are the anterior, posterior, anterolaterals, and posterolaterals.

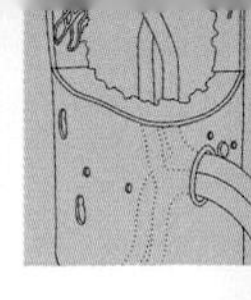

Vertebral Column (p. 133)

1. The bones of the adult vertebral column are the cervical vertebrae (7), thoracic vertebrae (12), lumbar vertebrae (5), the sacrum (5, fused), and the coccyx (4, fused).
2. The vertebral column contains normal curves that give strength, support, and balance.
3. The vertebrae are similar in structure, each consisting of a body, vertebral arch, and seven processes. Vertebrae in the different regions of the column vary in size, shape, and detail.

Thorax (p. 137)

1. The thoracic skeleton consists of the sternum, ribs, costal cartilages, and thoracic vertebrae.
2. The thoracic cage protects vital organs in the chest area.

Pectoral (Shoulder) Girdle (p. 139)

1. Each pectoral (shoulder) girdle consists of a clavicle and scapula.
2. Each attaches an upper limb to the trunk.

Upper Limb (p. 140)

1. There are 30 bones in each upper limb.
2. The upper limb bones include the humerus, ulna, radius, carpals, metacarpals, and phalanges.

Pelvic (Hip) Girdle (p. 143)

1. The pelvic (hip) girdle consists of two hip bones.
2. It attaches the lower limbs to the trunk at the sacrum.
3. Each hip bone consists of three fused components: ilium, pubis, and ischium.

Lower Limb (p. 144)

1. There are 30 bones in each lower limb.
2. The lower limb bones include the femur, patella, tibia, fibula, tarsals, metatarsals, and phalanges.
3. The bones of the foot are arranged in two arches, the longitudinal arch and the transverse arch, to provide support and leverage.

Comparison of Female and Male Skeletons (p. 146)

1. Male bones are generally larger and heavier than female bones and have more prominent markings for muscle attachment.
2. The female pelvis is adapted for pregnancy and childbirth. Differences in pelvic structure are listed in Table 6.4 on page 147.

Aging and the Skeletal System (p. 147)

1. The main effect of aging is a loss of calcium from bones, which may result in osteoporosis.
2. Another effect of aging is a decreased production of extracellular matrix proteins (mostly collagen fibers), which makes bones more brittle and thus more susceptible to fracture.

SELF-QUIZ

1. Match the following cell types to their functions:

____	**a.** chondroblasts	**A.** mature bone cells
____	**b.** osteoclasts	**B.** cells that form bone
____	**c.** chondrocytes	**C.** secrete cartilage matrix
____	**d.** osteocytes	**D.** mature cartilage cells
____	**e.** osteoblasts	**E.** involved in bone resorption

2. When trying to locate a foramen in a bone, you would look for

a. a large, rough projection
b. a ridge
c. a rounded projection
d. a shallow depression
e. an opening or hole

3. The ribs articulate with the

a. thoracic vertebrae
b. sacrum
c. cervical vertebrae
d. lumbar vertebrae
e. atlas and axis

4. Match the following:

____	**a.** run lengthwise through bone	**A.** lamellae
____	**b.** connect central canals with lacunae	**B.** lacunae
____	**c.** concentric rings of matrix	**C.** perforating (volkmann's) canal
____	**d.** connect nutrient arteries and nerves from the periosteum to the central canals	**D.** canaliculi
____	**e.** spaces that contain osteocytes	**E.** central (haversian) canal

5. The presence of an epiphyseal line in a long bone indicates that the bone

a. is undergoing resorption
b. has stopped growing in length
c. is growing in diameter
d. is still capable of growing in length
e. is broken

6. The hyoid bone is unique because it
   a. is the smallest bone in the skull
   b. can malform causing a cleft palate
   c. forms the paranasal sinuses
   d. is often broken when an individual falls forward
   e. does not articulate with any other bone

7. The bones that form the pectoral girdle are the
   a. clavicle and scapula b. scapula and sternum
   c. humerus and scapula d. clavicle and humerus
   e. coxal bones

8. The main hormone that regulates the Ca^{2+} balance between bone and blood is
   a. parathyroid hormone b. insulin c. testosterone
   d. insulinlike growth factors e. human growth hormone

9. Spongy bone differs from compact bone because spongy bone
   a. is made up of numerous osteons
   b. is found primarily in the diaphyses of long bones
   c. has latticework walls known as trabeculae
   d. contains few, small spaces known as lacunae
   e. has lamellae arranged in concentric rings

10. In which of the following individuals might you expect to find the smallest bone mass?
    a. 20-year-old male weightlifter
    b. 45-year-old female weightlifter
    c. 45-year-old male astronaut
    d. 80-year-old bedridden female
    e. 65-year-old bedridden male

11. Place the following steps of endochondral ossification in the correct order:
    1. Hyaline cartilage remains on the articular surfaces and epiphyseal plates
    2. Chondroblasts produce a growing hyaline cartilage model surrounded by the perichondrium
    3. Osteoblasts in perichondrium produce compact bone
    4. Secondary ossification centers form
    5. Primary ossification center and medullary cavity form

    a. 2, 3, 4, 5, 1 b. 2, 3, 5, 4, 1 c. 5, 2, 1, 3, 4
    d. 3, 2, 5, 4, 1 e. 5, 3, 2, 1, 4

12. Match each bone to its shape:

____	a. humerus	A. flat
____	b. carpus	B. irregular
____	c. vertebra	C. long
____	d. sternum	D. short

13. Where long bones form joints, the epiphyses are covered with
    a. yellow bone marrow b. osteoclasts c. periosteum
    d. endosteum e. hyaline cartilage

14. What substance in bone contributes to its tensile strength?
    a. red bone marrow
    b. collagen
    c. yellow bone marrow
    d. calcium phosphate
    e. loose fibrous connective tissue

15. The skeletal system is responsible for
    a. protecting internal organs from injury
    b. producing movement
    c. providing a supporting framework for the body
    d. hemopoiesis
    e. all of the above

16. For each of the following bones, place an AX in the blank if it belongs to the axial skeleton and an AP in the blank if it is part of the appendicular skeleton.

____	a. lacrimal	____	l. metatarsals	____	w. maxilla
____	b. clavicle	____	m. temporal	____	x. frontal
____	c. radius	____	n. metacarpals	____	y. inferior nasal concha
____	d. mandible	____	o. vomer		
____	e. patella	____	p. fibula		
____	f. carpals	____	q. palatine	____	z. humerus
____	g. scapula	____	r. hyoid	____	aa. ulna
____	h. sternum	____	s. tibia	____	bb. femur
____	i. phalanges	____	t. sphenoid	____	cc. ribs
____	j. tarsals	____	u. vertebrae	____	dd. occipital
____	k. ethmoid	____	v. coxal		

CRITICAL THINKING APPLICATIONS

1. J.R. was riding his motorcycle across the Big Span Bridge when he had a collision with a nearsighted sea gull. In the resulting crash, J.R. crushed his left leg, fracturing both leg bones; snapped the pointy distal end of his lateral forearm bone; and broke the most lateral and proximal bone in his wrist. The sea gull flew off when the ambulance arrived. Name the bones that J.R. broke.
2. While investigating her new baby brother, a 4-year-old girl discovers a soft spot on the baby's skull and announces that the baby needs to go back because "it's not finished yet." Explain the presence of soft spots in the infant's skull and the lack of soft spots in yours.
3. Old Grandma Olga is a tiny, stooped woman with a big sense of humor. Her favorite movie line is from *The Wizard of Oz* when the wicked witch says "I'm melting." "That's me," laughs Olga, "melting away, getting shorter every year." What is happening to Grandma Olga?
4. During the volleyball game, Kate jumped, twisted, spiked, scored, and screamed! She couldn't put any weight on her left leg and her left knee swelled rapidly to twice its usual size. X-rays revealed a fracture of the proximal tibia. In layman's terms, what is the location of Kate's fracture? What caused the rapid swelling? What are the body's requirements for bone healing?

ANSWERS TO FIGURE QUESTIONS

6.1 The articular cartilage reduces friction at joints; red bone marrow produces blood cells; and the endosteum lines the medullary cavity.

6.2 Because the central canals are the main blood supply to the osteocytes, their blockage would lead to death of osteocytes.

6.3 The flat bones of the skull and mandible develop by intramembranous ossification.

6.4 The epiphyseal lines are indications of growth zones that have ceased to function.

6.5 Heartbeat, respiration, nerve cell functioning, enzyme functioning, and blood clotting are all processes that depend on proper levels of calcium.

6.6 Axial skeleton: skull and vertebral column. Appendicular skeleton: clavicle, shoulder girdle, humerus, pelvic girdle, and femur.

6.7 The cranial bones are the frontal, parietal, occipital, sphenoid, ethmoid, and temporal bones.

6.8 The foramen magnum is the largest foramen in the skull.

6.9 Crista galli of ethmoid bone, frontal, parietal, temporal, occipital, temporal, parietal, frontal, and crista galli of ethmoid bone articulate in clockwise order with the sphenoid bone.

6.10 The perpendicular plate of the ethmoid bone forms the top part of the nasal septum.

6.11 The paranasal sinuses produce mucus and serve as resonating chambers for vocalization.

6.12 The thoracic and sacral curves are concave.

6.13 The vertebral foramina enclose the spinal cord, and the intervertebral foramina provide spaces for spinal nerves to exit the vertebral column.

6.14 The atlas and axis permit movement of the head to signify "no."

6.15 The lumbar vertebrae support more weight than the thoracic and cervical vertebrae.

6.16 The sacral foramina are passageways for nerves and blood vessels.

6.17 The true ribs are pairs 1 through 7; the false ribs are pairs 8 through 12; and the floating ribs are pairs 11 and 12.

6.18 A pectoral girdle consists of a clavicle and a scapula.

6.19 The glenoid cavity of the scapula articulates with the humerus.

6.20 The "elbow" part of the ulna is the olecranon.

6.21 The knuckles are the heads of the metacarpals.

6.22 The true pelvis surrounds the pelvic organs in the pelvic cavity.

6.23 The femur fits into the acetabulum.

6.24 The distal end of the femur articulates with the tibia and the patella.

6.25 The tibia is the weight-bearing bone of the leg.

6.26 The talus articulates with the tibia and fibula.

6.27 The arches are not rigid, yielding when weight is applied and springing back when weight is lifted to allow them to absorb the shock of walking and running.

6.28 A drug that inhibits the activity of osteoclasts might lessen the effects of osteoporosis.

chapter 7 JOINTS

did you know? ***For many years, people believed that exercise accelerated joint degeneration, and that people with arthritis should avoid physical activity. Scientists now believe that a sedentary lifestyle leads to loss of strength in muscles, tendons, ligaments, and other joint structures, which makes movement even more painful and difficult. When muscles and joints atrophy, the resulting weakness makes joints less stable, and more vulnerable to injury. Physical activity helps to strengthen joint structures and delay the progress of arthritis. Low- or non-impact activities such as strength training, swimming, and cycling can improve fitness, functional status, and quality of life for people with arthritis.***

Focus on Wellness, page 167

www.wiley.com/college/apcentral

Bones are too rigid to bend without being damaged. Fortunately, flexible connective tissues form joints that hold bones together while in most cases permitting some degree of movement. If you have ever damaged these areas, you know how difficult it is to walk with a cast over your knee or to turn a doorknob with a splint on your finger. A ***joint*** (also called an ***articulation***) is a point of contact between bones, between cartilage and bones, or between teeth and bones. When we say one bone articulates with another bone, we mean that the two bones form a joint. ***Arthrology*** (ar-THROL-ō-jē; *arthr-* = joint; *-logy* = study of) is the scientific study of joints. Many joints of the body permit movement. The study of motion of the human body is called ***kinesiology*** (ki-nē′-sē-OL-ō-jē; *kinesi-* = movement).

looking back to move ahead . . .

- **Collagen Fibers (page 84)**
- **Dense Regular Connective Tissue (page 86)**
- **Cartilage (page 89)**
- **Synovial Membranes (page 91)**
- **Divisions of the Skeletal System (page 124)**

JOINTS

OBJECTIVES • Describe how the structure of a joint determines its function.

• Describe the structural and functional classes of joints.

A joint's structure determines its combination of strength and flexibility. At one end of the spectrum are joints that permit no movement and are thus very strong, but inflexible. In contrast, other joints afford fairly free movement and are thus flexible but not as strong. In general, the closer the fit at the point of contact, the stronger the joint. At tightly fitted joints, movement is obviously more restricted. The looser the fit, the greater the movement. However, loosely fitted joints are prone to displacement of the articulating bones from their normal positions (dislocation). Movement at joints is also determined by (1) the shape of the articulating bones, (2) the flexibility (tension or tautness) of the ligaments that bind the bones together, and (3) the tension of associated muscles and tendons. Joint flexibility may also be affected by hormones. For example, toward the end of pregnancy, a hormone called relaxin increases the flexibility of the fibrocartilage of the pubic symphysis and loosens the ligaments between the sacrum and hip bone. These changes enlarge the pelvic outlet, which assists in delivery of the baby.

Joints are classified structurally, based on their anatomical characteristics, and functionally, based on the type of movement they permit.

The structural classification of joints is based on two criteria: (1) the presence or absence of a space between the articulating bones, called a synovial cavity, and (2) the type of connective tissue that holds the bones together. Structurally, joints are classified as one of the following types:

- **Fibrous joints** (FĪ-brus): There is no synovial cavity and the bones are held together by fibrous connective tissue that is rich in collagen fibers.
- **Cartilaginous joints** (kar-ti-LAJ-i-nus): There is no synovial cavity and the bones are held together by cartilage.
- **Synovial joints** (si-NŌ-vē-al): The bones forming the joint have a synovial cavity and are united by the dense irregular connective tissue of an articular capsule, and often by accessory ligaments.

The functional classification of joints relates to the degree of movement they permit. Functionally, joints are classified as one of the following types:

- ***Synarthrosis*** (sin′-ar-THRŌ-sis; *syn-* = together): An immovable joint. The plural is *synarthroses.*
- ***Amphiarthrosis*** (am′-fē-ar-THRŌ-sis; *amphi-* = on both sides): A slightly movable joint. The plural is *amphiarthroses.*
- ***Diarthrosis*** (dī′-ar-THRŌ-sis = movable joint): A freely movable joint. The plural is *diarthroses.* All diarthroses are synovial joints. They have a variety of shapes and permit several different types of movements.

The following sections present the joints of the body according to their structural classification. As we examine the structure of each type of joint, we will also explore its functional attributes.

CHECKPOINT

1. What factors determine movement at joints?

FIBROUS JOINTS

OBJECTIVE • Describe the structure and functions of the three types of fibrous joints.

Fibrous joints permit little or no movement. The three types of fibrous joints are (1) sutures, (2) syndesmoses, and (3) gomphoses.

1. A ***suture*** (SOO-cher; *sutur-* = seam) is a fibrous joint composed of a thin layer of dense fibrous connective tissue. Sutures unite the bones of the skull. An example is the coronal suture between the frontal and parietal bones (Figure 7.1a). The irregular, interlocking edges of sutures give them added strength and decrease their chance of fracturing. Because a suture is immovable, it is classified functionally as a synarthrosis.
2. A ***syndesmosis*** (sin′-dez-MŌ-sis; *syndesmo-* = band or ligament) is a fibrous joint in which the distance between the articulating bones and the amount of dense fibrous connective tissue is greater than in a suture (Figure 7.1b). One example of a syndesmosis is the distal articulation between the tibia and fibula where the anterior tibiofibular ligament connects the bones. Because it permits slight movement, a syndesmosis is classified functionally as an amphiarthrosis.
3. A ***gomphosis*** (gom-FŌ-sis; *gompho-* = a bolt or nail; plural is *gomphoses*) is a type of fibrous joint in which a cone-shaped peg fits into a socket. The only gomphoses in the human body are the articulations of the roots of the teeth with the sockets of the alveolar processes of the maxillae and mandible (Figure 7.1c). The dense fibrous connective tissue between the root of a tooth and its socket is the periodontal ligament. A gomphosis is classified functionally as a synarthrosis, an immovable joint.

CHECKPOINT

2. Which fibrous joints are synarthroses? Which are amphiarthroses?

Figure 7.1 Fibrous joints.

At a fibrous joint, the bones are held together by connective tissue containing many collagen fibers.

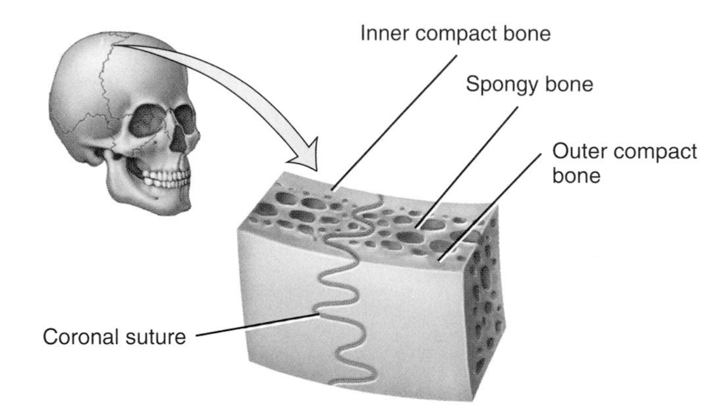

(a) Suture between skull bones

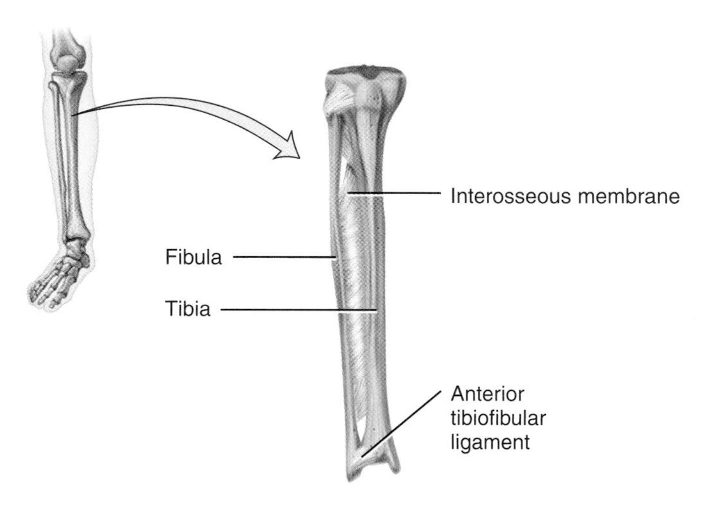

(b) Syndesmosis between distal tibia and fibula

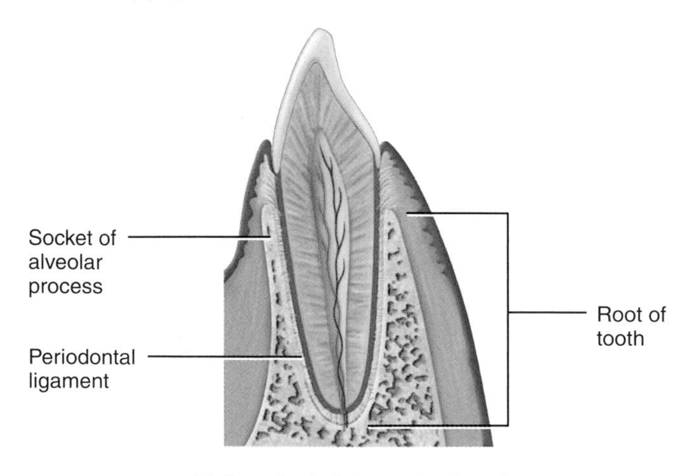

(c) Gomphosis between tooth and socket of alveolar process

Functionally, why are sutures classified as synarthroses and syndesmoses classified as amphiarthroses?

CARTILAGINOUS JOINTS

OBJECTIVE • Describe the structure and functions of the two types of cartilaginous joints.

Like a fibrous joint, a ***cartilaginous*** (car-ti-LAJ-i-nus) ***joint*** allows little or no movement. Here the articulating bones are tightly connected by either fibrocartilage or hyaline cartilage. The two types of cartilaginous joints are synchondroses and symphyses.

1. A ***synchondrosis*** (sin′-kon-DRŌ-sis; *chondro-* = cartilage) is a cartilaginous joint in which the connecting material is hyaline cartilage. An example of a synchondrosis is the epiphyseal plate that connects the epiphysis and diaphysis of an elongating bone (Figure 7.2a). Functionally, a synchondrosis is a synarthrosis, an immovable joint. When bone growth stops, bone replaces the hyaline cartilage.
2. A ***symphysis*** (SIM-fi-sis = growing together) is a cartilaginous joint in which the ends of the articulating bones are covered with hyaline cartilage, but the bones are connected by a broad, flat disc of fibrocartilage. The pubic symphysis between the anterior surfaces of the hip bones is one example of a symphysis (Figure 7.2b). This type of joint is also found at the intervertebral joints between bodies of vertebrae. Functionally, a symphysis is an amphiarthrosis, a slightly movable joint.

■ **CHECKPOINT**

3. Which cartilaginous joints are synarthroses? Which are amphiarthroses?

SYNOVIAL JOINTS

OBJECTIVE • Describe the structure of synovial joints.

Structure of Synovial Joints

Synovial joints (si-NŌ-vē-al) have certain characteristics that distinguish them from other joints. The unique characteristic of a synovial joint is the presence of a space called a ***synovial (joint) cavity*** between the articulating bones (Figure 7.3). The synovial cavity allows a joint to be freely movable. Hence, all synovial joints are classified functionally as diarthroses. The bones at a synovial joint are covered by ***articular cartilage***, which is hyaline cartilage. Articular cartilage reduces friction between bones in the joint during movement and helps to absorb shock.

A sleevelike ***articular capsule*** surrounds a synovial joint, encloses the synovial cavity, and unites the articulating bones.

Figure 7.2 Cartilaginous joints.

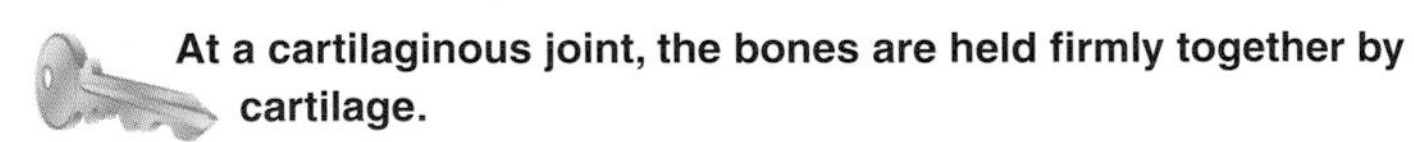

At a cartilaginous joint, the bones are held firmly together by cartilage.

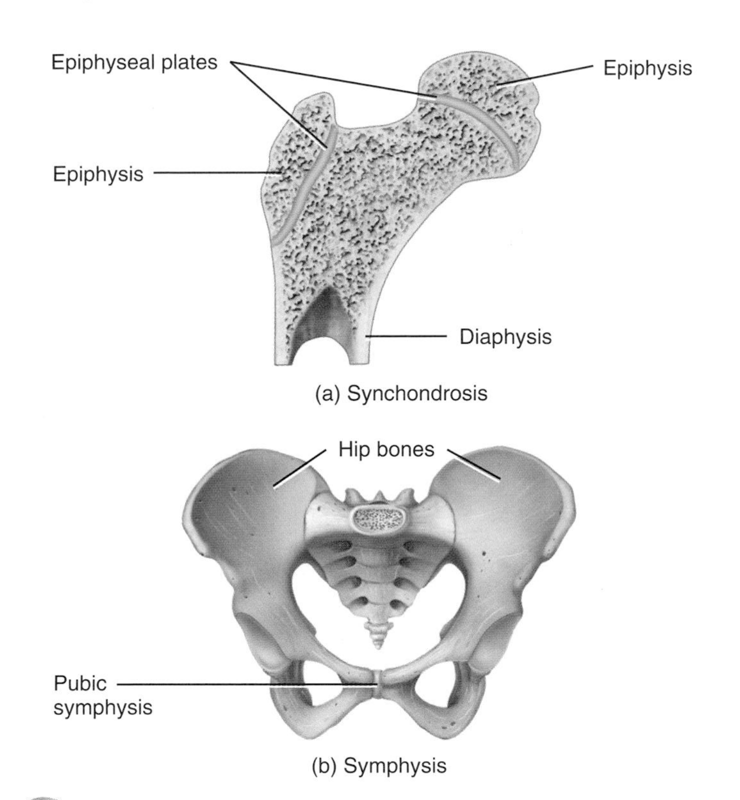

What is the structural difference between a synchondrosis and a symphysis?

The articular capsule is composed of two layers, an outer fibrous capsule and an inner synovial membrane (Figure 7.3). The outer layer, the ***fibrous capsule***, usually consists of dense irregular connective tissue that attaches to the periosteum of the articulating bones. The fibers of some fibrous capsules are arranged in parallel bundles that are highly adapted for resisting strains. Such fiber bundles are called ***ligaments*** (*liga-* = bound or tied) and are one of the main mechanical factors that hold bones close together in a synovial joint. The inner layer of the articular capsule, the ***synovial membrane***, is composed of areolar connective tissue with elastic fibers. At many synovial joints the synovial membrane includes accumulations of adipose tissue, called ***articular fat pads*** (see Figure 7.10c).

The synovial membrane secretes ***synovial fluid*** (*ov-* = egg), which forms a thin film over the surfaces within the articular capsule. This viscous, clear or pale yellow fluid was named for its similarity in appearance and consistency to uncooked egg white (albumin). Its several functions include reducing friction by lubricating the joint, and supplying nutrients to and removing metabolic wastes from the chondrocytes within articular cartilage. When a synovial joint is immobile for a time, the fluid is quite viscous (gel-like), but as joint movement increases, the fluid becomes less viscous. One of the benefits of a warm-up before exercise is that it stimulates the production and secretion of synovial fluid. More fluid means less stress on the joint during exercise.

Many synovial joints also contain ***accessory ligaments*** that lie outside and inside the articular capsule. Examples of accessory ligaments outside the articular capsule are the fibular (lateral) and tibial (medial) collateral ligaments of the knee joint (see Figure 7.10d). Examples of accessory ligaments inside the articular capsule are the anterior and posterior cruciate ligaments of the knee joint (see Figure 7.10d).

Inside some synovial joints, such as the knee, are pads of fibrocartilage that lie between the articular surfaces of the bones and are attached to the fibrous capsule. These pads are called ***articular discs*** or ***menisci*** (me-NIS-sī; singular is *meniscus*). Figure 7.10d depicts the lateral and medial menisci in the knee joint. By modifying the shape of the joint surfaces of the articulating bones, articular discs allow two bones of different shapes to fit more tightly. Articular discs also help to maintain the stability of the joint and direct the flow of synovial fluid to the areas of greatest friction.

Figure 7.3 Structure of a typical synovial joint. Note the two layers of the articular capsule: the fibrous capsule and the synovial membrane. Synovial fluid fills the synovial cavity, which is located between the synovial membrane and the hyaline articular cartilage.

The distinguishing feature of a synovial joint is the synovial (joint) cavity between the articulating bones.

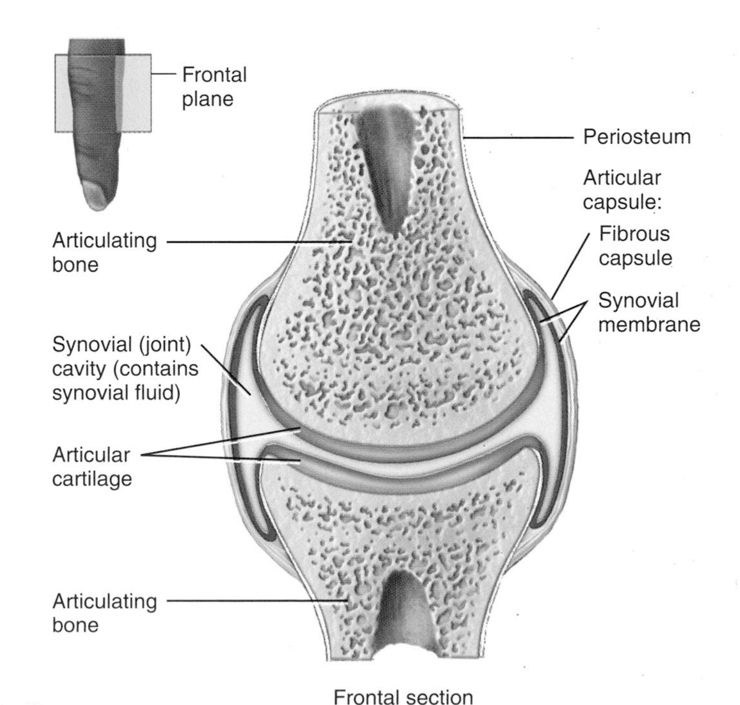

What is the functional classification of synovial joints?

The tearing of articular discs (menisci) in the knee, commonly called **torn cartilage**, occurs often among athletes. Such damaged cartilage will begin to wear and may precipitate arthritis unless it is surgically removed (meniscectomy). Surgical repair of the torn cartilage is required because of the avascular nature of cartilage and may be assisted by **arthroscopy** (ar-THROS-kō-pē; *-scopy* = observation), the visual examination of the interior of a joint, usually the knee, with an *arthroscope*, a lighted, pencil-thin instrument. Arthroscopy is used to determine the nature and extent of damage following knee injury and to monitor the progression of disease and the effects of therapy. In addition, the insertion of surgical instruments through the arthroscope or other incisions enables a physician to remove torn cartilage and repair damaged cruciate ligaments in the knee; to remodel poorly formed cartilage; to obtain tissue samples for analysis; and to perform surgery on other joints, such as the shoulder, elbow, ankle, and wrist.

The various movements of the body create friction between moving parts. Saclike structures called ***bursae*** (BER-sē = purses; singular is *bursa*) are strategically situated to reduce friction in some synovial joints, such as the shoulder and knee joints (see Figure 7.10c). Bursae are not strictly part of synovial joints, but do resemble joint capsules because their walls consist of connective tissue lined by a synovial membrane. They are also filled with a fluid similar to synovial fluid. Bursae are located between the skin and bone in places where skin rubs over bone. They are also found between tendons and bones, muscles and bones, and ligaments and bones. The fluid-filled bursal sacs cushion the movement of one body part over another.

An acute or chronic inflammation of a bursa, for example in the shoulder and knee, is called **bursitis**. The condition may be caused by trauma, by an acute or chronic infection (including syphilis and tuberculosis), or by rheumatoid arthritis (described on page 168). Repeated, excessive exertion of a joint often results in bursitis, with local inflammation and the accumulation of fluid. Symptoms include pain, swelling, tenderness, and limited movement. Treatment may include oral anti-inflammatory agents and injections of cortisol-like steroids.

■ **CHECKPOINT**

4. How does the structure of synovial joints classify them as diarthroses?
5. What are the functions of articular cartilage, the articular capsule, synovial fluid, articular discs, and bursae?

TYPES OF MOVEMENTS AT SYNOVIAL JOINTS

OBJECTIVE • Describe the types of movements that can occur at synovial joints.

Anatomists, physical therapists, and kinesiologists use specific terminology to designate specific types of movement that can occur at a synovial joint. These precise terms indicate the form of motion, the direction of movement, or the relationship of one body part to another during movement. Movements at synovial joints are grouped into four main categories: (1) gliding, (2) angular movements, (3) rotation, and (4) special movements. The last category includes movements that occur only at certain joints.

Gliding

Gliding is a simple movement in which relatively flat bone surfaces move back-and-forth and side-to-side relative to one another. This can be illustrated between the acromion of the scapula and clavicle by placing your upper limb at your side, rotating it about your head, and lowering it again (see Figure 7.7b). Gliding movements are limited in range due to the loose-fitting structure of the articular capsule and associated ligaments and bones.

Angular Movements

In ***angular movements***, there is an increase or a decrease in the angle between articulating bones. The principal angular movements are flexion, extension, hyperextension, abduction, adduction, and circumduction and are discussed with respect to the body in the anatomical position. In ***flexion*** (FLEK-shun = to bend), there is a decrease in the angle between articulating bones; in ***extension*** (eks-TEN-shun = to stretch out), there is an increase in the angle between articulating bones, often to restore a part of the body to the anatomical position after it has been flexed (Figure 7.4). Examples of flexion include bending the head toward the chest (Figure 7.4a); moving the humerus forward at the shoulder joint as in swinging the arms forward while walking (Figure 7.4b); moving the forearm toward the arm (Figure 7.4c); moving the palm toward the forearm (Figure 7.4d); moving the femur forward, as in walking (Figure 7.4e); and bending the knee (Figure 7.4f). Extension is simply the reverse of these movements.

Continuation of extension beyond the anatomical position is called ***hyperextension*** (*hyper-* = beyond or excessive). Examples of hyperextension include bending the head backward (Figure 7.4a); moving the humerus backward, as in swinging the arms backward while walking (Figure 7.4b); moving the palm backward at the wrist joint (Figure 7.4d);

Figure 7.4 Angular movements at synovial joints: flexion, extension, and hyperextension.

In angular movements, there is an increase or decrease in the angle between articulating bones.

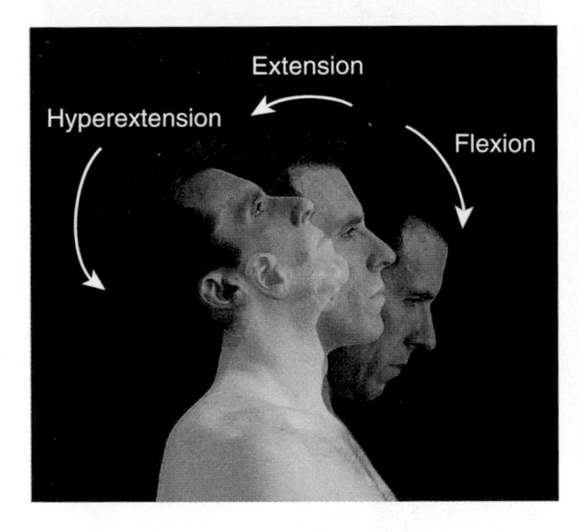

(a) Joints between atlas and occipital bone and between cervical vertebrae

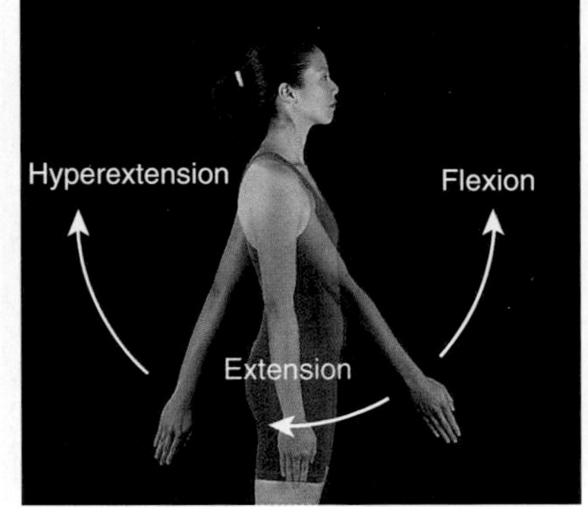

(b) Shoulder joint

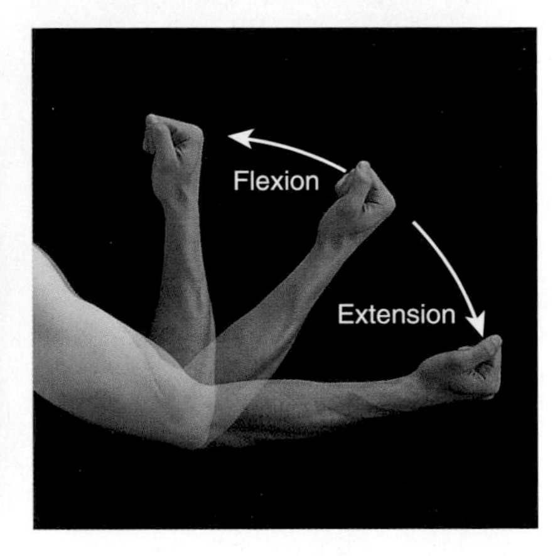

(c) Elbow joint

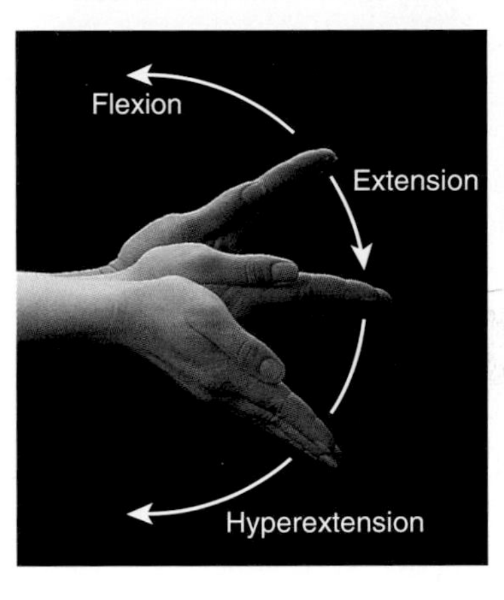

(d) Wrist joint

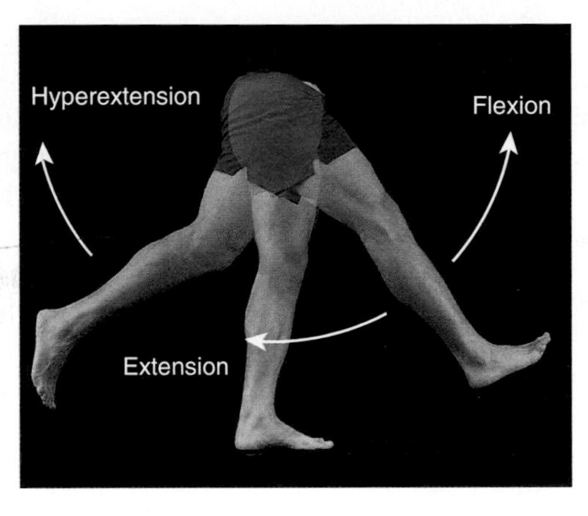

(e) Hip joint

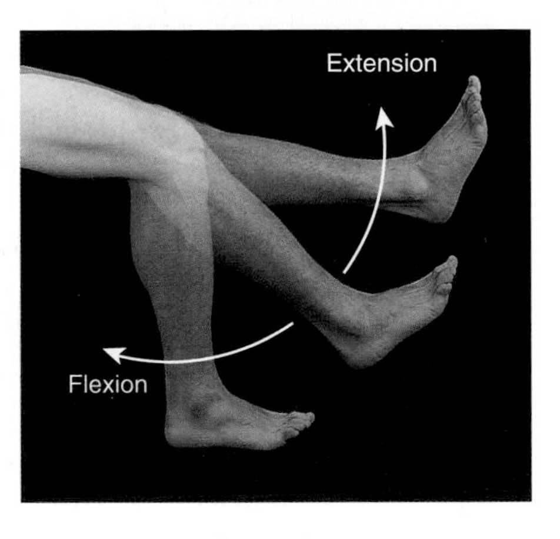

(f) Knee joint

What prevents hyperextension at some synovial joints?

and moving the femur backward, as in walking (Figure 7.4e). Hyperextension of other joints, such as the elbow, interphalangeal joints (fingers and toes), and knee joints, is usually prevented by the arrangement of ligaments and bones.

Abduction (ab-DUK-shun; *ab-* = away; *-duct* = to lead) is the movement of a bone away from the midline, and ***adduction*** (ad-DUK-shun; *ad-* = toward) is the movement of a bone toward the midline. Examples of abduction include lateral movement of the humerus upward (Figure 7.5a), lateral movement of the palm away from the body (Figure 7.5b), and lateral movement of the femur away from the body (Figure 7.5c). Movement in the opposite direction (medially) in each case produces adduction (Figure 7.5).

Circumduction (ser-kum-DUK-shun; *circ-* = circle) is movement of the distal end of a part of the body in a circle (Figure 7.6). Examples of joints that allow circumduction include the humerus at the shoulder joint (making a circle with your arm) and the femur at the hip joint (making a circle with your leg). Circumduction is more limited at the hip due to greater tension on the ligaments and muscles.

Rotation

In ***rotation*** (rō-TĀ-shun; *rota-* = to revolve) a bone revolves around its own longitudinal axis. An example is turning the head from side to side, as in signifying "no" (Figure 7.7a on page 163). In the limbs, rotation is defined relative to the midline. If the anterior surface of a bone of the limb is turned toward the midline, the movement is called *medial (internal) rotation*. You can medially rotate the humerus at the shoulder joint as follows: Starting in the anatomical position, flex your elbow and then draw your palm across the chest (Figure 7.7b). If the anterior surface of the

Figure 7.5 Angular movements at synovial joints: abduction and adduction.

Condyloid, saddle, and ball-and-socket joints permit abduction and adduction.

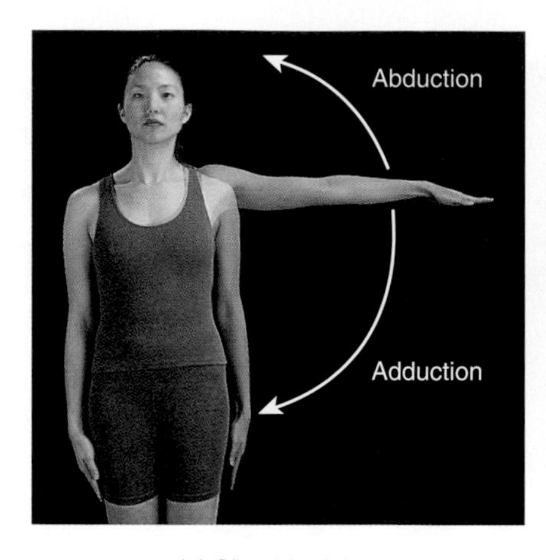

(a) Shoulder joint

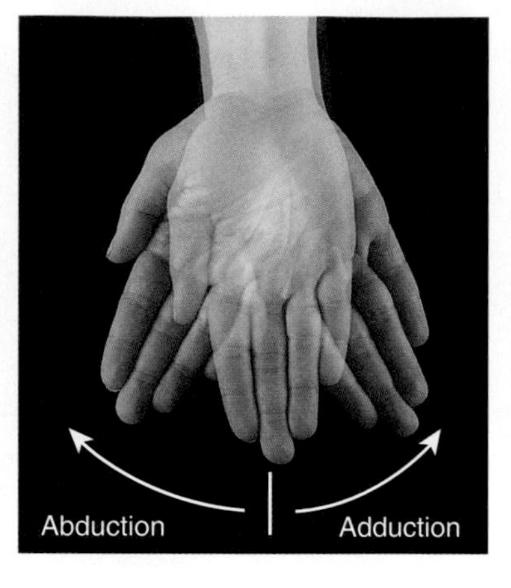

(b) Wrist joint

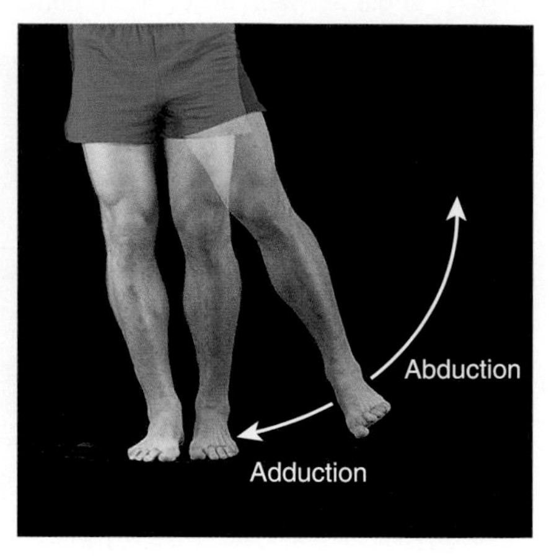

(c) Hip joint

One way to remember what adduction means is use of the phrase "adding your limb to your trunk." Why is this an effective learning device?

bone of a limb is turned away from the midline, the movement is called *lateral (external) rotation* (see Figure 7.7b).

Special Movements

The ***special movements*** that occur only at certain joints include elevation, depression, protraction, retraction, inversion, eversion, dorsiflexion, plantar flexion, supination, and pronation (Figure 7.8).

- ***Elevation*** (el′-e-VĀ-shun = to lift up) is the upward movement of a part of the body, such as closing the mouth to elevate the mandible (Figure 7.8a) or shrugging the shoulders to elevate the scapula.
- ***Depression*** (dē-PRESH-un = to press down) is the downward movement of a part of the body, such as opening the mouth to depress the mandible (Figure 7.8b) or returning shrugged shoulders to the anatomical position to depress the scapula.
- ***Protraction*** (prō-TRAK-shun = to draw forth) is the movement of a part of the body forward. You can protract your mandible by thrusting it outward (Figure 7.8c) or protract your clavicles by crossing your arms.
- ***Retraction*** (rē-TRAK-shun = to draw back) is the movement of a protracted part of the body back to the anatomical position (Figure 7.8d).
- ***Inversion*** (in-VER-zhun = to turn inward) is movement of the soles medially so that they face each other (Figure 7.8e).
- ***Eversion*** (ē-VER-zhun = to turn outward) is movement of the soles laterally so that they face away from each other (Figure 7.8f).

Figure 7.6 Angular movements at synovial joints: circumduction.

Circumduction is the movement of the distal end of a body part in a circle.

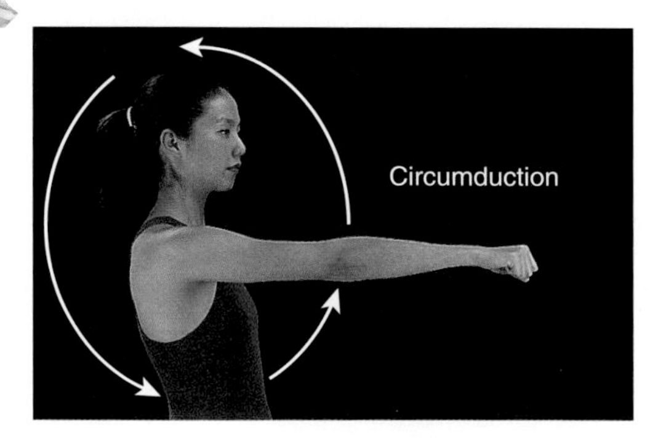

(a) Shoulder joint

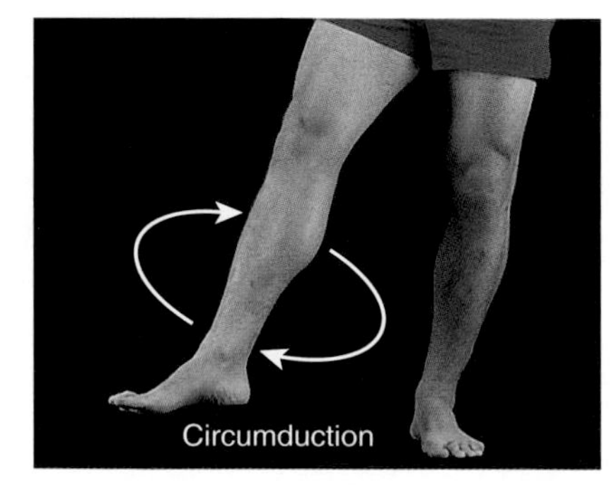

(b) Hip joint

List two joints where circumduction can occur.

Figure 7.7 **Rotation at synovial joints.**

In rotation, a bone revolves around its own longitudinal axis.

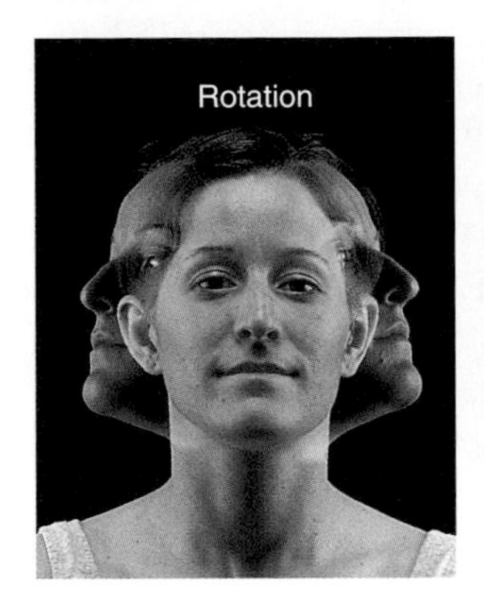

(a) Atlanto-axial joint

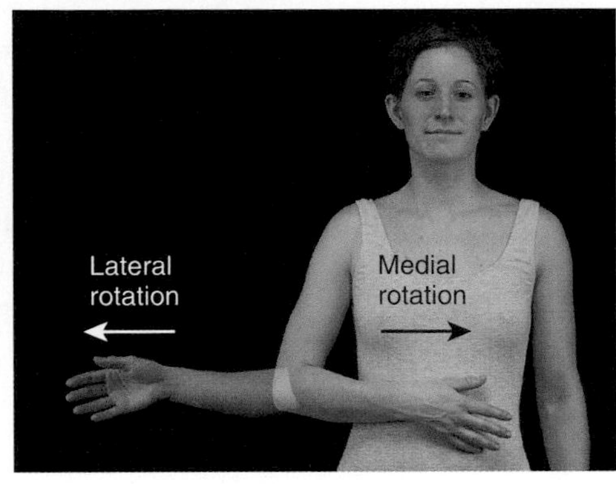

(b) Shoulder joint

How do medial and lateral rotation differ?

- ***Dorsiflexion*** (dor′-si-FLEK-shun) is bending of the foot in the direction of the dorsum (superior surface), as when you stand on your heels (Figure 7.8g).
- ***Plantar flexion*** involves bending of the foot in the direction of the plantar surface (Figure 7.8g), as when standing on your toes.
- ***Supination*** (soo′-pi-NĀ-shun) is movement of the forearm so that the palm is turned forward (Figure 7.8h). Supination of the palms is one of the defining features of the anatomical position (see Figure 1.4 on page 11).
- ***Pronation*** (prō-NĀ-shun) is movement of the forearm so that the palm is turned backward (Figure 7.8h).

■ **CHECKPOINT**

6. Define each of the movements at synovial joints just described and give an example of each.

Figure 7.8 **Special movements at synovial joints.**

Special movements occur only at certain synovial joints.

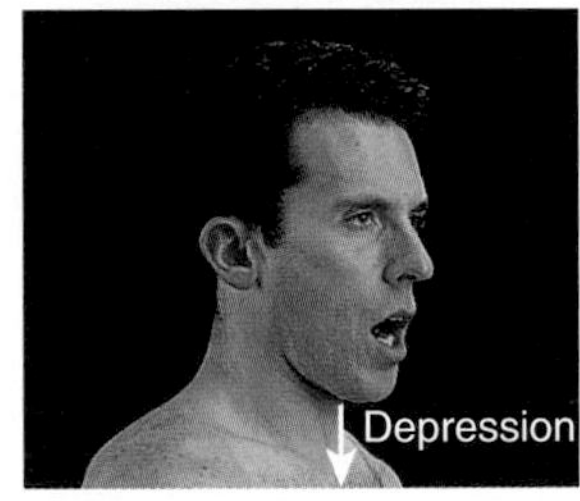

(a) Temporomandibular joint (b)

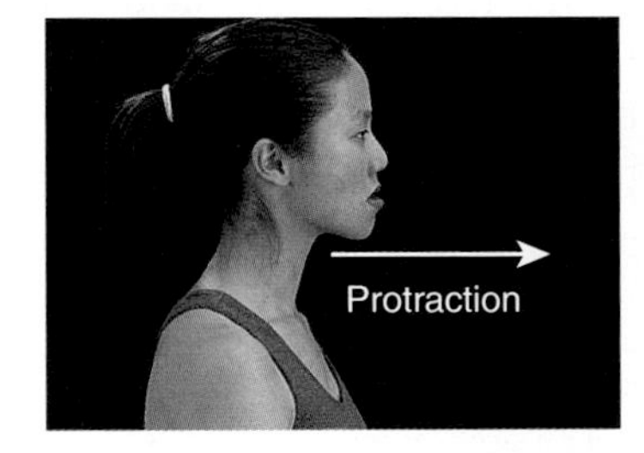

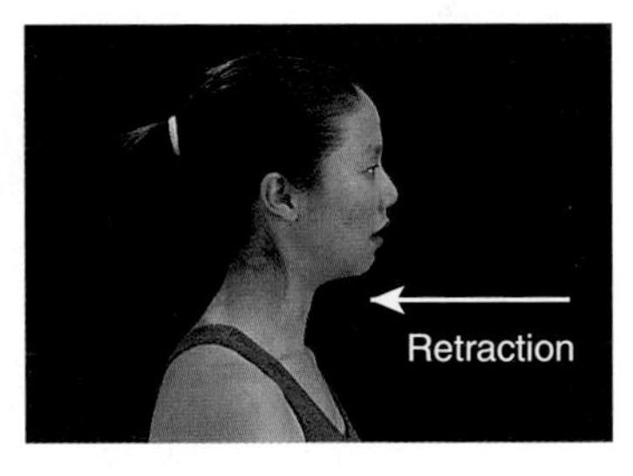

(c) Temporomandibular joint (d)

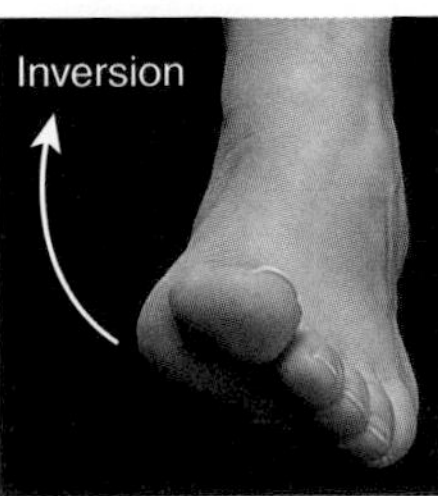

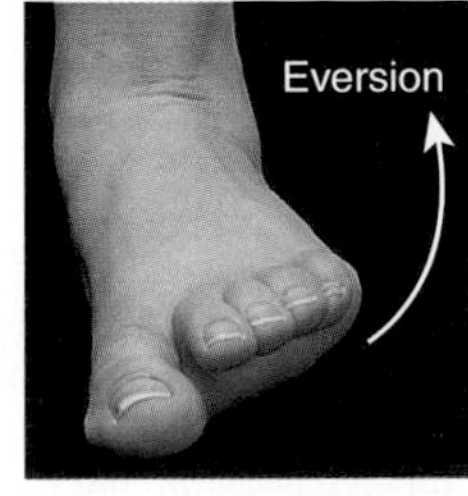

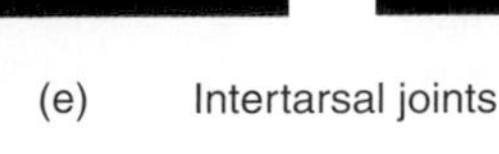

(e) Intertarsal joints (f)

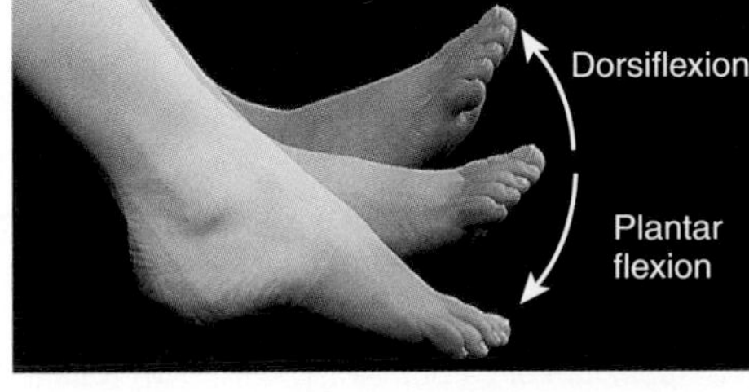

(g) Ankle joint

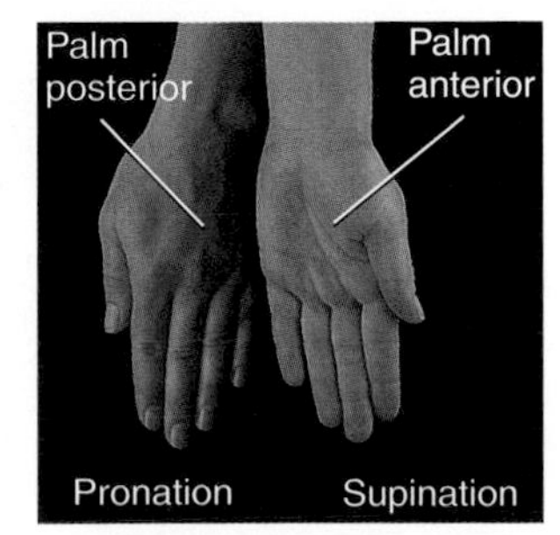

(h) Radioulnar joint

What movement of the shoulder girdle is involved in bringing the arms forward until the elbows touch?

TYPES OF SYNOVIAL JOINTS

OBJECTIVE • **Describe the six subtypes of synovial joints.**

Although all synovial joints have a similar structure, the shapes of the articulating surfaces vary and thus various types of movement are possible. Accordingly, synovial joints are divided into six subtypes: planar, hinge, pivot, condyloid, saddle, and ball-and-socket joints.

1. The articulating surfaces of bones in ***planar joints*** are flat or slightly curved (Figure 7.9a). Some examples of planar joints are the intercarpal (between carpal bones at the wrist), intertarsal (between tarsal bones at the ankle), sternoclavicular (between the sternum and the clavicle), and acromioclavicular (between the acromion of the scapula and the clavicle) joints. Planar joints primarily permit gliding movements.
2. In ***hinge joints***, the convex surface of one bone fits into the concave surface of another bone (Figure 7.9b).

Figure 7.9 Types of synovial joints. For each subtype, a drawing of the actual joint and a simplified diagram are shown.

Synovial joints are classified into subtypes on the basis of the shapes of the articulating bone surfaces.

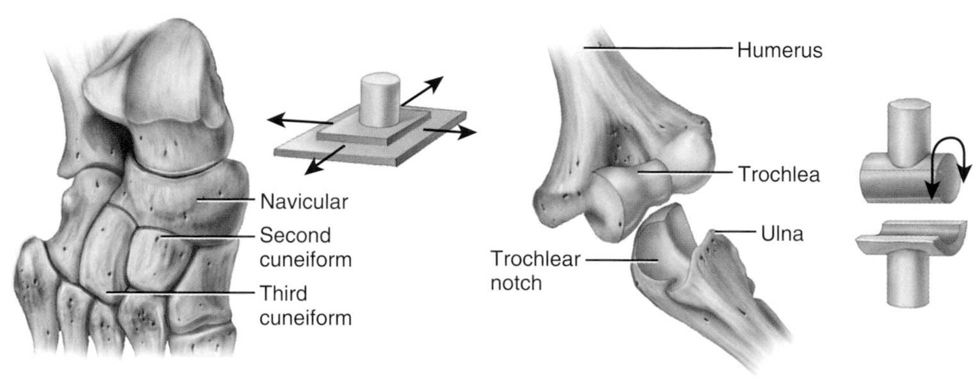

(a) Planar joint between the navicular and second and third cuneiforms of the tarsus in the foot

(b) Hinge joint between trochlea of humerus and trochlear notch of ulna at the elbow

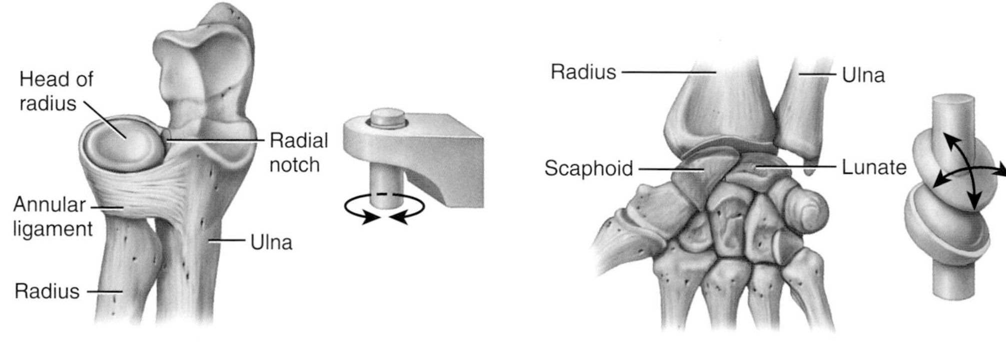

(c) Pivot joint between head of radius and radial notch of ulna

(d) Condyloid joint between radius and scaphoid and lunate bones of the carpus (wrist)

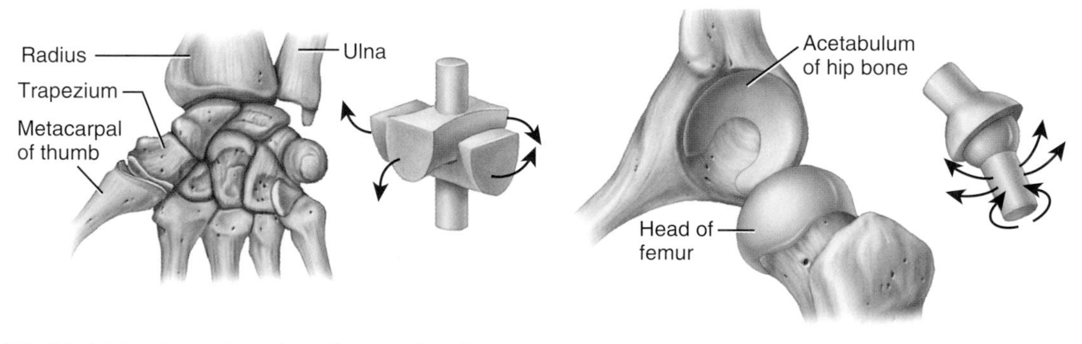

(e) Saddle joint between trapezium of carpus (wrist) and metacarpal of thumb

(f) Ball-and-socket joint between head of the femur and acetabulum of the hip bone

Which joints permit the greatest range of motion?

Examples of hinge joints are the knee, elbow, ankle, and interphalangeal joints (between the phalanges of the fingers and toes). As the name implies, hinge joints produce an angular, opening-and-closing motion like that of a hinged door. Hinge joints permit only flexion and extension.

3. In ***pivot joints***, the rounded or pointed surface of one bone articulates with a ring formed partly by another bone and partly by a ligament (Figure 7.9c). A pivot joint allows rotation around its own longitudinal axis. Examples of pivot joints are the atlantoaxial joint, in which the atlas rotates around the axis and permits you to turn your head from side to side as in signifying "no," and the radioulnar joints that allow you to move your palms forward and backward.
4. In ***condyloid joints*** (KON-di-loyd = knucklelike), the convex oval-shaped projection of one bone fits into the concave oval-shaped depression of another bone (Figure 7.9d). Examples are the wrist and metacarpophalangeal joints (between the metacarpals and phalanges) of the second through fifth digits. A condyloid joint permits flexion, extension, abduction, adduction, and circumduction.
5. In ***saddle joints***, the articular surface of one bone is saddle-shaped, and the articular surface of the other bone fits into the saddle like a rider sitting on a horse (Figure 7.9e). An example of a saddle joint is the carpometacarpal joint between the trapezium of the carpus and metacarpal of the thumb. Saddle joints permit flexion, extension, abduction, adduction, and circumduction.
6. In ***ball-and-socket joints***, the ball-like surface of one bone fits into a cuplike depression of another bone (Figure 7.9f). Ball-and-socket joints permit movement in several directions (flexion, extension, abduction, adduction, circumduction, and rotation); the only examples in the human body are the shoulder and hip joints.

■ **CHECKPOINT**

7. Where in the body can each subtype of synovial joint be found?

DETAILS OF A SYNOVIAL JOINT: THE KNEE JOINT

OBJECTIVE • Describe the principal structures and functions of the knee joint.

To give you an idea of the complexity of a synovial joint, we will examine some of the structural features of the knee joint, the largest and most complex joint in the body. Among the main structures of the knee joint are the following (Figure 7.10).

1. The articular capsule is strengthened by muscle tendons surrounding the joint.
2. The ***patellar ligament*** extends from the patella to the tibia and strengthens the anterior surface of the joint.
3. The ***oblique popliteal ligament*** (pop-LIT-ē-al) strengthens the posterior surface of the joint.
4. The ***arcuate popliteal ligament*** strengthens the lower lateral part of the posterior surface of the joint.
5. The ***tibial (medial) collateral ligament*** strengthens the medial aspect of the joint.
6. The ***fibular (lateral) collateral ligament*** strengthens the lateral aspect of the joint.
7. The ***anterior cruciate ligament (ACL)*** extends posteriorly and laterally from the tibia to the femur. The ACL is stretched or torn in about 70% of all serious knee injuries.
8. The ***posterior cruciate ligament (PCL)*** extends anteriorly and medially from the tibia to the femur. The ACL and PCL limit anterior and posterior movement of the femur and maintain the alignment of the femur with the tibia.
9. The menisci, fibrocartilage discs between the tibial and femoral condyles, help compensate for the irregular shapes of the articulating bones. The two menisci of the knee joint are the *medial meniscus*, a semicircular piece of fibrocartilage on the medial aspect of the knee, and the *lateral meniscus*, a nearly circular piece of fibrocartilage on the lateral aspect of the knee.
10. The bursae, saclike structures filled with fluid, help reduce friction.

■ **CHECKPOINT**

8. Which ligaments strengthen the posterior aspect of the knee joint?

AGING AND JOINTS

OBJECTIVE • Explain the effects of aging on joints.

Aging usually results in decreased production of synovial fluid in joints. In addition, the articular cartilage becomes thinner with age, and ligaments shorten and lose some of their flexibility. The effects of aging on joints are influenced by genetic factors and by wear and tear, and vary considerably from one person to another. Although degenerative changes in joints may begin as early as age 20, most changes do not occur until much later. By age 80, almost everyone de-

Figure 7.10 Structure of the right knee joint.

The knee joint is the largest and most complex joint in the body.

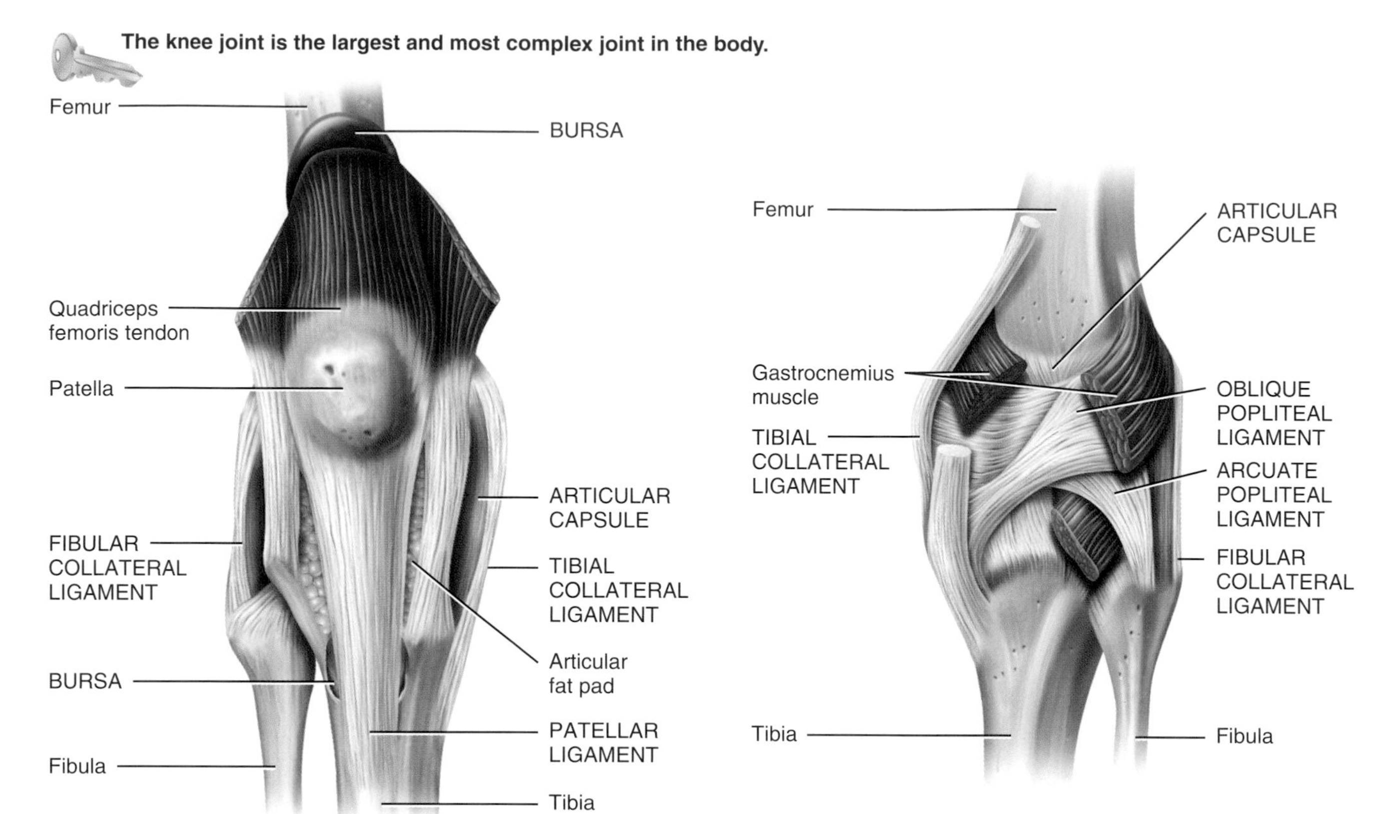

(a) Anterior superficial view

(b) Posterior deep view

Sagittal plane

Femur

LATERAL MENISCUS

Tibia

Gastrocnemius muscle

Quadriceps femoris tendon

Patella

Articular fat pad

BURSA

PATELLAR LIGAMENT

(c) Sagittal section

Femur

ANTERIOR CRUCIATE LIGAMENT (ACL)

LATERAL MENISCUS

FIBULAR COLLATERAL LIGAMENT

POSTERIOR CRUCIATE LIGAMENT (PCL)

MEDIAL MENISCUS

TIBIAL COLLATERAL LIGAMENT

Fibula

Tibia

(d) Anterior deep view (flexed)

What structures are damaged in the knee injury called torn cartilage?

Focus on Wellness

Joint Care—Prevent Repetitive Motion Injury

Diarthroses (freely movable joints) allow extensive movement. But the human body is not a machine, and diarthroses were not designed to withstand the repetition of a given motion over and over and over again, all day long. When you repeat the same motion for extended periods of time, you may overstress the joint or joints responsible for that motion and the associated soft-tissue structures, such as the articular capsule, ligaments, bursae, muscles, tendons, and nerves. Repeated episodes of mechanical stress can lead to the development of ***repetitive motion injuries***.

Repeat That?

Repetitive motion injuries are a type of ***cumulative trauma disorder (CTD)***, which is a group of disorders characterized by ongoing damage to soft tissues. Repetitive motion injuries are the most common type of CTD, but CTDs may also involve trauma due to exposure to cold or hot temperatures, certain types of lighting, vibration, and so forth. Repetitive motion injuries are similar in many ways to the ***overuse injuries*** that athletes often experience. Just as tennis players may develop epicondylitis (tennis elbow), so too may construction workers who perform repeated elbow flexion and extension in their work and students who spend hours a day using their computer mouse ("mouse elbow").

Repetitive motions alone may cause repetitive motion injuries. Risk increases when repetitive motions are coupled with poor posture and biomechanics, which put excess strain on joints. Joint stress also increases when a person must apply force with the motion, such as when gripping or lifting heavy objects. The joints at highest risk are those that are the weakest. Wrists, backs, elbows, shoulders, and necks are the most common sites of repetitive motion injury.

Repetitive motion injuries usually develop slowly over a long period of time. They typically begin with mild to moderate discomfort in the affected joints, especially at night. Other symptoms include swelling in the joint, muscle fatigue, numbness, and tingling. Symptoms may come and go at first, but then become constant. Symptoms of more advanced damage include more intense pain, muscle weakness, and nerve problems. If left untreated, repetitive motion injuries can be extremely painful. They also may severely limit a joint's range of motion. Fortunately, because they develop slowly, most repetitive motion injuries are discovered early enough to be successfully treated.

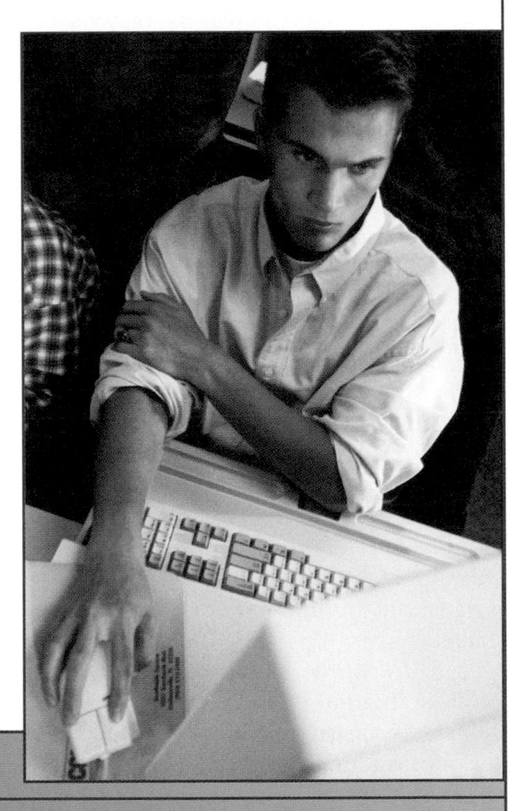

► **Think It Over . . .**

► ***Carpal tunnel syndrome is a repetitive motion injury in which pressure develops on the median nerve as it passes through the carpal tunnel, a narrow tunnel of bone and ligament at the wrist. Pressure on this nerve causes numbness, tingling, and pain in some or all of the fingers. What kind of workers do you think might be most at risk for the development of carpal tunnel syndrome?***

velops some type of degeneration in the knees, elbows, hips, and shoulders. It is also common for elderly individuals to develop degenerative changes in the vertebral column, resulting in a hunched-over posture and pressure on nerve roots. One type of arthritis, called osteoarthritis, is at least partially age related. Nearly everyone over age 70 has evidence of some osteoarthritic changes. Stretching and aerobic exercises that attempt to maintain full range of motion are helpful in minimizing the effects of aging. They help to maintain the effective functioning of ligaments, tendons, muscles, synovial fluid, and articular cartilage.

■ **CHECKPOINT**

9. Which joints show evidence of degeneration in nearly all individuals as aging progresses?

• • •

Now that you have a basic understanding of bones and joints, we will examine the structure and functions of muscular tissue and muscles. In this way you will understand how bones, joints, and muscles work together to produce various movements.

COMMON DISORDERS

Common Joint Injuries

Rotator cuff injury is a strain or tear in the rotator cuff muscles (see Figure 8.19 on page 205) and is a common injury among baseball pitchers and volleyball players, racket sports players, swimmers, and violinists, due to shoulder movements that involve vigorous circumduction. It also occurs as a result of wear and tear, aging, trauma, poor posture, improper lifting, and repetitive motions in certain jobs, such as placing items on a shelf above your head. Most often, there is tearing of the supraspinatus muscle tendon of the rotator cuff. This tendon is especially predisposed to wear-and-tear because of its location between the head of the humerus and acromion of the scapula, which compresses the tendon during shoulder movements.

A ***separated shoulder*** is an injury of the acromioclavicular joint, the joint formed by the acromion of the scapula and the acromial end of the clavicle. It most often happens with forceful trauma, as may happen when the shoulder strikes the ground in a fall.

Tennis elbow most commonly refers to pain at or near the lateral epicondyle of the humerus, usually caused by an improperly executed backhand. The extensor muscles strain or sprain, resulting in pain. ***Little-league elbow*** typically develops because of a heavy pitching schedule or throwing many curve balls, especially in youngsters. In this injury, the elbow may enlarge, fragment, or separate.

A ***dislocation of the radial head*** is the most common upper limb dislocation in children. In this injury, the head of the radius slides past or ruptures the ligament that forms a collar around the head of the radius at the proximal radioulnar joint. Dislocation is most apt to occur when a strong pull is applied to the forearm while it is extended and supinated, for instance while swinging a child around with outstretched arms.

The knee joint is the joint most vulnerable to damage because it is a mobile, weight-bearing joint and its stability depends almost entirely on its associated ligaments and muscles. Further, there is no correspondence of the articulating bones. A ***swollen knee*** may occur immediately or hours after an injury. The initial swelling is due to escape of blood from damaged blood vessels adjacent to areas involving rupture of the anterior cruciate ligament, damage to synovial membranes, torn menisci, fractures, or collateral ligament sprains. Delayed swelling is due to excessive production of synovial fluid, a condition commonly referred to as "water on the knee." A common type of knee injury in football is ***rupture of the tibial collateral ligaments***, often associated with tearing of the anterior cruciate ligament and medial meniscus (torn cartilage). Usually, a hard blow to the lateral side of the knee while the foot is fixed on the ground causes the damage. A ***dislocated knee*** refers to the displacement of the tibia relative to the femur. The most common type is dislocation anteriorly, resulting from hyperextension of the knee. A frequent consequence of a dislocated knee is damage to the popliteal artery.

Rheumatism and Arthritis

Rheumatism (ROO-ma-tizm) is any painful disorder of the supporting structures of the body—bones, ligaments, tendons, or muscles—that is not caused by infection or injury. ***Arthritis*** is a form of rheumatism in which the joints are swollen, stiff, and painful. It afflicts about 45 million people in the United States, and is the leading cause of physical disability among adults over age 65.

Rheumatoid arthritis (RA) is an autoimmune disease in which the immune system of the body attacks its own tissues—in this case, its own cartilage and joint linings. The primary symptom of RA is inflammation of the synovial membrane. RA is characterized by inflammation of the joint, which causes redness, warmth, swelling, pain, and loss of function.

Osteoarthritis (os′-tē-ō-ar-THRĪ-tis) is a degenerative joint disease in which joint cartilage is gradually lost. It results from a combination of aging, irritation of the joints, muscle weakness, and wear and abrasion. Commonly known as "wear-and-tear" arthritis, osteoarthritis is the most common type of arthritis. A major distinction between osteoarthritis and rheumatoid arthritis is that osteoarthritis strikes the larger joints (knees, hips) first, and rheumatoid arthritis first strikes smaller joints such as those in the fingers. A relatively new treatment for osteoarthritis of some joints is called *viscosupplementation*, in which hyaluronic acid is injected into a joint to improve lubrication. Results are usually as good as those involving corticosteroids.

In ***gouty arthritis*** (GOW-tē), sodium urate crystals are deposited in the soft tissues of the joints. The crystals irritate and erode the cartilage, causing inflammation, swelling, and acute pain. If the disorder is not treated, the ends of the articulating bones fuse, and the joint becomes immovable.

Joints that have been severely damaged by diseases such as arthritis, or by injury, may be replaced surgically with artifical joints in a procedure referred to as ***arthroplasty*** (AR-thrō-plas′-tē; *arthr-* = joint; *-plasty* = plastic repair of). Although most joints in the body can undergo arthroplasty, the ones most commonly replaced are the hips, knees, and shoulders. During the procedure, the ends of the damaged bones are removed and the metal, ceramic, or plastic components are fixed in place. The goals of arthroplasty are to relieve pain and increase range of motion.

Sprain and Strain

A ***sprain*** is the forcible wrenching or twisting of a joint that stretches or tears its ligaments but does not dislocate the bones. It occurs when the ligaments are stressed beyond their normal capacity. Sprains also may damage surrounding blood vessels, muscles, tendons, or nerves. Severe sprains may be so painful that the joint cannot be moved. There is considerable swelling, which results from hemorrhage of ruptured blood vessels. The ankle joint is most often sprained; the lower back is another frequent location for sprains. A ***strain*** is a stretched or partially torn muscle. It often occurs when a muscle contracts suddenly and powerfully—for example, in sprinters when they accelerate quickly.

MEDICAL TERMINOLOGY AND CONDITIONS

Arthralgia (ar-THRAL-jē-a; *arthr-* = joint; *-algia* = pain) Pain in a joint.

Bursectomy (bur-SEK-tō-mē; *-ectomy* = to cut out) Removal of a bursa.

Chondritis (kon-DRĪ-tis; *chondro-* = cartilage) Inflammation of cartilage.

Dislocation (*dis-* = apart) or ***luxation*** (luks-A-shun; *lux-* = dislocation) The displacement of a bone from a joint with tearing of ligaments, tendons, and articular capsules. A partial or incomplete dislocation is called a ***subluxation***.

Synovitis (sin′-ō-VĪ-tis) Inflammation of a synovial membrane in a joint

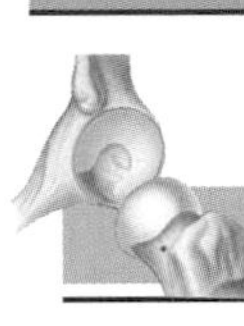

STUDY OUTLINE

Joints (p. 157)

1. A joint (articulation) is a point of contact between two bones, cartilage and bone, or teeth and bone.
2. A joint's structure determines its combination of strength and flexibility.
3. Structural classification is based on the presence or absence of a synovial cavity and the type of connecting tissue. Structurally, joints are classified as fibrous, cartilaginous, or synovial.
4. Functional classification of joints is based on the degree of movement permitted. Joints may be synarthroses (immovable), amphiarthroses (slightly movable), or diarthroses (freely movable).

Fibrous Joints (p. 157)

1. There is no joint cavity and the bones are held together by fibrous connective tissue in fibrous joints.
2. These joints include immovable sutures (found between skull bones), slightly movable syndesmoses (such as the distal joint between the tibia and fibula), and immovable gomphoses (roots of teeth in alveoli of the mandible and maxilla).

Cartilaginous Joints (p. 158)

1. There is no joint cavity and the bones are held together by cartilage in cartilaginous joints.
2. These joints include immovable synchondroses united by hyaline cartilage (epiphyseal plates) and slightly movable symphyses united by fibrocartilage (pubic symphysis).

Synovial Joints (p. 158)

1. A synovial joint contains a synovial cavity. All synovial joints are diarthroses.
2. Other characteristics of a synovial joint are the presence of articular cartilage and an articular capsule, made up of a fibrous capsule and a synovial membrane.
3. The synovial membrane secretes synovial fluid, which forms a thin, viscous film over the surfaces within the articular capsule.
4. Many synovial joints also contain accessory ligaments and articular discs.
5. Bursae are saclike structures, similar in structure to joint capsules, that reduce friction in joints such as the shoulder and knee joints.

Types of Movements at Synovial Joints (p. 160)

1. In a gliding movement, the nearly flat surfaces of bones move back-and-forth and side-to-side.
2. In angular movements, there is a change in the angle between bones. Examples are flexion–extension, hyperextension, abduction–adduction, and circumduction.
3. In rotation, a bone moves around its own longitudinal axis.
4. Special movements occur at specific synovial joints in the body. Examples are as follows: elevation–depression, protraction–retraction, inversion–eversion, dorsiflexion–plantar flexion, and supination–pronation.

Types of Synovial Joints (p. 164)

1. Types of synovial joints are planar, hinge, pivot, condyloid, saddle, and ball-and-socket.
2. In a planar joint, the articulating surfaces are flat; examples are joints between carpals and tarsals.
3. In a hinge joint, the convex surface of one bone fits into the concave surface of another; examples are the elbow, knee, and ankle joints.
4. In a pivot joint, a round or pointed surface of one bone fits into a ring formed by another bone and a ligament; examples are the atlantoaxial and radioulnar joints.
5. In a condyloid joint, an oval-shaped projection of one bone fits into an oval cavity of another; examples are the wrist joint and metacarpophalangeal joints for the second through fifth digits.
6. In a saddle joint, the articular surface of one bone is shaped like a saddle, and the other bone fits into the "saddle" like a rider on a horse; an example is the carpometacarpal joint between the trapezium and the metacarpal of the thumb.
7. In a ball-and-socket joint, the ball-shaped surface of one bone fits into the cuplike depression of another; examples are the shoulder and hip joints.

Details of a Synovial Joint: The Knee Joint (p. 165)

1. The knee joint is a diarthrosis that illustrates the complexity of this type of joint.
2. It contains an articular capsule, several ligaments within and around the outside of the joint, menisci, and bursae.

Aging and Joints (p. 165)

1. With aging, a decrease in synovial fluid, thinning of articular cartilage, and decreased flexibility of ligaments occur.
2. Most individuals experience some degeneration in the knees, elbows, hips, and shoulders due to the aging process.

SELF-QUIZ

1. A joint that has a ______ fit offers a great amount of movement and is ______ likely to become dislocated.
   a. tight, less b. tight, more c. loose, less
   d. loose, more e. flexible, less

2. An example of a fibrous joint in which the bones are immovable is a
   a. suture b. syndesmosis c. synovial d. symphysis
   e. synchondrosis

3. Pulling out a tooth would disarticulate which type of joint?
   a. symphysis b. synovial c. gomphosis
   d. cartilaginous e. suture

4. Which of the following is NOT a function of synovial fluid?
   a. It acts as a lubricant.
   b. It helps strengthen the joint.
   c. It removes microbes and debris from the joint.
   d. It provides nutrients to the tissues around the joints.
   e. It removes metabolic wastes.

5. Articular cartilage and bursae would most likely be found in which of the following?
   a. a gomphosis b. a suture c. the pubic symphysis
   d. the knee e. a synchondrosis

6. Which of the following structures provides flexibility to a joint while also preventing dislocation?
   a. bursae b. articular cartilage c. synovial fluid
   d. muscles e. articular capsule

7. The joints between the vertebrae and the joint between the hip bones are examples of which joint type?
   a. synovial b. symphysis c. fibrous
   d. synchondrosis e. suture

8. Match the following:

____	a. the joint between the atlas and axis	A. planar joint
____	b. allows gliding movements	B. hinge joint
____	c. the joint between the carpal and metacarpal of the thumb	C. ball-and-socket joint
____	d. hip joint	D. pivot joint
____	e. knee joint	E. saddle joint

9. Which of the following diarthrotic joints allows for the greatest degree of movement?
   a. ball-and-socket b. hinge c. condyloid
   d. pivot e. saddle

10. Moving the femur forward when walking is an example of
    a. abduction b. circumduction c. flexion
    d. gliding e. inversion

11. When a gymnast performs the "splits," the primary movement at the hip joint is
    a. rotation b. adduction c. extension d. gliding
    e. abduction

12. In the anatomical position, the palms are
    a. supinated b. flexed c. inverted d. pronated
    e. protracted

13. A fluid-filled sac found between skin and bone that helps reduce friction between the skin and bone is a
    a. meniscus b. bursa c. ligament
    d. articular capsule e. synovial membrane

14. Nodding your head "yes" in response to a question involves
    a. abduction and adduction b. circumduction
    c. extension and hyperextension d. rotation
    e. flexion and extension

15. Match the following:

____	a. movement of a bone around its own axis	A. rotation
____	b. movement away from the midline of the body	B. supination
____	c. turning the palm so it faces forward	C. depression
____	d. downward movement of a body part	D. adduction
____	e. movement toward the midline of the body	E. retraction
____	f. movement of the mandible or shoulder backward	F. pronation
____	g. turning the palm so it faces backward	G. abduction
____	h. upward movement of a body part	H. hyperextension
____	i. movement of the distal end of a body part in a circle	I. circumduction
____	j. movement beyond the plane of extension	J. elevation

CRITICAL THINKING APPLICATIONS

1. After your second A & P exam, you dropped to one knee, tipped your head back, raised one arm over your head, clenched your fist, pumped your arm up and down, and yelled "Yes!" Use the proper terms to describe the movements undertaken by the various joints.
2. Aunt Rosa's hip has been bothering her for years, and now she can hardly walk. Her doctor suggested a hip replacement. "It's one of those synonymous joints," Aunt Rosa explained. What type of joint is the hip joint? What types of movements can it perform?
3. Remember Kate, the volleyball player from Chapter 6? Her cast finally came off today. The orthopedist tested her knee's range of motion and declared that the ACL appeared to be intact. What is the ACL? How does the ACL contribute to the knee joint's stability?
4. Drew got slammed by a wave while bodysurfing. Now his arm feels useless, he's got an odd bump on his shoulder, and his shoulder really hurts! What happened to Drew's shoulder?

ANSWERS TO FIGURE QUESTIONS

7.1 Sutures are synarthroses because they are immovable; syndesmoses are classified as amphiarthroses because they are slightly movable.

7.2 Hyaline cartilage holds a synchondrosis together, and fibrocartilage holds a symphysis together.

7.3 Synovial joints are diarthroses, freely movable joints.

7.4 The arrangement of ligaments and bones prevents hyperextension at some synovial joints.

7.5 When you adduct your arm or leg, you bring it closer to the midline of the body, thus "adding" it to the trunk.

7.6 Circumduction can occur at the shoulder joint and at the hip joint.

7.7 The anterior surface of a bone or limb rotates toward the midline in medial rotation, and away from the midline in lateral rotation.

7.8 Bringing the arms forward until the elbows touch is an example of protraction.

7.9 Ball-and-socket joints permit the greatest range of movement.

7.10 In torn cartilage injuries of the knee, the menisci are damaged.

chapter 8

THE MUSCULAR SYSTEM

did you know? ***Strength training exercise results not only in stronger muscles, but in many other health benefits as well. Strength training helps to increase bone strength, increasing the deposition of bone minerals in young adults and helping to prevent, or at least slow, their loss in later life. By increasing muscle mass, strength training raises resting metabolic rate, the amount of energy expended at rest, so you can eat more food without gaining weight. Strength training helps to prevent back injury and injury from participation in sports and other physical activities. Psychological benefits include reductions in feelings of stress and fatigue.***

Focus on Wellness, page 190

www.wiley.com/college/apcentral

Movements such as throwing a ball, biking, and walking require an interaction between bones and muscles. To understand how muscles produce different movements, you will learn where the muscles attach on individual bones and the types of joints acted on by the contracting muscles. The bones, muscles, and joints together form an integrated system called the ***musculoskeletal system***. The scientific study of muscles is known as ***myology*** (mī-OL-ō-jē; *my-* = muscle; *-logy* = study of). The branch of medical science concerned with the prevention or correction of disorders of the musculoskeletal system is called ***orthopedics*** (or′-thō-PĒ-diks; *ortho-* = correct; *pedi* = child).

looking back to move ahead . . .

- Muscular Tissue (page 90)
- Adenosine Triphosphate (page 38)
- Divisions of the Skeletal System (page 124)
- Joints (page 157)
- Types of Movements at Synovial Joints (page 160)

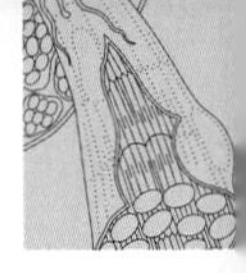

OVERVIEW OF MUSCULAR TISSUE

OBJECTIVE • Describe the types and functions of muscular tissue.

Types of Muscular Tissue

Muscular tissue constitutes about 40% to 50% of the total body weight and is composed of highly specialized cells. Recall from Chapter 4 that the three types of muscular tissue are skeletal, cardiac, and smooth. As its name suggests, most ***skeletal muscle tissue*** is attached to bones and moves parts of the skeleton. It is ***striated***; that is, *striations*, or alternating light and dark bands, are visible under a microscope. Because skeletal muscle can be made to contract and relax by conscious control, it is ***voluntary***. Due to the presence of a small number of cells that can undergo cell division, skeletal muscle has a limited capacity for regeneration.

Cardiac muscle tissue, found only in the heart, forms the bulk of the heart wall. The heart pumps blood through blood vessels to all parts of the body. Like skeletal muscle tissue, cardiac muscle tissue is ***striated***. However, unlike skeletal muscle tissue, it is ***involuntary***: Its contractions are not under conscious control. Cardiac muscle can regenerate under certain conditions. This will be explained in Chapter 15.

Smooth muscle tissue is located in the walls of hollow internal structures, such as blood vessels, airways, the stomach, and the intestines. It participates in internal processes such as digestion and the regulation of blood pressure. Smooth muscle is ***nonstriated*** (lacks striations) and ***involuntary*** (not under conscious control). Although smooth muscle tissue has considerable capacity to regenerate when compared with other muscle tissues, this capacity is limited when compared to other types of tissues, for example, epithelium.

Functions of Muscular Tissue

Through sustained contraction or alternating contraction and relaxation, muscular tissue has five key functions: producing body movements, stabilizing body positions, regulating organ volume, moving substances within the body, and generating heat.

1. **Producing body movements.** Body movements such as walking, running, writing, or nodding the head rely on the integrated functioning of skeletal muscles, bones, and joints.
2. **Stabilizing body positions.** Skeletal muscle contractions stabilize joints and help maintain body positions, such as standing or sitting. Postural muscles contract continuously when a person is awake; for example, sustained contractions of your neck muscles hold your head upright.
3. **Regulating organ volume.** Sustained contractions of ringlike bands of smooth muscles called *sphincters* prevent outflow of the contents of a hollow organ. Temporary storage of food in the stomach or urine in the urinary bladder is possible because smooth muscle sphincters close off the outlets of these organs.
4. **Moving substances within the body.** Cardiac muscle contractions pump blood through the body's blood vessels. Contraction and relaxation of smooth muscle in the walls of blood vessels helps adjust their diameter and thus regulate blood flow. Smooth muscle contractions also move food and other substances through the gastrointestinal tract, push gametes (sperm and oocytes) through the reproductive system, and propel urine through the urinary system. Skeletal muscle contractions aid the return of blood to the heart.
5. **Producing heat.** As muscular tissue contracts, it produces heat. Much of the heat released by muscles is used to maintain normal body temperature. Involuntary contractions of skeletal muscle, known as shivering, can help warm the body by greatly increasing the rate of heat production.

■ **CHECKPOINT**

1. What features distinguish the three types of muscular tissue?
2. What are the general functions of muscular tissue?

SKELETAL MUSCLE TISSUE

OBJECTIVES • Explain the relation of connective tissue components, blood vessels, and nerves to skeletal muscles.

• Describe the histology of a skeletal muscle fiber.

Each skeletal muscle is a separate organ composed of hundreds to thousands of cells, which are called ***muscle fibers*** because of their elongated shapes. Connective tissues surround muscle fibers and whole muscles, and blood vessels and nerves penetrate muscles.

Connective Tissue Components

Several connective tissue coverings are associated with skeletal muscle (Figure 8.1). The entire muscle is wrapped in ***epimysium*** (ep′-i-MĪZ-ē-um; *epi-* = upon). ***Perimysium*** (per′-i-MĪZ-ē-um; *peri-* = around) surrounds bundles of 10 to 100 or more muscle fibers called ***fascicles*** (FAS-i-kuls = little bundle). Finally, ***endomysium*** (en′-dō-MĪZ-ē-um; *endo-* = within) wraps each individual muscle fiber. Epimysium, perimysium, and endomysium extend beyond the muscle as a ***tendon***—a cord of dense regular connective tissue composed of parallel bundles of collagen fibers. Its

Figure 8.1 **Organization of skeletal muscle and its connective tissue coverings.**

A skeletal muscle consists of individual muscle fibers (cells) bundled into fascicles and surrounded by three connective tissue layers.

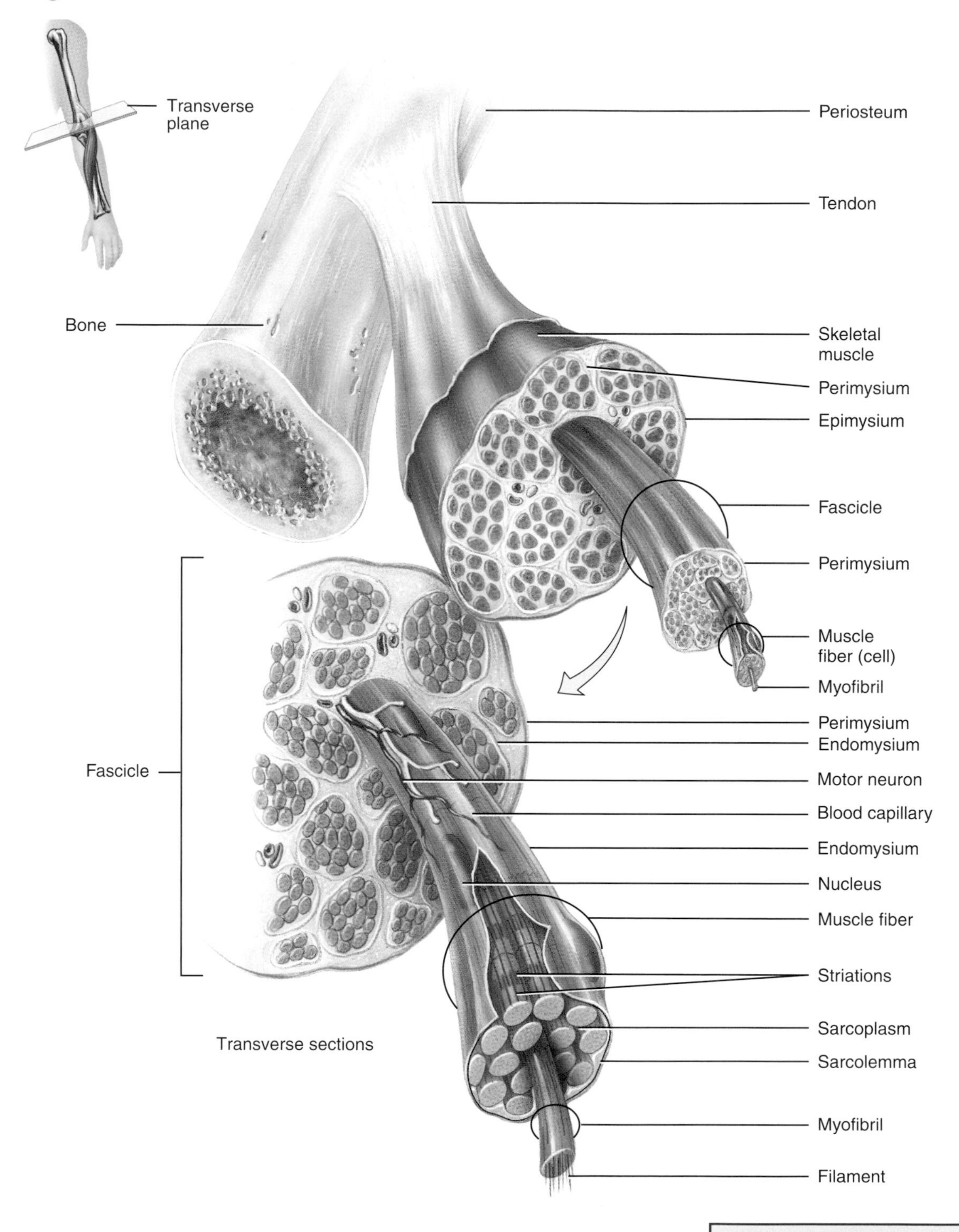

Functions of Muscles

1. Produces body motions.
2. Stabilizes body positions.
3. Regulates organ volume.
4. Moves substances within the body.
5. Produces heat.

Starting with the connective tissue that surrounds an individual muscle fiber (cell) and working toward the outside, list the connective tissue layers in order.

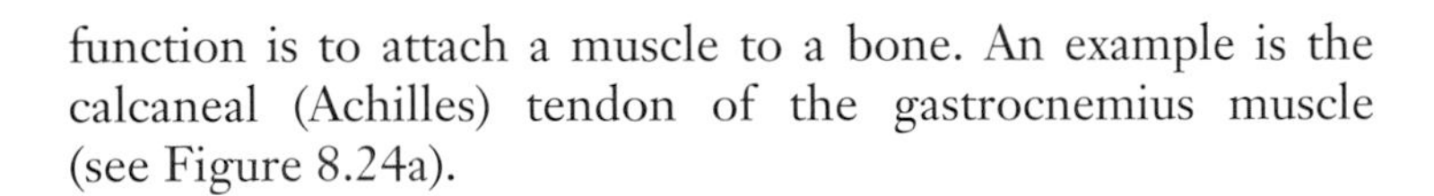

function is to attach a muscle to a bone. An example is the calcaneal (Achilles) tendon of the gastrocnemius muscle (see Figure 8.24a).

Nerve and Blood Supply

Skeletal muscles are well supplied with nerves and blood vessels (Figure 8.1), both of which are directly related to contraction, the chief characteristic of muscle. Muscle contraction also requires a good deal of ATP and therefore large amounts of nutrients and oxygen for ATP synthesis. Moreover, the waste products of these ATP-producing reactions must be eliminated. Thus, prolonged muscle action depends on a rich blood supply to deliver nutrients and oxygen and to remove wastes.

Generally, an artery and one or two veins accompany each nerve that penetrates a skeletal muscle. Within the endomysium, microscopic blood vessels called capillaries are distributed so that each muscle fiber is in close contact with one or more capillaries. Each skeletal muscle fiber also makes contact with the terminal portion of a neuron.

Histology

Microscopic examination of a skeletal muscle reveals that it consists of thousands of elongated, cylindrical cells called ***muscle fibers*** arranged parallel to one another (Figure 8.2a). Each muscle fiber is covered by a plasma membrane called the ***sarcolemma*** (*sarco-* = flesh; *-lemma* = sheath). ***Transverse tubules (T tubules)*** tunnel in from the surface toward the center of each muscle fiber. Multiple nuclei lie at the periphery of the fiber, next to the sarcolemma. The muscle fiber's cytoplasm, called ***sarcoplasm***, contains many mitochondria that produce large amounts of ATP during muscle contraction. Extending throughout the sarcoplasm is ***sarcoplasmic reticulum*** (sar′-kō-PLAZ-mik re-TIK-ū-lum), a network of fluid-filled membrane-enclosed tubules (similar to smooth endoplasmic reticulum) that stores calcium ions required for muscle contraction. Also in the sarcoplasm are numerous molecules of ***myoglobin*** (mī′-ō-GLŌ-bin), a reddish pigment similar to hemoglobin in blood. In addition to the characteristic color it lends to skeletal muscle, myoglobin stores oxygen until it is needed by mitochondria to generate ATP.

Extending along the entire length of the muscle fiber are cylindrical structures called ***myofibrils.*** Each myofibril, in turn, consists of two types of protein filaments called ***thin filaments*** and ***thick filaments*** (Figure 8.2b), which do not extend the entire length of a muscle fiber. Filaments overlap in specific patterns and form compartments called ***sarcomeres*** (*-meres* = parts), the basic functional units of striated muscle fibers (Figure 8.2b, c). Sarcomeres are separated from one another by zig-zagging zones of dense protein material called ***Z discs***. Within each sarcomere a darker area, called the ***A band***, extends the entire length of the thick filaments. At the center of each A band is a narrow ***H zone***, which contains only the thick filaments. At both ends of the A band, thick and thin filaments overlap. A lighter-colored area to either side of the A band, called the ***I band***, contains the rest of the thin filaments but no thick filaments. Each I band extends into two sarcomeres, divided in half by a Z disc (see Figure 8.4a). The alternating darker A bands and lighter I bands give the muscle fiber its striated appearance.

Thick filaments are composed of the protein ***myosin***, which is shaped like two golf clubs twisted together (Figure 8.3a on page 177). The *myosin tails* (golf club handles) are arranged parallel to each other, forming the shaft of the thick filament. The heads of the golf clubs project outward from the surface of the shaft. These projecting heads are referred to as *myosin heads*.

Thin filaments are anchored to the Z discs. Their main component is the protein ***actin***. Individual actin molecules join to form an actin filament that is twisted into a helix (Figure 8.3b). Each actin molecule contains *a myosin-binding site*, where a myosin head can attach. The thin filaments contain two other proteins, ***tropomyosin*** and ***troponin***. In a relaxed muscle, myosin is blocked from binding to actin because strands of tropomyosin cover the *myosin-binding sites* on actin. The tropomyosin strands, in turn, are held in place by troponin molecules.

Muscular atrophy (A-trō-fē; *a-* = without, *-trophy* = nourishment) is a wasting away of muscles. Individual muscle fibers decrease in size because of progressive loss of myofibrils. The atrophy that occurs if muscles are not used is termed *disuse atrophy*. Bedridden individuals and people with casts experience disuse atrophy because the number of nerve impulses to inactive muscle is greatly reduced. If the nerve supply to a muscle is disrupted or cut, the muscle undergoes *denervation atrophy*. In about 6 months to 2 years, the muscle will be one-quarter of its original size, and the muscle fibers will be replaced by fibrous connective tissue. The transition to connective tissue, when complete, cannot be reversed.

Muscular hypertrophy (hī-PER-trō-fē; *hyper-* = above or excessive) is an increase in muscle fiber diameter owing to the production of more myofibrils, mitochondria, sarcoplasmic reticulum, etc. It results from very forceful, repetitive muscular activity, such as strength training. Because hypertrophied muscles contain more myofibrils, they are capable of contractions that are more forceful.

■ **CHECKPOINT**

3. What type of connective tissue coverings are associated with skeletal muscle?
4. Why is a rich blood supply important for muscle contraction?
5. What is a sarcomere? What does a sarcomere contain?

Figure 8.2 **Organization of skeletal muscle from gross to molecular levels.**

The structural organization of a skeletal muscle from macroscopic to microscopic is as follows: skeletal muscle, fascicle (bundle of muscle fibers), muscle fiber, myofibril, and thin and thick filaments.

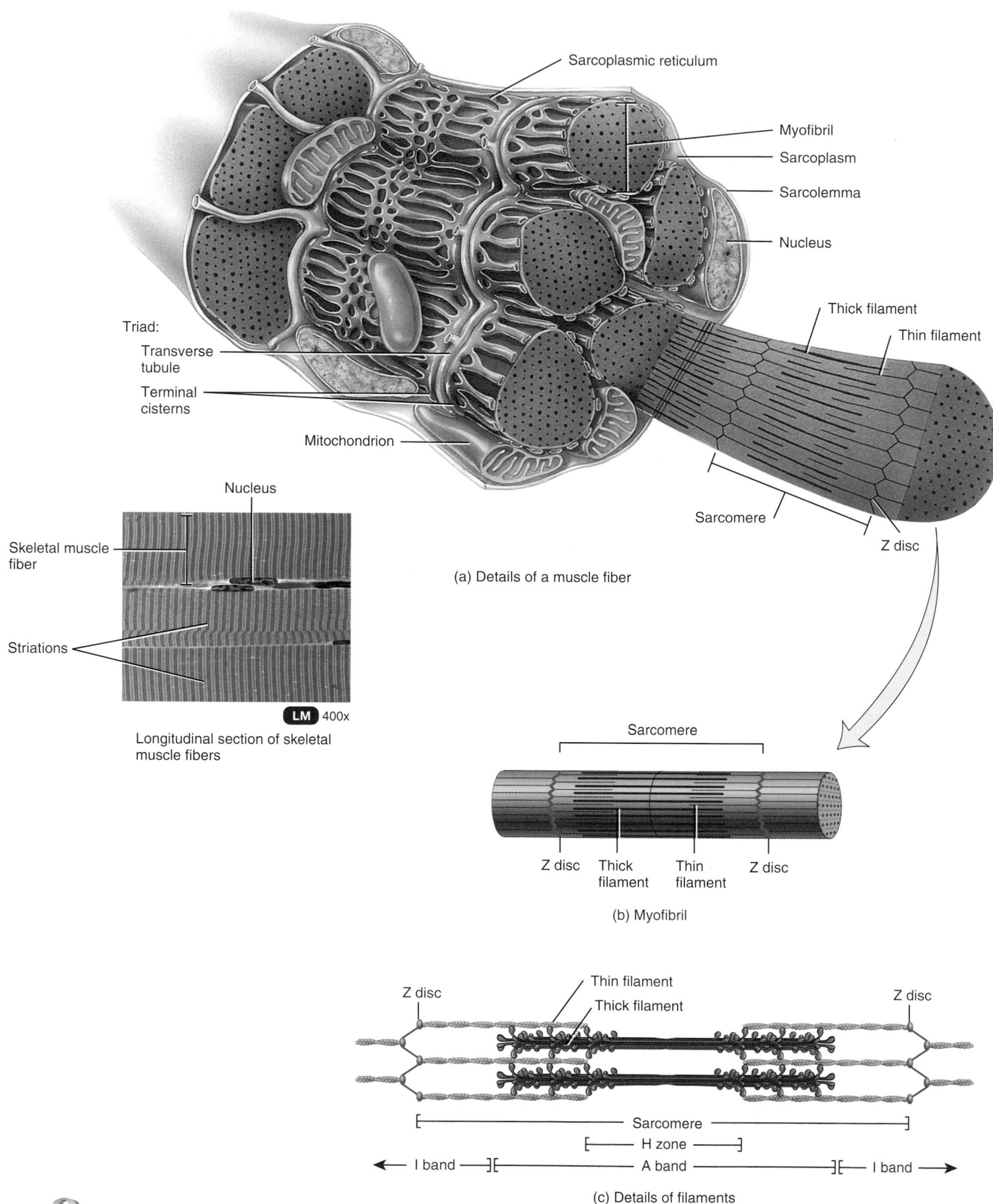

(a) Details of a muscle fiber

(b) Myofibril

(c) Details of filaments

Which filaments are part of the A band and I band?

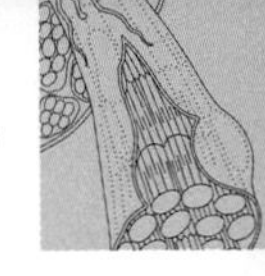

Figure 8.3 Detailed structure of filaments. (a) About 300 myosin molecules compose a thick filament. The myosin tails all point toward the center of the sarcomere. (b) Thin filaments contain actin, troponin, and tropomyosin.

Myofibrils contain thick and thin filaments.

Thick filament

Myosin tail

Myosin heads

(a) One thick filament (above) and a myosin molecule (below)

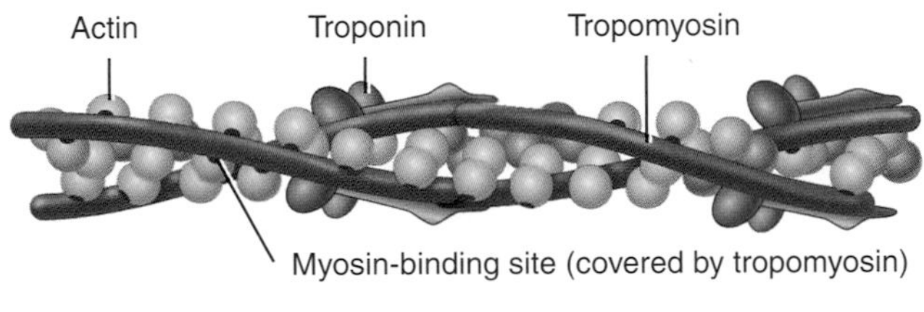

(b) Portion of a thin filament

What proteins are present in the A band and in the I band?

CONTRACTION AND RELAXATION OF SKELETAL MUSCLE

OBJECTIVE • **Explain how skeletal muscle fibers contract and relax.**

Sliding-Filament Mechanism

During muscle contraction, myosin heads of the thick filaments pull on the thin filaments, causing the thin filaments to slide toward the center of a sarcomere (Figure 8.4a, b). As the thin filaments slide, the I bands and H zones become narrower (Figure 8.4b) and eventually disappear altogether when the muscle is maximally contracted (Figure 8.4c).

The thin filaments slide past the thick filaments because the myosin heads move like the oars of a boat, pulling on the actin molecules of the thin filaments. Although the sarcomere shortens because of the increased overlap of thin and thick filaments, the lengths of the thin and thick filaments do not change. The sliding of filaments and shortening of sarcomeres in turn cause the shortening of the muscle fibers. This process, the ***sliding-filament mechanism*** of muscle contraction, occurs only when the level of calcium ions (Ca^{2+}) is high enough and ATP is available, for reasons you will see shortly.

Neuromuscular Junction

Before a skeletal muscle fiber can contract, it must be stimulated by an electrical signal called a ***muscle action potential*** delivered by its neuron called a ***motor neuron***. A single motor neuron along with all the muscle fibers it stimulates is called a ***motor unit***. Stimulation of one motor neuron causes all the muscle fibers in that motor unit to contract at the same time. Muscles that control small, precise movements, such as the muscles that move the eyes, have 10 to 20 muscle fibers per motor unit. Muscles of the body that are responsible for large, powerful movements, such as the biceps brachii in the arm and gastrocnemius in the leg, have as many as 2000 to 3000 muscle fibers in some motor units.

As the *axon* (long process) of a motor neuron enters a skeletal muscle, it divides into branches called *axon terminals* that approach—but do not touch—the sarcolemma of a muscle fiber (Figure 8.5a, b). The ends of the axon terminals

Figure 8.4 Sliding-filament mechanism of muscle contraction.

During muscle contraction, thin filaments move inward toward the H zone.

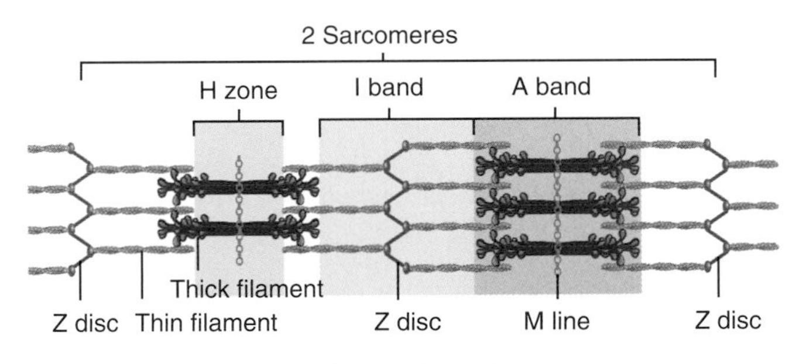

(a) Relaxed muscle

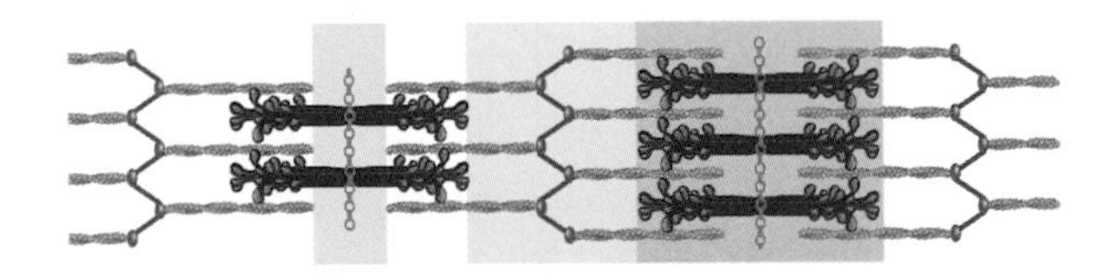

(b) Partially contracted muscle

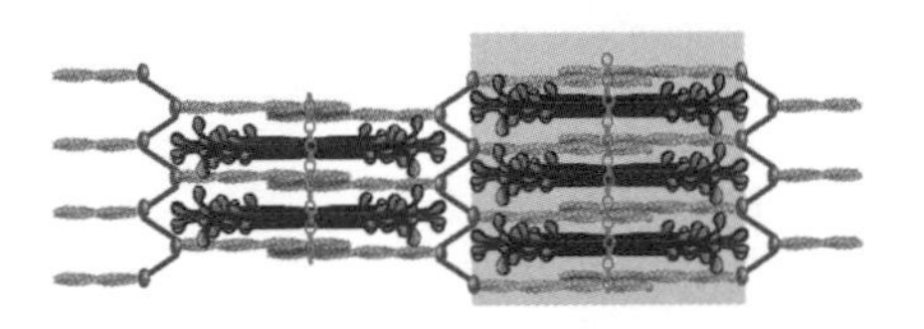

(c) Maximally contracted muscle

What happens to the I bands as muscle contracts? Do the lengths of the thick and thin filaments change during contraction?

enlarge into swellings known as *synaptic end bulbs*, which contain *synaptic vesicles* filled with a chemical *neurotransmitter*. The region of the sarcolemma near the axon terminal is called the ***motor end plate***. The space between the axon terminal and sarcolemma is the ***synaptic cleft***. The synapse formed between the axon terminals of a motor neuron and the motor end plate of a muscle fiber is known as the ***neuromuscular junction (NMJ)***. At the NMJ, a motor neuron excites a skeletal muscle fiber in the following way (Figure 8.5c):

1. **Release of acetylcholine.** Arrival of the nerve impulse at the synaptic end bulbs triggers release of the neurotransmitter ***acetylcholine (ACh)*** (as′-e-til-KŌ-lēn). ACh then diffuses across the synaptic cleft between the motor neuron and the motor end plate.
2. **Activation of ACh receptors.** Binding of ACh to its receptor in the motor end plate opens ion channels that allow small cations, especially sodium ions (Na^+), to flow across the membrane.

Figure 8.5 Neuromuscular junction.

A neuromuscular junction includes the axon terminal of a motor neuron plus the motor end plate of a muscle fiber.

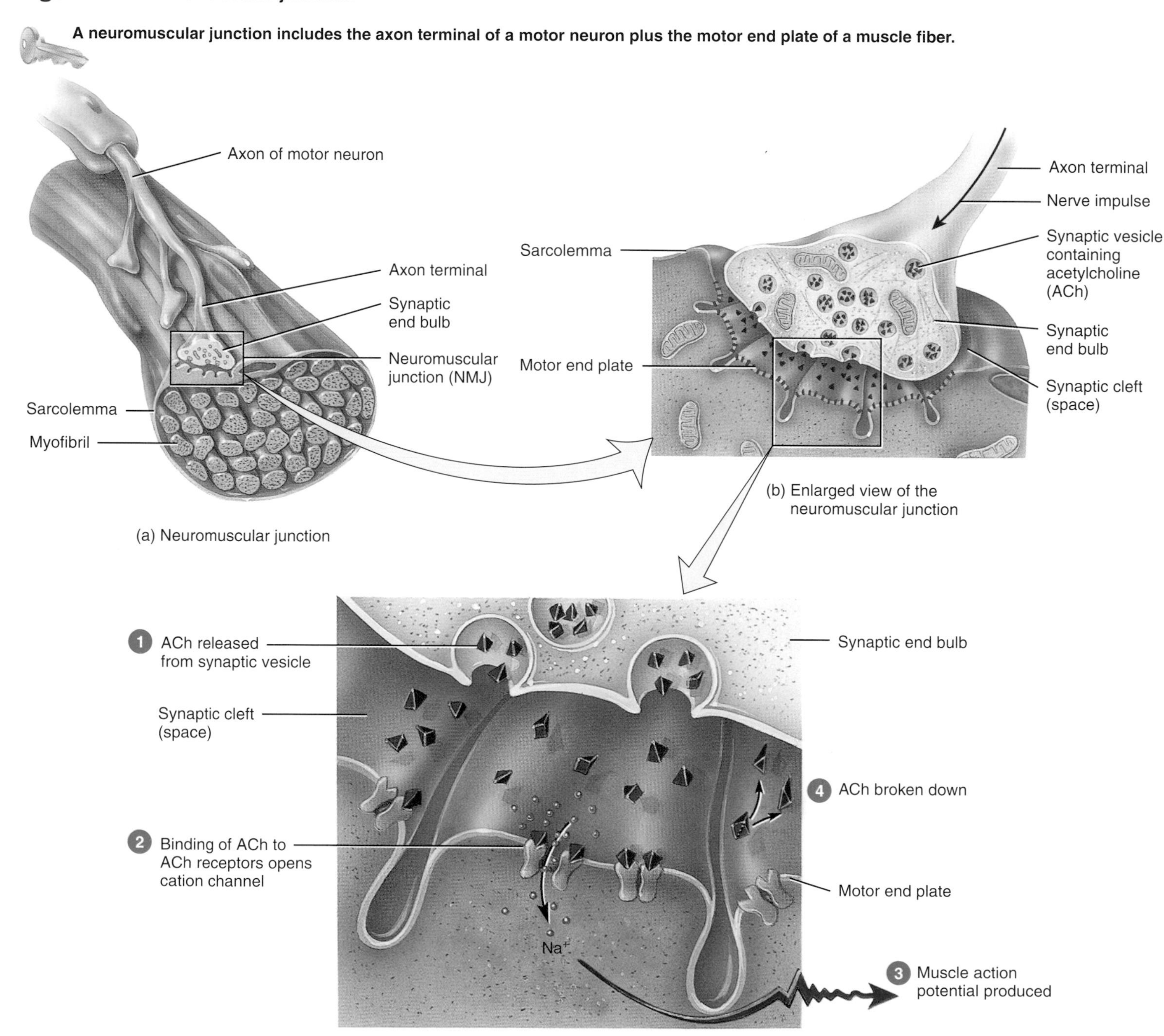

(c) Binding of acetylcholine to ACh receptors in the motor end plate

What is the motor end plate?

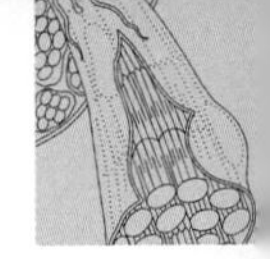

3. **Generation of muscle action potential.** The inflow of Na^+ (down its concentration gradient) generates a muscle action potential. The muscle action potential then travels along the sarcolemma and through the T tubules. Each nerve impulse normally elicits one muscle action potential. If another nerve impulse releases more acetylcholine, then steps 2 and 3 repeat. See chapter 9 for the details of nerve impulse generation.
4. **Breakdown of ACh.** The effect of ACh lasts only briefly because the neurotransmitter is rapidly broken down in the synaptic cleft by an enzyme called ***acetylcholinesterase (AChE)*** (as′-e-til-kō′-lin-ES-ter-ās).

Functioning of the NMJ can be altered by several toxins and drugs. Botulinum toxin, produced by the bacterium *Clostridium botulinum,* blocks release of ACh. As a result, muscle contraction does not occur. The bacteria proliferate in improperly canned foods, and their toxin is one of the most lethal chemicals known. A tiny amount can cause death by paralyzing the diaphragm, the main muscle that powers breathing. Yet, it is also the first bacterial toxin to be used as a medicine (Botox®). Injections of Botox into the affected muscles can help patients who have strabismus (crossed eyes) or blepharospasm (uncontrollable blinking). It is also used as a cosmetic treatment to relax muscles that cause facial wrinkles and to alleviate chronic back pain due to muscle spasms in the lumbar region.

Physiology of Contraction

Both Ca^{2+} and energy, in the form of ATP, are needed for muscle contraction. When a muscle fiber is relaxed (not contracting), there is a low concentration of Ca^{2+} in the sarcoplasm because the membrane of the sarcoplasmic reticulum contains Ca^{2+} active transport pumps that continually transport Ca^{2+} from the sarcoplasm into the sarcoplasmic reticulum (see Figure 8.7 7). However, when a muscle action potential travels along the sarcolemma and into the transverse tubule system, Ca^{2+} release channels open (see Figure 8.7 4), allowing Ca^{2+} to escape into the sarcoplasm. The Ca^{2+} binds to troponin molecules in the thin filaments, causing the troponin to change shape. This change in shape releases the troponin–tropomyosin complex from the myosin-binding sites on actin (see Figure 8.7 5).

Once the myosin-binding sites are uncovered, the ***contraction cycle***—the repeating sequence of events that causes the filaments to slide—begins, as shown in Figure 8.6:

1. **Splitting ATP.** The myosin heads contain ATPase, an enzyme that splits ATP into ADP (adenosine diphos-

Figure 8.6 The contraction cycle. Sarcomeres shorten through repeated cycles in which the myosin heads (crossbridges) attach to actin, rotate, and detach.

During the power stroke of contraction, crossbridges rotate and move the thin filaments past the thick filaments toward the center of the sarcomere.

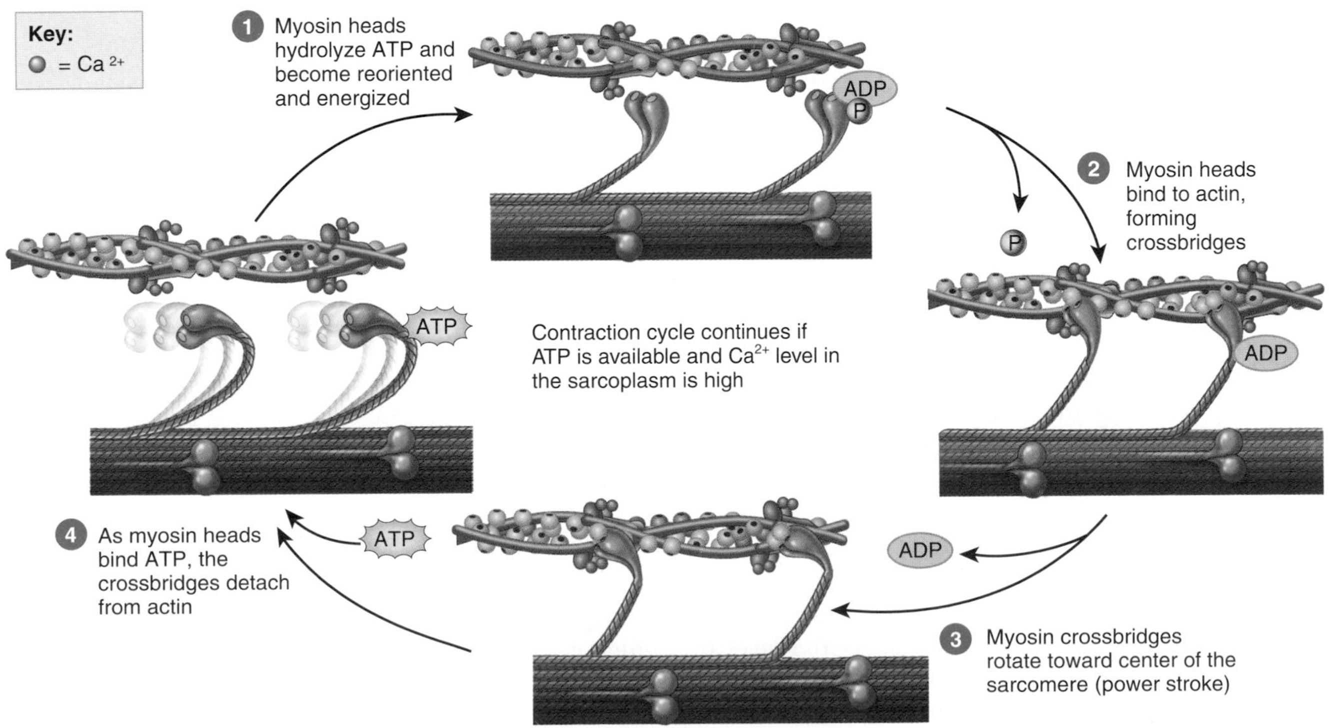

What causes crossbridges to detach from actin?

phate) and P (a phosphate group). This splitting reaction transfers energy to the myosin head, although ADP and P remain attached to it.

2. **Forming crossbridges.** The energized myosin heads attach to the myosin-binding sites on actin, and release the phosphate groups. When myosin heads attach to actin during contraction, they are referred to as ***crossbridges***.
3. **Power stroke.** After the crossbridges form, the ***power stroke*** occurs. During the power stroke, the crossbridge rotates or swivels and releases the ADP. The force produced as hundreds of crossbridges swivel slides the thin filament past the thick filament toward the center of the sarcomere.
4. **Binding ATP and detaching.** At the end of the power stroke, the crossbridges remain firmly attached to actin. When they bind another molecule of ATP, the myosin heads detach from actin.

As the myosin ATPase again splits ATP, the myosin head is reoriented and energized, ready to combine with another myosin-binding site farther along the thin filament. The contraction cycle repeats as long as ATP and Ca^{2+} are available in the sarcoplasm. At any one instant, some of the myosin heads are attached to actin, forming crossbridges and generating force, and other myosin heads are detached from actin and getting ready to bind again. During a maximal contraction, the sarcomere can shorten by as much as half its resting length.

After a person dies, Ca^{2+} begins to leak out of the sarcoplasmic reticulum and binds to troponin, causing the thin filaments to slide. ATP production has ceased, however, so the crossbridges cannot detach from actin. The resulting stiffness of the muscles is termed **rigor mortis**, rigidity of death. It begins 3–4 hours after death, lasts about 24 hours, and then disappears as enzymes from lysozymes digest crossbridges.

Relaxation

Two changes permit a muscle fiber to relax after it has contracted. First, the neurotransmitter acetylcholine is rapidly broken down by the enzyme acetylcholinesterase (AChE). When nerve action potentials cease, release of ACh stops, and AChE rapidly breaks down the ACh already present in the synaptic cleft. This ends the generation of muscle action potentials, and the Ca^{2+} release channels in the sarcoplasmic reticulum membrane close.

Second, calcium ions are rapidly transported from the sarcoplasm into the sarcoplasmic reticulum. As the level of Ca^{2+} in the sarcoplasm falls, the tropomyosin–troponin complex slides back over the myosin-binding sites on actin. Once the myosin-binding sites are covered, the thin filaments slip back to their relaxed positions. Figure 8.7 summarizes the events of contraction and relaxation in a muscle fiber.

Muscle Tone

Even when a whole muscle is not contracting, a small number of its motor units are involuntarily activated to produce a sustained contraction of their muscle fibers. This process results in ***muscle tone*** (*tonos* = tension). To sustain muscle tone, small groups of motor units are alternately active and inactive in a constantly shifting pattern. Muscle tone keeps skeletal muscles firm, but it does not result in a contraction strong enough to produce movement. For example, the tone of muscles in the back of the neck keeps the head upright and prevents it from slumping forward on the chest. Recall that skeletal muscle contracts only after it is activated by acetylcholine released by nerve impulses in its motor neurons. Hence, muscle tone is established by neurons in the brain and spinal cord that excite the muscle's motor neurons. When the motor neurons serving a skeletal muscle are damaged or cut, the muscle becomes ***flaccid*** (FLAS-id = flabby), a state of limpness in which muscle tone is lost.

■ **CHECKPOINT**

6. Explain how a skeletal muscle contracts and relaxes.
7. What is the importance of the neuromuscular junction?

METABOLISM OF SKELETAL MUSCLE TISSUE

OBJECTIVES • **Describe the sources of ATP and oxygen for muscle contraction.**
• **Define muscle fatigue and list its possible causes.**

Energy for Contraction

Unlike most cells of the body, skeletal muscle fibers often switch between virtual inactivity, when they are relaxed and using only a modest amount of ATP, and great activity, when they are contracting and using ATP at a rapid pace. However, the ATP present inside muscle fibers is enough to power contraction for only a few seconds. If strenuous exercise is to continue, additional ATP must be synthesized. Muscle fibers have three sources for ATP production: (1) creatine phosphate, (2) anaerobic cellular respiration, and (3) aerobic cellular respiration.

While at rest, muscle fibers produce more ATP than they need. Some of the excess ATP is used to make ***creatine phosphate***, an energy-rich molecule that is unique to muscle

Figure 8.7 Summary of the events of contraction and relaxation in a skeletal muscle fiber.

Acetylcholine released at the neuromuscular junction triggers a muscle action potential, which leads to muscle contraction.

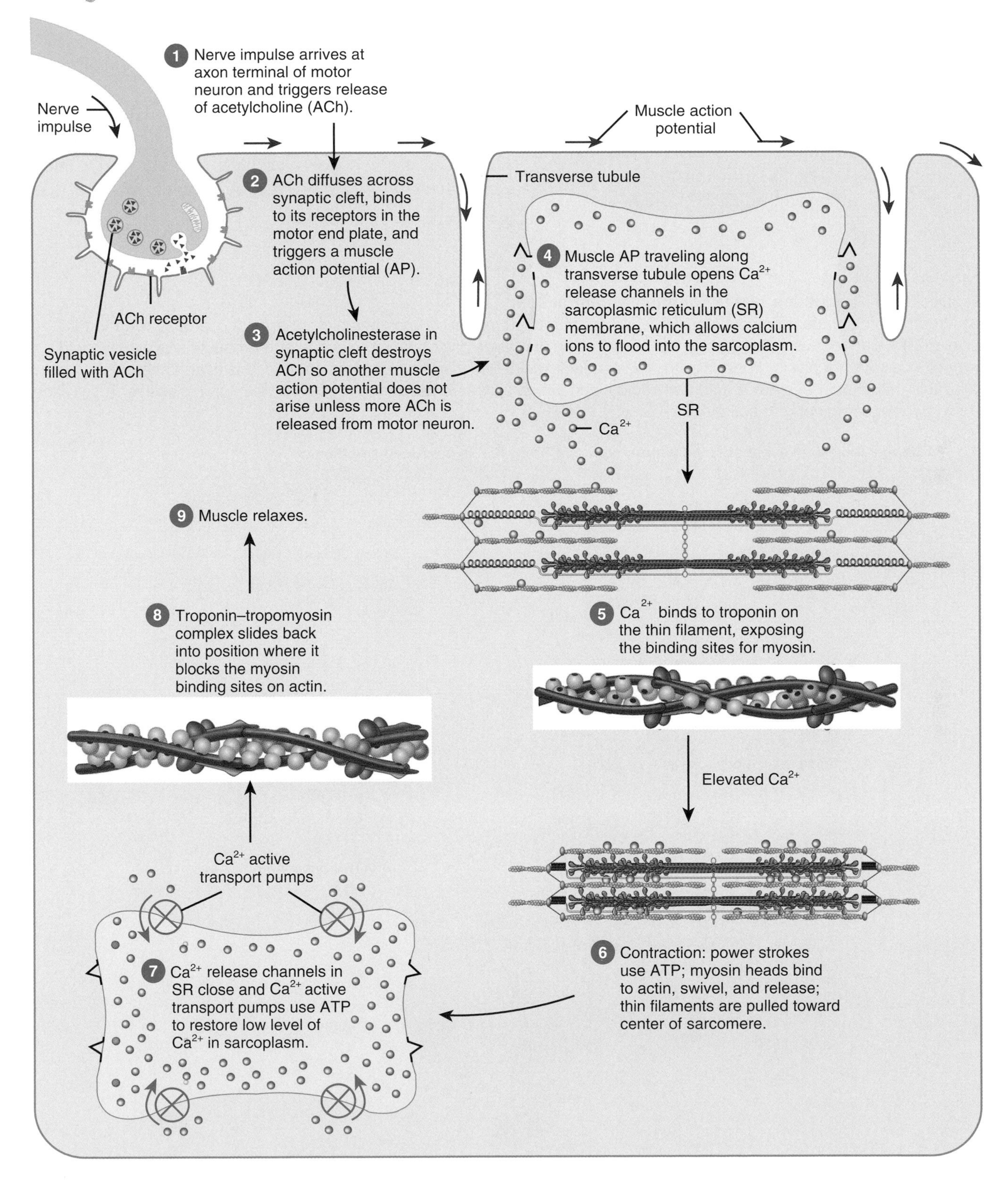

The power stroke occurs during which numbered step in this figure?

fibers (Figure 8.8a). One of ATP's high-energy phosphate groups is transferred to creatine, forming creatine phosphate and ADP (adenosine diphosphate). ***Creatine*** is a small, amino acid-like molecule that is synthesized in the liver, kidneys, and pancreas and derived from certain foods (milk, red meat, fish), then transported to muscle fibers. While muscle is contracting, the high-energy phosphate group can be transferred from creatine phosphate back to ADP, quickly forming new ATP molecules. Together, creatine phosphate and ATP provide enough energy for muscles to contract maximally for about 15 seconds. This energy is sufficient for short bursts of intense activity, for example, running a 100-meter dash.

Adults need to synthesize and ingest a total of about 2 grams of creatine daily to make up for the urinary loss of creatinine, the breakdown product of creatine. Some studies have demonstrated improved performance during intense exercise in subjects who had ingested creatine supplements. For example, college football players who received supplements of 15 grams per day for 28 days gained more muscle mass and had larger gains in lifting power and sprinting performance than the control subjects. Other studies, however, have failed to find a performance-enhancing effect of creatine supplementation. In addition,

Figure 8.8 Production of ATP for muscle contraction.
(a) Creatine phosphate, formed from ATP while the muscle is relaxed, transfers a high-energy phosphate group to ADP, forming ATP, during muscle contraction. (b) Breakdown of muscle glycogen into glucose and production of pyruvic acid from glucose via glycolysis produce both ATP and lactic acid. Because no oxygen is needed, this is an anaerobic pathway. (c) Within mitochondria, pyruvic acid, fatty acids, and amino acids are used to produce ATP via aerobic cellular respiration, an oxygen-requiring set of reactions.

During a long-term event such as a marathon race, most ATP is produced aerobically.

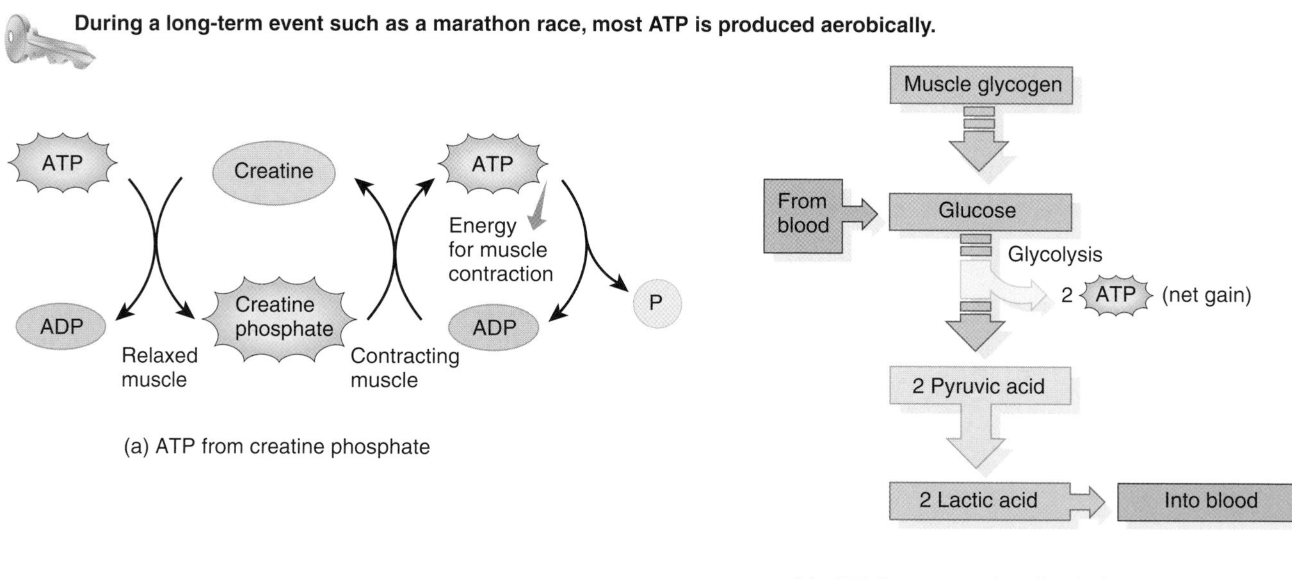

(a) ATP from creatine phosphate

(b) ATP from anaerobic glycolysis

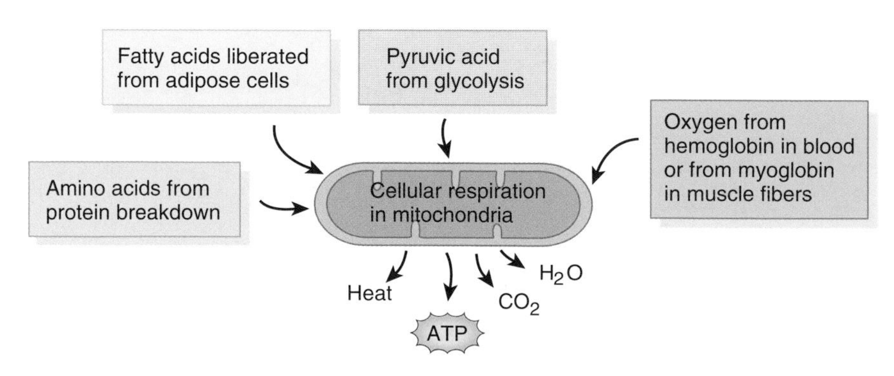

(c) ATP from aerobic cellular respiration

Where inside a skeletal muscle fiber are the events shown here occurring?

ingesting extra creatine decreases the body's own synthesis of creatine, and it is not known whether natural synthesis recovers after long-term creatine supplementation. Further research is needed to determine both the long-term safety and the value of **creatine supplementation.**

When muscle activity continues past the 15-second mark, the supply of creatine phosphate is depleted. The next source of ATP is *glycolysis*, a series of cytosolic reactions that produces 2 ATPs by breaking down a glucose molecule to pyruvic acid. Glucose passes easily from the blood into contracting muscle fibers and also is produced within muscle fibers by breakdown of glycogen (Figure 8.8b). When oxygen levels are low as a result of vigorous muscle activity, most of the pyruvic acid is converted to lactic acid, a process called ***anaerobic cellular respiration*** because it occurs without using oxygen. Anaerobic cellular respiration can provide enough energy for about 30 to 40 seconds of maximal muscle activity. Together, conversion of creatine phosphate and glycolysis can provide enough ATP to run a 400-meter race.

Muscle activity that lasts longer than half a minute depends increasingly on ***aerobic cellular respiration***, a series of oxygen-requiring reactions that produce ATP in mitochondria. Muscle fibers have two sources of oxygen: (1) oxygen that diffuses into them from the blood and (2) oxygen released by myoglobin in the sarcoplasm. ***Myoglobin*** is an oxygen-binding protein found only in muscle fibers. It binds oxygen when oxygen is plentiful and releases oxygen when it is scarce. If enough oxygen is present, pyruvic acid enters the mitochondria, where it is completely oxidized in reactions that generate ATP, carbon dioxide, water, and heat (Figure 8.8c). In comparison with anaerobic cellular respiration, aerobic cellular respiration yields much more ATP, about 36 molecules of ATP from each glucose molecule. In activities that last more than 10 minutes, aerobic cellular respiration provides most of the needed ATP.

Muscle Fatigue

The inability of a muscle to contract forcefully after prolonged activity is called ***muscle fatigue***. One important factor in muscle fatigue is lowered release of calcium ions from the sarcoplasmic reticulum, resulting in a decline of Ca^{2+} level in the sarcoplasm. Other factors that contribute to muscle fatigue include depletion of creatine phosphate, insufficient oxygen, depletion of glycogen and other nutrients, buildup of lactic acid and ADP, and failure of nerve impulses in the motor neuron to release enough acetylcholine.

Oxygen Consumption After Exercise

During prolonged periods of muscle contraction, increases in breathing and blood flow enhance oxygen delivery to muscular tissue. After muscle contraction has stopped, heavy breathing continues for a period of time, and oxygen consumption remains above the resting level. The term ***oxygen debt*** refers to the added oxygen, over and above the oxygen consumed at rest, that is taken into the body after exercise. This extra oxygen is used to "pay back" or restore metabolic conditions to the resting level in three ways: (1) to convert lactic acid back into glycogen stores in the liver, (2) to resynthesize creatine phosphate and ATP, and (3) to replace the oxygen removed from myoglobin.

The metabolic changes that occur *during exercise*, however, account for only some of the extra oxygen used *after exercise*. Only a small amount of resynthesis of glycogen occurs from lactic acid. Instead, glycogen stores are replenished much later from dietary carbohydrates. Much of the lactic acid that remains after exercise is converted back to pyruvic acid and used for ATP production via aerobic cellular respiration. Ongoing changes after exercise also boost oxygen use. First, the elevated body temperature after strenuous exercise increases the rate of chemical reactions throughout the body. Faster reactions use ATP more rapidly, and more oxygen is needed to produce ATP. Second, the heart and muscles used in breathing are still working harder than they were at rest, and thus they consume more ATP. Third, tissue repair processes are occurring at an increased pace. For these reasons, ***recovery oxygen uptake*** is a better term than oxygen debt for the elevated use of oxygen after exercise.

CHECKPOINT

8. What are the sources of ATP for muscle fibers?
9. What factors contribute to muscle fatigue?
10. Why is the term *recovery oxygen uptake* more accurate than *oxygen debt*?

CONTROL OF MUSCLE TENSION

OBJECTIVES • **Explain the three phases of a twitch contraction.**

- **Describe how the frequency of stimulation and motor unit recruitment affect muscle tension.**
- **Compare the three types of skeletal muscle fibers.**
- **Distinguish between isotonic and isometric contractions.**

The contraction that results from a single muscle action potential, a muscle twitch, has significantly smaller force than the maximum force or tension the fiber is capable of producing. The total tension that a *single* muscle fiber can produce depends mainly on the rate at which nerve impulses arrive at its neuromuscular junction. The number of impulses per second is the *frequency of stimulation*. When considering the contraction of a *whole* muscle, the total tension it can produce depends on the number of muscle fibers that are contracting in unison.

Twitch Contraction

A ***twitch contraction*** is a brief contraction of all the muscle fibers in a motor unit in response to a single action potential in its motor neuron. Figure 8.9 shows a recording of a muscle contraction, called a ***myogram***. Note that a brief delay, called the *latent period*, occurs between application of the stimulus (time zero on the graph) and the beginning of contraction. During the latent period, the muscle action potential sweeps over the sarcolemma and calcium ions are released from the sarcoplasmic reticulum. During the second phase, the *contraction period* (upward tracing), repetitive power strokes are occurring, generating tension or force of contraction. In the third phase, the *relaxation period* (downward tracing), power strokes cease because the level of Ca^{2+} in the sarcoplasm is decreasing to the resting level. (Recall that calcium ions are actively transported back into the sarcoplasmic reticulum.)

Frequency of Stimulation

If a second stimulus occurs before a muscle fiber has completely relaxed, the second contraction will be stronger than the first because the second contraction begins when the fiber is at a higher level of tension (Figure 8.10a, b). This phenomenon, in which stimuli arriving one after the other cause larger contractions, is called ***wave summation***. When a skeletal muscle fiber is stimulated at a rate of 20 to 30 times per second, it can only partially relax between stimuli. The result is a sustained but wavering contraction called ***unfused (incomplete) tetanus*** (*tetan-* = rigid, tense; Figure 8.10c). When a skeletal muscle fiber is stimulated at a higher rate of 80 to 100 times per second, it does not relax at all. The result is ***fused (complete) tetanus***, a sustained contraction in which individual twiches cannot be detected (Figure 8.10d).

Figure 8.9 Myogram of a twitch contraction. The arrow indicates the time at which the stimulus occurred.

A myogram is a record of a muscle contraction.

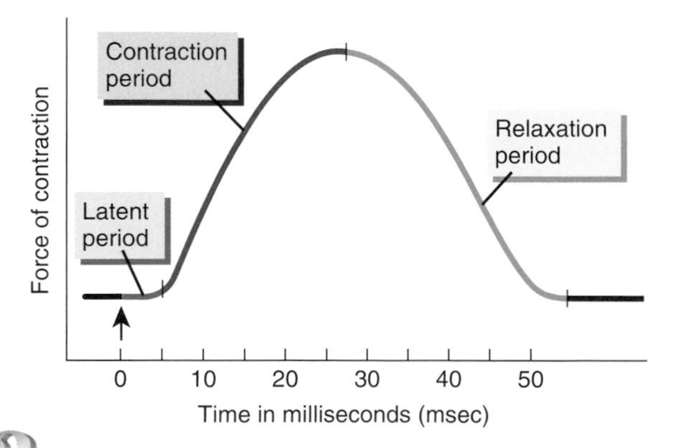

During which period do sarcomeres shorten?

Motor Unit Recruitment

The process in which the number of contracting motor units is increased is called ***motor unit recruitment***. Normally, the various motor neurons to a whole muscle fire *asynchronously* (at different times): While some motor units are contracting, others are relaxed. This pattern of motor unit activity delays muscle fatigue by allowing alternately contracting motor units to relieve one another, so that the contraction can be sustained for long periods.

Recruitment is one factor responsible for producing smooth movements rather than a series of jerky movements. Precise movements are brought about by small changes in muscle contraction. Typically, the muscles that produce precise movements are composed of small motor units. In this way, when a motor unit is recruited or turned off, only slight changes occur in muscle tension. On the other hand, large motor units are active where large tension is needed and precision is less important.

Types of Skeletal Muscle Fibers

Skeletal muscles contain three types of muscle fibers, which are present in varying proportions in different muscles of the body: The fiber types are (1) slow oxidative fibers, (2) fast oxidative–glycolytic fibers, and (3) fast glycolytic fibers.

Slow oxidative (SO) fibers or *red fibers* are small in diameter and appear dark red because they contain a large amount of myoglobin. Because they have many large mitochondria, SO fibers generate ATP mainly by aerobic cellular respiration, which is why they are called oxidative fibers. These fibers are said to be "slow" because the contraction cycle proceeds at a slower pace than in "fast" fibers. SO fibers are very resistant to fatigue and are capable of prolonged, sustained contractions.

Fast oxidative–glycolytic (FOG) fibers are intermediate in diameter between the other two types. Like slow oxidative fibers, they contain a large amount of myoglobin, and thus appear dark red. FOG fibers can generate considerable ATP by aerobic cellular respiration, which gives them a moderately high resistance to fatigue. Because their glycogen content is high, they also generate ATP by anaerobic glycolysis. These fibers are "fast" because they contract and relax more quickly than SO fibers.

Fast glycolytic (FG) fibers or *white fibers* are largest in diameter, contain the most myofibrils, and generate the most powerful and most rapid contractions. They have a low myoglobin content and few mitochondria. FG fibers contain large amounts of glycogen and generate ATP mainly by anaerobic glycolysis. They are used for intense movements of short duration, but they fatigue quickly. Strength-training

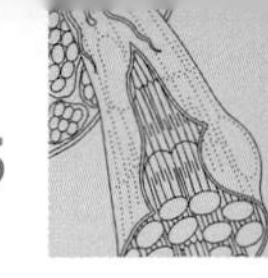

Figure 8.10 Myograms showing the effects of different frequencies of stimulation. (a) Single twitch. (b) When a second stimulus occurs before the muscle has relaxed, wave summation occurs, and the second contraction is stronger than the first. (The dashed line indicates the force of contraction expected in a single twitch.) (c) In unfused tetanus, the curve looks jagged due to partial relaxation of the muscle between stimuli. (d) In fused tetanus, the contraction force is steady and sustained.

Due to wave summation, the tension produced during a sustained contraction is greater than during a single twitch.

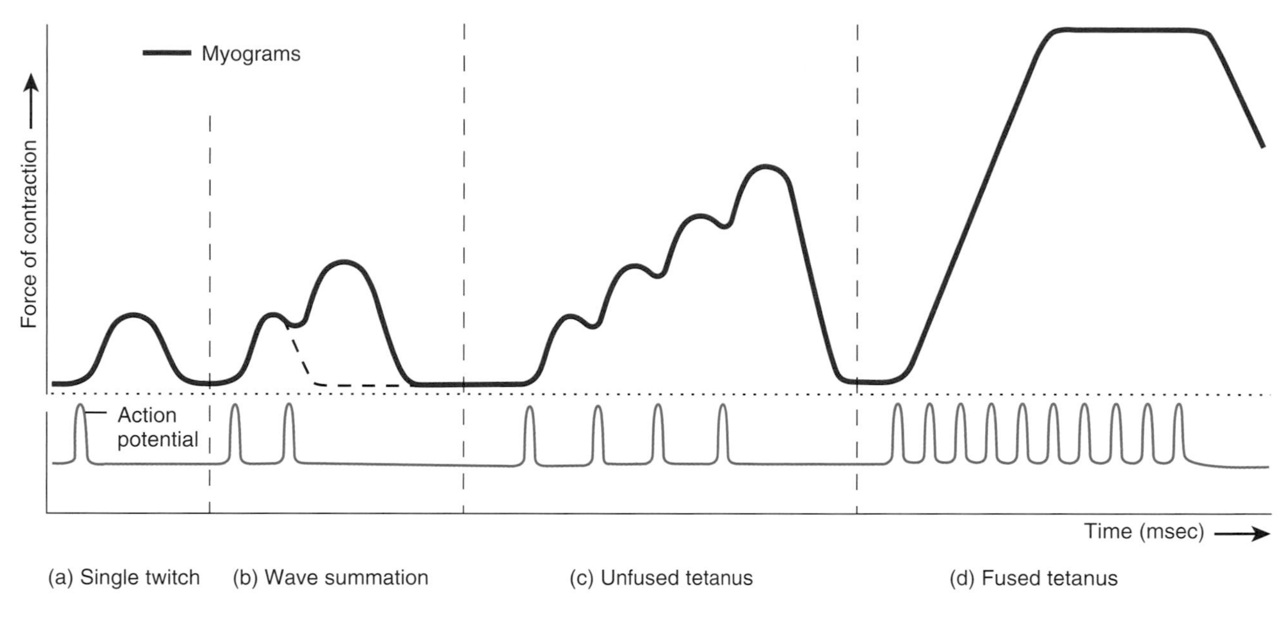

What frequency of stimulation is needed to produce fused tetanus?

programs that engage a person in activities requiring great strength for short times produce increases in the size, strength, and glycogen content of FG fibers.

Most skeletal muscles are a mixture of all three types of skeletal muscle fibers, about half of which are SO fibers. The proportions vary somewhat, depending on the action of the muscle, the person's training program, and genetic factors. For example, the continually active postural muscles of the neck, back, and legs have a high proportion of SO fibers. Muscles of the shoulders and arms, in contrast, are not constantly active but are used intermittently and briefly to produce large amounts of tension, such as in lifting and throwing. These muscles have a high proportion of FG fibers. Leg muscles, which not only support the body but are also used for walking and running, have large numbers of both SO and FOG fibers.

Even though most skeletal muscles are a mixture of all three types of skeletal muscle fibers, the skeletal muscle fibers of any given motor unit are all of the same type. The different motor units in a muscle are recruited in a specific order, depending on need. For example, if weak contractions suffice to perform a task, only SO motor units are activated. If more force is needed, the motor units of FOG fibers are also recruited. Finally, if maximal force is required, motor units of FG fibers are also called into action.

Isometric and Isotonic Contractions

Muscle contractions are classified as either isotonic or isometric. In an ***isotonic contraction*** (*iso-* = equal; *-tonic* = tension), the tension (force of contraction) developed by the muscle remains almost constant while the muscle changes its length. Isotonic contractions are used for body movements and for moving objects. For example, picking up a book from a table involves isotonic contractions of the biceps brachii muscle in the arm.

In an ***isometric contraction*** (*-metro* = measure or length), the tension generated is not enough to exceed the resistance of the object to be moved and the muscle does not change its length. Isometric contractions occur when you try to lift a box but the box does not move because it is too heavy. Isometric contractions are important for maintaining posture and supporting objects in a fixed position.

■ CHECKPOINT

11. Define the following terms: myogram, twitch contraction, wave summation, unfused tetanus, and fused tetanus.

12. What characteristics distinguish the three types of skeletal muscle fibers?

13. Provide examples of isometric and isotonic contractions.

EXERCISE AND SKELETAL MUSCLE TISSUE

OBJECTIVE • Describe the effects of exercise on skeletal muscle tissue.

The relative ratio of fast glycolytic (FG) and slow oxidative (SO) fibers in each muscle is genetically determined and helps account for individual differences in physical performance. For example, people with a higher proportion of FG fibers often excel in activities that require periods of intense activity, such as weight lifting or sprinting. People with higher percentages of SO fibers are better at activities that require endurance, such as long-distance running.

Although the total number of skeletal muscle fibers usually does not increase, the characteristics of those present can change to some extent. Various types of exercises can induce changes in the fibers in a skeletal muscle. Endurance-type (aerobic) exercises, such as running or swimming, cause a gradual transformation of some FG fibers into fast oxidative–glycolytic (FOG) fibers. The transformed muscle fibers show slight increases in diameter, number of mitochondria, blood supply, and strength. Endurance exercises also result in cardiovascular and respiratory changes that cause skeletal muscles to receive better supplies of oxygen and nutrients but do not increase muscle mass. By contrast, exercises that require great strength for short periods produce an increase in the size and strength of FG fibers. The increase in size is due to increased synthesis of thick and thin filaments. The overall result is muscle enlargement (hypertrophy), as evidenced by the bulging muscles of body builders.

■ **CHECKPOINT**

14. Explain how the characteristics of skeletal muscle fibers may change with exercise.

CARDIAC MUSCLE TISSUE

OBJECTIVE • Describe the structure and function of cardiac muscle tissue.

Most of the heart consists of ***cardiac muscle tissue***. Like skeletal muscle, cardiac muscle is also *striated*, but its action is *involuntary*: Its alternating cycles of contraction and relaxation are not consciously controlled. Cardiac muscle fibers often are branched; are shorter in length and larger in diameter than skeletal muscle fibers; and have a single, centrally located nucleus (see Figure 15.2b on page 366). Cardiac muscle fibers interconnect with one another by irregular transverse thickenings of the sarcolemma called ***intercalated discs*** (in-TER-ka-lāt-ed = to insert between). The intercalated discs hold the fibers together and contain *gap junctions*, which allow muscle action potentials to spread quickly from one cardiac muscle fiber to another.

A major difference between skeletal muscle and cardiac muscle is the source of stimulation. We have seen that skeletal muscle tissue contracts only when stimulated by acetylcholine released by a nerve impulse in a motor neuron. In contrast, the heart beats because some of the cardiac muscle fibers act as a pacemaker to initiate each cardiac contraction. The built-in or intrinsic rhythm of heart contractions is called ***autorhythmicity*** (aw′-tō-rith-MIS-i-tē). Several hormones and neurotransmitters can increase or decrease heart rate by speeding or slowing the heart's pacemaker.

Under normal resting conditions, cardiac muscle tissue contracts and relaxes an average of about 75 times a minute. Thus, cardiac muscle tissue requires a constant supply of oxygen and nutrients. The mitochondria in cardiac muscle fibers are larger and more numerous than in skeletal muscle fibers and produce most of the needed ATP via aerobic cellular respiration. In addition, cardiac muscle fibers can use lactic acid, released by skeletal muscle fibers during exercise, to make ATP.

■ **CHECKPOINT**

15. What are the major structural and functional differences between cardiac and skeletal muscle tissue?

SMOOTH MUSCLE TISSUE

OBJECTIVE • Describe the structure and function of smooth muscle tissue.

Smooth muscle tissue is found in many internal organs and blood vessels. Like cardiac muscle, smooth muscle is *involuntary*. Smooth muscle fibers are considerably smaller in length and diameter than skeletal muscle fibers and are tapered at both ends. Within each fiber is a single, oval, centrally located nucleus (Figure 8.11). In addition to thick and thin filaments, smooth muscle fibers also contain ***intermediate filaments***. Because the various filaments have no regular pattern of overlap, smooth muscle fibers lack alternating dark and light bands and thus appear *nonstriated*, or smooth.

In smooth muscle fibers, the thin filaments attach to structures called ***dense bodies***, which are functionally similar to Z discs in striated muscle fibers. Some dense bodies are dispersed throughout the sarcoplasm; others are attached to the sarcolemma. Bundles of intermediate filaments also attach to dense bodies and stretch from one dense body to another. During contraction, the sliding filament mechanism involving thick and thin filaments generates tension that is

Figure 8.11 Histology of smooth muscle tissue. A smooth muscle fiber is shown in the relaxed state (left) and the contracted state (right).

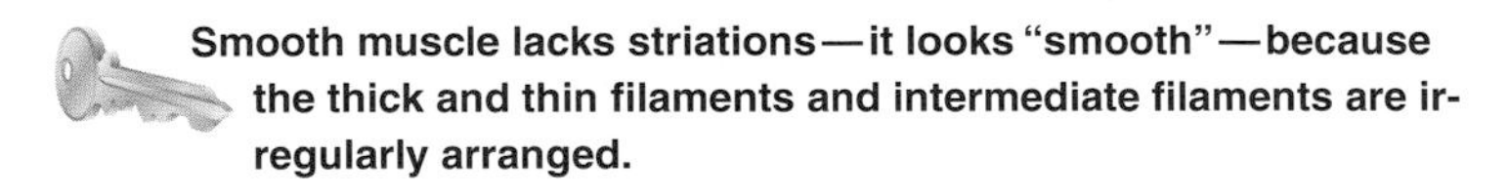

Smooth muscle lacks striations—it looks "smooth"—because the thick and thin filaments and intermediate filaments are irregularly arranged.

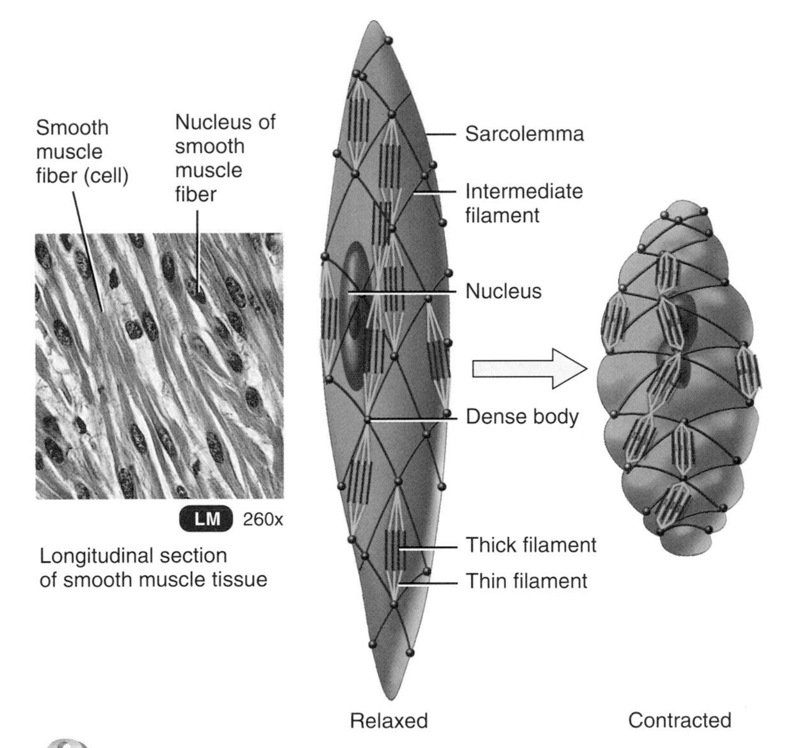

Which type of smooth muscle is found in the walls of hollow organs?

transmitted to intermediate filaments. These, in turn, pull on the dense bodies attached to the sarcolemma, causing a lengthwise shortening of the muscle fiber.

There are two kinds of smooth muscle tissue, visceral and multiunit. The more common type is ***visceral (single-unit) muscle tissue***. It is found in sheets that wrap around to form part of the walls of small arteries and veins and hollow viscera such as the stomach, intestines, uterus, and urinary bladder. The fibers in visceral muscle tissue are tightly bound together in a continuous network. Like cardiac muscle, visceral smooth muscle is autorhythmic. Because the fibers connect to one another by gap junctions, muscle action potentials spread throughout the network. When a neurotransmitter, hormone, or autorhythmic signal stimulates one fiber, the muscle action potential spreads to neighboring fibers, which then contract in unison, as a single unit.

The second kind of smooth muscle tissue, ***multiunit smooth muscle tissue***, consists of individual fibers, each with its own motor nerve endings. Unlike stimulation of a single visceral muscle fiber, which causes contraction of many adjacent fibers, stimulation of a single multiunit smooth muscle fiber causes contraction of that fiber only. Multiunit smooth muscle tissue is found in the walls of large arteries, in large airways to the lungs, in the arrector pili muscles attached to hair follicles, and in the internal eye muscles.

Compared with contraction in a skeletal muscle fiber, contraction in a smooth muscle fiber starts more slowly and lasts much longer. Calcium ions enter smooth muscle fibers slowly and also move slowly out of the muscle fiber when excitation declines, which delays relaxation. The prolonged presence of Ca^{2+} in the cytosol provides for ***smooth muscle tone***, a state of continued partial contraction. Smooth muscle tissue can thus sustain long-term tone, which is important in the walls of blood vessels and in the walls of organs that maintain pressure on their contents. Finally, smooth muscle can both shorten and stretch to a greater extent than other muscle types. Stretchiness permits smooth muscle in the wall of hollow organs such as the uterus, stomach, intestines, and urinary bladder to expand as their contents enlarge, while still retaining the ability to contract.

Most smooth muscle fibers contract or relax in response to nerve impulses from the autonomic (involuntary) nervous system. In addition, many smooth muscle fibers contract or relax in response to stretching; hormones; or local factors such as changes in pH, oxygen and carbon dioxide levels, temperature, and ion concentrations. For example, the hormone epinephrine, released by the adrenal medulla, causes relaxation of smooth muscle in the airways and in some blood vessel walls.

Table 8.1 presents a summary of the major characteristics of the three types of muscular tissue.

■ **CHECKPOINT**

16. How do visceral and multiunit smooth muscle differ?

17. What are the major structural and functional differences between smooth and skeletal muscle tissue?

AGING AND MUSCULAR TISSUE

OBJECTIVE • Explain the effects of aging on skeletal muscle.

Beginning at about 30 years of age, humans undergo a slow, progressive loss of skeletal muscle mass that is replaced largely by fibrous connective tissue and adipose tissue. In part, this decline is due to decreased levels of physical activity. Accompanying the loss of muscle mass is a decrease in maximal strength, a slowing of muscle reflexes, and a loss of flexibility. In some muscles, a selective loss of muscle fibers of a given type may occur. With aging, the relative number of slow oxidative fibers appears to increase. This could be due either to atrophy of the other fiber types or their conversion into slow oxidative fibers. Whether this is an effect of aging

Table 8.1 Summary of the Principal Features of Muscular Tissue

Characteristics	Skeletal Muscle	Cardiac Muscle	Smooth Muscle
Cell Appearance and Features	Long cylindrical fiber with many peripherally located nuclei; striated; unbranched	Branched cylindrical fiber, usually with one centrally located nucleus; intercalated discs join neighboring fibers; striated	Fiber is thickest in the middle, tapered at each end, has one centrally located nucleus; not striated
Location	Primarily attached to bones by tendons	Heart	Walls of hollow viscera, airways, blood vessels, iris and ciliary body of the eye, arrector pili of hair follicles
Fiber Diameter	Very large (10–100μm)*	Large (10–20 μm)	Small (3–8 μm)
Fiber Length	Very large (100 μm–30 cm)	Small (50–100 μm)	Intermediate (30–200 μm)
Sarcomeres	Yes	Yes	No
Transverse Tubules	Yes, aligned with each A–I band junction	Yes, aligned with each Z disc	No
Speed of Contraction	Fast	Moderate	Slow
Nervous Control	Voluntary	Involuntary	Involuntary
Capacity for Regeneration	Limited	Limited	Considerable compared with other muscle tissues, but limited compared with tissues such as epithelium

*1 micrometer (μm) = 1/25,000 of an inch.

itself or mainly reflects the more limited physical activity of older people is still an unresolved question. Nevertheless, aerobic activities and strength training programs are effective in older people and can slow or even reverse the age-associated decline in muscular performance.

■ **CHECKPOINT**

18. Why does muscle strength decrease with aging?

HOW SKELETAL MUSCLES PRODUCE MOVEMENT

OBJECTIVE • Describe how skeletal muscles cooperate to produce movement.

Now that you have a basic understanding of the structure and functions of muscular tissue, we will examine how skeletal muscles cooperate to produce various body movements.

Origin and Insertion

Based on the description of muscular tissue, we can define a ***skeletal muscle*** as an organ composed of several different types of tissues. These include skeletal muscle tissue, vascular tissue (blood vessels and blood), nervous tissue (motor neurons), and several types of connective tissues.

Skeletal muscles are not attached directly to bones; they produce movements by pulling on tendons, which, in turn, pull on bones. Most skeletal muscles cross at least one joint and are attached to the articulating bones that form the joint (Figure 8.12). When the muscle contracts, it draws one bone toward the other. The two bones do not move equally. One is held nearly in its original position; the attachment of a muscle (by means of a tendon) to the stationary bone is called the ***origin***. The other end of the muscle is attached by means of a tendon to the movable bone at a point called the ***insertion***. The fleshy portion of the muscle between the tendons of the origin and insertion is called the ***belly***. A good analogy is a spring on a door. The part of the spring attached to the door represents the insertion, the part attached to the frame is the origin, and the coils of the spring are the belly.

Tenosynovitis (ten′-ō-sin-ō-VĪ-tis), commonly known as **tendinitis**, is a painful inflammation of the tendons, tendon sheaths, and synovial membranes of joints. The tendons most often affected are at the wrists, shoulders, elbows (resulting in *tennis elbow*), finger joints (resulting in *trigger finger*), ankles, and feet. The affected sheaths some-

Figure 8.12 Relationship of skeletal muscles to bones. Skeletal muscles produce movements by pulling on tendons attached to bones.

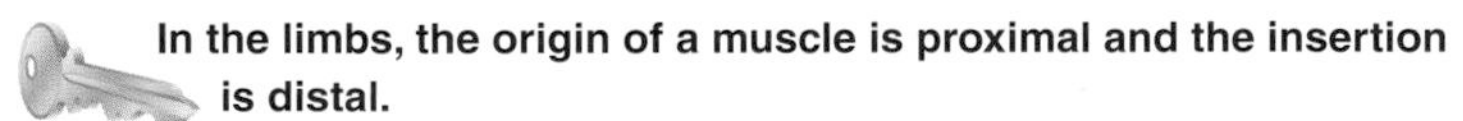

In the limbs, the origin of a muscle is proximal and the insertion is distal.

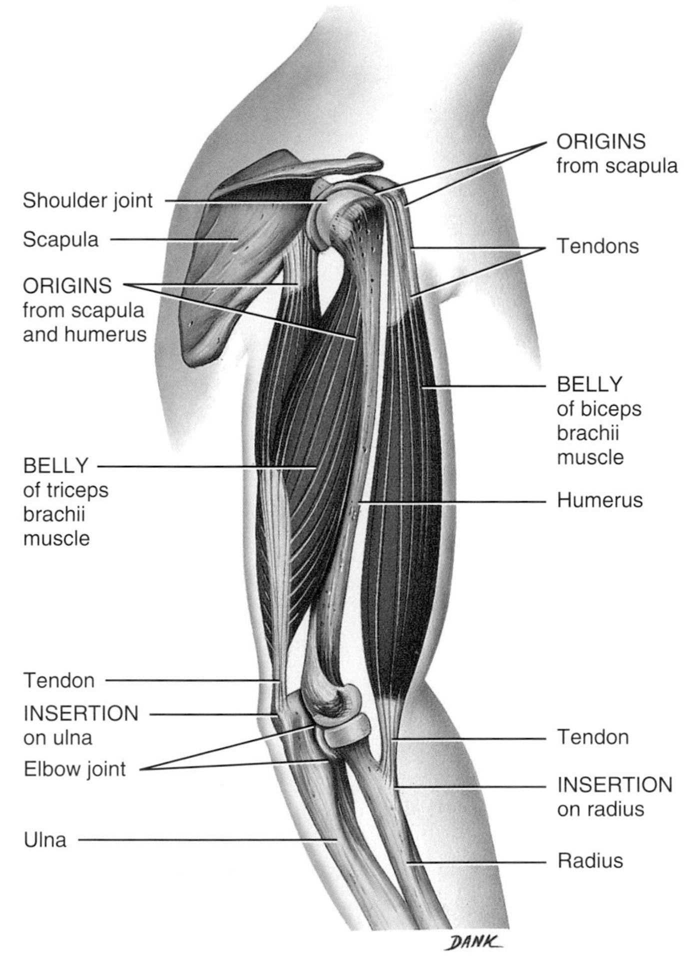

Origin and insertion of a skeletal muscle

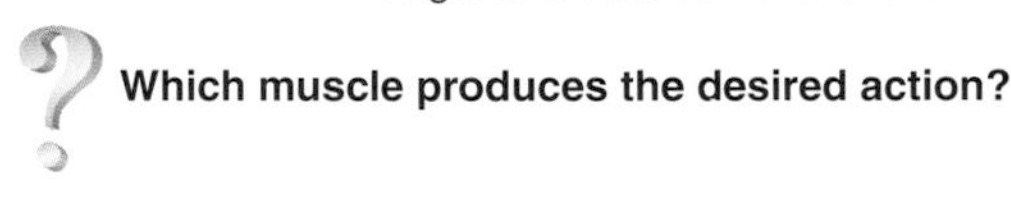

Which muscle produces the desired action?

times become visibly swollen due to fluid accumulation. The joint is tender, and movement of the body part often causes pain. Trauma, strain, or excessive exercise may cause tenosynovitis. For instance, tying shoelaces too tightly may cause tenosynovitis of the dorsum of the foot. Also, gymnasts are prone to developing the condition because of chronic, repetitive, and maximum hyperextension at the wrists.

Group Actions

Most movements occur because several skeletal muscles are acting in groups rather than individually. Also, most skeletal muscles are arranged in opposing pairs at joints, that is, flexors–extensors, abductors–adductors, and so on. A muscle that causes a desired action is referred to as the ***prime mover*** or ***agonist*** (= leader). Often, another muscle, called the ***antagonist*** (*ant-* = against), relaxes while the prime mover contracts. The antagonist has an effect opposite to that of the prime mover; that is, the antagonist stretches and yields to the movement of the prime mover. When you bend (flex) your elbow, the biceps brachii is the prime mover. While the biceps brachii is contracting, the triceps brachii, the antagonist, is relaxing (see Figure 8.20). Do not assume, however, that the biceps brachii is always the prime mover and the triceps brachii is always the antagonist. For example, when straightening (extending) the elbow, the triceps brachii serves as the prime mover and the biceps brachii functions as the antagonist. If the prime mover and antagonist contracted together with equal force, there would be no movement, as in an isometric contraction.

Most movements also involve muscles called ***synergists*** (SIN-er-gists; *syn-* = together; *erg-* = work), which help the prime mover function more efficiently by reducing unnecessary movement. Some muscles in a group also act as ***fixators***, stabilizing the origin of the prime mover so that the prime mover can act more efficiently. Under different conditions and depending on the movement, many muscles act at various times as prime movers, antagonists, synergists, or fixators.

■ CHECKPOINT

19. Distinguish between the origin and insertion of a skeletal muscle.

20. Explain why most body movements occur because several skeletal muscles act in groups rather than individually.

PRINCIPAL SKELETAL MUSCLES

OBJECTIVES • **List and describe the ways that skeletal muscles are named.**

• **Describe the location of skeletal muscles in various regions of the body and identify their functions.**

The names of most of the nearly 700 skeletal muscles are based on specific characteristics. Learning the terms used to indicate specific characteristics will help you remember the names of the muscles (Table 8.2 on page 191).

Exhibits 8.1 through 8.13 list the principal skeletal muscles of the body with their origins, insertions, and actions. (By no means have all the muscles of the body been included.) For each exhibit, an overview section provides a general orientation to the muscles and their functions or unique characteristics. To make it easier for you to learn to say the names of skeletal muscles and understand how they are named, we have provided phonetic pronunciations and word roots that

FOCUS ON WELLNESS

Effective Stretching Increases Muscle Flexibility

A certain degree of elasticity is an important attribute of skeletal muscles and their connective tissue attachments. Greater elasticity contributes to a greater degree of ***flexibility***, increasing the range of motion of a joint. A joint's ***range of motion (ROM)*** is the maximum ability to move the bones about the joint through an arc of a circle. For example, a person may normally be able to extend the knee joint from 30° when it is maximally flexed to 170° when fully extended. The ROM or degree of flexibility is then 170 − 30° = 140°. Physical therapists measure improvements in flexibility by increases in ROM.

Stretching It

When a relaxed muscle is physically stretched, its ability to lengthen is limited by connective tissue structures, such as fasciae. Regular stretching gradually lengthens these structures, but the process occurs very slowly. To see an improvement in flexibility, stretching exercises must be performed regularly—daily, if possible—for many weeks.

Tissues stretch best when slow, gentle force is applied at elevated tissue temperatures. An external source of heat, such as hot packs or ultrasound, can be used. But 10 or more minutes of muscular contraction is also a good way to raise muscle temperature. Exercise heats the muscle more deeply and thoroughly. That's where the term "warm-up" comes from. It's important to warm up *before* stretching, not vice versa. Stretching cold muscles does not increase flexibility and may even cause injury.

Just Relax . . .

The easiest and safest way to increase flexibility is with static stretching. A good static stretch is slow and gentle. After warming up, you get into a comfortable stretching position and relax. Continuing to relax and breathe deeply, you reach just a little farther, and a little farther, holding the stretch for at least 30 seconds. If you have difficulty relaxing, you know you have stretched too far. Ease up until you feel a stretch but no strain.

When stretching, it is important to relax. Sounds simple, right? But if you ever visit an exercise class, you'll notice some people who are all tense, rigid, and hunched up, because the stretching positions are uncomfortable. As a result, their muscles tighten up in protest. These people figure they'd better push a little harder, and they tense up even more. They are unintentionally activating the motor neurons that initiate muscular contraction in the very muscles they are supposed to be relaxing, which of course interferes with the muscle's ability to elongate and stretch.

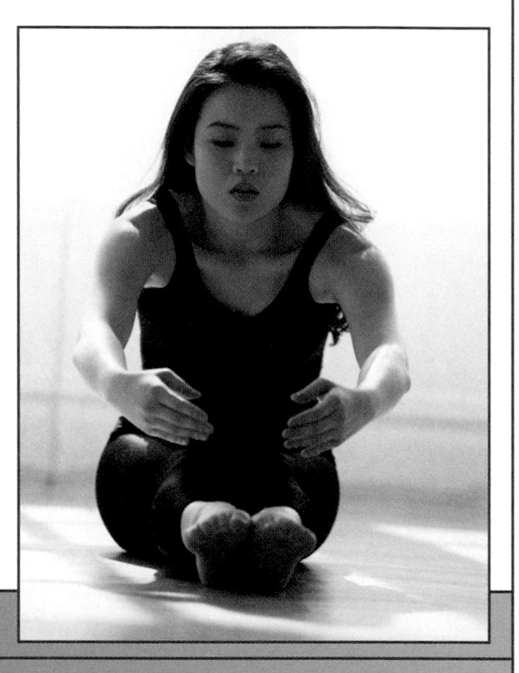

► **THINK IT OVER . . .**

► ***Using the information presented in this chapter, try to figure out which muscles are being stretched when you place one foot (keep that leg straight) up on a bar or chair.***

indicate how the muscles are named (refer also to Table 8.2). Once you have mastered the naming of the muscles, their actions will have more meaning and be easier to remember.

The muscles are divided into groups according to the part of the body on which they act. Figure 8.13 on pages 192–193 shows general anterior and posterior views of the muscular system. As you study groups of muscles in the following exhibits, refer to Figure 8.13 to see how each group is related to all others.

• • •

To appreciate the many ways that the muscular system contributes to homeostasis of other body systems, examine Focus on Homeostasis: The Muscular System on page 218. Next, in Chapter 9, we will see how the nervous system is organized, how neurons generate nerve impulses that activate muscle tissues as well as other neurons, and how synapses function.

Table 8.2 Characteristics Used to Name Skeletal Muscles

Name	Meaning	Example	Figure
***Direction*: Orientation of muscle fibers relative to the body's midline**			
Rectus	Parallel to midline	Rectus abdominis	8.16b
Transverse	Perpendicular to midline	Transverse abdominis	8.16b
Oblique	Diagonal to midline	External oblique	8.16a
***Size*: Relative size of the muscle**			
Maximus	Largest	Gluteus maximus	8.23b
Minimus	Smallest	Gluteus minimus	8.23b
Longus	Longest	Adductor longus	8.23a
Latissimus	Widest	Latissimus dorsi	8.13b
Longissimus	Longest	Longissimus muscles	8.22
Magnus	Large	Adductor magnus	8.23b
Major	Larger	Pectoralis major	8.13a
Minor	Smaller	Pectoralis minor	8.19a
Vastus	Great	Vastus lateralis	8.23a
***Shape*: Relative shape of the muscle**			
Deltoid	Triangular	Deltoid	8.13b
Trapezius	Trapezoid	Trapezius	8.13b
Serratus	Saw-toothed	Serratus anterior	8.13a
Rhomboid	Diamond-shaped	Rhomboid major	8.18b
Orbicularis	Circular	Orbicularis oculi	8.14
Pectinate	Comblike	Pectineus	8.23a
Piriformis	Pear-shaped	Piriformis	8.23b
Platys	Flat	Platysma	8.13a
Quadratus	Square	Quadratus lumborum	8.17b
Gracilis	Slender	Gracilis	8.23a
***Action*: Principal action of the muscle**			
Flexor	Decreases joint angle	Flexor carpi radialis	8.21a
Extensor	Increases joint angle	Extensor carpi ulnaris	8.21b
Abductor	Moves bone away from midline	Abductor pollicis longus	8.21b
Adductor	Moves bone closer to midline	Adductor longus	8.23a
Levator	Produces superior movement	Levator scapulae	8.18
Depressor	Produces inferior movement	Depressor labii inferioris	
Supinator	Turns palm anteriorly	Supinator	
Pronator	Turns palm posteriorly	Pronator teres	8.21a
Sphincter	Decreases size of opening	External anal sphincter	19.15b
Tensor	Makes a body part rigid	Tensor fasciae latae	8.23a
***Number of Origins*: Number of tendons of origin**			
Biceps	Two origins	Biceps brachii	8.20a
Triceps	Three origins	Triceps brachii	8.20b
Quadriceps	Four origins	Quadriceps femoris	8.23a
***Location*: Structure near which a muscle is found**			
Example: Temporalis, a muscle near the temporal bone (Figure 8.14).			
***Origin and Insertion*: Sites where muscle originates and inserts**			
Example: Brachioradialis, originating on the humerus and inserting on the radius (Figure 8.21a).			

Figure 8.13 Principal superficial skeletal muscles.

Most movements require contraction of several skeletal muscles acting in groups rather than individually.

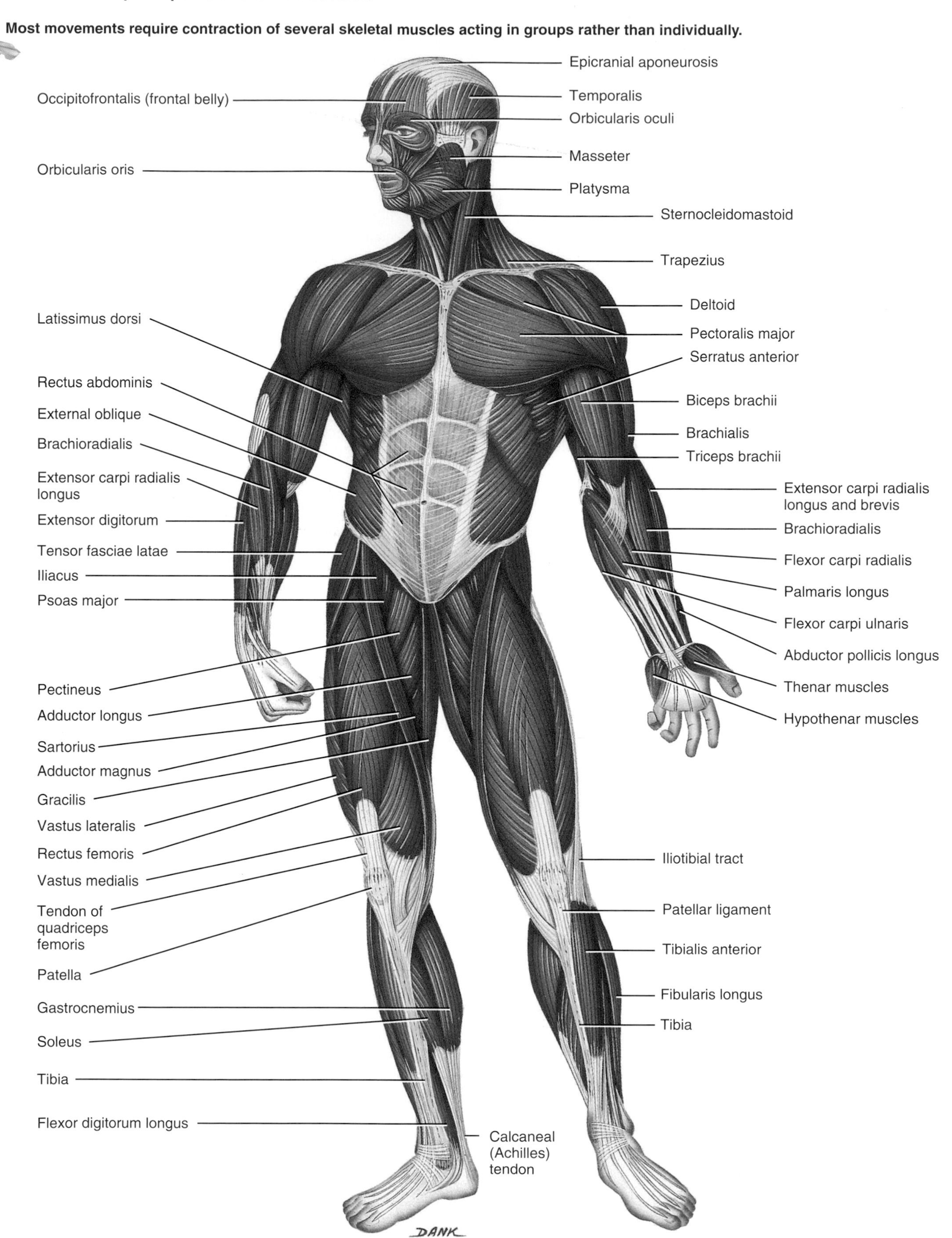

(a) Anterior view

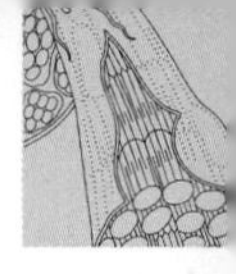

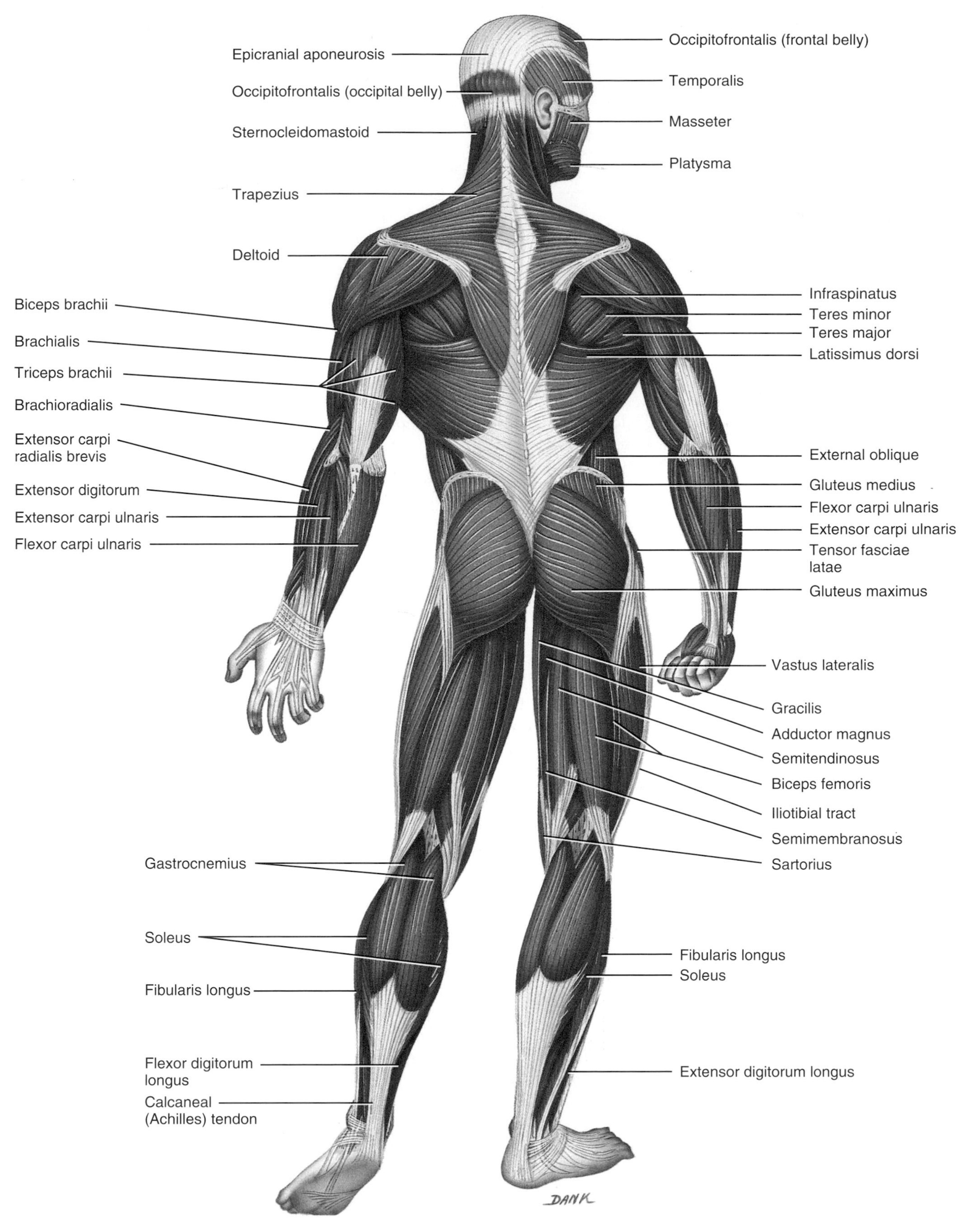

(b) Posterior view

Which is an example of a muscle named for the following characteristics: direction of fibers, shape, action, size, origin and insertion, location, and number of origins?

Exhibit 8.1 Muscles of Facial Expression *(Figure 8.14)*

OBJECTIVE • Describe the origin, insertion, and action of the muscles of facial expression.

Overview: The muscles of facial expression provide humans with the ability to express a wide variety of emotions, including displeasure, surprise, fear, and happiness. The muscles themselves lie within the layers of superficial fascia (connective tissue beneath the skin). As a rule, their origins are in the fascia or in the bones of the skull, with insertions into the skin. The "movable bone" in this case is the skin rather than a joint.

Bell's palsy, also known as **facial paralysis**, is a one-sided paralysis of the muscles of facial expression as a result of damage or disease of the facial (VII) nerve. Although the cause is unknown, a relationship between the herpes simplex virus and inflammation of the facial nerve has been suggested. In severe cases, the paralysis causes the entire side of the face to droop, and the person cannot wrinkle the forehead, close the eye, or pucker the lips on the affected side. Drooling and difficulty in swallowing also occur. Eighty percent of patients recover completely within a few weeks to a few months. For others, paralysis is permanent.

Relating muscles to movements: Arrange the muscles in this exhibit into two groups: (1) those that act on the mouth and (2) those that act on the eyes.

■ CHECKPOINT

What muscles would you use to show surprise, express sadness, show your upper teeth, pucker your lips, squint, and blow up a balloon?

Muscle	Origin	Insertion	Action
Occipitofrontalis (ok-sip-i-tō-frun-TĀ-lis)			
Frontal belly	Epicranial aponeurosis (ap′-ō-noo-RŌ-sis) (flat tendon that attaches to the frontalis and occipitalis muscles).	Skin superior to orbit.	Draws scalp forward, raises eyebrows, and wrinkles skin of forehead horizontally as in a look of surprise.
Occipital belly (*occipit-* = base of skull) (See Figure 8.13b.)	Occipital and temporal	Epicranial aponeurosis.	Draws scalp backward.
Orbicularis oris (or-bi′-kū-LAR-is OR-is; *-orb* = circular; *or* = mouth)	Muscle fibers surrounding opening of mouth.	Skin at corner of mouth.	Closes and protrudes lips, compresses lips against teeth, and shapes lips during speech.
Zygomaticus major (zī-gō-MA-ti-kus; *zygomatic* = cheek bone; *major* = greater)	Zygomatic bone.	Skin at angle of mouth and orbicularis oris.	Draws angle of mouth upward and outward, as in smiling or laughing.
Buccinator (BUK-si-nā′-tor; *bucia* = cheek)	Maxilla and mandible.	Orbicularis oris.	Presses cheeks against teeth and lips, as in whistling, blowing, and sucking; draws corner of mouth laterally; assists in mastication (chewing) by keeping food between the teeth (and not between teeth and cheeks).
Platysma (pla-TIZ-ma; *platys* = flat)	Fascia over deltoid and pectoralis major muscles.	Mandible, muscles around angle of mouth, and skin of lower face.	Draws outer part of lower lip downward and backward as in pouting; depresses mandible.
Orbicularis oculi (OK-ū-lī; *oculi* = eye)	Medial wall of orbit.	Circular path around orbit.	Closes eye.
Levator palpebrae superioris (le-VĀ-tor PAL-pe-brē soo-per′-ē-OR-is; *palpebrae* = eyelids) (see Figure 8.15.)	Roof of orbit.	Skin of upper eyelid.	Elevates upper eyelid (opens eye).

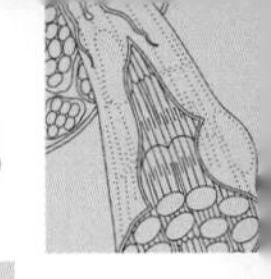

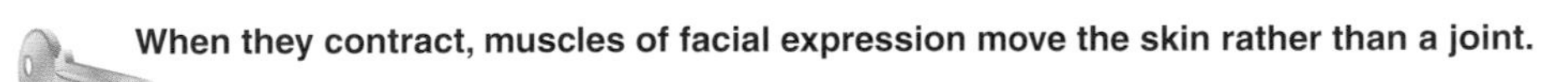

Figure 8.14 Muscles of facial expression. In this and subsequent figures in the chapter, the muscles indicated in all uppercase letters are the ones specifically referred to in the corresponding exhibit.

When they contract, muscles of facial expression move the skin rather than a joint.

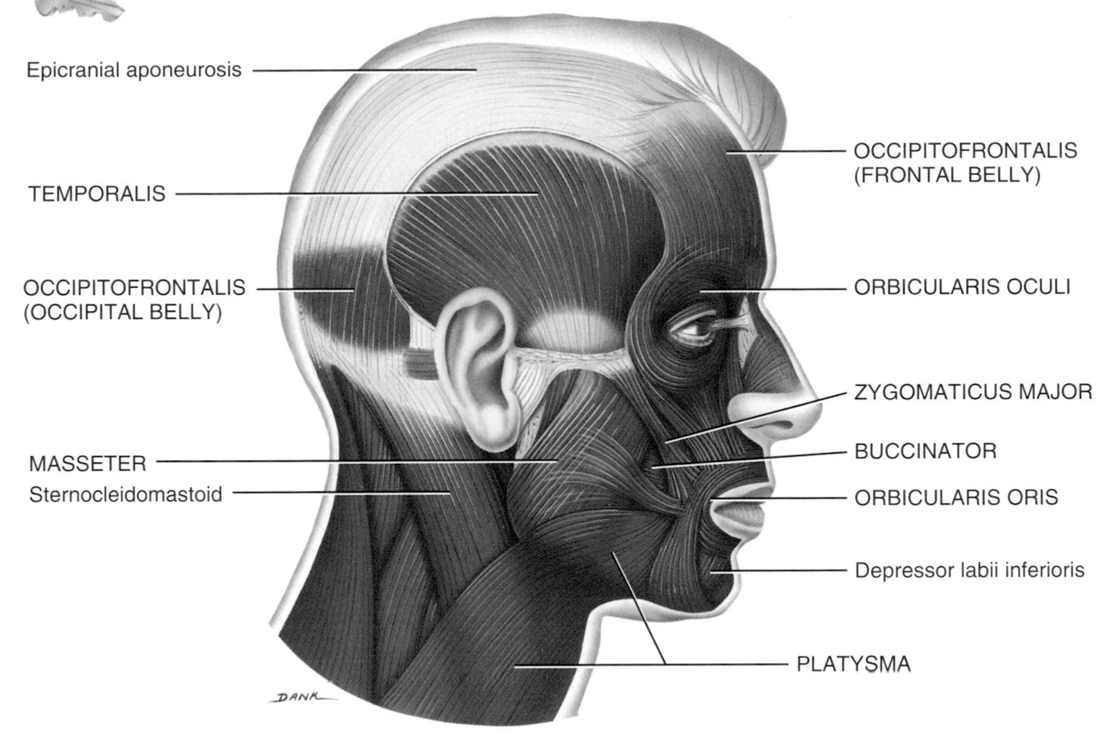

Right lateral superficial view

Which muscles of facial expression cause smiling, pouting, and squinting?

Exhibit 8.2 Muscles That Move the Mandible (Lower Jaw) *(See Figure 8.14)*

OBJECTIVE • Describe the origin, insertion, and action of the muscles that move the mandible.

Overview: Muscles that move the mandible (lower jaw) are also known as muscles of ***mastication*** (mas′-ti-KĀ-shun = to chew) because they are involved in biting and chewing. These muscles also assist in speech.

Relating muscles to movements: Arrange the muscles in this exhibit and the previous exhibit according to their actions on the mandible: (1) elevation, (2) depression, and (3) retraction. The same muscle may be mentioned more than once.

■ **CHECKPOINT**

What would happen if you lost tone in the masseter and temporalis muscles?

Muscle	Origin	Insertion	Action
Masseter (MA-se-ter; *maseter* = chewer) See Figure 8.14.	Maxilla and zygomatic arch.	Mandible.	Elevates mandible as in closing mouth.
Temporalis (tem′-por-Ā-lis; *tempori* = temples) See Figure 8.14.	Temporal bone.	Mandible.	Elevates and retracts (draws back) mandible.

Exhibit 8.3 Muscles That Move the Eyeballs: Extrinsic Muscles *(Figure 8.15)*

OBJECTIVE • Describe the origin, insertion, and action of the extrinsic muscles of the eyeballs.

Overview: Two types of muscles are associated with the eyeball, extrinsic and intrinsic. *Extrinsic muscles* originate outside the eyeball and are inserted on its outer surface (sclera). They move the eyeballs in various directions. *Intrinsic muscles* originate and insert entirely within the eyeball. They move structures within the eyeballs, such as the iris and the lens.

Movements of the eyeballs are controlled by three pairs of extrinsic muscles: (1) superior and inferior recti, (2) lateral and medial recti, and (3) superior and inferior obliques. Two pairs of rectus muscles move the eyeball in the direction indicated by their respective names: superior, inferior, lateral, and medial. One pair of muscles, the oblique muscles—superior and inferior—rotate the eyeball on its axis. The extrinsic muscles of the eyeballs are among the fastest contracting and most precisely controlled skeletal muscles of the body.

Strabismus is a condition in which the two eyes are not properly aligned. A lesion of the oculomotor (III) nerve, which controls the superior, inferior, and medial recti and the inferior oblique muscles, causes the eyeball to move laterally when at rest. The person cannot move the eyeball medially and inferiorly. A lesion in the abducens (VI) nerve, which innervates the lateral rectus muscle, causes the eyeball to move medially when at rest with inability to move the eyeball laterally.

Relating muscles to movements: Arrange the muscles in this exhibit according to their actions on the eyeballs: (1) elevation, (2) depression, (3) abduction, (4) adduction, (5) medial rotation, and (6) lateral rotation. The same muscle may be mentioned more than once.

■ **CHECKPOINT**

Which muscles contract and relax in each eye as you gaze to your left without moving your head?

Muscle	Origin	Insertion	Action
Superior rectus (REK-tus; *superior* = above; *rect-* = straight; here, muscle fibers that are parallel to long axis of eyeball)	Tendinous ring attached to bony orbit around optic foramen.	Superior and central part of eyeball.	Moves eyeball upward (elevation) and medially (adduction), and rotates it medially.
Inferior rectus *(inferior* = below)	Same as above.	Inferior and central part of eyeball.	Moves eyeball downward (depression) and medially (adduction), and rotates it medially.
Lateral rectus	Same as above.	Lateral side of eyeball.	Moves eyeball laterally (abduction).
Medial rectus	Same as above.	Medial side of eyeball.	Moves eyeball medially (adduction).
Superior oblique (ō-BLĒK; *oblique* = slanting; here, muscle fibers run diagonally to long axis of eyeball)	Same as above.	Eyeball between superior and lateral recti. The muscle moves through a ring of fibrocartilaginous tissue called the trochlea (*trochlea* = pulley).	Moves eyeball downward (depression) and laterally (abduction), and rotates it medially.
Inferior oblique	Maxilla.	Eyeball between inferior and lateral recti.	Moves eyeball upward (elevation) and laterally (abduction), and rotates it laterally.

Figure 8.15 Extrinsic muscles of the eyeballs.

The extrinsic muscles of the eyeball are among the fastest contracting and most precisely controlled skeletal muscles in the body.

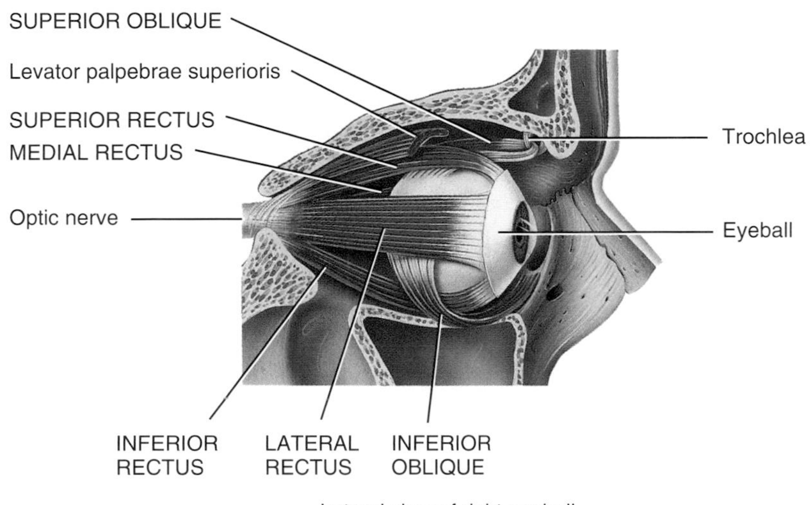

Lateral view of right eyeball

Which muscle passes through the trochlea?

Exhibit 8.4 Muscles That Act on the Anterior Abdominal Wall *(Figure 8.16)*

OBJECTIVE • Describe the origin, insertion, and action of the muscles that act on the anterior abdominal wall.

Overview: The anterior abdominal wall is composed of skin; fascia; and four pairs of muscles: rectus abdominis, external oblique, internal oblique, and transverse abdominis.

A **hernia** is a protrusion of an organ through a structure that normally contains it, which creates a lump that can be seen or felt through the skin's surface. The inguinal region is a weak area in the abdominal wall. It is often the site of an **inguinal hernia**, a rupture or separation of a portion of the inguinal area of the abdominal wall resulting in the protrusion of a part of the small intestine. Hernia is much more common in males than in females because the inguinal canals in males are larger to accommodate the spermatic cord and ilioinguinal nerve. Treatment of hernias most often involves surgery. The organ that protrudes is "tucked" back into the abdominal cavity and the defect in the abdominal muscles is repaired. In addition, a mesh is often applied to reinforce the area of weakness.

Relating muscles to movements: Arrange the muscles in this exhibit according to the following actions on the vertebral column: (1) flexion, (2) lateral flexion, (3) extension, and (4) rotation. The same muscle may be mentioned more than once.

■ **CHECKPOINT**

Which muscles do you contract when you "suck in your tummy," thereby compressing the anterior abdominal wall?

Muscle	Origin	Insertion	Action
Rectus abdominis (REK-tus ab-DOM-in-is; *rect-* = straight, fibers parallel to midline; *abdomin-* = abdomen)	Pubis and pubic symphysis.	Cartilage of fifth to seventh ribs and xiphoid process of sternum.	Flexes vertebral column, and compresses abdomen to aid in defecation, urination, forced expiration, and childbirth.
External oblique (ō-BLĒK; *external* = closer to surface; *oblique* = slanting; here, fibers that are diagonal to midline)	Lower eight ribs.	Crest of ilium and linea alba (a tough connective tissue band that runs from the xiphoid process of the sternum to the pubic symphysis).	Contraction of both external obliques compresses abdomen and flexes vertebral column; contraction of one side alone bends vertebral column laterally and rotates it.
Internal oblique (*internal* = farther from surface)	Ilium, inguinal ligament, and thoracolumbar fascia.	Cartilage of last three or four ribs and linea alba.	Contraction of both internal obliques compresses abdomen and flexes vertebral column; contraction of one side alone bends vertebral column laterally and rotates it.
Transverse abdominis (*transverse* = fibers that are perpendicular to midline)	Ilium, inguinal ligament, lumbar fascia, and cartilages of last six ribs.	Xiphoid process of sternum, linea alba, and pubis.	Compresses abdomen.

Figure 8.16 Muscles of the male anterolateral abdominal wall.

The inguinal ligament separates the thigh from the body wall.

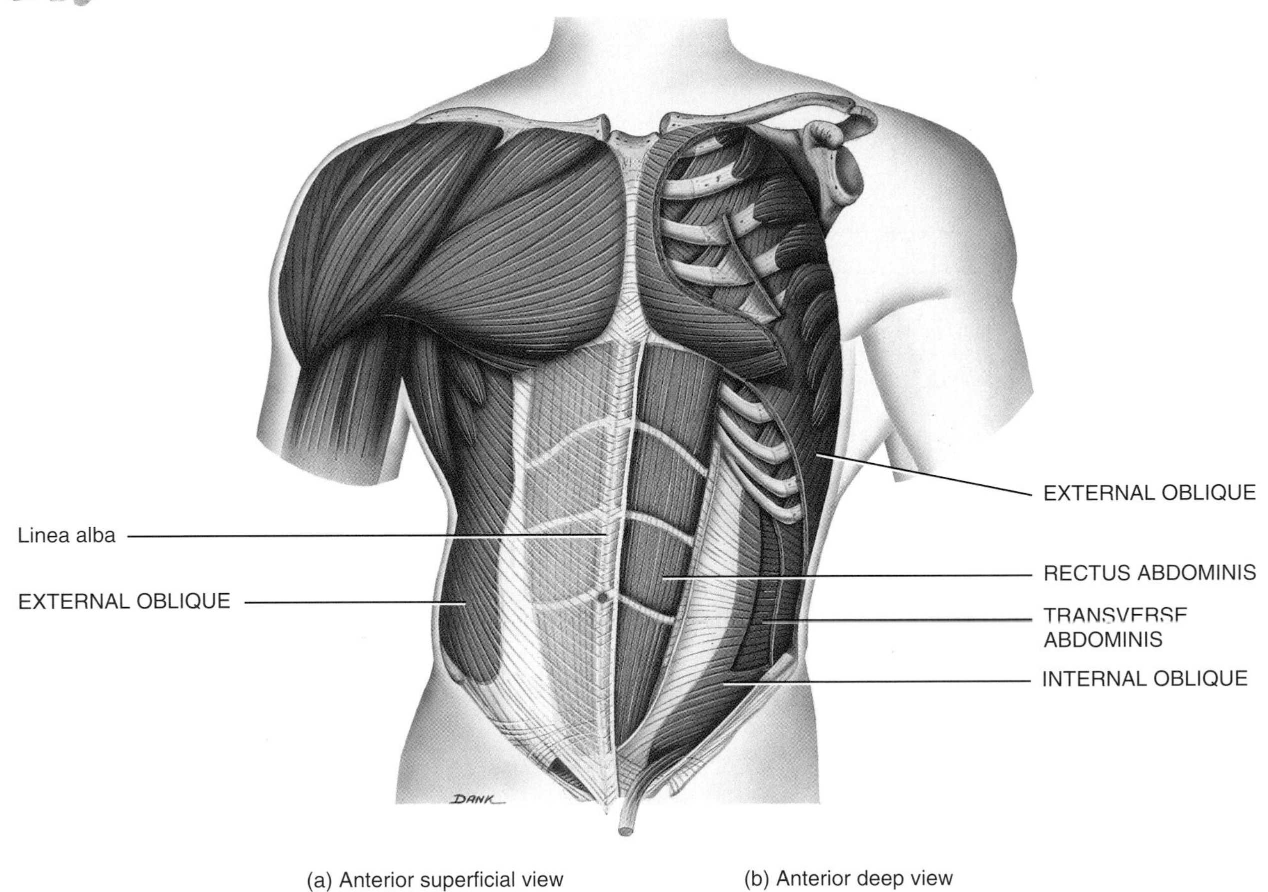

(a) Anterior superficial view (b) Anterior deep view

Which abdominal muscle aids in urination?

Exhibit 8.5 Muscles Used in Breathing *(Figure 8.17)*

OBJECTIVE • Describe the origin, insertion, and action of the muscles used in breathing.

Overview: The muscles described here alter the size of the thoracic cavity so that breathing can occur. Inhalation (breathing in) occurs when the thoracic cavity increases in size, and exhalation (breathing out) occurs when the thoracic cavity decreases in size.

The dome-shaped ***diaphragm*** is the most important muscle that powers quiet breathing. The ***external intercostals***, located between the ribs, assist the diaphragm during quiet breathing. The ***internal intercostals***, also between the ribs, run at right angles to the external intercostals.

Relating muscles to movements: Arrange the muscles in this exhibit according to the following actions on the size of the thorax: (1) increase in vertical dimension, (2) increase in lateral and anteroposterior dimensions, and (3) decrease in lateral and anteroposterior dimensions.

■ **CHECKPOINT**

What situations would require forceful breathing?

Muscle	Origin	Insertion	Action
Diaphragm (DĪ-a-fram; *dia* = across; *-phragm* = wall)	Xiphoid process of the sternum, costal cartilages of the inferior six ribs, lumbar vertebrae, and twelfth rib.	Central tendon.	Contraction of the diaphragm causes it to flatten and increases the vertical (top-to-bottom) dimension of the thoracic cavity, resulting in inhalation; relaxation of the diaphragm causes it to move superiorly and decreases the vertical dimension of the thoracic cavity, resulting in exhalation.
External intercostals (in′-ter-KOS-tals; *external* = closer to surface; *inter-* = between; *costa* = rib)	Inferior border of rib above.	Superior border of rib below.	Contraction elevates the ribs and increases the anteroposterior (front-to-back) and lateral (side-to-side) dimensions of the thoracic cavity, resulting in inhalation; relaxation depresses the ribs and decreases the anteroposterior and lateral dimensions of the thoracic cavity, resulting in exhalation.
Internal intercostals (*internal* = farther from surface)	Superior border of rib below.	Inferior border of rib above.	Contraction draws adjacent ribs together to further decrease the anteroposterior and lateral dimensions of the thoracic cavity during forced exhalation.

Figure 8.17 Muscles used in breathing.

The muscles used in breathing alter the size of the thoracic cavity.

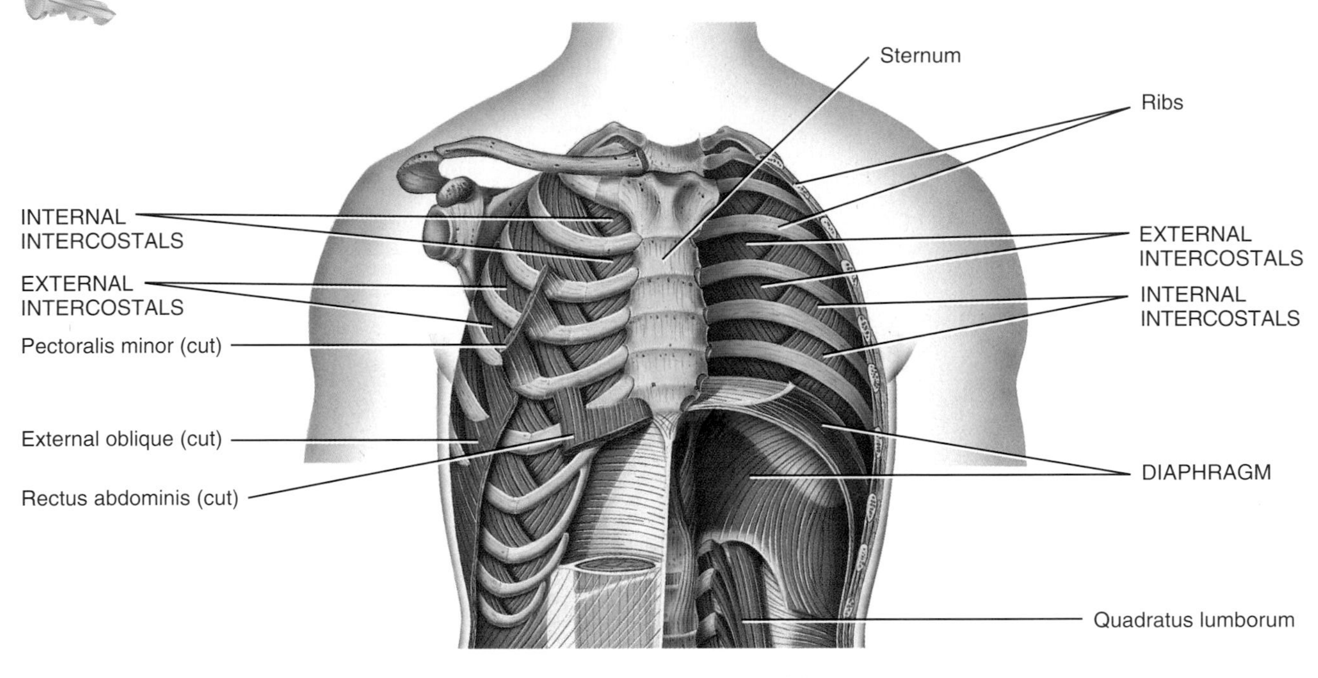

(a) Anterior superficial view (b) Anterior deep view

Which muscles contract during a normal quiet inhalation?

Exhibit 8.6 Muscles That Move the Pectoral (Shoulder) Girdle *(Figure 8.18)*

OBJECTIVE • Describe the origin, insertion, and action of the muscles that move the pectoral girdle.

Overview: Muscles that move the pectoral (shoulder) girdle originate on the axial skeleton and insert on the clavicle or scapula. The main action of the muscles is to hold the scapula in place so that it can function as a stable point of origin for most of the muscles that move the humerus (arm bone).

Relating muscles to movements: Arrange the muscles in this exhibit according to the following actions on the scapula: (1) depression, (2) elevation, (3) lateral and forward movement, and (4) medial and backward movement. The same muscle may be mentioned more than once.

■ CHECKPOINT

Which muscle in this exhibit not only moves the pectoral girdle but also assists in forced inhalation?

Muscle	Origin	Insertion	Action
Pectoralis minor (pek′-tor-Ā-lis; *pect-* = breast, chest, thorax; *minor* = lesser)	Third through fifth ribs.	Scapula.	Depresses scapula, moves it laterally and forward, and rotates it downward (movement of glenoid cavity upward); elevates third through fifth ribs during forced inhalation when scapula is fixed.
Serratus anterior (ser-Ā-tus; *serratus* = saw-toothed; *anterior* = before)	Upper eight or nine ribs.	Scapula.	Moves scapula laterally and forward, and rotates it upward (movement of glenoid cavity downward); elevates ribs when scapula is fixed; known as "boxer's muscle" because it is important in horizontal arm movements such as punching and pushing.
Trapezius (tra-PĒ-zē-us; *trapezi-* = trapezoid-shaped) (See also Figure 8.13b.)	Occipital bone and spines of seventh cervical and all thoracic vertebrae.	Clavicle and scapula.	Elevates clavicle; moves scapula medially and backward, rotates it upward, and elevates or depresses it; extends head.
Levator scapulae (le-VĀ-tor SKA-pū-lē; *levator* = to raise; *scapulae* = of the scapula)	Upper four or five cervical vertebrae.	Scapula.	Elevates scapula and rotates it downward.
Rhomboid major (rom-BOYD); *rhomboid* = rhomboid or diamond-shaped)	Spines of second to fifth thoracic vertebrae.	Scapula.	Elevates scapula, moves it medially and backward, and rotates it downward.

Figure 8.18 Muscles that move the pectoral (shoulder) girdle.

Muscles that move the pectoral girdle originate on the axial skeleton and insert on the clavicle or scapula.

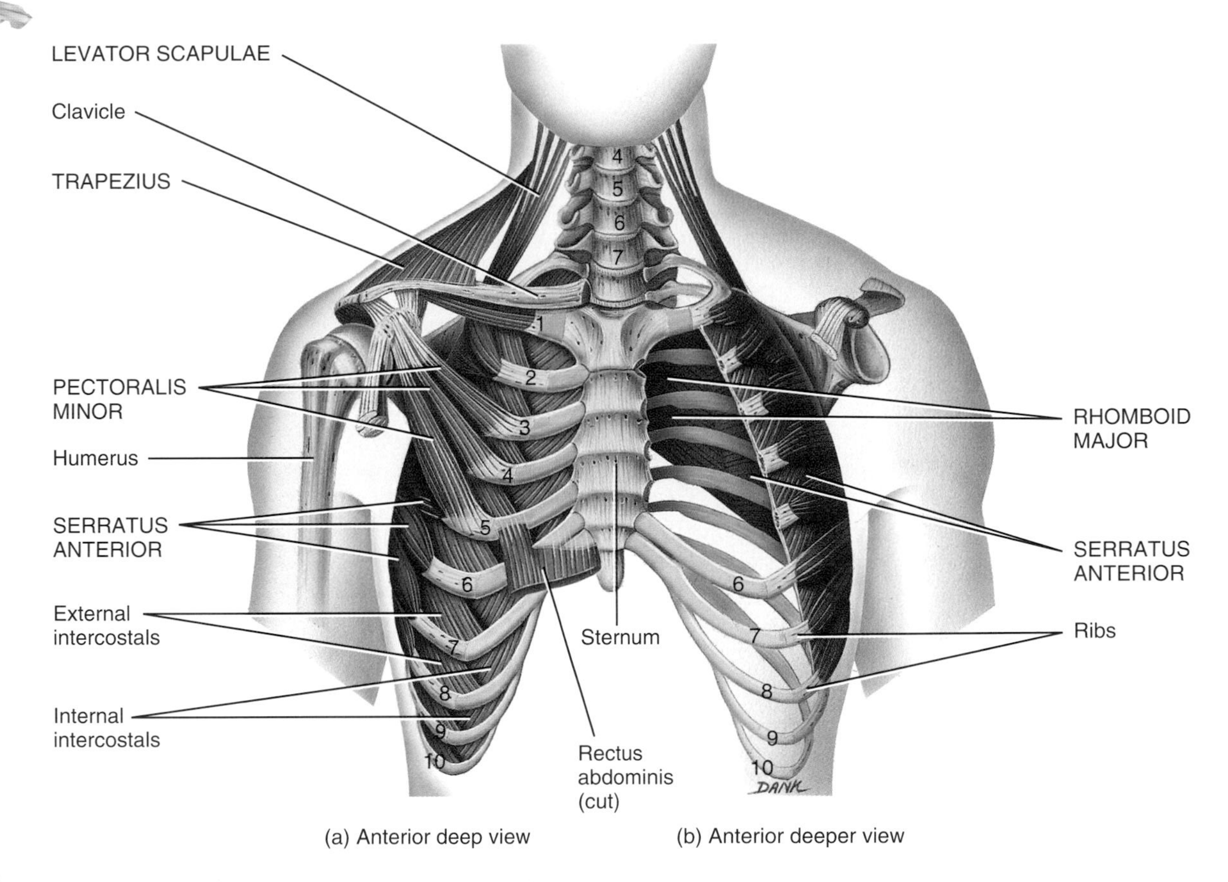

(a) Anterior deep view (b) Anterior deeper view

Which muscles originate on the ribs? The vertebrae?

Exhibit 8.7 Muscles That Move the Humerus (Arm Bone) *(Figure 8.19)*

OBJECTIVE • Describe the origin, insertion, and action of the muscles that move the humerus.

Overview: Of the nine muscles that cross the shoulder joint, only two of them (pectoralis major and latissimus dorsi) do not originate on the scapula.

The strength and stability of the shoulder joint are provided by four deep muscles of the shoulder and their tendons: subscapularis, supraspinatus, infraspinatus, and teres minor. The tendons are arranged in a nearly complete circle around the joint, like the cuff on a shirt sleeve. This arrangement is called the ***rotator cuff***.

One of the most common causes of shoulder pain and dysfunction in athletes is known as **impingement syndrome**. The repetitive movement of the arm over the head that is common in baseball, overhead racquet sports, lifting weights over the head, spiking a volleyball, and swimming puts these athletes at risk for developing this syndrome. It may also be caused by a direct blow or stretch injury. Continual pinching of the supraspinatus tendon as a result of overhead motions causes it to become inflamed and results in pain. If movement is continued despite the pain, the tendon may degenerate near the attachment to the humerus and ultimately may tear away from the bone (rotator cuff injury). Treatment consists of resting the injured tendons, strengthening the shoulder through exercise, and surgery if the injury is particularly severe.

Relating muscles to movements: Arrange the muscles in this exhibit according to the following actions on the humerus at the shoulder joint: (1) flexion, (2) extension, (3) abduction, (4) adduction, (5) medial rotation, and (6) lateral rotation. The same muscle may be mentioned more than once.

■ **CHECKPOINT**

What is the rotator cuff?

Muscle	Origin	Insertion	Action
Pectoralis major (pek′-tō-RĀ-lis; *pector-* = chest; *major* = greater) (See also Figure 8.13a.)	Clavicle, sternum, cartilages of second to sixth ribs.	Humerus.	Adducts and rotates arm medially at shoulder joint; flexes and extends arm at shoulder joint.
Latissimus dorsi (la-TIS-i-mus DOR-sī; *latissimus* = widest; *dorsi* = of the back) (See also Figure 8.13b.)	Spines of lower six thoracic vertebrae, lumbar vertebrae, sacrum and ilium, lower four ribs.	Humerus.	Extends, adducts, and rotates arm medially at shoulder joint; draws arm downward and backward.
Deltoid (DEL-toyd; *deltoid* = triangularly shaped) (See also Figure 8.13a, b.)	Clavicle and scapula.	Humerus.	Abducts, flexes, extends, and rotates arm at shoulder joint.
Subscapularis (sub-scap′-ū-LĀ-ris; *sub-* = below; *scapularis* = scapula)	Scapula.	Humerus.	Rotates arm medially at shoulder joint.
Supraspinatus (soo′-pra-spi-NĀ-tus; *supra-* = above; *spina-* = spine of scapula)	Scapula.	Humerus.	Assists deltoid muscle in abducting arm at shoulder joint.
Infraspinatus (in′-fra-spi-NĀ-tus; *infra-* = below) (See Figure 8.13b.)	Scapula.	Humerus.	Rotates arm laterally and adducts arm at shoulder joint.
Teres major (TE-rēz) (*teres* = long and round) (see Figure 8.13b.)	Scapula.	Humerus.	Extends arm at shoulder joint; assists in adduction and rotation of arm medially at shoulder joint.
Teres minor (See Figure 8.13b.)	Scapula.	Humerus.	Rotates arm laterally, extends and adducts arm at shoulder joint.
Coracobrachialis (kor′-a-kō-brā-kē-Ā-lis; *coraco* = coracoid process; *brachi-* = arm)	Scapula.	Humerus.	Flexes and adducts arm at shoulder joint.

Figure 8.19 Muscles that move the humerus (arm bone).

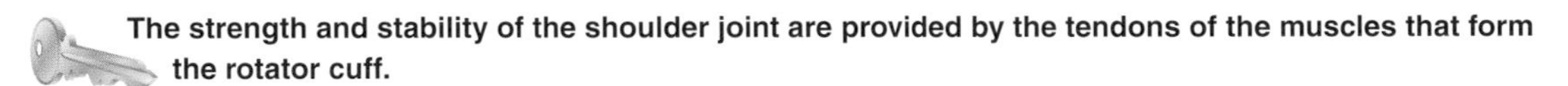

The strength and stability of the shoulder joint are provided by the tendons of the muscles that form the rotator cuff.

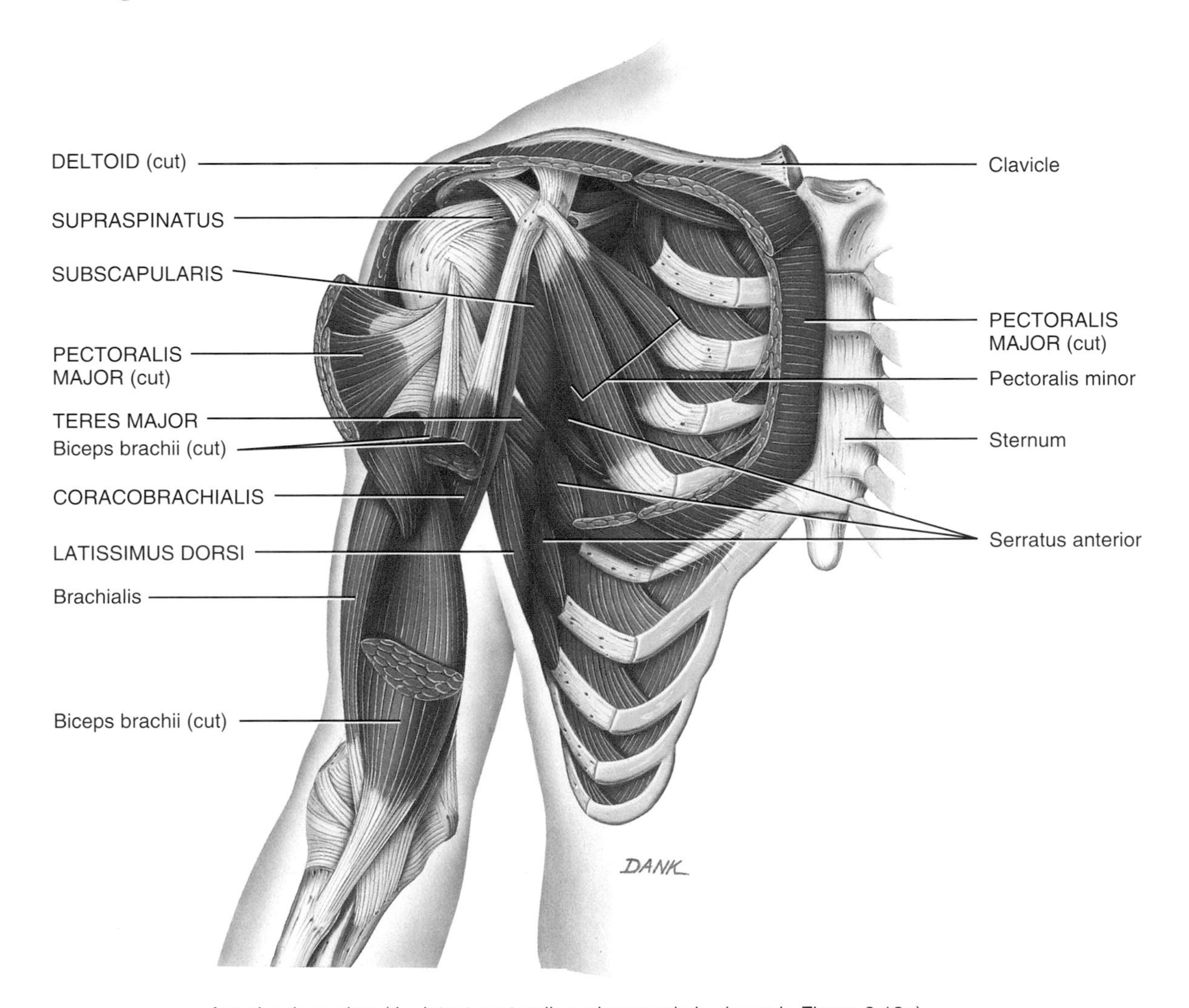

Anterior deep view (the intact pectoralis major muscle is shown in Figure 8.13a)

Of the nine muscles that cross the shoulder joint, which two muscles do not originate on the scapula?

Exhibit 8.8 Muscles That Move the Radius and Ulna (Forearm Bones) *(Figure 8.20)*

OBJECTIVE • Describe the origin, insertion, and action of the muscles that move the radius and ulna.

Overview: Recall that the elbow joint is a hinge joint, capable only of flexion and extension. The biceps brachii, brachialis, and brachioradialis are flexors of the elbow joint; the triceps brachii is an extensor. Other muscles that move the radius and ulna are concerned with supination and pronation. In the limbs, functionally related skeletal muscles and their associated blood vessels and nerves are grouped together by deep fascia into regions called ***compartments***. Thus, in the arm, the biceps brachii, brachialis, and coracobrachialis muscles constitute the ***anterior (flexor) compartment***; the triceps brachii muscle forms the ***posterior (extensor) compartment***.

Relating muscles to movements: Arrange the muscles in this exhibit according to the following actions: (1) flexion and extension of the elbow joint; (2) supination and pronation of the forearm; and (3) flexion and extension of the humerus. The same muscle may be mentioned more than once.

■ CHECKPOINT

Which muscles are in the anterior and posterior compartments of the arm?

Muscle	Origin	Insertion	Action
Biceps brachii (BĪ-ceps BRĀ-kē-ī; *biceps* = two heads of origin; *brachi* = of the arm)	Scapula.	Radius.	Flexes and supinates forearm at elbow joint; flexes arm at shoulder joint.
Brachialis (brā′-kē-Ā-lis)	Humerus.	Ulna.	Flexes forearm at elbow joint.
Brachioradialis (brā′-kē-ō-rā′-dē-Ā-lis; *radi-* = radius) (See Figure 8.21a.)	Humerus.	Radius.	Flexes forearm at elbow joint.
Triceps brachii (TRĪ-ceps BRĀ-kē-ī; *triceps* = three heads of origin)	Scapula and humerus.	Ulna.	Extends forearm at elbow joint; extends arm at shoulder joint.
Supinator (SOO-pi-nā-tor; *supination* = turning palm forward). (Not illustrated.)	Humerus and ulna.	Radius.	Supinates forearm.
Pronator teres (PRŌ-nā-tor TE-rēz; *pronation* = turning palm backward) (See Figure 8.21a.)	Humerus and ulna.	Radius.	Pronates forearm.

Figure 8.20 Muscles that move the radius and ulna (forearm bones).

The anterior arm muscles flex the forearm, but the posterior arm muscles extend it.

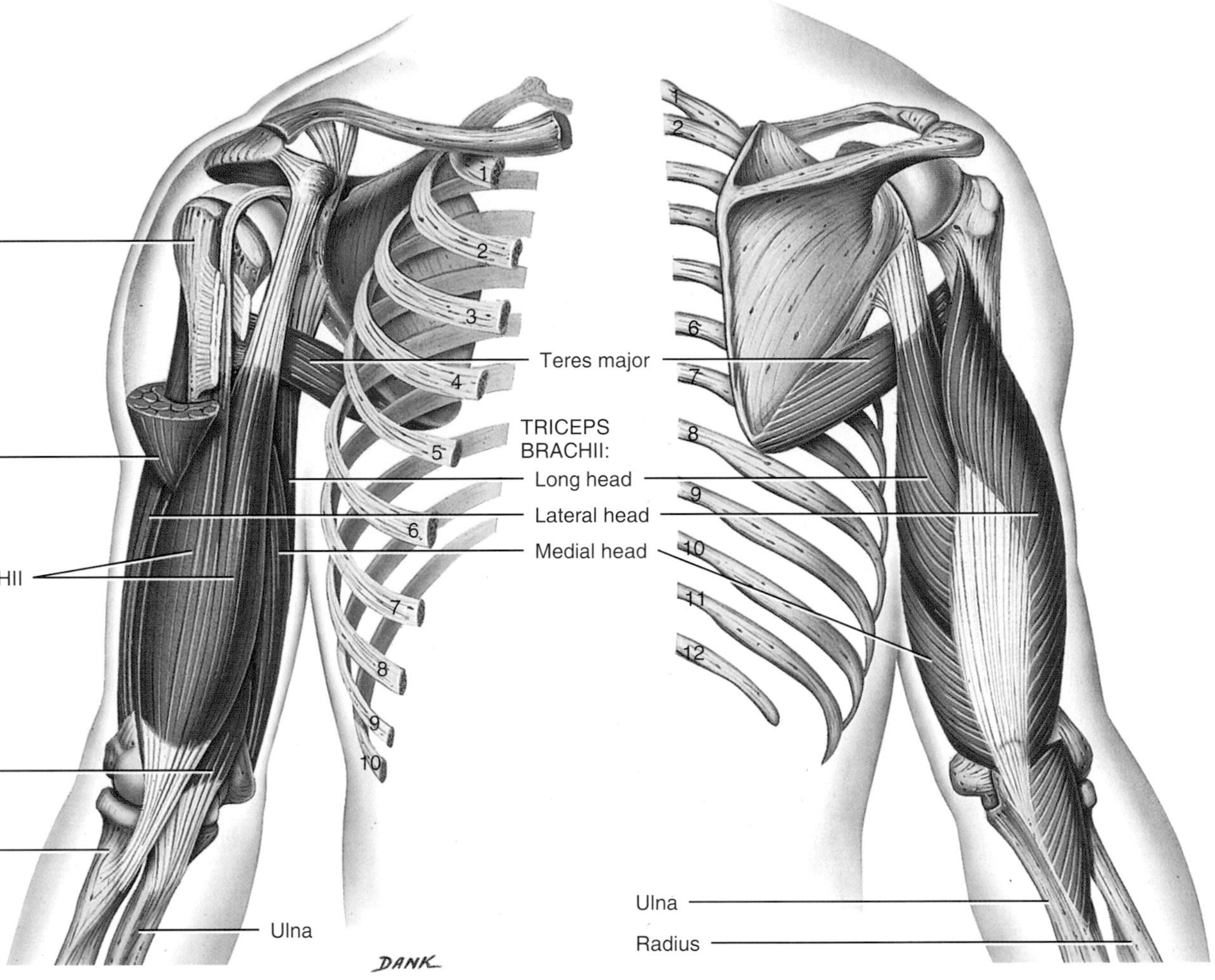

Humerus
Deltoid (cut)
BICEPS BRACHII
BRACHIALIS
Radius
Ulna
Teres major
TRICEPS BRACHII:
Long head
Lateral head
Medial head
Ulna
Radius

(a) Anterior view

(b) Posterior view

What is a compartment?

Exhibit 8.9 Muscles That Move the Wrist, Hand, and Fingers *(Figure 8.21)*

OBJECTIVE • **Describe the origin, insertion, and action of the muscles that move the wrist, hand, and fingers.**

Overview: Muscles that move the wrist, hand, and fingers are located on the forearm and are many and varied. Their names for the most part give some indication of their origin, insertion, or action. On the basis of location and function, the muscles are divided into two compartments. The ***anterior (flexor) compartment muscles*** originate on the humerus and typically insert on the carpals, metacarpals, and phalanges. The bellies of these muscles form the bulk of the proximal forearm. The ***posterior (extensor) compartment muscles*** arise on the humerus and insert on the metacarpals and phalanges.

The tendons of the muscles of the forearm that attach to the wrist or continue into the hand, along with blood vessels and nerves, are held close to bones by fascia. The tendons are also surrounded by tendon sheaths. At the wrist, the deep fascia is thickened into fibrous bands called ***retinacula*** (re-ti-NAK-ū-la; retinacul = a holdfast; singular is *retinaculum*). The ***flexor retinaculum*** is located over the palmar surface of the carpal bones. Through it pass the long flexor tendons of the fingers and wrist and the median nerve. The ***extensor retinaculum*** is located over the dorsal surface of the carpal bones. Through it pass the extensor tendons of the wrist and fingers.

The **carpal tunnel** is a narrow passageway formed anteriorly by the flexor retinaculum and posteriorly by the carpal bones. Through this tunnel pass the median nerve, the most superficial structure, and the long flexor tendons for the digits (Figure 8.21c). Structures within the carpal tunnel, especially the median nerve, are vulnerable to compression, and the resulting condition is called **carpal tunnel syndrome**. Compression of the median nerve leads to sensory changes over the lateral side of the hand and muscle weakness in the thenar eminence. This results in pain, numbness, and tingling of the fingers. The condition may be caused by inflammation of the digital tendon sheaths, fluid retention, excessive exercise, infection, trauma, and/or repetitive activities that involve flexion of the wrist, such as keyboarding, cutting hair, and playing a piano. Treatment may involve the use of nonsteroidal anti-inflammatory drugs (such as ibuprofen or aspirin), wearing a wrist splint, corticosteroid injections, or surgery to cut the flexor retinaculum and release pressure on the median nerve.

Relating muscles to movements: Arrange the muscles in this exhibit according to the following actions: (1) flexion, extension, abduction, and adduction of the wrist joint and (2) flexion and extension of the phalanges. The same muscle may be mentioned more than once.

■ **CHECKPOINT**

Which muscles and actions of the wrist, hand, and digits are used when writing?

Muscle	Origin	Insertion	Action
Anterior (Flexor) Compartment			
Flexor carpi radialis (FLEK-sor KAR-pē rā′-dē-Ā-lis; *flexor* = decreases angle at joint; *carpus* = wrist; *radi-* = radius)	Humerus.	Second and third metacarpals.	Flexes and abducts hand at wrist joint.
Flexor carpi ulnaris (ul-NAR-is; *ulnar-* = ulna)	Humerus and ulna.	Pisiform, hamate, and fifth metacarpal.	Flexes and adducts hand at wrist joint.
Palmaris longus (pal-MA-ris LON-gus; *palma* = palm; *longus* = long)	Humerus.	Flexor retinaculum.	Weakly flexes hand at wrist joint.
Flexor digitorum superficialis (soo′-per- fish′-ē-A-lis; digit = finger or toe; *superficialis* = closer to surface)	Humerus, ulna, and radius.	Middle phalanges.	Flexes hand at wrist joint; flexes phalanges of each finger.
Flexor digitorum profundus (di′-ji-TOR-um pro-FUN-dus *profundus* = deep). (Not illustrated.)	Ulna.	Bases of distal phalanges.	Flexes hand at wrist joint; flexes phalanges of each finger.
Posterior (Extensor) Compartment			
Extensor carpi radialis longus (eks-TEN-sor; *extensor* = increases angle at joint)	Humerus.	Second metacarpal.	Extends and abducts hand at wrist joint.
Extensor carpi ulnaris	Humerus and ulna.	Fifth metacarpal.	Extends and adducts hand at wrist joint.
Extensor digitorum	Humerus.	Second through fifth phalanges.	Extends hand at wrist joint; extends phalanges.

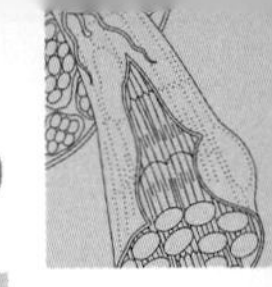

Figure 8.21 Muscles that move the wrist, hand, and fingers.

The anterior compartment muscles function as flexors, and the posterior compartment muscles function as extensors.

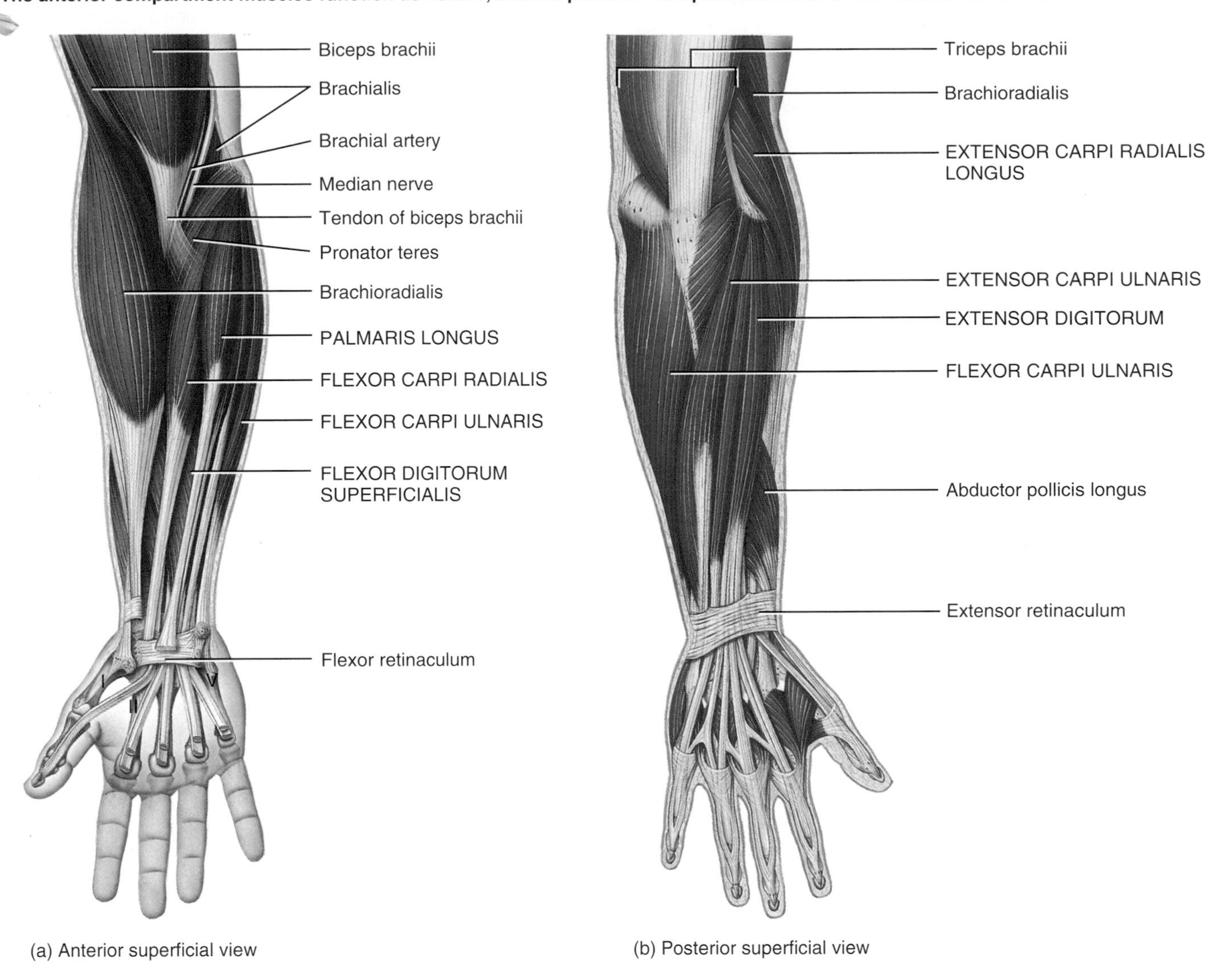

(a) Anterior superficial view

(b) Posterior superficial view

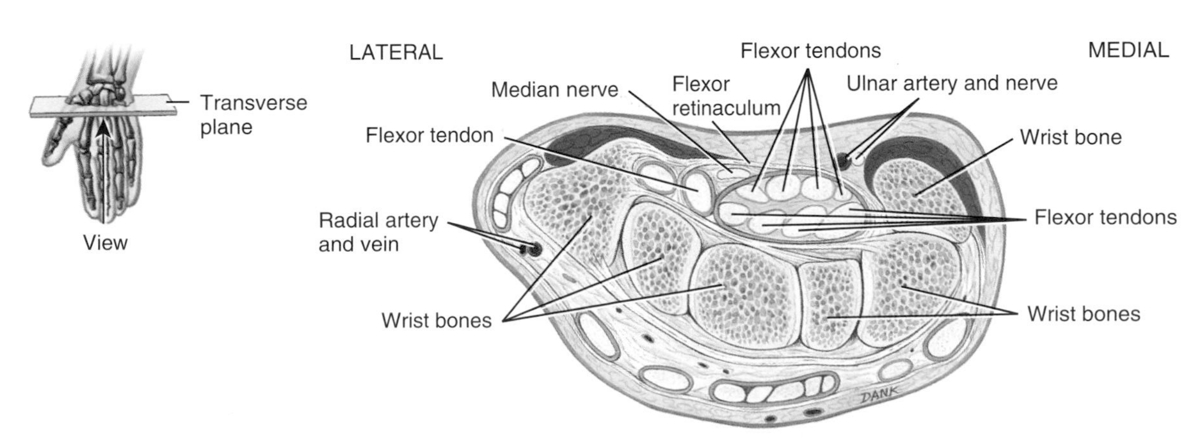

(c) Inferior view of transverse section

Which nerve is associated with the flexor retinaculum?

Exhibit 8.10 Muscles That Move the Vertebral Column (Backbone) *(Figure 8.22)*

OBJECTIVE • Describe the origin, insertion, and action of the muscles that move the vertebral column.

Overview: The ***erector spinae muscles*** form the largest muscular mass of the back, forming a prominent bulge on either side of the vertebral column (Figure 8.22). It consists of three groups of overlapping muscles: ***iliocostalis group*** (il′-ē-ō-kos-TĀ-lis), ***longissimus group*** (lon′-JI-si-mus), and ***spinalis group*** (spi-NĀ-lis). Other muscles that move the vertebral column include the ***sternocleidomastoid, quadratus lumborum, rectus abdominis*** (see Exhibit 8.4), ***psoas major*** (see Exhibit 8.11), and ***iliacus*** (see Exhibit 8.11).

Full flexion at the waist, as in touching your toes, overstretches the erector spinae muscles and muscles that are overstretched cannot contract effectively. Straightening up from such a position is therefore initiated by the hamstring muscles on the back of the thigh and the gluteus maximus muscles of the buttocks. The erector spinae muscles join in as the degree of flexion decreases. **Improperly lifting a heavy weight**, however, can strain the erector spinae muscles. The result can be painful muscle spasms, tearing of tendons and ligaments of the lower back, and rupturing of intervertebral discs. The lumbar muscles are adapted for maintaining posture, not for lifting. This is why it is important to kneel and use the powerful extensor muscles of the thighs and buttocks while lifting a heavy load.

Relating muscles to movements: Arrange the muscles in this exhibit according to the following actions on the vertebral column: (1) flexion and (2) extension.

■ **CHECKPOINT**

Which groups of muscles make up the erector spinae?

Muscle	Origin	Insertion	Action
Erector spinae (e-REK-tor SPI-nē; *erector* = raise; *spinae* = of the spine) (iliocostalis group, longissimus group, and spinalis group)	All ribs plus cervical, thoracic, and lumbar vertebrae.	Occipital bone, temporal bone, ribs, and vertebrae.	Extends head; extends and laterally flexes vertebral column.
Sternocleidomastoid (ster′-nō-klī-dō-MAS-toid; *sternum* = breastbone; *cleido-* = clavicle; *mastoid* = mastoid process of temporal bone) (See Figure 8.13b.)	Sternum and clavicle.	Temporal bone.	Contractions of both muscles flex cervical part of the vertebral column and flex the head; contraction of one muscle rotates head toward side opposite contracting muscle.
Quadratus lumborum (kwod-RĀ-tus lum-BOR-um; *quadratus* = four-sided; *lumbo* = lumbar region). (See Figure 8.17b.)	Ilium.	Twelfth rib and upper four lumbar vertebrae.	Contractions of both muscles extend lumbar part of the vertebral column; contraction of one muscle flexes lumbar part of vertebral column.

Figure 8.22 Major muscles that move the vertebral column (backbone).

The erector spinae muscles extend the vertebral column.

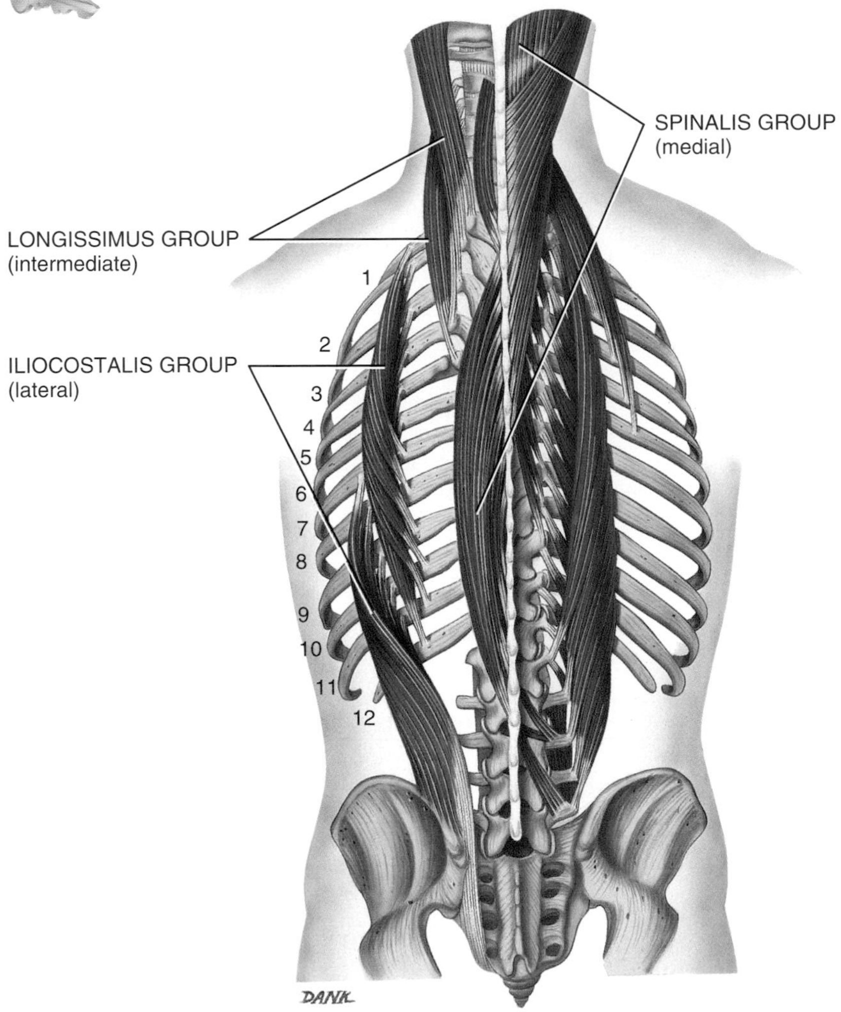

Posterior view of erector spinae muscles

Which muscles constitute the erector spinae?

Exhibit 8.11 Muscles That Move the Femur (Thigh Bone) *(Figure 8.23)*

OBJECTIVE • Describe the origin, insertion, and action of the muscles that move the femur.

Overview: Muscles of the lower limbs are larger and more powerful than those of the upper limbs to provide stability, locomotion, and maintenance of posture. In addition, muscles of the lower limbs often cross two joints and act equally on both. The majority of muscles that act on the femur originate on the pelvic (hip) girdle and insert on the femur. The anterior muscles are the psoas major and iliacus, together referred to as the ***iliopsoas*** (il′-ē-ō-SŌ-as). The remaining muscles (except for the pectineus, adductors, and tensor fasciae latae) are posterior muscles. Technically, the pectineus and adductors are components of the medial compartment of the thigh, but they are included in this exhibit because they act on the thigh. The tensor fasciae latae is laterally placed. The ***fascia lata*** is a deep fascia of the thigh that encircles the entire thigh. It is well developed laterally, where together with the tendons of the gluteus maximus and tensor fasciae latae it forms a structure called the ***iliotibial tract***. The tract inserts into the lateral condyle of the tibia.

The major muscles of the inner thigh function to move the legs medially. This muscle group is important in activities such as sprinting, hurdling, and horseback riding. A rupture or tear of one or more of these muscles can cause a **groin pull**. Groin pulls most often occur during sprinting or twisting, or from kicking a solid, perhaps stationary object. Symptoms of a groin pull may be sudden, or may not surface until the day after the injury, and include sharp pain in the inguinal region, swelling, bruising, or inability to contract the muscles. As with most strain injuries, treatment involves RICE therapy, which stands for *R*est, *I*ce, *C*ompression, and *E*levation. Ice should be applied immediately, and the injured part should be elevated and rested. An elastic bandage should be applied, if possible, to compress the injured tissue.

Relating muscles to movements: Arrange the muscles in this exhibit according to the following actions on the thigh at the hip joint: (1) flexion, (2) extension, (3) abduction, (4) adduction, (5) medial rotation, and (6) lateral rotation. The same muscle may be mentioned more than once.

■ **CHECKPOINT**

What forms the iliotibial tract?

Muscle	Origin	Insertion	Action
Psoas major (SŌ-as; *psoa* = a muscle of loin)	Lumbar vertebrae.	Femur.	Flexes and rotates thigh laterally at the hip joint; flexes vertebral column.
Iliacus (il′-ē-AK-us; *iliac* = ilium)	Ilium.	With psoas major into femur.	Flexes and rotates thigh laterally at the hip joint; flexes vertebral column.
Gluteus maximus (GLOO-tē-us MAK-si-mus; *glute-* = buttock; *maximus* = largest) (See also Figure 8.13b.)	Ilium, sacrum, coccyx, and aponeurosis of sacrospinalis.	Iliotibial tract of fascia lata and femur.	Extends and rotates thigh laterally at the hip joint.
Gluteus medius (MĒ-dē-us; *medi-* = middle) (See also Figure 8.13b.)	Ilium.	Femur.	Abducts and rotates thigh medially at the hip joint.
Tensor fasclae latae (TEN-sor FA-shē-ē LĀ-tē; *tensor* = makes tense; *fasciae-* = of the band; *lat-* = wide)	Ilium.	Tibia by means of the iliotibial tract.	Flexes and abducts thigh at the hip joint.
Adductor longus (LONG-us; *adductor* = moves part closer to midline; *longus* = long)	Pubis and pubic symphysis.	Femur.	Adducts, medially rotates, and flexes thigh at the hip joint.
Adductor magnus (MAG-nus; *magnus* = large)	Pubis and ischium.	Femur.	Adducts, flexes, medially rotates and extends thigh (anterior part flexes, posterior part extends) at the hip joint.
Piriformis (pir-i-FOR-mis; *piri-* = pear; *form-* = shape)	Sacrum.	Femur.	Rotates thigh laterally and abducts it at the hip joint.
Pectineus (pek-TIN-ē-us; *pectin-* = comb-shaped)	Pubis.	Femur.	Flexes and adducts thigh at the hip joint.

Figure 8.23 **Muscles that move the femur (thigh bone).**

Most muscles that move the femur originate on the pelvic (hip) girdle and insert on the femur.

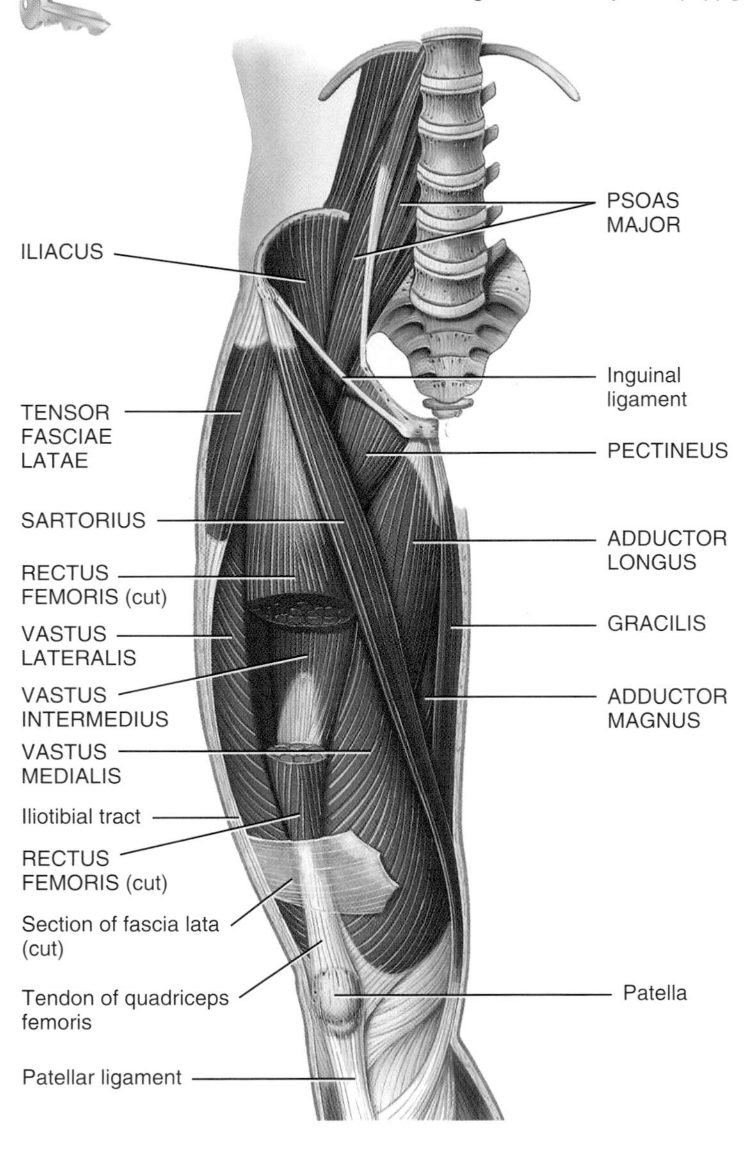

(a) Anterior superficial view

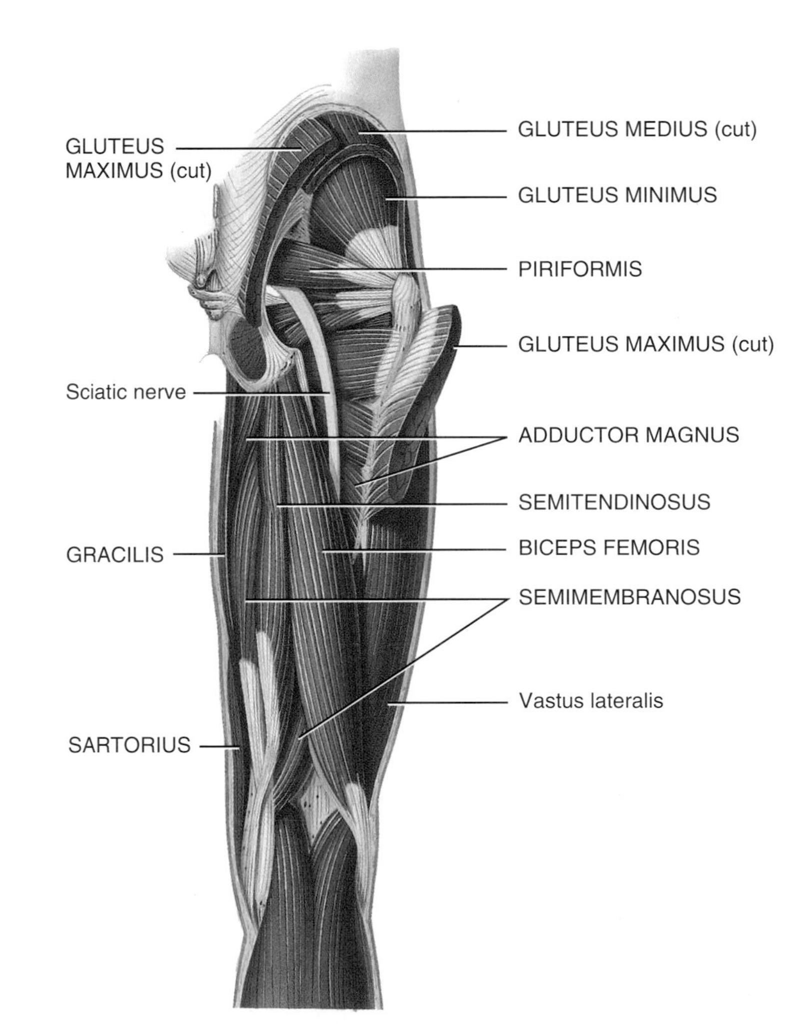

(b) Posterior superficial view

Which muscles are part of the quadriceps femoris? The hamstrings?

Exhibit 8.12 Muscles That Move the Femur (Thigh Bone) and Tibia and Fibula (Leg Bones) *(See Figure 8.23)*

OBJECTIVE • Describe the origin, insertion, and action of the muscles that move the femur, tibia, and fibula.

Overview: The muscles that move the femur, tibia, and fibula originate in the hip and thigh and are separated into compartments by deep fascia. The ***medial (adductor) compartment*** is so named because its muscles adduct the thigh. The adductor magnus, adductor longus, and pectineus muscles, components of the medial compartment, are included in Exhibit 8.11 because they act on the femur. The gracilis, the other muscle in the medial compartment, not only adducts the thigh but also flexes the leg. For this reason, it is included in this exhibit.

The ***anterior (extensor) compartment*** is so designated because its muscles act to extend the leg at the knee joint, and some also flex the thigh at the hip joint. It is composed of the quadriceps femoris and sartorius muscles. The quadriceps femoris muscle is the largest muscle in the body but has four distinct parts, usually described as four separate muscles (rectus femoris, vastus lateralis, vastus medialis, and vastus intermedius). The common tendon for the four muscles is the ***quadriceps tendon***, which attaches to the patella. The tendon continues below the patella as the ***patellar ligament*** and attaches to the tibial tuberosity. The sartorius muscle is the longest muscle in the body, extending from the ilium of the hip bone to the medial side of the tibia. It moves both the thigh and the leg.

The ***posterior (flexor) compartment*** is so named because its muscles flex the leg (but also extend the thigh). Included are the hamstrings (biceps femoris, semitendinosus, and semimembranosus), so named because their tendons are long and string-like in the popliteal area.

A strain or partial tear of the proximal hamstring muscles is referred to as **"pulled hamstrings"** or **hamstring strains**. They are common sports injuries in individuals who run very hard and/or are required to perform quick starts and stops. Sometimes the violent muscular exertion required to perform a feat tears off part of the tendinous origins of the hamstrings, especially the biceps femoris, from the ischial tuberosity. This injury is usually accompanied by a *contusion* (bruising) and tearing of some of the muscle fibers and rupture of blood vessels, producing a *hematoma* (collection of blood) and pain. Adequate training with good balance between the quadriceps femoris and hamstrings and stretching exercises before running or competing are important in preventing this injury.

Relating muscles to movements: Arrange the muscles in this exhibit according to the following actions on the thigh at the hip joint: (1) abduction, (2) adduction, (3) lateral rotation, (4) flexion, and (5) extension; and according to the following action on the leg: (1) flexion and (2) extension. The same muscle may be mentioned more than once.

■ CHECKPOINT

Which muscle tendons form the medial and lateral borders of the popliteal fossa?

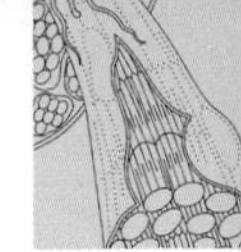

Muscle	Origin	Insertion	Action
Medial (Adductor) Compartment			
Adductor magnus (MAG-nus) **Adductor longus** (LONG-us) **Pectineus** (pek-TIN-ē-us)	See Exhibit 8.11.		
Gracilis (GRAS-i-lis; *gracilis* = slender)	Pubic symphysis.	Tibia.	Adducts and medially rotates thigh at hip joint; flexes leg at knee joint.
Medial (Extensor) Compartment			
Quadriceps femoris (KWOD-ri-seps FEM-or-is; *quadriceps* = four heads of origin; *femoris* = femur)			
Rectus femoris (REK-tus FEM-or-is; *rectus* = straight; here, fibers run parallel to midline)	Ilium.	Patella by means of quadriceps tendon and then tibial tuberosity by means of patellar ligament.	All four heads extend leg at knee joint; rectus femoris muscle alone also flexes thigh at hip joint.
Vastus lateralis (VAS-tus lat′-er-Ā-lis; *vast-* = large; *lateralis* = lateral)	Femur.		
Vastus medialis (mē′-dē-Ā-lis; *medialis* = medial)	Femur.		
Vastus intermedius (in′-ter-MĒ-dē-us; *intermedius* = middle)	Femur.		
Sartorius (sar-TOR-ē-us; *sartor-* = tailor; refers to cross-legged position of tailors) Longest muscle in the body.	Ilium.	Tibia.	Flexes leg at knee joint; flexes, abducts, and laterally rotates thigh at hip joint, thus crossing leg.
Posterior (Flexor) Compartment			
Hamstrings			
Biceps femoris (BĪ-ceps FEM-or-is; *biceps* = two heads of origin)	Ischium and femur.	Fibula and tibia.	Flexes leg at knee joint; extends thigh at hip joint.
Semitendinosus (sem′-ē-TEN-di-nō′-sus; *semi-* = half; *tendo-* = tendon)	Ischium.	Tibia.	Flexes leg at knee joint; extends thigh at hip joint.
Semimembranosus (sem′-ē-MEM-bra-nō′-sus; *membran-* = membrane)	Ischium.	Tibia.	Flexes leg at knee joint; extends thigh at hip joint.

Exhibit 8.13 Muscles That Move the Foot and Toes *(Figure 8.24)*

OBJECTIVE • Describe the origin, insertion, and action of the muscles that move the foot and toes.

Overview: Muscles that move the foot and toes are located in the leg. The muscles of the leg, like those of the thigh, are divided into three compartments by deep fascia. The ***anterior compartment*** consists of muscles that dorsiflex the foot. In a situation like that at the wrist, the tendons of the muscles of the anterior compartment are held firmly to the ankle bones by thickenings of deep fascia called the ***superior extensor retinaculum*** and ***inferior extensor retinaculum***. The ***lateral compartment*** contains muscles that plantar flex and evert the foot. The ***posterior compartment*** consists of superficial and deep muscles. The superficial muscles (gastrocnemius and soleus) share a common tendon of insertion, the calcaneal (Achilles) tendon, the strongest tendon of the body.

Shinsplint syndrome, or simply **shinsplints**, refers to pain or soreness along the medial, distal two-thirds of the tibia. It may be caused by tendinitis of the tibialis anterior or toe flexors, inflammation of the periosteum around the tibia, or stress fractures of the tibia. The tendinitis usually occurs when poorly conditioned runners run on hard or banked surfaces with poorly supportive running shoes or walking or running up and down hills. The condition may also occur as a result of vigorous activity of the legs following a period of relative inactivity. The muscles in the anterior compartment (mainly the tibialis anterior) can be strengthened to balance the stronger posterior compartment muscles.

Relating muscles to movements: Arrange the muscles in this exhibit according to the following actions on the foot: (1) dorsiflexion, (2) plantar flexion, (3) inversion, and (4) eversion; and according to the following actions on the toes: (1) flexion and (2) extension. The same muscle may be mentioned more than once.

■ **CHECKPOINT**

What is the function of the superior and inferior extensor retinaculum?

Figure 8.24 Muscles that move the foot and toes.

The superficial muscles of the posterior compartment share a common tendon of insertion, the calcaneal (Achilles) tendon, that inserts into the calcaneal bone of the ankle.

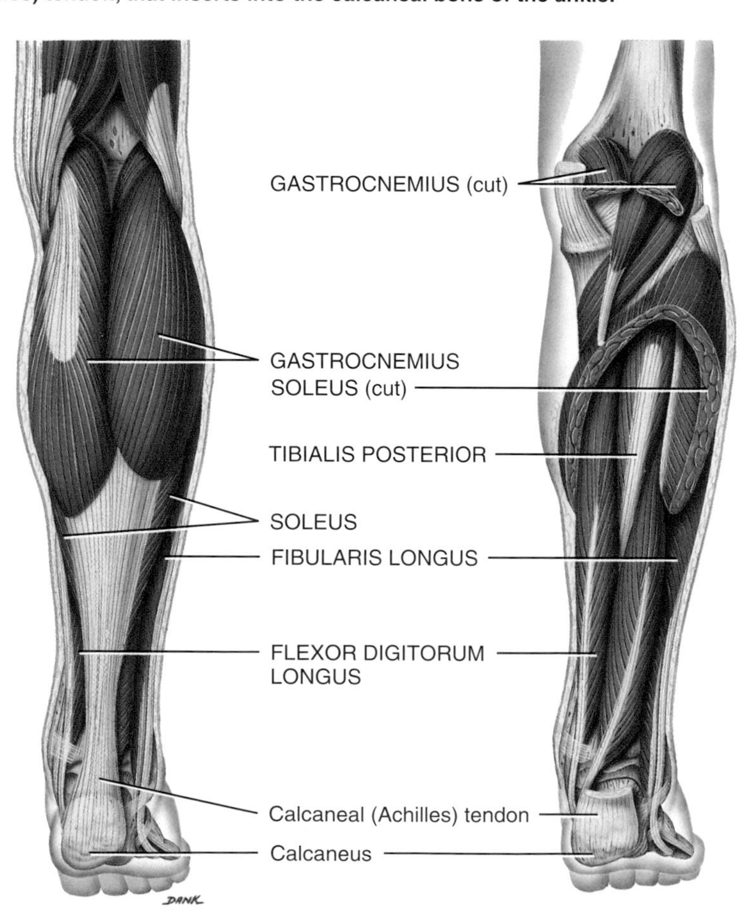

(a) Posterior superficial view

(b) Posterior deep view

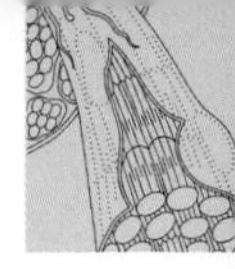

Muscle	Origin	Insertion	Action
Anterior Compartment			
Tibialis anterior (tib′-ē-Ā-lis; *tibialis* = tibia; *anterior* = front)	Tibia.	First metatarsal and first cuneiform.	Dorsiflexes and inverts foot.
Extensor digitorum longus (eks-TEN-sor di′-ji-TOR-um LON-gus; *extensor* = increases angle at joint; *digitorum* = finger or toe; *longus* = long)	Tibia and fibula.	Middle and distal phalanges of four outer toes.	Dorsiflexes and everts foot; extends toes.
Lateral Compartment			
Fibularis (Peroneus) longus (fib-ū-LAR-is LON-gus)	Fibula and tibia.	First metatarsal and first cuneiform.	Plantar flexes and everts foot.
Posterior Compartment			
Gastrocnemius (gas′-trok-NĒ-mē-us; *gastro-* = belly; *-cnem* = leg)	Femur.	Calcaneus by means of calcaneal (Achilles) tendon.	Plantar flexes foot; flexes leg at knee joint.
Soleus (SŌ-lē-us; *soleus* = a type of flatfish)	Fibula and tibia.	Calcaneus by means of calcaneal (Achilles) tendon.	Plantar flexes foot.
Tibialis posterior (*posterior* = back)	Tibia and fibula.	Second, third, and fourth metatarsals; navicular; all three cuneiforms, and cuboid.	Plantar flexes and inverts foot.
Flexor digitorum longus (FLEK-sor; *flexor* = decreases angle at joint)	Tibia.	Distal phalanges of four outer toes.	Plantar flexes foot; flexes toes.

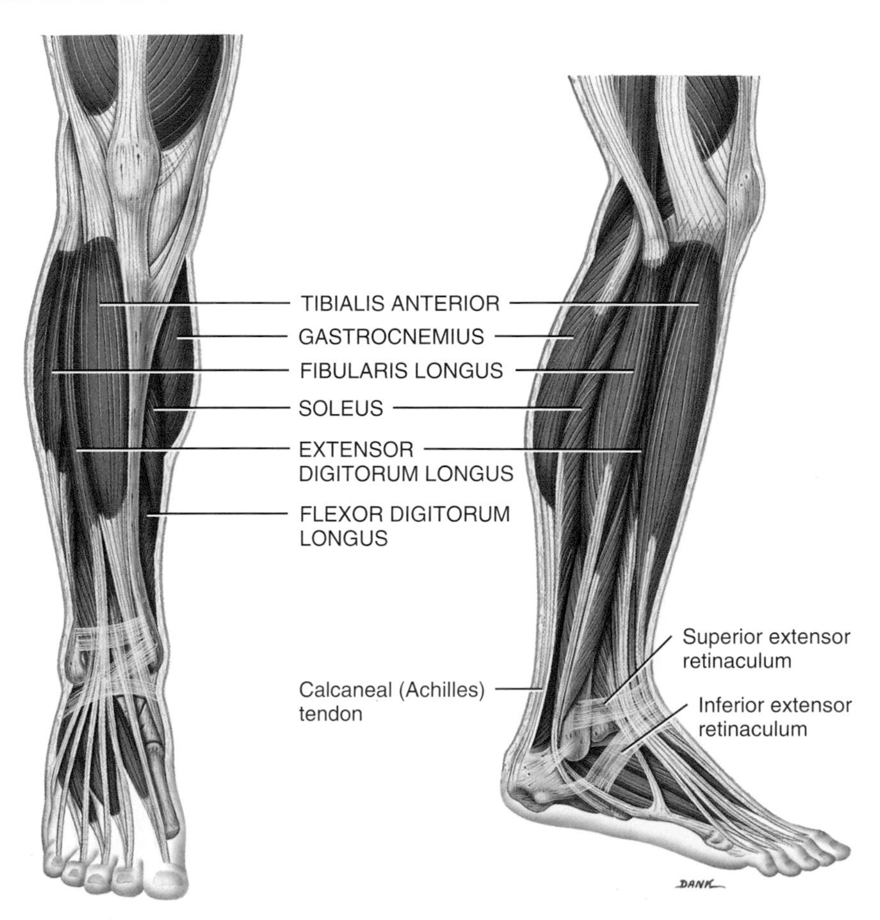

(c) Anterior superficial view

(d) Right lateral superficial view

Which muscle is primarily affected in shinsplint syndrome?

FOCUS ON HOMEOSTASIS

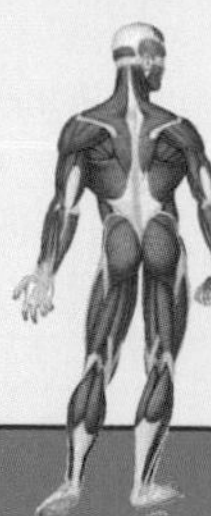

THE MUSCULAR SYSTEM

BODY SYSTEM		CONTRIBUTION OF THE MUSCULAR SYSTEM
For all body systems		The muscular system and muscular tissues produce body movements, stabilize body positions, move substances within the body, and produce heat that helps maintain normal body temperature.
Integumentary system	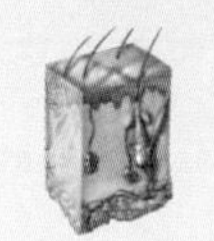	Pull of skeletal muscles on attachments to skin of face causes facial expressions; muscular exercise increases skin blood flow.
Skeletal system	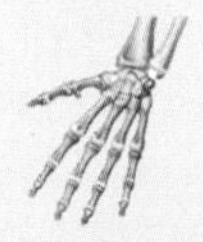	Skeletal muscle causes movement of body parts by pulling on attachments to bones; skeletal muscle provides stability for bones and joints.
Nervous system	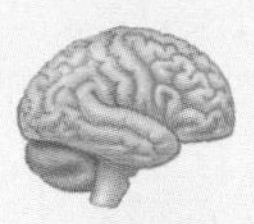	Smooth, cardiac, and skeletal muscles carry out commands for the nervous system; shivering—involuntary contraction of skeletal muscles that is regulated by the brain—generates heat to raise body temperature.
Endocrine system	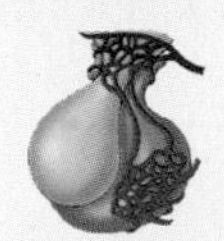	Regular activity of skeletal muscles (exercise) improves the action of some hormones, such as insulin; muscles protect some endocrine glands.
Cardiovascular system	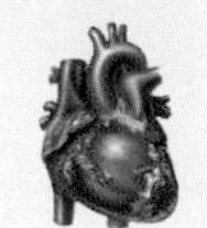	Cardiac muscle powers the pumping action of the heart; contraction and relaxation of smooth muscle in blood vessel walls help adjust the amount of blood flowing through various body tissues; contraction of skeletal muscles in the legs assists return of blood to the heart; regular exercise causes cardiac hypertrophy (enlargement) and increases the heart's pumping efficiency; lactic acid produced by active skeletal muscles may be used for ATP production by the heart.
Lymphatic system and immunity	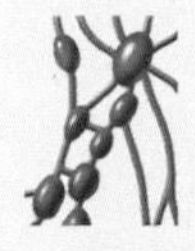	Skeletal muscles protect some lymph nodes and lymphatic vessels and promote the flow of lymph inside lymphatic vessels; exercise may increase or decrease some immune responses.
Respiratory system	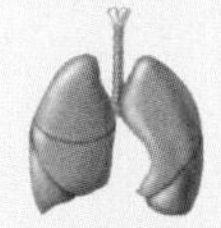	Skeletal muscles involved with breathing cause air to flow into and out of the lungs; smooth muscle fibers adjust the size of airways; vibrations in skeletal muscles of the larynx control air flowing past vocal cords, regulating voice production; coughing and sneezing, due to skeletal muscle contractions, help clear airways; regular exercise improves the efficiency of breathing.
Digestive system	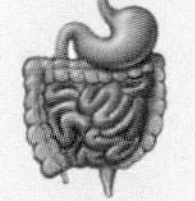	Skeletal muscles protect and support organs in the abdominal cavity; alternating contraction and relaxation of skeletal muscles power chewing and initiate swallowing; smooth muscle sphincters control the volume of organs of the gastrointestinal (GI) tract; smooth muscles in the walls of the GI tract mix and move its contents through the tract.
Urinary system	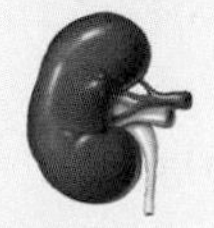	Skeletal muscle and smooth muscle sphincters and smooth muscle in the wall of the urinary bladder control whether urine is stored in the urinary bladder or voided (urination).
Reproductive systems	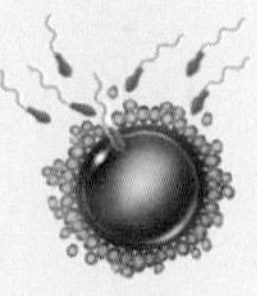	Skeletal and smooth muscle contractions eject semen; smooth muscle contractions propel oocytes through uterine tubes, help regulate flow of menstrual blood from the uterus, and force baby from the uterus during childbirth; during intercourse, skeletal muscle contractions are associated with orgasm and pleasurable sensations in both sexes.

COMMON DISORDERS

Skeletal muscle function may be abnormal due to disease or damage of any of the components of a motor unit: somatic motor neurons, neuromuscular junctions, or muscle fibers. The term ***neuromuscular disease*** encompasses problems at all three sites; the term ***myopathy*** (mī-OP-a-thē; *-pathy* = disease) signifies a disease or disorder of the skeletal muscle tissue itself.

Myasthenia Gravis

Myasthenia gravis (mī-as-THĒ-nē-a GRAV-is) is an autoimmune disease that causes chronic, progressive damage of the neuromuscular junction. In people with myasthenia gravis, the immune system inappropriately produces antibodies that bind to and block some ACh receptors, thereby decreasing the number of functional ACh receptors at the motor end plates of skeletal muscles (see Figure 8.5). Because 75% of patients with myasthenia gravis have hyperplasia or tumors of the thymus, it is possible that thymic abnormalities cause the disorder. As the disease progresses, more ACh receptors are lost. Thus, muscles become increasingly weaker, fatigue more easily, and may eventually cease to function.

Myasthenia gravis occurs in about 1 in 10,000 people and is more common in women, who typically are ages 20 to 40 at onset, than in men, who usually are ages 50 to 60 at onset. The muscles of the face and neck are most often affected. Initial symptoms include weakness of the eye muscles, which may produce double vision, and difficulty in swallowing. Later, the person has difficulty chewing and talking. Eventually the muscles of the limbs may become involved. Death may result from paralysis of the respiratory muscles, but often the disorder does not progress to this stage.

Muscular Dystrophy

The term ***muscular dystrophy*** refers to a group of inherited muscle-destroying diseases that cause progressive degeneration of skeletal muscle fibers. The most common form of muscular dystrophy is *DMD—Duchenne muscular dystrophy* (doo-SHĀN). Because the mutated gene is on the X chromosome, which males have only one of, DMD strikes boys almost exclusively. (Sex-linked inheritance is described in Chapter 24.) Worldwide, about 1 in every 3500 male babies—21,000 in all—are born with DMD each year. The disorder usually becomes apparent between the ages of 2 and 5, when parents notice the child falls often and has difficulty running, jumping, and hopping. By age 12 most boys with DMD are unable to walk. Respiratory or cardiac failure usually causes death between the ages of 20 and 30.

In DMD, the gene that codes for the protein dystrophin is mutated and little or no dystrophin is present (dystrophin provides structural reinforcement for the skeletal muscle fiber sarcolemma). Without the reinforcing effect of dystrophin, the sarcolemma easily tears during muscle contraction. Because their plasma membranes are damaged, muscle fibers slowly rupture and die.

Fibromyalgia

Fibromyalgia (*algia* = painful condition) is a painful, nonarticular rheumatic disorder that usually appears between the ages of 25 and 50. An estimated 3 million people in the United States suffer from fibromyalgia, which is 15 times more common in women than in men. The disorder affects the fibrous connective tissue components of muscles, tendons, and ligaments. A striking sign is pain that results from gentle pressure at specific "tender points." Even without pressure, there is pain, tenderness, and stiffness of muscles, tendons, and surrounding soft tissues. Besides muscle pain, those with fibromyalgia report severe fatigue, poor sleep, headaches, depression, and inability to carry out their daily activities. Often, a gentle aerobic fitness program is beneficial.

Abnormal Contractions of Skeletal Muscle

One kind of abnormal muscular contraction is a ***spasm***, a sudden involuntary contraction of a single muscle in a large group of muscles. A painful spasmodic contraction is known as a ***cramp***. A ***tic*** is a spasmodic twitching made involuntarily by muscles that are ordinarily under voluntary control. Twitching of the eyelid and facial muscles are examples of tics. A ***tremor*** is a rhythmic, involuntary, purposeless contraction that produces a quivering or shaking movement. A ***fasciculation*** is an involuntary, brief twitch of an entire motor unit that is visible under the skin; it occurs irregularly and is not associated with movement of the affected muscle. Fasciculations may be seen in multiple sclerosis (see page 237) or in amyotrophic lateral sclerosis (Lou Gehrig's disease). A ***fibrillation*** is a spontaneous contraction of a single muscle fiber that is not visible under the skin but can be recorded by electromyography. Fibrillations may signal destruction of motor neurons.

Running Injuries

Nearly 70% of those who jog or run sustain some type of running-related injury. Most such injuries are minor, but some are quite serious. In addition, untreated or inappropriately treated minor injuries may become chronic. Among runners, common sites of injury include the ankle, knee, calcaneal (Achilles) tendon, hip, groin, foot, and back. Of these, the knee often is the most severely injured area.

Running injuries are frequently related to faulty training techniques. This may involve improper (or lack of) warm-up routines, running too much, or running too soon after an injury. Or it might involve extended running on hard and/or uneven surfaces. Poorly constructed or worn-out running shoes can also contribute to injury, as can any biomechanical problem (such as a fallen arch) aggravated by running.

Most sports injuries should be treated initially with RICE therapy, which stands for Rest, Ice, Compression, and Elevation. Immediately apply ice, and rest and elevate the injured part. Then apply an elastic bandage, if possible, to compress the injured tissue. Continue using RICE for 2 to 3 days, and resist the temptation to apply heat, which may worsen the swelling. Follow-up treatment may include alternating moist heat and ice massage to enhance blood flow in the injured area. Sometimes it is helpful to take nonsteroidal anti-inflammatory drugs (NSAIDs) or to have local injections of corticosteroids.

During the recovery period, it is important to keep active using an alternative fitness program that does not worsen the original injury. This activity should be determined in consultation with a physician. Finally, careful exercise is needed to rehabilitate the injured area itself.

Effects of Anabolic Stereoids

The use of ***anabolic steroids*** by athletes has received widespread attention. These steroid hormones, similar to testosterone, are taken to increase muscle size and strength. The large doses needed to produce an effect, however, have damaging, sometimes even devastating side effects, including liver cancer, kidney damage, increased risk of heart disease, stunted growth, wide mood swings, and increased irritability and aggression. Additionally, females who take anabolic steroids may experience atrophy of the breasts and uterus, menstrual irregularities, sterility, facial hair growth, and deepening of the voice. Males may experience diminished testosterone secretion, atrophy of the testes, and baldness.

MEDICAL TERMINOLOGY AND CONDITIONS

Electromyography or ***EMG*** (e-lek′-trō-mī-OG-ra-fē; *electro-* = electricity; *myo-* = muscle; *-graphy* = to write) The recording and study of electrical changes that occur in muscular tissue.

Hypertonia (*hyper-* = above) Increased muscle tone, characterized by increased muscle stiffness and sometimes associated with a change in normal reflexes.

Hypotonia (*hypo-* = below) Decreased or lost muscle tone.

Muscle strain Tearing of a muscle because of forceful impact, accompanied by bleeding and severe pain. Also known as a *charley horse* or pulled muscle. It often occurs in contact sports and typically affects the quadriceps femoris muscle on the anterior surface of the thigh.

Myalgia (mī-AL-jē-a; *-algia* = painful condition) Pain in or associated with muscles.

Myoma (mī-Ō-ma; *-oma* = tumor) A tumor consisting of muscular tissue.

Myomalacia (mī′-ō-ma-LĀ-shē-a; *-malacia* = soft) Pathological softening of muscle tissue.

Myositis (mī′-ō-SĪ-tis; *-itis* = inflammation of) Inflammation of muscle fibers (cells).

Myotonia (mī′-ō-TŌ-nē-a; *-tonia* = tension) Increased muscular excitability and contractility, with decreased power of relaxation; tonic spasm of the muscle.

STUDY OUTLINE

Overview of Muscular Tissue (p. 173)

1. The three types of muscular tissue are skeletal muscle, cardiac muscle, and smooth muscle (summarized in Table 8.1 on page 188).
2. Skeletal muscle tissue is mostly attached to bones. It is striated and voluntary.
3. Cardiac muscle tissue forms most of the wall of the heart. It is striated and involuntary.
4. Smooth muscle tissue is located in viscera. It is nonstriated and involuntary.
5. Through contraction and relaxation, muscular tissue has five key functions: producing body movements, stabilizing body positions, regulating organ volume, moving substances within the body, and producing heat.

Skeletal Muscle Tissue (p. 173)

1. Connective tissue coverings associated with skeletal muscle include the epimysium, covering an entire muscle; perimysium, covering fascicles; and endomysium, covering individual muscle fibers.
2. Tendons are extensions of connective tissue beyond muscle fibers that attach the muscle to bone.
3. Skeletal muscles are well supplied with nerves and blood vessels, which provide nutrients and oxygen for contraction.
4. Skeletal muscle consists of muscle fibers (cells) covered by a sarcolemma that features tunnel-like extensions, the transverse tubules. The fibers contain sarcoplasm, multiple nuclei, many mitochondria, myoglobin, and sarcoplasmic reticulum.
5. Each fiber also contains myofibrils that contain thin and thick filaments. The filaments are arranged in functional units called sarcomeres.
6. Thin filaments are composed of actin, tropomyosin, and troponin; thick filaments consist of myosin.

Contraction and Relaxation of Skeletal Muscle (p. 177)

1. Muscle contraction occurs when myosin heads attach to and "walk" along the thin filaments at both ends of a sarcomere, progressively pulling the thin filaments toward the center of a sarcomere. As the thin filaments slide inward, the Z discs come closer together, and the sarcomere shortens.
2. The neuromuscular junction (NMJ) is the synapse between a motor neuron and a skeletal muscle fiber. The NMJ includes the axon terminals and synaptic end bulbs of a motor neuron plus the adjacent motor end plate of the muscle fiber sarcolemma.

3. A motor neuron and all of the muscle fibers it stimulates form a motor unit. A single motor unit may include as few as 10 or as many as 2000 muscle fibers.
4. When a nerve impulse reaches the synaptic end bulbs of a somatic motor neuron, it triggers the release of acetylcholine (ACh) from synaptic vesicles. ACh diffuses across the synaptic cleft and binds to ACh receptors, initiating a muscle action potential. Acetylcholinesterase then quickly destroys ACh.
5. An increase in the level of Ca^{2+} in the sarcoplasm, caused by the muscle action potential, starts the contraction cycle; as a decrease in the level of Ca^{2+} turns off the contraction cycle.
6. The contraction cycle is the repeating sequence of events that causes sliding of the filaments: (1) myosin ATPase splits ATP and becomes energized, (2) the myosin head attaches to actin forming a crossbridge, (3) the crossbridge generates force as it swivels or rotates toward the center of the sarcomere (power stroke), and (4) binding of ATP to myosin detaches myosin from actin. The myosin head again splits ATP, returns to its original position, and binds to a new site on actin as the cycle continues.
7. Ca^{2+} active transport pumps continually remove Ca^{2+} from the sarcoplasm into the sarcoplasmic reticulum (SR). When the level of Ca^{2+} in the sarcoplasm decreases, the troponin–tropomyosin complexes slide back over and cover the myosin-binding sites, and the muscle fiber relaxes.
8. Continual involuntary activation of a small number of motor units produces muscle tone, which is essential for maintaining posture.

Metabolism of Skeletal Muscle Tissue (p. 180)

1. Muscle fibers have three sources for ATP production: creatine phosphate, anaerobic cellular respiration, and aerobic cellular respiration.
2. The transfer of a high-energy phosphate group from creatine phosphate to ADP forms new ATP molecules. Together, creatine phosphate and ATP provide enough energy for muscles to contract maximally for about 15 seconds.
3. Glucose is converted to pyruvic acid in the reactions of glycolysis, which yield two ATPs without using oxygen. These anaerobic reactions can provide enough ATP for about 30 to 40 seconds of maximal muscle activity.
4. Muscular activity that lasts longer than half a minute depends on aerobic cellular respiration, mitochondrial reactions that require oxygen to produce ATP. Aerobic cellular respiration yields about 36 molecules of ATP from each glucose molecule.
5. The inability of a muscle to contract forcefully after prolonged activity is muscle fatigue.
6. Elevated oxygen use after exercise is called recovery oxygen uptake.

Control of Muscle Tension (p. 183)

1. A twitch contraction is a brief contraction of all the muscle fibers in a motor unit in response to a single action potential.
2. A record of a contraction is called a myogram. It consists of a latent period, a contraction period, and a relaxation period.
3. Wave summation is the increased strength of a contraction that occurs when a second stimulus arrives before the muscle has completely relaxed after a previous stimulus.
4. Repeated stimuli can produce unfused tetanus, a sustained muscle contraction with partial relaxation between stimuli; more rapidly repeating stimuli will produce fused tetanus, a sustained contraction without partial relaxation between stimuli.
5. Motor unit recruitment is the process of increasing the number of active motor units.
6. On the basis of their structure and function, skeletal muscle fibers are classified as slow oxidative (SO), fast oxidative–glycolytic (FOG), and fast glycolytic (FG) fibers.
7. Most skeletal muscles contain a mixture of all three fiber types; their proportions vary with the typical action of the muscle.
8. The motor units of a muscle are recruited in the following order: first SO fibers, then FOG fibers, and finally FG fibers.
9. In an isometric contraction, there is no change in the length of a muscle, but the muscle develops considerable tension. In an isotonic contraction, there is a change in the length of a muscle, but no change in its tension.

Exercise and Skeletal Muscle Tissue (p. 186)

1. Various types of exercises can induce changes in the fibers in a skeletal muscle. Endurance-type (aerobic) exercises cause a gradual transformation of some fast glycolytic (FG) fibers into fast oxidative–glycolytic (FOG) fibers.
2. Exercises that require great strength for short periods produce an increase in the size and strength of fast glycolytic (FG) fibers. The increase in size is due to increased synthesis of thick and thin filaments.

Cardiac Muscle Tissue (p. 186)

1. Cardiac muscle tissue, which is striated and involuntary, is found only in the heart.
2. Each cardiac muscle fiber usually contains a single centrally located nucleus and exhibits branching.
3. Cardiac muscle fibers are connected by means of intercalated discs, which hold the muscle fibers together and allow muscle action potentials to quickly spread from one cardiac muscle fiber to another.
4. Cardiac muscle tissue contracts when stimulated by its own autorhythmic fibers. Due to its continuous, rhythmic activity, cardiac muscle depends greatly on aerobic cellular respiration to generate ATP.

Smooth Muscle Tissue (p. 186)

1. Smooth muscle tissue is nonstriated and involuntary.
2. In addition to thin and thick filaments, smooth muscle fibers contain intermediate filaments and dense bodies.
3. Visceral (single-unit) smooth muscle is found in the walls of hollow viscera and of small blood vessels. Many visceral fibers form a network that contracts in unison.
4. Multiunit smooth muscle is found in large blood vessels, large airways to the lungs, arrector pili muscles, and the eye. The fibers contract independently rather than in unison.
5. The duration of contraction and relaxation is longer in smooth muscle than in skeletal muscle.

6. Smooth muscle fibers can be stretched considerably and still retain the ability to contract.
7. Smooth muscle fibers contract in response to nerve impulses, stretching, hormones, and local factors.

Aging and Muscular Tissue (p. 187)

1. Beginning at about 30 years of age, there is a slow, progressive loss of skeletal muscle, which is replaced by fibrous connective tissue and fat.
2. Aging also results in a decrease in muscle strength, slower muscle reflexes, and loss of flexibility.

How Skeletal Muscles Produce Movement (p. 188)

1. Skeletal muscles produce movement by pulling on tendons attached to bones.
2. The attachment to the stationary bone is the origin. The attachment to the movable bone is the insertion.
3. The prime mover (agonist) produces the desired action. The antagonist produces an opposite action. The synergist assists the prime mover by reducing unnecessary movement. The fixator stabilizes the origin of the prime mover so that it can act more efficiently.

Principal Skeletal Muscles (p. 189)

1. The principal skeletal muscles of the body are grouped according to region, as shown in Exhibits 8.1 through 8.13.
2. In studying muscle groups, refer to Figure 8.13 on pages 192–193 to see how each group is related to all others.
3. The names of most skeletal muscles indicate specific characteristics.
4. The major descriptive categories are direction of fibers, location, size, number of origins (or heads), shape, origin and insertion, and action (see Table 8.2 on page 191).

SELF-QUIZ

1. The characteristic of muscular tissue that allows it to return to its original shape after contraction is
   a. extensibility **b.** excitability **c.** fused tetanus **d.** contractility **e.** elasticity

2. Match the connective tissue coverings with their locations:
   ____ **a.** wraps an entire muscle
   ____ **b.** lies immediately under the skin
   ____ **c.** separates muscle into functional groups
   ____ **d.** surrounds each individual muscle fiber
   ____ **e.** divides muscle fibers into fascicles

   A. endomysium
   B. deep fascia
   C. perimysium
   D. epimysium
   E. superficial fascia

3. Which of the following statements about skeletal muscle tissue is NOT true?
   a. Skeletal muscle requires a large blood supply.
   b. Skeletal muscle fibers have many mitochondria.
   c. The arrangement of thick and thin filaments produces the striations in skeletal muscle tissue.
   d. Skeletal muscle fibers contain gap junctions that help conduct action potentials from one fiber to another.
   e. A skeletal muscle fiber has many nuclei.

4. Match the following:
   ____ **a.** network of tubules that stores calcium
   ____ **b.** pigment that stores oxygen
   ____ **c.** composed of myosin
   ____ **d.** composed of actin, tropomyosin, and troponin
   ____ **e.** tunnel-like extensions of sarcolemma

   A. thick filaments
   B. transverse tubules
   C. sarcoplasmic reticulum
   D. myoglobin
   E. thin filaments

5. The sarcolemma is the equivalent of the
   a. cytoplasm **b.** nucleus **c.** plasma membrane **d.** endoplasmic reticulum **e.** mitochondria

6. You begin an intensive weightlifting plan because you want to enter a weightlifting contest. During the activity of weightlifting, your skeletal muscles will obtain energy (ATP) primarily through
   a. anaerobic cellular respiration **b.** the complete breakdown of pyruvic acid in the mitochondria **c.** hyperplasia **d.** hypertrophy **e.** aerobic cellular respiration

7. Which of the following events of skeletal muscle contraction does NOT occur during the latent period?
   a. Sarcomeres shorten.
   b. Action potentials conduct into the T tubules.
   c. The concentration of calcium ions increases in the sarcoplasm.
   d. Myosin-binding sites on the thin filaments are exposed.
   e. Calcium release channels in the sarcoplasmic reticulum open.

8. For each of the following descriptions, indicate if it refers to skeletal muscle, cardiac muscle, or smooth muscle. Use the abbreviations SK for skeletal, CA for cardiac, and SM for smooth. The same response may be used more than once.
   ____ **a.** involuntary
   ____ **b.** multinucleated
   ____ **c.** striated
   ____ **d.** contain intercalated discs
   ____ **e.** elongated, cylindrical cells
   ____ **f.** voluntary
   ____ **g.** cells that taper at both ends
   ____ **h.** nonstriated
   ____ **i.** muscle fibers contractindividually
   ____ **j.** autorhythmic

9. When ATP in the sarcoplasm is exhausted, the muscle must rely on _____ to quickly produce more ATP from ADP for contraction.

   a. acetylcholine b. creatine phosphate c. lactic acid
   d. pyruvic acid e. acetylcholinesterase

10. A motor unit consists of

    a. a transverse tubule and its associated sarcomeres
    b. a motor neuron and all of the muscle fibers it stimulates
    c. a muscle and all of its motor neurons
    d. all of the filaments encased within a sarcomere
    e. the motor end plate and the transverse tubules

11. Thick filaments

    a. include actin, troponin, and tropomyosin
    b. compose the I band
    c. stretch the entire length of a sarcomere
    d. have binding sites for Ca^{2+}
    e. have myosin heads (crossbridges) used for the power stroke

12. The chemical that prevents the continuous stimulation of a muscle fiber is

    a. Ca^{2+} b. acetylcholinesterase c. ATP
    d. acetylcholine e. troponin–tropomyosin

13. Which of the following is NOT associated with muscle fatigue?

    a. depletion of creatine phosphate b. lack of oxygen
    c. decrease in Ca^{2+} levels in the sarcoplasm
    d. decrease in lactic acid levels e. lack of glycogen

14. All of the following may result in an increase in muscle size EXCEPT

    a. denervation atrophy b. weight training
    c. human growth hormone d. testosterone
    e. isotonic contraction

15. Skeletal muscles are named using several characteristics. Which characteristic is NOT used to name skeletal muscles?

    a. direction of fibers b. size c. speed of contraction
    d. location e. shape

16. Arrange the following in the correct order for skeletal muscle fiber contraction.

    1. Sarcoplasmic reticulum releases Ca^{2+}.
    2. Ca^{2+} combines with troponin.
    3. Acetylcholine is released from the axon terminal.
    4. Action potential travels into transverse tubules.
    5. Energized myosin heads (crossbridges) attach to actin.
    6. Thin filaments slide toward the center of the sarcomere.

    a. 3, 4, 1, 2, 5, 6
    b. 4, 3, 2, 1, 5, 6
    c. 1, 2, 3, 4, 5, 6
    d. 4, 1, 3, 5, 2, 6
    e. 3, 1, 4, 5, 2, 6

17. Your instructor asks you to pick up a box of books and carry them to the library in another building. You try to pick up the box, but the box is too heavy to move. Which of the following types of muscle contractions would you be utilizing?

    a. hypertonic b. isotonic only c. spastic
    d. isometric only e. isometric and isotonic

18. Match the following:

    ____ a. extends and laterally rotates thigh at the hip joint
    ____ b. adducts and medially rotates thigh at the hip joint
    ____ c. compresses abdomen and flexes vertebral column
    ____ d. flexes the neck
    ____ e. flexes and abducts wrist joint
    ____ f. extends phalanges
    ____ g. adducts and rotates arm medially at shoulder joint
    ____ h. extends leg at the knee and flexes thigh at hip joint
    ____ i. plantar flexes foot at ankle joint and flexes leg at knee joint
    ____ j. dorsiflexes and inverts foot
    ____ k. abducts, flexes, extends, and rotates arm at shoulder joint
    ____ l. elevates clavicle; depresses or elevates scapula
    ____ m. elevates mandible; closes mouth
    ____ n. wrinkles skin of forehead horizontally as in a look of surprise
    ____ o. extends, adducts, and rotates arm medially at shoulder joint

    A. trapezius
    B. flexor carpi radialis
    C. tibialis anterior
    D. adductor longus
    E. gluteus maximus
    F. quadriceps group
    G. rectus abdominis
    H. sternocleidomastoid
    I. frontal belly of occipitofrontalis
    J. gastrocnemius
    K. deltoid
    L. masseter
    M. extensor digitorum
    N. latissimus dorsi
    O. pectoralis major

19. Match the following:

____ a. extend from the thick filaments
____ b. contain myosin-binding site
____ c. dense area that separates sarcomeres
____ d. contain acetylcholine
____ e. striated zone of the sarcomere composed of thick and thin filaments
____ f. space between axon terminal and the sarcolemma
____ g. striated zone of the sarcomere composed of thin filaments only
____ h. region of sarcolemma near the adjoining axon terminal

A. I band
B. synaptic vesicles
C. myosin heads
D. Z discs
E. motor end plate
F. actin molecules
G. A band
H. synaptic cleft

20. Matching the following:

____ a. works with prime mover to reduce unnecessary movement
____ b. muscle in a group that produces desired movement
____ c. stationary end of a muscle
____ d. muscle that has an action opposite to that of another muscle
____ e. helps stabilize the origin of the prime mover
____ f. the end of a muscle attached to the movable bone

A. insertion
B. origin
C. synergist
D. antagonist
E. prime mover
F. fixator

CRITICAL THINKING APPLICATIONS

1. The newspaper reported several cases of botulism poisoning following a fund-raiser potluck dinner for the local clinic. The cause appeared to be three-bean salad "flavored" with the bacterium *Clostridium botulinum*. What would be the result of botulism poisoning on muscle function?

2. Ali's nephew was squealing with laughter. She was entertaining him by sticking her thumb in her pursed lips, raising her eyebrows, pumping her arm up and down, and puffing her cheeks in and out. Name the muscles Ali was using to maneuver her face.

3. When her cast finally came off after six long weeks, Kate thought she'd be all set to rejoin her volleyball team, but now her left thigh is only half the size of her right. Explain what happened to her thigh and what she needs to do to get back in the game.

4. The coach of the track team has his athletes crosstraining. They ran 10 miles on Monday, then on Tuesday they lifted weights. How do these types of exercise affect the muscles?

ANSWERS TO FIGURE QUESTIONS

8.1 In order from the inside toward the outside, the connective tissue layers are endomysium, perimysium, and epimysium.

8.2 The A band is composed of thick filaments in its center and overlapping thick and thin filaments at each end; the I band is composed of thin filaments.

8.3 A band: myosin, actin, troponin, and tropomyosin. I band: actin, troponin, and tropomyosin.

8.4 The I bands disappear. The lengths of the thick and thin filaments do not change.

8.5 The motor end plate is the region of the sarcolemma near the axon terminal.

8.6 Binding of ATP to the myosin heads detaches them from actin.

8.7 The power stroke occurs during step 6.

8.8 Glycolysis, exchange of phosphate between creatine phosphate and ADP, and glycogen breakdown occur in the cytosol. Oxidation of pyruvic acid, amino acids, and fatty acids (aerobic cellular respiration) occurs in the mitochondria.

8.9 Sarcomeres shorten during the contraction period.

8.10 Fused tetanus occurs when the frequency of stimulation reaches 80 to 100 stimuli per second.

8.11 The walls of hollow organs contain visceral (single-unit) smooth muscle.

8.12 The prime mover or agonist produces the desired action.

8.13 The following are some possible responses (there are other correct answers): direction of fibers—external oblique; shape—deltoid; action—extensor digitorum; size—gluteus maximus; origin and insertion—sternocleidomastoid; location—tibialis anterior; number of origins—biceps brachii.

8.14 Smiling—zygomaticus major; pouting—platysma; squinting—orbicularis oculi.

8.15 The superior oblique passes through the trochlea.

8.16 The rectus abdominis aids in urination.

8.17 The diaphragm and external intercostals contract during a normal quiet inhalation.

8.18 The pectoralis minor and serratus anterior have origins on the ribs; the trapezius, levator scapulae, and rhomboid major have origins on the vertebrae.

8.19 The pectoralis major and latissimus dorsi are muscles that cross the shoulder joint but do not originate on the scapula.

8.20 A compartment is a group of functionally related skeletal muscles in a limb, along with their blood vessels and nerves.

8.21 The median nerve is associated with the flexor retinaculum.

8.22 The iliocostalis, longissimus, and spinalis constitute the erector spinae.

8.23 Quadriceps femoris—rectus femoris, vastus lateralis, vastus medialis, and vastus intermedius; hamstrings—biceps femoris, semitendinosus, semimembranosus.

8.24 Shinsplint syndrome affects the tibialis anterior.

NERVOUS TISSUE chapter 9

did you know?

Depression is characterized by a mixture of psychological and physical symptoms, and is marked by changes in nervous system function. Depression is associated with imbalances in some of the chemicals that transmit messages between nerve cells. These chemicals are called neurotransmitters. Sometimes not enough of a neurotransmitter is produced. Other times the nerve cells do not respond to the neurotransmitter as they should. One of the neurotransmitters that plays an important role in depression is serotonin. Psychologists do not yet know whether the feelings of depression cause or are caused by neurotransmitter changes.

Focus on Wellness, page 236

www.wiley.com/college/apcentral

Together, all nervous tissues in the body comprise the ***nervous system***. Among the 11 body systems, the nervous system and the endocrine system play the most important roles in maintaining homeostasis. The nervous system, the subject of this and the next three chapters, can respond rapidly to help adjust body processes using nerve impulses. The endocrine system typically operates more slowly and exerts its influence on homeostasis by releasing hormones that the blood delivers to cells throughout the body. Besides helping maintain homeostasis, the nervous system is responsible for our perceptions, behaviors, and memories. It also initiates all voluntary movements. The branch of medical science that deals with the normal functioning and disorders of the nervous system is called ***neurology*** (noo-ROL-ō-jē; *neuro-* = nerve or nervous system; *-logy* = study of).

looking back to move ahead . . .

- Ion Channels (page 48)
- Sodium-potassium Pump (page 50)
- Nervous Tissue (page 90)
- Sensory Nerve Endings and Sensory Receptors in the Skin (page 100)
- Release of Acetylcholine at the Neuromuscular Junction (page 179)

OVERVIEW OF THE NERVOUS SYSTEM

OBJECTIVES • List the structures and basic functions of the nervous system.

• Describe the organization of the nervous system.

Structures of the Nervous System

The nervous system is an intricate, highly organized network of billions of neurons and even more neuroglia. The structures that make up the nervous system include the brain, cranial nerves and their branches, the spinal cord, spinal nerves and their branches, ganglia, enteric plexuses, and sensory receptors (Figure 9.1).

The skull encloses the ***brain***, which contains about 100 billion neurons. Twelve pairs (right and left) of ***cranial nerves***, numbered I through XII, emerge from the base of the brain. A ***nerve*** is a bundle of hundreds to thousands of axons plus associated connective tissue and blood vessels that lie outside the brain and spinal cord. Each nerve follows a defined path and serves a specific region of the body. For example, cranial nerve I carries signals for the sense of smell from the nose to the brain.

The ***spinal cord*** connects to the brain and is encircled by the bones of the vertebral column. It contains about 100 million neurons. Thirty-one pairs of ***spinal nerves*** emerge from the spinal cord, each serving a specific region on the right or left side of the body. ***Ganglia*** (GANG-lē-a = swelling or knot) are small masses of nervous tissue that are located outside the brain and spinal cord. Ganglia contain cell bodies of neurons and are closely associated with cranial and spinal nerves. In the walls of organs of the gastrointestinal tract are extensive networks of neurons, called ***enteric plexuses***, that help regulate the digestive system (Figure 9.1). ***Sensory receptors*** are either the dendrites of sensory neurons (such as sensory receptors in the skin) or separate, specialized cells that monitor changes in the internal or external environment (such as photoreceptors in the retina of the eye).

Functions of the Nervous System

The nervous system carries out a complex array of tasks, such as sensing various smells, producing speech, remembering past events, providing signals that control body movements,

Figure 9.1 Major structures of the nervous system.

The nervous system includes the brain, cranial nerves, spinal cord, spinal nerves, ganglia, enteric plexuses, and sensory receptors.

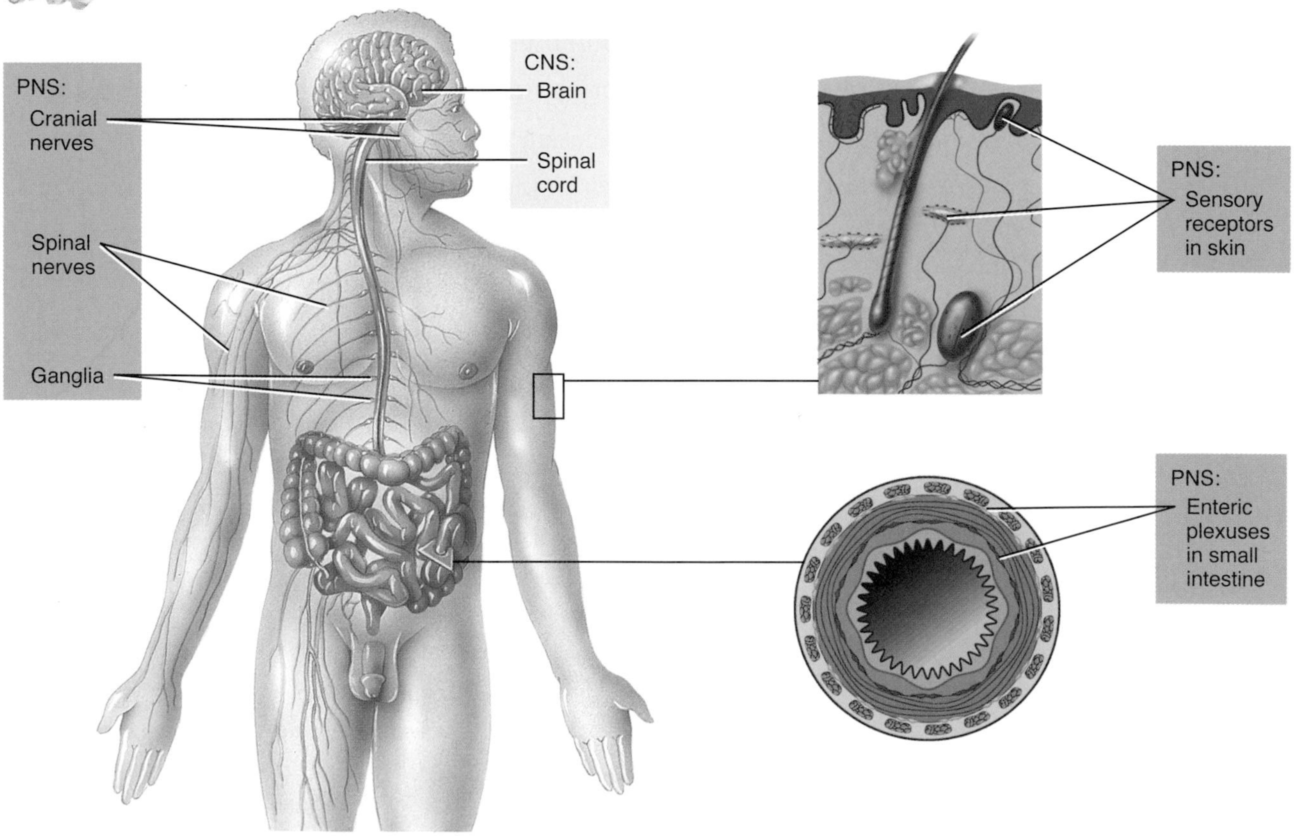

What is the total number of cranial and spinal nerves in your body?

and regulating the operation of internal organs. These diverse activities can be grouped into three basic functions: sensory, integrative, and motor.

- ***Sensory function.*** The sensory receptors *detect* many different types of stimuli, both within your body, such as an increase in blood temperature, and outside your body, such as a touch on your arm. ***Sensory*** or ***afferent neurons*** (AF-er-ent NOOR-onz; *af-* = toward; *-ferent* = carried) carry this sensory information into the brain and spinal cord through cranial and spinal nerves.
- ***Integrative function.*** The nervous system *integrates* (processes) sensory information by analyzing and storing some of it and by making decisions for appropriate responses. An important integrative function is ***perception***, the conscious awareness of sensory stimuli. Perception occurs in the brain. Many of the neurons that participate in integration are ***interneurons***, whose axons extend for only a short distance and contact nearby neurons in the brain or spinal cord. Interneurons comprise the vast majority of neurons in the body.
- ***Motor function.*** Once a sensory stimulus is perceived, the nervous system may elicit an appropriate motor response such as muscle contraction or gland secretion. The neurons that serve this function are ***motor*** or ***efferent neurons*** (EF-er-ent; *ef-* = away from). Motor neurons carry information from the brain toward the spinal cord or out of the brain and spinal cord to ***effectors*** (muscles and glands) through cranial and spinal nerves. Stimulation of the effectors by motor neurons causes muscles to contract and glands to secrete.

Organization of the Nervous System

The two main subdivisions of the nervous system are the ***central nervous system (CNS)***, which consists of the brain and spinal cord, and the ***peripheral*** (pe-RIF-er-al) ***nervous system (PNS)***, which includes all nervous tissue outside the CNS. The CNS integrates and correlates many different kinds of incoming sensory information. The CNS is also the source of thoughts, emotions, and memories. Most nerve impulses that stimulate muscles to contract and glands to secrete originate in the CNS. Structural components of the PNS are cranial nerves and their branches, spinal nerves and their branches, ganglia, and sensory receptors. Figure 9.2 shows the further functional subdivision of the PNS into a ***somatic nervous system (SNS)*** (*somat-* = body), an ***autonomic nervous system (ANS)*** (*auto-* = self; *-nomic* = law), and an ***enteric nervous system (ENS)*** (*enter-* = intestines). The somatic nervous system consists of (1) sensory neurons that convey information from somatic receptors in the head, body wall, and limbs and from receptors for the special senses of vision, hearing, taste, and smell to the CNS and (2) motor neurons that conduct impulses from the CNS to *skeletal muscles* only. Because these motor responses can be consciously controlled, the action of this part of the PNS is *voluntary*.

The ANS (the focus of Chapter 11) consists of (1) sensory neurons that convey information from autonomic sensory receptors, located primarily in visceral organs such as the stomach and lungs, to the CNS, and (2) motor neurons that conduct nerve impulses from the CNS to *smooth muscle*, *cardiac muscle*, and *glands*. Because its motor responses are not normally under conscious control, the action of the ANS is *invol-*

Figure 9.2 Organization of the nervous system. Subdivisions of the PNS are the somatic nervous system (SNS), the autonomic nervous system (ANS), and the enteric nervous system (ENS).

The two main subdivisions of the nervous system are (1) the central nervous system (CNS), consisting of the brain and spinal cord, and (2) the peripheral nervous system (PNS), consisting of all nervous tissue outside the CNS.

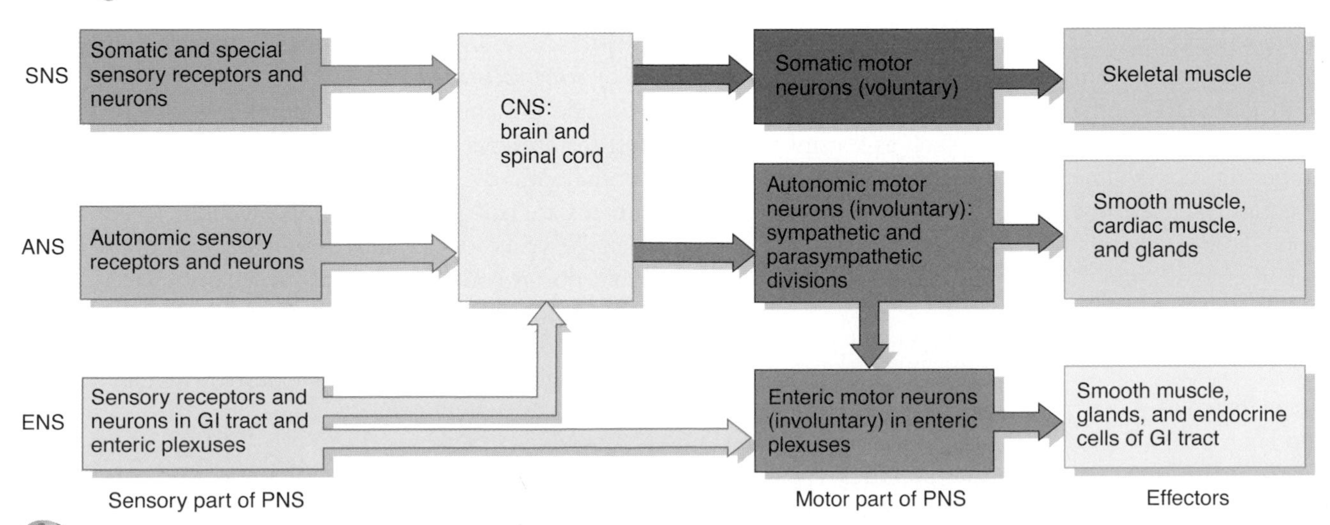

Which types of neurons carry input to the CNS and output from the CNS?

untary. The motor part of the ANS consists of two divisions, the *sympathetic division* and the *parasympathetic division*. With a few exceptions, effectors are innervated by both divisions, and usually the two divisions have opposing actions. For example, sympathetic neurons speed the heartbeat, and parasympathetic neurons slow it down. In general, the sympathetic division helps support exercise or emergency actions, so-called "fight-or-flight" responses, and the parasympathetic division takes care of "rest-and-digest" activities.

The enteric nervous system is the "brain of the gut," and its operation is involuntary. Its neurons extend most of the length of the gastrointestinal (GI) tract. Sensory neurons of the enteric nervous system monitor chemical changes within the GI tract and the stretching of its walls. Enteric motor neurons govern contraction of GI tract smooth muscle, secretions of the GI tract organs, such as acid secretion by the stomach, and activity of GI tract endocrine cells.

■ **CHECKPOINT**

1. What are the components of the CNS and PNS?
2. What kinds of problems would result from damage of sensory neurons, interneurons, and motor neurons?
3. What are the components and functions of the somatic, autonomic, and enteric nervous systems? Which subdivisions have involuntary actions?

HISTOLOGY OF NERVOUS TISSUE

OBJECTIVES • Contrast the histological characteristics and the functions of neurons and neuroglia.

• Distinguish between gray matter and white matter.

Nervous tissue consists of two types of cells: neurons and neuroglia. ***Neurons*** (nerve cells) are the basic information-processing units of the nervous system and are specialized for nerve impulse (action potential) conduction. They provide most of the unique functions of the nervous system, such as sensing, thinking, remembering, controlling muscle activity, and regulating glandular secretions. ***Neuroglia*** (noo-RŌG-lē-a; *glia* = glue) support, nourish, and protect the neurons and maintain homeostasis in the interstitial fluid that bathes neurons.

Neurons

Neurons usually have three parts: (1) a cell body, (2) dendrites, and (3) an axon (Figure 9.3). The ***cell body*** contains a nucleus surrounded by cytoplasm that includes typical organelles such as rough endoplasmic reticulum, lysosomes, mitochondria, and a Golgi complex. Most cellular molecules needed for a neuron's operation are synthesized in the cell body.

Two kinds of processes (extensions) emerge from the cell body of a neuron: multiple dendrites and a single axon. The cell body and the ***dendrites*** (= little trees) are the receiving or input parts of a neuron. Usually, dendrites are short, tapering, and highly branched, forming a tree-shaped array of processes that emerge from the cell body. The second type of process, the ***axon***, conducts nerve impulses toward another neuron, a muscle fiber, or a gland cell. An axon is a long, thin, cylindrical projection that often joins the cell body at a cone-shaped elevation called the *axon hillock* (= small hill). Nerve impulses usually arise at the axon hillock and then travel along the axon. Some axons have side branches called *axon collaterals*. The axon and axon collaterals end by dividing into many fine processes called *axon terminals*.

The site where two neurons or a neuron and an effector cell can communicate is termed a ***synapse***. The tips of most axon terminals swell into *synaptic end bulbs*. These bulb-shaped structures contain *synaptic vesicles*, tiny sacs that store chemicals called *neurotransmitters*. The neurotransmitter molecules released from synaptic vesicles are the means of communication at a synapse.

Myelination

The axons of most neurons are surrounded by a ***myelin sheath***, a many-layered covering composed of lipid and protein (Figure 9.3). Like insulation covering an electrical wire, the myelin sheath insulates the axon of a neuron and increases the speed of nerve impulse conduction. As you will learn shortly, Schwann cells in the PNS and oligodendrocytes in the CNS produce myelin sheaths by wrapping themselves around and around axons. Eventually, as many as 100 layers cover the axon, much as multiple layers of paper cover the cardboard tube in a roll of toilet paper. Gaps in the myelin sheath, called ***nodes of Ranvier*** (RON-vē-ā), appear at intervals along the axon (Figure 9.3). Axons with a myelin sheath are said to be ***myelinated***, and those without it are said to be ***unmyelinated.***

The amount of myelin increases from birth to maturity, and its presence greatly increases the speed of nerve impulse conduction. By the time a baby starts to talk, most myelin sheaths are partially formed, but myelination continues into the teenage years. An infant's responses to stimuli are neither as rapid nor as coordinated as those of an older child or an adult, in part because myelination is still in progress during infancy. Certain diseases, such as multiple sclerosis (see page 237) and Tay-Sachs disease (see page 56), destroy myelin sheaths.

Gray and White Matter

In a freshly dissected section of the brain or spinal cord, some regions look white and glistening, and others appear

Figure 9.3 Structure of a typical multipolar neuron. Arrows indicate the direction of information flow: dendrites → cell body → axon → axon terminals → synaptic end bulbs.

The basic parts of a neuron are several dendrites, a cell body, and a single axon.

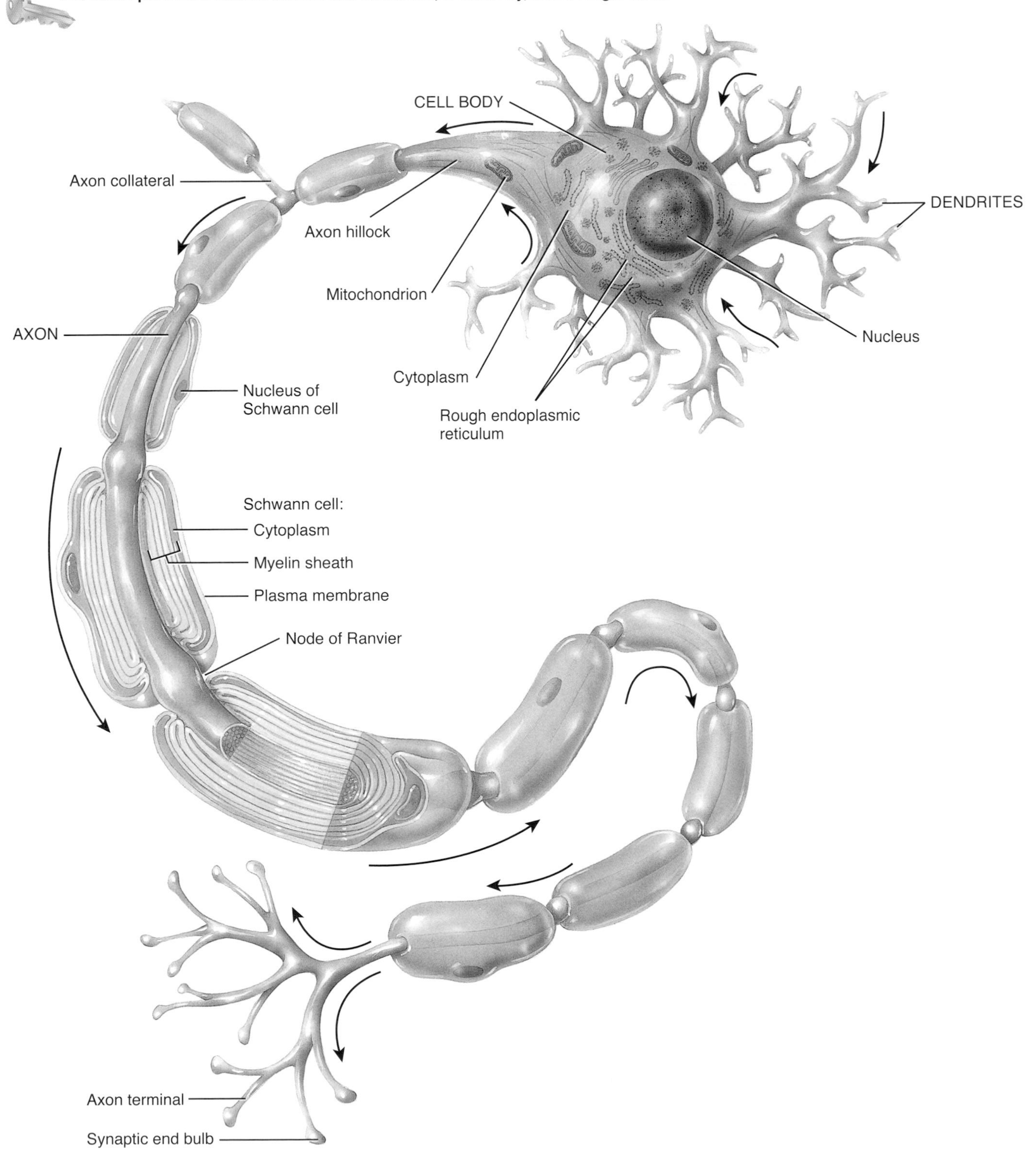

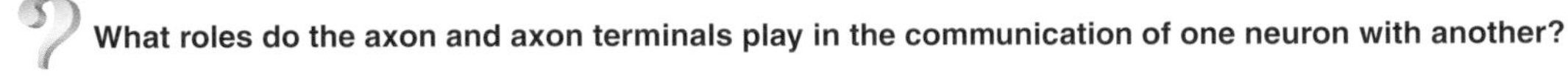

What roles do the axon and axon terminals play in the communication of one neuron with another?

gray. The ***white matter*** of nervous tissue consists primarily of myelinated axons of many neurons. The whitish color of myelin gives white matter its name. The ***gray matter*** of nervous tissue contains neuronal cell bodies, dendrites, unmyelinated axons, axon terminals, and neuroglia. It looks grayish, rather than white, because the cellular organelles impart a gray color and there is little or no myelin in these areas. Blood vessels are present in both white and gray matter.

In the spinal cord, the outer white matter surrounds an inner core of gray matter shaped like a butterfly or the letter H (see Figure 10.1 on page 243). In the brain, a thin shell of gray matter (cortex) covers the surface of the largest parts of the brain, the cerebrum and cerebellum (see Figures 10.10 and 10.11 on pages 254 and 256, respectively). When used to describe nervous tissue, a ***nucleus*** is a cluster of neuronal cell bodies within the CNS. (Recall that the term *ganglion* refers to a similar arrangement within the PNS). Many nuclei of gray matter lie deep within the brain. Much of the CNS white matter consists of ***tracts***, which are bundles of axons in the CNS that extend for some distance up or down the spinal cord or connect parts of the brain with each other and with the spinal cord. (Recall that the term *nerve* refers to a bundle of axons in the PNS).

Human neurons have very limited powers of **regeneration**, the capability to replicate or repair themselves. In the PNS, axons and dendrites may undergo repair if the cell body is intact and if the Schwann cells are functional. The Schwann cells on either side of an injured site multiply by mitosis, grow toward each other, and may form a **regeneration tube** across the injured area. The tube guides axonal regrowth from the proximal area across the injured area into the distal area previously occupied by the original axon. Regrowth is slow, in part, because many needed materials must be transported from their sites of synthesis in the cell body several inches or feet down the axon to the growth region. New axons cannot grow if the gap becomes filled with scar tissue. In the CNS, even when the cell body remains intact, a cut axon is usually not repaired. The presence of CNS myelin is one factor that actively inhibits regeneration of neurons.

Neuroglia

Neuroglia constitute about half the volume of the CNS. Their name derives from the idea of early histologists that they were the "glue" that held nervous tissue together. We now know that neuroglia are not merely passive bystanders but rather active participants in the operation of nervous tissue. Generally, neuroglia are smaller than neurons, and they are 5 to 50 times more numerous. In contrast to neurons, glia do not generate or conduct nerve impulses, and they can multiply and divide in the mature nervous system. In cases of injury or disease, neuroglia multiply to fill in the spaces formerly occupied by neurons. Brain tumors derived from glia, called ***gliomas***, often are highly malignant and grow rapidly. Of the six types of neuroglia, four—astrocytes, oligodendrocytes, microglia, and ependymal cells—are found only in the CNS. The remaining two types—Schwann cells and satellite cells—are present in the PNS. Table 9.1 shows the appearance of neuroglia and lists their funcions.

■ **CHECKPOINT**

4. What are the functions of the dendrites, cell body, axon, and synaptic end bulbs of a neuron?
5. Which cells produce myelin in nervous tissue, and what is the function of a myelin sheath?
6. What are the functions of neuroglia?

ACTION POTENTIALS

OBJECTIVE • Describe how a nerve impulse is generated and conducted.

Neurons communicate with one another by means of nerve action potentials, also called nerve impulses. Recall from Chapter 8 that a muscle fiber contracts in response to a muscle action potential. The generation of action potentials in both muscle fibers and neurons depends on two basic features of the plasma membrane: the existence of a resting membrane potential and the presence of specific types of ion channels. Many body cells exhibit a ***membrane potential***, a difference in the amount of electrical charge on the inside of the plasma membrane as compared to the outside. The membrane potential is like voltage stored in a battery. A cell that has a membrane potential is said to be ***polarized***. When muscle fibers and neurons are "at rest" (not conducting action potentials), the voltage across the plasma membrane is termed the ***resting membrane potential.***

If you connect the positive and negative terminals of a battery with a piece of metal (look in the battery compartment of your portable radio), an *electrical current* carried by electrons flows from the battery, allowing you to listen to your favorite music. In living tissues, the flow of *ions* (rather than electrons) constitutes electrical currents. The main sites where ions can flow across the membrane are through the pores of various types of ion channels.

Ion Channels

When they are open, ion channels allow specific ions to diffuse across the plasma membrane from where the ions are more concentrated to where they are less concentrated. Similarly, positively charged ions will move toward a negatively

Table 9.1 Neuroglia in the CNS and PNS

Type of Neuroglial Cell	Functions	Type of Neuroglial Cell	Functions
Central Nervous System			
Astrocytes (AS-trō-sītz; *astro-* = star; *-cyte* = cell)	Support neurons; protect neurons from harmful substances; help maintain proper chemical environment for generation of nerve impulses; assist with growth and migration of neurons during brain development; play a role in learning and memory; help form the blood–brain barrier.	**Oligodendrocytes** (OL-i-gō-den′-drō- sītz; *oligo-* = few; *dendro-* = tree)	Produce and maintain myelin sheath around several adjacent axons of CNS neurons.
Microglia (mī-KROG-lē-a; *micro-* = small)	Protect CNS cells from disease by engulfing invading microbes; migrate to areas of injured nerve tissue where they clear away debris of dead cells.	**Ependymal cells** (ep-EN-di-mal; *epen-* = above; *dym-* = garment)	Line ventricles of the brain (cavities filled with cerebrospinal fluid) and central canal of the spinal cord; form cerebrospinal fluid and assist in its circulation.
Peripheral Nervous System			
Schwann cells	Produce and maintain myelin sheath around a single axon of a PNS neuron; participate in regeneration of PNS axons.	**Satellite cells** (SAT-i-līt)	Support neurons in PNS ganglia and regulate exchange of materials between neurons and interstitial fluid.

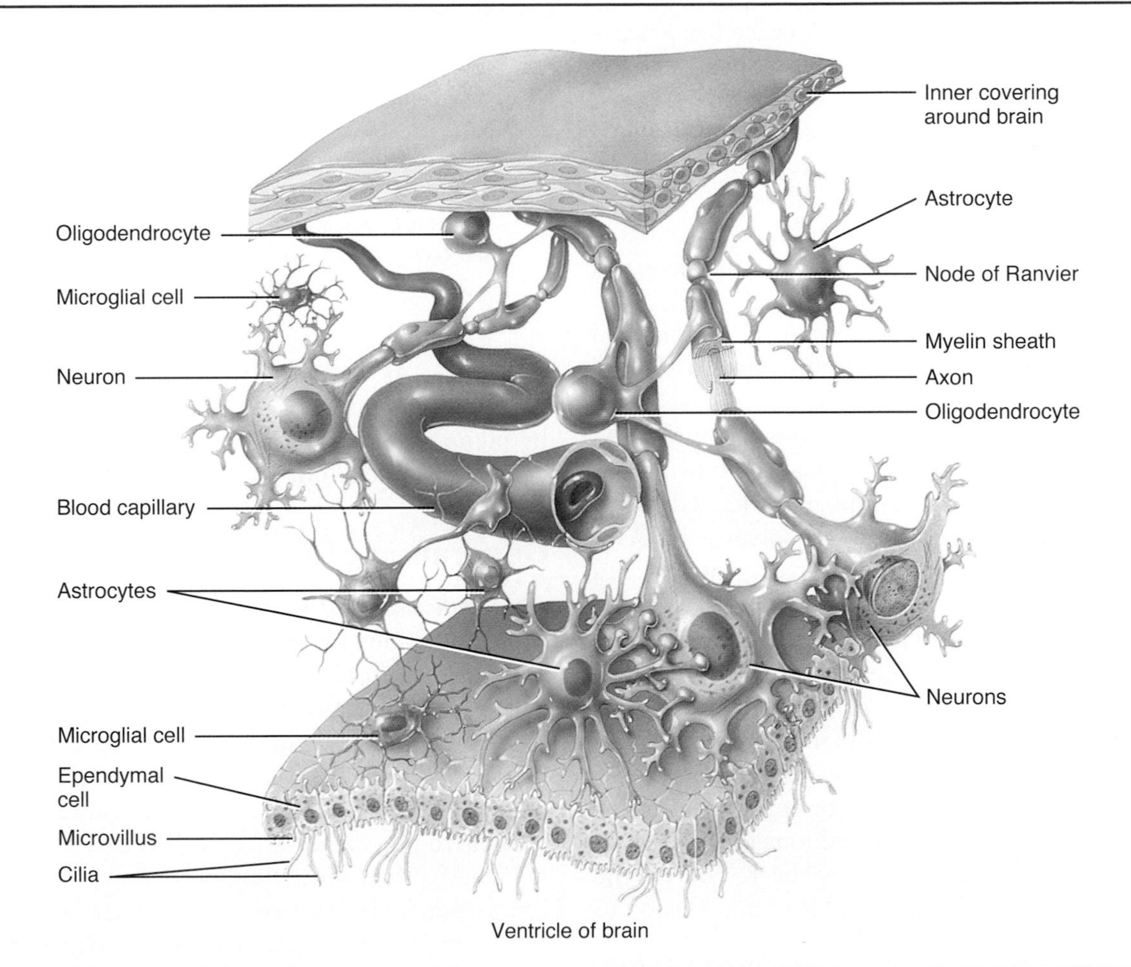

charged area, and negatively charged ions will move toward a positively charged area. As ions diffuse across a plasma membrane to equalize differences in charge or concentration, the result is a flow of current that can change the membrane potential.

Two different types of ion channels are leakage channels and gated channels. ***Leakage channels*** allow a small but steady stream of ions to leak across the membrane. Because plasma membranes typically have many more potassium ion (K^+) leakage channels than sodium ion (Na^+) leakage channels, the membrane's permeability to K^+ is much higher than its permeability to Na^+. ***Gated channels***, in contrast, open and close on command (see Figure 3.5 on page 48). ***Voltage-gated channels***—channels that open in response to a change in membrane potential—are used to generate and conduct action potentials.

Resting Membrane Potential

In a resting neuron, the outside surface of the plasma membrane has a positive charge and the inside surface has a negative charge. The separation of positive and negative electrical charges is a form of potential energy, which can be measured in volts. For example, two 1.5-volt batteries can power a portable radio. Voltages produced by cells typically are much smaller and are measured in millivolts (1 millivolt = 1 mV = 1/1000 volt). In neurons, the resting membrane potential is about −70 mV. The minus sign indicates that the inside of the membrane is negative relative to the outside.

The resting membrane potential arises from the unequal distributions of various ions in cytosol and interstitial fluid (Figure 9.4). Interstitial fluid is rich in sodium ions (Na^+) and chloride ions (Cl^-). Inside cells, the main positively charged ions in the cytosol are potassium ions (K^+), and the two dominant negatively charged ions are phosphates attached to organic molecules, such as the three phosphates in ATP (adenosine triphosphate), and amino acids in proteins. Because the concentration of K^+ is higher in cytosol and because plasma membranes have many K^+ leakage channels, potassium ions diffuse down their concentration gradient—out of cells into the interstitial fluid. As more and more positive potassium ions exit, the inside of the membrane becomes increasingly negative, and the outside of the membrane becomes increasingly positive. Another factor contributes to the negativity inside: Most negatively charged ions inside the cell are not free to leave. They cannot follow the K^+ out of the cell because they are attached either to large proteins or to other large molecules.

Membrane permeability to Na^+ is very low because there are only a few sodium leakage channels. Nevertheless, sodium ions do slowly diffuse inward, down their concentration gradient. Left unchecked, such inward leakage of Na^+ would eventually destroy the resting membrane potential. The small inward Na^+ leak and outward K^+ leak are offset by

Figure 9.4 The distribution of ions that produces the resting membrane potential.

The resting membrane potential is due to a small buildup of negatively charged ions, mainly organic phosphates (PO_4^{3-}) and proteins, in the cytosol just inside the membrane and an equal buildup of positively charged ions, mainly sodium ions (Na^+), in the interstitial fluid just outside the membrane.

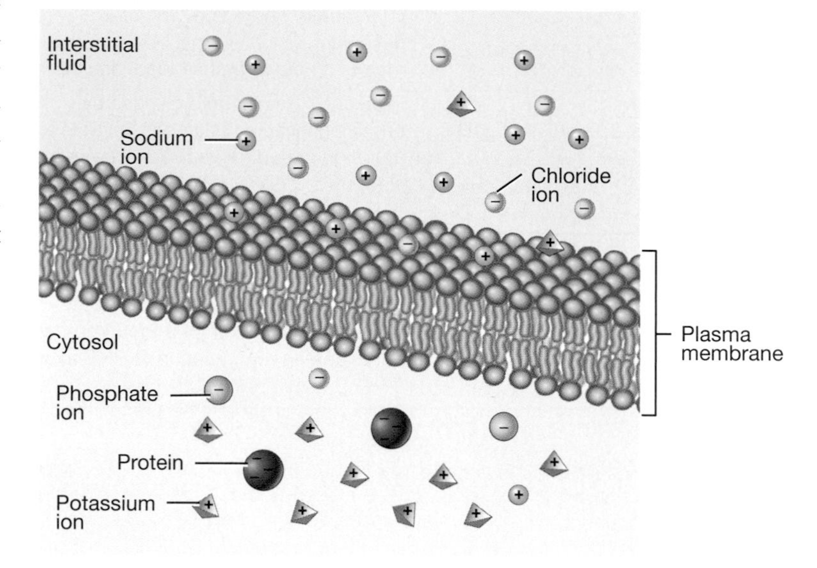

What is a typical value for the resting membrane potential of a neuron?

the sodium-potassium pumps (see Figure 3.9 on page 51). These pumps help maintain the resting membrane potential by pumping out Na^+ as fast as it leaks in. At the same time, the sodium-potassium pumps bring in K^+.

Generation of Action Potentials

An ***action potential (AP)*** or ***impulse*** is a sequence of rapidly occurring events that decrease and reverse the membrane potential and then eventually restore it to the resting state. The ability of muscle fibers and neurons to convert stimuli into action potentials is called ***electrical excitability***. A ***stimulus*** is anything in the cell's environment that can change the resting membrane potential. If a stimulus causes the membrane to depolarize to a critical level, called ***threshold*** (typically, about −55 mV), then an action potential arises (Figure 9.5). An action potential has two main phases: a depolarizing phase and a repolarizing phase. During the ***depolarizing phase***, the negative membrane potential becomes less negative, reaches zero, and then becomes positive. Then, during the ***repolarizing phase***, the membrane polarization is restored to its resting state of −70 mV (Figure 9.5). In neurons, the depolarizing and repolarizing phases of an action potential typically last about one millisecond (1/1000 sec).

During an action potential, depolarization to threshold briefly opens two types of voltage-gated ion channels. In neurons, these channels are present mainly in the plasma membrane of the axon and axon terminals. First, a threshold depolarization opens voltage-gated Na^+ channels. As these channels open, about 20,000 sodium ions rush into the cell, causing the depolarizing phase. The inflow of Na^+ causes the membrane potential to pass 0 mV and finally reach +30 mV (Figure 9.5). Second, the threshold depolarization also opens voltage-gated K^+ channels. The voltage-gated K^+ channels open more slowly, so their opening occurs at about the same time the voltage-gated Na^+ channels are automatically closing. As the K^+ channels open, potassium ions flow out of the cell, producing the repolarizing phase.

While the voltage-gated K^+ channels are open, outflow of K^+ may be large enough to cause an ***after-hyperpolarizing phase*** of the action potential (Figure 9.5). During hyperpolarization, the membrane potential becomes even *more negative* than the resting level. Finally, as K^+ channels close, the membrane potential returns to the resting level of −70 mV.

Action potentials arise according to the ***all-or-none principle***. As long as a stimulus is strong enough to cause depolarization to threshold, the voltage-gated Na^+ and K^+ channels open, and an action potential occurs. A much stronger stimulus cannot cause a larger action potential because the size of an action potential is always the same. A weak stimulus that fails to cause a threshold-level depolarization does not elicit an action potential. For a brief time after an action potential begins, a muscle fiber or neuron cannot generate another action potential. This time is called the *refractory period*.

Figure 9.5 Action potential (AP). When a stimulus depolarizes the membrane to threshold, an action potential is generated.

An action potential consists of depolarizing and repolarizing phases.

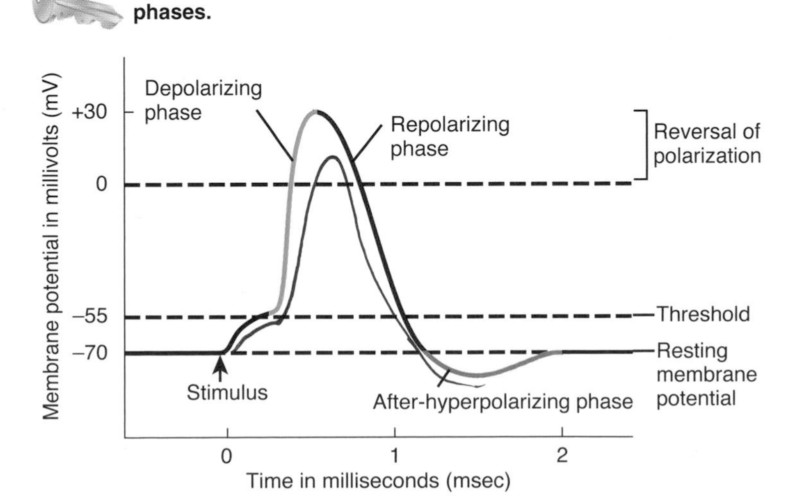

Which channels are open during depolarization? During repolarization?

Conduction of Nerve Impulses

To communicate information from one part of the body to another, nerve impulses must travel from where they arise, usually at the axon hillock, along the axon to the axon terminals (Figure 9.6). This type of impulse movement, which operates by positive feedback, is called ***conduction*** or ***propagation***. Depolarization to threshold at the axon hillock opens voltage-gated Na^+ channels. The resulting inflow of sodium ions depolarizes the adjacent membrane to threshold, which opens even more voltage-gated Na^+ channels, a positive feedback effect. Thus, a nerve impulse self-conducts along the axon plasma membrane. This situation is similar to pushing on the first domino in a long row: When the push on the first domino is strong enough, that domino falls against the second domino, and eventually the entire row topples.

The type of action potential conduction that occurs in unmyelinated axons (and in muscle fibers) is called ***continuous conduction***. In this case, each adjacent segment of the plasma membrane depolarizes to threshold and generates an action potential that depolarizes the next patch of the membrane (Figure 9.6a). Note that the impulse has traveled only a relatively short distance after 10 milliseconds (10 msec).

In myelinated axons, conduction is somewhat different. The voltage-gated Na^+ and K^+ channels are located primarily at the nodes of Ranvier, the gaps in the myelin sheath. When a nerve impulse conducts along a myelinated axon, current carried by Na^+ and K^+ flows through the interstitial fluid surrounding the myelin sheath and through the cytosol from one node to the next (Figure 9.6b). The nerve impulse at the first node generates ionic currents that open voltage-gated Na^+ channels at the second node and trigger a nerve impulse there. Then the nerve impulse from the second node generates an ionic current that opens voltage-gated Na^+ channels at the third node, and so on. Each node depolarizes and then repolarizes. Note the impulse has traveled much farther along the myelinated axon in Figure 9.6b in the same interval. Because current flows across the membrane only at the nodes, the impulse appears to leap from node to node as each nodal area depolarizes to threshold. This type of impulse conduction is called ***saltatory conduction*** (SAL-ta-tō-rē; *saltat-* = leaping).

The diameter of the axon and the presence or absence of a myelin sheath are the most important factors that determine the speed of nerve impulse conduction. Axons with large diameters conduct impulses faster than those with small diameters. Also, myelinated axons conduct impulses faster than do unmyelinated axons. Axons with the largest diameters are all myelinated and therefore capable of saltatory conduction. The smallest diameter axons are unmyelinated, so their conduction is continuous. Axons conduct impulses at higher speeds when warmed and at lower speeds when cooled. Pain resulting from tissue injury such as that caused by a minor burn can be reduced by the application of ice because cooling slows conduction of nerve impulses along the axons of pain-sensitive neurons.

Figure 9.6 Conduction of a nerve impulse after it arises at the axon hillock. Dotted lines indicate ionic current flow. (a) In continuous conduction along an unmyelinated axon, ionic currents flow across each adjacent portion of the plasma membrane. (b) In saltatory conduction along a myelinated axon, the nerve impulse at the first node generates ionic currents in the cytosol and interstitial fluid that open voltage-gated Na^+ channels at the second node, and so on at each subsequent node.

Unmyelinated axons exhibit continuous conduction, and myelinated axons exhibit saltatory conduction.

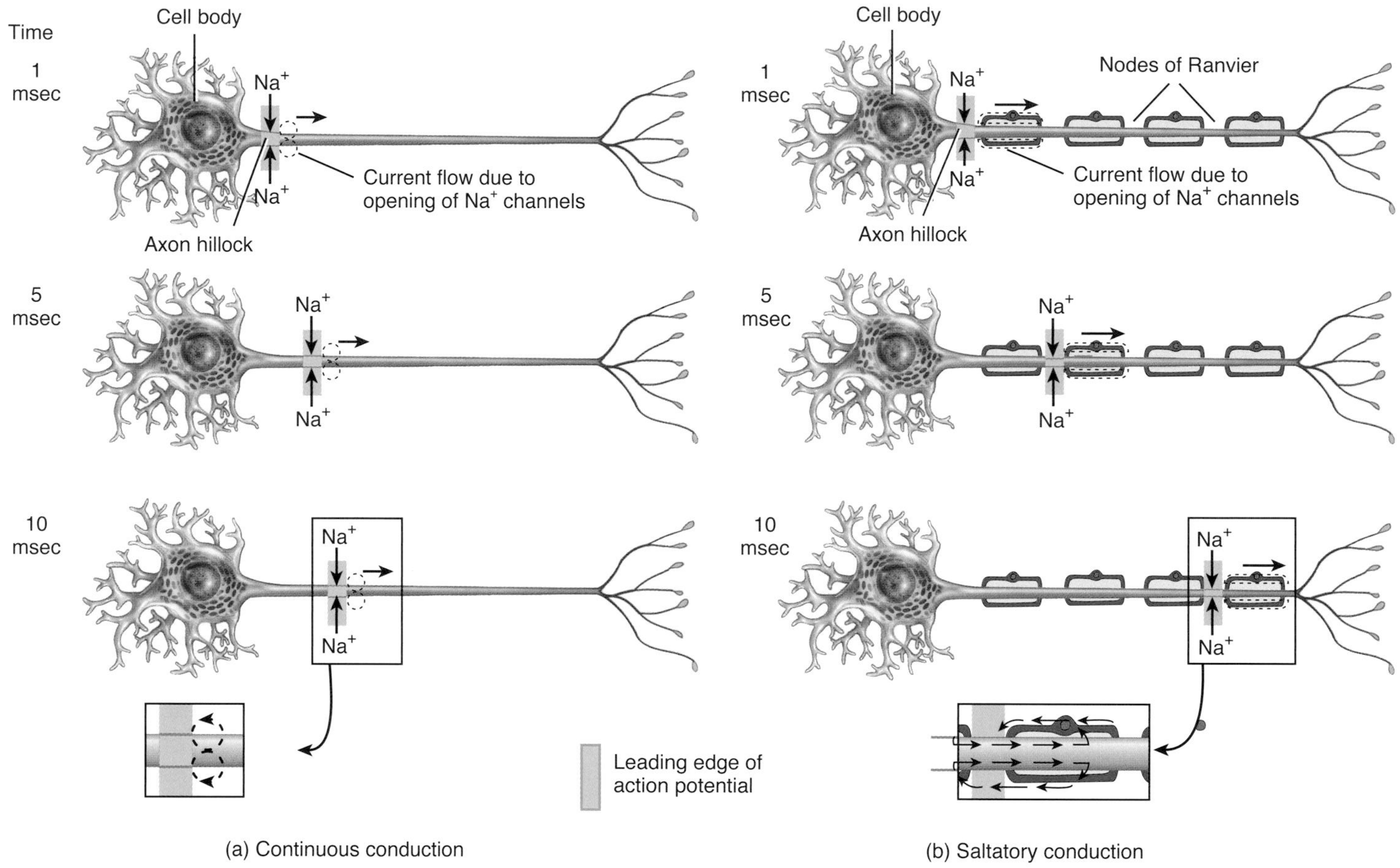

(a) Continuous conduction

(b) Saltatory conduction

What factors influence the speed of nerve impulse conduction?

Local anesthetics are drugs that block pain. Examples include procaine (Novocaine®) and Lidocaine, which may be used to produce anesthesia in the skin during suturing of a gash, in the mouth during dental work, or in the lower body during childbirth. These drugs act by blocking the opening of voltage-gated Na^+ channels. Nerve impulses cannot conduct past the obstructed region, so pain signals do not reach the CNS.

■ **CHECKPOINT**

7. What are the meanings of the terms resting membrane potential, depolarization, repolarization, nerve impulse, and refractory period?
8. How is saltatory conduction different from continuous conduction?

SYNAPTIC TRANSMISSION

OBJECTIVE • **Explain the events of synaptic transmission and the types of neurotransmitters used.**

Now that you know how action potentials arise and conduct along the axon of an individual neuron, we will explore how neurons communicate with one another. At synapses, neurons communicate with other neurons or with effectors by a series of events known as ***synaptic transmission***. In Chapter 8 we examined the events occurring at the neuromuscular junction, the synapse between a somatic motor neuron and a skeletal muscle fiber (see Figure 8.5 on page 178). Synapses between neurons operate in a similar way. The neuron sending the signal is called the ***presynaptic neuron*** (*pre-* = before), and the neuron receiving the message is called the ***postsynaptic neuron*** (*post-* = after).

Events at a Synapse

Although the presynaptic and postsynaptic neurons are in close proximity at a synapse, their plasma membranes do not touch. They are separated by the ***synaptic cleft***, a tiny space filled with interstitial fluid. Because nerve impulses cannot conduct across the synaptic cleft, an alternate, indirect form of communication occurs across this space. A typical synapse operates as follows (Figure 9.7):

1. A nerve impulse arrives at a synaptic end bulb of a presynaptic axon.
2. The depolarizing phase of the nerve impulse opens *voltage-gated Ca^{2+} channels*, which are present in the membrane of synaptic end bulbs. Because calcium ions are more concentrated in the interstitial fluid, Ca^{2+} flows into the synaptic end bulb through the opened channels.
3. An increase in the concentration of Ca^{2+} inside the synaptic end bulb triggers exocytosis of some of the synaptic vesicles, which releases thousands of neurotransmitter molecules into the synaptic cleft.
4. The neurotransmitter molecules diffuse across the synaptic cleft and bind to ***neurotransmitter receptors*** in the postsynaptic neuron's plasma membrane.
5. Binding of neurotransmitter molecules opens ion channels, which allows certain ions to flow across the membrane.
6. As ions flow through the opened channels, the voltage across the membrane changes. Depending on which ions

Figure 9.7 Synaptic transmission at a chemical synapse. Exocytosis of synaptic vesicles from a presynaptic neuron releases neurotransmitter molecules, which bind to receptors in the plasma membrane of the postsynaptic neuron.

At a chemical synapse, a presynaptic electrical signal (nerve impulse) is converted into a chemical signal (neurotransmitter release). The chemical signal is then converted back into an electrical signal (depolarization or hyperpolarization) in the postsynaptic cell.

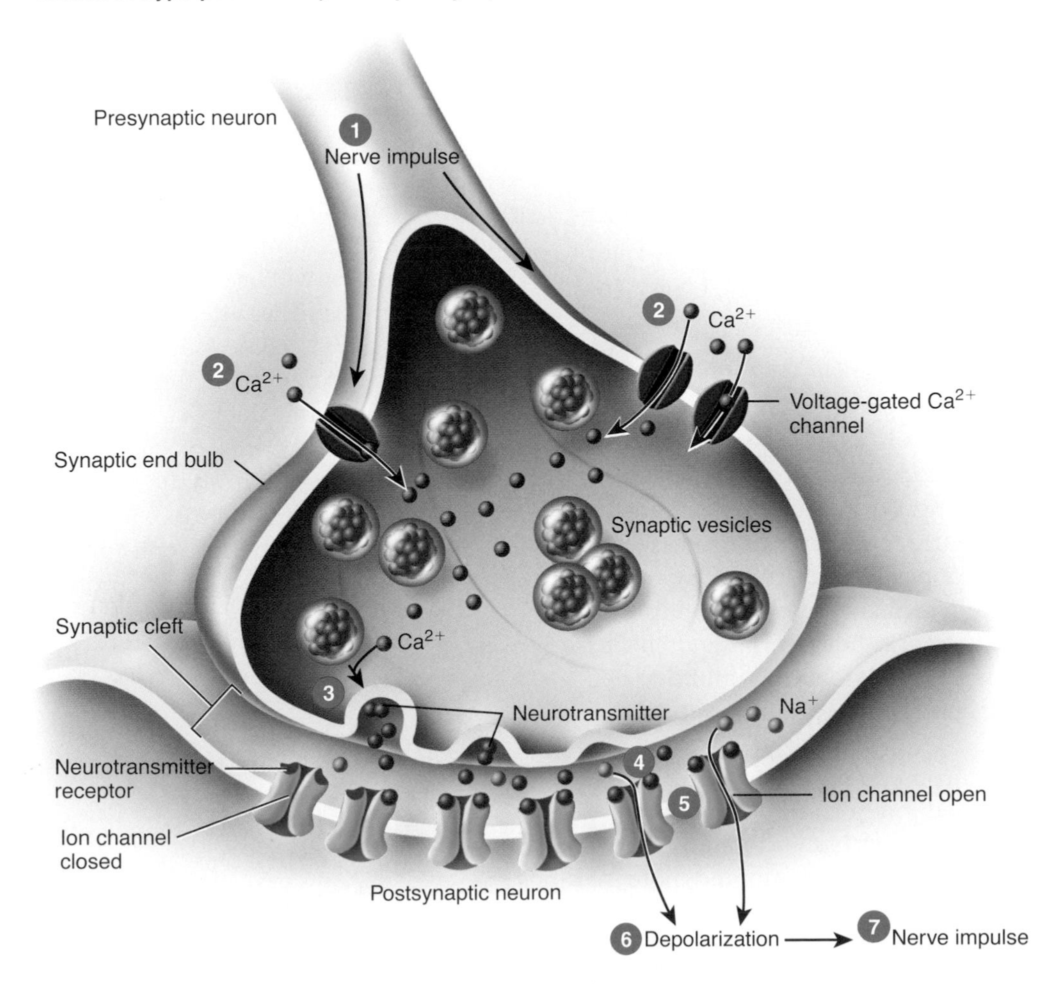

What causes the voltage-gated Ca^{2+} channels in synaptic end bulbs to open?

Focus on Wellness

Neurotransmitters—Why Food Affects Mood

Everyone who has enjoyed the soothing relaxation of a good meal has experienced the effect of food on mood. Neurons manufacture neurotransmitters from chemicals that come from food, so you could say that the story of the food–mood link begins with digestion. Many neurotransmitters are made from amino acids, which are the basic building blocks of proteins. Amino acids are made available when your body digests the protein in the food you eat. For example, the neurotransmitter serotonin is made from the amino acid tryptophan, and both dopamine and norepinephrine are synthesized from the amino acid tyrosine.

Mind-altering Food?

Regulation of neurotransmitter levels in the brain is quite complicated and depends not only on the availability of amino acid (and other) precursors, but also on competition of these precursors for entry into the brain. Consider serotonin, one of the neurotransmitters that appears to have an important effect on mood. Serotonin leads to feelings of relaxation and sleepiness.

Although serotonin is manufactured from the amino acid tryptophan, protein foods do not lead to higher levels of tryptophan in the blood or brain. This is because, after a high-protein meal, tryptophan must compete with more than 20 other amino acids for entry into the central nervous system, so its concentration in the brain remains relatively low. On the other hand, consumption of carbohydrate-rich foods, such as bread, pasta, potatoes, or sweets, is associated with an increase in the synthesis and release of serotonin in the brain. The result: Carbohydrates help us feel relaxed and sleepy.

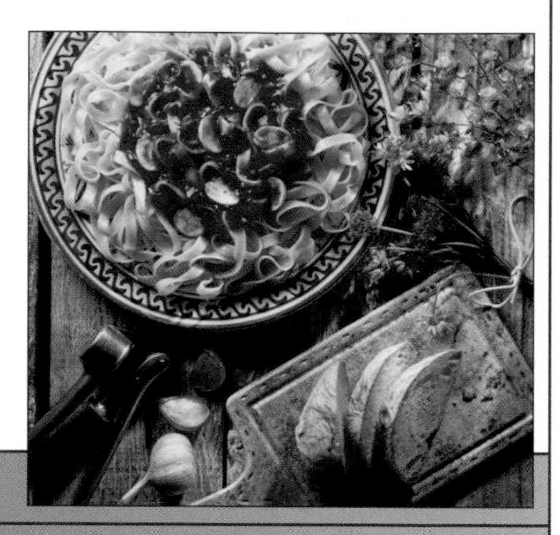

► **Think It Over . . .**

► ***Why might consuming a high-protein diet for several days lead to cravings for carbohydrate-rich foods?***

the channels admit, the voltage change may be a depolarization or a hyperpolarization.

7. If a depolarization occurs in the postsynaptic neuron and reaches threshold, then it triggers one or more nerve impulses.

At most synapses, only *one-way information transfer* can occur—from a presynaptic neuron to a postsynaptic neuron or to an effector, such as a muscle fiber or a gland cell. For example, synaptic transmission at a neuromuscular junction (NMJ) proceeds from a somatic motor neuron to a skeletal muscle fiber (but not in the opposite direction). Only synaptic end bulbs of presynaptic neurons can release neurotransmitters, and only the postsynaptic neuron's membrane has the correct receptor proteins to recognize and bind that neurotransmitter. As a result, nerve impulses move along their pathways in one direction.

When a postsynaptic neuron depolarizes, the effect is excitatory: If threshold is reached, one or more nerve impulses occur. By contrast, hyperpolarization has an inhibitory effect on the postsynaptic neuron: As the membrane potential moves farther away from threshold, nerve impulses are less likely to arise. A typical neuron in the CNS receives input from 1000 to 10,000 synapses. Some of this input is excitatory and some is inhibitory. The sum of all the excitatory and inhibitory effects at any given time determines whether one or more impulses will occur in the postsynaptic neuron.

A neurotransmitter affects the postsynaptic neuron, muscle fiber, or gland cell as long as it remains bound to its receptors. Thus, removal of the neurotransmitter is essential for normal synaptic function. Neurotransmitter is removed in three ways. (1) Some of the released neurotransmitter molecules diffuse away from the synaptic cleft. Once a neurotransmitter molecule is out of reach of its receptors, it can no longer exert an effect. (2) Some neurotransmitters are destroyed by enzymes. (3) Many neurotransmitters are actively transported back into the neuron that released them (reuptake). Others are transported into neighboring neuroglia (uptake).

Several therapeutically important drugs selectively block reuptake of specific neurotransmitters. For example, the drug fluoxetine (Prozac®) is a **selective serotonin reuptake inhibitor (SSRI)**. By blocking reuptake of serotonin, Prozac prolongs the activity of this neurotransmitter at synapses in the brain. SSRIs provide relief for those suffering from some forms of depression.

Neurotransmitters

About 100 substances are either known or suspected neurotransmitters. Most neurotransmitters are synthesized and loaded into synaptic vesicles in the synaptic end bulbs, close to their site of release. One of the best-studied neurotransmitters is ***acetylcholine (ACh)***, which is released by many PNS neurons and by some CNS neurons. ACh is an excitatory neurotransmitter at some synapses, such as the neuromuscular junction. It is also known to be an inhibitory neurotransmitter at other synapses. For example, parasympathetic neurons slow heart rate by releasing ACh at inhibitory synapses.

Several amino acids are neurotransmitters in the CNS. ***Glutamate*** and ***aspartate*** have powerful excitatory effects. Two other amino acids, ***gamma aminobutyric*** (GAM-ma am-i-nō-bū-TIR-ik) ***acid (GABA)*** and ***glycine***, are important inhibitory neurotransmitters. Antianxiety drugs such as diazepam (Valium®) enhance the action of GABA.

Some neurotransmitters are modified amino acids. These include norepinephrine, dopamine, and serotonin. ***Norepinephrine (NE)*** plays roles in arousal (awakening from deep sleep), dreaming, and regulating mood. Brain neurons containing the neurotransmitter ***dopamine (DA)*** are active during emotional responses, addictive behaviors, and pleasurable experiences. In addition, dopamine-releasing neurons help regulate skeletal muscle tone and some aspects of movement due to contraction of skeletal muscles. One form of schizophrenia is due to accumulation of excess dopamine. ***Serotonin*** is thought to be involved in sensory perception, temperature regulation, control of mood, appetite, and the onset of sleep.

Neurotransmitters consisting of amino acids linked by peptide bonds are called ***neuropeptides*** (noor-ō-PEP-tīds). The ***endorphins*** (en-DOR-fins) are neuropeptides that are the body's natural painkillers. Acupuncture may produce analgesia (loss of pain sensation) by increasing the release of endorphins. They have also been linked to improved memory and learning and to feelings of pleasure or euphoria.

An important newcomer to the ranks of recognized neurotransmitters is the simple gas ***nitric oxide (NO)***, which is different from all previously known neurotransmitters because it is not synthesized in advance and packaged into synaptic vesicles. Rather, it is formed on demand, diffuses out of cells that produce it and into neighboring cells, and acts immediately. Some research suggests that NO plays a role in learning and memory.

Substances naturally present in the body as well as drugs and toxins can **modify the effects of neurotransmitters** in several ways. Cocaine produces euphoria—intensely pleasurable feelings—by blocking reuptake of dopamine. This action allows dopamine to linger longer in synaptic clefts, producing excessive stimulation of certain brain regions. Isoproterenol (Isuprel®) can be used to dilate the airways during an asthma attack because it binds to and activates receptors for norepinephrine. Zyprexa®, a drug prescribed for schizophrenia, is effective because it binds to and blocks receptors for serotonin and dopamine.

■ **CHECKPOINT**

9. How are neurotransmitters removed after they are released from synaptic vesicles?

COMMON DISORDERS

Multiple Sclerosis

Multiple sclerosis (MS) is a disease that causes progressive destruction of myelin sheaths of neurons in the CNS. It afflicts about 2 million people worldwide and affects females twice as often as males. The condition's name describes the anatomical pathology: In *multiple* regions, the myelin sheaths deteriorate to *scleroses*, which are hardened scars or plaques. The destruction of myelin sheaths slows and then short-circuits conduction of nerve impulses.

The most common form of the condition is relapsing-remitting MS, which usually appears in early adulthood. The first symptoms may include a feeling of heaviness or weakness in the muscles, abnormal sensations, or double vision. An attack is followed by a period of remission during which the symptoms temporarily disappear. One attack follows another over the years. The result is a progressive loss of function interspersed with remission periods, during which symptoms abate.

MS is an autoimmune disease—the body's own immune system spearheads the attack. Although the trigger of MS is unknown, both genetic susceptibility and exposure to some environmental factor (perhaps a herpes virus) appear to contribute. Many patients with relapsing-remitting MS are treated with injections of beta interferon. This treatment lengthens the time between relapses, decreases the severity of relapses, and slows formation of new lesions in some cases. Unfortunately, not all MS patients can tolerate beta interferon, and therapy becomes less effective as the disease progresses.

Epilepsy

Epilepsy is a disorder characterized by short, recurrent, periodic attacks of motor, sensory, or psychological malfunction, although it almost never affects intelligence. The attacks, called *epileptic seizures*, afflict about 1% of the world's population. They are initiated by abnormal, synchronous electrical discharges from millions of neurons in the brain. As a result, lights, noise, or smells may be sensed when the eyes, ears, and nose have not been stimulated. In addition, the skeletal muscles of a person having a seizure may contract involuntarily. *Partial seizures* begin in a small focus on one side of the brain and produce milder symptoms; *generalized seizures* involve larger areas on both sides of the brain and loss of consciousness.

Epilepsy has many causes, including brain damage at birth (the most common cause); metabolic disturbances such as insufficient glucose or oxygen in the blood; infections; toxins; loss of blood or low blood pressure; head injuries; and tumors and abscesses of the brain. However, most epileptic seizures have no demonstrable cause.

Epileptic seizures often can be eliminated or alleviated by antiepileptic drugs, such as phenytoin, carbamazepine, and valproate sodium. An implantable device that stimulates the vagus (X) nerve also has produced dramatic results in reducing seizures in some patients whose epilepsy was not well-controlled by drugs.

MEDICAL TERMINOLOGY AND CONDITIONS

Demyelination (dē-mī-e-li-NĀ-shun) Loss or destruction of myelin sheaths around axons in the CNS or PNS.

Guillain-Barré Syndrome (GBS) (gē-an ba-RĀ) A demyelinating disorder in which macrophages remove myelin from PNS axons. It is a common cause of sudden paralysis and may result from the immune system's response to a bacterial infection. Most patients recover completely or partially, but about 15% remain paralyzed.

Neuropathy (noo-ROP-a-thē; *neuro-* = a nerve; *-pathy* = disease) Any disorder that affects the nervous system, but particularly a disorder of a cranial or spinal nerve.

STUDY OUTLINE

Overview of the Nervous System (p. 226)

1. Components of the nervous system include the brain, 12 pairs of cranial nerves and their branches, the spinal cord, 31 pairs of spinal nerves and their branches, sensory receptors, ganglia, and enteric plexuses.
2. Three basic functions of the nervous system are detecting stimuli (sensory function); analyzing, integrating, and storing sensory information (integrative function); and responding to integrative decisions (motor function).
3. Sensory (afferent) neurons provide input to the CNS; motor (efferent) neurons carry output from the CNS to effectors.
4. The two main subsystems of the nervous system are (1) the central nervous system (CNS), the brain and spinal cord, and (2) the peripheral nervous system (PNS), all nervous tissues outside the brain and spinal cord.
5. The PNS also is subdivided into the somatic nervous system (SNS), autonomic nervous system (ANS), and enteric nervous system (ENS).
6. The SNS consists of (1) sensory neurons that conduct impulses from somatic and special sense receptors to the CNS, and (2) motor neurons from the CNS to skeletal muscles.
7. The ANS contains (1) sensory neurons from visceral organs and (2) motor neurons in two divisions, sympathetic and parasympathetic, that convey impulses from the CNS to smooth muscle tissue, cardiac muscle tissue, and glands.
8. The ENS consists of neurons in two enteric plexuses that extend the length of the gastrointestinal (GI) tract; it monitors sensory changes and controls operation of the GI tract.

Histology of Nervous Tissue (p. 228)

1. Nervous tissue consists of two types of cells: neurons and neuroglia. Neurons are specialized for nerve impulse conduction and provide most of the unique functions of the nervous system, such as sensing, thinking, remembering, controlling muscle activity, and regulating glandular secretions. Neuroglia support, nourish, and protect the neurons and maintain homeostasis in the interstitial fluid that bathes neurons.
2. Most neurons have three parts. The dendrites are the main receiving or input region. Integration occurs in the cell body. The output part typically is a single axon, which conducts nerve impulses toward another neuron, a muscle fiber, or a gland cell.
3. Two types of neuroglia produce myelin sheaths: oligodendrocytes myelinate axons in the CNS, and Schwann cells myelinate axons in the PNS.
4. White matter primarily contains myelinated axons; gray matter contains neuronal cell bodies, dendrites, axon terminals, unmyelinated axons, and neuroglia.
5. In the spinal cord, gray matter forms an H-shaped inner core that is surrounded by white matter. In the brain, a thin, superficial shell of gray matter covers the cerebrum and cerebellum.
6. Neuroglia include astrocytes, oligodendrocytes, microglia, ependymal cells, Schwann cells, and satellite cells (see Table 9.1 on page 231).

Action Potentials (p. 230)

1. Neurons communicate with one another using nerve action potentials, also called nerve impulses.
2. Generation of action potentials depends on the existence of a resting membrane potential and the presence of voltage-gated channels for Na^+ and K^+.
3. A typical value for the resting membrane potential (difference in electrical charge across the plasma membrane) is −70 mV. A cell that exhibits a membrane potential is polarized.

4. The resting membrane potential arises due to an unequal distribution of ions on either side of the plasma membrane and a higher membrane permeability to K^+ than to Na^+. The level of K^+ is higher inside and the level of Na^+ is higher outside, a situation that is maintained by sodium-potassium pumps.
5. The ability of muscle fibers and neurons to respond to a stimulus and convert it into action potentials is called excitability.
6. During an action potential, voltage-gated Na^+ and K^+ channels open in sequence. Opening of voltage-gated Na^+ channels results in depolarization, the loss and then reversal of membrane polarization (from -70 mV to $+30$ mV). Then, opening of voltage-gated K^+ channels allows repolarization, recovery of the membrane potential to the resting level.
7. According to the all-or-none principle, if a stimulus is strong enough to generate an action potential, the impulse generated is of a constant size.
8. During the refractory period, another action potential cannot be generated.
9. Nerve impulse conduction that occurs as a step-by-step process along an unmyelinated axon is called continuous conduction. In saltatory conduction, a nerve impulse "leaps" from one node of Ranvier to the next along a myelinated axon.
10. Axons with larger diameters conduct impulses faster than those with smaller diameters; myelinated axons conduct impulses faster than unmyelinated axons.

Synaptic Transmission (p. 234)

1. Neurons communicate with other neurons and with effectors at synapses in a series of events known as synaptic transmission.
2. At a synapse, a neurotransmitter is released from a presynaptic neuron into the synaptic cleft and then binds to receptors on the postsynaptic plasma membrane.
3. An excitatory neurotransmitter depolarizes the postsynaptic neuron's membrane, brings the membrane potential closer to threshold, and increases the chance that one or more action potentials will arise. An inhibitory neurotransmitter hyperpolarizes the membrane of the postsynaptic neuron, thereby inhibiting action potential generation.
4. Neurotransmitter is removed in three ways: diffusion, enzymatic destruction, and reuptake by neurons or neuroglia.
5. Important neurotransmitters include acetylcholine, glutamate, aspartate, gamma amino butyric acid (GABA), glycine, norepinephrine, dopamine, serotonin, neuropeptides, and nitric oxide.

SELF-QUIZ

1. Which of the following are incorrectly matched?
   a. central nervous system: composed of the brain and spinal cord
   b. somatic nervous system: includes motor neurons to skeletal muscles
   c. sympathetic nervous system: includes motor neurons to skeletal, smooth, and cardiac muscles
   d. peripheral nervous system: includes cranial and spinal nerves
   e. autonomic nervous system: includes parasympathetic and sympathetic divisions
2. The portion of the nervous system that regulates the gastrointestinal (GI) tract is the
   a. somatic nervous system b. sympathetic division
   c. integrative division d. central nervous system
   e. enteric nervous system
3. Damage to dendrites would interfere with a neuron's ability to
   a. receive input b. make proteins c. conduct nerve impulses to another neuron d. release neurotransmitters
   e. form myelin
4. The type of cell that produces myelin sheaths around axons in the CNS is the
   a. astrocyte b. myelinocyte c. Schwann cell
   d. oligodendrocyte e. microglia
5. A bundle of axons in the CNS is
   a. a tract b. a nucleus c. a mixed nerve
   d. a ganglion e. an enteric plexus
6. Which of the following is NOT true concerning the repair of nervous tissue?
   a. If the cell body is not damaged, neurons in the PNS may be able to repair themselves.
   b. In the CNS, myelin inhibits neuronal regeneration.
   c. Injury to the CNS is usually permanent.
   d. Active Schwann cells contribute to the repair process in the PNS
   e. A regeneration tube forms across the injured area of a CNS neuron that undergoes repair.
7. In a resting neuron
   a. there is a high concentration of K^+ outside the cell
   b. negatively charged ions move freely through the plasma membrane
   c. the sodium-potassium pumps help maintain the low concentration of Na^+ inside the cell
   d. the outside surface of the plasma membrane has a negative charge
   e. the plasma membrane is highly permeable to Na^+
8. The depolarizing phase of a nerve impulse is caused by a
   a. rush of Na^+ into the neuron
   b. rush of Na^+ out of the neuron
   c. rush of K^+ into the neuron
   d. rush of K^+ out of the neuron
   e. pumping of K^+ into the neuron

9. If a stimulus is strong enough to generate an action potential, the impulse generated is of a constant size. A stronger stimulus cannot generate a larger impulse. This is known as
   a. the principle of polarization–depolarization
   b. saltatory conduction
   c. the all-or-none principle
   d. the principle of reflex action
   e. the absolute refractory period

10. Place the following events in the correct order of occurrence:
   1. Voltage-gated Na^+ channels open and permit Na^+ to rush inside the neuron.
   2. The Na^+/K^+ pump restores the ions to their original sites.
   3. A stimulus of threshold strength is applied to the neuron.
   4. The membrane polarization changes from negative (−55 mV) to positive (+30 mV).
   5. Voltage-gated K^+ channels open, and K^+ flows out of the neurons.

   a. 4, 1, 2, 3, 5 b. 4, 3, 1, 2, 5 c. 3, 1, 4, 2, 5 d. 5, 3, 1, 4, 2 e. 3, 1, 4, 5, 2

11. Saltatory conduction occurs
   a. in unmyelinated axons
   b. at the nodes of Ranvier
   c. in the smallest diameter axons
   d. in skeletal muscle fibers
   e. in cardiac muscle fibers

12. The speed of nerve impulse conduction is increased by
   a. cold b. a very strong stimulus c. small diameter of the axon d. myelination e. astrocytes

13. For a signal to be transmitted by means of a chemical synapse from a presynaptic neuron to a postsynaptic neuron,
   a. the presynaptic neuron must be touching the postsynaptic neuron
   b. the postsynaptic neuron must contain neurotransmitter receptors
   c. there must be gap junctions present between the two neurons
   d. the postsynaptic neuron needs to release neurotransmitters from its synaptic vesicles
   e. the neurons must be myelinated

14. What would happen at the postsynaptic neuron if the total inhibitory effects of the neurotransmitters were greater than the total excitatory effects?
   a. A nerve impulse would be generated.
   b. It would be easier to generate a nerve impulse when the next stimulus was received.
   c. The nerve impulse would be rerouted to another neuron.
   d. No nerve impulse would be generated.
   e. The neurotransmitter would be broken down more quickly.

15. Match the following neurotransmitters with their descriptions.

   ____ a. inhibitory amino acid in the CNS
   ____ b. a gaseous neurotransmitter that is not packaged into synaptic vesicles
   ____ c. excitatory amino acid in the CNS
   ____ d. body's natural painkillers
   ____ e. helps regulate mood
   ____ f. neurotransmitter that activates skeletal muscle fibers

   A. serotonin
   B. acetylcholine
   C. endorphins
   D. GABA
   E. nitric oxide
   F. glutamate

16. Match the following.

   ____ a. the portion of a neuron containing the nucleus
   ____ b. rounded structure at the distal end of an axon terminal
   ____ c. highly branched, input part of a neuron
   ____ d. sac in which neurotransmitter is stored
   ____ e. neuron located entirely within the CNS
   ____ f. long, cylindrical process that conducts impulses toward another neuron
   ____ g. produces myelin sheath in PNS
   ____ h. unmyelinated gap in the myelin sheath
   ____ i. substance that increases the speed of nerve impulse conduction
   ____ j. neuron that conveys information from a receptor to the CNS
   ____ k. neuron that conveys information from the CNS to an effector
   ____ l. bundle of many axons in the PNS
   ____ m. bundle of many axons in the CNS
   ____ n. group of cell bodies in the PNS
   ____ o. group of cell bodies in the CNS
   ____ p. substance used for communication at chemical synapses

   A. synaptic end bulb
   B. motor neuron
   C. sensory neuron
   D. dendrite
   E. interneuron
   F. nucleus
   G myelin sheath
   H. Schwann cell
   I. cell body
   J. node of Ranvier
   K. ganglion
   L. nerve
   M. neurotransmitter
   N. tract
   O. synaptic vesicle
   P. axon

CRITICAL THINKING APPLICATIONS

1. The buzzing of the alarm clock awoke Rodrigo. He stretched, yawned, and started to salivate as he smelled the brewing coffee. List the divisions of the nervous system that are involved in each of these activities.

2. Angelina just figured out that her A & P class actually starts at 10:00 A.M. and not at 10:15, which has been her arrival time since the beginning of the term. One of the other students remarks that Angelina's "gray matter is pretty thin." Should Angelina thank him?

3. Sarah really looks forward to the great feeling she has after going for a nice long run on the weekends. By the end of her run, she doesn't even feel the pain in her sore feet. Sarah read in a magazine that some kind of natural brain chemical was responsible for the "runner's high" that she feels. Are there such chemicals in Sarah's brain?

4. The pediatrician was trying to educate the anxious new parents of a six-month-old baby. "No, don't worry about him not walking yet. The myelination of the baby's nervous system is not finished yet." Explain what the pediatrician means by this reassurance.

ANSWERS TO FIGURE QUESTIONS

9.1 The total number of cranial and spinal nerves in your body is $(12 \times 2) + (31 \times 2) = 86$.

9.2 Sensory or afferent neurons carry input to the CNS. Motor or efferent neurons carry output from the CNS.

9.3 The axon conducts nerve impulses and transmits the message to another neuron or effector cell by releasing a neurotransmitter at its axon terminals.

9.4 A typical value for the resting membrane potential in a neuron is -70 mV.

9.5 Voltage-gated Na^+ channels are open during the depolarizing phase, and voltage-gated K^+ channels are open during the repolarizing phase of an action potential.

9.6 The two main factors that influence conduction speed of a nerve impulse are the axon diameter (larger axons conduct impulses more rapidly) and the presence or absence of a myelin sheath (myelinated axons conduct more rapidly than unmyelinated axons).

9.7 The depolarizing phase of the action potential opens the voltage-gated Ca^{2+} channels in synaptic end bulbs.

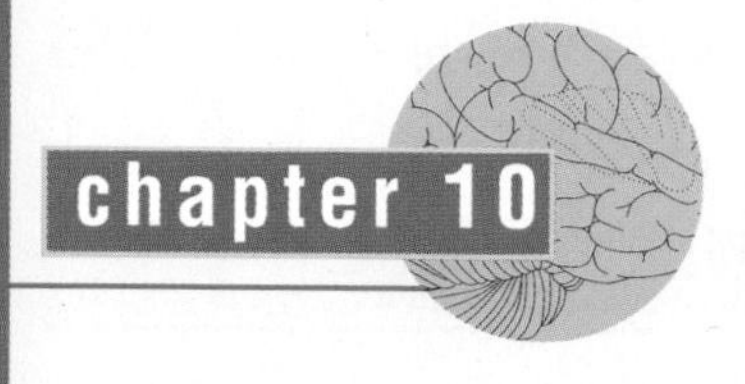

CENTRAL NERVOUS SYSTEM, SPINAL NERVES, AND CRANIAL NERVES

did you know? ***Athletes perform exercise training to stimulate physiological adaptations that lead to improved sports performance. Researchers now believe that a similar kind of training effect may occur in one of the most important organs in your body: the brain. Just as muscles respond to strength training by increasing in size and becoming stronger, so does the brain respond by increasing the number of neural pathways, the connections between neurons that allow you to think and to remember. Undergoing years of "mental exercise" may be one of the reasons that people with college degrees have a lower risk of developing Alzheimer disease.***

Focus on Wellness, page 257

www.wiley.com/college/apcentral

Now that you understand how the nervous system functions on the cellular level, in this chapter we will explore the structure and functions of the ***central nervous system* (CNS)**, which consists of the brain and spinal cord. We will also examine spinal nerves and cranial nerves, which are part of the peripheral nervous system (PNS) (see Figure 9.1 on page 226).

looking back to move ahead . . .

- **Skull and Hyoid Bone (page 125)**
- **Vertebral Column (page 133)**
- **Structures of the Nervous System (page 226)**
- **Structure of a Neuron (page 228)**
- **Gray and White Matter (page 228)**

SPINAL CORD STRUCTURE

OBJECTIVES • Describe how the spinal cord is protected.

• Describe the structure of the spinal cord.

Protection and Coverings: Vertebral Canal and Meninges

The spinal cord is located within the vertebral cavity of the vertebral column. Because the wall of the vertebral cavity is essentially a ring of bone, the cord is well protected. The vertebral ligaments, meninges, and cerebrospinal fluid provide additional protection.

The ***meninges*** (me-NIN-jēz) are three layers of connective tissue coverings that extend around the spinal cord and brain. The meninges that protect the spinal cord, the *spinal meninges* (Figure 10.1), are continuous with those that protect the brain, the *cranial meninges* (see Figure 10.7). The outermost of the three layers of the meninges is called the ***dura mater*** (DOO-ra MĀter = tough mother). Its tough, dense irregular connective tissue helps protect the delicate structures of the CNS. The tube of spinal dura mater extends to the second sacral vertebra, well beyond the spinal cord, which ends at about the level of the second lumbar vertebra. The spinal cord is also protected by a cushion of fat and connective tissue located in the ***epidural space***, a space between the dura mater and vertebral column.

The middle layer of the meninges is called the ***arachnoid mater*** (a-RAK-noyd; *arachn-* = spider; *-oid* = similar to) because the arrangement of its collagen and elastic fibers resembles a spider's web. The inner layer, the ***pia mater*** (PĒ-a MĀ-ter; *pia* = delicate), is a transparent layer of collagen and elastic fibers that adheres to the surface of the spinal cord and brain. It contains numerous blood vessels. Between the arachnoid mater and the pia mater is the ***subarachnoid space***, where cerebrospinal fluid circulates.

In a **spinal tap (lumbar puncture)**, a local anesthetic is given, and a long needle is inserted into the subarachnoid space. In adults, a spinal tap is normally performed between the third and fourth or fourth and fifth lumbar vertebrae. Because this region is inferior to the lowest portion of the spinal cord, it provides relatively safe access. The procedure is used to withdraw cerebrospinal fluid (CSF) for diagnostic purposes; to introduce antibiotics, contrast media for myelography, or anesthetics; to administer chemotherapy; to measure CSF pressure; and/or to evaluate the effects of treatment for diseases such as meningitis.

Figure 10.1 Spinal meninges.

Meninges are connective tissue coverings that surround the brain and spinal cord.

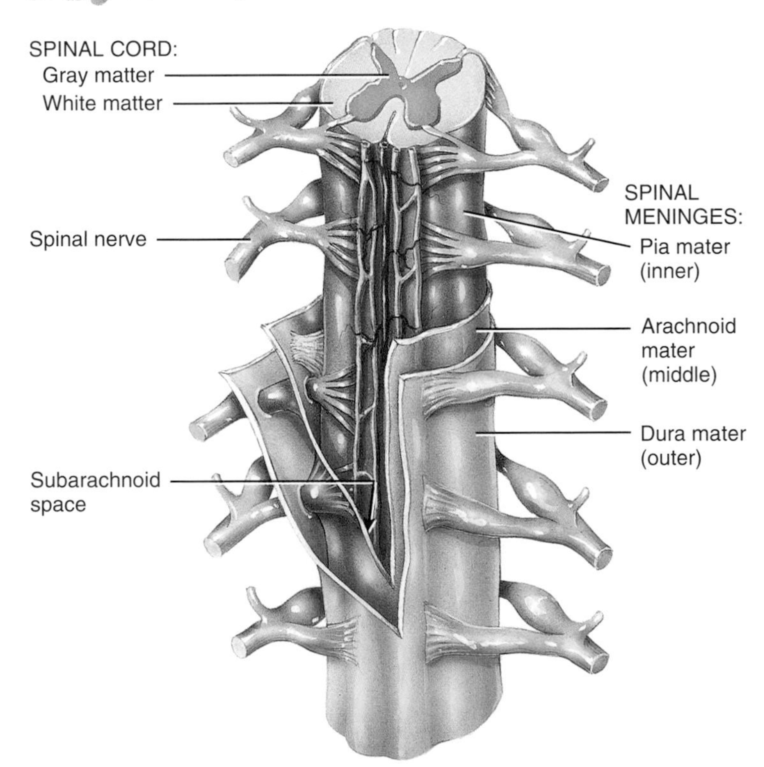

Anterior view and transverse section through spinal cord

In which meningeal space does cerebrospinal fluid circulate?

Gross Anatomy of the Spinal Cord

The length of the adult ***spinal cord*** ranges from 42 to 45 cm (16 to 18 in.). It extends from the lowest part of the brain, the medulla oblongata, to the upper border of the second lumbar vertebra in the vertebral column (Figure 10.2). Because the spinal cord is shorter than the vertebral column, nerves that arise from the lumbar, sacral, and coccygeal regions of the spinal cord do not leave the vertebral column at the same level they exit the cord. The roots of these spinal nerves angle down the vertebral cavity like wisps of flowing hair. They are appropriately named the ***cauda equina*** (KAW-da ē-KWĪ-na), meaning horse's tail. The spinal cord has two conspicuous enlargements: The ***cervical enlargement*** contains nerves that supply the upper limbs, and the ***lumbar enlargement*** contains nerves supplying the lower limbs. Each of 31 ***spinal segments*** of the spinal cord gives rise to a pair of spinal nerves (Figure 10.2).

Two grooves, the deep ***anterior median fissure*** and the shallow ***posterior median sulcus***, divide the spinal cord into right and left halves (see Figure 10.3). In the spinal cord, white matter surrounds a centrally located H-shaped mass of gray matter. In the center of the gray matter is the ***central canal***, a small space that extends the length of the cord and contains cerebrospinal fluid.

Figure 10.2 Spinal cord and spinal nerves. Selected nerves are labeled on the left side of the figure. Together, the lumbar and sacral plexuses are called the lumbosacral plexus.

The spinal cord extends from the base of the skull to the superior border of the second lumbar vertebra.

CERVICAL PLEXUS (C1–C5):
Phrenic nerve
BRACHIAL PLEXUS (C5–T1):
Musculocutaneous nerve
Axillary nerve
Median nerve
Radial nerve
Ulnar nerve
Intercostal (thoracic) nerves
LUMBAR PLEXUS (L1–L4):
Ilioinguinal nerve
Femoral nerve
Obturator nerve
SACRAL PLEXUS (L4–S4):
Superior gluteal nerve
Inferior gluteal nerve
Sciatic nerve
Pudendal nerve
C1
C2
C3
C4
C5
C6
C7
C8
T1
T2
T3
T4
T5
T6
T7
T8
T9
T10
T11
T12
L1
L2
L3
L4
L5
S1
S2
S3
S4
S5
Medulla oblongata of brain
Atlas (first cervical vertebra)
CERVICAL NERVES (8 pairs)
Cervical enlargement
First thoracic vertebra
THORACIC NERVES (12 pairs)
Lumbar enlargement
Second lumbar vertebra
LUMBAR NERVES (5 pairs)
Cauda equina
Ilium of hip bone
Sacrum
SACRAL NERVES (5 pairs)
COCCYGEAL NERVES (1 pair)

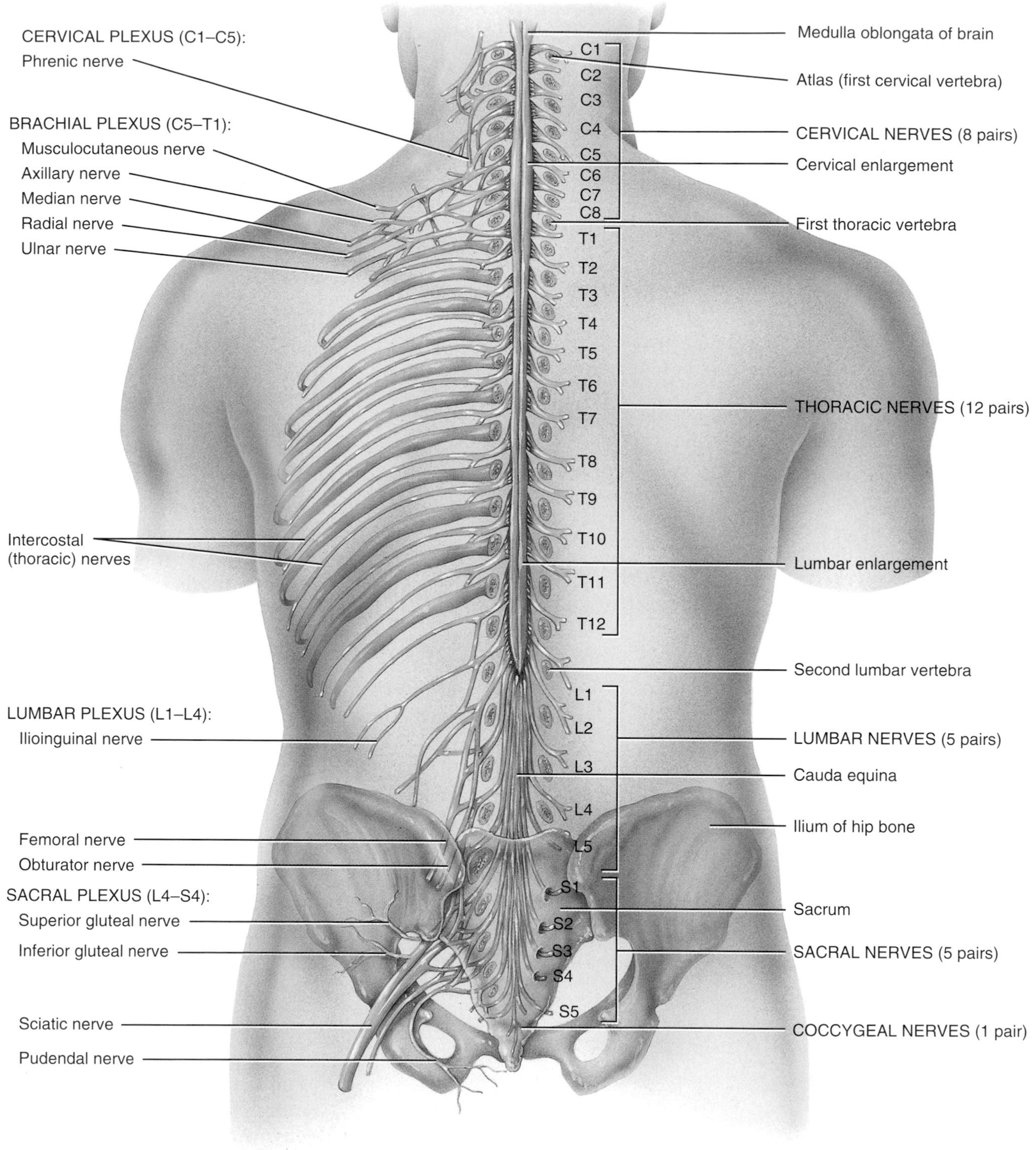

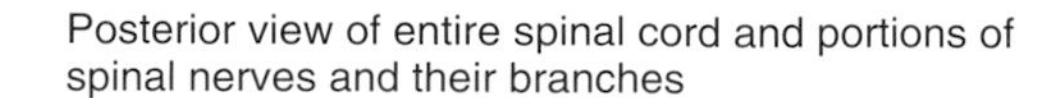
Posterior view of entire spinal cord and portions of spinal nerves and their branches

Are spinal nerves part of the CNS or the PNS?

Internal Structure of the Spinal Cord

The gray matter of the spinal cord contains neuronal cell bodies, dendrites, unmyelinated axons, axon terminals, and neuroglia. On each side of the spinal cord, the gray matter is subdivided into regions called ***horns***, named relative to their location: anterior, lateral, and posterior (Figure 10.3). The ***anterior (ventral) gray horns*** contain cell bodies of somatic motor neurons, which provide the nerve impulses that cause contraction of skeletal muscles. The ***posterior (dorsal) gray horns*** contain somatic and autonomic sensory neurons. Between the anterior and posterior gray horns are the ***lateral gray horns***, which are present only in the thoracic, upper lumbar, and sacral segments of the cord. The lateral horns contain the cell bodies of autonomic motor neurons that regulate the activity of smooth muscle, cardiac muscle, and glands.

The white matter of the spinal cord consists primarily of myelinated axons of neurons and is organized into regions called anterior, lateral, and posterior ***white columns***. Each column contains one or more ***tracts***, which are distinct bundles of axons having a common origin or destination and carrying similar information. ***Sensory (ascending) tracts*** consist of axons that conduct nerve impulses toward the brain. Tracts consisting of axons that carry nerve impulses down the spinal cord are called ***motor (descending) tracts***. Sensory and motor tracts of the spinal cord are continuous with sensory and motor tracts in the brain. Often, the name of a tract indicates its position in the white matter, where it begins and ends, and the direction of nerve impulse conduction. For example, the anterior spinothalamic tract is located in the *anterior* white column; it begins in the *spinal cord*, and it ends in the *thalamus* (a region of the brain) (see Figure 10.14b).

■ CHECKPOINT

1. How is the spinal cord protected?
2. What body regions are served by nerves from the cervical and lumbar enlargements?
3. Distinguish between a horn and a column in the spinal cord.

Figure 10.3 Internal structure of the spinal cord. Columns of white matter surround the gray matter.

The spinal cord conducts nerve impulses along tracts and serves as an integrating center for spinal reflexes.

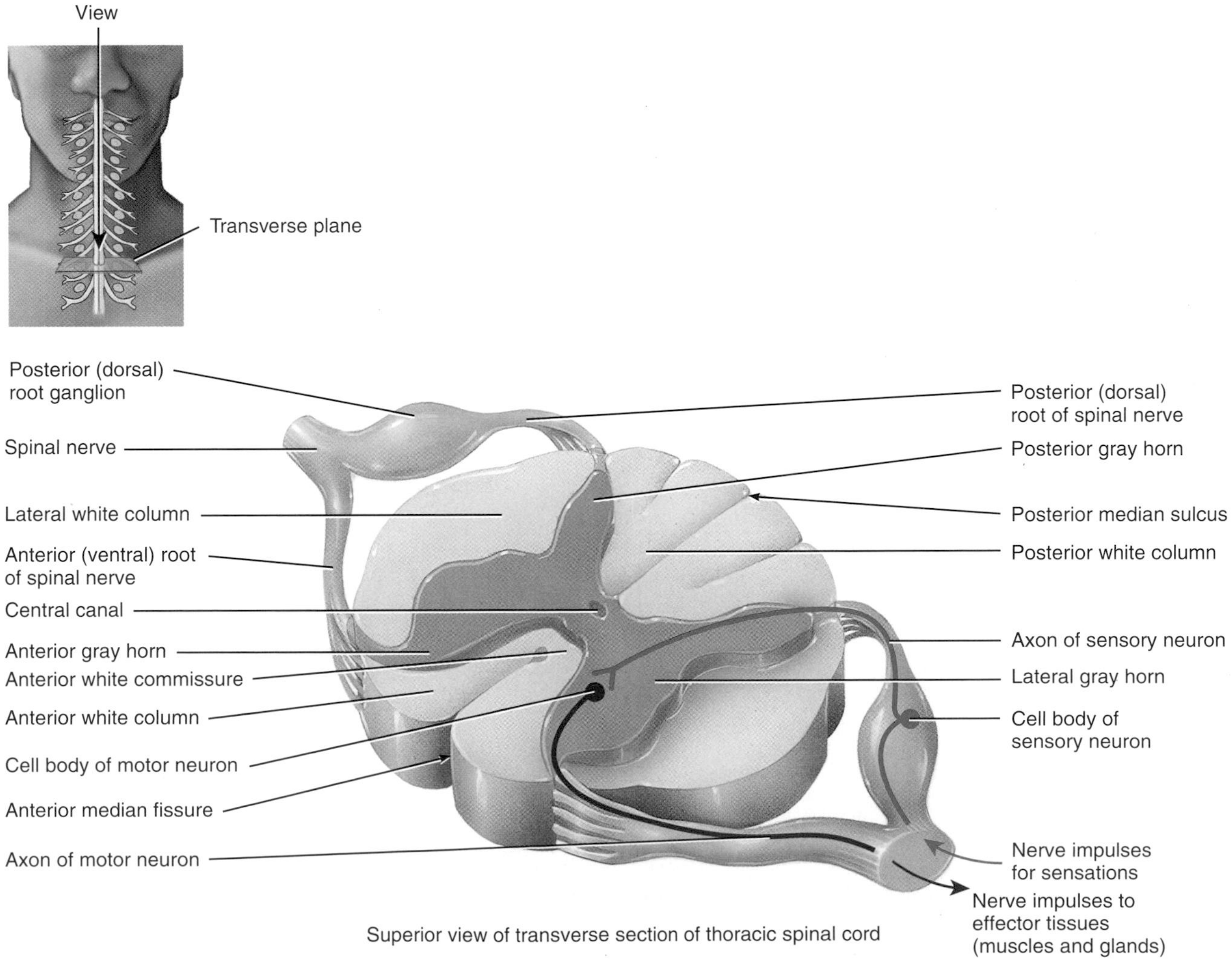

View

Transverse plane

Posterior (dorsal) root ganglion

Spinal nerve

Lateral white column

Anterior (ventral) root of spinal nerve

Central canal

Anterior gray horn

Anterior white commissure

Anterior white column

Cell body of motor neuron

Anterior median fissure

Axon of motor neuron

Posterior (dorsal) root of spinal nerve

Posterior gray horn

Posterior median sulcus

Posterior white column

Axon of sensory neuron

Lateral gray horn

Cell body of sensory neuron

Nerve impulses for sensations

Nerve impulses to effector tissues (muscles and glands)

Superior view of transverse section of thoracic spinal cord

What is the difference between a *horn* and a *column* in the spinal cord?

SPINAL NERVES

OBJECTIVE • Describe the composition, coverings, and distribution of spinal nerves.

Spinal nerves are the paths of communication between the spinal cord and the nerves that serve specific regions of the body. Two bundles of axons, called ***roots***, connect each spinal nerve to a segment of the cord (Figure 10.3). The ***posterior (dorsal) root*** contains only sensory axons, which conduct nerve impulses for sensations from the skin, muscles, and internal organs into the central nervous system. Each posterior root also has a swelling, the ***posterior (dorsal) root ganglion***, that contains the cell bodies of sensory neurons. The other point of attachment of a spinal nerve to the cord is the ***anterior (ventral) root***. It contains axons of ***somatic motor neurons***, which conduct nerve impulses from the CNS to skeletal muscles, and ***autonomic motor neurons***, which conduct impulses to smooth muscle, cardiac muscle, and glands. A spinal nerve thus contains both sensory and motor axons and therefore is a ***mixed nerve***.

Spinal nerves and the nerves that branch from them are part of the peripheral nervous system (PNS). They connect the CNS to sensory receptors, muscles, and glands in all parts of the body. The 31 pairs of spinal nerves are named and numbered according to the region and level of the vertebral column from which they emerge (see Figure 10.2). There are 8 pairs of cervical nerves, 12 pairs of thoracic nerves, 5 pairs of lumbar nerves, 5 pairs of sacral nerves, and 1 pair of coccygeal nerves. The first cervical pair emerges above the atlas. All other spinal nerves leave the vertebral column by passing through the ***intervertebral foramina***, the holes between vertebrae.

Spinal Nerve Coverings

Each spinal nerve (and cranial nerve) contains layers of protective connective tissue coverings (Figure 10.4). Individual axons, whether myelinated or unmyelinated, are wrapped in ***endoneurium*** (en′-dō-NOO-rē-um; *endo-* = within or inner). Groups of axons with their endoneurium are arranged in bundles, called ***fascicles***, each of which is wrapped in ***perineurium*** (per′-i-NOO-rē-um; *peri-* = around). The superficial covering over the entire nerve is the ***epineurium*** (ep′-i-NOO-rē-um; *epi-* = over). The dura mater of the spinal meninges fuses with the epineurium as a spinal nerve passes through the intervertebral foramen. Note the presence of many blood vessels, which nourish nerves, within the perineurium and epineurium.

Distribution of Spinal Nerves

Plexuses

A short distance after passing through its intervertebral foramen, a spinal nerve divides into several branches. Many of the spinal nerve branches do not extend directly to the body structures they supply. Instead, they form networks on either side of the body by joining with axons from adjacent nerves. Such a network is called a ***plexus*** (= braid or network). Emerging from the plexuses are nerves bearing names that are often descriptive of the general regions they supply or the course they take. Each of the nerves, in turn, may have several branches named for the specific structures they supply.

The major plexuses are the cervical plexus, brachial plexus, lumbar plexus, and sacral plexus (see Figure 10.2). The ***cervical plexus*** supplies the skin and muscles of the posterior head, neck, upper part of the shoulders, and the diaphragm. The phrenic nerves, which stimulate the diaphragm to contract, arise from the cervical plexus. Damage to the spinal cord above the origin of the phrenic nerves may cause respiratory failure. The ***brachial plexus*** constitutes the nerve supply for the upper limbs and several neck and shoulder muscles. Among the nerves that arise from the brachial plexus are the musculocutaneous, axillary, median, radial, and ulnar nerves. The ***lumbar plexus*** supplies the abdominal wall, external genitals, and part of the lower limbs. Arising from this plexus are the ilioinguinal, femoral, and obturator nerves. The ***sacral plexus*** supplies the buttocks, perineum,

Figure 10.4 Composition and connective tissue coverings of a spinal nerve.

Three layers of connective tissue wrappings protect axons: endoneurium surrounds individual axons, perineurium surrounds bundles of axons, and epineurium surrounds an entire nerve.

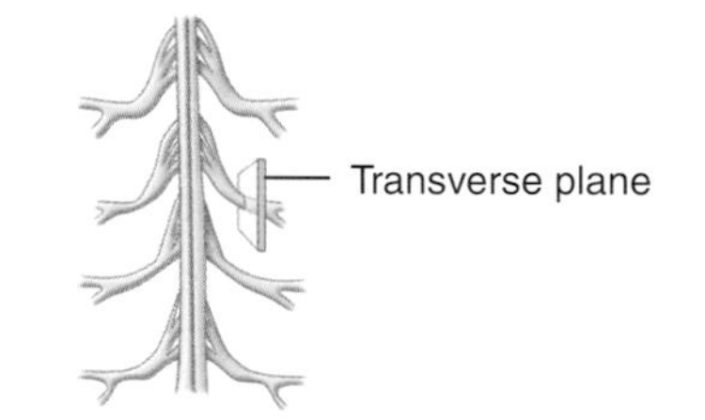

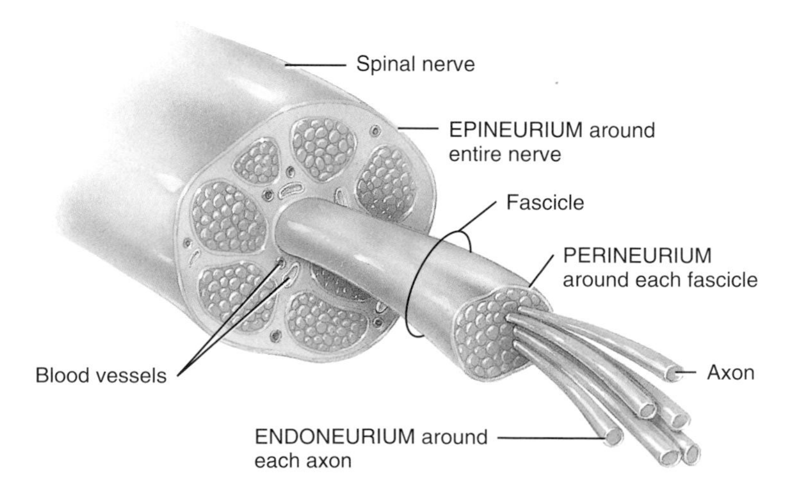

Transverse section showing the coverings of a spinal nerve

Why are all spinal nerves classified as mixed nerves?

and lower limbs. Among the nerves that arise from this plexus are the gluteal, sciatic, and pudendal nerves. The sciatic nerve is the longest nerve in the body.

Spinal nerves T2 to T11 do not form plexuses. They are known as ***intercostal nerves*** and extend directly to the structures they supply, including the muscles between ribs, abdominal muscles, and skin of the chest and back (see Figure 10.2).

■ **CHECKPOINT**

4. How do spinal nerves connect to the spinal cord?
5. Which regions of the body are supplied by plexuses, and which are served by intercostal nerves?

SPINAL CORD FUNCTIONS

OBJECTIVES • **Describe the functions of the spinal cord.** • **Describe the components of a reflex arc.**

The spinal cord white matter and gray matter have two major functions in maintaining homeostasis. (1) The white matter of the spinal cord consists of tracts that serve as highways for nerve impulse conduction. Along these highways, sensory impulses travel toward the brain and motor impulses travel from the brain toward skeletal muscles and other effector tissues. The route that nerve impulses follow from a neuron in one part of the body to other neurons elsewhere in the body is called a ***pathway***. After describing the functions of various regions of the brain, we will depict some important pathways that connect the spinal cord and brain (see Figures 10.14 and 10.15). (2) The gray matter of the spinal cord receives and integrates incoming and outgoing information and is a site for integration of reflexes. A ***reflex*** is a fast, involuntary sequence of actions that occurs in response to a particular stimulus. Some reflexes are inborn, such as pulling your hand away from a hot surface before you even feel that it is hot (a ***withdrawal reflex***). Other reflexes are learned or acquired, such as the many reflexes you learn while acquiring driving skills. When integration takes place in the spinal cord gray matter, the reflex is a ***spinal reflex***. By contrast, if integration occurs in the brain stem rather than the spinal cord, the reflex is a ***cranial reflex***. An example is the tracking movements of your eyes as you read this sentence.

The pathway followed by nerve impulses that produce a reflex is known as a ***reflex arc***. Using the ***patellar reflex*** (knee jerk reflex) as an example, the basic components of a reflex arc are as follows (Figure 10.5):

1. **Sensory receptor.** The distal end of a sensory neuron (or sometimes a separate receptor cell) serves as a ***sensory receptor***. Sensory receptors respond to a specific type of stimulus by generating one or more nerve impulses. In the patellar reflex, sensory receptors known as *muscle spindles* detect slight stretching of the quadriceps femoris (anterior thigh) muscle when the patellar (knee cap) ligament is tapped with a reflex hammer.
2. **Sensory neuron.** The nerve impulses conduct from the sensory receptor along the axon of a ***sensory neuron*** to its axon terminals, which are located in the CNS gray matter. Axon branches of the sensory neuron also relay nerve impulses to the brain, allowing conscious awareness that the reflex has occurred.
3. **Integrating center.** One or more regions of gray matter in the CNS act as an ***integrating center***. In the simplest type of reflex, such as the patellar reflex, the integrating center is a single synapse between a sensory neuron and a motor neuron. In other types of reflexes, the integrating center includes one or more interneurons.
4. **Motor neuron.** Impulses triggered by the integrating center pass out of the spinal cord (or brain stem, in the case of a cranial reflex) along a ***motor neuron*** to the part of the body that will respond. In the patellar reflex, the axon of the motor neuron extends to the quadriceps femoris muscle.
5. **Effector.** The part of the body that responds to the motor nerve impulse, such as a muscle or gland, is the ***effector***. Its action is a reflex. If the effector is skeletal muscle, the reflex is a ***somatic reflex***. If the effector is smooth muscle, cardiac muscle, or a gland, the reflex is an ***autonomic (visceral) reflex***. For example, the acts of swallowing, urinating, and defecating all involve autonomic reflexes. The patellar reflex is a somatic reflex because its effector is the quadriceps femoris muscle, which contracts and thereby relieves the stretching that initiated the reflex. In sum, the patellar reflex causes extension of the knee by contraction of the quadriceps femoris muscle in response to tapping the patellar ligament.

Damage or disease anywhere along a reflex arc can cause the reflex to be absent or abnormal. For example, **absence of the patellar reflex** could indicate damage of the sensory or motor neurons, or a spinal cord injury, in the lumbar region. Somatic reflexes generally can be tested simply by tapping or stroking the body surface. Most autonomic reflexes, by contrast, are not practical diagnostic tools because it is difficult to stimulate visceral receptors, which are deep inside the body. An exception is the pupillary light reflex, in which the pupils of both eyes decrease in diameter when either eye is exposed to light. Because the reflex arc includes synapses in lower parts of the brain, the **absence of a normal pupillary light reflex** may indicate brain damage or injury.

■ **CHECKPOINT**

6. What is the significance of the white matter tracts of the spinal cord?
7. How are somatic and autonomic reflexes similar and different?

Figure 10.5 Patellar reflex, showing general components of a reflex arc. The arrows show the direction of nerve impulse conduction.

A reflex is a fast, involuntary sequence of actions that occurs in response to a particular stimulus.

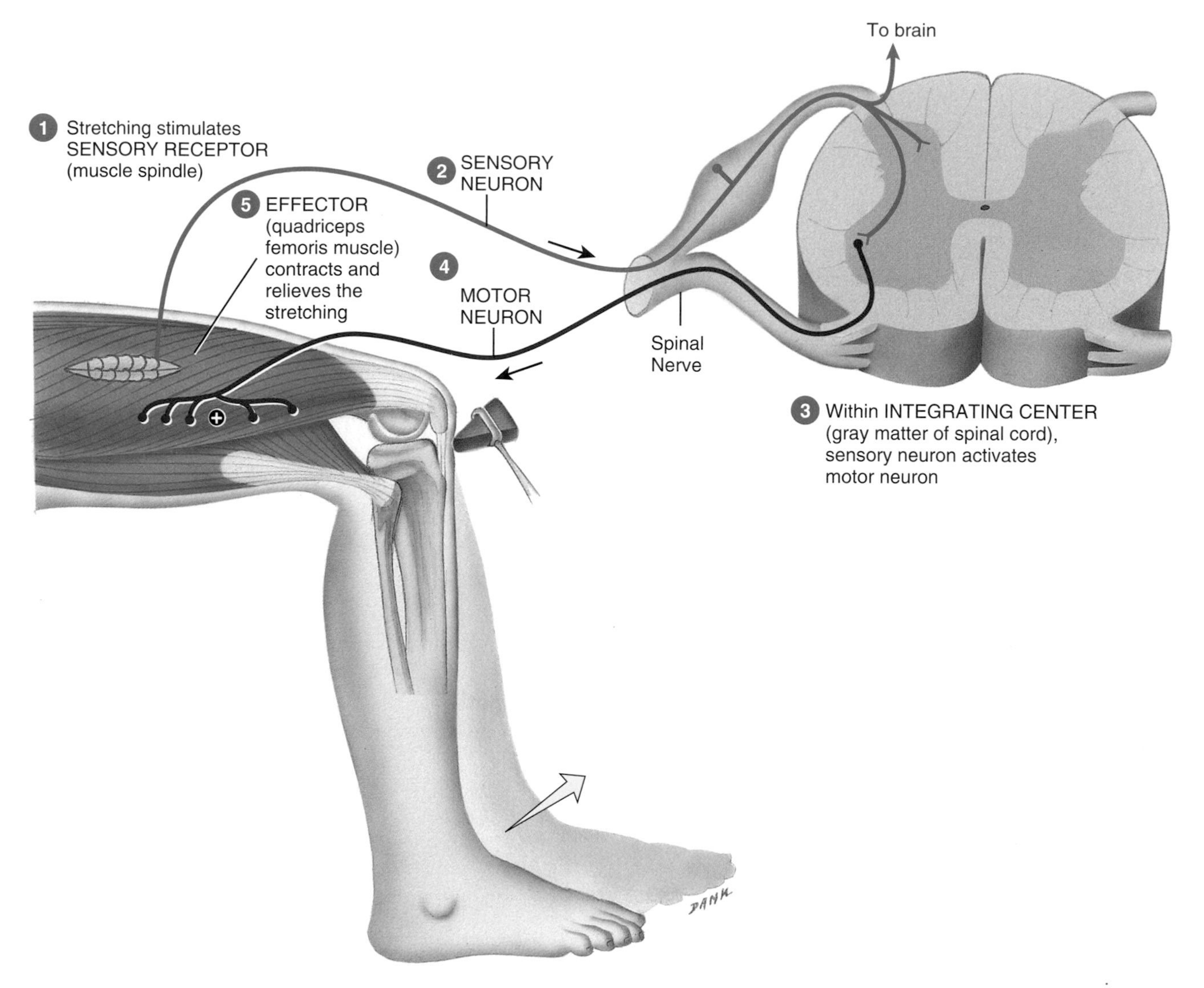

Which attachment of a spinal nerve to the spinal cord contains axons of sensory neurons? Which attachment contains axons of motor neurons?

BRAIN

OBJECTIVES • **Discuss how the brain is protected and supplied with blood.**

- **Name the major parts of the brain and explain the function of each part.**
- **Describe three somatic sensory and somatic motor pathways.**

Next, we will consider the major parts of the brain, how the brain is protected, and how it is related to the spinal cord and cranial nerves.

Major Parts and Protective Coverings

The ***brain*** is one of the largest organs of the body, consisting of about 100 billion neurons and 10–50 trillion neuroglia with a mass of about 1300 g (almost 3 lb). The four major parts are the brain stem, diencephalon, cerebrum, and cerebellum (Figure 10.6). The ***brain stem*** is continuous with the spinal cord and consists of the medulla oblongata, pons, and midbrain. Above the brain stem is the ***diencephalon*** (dī′-en-SEF-a-lon; *di-* = through; *-encephalon* = brain), consisting mostly of the thalamus, hypothalamus, and pineal gland. Supported on the diencephalon and brain stem and forming the bulk of the brain is the ***cerebrum*** (se-RĒ-brum = brain).

Figure 10.6 Brain. The pituitary gland is discussed together with the endocrine system in Chapter 13.

The four major parts of the brain are the brain stem, cerebellum, diencephalon, and cerebrum.

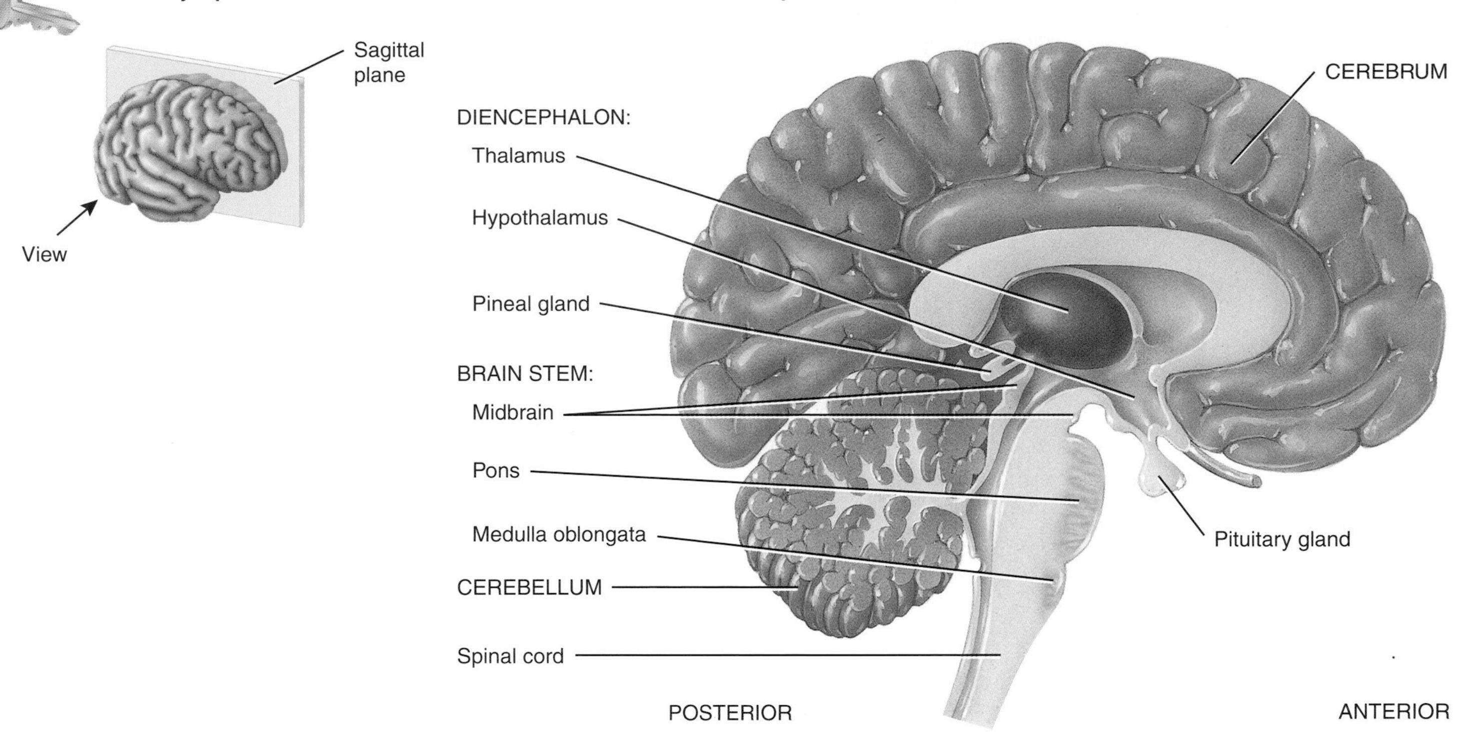

(a) Sagittal section, medial view

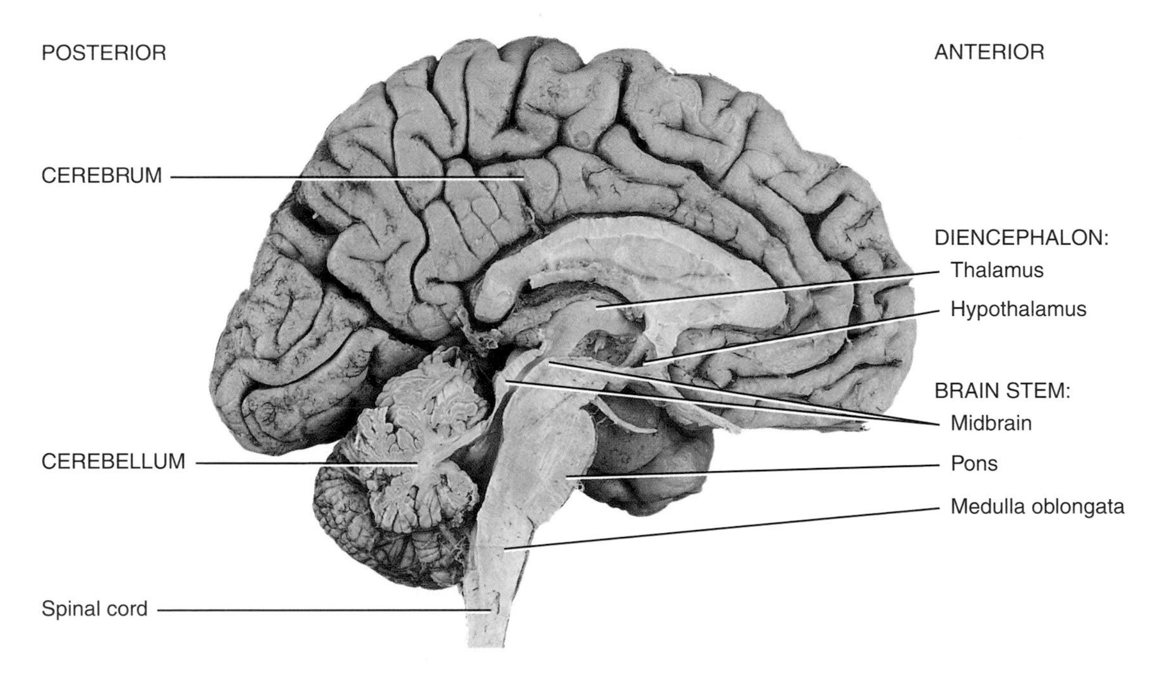

(b) Sagittal section, medial view

Which part of the brain attaches to the spinal cord?

The surface of the cerebrum is composed of a thin layer of gray matter, the *cerebral cortex* (*cortex* = rind or bark), beneath which lies the cerebral white matter. Posterior to the brain stem is the ***cerebellum*** (ser′-e-BEL-um = little brain).

As you learned earlier in the chapter, the brain is protected by the cranium and cranial meninges. The ***cranial meninges*** have the same names as the spinal meninges: the outermost ***dura mater***, middle ***arachnoid mater***, and innermost ***pia mater*** (Figure 10.7).

Brain Blood Supply and the Blood–Brain Barrier

Although the brain constitutes only about 2% of total body weight, it requires about 20% of the body's oxygen supply. If blood flow to the brain stops, even briefly, unconsciousness may result. Brain neurons that are totally deprived of oxygen for four or more minutes may be permanently injured. Blood supplying the brain also contains glucose, the main source of energy for brain cells. Because virtually no glucose is stored in the brain, the supply of glucose also must be continuous. If blood entering the brain has a low level of glucose, mental confusion, dizziness, convulsions, and loss of consciousness may occur.

The existence of a ***blood–brain barrier (BBB)*** protects brain cells from harmful substances and pathogens by preventing passage of many substances from blood into brain tissue. This barrier consists basically of very tightly sealed blood capillaries (microscopic blood vessels) in the brain. However, lipid-soluble substances such as oxygen, carbon dioxide, alcohol, and most anesthetic agents, easily cross the blood-brain barrier. Trauma, certain toxins, and inflammation can cause a breakdown of the blood–brain barrier.

Cerebrospinal Fluid

The spinal cord and brain are further protected against chemical and physical injury by ***cerebrospinal fluid (CSF)***. CSF is a clear, colorless liquid that carries oxygen, glucose, and other needed chemicals from the blood to neurons and neuroglia and removes wastes and toxic substances produced by brain and spinal cord cells. CSF circulates through the subarachnoid space (between the arachnoid mater and pia mater), around the brain and spinal cord, and through cavities in the brain known as ***ventricles*** (VEN-tri-kuls = little cavities). There are four ventricles: two ***lateral ventricles***, one ***third ventricle***, and one ***fourth ventricle*** (Figure 10.7). Openings connect them with one another, with the central canal of the spinal cord, and with the subarachnoid space.

The sites of CSF production are the ***choroid plexuses*** (KŌ-royd = membranelike), which are specialized networks of capillaries in the walls of the ventricles (Figure 10.7). Covering the choroid plexus capillaries are ependymal cells, which form cerebrospinal fluid from blood plasma by filtration and secretion. From the fourth ventricle, CSF flows into the central canal of the spinal cord and into the subarachnoid space around the surface of the brain and spinal cord. CSF is gradually reabsorbed into the blood through ***arachnoid villi***, which are fingerlike extensions of the arachnoid mater. The CSF drains primarily into a vein called the ***superior sagittal sinus*** (Figure 10.7). Normally, the volume of CSF remains constant at 80 to 150 mL (3 to 5 oz) because it is absorbed as rapidly as it is formed.

Abnormalities in the brain—tumors, inflammation, or developmental malformation—can interfere with the drainage of CSF from the ventricles into the subarachnoid space. When *excess* CSF accumulates in the ventricles, the CSF pressure rises. Elevated CSF pressure causes a condition called **hydrocephalus** (hī′-drō-SEF-a-lus; *hydro-* = water; *cephal-* = head). In a baby in whom the fontanels have not yet closed, the head bulges due to the increased pressure. If the condition persists, the fluid buildup compresses and damages the delicate nervous tissue. Hydrocephalus is relieved by draining the excess CSF. A neurosurgeon may implant a drain line, called a shunt, into the lateral ventricle to divert CSF into the superior vena cava or abdominal cavity, where it can be absorbed by the blood. In adults, hydrocephalus may occur after head injury, meningitis, or subarachnoid hemorrhage. This condition can quickly become life-threatening and requires immediate intervention; since the adult skull bones have already fused, nervous tissue damage occurs quickly.

Brain Stem

The brain stem is the part of the brain between the spinal cord and the diencephalon. It consists of three regions: (1) the medulla oblongata, (2) pons, and (3) midbrain. Extending through the brain stem is the reticular formation, a region where gray and white matter are intermingled.

Medulla Oblongata

The ***medulla oblongata*** (me-DOOL-la ob′-long-GA-ta), or simply ***medulla***, is a continuation of the spinal cord (see Figure 10.6). It forms the inferior part of the brain stem (Figure 10.8 on page 252). Within the medulla's white matter are all sensory (ascending) and motor (descending) tracts extending between the spinal cord and other parts of the brain.

The medulla also contains several nuclei, which are masses of gray matter where neurons form synapses with one another. Two major nuclei are the ***cardiovascular center***, which regulates the rate and force of the heartbeat and the diameter of blood vessels (see Figure 15.9 on page 377), and the ***medullary rhythmicity area***, which adjusts the basic rhythm of breathing (see Figure 18.12 on page 462). Nuclei associated with sensations of touch and vibration are located

Figure 10.7 Meninges and ventricles of the brain.

Cerebrospinal fluid (CSF) protects the brain and spinal cord and delivers nutrients from the blood to the brain and spinal cord; CSF also removes wastes from the brain and spinal cord to the blood.

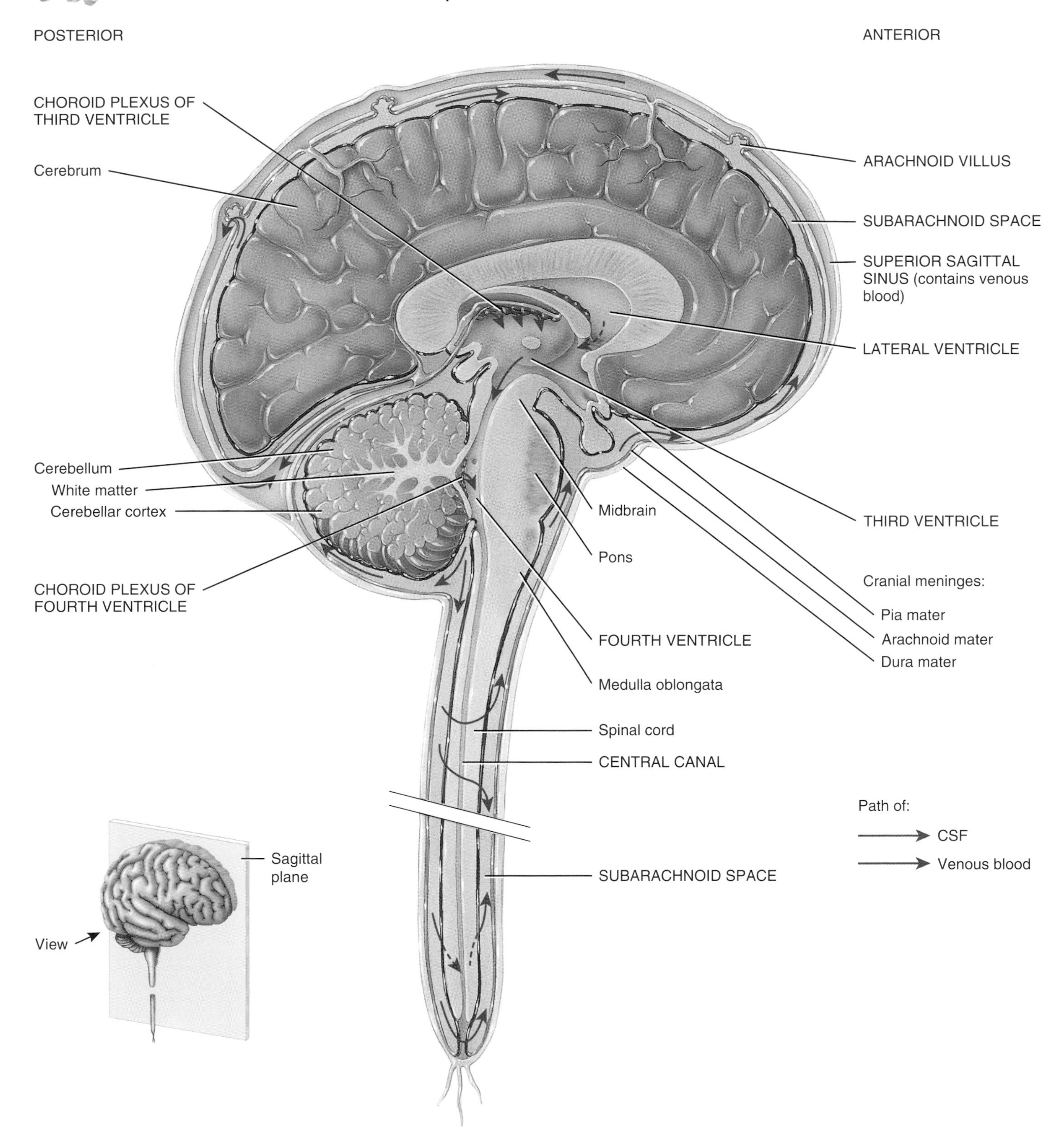

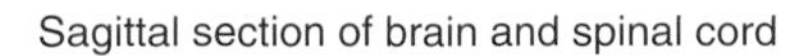
Sagittal section of brain and spinal cord

Where is CSF formed and absorbed?

Figure 10.8 Inferior aspect of the brain, showing the brain stem and cranial nerves.

The brain stem consists of the medulla oblongata, pons, and midbrain.

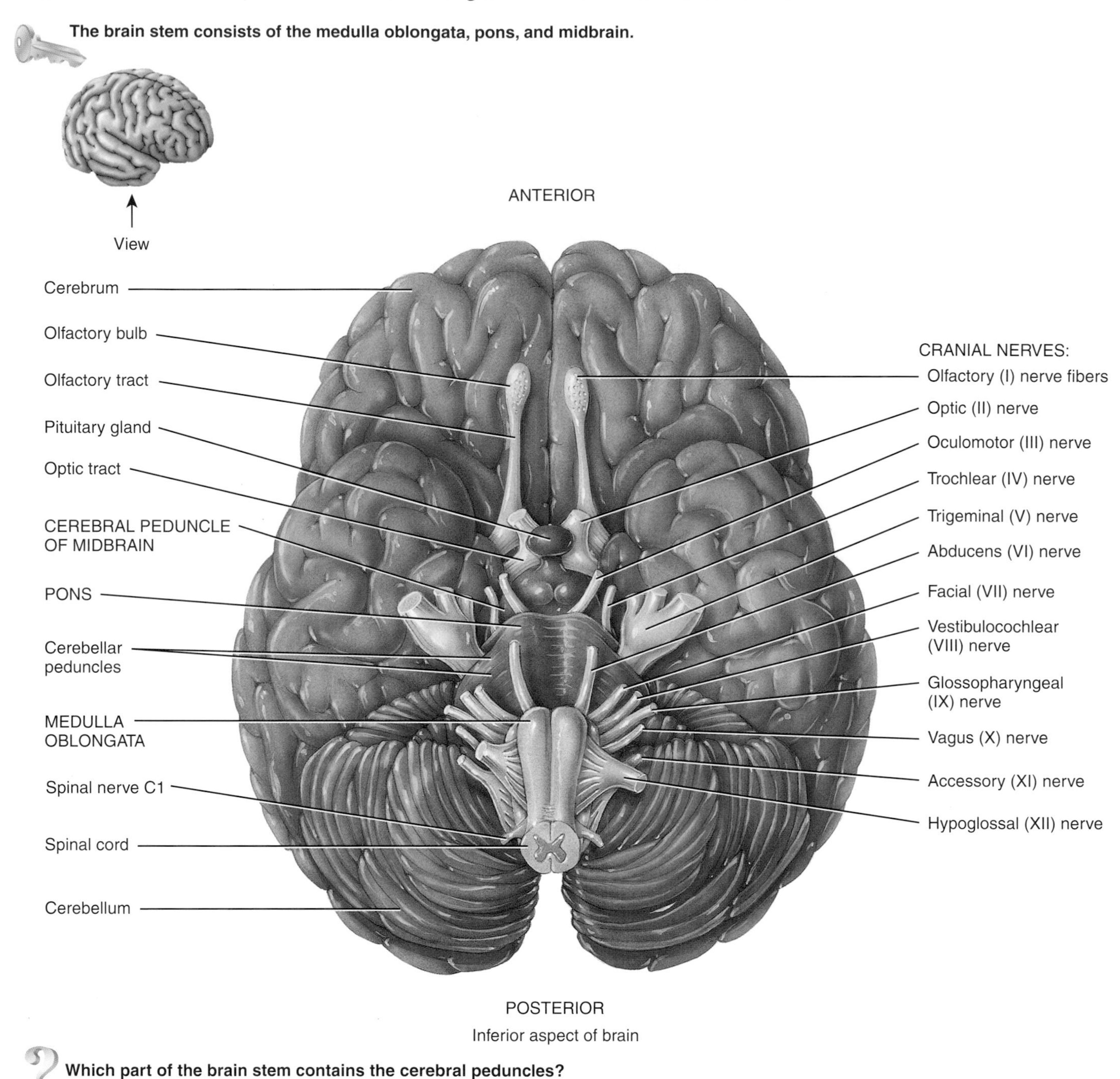

Inferior aspect of brain

Which part of the brain stem contains the cerebral peduncles?

in the posterior part of the medulla. Many ascending sensory axons form synapses in these nuclei (see Figure 10.14a). Other nuclei in the medulla control reflexes for swallowing, vomiting, coughing, hiccupping, and sneezing. Finally, the medulla contains nuclei associated with five pairs of cranial nerves (Figure 10.8): vestibulocochlear (VIII) nerves, glossopharyngeal (IX) nerves, vagus (X) nerves, accessory (XI) nerves (cranial portion), and hypoglossal (XII) nerves.

Given the many vital activities controlled by the medulla, it is not surprising that a hard blow to the back of the head or upper neck can be fatal. **Damage to the medullary rhythmicity area** is particularly serious and can rapidly lead to death. Symptoms of nonfatal injury to the medulla may include paralysis and loss of sensation on the opposite side of the body, and irregularities in breathing or heart rhythm.

Pons

The ***pons*** (= bridge) is above the medulla and anterior to the cerebellum (Figures 10.6, 10.7, and 10.8). Like the medulla, the pons consists of both nuclei and tracts. As its name implies, the pons is a bridge that connects parts of the brain with one another. These connections are bundles of axons. Some axons of the pons connect the right and left sides of the cerebellum. Others are part of ascending sensory tracts and descending motor tracts. Several nuclei in the pons are the sites where signals for voluntary movements that originate in the cerebral cortex are relayed into the cerebellum. Other nuclei in the pons help control breathing. The pons also contains nuclei associated with the following four pairs of cranial nerves (Figure 10.8): trigeminal (V) nerves, abducens (VI) nerves, facial (VII) nerves, and vestibulocochlear (VIII) nerves.

Midbrain

The ***midbrain*** connects the pons to the diencephalon (Figures 10.6, 10.7, and 10.8). The anterior part of the midbrain consists of a pair of large tracts called ***cerebral peduncles*** (pe-DUNK-kuls or PĒ-dung-kuls = little feet; Figure 10.9). They contain axons of motor neurons that conduct nerve impulses from the cerebrum to the spinal cord, medulla, and pons and axons of sensory neurons that extend from the medulla to the thalamus.

Nuclei of the midbrain include the ***substantia nigra*** (sub-STAN-shē-a = substance; NĪ-gra = black), which is large and darkly pigmented. Loss of these neurons is associated with Parkinson disease (see page 266). Also present are the right and left ***red nuclei***, which look reddish due to their rich blood supply and an iron-containing pigment in their neuronal cell bodies. Axons from the cerebellum and cerebral cortex form synapses in the red nuclei, which function with the cerebellum to coordinate muscular movements. Other nuclei in the midbrain are associated with two pairs of cranial nerves (see Figure 10.8): oculomotor (III) nerves and trochlear (IV) nerves.

The midbrain also contains nuclei that appear as four rounded bumps on the posterior surface. The two superior bumps are the ***superior colliculi*** (ko-LIK-ū-lī = little hills; singular is *colliculus*) (Figure 10.9). Several reflex arcs pass through the superior colliculi: tracking and scanning movements of the eyes and reflexes that govern movements of the eyes, head, and neck in response to visual stimuli. The two ***inferior colliculi*** are part of the auditory pathway, relaying impulses from the receptors for hearing in the ear to the thalamus. They also are reflex centers for the startle reflex, sudden movements of the head and body that occur when you are surprised by a loud noise.

Reticular Formation

In addition to the well-defined nuclei already described, much of the brain stem consists of small clusters of neuronal cell bodies (gray matter) intermingled with small bundles of myelinated axons (white matter). This region is known as the ***reticular formation*** (*ret-* = net) due to its netlike arrange-

Figure 10.9 Midbrain.

The midbrain connects the pons to the diencephalon.

View
Transverse plane
POSTERIOR
Superior colliculus
Reticular formation
Medial lemniscus
Oculomotor nucleus
Red nucleus
Substantia nigra
Corticospinal, corticopontine, and corticobulbar axons
Cerebral peduncle
Oculomotor (III) nerve
ANTERIOR

Transverse section of midbrain

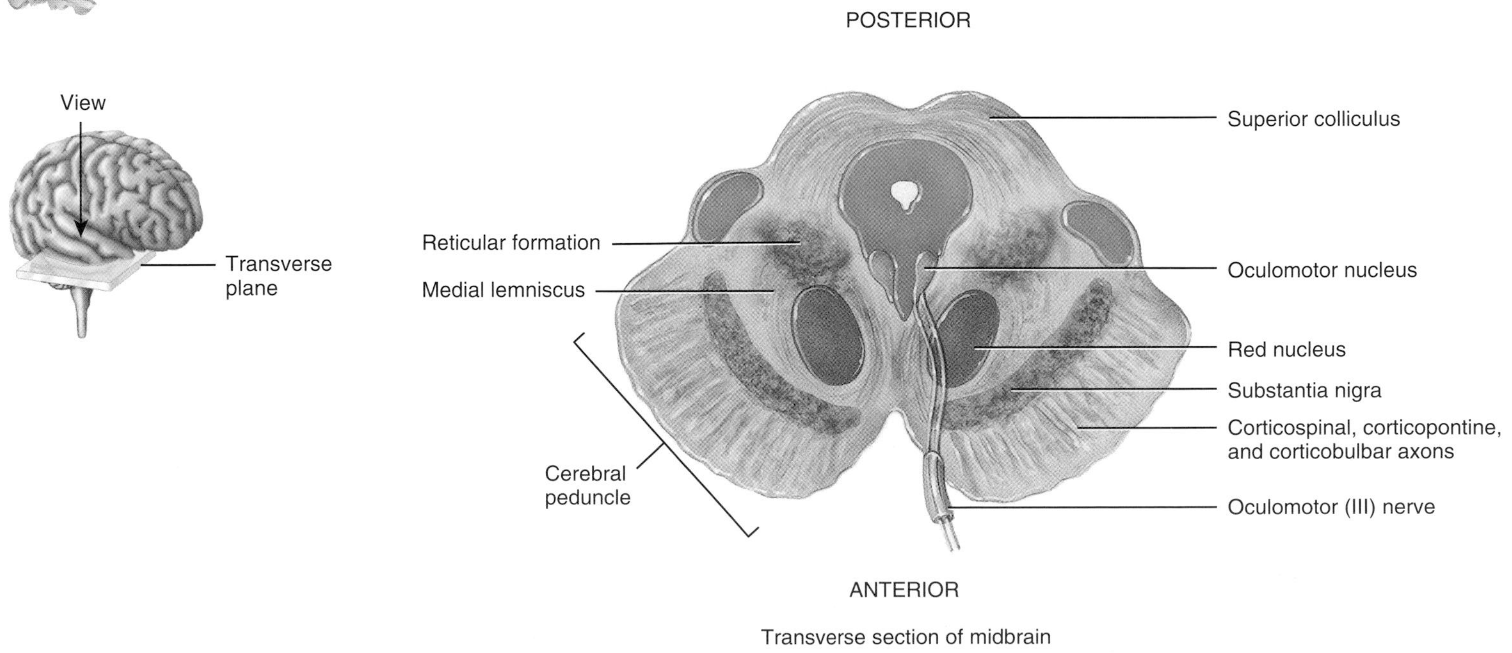

What functions are carried out by the superior colliculi?

ment of white matter and gray matter. Neurons within the reticular formation have both ascending (sensory) and descending (motor) functions.

The ascending part of the reticular formation is called the ***reticular activating system (RAS)*** , which consists of sensory axons that project to the cerebral cortex. When the RAS is stimulated, many nerve impulses pass upward to widespread areas of the cerebral cortex. The result is a state of wakefulness called ***consciousness***. The RAS helps maintain consciousness and is active during awakening from sleep. Inactivation of the RAS produces ***sleep***, a state of partial unconsciousness from which an individual can be aroused. The reticular formation's main descending function is to help regulate muscle tone, which is the slight degree of contraction in normal resting muscles.

■ CHECKPOINT

8. What is the significance of the blood–brain barrier?
9. What structures are the sites of CSF production, and where are they located?
10. Where are the medulla, pons, and midbrain located relative to one another?
11. What functions are governed by nuclei in the brain stem?
12. What are two important functions of the reticular formation?

Diencephalon

Major regions of the ***diencephalon*** include the thalamus, hypothalamus, and pineal gland (see Figure 10.6).

Thalamus

The ***thalamus*** (THAL-a-mus = inner chamber) consists of paired oval masses of gray matter, organized into nuclei, with interspersed tracts of white matter (Figure 10.10). Nuclei of the thalamus are important relay stations for sensory impulses that are conducting to the cerebral cortex from the spinal cord, brain stem, cerebellum, and other parts of the cerebrum.

The thalamus contributes to motor functions by transmitting information from the cerebellum and basal ganglia to motor areas of the cerebral cortex. It also relays nerve impulses between different areas of the cerebrum. The thalamus contributes to the regulation of autonomic activities and the maintenance of consciousness.

Hypothalamus

The ***hypothalamus*** (*hypo-* = under) is the small portion of the diencephalon that lies below the thalamus and above the pituitary gland (see Figures 10.6 and 10.10). Although its size is small, the hypothalamus controls many important body activities, most of them related to homeostasis. The chief functions of the hypothalamus are as follows:

Figure 10.10 Diencephalon: thalamus and hypothalamus. Also shown are the basal ganglia—the caudate nucleus, putamen, and globus pallidus.

The thalamus is the principal relay station for sensory impulses that reach the cerebral cortex from other parts of the brain and the spinal cord.

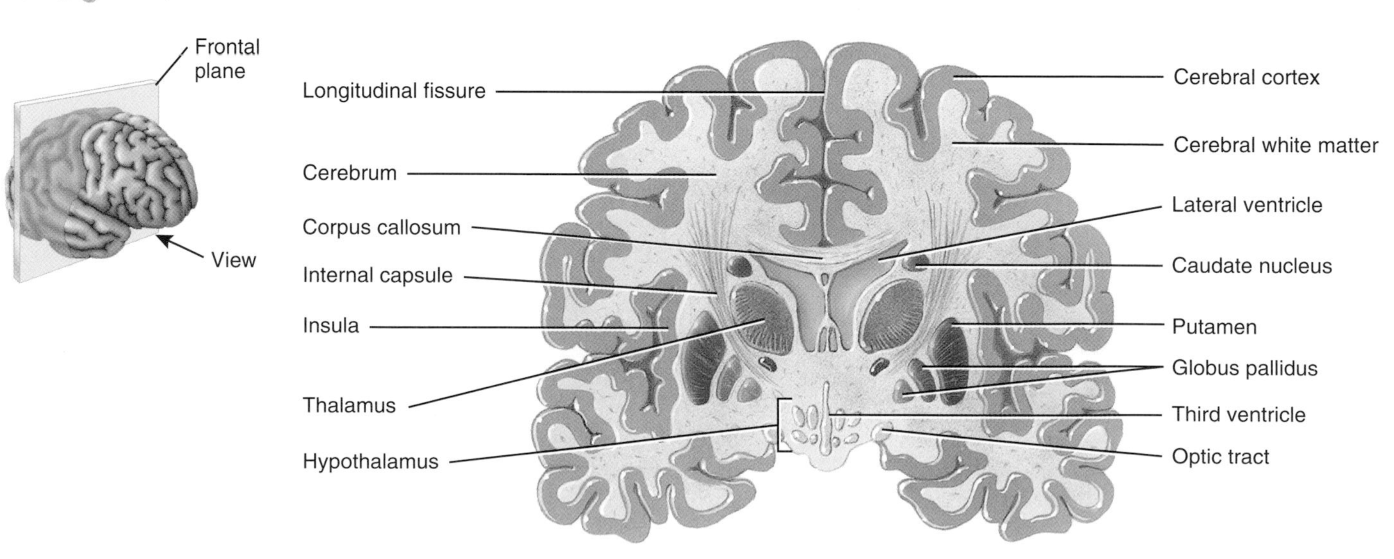

In which major part of the brain are the basal ganglia located, and what kind of tissue composes them?

1. **Control of the ANS**. The hypothalamus controls and integrates activities of the autonomic nervous system, which regulates contraction of smooth and cardiac muscle and the secretions of many glands. Through the ANS, the hypothalamus helps to regulate activities such as heart rate, movement of food through the gastrointestinal tract, and contraction of the urinary bladder.
2. **Control of the pituitary gland and production of hormones.** The hypothalamus controls the release of several hormones from the pituitary gland and thus serves as a primary connection between the nervous system and endocrine system. The hypothalamus also produces two hormones that are stored in the pituitary gland prior to their release.
3. **Regulation of emotional and behavioral patterns**. Together with the limbic system (described shortly), the hypothalamus regulates feelings of rage, aggression, pain, and pleasure, and the behavioral patterns related to sexual arousal.
4. **Regulation of eating and drinking**. The hypothalamus regulates eating behavior and also contains a ***thirst center***. When certain cells in the hypothalamus are stimulated by rising osmotic pressure of the interstitial fluid, they cause the sensation of thirst. The intake of water by drinking restores the osmotic pressure to normal, removing the stimulation and relieving the thirst.
5. **Control of body temperature**. If the temperature of blood flowing through the hypothalamus is above normal, the hypothalamus directs the autonomic nervous system to stimulate activities that promote heat loss. If, however, blood temperature is below normal, the hypothalamus generates impulses that promote heat production and retention.
6. **Regulation of circadian rhythms and states of consciousness**. The hypothalamus establishes patterns of awakening and sleep that occur on a circadian (daily) schedule.

Pineal Gland

The ***pineal gland*** (PĪN-ē-al = pinecone-like) is about the size of a small pea and protrudes from the posterior midline of the third ventricle (see Figure 10.6). Because the pineal gland secretes the hormone *melatonin*, it is part of the endocrine system. Melatonin promotes sleepiness and contributes to the setting of the body's biological clock.

Cerebellum

The ***cerebellum*** consists of two ***cerebellar hemispheres***, which are located posterior to the medulla and pons and below the cerebrum (see Figure 10.6). The surface of the cerebellum, called the ***cerebellar cortex***, consists of gray matter. Beneath the cortex is ***white matter*** that resembles the branches of a tree (see Figure 10.7). Deep within the white matter are masses of gray matter, the ***cerebellar nuclei***. The cerebellum attaches to the brain stem by bundles of axons called ***cerebellar peduncles*** (see Figure 10.8).

The cerebellum compares intended movements programmed by the cerebral cortex with what is actually happening. It constantly receives sensory impulses from muscles, tendons, joints, equilibrium receptors, and visual receptors. The cerebellum helps to smooth and coordinate complex sequences of skeletal muscle contractions. It regulates posture and balance and is essential for all skilled motor activities, from catching a baseball to dancing.

Damage to the cerebellum through trauma or disease disrupts muscle coordination, a condition called **ataxia** (*a-* = without; *-taxia* = order). Blindfolded people with ataxia cannot touch the tip of their nose with a finger because they cannot coordinate movement with their sense of where a body part is located. Another sign of ataxia is a changed speech pattern due to uncoordinated speech muscles. Cerebellar damage may also result in staggering or abnormal walking movements. People who consume too much alcohol show signs of ataxia because alcohol inhibits activity of the cerebellum. Alcohol overdose also suppresses the medullary rhythmicity area and may result in death.

Cerebrum

The ***cerebrum*** consists of the ***cerebral cortex*** (an outer rim of gray matter), an internal region of cerebral white matter, and gray matter nuclei deep within the white matter (Figure 10.10). The cerebrum provides us with the ability to read, write, and speak; to make calculations and compose music; to remember the past and plan for the future; and to create. During embryonic development, when there is a rapid increase in brain size, the gray matter of the cerebral cortex enlarges much faster than the underlying white matter. As a result, the cerebral cortex rolls and folds upon itself so that it can fit into the cranial cavity. The folds are called ***gyri*** (JĪ-rī = circles; singular is *gyrus*) (Figure 10.11). The deep grooves between folds are ***fissures***; the shallow grooves are ***sulci*** (SUL-sī = groove; singular is *sulcus*, SUL-kus). The ***longitudinal fissure*** separates the cerebrum into right and left halves called ***cerebral hemispheres***. The hemispheres are connected internally by the ***corpus callosum*** (kal-LŌ-sum; *corpus* = body; *callosum* = hard), a broad band of white matter containing axons that extend between the hemispheres (see Figure 10.10).

Each cerebral hemisphere has four lobes that are named after the bones that cover them: ***frontal lobe***, ***parietal lobe***, ***temporal lobe***, and ***occipital lobe*** (Figure 10.11). The ***central sulcus*** separates the frontal and parietal lobes. A major gyrus, the ***precentral gyrus***, is located immediately anterior to the central sulcus. The precentral gyrus contains the primary

Figure 10.11 Cerebrum. The inset in (a) indicates the differences among a gyrus, a sulcus, and a fissure. Because the insula cannot be seen externally, it has been projected to the surface in (b).

The cerebrum provides us with the ability to read, write, and speak; make calculations and compose music; remember the past and make future plans; and create.

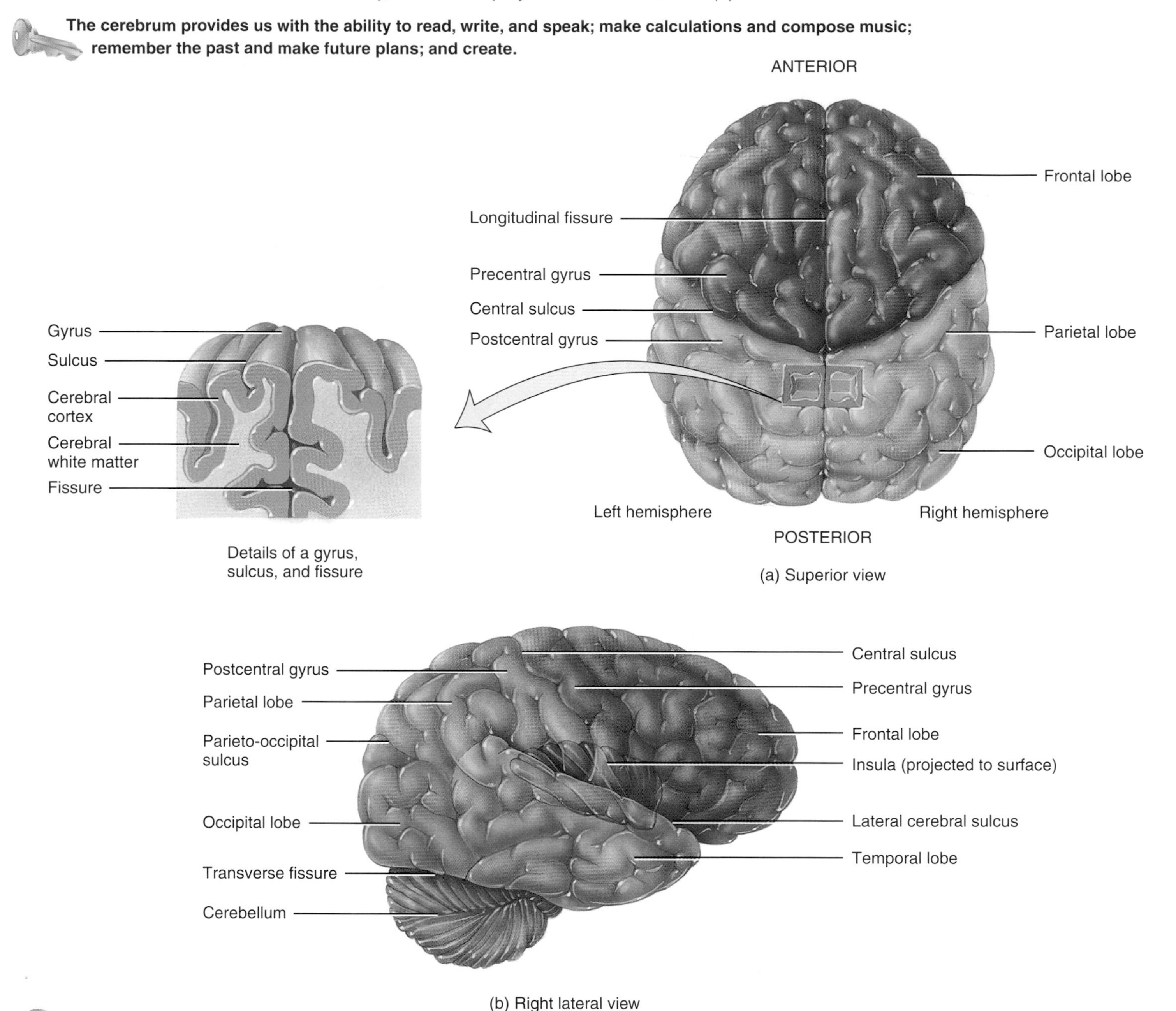

What structure separates the right and left cerebral hemispheres?

motor area of the cerebral cortex. The ***postcentral gyrus***, located immediately posterior to the central sulcus, contains the primary somatosensory area of the cerebral cortex, which is discussed shortly. A fifth part of the cerebrum, the ***insula***, cannot be seen at the surface of the brain because it lies within the lateral cerebral sulcus, deep to the parietal, frontal, and temporal lobes (see Figure 10.10).

The ***cerebral white matter*** consists of myelinated and unmyelinated axons that transmit impulses between gyri in the same hemisphere, from the gyri in one cerebral hemisphere to the corresponding gyri in the opposite cerebral hemisphere via the corpus callosum, and from the cerebrum to other parts of the brain and spinal cord.

Deep within each cerebral hemisphere are three nuclei (masses of gray matter) that are collectively termed the ***basal ganglia*** (see Figure 10.10). (Recall that "ganglion" usually means a collection of neuronal cell bodies *outside* the CNS. The name here is the one exception to that general rule.)

FOCUS ON WELLNESS

Coffee Nerves—The Health Risks of Caffeine

Caffeine has been enjoyed by people around the world since the beginning of history. Found naturally in over 60 plants, caffeine is the most widely consumed drug in North America, primarily as a component of coffee, tea, cola, and other beverages. It is also found in many over-the-counter drugs and in small amounts in chocolate. The average coffee drinker consumes 3 cups a day, with some people (including many college students) consuming 10 cups or more a day. How is all of this caffeine affecting our health?

The Java Jitters

Caffeine's most obvious effect is on the nervous system. Caffeine mimics the effects of the sympathetic division of the autonomic nervous system. In general, both caffeine and sympathetic arousal tend to wind you up. For example, they both make your heart beat faster and harder. (You will learn more about the autonomic nervous system in Chapter 11.)

The immediate effects of caffeine vary greatly from person to person. Some people find that any amount of caffeine causes undesirable symptoms such as muscle twitches, anxiety, increased blood pressure, an irregular heartbeat, digestive complaints, headache, and difficulty sleeping. Other people get one or more of these symptoms only if their caffeine consumption exceeds a certain threshold. How harmful is caffeine overload? If these symptoms are short-lived, and caffeine consumption is reduced or eliminated, no lasting harm seems to occur in otherwise healthy adults.

How Much Is Too Much?

While caffeine tolerance varies, studies suggest that long-term consumption of moderate amounts of caffeine probably poses little or no risk to long-term health. A moderate amount of caffeine is equivalent to that contained in two cups of coffee per day. This guideline does not apply to people who experience negative symptoms with caffeine; people should avoid caffeine in any amount that leads to unhealthy symptoms. Some animal studies suggest a link between caffeine and birth defects. Thus, the U.S. Food and Drug Administration recommends that women avoid or greatly reduce caffeine intake during pregnancy.

► THINK IT OVER . . .

► ***Why might your tolerance for caffeine go down during high-stress times, such as during final exam week? (Hint: Feelings of stress are associated with overactivation of the sympathetic division of the autonomic nervous system.)***

They are the *globus pallidus* (*globus* = ball; *pallidus* = pale), the *putamen* (pū-TĀ-men = shell), and the *caudate nucleus* (*caud-* = tail). A major function of the basal ganglia is to help initiate and terminate movements. They also help regulate the muscle tone required for specific body movements and control subconscious contractions of skeletal muscles, such as automatic arm swings while walking.

Damage to the basal ganglia results in uncontrollable shaking (tremor), muscular rigidity (stiffness), and involuntary muscle movements. Movement disruptions also are a hallmark of Parkinson disease (see page 266). In this disorder, neurons that extend from the substantia nigra to the putamen and caudate nucleus degenerate, causing the disruptions.

Limbic System

Encircling the upper part of the brain stem and the corpus callosum is a ring of structures on the inner border of the cerebrum and floor of the diencephalon that constitutes the ***limbic system*** (*limbic* = border) (Figure 10.12). The limbic system is sometimes called the "emotional brain" because it plays a primary role in a range of emotions, including pain, pleasure, docility, affection, and anger. Although behavior is a function of the entire nervous system, the limbic system controls most of its involuntary aspects related to survival. Animal experiments suggest that it has a major role in controlling the overall pattern of behavior. Together with parts of the cerebrum, the limbic system also functions in memory; damage to the limbic system causes memory impairment.

Figure 10.12 The limbic system. The components of the limbic system are shaded green.

The limbic system governs emotional aspects of behavior.

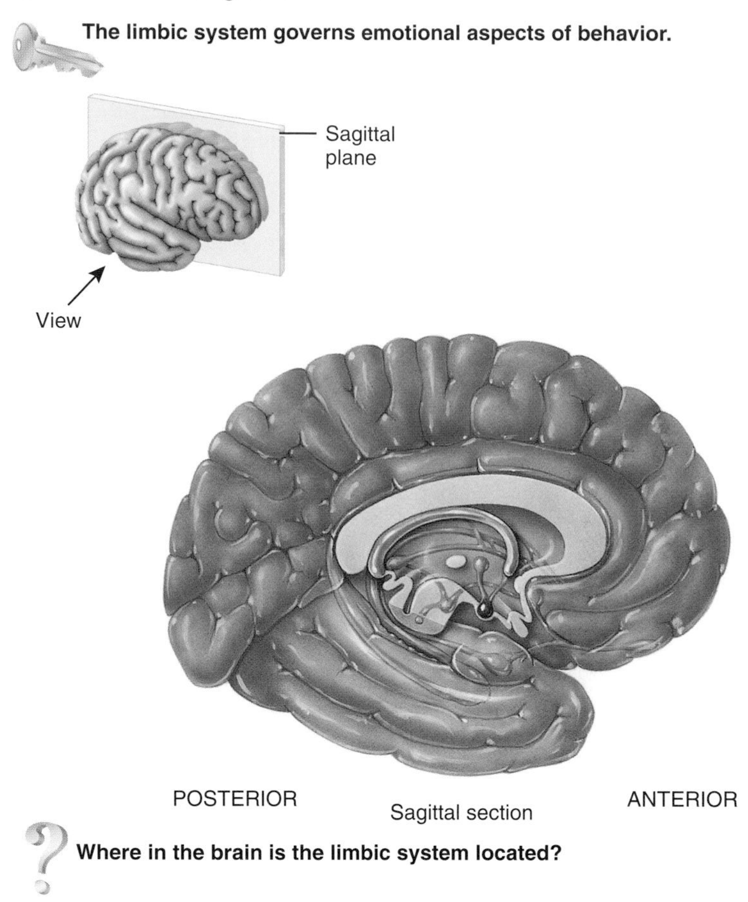

Where in the brain is the limbic system located?

Functional Areas of the Cerebral Cortex

Specific types of sensory, motor, and integrative signals are processed in certain regions of the cerebral cortex (Figure 10.13). Generally, ***sensory areas*** receive sensory information and are involved in ***perception***, the conscious awareness of a sensation; ***motor areas*** initiate movements; and ***association areas*** deal with more complex integrative functions such as memory, emotions, reasoning, will, judgment, personality traits, and intelligence.

SENSORY AREAS Sensory input to the cerebral cortex flows mainly to the posterior half of the cerebral hemispheres, to regions behind the central sulci. In the cerebral cortex, primary sensory areas have the most direct connections with peripheral sensory receptors.

The *primary somatosensory area* (sō′-mat-ō-SEN-sō-rē) is posterior to the central sulcus of each cerebral hemisphere in the postcentral gyrus of the parietal lobe (Figure 10.13). It receives nerve impulses for touch, proprioception (joint and muscle position), pain, itching, tickle, and temperature and is involved in the perception of these sensations. The primary somatosensory area allows you to pinpoint where sensations originate, so that you know exactly where on your body to swat that mosquito. The *primary visual area*, located in the occipital lobe, receives visual information and is involved in visual perception. The *primary auditory area*, located in the temporal lobe, receives information for sound and is involved in auditory perception. The *primary gustatory area*, located at the base of the postcentral gyrus, receives impulses for taste and is involved in gustatory perception. The *primary olfactory area*, located on the medial aspect of the temporal lobe (and thus is not visible in Figure 10.13), receives impulses for smell and is involved in olfactory perception.

MOTOR AREAS Motor output from the cerebral cortex flows mainly from the anterior part of each hemisphere. Among the most important motor areas are the primary motor area and Broca's speech area (Figure 10.13). The *primary motor area* is located in the precentral gyrus of the frontal lobe in each hemisphere. Each region in the primary motor area controls voluntary contractions of specific muscles on the opposite side of the body. *Broca's speech area* (BRO-kaz) is located in the frontal lobe close to the lateral cerebral sulcus. Speaking and understanding language are complex activities that involve several sensory, association, and motor areas of the cortex. In 97% of the population, these language areas are localized in the *left* hemisphere. Neural connections between Broca's speech area, the premotor area, and primary motor area activate muscles needed for speaking and breathing muscles.

ASSOCIATION AREAS The association areas of the cerebrum consist of some motor and sensory areas, plus large areas on the lateral surfaces of the occipital, parietal, and temporal lobes and on the frontal lobes anterior to the motor areas. Tracts connect association areas to one another. The *somatosensory association area*, just posterior to the primary somatosensory area, integrates and interprets somatic sensations such as the exact shape and texture of an object. Another role of the somatosensory association area is the storage of memories of past sensory experiences, enabling you to compare current sensations with previous experiences. For example, the somatosensory association area allows you to recognize objects such as a pencil and a paperclip simply by touching them. The *visual association area*, located in the occipital lobe, relates present and past visual experiences and is essential for recognizing and evaluating what is seen. The *auditory association area*, located below the primary auditory area in the temporal cortex, allows you to recognize a particular sound as speech, music, or noise.

Wernicke's area, a broad region in the *left* temporal and parietal lobes, interprets the meaning of speech by recognizing spoken words. It is active as you translate words into thoughts. The regions in the *right* hemisphere that correspond to Broca's and Wernicke's areas in the left hemisphere also contribute to verbal communication by adding emotional content, for instance, anger or joy, to spoken words. The *common integrative area* receives and interprets nerve impulses from the somatosensory, visual, and auditory association areas, and from the primary gustatory area, primary ol-

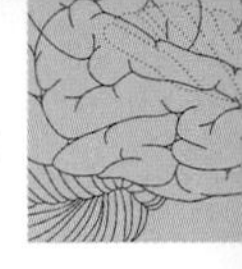

Figure 10.13 Functional areas of the cerebrum. Broca's speech area and Wernicke's area are in the left cerebral hemisphere of most people; they are shown here to indicate their relative locations.

Particular areas of the cerebral cortex process sensory, motor, and integrative signals.

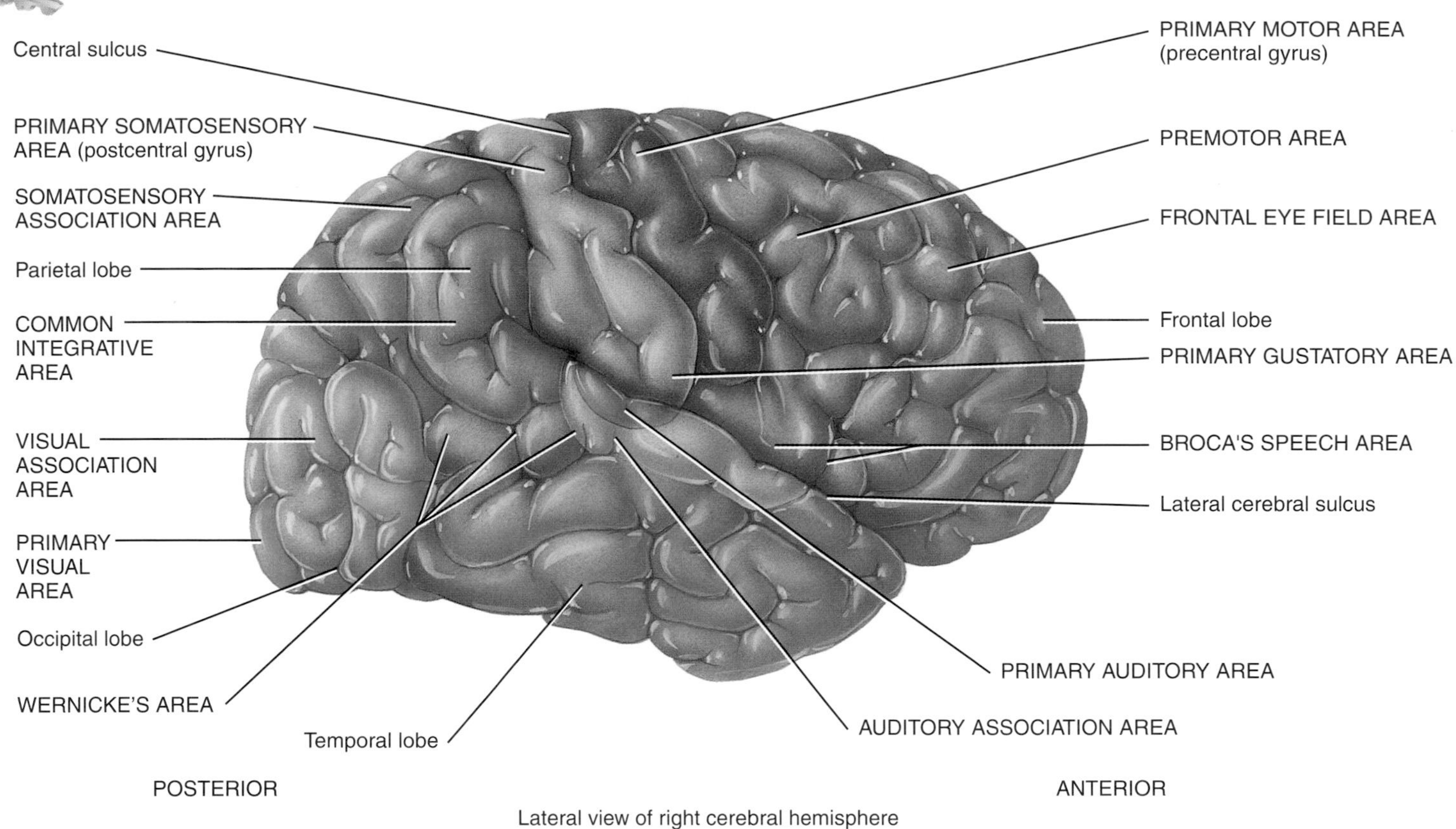

Lateral view of right cerebral hemisphere

Which part of the cerebrum localizes exactly where somatic sensations occur?

factory area, the thalamus, and parts of the brain stem. The *premotor area*, immediately anterior to the primary motor area, generates nerve impulses that cause a specific group of muscles to contract in a specific sequence, for example, to write a word. The *frontal eye field area* in the frontal cortex controls voluntary scanning movements of the eyes, such as those that occur while you are reading this sentence.

Injury to language areas of the cerebral cortex results in **aphasia** (a-FĀ-zē-a; *a-* = without; *-phasia* = speech), an inability to use or comprehend words. Damage to Broca's speech area results in *nonfluent aphasia*, an inability to properly form words. People with nonfluent aphasia know what they wish to say but cannot properly speak the words. Damage to Wernicke's area, the common integrative area or auditory association area, results in *fluent aphasia*, characterized by faulty understanding of spoken or written words. A person experiencing this type of aphasia may produce strings of words that have no meaning ("word salad"). For example, someone with fluent aphasia might say, "I rang car porch dinner light river pencil."

Somatic Sensory and Somatic Motor Pathways

Somatic sensory information from the body ascends to the primary somatosensory area via two main somatic sensory pathways: (1) the posterior column–medial lemniscus pathway and (2) the spinothalamic pathways. By contrast, nerve impulses that cause contraction of skeletal muscles descend along many pathways that originate mainly in the primary motor area of the brain and in the brain stem.

Somatic sensory pathways relay information from somatic sensory receptors to the primary somatosensory area in the cerebral cortex. The pathways consist of thousands of sets of three neurons (Figure 10.14).

Nerve impulses for conscious awareness of the position of muscles and joints (proprioception) and for most touch sensations ascend to the cortex along the ***posterior column–medial lemniscus pathway*** (Figure 10.14a). The name of the pathway comes from the names of two white matter tracts that convey the impulses: the posterior column of the spinal cord and the medial lemniscus of the brain stem. Impulses conducted along the posterior column–medial lemniscus pathway give rise to three main types of sensations:

Figure 10.14 Somatic sensory pathways. Circles represent cell bodies and dendrites, lines represent axons, and Y-shaped forks represent axon terminals. Arrows indicate the direction of nerve impulse conduction. (a) In the posterior column–medial lemniscus pathway, the first-order neuron in the pathway ascends to the medulla oblongata via the posterior column (white matter located on the posterior side of the spinal cord). In the medulla, it synapses with a second-order neuron, which then extends through the medial lemniscus to the thalamus on the opposite side. The third-order neuron extends from the thalamus to the cerebral cortex. (b) In the anterolateral pathway, the first-order neuron synapses with a second-order neuron in the spinal cord gray matter. The second-order neuron extends to the thalamus on the opposite side, and the third-order neuron extends from the thalamus to the cerebral cortex.

Nerve impulses for somatic sensations conduct to the primary somatosensory area (postcentral gyrus) of the cerebral cortex.

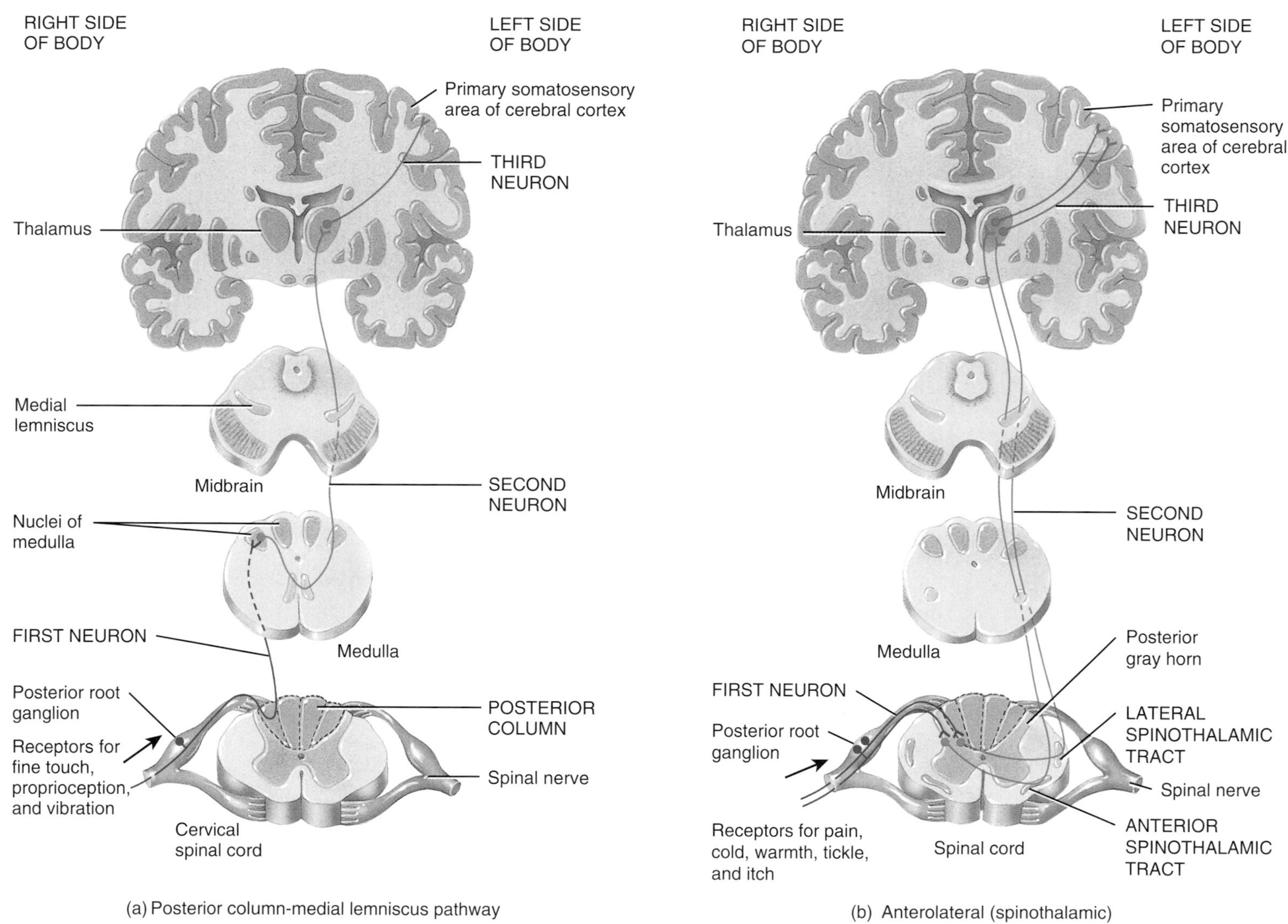

(a) Posterior column-medial lemniscus pathway

(b) Anterolateral (spinothalamic) pathways

Which somatic sensations could be lost due to damage of the spinothalamic tracts?

- ***Fine touch*** is the ability to recognize what point on the body is touched plus the shape, size, and texture of the source of stimulation.
- ***Proprioception*** is the awareness of the precise position of body parts, and *kinesthesia* is the awareness of directions of movement.
- ***Vibratory sensations*** arise when rapidly fluctuating touch stimuli are present.

The ***spinothalamic pathways*** (spī-nō-tha-LAM-ik) begin in two spinal cord tracts—the ***anterior spinothalamic tract*** and the ***lateral spinothalamic tract*** (Figure 10.14b). These tracts relay impulses for pain, thermal (hot and cold temperature), tickle, and itch sensations.

Neurons in the brain and spinal cord coordinate all voluntary and involuntary movements. Ultimately, all ***somatic motor pathways*** that control movement converge on neurons known as ***lower motor neurons*** (Figure 10.15). The axons of lower motor neurons extend out of the brain stem to stimulate skeletal muscles in the head and out of the spinal cord to stimulate skeletal muscles in the limbs and trunk.

Lower motor neurons receive their instructions from many other neurons in the brain and spinal cord.

1. Nearby *local interneurons* help coordinate rhythmic activity in specific muscle groups, such as alternating flexion and extension of the lower limbs during walking.
2. Local interneurons and lower motor neurons receive input from ***upper motor neurons*** (Figure 10.15). Upper motor neurons plan, initiate, and direct sequences of voluntary movements. Two major tracts that conduct nerve impulses from upper motor neurons in the cerebral cortex are the ***lateral corticospinal tract*** and ***anterior corticospinal tract***. Notice that axons of upper motor neurons from one cerebral hemisphere cross over and synapse with lower motor neurons in the other side of the spinal cord (Figure 10.15).
3. The *basal ganglia* communicate with motor areas of the cerebral cortex, thalamus, and substantia nigra. These connections help initiate and terminate movements, suppress unwanted movements, and establish a normal level of muscle tone.
4. Neurons connect the *cerebellum* with motor areas of the cerebral cortex and the brain stem. The cerebellum coordinates body movements and helps maintain normal posture and balance.

Damage or disease of *lower* motor neurons produces **flaccid paralysis** of muscles on the same side of the body: the muscles lack voluntary control and reflexes, muscle tone is decreased or lost, and the muscle remains flaccid (limp). Injury or disease of *upper* motor neurons causes **spastic paralysis** of muscles on the opposite side of the body. In this condition muscle tone is increased, reflexes are exaggerated, and pathological reflexes appear.

Hemispheric Lateralization

Although the brain is quite symmetrical, there are subtle anatomical differences between the two hemispheres. They are also functionally different in some ways, with each hemisphere specializing in certain functions. This functional asymmetry is termed ***hemispheric lateralization***.

As you have seen, the left hemisphere receives sensory signals from and controls the right side of the body, and the right hemisphere receives sensory signals from and controls the left side of the body. In addition, the left hemisphere is more important for spoken and written language, numerical and scientific skills, ability to use and understand sign language, and reasoning in most people. Patients with damage

Figure 10.15 Somatic motor pathways. Shown here are the two most direct pathways whereby signals initiated by the primary motor area in one hemisphere control skeletal muscles on the opposite side of the body. Circles represent cell bodies and dendrites, lines represent axons, and Y-shaped forks represent axon terminals.

Lower motor neurons stimulate skeletal muscles to produce movements.

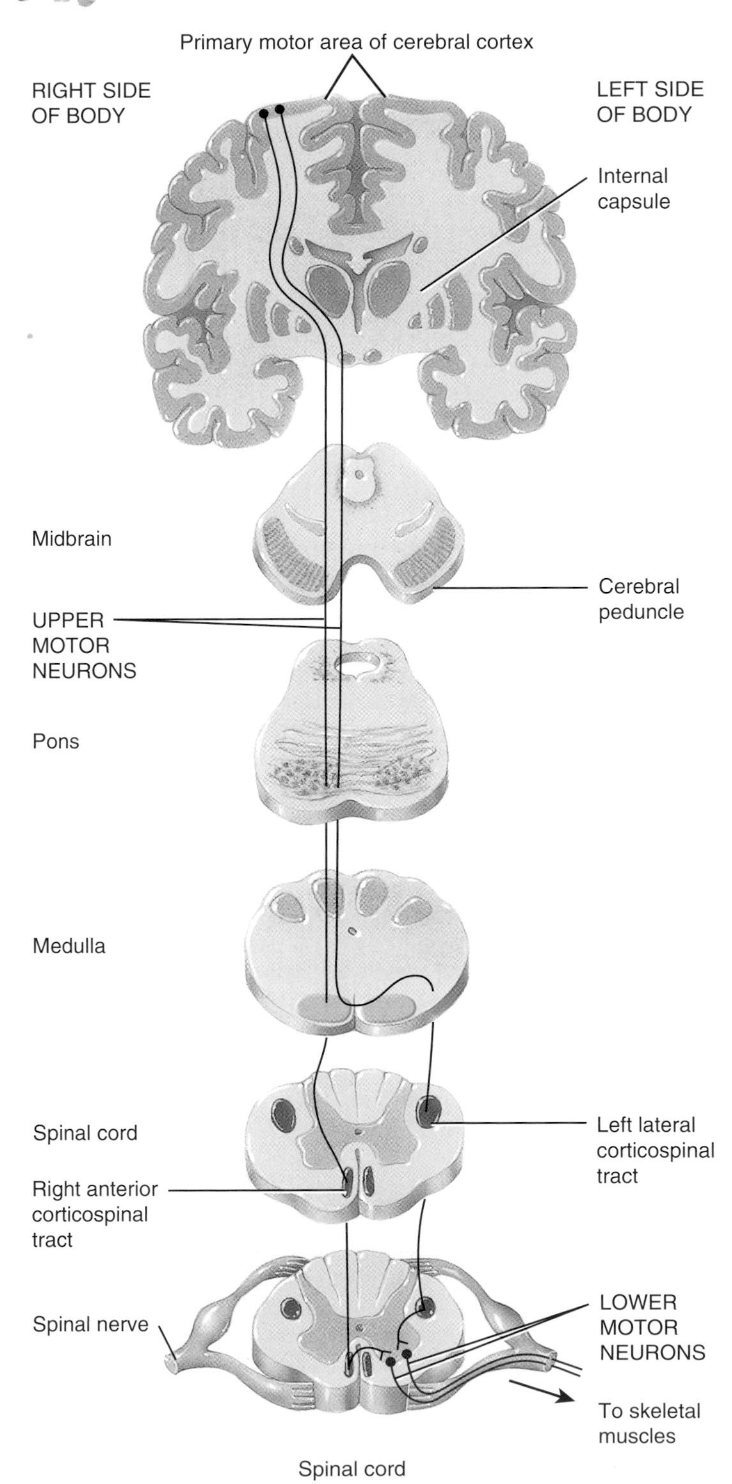

What two spinal cord tracts conduct impulses along axons of upper motor neurons?

in the left hemisphere, for example, often have difficulty speaking. The right hemisphere is more important for musical and artistic awareness; spatial and pattern perception; recognition of faces and emotional content of language; and for generating mental images of sight, sound, touch, taste, and smell.

Memory

Without memory, we would repeat mistakes and be unable to learn. Similarly, we would not be able to repeat our successes or accomplishments, except by chance. ***Memory*** is the process by which information acquired through learning is stored and retrieved. For an experience to become part of memory, it must produce structural and functional changes in the brain. The parts of the brain known to be involved with memory include the association areas of the frontal, parietal, occipital, and temporal lobes; parts of the limbic system; and the diencephalon. Memories for motor skills, such as how to serve a tennis ball, are stored in the basal ganglia and cerebellum as well as in the cerebral cortex.

Electroencephalogram (EEG)

At any instant, brain neurons are generating millions of nerve impulses. Taken together, these electrical signals are called ***brain waves***. Brain waves generated by neurons close to the brain surface, mainly neurons in the cerebral cortex, can be detected by metal electrodes placed on the forehead and scalp. A record of such waves is called an ***electroencephalogram*** (e-lek′-trō-en-SEF-a-lō-gram) or ***EEG***. Electroencephalograms are useful for studying normal brain functions, such as changes that occur during sleep. Neurologists also use them to diagnose a variety of brain disorders, such as epilepsy, tumors, metabolic abnormalities, sites of trauma, and degenerative diseases.

Table 10.1 summarizes the principal parts of the brain and their functions.

Table 10.1 Summary of Functions of Principal Parts of the Brain

Part	Function	Part	Function
Brain Stem	***Medulla oblongata:*** Relays motor and sensory impulses between other parts of the brain and the spinal cord. Reticular formation (also in pons, midbrain, and diencephalon) functions in consciousness and arousal. Vital centers regulate heartbeat, breathing (together with pons), and blood vessel diameter. Other centers coordinate swallowing, vomiting, coughing, sneezing, and hiccupping. Contains nuclei of origin for cranial nerves VIII, IX, X, XI, and XII.	**Diencephalon**	***Thalamus:*** Relays almost all sensory impulses to the cerebral cortex. Provides crude perception of touch, pressure, pain, and temperature. Also functions in cognition and awareness. ***Hypothalamus:*** Controls and integrates activities of the autonomic nervous system and pituitary gland. Regulates emotional and behavioral patterns and circadian rhythms. Controls body temperature and regulates eating and drinking behavior. Helps maintain waking state and establishes patterns of sleep. ***Pineal gland:*** Secretes the hormone melatonin.
	Pons: Relays impulses from one side of the cerebellum to the other and between the medulla and midbrain. Contains nuclei of origin for cranial nerves V, VI, VII, and VIII. Together with the medulla, helps control breathing.	**Cerebellum**	Compares intended movements with what is actually happening to coordinate complex, skilled movements. Regulates posture and balance.
	Midbrain: Relays motor impulses from the cerebral cortex to the pons and sensory impulses from the spinal cord to the thalamus. Most of substantia nigra and red nucleus contribute to control of movement. Contains nuclei of origin for cranial nerves III and IV.	**Cerebrum**	Sensory areas are involved in the perception of sensory information, motor areas control muscular movement, and association areas deal with more complex integrative functions such as memory, personality traits, and intelligence. Basal ganglia coordinate automatic muscle movements and help regulate muscle tone. Limbic system functions in emotional aspects of behavior related to survival.

■ **CHECKPOINT**

13. Why is the hypothalamus considered part of both the nervous system and the endocrine system?

14. What are the functions of the cerebellum and basal ganglia?

15. Where are the primary somatosensory area and primary motor area located in the brain? What are their functions?

16. What areas of the cerebral cortex are needed for normal language abilities?

17. Compare and contrast the posterior column–medial lemniscus pathway and the spinothalamic pathways.

CRANIAL NERVES

OBJECTIVE • **Identify the 12 pairs of cranial nerves by name and number and give the functions of each.**

The 12 pairs of ***cranial nerves***, like spinal nerves, are part of the peripheral nervous system. The cranial nerves are designated with roman numerals and with names (see Figure 10.8). The roman numerals indicate the order (anterior to posterior) in which the nerves emerge from the brain. The names indicate the distribution or function.

Cranial nerves emerge from the nose (cranial nerve I), the eyes (cranial nerve II), the inner ear (cranial nerve VIII), the brain stem (cranial nerves III–XII), and the spinal cord (part of cranial nerve XI). Two cranial nerves (cranial nerves I and II) contain only sensory axons and thus are *sensory nerves*. The rest are *mixed nerves* because they contain axons of both sensory and motor neurons. Cranial nerves III, IV, VI, XI, and XII are mainly motor. A few of their axons are sensory axons from muscle proprioceptors, but most of their axons are motor neurons that innervate skeletal muscles. Cranial nerves III, VII, IX, and X include both somatic and autonomic motor axons. The somatic axons stimulate skeletal muscles; the autonomic axons, which are part of the parasympathetic division, go to glands, smooth muscle, and cardiac muscle.

Table 10.2 lists the cranial nerves, along with their components (sensory or mixed) and functions.

■ **CHECKPOINT**

18. What is the difference between a mixed cranial nerve and a sensory cranial nerve?

Table 10.2 Summary of Cranial Nerves (see Figure 10.8)

Number	Name*	Components	Function
I	**Olfactory nerve** (ol-FAK-tō-rē; *olfact-* = to smell)	***Sensory:*** Axons in the lining of the nose.	Smell.
II	**Optic nerve** (OP-tik; *opti-* = the eye, vision)	***Sensory:*** Axons from the retina of the eye.	Vision.
III	**Oculomotor nerve** (ok′-ū-lō-MŌ-tor; *oculo-* = eye; *-motor* = mover)	***Sensory part:*** Axons from proprioceptors in the eyeball muscles. ***Motor part:*** Axons of somatic motor neurons that stimulate muscles of upper eyelid and four muscles that move the eyeballs plus axons of parasympathetic neurons that pass to two smooth muscles—the ciliary muscle of the eyeball and the sphincter muscle of the iris.	Muscle sense (proprioception). Movement of eyelid and eyeball; alters lens for near vision and constricts pupil.
IV	**Trochlear nerve** (TRŌK-lē-ar; *trochle-* = a pulley)	***Sensory part:*** Axons from proprioceptors in the superior oblique muscles (muscles that move the eyeballs). ***Motor part:*** Axons of somatic motor neurons that stimulate the superior oblique muscles.	Muscle sense (proprioception). Movement of the eyeball.
V	**Trigeminal nerve** (trī-JEM-i-nal = triple, for its three branches)	***Sensory part:*** Consists of three branches: the *ophthalmic nerve* contains axons from the scalp and forehead skin; the *maxillary nerve* contains axons from the lower eyelid, nose, upper teeth, upper lip, and pharynx; and the *mandibular nerve* contains axons from the tongue, lower teeth, and the lower side of the face. ***Motor part:*** Axons of somatic motor neurons that stimulate muscles used in chewing.	Touch, pain, and temperature sensations and muscle sense (proprioception). Chewing.

(Continues)

Table 10.2 Summary of Cranial Nerves (Continued)

Number	Name*	Components	Function
VI	**Abducens nerve** (ab-DOO-senz; *ab-* = away; *-ducens* = to lead)	***Sensory part:*** Axons from proprioceptors in lateral rectus muscles (muscles that move the eyeballs).	Muscle sense (proprioception).
		Motor part: Axons of somatic motor neurons that stimulate the lateral rectus muscles.	Movement of eyeball.
VII	**Facial nerve** (FĀ-shal = face)	***Sensory part:*** Axons from taste buds on tongue and axons from proprioceptors in muscles of face and scalp.	Taste and muscle sense (proprioception).
		Motor part: Axons of somatic motor neurons that stimulate facial, scalp, and neck muscles plus parasympathetic axons that stimulate lacrimal (tear) glands and salivary glands.	Facial expressions; secretion of tears and saliva.
VIII	**Vestibulocochlear nerve** (vest-tib-ū-lō-KOK-lē-ar; *vestibulo-* = small cavity; *-cochlear* = a spiral, snail-like)	***Vestibular branch, sensory part:*** Axons from semicircular canals, saccule, and utricle (organs of equilibrium).	Equilibrium.
		Vestibular branch, motor part: Axons that synapse with sensory receptors (hair cells) for equilibrium.	Adjusts sensitivity of hair cells.
		Cochlear branch, sensory part: Axons from spiral organ (organ of hearing).	Hearing.
		Cochlear branch, motor part: Axons that synapse with sensory receptors (hair cells) for hearing.	Modifies responses of hair cells.
IX	**Glossopharyngeal nerve** (glos′-ō-fa-RIN-jē-al; *glosso-* = tongue; *-pharyngeal* = throat)	***Sensory portion:*** Axons from taste buds and somatic sensory receptors on part of tongue, from proprioceptors in some swallowing muscles, and from stretch receptors in carotid sinus and chemoreceptors in carotid body.	Taste and somatic sensations (touch, pain, and temperature) from tongue; muscle sense (proprioception); monitoring blood pressure; monitoring oxygen and carbon dioxide in blood for regulation of breathing.
		Motor portion: Axons of somatic motor neurons that stimulate swallowing muscles of throat plus parasympathetic axons that stimulate a salivary gland.	Swallowing; secretion of saliva.
X	**Vagus nerve** (VĀ-gus; *vagus* = vagrant or wandering)	***Sensory portion:*** Axons from proprioceptors in muscles of neck and throat, from stretch receptors and chemoreceptors in carotid sinus and carotid body, from chemoreceptors in aortic body, and from visceral sensory receptors in most organs of the thoracic and abdominal cavities.	Somatic sensations (touch, pain, temperature) from throat and pharynx; monitoring of blood pressure; monitoring of oxygen and carbon dioxide in blood for regulation of breathing; sensations from visceral organs in thorax and abdomen.
		Motor portion: Axons of somatic motor neurons that stimulate skeletal muscles of the throat and neck plus parasympathetic axons that supply smooth muscle in the airways, esophagus, stomach, small intestine, most of the large intestine, and gallbladder; cardiac muscle in the heart; and glands of the gastrointestinal tract.	Swallowing, coughing, and voice production; smooth muscle contraction and relaxation in organs of the gastrointestinal tract; slowing of the heart rate; secretion of digestive fluids.

Number	Name*	Components	Function
XI	**Accessory nerve** (ak-SES-ō-re = assisting)	***Sensory part:*** Axons from proprioceptors in muscles of throat and voice box.	Muscle sense (proprioception).
		Motor part: Axons of somatic motor neurons that stimulate muscles of the throat and neck.	Swallowing and movements of head and shoulders.
XII	**Hypoglossal nerve** (hi′-pō-GLOS-al; *hypo-* = below; *-glossal* = tongue)	***Sensory part:*** Axons from proprioceptors in tongue muscles.	Muscle sense (proprioception).
		Motor part: Axons of somatic motor neurons that stimulate muscles of tongue.	Movement of tongue during speech and swallowing.

*A mnemonic device that can be used to remember the names of the nerves is: "**O**h, **o**h, **o**h, **t**o **t**ouch **a**nd **f**eel **v**ery **g**reen **v**egetables—**AH**!" Each boldfaced letter corresponds to the first letter of a pair of cranial nerves.

AGING AND THE NERVOUS SYSTEM

OBJECTIVE • Describe the effects of aging on the nervous system.

The brain grows rapidly during the first few years of life. Growth is due mainly to an increase in the size of neurons already present, the proliferation and growth of neuroglia, the development of dendritic branches and synaptic contacts, and continuing myelination of axons. From early adulthood onward, brain mass declines. By the time a person reaches age 80, the brain weighs about 7% less than it did in young adulthood. Although the number of neurons present does not decrease very much, the number of synaptic contacts declines. Associated with the decrease in brain mass is a decreased capacity for sending nerve impulses to and from the brain. As a result, processing of information diminishes. Conduction velocity decreases, voluntary motor movements slow down, and reflex times increase.

■ **CHECKPOINT**

19. How is brain mass related to age?

COMMON DISORDERS

Spinal Cord Injury

Most spinal cord injuries are due to trauma as a result of factors such as automobile accidents, falls, contact sports, diving, or acts of violence. The effects of the injury depend on the extent of direct trauma to the spinal cord or compression of the cord by fractured or displaced vertebrae or blood clots. Although any segment of the spinal cord may be involved, most common sites of injury are in the cervical, lower thoracic, and upper lumbar regions. Depending on the location and extent of spinal cord damage, paralysis may occur. ***Monoplegia*** (*mono-* = one; *-plegia* = blow or strike) is paralysis of one limb only. ***Diplegia*** (*di-* = two) is paralysis of both upper limbs or both lower limbs. ***Paraplegia*** (*para-* = beyond) is paralysis of both lower limbs. ***Hemiplegia*** (*hemi-* = half) is paralysis of the upper limb, trunk, and lower limb on one side of the body, and ***quadriplegia*** (*quad-* = four) is paralysis of all four limbs.

Shingles

Shingles is an acute infection of the peripheral nervous system caused by *herpes zoster* (HER-pēz ZOS-ter), the virus that also causes chickenpox. After a person recovers from chickenpox, the virus retreats to a posterior root ganglion. If the virus is reactivated, it may leave the ganglion and travel down sensory axons to the skin. The result is pain, discoloration of the skin, and a characteristic line of skin blisters. The line of blisters marks the distribution of the particular sensory nerve belonging to the infected posterior root ganglion.

Amyotrophic Lateral Sclerosis

Amyotrophic lateral sclerosis (ALS) (ā′-mī-ō′-TROF-ik; *a-* = without; *myo-* = muscle; *trophic* = nourishment) is a progressive degenerative disease that attacks motor areas of the cerebral cortex, axons of upper motor neurons, and lower motor neuron cell bodies. ALS is commonly known as *Lou Gehrig's disease* after the New York Yankees baseball player who died of it at age 37 in 1941. ALS causes progressive muscle weakness and atrophy. ALS often begins in sections of the spinal cord that serve the hands and arms but rapidly spreads to involve the whole body and face, without affecting intellect or sensations. Death typically occurs in 2 to 5 years. ALS may be caused by the buildup in the synaptic cleft of the neurotransmitter glutamate released by motor neurons. The excess glutamate causes motor neurons to malfunction and eventually die. The drug riluzole, which is used to treat ALS, reduces damage to motor neurons by decreasing the release of glutamate. Other factors implicated in the development of ALS include damage to motor neurons by free radicals, autoimmune responses, viral infection, deficiency of nerve growth factor, apoptosis (programmed cell death), environmental toxins, and trauma.

Cerebrovascular Accident

The most common brain disorder is a ***cerebrovascular accident (CVA)***, also called a ***stroke*** or ***brain attack***. CVAs affect 500,000 people a year in the United States and represent the third leading cause of death, behind heart attacks and cancer. A CVA is characterized by abrupt onset of persisting symptoms, such as paralysis or loss of sensation, that arise from destruction of brain tissue. Common causes of CVAs are hemorrhage from a blood vessel in the pia mater or brain, blood clots, and formation of cholesterol-containing atherosclerotic plaques that block brain blood flow. The risk factors implicated in CVAs are high blood pressure, high blood cholesterol, heart disease, narrowed carotid arteries, transient ischemic attacks (discussed next), diabetes, smoking, obesity, and excessive alcohol intake.

Transient Ischemic Attack

A ***transient ischemic attack (TIA)*** is an episode of temporary cerebral dysfunction caused by impaired blood flow to part of the brain. Symptoms include dizziness, weakness, numbness, or paralysis in a limb or in one side of the body; drooping of one side of the face; headache; slurred speech or difficulty understanding speech; and a partial loss of vision or double vision. Sometimes nausea or vomiting also occurs. The onset of symptoms is sudden and reaches maximum intensity almost immediately. A TIA usually persists for 5 to 10 minutes and only rarely lasts as long as 24 hours. It leaves no persistent neurological deficits. The causes of TIAs include blood clots, atherosclerosis, and certain blood disorders.

Poliomyelitis

Poliomyelitis, or simply ***polio***, is caused by a virus called poliovirus. The onset of the disease is marked by fever, severe headache, a stiff neck and back, deep muscle pain and weakness, and loss of certain somatic reflexes. In its most serious form, the virus produces paralysis by destroying cell bodies of motor neurons, specifically those in the anterior horns of the spinal cord and in the nuclei of the cranial nerves. Polio can cause death from respiratory or heart failure if the virus invades neurons in vital centers that control breathing and heart functions in the brain stem. Even though polio vaccines have virtually eradicated polio in the United States, outbreaks of polio continue throughout the world. Due to international travel, polio could easily be reintroduced into North America if individuals are not vaccinated appropriately.

Several decades after suffering a severe attack of polio and following their recovery from it, some individuals develop a condition called ***post-polio syndrome***. This neurological disorder is characterized by progressive muscle weakness, extreme fatigue, loss of function, and pain, especially in muscles and joints. Post-polio syndrome seems to involve a slow degeneration of motor neurons that innervate muscle fibers. Triggering factors appear to be a fall, a minor accident, surgery, or prolonged bed rest. Possible causes include overuse of surviving motor neurons over time, smaller motor neurons because of the initial infection by the virus, reactivation of dormant polio viruses, immune-mediated responses, hormone deficiencies, and environmental toxins. Treatment consists of muscle-strengthening exercises, administration of drugs to enhance the action of acetylcholine in stimulating muscle contraction, and administration of nerve growth factors to stimulate both nerve and muscle growth.

Parkinson Disease

Parkinson disease (PD) is a progressive disorder of the CNS that typically affects its victims around age 60. Neurons that extend from the substantia nigra to the putamen and caudate nucleus, where they release the neurotransmitter dopamine (DA), degenerate in PD. The cause of PD is unknown, but toxic environmental chemicals, such as pesticides, herbicides, and carbon monoxide, are suspected contributing agents. Only 5% of PD patients have a family history of the disease.

In PD patients, involuntary skeletal muscle contractions often interfere with voluntary movement. For instance, the muscles of the upper limb may alternately contract and relax, causing the hand to shake. This shaking, called ***tremor***, is the most common symptom of PD. Also, muscle tone may increase greatly, causing rigidity of the involved body part. Rigidity of the facial muscles gives the face a masklike appearance. The expression is characterized by a wide-eyed, unblinking stare and a slightly open mouth with uncontrolled drooling.

Motor performance is also impaired by ***bradykinesia*** (*brady-* = slow), slowness of movements. Activities such as shaving, cutting food, and buttoning a shirt take longer and become increasingly more difficult as the disease progresses. Muscular movements also exhibit ***hypokinesia*** (*hypo-* = under), decreasing range of motion. For example, words are written smaller, letters are poorly formed, and eventually handwriting becomes illegible. Often, walking is impaired; steps become shorter and shuffling, and arm swing diminishes. Even speech may be affected.

Alzheimer Disease

Alzheimer disease (ALTZ-hī-mer) or ***AD*** is a disabling senile dementia, the loss of reasoning and ability to care for oneself, that afflicts about 11% of the population over age 65. In the United States, AD afflicts about 4 million people and claims over 100,000 lives a year. The cause of most AD cases is still unknown, but evidence suggests it is due to a combination of genetic factors, environmental or lifestyle factors, and the aging process. Mutations in three different genes (coding for presenilin-1, presenilin-2, and amyloid precursor protein) lead to early-onset forms of AD in afflicted families but account for less than 1% of all cases. An environmental risk factor for developing AD is a history of head injury. A similar dementia occurs in boxers, probably caused by repeated blows to the head.

Individuals with AD initially have trouble remembering recent events. They then become confused and forgetful, often repeating questions or getting lost while traveling to previously familiar places. Disorientation increases; memories of past events disappear; and episodes of paranoia, hallucination, or violent changes in mood may occur. As their minds continue to deteriorate, AD patients lose their ability to read, write, talk, eat, or walk. At autopsy, brains of AD victims show three distinct structural abnormalities: (1) loss of neurons that liberate acetylcholine from a brain region called the nucleus basalis, located below the globus pallidus; (2) beta-amyloid plaques, clusters of abnormal proteins deposited outside neurons; and (3) neurofibrillary tangles, abnormal bundles of protein filaments inside neurons in affected brain regions. A person with AD usually dies of some complication that afflicts bedridden patients, such as pneumonia.

MEDICAL TERMINOLOGY AND CONDITIONS

Analgesia (an′-al-JĒ-zē-a; *an-* = without; *-algesia* = painful condition) Pain relief.

Anesthesia (an′-es-THĒ-zē-a; *-esthesia* = feeling) Loss of sensation.

Consciousness (KON-shus-nes) A state of wakefulness in which an individual is fully alert, aware, and oriented, partly as a result of feedback between the cerebral cortex and reticular activating system.

Dementia (de-MEN-shē-a; *de-* = away from; *-mentia* = mind) Permanent or progressive general loss of intellectual abilities, including impairment of memory, judgment, and abstract thinking, and changes in personality.

Encephalitis (en′-sef-a-LĪ-tis) An acute inflammation of the brain caused by either a direct attack by any of several viruses or an allergic reaction to any of the many viruses that are normally harmless to the central nervous system. If the virus affects the spinal cord as well, the condition is called ***encephalomyelitis***.

Epidural block Injection of an anesthetic drug into the epidural space, the space between the dura mater and the vertebral column, to cause a temporary loss of sensation. Such injections in the lower lumbar region are used to control pain during childbirth.

Meningitis (men-in-JĪ-tis) Inflammation of the meninges.

Nerve block Loss of sensation due to injection of a local anesthetic; an example is local dental anesthesia.

Neuralgia (noo-RAL-jē-a; *neur-* = nerve; *-algia* = pain) Attacks of pain along the entire length or a branch of a peripheral sensory nerve.

Neuritis (*neur-* = nerve; *-itis* = inflammation) Inflammation of one or several nerves, resulting from irritation caused by bone fractures, contusions, or penetrating injuries. Additional causes include infections; vitamin deficiency (usually thiamine); and poisons such as carbon monoxide, carbon tetrachloride, heavy metals, and some drugs.

Reye (RĪ) ***syndrome*** Occurs after a viral infection, particularly chickenpox or influenza, most often in children or teens who have taken aspirin; characterized by vomiting and brain dysfunction (disorientation, lethargy, and personality changes) that may progress to coma and death.

Sciatica (sī-AT-i-ka) A type of neuritis characterized by severe pain along the path of the sciatic nerve or its branches; may be caused by a slipped disc, pelvic injury, osteoarthritis of the backbone, or pressure from an expanding uterus during pregnancy.

STUDY OUTLINE

Spinal Cord Structure (p. 243)

1. The spinal cord is protected by the vertebral column, meninges, and cerebrospinal fluid.
2. The meninges are three connective tissue coverings of the spinal cord and brain: dura mater, arachnoid mater, and pia mater.
3. Removal of cerebrospinal fluid from the subarachnoid space is called a spinal tap. The procedure is used to remove CSF and to introduce antibiotics, anesthetics, and chemotherapy.
4. The spinal cord extends from the lowest part of the brain, the medulla oblongata, to the upper border of the second lumbar vertebra in the vertebral column.
5. The spinal cord contains cervical and lumbar enlargements that serve as points of origin for nerves to the limbs.
6. The roots of the nerves arising from the lumbar, sacral, and coccygeal regions of the cord are called the cauda equina.
7. The gray matter in the spinal cord is divided into horns and the white matter into columns. Parts of the spinal cord observed in cross section are the central canal; anterior, posterior, and lateral gray horns; anterior, posterior, and lateral white columns; and sensory (ascending) and motor (descending) tracts.

Spinal Nerves (p. 246)

1. The 31 pairs of spinal nerves are named and numbered according to the region and level of the spinal cord from which they emerge.
2. There are 8 pairs of cervical, 12 pairs of thoracic, 5 pairs of lumbar, 5 pairs of sacral, and 1 pair of coccygeal nerves.
3. Spinal nerves are attached to the spinal cord by means of a posterior root and an anterior root.
4. All spinal nerves are mixed nerves containing sensory and motor axons.
5. Branches of spinal nerves, except for T2 to T11, form networks of nerves called plexuses. Nerves T2 to T11 do not form plexuses and are called intercostal nerves.

6. The major plexuses are the cervical, brachial, lumbar, and sacral plexuses.

Spinal Cord Functions (p. 247)

1. The spinal cord white matter and gray matter have two major functions in maintaining homeostasis. The white matter serves as highways for nerve impulse conduction. The gray matter receives and integrates incoming and outgoing information and is a site for integration of reflexes.
2. A reflex is a fast, involuntary sequence of actions that occurs in response to a particular stimulus. The basic components of a reflex arc are a receptor, a sensory neuron, an integrating center, a motor neuron, and an effector.

Brain (p. 248)

1. The major parts of the brain are the brain stem, diencephalon, cerebellum, and cerebrum (see Table 10.1 on page 262). The brain stem consists of the medulla oblongata, pons, and midbrain. The diencephalon consists of the thalamus, hypothalamus, and pineal gland.
2. The brain is well supplied with oxygen and nutrients. Any interruption of the oxygen supply to the brain can weaken, permanently damage, or kill brain cells. Glucose deficiency may produce dizziness, convulsions, and unconsciousness.
3. The blood–brain barrier (BBB) limits the passage of certain material from the blood into the brain.
4. The brain is protected by cranial bones, meninges, and cerebrospinal fluid.
5. The cranial meninges are continuous with the spinal meninges and are named dura mater, arachnoid mater, and pia mater.
6. Cerebrospinal fluid is formed in the choroid plexuses and circulates continually through the subarachnoid space, ventricles, and central canal.
7. Cerebrospinal fluid protects by serving as a shock absorber. It also delivers nutritive substances from the blood and removes wastes.
8. The medulla oblongata, or medulla, is continuous with the upper part of the spinal cord. It contains regions for regulating heart rate, diameter of blood vessels, breathing, swallowing, coughing, vomiting, sneezing, and hiccupping. Cranial nerves VIII–XII originate at the medulla.
9. The pons links parts of the brain with one another; it relays impulses for voluntary skeletal movements from the cerebral cortex to the cerebellum, and it contains two regions that control breathing. Cranial nerves V–VII and part of VIII originate at the pons.
10. The midbrain is between the pons and diencephalon. It conveys motor impulses from the cerebrum to the cerebellum and spinal cord, sends sensory impulses from the spinal cord to the thalamus, and mediates auditory and visual reflexes. It also contains nuclei associated with cranial nerves III and IV.
11. The reticular formation is a netlike arrangement of gray and white matter extending throughout the brain stem that alerts the cerebral cortex to incoming sensory signals and helps regulate muscle tone.
12. The thalamus contains nuclei that serve as relay stations for sensory impulses to the cerebral cortex. It also contributes to motor functions by transmitting information from the cerebellum and basal ganglia to motor areas of the cerebral cortex.
13. The hypothalamus is inferior to the thalamus. It controls the autonomic nervous system, secretes hormones, functions in rage and aggression, governs body temperature, regulates food and fluid intake, and establishes circadian rhythms.
14. The cerebellum occupies the inferior and posterior aspects of the cranial cavity. It attaches to the brain stem by cerebellar peduncles. It coordinates movements and helps maintain normal muscle tone, posture, and balance.
15. The cerebrum is the largest part of the brain. Its cortex contains gyri (convolutions), fissures, and sulci. The cerebral lobes are frontal, parietal, temporal, and occipital.
16. The white matter is deep to the cortex and consists of myelinated and unmyelinated axons extending to other CNS regions.
17. The basal ganglia are several groups of nuclei in each cerebral hemisphere. They help control automatic movements of skeletal muscles and help regulate muscle tone.
18. The limbic system encircles the upper part of the brain stem and the corpus callosum. It functions in emotional aspects of behavior and memory.
19. The sensory areas of the cerebral cortex receive and perceive sensory information. The motor areas govern muscular movement. The association areas are concerned with emotional and intellectual processes.
20. Somatic sensory pathways from receptors to the cerebral cortex involve sets of three neurons. The posterior column–medial lemniscus pathway relays nerve impulses for sensations of fine touch, proprioception, and vibrations. The lateral and anterior spinothalamic tracts relay impulses for pain, thermal, tickle, and itch sensations.
21. All somatic motor pathways that control movement converge on lower motor neurons. Input to lower motor neurons comes from local interneurons, upper motor neurons, basal ganglia neurons, and cerebellar neurons.
22. Subtle anatomical differences exist between the two cerebral hemispheres, and each has some unique functions.
23. Memory, the ability to store and recall thoughts, involves persistent changes in the brain.
24. Brain waves generated by the cerebral cortex are recorded as an electroencephalogram (EEG), which may be used to diagnose epilepsy, infections, and tumors.

Cranial Nerves (p. 263)

1. Twelve pairs of cranial nerves emerge from the brain.
2. Like spinal nerves, cranial nerves are part of the PNS. See Table 10.2 on pages 263–265 for the names, components, and functions of each of the cranial nerves.

Aging and the Nervous System (p. 265)

1. The brain grows rapidly during the first few years of life.
2. Age-related effects involve loss of brain mass and decreased capacity for sending nerve impulses.

SELF-QUIZ

1. Which sequence best represents a reflex arc from the stimulus to the response?
   1. effector
   2. integrating center
   3. motor neuron
   4. receptor
   5. sensory neuron

   a. 3, 1, 4, 5, 2 b. 1, 5, 2, 3, 4 c. 4, 3, 2, 5, 1
   d. 5, 2, 3, 4, 1 e. 4, 5, 2, 3, 1

2. Which of the following would carry sensory nerve impulses?
   a. anterior spinothalamic tract b. anterior root
   c. lateral corticospinal tract d. direct pathways
   e. pyramids

3. An inability to distinguish keys in your pocket by touch could indicate damage to the
   a. gray matter of the cerebellum
   b. lateral spinothalamic tract
   c. posterior column–medial lemniscus pathway
   d. anterior ramus
   e. primary motor cortex

4. Carpal tunnel syndrome is due to damage to a nerve in the
   a. lumbar plexus b. cervical plexus c. brachial plexus
   d. cauda equina e. sacral plexus

5. A needle used in a spinal tap would penetrate (in order):
   1. arachnoid 2. dura mater 3. epidural space
   4. subarachnoid space

   a. 1, 2, 3, 4 b. 2, 3, 1, 4 c. 3, 1, 4, 2 d. 3, 2, 1, 4
   e. 4, 1, 2, 3

6. The diencephalon is composed of the
   a. medulla, pons, and hypothalamus
   b. midbrain, hypothalamus, and thalamus
   c. cerebellum and midbrain
   d. medulla, pons, and midbrain
   e. hypothalamus and thalamus

7. Which of the following statements about the blood supply to the brain is NOT true?
   a. The brain needs a constant supply of glucose delivered by the blood.
   b. The structure of the brain capillaries allows selective passage of certain materials from the blood into the brain.
   c. The glucose brought to the brain can be stored for future use
   d. Brain neurons that are totally deprived of oxygen for four minutes or more may be permanently injured.
   e. The brain requires about 20% of the body's oxygen supply.

8. After a car accident, Joe exhibits severe dizziness, difficulty in walking, and slurred speech. He may have damaged his
   a. cerebellum b. pons c. reticular activating system
   d. fifth cranial nerve e. midbrain

9. Which of the following is NOT a function of cerebrospinal fluid?
   a. protection b. circulation c. conduction of nerve impulses d. nutrition e. shock absorption

10. Which part of the brain contains the centers that control the heart rate and breathing rhythm?
    a. medulla b. midbrain c. cerebellum
    d. thalamus e. pons

11. The part of the brain that serves as a link between the nervous and endocrine systems is the
    a. reticular formation b. hypothalamus c. pons
    d. brain stem e. cerebellum

12. Which of the following is NOT a function of the hypothalamus?
    a. regulates food intake b. controls body temperature
    c. regulates feelings of rage and aggression
    d. helps establish sleep patterns
    e. allows crude interpretation of pain and pressure

13. The part(s) of the brain concerned with memory, reasoning, judgment, and intelligence is (are) the
    a. sensory areas b. limbic system c. motor areas
    d. cerebellum e. association areas

14. A broad band of white matter that connects the two cerebral hemispheres is the
    a. corpus callosum b. gyrus c. insula
    d. ascending tract e. basal ganglia

15. The ringing of your alarm clock in the morning wakes you up by stimulating the
    a. thalamus b. reticular activating system
    c. Broca's area d. basal ganglia e. spinal cord

16. Match the following functions to the primary lobe in which they are located:

    ___ a. contains primary visual area that allows interpretation of shape and color
    ___ b. receives impulses for smell
    ___ c. contains primary motor area that controls muscle movement
    ___ d. receives sensory impulses for touch, pain, and temperature

    A. frontal lobe
    B. parietal lobe
    C. occipital lobe
    D. temporal lobe

17. When entering a restaurant, you are bombarded with many different sensory stimuli. The part of the brain that combines all of those sensory inputs so that you can respond appropriately is the
    a. somatosensory association area
    b. common integrative area
    c. premotor area
    d. Wernicke's area
    e. hypothalamus

18. Which cranial nerves contain only sensory fibers?
a. olfactory, optic, and glossopharyngeal
b. optic and oculomotor
c. optic and trochlear
d. optic and olfactory
e. vagus and facial

19. Which two of the following cranial nerves are NOT involved in controlling movement of the eyeball?
a. oculomotor
b. trochlear
c. facial
d. abducens
e. trigeminal

20. Match the following:

___ **a.** organization of white matter in the spinal cord	**A.** longitudinal fissure
___ **b.** absorb cerebrospinal fluid	**B.** sulci
___ **c.** extension of nerves beyond the end of the spinal cord	**C.** ventricles
___ **d.** folds of the cerebral cortex	**D.** anterior median fissure
___ **e.** contains the sensory fibers of a spinal nerve	**E.** central canal
___ **f.** contains the motor fibers of a spinal nerve	**F.** posterior (dorsal) root
___ **g.** separates the cerebrum into right and left halves	**G.** columns
___ **h.** divides spinal cord into right and left sides	**H.** arachnoid villi
___ **i.** brain cavities where CSF circulates	**I.** anterior (ventral) root
___ **j.** shallow grooves in the cerebrum	**J.** gyri
___ **k.** contains CSF in the spinal cord	**K.** cauda equina

CRITICAL THINKING APPLICATIONS

1. After a few days of using her new crutches, Kate's arms and hands felt tingly and numb. The physical therapist said Kate had a case of "crutch palsy" from improper use of her crutches. Kate had been leaning her armpits on the crutches while hobbling along. What caused the numbness in her arms and hands?
2. Dennis was a little nervous. It was his first visit to the dentist in 10 years. "You won't feel a thing," said the dentist as she injected several doses of "numbing" medication. While having lunch right after the visit, soup drips down Dennis' chin because he still doesn't feel a thing in his lower lip and right upper lip. What happened to Dennis?
3. An elderly relative suffered a stroke and now has difficulty with the movement of her right upper limb. She is also working with a therapist due to some speech problems. What areas of the brain were damaged by the stroke?
4. Lynn flicked on the light when she heard her husband's yell. Kyle was bouncing on his left foot while holding his right foot in his hand. A pin was sticking out of the bottom of his foot. Explain Kyle's response to stepping on the pin.

ANSWERS TO FIGURE QUESTIONS

10.1 CSF circulates in the subarachnoid space.

10.2 Spinal nerves are part of the PNS (peripheral nervous system).

10.3 A horn is an area of gray matter, and a column is a region of white matter in the spinal cord.

10.4 All spinal nerves are mixed (have sensory and motor components) because the posterior root containing sensory axons and the anterior root containing motor axons unite to form the spinal nerve.

10.5 Axons of sensory neurons are part of the posterior root, and axons of motor neurons are part of the anterior root.

10.6 The medulla oblongata of the brain attaches to the spinal cord.

10.7 CSF is formed in the choroid plexuses and is reabsorbed through arachnoid villi into blood in the superior sagittal sinus.

10.8 The midbrain contains the cerebral peduncles.

10.9 The superior colliculi govern eye movements for tracking moving images and scanning stationary images and are responsible for reflexes that govern movements of the eyes, head, and neck in response to visual stimuli.

10.10 The basal ganglia are located in the cerebrum and are composed of gray matter.

10.11 The longitudinal fissure separates the right and left cerebral hemispheres.

10.12 The limbic system is located on the inner border of the cerebrum and floor of the diencephalon.

10.13 The primary somatosensory area localizes somatic sensations.

10.14 Damage to the spinothalamic tracts could produce loss of pain, thermal, tickle, and itch sensations.

10.15 In the spinal cord, the lateral and anterior corticospinal tracts conduct impulses along axons of upper motor neurons.

chapter 11

AUTONOMIC NERVOUS SYSTEM

did you know?

The "fight-or-flight" response of the sympathetic nervous system is very helpful when you encounter a snarling dog or need to escape from a burning building. But when the emergency is over, your parasympathetic nervous system needs time to help your body relax and recover. What happens when stress builds up, and no recovery occurs? When your days are filled with negative stress and an overactivated sympathetic nervous system, stress-related health problems may develop. Chronic, unrelenting, overwhelming stress interferes with the body's ability to maintain homeostasis and health. Learning relaxation and stress reduction skills can reduce the harmful effects of stress upon the body.

Focus on Wellness, page 278

www.wiley.com/college/apcentral

The part of the nervous system that regulates smooth muscle, cardiac muscle, and certain glands is the ***autonomic nervous system (ANS)***. Recall that together the ANS and somatic nervous system compose the peripheral nervous system; see Figure 9.1 on page 226. The ANS was originally named *autonomic* (*auto-* = self; *-nomic* = law) because it was thought to function in a self-governing manner. Although the ANS usually does operate without conscious control from the cerebral cortex, it is regulated by other brain regions, mainly the hypothalamus and brain stem. In this chapter, we compare the structural and functional features of the somatic and autonomic nervous systems. Then we discuss the anatomy of the motor portion of the ANS and compare the organization and actions of its two major branches, the sympathetic and parasympathetic divisions.

looking back to move ahead . . .

- **Structures of the Nervous System (page 226)**
- **Sensory and Motor Components of the ANS and ANS Effectors (page 227)**

COMPARISON OF SOMATIC AND AUTONOMIC NERVOUS SYSTEMS

OBJECTIVE • Compare the main structural and functional differences between the somatic and autonomic parts of the nervous system.

As you learned in Chapter 10, the somatic nervous system includes both sensory and motor neurons. The sensory neurons convey input from receptors for the special senses (vision, hearing, taste, smell, and equilibrium, described in Chapter 12) and from receptors for somatic senses (pain, temperature, touch, and proprioceptive sensations). All these sensations normally are consciously perceived. In turn, somatic motor neurons synapse with skeletal muscle—the effector tissue of the somatic nervous system—and produce conscious, voluntary movements. When a somatic motor neuron stimulates a skeletal muscle, the muscle contracts. If somatic motor neurons cease to stimulate a muscle, the result is a paralyzed, limp muscle that has no muscle tone. In addition, even though we are generally not conscious of breathing, the muscles that generate breathing movements are skeletal muscles controlled by somatic motor neurons. If the respiratory motor neurons become inactive, breathing stops.

The main input to the ANS comes from ***autonomic sensory neurons***. These neurons are associated with sensory receptors that monitor internal conditions, such as blood CO_2 level or the degree of stretching in the walls of internal organs or blood vessels. When the viscera are functioning properly, these sensory signals usually are not consciously perceived.

Autonomic motor neurons regulate ongoing activities in their effector tissues, which are cardiac muscle, smooth muscle, and glands, by both excitation and inhibition. Unlike skeletal muscle, these tissues often function to some extent even if their nerve supply is damaged. The heart continues to beat, for instance, when it is removed for transplantation into another person. Examples of autonomic responses are changes in the diameter of the pupil, dilation and constriction of blood vessels, and changes in the rate and force of the heartbeat. Because most autonomic responses cannot be consciously altered or suppressed to any great degree, they are the basis for polygraph ("lie detector") tests. However, practitioners of yoga or other techniques of meditation and those who employ biofeedback methods may learn how to modulate ANS activities. For example, they may be able to voluntarily decrease their heart rate or blood pressure.

Figure 11.1 compares somatic and autonomic motor neurons. The axon of a somatic motor neuron extends all the way from the CNS to the skeletal muscle fibers that it stimulates (Figure 11.1a). By contrast, autonomic motor pathways consist of sets of *two* motor neurons (Figure 11.1b). The first neuron, called the ***preganglionic neuron***, has its cell body in the CNS, either in the lateral gray horn of the spinal cord or in a nucleus of the brainstem. Its axon extends from the CNS via a cranial or a spinal nerve to an *autonomic ganglion*, where it synapses with the second neuron. (Recall that a ganglion is a collection of neuronal cell bodies usually outside the CNS.) The second neuron, the ***postganglionic neuron***, lies entirely in the peripheral nervous system. Its cell body is located in an autonomic ganglion, and its axon extends from the ganglion to the effector (smooth muscle, cardiac muscle, or a gland). The effect of the postganglionic neuron on the effector may be either excitation (causing contraction of smooth or cardiac muscle or increasing secretions of glands) or inhibition (causing relaxation of smooth or cardiac muscle or decreasing secretions of glands). In contrast, a single somatic motor neuron extends from the CNS and always excites its effector (causing contraction of skeletal muscle) (Figure 11.1a). Another difference between autonomic and somatic motor neurons is that all somatic motor neurons release acetylcholine (ACh) as their neurotransmitter. Some autonomic motor neurons release ACh; others release norepinephrine (NE).

The output (motor) part of the ANS has two main branches: the ***sympathetic division*** and the ***parasympathetic division***. Most organs have ***dual innervation***; that is, they receive impulses from both sympathetic and parasympathetic neurons. In general, nerve impulses from one division stimulate the organ to increase its activity (excitation), whereas impulses from the other division decrease the organ's activity (inhibition). For example, an increased rate of nerve impulses from the sympathetic division increases heart rate, and an increased rate of nerve

Table 11.1 Comparison of Somatic and Autonomic Nervous Systems

Property	Somatic	Autonomic
Effectors	Skeletal muscles.	Cardiac muscle, smooth muscle, and glands.
Type of control	Mainly voluntary.	Mainly involuntary.
Neural pathway	One motor neuron extends from CNS and synapses directly with a skeletal muscle fiber.	One motor neuron extends from the CNS and synapses with another motor neuron in a ganglion; the second motor neuron synapses with an autonomic effector.
Neurotransmitter	Acetylcholine.	Acetylcholine or norepinephrine.
Action of neurotransmitter on effector	Always excitatory (causing contraction of skeletal muscle).	May be excitatory (causing contraction of smooth muscle, increased heart rate, increased force of heart contraction, or increased secretions from glands) or inhibitory (causing relaxation of smooth muscle, decreased heart rate, or decreased secretions from glands).

Figure 11.1 Comparison of somatic and autonomic motor neuron pathways to their effector tissues.

Stimulation by the autonomic motor neurons can either excite or inhibit smooth muscle, cardiac muscle, and glands. Stimulation by somatic motor neurons always causes contraction of skeletal muscle.

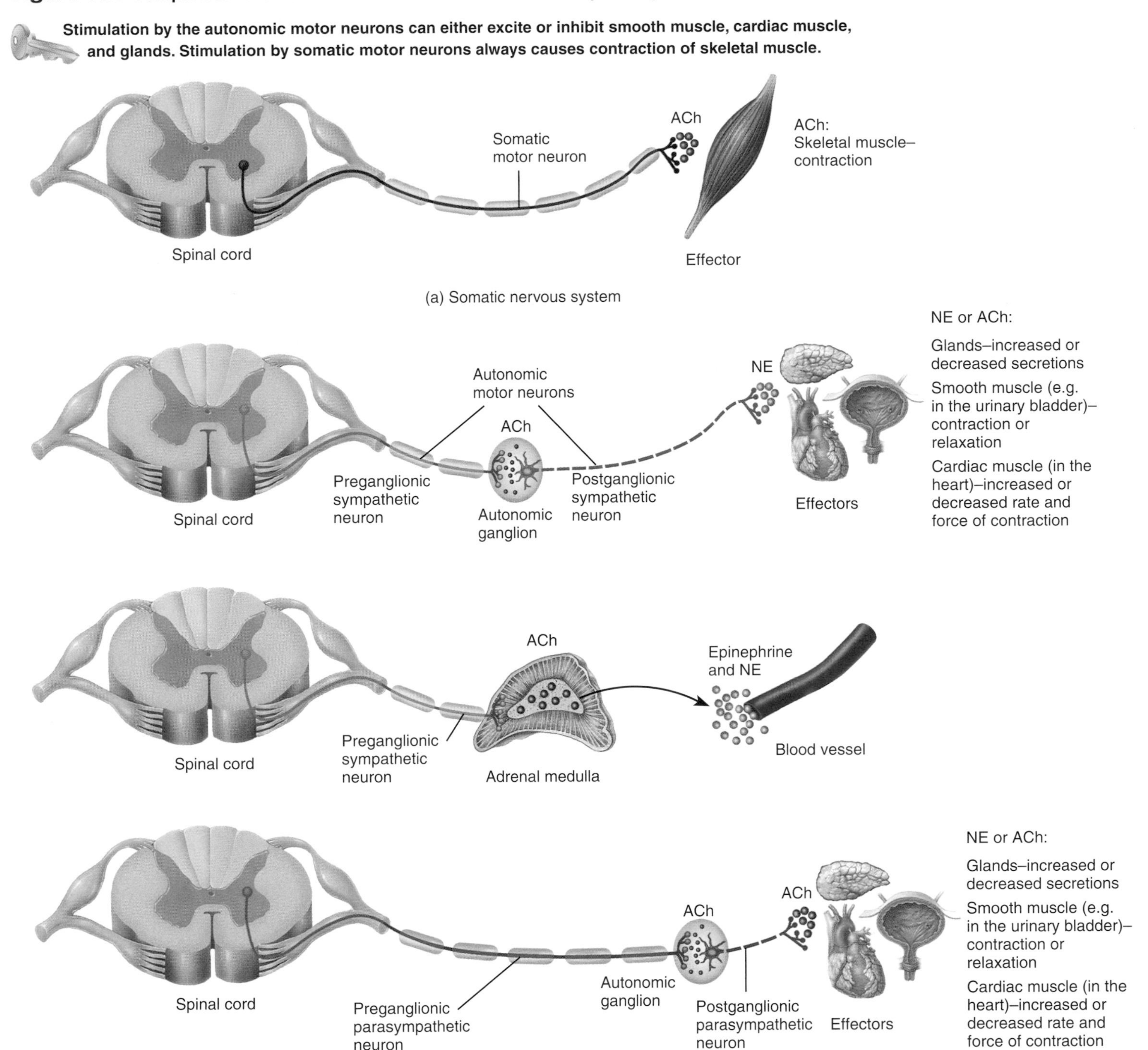

What does "dual innervation" mean?

impulses from the parasympathetic division decreases heart rate. Table 11.1 on page 272 summarizes the similarities and differences between the somatic and autonomic nervous systems.

■ **CHECKPOINT**

1. Why is the autonomic nervous system so named?
2. What are the main input and output components of the autonomic nervous system?

STRUCTURE OF THE AUTONOMIC NERVOUS SYSTEM

OBJECTIVE • Identify the structural features of the autonomic nervous system.

We will now examine the structure of preganglionic neurons, ganglia, and postganglionic neurons and how they relate to the activities of the autonomic nervous system.

Organization of the Sympathetic Division

The sympathetic division of the ANS is also called the *thoracolumbar division* (thōr′-a-kō-LUM-bar) because the outflow of sympathetic nerve impulses comes from the thoracic and lumbar segments of the spinal cord (Figure 11.2). The sympathetic preganglionic neurons have their cell bodies in the 12 thoracic and the first two lumbar segments of the spinal cord. The preganglionic axons emerge from the spinal cord through the anterior root of a spinal nerve along with axons of somatic motor neurons. After exiting the cord, the sympathetic preganglionic axons extend to a sympathetic ganglion.

In the sympathetic ganglia, sympathetic preganglionic neurons synapse with postganglionic neurons. Because the sympathetic trunk ganglia are near the spinal cord, most sympathetic preganglionic axons are short. ***Sympathetic trunk ganglia*** lie in two vertical rows, one on either side of the vertebral column (Figure 11.2). Most postganglionic axons emerging from sympathetic trunk ganglia supply organs above the diaphragm. Other sympathetic ganglia, the ***prevertebral ganglia***, lie anterior to the vertebral column and close to the large abdominal arteries. These include the *celiac ganglion* (SĒ-lē-ak), the *superior mesenteric ganglion*, and the *inferior mesenteric ganglion*. In general, postganglionic axons emerging from the prevertebral ganglia innervate organs below the diaphragm.

Once the axon of a preganglionic neuron of the sympathetic division enters a sympathetic trunk ganglion, it may follow one of four paths:

1. It may synapse with postganglionic neurons in the sympathetic trunk ganglion it first reaches.
2. It may ascend or descend to a higher or lower sympathetic trunk ganglion before synapsing with postganglionic neurons.
3. It may continue, without synapsing, through the sympathetic trunk ganglion to end at a prevertebral ganglion and synapse with postganglionic neurons there.
4. It may extend to and terminate in the adrenal medulla.

A single sympathetic preganglionic axon has many branches and may synapse with 20 or more postganglionic neurons. Thus, nerve impulses that arise in a single preganglionic neuron may activate many different postganglionic neurons that in turn synapse with several autonomic effectors. This pattern helps explain why sympathetic responses can affect organs throughout the body almost simultaneously.

Most postganglionic axons leaving the cervical sympathetic trunk ganglia serve the head. They are distributed to sweat glands, smooth muscles of the eye, blood vessels of the face, nasal mucosa, and salivary glands. A few postganglionic axons from the cervical sympathetic trunk ganglia supply the heart. In the thoracic region, postganglionic axons from the sympathetic trunk serve the heart, lungs, and bronchi. Some axons from thoracic levels also supply sweat glands, blood vessels, and smooth muscles of hair follicles in the skin. In the abdomen, axons of postganglionic neurons leaving the prevertebral ganglia follow the course of various arteries to abdominal and pelvic autonomic effectors.

The sympathetic division of the ANS also includes part of the adrenal glands (Figure 11.2). The inner part of the adrenal gland, the ***adrenal medulla*** (me-DUL-a), develops from the same embryonic tissue as the sympathetic ganglia, and its cells are similar to sympathetic postganglionic neurons. Rather than extending to another organ, however, these cells release hormones into the blood. Upon stimulation by sympathetic preganglionic neurons, cells of the adrenal medulla release a mixture of hormones—about 80% ***epinephrine*** and 20% ***norepinephrine***. These hormones circulate throughout the body and intensify responses elicited by sympathetic postganglionic neurons.

In **Horner's syndrome**, sympathetic stimulation of one side of the face is lost due to an inherited mutation, an injury, or a disease that affects sympathetic outflow through the superior cervical ganglion. Symptoms occur in the head on the affected side and include drooping of the upper eyelid, constricted pupil, and lack of sweating.

Organization of the Parasympathetic Division

The parasympathetic division is also called the *craniosacral division* (krā′-nē-ō-SĀ-kral) because the outflow of parasympathetic nerve impulses comes from cranial nerve nuclei and sacral segments of the spinal cord. The cell bodies of parasympathetic preganglionic neurons are located in the nuclei of four cranial nerves (III, VII, IX, and X) in the brain stem and in the second through fourth sacral segments of the spinal cord (S2, S3, and S4) (Figure 11.3 on page 276). Parasympathetic preganglionic axons emerge from the CNS as part of a cranial nerve or as part of the anterior root of a spinal nerve. Axons of the vagus (X) nerve carry nearly 80% of the total parasympathetic outflow. In the thorax, axons of the vagus nerve extend to ganglia in the heart and the airways of the lungs. In the abdomen, axons of the vagus nerve extend to ganglia in the liver, stomach, pancreas, small intestine, and part of the large intestine. Parasympathetic preganglionic axons exit the sacral spinal cord in the anterior roots

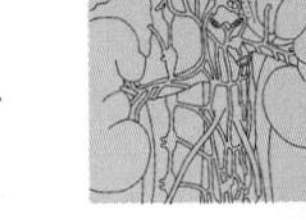

Figure 11.2 Structure of the sympathetic division of the autonomic nervous system. Although some innervated structures are diagrammed only for one side of the body, the sympathetic division actually innervates tissues and organs on both sides.

Cell bodies of sympathetic preganglionic neurons are located in the gray matter in the 12 thoracic and first two lumbar segments of the spinal cord.

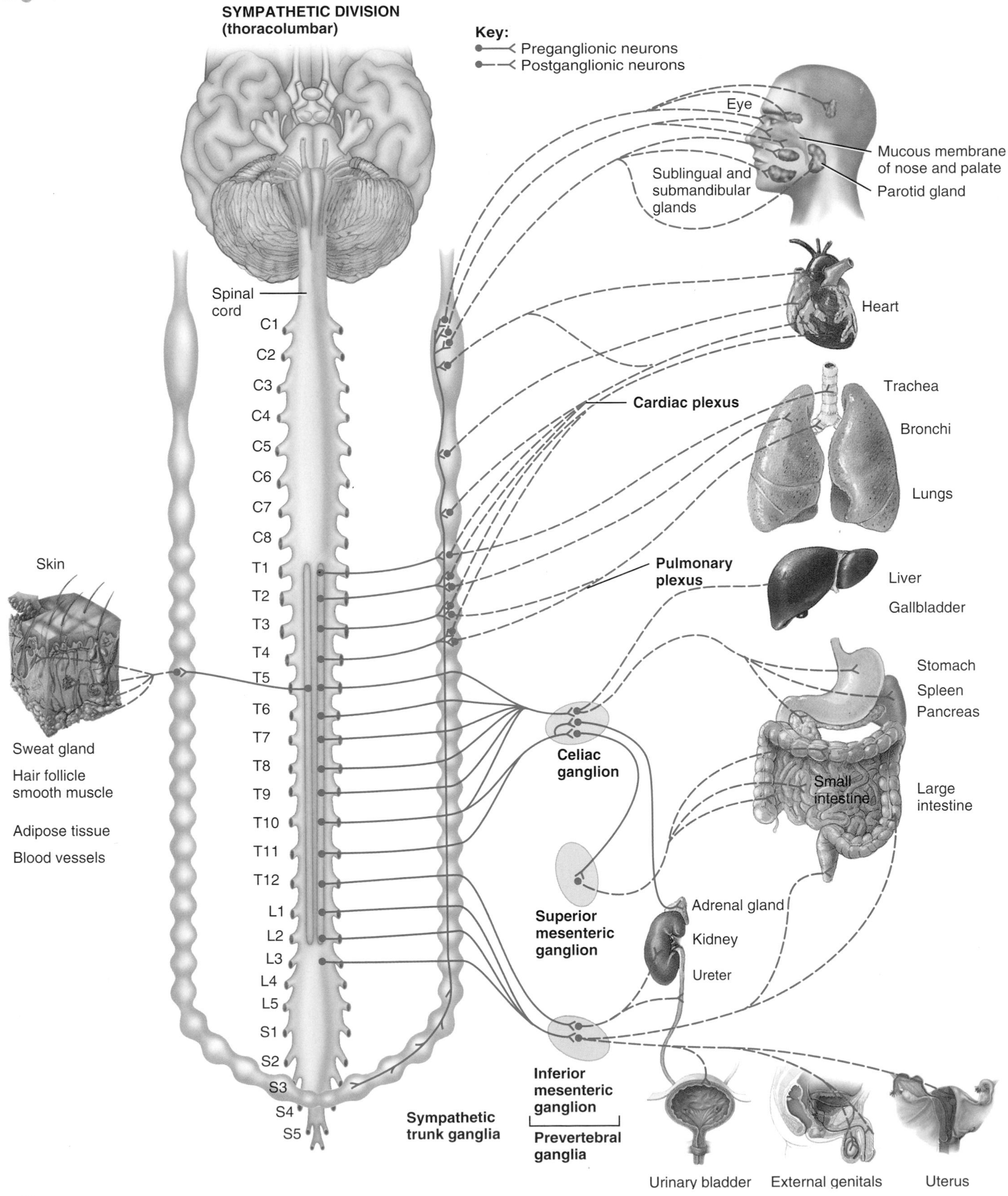

Which neurons synapse in a sympathetic trunk ganglion?

Figure 11.3 Structure of the parasympathetic division of the autonomic nervous system. Although some innervated structures are diagrammed on one side of the body, the parasympathetic division actually innervates organs on both sides.

Cell bodies of parasympathetic preganglionic neurons are located in brain stem nuclei and in the gray matter in the second through fourth sacral segments of the spinal cord.

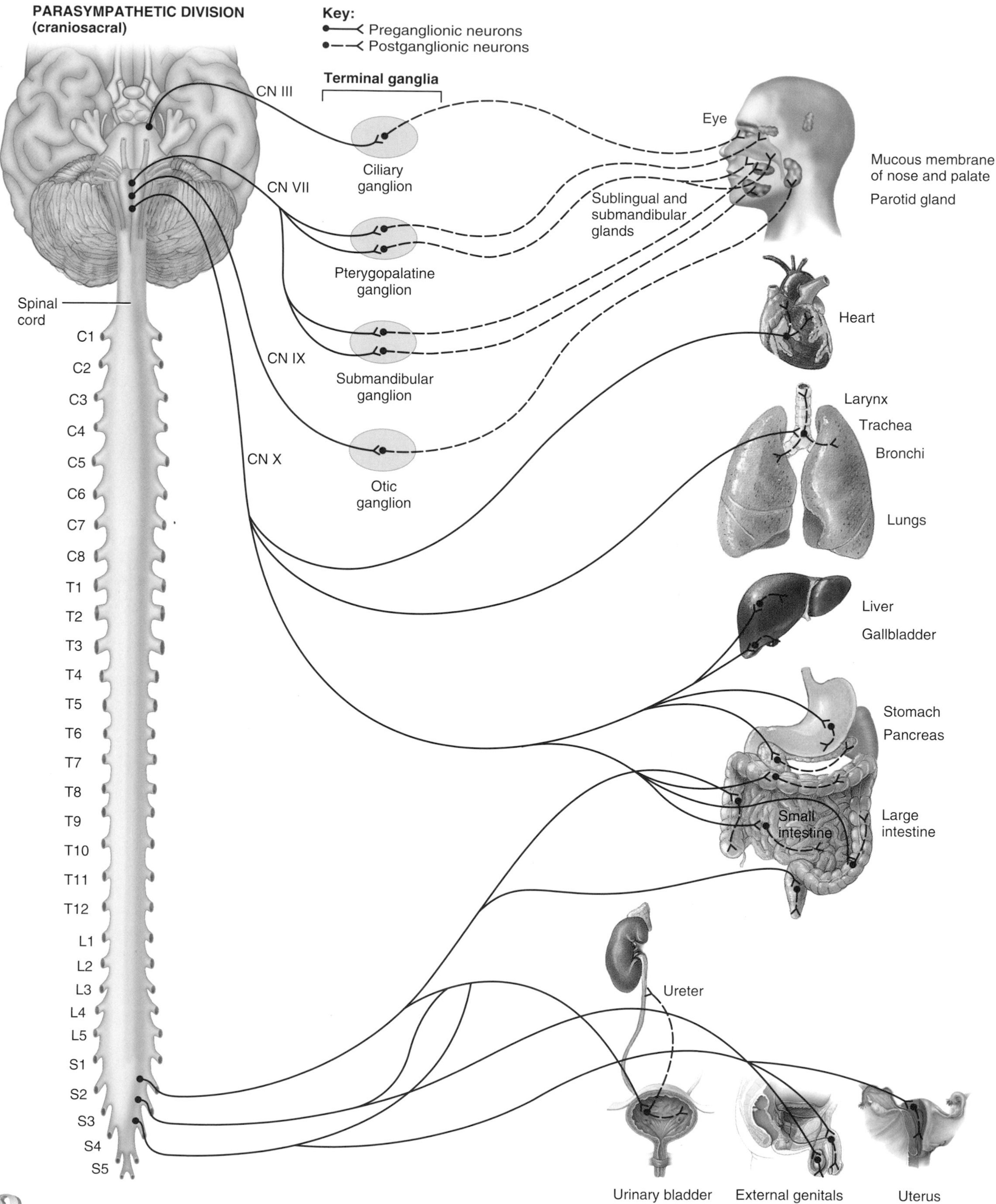

Which division, sympathetic or parasympathetic, has longer preganglionic axons? *(Hint: Compare Figures 11.2 and 11.3.)*

of the second through fourth sacral nerves. The axons then extend to ganglia in the walls of the colon, ureters, urinary bladder, and reproductive organs.

Preganglionic axons of the parasympathetic division synapse with postganglionic neurons in ***terminal ganglia***, which are located close to or actually within the wall of the innervated organ. Terminal ganglia in the head receive preganglionic axons from the oculomotor (III), facial (VII), or glossopharyngeal (IX) cranial nerves and supply structures in the head (Figure 11.3). Axons in the vagus (X) nerve extend to many terminal ganglia in the thorax and abdomen. Because the axons of parasympathetic preganglionic neurons extend from the brain stem or sacral spinal cord to a terminal ganglion in an innervated organ, they are longer than most of the axons of sympathetic preganglionic neurons (compare Figures 11.2 and 11.3).

In contrast to the preganglionic axons, most parasympathetic postganglionic axons are very short because the terminal ganglia lie in the walls of their autonomic effectors. In the ganglion, the preganglionic neuron usually synapses with only four or five postganglionic neurons, all of which supply the same effector. Thus, parasympathetic responses are localized to a single effector.

> A **megacolon** (*mega-* = big) is an abnormally large colon. In congenital megacolon, parasympathetic nerves to the distal segment of the colon do not develop properly. Loss of motor function in the segment causes massive dilation of the normal proximal colon. The condition results in extreme constipation, abdominal distension, and occasionally, vomiting. Surgical removal of the affected segment of the colon corrects the disorder.

■ **CHECKPOINT**

3. Describe the locations of sympathetic trunk ganglia, prevertebral ganglia, and terminal ganglia. Which types of autonomic neurons synapse in each type of ganglion?
4. How can the sympathetic division produce simultaneous effects throughout the body, when parasympathetic effects typically are localized to specific organs?

FUNCTIONS OF THE AUTONOMIC NERVOUS SYSTEM

OBJECTIVE • Describe the functions of the sympathetic and parasympathetic divisions of the autonomic nervous system.

ANS Neurotransmitters

Neurotransmitters are chemical substances released by neurons at synapses. Autonomic neurons release neurotransmitters at synapses between neurons (preganglionic to postganglionic) and at synapses with autonomic effectors (smooth muscle, cardiac muscle, and glands). Some ANS neurons release acetylcholine; others release norepinephrine.

ANS neurons that release *acetylcholine* include (1) all sympathetic and parasympathetic preganglionic neurons, (2) all parasympathetic postganglionic neurons, and (3) a few sympathetic postganglionic neurons. Because acetylcholine is quickly inactivated by the enzyme *acetylcholinesterase (AChE)*, parasympathetic effects are short-lived and localized.

Most sympathetic postganglionic neurons release the neurotransmitter *norepinephrine (NE)*. Because norepinephrine is inactivated much more slowly than acetylcholine and because the adrenal medulla also releases epinephrine and norepinephrine into the bloodstream, the effects of activation of the sympathetic division are longer lasting and more widespread than those of the parasympathetic division. For instance, your heart continues to pound for several minutes after a near miss at a busy intersection due to the long-lasting effects of the sympathetic division.

Activities of the ANS

As noted earlier, most body organs receive instructions from both divisions of the ANS, which typically work in opposition to one another. The balance between sympathetic and parasympathetic activity or "tone" is regulated by the hypothalamus. Typically, the hypothalamus turns up sympathetic tone at the same time it turns down parasympathetic tone, and vice versa. A few structures receive only sympathetic innervation—sweat glands, arrector pili muscles attached to hair follicles in the skin, the kidneys, the spleen, most blood vessels, and the adrenal medullae (see Figure 11.2). In these structures there is no opposition from the parasympathetic division. Still, an increase in sympathetic tone has one effect, and a decrease in sympathetic tone produces the opposite effect.

Sympathetic Activities

During physical or emotional stress, high sympathetic tone favors body functions that can support vigorous physical activity and rapid production of ATP. At the same time, the sympathetic division reduces body functions that favor the storage of energy. Besides physical exertion, a variety of emotions—such as fear, embarrassment, or rage—stimulate the sympathetic division. Visualizing body changes that occur during "E situations" (exercise, emergency, excitement, embarrassment) will help you remember most of the sympathetic responses. Activation of the sympathetic division and release of hormones by the adrenal medullae result in a series of physiological responses collectively called the ***fight-or-flight response***, in which the following occur:

1. The pupils of the eyes dilate.
2. Heart rate, force of heart contraction, and blood pressure increase.

Focus on Wellness

Mind-Body Exercise—An Antidote to Stress

When we think of exercise, we usually think of toning up our muscles and maybe our hearts. But when some people think of exercise, their focus is on toning up neural input from the parasympathetic division of the autonomic nervous system. As you learned in this chapter, activation of the parasympathetic division helps restore homeostasis in many systems and is associated with feelings of relaxation.

Mind-Body Harmony

Mind-body exercise refers to exercise systems such as tai chi, hatha yoga, and many forms of the martial arts that couple muscular activity with an internally directed focus. These exercise systems exercise the mind as well as the body. Their internally directed focus usually includes an awareness of breathing, energy, and other physical sensations.

Practitioners often refer to this internal awareness as "mindful," meaning that the exerciser is open to physical and emotional sensations with an understanding, nonjudgmental attitude. A mindful attitude is typical of many kinds of meditation and relaxation practices. For example, when practicing a yoga pose, you would think something like "Deep, steady breathing; relax into the pose; shoulders pulling back, neck lengthening," rather than "That girl next to me sure is flexible; I'm really a failure at this stuff." Of course, in real life such external thoughts do sneak in, but we can redirect our attention back to a more neutral, nonjudgmental style.

Mind-Body Benefits

People practicing mind-body activities reap benefits from both the physical and mental activity. Hatha yoga, tai chi, and the martial arts increase muscular strength and flexibility, posture, balance, and coordination, and if performed vigorously, they can even improve cardiovascular health and endurance to some extent. In addition, the stress relief provided by the activity extends into both physical and psychological realms. Feelings of mental relaxation and emotional well-being translate into better resting blood pressure, a healthier immune system, and more relaxed muscles. Less stress can also mean an improvement in health habits. Those who practice mind-body exercise often improve their eating habits and reduce harmful behaviors such as cigarette smoking.

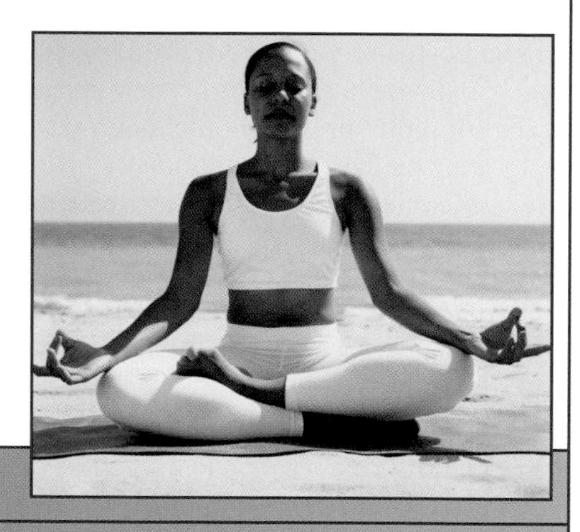

▶ **Think It Over . . .**

▶ ***How could you make walking more of a mind-body activity?***

3. The airways dilate, allowing faster movement of air into and out of the lungs.
4. The blood vessels that supply nonessential organs such as the kidneys and gastrointestinal tract constrict, which reduces blood flow through these tissues. The result is a slowing of urine formation and digestive activities, which are not essential during exercise.
5. Blood vessels that supply organs involved in exercise or fighting off danger—skeletal muscles, cardiac muscle, liver, and adipose tissue—dilate, which allows greater blood flow through these tissues.
6. Liver cells break down glycogen to glucose, and adipose cells break down triglycerides to fatty acids and glycerol, providing molecules that can be used by body cells for ATP production.
7. Release of glucose by the liver increases blood glucose level.
8. Processes that are not essential for meeting the stressful situation are inhibited. For example, muscular movements of the gastrointestinal tract and digestive secretions decrease or even stop.

Parasympathetic Activities

In contrast to the "fight-or-flight" activities of the sympathetic division, the parasympathetic division enhances "rest-and-digest" activities. Parasympathetic responses support body functions that conserve and restore body energy during times of rest and recovery. In the quiet intervals between periods of exercise, parasympathetic impulses to the digestive glands and the smooth muscle of the gastrointestinal tract predominate over sympathetic impulses. This allows energy-supplying food to be digested and absorbed. At the same time, parasympathetic responses reduce body functions that support physical activity.

The acronym *SLUDD* can be helpful in remembering five parasympathetic responses. It stands for salivation (S), lacrimation (L), urination (U), digestion (D), and defecation (D). Mainly the parasympathetic division stimulates all of these activities. Besides the increasing SLUDD responses, other important parasympathetic responses are "three decreases": decreased heart rate, decreased diameter of airways, and decreased diameter (constriction) of the pupils.

Table 11.2 lists the responses of glands, cardiac muscle, and smooth muscle to stimulation by the sympathetic and parasympathetic divisions of the ANS.

Table 11.2 Functions of the Autonomic Nervous System

Effector	Effect of Sympathetic Stimulation	Effect of Parasympathetic Stimulation
Glands		
Sweat	Increased sweating.	No known effect.
Lacrimal (tear)	Slight secretion of tears.	Secretion of tears.
Adrenal medulla	Secretion of epinephrine and norepinephrine.	No known effect.
Pancreas	Inhibition of secretion of digestive enzymes and insulin (hormone that lowers blood glucose level); secretion of glucagon (hormone that raises blood glucose level).	Secretion of digestive enzymes and insulin.
Posterior pituitary	Secretion of antidiuretic hormone (ADH).	No known effect.
Liver*	Breakdown of glycogen into glucose, synthesis of new glucose, and release of glucose into the blood; decreases bile secretion.	Promotes synthesis of glycogen; increases bile secretion.
Adipose tissue*	Breakdown of triglycerides and release of fatty acids into blood.	No known effect.
Cardiac Muscle		
Heart	Increased heart rate and increased force of atrial and ventricular contraction.	Decreased heart rate and decreased force of atrial contraction.
Smooth Muscle		
Radial muscle of iris of eye	Dilation of the pupil.	No known effect.
Circular muscle of iris of eye	No known effect.	Constriction of the pupil.
Ciliary muscle of eye	Relaxation to adjust shape of lens for distant vision.	Contraction to adjust shape of lens for close vision.
Gallbladder and ducts	Storage of bile in the gallbladder.	Release of bile into the small intestine.
Stomach and intestines	Decreased motility (movement); contraction of sphincters.	Increased motility; relaxation of sphincters.
Lungs (smooth muscle of bronchi)	Widening of the airways (bronchodilation).	Narrowing of the airways (bronchoconstriction).
Urinary bladder	Relaxation of muscular wall; contraction of internal sphincter.	Contraction of muscular wall; relaxation of internal sphincter.
Spleen	Contraction and discharge of stored blood into general circulation.	No known effect.
Smooth muscle of hair follicles	Contraction that results in erection of hairs, producing "goose bumps."	No known effect.
Uterus	Inhibits contraction in nonpregnant women; stimulates contraction in pregnant women.	Minimal effect.
Sex organs	In men, causes ejaculation of semen.	Vasodilation; erection of clitoris (women) and penis (men).
Salivary glands (arterioles)	Decreases secretion of saliva.	Stimulates secretion of saliva.
Gastric glands and intestinal glands (arterioles)	Inhibits secretion.	Promotes secretion.
Kidney (arterioles)	Decreases production of urine.	No known effect.
Skeletal muscle (arterioles)	Vasodilation in most, which increases blood flow.	No known effect.
Heart (coronary arterioles)	Vasodilation in most, which increases blood flow.	Causes slight constriction, which decreases blood flow.

*Listed with glands because they release substances into the blood.

Dysautonomia (dis-aw-tō-NŌ-mē-a; *dys-* = difficult; *autonomia* = self-governing) is an inherited disorder in which the autonomic nervous system functions abnormally. Symptoms include reduced tear gland secretions, poor vasomotor control, motor incoordination, skin blotching, absence of pain sensation, difficulty swallowing, decreased reflex responses, excessive vomiting, and emotional instability.

■ **CHECKPOINT**

5. What are some examples of the opposite effects of the sympathetic and parasympathetic divisions of the autonomic nervous system?
6. What happens during the fight-or-flight response?
7. Why is the parasympathetic division of the ANS considered the rest-and-digest division?

• • •

Now that we have discussed the structure and function of the nervous system, we will next consider in Chapter 12 how sensory information is relayed to the nervous system and how the nervous system responds to it.

COMMON DISORDERS

Autonomic Dysreflexia

Autonomic dysreflexia is an exaggerated response of the sympathetic division of the ANS that occurs in about 85% of individuals with spinal cord injury at or above the level of T6. The condition occurs due to interruption of the control of ANS neurons by higher centers. When certain sensory impulses, such as those resulting from stretching of a full urinary bladder, are unable to ascend the spinal cord, mass stimulation of the sympathetic nerves below the level of injury occurs. Among the effects of increased sympathetic activity is severe vasoconstriction, which elevates blood pressure. In response, the cardiovascular center in the medulla oblongata (1) increases parasympathetic output via the vagus nerve, which decreases heart rate, and (2) decreases sympathetic output, which causes dilation of blood vessels above the level of the injury.

Autonomic dysreflexia is characterized by a pounding headache; severe high blood pressure (hypertension); flushed, warm skin with profuse sweating above the injury level; pale, cold, and dry skin below the injury level; and anxiety. It is an emergency condition that requires immediate intervention. If untreated, autonomic dysreflexia can cause seizures, stroke, or heart attack.

Raynaud Phenomenon

In ***Raynaud phenomenon*** (rā-NŌ), the fingers and toes become ischemic (lack blood) after exposure to cold or with emotional stress. The condition is due to excessive sympathetic stimulation of smooth muscle in the arterioles of the fingers and toes. When the arterioles constrict in response to sympathetic stimulation, blood flow is greatly diminished. Symptoms are colorful—red, white, and blue. Fingers and toes may look white due to blockage of blood flow or look blue (cyanotic) due to deoxygenated blood in capillaries. With rewarming after cold exposure, the arterioles may dilate, causing the fingers and toes to look red. The disorder is most common in young women and occurs more often in cold climates.

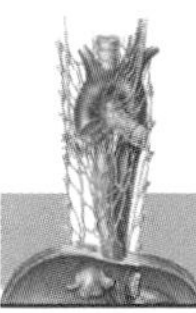

STUDY OUTLINE

Comparison of Somatic and Autonomic Nervous Systems (p. 272)

1. The part of the nervous system that regulates smooth muscle, cardiac muscle, and certain glands is the autonomic nervous system (ANS). The ANS usually operates without conscious control from the cerebral cortex, but other brain regions, mainly the hypothalamus and brain stem, regulate it.
2. The axons of somatic motor neurons extend from the CNS and synapse directly with an effector (skeletal muscle). Autonomic motor pathways consist of two motor neurons. The axon of the first motor neuron extends from the CNS and synapses in a ganglion with the second motor neuron; the second neuron synapses with an effector (smooth muscle, cardiac muscle, or a gland).
3. The output (motor) portion of the ANS has two divisions: sympathetic and parasympathetic. Most body organs receive dual innervation; usually one ANS division causes excitations and the other causes inhibition.

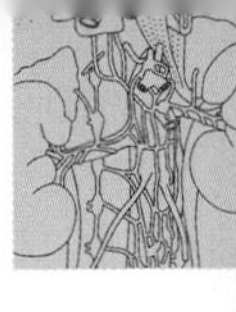

4. Somatic motor neurons release acetylcholine (ACh), and autonomic motor neurons release either acetylcholine or norepinephrine (NE).
5. Somatic nervous system effectors are skeletal muscles; ANS effectors include cardiac muscle, smooth muscle, and glands.
6. Table 11.1 on page 272 compares the somatic and autonomic nervous systems.

Structure of the Autonomic Nervous System (p. 273)

1. The sympathetic division of the ANS is also called the thoracolumbar division because the outflow of sympathetic nerve impulses comes from the thoracic and lumbar segments of the spinal cord. Cell bodies of sympathetic preganglionic neurons are in the 12 thoracic and the first two lumbar segments of the spinal cord.
2. Sympathetic ganglia are classified as sympathetic trunk ganglia (lateral to the vertebral column) or prevertebral ganglia (anterior to the vertebral column).
3. A single sympathetic preganglionic axon may synapse with 20 or more postganglionic neurons. Sympathetic responses can affect organs throughout the body almost simultaneously.
4. The parasympathetic division is also called the craniosacral division because the outflow of parasympathetic nerve impulses comes from cranial nerve nuclei and sacral segments of the spinal cord. The cell bodies of parasympathetic preganglionic neurons are located in the nuclei of cranial nerves III, VII, IX, and X in the brain stem and in three sacral segments of the spinal cord (S2, S3, and S4).
5. Parasympathetic ganglia are called terminal ganglia and are located near or within autonomic effectors. Parasympathetic terminal ganglia are close to or in the walls of their autonomic effectors, so most parasympathetic postganglionic axons are very short. In the ganglion, the preganglionic neuron usually synapses with only four or five postganglionic neurons, all of which supply the same effector. Thus, parasympathetic responses are localized to a single effector.

Functions of the Autonomic Nervous System (p. 277)

1. Some ANS neurons release acetylcholine, and others release norepinephrine; the result is excitation in some cases and inhibition in others.
2. ANS neurons that release acetylcholine include (1) all sympathetic and parasympathetic preganglionic neurons, (2) all parasympathetic postganglionic neurons, and (3) a few sympathetic postganglionic neurons.
3. Most sympathetic postganglionic neurons release the neurotransmitter norepinephrine (NE). The effects of NE are longer-lasting and more widespread than those of acetylcholine.
4. Activation of the sympathetic division causes widespread responses and is referred to as the fight-or-flight response. Activation of the parasympathetic division produces more restricted responses that typically are concerned with rest-and-digest activities.
5. Table 11.2 on page 279 summarizes the main functions of the sympathetic and parasympathetic divisions of the ANS.

SELF-QUIZ

1. In comparing the somatic nervous system with the autonomic nervous system, which of the following statements is true?
   a. The autonomic nervous system controls involuntary movements in skeletal muscle.
   b. The somatic nervous system controls voluntary activity in glands and smooth muscle.
   c. The autonomic nervous system controls involuntary activity in cardiac muscle, smooth muscle, and glands.
   d. The autonomic nervous system produces voluntary activity in smooth muscle and glands.
   e. The somatic nervous system controls involuntary movements, in smooth muscle, cardiac muscle, and glands.
2. Neurons in the autonomic nervous system include
   a. two motor neurons and one ganglion
   b. one motor neuron and two ganglia
   c. two motor neurons and two ganglia
   d. one motor and one sensory neuron, and no ganglia
   e. one motor and one sensory neuron, and one ganglion
3. Which statement is NOT true?
   a. Most sympathetic postganglionic neurons release norepinephrine.
   b. Parasympathetic preganglionic neurons release acetylcholine.
   c. Sympathetic effects are more localized and short-lived than parasympathetic effects.
   d. The effects from norepinephrine tend to be long-lasting.
   e. Branches of a single postganglionic neuron in the sympathetic division extend to many organs.
4. Which of the following pairs is mismatched?
   a. acetylcholine, parasympathetic nervous system
   b. fight-or-flight, sympathetic nervous system
   c. conserves body energy, parasympathetic nervous system
   d. rest-and-digest, parasympathetic nervous system
   e. norepinephrine, parasympathetic nervous system

5. Which of the following statements is NOT true concerning the autonomic nervous system?
   a. Most autonomic responses cannot be consciously cotrolled.
   b. In general, if the sympathetic division increases the activity in a specific organ, then the parasympathetic division decreases the activity of that organ.
   c. Sensory receptors monitor internal body conditions.
   d. Sensory neurons include pre- and postganglionic neurons.
   e. Most visceral effectors receive dual innervation.

6. Which part of the central nervous system contains centers that regulate the autonomic nervous system?
   a. hypothalamus b. cerebellum c. spinal cord
   d. basal ganglia e. thalamus

7. Place the following structures in the correct order as they relate to an autonomic nervous system response from receipt of the stimulus to response:

   1. visceral effector 2. centers in the CNS 3. autonomic ganglion 4. receptor and autonomic sensory neuron 5. preganglionic neuron 6. postganglionic neuron

   a. 4, 5, 2, 3, 6, 1 b. 5, 6, 2, 3, 1, 4 c. 1, 6, 3, 5, 2, 4
   d. 4, 2, 5, 3, 6, 1 e. 2, 4, 5, 6, 3, 1

8. Which of the following activities would NOT be monitored by autonomic sensory neurons?
   a. carbon dioxide levels in the blood
   b. hearing and equilibrium
   c. blood pressure
   d. stretching of the walls of visceral organs
   e. nausea from damaged viscera

9. The autonomic ganglia associated with the parasympathetic division are the
   a. trunk ganglia b. prevertebral ganglia c. posterior root ganglia d. terminal ganglia e. basal ganglia

10. Which of these statements about the parasympathetic division of the autonomic nervous system is NOT true? The parasympathetic division
    a. arises from the cranial nerves in the brain stem and sacral spinal cord segments
    b. is concerned with conserving and restoring energy
    c. uses acetylcholine as its neurotransmitter
    d. has ganglia near or within visceral effectors
    e. initiates responses in preganglionic neurons that synapse with 20 or more postganglionic neurons

11. Which nerve carries most of the parasympathetic output from the brain?
    a. spinal b. vagus c. oculomotor d. facial
    e. glossopharyngeal

12. Which of the following would NOT be affected by the autonomic nervous system?
    a. heart b. intestines c. urinary bladder
    d. skeletal muscle e. reproductive organs

13. Which of the following neurons release norepinephrine?
    a. somatic motor neurons
    b. sympathetic postganglionic neurons
    c. sympathetic preganglionic neurons
    d. parasympathetic postganglionic neurons
    e. parasympathetic preganglionic neurons

14. Match the following:

    ___ a. cluster of cell bodies outside the CNS
    ___ b. cell body is in ganglion; unmyelinated axon extends to effector
    ___ c. cell body lies inside the CNS; myelinated axon extends to ganglion
    ___ d. their postganglionic axons innervate organs below the diaphragm
    ___ e. their postganglionic axons supply organs above the diaphragm
    ___ f. contain the cell bodies and dendrites of parasympathetic postganglionic neurons

    A. sympathetic trunk ganglia
    B. prevertebral ganglia
    C. ganglion
    D. terminal ganglia
    E. preganglionic neuron
    F. postganglionic neuron

15. For each of the following, place a P if it refers to increased activity of the parasympathetic division or an S if it refers to increased activity of the sympathetic division.
    ___ a. dilates pupils
    ___ b. decreases heart rate
    ___ c. causes bronchoconstriction
    ___ d. stimulates breakdown of triglycerides
    ___ e. inhibits secretion of digestive enzymes and insulin
    ___ f. stimulates the gastrointestinal tract
    ___ g. occurs during exercise
    ___ h. causes release of glucose from the liver
    ___ i. dilates blood vessels to cardiac muscle

CRITICAL THINKING APPLICATIONS

1. It's Thanksgiving and you've just eaten a huge turkey dinner with all the trimmings. Now you're going to watch the big game on TV, if you can make it to the couch! Which division of the nervous system will be handling your body's post-dinner activities? Give examples of some organs and the effects on their functions.
2. Anthony wanted a toy on the top of the bookcase, so he climbed up the shelves. His mother ran in when she heard the crash and lifted the heavy bookcase with one arm while pulling her son out with the other. Later that day, she could not lift the bookcase back into position by herself. How do you explain the temporary "supermom" effect?
3. Taylor was watching a scary late-night horror movie when she heard a door slam and a cat's yowl. The hair rose on her arms and she was covered with goose bumps. Trace the pathway taken by the impulses from her CNS to her arms.
4. In the novel *The Hitchhiker's Guide to the Galaxy*, the character Zaphod Beebleborox has two heads and therefore two brains. Is this what is meant by dual innervation? Explain.

ANSWERS TO FIGURE QUESTIONS

11.1 Dual innervation means that an organ receives impulses from both the sympathetic and parasympathetic divisions of the ANS.

11.2 In the sympathetic trunk ganglia, sympathetic preganglionic axons form synapses with cell bodies and dendrites of sympathetic postganglionic neurons.

11.3 Most parasympathetic preganglionic axons are longer than most sympathetic preganglionic axons because parasympathetic ganglia are located in the walls of visceral organs, while most sympathetic ganglia are close to the spinal cord in the sympathetic trunk.

chapter 12

SOMATIC SENSES AND SPECIAL SENSES

did you know?

Some things improve with age, but hearing is not one of them. Damage to the hair cells that convert sound waves into nerve impulses accumulates over a lifetime, and by the time hearing loss is discovered, irreversible damage has already occurred. Exposure to excessive noise is the most common cause of hair cell damage. Damage increases with both the intensity and duration of exposure. The hair cells appear to be less traumatized by short periods of loud noise, such as a fire alarm going off, than by chronic exposure to moderately loud noise, such as the noise of vacuum cleaners, power tools, engines, and loud music.

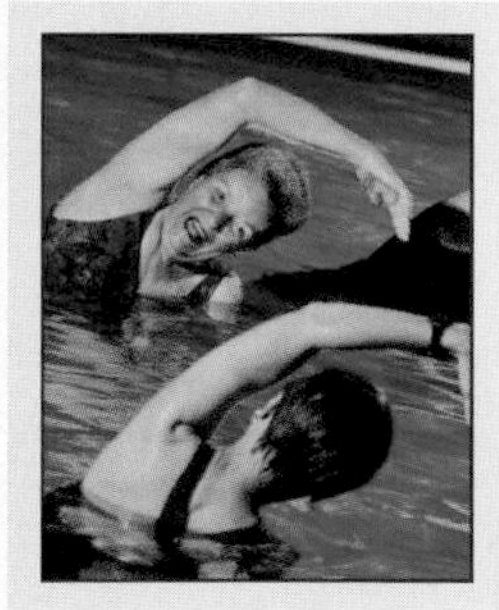

Focus on Wellness, page 289

www.wiley.com/college/apcentral

Consider what would happen if you could not feel the pain of a hot pot handle or an inflamed appendix, or if you could not see an oncoming car, hear a baby's cry, smell smoke, taste your favorite dessert, or maintain your balance on a flight of stairs. In short, if you could not "sense" your environment and make the necessary homeostatic adjustments, you could not survive very well on your own.

looking back to move ahead . . .

- **Sensory Nerve Endings and Sensory Receptors in the Skin (page 100)**
- **Somatic Sensory Pathways (page 259)**

OVERVIEW OF SENSATIONS

OBJECTIVE • Define a sensation and describe the conditions needed for a sensation to occur.

Most of us are aware of sensory input to the central nervous system (CNS) from structures associated with smell, taste, vision, hearing, and balance. These five senses are known as the ***special senses***. The other senses are termed ***general senses*** and include both somatic senses and visceral senses. ***Somatic senses*** (*somat-* = of the body) include tactile sensations (touch, pressure, and vibration); thermal sensations (warm and cold); pain sensations; and proprioceptive sensations (joint and muscle position sense and movements of the limbs and head). ***Visceral senses*** provide information about conditions within internal organs.

Definition of Sensation

Sensation is the conscious or subconscious awareness of changes in the external or internal environment. For a sensation to occur, four conditions must be satisfied:

1. A *stimulus*, or change in the environment, capable of activating certain sensory neurons, must occur. A stimulus that activates a sensory receptor may be in the form of light, heat, pressure, mechanical energy, or chemical energy.
2. A *sensory receptor* must convert the stimulus to an electrical signal, which ultimately produces one or more nerve impulses if it is large enough.
3. The nerve impulses must be *conducted* along a neural pathway from the sensory receptor to the brain.
4. A region of the brain must receive and *integrate* the nerve impulses into a sensation.

Characteristics of Sensations

As you have learned in Chapter 10, ***perception*** is the conscious awareness and interpretation of sensations and is primarily a function of the cerebral cortex. You seem to see with your eyes, hear with your ears, and feel pain in an injured part of your body. This is because sensory nerve impulses from each part of the body arrive in a specific region of the cerebral cortex, which interprets the sensation as coming from the stimulated sensory receptors. A given sensory neuron carries information for one type of sensation only. Neurons relaying impulses for touch, for example, do not also conduct impulses for pain. The specialization of sensory neurons enables nerve impulses from the eyes to be perceived as sight and those from the ears to be perceived as sounds.

A characteristic of most sensory receptors is ***adaptation***, a decrease in the strength of a sensation during a prolonged stimulus. Adaptation is caused in part by a decrease in the responsiveness of sensory receptors. As a result of adaptation, the perception of a sensation may fade or disappear even though the stimulus persists. For example, when you first step into a hot shower, the water may feel very hot, but soon the sensation decreases to one of comfortable warmth even though the stimulus (the high temperature of the water) does not change. Receptors vary in how quickly they adapt. Receptors associated with pressure, touch, and smell adapt rapidly. Slowly adapting receptors monitor stimuli associated with pain, body position, and the chemical composition of the blood.

Types of Sensory Receptors

Both structural and functional characteristics of sensory receptors can be used to group them into different classes (Table 12.1). Structurally, the simplest are *free nerve endings*,

Table 12.1 Classification of Sensory Receptors

Basis of Classification	Description
Structure	
Free nerve endings	Bare dendrites are associated with pain, thermal, tickle, itch, and some touch sensations.
Encapsulated nerve endings	Dendrites enclosed in a connective tissue capsule, such as a corpuscle of touch.
Separate cells	Receptor cell synapses with first-order neuron; located in the retina of the eye (photoreceptors), inner ear (hair cells), and taste buds of the tongue (gustatory receptor cells).
Function	
Mechanoreceptors	Detect mechanical pressure; provide sensations of touch, pressure, vibration, proprioception, and hearing and equilibrium; also monitor stretching of blood vessels and internal organs.
Thermoreceptors	Detect changes in temperature.
Nociceptors	Respond to painful stimuli resulting from physical or chemical damage to tissue.
Photoreceptors	Detect light that strikes the retina of the eye.
Chemoreceptors	Detect chemicals in mouth (taste), nose (smell), and body fluids.
Osmoreceptors	Sense the osmotic pressure of body fluids.

which are bare dendrites that lack any structural specializations at their ends that can be seen under a light microscope (Figure 12.1). Receptors for pain, thermal, tickle, itch, and some touch sensations are free nerve endings. Receptors for other somatic and visceral sensations, such as touch, pressure, and vibration, have *encapsulated nerve endings*. Their dendrites are enclosed in a connective tissue capsule with a distinctive microscopic structure. Still other sensory receptors consist of specialized, *separate cells* that synapse with sensory neurons, for example, hair cells in the inner ear.

Another way to group sensory receptors is functionally—according to the type of stimulus they detect. Most stimuli are in the form of mechanical energy, such as sound waves or pressure changes; electromagnetic energy, such as light or heat; or chemical energy, such as in a molecule of glucose.

- *Mechanoreceptors* are sensitive to mechanical stimuli such as the deformation, stretching, or bending of cells. Mechanoreceptors provide sensations of touch, pressure, vibration, proprioception, and hearing and equilibrium. They also monitor the stretching of blood vessels and internal organs.
- *Thermoreceptors* detect changes in temperature.
- *Nociceptors* respond to painful stimuli resulting from physical or chemical damage to tissue.
- *Photoreceptors* detect light that strikes the retina of the eye.
- *Chemoreceptors* detect chemicals in the mouth (taste), nose (smell), and body fluids.
- *Osmoreceptors* detect the osmotic pressure of body fluids.

CHECKPOINT

1. Which senses are "special senses"?
2. How is a sensation different from a perception?

SOMATIC SENSES

OBJECTIVES • **Describe the location and function of the receptors for tactile, thermal, and pain sensations.**
• **Identify the receptors for proprioception and describe their functions.**

Somatic sensations arise from stimulation of sensory receptors in the skin, mucous membranes, muscles, tendons, and

Figure 12.1 Structure and location of sensory receptors in the skin and subcutaneous layer.

The somatic sensations of touch, pressure, vibration, warmth, cold, and pain arise from sensory receptors in the skin, subcutaneous layer, and mucous membranes.

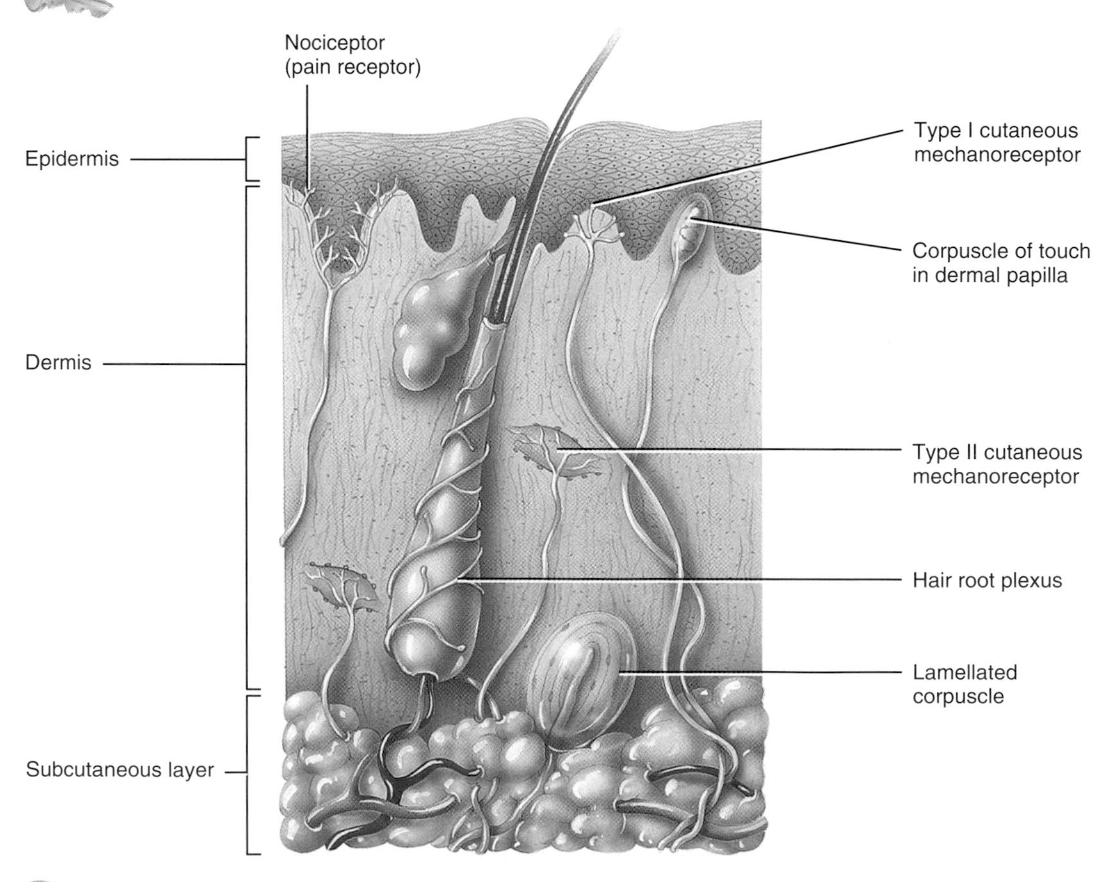

Which receptors are especially abundant in the fingertips, palms, and soles?

joints. The sensory receptors for somatic sensations are distributed unevenly. Some parts of the body surface are densely populated with receptors, and other parts contain only a few. The areas with the largest numbers of sensory receptors are the tip of the tongue, the lips, and the fingertips.

Tactile Sensations

The ***tactile sensations*** (TAK-tīl; *tact-* = touch) are touch, pressure, vibration, itch, and tickle. Itch and tickle sensations are detected by free nerve endings. All other tactile sensations are detected by a variety of encapsulated mechanoreceptors (Table 12.1). Tactile receptors in the skin or subcutaneous layer include corpuscles of touch, hair root plexuses, type I and II cutaneous mechanoreceptors, lamellated corpuscles, and free nerve endings (Figure 12.1).

Touch

Sensations of ***touch*** generally result from stimulation of tactile receptors in the skin or subcutaneous layer. There are two types of rapidly adapting touch receptors. ***Corpuscles of touch***, or ***Meissner corpuscles*** (MĪS-ner), are located in the dermal papillae of hairless skin. Each corpuscle is an egg-shaped mass of dendrites enclosed by a capsule of connective tissue. They are abundant in the fingertips, hands, eyelids, tip of the tongue, lips, nipples, soles, clitoris, and tip of the penis. ***Hair root plexuses*** consist of free nerve endings wrapped around hair follicles in hairy skin. Hair root plexuses detect movements on the surface of the skin that disturb hairs. For example, an insect landing on a hair causes movement of the hair shaft that stimulates the free nerve endings.

There are also two types of slowly adapting touch receptors. ***Type I cutaneous mechanoreceptors***, also known as ***Merkel discs***, are saucer-shaped, flattened free nerve endings that make contact with Merkel cells of the stratum basale; they are plentiful in the fingertips, hands, lips, and external genitalia. ***Type II cutaneous mechanoreceptors***, or ***Ruffini corpuscles***, are elongated, encapsulated receptors located deep in the dermis, and in ligaments and tendons as well. Present in the hands and abundant on the soles, they are most sensitive to stretching that occurs as digits or limbs are moved.

Pressure and Vibration

Pressure is a sustained sensation that is felt over a larger area than touch. Receptors that contribute to sensations of pressure include corpuscles of touch, type I mechanoreceptors, and lamellated corpuscles. ***Lamellated***, or ***pacinian*** (pa-SIN-ē-an), ***corpuscles*** are large oval structures composed of a multilayered connective tissue capsule that encloses a nerve ending (Figure 12.1). Like corpuscles of touch, lamellated corpuscles adapt rapidly. They are widely distributed in the body: in the dermis and subcutaneous layer; in tissues that underlie mucous and serous membranes; around joints, tendons, and muscles; in the periosteum; and in the mammary glands, external genitalia, and certain viscera, such as the pancreas and urinary bladder.

Sensations of ***vibration*** result from rapidly repetitive sensory signals from tactile receptors. The receptors for vibration sensations are corpuscles of touch and lamellated corpuscles. Corpuscles of touch can detect lower-frequency vibrations; lamellated corpuscles detect higher-frequency vibrations.

Itch and Tickle

The ***itch*** sensation results from stimulation of free nerve endings by certain chemicals, such as bradykinin, often as a result of a local inflammatory response. Receptors for the ***tickle*** sensation are thought to be free nerve endings and lamellated corpuscles. This intriguing sensation typically arises only when someone else touches you, not when you touch yourself. The explanation of this puzzle seems to lie in the nerve impulses that conduct to and from the cerebellum when you are moving your fingers and touching yourself that don't occur when someone else is tickling you.

Patients who have had a limb amputated may still experience sensations such as itching, pressure, tingling, or pain as if the limb were still there. This phenomenon is called **phantom limb sensation.** One explanation for phantom limb sensations is that the cerebral cortex interprets impulses arising in the proximal portions of sensory neurons that previously carried impulses from the limb as coming from the nonexistent (phantom) limb. Another explanation for phantom limb sensations is that neurons in the brain that previously received sensory impulses from the missing limb are still active, giving rise to false sensory perceptions.

Thermal Sensations

Thermoreceptors are free nerve endings. Two distinct ***thermal sensations***—coldness and warmth—are mediated by different receptors. Temperatures between 10° and 40°C (50–105°F) activate *cold receptors*, which are located in the epidermis. *Warm receptors* are located in the dermis and are activated by temperatures between 32° and 48°C (90–118°F). Cold and warm receptors both adapt rapidly at the onset of a stimulus but continue to generate nerve impulses more slowly throughout a prolonged stimulus. Temperatures below 10°C and above 48°C stimulate mainly nociceptors, rather than thermoreceptors, producing painful sensations.

Pain Sensations

The sensory receptors for pain, called ***nociceptors*** (nō′-sē-SEP-tors; *noci-* = harmful), are free nerve endings (Figure 12.1). Nociceptors are found in practically every tissue of the body except the brain, and they respond to several types of stimuli. Excessive stimulation of sensory receptors, excessive stretching of a structure, prolonged muscular contractions, inadequate blood flow to an organ, or the presence of certain chemical substances can all produce the sensation of pain. Pain may persist even after a pain-producing stimulus is removed because pain-causing chemicals linger and because nociceptors exhibit very little adaptation. The lack of adaptation of nociceptors serves a protective function: If there were adaptation to painful stimuli, irreparable tissue damage could result.

There are two types of pain: fast and slow. The perception of ***fast pain*** occurs very rapidly, usually within 0.1 second after a stimulus is applied. This type of pain is also known as acute, sharp, or pricking pain. The pain felt from a needle puncture or knife cut to the skin are examples of fast pain. Fast pain is not felt in deeper tissues of the body. The perception of ***slow pain*** begins a second or more after a stimulus is applied. It then gradually increases in intensity over a period of several seconds or minutes. This type of pain, which may be excruciating, is also referred to as chronic, burning, aching, or throbbing pain. Slow pain can occur both in the skin and in deeper tissues or internal organs. An example is the pain associated with a toothache.

Fast pain is very precisely localized to the stimulated area. For example, if someone pricks you with a pin, you know exactly which part of your body was stimulated. Somatic slow pain is well localized but more diffuse (involves large areas); it usually appears to come from a larger area of the skin. In many instances of visceral pain, the pain is felt in or just deep to the skin that overlies the stimulated organ, or in a surface area far from the stimulated organ. This phenomenon is called ***referred pain*** (Figure 12.2). In general, the visceral organ involved and the area in which the pain is referred are served by the same segment of the spinal cord. For example, sensory neurons from the heart, the skin over the heart, and the skin along the medial aspect of the left arm enter spinal cord segments T1 to T5. Thus, the pain of a heart attack typically is felt in the skin over the heart and along the left arm.

Some pain sensations occur out of proportion to minor damage or persist chronically for no obvious reason. In such cases, **analgesia** (*an-* = without; *-algesia* = pain) or pain relief is needed. Analgesic drugs such as aspirin and ibuprofen (for example, Advil®) block formation of some chemicals that stimulate nociceptors. Local anesthetics, such as Novocaine®, provide short-term pain relief by blocking conduction of nerve impulses. Morphine and other opiate drugs alter the quality of pain perception in the brain; pain is still sensed, but it is no longer perceived as so unpleasant.

Figure 12.2 Distribution of referred pain. The colored parts of the diagrams indicate skin areas to which visceral pain is referred.

Nociceptors are present in almost every tissue of the body.

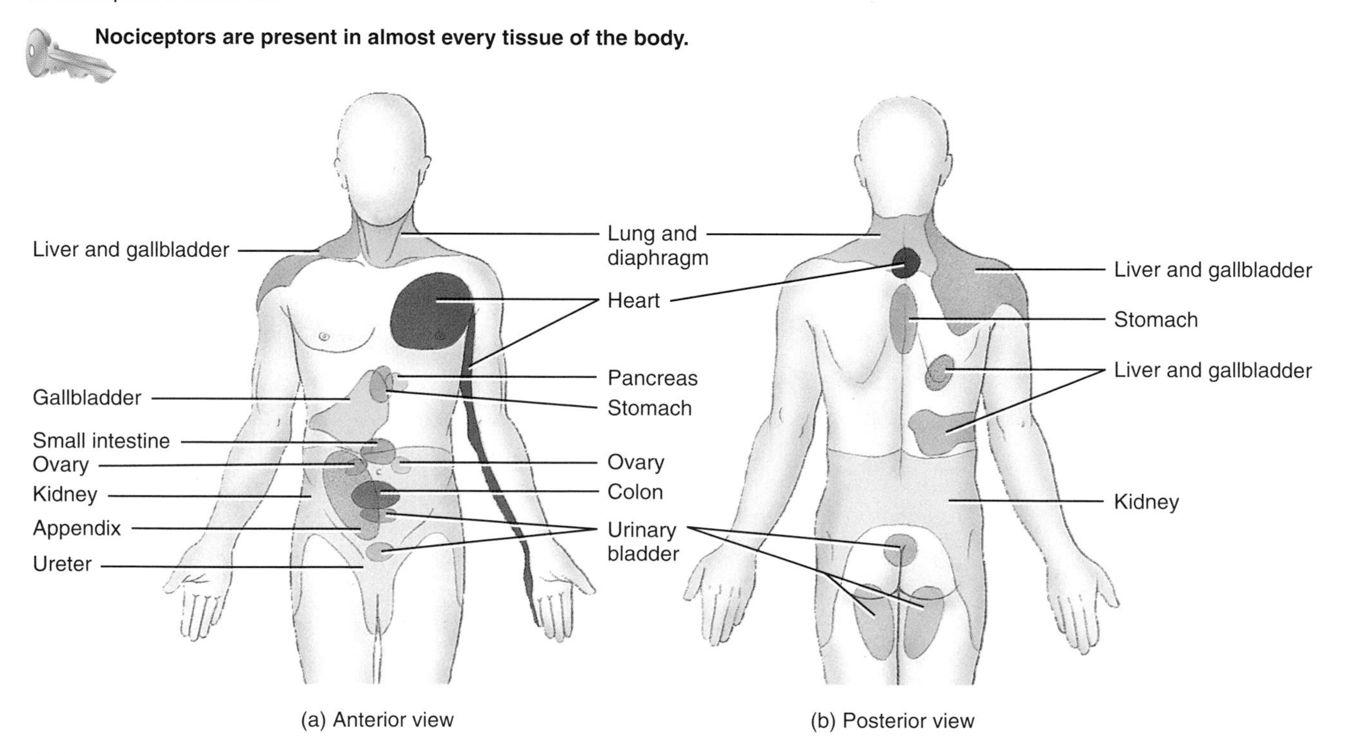

Which visceral organ has the broadest area for referred pain?

FOCUS ON WELLNESS

Pain Management—Sensation Modulation

Pain is a useful sensation when it alerts us to an injury that needs attention. We pull our finger away from a hot stove, we take off shoes that are too tight, and we rest an ankle that has been sprained. We do what we can to help the injury heal and meanwhile take over-the-counter or prescription painkillers until the pain goes away.

Pain that persists for longer than two or three months despite appropriate treatment is known as ***chronic pain***. The most common forms of chronic pain are low back pain and headache. Cancer, arthritis, fibromyalgia, and many other disorders are associated with chronic pain. People experiencing chronic pain often experience chronic frustration as they are sent from one specialist to another in search of a diagnosis.

The goal of pain management programs, developed to help people with chronic pain, is to decrease pain as much as possible, and then help patients learn to cope with whatever pain remains. Because no single treatment works for everyone, pain management programs typically offer a wide variety of treatments from surgery and nerve blocks to acupuncture and exercise therapy. Following are some of the therapies that complement medical and surgical treatment for the management of chronic pain.

Counseling

Pain used to be regarded as a purely physical response to physical injury. Psychological factors are now understood to serve as important mediators in the perception of pain. Feelings such as fear and anxiety strengthen the pain perceptions. Pain may be used to avoid certain situations, or to gain attention. Depression and associated symptoms such as sleep disturbances can contribute to chronic pain. Psychological counseling techniques can help people with chronic pain confront issues that may be worsening their pain.

Relaxation and Meditation

Relaxation and meditation techniques may reduce pain by decreasing anxiety and giving people a sense of personal control. Some of these techniques include deep breathing, visualization of positive images, and muscular relaxation. Others encourage people to become more aware of thoughts and situations that increase or decrease pain or provide a mental distraction from the sensations of pain.

Exercise

People with chronic pain tend to avoid movement because it hurts. Inactivity causes muscles and joint structures to atrophy, which may eventually cause the pain to worsen. Regular exercise and improved fitness help to relieve pain. Why? Exercise stimulates the production of endorphins, chemicals produced by the body to relieve pain. It also improves self-confidence, can serve as a distraction from pain, and improves sleep quality, which is often a problem for people with chronic pain.

► THINK IT OVER . . .

► *In what part of the nervous system do relaxation techniques have their effect?*

Proprioceptive Sensations

Proprioceptive sensations (prō-prē-ō-SEP-tive; *proprio-* = one's own) allow us to know where our head and limbs are located and how they are moving even if we are not looking at them, so that we can walk, type, or dress without using our eyes. ***Kinesthesia*** (kin′-es-THĒ-zē-a; *kin-* = motion; *-esthesia* = perception) is the perception of body movements. Proprioceptive sensations arise in receptors termed ***proprioceptors***. Proprioceptors are located in skeletal muscles (muscle spindles), in tendons (tendon organs), in and around synovial joints (joint kinesthetic receptors), and in the inner ear (hair cells). Those proprioceptors embedded in muscles, tendons, and synovial joints inform us of the degree to which muscles are contracted, the amount of tension on tendons, and the positions of joints. Hair cells of the inner ear monitor the orientation of the head relative to the ground and head position during movements. Proprioceptive sensations also allow us to estimate the weight of objects and determine the muscular effort necessary to perform a task. For example, as you pick up a bag you quickly realize whether it contains popcorn or books, and you then exert the correct amount of effort needed to lift it.

Nerve impulses for conscious proprioception pass along sensory tracts in the spinal cord and brain stem and are relayed to the primary somatosensory area (postcentral gyrus) in the parietal lobe of the cerebral cortex (see Figure 10.13 on page 259). Proprioceptive impulses also pass to the cerebellum, where they contribute to the cerebellum's role in coordinating skilled movements. Because proprioceptors adapt slowly and only slightly, the brain continually receives nerve impulses related to the position of different body parts and makes adjustments to ensure coordination.

■ **CHECKPOINT**

3. Why is it beneficial to your well-being that nociceptors and proprioceptors exhibit very little adaptation?
4. Which somatic sensory receptors detect touch sensations?
5. What is referred pain, and how is it useful in diagnosing internal disorders?

SPECIAL SENSES

Receptors for the special senses—smell, taste, sight, hearing, and equilibrium—are housed in complex sensory organs such as the eyes and ears. Like the general senses, the special senses allow us to detect changes in our environment. ***Ophthalmology*** (of′-thal-MOL-ō-jē; *ophthalmo-* = eye; *-logy* = study of) is the science that deals with the eye and its disorders. The other special senses are, in large part, the concern of ***otorhinolaryngology*** (ō′-tō-rī′-nō-lar′-in-GOL-ō-jē; *oto-* = ear; *rhino-* = nose; *laryngo-* = larynx), the science that deals with the ears, nose, and throat and their disorders.

OLFACTION: SENSE OF SMELL

OBJECTIVE • Describe the receptors for olfaction and the olfactory pathway to the brain.

The nose contains 10–100 million receptors for the sense of smell, or ***olfaction*** (ol-FAK-shun; *olfact-* = smell). Because some nerve impulses for smell and taste propagate to the limbic system, certain odors and tastes can evoke strong emotional responses or a flood of memories.

Structure of the Olfactory Epithelium

The olfactory epithelium occupies the upper portion of the nasal cavity (Figure 12.3a) and consists of three types of cells: olfactory receptors, supporting cells, and basal stem cells (Figure 12.3b). ***Olfactory receptors*** are the first-order neurons of the olfactory pathway. Several cilia called ***olfactory hairs*** project from a knob-shaped tip on each olfactory receptor. The olfactory hairs are the parts of the olfactory receptor that respond to inhaled chemicals. Chemicals that have an odor and can therefore stimulate the olfactory hairs are called ***odorants***. The axons of olfactory receptors extend from the olfactory epithelium to the olfactory bulb. ***Supporting cells*** are columnar epithelial cells of the mucous membrane lining the nose. They provide physical support, nourishment, and electrical insulation for the olfactory receptors, and they help detoxify chemicals that come in contact with the olfactory epithelium. ***Basal cells*** are stem cells located between the bases of the supporting cells and continually undergo cell division to produce new olfactory receptors, which live for only a month or so before being replaced. This process is remarkable because olfactory receptors are neurons, and in general, mature neurons are not replaced. ***Olfactory glands*** produce mucus that moistens the surface of the olfactory epithelium and serves as a solvent for inhaled odorants.

Stimulation of Olfactory Receptors

Many attempts have been made to distinguish among and classify "primary" sensations of smell. Genetic evidence now suggests the existence of hundreds of primary odors. Our ability to recognize about 10,000 different odors probably depends on patterns of activity in the brain that arise from activation of many different combinations of olfactory receptors. Olfactory receptors react to odorant molecules by producing an electrical signal that triggers one or more nerve impulses. Adaptation (decreasing sensitivity) to odors occurs rapidly. Olfactory receptors adapt by about 50% in the first second or so after stimulation and very slowly thereafter.

The Olfactory Pathway

On each side of the nose, about 40 bundles of the slender, unmyelinated axons of olfactory receptors extend through about 20 holes in the cribriform plate of the ethmoid bone (Figure 12.3b). These bundles of axons collectively form the right and left ***olfactory (I) nerves***. The olfactory nerves terminate in the brain in paired masses of gray matter called the ***olfactory bulbs***, which are located below the frontal lobes of the cerebrum. Within the olfactory bulbs, the axon terminals of olfactory receptors—the first-order neurons—form synapses with the dendrites and cell bodies of second-order neurons in the olfactory pathway.

The axons of the neurons extending from the olfactory bulb form the ***olfactory tract***. Some of the axons of the olfactory tract project to the *primary olfactory area* in the temporal lobe of the cerebral cortex (see Figure 10.13 on page 259),

Figure 12.3 Olfactory epithelium and olfactory receptors. (a) Location of olfactory epithelium in the nasal cavity. (b) Anatomy of olfactory receptors, whose axons extend through the cribriform plate to the olfactory bulb.

The olfactory epithelium consists of olfactory receptors, supporting cells, and basal cells.

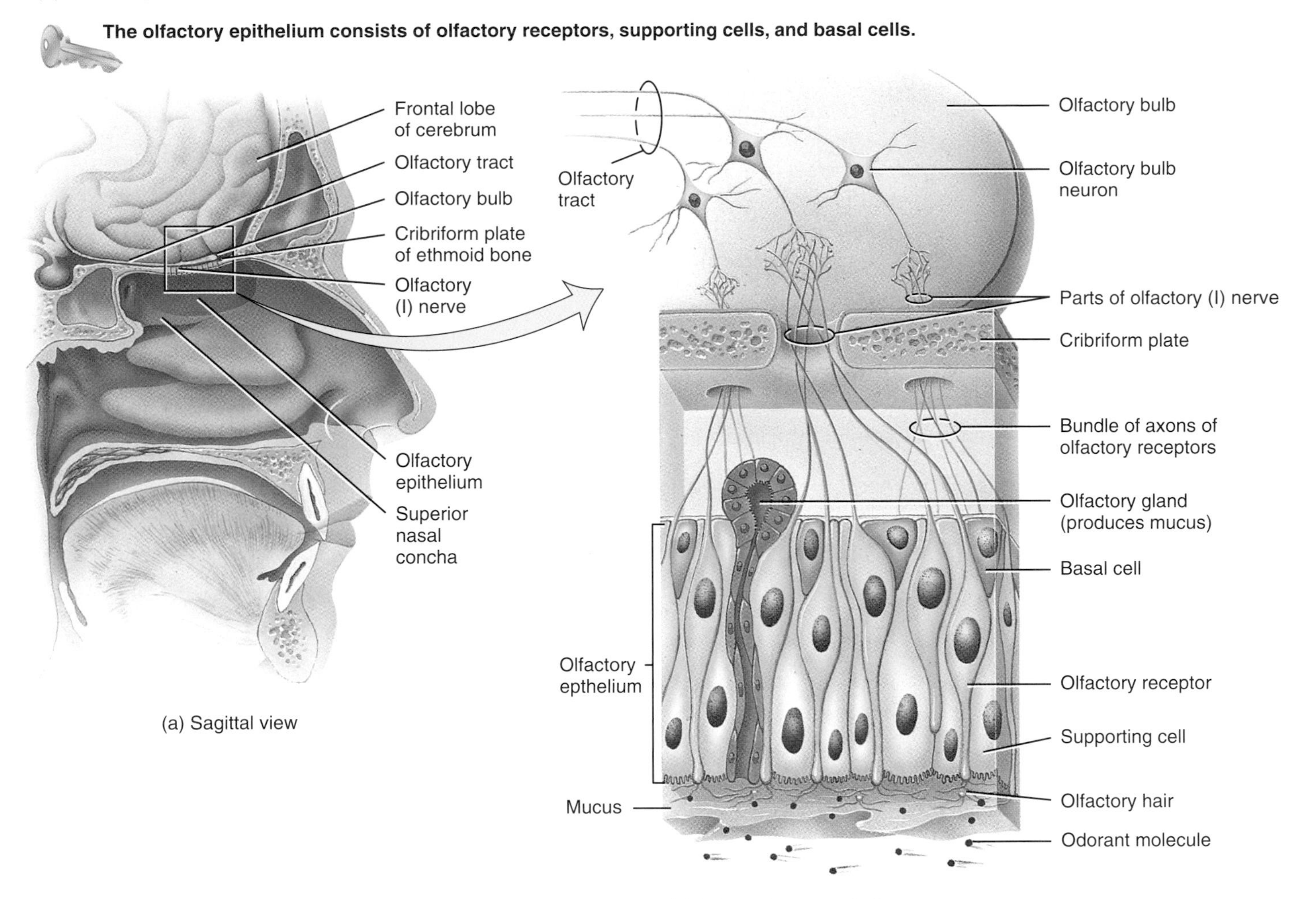

What is the function of basal stem cells?

where conscious awareness of smell begins. Other axons of the olfactory tract project to the limbic system and hypothalamus; these connections account for emotional and memory-evoked responses to odors. Examples include sexual excitement upon smelling a certain perfume or nausea upon smelling a food that once made you violently ill.

Hyposmia (hī-POZ-mē-a; *hypo-* = below; *-osmia* = smell, odor), a reduced ability to smell, affects half of those over age 65 and 75% of those over age 80. With aging the sense of smell deteriorates. Hyposmia also can be caused by neurological changes, such as a head injury, Alzheimer disease, or Parkinson disease; certain drugs, such as antihistamines, analgesics, or steroids; and the damaging effects of smoking.

■ **CHECKPOINT**

6. What functions are carried out by the three types of cells in the olfactory epithelium?
7. Define the following terms: olfactory nerve, olfactory bulb, and olfactory tract.

GUSTATION: SENSE OF TASTE

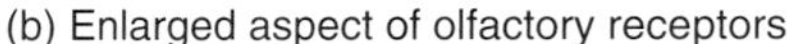

OBJECTIVE • Describe the receptors for gustation and the gustatory pathway to the brain.

Taste or ***gustation*** (gus-TĀ-shun; *gust-* = taste) is much simpler than olfaction because only five primary tastes can be distinguished: *sour, sweet, bitter, salty*, and *umami* (ū-MAM-ē).

The umami taste is described as "meaty" or "savory." All other flavors, such as chocolate, pepper, and coffee, are combinations of the five primary tastes, plus the accompanying olfactory and tactile (touch) sensations. Odors from food can pass upward from the mouth into the nasal cavity, where they stimulate olfactory receptors. Because olfaction is much more sensitive than taste, a given concentration of a food substance may stimulate the olfactory system thousands of times more strongly than it stimulates the gustatory system. When you have a cold or are suffering from allergies and cannot taste your food, it is actually olfaction that is blocked, not taste.

Structure of Taste Buds

The receptors for taste sensations are located in the ***taste buds*** (Figure 12.4). Most of the nearly 10,000 taste buds of a young adult are on the tongue, but some are also found on the roof of the mouth, pharynx (throat), and epiglottis (cartilage lid over the voice box). The number of taste buds declines with age. Taste buds are found in elevations on the tongue called ***papillae*** (pa-PIL-ē; singular is *papilla*), which provide a rough texture to the upper surface of the tongue (Figure 12.4 a,b). *Vallate papillae* (VAL-āt = wall-like) form an inverted V-shaped row at the back of the tongue. *Fungiform papillae* (FUN-ji-form = mushroomlike) are

Figure 12.4 The relationship of gustatory receptors in taste buds to tongue papillae.

Gustatory (taste) receptor cells are located in taste buds.

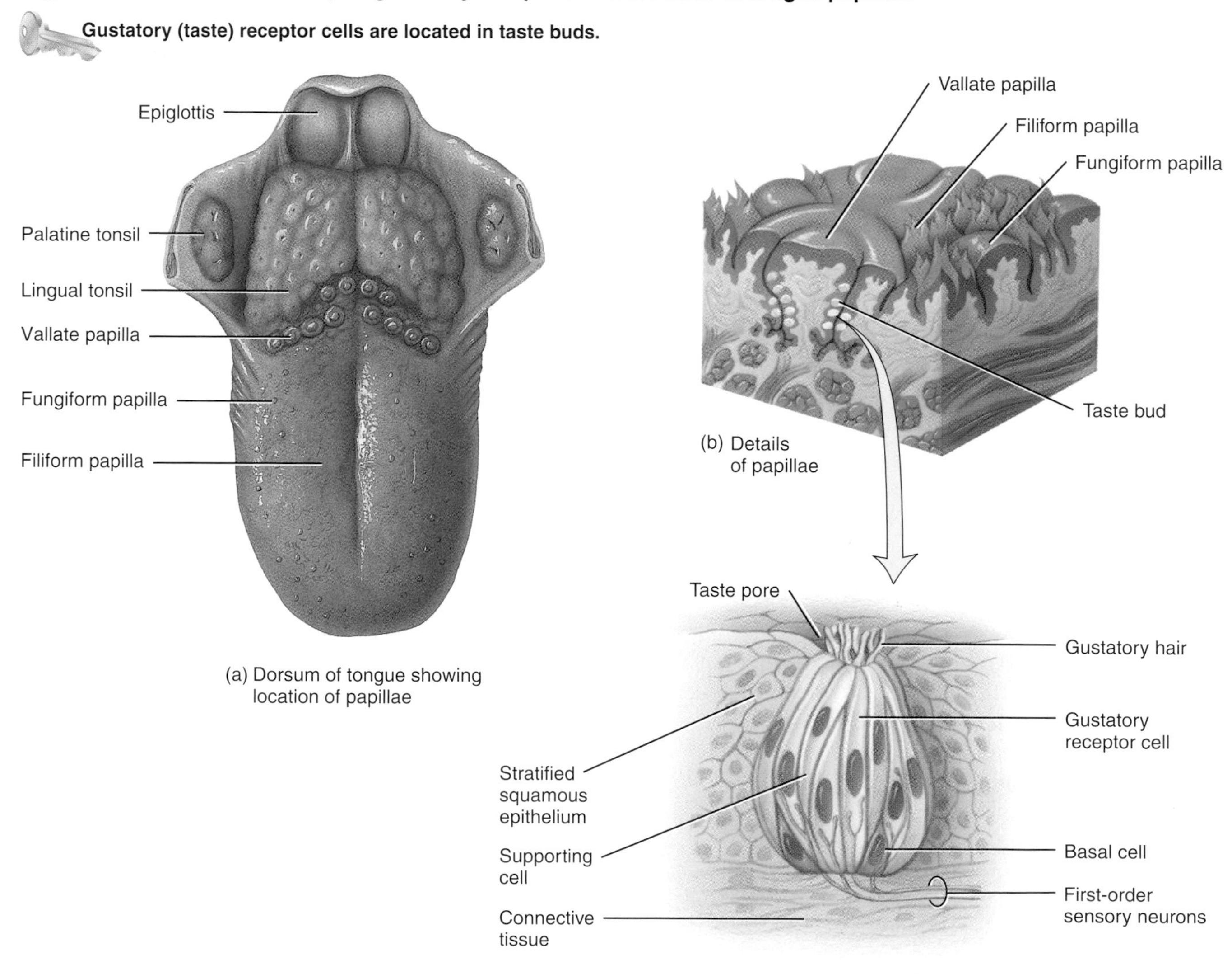

In order, from the tongue to the brain, what structures form the gustatory pathway?

mushroom-shaped elevations scattered over the entire surface of the tongue. In addition, the entire surface of the tongue has *filiform papillae* (FIL-i-form = threadlike), which contain touch receptors but no taste buds.

Each ***taste bud*** is an oval body consisting of three types of epithelial cells: supporting cells, gustatory receptor cells, and basal cells (Figure 12.4c). The ***supporting cells*** surround about 50 ***gustatory receptor cells***. A single, long ***gustatory hair*** projects from each gustatory receptor cell to the external surface through the ***taste pore***, an opening in the taste bud. ***Basal cells*** are stem cells that produce supporting cells, which then develop into gustatory receptor cells that have a life span of about 10 days. The gustatory receptor cells are separate receptor cells. They do not have an axon (like olfactory receptors) but rather synapse with dendrites of the first-order sensory neurons of the gustatory pathway.

Stimulation of Gustatory Receptors

Chemicals that stimulate gustatory receptor cells are known as ***tastants***. Once a tastant is dissolved in saliva, it can enter taste pores and make contact with the plasma membrane of the gustatory hairs. The result is an electrical signal that stimulates release of neurotransmitter molecules from the gustatory receptor cell. Nerve impulses are triggered when these neurotransmitter molecules bind to their receptors on the dendrites of the first-order sensory neuron. The dendrites branch profusely and contact many gustatory receptors in several taste buds. Individual gustatory receptor cells may respond to more than one of the five primary tastes. Complete adaptation (loss of sensitivity) to a specific taste can occur in 1 to 5 minutes of continuous stimulation.

If all tastants cause release of neurotransmitter from many gustatory receptor cells, why do foods taste different? The answer to this question is thought to lie in the patterns of nerve impulses in groups of first-order taste neurons that synapse with the gustatory receptor cells. Different tastes arise from activation of different groups of taste neurons. In addition, although each individual gustatory receptor cell responds to more than one of the five primary tastes, it may respond more strongly to some tastants than to others.

The Gustatory Pathway

Three cranial nerves contain axons of first-order gustatory neurons that innervate the taste buds. The facial (VII) nerve and glossopharyngeal (IX) nerve serve the tongue; the vagus (X) nerve serves the throat and epiglottis. From taste buds, impulses propagate along these cranial nerves to the medulla oblongata. From the medulla, some axons carrying taste signals project to the limbic system and the hypothalamus, and others project to the thalamus. Taste signals that project from the thalamus to the *primary gustatory area* in the parietal lobe of the cerebral cortex (see Figure 10.13 on page 259) give rise to the conscious perception of taste.

Probably because of taste projections to the hypothalamus and limbic system, there is a strong link between taste and pleasant or unpleasant emotions. Sweet foods evoke reactions of pleasure while bitter ones cause expressions of disgust, even in newborn babies. This phenomenon is the basis for **taste aversion**, in which people and animals quickly learn to avoid a food if it upsets the digestive system. Because the drugs and radiation treatments used to combat cancer often cause nausea and gastrointestinal upset regardless of what foods are consumed, cancer patients may lose their appetite because they develop taste aversions for most foods.

CHECKPOINT

8. How do olfactory receptors and gustatory receptor cells differ in structure and function?
9. Compare the olfactory and gustatory pathways.

VISION

OBJECTIVES • **Describe the accessory structures of the eye, the layers of the eyeball, the lens, the interior of the eyeball, image formation, and binocular vision.**

• **Describe the receptors for vision and the visual pathway to the brain.**

More than half the sensory receptors in the human body are located in the eyes, and a large part of the cerebral cortex is devoted to processing visual information. In this section of the chapter, we examine the accessory structures of the eye, the eyeball itself, the formation of visual images, the physiology of vision, and the visual pathway from the eye to the brain.

Accessory Structures of the Eye

The ***accessory structures*** of the eye are the eyebrows, eyelashes, eyelids, extrinsic muscles that move the eyeballs, and lacrimal (tear-producing) apparatus. The ***eyebrows*** and ***eyelashes*** help protect the eyeballs from foreign objects, perspiration, and direct rays of the sun (Figure 12.5). The upper and lower ***eyelids*** shade the eyes during sleep, protect the eyes from excessive light and foreign objects, and spread lubricating secretions over the eyeballs (by blinking). Six extrinsic eye muscles cooperate to move each eyeball right, left, up, down, and diagonally: the *superior rectus*, *inferior rectus*, *lateral rectus*, *medial rectus*, *superior oblique*, and *inferior*

Figure 12.5 Accessory structures of the eye.

Accessory structures of the eye are the eyebrows, eyelashes, eyelids, extrinsic eye muscles, and the lacrimal apparatus.

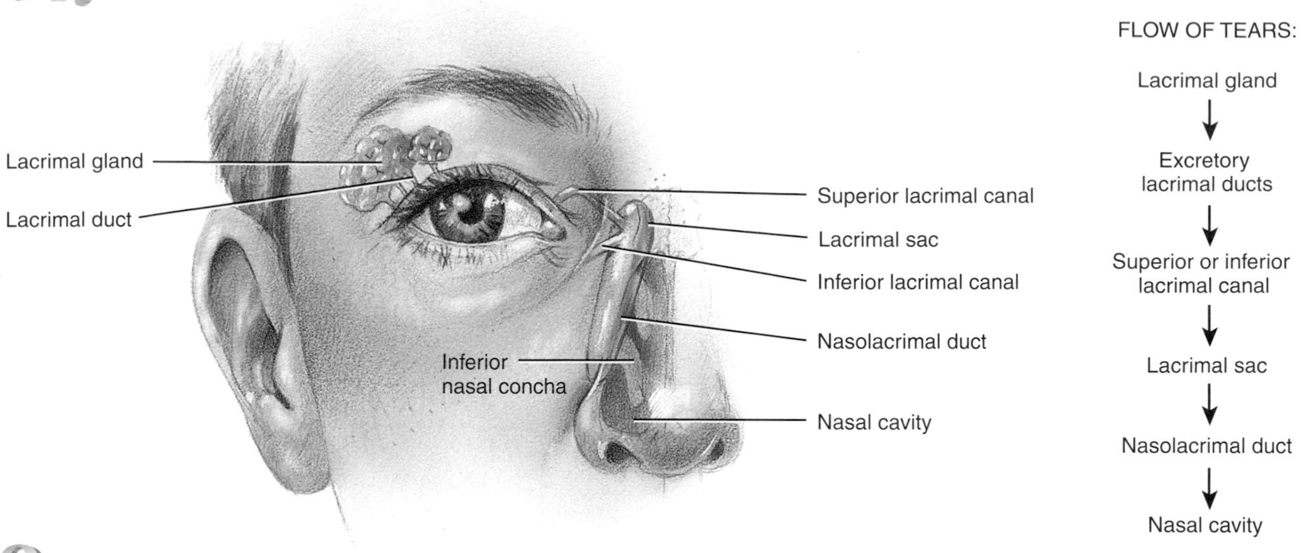

What are the functions of tears?

oblique. Neurons in the brain stem and cerebellum coordinate and synchronize the movements of the eyes.

The ***lacrimal apparatus*** (*lacrima* = tear) is a group of glands, ducts, canals, and sacs that produce and drain ***lacrimal fluid*** or ***tears*** (Figure 12.5). The right and left ***lacrimal glands*** are each about the size and shape of an almond. They secrete tears through the ***lacrimal ducts*** onto the surface of the upper eyelid. Tears then pass over the surface of the eyeball toward the nose into two ***lacrimal canals*** and a ***nasolacrimal duct***, which allow the tears to drain into the nasal cavity.

Tears are a watery solution containing salts, some mucus, and a bacteria-killing enzyme called ***lysozyme***. Tears clean, lubricate, and moisten the portion of the eyeball exposed to the air to prevent it from drying. Normally, tears are cleared away by evaporation or by passing into the nasal cavity as fast as they are produced. If, however, an irritating substance makes contact with the eye, the lacrimal glands are stimulated to oversecrete and tears accumulate. This protective mechanism dilutes and washes away the irritant. Only humans express emotions, both happiness and sadness, by ***crying***. In response to parasympathetic stimulation, the lacrimal glands produce excessive tears that may spill over the edges of the eyelids and even fill the nasal cavity with fluid. This is how crying produces a runny nose.

Layers of the Eyeball

The adult ***eyeball*** measures about 2.5 cm (1 inch) in diameter and is divided into three layers: fibrous tunic, vascular tunic, and retina (Figure 12.6).

Fibrous Tunic

The ***fibrous tunic*** is the outer coat of the eyeball. It consists of an anterior cornea and a posterior sclera. The ***cornea*** (KOR-nē-a) is a transparent fibrous coat that covers the colored iris. Because it is curved, the cornea helps focus light rays onto the retina. The ***sclera*** (SKLER-a = hard), the "white" of the eye, is a coat of dense connective tissue that covers all of the entire eyeball except the cornea. The sclera gives shape to the eyeball, makes it more rigid, and protects its inner parts. An epithelial layer called the ***conjunctiva*** (kon-junk-TĪ-va) covers the sclera but not the cornea and lines the inner surface of the eyelids.

Vascular Tunic

The ***vascular tunic*** is the middle layer of the eyeball and is composed of the choroid, ciliary body, and iris. The ***choroid*** (KŌ-royd) is a thin membrane that lines most of the internal surface of the sclera. It contains many blood vessels that help nourish the retina. The choroid also contains melanocytes that produce the pigment melanin, which causes this layer to appear dark brown in color. Melanin in the choroid absorbs stray light rays, which prevents reflection and scattering of light within the eyeball. As a result, the image cast on the retina by the cornea and lens remains sharp and clear.

At the front of the eye, the choroid becomes the ***ciliary body*** (SIL-ē-ar′-ē). The ciliary body consists of the *ciliary processes*, folds on the inner surface of the ciliary body whose capillaries secrete a fluid called aqueous humor, and the *ciliary muscle*, a smooth muscle that alters the shape of the lens for viewing objects up close or at a distance. The ***lens***, a transparent structure that focuses light rays onto the retina, is

Figure 12.6 Structure of the eyeball.

The wall of the eyeball consists of three layers: the fibrous tunic, the vascular tunic, and the retina.

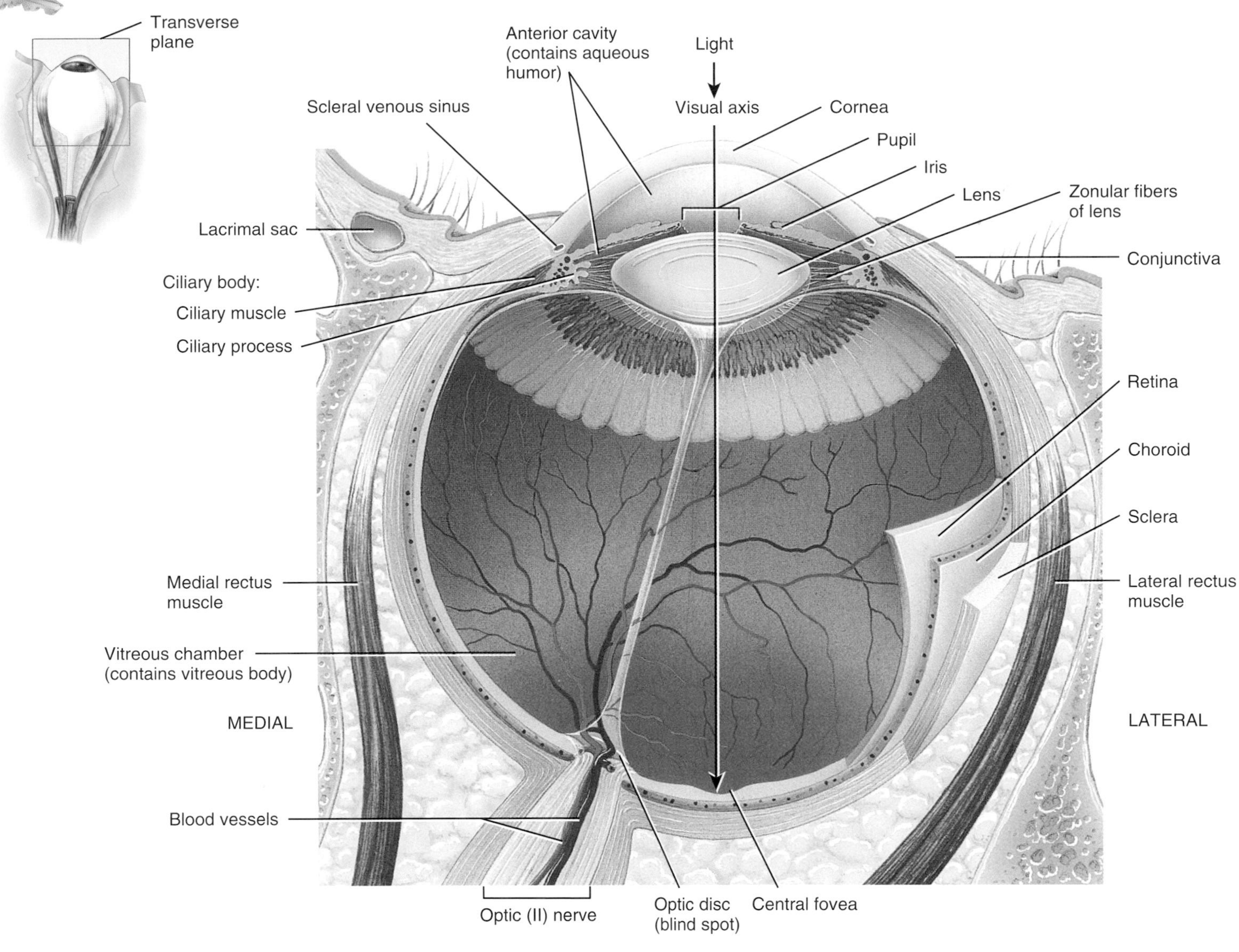

Superior view of transverse section of right eyeball

What are the components of the fibrous tunic and vascular tunic?

constructed of many layers of elastic protein fibers. *Zonular fibers* attach the lens to the ciliary muscle and hold the lens in position.

The ***iris*** (= colored circle) is the colored part of the eyeball. It includes both circular and radial smooth muscle fibers. The hole in the center of the iris, through which light enters the eyeball, is the ***pupil*** (Figure 12.7). The smooth muscle of the iris regulates the amount of light passing through the lens. When the eye is stimulated by bright light, the parasympathetic division of the autonomic nervous system (ANS) causes contraction of the circular muscles of the iris, which decreases the size of the pupil (constriction). When the eye must adjust to dim light, the sympathetic division of the ANS causes the radial muscles to contract, which increases the size of the pupil (dilation).

Using an **ophthalmoscope**, (of-THAL-mō-skōp; *ophthalmos-* = eye; *-skopeo* = to examine), an observer can peer through the pupil and see a magnified image of the retina and the blood vessels that cross it. The surface of the retina is the only place in the body where blood vessels can be viewed directly and examined for pathological changes, such as those that occur with hypertension or diabetes mellitus.

Figure 12.7 Responses of the pupil to light of varying brightness.

Contraction of the circular muscles causes constriction of the pupil; contraction of the radial muscles causes dilation of the pupil.

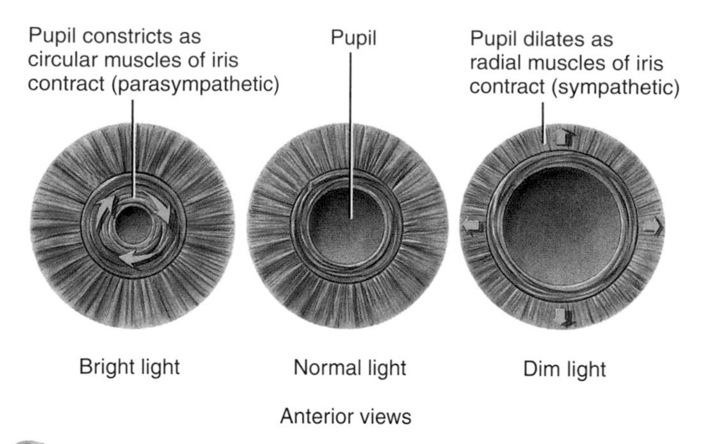

Which division of the autonomic nervous system causes pupillary constriction? Which causes pupillary dilation?

Retina

The third and inner coat of the eyeball, the ***retina***, lines the posterior three-quarters of the eyeball and is the beginning of the visual pathway (Figure 12.8). Two layers comprise the retina: the neural layer and the pigmented layer. The *neural layer* of the retina is a multilayered outgrowth of the brain. Three distinct layers of retinal neurons—the ***photoreceptor layer***, the ***bipolar cell layer***, and the ***ganglion cell layer***—are separated by two zones, the outer and inner synaptic layers, where synaptic contacts are made. Note that light passes through the ganglion and bipolar cell layers and both synaptic layers before it reaches the photoreceptor layer.

The *pigmented layer* of the retina is a sheet of melanin-containing epithelial cells located between the choroid and the neural part of the retina. The melanin in the pigmented layer of the retina, like in the choroid, also helps to absorb stray light rays.

Photoreceptors are specialized cells that begin the process by which light rays are ultimately converted to nerve impulses. There are two types of photoreceptors: rods and cones. ***Rods*** allow us to see shades of gray in dim light, such as moonlight. Brighter lights stimulate the ***cones***, giving rise to highly acute, color vision. Three types of cones are present

Figure 12.8 Microscopic structure of the retina. The downward blue arrow at left indicates the direction of the signals passing through the neural layer of the retina. Eventually, nerve impulses arise in ganglion cells and propagate along their axons, which make up the optic (II) nerve.

In the retina, visual signals pass from photoreceptors to bipolar cells to ganglion cells.

What are the two types of photoreceptors, and how do their functions differ?

in the retina: (1) *blue cones*, which are sensitive to blue light, (2) *green cones*, which are sensitive to green light; and (3) *red cones*, which are sensitive to red light. Color vision results from the stimulation of various combinations of these three types of cones. Just as an artist can obtain almost any color by mixing them on a palette, the cones can code for different colors by differential stimulation. There are about 6 million cones and 120 million rods. Cones are most densely concentrated in the ***central fovea***, a small depression in the center of the ***macula lutea*** (MAK-ū-la LOO-tē-a), or yellow spot, in the exact center of the retina. The central fovea is the area of highest *visual acuity* or *resolution* (sharpness of vision) because of its high concentration of cones. The main reason that you move your head and eyes while looking at something, such as the words of this sentence, is to place images of interest on your fovea. Rods are absent from the central fovea and macula lutea and increase in numbers toward the periphery of the retina.

From photoreceptors, information flows through the outer synaptic layer to the bipolar cells of the bipolar cell layer, and then from bipolar cells through the inner synaptic layer to the ganglion cells of the ganglion cell layer. Between 6 and 600 rods synapse with a single bipolar cell in the outer synaptic layer; a cone usually synapses with just one bipolar cell. The convergence of many rods onto a single bipolar cell increases the light sensitivity of rod vision but slightly blurs the image that is perceived. Cone vision, although less sensitive, has higher acuity because of the one-to-one synapses between cones and their bipolar cells. The axons of the ganglion cells extend posteriorly to a small area of the retina called the ***optic disc (blind spot)***, where they all exit as the optic (II) nerve (see Figure 12.6). Because the optic disc contains no rods or cones, we cannot see an image that strikes the blind spot. Normally, you are not aware of having a blind spot, but you can easily demonstrate its presence. Cover your left eye and gaze directly at the cross below. Then increase or decrease the distance between the book and your eye. At some point, the square will disappear as its image falls on the blind spot.

Interior of the Eyeball

The lens divides the interior of the eyeball into two cavities, the anterior cavity and the vitreous chamber. The ***anterior cavity*** lies anterior to the lens and is filled with ***aqueous humor*** (Ā-kwē-us HŪ-mor; *aqua* = water), a watery fluid similar to cerebrospinal fluid. Blood capillaries of the ciliary processes of the ciliary body secrete aqueous humor into the anterior cavity. It then drains into the *scleral venous sinus (canal of Schlemm)*, an opening where the sclera and cornea meet (see Figure 12.6), and reenters the blood. The aqueous humor helps maintain the shape of the eye and nourishes the lens and cornea, neither of which has blood vessels. Normally, aqueous humor is completely replaced about every 90 minutes.

Behind the lens is the second, and larger, cavity of the eyeball, the ***vitreous chamber***. It contains a clear, jellylike substance called the ***vitreous body***, which forms during embryonic life and is not replaced thereafter. This substance helps prevent the eyeball from collapsing and holds the retina flush against the choroid.

The pressure in the eye, called ***intraocular pressure***, is produced mainly by the aqueous humor with a smaller contribution from the vitreous body. Intraocular pressure maintains the shape of the eyeball and keeps the retina smoothly pressed against the choroid so the retina is well nourished and forms clear images. Normal intraocular pressure (about 16 mm Hg) is maintained by a balance between production and drainage of the aqueous humor.

Table 12.2 summarizes the structures of the eyeball.

Image Formation and Binocular Vision

In some ways the eye is like a camera: Its optical elements focus an image of some object on a light-sensitive "film"—the retina—while ensuring the correct amount of light makes the proper "exposure." To understand how the eye forms clear images of objects on the retina, we must examine three processes: (1) the refraction or bending of light by the lens and cornea, (2) the change in shape of the lens, and (3) constriction or narrowing of the pupil.

Refraction of Light Rays

When light rays traveling through a transparent substance (such as air) pass into a second transparent substance with a different density (such as water), they bend at the junction between the two substances. This bending is called ***refraction*** (Figure 12.9a on page 299). About 75% of the total refraction of light occurs at the cornea. Then, the lens of the eye further refracts the light rays so that they come into exact focus on the retina.

Images focused on the retina are inverted (upside down) (Figure 12.9b, c). They also undergo right-to-left reversal; that is, light from the right side of an object strikes the left side of the retina, and vice versa. The reason the world does not look inverted and reversed is that the brain "learns" early in life to coordinate visual images with the orientations of objects. The brain stores the inverted and reversed images we acquire when we first reach for and touch objects and interprets those visual images as being correctly oriented in space.

When an object is more than 6 meters (20 ft) away from the viewer, the light rays reflected from the object are nearly parallel to one another, and the curvatures of the cornea and lens exactly focus the image on the retina (Figure 12.9b). However, light rays from objects closer than 6 meters are divergent rather than parallel (Figure 12.9c). The rays must be refracted more if they are to be focused on the retina. This

Table 12.2 Summary of the Structures of the Eyeball and Their Functions

Structure	Function
Fibrous tunic Cornea Sclera	***Cornea:*** Admits and refracts (bends) light. ***Sclera:*** Provides shape and protects inner parts.
Vascular tunic Iris Ciliary body Choroid	***Iris:*** Regulates the amount of light that enters eyeball. ***Ciliary body:*** Secretes aqueous humor and alters the shape of the lens for near or far vision (accommodation). ***Choroid:*** Provides blood supply and absorbs scattered light.
Retina Retina	Receives light and converts it into nerve impulses. Provides output to brain via axons of ganglion cells, which form the optic (II) nerve.
Lens  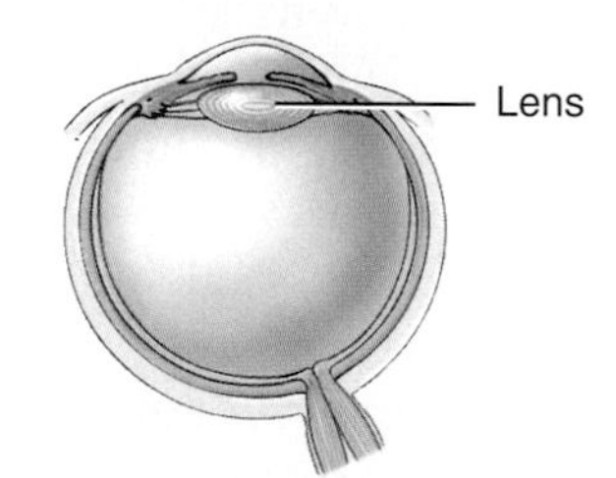	Refracts light.
Anterior cavity 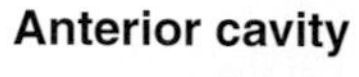 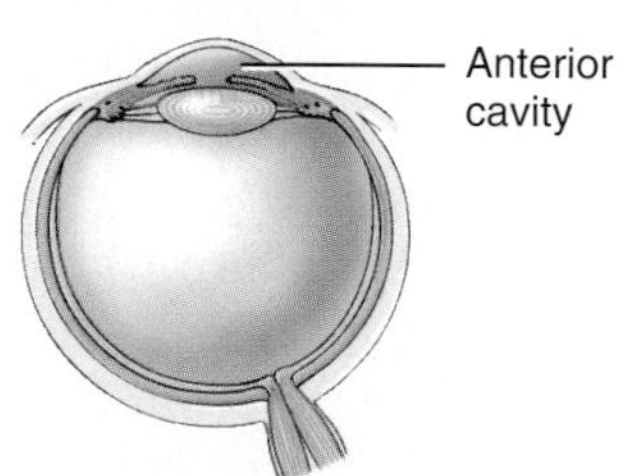	Contains aqueous humor that helps maintain the shape of the eyeball and supplies oxygen and nutrients to the lens and cornea.
Vitreous chamber 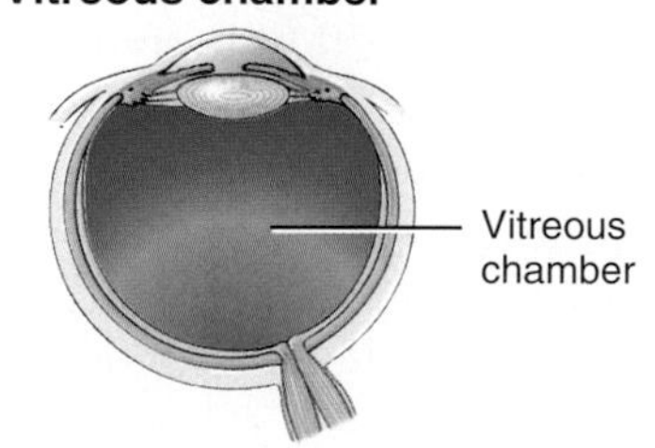	Contains the vitreous body, which helps maintain the shape of eyeball and keeps the retina attached to the choroid.

Figure 12.9 Refraction of light rays and accommodation.

Refraction is the bending of light rays.

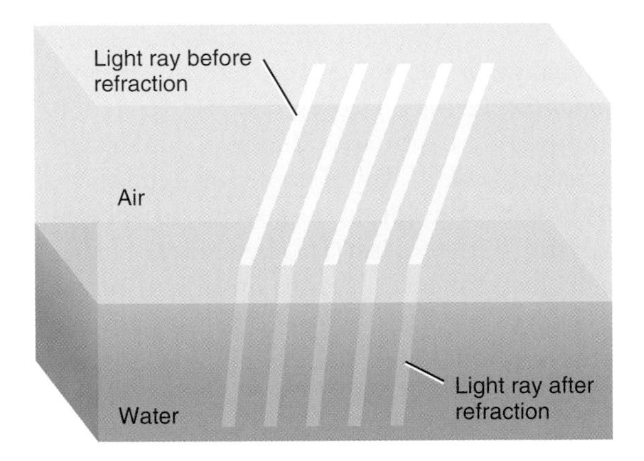

(a) Refraction of light rays

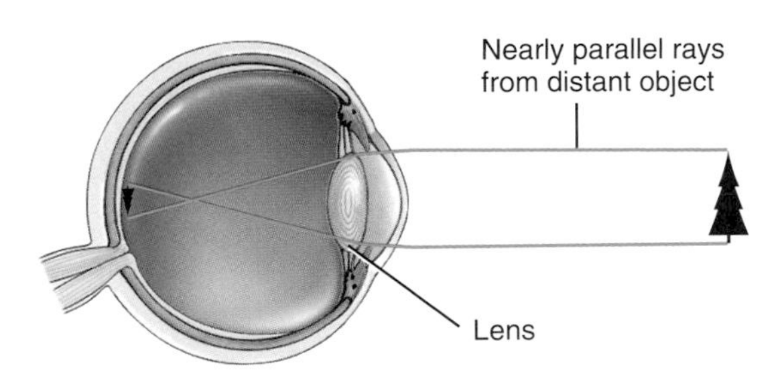

(b) Viewing distant object

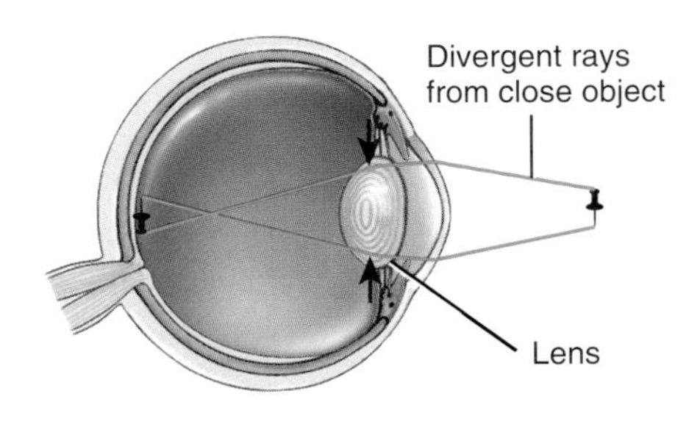

(c) Accommodation

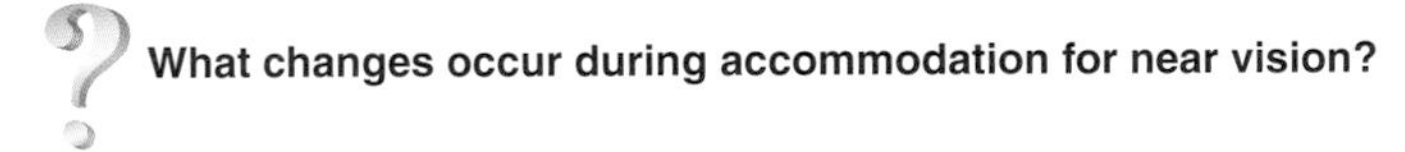

additional refraction is accomplished by changes in the shape of the lens.

Accommodation

A surface that curves outward, like the surface of a ball, is said to be *convex*. The convex surface of a lens refracts incoming light rays toward each other, so that they eventually intersect. The lens of the eye is convex on both its anterior and posterior surfaces, and its ability to refract light increases as its curvature becomes greater. When the eye is focusing on a close object, the lens becomes more convex and refracts the light rays more. This increase in the curvature of the lens for near vision is called ***accommodation*** (Figure 12.9c).

When you are viewing distant objects, the ciliary muscle of the ciliary body is relaxed and the lens is fairly flat because it is stretched in all directions by taut zonular fibers. When you view a close object, the ciliary muscle contracts, which pulls the ciliary process and choroid forward toward the lens. This action releases tension on the lens, allowing it to become rounder (more convex), which increases its focusing power and causes greater convergence of the light rays.

The normal eye, known as an ***emmetropic eye*** (em′-e-TROP-ik), can sufficiently refract light rays from an object 6 m (20 ft) away so that a clear image is focused on the retina (Figure 12.10a). Many people, however, lack this ability

Figure 12.10 Normal and abnormal refraction in the eyeball. (a) In the normal (emmetropic) eye, light rays from an object are bent sufficiently by the cornea and lens to focus on the central fovea. (b) In the nearsighted (myopic) eye, the image is focused in front of the retina. (c) Correction is by use of a concave lens that diverges entering light rays so that they have to travel further through the eyeball. (d) In the farsighted (hyperopic) eye, the image is focused behind the retina. (e) Correction is by a convex lens that causes entering light rays to converge.

In uncorrected myopia, distant objects can't be seen clearly; in uncorrected hyperopia, nearby objects can't be seen clearly.

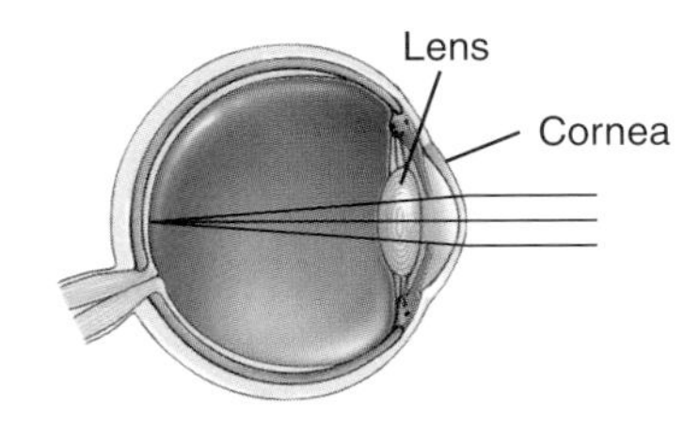

(a) Normal (emmetropic) eye

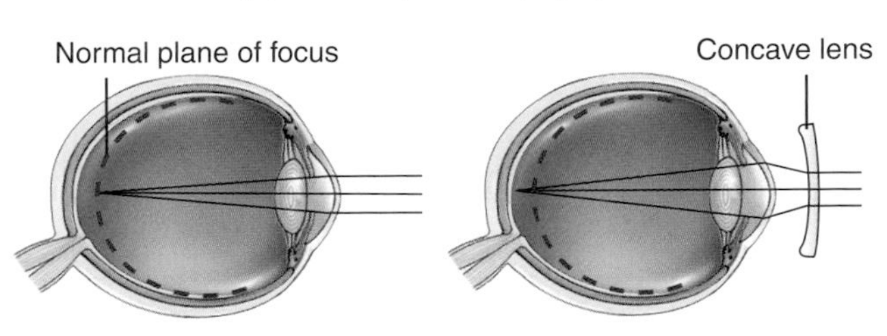

(b) Nearsighted (myopic) eye, uncorrected

(c) Nearsighted (myopic) eye, corrected

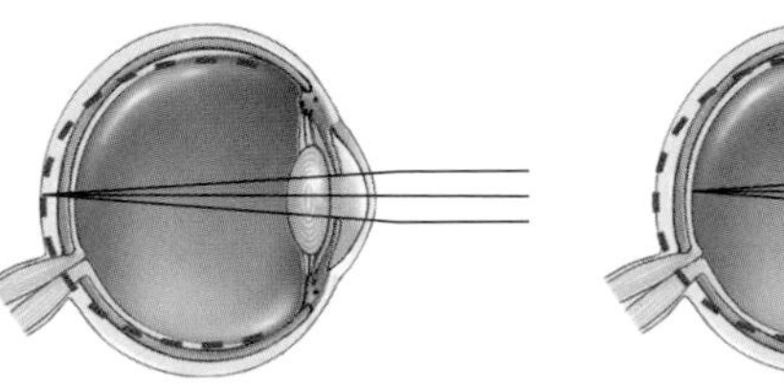

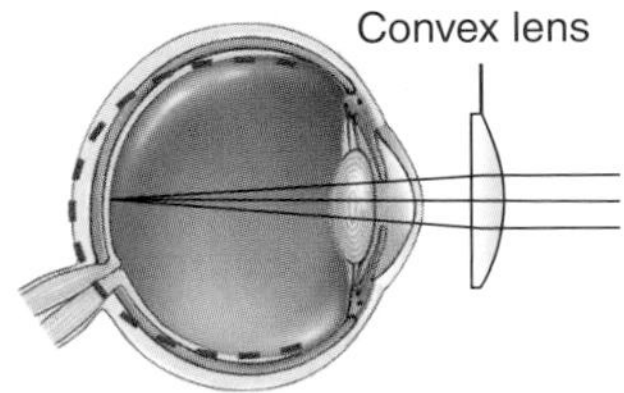

(d) Farsighted (hyperopic) eye, uncorrected

(e) Farsighted (hyperopic) eye, corrected

What is presbyopia?

because of refraction abnormalities. Among these abnormalities are ***myopia*** (mī-Ō-pē-a), or nearsightedness, which occurs when the eyeball is too long relative to the focusing power of the cornea and lens. Myopic individuals can see nearby objects clearly, but not distant objects. In ***hyperopia*** (hī-per-Ō-pē-a) or farsightedness, also known as ***hypermetropia*** (hī′-per-me-TRŌ-pē-a), the eyeball length is short relative to the focusing power of the cornea and lens. Hyperopic individuals can see distant objects clearly, but not nearby objects. Figure 12.10b–e illustrates these conditions and shows how they are corrected. Another refraction abnormality is ***astigmatism*** (a-STIG-ma-tizm), in which either the cornea or the lens has an irregular curvature.

> With aging, the lens loses some of its elasticity so its ability to accommodate decreases. At about age 40, people who have not previously worn glasses begin to require them for close vision, such as reading. This condition is called **presbyopia** (prez′-bē-Ō-pē-a; *presby-* = old; *-opia* = pertaining to the eye or vision).

Constriction of the Pupil

Constriction of the pupil is a narrowing of the diameter of the hole through which light enters the eye due to contraction of the circular muscles of the iris. This autonomic reflex occurs simultaneously with accommodation and prevents light rays from entering the eye through the periphery of the lens. Light rays entering at the periphery of the lens would not be brought to focus on the retina and would result in blurred vision. The pupil, as noted earlier, also constricts in bright light to limit the amount of light that strikes the retina.

Convergence

In humans, both eyes focus on only one set of objects, a characteristic called ***binocular vision***. This feature of our visual system allows the perception of depth and an appreciation of the three-dimensional nature of objects. When you stare straight ahead at a distant object, the incoming light rays are aimed directly at the pupils of both eyes and are refracted to comparable spots on the two retinas. As you move closer to the object, your eyes must rotate toward the nose if the light rays from the object are to strike comparable points on both retinas. ***Convergence*** is the name for this automatic movement of the two eyeballs toward the midline, which is caused by the coordinated action of the extrinsic eye muscles. The nearer the object, the greater the convergence needed to maintain binocular vision.

Stimulation of Photoreceptors

After an image is formed on the retina by refraction, accommodation, constriction of the pupil, and convergence, light rays must be converted into neural signals. The initial step in this process is the absorption of light rays by the rods and cones of the retina. To understand how absorption occurs, it is necessary to understand the role of photopigments.

A ***photopigment*** is a substance that can absorb light and undergo a change in structure. The photopigment in rods is called ***rhodopsin*** (*rhodo-* = rose; *-opsin* = related to vision) and is composed of a protein called *opsin* and a derivative of vitamin A called *retinal.* Any amount of light in a darkened room causes some rhodopsin molecules to split into opsin and retinal and initiate a series of chemical changes in the rods. When the light level is dim, opsin and retinal recombine into rhodopsin as fast as rhodopsin is split apart. Rods usually are nonfunctional in daylight, however, because rhodopsin is split apart faster than it can be reformed. After going from bright sunlight into a dark room, it takes about 40 minutes before the rods function maximally.

Cones function in bright light and provide color vision. As in rods, absorption of light rays causes breakdown of photopigment molecules. The photopigments in cones also contain retinal, but there are three different opsin proteins—one in each of the three types of cones. The cone photopigments reform much more quickly than the rod photopigment.

> The complete loss of cone vision causes a person to become legally blind. In contrast, a person who loses rod vision mainly has difficulty seeing in dim light and thus should not, for example, drive at night. Prolonged vitamin A deficiency and the resulting below-normal amount of rhodopsin may cause **night blindness**, an inability to see well at low light levels. An individual with an absence or deficiency of one of the three types of cones from the retina cannot distinguish some colors from others and is said to be **colorblind**. In the most common type, *red–green color blindness*, either red cones or green cones are missing. Thus, the person cannot distinguish between red and green. The inheritance of color blindness is illustrated in Figure 24.12 on page 606.

The Visual Pathway

After stimulation by light, the rods and cones trigger electrical signals in bipolar cells. Bipolar cells transmit both excitatory and inhibitory signals to ganglion cells. The ganglion cells become depolarized and generate nerve impulses. The axons of the ganglion cells exit the eyeball as the ***optic (II) nerve*** (Figure 12.11) and extend posteriorly to the ***optic chiasm*** (KĪ-azm = a crossover, as in the letter X). In the optic chiasm, about half of the axons from each eye cross to the opposite side of the brain. After passing the optic chiasm, the axons, now part of the ***optic tract***, terminate in the thalamus. Here they synapse with neurons whose axons project to the

Figure 12.11 Visual pathway.

At the optic chiasm, half of the retinal ganglion cell axons from each eye cross to the opposite side of the brain.

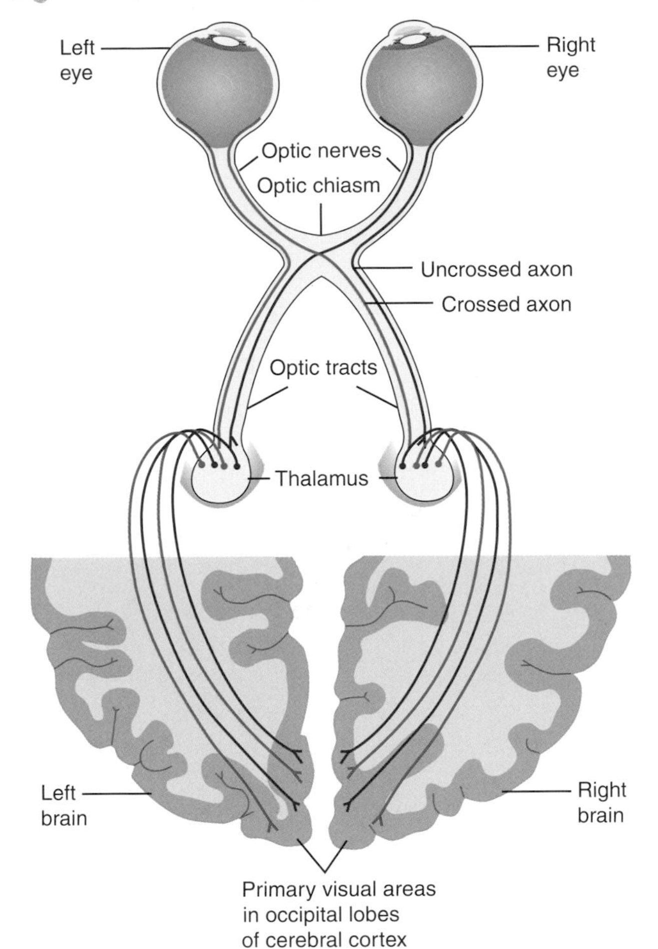

What is the correct order of structures that carry nerve impulses from the retina to the occipital lobe?

primary visual areas in the occipital lobes of the cerebral cortex (see Figure 10.13 on page 259). Because of crossing at the optic chiasm, the right side of the brain receives signals from both eyes for interpretation of visual sensations from the left side of an object, and the left side of the brain receives signals from both eyes for interpretation of visual sensations from the right side of an object.

■ CHECKPOINT

10. How does the shape of the lens change during accomodation?
11. How do photopigments respond to light?
12. By what pathway do nerve impulses triggered by an object in the left half of the visual field of the left eye reach the primary visual area of the cerebral cortex?

HEARING AND EQUILIBRIUM

OBJECTIVES • **Describe the structures of the outer, middle, and inner ear.**

• **Describe the receptors for hearing and equilibrium and their pathways to the brain.**

The ear is a marvelously sensitive structure. Its sensory receptors can convert sound vibrations into electrical signals 1000 times faster than photoreceptors can respond to light. Beside receptors for sound waves, the ear also contains receptors for equilibrium (balance).

Structure of the Ear

The ear is divided into three main regions: the outer ear, which collects sound waves and channels them inward; the middle ear, which conveys sound vibrations to the oval window; and the inner ear, which houses the receptors for hearing and equilibrium.

Outer Ear

The ***outer ear*** collects sound waves and passes them inward (Figure 12.12). It consists of an auricle, external auditory canal, and eardrum. The ***auricle***, the part of the ear that you can see, is a skin-covered flap of elastic cartilage shaped like the flared end of a trumpet. It plays a small role in collecting sound waves and directing them toward the ***external auditory canal*** (*audit-* = hearing), a curved tube that extends from the auricle and directs sound waves toward the eardrum. The canal contains a few hairs and ***ceruminous glands*** (se-ROO-mi-nus; *cer-* = wax), which secrete ***cerumen*** (se-ROO-men) (earwax). The hairs and cerumen help prevent foreign objects from entering the ear. The ***eardrum***, also called the ***tympanic membrane*** (tim-PAN-ik; *tympan-* = adrum), is a thin, semitransparent partition between the external auditory canal and the middle ear. Sound waves cause the eardrum to vibrate. Tearing of the tympanic membrane, due to trauma or infection, is called a ***perforated eardrum***.

Middle Ear

The ***middle ear*** is a small, air-filled cavity between the eardrum and inner ear (Figure 12.12). An opening in the anterior wall of the middle ear leads directly into the ***auditory tube***, commonly known as the ***eustachian tube***, which connects the middle ear with the upper part of the throat. When the auditory tube is open, air pressure can equalize on both sides of the eardrum. Otherwise, abrupt changes in air pressure on one side of the eardrum might cause it to rupture. During swallowing and yawning, the tube opens, which explains why yawning can help equalize the pressure changes that occur while flying in an airplane.

Figure 12.12 Structure of the ear.

The ear has three principal regions: the outer ear, the middle ear, and the inner ear.

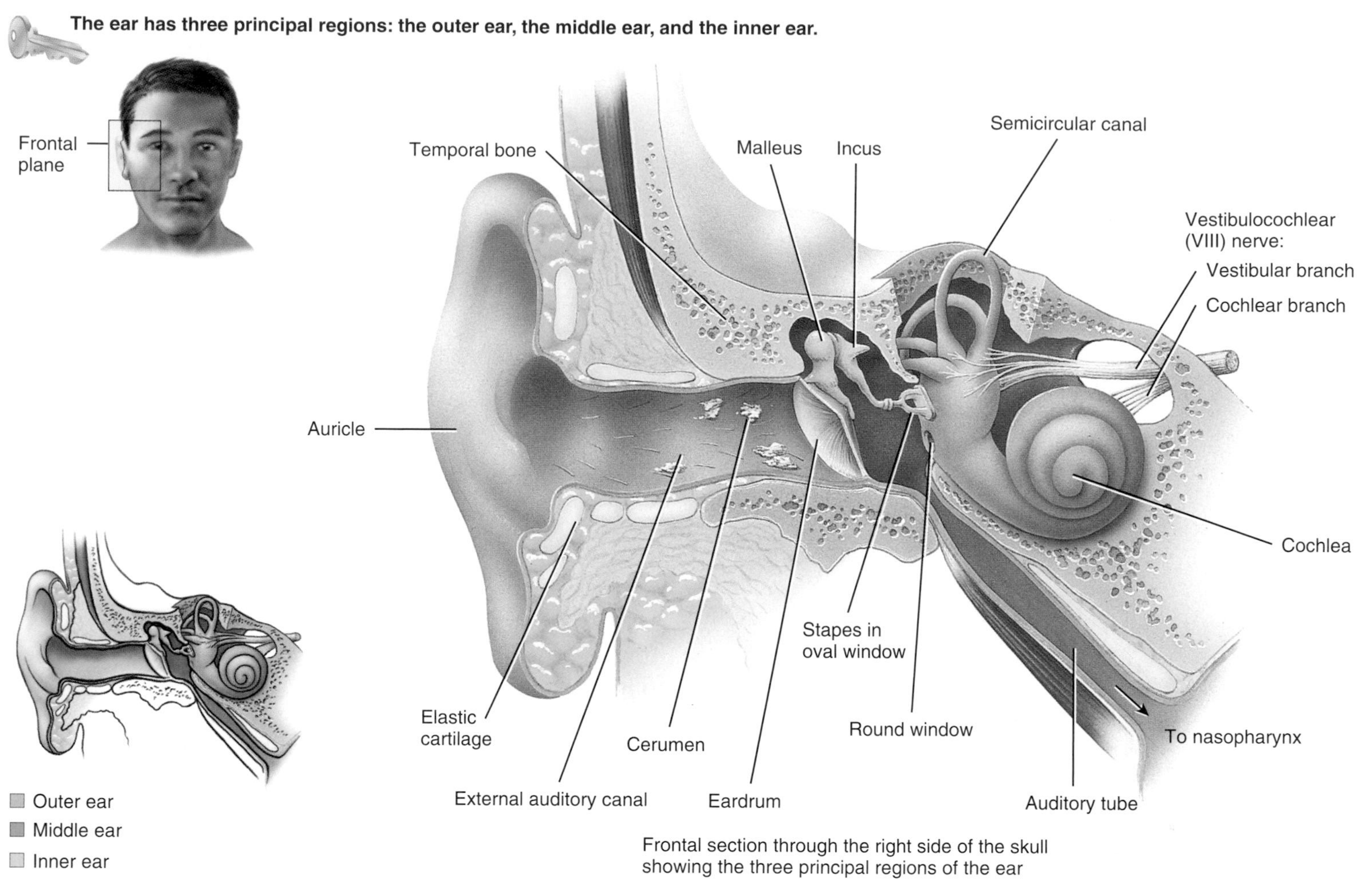

Frontal section through the right side of the skull showing the three principal regions of the ear

Where are the receptors for hearing and equilibrium located?

Extending across the middle ear and attached to it by means of ligaments are three tiny bones called ***auditory ossicles*** (OS-si-kuls) that are named for their shapes: the ***malleus*** (MAL-ē-us), ***incus*** (ING-kus), and ***stapes*** (STĀ-pēz), commonly called the hammer, anvil, and stirrup (Figure 12.12). Equally tiny skeletal muscles control the amount of movement of these bones to prevent damage by excessively loud noises. The stapes fits into a small opening in the thin bony partition between the middle and inner ear called the ***oval window***, where the inner ear begins.

Inner Ear

The ***inner ear*** is divided into the outer bony labyrinth and inner membranous labyrinth (Figure 12.13). The ***bony labyrinth*** (LAB-i-rinth) is a series of cavities in the temporal bone, including the cochlea, vestibule, and semicircular canals. The cochlea is the sense organ for hearing, and the vestibule and semicircular canals are the sense organs for equilibrium and balance. The bony labyrinth contains a fluid called *perilymph*. This fluid surrounds the inner ***membranous labyrinth***, a series of sacs and tubes with the same general shape as the bony labyrinth. The membranous labyrinth contains a fluid called *endolymph*.

The ***vestibule*** (VES-ti-būl) is the oval-shaped middle part of the bony labyrinth. The membranous labyrinth in the vestibule consists of two sacs called the ***utricle*** (Ū-tri-kl = little bag) and ***saccule*** (SAK-ūl = little sac). Behind the vestibule are the three bony ***semicircular canals***. The anterior and posterior semicircular canals are both vertical, and the lateral canal is horizontal. One end of each canal enlarges into a swelling called the ***ampulla*** (am-POOL-la = little jar). The portions of the membranous labyrinth that lie inside the bony semicircular canals are called the ***semicircular ducts***, which connect with the utricle of the vestibule.

A transverse section through the ***cochlea*** (KOK-lē-a = snail's shell), a bony spiral canal that resembles a snail's shell,

Figure 12.13 Details of the right inner ear. (a) Relationship of the scala tympani, cochlear duct, and scala vestibuli. The arrows indicate the transmission of sound waves. (b) Details of the spiral organ (organ of Corti).

The three channels in the cochlea are the scala vestibuli, scala tympani, and cochlear duct.

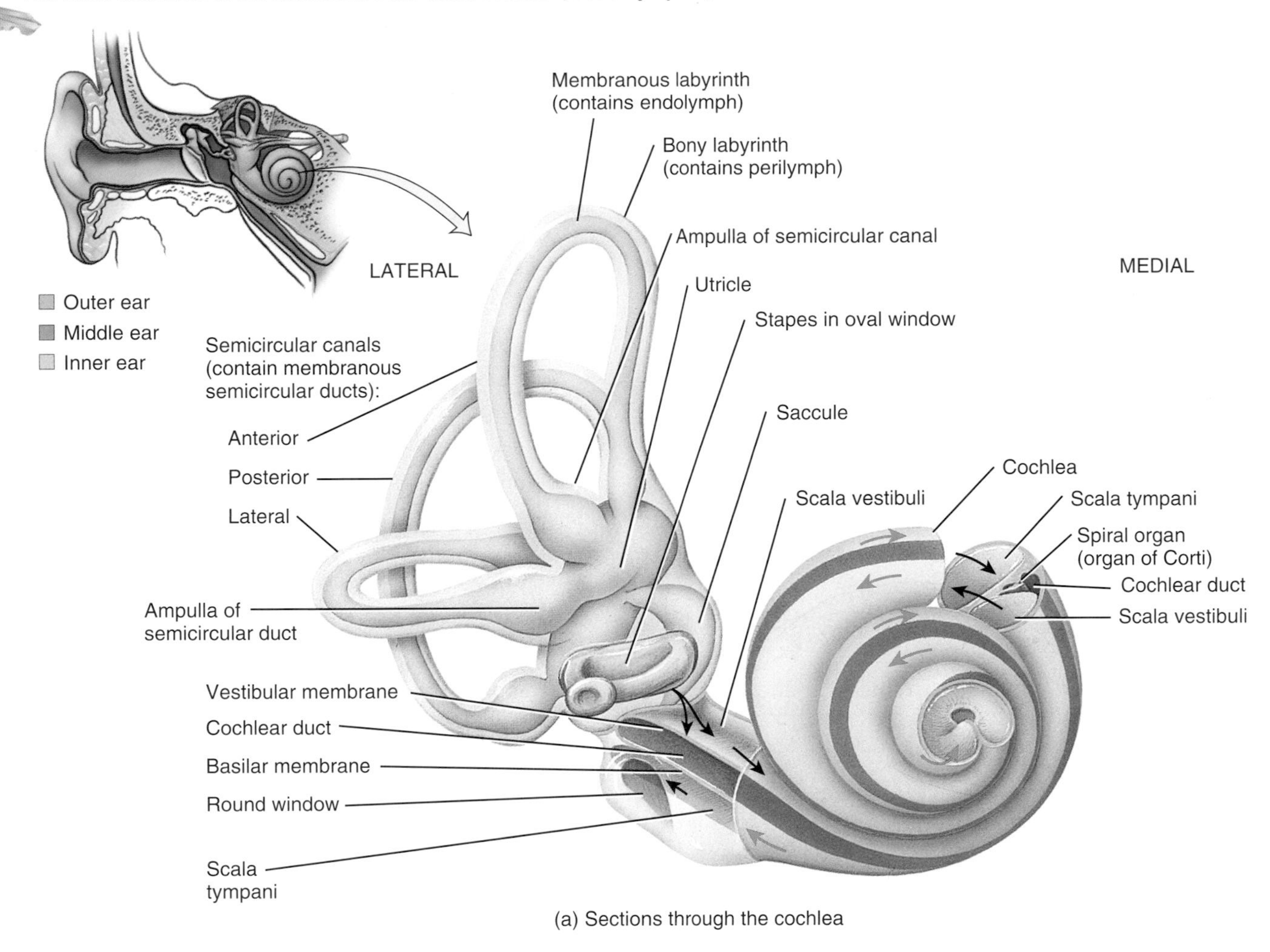

(a) Sections through the cochlea

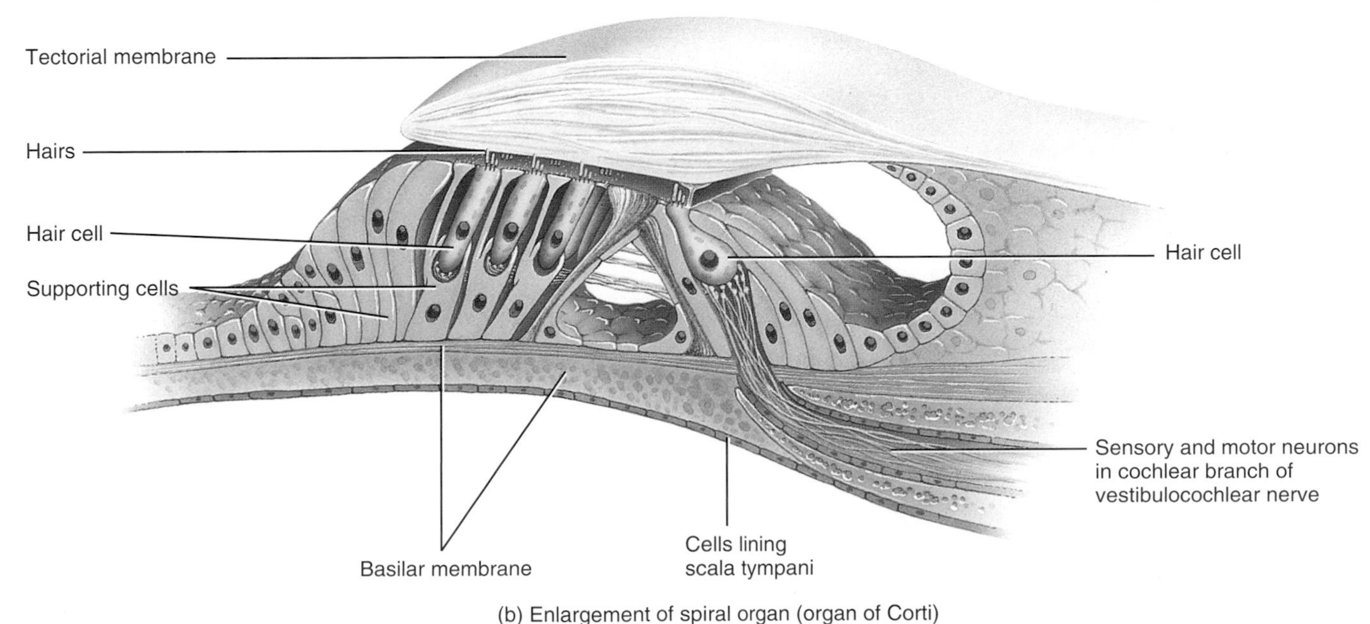

(b) Enlargement of spiral organ (organ of Corti)

What structures separate the outer ear from the middle ear? The middle ear from the inner ear?

shows that it is divided into three channels: cochlear duct, scala vestibuli, and scala tympani. The *cochlear duct* is a continuation of the membranous labyrinth into the cochlea; it is filled with endolymph. The channel above the cochlear duct is the *scala vestibuli*, which ends at the oval window. The channel below the cochlear duct is the *scala tympani*, which ends at the ***round window*** (a membrane-covered opening directly below the oval window). Both the scala vestibuli and scala tympani are part of the bony labyrinth of the cochlea and are filled with perilymph. The scala vestibuli and scala tympani are completely separated, except for an opening at the apex of the cochlea. Between the cochlear duct and the scala vestibuli is the *vestibular membrane*. Between the cochlear duct and scala tympani is the *basilar membrane*.

Resting on the basilar membrane is the ***spiral organ (organ of Corti)***, the organ of hearing (Figure 12.13b). The spiral organ consists of *supporting cells* and *hair cells*. The hair cells, the receptors for auditory sensations, have long processes at their free ends that extend into the endolymph of the cochlear duct. The hair cells form synapses with sensory and motor neurons in the cochlear branch of the vestibulocochlear (VIII) nerve. The *tectorial membrane*, a flexible gelatinous membrane, covers the hair cells.

Physiology of Hearing

The events involved in stimulation of hair cells by sound waves are as follows (Figure 12.14):

1. The auricle directs sound waves into the external auditory canal.
2. Sound waves striking the eardrum cause it to vibrate. The distance and speed of its movement depend on the intensity and frequency of the sound waves. More intense (louder) sounds produce larger vibrations. The eardrum vibrates slowly in response to low-frequency (low-pitched) sounds and rapidly in response to high-frequency (high-pitched) sounds.
3. The central area of the eardrum connects to the malleus, which also starts to vibrate. The vibration is transmitted from the malleus to the incus and then to the stapes.
4. As the stapes moves back and forth, it pushes the oval window in and out.

Figure 12.14 Physiology of hearing shown in the right ear. The numbers correspond to the events listed in the text. The cochlea has been uncoiled in order to more easily visualize the transmission of sound waves and their subsequent distortion of the vestibular and basilar membranes of the cochlear duct.

Sound waves originate from vibrating objects.

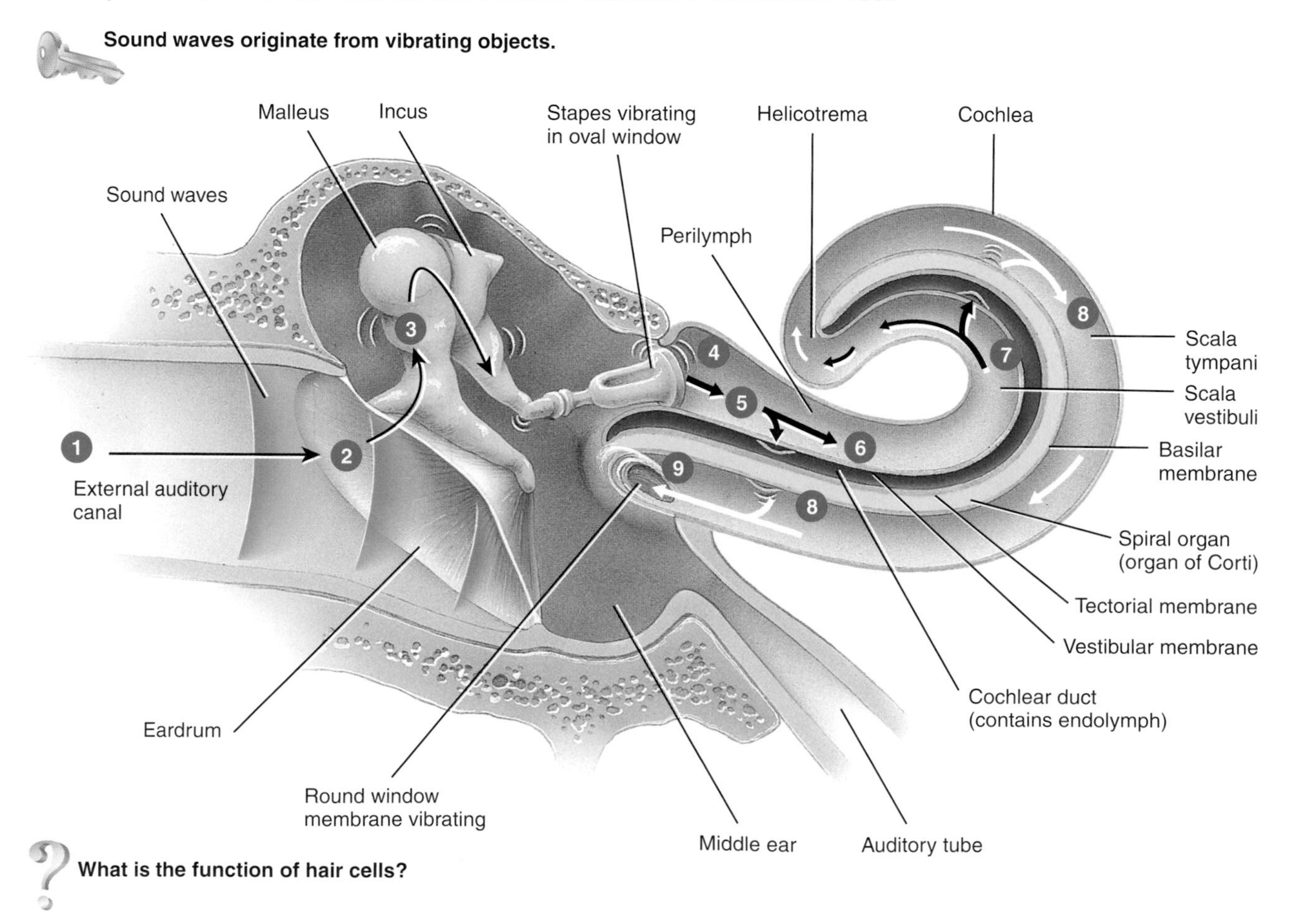

What is the function of hair cells?

5. The movement of the oval window sets up fluid pressure waves in the perilymph of the cochlea. As the oval window bulges inward, it pushes on the perilymph of the scala vestibuli.
6. The fluid pressure waves are transmitted from the scala vestibuli to the scala tympani and eventually to the membrane covering the round window, causing it to bulge outward into the middle ear. (See 9 in the figure.)
7. As the pressure waves deform the walls of the scala vestibuli and scala tympani, they also push the vestibular membrane back and forth, creating pressure waves in the endolymph inside the cochlear duct.
8. The pressure waves in the endolymph cause the basilar membrane to vibrate, which moves the hair cells of the spiral organ against the tectorial membrane. Bending of their hairs stimulates the hair cells to release neurotransmitter molecules at synapses with sensory neurons that are part of the vestibulocochlear (VIII) nerve (see Figure 12.13b). Then, the sensory neurons generate nerve impulses that conduct along the vestibulocochlear (VIII) nerve.

Sound waves of various frequencies cause certain regions of the basilar membrane to vibrate more intensely than other regions. Each segment of the basilar membrane is "tuned" for a particular pitch. Because the membrane is narrower and stiffer at the base of the cochlea (closer to the oval window), high-frequency (high-pitched) sounds induce maximal vibrations in this region. Toward the apex of the cochlea, the basilar membrane is wider and more flexible; low-frequency (low-pitched) sounds cause maximal vibration of the basilar membrane there. Loudness is determined by the intensity of sound waves. High-intensity sound waves cause larger vibrations of the basilar membrane, which leads to a higher frequency of nerve impulses reaching the brain. Louder sounds also may stimulate a larger number of hair cells.

Besides its role in detecting sounds, the cochlea has the surprising ability to *produce* sounds, which are called **otoacoustic emissions**. These sounds arise from vibrations of the hair cells themselves, caused in part by signals from *motor* neurons that synapse with the hair cells. A sensitive microphone placed next to the eardrum can pick up these very-low-volume sounds. Detection of otoacoustic emissions is a fast, inexpensive, and noninvasive way to screen newborns for hearing defects. In deaf babies, otoacoustic emissions are not produced or are greatly reduced in size.

Auditory Pathway

Sensory neurons in the cochlear branch of each vestibulocochlear (VIII) nerve terminate in the medulla oblongata on the same side of the brain. From the medulla, axons ascend to the midbrain, then to the thalamus, and finally to the primary auditory area in the temporal lobe (see Figure 10.13 on page 259). Because many auditory axons cross to the opposite side, the right and left primary auditory areas receive nerve impulses from both ears.

Physiology of Equilibrium

You learned about the anatomy of the inner ear structures for equilibrium in the previous section. In this section we will cover the physiology of balance, or how you are able to stay on your feet after tripping over your roommate's shoes. There are two types of ***equilibrium*** (balance). One kind, called ***static equilibrium***, refers to the maintenance of the position of the body (mainly the head) relative to the force of gravity. The second kind, ***dynamic equilibrium***, is the maintenance of body position (mainly the head) in response to sudden movements such as rotation, acceleration, and deceleration. Collectively, the receptor organs for equilibrium, which include the saccule, utricle, and membranous semicircular ducts, are called the ***vestibular apparatus*** (ves-TIB-ū-lar).

Static Equilibrium

The walls of both the utricle and the saccule contain a small, thickened region called a ***macula*** (MAK-ū-la; *macula* = spot). The two maculae (plural), which are perpendicular to one another, are the receptors for static equilibrium. The maculae provide sensory information on the position of the head in space and help maintain appropriate posture and balance. The maculae also contribute to some aspects of dynamic equilibrium by detecting linear acceleration and deceleration, such as the sensations you feel while in an elevator or a car that is speeding up or slowing down.

The two maculae consist of two kinds of cells: ***hair cells***, which are the sensory receptors, and ***supporting cells*** (Figure 12.15). The hairs of the hair cells protrude into a thick, jellylike substance called the ***otolithic membrane***. A layer of dense calcium carbonate crystals, called ***otoliths*** (*oto-* = ear; *-liths* = stones), extends over the entire surface of the otolithic membrane. If you tilt your head forward, gravity pulls the membrane (and the otoliths) so it slides over the hair cells in the direction of the tilt. This stimulates the hair cells and triggers nerve impulses that conduct along the ***vestibular branch*** of the vestibulocochlear (VIII) nerve (see Figure 12.12).

Dynamic Equilibrium

The three membranous semicircular ducts lie at right angles to one another in three planes (see Figure 12.13a). The positioning permits detection of rotational acceleration or deceleration. The dilated portion of each duct, the ampulla, contains a small elevation called the ***crista*** (KRIS-ta = crest;

Figure 12.15 Location and structure of receptors in the maculae of the right ear. Both sensory neurons (blue) and motor neurons (red) synapse with the hair cells.

Movements of the otolithic membrane stimulate the hair cells.

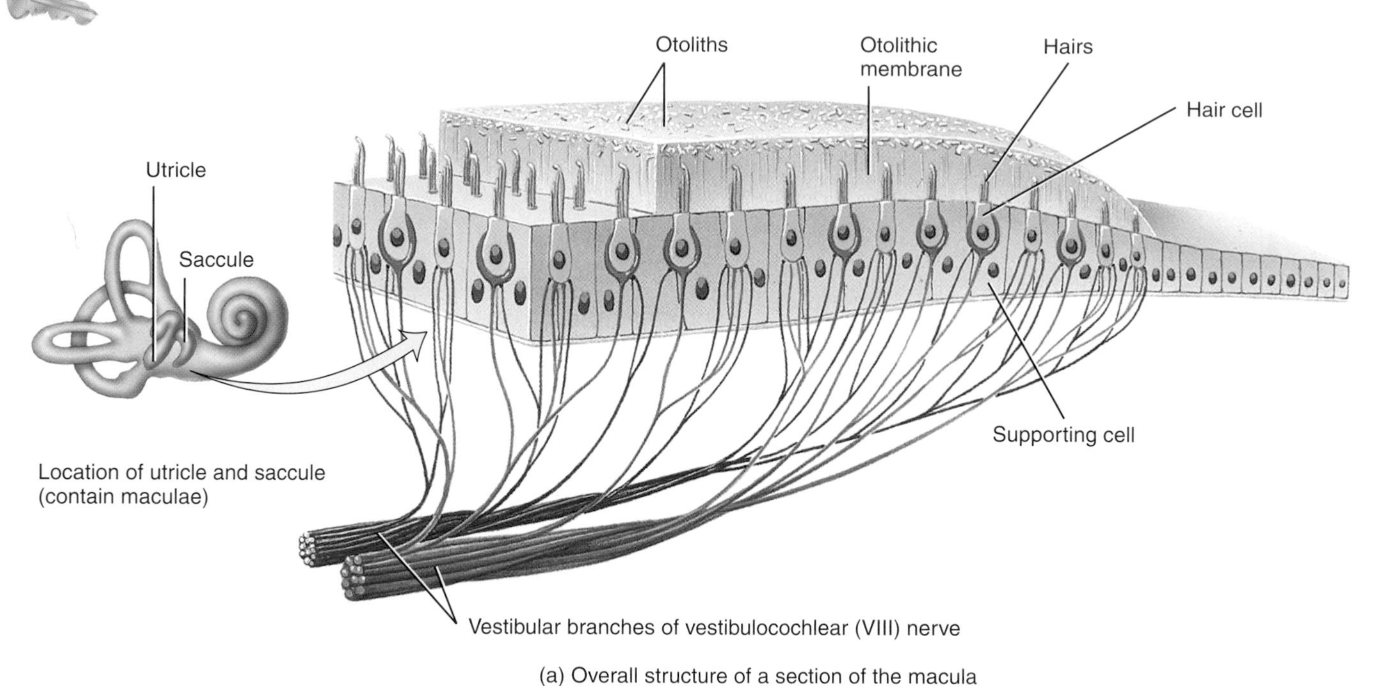

(a) Overall structure of a section of the macula

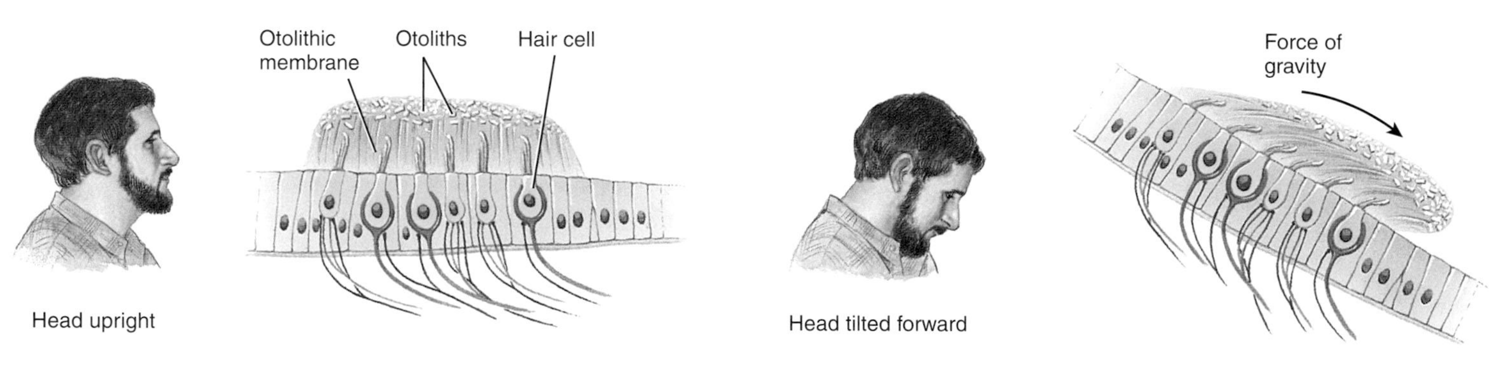

(b) Position of macula with head upright (left) and tilted forward (right)

What is the function of the maculae?

plural is *cristae*) (Figure 12.16). Each *crista* contains a group of ***hair cells*** and ***supporting cells***. Covering the crista is a mass of gelatinous material called the ***cupula*** (KŪ-pū-la). When the head moves, the attached membranous semicircular ducts and hair cells move with it. However, the endolymph within the membranous semicircular ducts is not attached and lags behind due to its inertia. As the moving hair cells drag along the stationary endolymph, the hairs bend. Bending of the hairs causes electrical signals in the hair cells. In turn, these signals trigger nerve impulses in sensory neurons that are part of the vestibular branch of the vestibulocochlear (VIII) nerve.

Equilibrium Pathways

Most of the vestibular branch axons of the vestibulocochlear (VIII) nerve enter the brain stem and then extend to the medulla or the cerebellum, where they synapse with the next neurons in the equilibrium pathways. From the medulla, some axons conduct nerve impulses along the cranial nerves that control eye movements and head and neck movements. Other axons form a spinal cord tract that conveys impulses for regulation of muscle tone in response to head movements. Various pathways among the medulla, cerebellum, and cerebrum enable the cerebellum to play a key role in maintaining equilibrium. The cerebellum continuously re-

Figure 12.16 Location and structure of the membranous semicircular ducts of the right ear. Both sensory neurons (blue) and motor neurons (red) synapse with the hair cells. The ampullary nerves are branches of the vestibular division of the vestibulocochlear (VIII) nerve.

The positions of the membranous semicircular ducts permit detection of rotational movements.

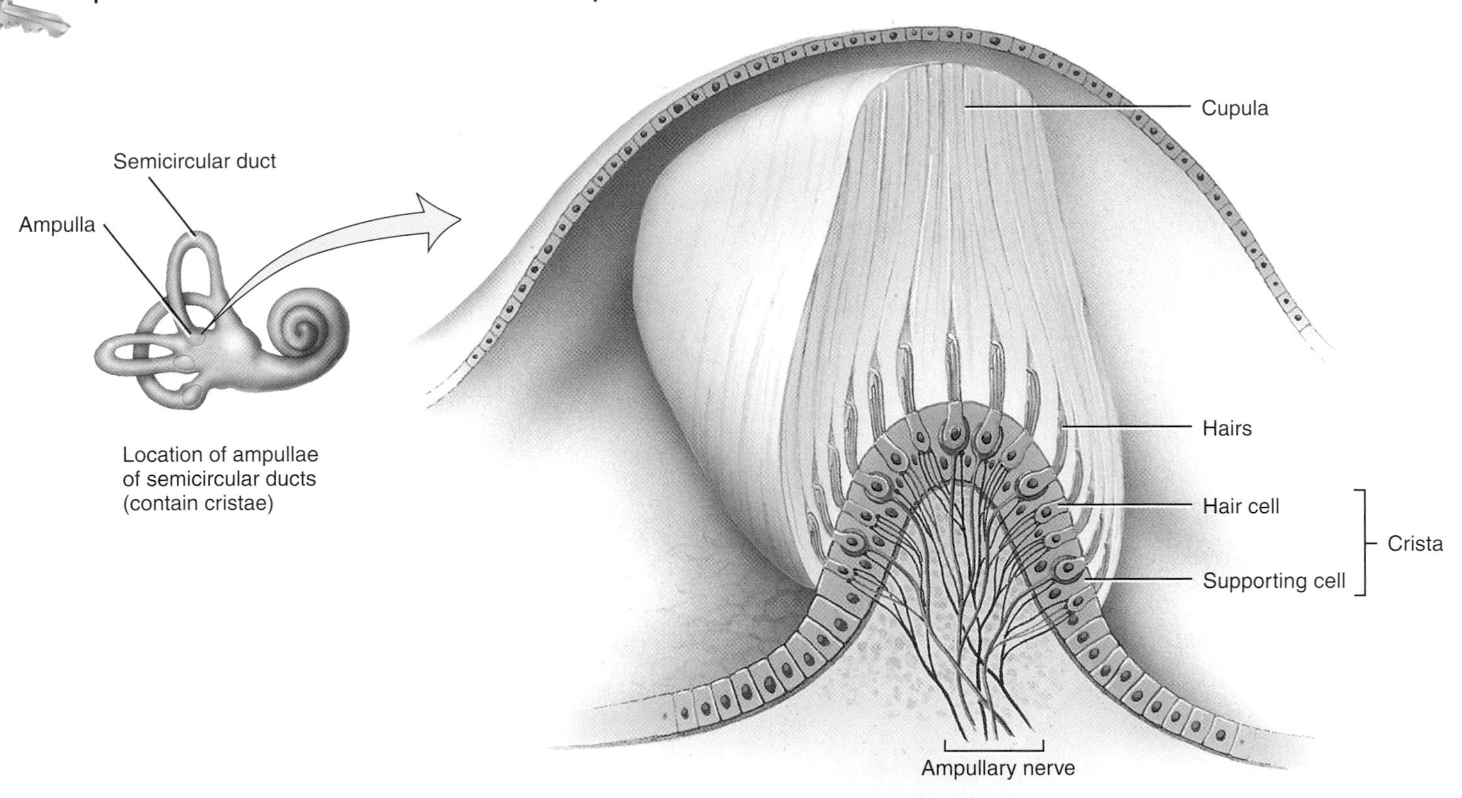

(a) Details of a crista

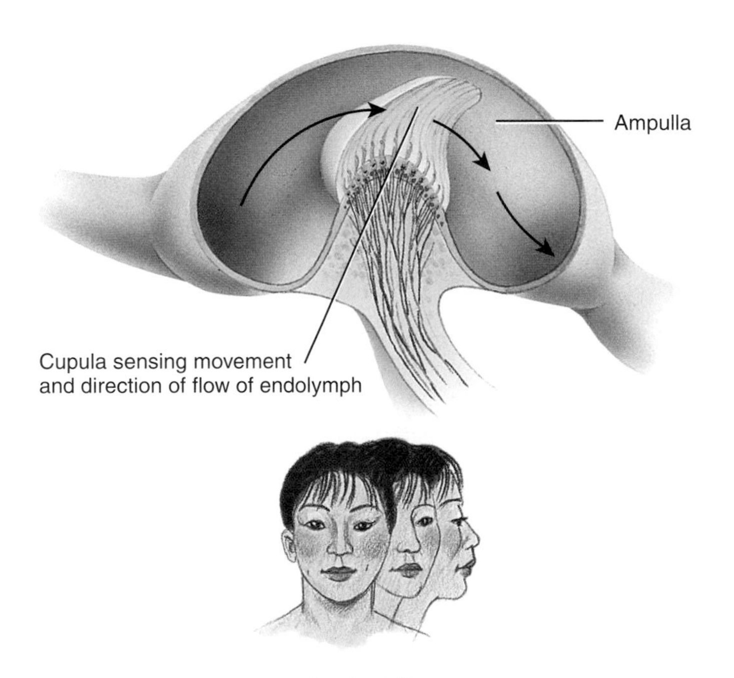

With which type of equilibrium are the membranous semicircular ducts associated?

ceives sensory information from the utricle and saccule. In response, the cerebellum makes adjustments to the signals going from the motor cortex to specific skeletal muscles to maintain equilibrium.

Table 12.3 summarizes the structures of the ear related to hearing and equilibrium.

■ CHECKPOINT

13. What is the pathway for auditory impulses from the cochlea to the cerebral cortex?

14. Compare the function of the maculae in maintaining static equilibrium with the role of the cristae in maintaining dynamic equilibrium.

• • •

Now that our exploration of the nervous system and sensations is completed, you can appreciate the many ways that the nervous system contributes to homeostasis of other body systems by examining Focus on Homeostasis: The Nervous System on page 309. Next, in Chapter 13, we will see how the hormones released by the endocrine system also help maintain homeostasis of many body processes.

Table 12.3 Summary of Structures of the Ear Related to Hearing and Equilibrium

Regions of the Ear and Key Structures	Functions
Outer Ear 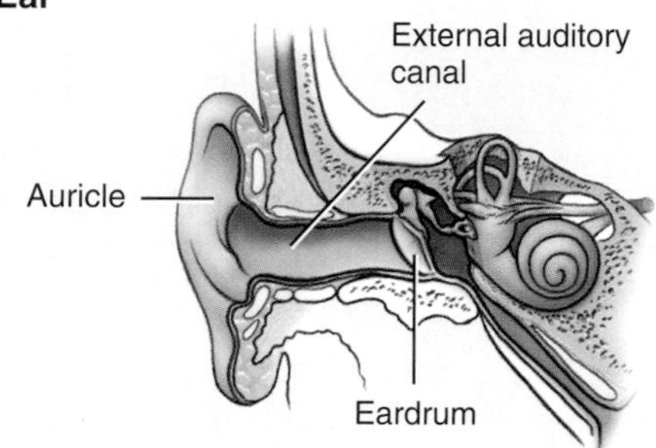	***Auricle:*** Collects sound waves. ***External auditory canal:*** Directs sound waves to the eardrum. ***Eardrum (tympanic membrane):*** Sound waves cause it to vibrate, which, in turn, causes the malleus to vibrate.
Middle Ear 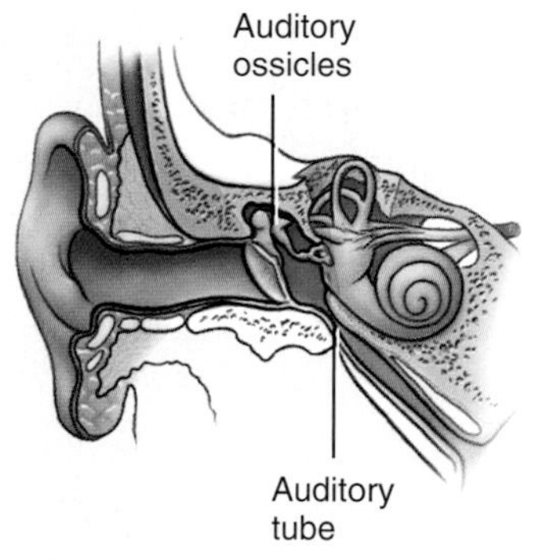	***Auditory ossicles:*** Transmit and amplify vibrations from tympanic membrane to oval window. ***Auditory (eustachian) tube:*** Equalizes air pressure on both sides of the tympanic membrane.
Inner Ear 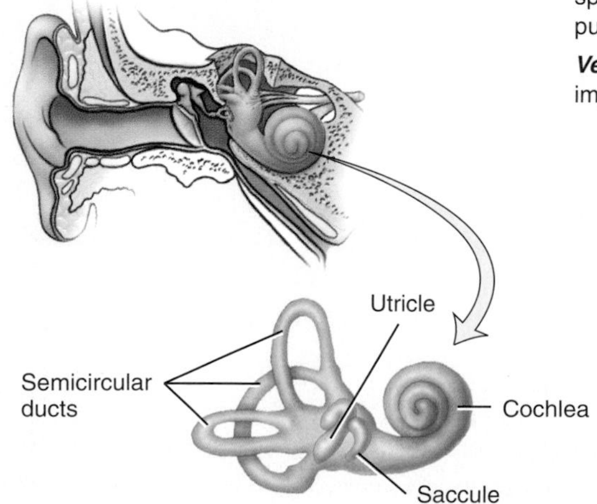	***Cochlea:*** Contains a series of fluids, channels, and membranes that transmit vibrations to the spiral organ (organ of Corti), the organ of hearing; hair cells in the spiral organ trigger nerve impulses in the cochlear branch of the vestibulocochlear (VIII) nerve. ***Vestibular apparatus:*** Includes semicircular ducts, utricle, and saccule, which generate nerve impulses that propagate along the vestibular branch of the vestibulocochlear (VIII) nerve. ***Semicircular ducts:*** Contain cristae, sites of hair cells for dynamic equilibrium. ***Utricle:*** Contains macula, site of hair cells for static and dynamic equilibrium. ***Saccule:*** Contains macula, site of hair cells for static and dynamic equilibrium.

FOCUS ON HOMEOSTASIS

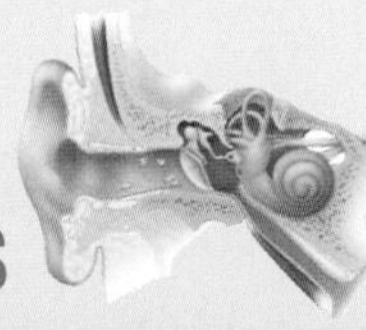

The Nervous System

Body System		Contribution of the Nervous System
For all body systems		Together with hormones from the endocrine system, nerve impulses provide communication and regulation of most body tissues.
Integumentary system	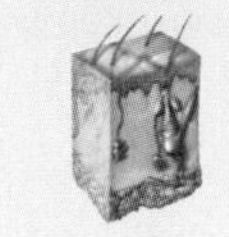	Sympathetic nerves of the autonomic nervous system (ANS) control contraction of smooth muscles attached to hair follicles and secretion of perspiration from sweat glands.
Skeletal system	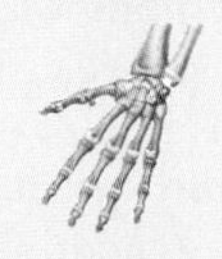	Nociceptors (pain receptors) in bone tissue warn of bone trauma or damage.
Muscular system		Somatic motor neurons receive instructions from motor areas of the brain and stimulate contraction of skeletal muscles to bring about body movements. The basal ganglia and reticular formation set the level of muscle tone. The cerebellum coordinates skilled movements.
Endocrine system		The hypothalamus regulates secretion of hormones from the anterior and posterior pituitary. The ANS regulates secretion of hormones from the adrenal medulla and pancreas.
Cardiovascular system	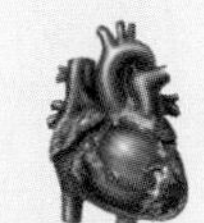	The cardiovascular center in the medulla oblongata provides nerve impulses to the ANS that govern heart rate and the forcefulness of the heartbeat. Nerve impulses from the ANS also regulate blood pressure and blood flow through blood vessels.
Lymphatic system and immunity	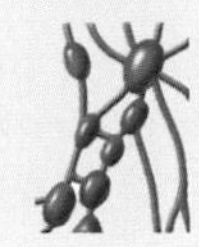	Certain neurotransmitters help regulate immune responses. Activity in the nervous system may increase or decrease immune responses.
Respiratory system	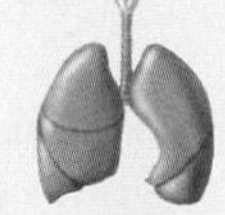	Respiratory areas in the brain stem control breathing rate and depth. The ANS helps regulate the diameter of airways.
Digestive system	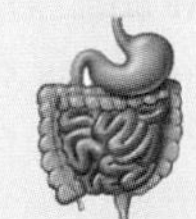	The ANS and enteric nervous system (ENS) help regulate digestion. The parasympathetic division of the ANS stimulates many digestive processes.
Urinary system	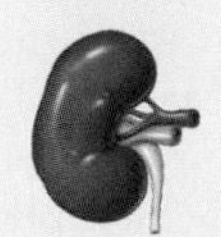	The ANS helps regulate blood flow to kidneys, thereby influencing the rate of urine formation; brain and spinal cord centers govern emptying of urinary bladder.
Reproductive systems	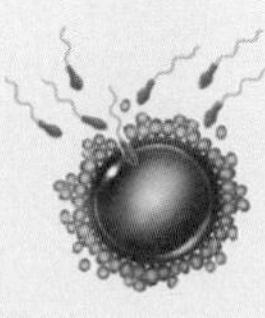	The hypothalamus and limbic system govern a variety of sexual behaviors; the ANS brings about erection of the penis in males and the clitoris in females and ejaculation of semen in males. The hypothalamus regulates release of anterior pituitary hormones that control the gonads (ovaries and testes). Nerve impulses elicited by touch stimuli from suckling infant cause release of oxytocin and milk ejection in nursing mothers.

COMMON DISORDERS

Cataracts

A common cause of blindness is a loss of transparency of the lens known as a ***cataract*** (CAT-a-rakt). The lens becomes cloudy (less transparent) due to changes in the structure of the lens proteins. Cataracts often occur with aging but may also be caused by injury, excessive exposure to ultraviolet rays, certain medications (such as long-term use of steroids), or complications of other diseases (for example, diabetes). People who smoke also have increased risk of developing cataracts. Fortunately, sight can usually be restored by surgical removal of the old lens and implantation of an artificial one.

Glaucoma

In ***glaucoma*** (glaw-KŌ-ma), the most common cause of blindness in the United States, a buildup of aqueous humor within the anterior cavity causes an abnormally high intraocular pressure. Persistent pressure results in a progression from mild visual impairment to irreversible destruction of the retina, damage to the optic nerve, and blindness. Because glaucoma is painless, and because the other eye initially compensates to a large extent for the loss of vision, a person may experience considerable retinal damage and loss of vision before the condition is diagnosed.

Deafness

Deafness is significant or total hearing loss. *Sensorineural deafness* is caused by either impairment of hair cells in the cochlea or damage of the cochlear branch of the vestibulocochlear nerve. This type of deafness may be caused by atherosclerosis, which reduces blood supply to the ears; repeated exposure to loud noise, which destroys hair cells of the spiral organ; or certain drugs such as aspirin and streptomycin. *Conduction deafness* is caused by impairment of the outer and middle ear mechanisms for transmitting sounds to the cochlea. It may be caused by otosclerosis, the deposition of new bone around the oval window; impacted cerumen; injury to the eardrum; or aging, which often results in thickening of the eardrum and stiffening of the joints of the auditory ossicles.

Ménière's Disease

Ménière's disease (men′-ē-ĀRZ) results from an increased amount of endolymph that enlarges the membranous labyrinth. Among the symptoms are fluctuating hearing loss (caused by distortion of the basilar membrane of the cochlea) and roaring tinnitus (ringing). Vertigo (a sensation of spinning or whirling) is characteristic of Ménière's disease. Almost total destruction of hearing may occur over a period of years.

Otitis Media

Otitis media is an acute infection of the middle ear caused primarily by bacteria and associated with infections of the nose and throat. Symptoms include pain; malaise (discomfort or uneasiness); fever; and a reddening and outward bulging of the eardrum, which may rupture unless prompt treatment is received (this may involve draining pus from the middle ear). Bacteria from the nasopharynx passing into the auditory tube is the primary cause of all middle ear infections. Children are more susceptible than adults to middle ear infections because their auditory tubes are almost horizontal, which decreases drainage.

MEDICAL TERMINOLOGY AND CONDITIONS

Age-related macular disease (AMD) Degeneration of the macula lutea of the retina in persons 50 years of age and older.

Anosmia (an-OZ-mē-a; *a-* = without; *-osmi* = smell, odor) Total lack of the sense of smell.

Cochlear implant A device that translates sounds into electrical signals that can be interpreted by the brain. It is especially useful for people with deafness caused by damage to hair cells in the cochlea.

Conjunctivitis (pinkeye) An inflammation of the conjunctiva; when caused by bacteria such as pneumococci, staphylococci, or *Hemophilus influenzae*, it is very contagious and more common in children. May also be caused by irritants, such as dust, smoke, or pollutants in the air, in which case it is not contagious.

Detached retina Detachment of the neural portion of the retina from the pigment epithelium due to trauma, disease, or age-related degeneration. The result is distorted vision and blindness.

LASIK Surgery with a laser to correct the curvature of the cornea for conditions such as nearsightedness, farsightedness, and astigmatism.

Nystagmus (nis-TAG-mus; *nystagm-* = nodding or drowsy) A rapid involuntary movement of the eyeballs, possibly caused by a disease of the central nervous system. It is associated with conditions that cause vertigo.

Otalgia (ō-TAL-jē-a; *ot-* = ear; *-algia* = pain) Earache.

Retinoblastoma (ret-i-nō-blas-TŌ-ma; *-oma* = tumor) A tumor arising from immature retinal cells; it accounts for 2% of childhood cancers.

Scotoma (skō-TŌ-ma = darkness) An area of reduced or lost vision in the visual field.

Strabismus (stra-BIZ-mus) An imbalance in the extrinsic eye muscles that causes a misalignment of one eye so that its line of vision is not parallel with that of the other eye (cross-eyes) and both eyes are not pointed at the same object at the same time; the condition produces a squint.

Tinnitus (ti-NĪ-tus) A ringing, roaring, or clicking in the ears.

Trachoma (tra-KŌ-ma) A serious form of conjunctivitis and the greatest single cause of blindness in the world. It is caused by the bacterium *Chlamydia trachomatis*. The disease produces an excessive growth of subconjunctival tissue and invasion of blood vessels into the cornea, which progresses until the entire cornea is opaque, causing blindness.

Vertigo (VER-ti-gō = dizziness) A sensation of spinning or movement in which the world seems to revolve or the person seems to revolve in space.

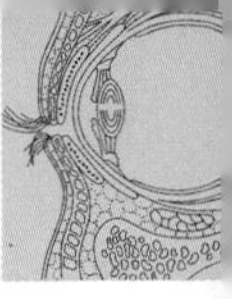

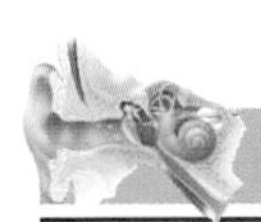

STUDY OUTLINE

Overview of Sensations (p. 285)

1. Sensation is the conscious or subconscious awareness of external and internal stimuli.
2. Two general classes of senses are (1) general senses, which include somatic senses and visceral senses, and (2) special senses, which include smell, taste, vision, hearing, and equilibrium (balance).
3. The conditions for a sensation to occur are reception of a stimulus by a sensory receptor, conversion of the stimulus into one or more nerve impulses, conduction of the impulses to the brain, and integration of the impulses by a region of the brain.
4. Sensory impulses from each part of the body arrive in specific regions of the cerebral cortex.
5. Adaptation is a decrease in sensation during a prolonged stimulus. Some receptors are rapidly adapting; others are slowly adapting.
6. Receptors can be classified structurally by their microscopic features as free nerve endings, encapsulated nerve endings, or separate cells. Functionally, receptors are classified by the type of stimulus they detect as mechanoreceptors, thermoreceptors, nociceptors, photoreceptors, osmoreceptors, and chemoreceptors.

Somatic Senses (p. 286)

1. Somatic sensations include tactile sensations (touch, pressure, vibration, itch, and tickle), thermal sensations (heat and cold), pain sensations, and proprioceptive sensations (joint and muscle position sense and movements of the limbs). Receptors for these sensations are located in the skin, mucous membranes, muscles, tendons, and joints.
2. Receptors for touch include corpuscles of touch (Meissner corpuscles), hair root plexuses, type I cutaneous mechanoreceptors (Merkel discs), and type II cutaneous mechanoreceptors (Ruffini corpuscles). Receptors for pressure and vibration are lamellated (pacinian) corpuscles. Tickle and itch sensations result from stimulation of free nerve endings.
3. Thermoreceptors, free nerve endings in the epidermis and dermis, adapt to continuous stimulation.
4. Nociceptors are free nerve endings that are located in nearly every body tissue; they provide pain sensations.
5. Proprioceptors inform us of the degree to which muscles are contracted, the amount of tension present in tendons, the positions of joints, and the orientation of the head.

Olfaction: Sense of Smell (p. 290)

1. The olfactory epithelium in the upper portion of the nasal cavity contains olfactory receptors, supporting cells, and basal stem cells.
2. Individual olfactory receptors respond to hundreds of different odorant molecules by producing an electrical signal that triggers one or more nerve impulses. Adaptation (decreasing sensitivity) to odors occurs rapidly.
3. Axons of olfactory receptors form the olfactory nerves, which convey nerve impulses to the olfactory bulbs. From there, impulses conduct via the olfactory tracts to the limbic system, hypothalamus, and cerebral cortex (temporal lobe).

Gustation: Sense of Taste (p. 291)

1. The receptors for gustation, the gustatory receptor cells, are located in taste buds.
2. To be tasted, substances must be dissolved in saliva.
3. The five primary tastes are salty, sweet, sour, bitter, and umami.
4. Gustatory receptor cells trigger impulses in cranial nerves VII (facial), IX (glossopharyngeal), and X (vagus). Impulses for taste conduct to the medulla oblongata, limbic system, hypothalamus, thalamus, and the primary gustatory area in the parietal lobe of the cerebral cortex.

Vision (p. 293)

1. Accessory structures of the eyes include the eyebrows, eyelids, eyelashes, the lacrimal apparatus (which produces and drains tears), and extrinsic eye muscles (which move the eyes).
2. The eyeball has three layers: (a) fibrous tunic (sclera and cornea), (b) vascular tunic (choroid, ciliary body, and iris), and (c) retina.
3. The retina consists of a neural layer (photoreceptor layer, bipolar cell layer, and ganglion cell layer) and a pigmented layer (a sheet of melanin-containing epithelial cells).
4. The anterior cavity contains aqueous humor; the vitreous chamber contains the vitreous body.
5. Image formation on the retina involves refraction of light rays by the cornea and lens, which focus an inverted image on the central fovea of the retina.
6. For viewing close objects, the lens increases its curvature (accommodation), and the pupil constricts to prevent light rays from entering the eye through the periphery of the lens.
7. Improper refraction may result from myopia (nearsightedness), hypermetropia (farsightedness), or astigmatism (irregular curvature of the cornea or lens).
8. Movement of the eyeballs toward the nose to view an object is called convergence.
9. The first step in vision is the absorption of light rays by photopigments in rods and cones (photoreceptors). Stimulation of the rods and cones then activates bipolar cells, which in turn activate the ganglion cells.
10. Nerve impulses arise in ganglion cells and conduct along the optic nerve, through the optic chiasm and optic tract to the thalamus. From the thalamus, the optic radiations extend to the primary visual area in the occipital lobe of the cerebral cortex.

Hearing and Equilibrium (p. 301)

1. The outer ear consists of the auricle, external auditory canal, and eardrum.
2. The middle ear consists of the auditory (eustachian) tube, auditory ossicles, oval window, and round window.

3. The inner ear consists of the bony labyrinth and membranous labyrinth. The inner ear contains the spiral organ (organ of Corti), the organ of hearing.
4. Sound waves enter the external auditory canal, strike the eardrum, pass through the ossicles, strike the oval window, set up pressure waves in the perilymph, strike the vestibular membrane and scala tympani, increase pressure in the endolymph, vibrate the basilar membrane, and stimulate hair cells in the spiral organ.
5. Hair cells release neurotransmitter molecules that can initiate nerve impulses in sensory neurons.
6. Sensory neurons in the cochlear branch of the vestibulocochlear nerve terminate in the medulla oblongata. Auditory signals then pass to the midbrain, thalamus, and temporal lobes.
7. Static equilibrium is the orientation of the body relative to the pull of gravity. The maculae of the utricle and saccule are the sense organs of static equilibrium.
8. Dynamic equilibrium is the maintenance of body position in response to rotation, acceleration, and deceleration. The maculae of the utricle and saccule and the cristae in the membranous semicircular ducts are the sense organs of dynamic equilibrium.
9. Most vestibular branch axons of the vestibulocochlear nerve enter the brain stem and terminate in the medulla and pons; other axons extend to the cerebellum.

SELF-QUIZ

1. You enter a sauna and it feels awfully hot, but soon the temperature feels comfortably warm. What have you have experienced?
   a. damage to your thermoreceptors
   b. sensory adaptation
   c. a change in the temperature of the sauna
   d. inactivation of your thermoreceptors
   e. damage to the parietal lobe
2. The lacrimal glands produce ________, which drain(s) into the ________.
   a. tears; anterior cavity
   b. tears; nasal cavity
   c. aqueous humor; anterior chamber
   d. aqueous humor; anterior cavity
   e. aqueous humor; scleral venous sinus
3. The spiral organ (organ of Corti)
   a. contains hair cells
   b. is responsible for equilibrium
   c. is filled with perilymph
   d. is another name for the auditory (eustachian) tube
   e. transmits auditory nerve impulses to the brain
4. Equilibrium and the activities of muscles and joints are monitored by
   a. olfactory receptors b. nociceptors
   c. tactile receptors d. proprioceptors
   e. thermoreceptors
5. In the retina, cone photoreceptors
   a. are more numerous than rods
   b. contain the photopigment rhodopsin
   c. are more sensitive to low light level than are rods
   d. reform their photopigments more slowly than do rods
   e. provide higher acuity vision than do rods
6. Which of the following is NOT required for a sensation to occur?
   a. the presence of a stimulus
   b. a receptor specialized to detect a stimulus
   c. the presence of slowly adapting receptors
   d. a sensory neuron to conduct impulses
   e. a region of the brain for integration of the nerve impulse
7. Match each receptor with its function.

___a. color vision	A. lamellated (pacinian) corpuscle
___b. taste	B. type I cutaneous mechanoreceptor
___c. smell	C. rod photoreceptor
___d. dynamic equilibrium	D. nociceptor
___e. vision in dim light	E. gustatory receptor cell
___f. stretch in a muscle	F. olfactory receptor
___g. static equilibrium	G. muscle spindle
___h. pressure	H. maculae
___i. touch	I. cristae
___j. detection of pain	J. cone photoreceptors

8. For taste to occur
   a. the mouth must be dry
   b. the chemical must be in contact with the basal cells
   c. filiform papillae must be stimulated
   d. the limbic system needs to be activated
   e. the gustatory hair must be stimulated by the dissolved chemical
9. Which of the following characteristics of taste is NOT true?
   a. Olfaction can affect taste.
   b. Three cranial nerves conduct the impulses for taste to the brain.

c. Taste adaptation occurs quickly.
d. Humans can recognize about 10 primary tastes.
e. Taste receptors are located in taste buds on the tongue, on the roof of the mouth, in the throat, and in the epiglottis.

10. You are seated at your desk and drop your pencil. As you lean over to retrieve it, what is occurring in your inner ear?
   a. The hair cells on the macula are responding to changes in static equilibrium.
   b. The hair cells in the cochlea are responding to changes in dynamic equilibrium.
   c. The cristae of each semicircular duct are responding to changes in dynamic equilibrium.
   d. The cochlear branch of the vestibulocochlear (VIII) nerve begins to transmit nerve impulses to the brain.
   e. The auditory (eustachian) tube makes adjustments for varying air pressures.

11. Kinesthesia is the
   a. perception of body movements
   b. ability to identify an object by feeling it
   c. sensation of weightlessness that occurs in outer space
   d. decrease in sensitivity of receptors to a prolonged stimulus
   e. movement of body parts in a rhythmic manner

12. Which of the following is NOT true about nociceptors?
   a. They respond to stimuli that may cause tissue damage.
   b. They consist of free nerve endings.
   c. They can be activated by excessive stimuli from other sensations.
   d. They are found in virtually every body tissue except the brain.
   e. They adapt very rapidly.

13. Which of the following is NOT a function of tears?
   a. moisten the eye b. wash away eye irritants
   c. destroy certain bacteria d. lubricate the eye
   e. provide nutrients to the cornea

14. Transmission of vibrations (sound waves) from the tympanic membrane to the oval window is accomplished by
   a. neurons b. the tectorial membrane
   c. the auditory ossicles d. the endolymph
   e. the auditory (Eustachian) tube

15. Match the following:

___a. focuses light rays onto the retina	A. sclera
___b. regulates the amount of light entering the eye	B. choroid
___c. contains aqueous humor	C. lacrimal glands
___d. contains blood vessels that help nourish the retina	D. lens
___e. produce tears	E. retina
___f. dense connective tissue that provides shape to the eye	F. iris
___g. contains photoreceptors	G. anterior cavity

16. Which of the following structures refracts light rays entering the eye?
   a. cornea b. sclera c. pupil d. retina
   e. conjunctiva

17. Your 45-year-old neighbor has recently begun to have difficulty reading the morning newspaper. You explain that this condition is known as ________ and is due to ________.
   a. myopia, inability of his eyes to properly focus light on his retinas
   b. night blindness, a vitamin A deficiency
   c. binocular vision, the eyes focusing on two different objects
   d. astigmatism, an irregularity in the curvature of the lens
   e. presbyopia, the loss of elasticity in the lens

18. Damage to cells in the central fovea would interfere with
   a. dynamic equilibrium b. accommodation
   c. visual acuity d. ability to see in dim light
   e. intraocular pressure

19. Place the following events concerning the visual pathway in the correct order:
   1. Nerve impulses exit the eye via the optic nerve.
   2. Optic tract axons terminate in the thalamus.
   3. Light reaches the retina.
   4. Rods and cones are stimulated.
   5. Synapses occur in the thalamus and continue to the primary visual area in the occipital lobe.
   6. Ganglion cells generate nerve impulses.

   a. 4, 1, 2, 5, 6, 3 b. 5, 4, 1, 3, 2, 6 c. 3, 4, 6, 1, 5, 2
   d. 3, 4, 6, 1, 2, 5 e. 3, 4, 5, 6, 1, 2

20. Place the following events of the auditory pathway in the correct order:
   1. Hair cells in the spiral organ bend as they rub against the tectorial membrane.
   2. Movement in the oval window begins movement in the perilymph.
   3. Nerve impulses exit the ear via the vestibulocochlear (VIII) nerve.
   4. The eardrum and auditory ossicles transmit vibrations from sound waves.
   5. Pressure waves from the perilymph cause bulging of the round window and formation of pressure waves in the endolymph.

   a. 4, 2, 5, 1, 3 b. 4, 5, 2, 3, 1 c. 5, 3, 2, 4, 1
   d. 3, 4, 5, 1, 2 e. 2, 4, 1, 5, 3

CRITICAL THINKING APPLICATIONS

1. When you first enter a coffee shop, the aroma of fresh java is wonderfully strong and full-bodied. After several minutes waiting in line, the odor is barely noticeable. Has something happened to the coffee or has something happened to you?
2. Cliff works the night shift and sometimes falls asleep in A&P class. What is the effect on the structures in his internal ear when his head falls backward as he slumps in his seat?
3. A medical procedure used to improve visual acuity involves shaving of a thin layer off the cornea. How could this procedure improve vision?
4. The optometrist put drops in Latasha's eyes during her eye exam. When Latasha looked in the mirror after the exam, her pupils were very large and her eyes were sensitive to the bright light. How did the eye drops produce this effect on Latasha's eyes?

ANSWERS TO FIGURE QUESTIONS

12.1 Corpuscles of touch (Meissner corpuscles) are abundant in the fingertips, palms, and soles.

12.2 The kidneys have the broadest area for referred pain.

12.3 Basal stem cells undergo cell division to produce new olfactory receptors.

12.4 The gustatory pathway: gustatory receptor cells → cranial nerves VII, IX, and X → medulla oblongata → thalamus → primary gustatory area in the parietal lobe of the cerebral cortex.

12.5 Tears clean, lubricate, and moisten the eyeball.

12.6 The fibrous tunic consists of the cornea and sclera; the vascular tunic consists of the choroid, ciliary body, and iris.

12.7 The parasympathetic division of the autonomic nervous system causes pupillary constriction; the sympathetic division causes pupillary dilation.

12.8 The two types of photoreceptors are rods and cones. Rods provide black-and-white vision in dim light; cones provide high visual acuity and color vision in bright light.

12.9 During accommodation, the ciliary muscle contracts, zonular fibers slacken, and the lens becomes more rounded (convex) and refracts light more.

12.10 Presbyopia is the loss of elasticity in the lens that occurs with aging.

12.11 Structures carrying visual impulses from the retina: axons of ganglion cells → optic (II) nerve → optic chiasm → optic tract → thalamus → primary visual area in occipital lobe of the cerebral cortex.

12.12 The receptors for hearing and equilibrium are located in the inner ear: cochlea (hearing) and semicircular ducts (equilibrium).

12.13 The eardrum (tympanic membrane) separates the outer ear from the middle ear. The oval and round windows separate the middle ear from the inner ear.

12.14 Hair cells convert a mechanical force (stimulus) into an electrical signal (depolarization and repolarization of the hair cell membrane).

12.15 The maculae are the receptors for static equilibrium and also contribute to dynamic equilibrium.

12.16 The membranous semicircular ducts function in dynamic equilibrium.

THE ENDOCRINE SYSTEM chapter 13

did you know? ***The high blood sugar levels seen in Type 2 diabetes occur because the body's ability to respond to the important hormone insulin is impaired. Researchers do not know exactly why cells become sluggish in their response to insulin, but they do know that obesity and a sedentary lifestyle increase a person's risk for developing this disorder. The number of young people diagnosed with Type 2 diabetes has increased almost tenfold since the 1980s, probably because of increasing rates of childhood obesity. Before that time, Type 2 diabetes was so rarely seen in children and adolescents that it was called "adult-onset diabetes."***

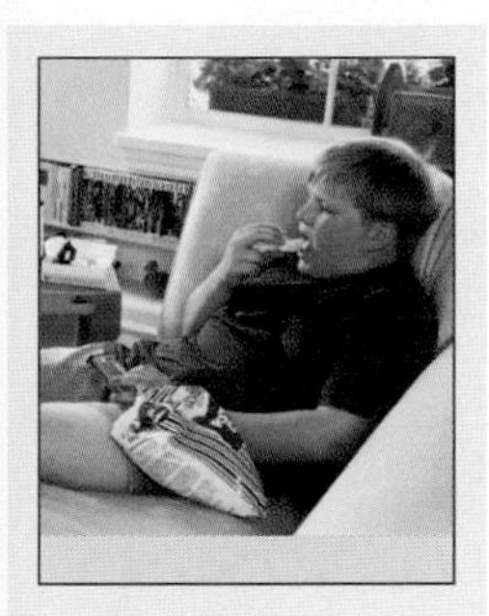

Focus on Wellness, page 332

www.wiley.com/college/apcentral

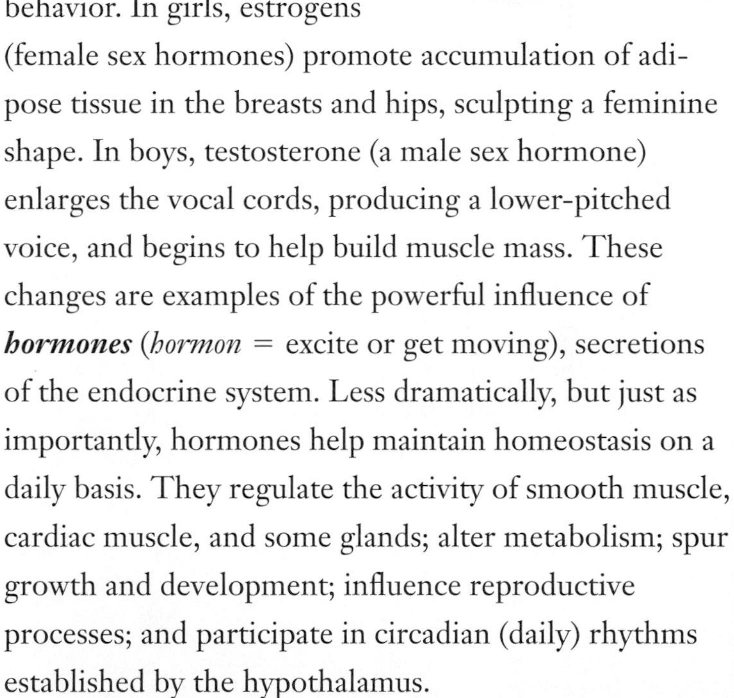

As they mature, boys and girls develop striking differences in physical appearance and behavior. In girls, estrogens (female sex hormones) promote accumulation of adipose tissue in the breasts and hips, sculpting a feminine shape. In boys, testosterone (a male sex hormone) enlarges the vocal cords, producing a lower-pitched voice, and begins to help build muscle mass. These changes are examples of the powerful influence of ***hormones*** (*hormon* = excite or get moving), secretions of the endocrine system. Less dramatically, but just as importantly, hormones help maintain homeostasis on a daily basis. They regulate the activity of smooth muscle, cardiac muscle, and some glands; alter metabolism; spur growth and development; influence reproductive processes; and participate in circadian (daily) rhythms established by the hypothalamus.

looking back to move ahead . . .

- **Steroids (page 35)**
- **The Plasma Membrane (page 46)**
- **Neurons (page 228)**
- **Negative and Positive Feedback Systems (page 8)**

INTRODUCTION

OBJECTIVE • List the components of the endocrine system.

The ***endocrine system*** consists of several endocrine glands plus many hormone-secreting cells in organs that have functions besides secreting hormones. In contrast to the nervous system, which controls body activities through the release of neurotransmitters at synapses, the endocrine system releases hormones into interstitial fluid and then into the bloodstream. The circulating blood then delivers hormones to virtually all cells throughout the body, and cells that recognize a particular hormone will respond. The nervous system and endocrine system often work together. For example, certain parts of the nervous system stimulate or inhibit the release of hormones by the endocrine system. Typically, the endocrine system acts more slowly than the nervous system, which often produces an effect within a fraction of a second. Moreover, the effects of hormones linger until they are cleared from the blood. The liver inactivates some hormones, and the kidneys excrete others in the urine.

As you learned in Chapter 4, two types of glands are present in the body: exocrine glands and endocrine glands. *Exocrine glands* secrete their products into *ducts* that carry the secretions into a body cavity, into the lumen of an organ, or onto the outer surface of the body. Sweat glands are one example of exocrine glands. The cells of *endocrine glands*, by contrast, secrete their products (hormones) into *interstitial fluid*, the fluid that surrounds tissue cells. Then, the hormones diffuse into blood capillaries, and blood carries them throughout the body.

The endocrine glands include the pituitary, thyroid, parathyroid, adrenal, and pineal glands (Figure 13.1). In addition, several organs and tissues are not exclusively classified as endocrine glands but contain cells that secrete hormones. These include the hypothalamus, thymus, pancreas, ovaries, testes, kidneys, stomach, liver, small intestine, skin, heart,

Figure 13.1 Location of endocrine glands and other organs that contain endocrine cells. Some nearby structures are shown for orientation (trachea, lungs, scrotum, and uterus).

Endocrine glands secrete hormones, which circulating blood delivers to target tissues.

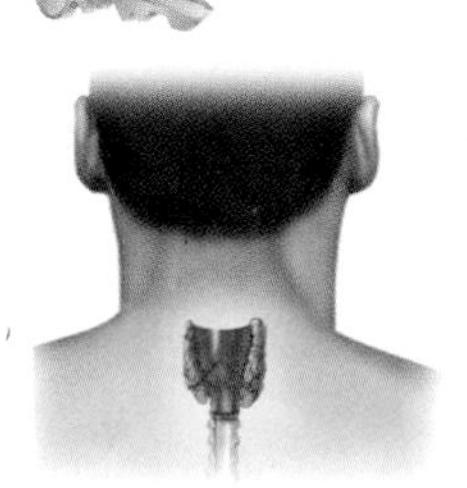

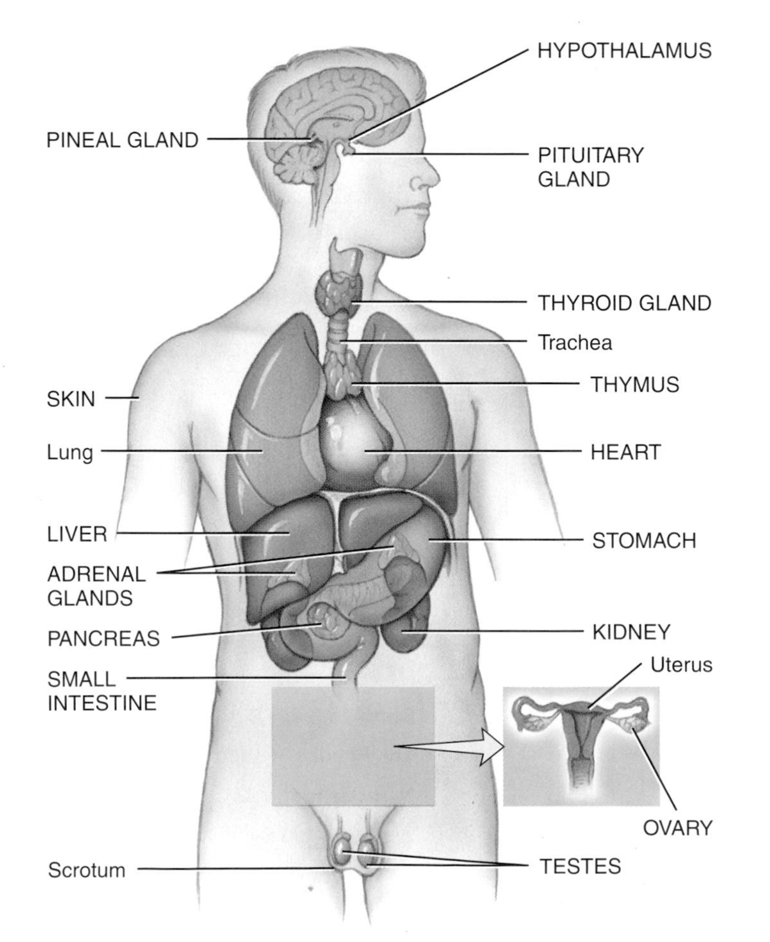

Functions of Hormones

1. Help regulate:
   - Chemical composition and volume of interstitial fluid
   - Metabolism and energy balance
   - Contraction of smooth and cardiac muscle fibers
   - Glandular secretions
   - Some immune system activities
2. Control growth and development.
3. Regulate operation of reproductive systems.
4. Help establish circadian rhythms.

What is the basic difference between endocrine glands and exocrine glands?

adipose tissue, and placenta. ***Endocrinology*** (en′-dō-kri-NOL-ō-jē; *endo-* = within; *-crino* = to secrete; *-logy* = study of) is the scientific and medical specialty concerned with hormonal secretions and the diagnosis and treatment of disorders of the endocrine system.

■ **CHECKPOINT**

1. Why are organs such as the kidneys, stomach, heart, and skin considered part of the endocrine system?

HORMONE ACTION

OBJECTIVES • Define target cells and describe the role of hormone receptors.

• Describe the two general mechanisms of action of hormones.

Target Cells and Hormone Receptors

Although a given hormone travels throughout the body in the blood, it affects only specific ***target cells***. Hormones, like neurotransmitters, influence their target cells by chemically binding to specific protein ***receptors***. Only the target cells for a given hormone have receptors that bind and recognize that hormone. For example, thyroid-stimulating hormone (TSH) binds to receptors on cells of the thyroid gland, but it does not bind to cells of the ovaries because ovarian cells do not have TSH receptors. Generally, a target cell has 2000 to 100,000 receptors for a particular hormone.

The drug **RU486 (mifepristone)** can be used to induce an abortion. It binds to the receptors for progesterone (a female sex hormone) and prevents progesterone from exerting its normal effects. When RU486 is given to a pregnant woman, the conditions needed for embryonic development are lost, and the embryo is sloughed off along with the lining of the uterus. This example illustrates an important principle: If a hormone is prevented from interacting with its receptors, the hormone cannot perform its normal functions.

Chemistry of Hormones

Chemically, some hormones are soluble in lipids (fats) and others are soluble in water. The lipid-soluble hormones include steroid hormones, thyroid hormones, and nitric oxide. ***Steroid hormones*** are made from cholesterol. The two ***thyroid hormones*** (T_3 and T_4) are made by attaching iodine atoms to the amino acid tyrosine. The gas ***nitric oxide (NO)*** functions as both a hormone and a neurotransmitter.

Most of the water-soluble hormones are made from amino acids. For instance, the amino acid tyrosine is modified to form the hormones epinephrine and norepinephrine (which are also neurotransmitters). Other water-soluble hormones consist of short chains of amino acids (peptide hormones), such as antidiuretic hormone (ADH) and oxytocin, or longer chains of amino acids (protein hormones), for instance, insulin and human growth hormone.

Mechanisms of Hormone Action

The response to a hormone depends on both the hormone and the target cell. Various target cells respond differently to the same hormone. Insulin, for example, stimulates the synthesis of glycogen in liver cells but the synthesis of triglycerides in adipose cells. To exert an effect, a hormone first must "announce its arrival" to a target cell by binding to its receptors. The receptors for lipid-soluble hormones are located inside target cells, and the receptors for water-soluble hormones are part of the plasma membrane of target cells.

Action of Lipid-Soluble Hormones

Lipid-soluble hormones diffuse through the lipid bilayer of the plasma membrane and bind to their receptors *within* target cells. They exert their effects in the following way (Figure 13.2):

Figure 13.2 Mechanism of action of lipid-soluble hormones.

Lipid-soluble hormones bind to their receptors inside target cells.

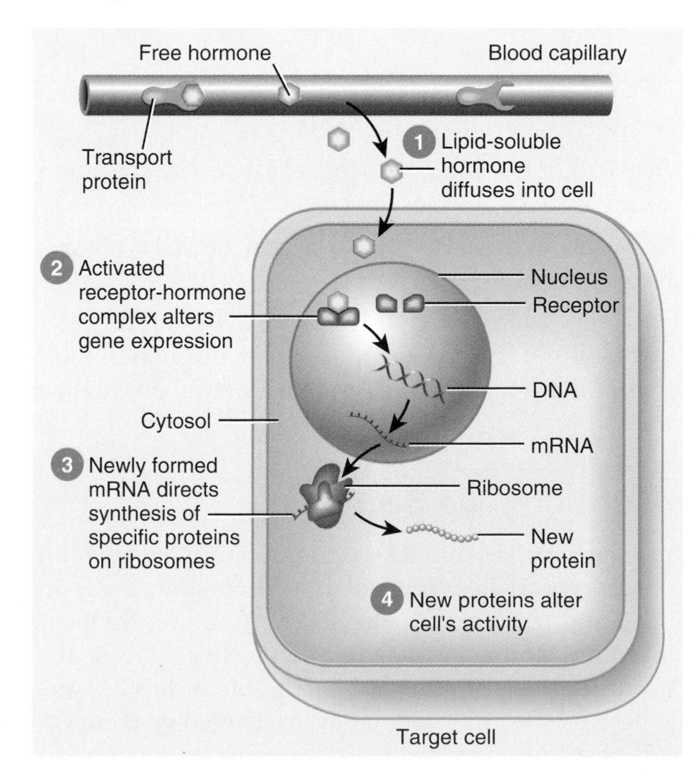

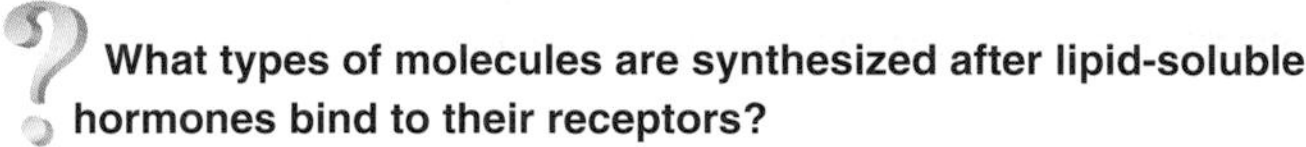

What types of molecules are synthesized after lipid-soluble hormones bind to their receptors?

1. A lipid-soluble hormone detaches from its transport protein in the bloodstream. Then, the free hormone diffuses from blood into interstitial fluid, and through the plasma membrane into a cell.
2. The hormone binds to and activates receptors within the cell. The activated receptor–hormone complex then alters gene expression: It turns specific genes on or off.
3. As the DNA is transcribed, new messenger RNA (mRNA) forms, leaves the nucleus, and enters the cytosol. There it directs synthesis of a new protein, often an enzyme, on the ribosomes.
4. The new proteins alter the cell's activity and cause the responses typical of that specific hormone.

Action of Water-soluble Hormones

Because most amino acid–based hormones are not lipid-soluble, they cannot diffuse through the lipid bilayer of the plasma membrane. Instead, water-soluble hormones bind to receptors that protrude from the target cell surface. When a water-soluble hormone binds to its receptor at the outer surface of the plasma membrane, it acts as the ***first messenger***. The first messenger (the hormone) then causes production of a ***second messenger*** inside the cell, where specific hormone-stimulated responses take place. One common second messenger is ***cyclic AMP (cAMP)***, which is synthesized from ATP.

Water-soluble hormones exert their effects as follows (Figure 13.3):

1. A water-soluble hormone (the first messenger) diffuses from the blood and binds to its receptor in a target cell's plasma membrane.
2. As a result of the binding, a reaction starts inside the cell that converts ATP into cyclic AMP.
3. Cyclic AMP (the second messenger) causes the activation of several proteins (such as enzymes).
4. Activated proteins cause reactions that produce physiological responses.
5. After a brief time, cyclic AMP is inactivated. Thus, the cell's response is turned off unless new hormone molecules continue to bind to their receptors in the plasma membrane.

Figure 13.3 Mechanism of action of water-soluble hormones.

Water-soluble hormones bind to receptors embedded in the plasma membrane of target cells.

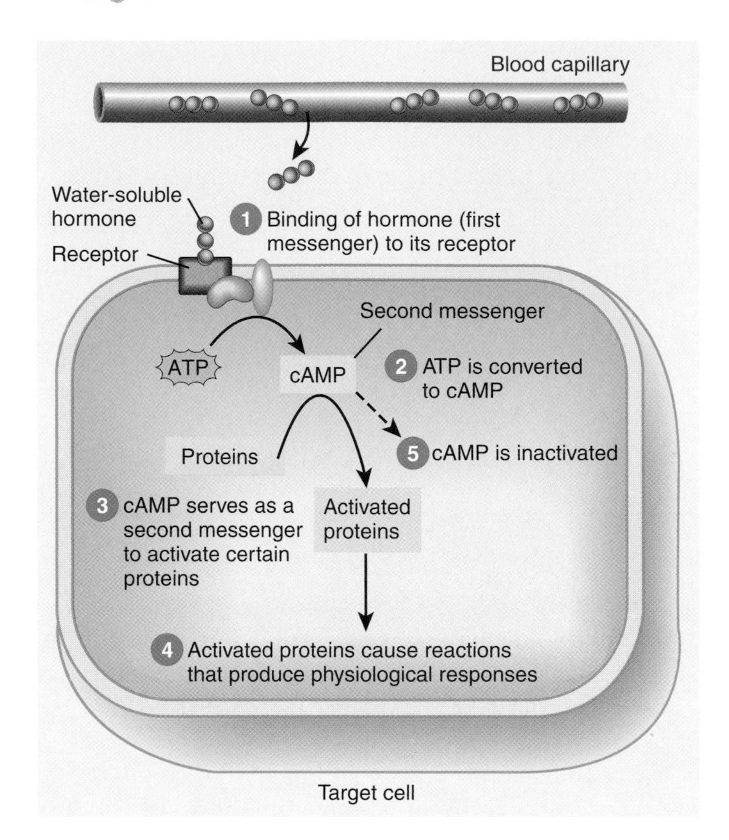

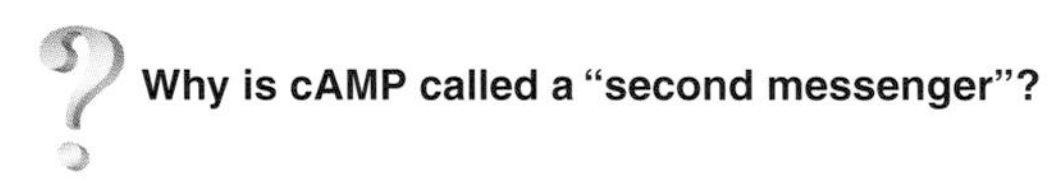

Control of Hormone Secretions

The release of most hormones occurs in short bursts, with little or no secretion between bursts. When stimulated, an endocrine gland releases its hormone in more frequent bursts, increasing the concentration of the hormone in the blood. In the absence of stimulation, the blood level of the hormone decreases as the hormone is inactivated or excreted. Regulation of secretion normally prevents overproduction or underproduction of any given hormone.

Hormone secretion is regulated by (1) signals from the nervous system, (2) chemical changes in the blood, and (3) other hormones. For example, nerve impulses to the adrenal medullae regulate the release of epinephrine and norepinephrine; blood Ca^{2+} level regulates the secretion of parathyroid hormone; and a hormone from the anterior pituitary (ACTH) stimulates the release of cortisol by the adrenal cortex. Most systems that regulate secretion of hormones work by negative feedback, but a few operate by positive feedback. For example, during childbirth, the hormone oxytocin stimulates contractions of the uterus, and uterine contractions, in turn, stimulate more oxytocin release, a positive feedback effect.

■ **CHECKPOINT**

2. Chemically, what types of molecules are hormones?
3. What are the general ways in which blood hormone levels are regulated?

HYPOTHALAMUS AND PITUITARY GLAND

OBJECTIVES • **Describe the locations of and relationship between the hypothalamus and the pituitary gland.**
• **Describe the functions of each hormone secreted by the pituitary gland.**

For many years, the ***pituitary gland*** (pi-TOO-i-tār-ē) or *hypophysis* (hī-POF-i-sis) was considered the "master" endocrine gland because it secretes several hormones that control other endocrine glands. We now know that the pituitary gland itself has a master—the ***hypothalamus***. This small region of the brain is the major link between the nervous and endocrine systems. Cells in the hypothalamus synthesize at least nine hormones, and the pituitary gland secretes seven. Together, these hormones play important roles in the regulation of virtually all aspects of growth, development, metabolism, and homeostasis.

The pituitary gland is about the size of a small grape and has two lobes: a larger ***anterior pituitary*** or *anterior lobe* and a smaller ***posterior pituitary*** or *posterior lobe* (Figure 13.4). Both lobes of the pituitary gland rest in the *hypophyseal fossa*, a cup-shaped depression in the sphenoid bone. A stalklike structure,

Figure 13.4 The pituitary gland and its blood supply. As shown in the inset, to the right, releasing and inhibiting hormones synthesized by hypothalamic neurosecretory cells diffuse into capillaries of the hypothalamus and are carried by the hypophyseal portal veins to the anterior pituitary.

Hypothalamic releasing and inhibiting hormones are an important link between the nervous and endocrine systems.

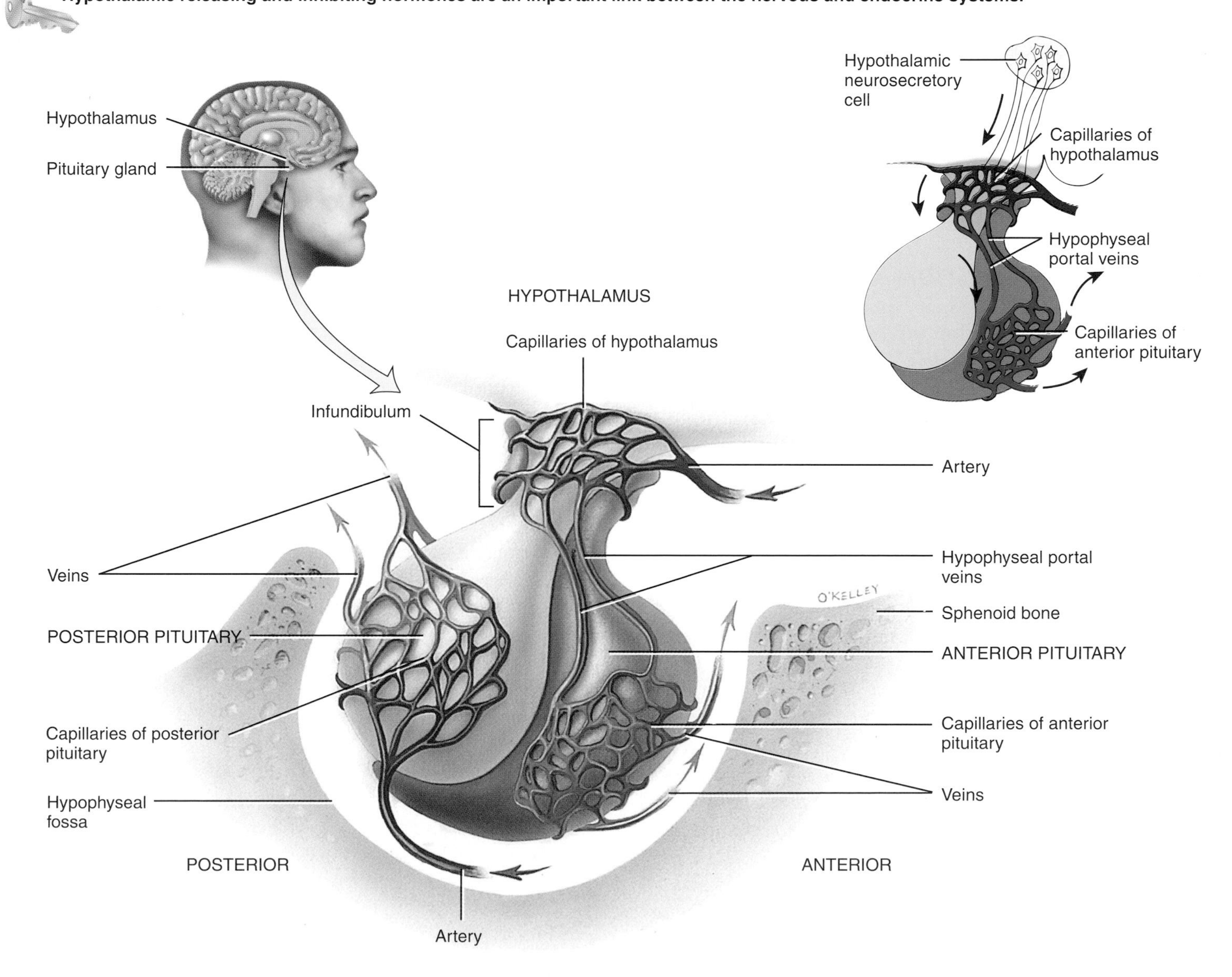

Which lobe of the pituitary gland does not synthesize the hormones it releases? Where are its hormones produced?

the ***infundibulum***, attaches the pituitary gland to the hypothalamus. Within the infundibulum, blood vessels termed ***hypophyseal portal veins*** (hī′-pō-FIZ-ē-al) connect capillaries in the hypothalamus to capillaries in the anterior pituitary. Axons of hypothalamic neurons called ***neurosecretory cells*** end near the capillaries of the hypothalamus (inset on right in Figure 13.4), where they release several hormones into the blood.

Anterior Pituitary Hormones

The anterior pituitary synthesizes and secretes hormones that regulate a wide range of bodily activities, from growth to reproduction. Secretion of anterior pituitary hormones is stimulated by ***releasing hormones*** and suppressed by ***inhibiting hormones***, both produced by neurosecretory cells of the hypothalamus. The hypophyseal portal veins deliver the hypothalamic releasing and inhibiting hormones from the hypothalamus to the anterior pituitary (Figure 13.4). This direct route allows the releasing and inhibiting hormones to act quickly on cells of the anterior pituitary before the hormones are diluted or destroyed in the general circulation. Those anterior pituitary hormones that act on other endocrine glands are called ***tropic hormones*** (TRŌ-pik) or ***tropins***.

Human Growth Hormone and Insulinlike Growth Factors

Human growth hormone (hGH) is the most abundant anterior pituitary hormone. The main function of hGH is to promote synthesis and secretion of small protein hormones called ***insulinlike growth factors (IGFs)*** or ***somatomedins***. IGFs are so named because some of their actions are similar to those of insulin. In response to hGH, cells in the liver, skeletal muscles, cartilage, bones, and other tissues secrete IGFs, which may either enter the bloodstream or act locally. IGFs stimulate protein synthesis, help maintain muscle and bone mass, and promote healing of injuries and tissue repair. They also enhance breakdown of triglycerides (fats), which releases fatty acids into the blood, and breakdown of liver glycogen, which releases glucose into the blood. Cells throughout the body can use the released fatty acids and glucose for the production of ATP.

The anterior pituitary releases hGH in bursts that occur every few hours, especially during sleep. Two hypothalamic hormones control secretion of hGH: *Growth hormone-releasing hormone (GHRH)* promotes secretion of human growth hormone, and *growth hormone-inhibiting hormone (GHIH)* suppresses it. Blood glucose level is a major regulator of GHRH and GHIH secretion. Low blood glucose level (hypoglycemia) stimulates the hypothalamus to secrete GHRH. By means of negative feedback, an increase in blood glucose concentration above the normal level (hyperglycemia) inhibits release of GHRH. By contrast, hyperglycemia stimulates the hypothalamus to secrete GHIH and hypoglycemia inhibits release of GHIH.

Thyroid-Stimulating Hormone

Thyroid-stimulating hormone (TSH) stimulates the synthesis and secretion of thyroid hormones by the thyroid gland. *Thyrotropin-releasing hormone (TRH)* from the hypothalamus controls TSH secretion. Release of TRH, in turn, depends on blood levels of thyroid hormones, which inhibit secretion of TRH via negative feedback. There is no thyrotropin-inhibiting hormone.

Follicle-Stimulating Hormone and Luteinizing Hormone

In females, the ovaries are the targets for ***follicle-stimulating hormone (FSH)*** and ***luteinizing hormone (LH)***. Each month FSH initiates the development of several ovarian follicles and LH triggers ovulation (described in Chapter 23). After ovulation, LH stimulates formation of the corpus luteum in the ovary and the secretion of progesterone (another female sex hormone) by the corpus luteum. FSH and LH also stimulate follicular cells to secrete estrogens. In males, FSH stimulates sperm production in the testes, and LH stimulates the testes to secrete testosterone. *Gonadotropin-releasing hormone (GnRH)* from the hypothalamus stimulates release of FSH and LH. The release of GnRH, FSH, and LH is suppressed by estrogens in females and by testosterone in males through negative feedback systems. There is no gonadotropin-inhibiting hormone.

Prolactin

Prolactin (PRL), together with other hormones, initiates and maintains milk production by the mammary glands. Ejection of milk from the mammary glands depends on the hormone oxytocin, which is released from the posterior pituitary. The function of prolactin is unknown in males, but prolactin hypersecretion causes erectile dysfunction (impotence, the inability to have an erection of the penis). In females, *prolactin-inhibiting hormone (PIH)* suppresses release of prolactin most of the time. Each month, just before menstruation begins, the secretion of PIH diminishes and the blood level of prolactin rises, but not enough to stimulate milk production. As the menstrual cycle begins anew, PIH is again secreted and the prolactin level drops. During pregnancy, very high levels of estrogens promote secretion of *prolactin-releasing hormone (PRH)*, which in turn stimulates release of prolactin.

Adrenocorticotropic Hormone

Adrenocorticotropic hormone (ACTH) or *corticotropin* controls the production and secretion of hormones called glucocorticoids by the cortex (outer portion) of the adrenal glands. Corticotropin-releasing hormone (CRH) from the hypothalamus stimulates secretion of ACTH. Stress-related stimuli, such as low blood glucose or physical trauma, and interleukin-1, a substance produced by macrophages, also stimulate release of ACTH. Glucocorticoids cause negative feedback inhibition of both CRH and ACTH release.

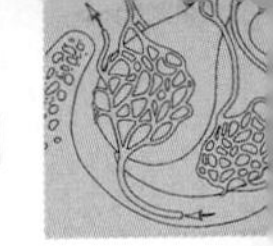

Melanocyte-Stimulating Hormone

There is little circulating melanocyte-stimulating hormone (MSH) in humans. Although an excessive amount of MSH causes darkening of the skin, the function of normal levels of MSH is unknown. The presence of MSH receptors in the brain suggests it may influence brain activity. Excessive corticotropin-releasing hormone (CRH) can stimulate MSH release, and dopamine inhibits MSH release.

Posterior Pituitary Hormones

The ***posterior pituitary*** contains the axons and axon terminals of more than 10,000 neurosecretory cells whose cell bodies are in the hypothalamus (Figure 13.5). Although the posterior pituitary does not *synthesize* hormones, it does *store* and *release* two hormones. In the hypothalamus, the hormones ***oxytocin*** (ok′-sē-TŌ-sin; *oxytoc-* = quick birth) and ***antidiuretic hormone (ADH)*** are synthesized and packaged

Figure 13.5 Hypothalamic neurosecretory cells synthesize oxytocin and antidiuretic hormone. Their axons extend from the hypothalamus to the posterior pituitary. Nerve impulses trigger release of the hormones from vesicles in the axon terminals in the posterior pituitary.

Oxytocin and antidiuretic hormone are synthesized in the hypothalamus and released into capillaries of the posterior pituitary.

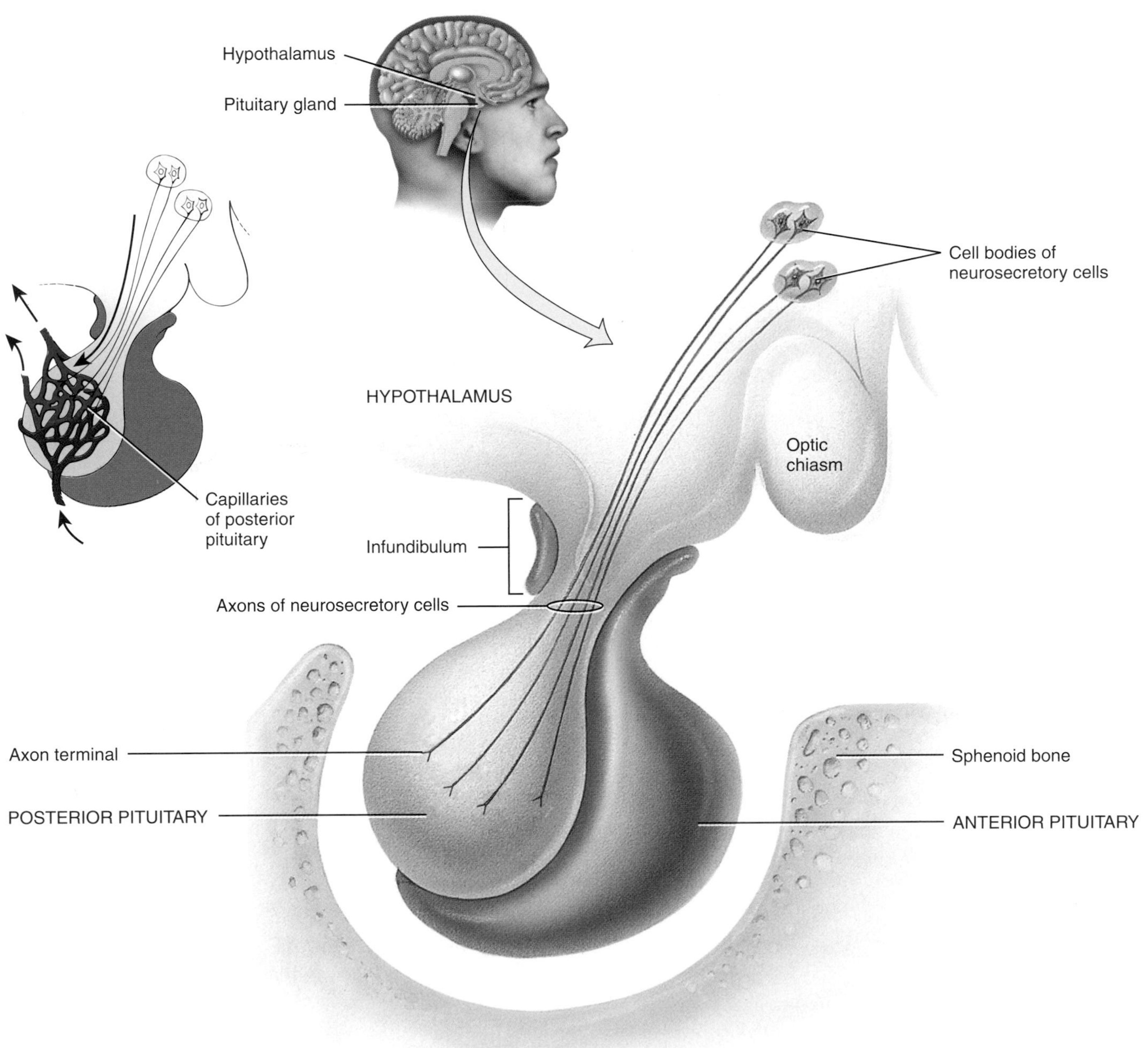

Where are the target cells of oxytocin located?

into secretory vesicles within the cell bodies of different neurosecretory cells. The vesicles then move down the axons to the axon terminals in the posterior pituitary. Nerve impulses that arrive at the axon terminals trigger release of these hormones into the capillaries of the posterior pituitary.

Oxytocin

During and after delivery of a baby, oxytocin has two target tissues: the mother's uterus and breasts. During delivery, oxytocin enhances contraction of smooth muscle cells in the wall of the uterus; after delivery, it stimulates milk ejection ("letdown") from the mammary glands in response to the mechanical stimulus provided by a suckling infant. Together, milk production and ejection constitute *lactation*. The function of oxytocin in males and in nonpregnant females is not clear. Experiments with animals have suggested actions within the brain that foster parental caretaking behavior toward young offspring. Oxytocin also may be partly responsible for the feelings of sexual pleasure during and after intercourse.

Years before oxytocin was discovered, midwives commonly let a first-born twin nurse at the mother's breast to speed the birth of the second child. Now we know why this practice is helpful—it stimulates release of oxytocin. Even after a single birth, nursing promotes expulsion of the placenta (afterbirth) and helps the uterus regain its smaller size. **Synthetic oxytocin (Pitocin®)** often is given to induce labor or to increase uterine tone and control hemorrhage just after giving birth.

Antidiuretic Hormone

An *antidiuretic* (*anti-* = against; *diuretic* = urine-producing agent) is a substance that decreases urine production. ***Antidiuretic hormone (ADH)*** causes the kidneys to retain more water, thus decreasing urine volume. In the absence of ADH, urine output increases more than tenfold, from the normal 1–2 liters to about 20 liters a day. ADH also decreases the water lost through sweating and causes constriction of arterioles. This hormone's other name, ***vasopressin*** (*vaso-* = vessel; *pressin-* = pressing or constricting), reflects its effect on increasing blood pressure.

The amount of ADH secreted varies with blood osmotic pressure and blood volume. Blood osmotic pressure is proportional to the concentration of solutes in the blood plasma. When body water is lost faster than it is taken in, a condition termed *dehydration*, the blood volume falls and blood osmotic pressure rises. Figure 13.6 shows regulation of ADH secretion and the actions of ADH on its target tissues.

1. High blood osmotic pressure—due to dehydration or a drop in blood volume because of hemorrhage, diarrhea, or excessive sweating—stimulates ***osmoreceptors***, neurons in the hypothalamus that monitor blood osmotic pressure.

Figure 13.6 Regulation of secretion and actions of antidiuretic hormone (ADH).

ADH acts to retain body water and increase blood pressure.

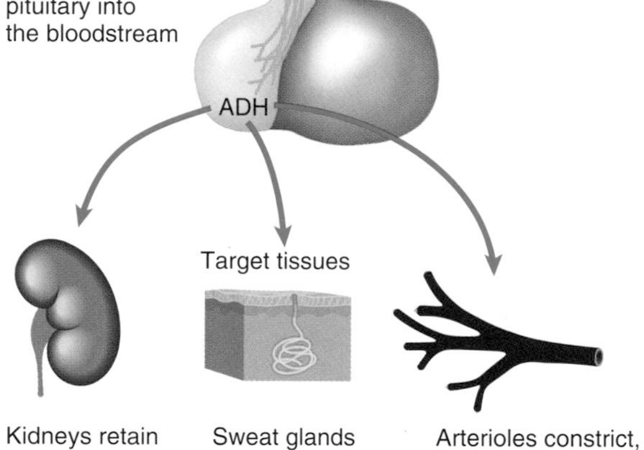

What effect would drinking a large glass of water have on the osmotic pressure of your blood, and how would the level of ADH change in your blood?

2. Osmoreceptors activate the hypothalamic neurosecretory cells that synthesize and release ADH.
3. When neurosecretory cells receive excitatory input from the osmoreceptors, they generate nerve impulses that cause the release of ADH in the posterior pituitary. The ADH then diffuses into blood capillaries of the posterior pituitary.
4. The blood carries ADH to three target tissues: the kidneys, sweat glands, and smooth muscle in blood vessel walls. The kidneys respond by retaining more water, which decreases urine output. Secretory activity of sweat glands decreases, which lowers the rate of water loss by

perspiration from the skin. Smooth muscle in the walls of arterioles (small arteries) contracts in response to high levels of ADH, which constricts (narrows) the lumen of these blood vessels and increases blood pressure.

5. Low blood osmotic pressure or increased blood volume inhibits the osmoreceptors.
6. Inhibition of osmoreceptors reduces or stops ADH secretion. The kidneys then retain less water by forming a larger volume of urine, secretory activity of sweat glands increases, and arterioles dilate. The blood volume and osmotic pressure of body fluids return to normal.

Secretion of ADH can also be altered in other ways. Pain, stress, trauma, anxiety, acetylcholine, nicotine, and drugs such as morphine, tranquilizers, and some anesthetics stimulate ADH secretion. Alcohol inhibits ADH secretion, thereby increasing urine output. The resulting dehydration may cause both the thirst and the headache typical of a hangover.

Table 13.1 lists the pituitary gland hormones and summarizes their actions.

■ **CHECKPOINT**

4. In what respect is the pituitary gland actually two glands?
5. How do hypothalamic releasing and inhibiting hormones influence secretions of anterior pituitary hormones?

THYROID GLAND

OBJECTIVE • **Describe the location, hormones, and functions of the thyroid gland.**

The butterfly-shaped ***thyroid gland*** is located just below the larynx (voice box). It is composed of right and left lobes, one on either side of the trachea (Figure 13.7a).

Microscopic spherical sacs called ***thyroid follicles*** (Figure 13.7b) make up most of the thyroid gland. The wall of each thyroid follicle consists primarily of cells called ***follicular cells***, which produce two hormones: ***thyroxine*** (thī-ROK-sēn), also called ***T_4*** because it contains four atoms of iodine, and ***triiodothyronine*** (trī-ī′-ō-dō-THĪ-rō-nēn) (***T_3***), which contains three atoms of iodine. T_3 and T_4 are also known as ***thyroid hormones***. The central cavity of each thyroid follicle contains stored thyroid hormones. As T_4 circulates in the blood and enters cells throughout the body, most of it is converted to T_3 by removal of one iodine atom.

A smaller number of cells called ***parafollicular cells*** lie between the follicles (Figure 13.7b). They produce the hormone calcitonin.

Actions of Thyroid Hormones

Because most body cells have receptors for thyroid hormones, T_3 and T_4 exert their effects throughout the body.

Table 13.1 Summary of Pituitary Gland Hormones and Their Actions

Hormone	Actions
Anterior Pituitary Hormones	
Human growth hormone (hGH)	Stimulates liver, muscle, cartilage, bone, and other tissues to synthesize and secrete insulinlike growth factors (IGFs). IGFs promote growth of body cells, protein synthesis, tissue repair, breakdown of triglycerides, and elevation of blood glucose level.
Thyroid-stimulating hormone (TSH)	Stimulates synthesis and secretion of thyroid hormones by the thyroid gland.
Follicle-stimulating hormone (FSH)	In females, initiates development of oocytes and induces secretion of estrogens by the ovaries. In males, stimulates testes to produce sperm.
Luteinizing hormone (LH)	In females, stimulates secretion of estrogens and progesterone, ovulation, and formation of corpus luteum. In males, stimulates testes to produce testosterone.
Prolactin (PRL)	In females, stimulates milk production by the mammary glands.
Adrenocorticotropic hormone (ACTH), also known as **corticotropin**	Stimulates secretion of glucocorticoids (mainly cortisol) by the adrenal cortex.
Melanocyte-stimulating hormone (MSH)	Exact role in humans is unknown but may influence brain activity. When present in excess, can cause darkening of skin.
Posterior Pituitary Hormones	
Oxytocin	Stimulates contraction of smooth muscle cells of uterus during childbirth. Stimulates milk ejection from the mammary glands.
Antidiuretic hormone (ADH), also known as **vasopressin**	Conserves body water by decreasing urine output. Decreases water loss through sweating. Raises blood pressure by constricting (narrowing) arterioles.

Figure 13.7 **Location and histology of the thyroid gland.**

Thyroid hormones regulate (1) oxygen use and basal metabolic rate, (2) cellular metabolism, and (3) growth and development.

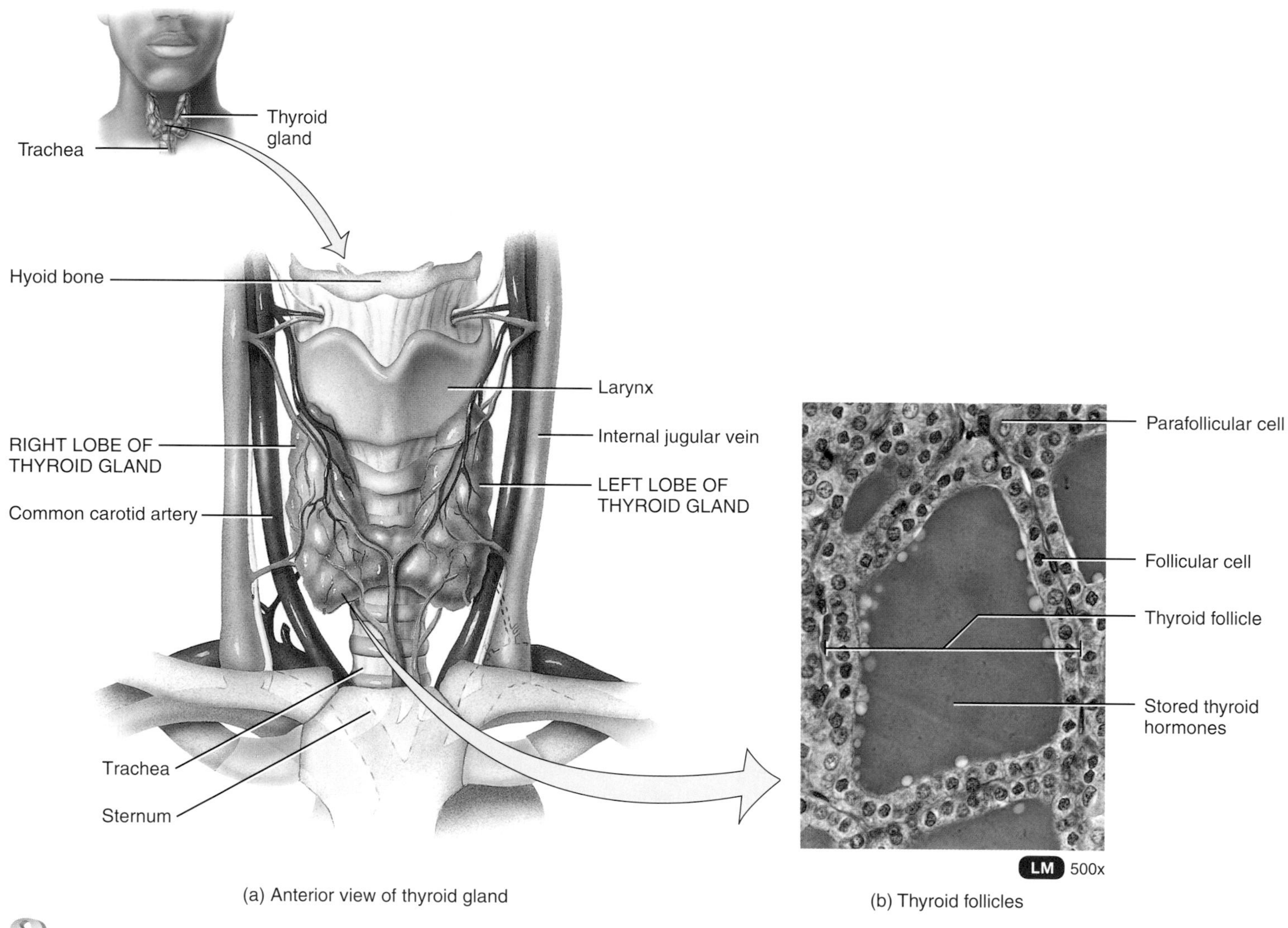

(a) Anterior view of thyroid gland

(b) Thyroid follicles

Which cells secrete T_3 and T_4? Which cells secrete calcitonin?

Thyroid hormones increase *basal metabolic rate (BMR)*, the rate of oxygen consumption under standard or basal conditions (awake, at rest, and fasting). The BMR rises due to increased synthesis and use of ATP. As cells use more oxygen to produce the ATP, more heat is given off, and body temperature rises. In this way, thyroid hormones play an important role in the maintenance of normal body temperature. The thyroid hormones also stimulate protein synthesis, increase the use of glucose and fatty acids for ATP production, increase the breakdown of triglycerides, and enhance cholesterol excretion, thus reducing blood cholesterol level. Together with human growth hormone and insulin, thyroid hormones stimulate body growth, particularly the growth of the nervous and skeletal systems.

Excess secretion of thyroid hormones is known as **hyperthyroidism**. Symptoms of hyperthyroidism include increased heart rate and more forceful heartbeats, increased blood pressure, and increased nervousness.

Control of Thyroid Hormone Secretion

Thyrotropin-releasing hormone (TRH) from the hypothalamus and thyroid-stimulating hormone (TSH) from the anterior pituitary stimulate synthesis and release of thyroid hormones, as shown in Figure 13.8:

1. Low blood level of thyroid hormones or low metabolic rate stimulate the hypothalamus to secrete TRH.
2. TRH is carried to the anterior pituitary, where it stimulates secretion of thyroid-stimulating hormone (TSH).
3. TSH stimulates thyroid follicular cell activity, including thyroid hormone synthesis and secretion, and growth of the follicular cells.
4. The thyroid follicular cells release thyroid hormones into the blood until the metabolic rate returns to normal.
5. An elevated level of thyroid hormones inhibits release of TRH and TSH (negative feedback).

Conditions that increase ATP demand—a cold environment, low blood glucose, high altitude, and pregnancy—also increase secretion of the thyroid hormones.

Calcitonin

The hormone produced by the parafollicular cells of the thyroid gland is ***calcitonin (CT)*** (kal-si-TŌ-nin). Calcitonin can decrease the level of calcium in the blood by inhibiting the action of osteoclasts, the cells that break down bone. The secretion of calcitonin is controlled by a negative feedback system (see Figure 13.10). Calcitonin's importance in normal physiology is unclear because it can be present in excess or completely absent without causing clinical symptoms.

Figure 13.8 Regulation of secretion of thyroid hormones.

TSH promotes release of thyroid hormones.

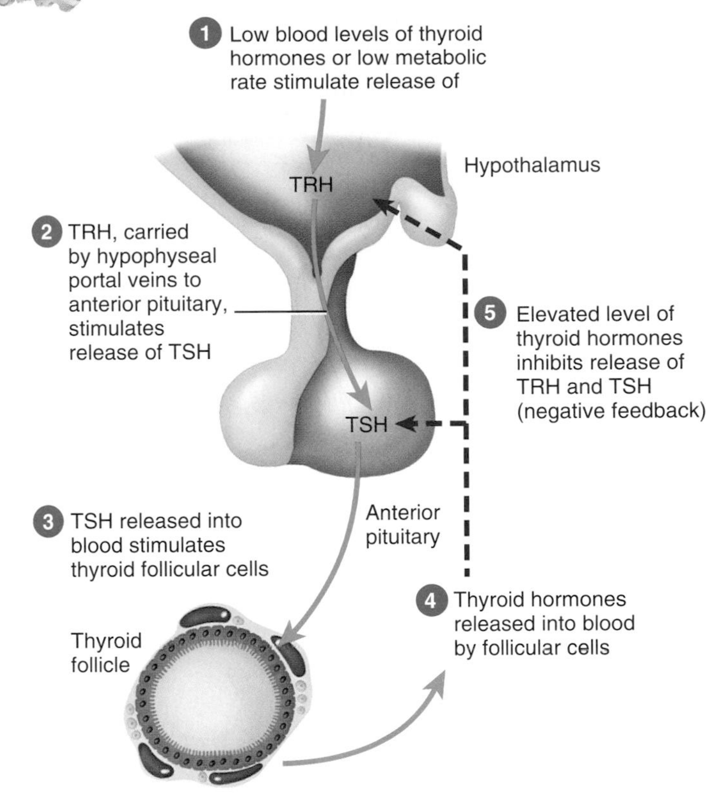

What is the effect of thyroid hormones on metabolic rate?

Miacalcin®, a calcitonin extract from salmon, is an effective treatment for osteoporosis, a disorder in which the pace of bone breakdown exceeds the pace of bone rebuilding. It inhibits breakdown of bone and accelerates uptake of calcium and phosphates into bone.

CHECKPOINT

6. How is the secretion of T_3 and T_4 regulated?
7. What are the actions of the thyroid hormones and calcitonin?

PARATHYROID GLANDS

OBJECTIVE • Describe the location, hormones, and functions of the parathyroid glands.

The ***parathyroid glands*** (*para-* = beside) are small, round masses of glandular tissue that are partially embedded in the posterior surface of the thyroid gland (Figure 13.9). Usually, one superior and one inferior parathyroid gland are attached to each thyroid lobe. Within the parathyroid glands are secretory cells called ***chief cells*** that release ***parathyroid hormone (PTH)***.

PTH is the major regulator of the levels of calcium (Ca^{2+}), magnesium (Mg^{2+}), and phosphate (HPO_4^{2-}) ions in the blood. PTH increases the number and activity of osteoclasts, which break down bone extracellular matrix and release Ca^{2+} and HPO_4^{2-} into the blood. PTH also produces three changes in the kidneys. First, it slows the rate at which Ca^{2+} and Mg^{2+} are lost from blood into the urine. Second, it increases loss of HPO_4^{2-} from blood in urine. Because more is lost in the urine than is gained from the bones, PTH decreases blood HPO_4^{2-} level and increases blood Ca^{2+} and Mg^{2+} levels. Third, PTH promotes formation of the hormone ***calcitriol***, the active form of vitamin D. Calcitriol acts on the gastrointestinal tract to increase the rate of Ca^{2+}, Mg^{2+}, and HPO_4^{2-} absorption from foods into the blood.

The blood calcium level directly controls the secretion of calcitonin and parathyroid hormone via negative feedback, and the two hormones have opposite effects on blood Ca^{2+} level (Figure 13.10).

1. A higher-than-normal level of calcium ions (Ca^{2+}) in the blood stimulates parafollicular cells of the thyroid gland to release more calcitonin.
2. CT inhibits the activity of osteoclasts, thereby decreasing blood Ca^{2+} level.
3. A lower-than-normal level of Ca^{2+} in the blood stimulates chief cells of the parathyroid gland to release more PTH.
4. PTH increases the number and activity of osteoclasts, which break down bone and release Ca^{2+} into the blood. PTH also slows loss of Ca^{2+} in the urine. Both actions of PTH raise the blood level of Ca^{2+}.
5. PTH also stimulates the kidneys to release calcitriol, the active form of vitamin D.
6. Calcitriol stimulates increased absorption of Ca^{2+} from foods in the gastrointestinal tract, which helps increase the blood level of Ca^{2+}.

Figure 13.9 Location of the parathyroid glands.

The four parathyroid glands are attached to the posterior surface of the thyroid gland.

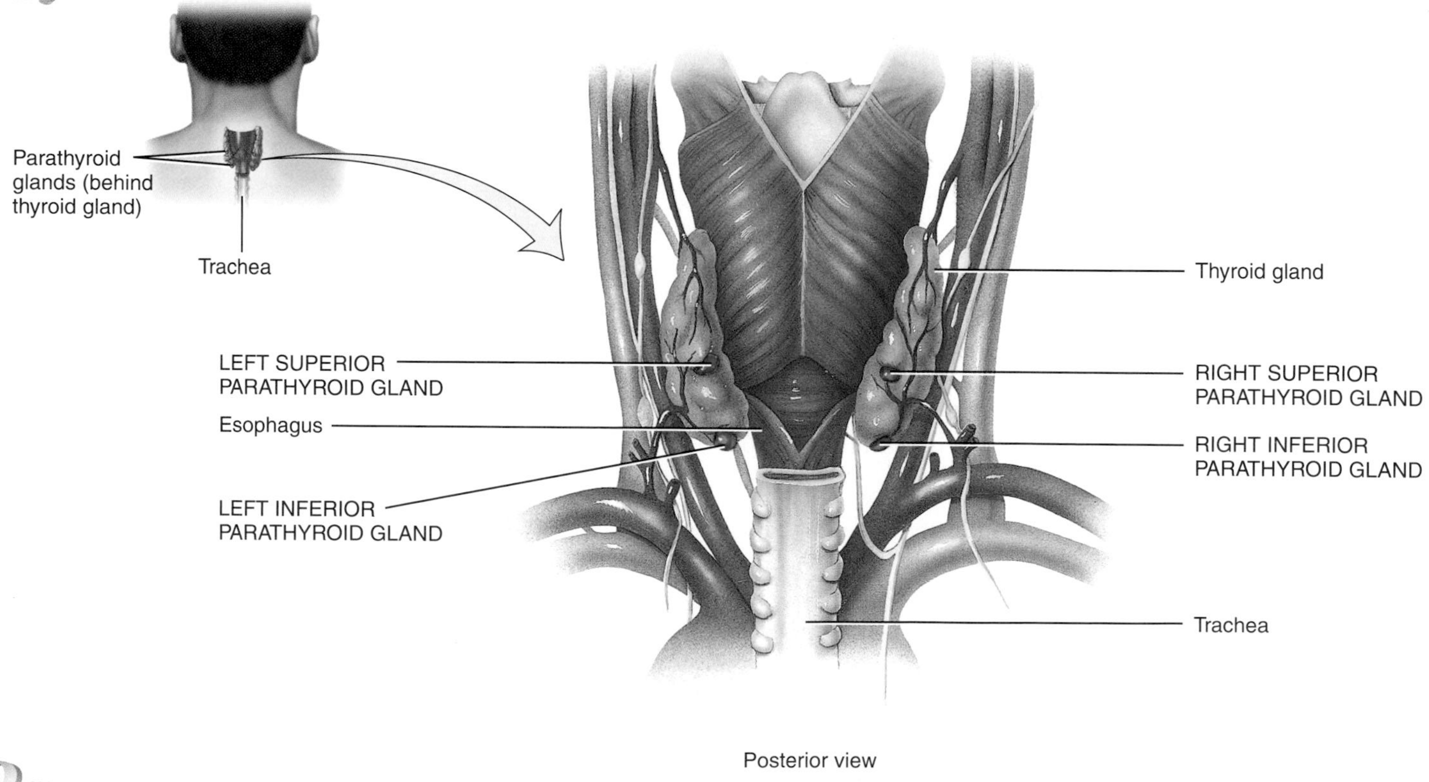

What effect does parathyroid hormone have on osteoclasts?

Figure 13.10 The roles of calcitonin (green arrows), parathyroid hormone (purple arrows), and calcitriol (orange arrows) in homeostasis of blood calcium level.

PTH and calcitonin have opposite effects on the level of calcium ions (Ca^{2+}) in the blood.

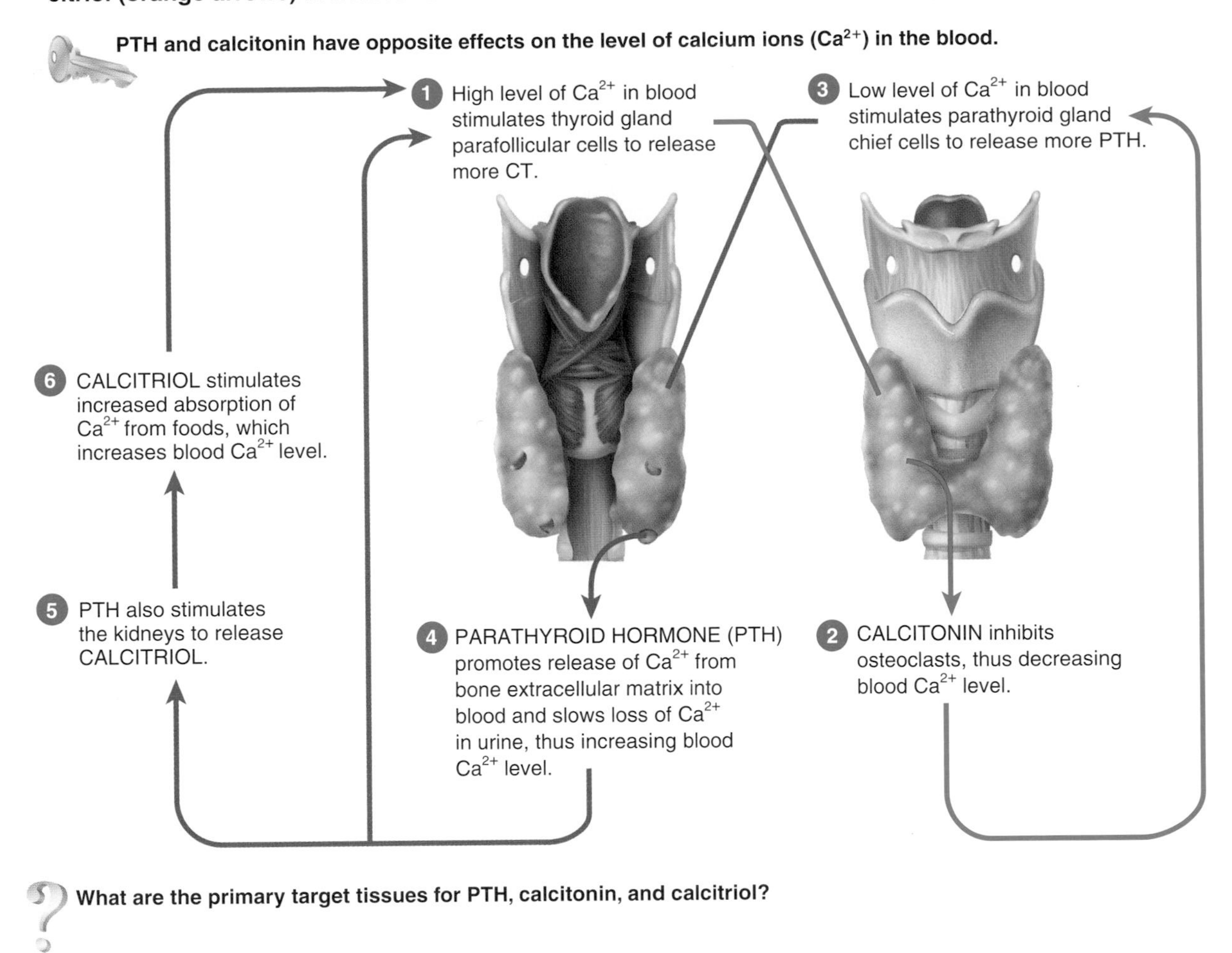

What are the primary target tissues for PTH, calcitonin, and calcitriol?

■ **CHECKPOINT**

8. How is secretion of PTH regulated?
9. In what ways are the actions of PTH and calcitriol similar and different?

PANCREATIC ISLETS

OBJECTIVE • Describe the location, hormones, and functions of the pancreatic islets.

The ***pancreas*** (*pan-* = all; *-creas* = flesh) is a flattened organ located in the curve of the duodenum, the first part of the small intestine (Figure 13.11a). It has both endocrine functions, discussed in this chapter, and exocrine functions, discussed in Chapter 19. The endocrine part of the pancreas consists of clusters of cells called ***pancreatic islets*** or ***islets of Langerhans*** (LAHNG-er-hanz). Some of the islet cells, the *alpha cells*, secrete the hormone ***glucagon*** (GLOO-ka-gon), and other islet cells, the *beta cells*, secrete ***insulin*** (IN-soo-lin). The islets also contain abundant blood capillaries and are surrounded by cells that form the exocrine part of the pancreas (Figure 13.11b, c).

Actions of Glucagon and Insulin

The main action of glucagon is to increase blood glucose level when it falls below normal, which provides neurons with glucose for ATP production. Insulin, by contrast, helps glucose move into cells, especially muscle fibers, which lowers blood glucose level when it is too high. The level of blood glucose controls secretion of both glucagon and insulin via negative feedback. Figure 13.12 on page 329 shows the conditions that stimulate the pancreatic islets to secrete their hormones, the ways in which glucagon and insulin produce their effects on blood glucose level, and the negative feedback control of hormone secretion.

Figure 13.11 Location and histology of the pancreas.

Hormones released by pancreatic islets regulate blood glucose level.

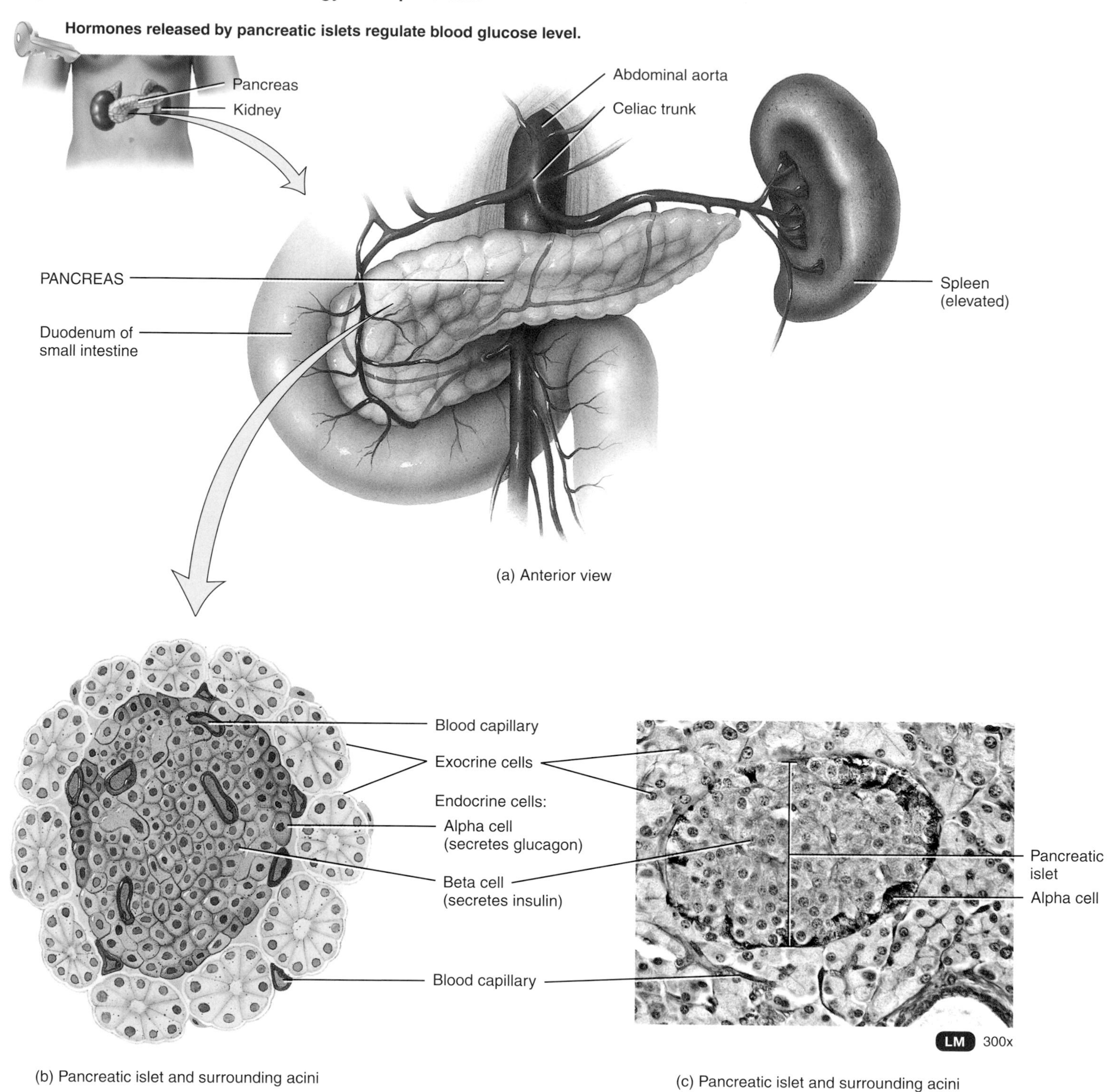

(b) Pancreatic islet and surrounding acini

(c) Pancreatic islet and surrounding acini

Is the pancreas an exocrine gland or an endocrine gland?

Figure 13.12 Regulation of blood glucose level by negative feedback systems involving glucagon (blue arrows) and insulin (orange arrows).

Low blood glucose stimulates secretion of glucagon, and high blood glucose stimulates secretion of insulin.

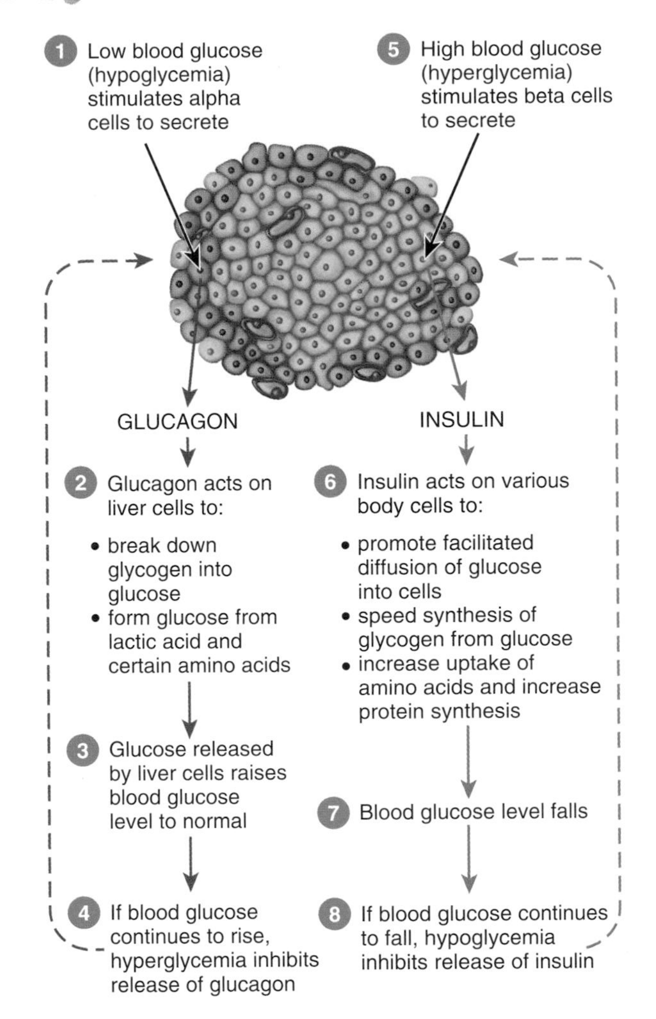

Why is glucagon sometimes called an "anti-insulin" hormone?

1. Low blood glucose level (hypoglycemia) stimulates secretion of glucagon.
2. Glucagon acts on liver cells to promote breakdown of glycogen into glucose and formation of glucose from lactic acid and certain amino acids.
3. As a result, the liver releases glucose into the blood more rapidly, and blood glucose level rises.
4. If blood glucose continues to rise, high blood glucose level (hyperglycemia) inhibits release of glucagon by alpha cells (negative feedback).
5. At the same time, high blood glucose level stimulates secretion of insulin.
6. Insulin acts on various cells in the body to promote facilitated diffusion of glucose into cells, especially skeletal muscle fibers; to speed synthesis of glycogen from glucose; to increase uptake of amino acids by cells; and to increase protein synthesis.
7. As a result, blood glucose level falls.
8. If blood glucose level drops below normal, low blood glucose inhibits release of insulin by beta cells (negative feedback).

In addition to affecting glucose metabolism, insulin promotes the uptake of amino acids into body cells and increases the synthesis of proteins and fatty acids within cells. Therefore, insulin is an important hormone when tissues are developing, growing, or being repaired.

Release of insulin and glucagon is also regulated by the autonomic nervous system (ANS). The parasympathetic division of the ANS stimulates secretion of insulin, for instance, during digestion and absorption of a meal. The sympathetic division of the ANS, by contrast, stimulates secretion of glucagon, as happens during exercise.

■ **CHECKPOINT**

10. What are the functions of insulin?

11. How are blood levels of glucagon and insulin controlled?

ADRENAL GLANDS

OBJECTIVE • Describe the location, hormones, and functions of the adrenal glands.

There are two ***adrenal glands***, one lying atop each kidney (Figure 13.13). Each adrenal gland has regions that produce different hormones: the outer ***adrenal cortex***, which makes up 85% of the gland, and the inner ***adrenal medulla***.

Adrenal Cortex Hormones

The adrenal cortex consists of three zones, each of which synthesizes and secretes different steroid hormones. The outer zone releases hormones called mineralocorticoids because they affect mineral homeostasis. The middle zone releases hormones called glucocorticoids because they affect glucose homeostasis. The inner zone releases androgens (steroid hormones that have masculinizing effects).

Figure 13.13 Location and histology of the adrenal glands.

The adrenal cortex secretes steroid hormones, and the adrenal medulla secretes epinephrine and norepinephrine.

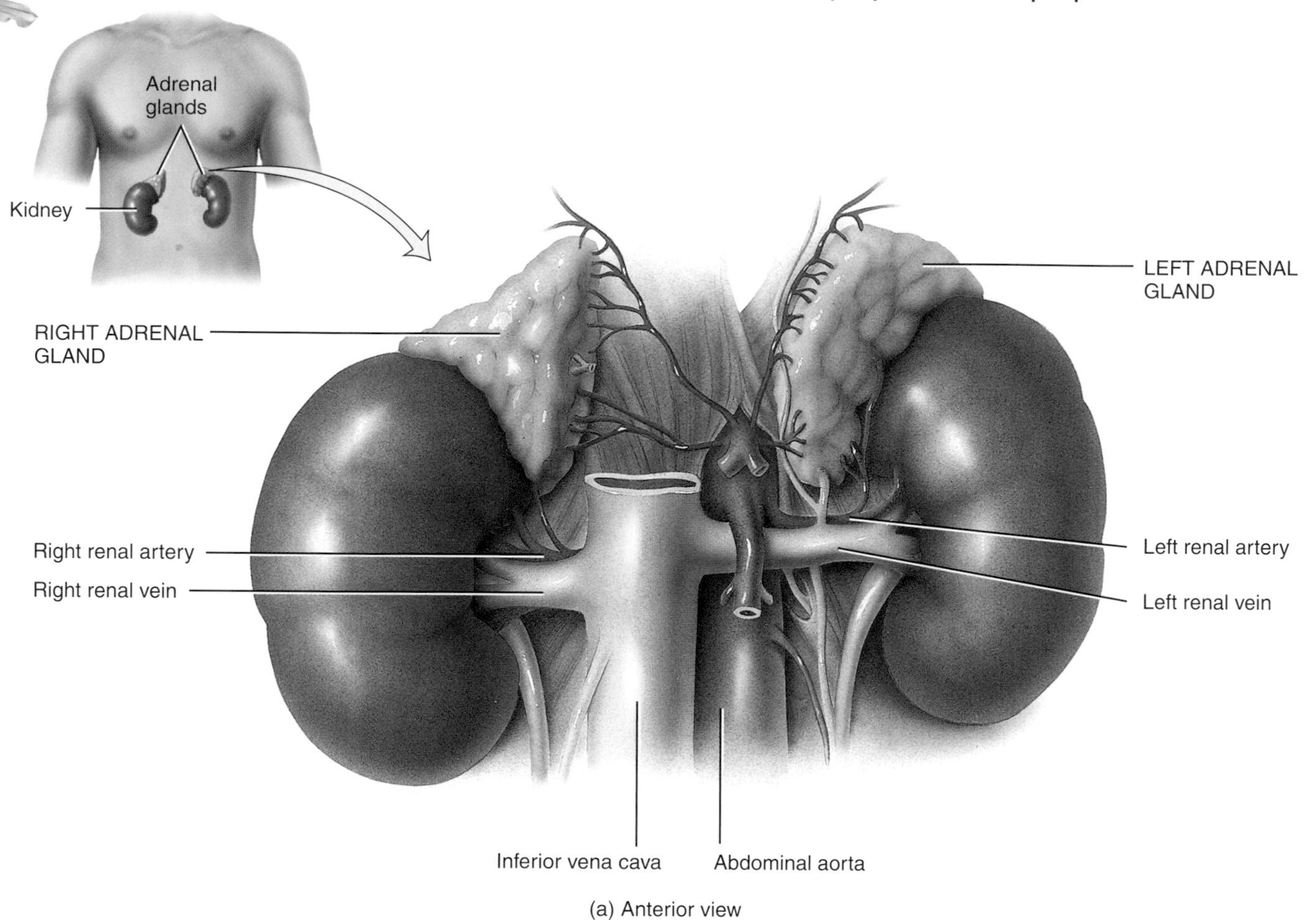

(a) Anterior view

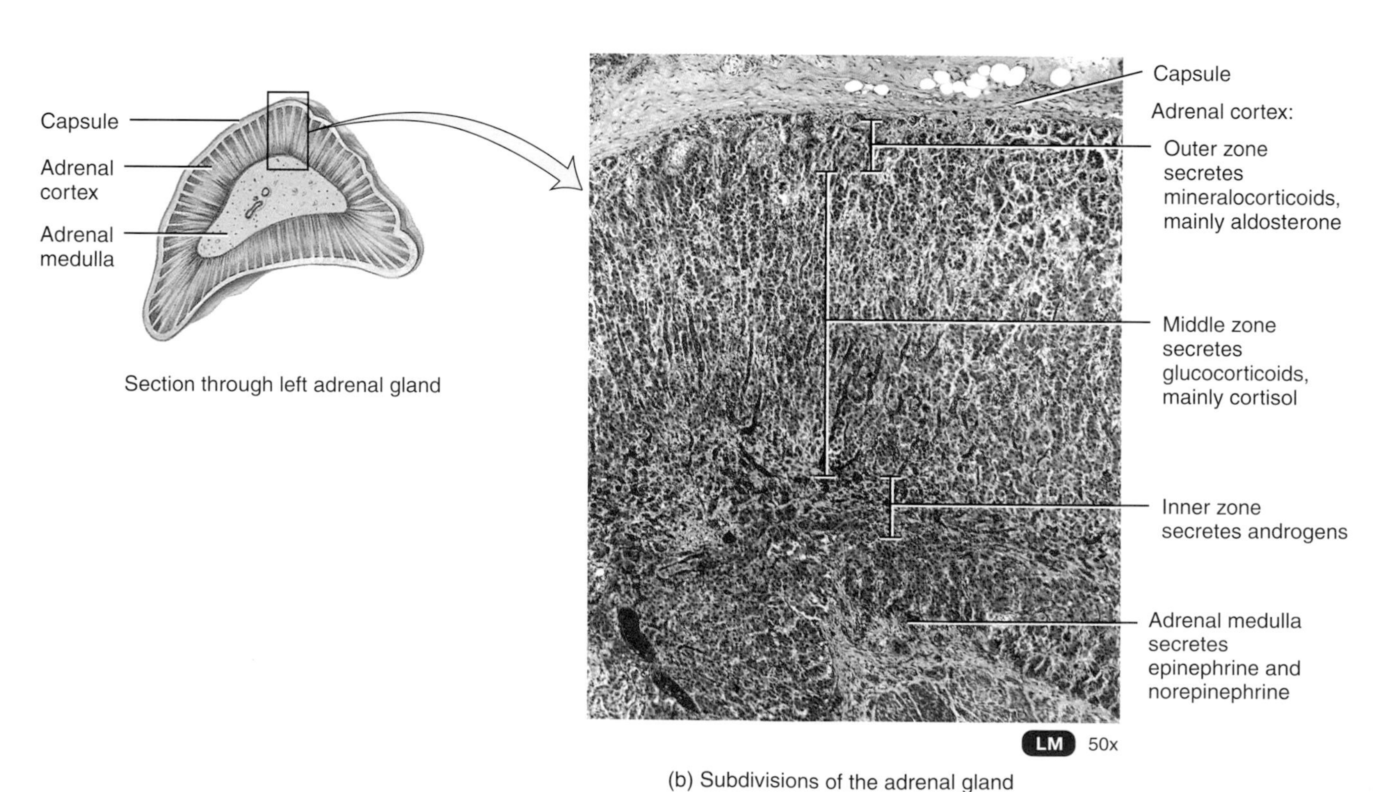

(b) Subdivisions of the adrenal gland

What hormones are secreted by the three zones of the adrenal cortex?

Mineralocorticoids

Aldosterone (al-DO-ster-ōn) is the major ***mineralocorticoid*** (min′-er-al-ō-KOR-ti-koyd). It regulates homeostasis of two mineral ions, namely, sodium ions (Na^+) and potassium ions (K^+). Aldosterone increases reabsorption of Na^+ from the urine into the blood, and it stimulates excretion of K^+ into the urine. It also helps adjust blood pressure and blood volume, and promotes excretion of H^+ in the urine. Such removal of acids from the body can help prevent acidosis (blood pH below 7.35).

Secretion of aldosterone occurs as part of the ***renin–angiotensin–aldosterone pathway*** (RĒ-nin an′-jē-ō-TEN-sin) (Figure 13.14). Conditions that initiate this pathway include dehydration, Na^+ deficiency, or hemorrhage, which decrease blood volume and blood pressure. Lowered blood pressure stimulates the kidneys to secrete the enzyme ***renin***, which promotes a reaction in the blood that forms *angiotensin I*. As blood flows through the lungs, another enzyme called ***angiotensin converting enzyme (ACE)*** converts inactive angiotensin I into the active hormone ***angiotensin II***. Angiotensin II stimulates the adrenal cortex to secrete aldosterone. Aldosterone, in turn, acts on the kidneys to promote the return of Na^+ and water to the blood. As more water returns to the blood (and less is lost in the urine), blood volume increases. As blood volume increases, blood pressure increases to normal.

Figure 13.14 The renin–angiotensin–aldosterone pathway.

Aldosterone helps regulate blood volume, blood pressure, and levels of Na^+ and K^+ in the blood.

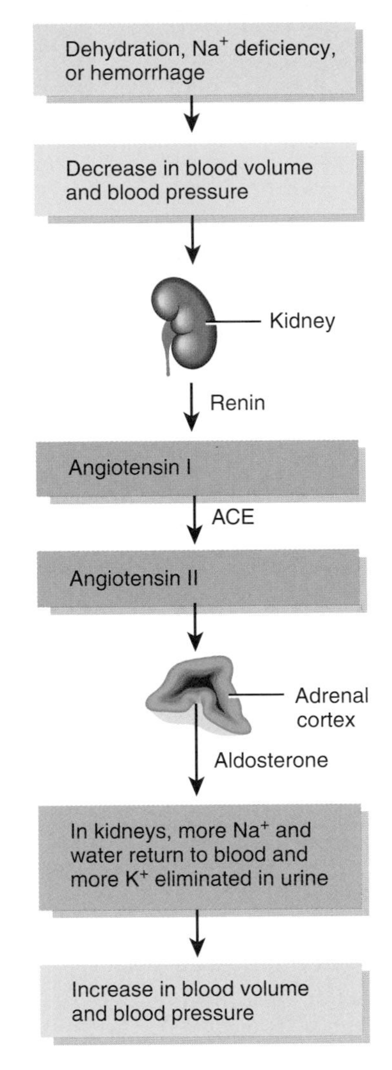

Could a drug that *blocks* the action of the enzyme ACE be used to raise or to lower blood pressure?

Glucocorticoids

The most abundant ***glucocorticoid*** (gloo′-kō-KOR-ti-koyd; *gluco-* = sugar; *cortic-* = the bark, shell) is ***cortisol***. Cortisol and other glucocorticoids have the following actions:

- **Protein breakdown**. Glucocorticoids increase the rate of protein breakdown, mainly in muscle fibers, and thus increase the liberation of amino acids into the bloodstream. The amino acids may be used by body cells for synthesis of new proteins or for ATP production.
- **Glucose formation**. Upon stimulation by glucocorticoids, liver cells may convert certain amino acids or lactic acid to glucose, which neurons and other cells can use for ATP production.
- **Breakdown of triglycerides**. Glucocorticoids stimulate the breakdown of triglycerides in adipose tissue. The fatty acids thus released into the blood can be used for ATP production by many body cells.
- **Anti-inflammatory effects**. Glucocorticoids inhibit white blood cells that participate in inflammatory responses. They are often used in the treatment of chronic inflammatory disorders such as rheumatoid arthritis. Unfortunately, glucocorticoids also retard tissue repair, which slows wound healing.
- **Depression of immune responses**. High doses of glucocorticoids depress immune responses. For this reason, glucocorticoids are prescribed for organ transplant recipients to decrease the risk of tissue rejection by the immune system.

The control of secretion of cortisol (and other glucocorticoids) occurs by negative feedback. A low blood level of cortisol stimulates neurosecretory cells in the hypothalamus to secrete *corticotropin-releasing hormone (CRH)*. The hypophyseal portal veins carry CRH to the anterior pituitary, where it stimulates release of *ACTH* (adrenocorticotropic hormone). ACTH, in turn, stimulates cells of the adrenal cortex to secrete cortisol. As the level of cortisol rises, it ex-

FOCUS ON WELLNESS

Insulin Resistance—A Metabolic Medley

One of the most common endocrine disorders, type 2 diabetes, is characterized by high levels of insulin in the blood. Insulin levels are high due to **insulin resistance**, a condition in which insulin receptors do not respond properly to insulin. Despite plentiful insulin, blood glucose level remains high, because the receptors are not letting insulin help the glucose across the membrane and into the cells.

A Little Riddle About Fat in the Middle

Many people who develop type 2 diabetes also develop hypertension (high blood pressure) and high blood cholesterol. They also tend to be overweight and sedentary. This cluster of disorders—termed ***metabolic syndrome***—may be related to excess adipose tissue around the abdominal viscera.

Why is abdominal fat riskier than other adipose tissue? Adipocytes (fat cells) in the abdominal region are metabolically "more active" than lower-body fat cells, and they are more responsive to hormones such as epinephrine. This means they release fatty acids into the bloodstream more readily, which, in the abdominal area, flows to the liver. The liver takes up the fatty acids and produces triglycerides that are packaged into very-low-density lipoprotein (VLDL) particles. Later, the VLDLs are converted into low-density lipoprotein (LDL) particles. Higher levels of LDLs are associated with the formation of artery-clogging atherosclerotic plaques.

The elevation in triglycerides may disrupt blood sugar regulation and trigger a rise in insulin. Elevated insulin levels in turn stimulate the sympathetic nervous system, which increases blood pressure. And there you have it, all in one package: high blood sugar, high blood lipids, hypertension, and abdominal obesity, a package that significantly increases artery disease risk.

Smoking, alcohol consumption, poor diet, and a sedentary lifestyle predispose a person to the development of type 2 diabetes. Both exercise and weight loss (in people who are overweight) increase the sensitivity of insulin receptors and improve transport of glucose into body cells.

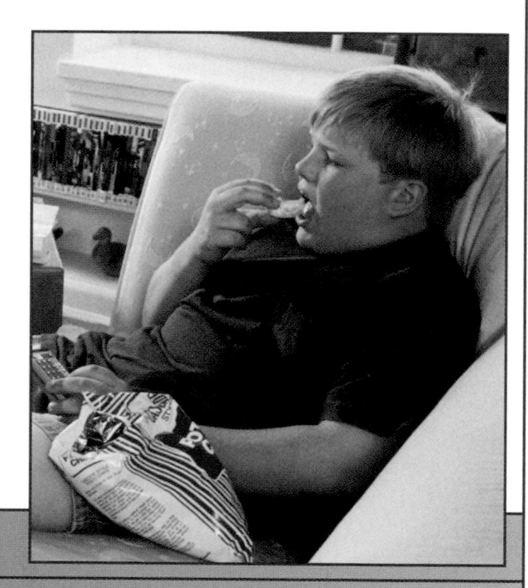

► **THINK IT OVER . . .**

► ***Why do you think extra fat tends to be lost more easily in the abdominal region than in the hips and thighs? Are you at risk for metabolic syndrome? Use a tape measure to take the circumferences of your hips and waist. A waist–hips ratio greater than 1.0 for men or 0.8 for women is considered risky.***

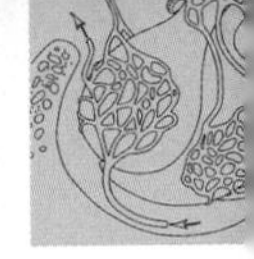

erts negative feedback inhibition both on the anterior pituitary to reduce release of ACTH and on the hypothalamus to reduce release of CRH.

Androgens

In both males and females, the adrenal cortex secretes small amounts of weak androgens. After puberty in males, androgens are also released in much greater quantity by the testes. Thus, the amount of androgens secreted by the adrenal gland in males is usually so low that their effects are insignificant. In females, however, adrenal androgens play important roles: They contribute to libido (sex drive) and are converted into estrogens (feminizing sex steroids) by other body tissues. After menopause, when ovarian secretion of estrogens ceases, all female estrogens come from conversion of adrenal androgens. Adrenal androgens also stimulate growth of axillary (armpit) and pubic hair in boys and girls and contribute to the growth spurt before puberty. Although control of adrenal androgen secretion is not fully understood, the main hormone that stimulates its secretion is ACTH.

> **Congenital adrenal hyperplasia (CAH)** is a genetic disorder in which one or more enzymes needed for the production of cortisol are absent. Because the cortisol level is low, secretion of ACTH by the anterior pituitary is high due to lack of negative feedback inhibition. ACTH, in turn, stimulates growth and secretory activity of the adrenal cortex. As a result, both adrenal glands are enlarged. However, certain steps leading to synthesis of cortisol are blocked. Thus, precursor molecules build up, and some of these are weak androgens that can be converted to testosterone. The result is **virilism**, or masculinization. In a female, virile characteristics include growth of a beard, development of a much deeper voice and a masculine distribution of body hair, growth of the clitoris so it may resemble a penis, atrophy of the breasts, and increased muscularity that produces a masculine physique. In males, virilism causes the same characteristics as in females, plus rapid development of the male sexual organs and emergence of male sexual desires.

Adrenal Medulla Hormones

The innermost region of each adrenal gland, the adrenal medulla, consists of sympathetic postganglionic cells of the autonomic nervous system (ANS) that are specialized to secrete hormones. The two main hormones of the adrenal medullae are ***epinephrine*** and ***norepinephrine (NE)***, also called adrenaline and noradrenaline.

In stressful situations and during exercise, impulses from the hypothalamus stimulate sympathetic preganglionic neurons, which in turn stimulate the cells of the adrenal medullae to secrete epinephrine and norepinephrine. These two hormones greatly augment the fight-or-flight response (see page 335). By increasing heart rate and force of contraction, epinephrine and norepinephrine increase the pumping output of the heart, which increases blood pressure. They also increase blood flow to the heart, liver, skeletal muscles, and adipose tissue; dilate airways to the lungs; and increase blood levels of glucose and fatty acids. Like the glucocorticoids of the adrenal cortex, epinephrine and norepinephrine also help the body resist stress.

■ **CHECKPOINT**

12. How do the adrenal cortex and adrenal medulla compare with regard to their location and histology?

13. How is secretion of adrenal cortex hormones regulated?

OVARIES AND TESTES

OBJECTIVE • Describe the location, hormones, and functions of the ovaries and testes.

Gonads are the organs that produce gametes—sperm in males and oocytes in females. The female gonads, the ***ovaries***, are paired oval bodies located in the pelvic cavity. They produce the female sex hormones ***estrogens*** and ***progesterone***. Along with FSH and LH from the anterior pituitary, the female sex hormones regulate the menstrual cycle, maintain pregnancy, and prepare the mammary glands for lactation. They also help establish and maintain the feminine body shape.

The ovaries also produce ***inhibin***, a protein hormone that inhibits secretion of follicle-stimulating hormone (FSH). During pregnancy, the ovaries and placenta produce a peptide hormone called ***relaxin***, which increases the flexibility of the pubic symphysis during pregnancy and helps dilate the uterine cervix during labor and delivery. These actions enlarge the birth canal, which helps ease the baby's passage.

The male gonads, the ***testes***, are oval glands that lie in the scrotum. They produce ***testosterone***, the primary androgen or male sex hormone. Testosterone regulates production of sperm and stimulates the development and maintenance of masculine characteristics such as beard growth and deepening of the voice. The testes also produce inhibin, which inhibits secretion of FSH. The detailed structure of the ovaries and testes and the specific roles of sex hormones will be discussed in Chapter 23.

> **Seasonal affective disorder (SAD)** is a type of depression that afflicts some people during the winter months, when day length is short. It is thought to be due, in part, to overproduction of melatonin. Bright light therapy—repeated exposure to artificial light—can provide relief.

■ **CHECKPOINT**

14. Why are the ovaries and testes included among the endocrine glands?

PINEAL GLAND

OBJECTIVE • Describe the location, hormone, and functions of the pineal gland.

The ***pineal gland*** (PĪN-ē-al = pinecone shape) is a small endocrine gland attached to the roof of the third ventricle of the brain at the midline (see Figures 13.1 and 10.6 on pages 316 and 249, respectively). One hormone secreted by the pineal gland is ***melatonin***, which contributes to setting the body's biological clock. More melatonin is released in darkness and during sleep; less melatonin is liberated in strong sunlight. In animals that breed during specific seasons, melatonin inhibits reproductive functions. Whether melatonin influences human reproductive function, however, is still unclear. Melatonin levels are higher in children and decline with age into adulthood, but there is no evidence that changes in melatonin secretion correlate with the onset of puberty and sexual maturation.

■ **CHECKPOINT**

15. What is the relationship between melatonin secretion and sleep?

OTHER HORMONES

OBJECTIVE • List the hormones secreted by cells in tissues and organs other than endocrine glands, and describe their functions.

Hormones from Other Endocrine Cells

Some tissues and organs other than those already described contain endocrine cells that secrete hormones. Table 13.2 summarizes these hormones and their actions.

Table 13.2 Summary of Hormones Produced by Endocrine Cells

Source and Hormone	Actions
Thymus	
Thymosin	Promotes the maturation of T cells (a type of white blood cell that destroys microbes and foreign substances) and may retard the aging process (discussed in Chapter 17).
Gastrointestinal Tract	
Gastrin	Promotes secretion of gastric juice and increases movements of the stomach (discussed in Chapter 19).
Glucose-dependent insulinotropic peptide (GIP)	Stimulates release of insulin by pancreatic beta cells (discussed in Chapter 19).
Secretin	Stimulates secretion of pancreatic juice and bile (discussed in Chapter 19).
Cholecystokinin (CCK)	Stimulates secretion of pancreatic juice, regulates release of bile from the gallbladder, and brings about a feeling of fullness after eating (discussed in Chapter 19).
Kidney	
Erythropoietin (EPO)	Increases rate of red blood cell production (discussed in Chapter 14).
Heart	
Atrial natriuretic peptide (ANP)	Decreases blood pressure (discussed in Chapter 16).
Adipose Tissue	
Leptin	Suppresses appetite and may increase the activity of FSH and LH (discussed in Chapter 20).
Placenta	
Human chorionic gonadotropin (hCG)	Stimulates the ovary to continue production of estrogens and progesterone during pregnancy (discussed in Chapter 24).

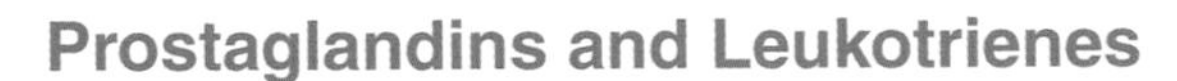

Prostaglandins and Leukotrienes

Two families of molecules derived from fatty acids, the ***prostaglandins*** (pros′-ta-GLAN-dins), or ***PGs***, and the ***leukotrienes*** (loo-kō-TRĪ-ēns), or ***LTs***, act locally as hormones in most tissues of the body. Virtually all body cells except red blood cells release these local hormones in response to chemical and mechanical stimuli. Because the PGs and LTs act close to their sites of release, they appear in only tiny quantities in the blood.

Leukotrienes stimulate movement of white blood cells and mediate inflammation. The prostaglandins alter smooth muscle contraction, glandular secretions, blood flow, reproductive processes, platelet function, respiration, nerve impulse transmission, fat metabolism, and immune responses. PGs also have roles in inflammation, promoting fever, and intensifying pain.

Aspirin and related **nonsteroidal anti-inflammatory drugs (NSAIDs)**, such as ibuprofen (Motrin®), inhibit a key enzyme in prostaglandin synthesis without affecting synthesis of leukotrienes. They are used to treat a wide variety of inflammatory disorders, from rheumatoid arthritis to tennis elbow.

■ **CHECKPOINT**

16. What hormones are secreted by the gastrointestinal tract, placenta, kidneys, skin, adipose tissue, and heart?

17. What are some functions of prostaglandins and leukotrienes?

THE STRESS RESPONSE

OBJECTIVE • Describe how the body responds to stress.

It is impossible to remove all stress from our everyday lives. Any stimulus that produces a stress response is called a ***stressor***. A stressor may be almost any disturbance—heat or cold, environmental poisons, toxins given off by bacteria, heavy bleeding from a wound or surgery, or a strong emotional reaction. Stressors may be pleasant or unpleasant, and they vary among people and even within the same person at different times. When homeostatic mechanisms are successful in counteracting stress, the internal environment remains within normal physiological limits. If stress is extreme, unusual, or long lasting, it elicits the ***stress response***, a sequence of bodily changes that can progress through three stages: (1) an initial fight-or-flight response, (2) a slower resistance reaction, and eventually (3) exhaustion.

The ***fight-or-flight response***, initiated by nerve impulses from the hypothalamus to the sympathetic division of the autonomic nervous system (ANS), including the adrenal medullae, quickly mobilizes the body's resources for immediate physical activity. It brings huge amounts of glucose and oxygen to the organs that are most active in warding off danger: the brain, which must become highly alert; the skeletal muscles, which may have to fight off an attacker or flee; and the heart, which must work vigorously to pump enough blood to the brain and muscles. Reduction of blood flow to the kidneys, however, promotes the release of renin, which sets into motion the renin–angiotensin–aldosterone pathway (see Figure 13.14). Aldosterone causes the kidneys to retain Na^+, which leads to water retention and elevated blood pressure. Water retention also helps preserve body fluid volume in the case of severe bleeding.

The second stage in the stress response is the ***resistance reaction***. Unlike the short-lived fight-or-flight response, which is initiated by nerve impulses from the hypothalamus, the resistance reaction is initiated in large part by hypothalamic releasing hormones and is a longer-lasting response. The hormones involved are corticotropin-releasing hormone (CRH), growth hormone-releasing hormone (GHRH), and thyrotropin-releasing hormone (TRH).

CRH stimulates the anterior pituitary to secrete ACTH, which in turn stimulates the adrenal cortex to release more cortisol. Cortisol then stimulates release of glucose by liver cells, breakdown of triglycerides into fatty acids, and catabolism of proteins into amino acids. Tissues throughout the body can use the resulting glucose, fatty acids, and amino acids to produce ATP or to repair damaged cells. Cortisol also reduces inflammation. A second hypothalamic releasing hormone, GHRH, causes the anterior pituitary to secrete human growth hormone (hGH). Acting via insulinlike growth factors, hGH stimulates breakdown of triglycerides and glycogen. A third hypothalamic releasing hormone, TRH, stimulates the anterior pituitary to secrete thyroid-stimulating hormone (TSH). TSH promotes secretion of thyroid hormones, which stimulate the increased use of glucose for ATP production. The combined actions of hGH and TSH thereby supply additional ATP for metabolically active cells.

The resistance stage helps the body continue fighting a stressor long after the fight-or-flight response dissipates. Generally, it is successful in seeing us through a stressful episode, and our bodies then return to normal. Occasionally, however, the resistance stage fails to combat the stressor: The resources of the body may eventually become so depleted that they cannot sustain the resistance stage, and ***exhaustion*** ensues. Prolonged exposure to high levels of cortisol and other hormones involved in the resistance reaction causes wasting of muscles, suppression of the immune system, ulceration of the gastrointestinal tract, and failure of pancreatic beta cells. In addition, pathological changes may occur because resistance reactions persist after the stressor has been removed.

Although the exact role of stress in human diseases is not known, it is clear that stress can temporarily inhibit certain components of the immune system. Stress-related disorders include gastritis, ulcerative colitis, irritable bowel syndrome, hypertension, asthma, rheumatoid arthritis, migraine headaches, anxiety, and depression. People under stress also are at a greater risk of developing a chronic disease or dying prematurely.

Posttraumatic stress disorder (PTSD) may develop in someone who has experienced, witnessed, or learned about a physically or psychologically distressing event. The immediate cause of PTSD appears to be the specific stressors associated with the events. Among the stressors are terrorism, hostage taking, imprisonment, serious accidents, torture, sexual or physical abuse, violent crimes, and natural disasters. In the United States, PTSD affects 10% of females and 5% of males. Symptoms of PTSD include reliving the event through nightmares or flashbacks; loss of interest and lack of motivation; poor concentration; irritability; and insomnia.

■ **CHECKPOINT**

18. What is the role of the hypothalamus during stress?

19. How are stress and immunity related?

AGING AND THE ENDOCRINE SYSTEM

OBJECTIVE • Describe the effects of aging on the endocrine system.

Although some endocrine glands shrink as we get older, their performance may or may not be compromised. Production of human growth hormone by the anterior pituitary decreases, which is one cause of muscle atrophy as aging proceeds. The thyroid gland often decreases its output of thyroid hormones with age, causing a decrease in metabolic rate, an increase in body fat, and hypothyroidism, which is seen more often in older people. Because there is less negative feedback (lower levels of thyroid hormones), the level of thyroid-stimulating hormone increases with age.

With aging, the blood level of PTH rises, perhaps due to inadequate dietary intake of calcium. In a study of older women who took 2,400 mg/day of supplemental calcium, blood levels of PTH were as low as those in younger women. Both calcitriol and calcitonin levels are lower in older persons. Together, the rise in PTH and the fall in calcitonin heighten the age-related decrease in bone mass that leads to osteoporosis and increased risk of fractures.

The adrenal glands contain increasingly more fibrous tissue and produce less cortisol and aldosterone with advancing age. However, production of epinephrine and norepinephrine remains normal. The pancreas releases insulin more slowly with age, and receptor sensitivity to glucose declines. As a result, blood glucose levels in older people increase faster and return to normal more slowly than in younger individuals.

The thymus is largest in infancy. After puberty, its size begins to decrease, and thymic tissue is replaced by adipose and areolar connective tissue. In older adults, the thymus has atrophied significantly. However, it still produces new T cells for immune responses.

The ovaries decrease in size with age, and they no longer respond to gonadotropins. The resultant decreased output of estrogens leads to conditions such as osteoporosis, high blood cholesterol, and atherosclerosis. FSH and LH levels are high due to less negative feedback inhibition of estrogens. Although testosterone production by the testes decreases with age, the effects are not usually apparent until very old age, and many elderly males can still produce active sperm in normal numbers.

■ **CHECKPOINT**

20. Which hormone is related to the muscle atrophy that occurs with aging?

• • •

To appreciate the many ways the endocrine system contributes to homeostasis of other body systems, examine Focus on Homeostasis: The Endocrine System on page 339. Next, in Chapter 14, we will begin to explore the cardiovascular system, starting with a description of the composition and functions of the blood.

COMMON DISORDERS

Disorders of the endocrine system often involve either ***hyposecretion*** (*hypo-* = too little or under), inadequate release of a hormone, or ***hypersecretion*** (*hyper-* = too much or above), excessive release of a hormone. In other cases, the problem is faulty hormone receptors or an inadequate number of receptors.

Pituitary Gland Disorders

Several disorders of the anterior pituitary involve human growth hormone (hGH). Undersecretion of hGH during the growth years slows bone growth, and the epiphyseal plates close before normal height is reached. This condition is called ***pituitary dwarfism.*** Other organs of the body also fail to grow, and the body proportions are childlike.

Oversecretion of hGH during childhood results in ***giantism (gigantism)***, an abnormal increase in the length of long bones. The person grows to be very tall, but body proportions are about normal. Figure 13.15a shows identical twins; one brother developed giantism due to a pituitary tumor. Oversecretion of hGH during adulthood is called ***acromegaly*** (ak′-rō-MEG-a-lē). Although hGH cannot produce further lengthening of the long bones because the epiphyseal plates are already closed, the bones of the hands, feet, cheeks, and jaws thicken and other tissues enlarge (Figure 13.15b).

The most common abnormality of the posterior pituitary is ***diabetes insipidus*** (dī-a-BĒ-tēs in-SIP-i-dus; *diabetes* = overflow; *insipidus* = tasteless). This disorder is due to defects in antidiuretic hormone (ADH) receptors or an inability to secrete ADH. Usually the disorder is caused by a brain tumor, head trauma, or brain surgery that damages the posterior pituitary or the hypothalamus. A common symptom is excretion of large volumes of urine, with resulting dehydration and thirst. Because so much water is lost in the urine, a person with diabetes insipidus may die of dehydration if deprived of water for only a day or so.

Thyroid Gland Disorders

Thyroid gland disorders affect all major body systems and are among the most common endocrine disorders. ***Congenital hypothyroidism***, hyposecretion of thyroid hormones that is present at birth,

Figure 13.15 Photographs of people with various endocrine disorders.

Disorders of the endocrine system often involve hyposecretion or hypersecretion of various hormones.

(a) A 22-year old man with pituitary giantism shown beside his identical twin

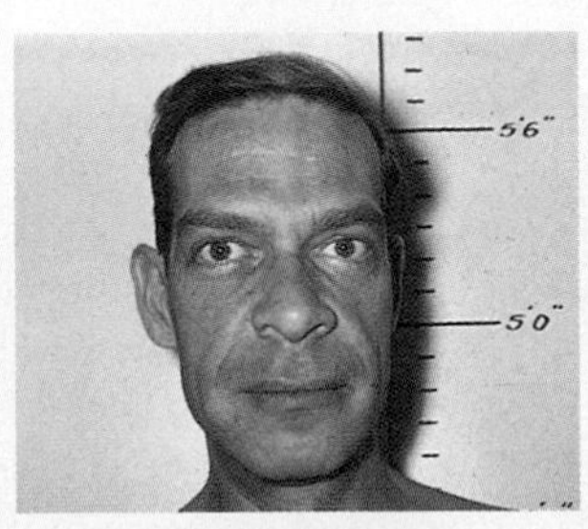

(b) Acromegaly (excess hGH during adulthood)

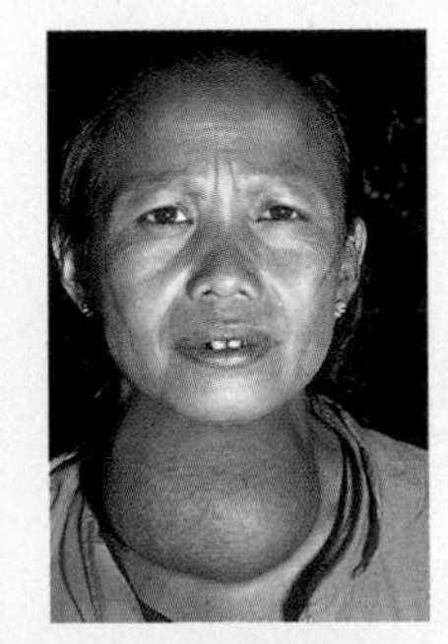

(c) Goiter (enlargement of thyroid gland)

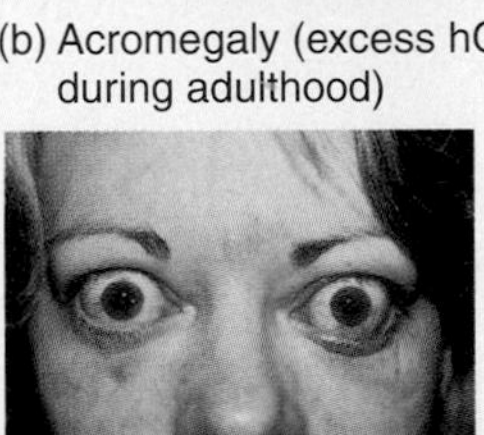

(d) Exophthalmos (excess thyroid hormones, as in Graves disease)

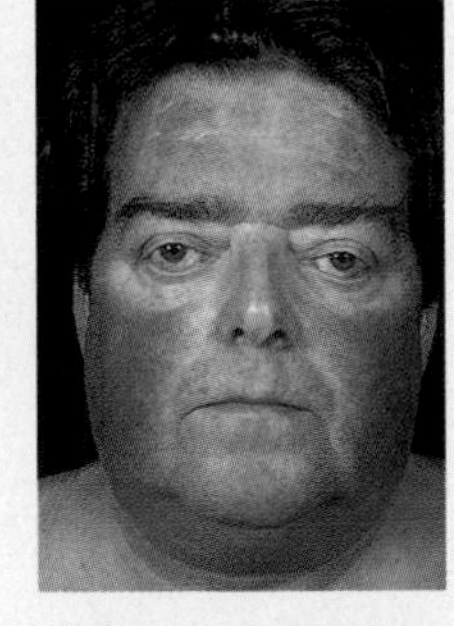

(e) Cushing's syndrome (excess glucocorticoids)

Which of the disorders shown here is due to antibodies that mimic the action of TSH?

has devastating consequences if not treated promptly. Previously termed *cretinism*, this condition causes severe mental retardation. At birth, the baby typically is normal because lipid-soluble maternal thyroid hormones crossed the placenta during pregnancy and allowed normal development. Most states require testing of all newborns to ensure adequate thyroid function. If congenital hypothyroidism exists, oral thyroid hormone treatment must be started soon after birth and continued for life.

Hypothyroidism during the adult years produces ***myxedema*** (mix-e-DĒ-ma), which occurs about five times more often in females than in males. A hallmark of this disorder is edema (accumulation of interstitial fluid) that causes the facial tissues to swell and look puffy. A person with myxedema has a slow heart rate, low body temperature, sensitivity to cold, dry hair and skin, muscular weakness, general lethargy, and a tendency to gain weight easily.

The most common form of hyperthyroidism is ***Graves disease***, which also occurs much more often in females than in males, usually before age 40. Graves disease is an autoimmune disorder in which the person produces antibodies that mimic the action of thyroid-stimulating hormone (TSH). The antibodies continually stimulate the thyroid gland to grow and produce thyroid hormones. Thus, the thyroid gland may enlarge to two to three times its normal size, a condition called ***goiter*** (GOY-ter; *guttur* = throat) (Figure 13.15c). Goiter also occurs in other thyroid diseases and if dietary intake of iodine is inadequate. Graves patients often have a peculiar edema behind the eyes, called *exophthalmos* (ek′-sof-THAL-mos), which causes the eyes to protrude (Figure 13.15d).

Parathyroid Gland Disorders

Hypoparathyroidism—too little parathyroid hormone—leads to a deficiency of Ca^{2+}, which causes neurons and muscle fibers to depolarize and produce action potentials spontaneously. This leads to twitches, spasms, and ***tetany*** (maintained contraction) of skeletal muscle. The leading cause of hypoparathyroidism is accidental damage to the parathyroid glands or to their blood supply during surgery to remove the thyroid gland.

Adrenal Gland Disorders

Hypersecretion of cortisol by the adrenal cortex produces ***Cushing's syndrome***. The condition is characterized by the breakdown of muscle proteins and redistribution of body fat, resulting in spindly arms and legs accompanied by a rounded "moon face" (Figure 13.15e), "buffalo hump" on the back, and pendulous (hanging) abdomen. The elevated level of cortisol causes hyperglycemia, osteoporosis, weakness, hypertension, increased susceptibility to infection, decreased resistance to stress, and mood swings.

Hyposecretion of glucocorticoids and aldosterone causes ***Addison's disease***. Symptoms include mental lethargy, anorexia, nausea and vomiting, weight loss, hypoglycemia, and muscular weakness. Loss of aldosterone leads to elevated potassium and decreased sodium in the blood; low blood pressure; dehydration; and decreased cardiac output, cardiac arrhythmias, and even cardiac arrest. The skin may have a "bronzed" appearance that often is mistaken for a suntan. Such was true in the case of President John F. Kennedy, whose Addison's disease was known to only a few while he was alive.

Usually benign tumors of the adrenal medulla, called ***pheochromocytomas*** (fē′-ō-krō′-mō-sī-TŌ-mas; *pheo-* = dusky; *chromo-* = color; *cyto-* = cell), cause oversecretion of epinephrine and norepinephrine. The result is a prolonged version of the fight-or-flight response: rapid heart rate, headache, high blood pressure, high levels of glucose in blood and urine, an elevated basal metabolic rate (BMR), flushed face, nervousness, sweating, and decreased gastrointestinal motility.

Pancreatic Islet Disorders

The most common endocrine disorder is ***diabetes mellitus*** (MEL-i-tus; *melli-* = honey sweetened), caused by an inability either to produce or to use insulin. Diabetes mellitus is the fourth leading cause of death by disease in the United States, primarily because of its damage to the cardiovascular system. Because insulin is unavailable to aid the movement of glucose into body cells, blood glucose level is high and glucose "spills" into the urine (glucosuria). Hallmarks of diabetes mellitus are the three "polys"; *polyuria*, excessive urine production due to an inability of the kidneys to reabsorb water; *polydipsia*, excessive thirst; and *polyphagia*, excessive eating.

Both genetic and environmental factors contribute to onset of the two types of diabetes mellitus—Type 1 and Type 2—but the exact mechanisms are still unknown. In ***Type 1 diabetes*** insulin level is low because the person's immune system destroys the pancreatic beta cells. Most commonly, Type 1 diabetes develops in people younger than age 20, though it persists throughout life. By the time symptoms arise, 80–90% of the islet beta cells have been destroyed.

Because insulin is not present to aid the entry of glucose into body cells, most cells use fatty acids to produce ATP. Stores of triglycerides in adipose tissue are broken down to fatty acids and glycerol. The byproducts of fatty acid breakdown—organic acids called ketones or ketone bodies—accumulate. Buildup of ketones causes blood pH to fall, a condition known as ***ketoacidosis***. Unless treated quickly, ketoacidosis can cause death.

Type 2 diabetes is much more common than Type 1. It most often occurs in people who are over 35 and overweight. The high glucose levels in the blood often can be controlled by diet, exercise, and weight loss. Sometimes, an antidiabetic drug such as *glyburide* (Diabeta®) is used to stimulate secretion of insulin by pancreatic beta cells. Although some Type 2 diabetics need insulin, many have a sufficient amount (or even a surplus) of insulin in the blood. For these people, diabetes arises not from a shortage of insulin but because target cells become less sensitive to it.

Hyperinsulinism most often results when a diabetic injects too much insulin. The main symptom is ***hypoglycemia***, decreased blood glucose level, which occurs because the excess insulin stimulates too much uptake of glucose by body cells. When blood glucose falls, neurons are deprived of the steady supply of glucose they need to function effectively. Severe hypoglycemia leads to mental disorientation, convulsions, unconsciousness, and shock and is termed ***insulin shock***. Death can occur quickly unless blood glucose is raised.

FOCUS ON HOMEOSTASIS

THE ENDOCRINE SYSTEM

BODY SYSTEM		CONTRIBUTION OF THE ENDOCRINE SYSTEM
For all body systems		Together with the nervous system, hormones of the endocrine system regulate the activity and growth of target cells throughout the body. Several hormones regulate metabolism, uptake of glucose, and molecules used for ATP production by body cells.
Integumentary system	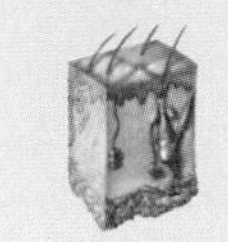	Androgens stimulate the growth of axillary and pubic hair and activation of sebaceous glands. Melanocyte-stimulating hormone (MSH) can cause darkening of the skin.
Skeletal system	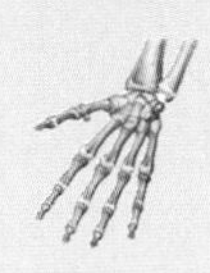	Human growth hormone (hGH) and insulinlike growth factors (IGFs) stimulate bone growth. Estrogens cause closure of epiphyseal plates at the end of puberty and help maintain bone mass in adults. Parathyroid hormone (PTH) promotes the release of calcium and other minerals from bone extracellular matrix into the blood. Thyroid hormones are needed for normal development and growth of the skeleton.
Muscular system	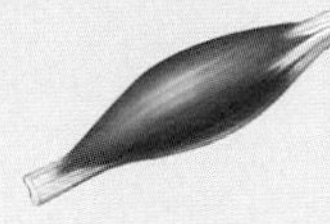	Epinephrine and norepinephrine help increase blood flow to exercising muscles. PTH maintains the proper level of Ca^{2+} in blood and interstitial fluid, which is needed for muscle contraction. Glucagon, insulin, and other hormones regulate metabolism in muscle fibers. IGFs, thyroid hormones, and insulin stimulate protein synthesis and thereby help maintain muscle mass.
Nervous system	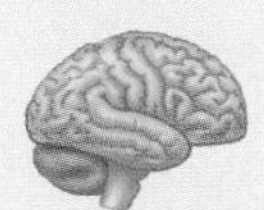	Several hormones, especially thyroid hormones, insulin, and IGFs, influence growth and development of the nervous system.
Cardiovascular system	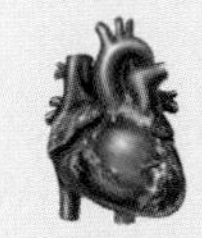	Erythropoietin (EPO) promotes the production of red blood cells. Aldosterone and antidiuretic hormone (ADH) increase blood volume. Epinephrine and norepinephrine increase the heart's rate and force of contraction. Several hormones elevate blood pressure during exercise and other stresses.
Lymphatic system and immunity	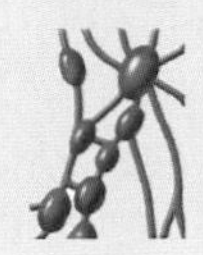	Glucocorticoids such as cortisol depress inflammation and immune responses. Hormones from the thymus promote maturation of T cells, a type of white blood cell that participates in immune responses.
Respiratory system	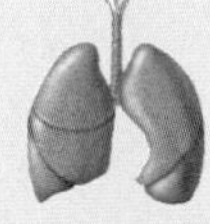	Epinephrine and norepinephrine dilate (widen) the airways during exercise and other stresses. Erythropoietin regulates the amount of oxygen carried in the blood by adjusting the number of red blood cells.
Digestive system	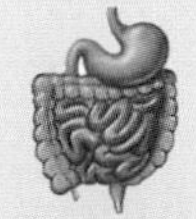	Epinephrine and norepinephrine depress activity of the digestive system. Gastrin, cholecystokinin, secretin, and GIP help regulate digestion. Calcitriol promotes absorption of dietary calcium. Leptin suppresses appetite.
Urinary system	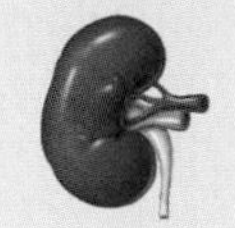	ADH, aldosterone, and atrial natriuretic peptide (ANP) adjust the rate of loss of water and ions in the urine, thereby regulating blood volume and ion levels in the blood.
Reproductive systems	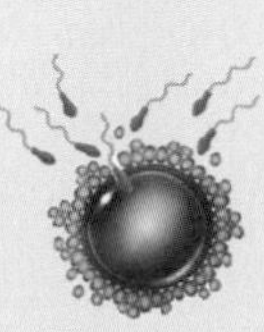	Hypothalamic releasing and inhibiting hormones, follicle-stimulating hormone (FSH), and luteinizing hormone (LH) regulate the development, growth, and secretions of the gonads (ovaries and testes). Estrogens and testosterone contribute to the development of oocytes and sperm and stimulate the development of sexual characteristics. Prolactin promotes milk production the in mammary glands. Oxytocin causes contraction of the uterus and ejection of milk from the mammary glands.

MEDICAL TERMINOLOGY AND CONDITIONS

Gynecomastia (gī-ne′-kō-MAS-tē-a; *gyneco-* = woman; *mast-* = breast) Excessive development of mammary glands in a male. Sometimes a tumor of the adrenal gland may secrete sufficient amounts of estrogen to cause the condition.

Hirsutism (HER-soo-tizm; *hirsut-* = shaggy) Presence of excessive bodily and facial hair in a male pattern, especially in women; may be due to excess androgen production caused by tumors or drugs.

Thyroid crisis (storm) A severe state of hyperthyroidism that can be life-threatening. It is characterized by high body temperature, rapid heart rate, high blood pressure, gastrointestinal symptoms (abdominal pain, vomiting, diarrhea), agitation, tremors, confusion, seizures, and possibly coma.

Virilizing adenoma (*aden-* = gland; *-oma* = tumor) Tumor of the adrenal gland that liberates excessive androgens, causing virilism (masculinization) in females. Occasionally, adrenal tumor cells liberate estrogens to the extent that a male patient develops gynecomastia. Such a tumor is called a ***feminizing adenoma.***

STUDY OUTLINE

Introduction (p. 316)

1. The nervous system controls homeostasis through the release of neurotransmitters; the endocrine system uses hormones.
2. The nervous system causes muscles to contract and glands to secrete; the endocrine system affects virtually all body tissues.
3. Exocrine glands (sweat, oil, mucous, digestive) secrete their products through ducts into body cavities or onto body surfaces.
4. The endocrine system consists of endocrine glands and several organs that contain endocrine tissue.

Hormone Action (p. 317)

1. Endocrine glands secrete hormones into interstitial fluid. Then, the hormones diffuse into the blood.
2. Hormones affect only specific target cells that have the proper receptors to bind a given hormone.
3. Chemically, hormones are either lipid-soluble (steroids, thyroid hormones, and nitric oxide) or water-soluble (modified amino acids, peptides, and proteins).
4. Lipid-soluble hormones affect cell function by altering gene expression.
5. Water-soluble hormones alter cell function by activating plasma membrane receptors, which elicit production of a second messenger that activates various proteins inside the cell.
6. Hormone secretion is controlled by signals from the nervous system, chemical changes in the blood, and other hormones.

Hypothalamus and Pituitary Gland (p. 319)

1. The pituitary gland is attached to the hypothalamus and consists of two lobes: the anterior pituitary and the posterior pituitary.
2. Hormones of the pituitary gland are controlled by inhibiting and releasing hormones produced by the hypothalamus. The hypophyseal portal veins carry hypothalamic releasing and inhibiting hormones from the hypothalamus to the anterior pituitary.
3. The anterior pituitary consists of cells that produce human growth hormone (hGH), prolactin (PRL), thyroid-stimulating hormone (TSH), follicle-stimulating hormone (FSH), luteinizing hormone (LH), adrenocorticotropic hormone (ACTH), and melanocyte-stimulating hormone (MSH).
4. Human growth hormone (hGH) stimulates body growth through insulinlike growth factors (IGFs) and is controlled by growth hormone-releasing hormone (GHRH) and growth hormone-inhibiting hormone (GHIH).
5. TSH regulates thyroid gland activities and is controlled by thyrotropin-releasing hormone (TRH).
6. FSH and LH regulate activities of the gonads—ovaries and testes—and are controlled by gonadotropin-releasing hormone (GnRH).
7. PRL helps stimulate milk production. Prolactin-inhibiting hormone (PIH) suppresses release of prolactin. Prolactin-releasing hormone (PRH) stimulates a rise in prolactin level during pregnancy.
8. ACTH regulates activities of the adrenal cortex and is controlled by corticotropin-releasing hormone (CRH).
9. The posterior pituitary contains axon terminals of neurosecretory cells whose cell bodies are in the hypothalamus.
10. Hormones made in the hypothalamus and released in the posterior pituitary include oxytocin, which stimulates contraction of the uterus and ejection of milk from the breasts, and antidiuretic hormone (ADH), which stimulates water reabsorption by the kidneys and constriction of arterioles.
11. Oxytocin secretion is stimulated by uterine stretching and by suckling during nursing; ADH secretion is controlled by the osmotic pressure of the blood and blood volume.

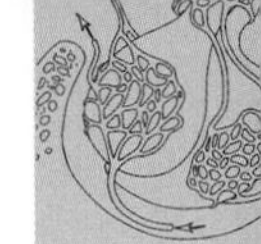

12. Table 13.1 on page 323 summarizes the hormones of the anterior and posterior pituitary.

Thyroid Gland (p. 323)

1. The thyroid gland is located below the larynx.
2. It consists of thyroid follicles composed of follicular cells, which secrete the thyroid hormones thyroxine (T_4) and triiodothyronine (T_3), and parafollicular cells, which secrete calcitonin.
3. Thyroid hormones regulate oxygen use and metabolic rate, cellular metabolism, and growth and development. Secretion is controlled by TRH from the hypothalamus and thyroid-stimulating hormone (TSH) from the anterior pituitary.
4. Calcitonin (CT) can lower the blood level of calcium; its secretion is controlled by the level of calcium in the blood.

Parathyroid Glands (p. 325)

1. The parathyroid glands are embedded on the posterior surfaces of the thyroid.
2. Parathyroid hormone (PTH) regulates the homeostasis of calcium, magnesium, and phosphate by increasing blood calcium and magnesium levels and decreasing blood phosphate level. PTH secretion is controlled by the level of calcium in the blood.

Pancreatic Islets (p. 327)

1. The pancreas lies in the curve of the duodenum. It has both endocrine and exocrine functions.
2. The endocrine portion consists of pancreatic islets or islets of Langerhans, which are made up of alpha and beta cells.
3. Alpha cells secrete glucagon and beta cells secrete insulin.
4. Glucagon increases blood glucose level, and insulin decreases blood glucose level. Secretion of both hormones is controlled by the level of glucose in the blood.

Adrenal Glands (p. 329)

1. The adrenal glands are located above the kidneys. They consist of an outer cortex and inner medulla.
2. The adrenal cortex is divided into three zones: The outer zone secretes mineralocorticoids; the middle zone secretes glucocorticoids; and the inner zone secretes androgens.
3. Mineralocorticoids (mainly aldosterone) increase sodium and water reabsorption and decrease potassium reabsorption. Secretion is controlled by the renin–angiotensin–aldosterone pathway.
4. Glucocorticoids (mainly cortisol) promote normal metabolism, help resist stress, and decrease inflammation. Secretion is controlled by ACTH.
5. Androgens secreted by the adrenal cortex stimulate growth of axillary and pubic hair, aid the prepubertal growth spurt, and contribute to libido.
6. The adrenal medullae secrete epinephrine and norepinephrine (NE), which are released under stress.

Ovaries and Testes (p. 333)

1. The ovaries are located in the pelvic cavity and produce estrogens, progesterone, and inhibin. These sex hormones regulate the menstrual cycle, maintain pregnancy, and prepare the mammary glands for lactation. They also help establish and maintain the feminine body shape.
2. The testes lie inside the scrotum and produce testosterone and inhibin. Testosterone regulates production of sperm and stimulates the development and maintenance of masculine characteristics such as beard growth and deepening of the voice.

Pineal Gland (p. 334)

1. The pineal gland, attached to the roof of the third ventricle in the brain, secretes melatonin, which contributes to setting the body's biological clock.

Other Hormones (p. 334)

1. Body tissues other than those normally classified as endocrine glands contain endocrine tissue and secrete hormones. These include the gastrointestinal tract, placenta, kidneys, skin, and heart. (See Table 13.2 on page 334.)
2. Prostaglandins and leukotrienes act locally in most body tissues.

The Stress Response (p. 335)

1. Stressors include surgical operations, poisons, infections, fever, and strong emotional responses.
2. If stress is extreme, it triggers the stress response, which occurs in three stages: the fight-or-flight response, resistance reaction, and exhaustion.
3. The fight-or-flight response is initiated by nerve impulses from the hypothalamus to the sympathetic division of the autonomic nervous system and the adrenal medullae. This response rapidly increases circulation and promotes ATP production.
4. The resistance reaction is initiated by releasing hormones secreted by the hypothalamus. Resistance reactions are longer lasting and accelerate breakdown reactions to provide ATP for counteracting stress.
5. Exhaustion results from depletion of body resources during the resistance stage.
6. Stress may trigger certain diseases by inhibiting the immune system.

Aging and the Endocrine System (p. 336)

1. Although some endocrine glands shrink as we get older, their performance may or may not be compromised.
2. Production of human growth hormone, thyroid hormones, cortisol, aldosterone, and estrogens decrease with advancing age.
3. With aging, the blood levels of TSH, LH, FSH, and PTH rise.
4. The pancreas releases insulin more slowly with age, and receptor sensitivity to glucose declines.
5. After puberty, thymus size begins to decrease, and thymic tissue is replaced by adipose and areolar connective tissue.

SELF-QUIZ

1. Which of the following is NOT true concerning hormones?
   a. Responses to hormones are generally slower and longer lasting than the responses stimulated by the nervous system.
   b. Hormones are generally controlled by negative feedback systems.
   c. The hypothalamus inhibits the release of some hormones.
   d. Most hormones are released steadily throughout the day.
   e. Hormone secretion is determined by the body's need to maintain homeostasis.

2. Which of the following statements concerning hormone action is NOT true?
   a. Hormones bring about changes in metabolic activities of cells
   b. Target cells must have receptors for a hormone.
   c. Lipid-soluble hormones may directly enter target cells and activate the genes.
   d. A hormone that attaches to a membrane receptor is termed the first messenger.
   e. ATP is a common second messenger in target cells.

3. Which of the following statements is NOT true?
   a. The secretion of hormones by the anterior pituitary is controlled by hypothalamic releasing hormones.
   b. The pituitary is attached to the hypothalamus by the infundibulum.
   c. Hypophyseal portal veins connect the posterior pituitary to the hypothalamus.
   d. The anterior pituitary constitutes the majority of the pituitary gland.
   e. The posterior pituitary releases hormones produced by neurosecretory cells of the hypothalamus.

4. The hormone that promotes milk release from the mammary glands and that stimulates the uterus to contract is
   a. oxytocin b. prolactin c. relaxin
   d. calcitonin e. follicle-stimulating hormone

5. The gland that prepares the body to react to stress by releasing epinephrine is the
   a. posterior pituitary b. anterior pituitary c. pineal
   d. adrenal e. pancreas

6. To help prevent rejection, organ transplant patients could be given
   a. glucocorticoids
   b. calcitonin
   c. mineralocorticoids
   d. thymopoietin
   e. melanocyte-stimulating hormone

7. A female who is sluggish, gaining weight, and has a low body temperature may be having problems with her
   a. pancreas b. parathyroid glands c. adrenal medullae
   d. ovaries e. thyroid gland

8. Destruction of the alpha cells of the pancreas might result in
   a. hypoglycemia b. seasonal affective disorder
   c. acromegaly d. hyperglycemia
   e. decreased urine output

9. Which of the following is NOT true concerning human growth hormone (hGH) and insulinlike growth factors?
   a. They stimulate protein synthesis.
   b. They have one primary target tissue in the body.
   c. They stimulate skeletal muscle growth.
   d. Hyposecretion in childhood results in dwarfism.
   e. Hypoglycemia can stimulate the release of hGH from the pituitary gland.

10. Follicle-stimulating hormone (FSH) acts on ______ and luteinizing hormone (LH) acts on ______.
    a. the ovaries, the testes b. the testes, the ovaries
    c. the ovaries and testes, the ovaries and testes
    d. the ovaries, the mammary glands
    e. the ovaries and uterus, the testes

11. An injection of adrenocorticotropic hormone (ACTH) would
    a. stimulate the ovaries
    b. influence thyroid gland activity
    c. stimulate the release of cortisol
    d. cause uterine contractions
    e. decrease urine output

12. Which of the following is NOT true concerning glucocorticoids?
    a. They help to control electrolyte balance.
    b. They help provide resistance to stress.
    c. They help promote normal metabolism.
    d. They are anti-inflammatory hormones.
    e. They provide the body with energy.

13. Mineralocorticoids
    a. help prevent the loss of potassium from the body
    b. are secreted based on the renin–angiotensin–aldosterone pathway
    c. increase the rate of sodium loss in the urine
    d. are involved in lowering the body's blood pressure
    e. increase water loss from the body by increasing urine production

14. A lack of iodine in the diet affects the production of which hormone?
    a. calcitonin
    b. parathyroid hormone
    c. aldosterone
    d. thyroxine
    e. glucagon

15. Which of the following hormones with opposite effects are correctly paired?
    a. parathyroid hormone, thyroid hormones
    b. parathyroid hormone, calcitonin
    c. oxytocin, glucocorticoids
    d. aldosterone, oxytocin
    e. thyroid hormones, thymosin

16. The hormone that normally functions as part of a positive feedback cycle is
    a. cortisol
    b. testosterone
    c. oxytocin
    d. insulin
    e. thyroxine

17. Match the following:

___a. produce thyroid hormones	A. posterior pituitary
___b. secrete insulin	B. adrenal cortex
___c. release hormones into capillaries of the posterior pituitary	C. follicular cells
___d. store oxytocin	D. alpha cells
___e. secrete glucagon	E. parafollicular cells
___f. produce calcitonin	F. beta cells
___g. secrete steroid hormones	G. neurosecretory cells

18. In a dehydrated person, you would expect to see an increased release of
    a. parathyroid hormone
    b. aldosterone
    c. insulin
    d. melatonin
    e. inhibin

19. Match the following:

___a. diabetes insipidus	A. hypersecretion of glucocorticoids
___b. diabetes mellitus	B. hyposecretion of antidiuretic hormone
___c. myxedema	C. hyposecretion of insulin
___d. Cushing's syndrome	D. hyposecretion of parathyroid hormone
___e. Addison's disease	E. hyposecretion of thyroid hormone
___f. tetany	F. hyposecretion of glucocorticoids

20. For each of the following, indicate at which stage they would occur as part of the stress response. Use **F** to indicate fight-or-flight response, **R** to indicate resistance reaction, and **E** to indicate exhaustion.
    ___ a. initiated by hypothalamic releasing hormones
    ___ b. initiated by the sympathetic division of the autonomic nervous system
    ___ c. immediately prepares the body for action
    ___ d. increases cortisol release
    ___ e. short-lived response
    ___ f. body resources become depleted
    ___ g. increased release of many hormones that ensure a continued ATP supply
    ___ h. failure of pancreatic beta cells
    ___ i. nonessential body functions inhibited

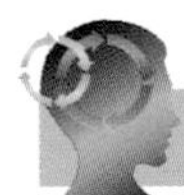

CRITICAL THINKING APPLICATIONS

1. Patrick was diagnosed with diabetes mellitus on his 8th birthday. His 65-year-old aunt was just diagnosed with diabetes also. Patrick is having a hard time understanding why he needs injections, while his aunt controls her blood sugar with diet and oral medication. Why is his aunt's treatment different from his?

2. Eddie, the tallest man in the world, suffered from an oversecretion of a pituitary hormone his entire life. Eddie died in early adulthood due to the effects of his condition. Name Eddie's condition and explain its cause.

3. Melatonin has been suggested as a possible aid for sleeping problems due to jet lag and rotating work schedules (shift work). It may also be involved in seasonal affective disorder (SAD). Explain how melatonin may affect sleeping.

4. Brian is in a 50-mile bike-a-thon on a hot summer day. He's breathing dust at the back of the pack, he's sweating profusely, and now he's lost his water bottle. Brian is not having a good time. How will his hormones respond to decreased intake of water and the stress of the situation?

ANSWERS TO FIGURE QUESTIONS

13.1 Secretions of endocrine glands diffuse into interstitial fluid and then into the blood; exocrine secretions flow into ducts that lead into body cavities or to the body surface.

13.2 RNA molecules are synthesized when genes are expressed (transcribed), and then mRNA codes for the synthesis of protein molecules.

13.3 It brings the message of the first messenger, the water-soluble hormone, into the cell.

13.4 The posterior pituitary releases hormones synthesized in the hypothalamus.

13.5 Oxytocin's target cells are in the uterus and mammary glands.

13.6 Absorption of a large glass of water in the intestines would decrease the osmotic pressure (concentration of solutes) of your blood plasma, turning off secretion of ADH and decreasing the ADH level in your blood.

13.7 Follicular cells secrete T_3 and T_4; parafollicular cells secrete calcitonin.

13.8 Thyroid hormones increase metabolic rate.

13.9 PTH increases the number and activity of osteoclasts.

13.10 Target tissues for PTH are bone and kidneys; the target tissue for calcitonin is bone; the target tissue for calcitriol is the gastrointestinal (GI) tract.

13.11 The pancreas is both an endocrine and an exocrine gland.

13.12 Glucagon is considered an "anti-insulin" hormone because it has several effects that are opposite to those of insulin.

13.13 The outer zone of the adrenal cortex secretes mineralocorticoids, the middle zone secretes glucocorticoids, and the inner zone secretes adrenal androgens.

13.14 Because drugs that block ACE lower blood pressure, they are used to treat high blood pressure (hypertension).

13.15 In Graves disease, antibodies are produced that mimic the action of TSH.

THE CARDIOVASCULAR SYSTEM: BLOOD

chapter 14

did you know?

The American Red Cross calls blood donation "The Gift of Life." But many people receiving a blood donation, or transfusion, worry about the safety of the blood they will receive. Many viruses can be transmitted easily from donor to patient through a transfusion. Because the blood supply is carefully screened for the presence of viruses such as HIV (the virus that causes AIDS) and the viruses that cause Hepatitis B and Hepatitis C, the risk of receiving unsafe blood is extremely low. For example, the risk of receiving a unit of HIV-positive blood is 1 in 1.5 million.

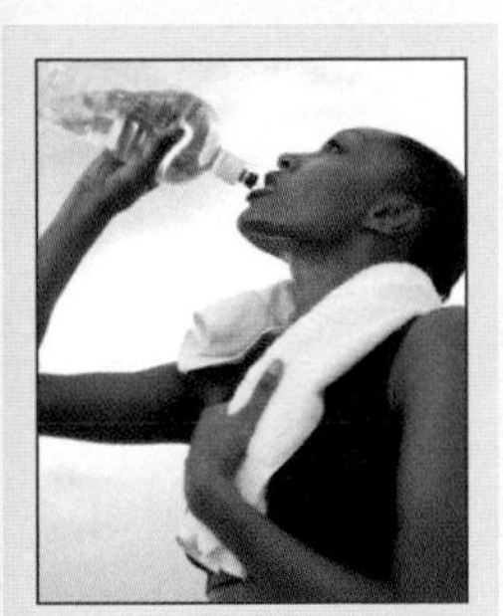

Focus on Wellness, page 359

www.wiley.com/college/apcentral

The ***cardiovascular system*** (*cardio-* = heart; *-vascular* = blood or blood vessels) consists of three interrelated components: blood, the heart, and blood vessels. The focus of this chapter is blood; the next two chapters will cover the heart and blood vessels, respectively.

Functionally, the cardiovascular system transports substances to and from body cells. To perform its functions, blood must circulate throughout the body. The heart serves as the pump for circulation, and blood vessels carry blood from the heart to body cells and from body cells back to the heart.

The branch of science concerned with the study of blood, blood-forming tissues, and the disorders associated with them is ***hematology*** (hēm-a-TOL-ō-jē; *hemo-* or *hemato-* = blood; *-logy* = study of).

looking back to move ahead . . .

- **Blood Tissue (page 90)**
- **Positive Feedback System (page 8)**
- **Phagocytosis (page 51)**

FUNCTIONS OF BLOOD

OBJECTIVE • List and describe the functions of blood.

Blood, a liquid connective tissue, has three general functions: transportation, regulation, and protection.

1. **Transportation**. Blood transports oxygen from the lungs to cells throughout the body and carbon dioxide (a waste product of cellular respiration; see Chapter 20) from the cells to the lungs. It also carries nutrients from the gastrointestinal tract to body cells, heat and waste products away from cells, and hormones from endocrine glands to other body cells.
2. **Regulation**. Blood helps regulate the pH of body fluids. The heat-absorbing and coolant properties of the water in blood plasma (see page 30) and its variable rate of flow through the skin help adjust body temperature. Blood osmotic pressure also influences the water content of cells.
3. **Protection**. Blood clots (becomes gel-like) in response to an injury, which protects against its excessive loss from the cardiovascular system. In addition, white blood cells protect against disease by carrying on phagocytosis and producing proteins called antibodies. Blood contains additional proteins, called interferons and complement, that also help protect against disease.

■ **CHECKPOINT**

1. Name several substances transported by blood.
2. How is blood protective?

COMPONENTS OF WHOLE BLOOD

OBJECTIVE • Discuss the formation, components, and functions of whole blood.

Blood is denser and more viscous (thicker) than water. The temperature of blood is about 38°C (100.4°F). Its pH is slightly alkaline, ranging from 7.35 to 7.45. Blood constitutes about 8% of the total body weight. The blood volume is 5 to 6 liters (1.5 gal) in an average-sized adult male and 4 to 5 liters (1.2 gal) in an average-sized adult female. The difference in volume is due to differences in body size.

Whole blood is composed of two portions: (1) *blood plasma*, a liquid that contains dissolved substances, and (2) *formed elements*, which are cells and cell fragments. If a sample of blood is centrifuged (spun at high speed) in a small glass tube, the cells sink to the bottom of the tube and the lighter-weight blood plasma forms a layer on top (Figure 14.1a). Blood is about 45% formed elements and 55% plasma. Normally, more than 99% of the formed elements are red blood cells (RBCs). The percentage of total blood volume occupied by red blood cells is termed the ***hematocrit*** (he-MAT-ō-krit). Pale, colorless white blood cells (WBCs) and platelets occupy less than 1% of total blood volume. They form a very thin layer, called the *buffy coat*, between the packed RBCs and blood plasma in centrifuged blood. Figure 14.1b shows the composition of blood plasma and the numbers of the various types of formed elements in blood.

Blood Plasma

When the formed elements are removed from blood, a straw-colored liquid called ***blood plasma*** (or simply ***plasma***) remains. Plasma is about 91.5% water, 7% proteins, and 1.5% solutes other than proteins. Proteins in the blood, the *plasma proteins*, are synthesized mainly by the liver. The most plentiful plasma proteins are the ***albumins***, which account for about 54% of all plasma proteins. Among other functions, albumins help maintain proper blood osmotic pressure, which is an important factor in the exchange of fluids across capillary walls. ***Globulins***, which compose 38% of plasma proteins, include ***antibodies***, defensive proteins produced during certain immune responses. ***Fibrinogen*** makes up about 7% of plasma proteins and is a key protein in formation of blood clots. Other solutes in plasma include electrolytes, nutrients, gases, regulatory substances such as enzymes and hormones, vitamins, and waste products.

Formed Elements

The ***formed elements*** of the blood are the following (see Figure 14.2 on page 348):

I. Red blood cells
II. White blood cells
   A. Granular leukocytes (contain conspicuous granules that are visible under a light microscope after staining)
      1. Neutrophils
      2. Eosinophils
      3. Basophils
   B. Agranular leukocytes (no granules are visible under a light microscope after staining)
      1. T and B lymphocytes and natural killer cells
      2. Monocytes
III. Platelets

Formation of Blood Cells

The process by which the formed elements of blood develop is called ***hemopoiesis*** (hē-mō-poy-Ē-sis; *-poiesis* = making).

Figure 14.1 Components of blood in a normal adult.

Blood is a connective tissue that consists of blood plasma (liquid) plus formed elements: red blood cells, white blood cells, and platelets.

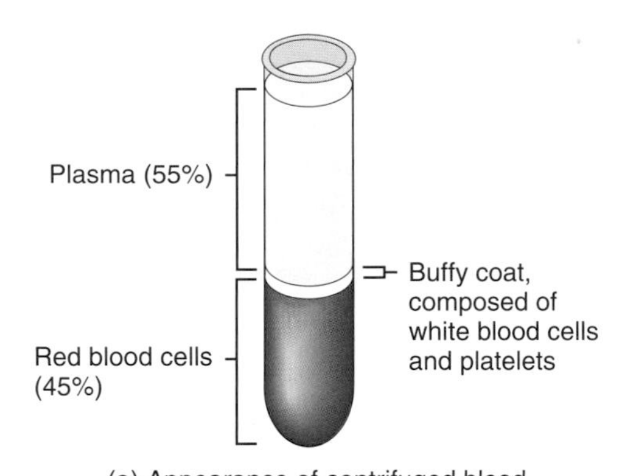

(a) Appearance of centrifuged blood

Functions of Blood

1. Transport of oxygen, carbon dioxide, nutrients, hormones, heat, and wastes.
2. Regulation of pH, body temperature, and water content of cells.
3. Protection against blood loss through clotting
4. Protection against disease through platelets; phagocytic white blood cells; and proteins such as antibodies, complement, and interferons.

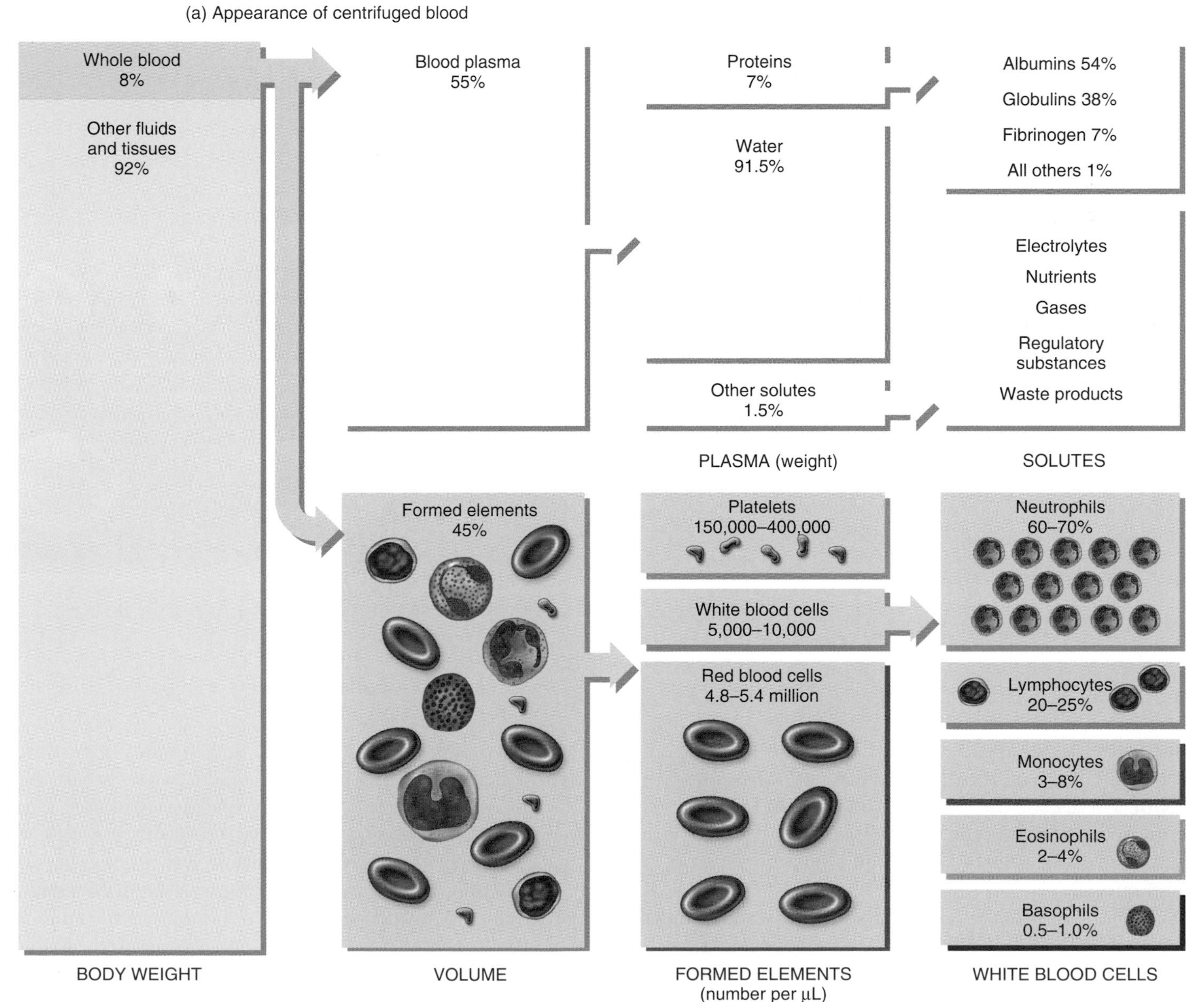

(b) Components of blood

Which formed elements of blood are most numerous?

Figure 14.2 Origin, development, and structure of blood cells. Some of the generations of some cell lines have been omitted.

Blood cell production, called hemopoiesis, occurs in red bone marrow after birth.

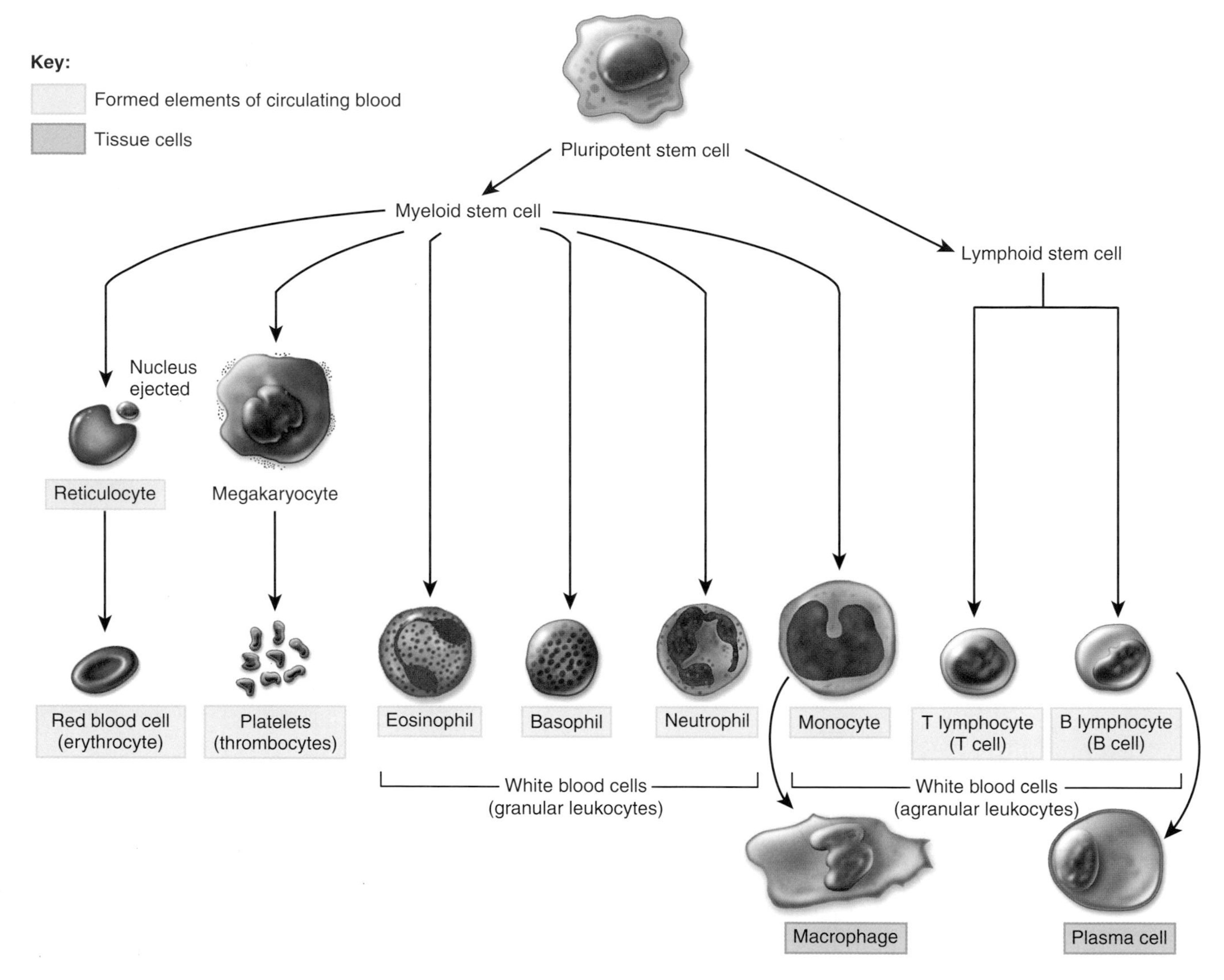

(a) Origin of blood cells from pluripotent stem cells

Before birth, hemopoiesis first occurs in the yolk sac of an embryo and later in the liver, spleen, thymus, and lymph nodes of a fetus. In the last three months before birth, red bone marrow becomes the primary site of hemopoiesis and continues as the source of blood cells after birth and throughout life.

Red bone marrow is a highly vascularized connective tissue located in the microscopic spaces between trabeculae of spongy bone tissue. It is present chiefly in bones of the axial skeleton, pectoral and pelvic girdles, and the proximal epiphyses of the humerus and femur. About 0.05–0.1% of red bone marrow cells are cells called ***pluripotent stem cells*** (ploo-RIP-ō-tent; *pluri-* = several). Pluripotent stem cells are cells that have the capacity to develop into many different types of cells (Figure 14.2a).

In response to stimulation by specific hormones, pluripotent stem cells generate two other types of stem cells which have the capacity to develop into fewer types of cells: *myeloid stem cells* and *lymphoid stem cells* (Figure 14.2a). Myeloid stem cells begin their development in red bone marrow and differentiate into several types of cells from which red blood cells, platelets, eosinophils, basophils, neutrophils, and monocytes develop. Lymphoid stem cells begin their development in red bone marrow but complete it in lymphatic tissues. They differentiate into cells from which the T and B lymphocytes develop.

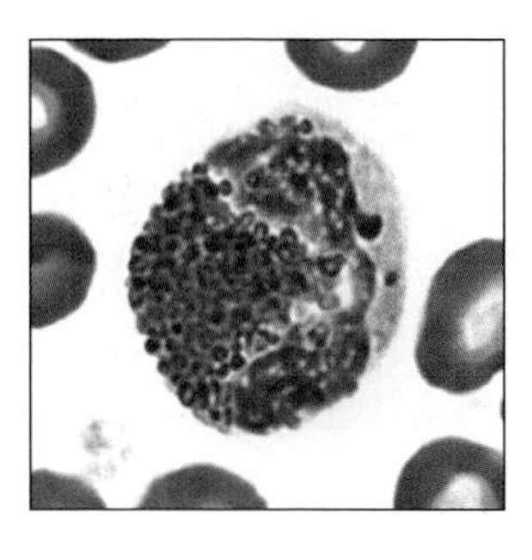

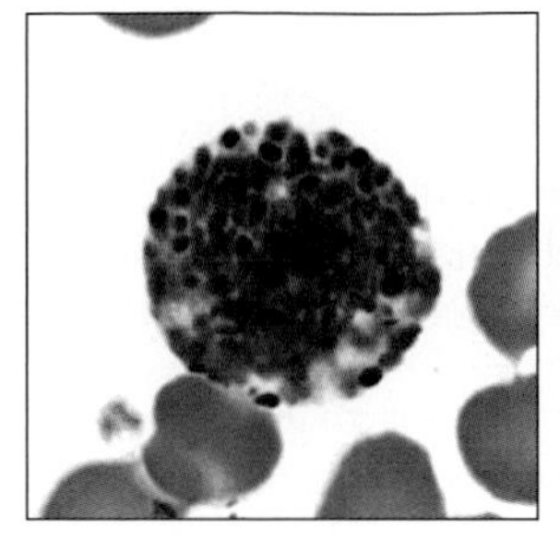

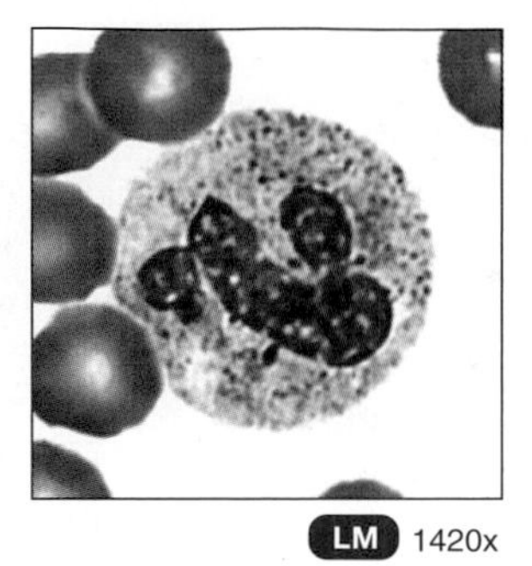

Eosinophil Basophil Neutrophil

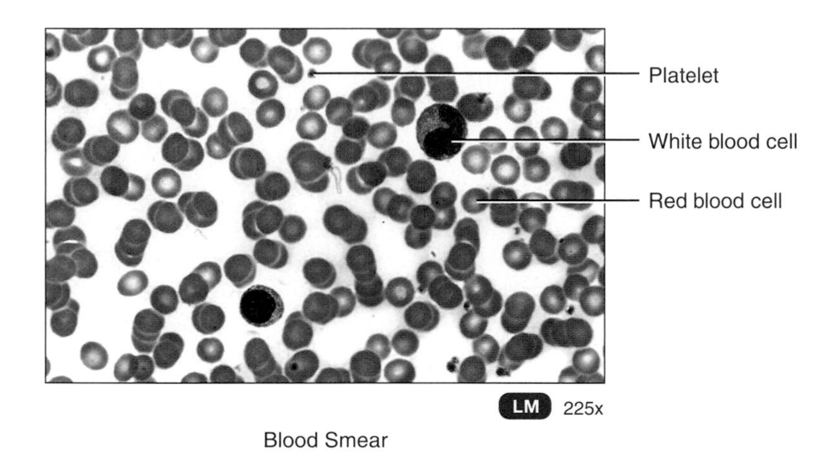

Blood Smear

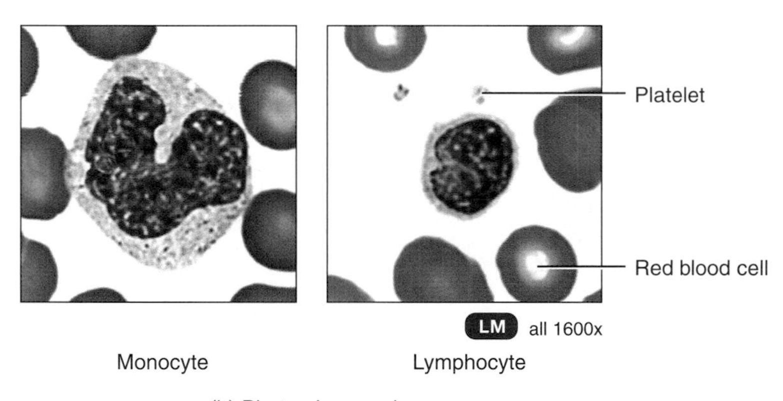

Monocyte Lymphocyte

(b) Photomicrographs

What percentage of body weight is made up of blood?

Red Blood Cells

RBC STRUCTURE ***Red blood cells (RBCs)*** or ***erythrocytes*** (e-RITH-rō-sīts; *erythro-* = red; *-cyte* = cell) contain the oxygen-carrying protein ***hemoglobin***, which is a pigment that gives whole blood its red color. Hemoglobin also transports about 23% of the carbon dioxide in the blood. A healthy adult male has about 5.4 million red blood cells per microliter (μL) of blood, and a healthy adult female has about 4.8 million. (One drop of blood is about 50μL.) Again this difference reflects differences in body size. To maintain normal numbers of RBCs, new mature cells must enter the circulation at the astonishing rate of at least 2 million per second, a pace that balances the equally high rate of RBC destruction. RBCs are biconcave (concave on both sides) discs averaging about 8μm* in diameter. Mature RBCs lack a nucleus and other organelles and can neither reproduce nor carry on extensive metabolic activities. However, all of their internal space is available for oxygen and carbon dioxide transport. Essentially, RBCs consist of a selectively permeable plasma membrane, cytosol, and hemoglobin.

*1μm = 1/25,000 of an inch or 1/10,000 of a centimeter (cm), which is 1/1000 of a millimeter (mm).

Since a biconcave disc has a much greater surface area for its volume (compared to a sphere or a cube), this shape provides a large surface area for the diffusion of gas molecules into and out of a RBC.

> Delivery of oxygen to muscles is a limiting factor in muscular feats. As a result, increasing the oxygen-carrying capacity of the blood enhances athletic performance, especially in endurance events. Because RBCs are the main transport vehicle for oxygen, athletes have tried several means of increasing their RBC count, causing **induced polycythemia**, to gain a competitive edge. Athletes have enhanced their RBC production by injecting *Epoetin alfa* (*Procrit®* or *Epogen®*), a drug that is used to treat anemia by stimulating the production of RBCs by red bone marrow. Practices that increase the number of RBCs are dangerous because they raise the viscosity of the blood, which increases the resistance to blood flow and makes the blood more difficult for the heart to pump. Increased viscosity also contributes to high blood pressure and increased risk of stroke. During the 1980s, at least 15 competitive cyclists died from heart attacks or strokes linked to suspected use of Epoetin alfa. Although the International Olympics Committee bans Epoetin alfa use, enforcement is difficult because the drug is identical to naturally occurring EPO.

RBC LIFE CYCLE Red blood cells live only about 120 days because of wear and tear on their plasma membranes as they squeeze through blood capillaries. Worn-out red blood cells are removed from circulation as follows (Figure 14.3).

1. Macrophages in the spleen, liver, and red bone marrow phagocytize ruptured and worn-out red blood cells, splitting apart the heme and globin portions of hemoglobin.
2. The protein globin is broken down into amino acids, which can be reused by body cells to synthesize other proteins.
3. Iron removed from the heme portion associates with the plasma protein ***transferrin*** (trans-FER-in; *trans-* = across; *ferr-* = iron), which acts as a transporter.
4. The iron-transferrin complex is then carried to red bone marrow, where RBC precursor cells use it in hemoglobin synthesis. Iron is needed for the heme portion of the hemoglobin molecule, and amino acids are needed for the globin portion. Vitamin B_{12} is also needed for synthesis of hemoglobin. (The lining of the stomach must produce a protein called *intrinsic factor* for absorption of dietary vitamin B_{12} from the GI tract into the blood.)
5. Erythropoiesis in red bone marrow results in the production of red blood cells, which enter the circulation.

Figure 14.3 Formation and destruction of red blood cells, and the recycling of hemoglobin components.

The rate of RBC formation by red bone marrow equals the rate of RBC destruction by macrophages.

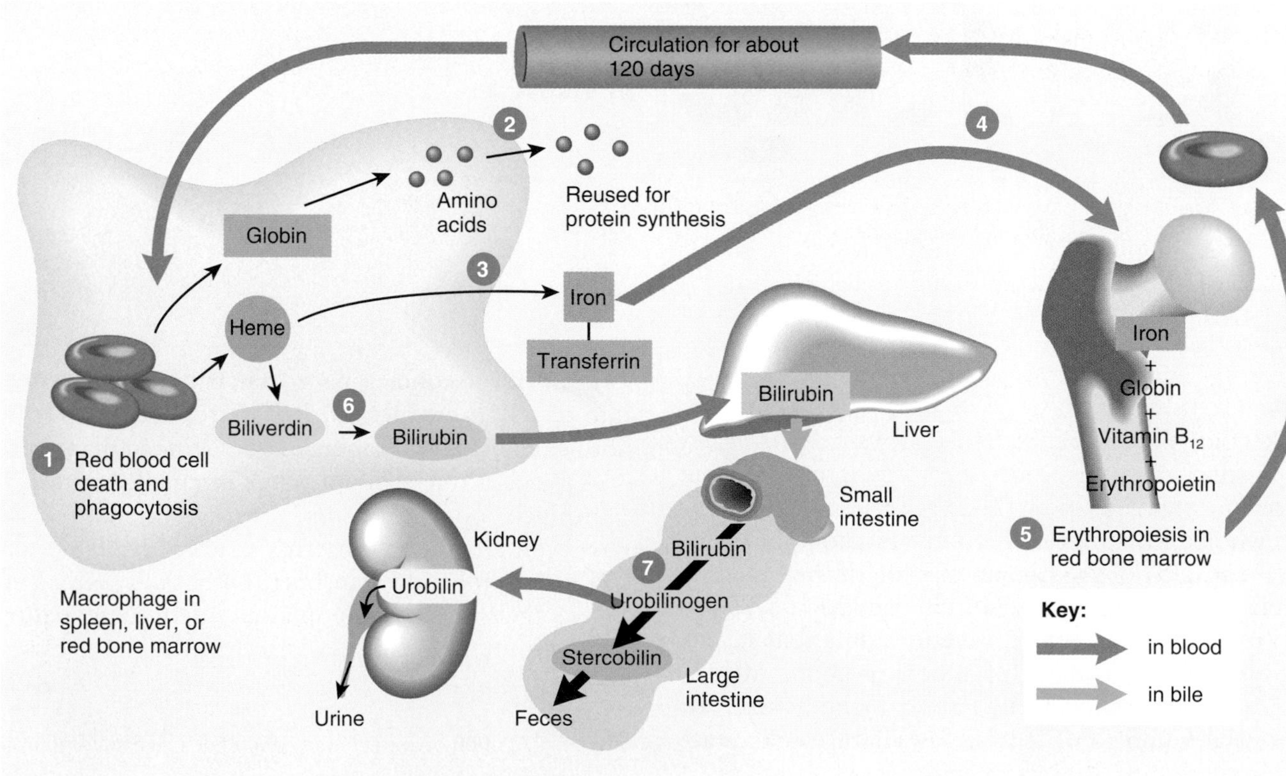

What substance is responsible for the brown color of feces?

6. When iron is removed from heme, the non-iron portion of heme is converted to ***biliverdin*** (bil′-i-VER-din), a green pigment, and then into ***bilirubin*** (bil′-i-ROO-bin), a yellow-orange pigment. Bilirubin enters the blood and is transported to the liver. Within the liver, bilirubin is secreted by liver cells into bile, which passes into the small intestine and then into the large intestine.
7. In the large intestine, bacteria convert bilirubin into ***urobilinogen*** (ūr-ō-bī-LIN-ō-jen). Some urobilinogen is absorbed back into the blood, converted to a yellow pigment called ***urobilin*** (ūr-ō-BĪ-lin), and excreted in urine. Most urobilinogen is eliminated in feces in the form of a brown pigment called ***stercobilin*** (ster′-kō-BĪ-lin), which gives feces its characteristic color.

Because free iron ions bind to and damage molecules in cells or in the blood, transferrin acts as a protective "protein escort" during transport of iron ions. As a result, plasma contains virtually no free iron.

RBC PRODUCTION The formation of blood cells in general is called hemopoiesis; the formation of just RBCs is termed ***erythropoiesis*** (e-rith′-rō-poy-Ē-sis). Near the end of erythropoiesis, an RBC precursor ejects its nucleus and becomes a ***reticulocyte*** (re-TIK-ū-lō-sīt; see Figure 14.2a). Loss of the nucleus causes the center of the cell to indent, producing the RBC's distinctive biconcave shape. Reticulocytes, which are about 34% hemoglobin and retain some mitochondria, ribosomes, and endoplasmic reticulum, pass from red bone marrow into the bloodstream. Reticulocytes usually develop into mature RBCs within 1 to 2 days after their release from bone marrow.

Normally, erythropoiesis and destruction of RBCs proceed at the same pace. If the oxygen-carrying capacity of the blood falls because erythropoiesis is not keeping up with RBC destruction, RBC production increases (Figure 14.4). The controlled condition in this particular negative feedback loop is the amount of oxygen delivered to the kidneys (and thus to body tissues in general). ***Hypoxia*** (hī-POKS-ē-a), a deficiency of oxygen, stimulates increased release of ***erythropoietin*** (e-rith′-rō-POY-ē-tin), or ***EPO***, a hormone made by the kidneys. EPO circulates through the blood to the red bone marrow, where it stimulates erythropoiesis. The larger the number of RBCs in the blood, the higher the oxygen delivery to the tissues (Figure 14.4). A person with prolonged hypoxia may develop a life-threatening condition called ***cyanosis*** (sī′-a-NŌ-sis), characterized by a bluish-purple skin coloration most easily seen in the nails and mucous membranes. Oxygen delivery may fall due to ***anemia*** (a lower-than-normal number of RBCs or reduced quantity of hemoglobin) or circulatory problems that reduce blood flow to tissues.

A test that measures the rate of erythropoiesis is called a reticulocyte count. This and several other tests related to red blood cells are explained in Table 14.1.

Figure 14.4 Negative feedback regulation of erythropoiesis (red blood cell formation).

The main stimulus for erythropoiesis is hypoxia, a decrease in the oxygen-carrying capacity of the blood.

Some stimulus disrupts homeostasis by
Decreasing
Oxygen delivery to kidneys (and other tissues)
Receptors
Kidney cells that secrete erythropoietin detect low oxygen level
Input
Increased erythropoietin secreted into blood
Control center
Proerythroblasts in red bone marrow mature more quickly into reticulocytes
Output
More reticulocytes enter circulating blood
Effectors
Larger number of RBCs in circulation
Increased oxygen delivery to tissues
Return to homeostasis when oxygen delivery to kidneys increases to normal

What is the term for cellular oxygen deficiency?

Premature newborns often exhibit anemia, due in part to inadequate production of erythropoietin. During the first weeks after birth, the liver, not the kidneys, produces most EPO. Because the liver is less sensitive than the kidneys to hypoxia, newborns have a smaller EPO response to anemia than do adults. In addition, in infants, fetal hemoglobin is converted into adult hemoglobin; since fetal hemoglobin carries up to 30% more oxygen, the loss of fetal hemoglobin makes the anemia worse.

Table 14.1 Obtaining Blood Samples and Common Medical Tests Involving Blood

I. Obtaining Blood Samples

A. Venipuncture. This most frequently used procedure involves withdrawal of blood from a vein using a sterile needle and syringe. (Veins are used instead of arteries because they are closer to the skin, more readily accessible, and contain blood at a much lower pressure.) A commonly used vein is the median cubital vein in front of the elbow (see Figure 16.14 on page 407). A tourniquet is wrapped around the arm, which stops blood flow through the veins and makes the veins below the tourniquet stand out.

B. Fingerstick. Using a sterile needle or lancet, a drop or two of capillary blood is taken from a finger, earlobe, or heel.

C. Arterial stick. Sample is most often taken from radial artery in the wrist or femoral artery in the thigh (see Figure 16.9 on page 397).

II. Testing Blood Samples

A. Reticulocyte count (indicates the rate of erythropoiesis)

Normal value: 0.5% to 1.5%.

Abnormal values: A high reticulocyte count might indicate the presence of bleeding or hemolysis (rupture of erythrocytes), or it may be the response of someone who is iron deficient. Low reticulocyte count in the presence of anemia might indicate a malfunction of the red bone marrow, owing to a nutritional deficiency, pernicious anemia, or leukemia.

B. Hematocrit (the percentage of red blood cells in blood). A hematocrit of 40 means that 40% of the volume of blood is composed of RBCs.

Normal values:

Females: 38 to 46 (average 42)
Males: 40 to 54 (average 47)

Abnormal values: The test is used to diagnose anemia, polycythemia (an increased percentage of red blood cells), and abnormal states of hydration. Anemia may vary from mild (hematocrit of 35) to severe (hematocrit of less than 15). Athletes often have a higher-than-average hematocrit, and the average hematocrit of persons living at high altitude is greater than that of persons living at sea level.

C. Differential white blood cell count (the percentage of each type of white blood cells in a sample of 100 WBCs)

Normal values:

Type of WBC	Percentage
neutrophils	60–70
eosinophils	2–4
basophils	0.5–1
lymphocytes	20–25
monocytes	3–8

Abnormal values: A high neutrophil count might result from bacterial infections, burns, stress, or inflammation; a low neutrophil count might be caused by radiation, certain drugs, vitamin B_{12} deficiency, or systemic lupus erythematosus (SLE) (see page 93). A high eosinophil count could indicate allergic reactions, parasitic infections, autoimmune disease, or adrenal insufficiency; a low eosinophil count could be caused by certain drugs, stress, or Cushing's syndrome. Basophils could be elevated in some types of allergic responses, leukemias, cancers, and hyperthyroidism; decreases in basophils could occur during pregnancy, ovulation, stress, and hyperthyroidism. High lymphocyte counts could indicate viral infections, immune diseases, and some leukemias; low lymphocyte counts might occur as a result of prolonged severe illness, high steroid levels, and immunosuppression. A high monocyte count could result from certain viral or fungal infections, tuberculosis (TB), some leukemias, and chronic diseases; low monocyte levels rarely occur.

D. Complete blood count (CBC) (provides information about the formed elements in blood)*

Normal values:

RBC count	About 5.4 million per μL in males About 4.8 million per μL in females
Hemoglobin	14–18 g/dl in adult males 12–16 g/dl in adult females
Hematocrit	See B
WBC count	5,000–10,000 per μL
Differential white blood count	See C
Platelet count	150,000–400,000 μL

Abnormal values: Increased RBC count, hemoglobin, and hematocrit occur in polycythemia, congenital heart disease, and hypoxia; decreased RBC count, hemoglobin, and hematocrit occur in hemorrhage and certain types of anemia. Increased WBC counts may indicate acute or chronic infections, trauma, leukemia, or stress (see also above under differential white blood cell count). Decreased WBC counts could indicate anemia and viral infections (see also above under differential white blood cell count). High platelet counts may indicate cancer, trauma, or cirrhosis. Low platelet counts could indicate anemia, allergic conditions, or hemorrhage.

*Not all components of a CBC have been included.

White Blood Cells

WBC STRUCTURE AND TYPES Unlike red blood cells, ***white blood cells (WBCs)*** or ***leukocytes*** (LOO-kō-sīts; *leuko-* = white) have nuclei and do not contain hemoglobin. WBCs are classified as either granular or agranular, depending on whether they contain chemical-filled cytoplasmic granule (vesicles) that are made visible by staining when viewed through a light microscope. The *granular leukocytes* include ***neutrophils*** (NOO-trō-fils), ***eosinophils*** (ē′-ō-SIN-ō-fils), and ***basophils*** (BĀ-sō-fils). The *agranular leukocytes* include ***lymphocytes*** and ***monocytes*** (MON-ō-sīts). (See Table 14.2 for the sizes and microscopic characteristics of WBCs.)

WBC FUNCTIONS The skin and mucous membranes of the body are continuously exposed to microbes (microscopic organisms), such as bacteria, some of which are capable of invading deeper tissues and causing disease. Once microbes enter the body, some WBCs combat them by ***phagocytosis***, and others produce antibodies. Neutrophils respond first to bacterial invasion, carrying on phagocytosis and releasing enzymes such as lysozyme that destroy certain bacteria. Monocytes take longer to reach the site of infection than neutrophils, but they eventually arrive in larger numbers. Monocytes that migrate into infected tissues develop into cells called ***wandering macrophages*** (*macro-* = large; *-phages* = eaters), which can phagocytize many more microbes than neutrophils. They also clean up cellular debris following an infection.

Eosinophils leave the capillaries and enter interstitial fluid. They release enzymes that combat inflammation in allergic reactions. Eosinophils also phagocytize antigen–antibody complexes and are effective against certain parasitic worms. A high eosinophil count often indicates an allergic condition or a parasitic infection.

Basophils are also involved in inflammatory and allergic reactions. They leave capillaries, enter tissues, and can liberate heparin, histamine, and serotonin. These substances intensify the inflammatory reaction and are involved in allergic reactions.

Three types of lymphocytes—B cells, T cells, and natural killer (NK) cells—are the major combatants in immune responses, which are described in detail in Chapter 17. B cells develop into plasma cells, which produce antibodies that help destroy bacteria and inactivate their toxins. T cells attack viruses, fungi, transplanted cells, cancer cells, and some bacteria. Natural killer cells attack a wide variety of infectious microbes and certain spontaneously arising tumor cells.

White blood cells and other nucleated body cells have proteins, called ***major histocompatibility (MHC) antigens***, protruding from their plasma membrane into the extracellular fluid. These "cell identity markers" are unique for each person (except identical twins). Although RBCs (which do not possess nuclei) possess blood group antigens, they lack the MHC antigens. An incompatible tissue transplant is rejected by the recipient due, in part, to differences in donor and recipient MHC antigens. The MHC antigens are used to type tissues to identify compatible donors and recipients and thus reduce the chance of tissue rejection.

WBC LIFE SPAN Red blood cells outnumber white blood cells about 700 to 1. There are normally about 5000 to 10,000 WBCs per μL of blood. Bacteria have continuous access to the body through the mouth, nose, and pores of the skin. Furthermore, many cells, especially those of epithelial tissue, age and die daily, and their remains must be removed. However, a WBC can phagocytize only a certain amount of material before it interferes with the WBC's own metabolic activities. Thus, the life span of most WBCs is only a few days. During a period of infection, many WBCs live only a few hours. However, some B and T cells remain in the body for years.

Leukocytosis (loo′-kō-sī-TŌ-sis), an increase in the number of WBCs, is a normal, protective response to stresses such as invading microbes, strenuous exercise, anesthesia, and surgery. Leukocytosis usually indicates an inflammation or infection. Because each type of white blood cell plays a different role, determining the percentage of each type in the blood assists in diagnosing the condition. This test, called a ***differential white blood cell count***, measures the number of each kind of white cell in a sample of 100 white blood cells (see Table 14.1). An abnormally low level of white blood cells (below 5000 cells/μL), called ***leukopenia*** (loo′-kō-PĒ-nē-a), is never beneficial; it may be caused by exposure to radiation, shock, and certain chemotherapeutic agents.

WBC PRODUCTION Leukocytes develop in red bone marrow. As shown in Figure 14.2a, monocytes and granular leukocytes develop from a myeloid stem cell. T and B cells develop from a lymphoid stem cell.

Platelets

Pluripotent stem cells also differentiate into cells that produce platelets (see Figure 14.2a). Some myeloid stem cells develop into cells called *megakaryoblasts*, which in turn transform into megakaryocytes, huge cells that splinter into 2000–3000 fragments in the red bone marrow and then enter the bloodstream. Each fragment, enclosed by a piece of the megakaryocyte cell membrane, is a ***platelet***. Between 150,000 and 400,000 platelets are present in each μL of blood. Platelets are disc-shaped, have a diameter of 2–4 μm, and exhibit many vesicles but no nucleus. When blood vessels are damaged, platelets help stop blood loss by forming a platelet plug. Their vesicles also contain chemicals that promote blood clotting (both processes are described shortly). After their short life span of 5–9 days, platelets are removed by macrophages in the spleen and liver.

A **bone marrow transplant** is the replacement of cancerous or abnormal red bone marrow with healthy red bone marrow in order to establish normal blood cell counts. The defective red bone marrow is destroyed by high doses of chemotherapy and whole body radiation just before the transplant takes place. These treatments kill the cancer cells and destroy the patient's immune system in order to decrease the chance of transplant rejection. The red bone marrow from a donor is usually removed from the hip bone under general anesthesia with a syringe and is then injected into the recipient's vein, much like a blood transfusion. The injected marrow migrates to the recipient's red bone marrow cavities, and the stem cells in the marrow multiply. If all goes well, the recipient's red bone marrow is replaced entirely by healthy, noncancerous cells.

Bone marrow transplants have been used to treat aplastic anemia, certain types of leukemia, severe combined immunodeficiency disease (SCID), Hodgkin's disease, non-Hodgkin's lymphoma, multiple myeloma, thalassemia, sickle-cell disease, breast cancer, ovarian cancer, testicular cancer, and hemolytic anemia. However, there are some drawbacks. Since the recipient's white blood cells have been completely destroyed by chemotherapy and radiation, the patient is extremely vulnerable to infection. (It takes about 2–3 weeks for transplanted bone marrow to produce enough white blood cells to protect against infection.) In addition, transplanted red bone marrow may produce T lymphocytes that attack the recipient's tissues. Another drawback is that patients must take immunosuppressive drugs for life. Because these drugs reduce the level of immune system activity, they increase the risk of infection.

Table 14.2 presents a summary of the formed elements in blood.

■ **CHECKPOINT**

3. Briefly outline the process of hemopoiesis.
4. What is erythropoiesis? How does erythropoiesis affect hematocrit? What factors speed up and slow down erythropoiesis?
5. What functions do neutrophils, eosinophils, basophils, monocytes, B cells, T cells, and natural killer cells perform?
6. How are leukocytosis and leukopenia different? What is a differential white blood cell count?

HEMOSTASIS

OBJECTIVE • Describe the various mechanisms that prevent blood loss.

Hemostasis (hē′-mō-STĀ-sis; *-stasis* = standing still) is a sequence of responses that stops bleeding when blood vessels are injured. (Be sure not to confuse the two words *hemostasis* and *homeostasis*.) The hemostatic response must be quick, localized to the region of damage, and carefully controlled. Three mechanisms can reduce loss of blood from blood vessels: (1) vascular spasm, (2) platelet plug formation, and (3) blood clotting (coagulation). When successful, hemostasis prevents ***hemorrhage*** (HEM-or-ij; *-rhage* = burst forth), the loss of a large amount of blood from the vessels. Hemostasis can prevent hemorrhage from smaller blood vessels, but extensive hemorrhage from larger vessels usually requires medical intervention.

Vascular Spasm

When a blood vessel is damaged, the smooth muscle in its wall contracts immediately, a response called a ***vascular spasm***. Vascular spasm reduces blood loss for several minutes to several hours, during which time the other hemostatic mechanisms begin to operate. The spasm is probably caused by damage to the smooth muscle and by reflexes initiated by pain receptors. As platelets accumulate at the damaged site, they release chemicals that enhance vasoconstriction (narrowing of a blood vessel), thus maintaining the vascular spasm.

Platelet Plug Formation

When platelets come into contact with parts of a damaged blood vessel, their characteristics change drastically and they quickly come together to form a platelet plug that helps fill the gap in the injured blood vessel wall. Platelet plug formation occurs as follows.

Initially, platelets contact and stick to parts of a damaged blood vessel, such as collagen fibers. Then, they interact with one another and begin to liberate the chemicals. The chemicals activate nearby platelets and sustain the vascular spasm, which decreases blood flow through the injured vessel. The release of platelet chemicals makes other platelets in the area sticky, and the stickiness of the newly recruited and activated platelets causes them to stick to the originally activated platelets. Eventually, a large number of platelets forms a mass called a ***platelet plug***. A platelet plug can stop blood loss completely if the hole in a blood vessel is small enough.

Blood Clotting

Normally, blood remains in its liquid form as long as it stays within its vessels. If it is withdrawn from the body, however, it thickens and forms a gel. Eventually, the gel separates from the liquid. The straw-colored liquid, called ***serum***, is simply plasma minus the clotting proteins. The gel is called a ***clot*** and consists of a network of insoluble protein fibers called ***fibrin*** in which the formed elements of blood are trapped (see Figure 14.5).

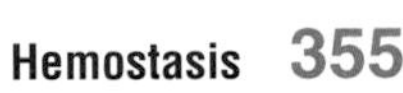

Table 14.2 Summary of Formed Elements in Blood

Name and Appearance	Number	Characteristics*	Functions
Red Blood Cells (RBCs) or Erythrocytes	4.8 million/μL in females; 5.4 million/μL in males.	7–8 μm diameter, biconcave discs, without a nucleus; live for about 120 days.	Hemoglobin within RBCs transports most of the oxygen and part of the carbon dixoide in the blood.
White Blood Cells (WBCs) or Leukocytes	5000–10,000/μL	Most live for a few hours to a few days.†	Combat pathogens and other foreign substances that enter the body.
Granular Leukocytes			
Neutrophils	60–70% of all WBCs.	10–12 μm diameter; nucleus has 2–5 lobes connected by thin strands of chromatin; cytoplasm has very fine, pale lilac granules.	Phagocytosis. Destruction of bacteria with lysozyme, defensins, and strong oxidants, such as superoxide anion, hydrogen peroxide, and hypochlorite anion.
Eosinophils	2–4% of all WBCs.	10–12 μm diameter; nucleus has 2 or 3 lobes; large, red-orange granules fill the cytoplasm.	Combat the effects of histamine in allergic reactions, phagocytize antigen–antibody complexes, and destroy certain parasitic worms.
Basophils	0.5–1% of all WBCs.	8–10 μm diameter; nucleus has 2 lobes; large cytoplasmic granules appear deep blue-purple.	Liberate heparin, histamine, and serotonin in allergic reactions that intensify the overall inflammatory response.
Agranular Leukocytes			
Lymphocytes (T cells, B cells, and **natural killer cells)**	20–25% of all WBCs.	Small lymphocytes are 6–9 μm in diameter; large lymphocytes are 10–14 μm in diameter; nucleus is round or slightly indented; cytoplasm forms a rim around the nucleus that looks sky blue; the larger the cell, the more cytoplasm is visible.	Mediate immune responses, including antigen–antibody reactions. B cells develop into plasma cells, which secrete antibodies. T cells attack invading viruses, cancer cells, and transplanted tissue cells. Natural killer cells attack a wide variety of infectious microbes and certain spontaneously arising tumor cells.
Monocytes	3–8% of all WBCs.	12–20 μm diameter; nucleus is kidney shaped or horseshoe shaped; cytoplasm is blue-gray and has foamy appearance.	Phagocytosis (after transforming into fixed or wandering macrophages).
Platelets	150,000–400,000/μL.	2–4 μm diameter cell fragments that live for 5–9 days; contain many vesicles but no nucleus.	Form platelet plug in hemostasis; release chemicals that promote vascular spasm and blood clotting.

*Colors are those seen when using Wright's stain.

†Some lymphocytes, called T and B memory cells, can live for many years once they are established.

The process of clot formation, called ***clotting (coagulation)***, is a series of chemical reactions that culminates in the formation of fibrin threads. If blood clots too easily, the result can be ***thrombosis***, clotting in an unbroken blood vessel. If the blood takes too long to clot, hemorrhage can result.

Clotting is a complex process in which various chemicals known as ***clotting factors*** activate each other. Clotting (coagulation) factors include calcium ions (Ca^{2+}), several enzymes that are made by liver cells and released into the blood, and various molecules associated with platelets or released by damaged tissues. Many clotting factors are identified by Roman numerals. Clotting occurs in three stages (Figure 14.5):

Figure 14.5 Blood clotting.

During blood clotting, the clotting factors activate each other, resulting in a cascade of reactions that includes positive feedback cycles.

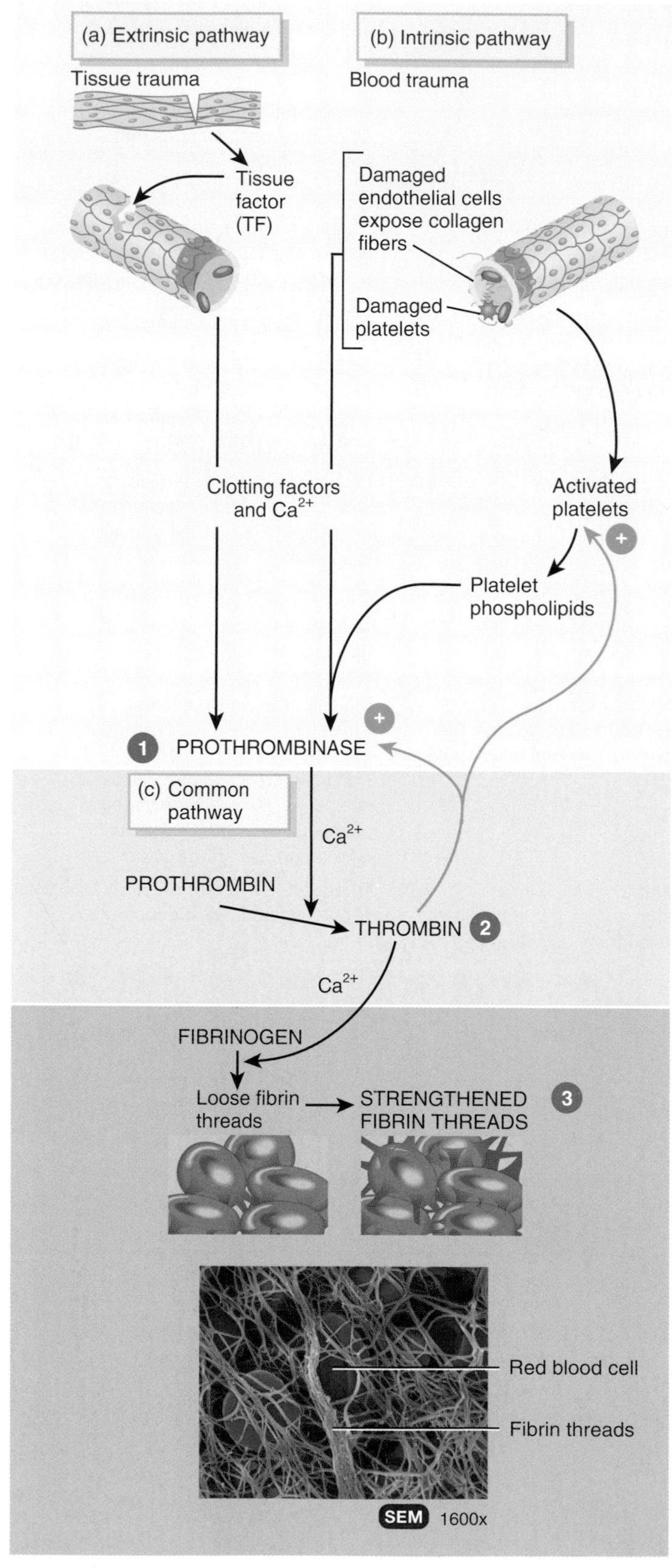

What is the outcome of stage 1 of clotting?

1. ***Prothrombinase*** is formed.
2. Prothrombinase converts ***prothrombin*** (a plasma protein formed by the liver with the help of vitamin K) into the enzyme ***thrombin***.
3. Thrombin converts soluble ***fibrinogen*** (another plasma protein formed by the liver) into insoluble fibrin. Fibrin forms the threads of the clot. (Cigarette smoke contains substances that interfere with fibrin formation.)

Prothrombinase can be formed in two ways, by either the extrinsic or the intrinsic pathway of blood clotting (Figure 14.5). The ***extrinsic pathway*** of blood clotting occurs rapidly, within seconds. It is so-named because damaged tissue cells release a tissue protein called ***tissue factor (TF)*** into the blood from *outside* (extrinsic to) blood vessels (Figure 14.5a). Following several additional reactions that require calcium ions (Ca^{2+}) and several clotting factors, tissue factor is eventually converted into prothrombinase. This completes the extrinsic pathway.

The ***intrinsic pathway*** of blood clotting (Figure 14.5b) is more complex than the extrinsic pathway, and it occurs more slowly, usually requiring several minutes. The intrinsic pathway is so-named because its activators are either in direct contact with blood or contained *within* (intrinsic to) the blood. If endothelial cells lining the blood vessels become roughened or damaged, blood can come in contact with collagen fibers in the adjacent connective tissue. Such contact activates clotting factors. In addition, trauma to endothelial cells activates platelets, causing them to release phospholipids that can also activate certain clotting factors. After several additional reactions that require Ca^{2+} and several clotting factors, prothrombinase is formed. Once formed, thrombin activates more platelets, resulting in the release of more platelet phospholipids, an example of a positive feedback cycle.

Clot formation occurs locally; it does not extend beyond the wound site into the general circulation. One reason for this is that fibrin has the ability to absorb and inactivate up to nearly 90% of the thrombin formed from prothrombin. This helps stop the spread of thrombin into the blood and thus inhibits clotting except at the wound.

Clot Retraction and Blood Vessel Repair

Once a clot is formed, it plugs the ruptured area of the blood vessel and thus stops blood loss. ***Clot retraction*** is the consolidation or tightening of the fibrin clot. The fibrin threads attached to the damaged surfaces of the blood vessel gradually contract as platelets pull on them. As the clot retracts, it pulls the edges of the damaged vessel closer together, decreasing the risk of further damage. Permanent repair of the blood vessel can then take place. In time, fibroblasts form connective tissue in the ruptured area, and new endothelial cells repair the vessel lining.

Hemostatic Control Mechanisms

Many times a day little clots start to form, often at a site of minor roughness inside a blood vessel. Usually, small, inappropriate clots dissolve in a process called ***fibrinolysis*** (fī′-bri-NOL-i-sis). When a clot is formed, an inactive plasma enzyme called ***plasminogen*** is incorporated into the clot. Both body tissues and blood contain substances that can activate plasminogen to ***plasmin***, an active plasma enzyme. Once plasmin is formed, it can dissolve the clot by digesting fibrin threads. Plasmin also dissolves clots at sites of damage once the damage is repaired.

Patients who are at increased risk of forming blood clots may receive an **anticoagulant drug**, a substance that delays, suppresses, or prevents blood clotting. Examples are heparin or warfarin. *Heparin*, an anticoagulant that is produced by mast cells and basophils, inhibits the conversion of prothrombin to thrombin, thereby preventing blood clot formation. Heparin extracted from animal tissues is often used to prevent clotting during hemodialysis and after open heart surgery. *Coumadin® (warfarin sodium)* acts as an antagonist to vitamin K and thus blocks synthesis of four clotting factors. To prevent clotting in donated blood, blood banks and laboratories often add a substance that removes Ca^{2+}, for example, CPD (citrate phosphate dextrose).

Clotting in Blood Vessels

Despite fibrinolysis and the action of anticoagulants, blood clots sometimes form within blood vessels. The endothelial surfaces of a blood vessel may be roughened as a result of *atherosclerosis* (accumulation of fatty substances on arterial walls), trauma, or infection. These conditions also make the platelets that are attracted to the rough spots more sticky. Clots may also form in blood vessels when blood flows too slowly, allowing clotting factors to accumulate in high enough concentrations to initiate a clot.

Clotting in an unbroken blood vessel is called ***thrombosis*** (*thromb-* = clot; *-osis* = a condition of). The clot itself, called a ***thrombus***, may dissolve spontaneously. If it remains intact, however, the thrombus may become dislodged and be swept away in the blood. A blood clot, bubble of air, fat from broken bones, or a piece of debris transported by the bloodstream is called an ***embolus*** (*em-* = in; *-bolus* = a mass; plural is *emboli*). Because emboli often form in veins, where blood flow is slower, the most common site for the embolus to become lodged is in the lungs, a condition called ***pulmonary embolism***. Massive emboli in the lungs may result in right ventricular failure and death in a few minutes or hours. An embolus that breaks away from an arterial wall may lodge in a smaller-diameter artery downstream. If it blocks blood flow to the brain, kidney, or heart, the embolus can cause a stroke, kidney failure, or heart attack, respectively.

In patients with heart and blood vessel disease, the events of hemostasis may occur even without external injury to a blood vessel. At low doses, **aspirin** inhibits vasoconstriction and platelet aggregation. It also reduces the chance of thrombus formation. Due to these effects, aspirin reduces the risk of transient ischemic attacks (TIA), strokes, myocardial infarction, and blockage of peripheral arteries.

Thrombolytic agents are chemical substances that are injected into the body to dissolve blood clots that have already formed to restore circulation. They either directly or indirectly activate plasminogen. The first thrombolytic agent, approved in 1982 for dissolving clots in the coronary arteries of the heart, was **streptokinase**, which is produced by streptococcal bacteria. A genetically engineered version of human **tissue plasminogen activator (tPA)** is now used to treat both heart attacks and brain attacks (strokes) that are caused by blood clots.

■ **CHECKPOINT**

7. What is hemostasis?
8. How do vascular spasm and platelet plug formation occur?
9. What is fibrinolysis? Why does blood rarely remain clotted inside blood vessels?

BLOOD GROUPS AND BLOOD TYPES

OBJECTIVE • **Describe the ABO and Rh blood groups.**

The surfaces of red blood cells contain a genetically determined assortment of ***antigens*** composed of glycolipids and glycoproteins called ***agglutinogens*** (ag′-loo-TIN-ō-jenz). Based on the presence or absence of various antigens, blood is categorized into different ***blood groups***. Within a given blood group there may be two or more different ***blood types***. There are at least 24 blood groups and more than 100 antigens that can be detected on the surface of red blood cells. Here we discuss two major blood groups: ABO and Rh.

ABO Blood Group

The ***ABO blood group*** is based on two antigens called ***A*** and ***B*** (Figure 14.6). People whose RBCs display only antigen A have type A blood. Those who have only antigen B are type B. Individuals who have both A and B antigens are type AB, and those who have neither antigen A nor B are type O. In about 80% of the population, soluble antigens of the ABO type appear in saliva and other body fluids, in which case blood type can be identified from a sample of saliva. The incidence of ABO blood types varies among different population groups, as indicated in Table 14.3.

Figure 14.6 Antigens and antibodies involved in the ABO blood grouping system.

Your plasma does not contain antibodies that could react with the antigens on your red blood cells.

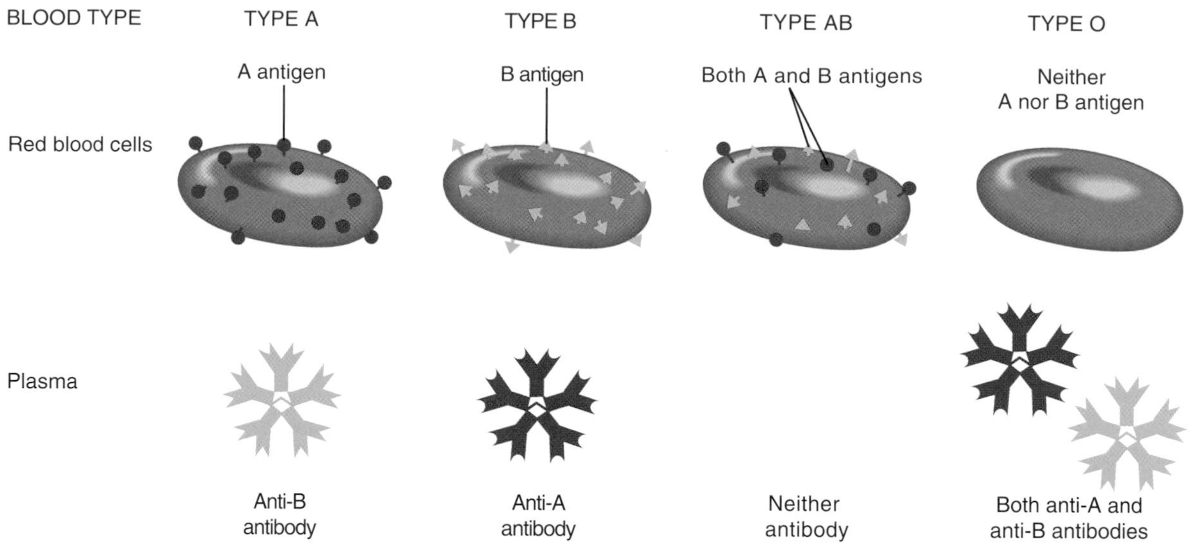

Which antibodies are found in type O blood?

In addition to antigens on RBCs, blood plasma usually contains ***antibodies*** or ***agglutinins*** (a-GLOO-ti-nins) that react with the A or B antigens if the two are mixed. These are the ***anti-A antibody***, which reacts with antigen A, and the ***anti-B antibody***, which reacts with antigen B. The antibodies present in each of the four ABO blood types are also shown in Figure 14.6. You do not have antibodies that react with your own antigens, but you do have antibodies for any antigens that your RBCs lack. For example, if you have type A blood, it means that you have A antigens on the surfaces of your RBCs, but anti-B antibodies in your blood plasma. If you had anti-A antibodies in your blood plasma, they would attack your RBCs.

Rh Blood Group

The ***Rh blood group*** is so named because the Rh antigen was first found in the blood of the rhesus monkey. People whose RBCs have the Rh antigen are designated Rh^+ (Rh positive); those who lack the Rh antigen are designated Rh^- (Rh negative). The percentages of Rh^+ and Rh^- individuals in various populations are shown in Table 14.3. Under normal circumstances, plasma does not contain anti-Rh antibodies. If an Rh^- person receives an Rh^+ blood transfusion, however, the immune system starts to make anti-Rh antibodies that do remain in the blood.

Table 14.3 Blood Types in the United States

	Blood Type (percentage)				
Population Group	**O**	**A**	**B**	**AB**	**Rh^+**
European American	45	40	11	4	85
African American	49	27	20	4	95
Korean	32	28	30	10	100
Japanese	31	38	21	10	100
Chinese	42	27	25	6	100
Native American	79	16	4	1	100

Transfusions

Despite the differences in RBC antigens, blood is the most easily shared of human tissues, saving many thousands of lives every year through transfusions. A ***transfusion*** (trans-FŪ-zhun) is the transfer of whole blood or blood components (red blood cells only or plasma only) into the bloodstream. Most often a transfusion is given to alleviate anemia or when blood volume is low, for example, after a severe hemorrhage.

In an incompatible blood transfusion, antibodies in the recipient's plasma bind to the antigens on the donated RBCs. When these antigen–antibody complexes form, they cause hemolysis and release hemoglobin into the plasma. Consider what happens if a person with type A blood receives a transfusion of type B blood. In this situation, two things can happen. First, the anti-B antibodies in the recipient's plasma can bind to the B antigens on the donor's RBCs, causing hemolysis. Second, the anti-A antibodies in the donor's plasma can bind to the A antigens on the recipient's RBCs. The second reaction is usually not serious because the donor's anti-A

FOCUS ON WELLNESS

Lifestyle and Blood Circulation—Let It Flow

Many people fear cholesterol, an evil substance that silently accumulates within artery walls, year after year, until it eventually kills its victim by shutting off blood flow to an important organ such as the heart or brain. But cholesterol is not the only villain in this atherosclerosis melodrama. Cholesterol contributes to the formation of arterial plaques, but the antagonist delivering the final blow is often a blood clot that forms in a blood vessel and subsequently blocks a narrowed artery, cutting off circulation to the tissues downstream. Fortunately, many of the things you can do to keep your arteries healthy also reduce your risk of blood clots.

Quit Smoking

If you need yet another reason to quit smoking, here it is: Smoking increases blood fibrinogen levels. Increased fibrinogen levels are associated with increased clotting risk. High fibrinogen levels increase platelet aggregation and fibrin deposition, contributing to both clotting and plaque deposition.

Exercise Regularly

Regular physical activity increases plasma volume. An increase in plasma volume means that the blood is more dilute, or "thinner," with a lower percentage of red blood cells and less fibrinogen, and consequently a reduced risk of blood clotting. Several studies have shown that vigorous exercise also reduces platelet stickiness and enhances fibrinolytic activity. These effects may help to explain why active people are at lower risk for heart disease and stroke. A sedentary lifestyle, by contrast, leads to increased clotting risk: Blood thickens as plasma volume decreases. Sedentary people have stickier platelets, which together with higher levels of fibrinogen are more likely to form blood clots.

Cope Effectively with Stress

Prolonged mental stress impairs fibrinolysis by decreasing the activity of tissue plasminogen activator (tPA), which helps break down fibrinogen.

Eat a Heart-healthy Diet

People with high blood cholesterol levels exhibit disturbances in coagulation, fibrinolysis, and platelet behavior. Lowering blood lipid levels by diet or drug therapy seems to reverse these disturbances and may be one way that a heart-healthy lifestyle reduces heart disease risk. An interesting study from Denmark found that volunteers who stuck to a low-fat, high-fiber diet showed increased fibrinolytic activity and thus a reduced risk of blood clot formation.

A moderate alcohol intake (one to two drinks per day) has been associated with a reduced heart disease risk. This risk reduction may be due in part to the increase in tPA level observed in moderate drinkers.

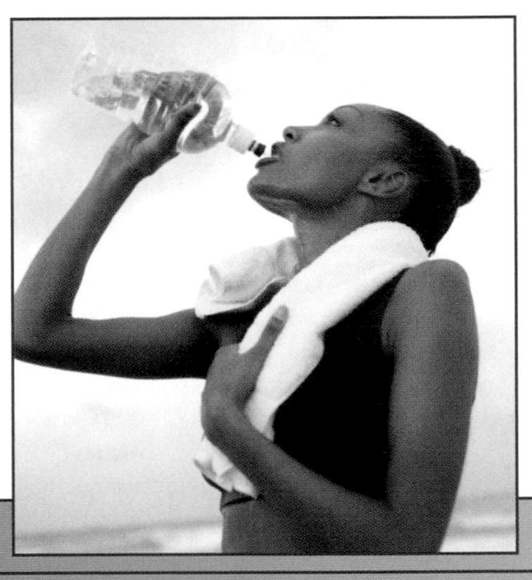

► **THINK IT OVER . . .**

► ***Why are people at risk for clot formation told to avoid sitting for extended periods of time, such as on long airplane flights or car rides?***

antibodies become so diluted in the recipient's plasma that they do not cause any significant hemolysis of the recipient's RBCs.

People with type AB blood do not have any anti-A or anti-B antibodies in their plasma. They are sometimes called "universal recipients" because theoretically they can receive blood from donors of all four ABO blood types. People with type O blood have neither A nor B antigens on their RBCs and are sometimes called "universal donors." Theoretically, because there are no antigens on their RBCs for antibodies to attack, they can donate blood to all four ABO blood types. Type O persons requiring blood may receive only type O blood, as they have antibodies to both A and B antigens in their plasma. In practice, use of the terms *universal recipient* and *universal donor* is misleading and dangerous. Blood contains antigens and antibodies other than those associated with the ABO system, and they can cause transfusion problems. Thus, blood should always be carefully matched before transfusion.

■ CHECKPOINT

10. What is the basis for distinguishing the various blood groups?

11. What precautions must be taken before giving a blood transfusion?

• • •

We will next direct our attention to the heart, the second major component of the cardiovascular system.

COMMON DISORDERS

Anemia

Anemia is a condition in which the oxygen-carrying capacity of blood is reduced. Many types of anemia exist; all are characterized by reduced numbers of RBCs or a decreased amount of hemoglobin in the blood. The person feels fatigued and is intolerant of cold, both of which are related to lack of oxygen needed for ATP and heat production. Also, the skin appears pale, due to the low content of red-colored hemoglobin circulating in skin blood vessels. Among the most important types of anemia are the following:

- *Iron-deficiency anemia*, the most prevalent kind of anemia, is caused by inadequate absorption of iron, excessive loss of iron, or insufficient intake of iron. Women are at greater risk for iron-deficiency anemia due to monthly menstrual blood loss.
- *Pernicious anemia* is caused by insufficient hemopoiesis resulting from an inability of the stomach to produce intrinsic factor (needed for absorption of dietary vitamin B_{12}).
- *Hemorrhagic anemia* is due to an excessive loss of RBCs through bleeding resulting from large wounds, stomach ulcers, or especially heavy menstruation.
- In *hemolytic anemia*, RBC plasma membranes rupture prematurely. The condition may result from inherited defects or from outside agents such as parasites, toxins, or antibodies from incompatible transfused blood.
- *Thalassemia* (thal′-a-SĒ-mē-a) is a group of hereditary hemolytic anemias in which there is an abnormality in one or more of the four polypeptide chains of the hemoglobin molecule. Thalassemia occurs primarily in populations from countries bordering the Mediterranean Sea.
- *Aplastic anemia* results from destruction of the red bone marrow caused by toxins, gamma radiation, and certain medications that inhibit enzymes needed for hemopoiesis.

Sickle Cell Disease

The RBCs of a person with ***sickle cell disease (SCD)*** contain Hb-S, an abnormal kind of hemoglobin. When Hb-S gives up oxygen to the interstitial fluid, it forms long, stiff, rodlike structures that bend the erythrocyte into a sickle shape. The sickled cells rupture easily. Even though the loss of RBCs stimulates erythropoiesis, it cannot keep pace with hemolysis; hemolytic anemia is the result. Prolonged oxygen reduction may eventually cause extensive tissue damage.

Hemophilia

Hemophilia (hē-mō-FIL-ē-a; *-philia* = loving) is an inherited deficiency of clotting in which bleeding may occur spontaneously or after only minor trauma. Different types of hemophilia are due to deficiencies of different blood clotting factors and exhibit varying degrees of severity. Hemophilia is characterized by spontaneous or traumatic subcutaneous and intramuscular hemorrhaging, nosebleeds, blood in the urine, and hemorrhages in joints that produce pain and tissue damage. Treatment involves transfusions of fresh plasma or concentrates of the deficient clotting factor to relieve the tendency to bleed.

Hemolytic Disease of the Newborn

Hemolytic disease of the newborn (HDN) is a problem that results from Rh incompatibility between a mother and her fetus. Normally, no direct contact occurs between maternal and fetal blood while a woman is pregnant. However, if a small amount of Rh^+ blood leaks from the fetus through the placenta into the bloodstream of an Rh^- mother, her body starts to make anti-Rh antibodies. Because the greatest possibility of fetal blood transfer occurs at delivery, the first-born baby typically is not affected. If the mother becomes pregnant again, however, her anti-Rh antibodies, made after delivery of the first baby, can cross the placenta and enter the bloodstream of the fetus. If the fetus is Rh^-, there is no problem, because Rh^- blood does not have the Rh antigen. If, however, the fetus is Rh^+, life-threatening *hemolysis* (rupture of RBCs) is likely to occur in the fetal blood. By contrast, ABO incompatibility between a mother and her fetus rarely causes problems because the anti-A and anti-B antibodies do not cross the placenta.

HDN is prevented by giving all Rh^- women an injection of anti-Rh antibodies called anti-Rh gamma globulin (RhoGAM) soon after every delivery, miscarriage, or abortion. These antibodies destroy any Rh antigens that are present so the mother doesn't produce her own antibodies to them. In the case of an Rh^+ mother, there are no complications, because she cannot make anti-Rh antibodies.

Leukemia

The term ***leukemia*** (loo-KĒ-mē-a; *leuko-* = white) refers to a group of red bone marrow cancers in which abnormal white blood cells multiply uncontrollably. The accumulation of the cancerous white blood cells in red bone marrow interferes with the production of red blood cells, white blood cells, and platelets. As a result, the oxygen-carrying capacity of the blood is reduced, an individual is more susceptible to infection, and blood clotting is abnormal. In most leukemias, the cancerous white blood cells spread to the lymph nodes, liver, and spleen, causing them to enlarge. All leukemias produce the usual symptoms of anemia (fatigue, intolerance to cold, and pale skin). In addition, weight loss, fever, night sweats, excessive bleeding, and recurrent infections may also occur.

MEDICAL TERMINOLOGY AND CONDITIONS

Autologous preoperative transfusion (aw-TOL-o-gus trans-FŪ-zhun; *auto-* = self) Donating one's own blood in preparation for surgery; can be done up to six weeks before elective surgery. Also called ***predonation***.

Blood bank A facility that collects and stores a supply of blood for future use by the donor or others. Because blood banks have now assumed additional and diverse functions (immunohematology reference work, continuing medical education, bone and tissue storage, and clinical consultation), they are more appropriately referred to as *centers of transfusion medicine*.

Cyanosis (sī-a-NŌ-sis; *cyano-* = blue) Slightly bluish/dark-purple skin discoloration, most easily seen in the nail beds and mucous membranes, due to an increased quantity of reduced hemoglobin (hemoglobin not combined with oxygen) in systemic blood.

Hemochromatosis (hē′-mō-krō′-ma-TŌ-sis; *chroma* = color) Disorder of iron metabolism characterized by excess deposits of iron in tissues (especially the liver, heart, pituitary gland, gonads, and pancreas) that result in discoloration (bronzing) of the skin, cirrhosis, diabetes mellitus, and bone and joint abnormalities.

Jaundice (*jaund-* = yellow) An abnormal yellowish discoloration of the sclerae of the eyes, skin, and mucous membranes due to excess bilirubin (yellow-orange pigment) in the blood that is produced when the heme pigment in aged red blood cells is broken down.

Phlebotomist (fle-BOT-ō-mist; *phlebo-* = vein; *-tom* = cut) A technician who specializes in withdrawing blood.

Polycythemia (pol′-ē-sī-THĒ-mē-a) An abnormal increase in the number of red blood cells in which hematocrit is above 55%, the upper limit of normal.

Septicemia (sep′-ti-SĒ-mē-a; *septic-* = decay; *-emia* = condition of blood) An accumulation of toxins or disease-causing bacteria in the blood. Also called ***blood poisoning***.

Thrombocytopenia (throm′-bō-sī′-tō-PĒ-nē-a; *-penia* = poverty) Very low platelet count that results in a tendency to bleed from capillaries.

STUDY OUTLINE

Functions of Blood (p. 346)

1. Blood transports oxygen, carbon dioxide, nutrients, wastes, and hormones.
2. It helps to regulate pH, body temperature, and water content of cells.
3. It prevents blood loss through clotting and combats microbes and toxins through the action of certain phagocytic white blood cells or specialized plasma proteins.

Components of Whole Blood (p. 346)

1. Physical characteristics of whole blood include a viscosity greater than that of water, a temperature of 38°C (100.4°F), and a pH range between 7.35 and 7.45.
2. Blood constitutes about 8% of body weight in an adult.
3. Blood consists of 55% plasma and 45% formed elements.
4. The formed elements in blood include red blood cells (erythrocytes), white blood cells (leukocytes), and platelets. Hematocrit is the percentage of red blood cells in whole blood.
5. Plasma contains 91.5% water, 7% proteins, and 1.5% solutes other than proteins.
6. Principal solutes include proteins (albumins, globulins, fibrinogen), nutrients, hormones, respiratory gases, electrolytes, and waste products.
7. Hemopoiesis, the formation of blood cells from pluripotent stem cells, occurs in red bone marrow.
8. Red blood cells (RBCs) are biconcave discs without nuclei that contain hemoglobin.
9. The function of the hemoglobin in red blood cells is to transport oxygen.
10. Red blood cells live about 120 days. A healthy male has about 5.4 million RBCs/μL of blood and a healthy female has about 4.8 million RBCs/μL.
11. After phagocytosis of aged red blood cells by macrophages, hemoglobin is recycled.
12. RBC formation, called erythropoiesis, occurs in adult red bone marrow. It is stimulated by hypoxia, which stimulates release of erythropoietin by the kidneys.
13. A reticulocyte count is a diagnostic test that indicates the rate of erythropoiesis.
14. White blood cells (WBCs) are nucleated cells. The two principal types are granular leukocytes (neutrophils, eosinophils, basophils) and agranular leukocytes (lymphocytes and monocytes).
15. The general function of WBCs is to combat inflammation and infection. Neutrophils and macrophages (which develop from monocytes) do so through phagocytosis.
16. Eosinophils combat inflammation in allergic reactions, phagocytize antigen–antibody complexes, and combat parasitic worms; basophils liberate heparin, histamine, and serotonin in allergic reactions that intensify the inflammatory response.
17. B cells (lymphocytes) are effective against bacteria and other toxins. T cells (lymphocytes) are effective against viruses, fungi, and cancer cells. Natural killer cells attack microbes and tumor cells.
18. White blood cells usually live for only a few hours or a few days. Normal blood contains 5000 to 10,000 WBCs/μL.
19. Platelets are disc-shaped cell fragments without nuclei.
20. Platelets are formed from megakaryocytes and take part in hemostasis by forming a platelet plug.
21. Normal blood contains 150,000 to 400,000 platelets/μL.

Hemostasis (p. 354)

1. Hemostasis, the stoppage of bleeding, involves vascular spasm, platelet plug formation, and blood clotting.
2. In vascular spasm, the smooth muscle of a blood vessel wall contracts.
3. Platelet plug formation is the aggregation of platelets to stop bleeding.
4. A clot is a network of insoluble protein fibers (fibrin) in which formed elements of blood are trapped. The chemicals involved in clotting are known as clotting factors.
5. Blood clotting involves a series of reactions that may be divided into three stages: formation of prothrombinase by either the extrinsic or intrinsic pathway, conversion of prothrombin into thrombin, and conversion of soluble fibrinogen into insoluble fibrin.
6. Normal coagulation involves clot retraction (tightening of the clot) and fibrinolysis (dissolution of the clot).
7. Anticoagulants (for example, heparin) prevent clotting.
8. Clotting in an unbroken blood vessel is called thrombosis. A thrombus that moves from its site of origin is called an embolus.

Blood Groups and Blood Types (p. 357)

1. In the ABO system, the antigens on RBCs, called A and B, determine blood type. Plasma contains antibodies termed anti-A and anti-B antibodies.
2. In the Rh system, individuals whose erythrocytes have Rh antigens are classified as Rh^+. Those who lack the antigen are Rh^-.

SELF-QUIZ

1. A hematocrit is
   a. used to measure the quantity of the five types of white blood cells
   b. essential for determining a person's blood type
   c. the percentage of red blood cells in whole blood
   d. also known as a platelet count
   e. involved in blood clotting

2. Match the following:
   ____ a. involved in certain immune responses
   ____ b. develop into mature red blood cells
   ____ c. required for vitamin B_{12} absorption
   ____ d. most abundant plasma protein
   ____ e. blood after formed elements are removed
   ____ f. plasma without clotting proteins
   ____ g. needed for blood clotting

   A. albumin
   B. fibrinogen
   C. intrinsic factor
   D. immunoglobulins
   E. plasma
   F. serum
   G. reticulocytes

3. In adults, erythropoiesis takes place in
   a. the liver b. yellow bone marrow c. red bone marrow
   d. lymphatic tissue e. the kidneys

4. Which of the following pigments contributes to the yellow color in urine?
   a. hemoglobin b. stercobilin c. biliverdin
   d. urobilin e. bilirubin

5. Which of the following statements is NOT true about red blood cells?
   a. The production of red blood cells is known as erythropoiesis.
   b. Red blood cells originate from pluripotent stem cells.
   c. Hypoxia increases the production of red blood cells.
   d. The liver takes part in the destruction and recycling of red blood cell components.
   e. Red blood cells have a lobed nucleus and granular cytoplasm.

6. A primary function of red blood cells is to
   a. maintain blood volume
   b. help blood clot
   c. provide immunity against some diseases
   d. clean up debris following infection
   e. deliver oxygen to the cells of the body

7. If a differential white blood cell count indicated higher than normal numbers of basophils, what may be occurring in the body?
   a. chronic infection b. allergic reaction c. leukopenia
   d. initial response to invading bacteria e. hemostasis

8. In a person with blood type A, the antibodies that would normally be present in the plasma is (are)
   a. anti-A antibody
   b. anti-B antibody
   c. both anti-A and anti-B antibodies
   d. neither anti-A nor anti-B
   e. anti-O antibodies

9. Hemolytic disease of the newborn (HDN) may occur in the fetus of a second pregnancy if
   a. the mother is Rh^+ and the baby is Rh^-
   b. the mother is Rh^+ and the baby is Rh^+
   c. the mother is Rh^- and the baby is Rh^-
   d. the mother is Rh^- and the baby is Rh^+
   e. the father is Rh^- and the mother is Rh^+

10. Place the following steps of hemostasis in the correct order.
    1. clot retraction 2. prothrombinase formed
    3. fibrinolysis by plasmin 4. vascular spasm
    5. conversion of prothrombin into thrombin
    6. platelet plug formation
    7. conversion of fibrinogen into fibrin

    a. 4, 6, 2, 5, 7, 1, 3 b. 5, 4, 7, 6, 2, 3, 1
    c. 2, 5, 6, 7, 1, 4, 3 d. 4, 6, 5, 2, 7, 1, 3
    e. 4, 2, 6, 5, 3, 7, 1

11. Which of the following is NOT a normal component of blood plasma?
   a. albumins b. fibrinogen c. hemoglobin
   d. immunoglobulins e. water

12. How does aspirin prevent thrombosis?
   a. It inhibits platelet aggregation.
   b. It interferes with Ca^{2+} absorption.
   c. It inhibits the conversion of prothrombin to thrombin.
   d. It acts as an enzyme to dissolve the thrombus.
   e. It prevents the accumulation of fatty substances on blood vessel walls.

13. Match the following:

____	a. become wandering macrophages	A. neutrophils
____	b. produce antibodies	B. eosinophils
____	c. are involved in allergic reactions	C. basophils
____	d. first to respond to bacterial invasion	D. lymphocytes
____	e. destroy antigen–antibody complexes; combat inflammation	E. monocytes

14. Hemostasis is
   a. maintenance of a steady state in the body
   b. an abnormal increase in leukocytes
   c. a hereditary condition in which spontaneous hemorrhaging occurs
   d. an anticoagulant produced by some leukocytes
   e. a series of events that stop bleeding

15. Which of the following are mismatched?
   a. white blood cell count below 5000 cells/μL, leukopenia
   b. red blood cell count of 250,000 cells/μL, normal adult male
   c. white blood cell count above 10,000 cells/μL, leukocytosis
   d. platelet count of 300,000 cells/μL, normal adult
   e. pH 7.4, normal blood

16. An individual with type A blood has ________ in the plasma membranes of red blood cells.
   a. antigen A b. antigen B c. major histocompatibility antigen A d. antigen A and antigen Rh
   e. antigen B and antigen Rh

17. Mrs. Smith arrives at a health clinic with her ill daughter Beth. It is suspected that Beth has recently developed a bacterial infection. It is likely that Beth's leukocyte count will be ________cells/μL of blood, a condition known as ________. A differential white blood cell count shows an abnormally high percentage of ________.
   a. 20,000, leukopenia, neutrophils
   b. 5000, leukocytosis, monocytes
   c. 7000, leukocytosis, basophils
   d. 2000, leukopenia, platelets
   e. 20,000, leukocytosis, neutrophils

18. Clot retraction
   a. draws torn edges of the damaged vessel closer together
   b. dissolves clots c. is also known as the intrinsic pathway
   d. involves the formation of fibrin from fibrinogen
   e. helps prevent the formation of an embolus

19. Persons with blood type AB are sometimes referred to as universal recipients because their blood
   a. lacks A and B antigens b. lacks anti-A and anti-B antibodies c. possesses type O antigens and anti-O antibodies
   d. has natural immunity to disease
   e. contains A and B antigens

20. A thrombus that is being transported by the bloodstream is called
   a. a plasma protein b. a platelet c. an embolus
   d. a wandering macrophage e. a reticulocyte

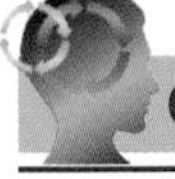

CRITICAL THINKING APPLICATIONS

1. Biliary atresia is a condition in which the ducts that transport bile out of the liver do not function properly. The whites of the eyes in a baby with this condition have a yellow color. What is the name of the yellow color and what is its cause?

2. A woman with blood type Rh^+ is married to a man with blood type Rh^- and is pregnant with their second child. What is the chance the baby will have hemolytic disease of the newborn (HDN)?

3. The school nurse sighed, "I just can't get used to the blue nail polish the kids are wearing. I keep thinking there's a medical problem." What type of problem might result in blue fingernails?

4. Very small numbers of pluripotent stem cells occur normally in blood. If these cells could be isolated and grown in sufficient numbers, what medically useful products could they produce?

ANSWERS TO FIGURE QUESTIONS

14.1 Red blood cells are the most numerous formed element in blood.

14.2 Blood makes up 8% of body weight.

14.3 Stercobilin is responsible for the brown color of feces.

14.4 Hypoxia means cellular oxygen deficiency.

14.5 Prothrombinase is formed during stage 1 of clotting.

14.6 Type O blood has anti-A and anti-B antibodies.

chapter 15

THE CARDIOVASCULAR SYSTEM: HEART

did you know? ***What is a "heart-healthy diet" and how does it help your heart? A heart-healthy diet is one that is low in saturated fats, high in fruits and vegetables, and contains plenty of fiber. A heart-healthy diet encourages the consumption of fish but warns against too much sugar and salt. A heart-healthy diet is actually an "artery-healthy diet" because it is associated with health improvements that reduce the risk of artery disease: better blood cholesterol levels, better blood pressure, and less obesity. Coronary artery disease is the leading cause of death from heart disease, so by keeping the heart's arteries healthy, the heart stays healthy as well.***

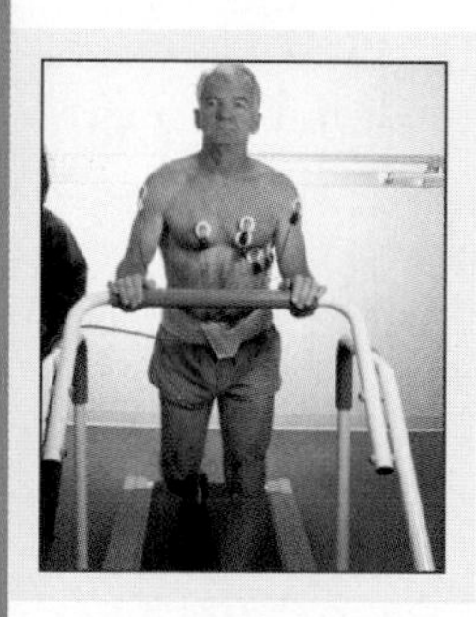

Focus on Wellness, page 376

In the last chapter we examined the composition and functions of blood. For blood to reach body cells and exchange materials with them, it must be constantly pumped by the heart through the body's blood vessels. The heart beats about 100,000 times every day, which adds up to about 35 million beats in a year. The left side of the heart pumps blood through an estimated 100,000 km (60,000 mi) of blood vessels. The right side of the heart pumps blood through the lungs, enabling blood to pick up oxygen and unload carbon dioxide. Even while you are sleeping, your heart pumps 30 times its own weight each minute, which amounts to about 5 liters (5.3 qt) to the lungs and the same volume to the rest of the body. At this rate, the heart pumps more than 14,000 liters (3,600 gal) of blood in a day, or 10 million liters (2.6 million gal) in a year. You don't spend all your time sleeping, however, and your heart pumps more vigorously when you are active. Thus, the actual blood volume the heart pumps in a single day is much larger.

The scientific study of the normal heart and the diseases associated with it is ***cardiology*** (kar′-dē-OL-ō-jē; *cardio-* = heart; *-logy* = study of). This chapter explores the design of the heart and the unique properties that permit it to pump for a lifetime without a moment of rest.

looking back to move ahead . . .

- **Functions of Blood (page 346)**
- **Membranes (page 90)**
- **Muscular Tissue (page 90)**
- **Cardiac Muscle Tissue (page 186)**
- **Free Radicals (page 25)**
- **ANS Neurotransmitters (page 277)**

STRUCTURE AND ORGANIZATION OF THE HEART

OBJECTIVES • Describe the location of the heart and the structure and functions of the pericardium.

- **Describe the layers of the heart wall and the chambers of the heart.**
- **Identify the major blood vessels that enter and exit the heart.**
- **Describe the structure and functions of the valves of the heart.**

Location and Coverings of the Heart

The ***heart*** is situated between the two lungs in the thoracic cavity, with about two-thirds of its mass lying to the left of the body's midline (Figure 15.1). Your heart is about the size of your closed fist. The pointed end, the *apex*, is formed by the tip of the left ventricle, a lower chamber of the heart, and rests on the diaphragm. The *base* of the heart is its posterior surface. It is formed by the atria (upper chambers of the heart), mostly the left atrium, into which the four pulmonary veins open, and a portion of the right atrium that receives the superior and inferior vena cavae (see Figure 15.3b). The base lies opposite the apex.

The membrane that surrounds and protects the heart and holds it in place is the ***pericardium*** (*peri-* = around). It consists of two parts: the fibrous pericardium and the serous pericardium (Figure 15.2). The outer ***fibrous pericardium*** is a tough, inelastic, dense irregular connective tissue. It prevents overstretching of the heart, provides protection, and anchors the heart in place.

The inner ***serous pericardium*** is a thinner, more delicate membrane that forms a double layer around the heart. The outer ***parietal layer*** of the serous pericardium is fused to the fibrous pericardium, and the inner ***visceral layer*** of the serous pericardium, also called the ***epicardium*** (*epi-* = on top of), adheres tightly to the surface of the heart. Between the parietal and visceral layers of the serous pericardium is a thin film of fluid. This fluid, known as ***pericardial fluid***, reduces friction between the membranes as the heart moves. The ***pericardial cavity*** is the space that contains the pericardial fluid.

Figure 15.1 Position of the heart and associated blood vessels in the thoracic cavity. In this and subsequent illustrations, vessels that carry oxygenated blood are colored red; vessels that carry deoxygenated blood are colored blue.

The heart is located between the lungs, with about two-thirds of its mass to the left of the midline.

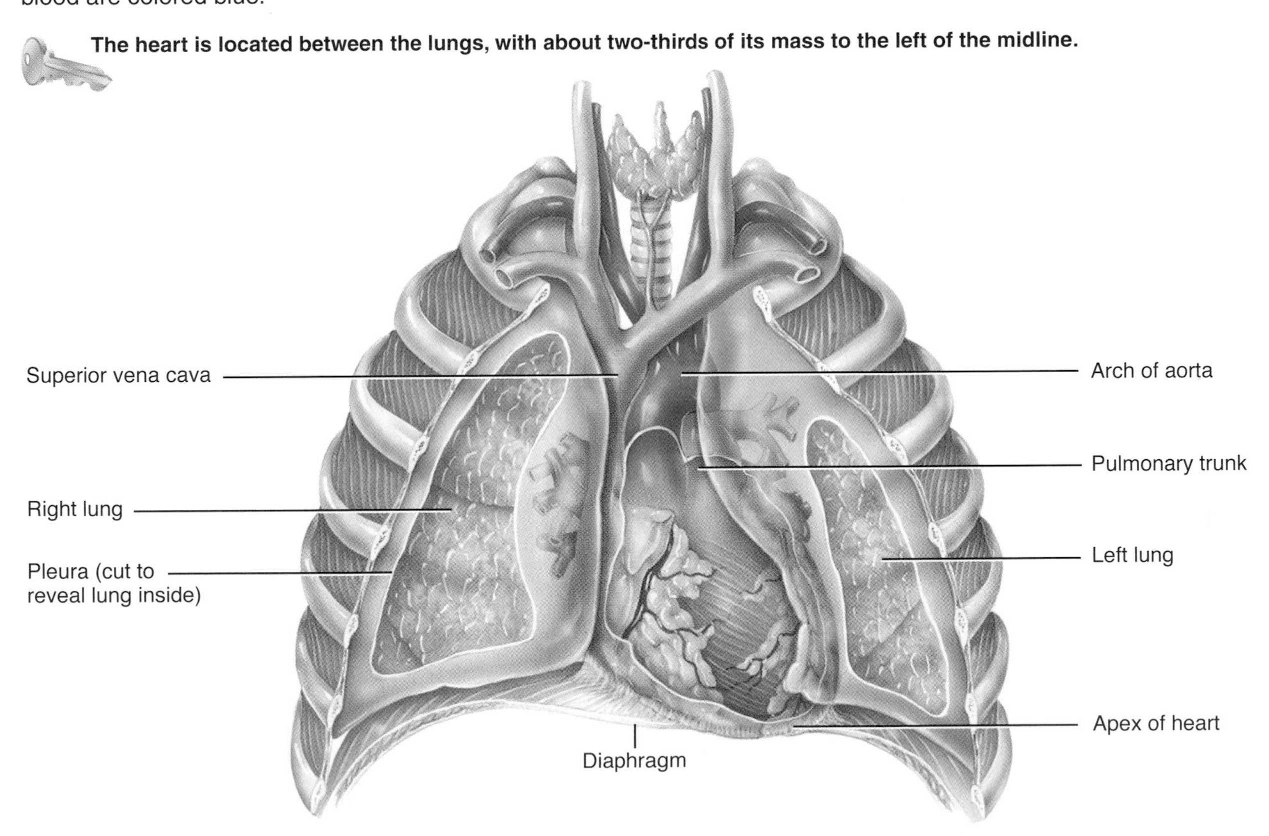

Anterior view of the heart in the thoracic cavity

What forms the base of the heart?

Figure 15.2 Pericardium and heart wall.

The pericardium is a sac that surrounds and protects the heart.

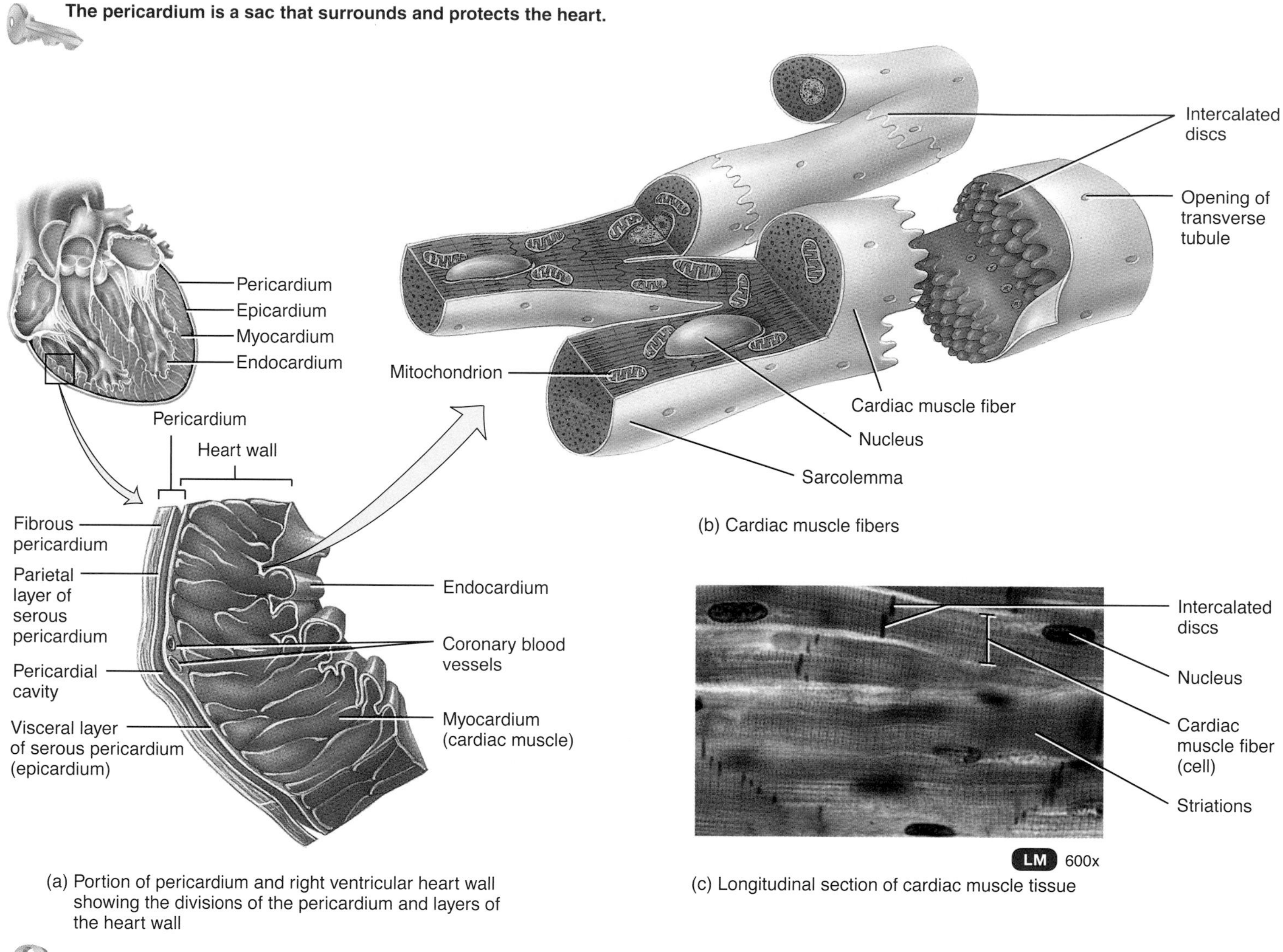

(a) Portion of pericardium and right ventricular heart wall showing the divisions of the pericardium and layers of the heart wall

(b) Cardiac muscle fibers

(c) Longitudinal section of cardiac muscle tissue

Which layer is both a part of the pericardium and a part of the heart wall?

Inflammation of the pericardium is called **pericarditis** (per′-i-kar-DĪ-tis). In one form of this condition, there is a buildup of pericardial fluid. If a great deal of fluid accumulates, this is a life-threatening condition because the fluid compresses the heart, a condition called *cardiac tamponade* (tam′-pon-ĀD). As a result of the compression, ventricular filling is decreased, cardiac output is reduced, venous return to the heart is diminished, blood presure falls, and breathing is difficult.

Heart Wall

The wall of the heart (Figure 15.2a) is composed of three layers: epicardium (external layer), myocardium (middle layer), and endocardium (inner layer). The ***epicardium***, which is also known as the visceral layer of serous pericardium, is the thin, transparent outer layer of the wall. It is composed of mesothelium and connective tissue.

The ***myocardium*** (*myo-* = muscle) consists of cardiac muscle tissue, which constitutes the bulk of the heart. This tissue is found only in the heart and is specialized in structure and function. The myocardium is responsible for the pumping action of the heart. Cardiac muscle fibers (cells) are involuntary, striated, and branched, and the tissue is arranged in interlacing bundles of fibers (Figure 15.2b).

Cardiac muscle fibers form two separate networks—one atrial and one ventricular. Each cardiac muscle fiber connects with other fibers in the networks by thickenings of the sarcolemma (plasma membrane) called ***intercalated discs***. Within the discs are ***gap junctions*** that allow action potentials to conduct from one cardiac muscle fiber to the next. The intercalated discs also link cardiac muscle fibers to one

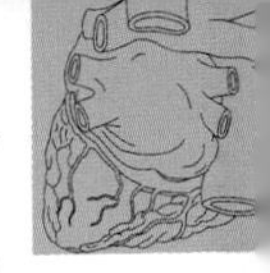

another so they do not pull apart. Each network contracts as a functional unit, so the atria contract separately from the ventricles. In response to a single action potential, cardiac muscle fibers develop a prolonged contraction, 10–15 times longer than the contraction observed in skeletal muscle fibers. Also, the refractory period of a cardiac fiber lasts longer than the contraction itself. Thus, another contraction of cardiac muscle cannot begin until relaxation is well underway. For this reason, tetanus (maintained contraction) cannot occur in cardiac muscle tissue.

The ***endocardium*** (*endo-* = within) is a thin layer of simple squamous epithelium that lines the inside of the myocardium and covers the valves of the heart and the tendons attached to the valves. It is continuous with the epithelial lining of the large blood vessels.

The heart of a heart attack survivor often has regions of infarcted (dead) cardiac muscle tissue that typically are replaced with noncontractile fibrous scar tissue over time. Our inability to repair damage from a heart attack has been attributed to a lack of stem cells in cardiac muscle and to the absence of mitosis in mature cardiac muscle fibers. A recent study of heart transplant recipients by American and Italian scientists, however, provides evidence for significant **replacement of heart cells**. The researchers studied men who had received a heart from a female, and then looked for the presence of a Y chromosome in heart cells. (All female cells except gametes have two X chromosomes and lack the Y chromosome.) Several years after the transplant surgery, between 7% and 16% of the heart cells in the transplanted tissue, including cardiac muscle fibers and endothelial cells in coronary arterioles and capillaries, had been replaced by the recipient's own cells, as evidenced by the presence of a Y chromosome. The study also revealed cells with some of the characteristics of stem cells in both transplanted hearts and control hearts. Evidently, stem cells can migrate from the blood into the heart and differentiate into functional muscle and endothelial cells. The hope is that researchers can learn how to "turn on" such regeneration of heart cells to treat people with heart failure or cardiomyopathy (diseased heart).

Chambers of the Heart

The heart contains four chambers (Figure 15.3). The two upper chambers are the ***atria*** (= entry halls or chambers), and the two lower chambers are the ***ventricles*** (= little bellies). Between the right atrium and left atrium is a thin partition called the ***interatrial septum*** (*inter-* = between; *septum* = a

Figure 15.3 Structure of the heart.

The four chambers of the heart are the two upper atria and two lower ventricles.

(a) Anterior external view showing surface features

(Continues)

Figure 15.3 *(Continued)*

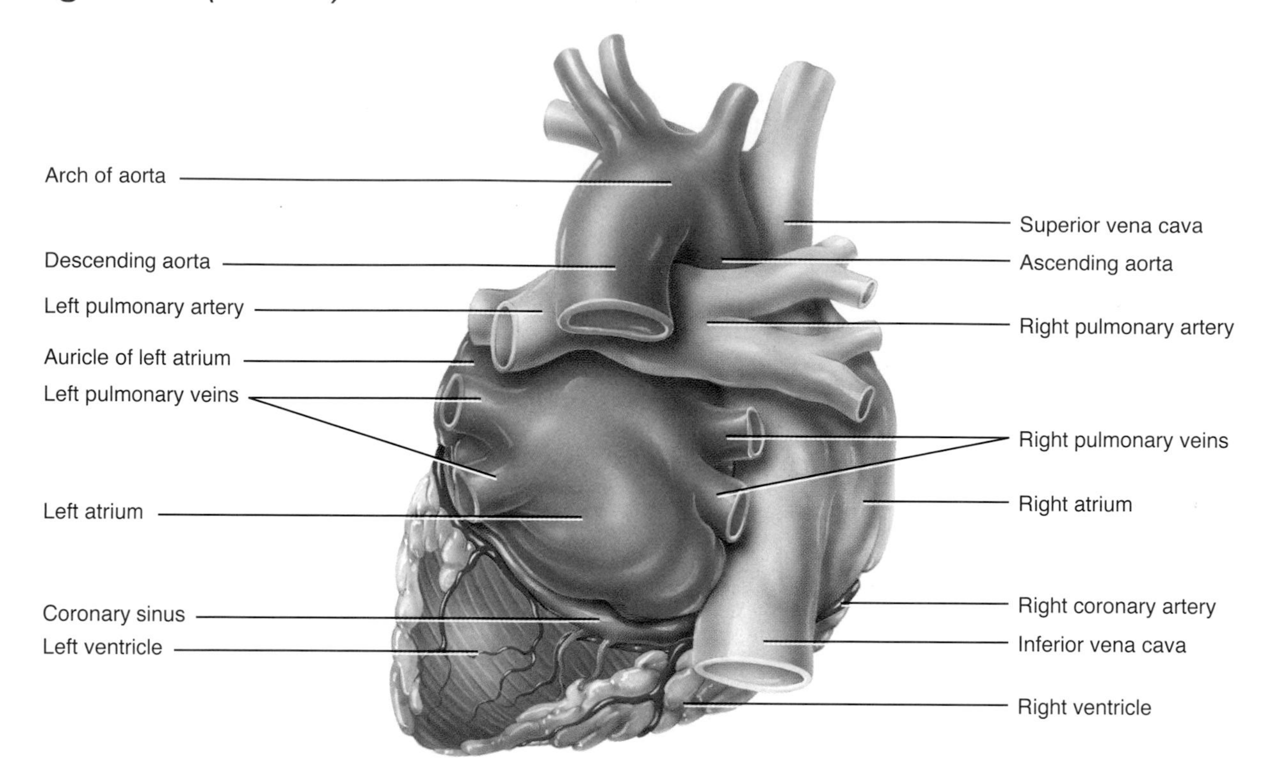

(b) Posterior external view showing surface features

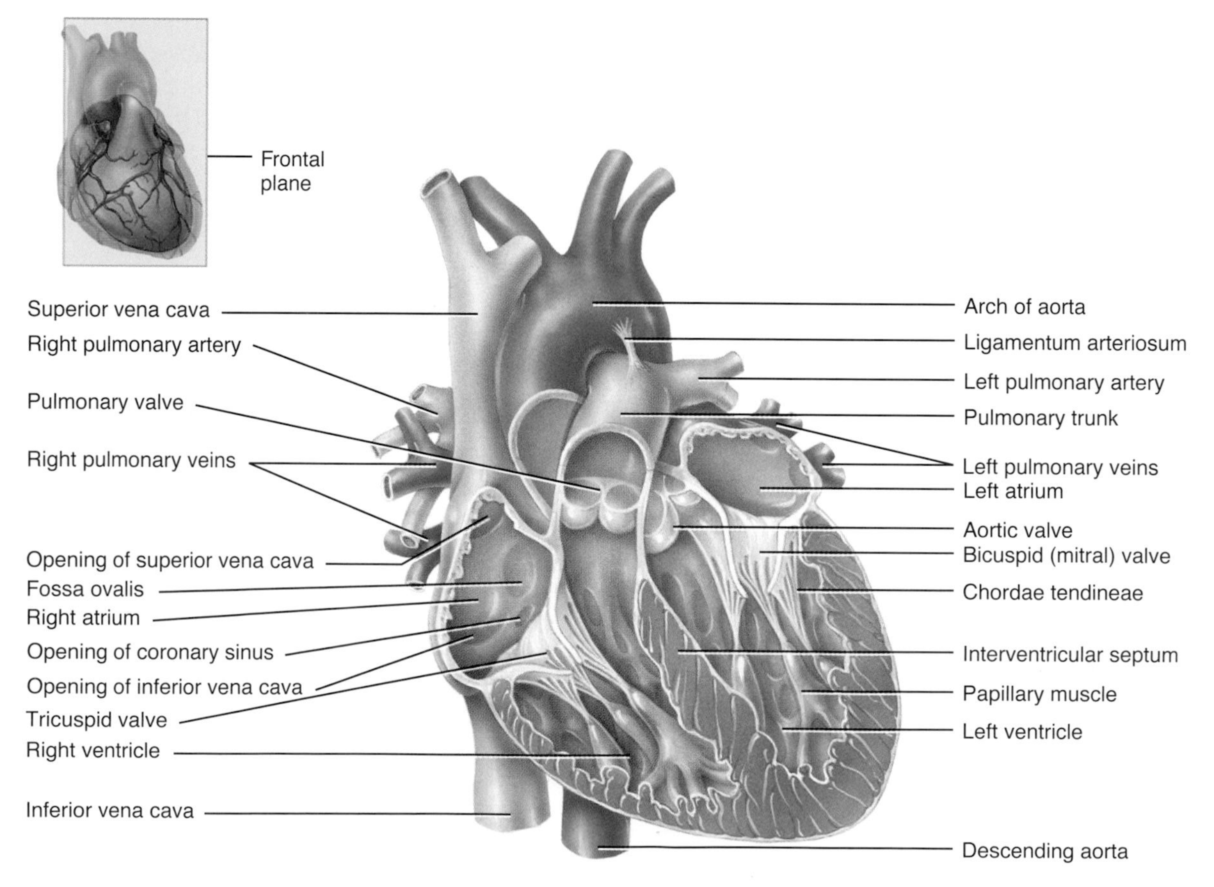

(c) Anterior view of frontal section showing internal anatomy

Through which type of vessel does blood flow away from the heart?

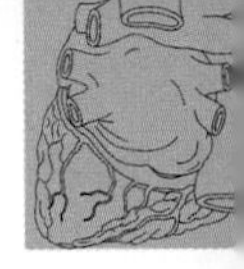

dividing wall or partition); a prominent feature of this septum is an oval depression called the ***fossa ovalis***. It is the remnant of the ***foramen ovale***, an opening in the fetal heart that directs blood from the right to left atrium in order to bypass the nonfunctioning fetal lungs. The foramen ovale normally closes soon after birth. An ***interventricular septum*** separates the right ventricle from the left ventricle (Figure 15.3c). On the anterior surface of each atrium is a wrinkled pouchlike structure called an ***auricle*** (OR-i-kul; *auri-* = ear), so named because of its resemblance to a dog's ear. Each auricle slightly increases the capacity of an atrium so that it can hold a greater volume of blood.

The thickness of the myocardium of the chambers varies according to the amount of work each chamber has to perform. The walls of the atria are thin compared to those of the ventricles because the atria need only enough cardiac muscle tissue to deliver blood into the ventricles (Figure 15.3c). The right ventricle pumps blood only to the lungs (pulmonary circulation); the left ventricle pumps blood to all other parts of the body (systemic circulation). The left ventricle must work harder than the right ventricle to maintain the same rate of blood flow, so the muscular wall of the left ventricle is considerably thicker than the wall of the right ventricle to overcome the greater pressure.

Great Vessels of the Heart

The right atrium receives ***deoxygenated blood*** (oxygen-poor blood that has given up some of its oxygen to cells) through three ***veins***, blood vessels that return blood to the heart. The ***superior vena cava*** (VĒ-na CĀ-va; *vena* = vein; *cava* = hollow, a cave) brings blood mainly from parts of the body above the heart; the ***inferior vena cava*** brings blood mostly from parts of the body below the heart; and the ***coronary sinus*** drains blood from most of the vessels supplying the wall of the heart (Figure 15.3b, c). The right atrium then delivers the deoxygenated blood into the right ventricle, which pumps it into the ***pulmonary trunk.*** The pulmonary trunk divides into a ***right*** and ***left pulmonary artery***, each of which carries blood to the corresponding lung. ***Arteries*** are blood vessels that carry blood away from the heart. In the lungs, the deoxygenated blood unloads carbon dioxide and picks up oxygen. This ***oxygenated blood*** (oxygen-rich blood that has picked up oxygen as it flows through the lungs) then enters the left atrium via four ***pulmonary veins***. The blood then passes into the left ventricle, which pumps the blood into the ***ascending aorta***. From here the oxygenated blood is carried to all parts of the body.

Between the pulmonary trunk and arch of the aorta is a structure called the ***ligamentum arteriosum***. It is the remnant of the ***ductus arteriosus***, a blood vessel in fetal circulation that allows most blood to bypass the nonfunctional fetal lungs (see page 412).

Valves of the Heart

As each chamber of the heart contracts, it pushes a volume of blood into a ventricle or out of the heart into an artery. To prevent the blood from flowing backward, the heart has four ***valves*** composed of dense connective tissue covered by endothelium. These valves open and close in response to pressure changes as the heart contracts and relaxes.

As their names imply, ***atrioventricular (AV) valves*** lie between the atria and ventricles (Figure 15.3c). The atrioventricular valve between the right atrium and right ventricle is called the ***tricuspid valve*** because it consists of three cusps (leaflets). The pointed ends of the cusps project into the ventricle. Tendonlike cords, called ***chordae tendineae*** (KOR-dē ten-DIN-ē-ē; *chord-* = cord; *tend-* = tendon), connect the pointed ends to ***papillary muscles*** (*papill-* = nipple), cardiac muscle projections located on the inner surface of the ventricles. The chordae tendineae prevent the valve cusps from pushing up into the atria when the ventricles contract.

The atrioventricular valve between the left atrium and left ventricle is called the ***bicuspid (mitral) valve.*** It has two cusps that work in the same way as the cusps of the tricuspid valve. For blood to pass from an atrium to a ventricle, an atrioventricular valve must open.

The opening and closing of the valves are due to pressure differences across the valves. When blood moves from an atrium to a ventricle, the valve is pushed open, the papillary muscles relax, and the chordae tendineae slacken (Figure 15.4a). When a ventricle contracts, the pressure of the ventricular blood drives the cusps upward until their edges meet and close the opening (Figure 15.4b). At the same time, contraction of the papillary muscles and tightening of the chordae tendineae help prevent the cusps from swinging upward into the atrium.

Near the origin of the pulmonary trunk and aorta are ***semilunar valves*** called the ***pulmonary valve*** and the ***aortic valve*** that prevent blood from flowing back into the heart (see Figure 15.3c). The pulmonary valve lies in the opening where the pulmonary trunk leaves the right ventricle. The aortic valve is situated at the opening between the left ventricle and the aorta. Each valve consists of three semilunar (half-moon-shaped) cusps that attach to the artery wall. Like the atrioventricular valves, the semilunar valves permit blood to flow in one direction only—in this case, from the ventricles into the arteries.

When the ventricles contract, pressure builds up within them. The semilunar valves open when pressure in the ventricles exceeds the pressure in the arteries, permitting ejection of blood from the ventricles into the pulmonary trunk and aorta (see Figure 15.4d). As the ventricles relax, blood starts to flow back toward the heart. This back-flowing blood fills the valve cusps, which tightly closes the semilunar valves (see Figure 15.4c).

Figure 15.4 Atrioventricular (AV) valves. The bicuspid and tricuspid valves operate in a similar manner.

Heart valves open and close in response to pressure changes as the heart contracts and relaxes.

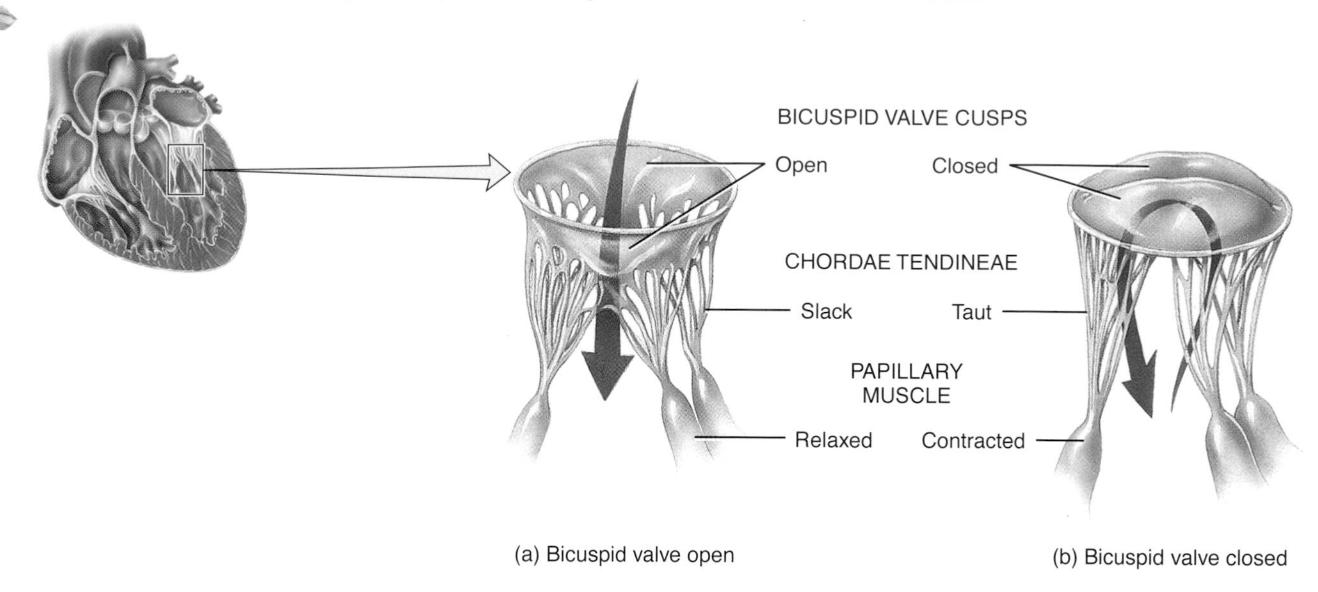

(a) Bicuspid valve open

(b) Bicuspid valve closed

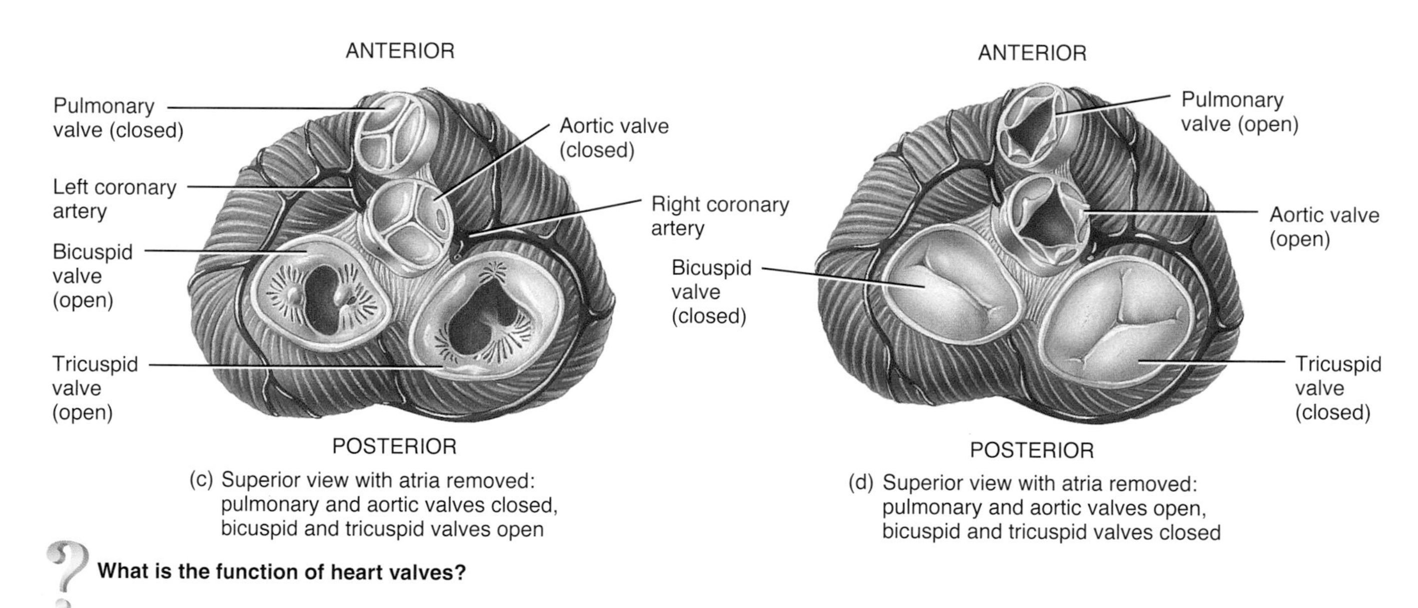

(c) Superior view with atria removed: pulmonary and aortic valves closed, bicuspid and tricuspid valves open

(d) Superior view with atria removed: pulmonary and aortic valves open, bicuspid and tricuspid valves closed

What is the function of heart valves?

When heart valves operate normally, they open fully and close completely at the proper times. A narrowing of a heart valve opening that restricts blood flow is known as **stenosis** (ste-NŌ-sis = a narrowing); failure of a valve to close completely is termed **insufficiency** or **incompetence**. In **mitral stenosis**, scar formation or a congenital defect causes narrowing of the mitral valve. One cause of **mitral insufficiency**, in which there is backflow of blood from the left ventricle into the left atrium, is **mitral valve prolapse (MVP)**. In MVP, one or both cusps of the mitral valve protrude into the left atrium during ventricular contraction. Mitral valve prolapse is one of the most common valvular disorders, affecting as much as 30% of the population. It is more prevalent in women than in men, and does not always pose a serious threat. In **aortic stenosis**, the aortic valve is narrowed, and in **aortic insufficiency**, there is backflow of blood from the aorta into the left ventricle.

If a heart valve cannot be repaired surgically, then the valve must be replaced. Tissue (biologic) valves may be provided by human donors or pigs; sometimes mechanical (artificial) valves made of plastic or metal are used. The aortic valve is the most commonly replaced heart valve.

■ CHECKPOINT

1. Identify the location of the heart.
2. Describe the various layers of the pericardium and the heart wall.
3. How do atria and ventricles differ in structure and function?
4. Which blood vessels that enter and exit the heart carry oxygenated blood? Which carry deoxygenated blood?
5. In correct sequence, which heart chambers, heart valves, and blood vessels would a drop of blood encounter from the time it flows out of the right atrium until it reaches the aorta?

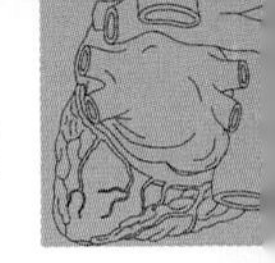

BLOOD FLOW AND BLOOD SUPPLY OF THE HEART

OBJECTIVES • **Explain how blood flows through the heart.**
• **Describe the clinical importance of the blood supply of the heart.**

Blood Flow Through the Heart

Blood flows through the heart from areas of higher blood pressure to areas of lower blood pressure. As the walls of the atria contract, the pressure of the blood within them increases. This increased blood pressure forces the AV valves open, allowing atrial blood to flow through the AV valves into the ventricles.

After the atria are finished contracting, the walls of the ventricles contract, increasing ventricular blood pressure and pushing blood through the semilunar valves into the pulmonary trunk and aorta. At the same time, the shape of the AV valve cusps causes them to be pushed shut, preventing backflow of ventricular blood into the atria. Figure 15.5 summarizes the flow of blood through the heart.

Figure 15.5 Blood flow through the heart.

The right and left coronary arteries deliver blood to the heart; the coronary veins drain blood from the heart into the coronary sinus.

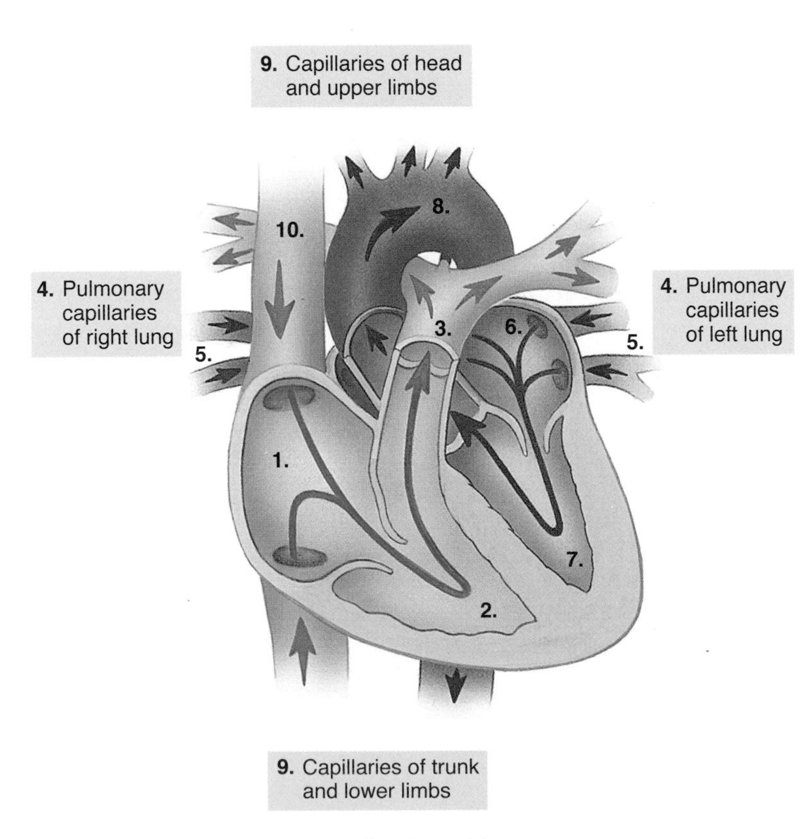

(a) Path of blood flow through heart

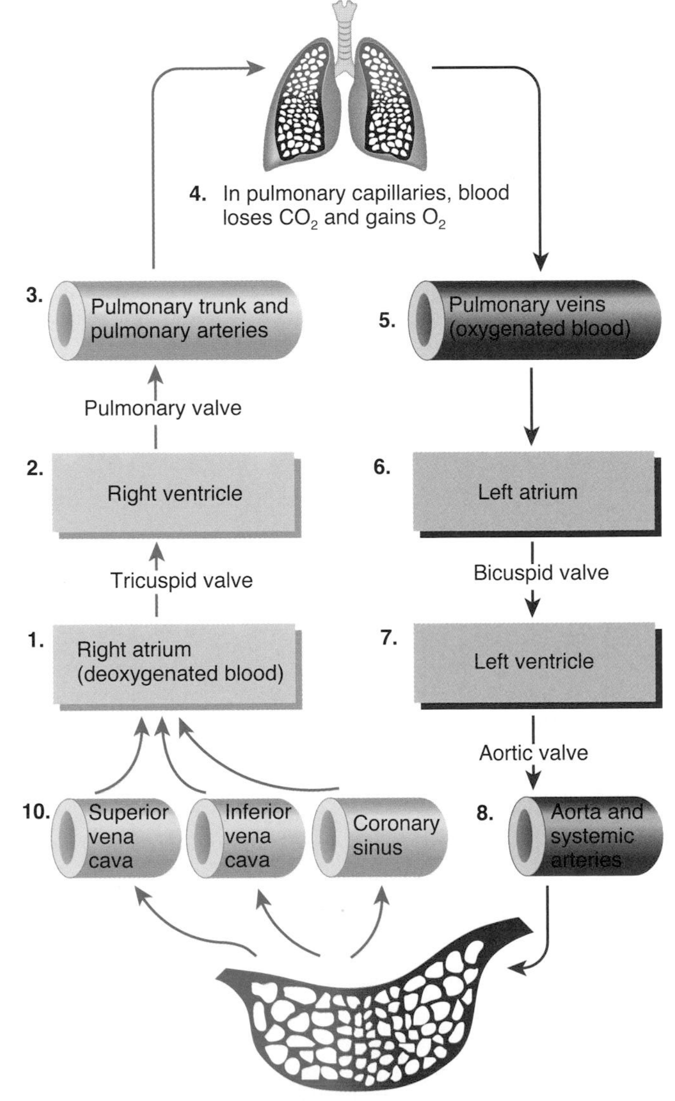

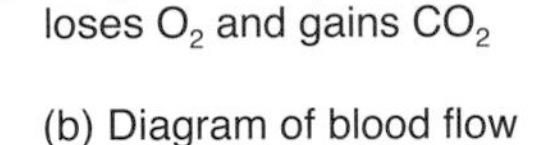
9. In systemic capillaries, blood loses O_2 and gains CO_2

(b) Diagram of blood flow

Which veins deliver deoxygenated blood into the right atrium?

Blood Supply of the Heart

The wall of the heart, like any other tissue, has its own blood vessels. The flow of blood through the numerous vessels in the myocardium is called ***coronary (cardiac) circulation***. The principal coronary vessels are the ***left*** and ***right coronary arteries***, which originate as branches of the ascending aorta (see Figure 15.3a). Each artery branches and then branches again to deliver oxygen and nutrients throughout the heart muscle. Most of the deoxygenated blood, which carries carbon dioxide and wastes, is collected by a large vein on the posterior surface of the heart, the ***coronary sinus*** (see Figure 15.3b), which empties into the right atrium.

Most parts of the body receive blood from branches of more than one artery, and where two or more arteries supply the same region, they usually connect. These connections, called ***anastomoses*** (a-nas′-tō-MŌ-sēs), provide alternate routes for blood to reach a particular organ or tissue. The myocardium contains many anastomoses that connect branches of a given coronary artery or extend between branches of different coronary arteries. They provide detours for arterial blood if a main route becomes obstructed. Thus, heart muscle may receive sufficient oxygen even if one of its coronary arteries is partially blocked.

When blockage of a coronary artery deprives the heart muscle of oxygen, **reperfusion**, the reestablishment of blood flow, may damage the tissue further. This surprising effect is due to the formation of oxygen **free radicals** from the reintroduced oxygen. Free radicals are electrically charged molecules that have an unpaired electron. Such molecules are unstable and highly reactive. They cause chain reactions that lead to cellular damage and death. To counter the effects of oxygen free radicals, body cells produce enzymes that convert free radicals to less reactive substances. In addition, some nutrients, such as vitamin E, vitamin C, beta-carotene, zinc, and selenium, are antioxidants, which remove oxygen free radicals. Drugs that lessen reperfusion damage after a heart attack or stroke are currently under development.

■ CHECKPOINT

6. Describe the main force that causes blood to flow through the heart.
7. Why is it that blood flowing through the chambers within the heart cannot supply sufficient oxygen or remove enough carbon dioxide from the myocardium?

CONDUCTION SYSTEM OF THE HEART

OBJECTIVE • Explain how each heartbeat is initiated and maintained.

About 1% of the cardiac muscle fibers are different from all others because they can generate action potentials over and over and do so in a rhythmical pattern. These cells have two important functions: They act as a ***pacemaker***, setting the rhythm for the entire heart, and they form the ***conduction system***, the route for action potentials throughout the heart muscle. The conduction system ensures that cardiac chambers are stimulated to contract in a coordinated manner, which makes the heart an effective pump. Cardiac action potentials pass through the following components of the conduction system (Figure 15.6):

1. Normally, cardiac excitation begins in the ***sinoatrial (SA) node***, located in the right atrial wall just inferior to the opening of the superior vena cava. An action potential spontaneously arises in the SA node and then conducts throughout both atria via gap junctions in the intercalated discs of atrial fibers (see Figure 15.2b). Following the action potential, the atria contract.
2. By conducting along atrial muscle fibers, the action potential also reaches the ***atrioventricular (AV) node***, located in the interatrial septum, just anterior to the opening of the coronary sinus. At the AV node, the action potential slows considerably, providing time for the atria to empty their blood into the ventricles.
3. From the AV node, the action potential enters the ***atrioventricular (AV) bundle*** (also known as the ***bundle of His***), in the interventricular septum. The AV bundle is the only site where action potentials can conduct from the atria to the ventricles.
4. After conducting along the AV bundle, the action potential then enters both the ***right*** and ***left bundle branches*** that course through the interventricular septum toward the apex of the heart.
5. Finally, large-diameter ***Purkinje fibers*** (pur-KIN-jē) rapidly conduct the action potential, first to the apex of the ventricles and then upward to the remainder of the ventricular myocardium. Then, a fraction of a second after the atria contract, the ventricles contract.

Figure 15.6 Conduction system of the heart. The SA node, located in the right atrial wall, is the heart's pacemaker, initiating cardiac action potentials that cause contraction of the heart's chambers. The arrows indicate the flow of action potentials through the atria. The route of action potentials through the numbered components of the conduction system is described in the text.

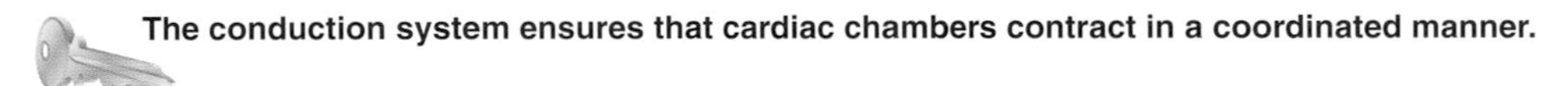

The conduction system ensures that cardiac chambers contract in a coordinated manner.

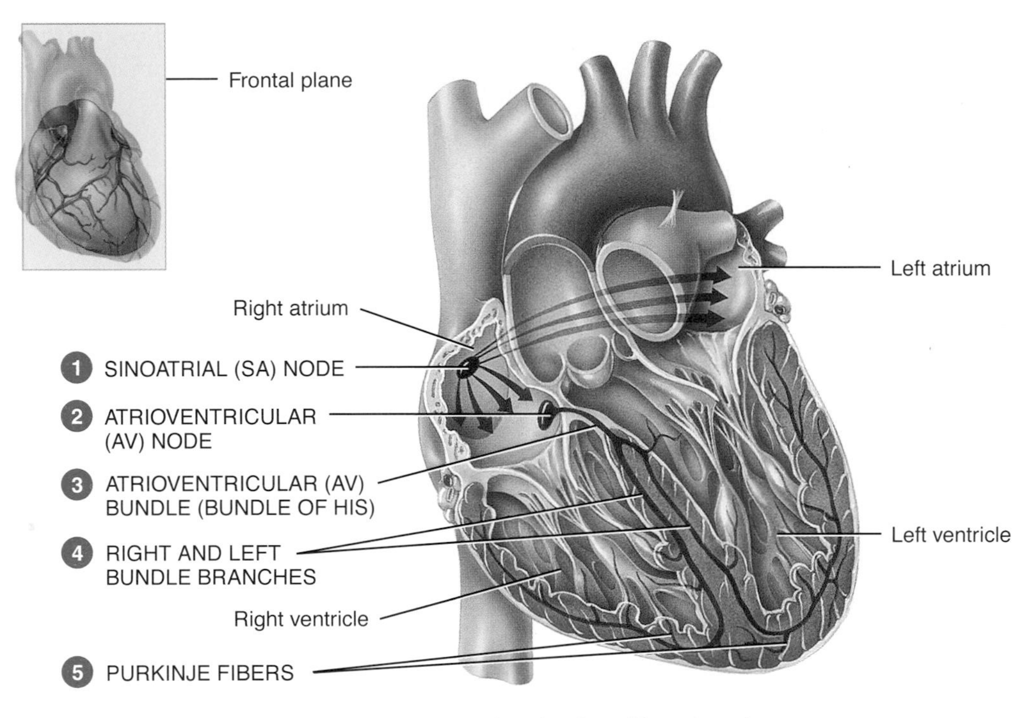

Which component of the conduction system provides the only route for action potentials to conduct between the atria and the ventricles?

The SA node initiates action potentials about 100 times per minute, faster than any other region of the conducting system. Thus, the SA node sets the rhythm for contraction of the heart—it is the *pacemaker* of the heart. Various hormones and neurotransmitters can speed or slow pacing of the heart by SA node fibers. In a person at rest, for example, acetylcholine released by the parasympathetic division of the ANS typically slows SA node pacing to about 75 action potentials per minute, causing 75 heartbeats per minute. If the SA node becomes diseased or damaged, the slower AV node fibers can become the pacemaker. With pacing by the AV node, however, heart rate is slower, only 40 to 60 beats/min. If the activity of both nodes is suppressed, the heartbeat may still be maintained by the AV bundle, a bundle branch, or Purkinje fibers. These fibers generate action potentials very slowly, about 20 to 35 times per minute. At such a low heart rate, blood flow to the brain is inadequate.

When the heart rate is too low, normal heart rhythm can be restored and maintained by surgically implanting an **artificial pacemaker**, a device that sends out small electrical currents to stimulate the heart to contract. A pacemaker consists of a battery and impulse generator and is usually implanted beneath the skin just inferior to the clavicle. The pacemaker is connected to one or two flexible wires (leads) that are threaded through the superior vena cava and then passed into the right atrium and right ventricle. Many of the newer pacemakers, called *activity-adjusted pacemakers*, automatically speed up the heartbeat during exercise.

■ CHECKPOINT

8. Describe the path of an action potential through the conduction system.

ELECTROCARDIOGRAM

OBJECTIVE • Describe the meaning and diagnostic value of an electrocardiogram.

Conduction of action potentials through the heart generates electrical currents that can be picked up by electrodes placed on the skin. A recording of the electrical changes that accompany the heartbeat is called an ***electrocardiogram*** (e-lek′-trō-KAR-dē-ō-gram), which is abbreviated as either ***ECG*** or ***EKG***.

Three clearly recognizable waves accompany each heartbeat. The first, called the ***P wave***, is a small upward deflection on the ECG (Figure 15.7); it represents atrial depolarization, the depolarizing phase of the cardiac action potential as it spreads from the SA node throughout both atria. Depolarization causes contraction. Thus, a fraction of a second after the P wave begins, the atria contract. The second wave, called the ***QRS complex***, begins as a downward deflection (Q); continues as a large, upright, triangular wave (R); and ends as a downward wave (S). The QRS complex represents the onset of ventricular depolarization, as the cardiac action potential spreads through the ventricles. Shortly after the QRS complex begins, the ventricles start to contract. The third wave is the ***T wave***, a dome-shaped upward deflection that indicates ventricular repolarization and occurs just before the ventricles start to relax. Repolarization of the atria is not usually evident in an ECG because it is masked by the larger QRS complex.

Variations in the size and duration of the waves of an ECG are useful in diagnosing abnormal cardiac rhythms and conduction patterns and in following the course of recovery from a heart attack. An ECG can also reveal the presence of a living fetus.

Figure 15.7 Normal electrocardiogram (ECG) of a single heartbeat. P wave = atrial depolarization; QRS complex = onset of ventricular depolarization; T wave = ventricular repolarization.

An electrocardiogram is a recording of the electrical activity that initiates each heartbeat.

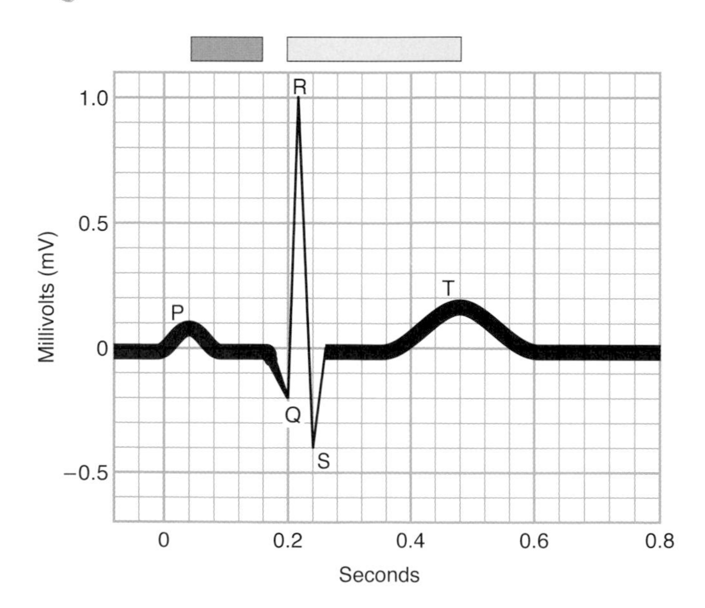

What event occurs in response to atrial depolarization?

■ **CHECKPOINT**

9. What is the significance of the P wave, QRS complex, and T wave?

THE CARDIAC CYCLE

OBJECTIVE • Describe the phases of the cardiac cycle.

A single ***cardiac cycle*** includes all the events associated with one heartbeat. In a normal cardiac cycle, the two atria contract while the two ventricles relax; then, while the two ventricles contract, the two atria relax. The term ***systole*** (SIS-tō-lē = contraction) refers to the phase of contraction; ***diastole*** (dī-AS-tō-lē = dilation or expansion) refers to the phase of relaxation. A cardiac cycle consists of systole and diastole of both atria plus systole and diastole of both ventricles.

For the purposes of our discussion, we will divide the *cardiac cycle* into three phases (Figure 15.8):

1. ***Relaxation period.*** The relaxation period begins at the end of a cardiac cycle when the ventricles start to relax and all four chambers are in diastole. Repolarization of the ventricular muscle fibers (T wave in the ECG) initiates relaxation. As the ventricles relax, pressure within them drops. When ventricular pressure drops below atrial pressure, the AV valves open and ventricular filling begins. About 75% of the ventricular filling occurs after the AV valves open and before the atria contract.
2. ***Atrial systole (contraction).*** An action potential from the SA node causes atrial depolarization, noted as the P wave in the ECG. Atrial systole follows the P wave, which marks the end of the relaxation period. As the atria contract, they force the last 25% of the blood into the ventricles. At the end of atrial systole, each ventricle contains about 130 mL of blood. The AV valves are still open and the semilunar valves are still closed.
3. ***Ventricular systole (contraction).*** The QRS complex in the ECG indicates ventricular depolarization, which leads to contraction of the ventricles. Ventricular contraction pushes blood against the AV valves, forcing them shut. As ventricular contraction continues, pressure inside the chambers quickly rises. When left ventricular

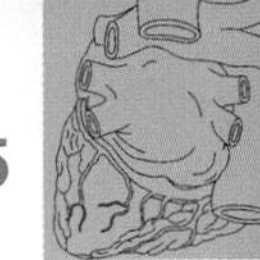

Figure 15.8 Cardiac cycle.

A cardiac cycle is composed of all the events associated with one heartbeat.

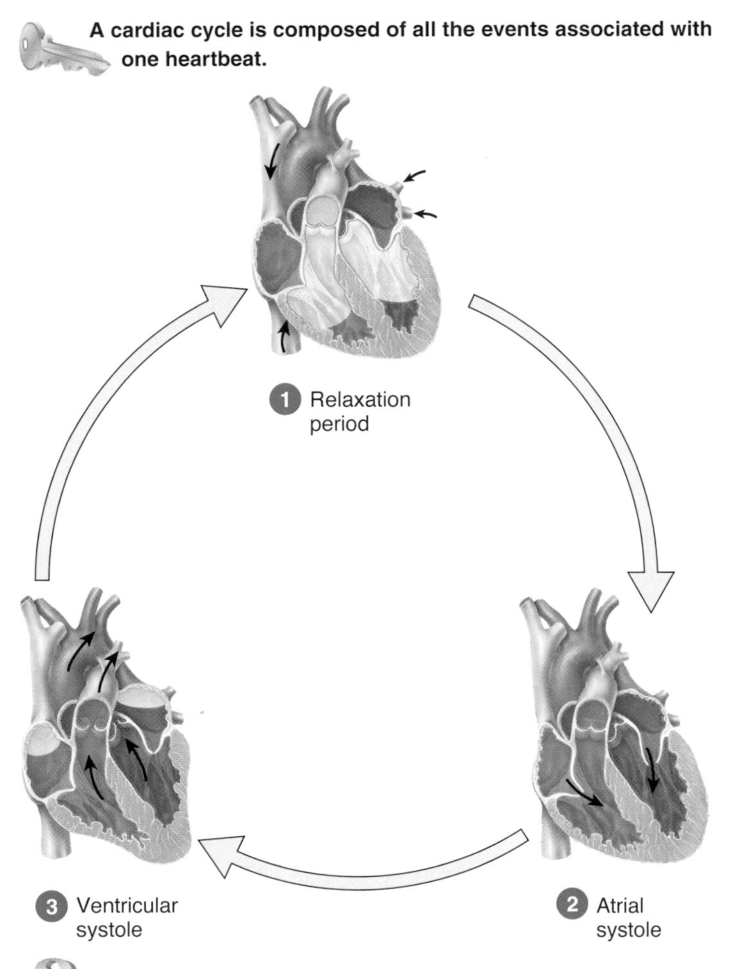

What is the term used for the contraction phase of the cardiac cycle? The relaxation phase?

pressure surpasses aortic pressure and right ventricular pressure rises above the pressure in the pulmonary trunk, both semilunar valves open, and ejection of blood from the heart begins. Ejection continues until the ventricles start to relax. At rest, the volume of blood ejected from each ventricle during ventricular systole is about 70 mL (a little more than 2 oz.). When the ventricles begin to relax, ventricular pressure drops, the semilunar valves close, and another relaxation period begins.

At rest, each cardiac cycle lasts about 0.8 sec. In one complete cycle, the first 0.4 sec of the cycle is the relaxation period, when all four chambers are in diastole. Then, the atria are in systole for 0.1 sec and in diastole for the next 0.7 sec. After atrial systole, the ventricles are in systole for 0.3 sec and in diastole for 0.5 sec. When the heart beats faster, for instance during exercise, the relaxation period is shorter.

Heart Sounds

The sound of the heartbeat comes primarily from turbulence in blood flow created by the closure of the valves, not from the contraction of the heart muscle. The first sound, ***lubb***, is a long, booming sound from the AV valves closing after ventricular systole begins. The second sound, a short, sharp sound, ***dupp***, is from the semilunar valves closing at the end of ventricular systole. There is a pause during the relaxation period. Thus, the cardiac cycle is heard as: lubb, dupp, pause; lubb, dupp, pause; lubb, dupp, pause.

Heart sounds provide valuable information about the mechanical operation of the heart. A **heart murmur** is an abnormal sound consisting of a clicking, rushing, or gurgling noise that is heard before, between, or after the normal heart sounds, or that may mask the normal heart sounds. Heart murmurs in children are extremely common and usually do not represent a health condition. These types of heart murmurs often subside or disappear with growth. Although some heart murmurs in adults are innocent, most often a murmur indicates a valve disorder.

■ **CHECKPOINT**

10. Explain the events that occur during each of the three phases of the cardiac cycle.

11. What causes the heart sounds?

CARDIAC OUTPUT

OBJECTIVE • Define cardiac output, explain how it is calculated, and describe how it is regulated.

The volume of blood ejected per minute from the left ventricle into the aorta is called the ***cardiac output (CO)***. (Note that the same amount of blood is also ejected from the right ventricle into the pulmonary trunk.) Cardiac output is determined by (1) the ***stroke volume (SV)***, the amount of blood ejected by the left ventricle during each beat (contraction), and (2) ***heart rate (HR)***, the number of heartbeats per minute. In a resting adult, stroke volume averages 70 mL, and heart rate is about 75 beats per minute. Thus the average cardiac output in a resting adult is

$$\begin{aligned}\text{Cardiac output} &= \text{stroke volume} \times \text{heart rate}\\ &= 70\ \text{mL/beat} \times 75\ \text{beats/min}\\ &= 5250\ \text{mL/min or } 5.25\ \text{liters/min}\end{aligned}$$

Factors that increase stroke volume or heart rate, such as exercise, increase cardiac output.

Regulation of Stroke Volume

Although some blood is always left in the ventricles at the end of their contraction, a healthy heart pumps out the blood that has entered its chambers during the previous diastole.

Focus on Wellness

Sudden Cardiac Death During Exercise — What's the Risk?

Regular physical activity helps keep hearts and arteries healthy. But very rarely, strenuous activity can precipitate a heart attack. In adults of middle age and older, a heart attack, or myocardial infarction, typically occurs because a blood clot lodges in an artery of the heart already narrowed by atherosclerotic plaque. During exercise, heart rate and blood pressure increase. Under this stress, an unstable plaque may rupture, stimulating the clotting process as the body tries to repair the damaged artery.

How Risky Is Exercise?

Many studies have been conducted in an attempt to quantify the risk imposed by strenuous exercise. Researchers have concluded that, in general, risk of heart attack is about two to six times higher during strenuous exercise than during light physical activity or rest. The statistical risk of heart attack varies considerably depending on a person's exercise history. Risk is lowest for those who exercise regularly and highest for people unaccustomed to exercise. Risk during exercise also rises with the number and severity of other cardiovascular risk factors. For example, people already diagnosed with heart disease are 10 times as likely to have a heart attack during exercise as apparently healthy individuals. Though this may sound discouraging, the risk can be seen from another point of view. Overall, the incidence of death during physical activity is very low, about 6 deaths per 100,000 middle-aged men per year.

Exercise is also safe for most people recovering from heart attacks. In 167 supervised cardiac rehabilitation exercise programs, the incidence of heart attack in 51,000 patients was 1 in 294,000 person-hours (number of people multiplied by number of hours each exercised). Incidence of death was only 1 in 784,000 person-hours.

Who Is at Risk?

Although these figures illustrate that risk of heart attack and death are quite low, if the death occurs to someone you love, it happens 100% and is still a tragic loss. People with diagnosed or suspected artery disease are most at risk and should check with their physicians before starting an exercise program. Risk can be reduced by exercising regularly (several times per week) at a low to moderate intensity, and by heeding any warning signs of cardiovascular disease, such as chest pain or pressure, abnormal heart rhythms, or dizziness.

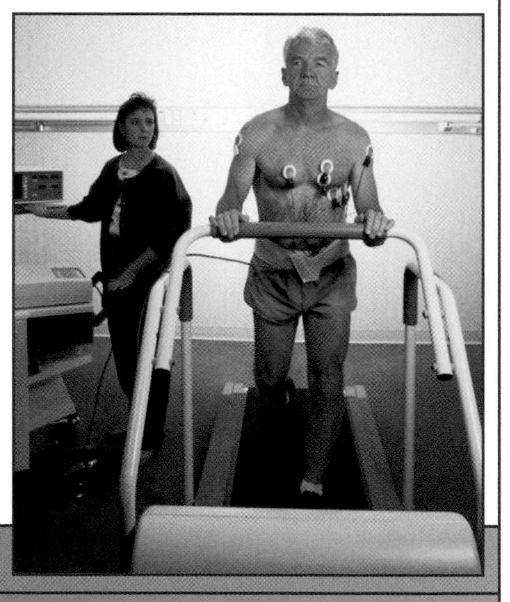

► **Think It Over . . .**

► ***Why do you think exercise stress tests are used to help diagnose coronary artery disease?***

The more blood that returns to the heart during diastole, the more blood that is ejected during the next systole. Three factors regulate stroke volume and ensure that the left and right ventricles pump equal volumes of blood:

1. **The degree of stretch in the heart before it contracts.** Within limits, the more the heart is stretched as it fills during diastole, the greater the force of contraction during systole, a relationship known as the ***Frank–Starling law of the heart.*** The situation is somewhat like stretching a rubber band: The more you stretch the heart, the more forcefully it contracts. In other words, within physiological limits, the heart pumps all the blood it receives. If the left side of the heart pumps a little more blood than the right side, a larger volume of blood returns to the right ventricle. On the next beat the right ventricle contracts more forcefully, and the two sides are again in balance.
2. **The forcefulness of contraction of individual ventricular muscle fibers.** Even at a constant degree of stretch, the heart can contract more or less forcefully when certain substances are present. Stimulation of the sympathetic division of the autonomic nervous system (ANS), hormones such as epinephrine and norepinephrine, increased Ca^{2+} level in the interstitial fluid, and the drug digitalis all increase the force of contraction of cardiac muscle fibers. In contrast, inhibition of the sympathetic division of the ANS, anoxia, acidosis, some anesthetics, and increased K^+ level in the extracellular fluid decrease contraction force.
3. **The pressure required to eject blood from the ventricles.** The semilunar valves open and ejection of blood from the heart begins when pressure in the right ventricle exceeds the pressure in the pulmonary trunk and

when the pressure in the left ventricle exceeds the pressure in the aorta. When the required pressure is higher than normal, the valves open later than normal, stroke volume decreases, and more blood remains in the ventricles at the end of systole.

In **congestive heart failure (CHF),** the heart is a failing pump. It pumps blood less and less effectively, leaving more blood in the ventricles at the end of each cycle. The result is a positive feedback cycle: Less-effective pumping leads to even lower pumping capability. Often, one side of the heart starts to fail before the other. If the left ventricle fails first, it can't pump out all the blood it receives, and blood backs up in the lungs. The result is *pulmonary edema,* fluid accumulation in the lungs that can lead to suffocation. If the right ventricle fails first, blood backs up in the systemic blood vessels. In this case, the resulting *peripheral edema* is usually most noticeable as swelling in the feet and ankles. Common causes of CHF are coronary artery disease (see page 379), long-term high blood pressure, myocardial infarctions, and valve disorders.

Regulation of Heart Rate

Adjustments to the heart rate are important in the short-term control of cardiac output and blood pressure. If left to itself, the sinoatrial node would set a constant heart rate of about 100 beats/min. However, tissues require different volumes of blood flow under different conditions. During exercise, for example, cardiac output rises to supply working tissues with increased amounts of oxygen and nutrients. The most important factors in the regulation of heart rate are the autonomic nervous system and the hormones epinephrine and norepinephrine, released by the adrenal glands.

Autonomic Regulation of Heart Rate

The nervous system regulation of the heart originates in the ***cardiovascular (CV) center*** in the medulla oblongata. This region of the brain stem receives input from a variety of sensory receptors and from higher brain centers, such as the limbic system and cerebral cortex. The cardiovascular center then directs appropriate output by increasing or decreasing the frequency of nerve impulses sent out to both the sympathetic and parasympathetic branches of the ANS (Figure 15.9).

Arising from the CV center are sympathetic neurons that reach the heart via ***cardiac accelerator nerves.*** They innervate the conduction system, atria, and ventricles. The norepinephrine released by cardiac accelerator nerves increases the heart rate. Also arising from the CV center are parasympathetic neurons that reach the heart via the ***vagus (X) nerves.*** These parasympathetic neurons extend to the conduction system and atria. The neurotransmitter they release—acetyl-

Figure 15.9 Autonomic nervous system regulation of heart rate.

The cardiovascular center in the medulla oblongata controls both sympathetic and parasympathetic nerves that innervate the heart.

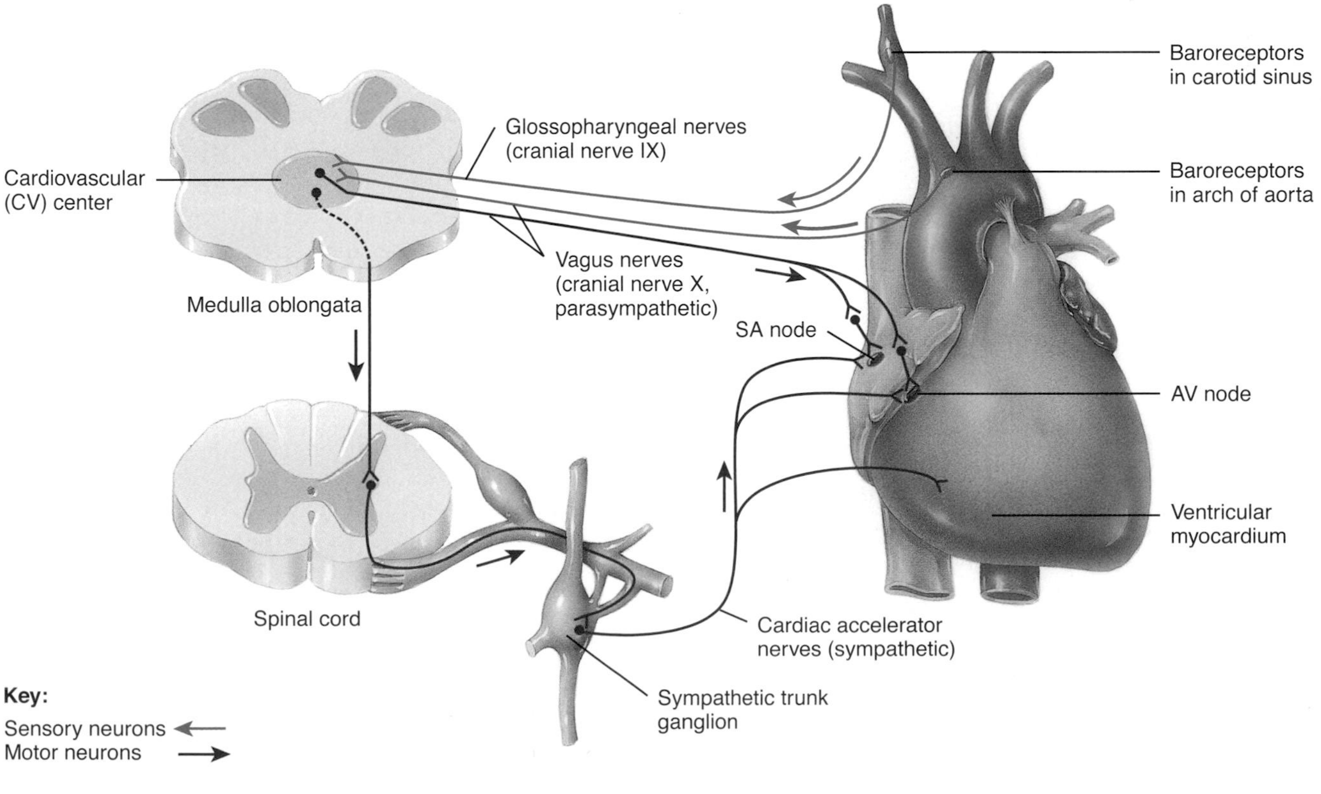

What effect does acetylcholine, released by parasympathetic nerves, have on heart rate?

choline (ACh)—decreases the heart rate by slowing the pacemaking activity of the SA node.

Several types of sensory receptors provide input to the cardiovascular center. For example, ***baroreceptors*** (*baro-* = pressure), neurons sensitive to blood pressure changes, are strategically located in the arch of the aorta and carotid arteries (arteries in the neck that supply blood to the brain). If there is an increase in blood pressure, the baroreceptors send nerve impulses along sensory neurons that are part of the glossopharyngeal (IX) and vagus (X) nerves to the CV center (Figure 15.9). The cardiovascular center responds by putting out more nerve impulses along the parasympathetic (motor) neurons that are also part of the vagus (X) nerves. The resulting decrease in heart rate lowers cardiac output and thus lowers blood pressure. If blood pressure falls, baroreceptors do not stimulate the cardiovascular center. As a result of this lack of stimulation, heart rate increases, cardiac output increases, and blood pressure increases to the normal level. ***Chemoreceptors***, neurons sensitive to chemical changes in the blood, detect changes in blood levels of chemicals such as O_2, CO_2, and H^+. Their relationship to the cardiovascular center is considered in Chapter 16 with regard to blood pressure.

Chemical Regulation of Heart Rate

Certain chemicals influence both the basic physiology of cardiac muscle and its rate of contraction. Chemicals with major effects on the heart fall into one of two categories:

1. **Hormones.** Epinephrine and norepinephrine (from the adrenal medullae) enhance the heart's pumping effectiveness by increasing both heart rate and contraction force. Exercise, stress, and excitement cause the adrenal medullae to release more hormones. Thyroid hormones also increase heart rate. One sign of hyperthyroidism (excessive levels of thyroid hormone) is tachycardia (elevated resting heart rate).
2. **Ions.** Elevated blood levels of K^+ or Na^+ decrease heart rate and contraction force. A moderate increase in extracellular and intracellular Ca^{2+} level increases heart rate and contraction force.

Other Factors in Heart Rate Regulation

Age, gender, physical fitness, and body temperature also influence resting heart rate. A newborn baby is likely to have a resting heart rate over 120 beats per minute; the rate then declines throughout childhood to the adult level of 75 beats per minute. Adult females generally have slightly higher resting heart rates than adult males, although regular exercise tends to bring resting heart rate down in both sexes. As adults age, their heart rates may increase.

Increased body temperature, such as occurs during fever or strenuous exercise, increases heart rate by causing the SA node to discharge more rapidly. Decreased body temperature decreases heart rate and force of contraction. During surgical repair of certain heart abnormalities, it is helpful to slow a patient's heart rate by deliberately cooling the body.

■ **CHECKPOINT**

12. Describe how stroke volume is regulated.
13. How does the autonomic nervous system help regulate heart rate?

EXERCISE AND THE HEART

OBJECTIVE • **Explain the relationship between exercise and the heart.**

Regardless of the current level, a person's cardiovascular fitness can be improved at any age with regular exercise. Some types of exercise are more effective than others for improving the health of the cardiovascular system. ***Aerobics***, any activity that works large body muscles for at least 20 minutes, elevates cardiac output and accelerates metabolic rate. Three to five such sessions a week are usually recommended for improving the health of the cardiovascular system. Brisk walking, running, bicycling, cross-country skiing, and swimming are examples of aerobic activities.

Sustained exercise increases the oxygen demand of the muscles. Whether the demand is met depends mainly on the adequacy of cardiac output and proper functioning of the respiratory system. After several weeks of training, a healthy person increases maximal cardiac output, thereby increasing the maximal rate of oxygen delivery to the tissues. Oxygen delivery also rises because skeletal muscles develop more capillary networks in response to long-term training.

During strenuous activity, a well-trained athlete can achieve a cardiac output double that of a sedentary person, in part because training causes hypertrophy (enlargement) of the heart. Even though the heart of a well-trained athlete is larger, *resting* cardiac output is about the same as in a healthy untrained person, because stroke volume is increased while heart rate is decreased. The resting heart rate of a trained athlete often is only 40–60 beats per minute (*resting bradycardia*). Regular exercise also helps to reduce blood pressure, anxiety, and depression; control weight; and increase the body's ability to dissolve blood clots by increasing fibrinolytic activity.

■ **CHECKPOINT**

14. What is aerobic exercise? Why are aerobic exercises beneficial?

• • •

The heart is the blood pump for the cardiovascular system, but it is the blood vessels that distribute blood to all parts of the body and collect blood from them. In the next chapter we will see how blood vessels accomplish this.

COMMON DISORDERS

Coronary Artery Disease

Coronary artery disease (CAD) affects about 7 million people and causes nearly 750,000 deaths in the United States each year. CAD is defined as the effects of the accumulation of atherosclerotic plaques (described shortly) in coronary arteries that lead to a reduction in blood flow to the myocardium. Some individuals have no signs or symptoms, others experience angina pectoris (chest pain), and still others suffer a heart attack.

People who possess combinations of certain risk factors are more likely to develop CAD. *Risk factors* (characteristics, symp–toms, or signs that are statistically associated with a greater chance of developing a disease) include smoking, high blood pressure, diabetes, high cholesterol levels, obesity, "type A" personality, sedentary lifestyle, and a family history of CAD. Most of these can be modified by changing diet and other habits or can be controlled by taking medications. However, other risk factors are unmodifiable—that is, beyond our control—including genetic predisposition (family history of CAD at an early age), age, and gender. For example, adult males are more likely than adult females to develop CAD; after age 70 the risks are roughly equal. Smoking is undoubtedly the number-one risk factor in all CAD-associated diseases, roughly doubling the risk of morbidity and mortality.

In recent years, a number of new risk factors (all modifiable) have been identified as significant predictors of CAD. *C-reactive proteins (CRPs)* are proteins produced by the liver or present in blood in an inactive form that are converted to an active form during inflammation. CRPs may play a direct role in the development of atherosclerosis by promoting the uptake of LDLs by macrophages. *Lipoprotein (a)* is an LDL-like particle that binds to endothelial cells, macrophages, and blood platelets, may promote the proliferation of smooth muscle fibers, and inhibits the breakdown of blood clots. *Fibrinogen* is a glycoprotein involved in blood clotting that may help regulate cellular proliferation, vasoconstriction, and platelet aggregation. *Homocysteine* is an amino acid that may induce blood vessel damage by promoting platelet aggregation and smooth muscle fiber proliferation.

Atherosclerosis (ath′-er-ō-skler-ō-sis) is a progressive disease characterized by the formation in the walls of large- and medium-sized arteries of lesions called ***atherosclerotic plaques*** (Figure 15.10).

To understand how atherosclerotic plaques develop, you will need to know about molecules produced by the liver and small intestine called ***lipoproteins***. These spherical particles consist of an inner core of triglycerides and other lipids and an outer shell of proteins, phospholipids, and cholesterol. Two major lipoproteins are ***low-density lipoproteins*** or ***LDLs*** and ***high-density lipoproteins*** or ***HDLs***. LDLs transport cholesterol from the liver to body cells for use in cell membrane repair and the production of steroid hormones and bile salts. However, excessive amounts of LDLs promote atherosclerosis, so the cholesterol in these particles is known as "bad cholesterol." HDLs, on the other hand, remove excess cholesterol from body cells and transport it to the liver for elimination. Because HDLs decrease blood cholesterol level, the cholesterol in HDLs is known as "good cholesterol." Basically, you want your LDL to be low and your HDL to be high.

It has recently been learned that inflammation, a defensive response of the body to tissue damage, plays a key role in the development of atherosclerotic plaques. As a result of the damage, blood

Figure 15.10 Photomicrographs of a transverse section of (a) a normal artery and (b) one partially obstructed by an atherosclerotic plaque.

Atherosclerosis is a progressive disease caused by the formation of atherosclerotic plaques.

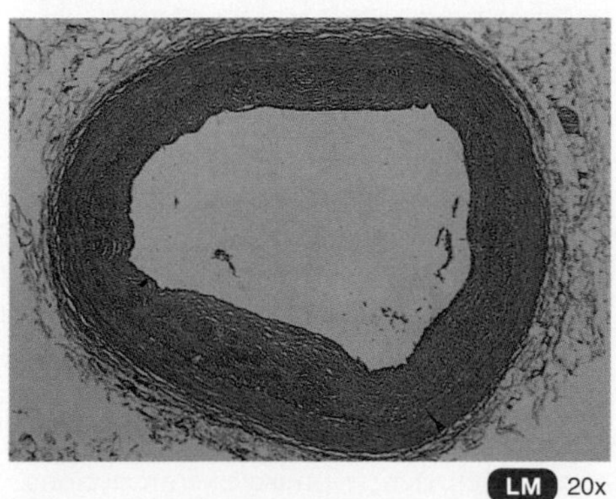

LM 20x

(a) Normal artery

Partially obstructed space through which blood flows

Atherosclerotic plaque

LM 20x

(b) Obstructed artery

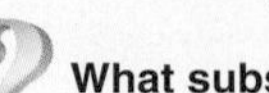

What substances are part of an atherosclerotic plaque?

vessels dilate and increase their permeability. The formation of atherosclerotic plaques begins when excess LDLs from the blood accumulate in the artery wall and undergo oxidation. In response, endothelial and smooth muscle cells of the artery secrete substances that attract monocytes from the blood and convert them into macrophages. The macrophages then ingest and become so filled with the oxidized LDL particles that they have a foamy appearance when viewed microscopically (***foam cells***). Together with T cells (lymphocytes), foam cells form a ***fatty streak***, the beginning of an atherosclerotic plaque. Following fatty streak formation, smooth muscle cells of the artery migrate to the top of the atherosclerotic plaque, forming a cap over it and thus walling it off from the blood.

Because most atherosclerotic plaques expand away from the bloodstream rather than into it, blood can flow through an artery with relative ease, often for decades. Most heart attacks occur when the cap over the plaque breaks open in response to chemicals produced by foam cells, causing a clot to form. If the clot in a coronary artery is large enough, it can significantly decrease or stop the flow of blood and result in a heart attack.

Treatment options for CAD include drugs (antihypertensive drugs, nitroglycerin, beta-blockers, and cholesterol-lowering and clot-dissolving agents) and various surgical and nonsurgical procedures designed to increase the blood supply to the heart.

Myocardial Ischemia and Infarction

Partial obstruction of blood flow in the coronary arteries may cause ***myocardial ischemia*** (is-KĒ-mē-a; *ische-* = to obstruct; *-emia* = in the blood), a condition of reduced blood flow to the myocardium. Usually, ischemia causes ***hypoxia*** (reduced oxygen supply), which may weaken cells without killing them. ***Angina pectoris*** (an-JĪ-na or AN-ji-na PEK-tō-ris), which literally means "strangled chest," is a severe pain that usually accompanies myocardial ischemia. Typically, sufferers describe it as a tightness or squeezing sensation, as though the chest were in a vise. The pain associated with angina pectoris is often referred to the neck, chin, or down the left arm to the elbow. ***Silent myocardial ischemia***, ischemic episodes without pain, is particulary dangerous because the person has no forewarning of an impending heart attack.

A complete obstruction to blood flow in a coronary artery may result in a ***myocardial infarction*** (in-FARK-shun), or ***MI***, commonly called a *heart attack*. *Infarction* means the death of an area of tissue because of interrupted blood supply. Because the heart tissue distal to the obstruction dies and is replaced by noncontractile scar tissue, the heart muscle loses some of its strength. Depending on the size and location of the infarcted (dead) area, an infarction may disrupt the conduction system of the heart and cause sudden death by triggering ventricular fibrillation. Treatment for a myocardial infarction may involve injection of a thrombolytic (clot-dissolving) agent such as streptokinase or tPA, plus heparin (an anticoagulant), or performing coronary angioplasty or coronary artery bypass grafting. Fortunately, heart muscle can remain alive in a resting person if it receives as little as 10–15% of its normal blood supply.

Congenital Defects

A defect that exists at birth (and usually before) is a ***congenital defect***. Among the several congenital defects that affect the heart are the following:

- In ***patent ductus arteriosus***, the ductus arteriosus (temporary blood vessel) between the aorta and the pulmonary trunk, which normally closes shortly after birth, remains open (see Figure 16.17 on page 413). Closure of the ductus arteriosus leaves a remnant called the ligamentum arteriosum (see Figure 15.3a).
- ***Atrial septal defect (ASD)*** is caused by incomplete closure of the interatrial septum. The most common type involves the foramen ovale, which normally closes shortly after birth (see Figure 16.17 on page 413).
- ***Ventricular septal defect (VSD)*** is caused by an incomplete closure of the interventricular septum.
- ***Valvular stenosis*** is a narrowing of one of the valves associated with blood flow through the heart.
- ***Tetralogy of Fallot*** (te-TRAL-Ō-jē of fa-LŌ) is a combination of four defects: an interventricular septal defect, an aorta that emerges from both ventricles instead of from the left ventricle only, a narrowed pulmonary semilunar valve, and an enlarged right ventricle.

Some congenital heart defects are being surgically corrected prior to birth in order to prevent complications at the time of birth and following the birth of an infant.

Arrhythmias

The usual rhythm of heartbeats, established by the SA mode, is called ***normal sinus rhythm***. The term ***arrhythmia*** (a-RITH-mē-a; *a-* = without) or ***dysrhythmia*** refers to an abnormal rhythm as a result of a defect in the conduction system of the heart. The heart may beat irregularly, too fast, or too slowly. Symptoms include chest pain, shortness of breath, lightheadedness, dizziness, and fainting. Arrhythmias may be caused by factors that stimulate the heart, such as stress, caffeine, alcohol, nicotine, cocaine, and certain drugs that contain caffeine or other stimulants. Arrhythmias may also be caused by a congenital defect, coronary artery disease, myocardial infarction, hypertension, defective heart valves, rheumatic heart disease, hyperthyroidism, and potassium deficiency.

One serious arrhythmia is called a ***heart block***. The most common heart block occurs in the atrioventricular node, which conducts impulses from the atria to the ventricles. This disturbance is called ***atrioventricular (AV) block***.

In ***atrial flutter***, the atrial rhythm averages between 240 and 360 beats per minute. The condition is essentially rapid atrial contractions accompanied by AV block. ***Atrial fibrillation*** is an uncoordinated contraction of the atrial muscles. When the muscle fibrillates, the muscle fibers of the atrium quiver individually instead of contracting together, canceling out the pumping of the atrium. ***Ventricular fibrillation (VF)*** is characterized by uncoordinated haphazard ventricular muscle contractions. Ventricular ejection ceases, and circulatory failure and death occur.

MEDICAL TERMINOLOGY AND CONDITIONS

Angiocardiography (an′-jē-ō-kar′-dē-OG-ra-fē; *angio-* = vessel; *cardio-* = heart) X-ray examination of the heart and great blood vessels after injection of a radiopaque dye into the bloodstream.

Cardiac arrest (KAR-dē-ak a-REST) A clinical term meaning cessation of an effective heartbeat. The heart may be completely stopped or in ventricular fibrillation.

Cardiac catheterization (kath′-e-ter-i-ZĀ-shun) Procedure that is used to visualize the heart's coronary arteries, chambers, valves, and great vessels. It may also be used to measure pressure in the heart and blood vessels; to assess cardiac output; and to measure the flow of blood through the heart and blood vessels, the oxygen content of blood, and the status of the heart valves and conduction system. The basic procedure involves inserting a catheter into a peripheral vein (for right heart catheterization) or artery (for left heart catheterization) and guiding it under fluoroscopy (x-ray observation).

Cardiac rehabilitation (rē-ha-bil-i-TĀ-shun) A supervised program of progressive exercise, psychological support, education, and training to enable a patient to resume normal activities following a myocardial infarction.

Cardiomegaly (kar′-dē-ō-MEG-a-lē; *mega-* = large) Heart enlargement.

Cor pulmonale (CP) (kor pul-mōn-ALE; *cor-* = heart; *pulmon-* = lung) Right ventricular hypertrophy caused by hypertension (high blood pressure) in the pulmonary circulation.

Cardiopulmonary resuscitation (kar′-dē-ō-PUL-mō-ner-ē re-sus′-i-TĀ-shun) (***CPR***) The artificial establishment of normal or near-normal respiration and circulation. The ***ABCs*** of cardiopulmonary resuscitation are ***Airway, Breathing***, and ***Circulation***, meaning the rescuer must establish an airway, provide artificial ventilation if breathing has stopped, and reestablish circulation if there is inadequate cardiac action.

Palpitation (pal′-pi-TĀ-shun) A fluttering of the heart or abnormal rate or rhythm of the heart.

Paroxysmal tachycardia (par′-ok-SIZ-mal tak′-e-KAR-dē-a) A period of rapid heartbeats that begins and ends suddenly.

Rheumatic fever (roo-MAT-ik) An acute systemic inflammatory disease that usually occurs after a streptococcal infection of the throat. The bacteria trigger an immune response in which antibodies that are produced to destroy the bacteria attack and inflame the connective tissues in joints, heart valves, and other organs. Even though rheumatic fever may weaken the entire heart wall, most often it damages the bicuspid (mitral) and aortic valves.

Sudden cardiac death The unexpected cessation of circulation and breathing due to an underlying heart disease such as ischemia, myocardial infarction, or a disturbance in cardiac rhythm.

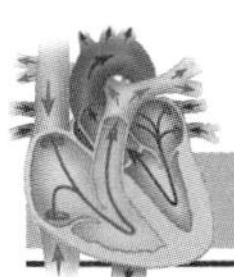

STUDY OUTLINE

Structure and Organization of the Heart (p. 365)

1. The heart is situated between the lungs, with about two-thirds of its mass to the left of the midline.
2. The pericardium consists of an outer fibrous layer and an inner serous pericardium.
3. The serous pericardium is composed of a parietal layer and a visceral layer.
4. Between the parietal and visceral layers of the serous pericardium is the pericardial cavity, a space filled with pericardial fluid that reduces friction between the two membranes.
5. The wall of the heart has three layers: epicardium, myocardium, and endocardium.
6. The chambers include two upper atria and two lower ventricles.
7. The blood flows through the heart from the superior and inferior venae cavae and the coronary sinus to the right atrium, through the tricuspid valve to the right ventricle, and through the pulmonary trunk to the lungs.
8. From the lungs, blood flows through the pulmonary veins into the left atrium, through the bicuspid valve to the left ventricle, and out through the aorta.
9. Four valves prevent the backflow of blood in the heart.
10. Atrioventricular (AV) valves, between the atria and their ventricles, are the tricuspid valve on the right side of the heart and the bicuspid (mitral) valve on the left.
11. The atrioventricular valves, chordae tendineae, and their papillary muscles stop blood from flowing back into the atria.
12. Each of the two arteries that leave the heart has a semilunar valve.

Blood Flow and Blood Supply of the Heart (p. 371)

1. Blood flows through the heart from areas of higher pressure to areas of lower pressure.
2. The pressure is related to the size and volume of a chamber.
3. The movement of blood through the heart is controlled by the opening and closing of the valves and the contraction and relaxation of the myocardium.
4. Coronary circulation delivers oxygenated blood to the myocardium and removes carbon dioxide from it.
5. Deoxygenated blood returns to the right atrium via the coronary sinus.

6. Malfunctions of this system can result in angina pectoris or myocardial infarction (MI).

Conduction System of the Heart (p. 372)

1. The conduction system consists of specialized cardiac muscle tissue that generates and distributes action potentials.
2. Components of this system are the sinoatrial (SA) node (pacemaker), atrioventricular (AV) node, atrioventricular (AV) bundle (bundle of His), bundle branches, and Purkinje fibers.

Electrocardiogram (p. 374)

1. The record of electrical changes during each cardiac cycle is referred to as an electrocardiogram (ECG).
2. A normal ECG consists of a P wave (depolarization of atria), QRS complex (onset of ventricular depolarization), and T wave (ventricular repolarization).
3. The ECG is used to diagnose abnormal cardiac rhythms and conduction patterns.

The Cardiac Cycle (p. 374)

1. A cardiac cycle consists of systole (contraction) and diastole (relaxation) of the chambers of the heart.
2. The phases of the cardiac cycle are (a) the relaxation period, (b) atrial systole, and (c) ventricular systole.
3. A complete cardiac cycle takes 0.8 sec at an average heartbeat of 75 beats per minute.
4. The first heart sound (lubb) represents the closing of the atrioventricular valves. The second sound (dupp) represents the closing of semilunar valves.

Cardiac Output (p. 375)

1. Cardiac output (CO) is the amount of blood ejected by the left ventricle into the aorta each minute: CO = stroke volume × beats per minute.
2. Stroke volume (SV) is the amount of blood ejected by a ventricle during ventricular systole. It is related to stretch on the heart before it contracts, forcefulness of contraction, and the amount of pressure required to eject blood from the ventricles.
3. Nervous control of the cardiovascular system originates in the cardiovascular center in the medulla oblongata.
4. Sympathetic impulses increase heart rate and force of contraction; parasympathetic impulses decrease heart rate.
5. Heart rate is affected by hormones (epinephrine, norepinephrine, thyroid hormones), ions (Na^+, K^+, Ca^{2+}), age, gender, physical fitness, and body temperature.

Exercise and the Heart (p. 378)

1. Sustained exercise increases oxygen demand on muscles.
2. Among the benefits of aerobic exercise are increased maximal cardiac output, decreased blood pressure, weight control, and increased ability to dissolve clots.

SELF-QUIZ

1. Match the following:

   ____ a. valve between the left atrium and left ventricle
   ____ b. valve between the right atrium and right ventricle
   ____ c. chamber that pumps blood to the lungs
   ____ d. chamber that pumps blood into aorta
   ____ e. chamber that receives oxygenated blood from lungs
   ____ f. chamber that receives deoxygenated blood from body
   ____ g. valve between the left ventricle and aorta
   ____ h. valve between right ventricle and pulmonary trunk

   A. aortic valve
   B. right atrium
   C. left atrium
   D. bicuspid (mitral) valve
   E. pulmonary valve
   F. right ventricle
   G. left ventricle
   H. tricuspid valve

2. Which of the following statements describes the pericardium?
   a. It is a layer of nervous tissue.
   b. It lines the inside of the myocardium.
   c. It is continuous with the epithelial lining of the large blood vessels.
   d. It is responsible for the contraction of the heart.
   e. It is a membrane that surrounds and protects the heart.

3. Which blood vessel delivers deoxygenated blood from the head and neck to the heart?
   a. pulmonary vein b. thoracic aorta
   c. pulmonary artery d. inferior vena cava
   e. superior vena cava

4. An embolus originating in the coronary sinus would first enter the
   a. right atrium b. pulmonary veins c. left atrium
   d. right ventricle e. aorta

5. The chordae tendineae and papillary muscles of the heart
   a. are responsible for connecting cardiac muscle fibers for the spread of action potentials
   b. can develop self-excitability and stimulate contraction
   c. help prevent the atrioventricular valves from protruding into the atria when the ventricles contract
   d. help anchor and protect the heart
   e. form the cusps (flaps) of the heart valves

6. Which chamber of the heart has the thickest layer of myocardium?
   a. right ventricle b. right atrium c. left ventricle
   d. left atrium e. coronary sinus

7. The normal "pacemaker" of the heart is the
   a. sinoatrial (SA) node b. atrioventricular (AV) node
   c. Purkinje fibers d. atrioventricular (AV) bundle
   e. right bundle branch

8. In normal heart action,
   a. the right atrium and ventricle contract, followed by the contraction of the left atrium and ventricle
   b. the order of contraction is right atrium, then right ventricle, then left atrium, then left ventricle
   c. the two atria contract together, and then the two ventricles contract together
   d. the right atrium and left ventricle contract, followed by the contraction of the left atrium and right ventricle
   e. all four chambers of the heart contract and then relax simultaneously

9. Heart sounds are produced by
   a. contraction of the myocardium
   b. closure of the heart valves
   c. the flow of blood in the coronary arteries
   d. the flow of blood in the ventricles
   e. the transmission of action potentials through the conduction system

10. Heart rate and strength of contraction are controlled by the cardiovascular center, which is located in the
    a. cerebrum b. pons c. right atrium d. medulla
    e. atrioventricular node

11. The portion of the ECG that corresponds to atrial depolarization is the
    a. R peak b. space between the T wave and P wave
    c. T wave d. P wave e. QRS complex

12. The opening of the semilunar valves is due to the pressure in the
    a. ventricles exceeding the pressure in the aorta and pulmonary trunk
    b. ventricles exceeding the pressure in the atria
    c. atria exceeding the pressure in the ventricles
    d. atria exceeding the pressure in the aorta and pulmonary trunk
    e. aorta and pulmonary trunk exceeding the pressure in the ventricles

13. On the anterior surface of each atrium is a wrinkled pouchlike structure called a(n)
    a. anterior interventricular sulcus b. coronary sulcus
    c. auricle d. interatrial septum
    e. posterior interventricular sulcus

14. The Frank–Starling law of the heart
    a. is important in maintaining equal blood output from both ventricles
    b. is used in reference to the force of contraction of the atria
    c. results in a decreased heart rate
    d. causes blood to accumulate in the lungs
    e. is related to the stretching of the cardiac muscle cells in the atria

15. Which of the following sequences best represents the pathway of an action potential through the heart's conduction system?
    1. sinoatrial (SA) node 2. Purkinje fibers
    3. atrioventricular (AV) bundle
    4. atrioventricular (AV) node
    5. right and left bundle branches
    a. 1, 4, 3, 2, 5 b. 4, 1, 3, 5, 2 c. 3, 4, 1, 2, 5
    d. 1, 4, 3, 5, 2 e. 2, 5, 3, 4, 1

16. Which of the following is NOT true concerning ventricular filling during the cardiac cycle?
    a. The atrioventricular (AV) valves are open.
    b. The ventricles fill to 75% of their capacity before the atria contract.
    c. The remaining 25% of the ventricular blood is forced into the ventricles when the atria contract.
    d. The semilunar valves are open.
    e. Ventricular filling begins when the ventricular pressure drops below the atrial pressure, causing the AV valves to open.

17. Cardiac output
    a. equals stroke volume (SV) × blood pressure (BP)
    b. equals stroke volume (SV) × heart rate (HR)
    c. is calculated using the formula for the Frank–Starling law of the heart
    d. is about 70 mL in the average adult male
    e. equals blood pressure (BP) × heart rate (HR)

18. Most heart problems are due to
    a. old age b. leakages at the valves
    c. problems in the coronary circulation
    d. the failure of the conduction system
    e. infections in the heart coverings

19. Using the situations that follow, indicate if the heart rate would speed up **(A)** or slow down **(B)**.
    ____ a. sympathetic stimulation of the sinoatrial (SA) node
    ____ b. decrease in blood pressure
    ____ c. fever
    ____ d. parasympathetic stimulation of the heart's conduction system
    ____ e. release of epinephrine
    ____ f. elevated K^+ level
    ____ g. release of acetylcholine
    ____ h. strenuous exercise
    ____ i. stimulation by the vagus (X) nerve
    ____ j. fear, anger, stress
    ____ k. cooling the body
    ____ l. hypoxia
    ____ m. excessive thyroid hormones

20. Match the following:

____ a. may cause a heart murmur	A. pericarditis
____ b. heart compression	B. mitral valve prolapse
____ c. inflammation of heart covering	C. myocardial infarction
____ d. heart chamber contraction	D. angina pectoris
____ e. chest pain from ischemia	E. diastole
____ f. heart attack	F. systole
____ g. heart chamber relaxation	G. cardiac tamponade

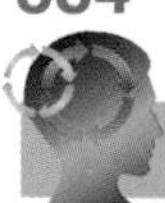

CRITICAL THINKING APPLICATIONS

1. Your uncle had an artificial pacemaker inserted after his last bout with heart trouble. What is the function of a pacemaker? For which heart structure does the pacemaker substitute?
2. Nikos was strolling across a four-lane highway when a car suddenly appeared out of nowhere. As he finished sprinting across the road, he felt his heart racing. Trace the route of the signal from his brain to his heart.
3. Jean-Claude, a member of the college's cross-country ski team, volunteered to have his heart function evaluated by the exercise physiology class. His resting pulse rate was 40 beats per minute. Assuming that he has an average cardiac output (CO), determine Jean-Claude's stroke volume (SV). Next, Jean-Claude rode an exercise bike until his heart rate had risen to 60 beats per minute. Assuming that his SV stayed constant, calculate Jean-Claude's CO during this moderate exercise.
4. Rosa's great Aunt Frieda likes to say that she has complaining feet and a mumbling heart. Aunt Frieda's physician uses the terms "edema" and "murmur." Explain Aunt Frieda's medical condition.

ANSWERS TO FIGURE QUESTIONS

15.1 The base of the heart consists mainly of the left atrium.

15.2 The visceral layer of the serous pericardium is also part of the heart wall (epicardium).

15.3 Blood flows away from the heart in arteries.

15.4 Heart valves prevent the backflow of blood.

15.5 The superior vena cava, inferior vena cava, and coronary sinus deliver deoxygenated blood into the right atrium.

15.6 The only electrical connection between the atria and the ventricles is the atrioventricular (AV) bundle.

15.7 Atrial depolarization causes contraction of the atria.

15.8 The contraction phase is called systole; the relaxation phase is called diastole.

15.9 Acetylcholine decreases heart rate.

15.10 Fatty substances, cholesterol, and smooth muscle fibers make up atherosclerotic plaques.

THE CARDIOVASCULAR SYSTEM: BLOOD VESSELS AND CIRCULATION

did you know?

Physical activity helps to protect the cardiovascular system in many ways. It improves blood cholesterol levels and blood sugar regulation. Regular exercise leads to increased output from the parasympathetic nervous system (not during exercise, but during the rest of the day), which leads to a lower resting heart rate and lower resting blood pressure. Physical activity burns calories, thus helping to prevent obesity. People who exercise regularly have lower rates of inflammation markers, such as C-reactive protein. Less inflammation suggests lower risk of artery disease. To maximize cardiovascular health, researchers recommend about an hour a day of brisk activity.

Focus on Wellness, page 410

www.wiley.com/college/apcentral

The cardiovascular system contributes to the homeostasis of other body systems by transporting and distributing blood throughout the body to deliver materials such as oxygen, nutrients, and hormones and to carry away wastes. This transport is accomplished by blood vessels, which form closed circulatory routes for blood to travel from the heart to body organs and back again. In Chapters 14 and 15 we discussed the composition and functions of blood and the structure and function of the heart. In this chapter, we examine the structure and functions of the different types of blood vessels that carry the blood to and from the heart, as well as factors that contribute to blood flow and regulation of blood pressure.

looking back to move ahead . . .

- **Diffusion (page 47)**
- **Medulla Oblongata (page 250)**
- **Antidiuretic Hormone (page 322)**
- **Mineralocorticoids (page 331)**
- **Great Vessels of the Heart (page 369)**

BLOOD VESSEL STRUCTURE AND FUNCTION

OBJECTIVES • Compare the structure and function of the different types of blood vessels.

- **Describe how substances enter and leave the blood in capillaries.**
- **Explain how venous blood returns to the heart.**

There are five types of blood vessels: arteries, arterioles, capillaries, venules, and veins. ***Arteries*** (AR-ter-ēz) carry blood *away from the heart* to body tissues. Two large arteries—the aorta and the pulmonary trunk—emerge from the heart and branch out into medium-sized arteries that serve various regions of the body. These medium-sized arteries then divide into small arteries, which, in turn, divide into still smaller arteries called ***arterioles*** (ar-TER-ē-ōls). Arterioles within a tissue or organ branch into numerous microscopic vessels called ***capillaries*** (KAP-i-lar′-ēz). Groups of capillaries within a tissue reunite to form small veins called ***venules*** (VEN-ūls). These, in turn, merge to form progressively larger vessels called veins. ***Veins*** (VĀNZ) are the blood vessels that convey blood from the tissues *back to the heart*.

At any one time, systemic veins and venules contain about 64% of the total volume of blood in the system, systemic arteries and arterioles about 13%, systemic capillaries about 7%, pulmonary blood vessels about 9%, and the heart chambers about 7%. Because veins contain so much of the blood, certain veins function as ***blood reservoirs***. The main blood reservoirs are the veins of the abdominal organs (especially the liver and spleen) and the skin. Blood can be diverted quickly from its reservoirs to other parts of the body, for example, to skeletal muscles to support increased muscular activity.

Arteries and Arterioles

The walls of arteries have three layers of tissue surrounding a hollow space, the ***lumen***, through which the blood flows (Figure 16.1a). The inner layer is composed of ***endothelium***, a type of simple squamous epithelium; a basement membrane; and an elastic tissue called the internal elastic lamina. The middle layer consists of smooth muscle and elastic tissue. The outer layer is composed mainly of elastic and collagen fibers.

Sympathetic fibers of the autonomic nervous system innervate vascular smooth muscle. An increase in sympathetic stimulation typically causes the smooth muscle to contract, squeezing the vessel wall and narrowing the lumen. Such a decrease in the diameter of the lumen of a blood vessel is called ***vasoconstriction***. In contrast, when sympathetic stimulation decreases, or in the presence of certain chemicals (such as nitric oxide or lactic acid), smooth muscle fibers relax. The resulting increase in lumen diameter is called ***vasodilation***. Additionally, when an artery or arteriole is damaged, its smooth muscle contracts, producing vascular spasm of the vessel. Such a vasospasm limits blood flow through the damaged vessel and helps reduce blood loss if the vessel is small.

The largest-diameter arteries contain a high proportion of elastic fibers in their middle layer, and their walls are relatively thin in proportion to their overall diameter. Such arteries are called ***elastic arteries***. These arteries help propel blood onward while the ventricles are relaxing. As blood is ejected from the heart into elastic arteries, their highly elastic walls stretch, accommodating the surge of blood. Then, while the ventricles are relaxing, the elastic fibers in the artery walls recoil, which forces blood onward toward the smaller arteries. Examples include the aorta and the brachiocephalic, common carotid, subclavian, vertebral, pulmonary, and common iliac arteries. Medium-sized arteries, on the other hand, contain more smooth muscle and fewer elastic fibers than elastic arteries. Such arteries are called ***muscular arteries*** and are capable of greater vasoconstriction and vasodilation to adjust the rate of blood flow. Examples include the brachial artery (arm) and radial artery (forearm).

An ***arteriole*** (= small artery) is a very small, almost microscopic, artery that delivers blood to capillaries. The smallest arterioles consist of little more than a layer of endothelium covered by a few smooth muscle fibers (see Figure 16.2a). Arterioles play a key role in regulating blood flow from arteries into capillaries. During vasoconstriction, blood flow from the arterioles to the capillaries is restricted; during vasodilation, the flow is significantly increased. A change in diameter of arterioles can also significantly alter blood pressure; vasodilation decreases blood pressure and vasoconstriction increases blood pressure.

Capillaries

Capillaries (*capillar-* = hairlike) are microscopic vessels that connect arterioles to venules (Figure 16.1c). Capillaries are present near almost every body cell, and they are known as *exchange vessels* because they permit the exchange of nutrients and wastes between the body's cells and the blood. The number of capillaries varies with the metabolic activity of the tissue they serve. Body tissues with high metabolic requirements, such as muscles, the liver, the kidneys, and the nervous system, have extensive capillary networks. Tissues with lower metabolic requirements, such as tendons and ligaments, contain fewer capillaries. A few tissues—all covering and lining epithelia, the cornea and lens of the eye, and cartilage—lack capillaries completely.

Figure 16.1 Comparative structure of blood vessels. The relative size of the capillary in (c) is enlarged for emphasis. Note the valve in the vein.

Arteries carry blood away from the heart to tissues. Veins carry blood from tissues back to the heart.

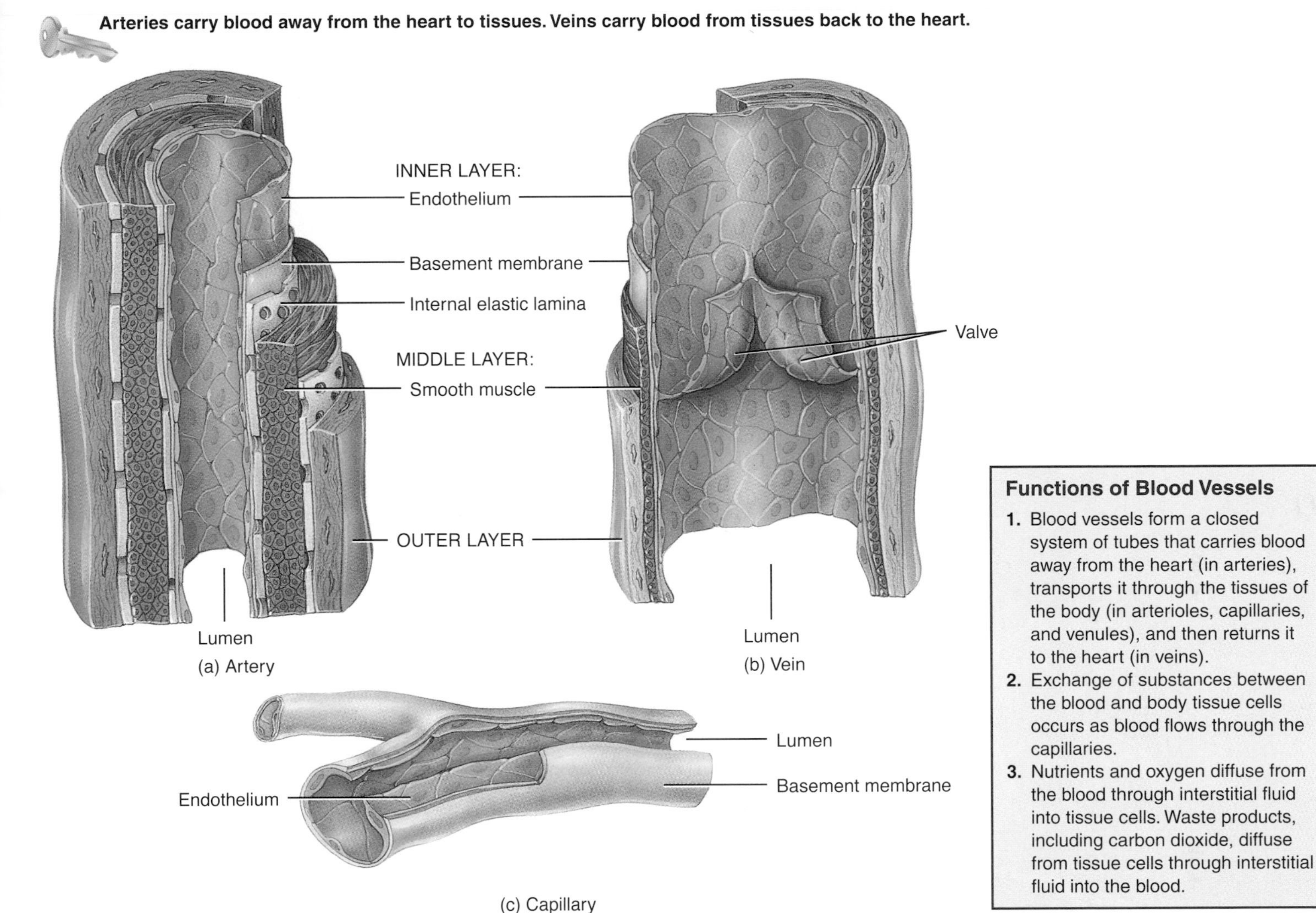

Would you expect a femoral artery or a femoral vein to have the thicker wall? A wider lumen?

Structure of Capillaries

A capillary consists of a layer of endothelium that is surrounded by basement membrane (Figure 16.1c). Because capillary walls are very thin, many substances easily pass through them to reach tissue cells from the blood or to enter the blood from interstitial fluid. The walls of all other blood vessels are too thick to permit the exchange of substances between blood and interstitial fluid. Depending on how tightly their endothelial cells are joined, different types of capillaries have varying degrees of permeability.

In some regions, capillaries link arterioles to venules directly. In other places, they form extensive branching networks (Figure 16.2). Blood flows through only a small part of a tissue's capillary network when metabolic needs are low. But when a tissue becomes active, the entire capillary network fills with blood. The flow of blood in capillaries is regulated by smooth muscle fibers in arteriole walls and by ***precapillary sphincters***, rings of smooth muscle at the point where capillaries branch from arterioles (Figure 16.2a). When precapillary sphincters relax, more blood flows into the connected capillaries; when precapillary sphincters contract, less blood flows through their capillaries.

Capillary Exchange

Because of the small diameter of capillaries, blood flows more slowly through them than through larger blood vessels. The

Figure 16.2 Capillaries. Because red blood cells and capillaries are nearly the same size, red blood cells squeeze through capillaries in single file.

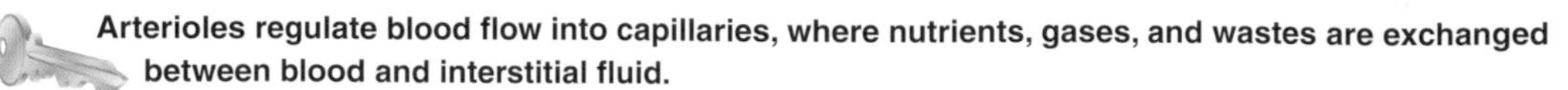

Arterioles regulate blood flow into capillaries, where nutrients, gases, and wastes are exchanged between blood and interstitial fluid.

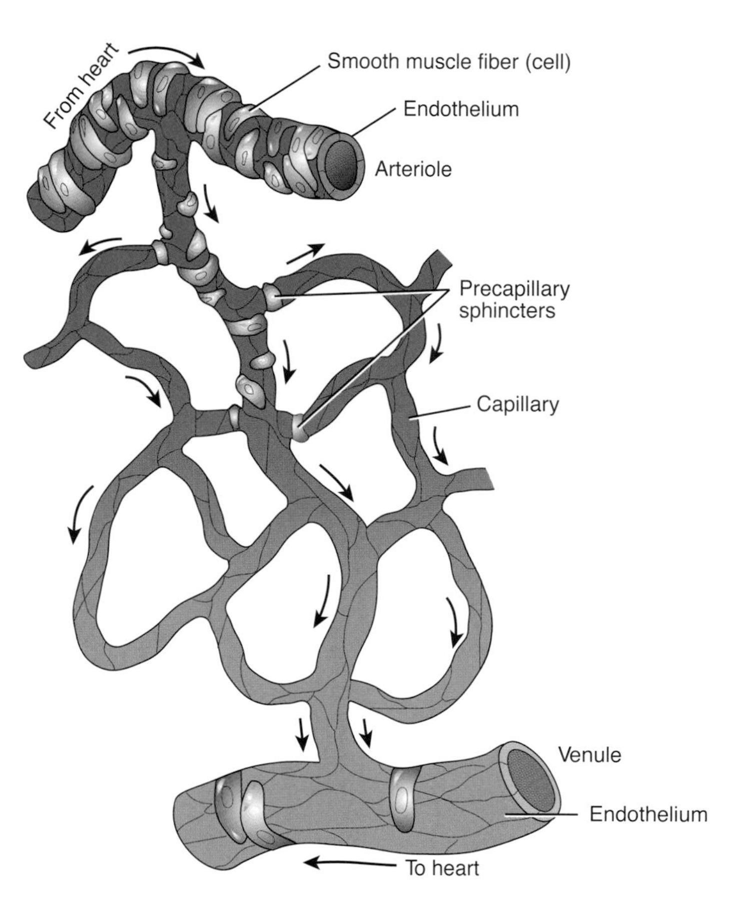

(a) Details of a capillary network

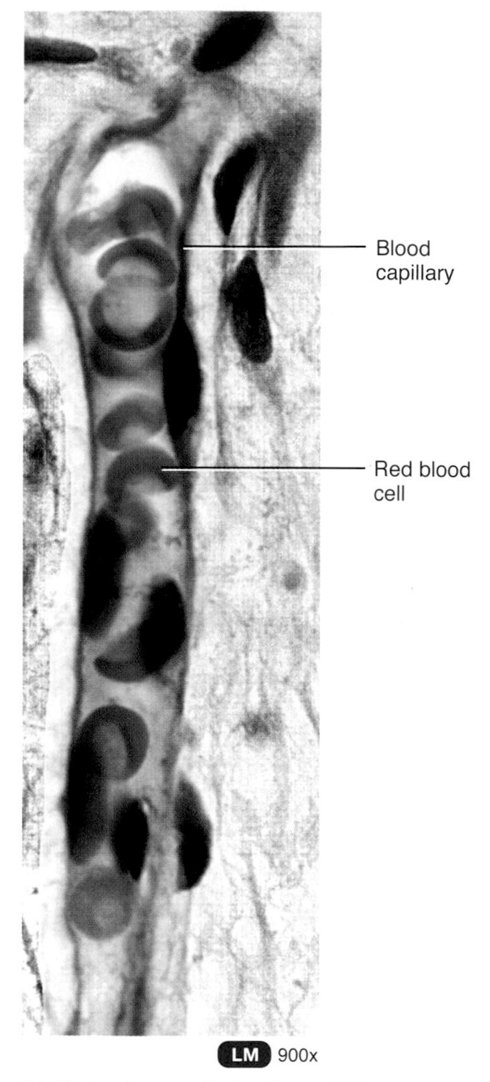

(b) Photomicrograph showing red blood cells squeezing through a blood capillary

Why do metabolically active tissues have extensive capillary networks?

slow flow aids the prime mission of the entire cardiovascular system: to keep blood flowing through capillaries so that ***capillary exchange***—the movement of substances into and out of capillaries—can occur.

Capillary blood pressure, the pressure of blood against the walls of capillaries, "pushes" fluid out of capillaries into interstitial fluid. An opposing pressure, termed ***blood colloid osmotic pressure***, "pulls" fluid into capillaries. (Recall that osmotic pressure is the pressure of a fluid due to its solute concentration. The higher the solute concentration, the greater the osmotic pressure.) Most solutes are present in nearly equal concentrations in blood and interstitial fluid. But the presence of proteins in plasma and their virtual absence in interstitial fluid gives blood the higher osmotic pressure. Blood colloid osmotic pressure is osmotic pressure due mainly to plasma proteins.

Capillary blood pressure is higher than blood colloid osmotic pressure for about the first half of the length of a typical capillary. Thus, water and solutes flow out of the blood capillary into the surrounding interstitial fluid, a movement called ***filtration*** (Figure 16.3). Because capillary blood pressure de-

Figure 16.3 Capillary exchange.

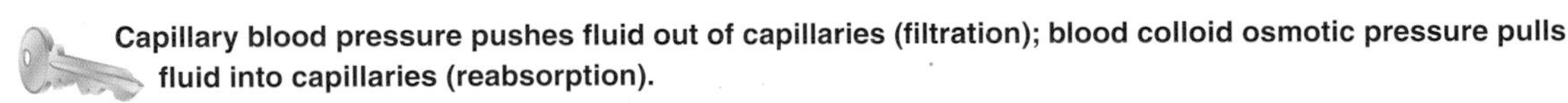

Capillary blood pressure pushes fluid out of capillaries (filtration); blood colloid osmotic pressure pulls fluid into capillaries (reabsorption).

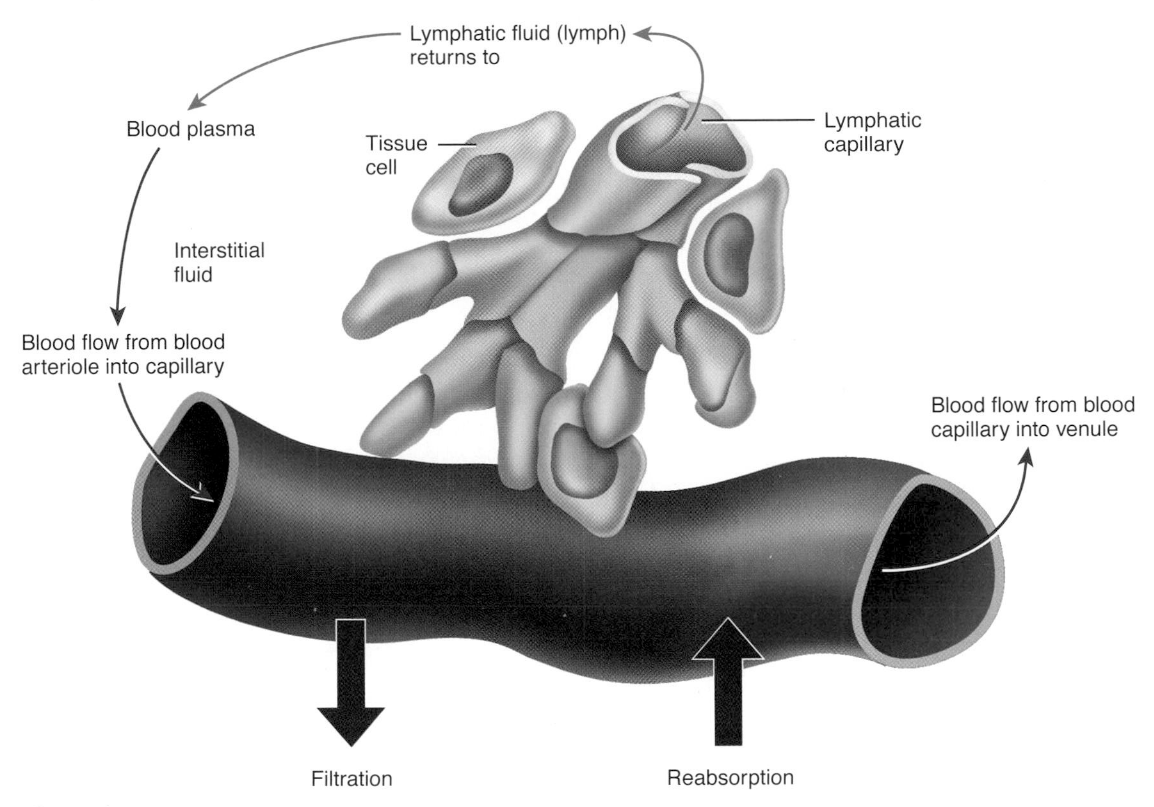

What happens to excess filtered fluid and proteins that are not reabsorbed?

creases progressively as blood flows along a capillary, at about the capillary's midpoint, blood pressure drops below blood colloid osmotic pressure. Then, water and solutes move from interstitial fluid into the blood capillary, a process termed ***reabsorption***. Normally, about 85% of the filtered fluid is reabsorbed. The excess filtered fluid and the few plasma proteins that do escape enter lymphatic capillaries and eventually are returned by the lymphatic system to the cardiovascular system.

Localized changes in each capillary network can regulate vasodilation and vasoconstriction. When vasodilators are released by tissue cells, they cause dilation of nearby arterioles and relaxation of precapillary sphincters. Then, blood flow into the capillary networks increases, and O_2 delivery to the tissue rises. Vasoconstrictors have the opposite effect. The ability of a tissue to automatically adjust its blood flow to match its metabolic demands is called ***autoregulation***.

Venules and Veins

When several capillaries unite, they form venules. Venules receive blood from capillaries and empty blood into veins, which return blood to the heart.

Structure of Venules and Veins

Venules (= little veins) are similar in structure to arterioles; their walls are thinner near the capillary end and thicker as they progress toward the heart. ***Veins*** are structurally similar to arteries, but their middle and inner layers are thinner (see Figure 16.1b). The outer layer of veins is the thickest layer. The lumen of a vein is wider than that of a corresponding artery.

In some veins, the inner layer folds inward to form ***valves*** that prevent the backflow of blood. In people with weak venous valves, gravity forces blood backward through the valve. This increases venous blood pressure, which pushes the vein's wall outward. After repeated overloading, the walls lose their elasticity and become stretched and flabby, a condition called ***varicose veins***.

By the time blood leaves the capillaries and moves into veins, it has lost a great deal of pressure. This can be observed in the blood leaving a cut vessel: Blood flows from a cut vein slowly and evenly, whereas it gushes out of a cut artery in rapid spurts. When a blood sample is needed, it is usually collected from a vein because pressure is low in veins and many of them are close to the skin surface.

Venous Return

Venous return, the volume of blood flowing back to the heart through systemic veins, occurs due to pressure generated in three ways: (1) contractions of the heart, (2) the skeletal muscle pump, and (3) the respiratory pump. Blood pressure is generated by contraction of the heart's ventricles and is measured in millimeters of mercury, abbreviated mm Hg. The pressure difference from venules (averaging about 16 mm Hg) to the right atrium (0 mm Hg), although small, normally is sufficient to cause venous return to the heart. When you stand, the pressure pushing blood up the veins in your lower limbs is barely enough to overcome the force of gravity pushing it back down.

The ***skeletal muscle pump*** operates as follows (Figure 16.4):

1. While you are standing at rest, both the venous valve closer to the heart and the one farther from the heart in this part of the leg are open, and blood flows upward toward the heart.
2. Contraction of leg muscles, such as when you stand on tiptoes or take a step, compresses the vein. The compression pushes blood through the valve closer to the heart, an action called *milking*. At the same time, the valve farther from the heart in the uncompressed segment of the vein closes as some blood is pushed against it. People who are immobilized through injury or disease lack these contractions of leg muscles. As a result, their venous return is slower and they may develop circulation problems.
3. Just after muscle relaxation, pressure falls in the previously compressed section of vein, which causes the valve closer to the heart to close. The valve farther from the heart now opens because blood pressure in the foot is higher than in the leg, and the vein fills with blood from the foot.

The ***respiratory pump*** is also based on alternating compression and decompression of veins. During inhalation (breathing in) the diaphragm moves downward, which causes a decrease in pressure in the thoracic cavity and an increase in pressure in the abdominal cavity. As a result, abdominal veins are compressed, and a greater volume of blood moves from the compressed abdominal veins into the decompressed thoracic veins and then into the right atrium. When the pressures reverse during exhalation (breathing out), the valves in the veins prevent backflow of blood from the thoracic veins to the abdominal veins.

■ **CHECKPOINT**

1. How do arteries, capillaries, and veins differ in function?
2. Distinguish between filtration and reabsorption.
3. What factors contribute to blood flow back to the heart?

Figure 16.4 Action of the skeletal muscle pump in returning blood to the heart. Steps are described in the text.

Milking refers to skeletal muscle contractions that drive venous blood toward the heart.

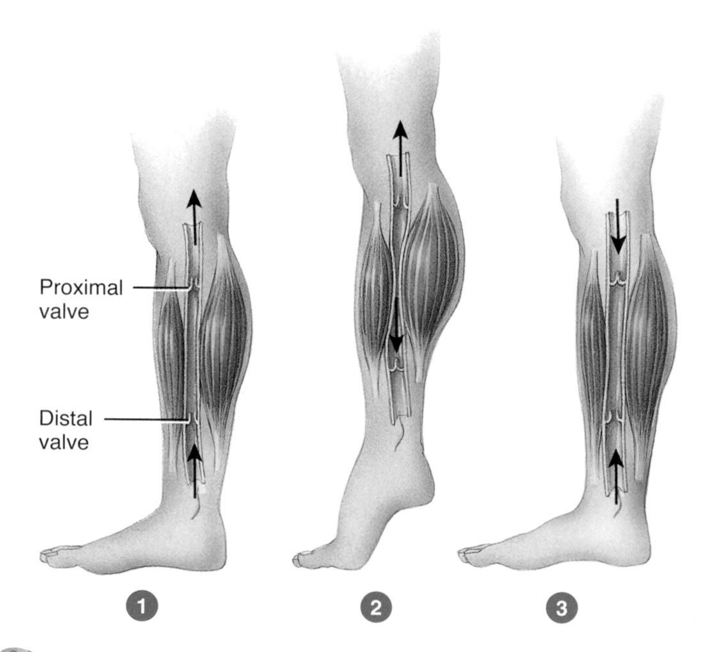

What mechanisms, besides cardiac contractions, act as pumps to boost venous return?

BLOOD FLOW THROUGH BLOOD VESSELS

OBJECTIVES • **Define blood pressure and describe how it varies throughout the systemic circulation.**

- **Identify the factors that affect blood pressure and vascular resistance.**
- **Describe how blood pressure and blood flow are regulated.**

We saw in Chapter 15 that cardiac output (CO) depends on stroke volume and heart rate. Two other factors influencing cardiac output and the proportion of blood that flows through specific circulatory routes are blood pressure and vascular resistance.

Blood Pressure

As you have just learned, blood flows from regions of higher pressure to regions of lower pressure; the greater the pressure difference, the greater the blood flow. Contraction of the ventricles generates ***blood pressure (BP)***, the pressure exerted by blood on the walls of a blood vessel. BP is highest in the aorta and large systemic arteries, where in a resting, young adult, it rises to about 110 mm Hg during systole (contraction) and drops to about 70 mm Hg during diastole (relaxation). Blood pressure falls progressively as the distance from the left ventricle increases (Figure 16.5), to about 35 mm Hg as blood passes into systemic capillaries. At the venous end of capillaries, blood pressure drops to about 16 mm Hg. Blood pressure continues to drop as blood enters systemic venules and then veins, and it reaches 0 mm Hg as blood returns to the right atrium.

Blood pressure depends in part on the total volume of blood in the cardiovascular system. The normal volume of blood in an adult is about 5 liters (5.3 qt). Any decrease in this volume, as from hemorrhage, decreases the amount of blood that is circulated through the arteries. A modest decrease can be compensated by homeostatic mechanisms that help maintain blood pressure, but if the decrease in blood volume is greater than 10% of total blood volume, blood pressure drops, with potentially life-threatening results. Conversely, anything that increases blood volume, such as water retention in the body, tends to increase blood pressure.

Figure 16.5 Blood pressure changes as blood flows through the systemic circulation. The dashed line is the mean (average) pressure in the aorta, arteries, and arterioles.

Blood pressure falls progressively as blood flows from systemic arteries through capillaries and back to the right atrium. The greatest drop in blood pressure occurs in the arterioles.

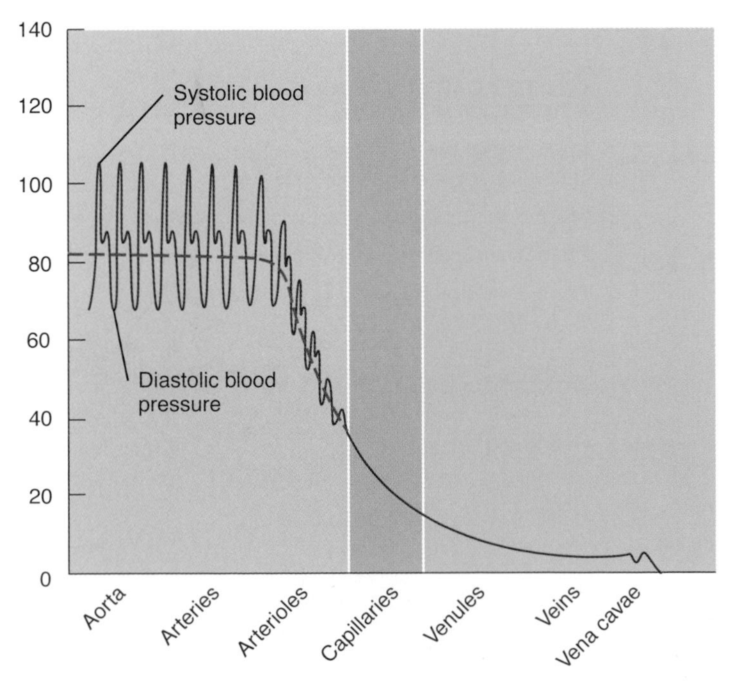

What is the relationship between blood pressure and blood flow?

Resistance

Vascular resistance is the opposition to blood flow due to friction between blood and the walls of blood vessels. An increase in vascular resistance increases blood pressure; a decrease in vascular resistance has the opposite effect. Vascular resistance depends on (1) size of the blood vessel lumen, (2) blood viscosity, and (3) total blood vessel length.

1. **Size of the lumen**. The smaller the lumen of a blood vessel, the greater its resistance to blood flow. Vasoconstriction narrows the lumen, and vasodilation widens it. Normally, moment-to-moment fluctuations in blood flow through a given tissue are due to vasoconstriction and vasodilation of the tissue's arterioles. As arterioles dilate, resistance decreases, and blood pressure falls. As arterioles constrict, resistance increases, and blood pressure rises.
2. **Blood viscosity**. The viscosity (thickness) of blood depends mostly on the ratio of red blood cells to plasma (fluid) volume, and to a smaller extent on the concentration of proteins in plasma. The higher the blood's viscosity, the higher the resistance. Any condition that increases the viscosity of blood, such as dehydration or polycythemia (an unusually high number of red blood cells), thus increases blood pressure. A depletion of plasma proteins or red blood cells, as a result of anemia or hemorrhage, decreases viscosity and thus decreases blood pressure.
3. **Total blood vessel length**. Resistance to blood flow increases when the total length of all blood vessels in the body increases. The longer the blood vessel, the greater the contact between the vessel wall and the blood. The greater the contact between the vessel wall and the blood, the greater the friction. An estimated 400 miles of additional blood vessels develop for each extra pound of fat, one reason why overweight individuals may have higher blood pressure.

Regulation of Blood Pressure and Blood Flow

Several interconnected negative feedback systems control blood pressure and blood flow by adjusting heart rate, stroke volume, vascular resistance, and blood volume. Some systems allow rapid adjustments to cope with sudden changes, such as the drop in blood pressure in the brain that occurs when you stand up; others provide long-term regulation. The body may also require adjustments to the distribution of blood flow. During exercise, for example, a greater percentage of blood flow is diverted to skeletal muscles.

Role of the Cardiovascular Center

In Chapter 15 we noted how the ***cardiovascular (CV) center*** in the medulla oblongata helps regulate heart rate and stroke volume. The CV center also controls the neural and hormonal negative feedback systems that regulate blood pressure and blood flow to specific tissues.

INPUT The cardiovascular center receives input from higher brain regions: the cerebral cortex, limbic system, and hypothalamus (Figure 16.6). For example, even before you start to run a race, your heart rate may increase due to nerve impulses conveyed from the limbic system to the CV center. If your body temperature rises during a race, the hypothalamus sends nerve impulses to the CV center. The resulting vasodilation of skin blood vessels allows heat to dissipate more rapidly from the surface of the skin.

The CV center also receives input from three main types of sensory receptors: proprioceptors, baroreceptors, and chemoreceptors. *Proprioceptors*, which monitor movements of joints and muscles, provide input to the cardiovascular center during physical activity, such as playing tennis, and cause the rapid increase in heart rate at the beginning of exercise.

Baroreceptors (pressure receptors) are located in the aorta, internal carotid arteries (arteries in the neck that supply blood to the brain), and other large arteries in the neck and chest. They send impulses continuously to the cardiovascular center to help regulate blood pressure. If blood pressure falls, the baroreceptors are stretched less, and they send nerve impulses at a slower rate to the cardiovascular center (Figure 16.7). In response, the cardiovascular center decreases parasympathetic stimulation of the heart and increases sympathetic stimulation of the heart. As the heart beats faster and more forcefully, and as vascular resistance increases, blood pressure increases to the normal level.

By contrast, when an increase in pressure is detected, the baroreceptors send impulses at a faster rate. The cardiovascular center responds by increasing parasympathetic stimulation and decreasing sympathetic stimulation. The resulting decreases in heart rate and force of contraction lower cardiac output, and vasodilation lowers vascular resistance. Decreased cardiac output and decreased vascular resistance both lower blood pressure.

Moving from a prone (lying down) to an erect position decreases blood pressure and blood flow in the head and upper part of the body. The drop in pressure, however, is quickly counteracted by the **baroreceptor reflexes.** Sometimes these reflexes operate more slowly than normal, especially in older people. As a result, a person can faint due to reduced brain blood flow upon standing up too quickly.

CHEMORECEPTOR REFLEXES *Chemoreceptors* (chemical receptors) that monitor blood levels of O_2, CO_2, and H^+ are located in the two ***carotid bodies*** in the common carotid ar-

Figure 16.6 The cardiovascular (CV) center. Located in the medulla oblongata, the CV center receives input from higher brain centers, proprioceptors, baroreceptors, and chemoreceptors. It provides output to both the sympathetic and parasympathetic divisions of the autonomic nervous system.

The cardiovascular center is the main region for the nervous system regulation of heart rate, force of heart contractions, and vasodilation or vasoconstriction of blood vessels.

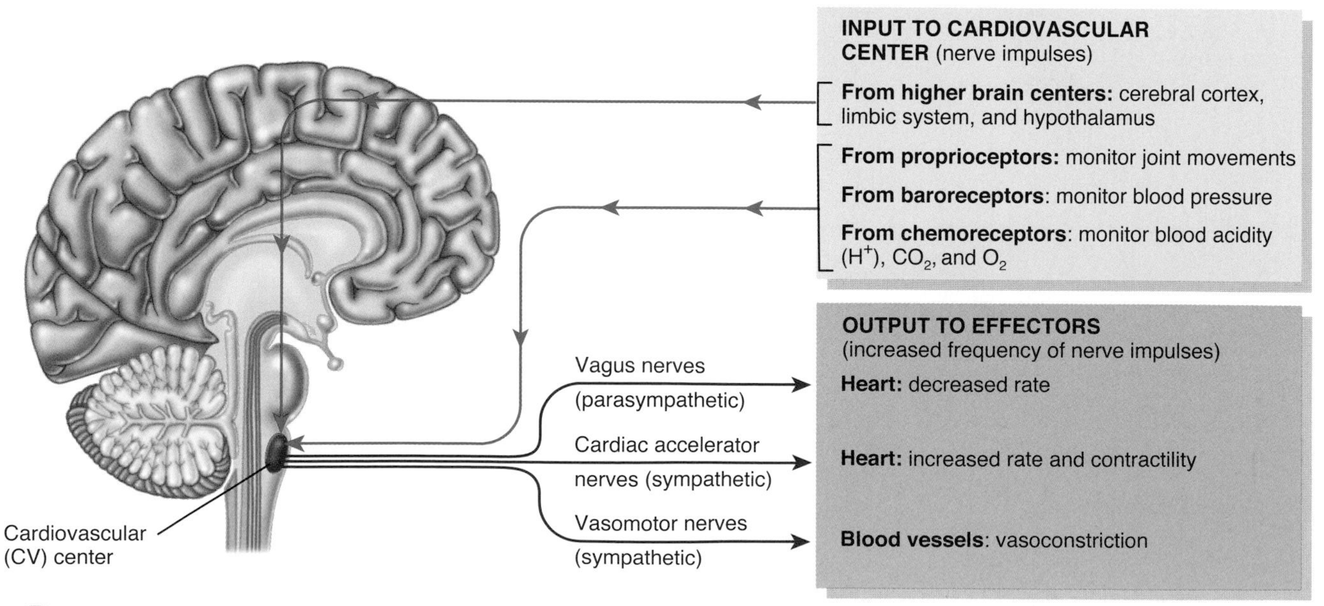

How does vasoconstriction affect vascular resistance and blood flow?

Figure 16.7 Negative feedback regulation of blood pressure via baroreceptor reflexes.

The baroreceptor reflex is a neural mechanism for rapid regulation of blood pressure.

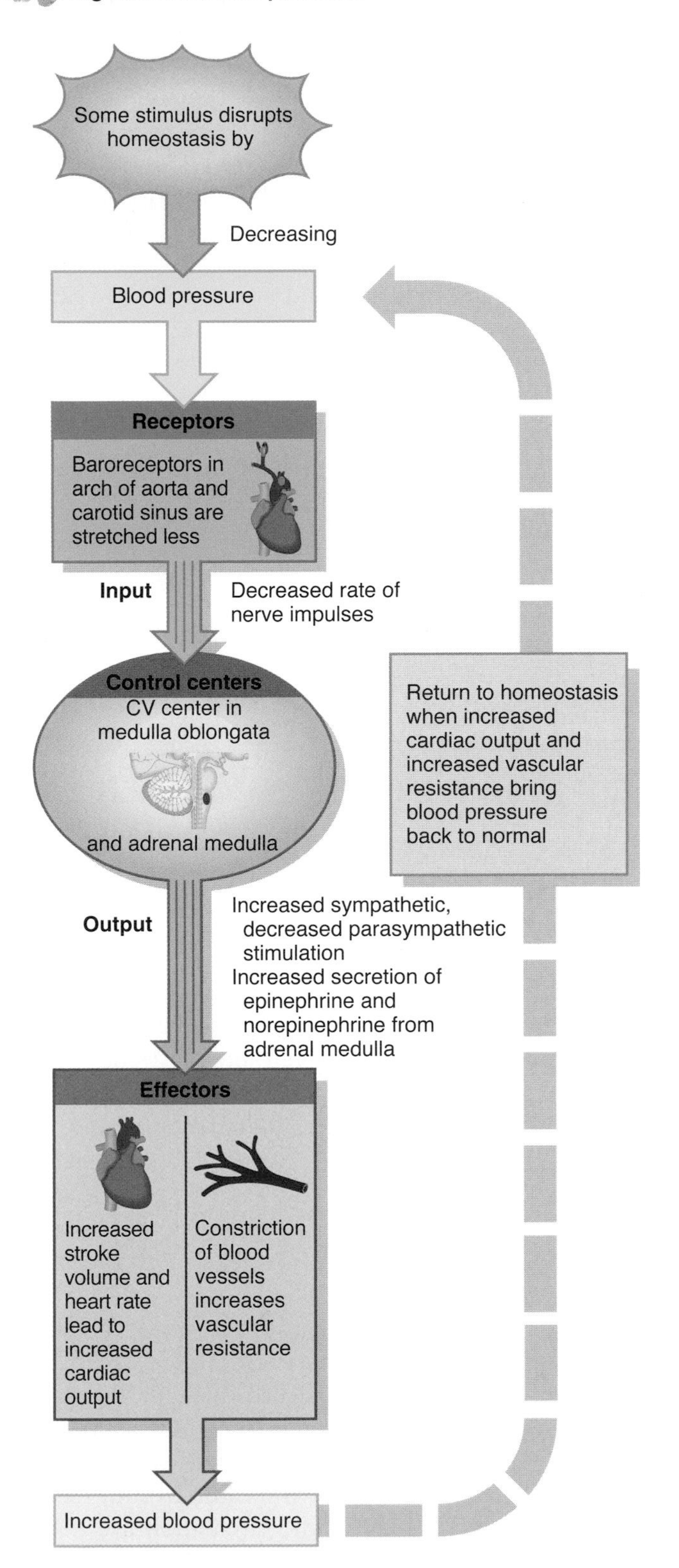

Does this negative feedback cycle happen when you lie down or when you stand up?

teries and in the ***aortic body*** in the arch of the aorta. *Hypoxia* (lowered O_2 availability), *acidosis* (an increase in H^+ concentration), or *hypercapnia* (excess CO_2) stimulates the chemoreceptors to send impulses to the cardiovascular center. In response, the CV center increases sympathetic stimulation of arterioles and veins, producing vasoconstriction and an increase in blood pressure.

OUTPUT Output from the cardiovascular center flows along sympathetic and parasympathetic fibers of the ANS (see Figure 16.6). An increase in sympathetic stimulation increases heart rate and the forcefulness of contraction, whereas a decrease in sympathetic stimulation decreases heart rate and contraction force. The vasomotor region of the cardiovascular center also sends impulses to arterioles throughout the body. The result is a moderate state of vasoconstriction, called ***vasomotor tone***, that sets the resting level of vascular resistance. Sympathetic stimulation of most veins results in movement of blood out of venous blood reservoirs, which increases blood pressure.

Hormonal Regulation of Blood Pressure and Blood Flow

Several hormones help regulate blood pressure and blood flow by altering cardiac output, changing vascular resistance, or adjusting the total blood volume.

1. **Renin–angiotensin–aldosterone (RAA) system.** When blood volume falls or blood flow to the kidneys decreases, certain cells in the kidneys secrete the enzyme *renin* into the bloodstream (see Figure 13.14 on page 331). Together, renin and angiotensin converting enzyme (ACE) produce the active hormone *angiotensin II*, which raises blood pressure by causing vasoconstriction. Angiotensin II also stimulates secretion of *aldosterone*, which increases reabsorption of sodium ions (Na^+) and water by the kidneys. The water reabsorption increases total blood volume, which increases blood pressure.
2. **Epinephrine and norepinephrine.** In response to sympathetic stimulation, the adrenal medulla releases epinephrine and norepinephrine. These hormones increase cardiac output by increasing the rate and force of heart contractions; they also cause vasoconstriction of arterioles and veins in the skin and abdominal organs.
3. **Antidiuretic hormone (ADH).** ADH is produced by the hypothalamus and released from the posterior pituitary in response to dehydration or decreased blood volume. Among other actions, ADH causes vasoconstriction, which increases blood pressure. For this reason ADH is also called ***vasopressin***.
4. **Atrial natriuretic peptide (ANP).** Released by cells in the atria of the heart, ANP lowers blood pressure by causing vasodilation and by promoting the loss of salt and water in the urine, which reduces blood volume.

■ **CHECKPOINT**

4. What two factors influence cardiac output?
5. Describe how blood pressure decreases as distance from the left ventricle increases.
6. What factors determine vascular resistance?
7. Explain the role of the cardiovascular center, reflexes, and hormones in regulating blood pressure.

CHECKING CIRCULATION

OBJECTIVE • Explain how pulse and blood pressure are measured.

Pulse

The alternate expansion and elastic recoil of an artery after each contraction and relaxation of the left ventricle is called a ***pulse***. The pulse is strongest in the arteries closest to the heart. It becomes weaker as it passes through the arterioles, and it disappears altogether in the capillaries. The radial artery at the wrist is most commonly used to feel the pulse. Other sites where the pulse may be felt include the brachial artery along the medial side of the biceps brachii muscle; the common carotid artery, next to the voice box, which is usually monitored during cardiopulmonary resuscitation; the popliteal artery behind the knee; and the dorsal artery of the foot above the instep of the foot.

The pulse rate normally is the same as the heart rate, about 75 beats per minute at rest. ***Tachycardia*** (tak′-i-KAR-dē-a; *tachy-* = fast) is a rapid resting heart or pulse rate over 100 beats/min. ***Bradycardia*** (brād′-i-KAR-dē-a; *brady-* = slow) indicates a slow resting heart or pulse rate under 50 beats/min.

Measurement of Blood Pressure

In clinical use, the term ***blood pressure*** usually refers to the pressure in arteries generated by the left ventricle during systole and the pressure remaining in the arteries when the ventricle is in diastole. Blood pressure is usually measured in the brachial artery in the left arm (see Figure 16.10a). The device used to measure blood pressure is a ***sphygmomanometer*** (sfig′-mō-ma-NOM-e-ter; *sphygmo-* = pulse; *manometer* = instrument used to measure pressure). When the pressure cuff is inflated above the blood pressure attained during systole, the artery is compressed so that blood flow stops. The technician places a stethoscope below the cuff over the brachial artery and then slowly deflates the cuff. When the cuff is deflated enough to allow the artery to open, a spurt of blood passes through, resulting in the first sound heard through the stethoscope. This sound corresponds to ***systolic blood pressure (SBP)***—the force with which blood is pushing against arterial walls during ventricular contraction. As the cuff is deflated further, the sounds suddenly become faint. This level, called the ***diastolic blood pressure (DBP)***, represents the force exerted by the blood remaining in arteries during ventricular relaxation.

The normal blood pressure of a young adult male is less than 120 mm Hg systolic and less than 80 mm Hg diastolic, reported, for example, as "110 over 70" and written as 110/70. In young adult females, the pressures are 8 to 10 mm Hg less. People who exercise regularly and are in good physical condition may have even lower blood pressures.

■ **CHECKPOINT**

8. What causes pulse?
9. Distinguish between systolic and diastolic blood pressure.

CIRCULATORY ROUTES

OBJECTIVE • Compare the major routes that blood takes through various regions of the body.

Blood vessels are organized into ***circulatory routes*** that carry blood throughout the body (Figure 16.8). As noted earlier, the two main circulatory routes are the systemic circulation and the pulmonary circulation.

Systemic Circulation

The ***systemic circulation*** includes the arteries and arterioles that carry blood containing oxygen and nutrients from the left ventricle to systemic capillaries throughout the body, plus the veins and venules that carry blood containing carbon dioxide and wastes to the right atrium. Blood leaving the aorta and traveling through the systemic arteries is a bright red color. As it moves through the capillaries, it loses some of its oxygen and takes on carbon dioxide, so that the blood in the systemic veins is a dark red color.

All systemic arteries branch from the ***aorta***, which arises from the left ventricle of the heart (see Figure 16.9). Deoxygenated blood returns to the heart through the systemic veins. All the veins of the systemic circulation empty into the ***superior vena cava***, ***inferior vena cava***, or the ***coronary sinus***, which, in turn, empty into the right atrium. The principal blood vessels of the systemic circulation are described and illustrated in Exhibits 16.1 through 16.7 and Figures 16.9 through 16.15 starting on page 396.

Figure 16.8 Circulatory routes. Red arrows indicate hepatic portal circulation. The details of the pulmonary circulation are shown here, and the details of the hepatic portal circulation are shown in Figure 16.16.

Blood vessels are organized into routes that deliver blood to various tissues of the body.

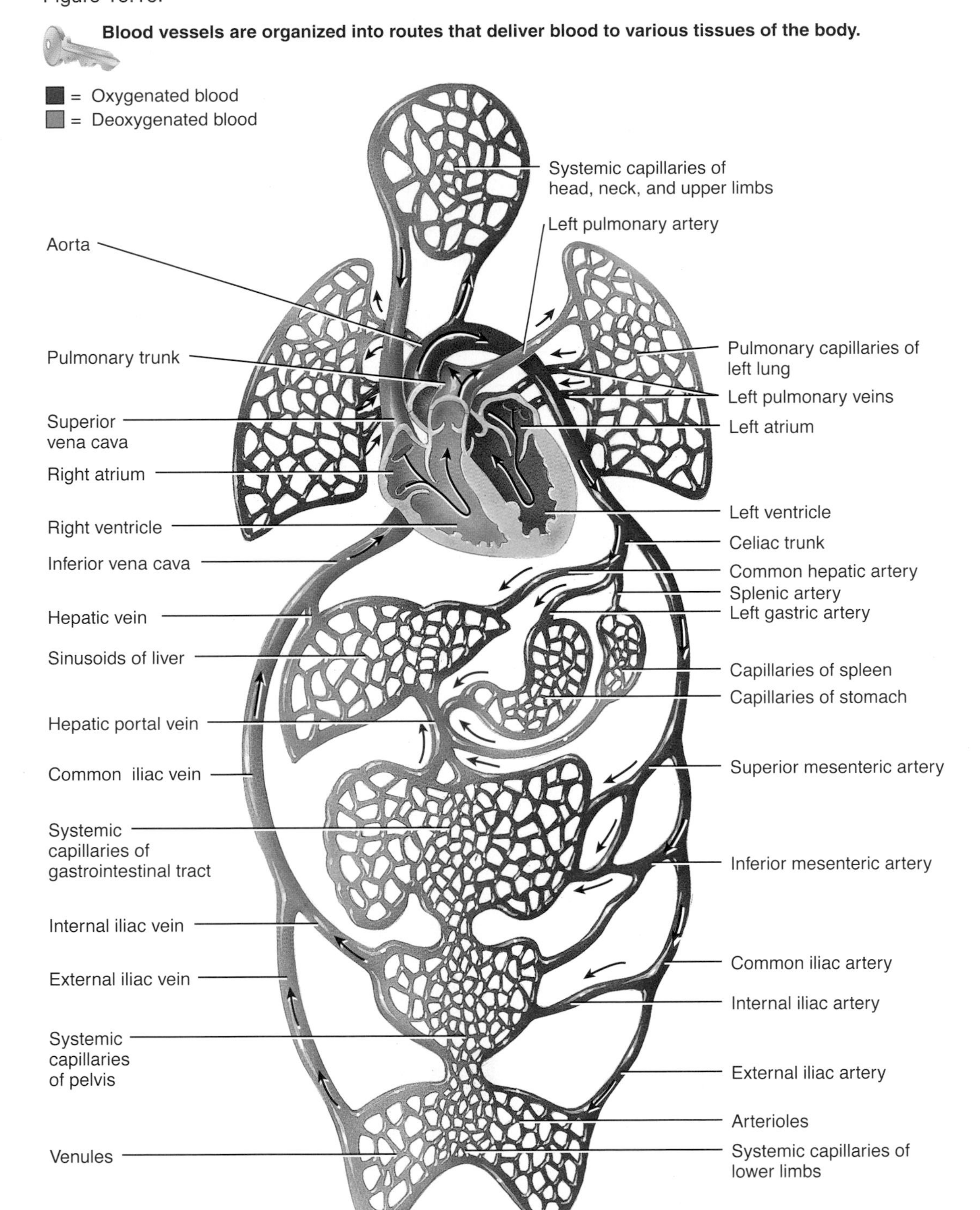

What are the two principal circulatory routes?

Exhibit 16.1 The Aorta and Its Branches *(Figure 16.9)*

OBJECTIVE • Identify the four principal divisions of the aorta and locate the major arterial branches arising from each.

• The ***aorta*** (*aortae* = to lift up), the largest artery of the body, is 2 to 3 cm (about 1 in.) in diameter. Its four principal divisions are the ascending aorta, arch of the aorta, thoracic aorta, and abdominal aorta. The ***ascending aorta*** emerges from the left ventricle posterior to the pulmonary trunk. It gives off two coronary artery branches that supply the myocardium of the heart. Then it turns to the left, forming the ***arch of the aorta***. Branches of the arch of the aorta are described in Exhibit 16.2. The part of the aorta between the arch of the aorta and the diaphragm, the ***thoracic aorta***, is about 20 cm (8 in.) long. The part of the aorta between the diaphragm and the common iliac arteries is the ***abdominal aorta*** (ab-DOM-i-nal). The main branches of the abdominal aorta are the ***celiac trunk***, the ***superior mesenteric artery***, and the ***inferior mesenteric artery***. The abdominal aorta divides at the level of the fourth lumbar vertebra into two ***common iliac arteries***, which carry blood to the lower limbs.

■ **CHECKPOINT**

What general regions do each of the four principal divisions of the aorta supply?

Division and Branches	Region Supplied
Ascending Aorta	
Right and left coronary arteries	Heart.
Arch of the Aorta (see Exhibit 16.2)	
Brachiocephalic trunk (brā′-kē-ō-se-FAL-ik)	
Right common carotid artery (ka-ROT-id)	Right side of head and neck.
Right subclavian artery (sub-KLĀ-vē-an)	Right upper limb.
Left common carotid artery	Left side of head and neck.
Left subclavian artery	Left upper limb.
Thoracic Aorta (*thorac-* = chest)	
Bronchial arteries (BRONG-kē-al)	Bronchi of lungs.
Esophageal arteries (e-sof′-a-JĒ-al)	Esophagus.
Posterior intercostal arteries (in′-ter-KOS-tal)	Intercostal and chest muscles.
Superior phrenic arteries (FREN-ik)	Superior and posterior surfaces of diaphragm.
Abdominal Aorta	
Inferior phrenic arteries (FREN-ik)	Inferior surface of diaphragm.
Celiac trunk (SĒ-lē-ak)	
Common hepatic artery (he-PAT-ik)	Liver.
Left gastric artery (GAS-trik)	Stomach and esophagus.
Splenic artery (SPLĒN-ik)	Spleen, pancreas, and stomach.
Superior mesenteric artery (MES-en-ter′-ik)	Small intestine, cecum, ascending and transverse colons, and pancreas.
Suprarenal arteries (soo′-pra-RĒ-nal)	Adrenal (suprarenal) glands.
Renal arteries (RĒ-nal)	Kidneys.
Gonadal arteries (gō-NAD-al)	
Testicular arteries (tes-TIK-ū-lar)	Testes (male).
Ovarian arteries (ō-VAR-ē-an)	Ovaries (female).
Inferior mesenteric artery	Transverse, descending, and sigmoid colons; rectum.
Common iliac arteries (IL-ē-ak)	
External iliac arteries	Lower limbs.
Internal iliac arteries	Uterus (female), prostate (male), muscles of buttocks, and urinary bladder.

Figure 16.9 Aorta and its principal branches.

All systemic arteries branch from the aorta.

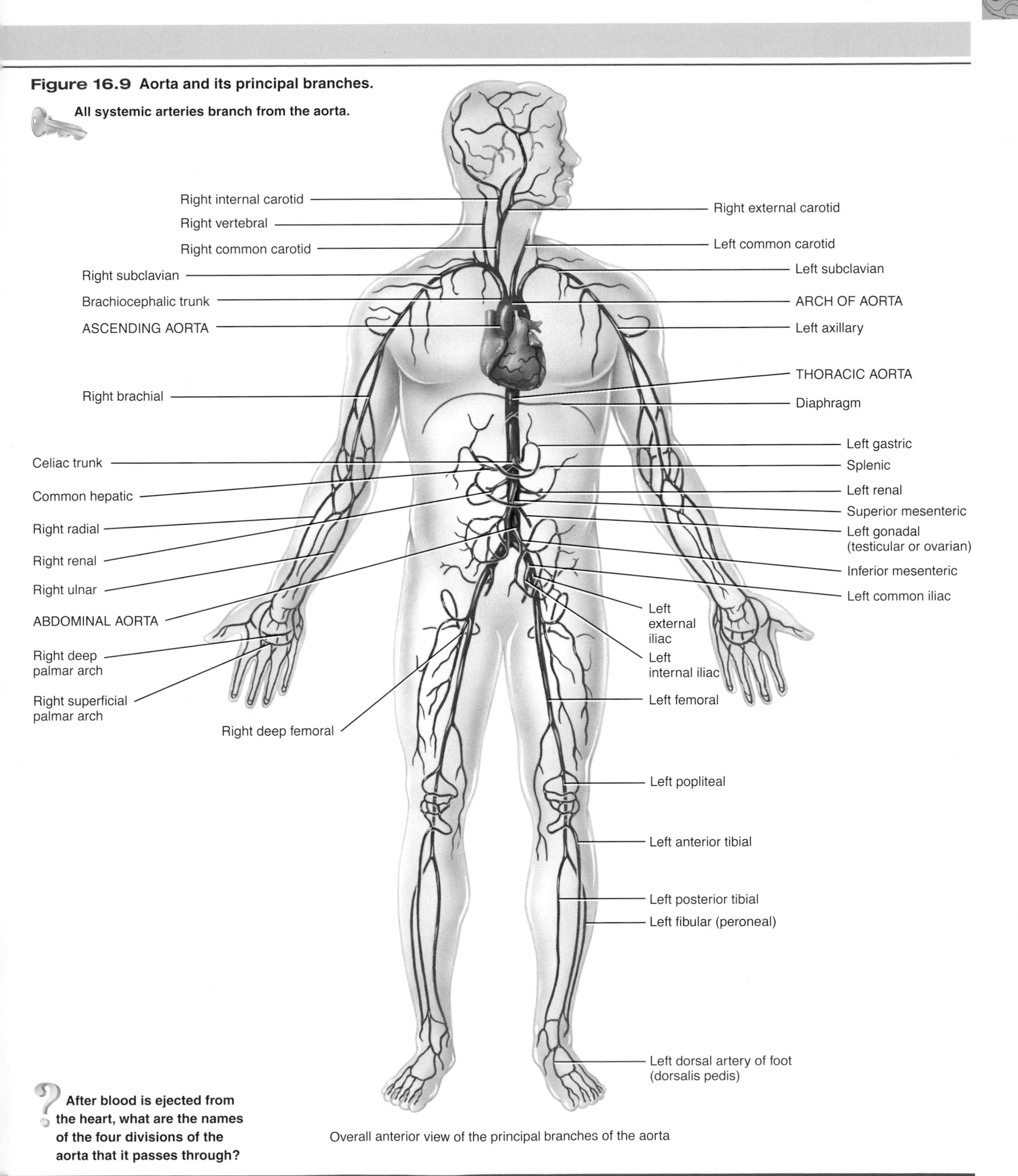

Overall anterior view of the principal branches of the aorta

After blood is ejected from the heart, what are the names of the four divisions of the aorta that it passes through?

Exhibit 16.2 The Arch of the Aorta *(Figure 16.10)*

OBJECTIVE • Identify the three arteries that branch from the arch of the aorta.

• The ***arch of the aorta***, the continuation of the ascending aorta, is 4 to 5 cm (almost 2 in.) in length. It has three branches. In order, as they emerge from the arch of the aorta, the three branches are the brachiocephalic trunk, the left common carotid artery, and the left subclavian artery.

■ **CHECKPOINT**

What general regions do the arteries that arise from the arch of the aorta supply?

Artery	Description and Region Supplied
Brachiocephalic Trunk (brā′-kē-ō-se-FAL-ik; *brachio-* = arm; *-cephalic* = head)	The ***brachiocephalic trunk*** divides to form the right subclavian artery and right common carotid artery (Figure 16.10a).
Right subclavian artery (sub-KLĀ-vē-an)	The ***right subclavian artery*** extends from the brachiocephalic trunk and then passes into the armpit (axilla). The general distribution of the artery is to the brain and spinal cord, neck, shoulder, and chest.
Axillary artery (AK-si-ler-ē = armpit)	The continuation of the right subclavian artery into the axilla is called the ***axillary artery***. Its general distribution is to the shoulder.
Brachial artery (BRĀ-kē-al = arm)	The ***brachial artery***, which provides the main blood supply to the arm, is the continuation of the axillary artery into the arm. It is commonly used to measure blood pressure. Just below the bend in the elbow, the brachial artery divides into the radial artery and ulnar artery.
Radial artery (RĀ-dē-al = radius)	The ***radial artery*** is a direct continuation of the brachial artery. It passes along the lateral (radial) aspect of the forearm and then through the wrist and hand; it is a common site for measuring radial pulse.
Ulnar artery (UL-nar = ulna)	The ***ulnar artery*** passes along the medial (ulnar) aspect of the forearm and then into the wrist and hand.
Superficial palmar arch (*palma* = palm)	The ***superficial palmar arch*** is formed mainly by the ulnar artery and extends across the palm. It gives rise to blood vessels that supply the palm and the fingers.
Deep palmar arch	The ***deep palmar arch*** is formed mainly by the radial artery. The arch extends across the palm and gives rise to blood vessels that supply the palm.
Vertebral artery (VER-te-bral)	Before passing into the axilla, the right subclavian artery gives off a major branch to the brain called the ***right vertebral artery*** (Figure 16.10b). The right vertebral artery passes through the foramina of the transverse processes of the cervical vertebrae and enters the skull through the foramen magnum to reach the inferior surface of the brain. Here it unites with the left vertebral artery to form the ***basilar artery*** (BAS-i-lar). The vertebral artery supplies the posterior portion of the brain with blood. The basilar artery supplies the cerebellum and pons of the brain and the internal ear.
Right common carotid artery (ka-ROT-id)	The ***right common carotid artery*** begins at the branching of the brachiocephalic trunk and supplies structures in the head (Figure 16.10b). Near the larynx (voice box), it divides into the right external and right internal carotid arteries.
External carotid artery	The ***external carotid artery*** supplies structures *external* to the skull.
Internal carotid artery	The ***internal carotid artery*** supplies structures *internal* to the skull such as the eyeball, ear, most of the cerebrum of the brain, and pituitary gland. Inside the cranium, the internal carotid arteries along with the basilar artery form an arrangement of blood vessels at the base of the brain near the hypophyseal fossa called the ***cerebral arterial circle (circle of Willis)***. From this circle (Figure 16.10c) arise arteries supplying most of the brain. The cerebral arterial circle is formed by the union of the ***anterior cerebral arteries*** (branches of internal carotids) and ***posterior cerebral arteries*** (branches of basilar artery). The ***posterior cerebral arteries*** are connected with the internal carotid arteries by the ***posterior communicating arteries*** (ko-MŪ-ni-kā′-ting). The anterior cerebral arteries are connected by the ***anterior communicating artery***. The ***internal carotid arteries*** are also considered part of the cerebral arterial circle. The functions of the cerebral arterial circle are to equalize blood pressure to the brain and provide alternate routes for blood flow to the brain, should the arteries become damaged.
Left Common Carotid Artery	Divides into basically the same branches with the same names as the right common carotid artery.
Left Subclavian Artery	Divides into basically the same branches with the same names as the right subclavian artery.

Figure 16.10 Arch of the aorta and its branches.

The arch of the aorta is the continuation of the ascending aorta.

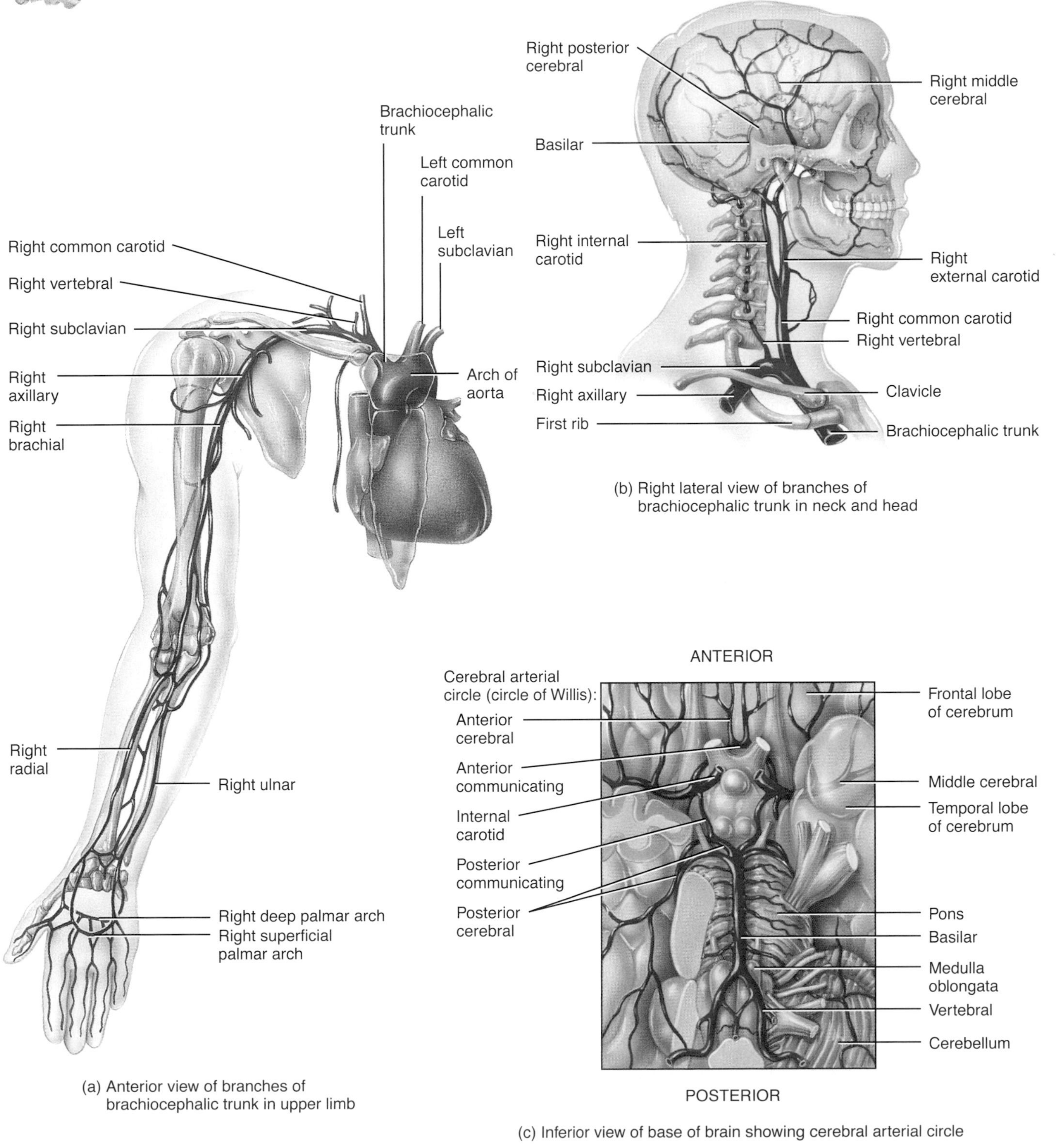

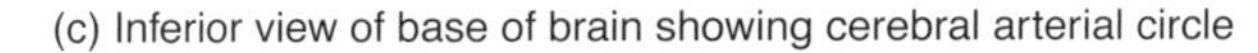

(a) Anterior view of branches of brachiocephalic trunk in upper limb

(b) Right lateral view of branches of brachiocephalic trunk in neck and head

(c) Inferior view of base of brain showing cerebral arterial circle

What are the three major branches of the arch of the aorta, in order of their origination?

Exhibit 16.3 Arteries of the Pelvis and Lower Limbs *(Figure 16.11)*

OBJECTIVE • Identify the two major branches of the common iliac arteries.

• The abdominal aorta ends by dividing into the right and left ***common iliac arteries***. These, in turn, divide into the ***internal*** and ***external iliac arteries***. In sequence, the external iliacs become the ***femoral arteries*** in the thighs, the ***popliteal arteries*** posterior to the knee, and the ***anterior*** and ***posterior tibial arteries*** in the legs.

■ **CHECKPOINT**

What general regions do the internal and external iliac arteries supply?

Artery	Description and Region Supplied
Common iliac arteries (IL-ē-ak = ilium)	At about the level of the fourth lumbar vertebra, the abdominal aorta divides into the right and left ***common iliac arteries***. Each gives rise to two branches: internal iliac and external iliac arteries. The general distribution of the common iliac arteries is to the pelvis, external genitals, and lower limbs.
Internal iliac arteries	The ***internal iliac arteries*** are the primary arteries of the pelvis. They supply the pelvis, buttocks, external genitals, and thigh.
External iliac arteries	The ***external iliac arteries*** supply the lower limbs.
Femoral arteries (FEM-o-ral = thigh)	The ***femoral arteries***, continuations of the external iliacs, supply the lower abdominal wall, groin, external genitals, and muscles of the thigh.
Popliteal arteries (pop′-li-TĒ-al = posterior surface of the knee)	The ***popliteal arteries***, continuations of the femoral arteries, supply muscles and the skin on the posterior of the legs; muscles of the calf; knee joint; femur; patella; and fibula.
Anterior tibial arteries (TIB-ē-al = shin bone)	The ***anterior tibial arteries***, which branch from the popliteal arteries, supply the knee joints, anterior muscles of the legs, skin on the anterior of the legs, and ankle joints. At the ankles, the anterior tibial arteries become the ***dorsal arteries of the foot (dorsalis pedis arteries)***, which supply the muscles, skin, and joints on the dorsal aspects of the feet. The dorsal arteries of the foot give off branches that supply the feet and toes.
Posterior tibial arteries	The ***posterior tibial arteries***, the direct continuations of the popliteal arteries, distribute to the muscles, bones, and joints of the leg and foot. Major branches of the posterior tibial arteries are the ***fibular (peroneal) arteries***, which supply the leg and ankle. Branching of the posterior tibial arteries gives rise to the medial and lateral plantar arteries. The ***medial plantar arteries*** (PLAN-tar = sole) supply the muscles and skin of the feet and toes. The ***lateral plantar arteries*** supply the feet and toes.

Figure 16.11 **Arteries of the pelvis and right lower limb.**

The internal iliac arteries carry most of the blood supply to the pelvis, buttocks, external genitals, and thigh.

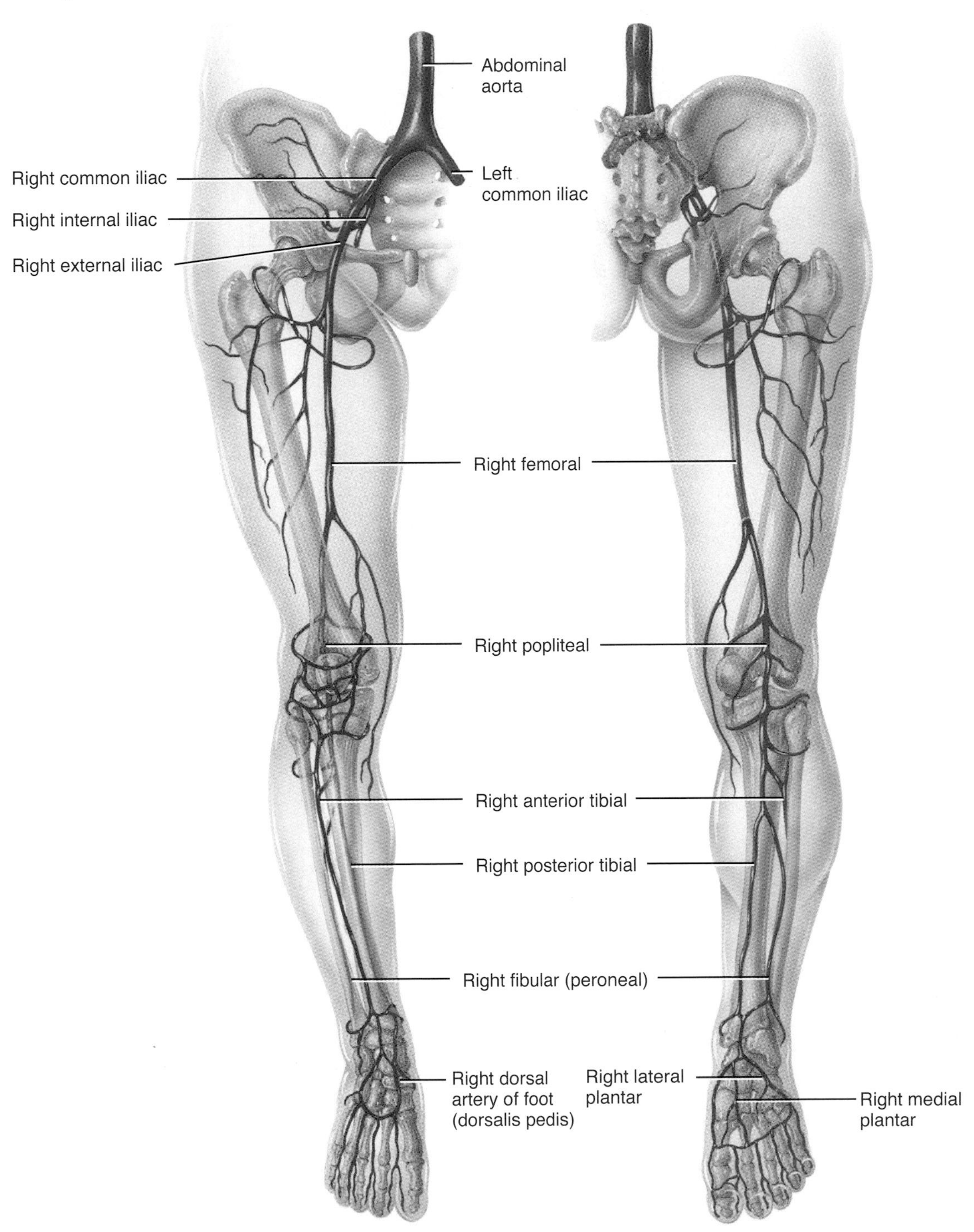

At what point does the abdominal aorta divide into the common iliac arteries?

Exhibit 16.4 Veins of the Systemic Circulation *(Figure 16.12)*

OBJECTIVE • Identify the three systemic veins that return deoxygenated blood to the heart.

• Arteries distribute blood to various parts of the body, and veins drain blood away from them. For the most part, arteries are deep. Veins may be ***superficial*** (located just beneath the skin) or ***deep***. Deep veins generally travel alongside arteries and usually bear the same name. Because there are no large superficial arteries, the names of superficial veins do not correspond to those of arteries. Superficial veins are clinically important as sites for withdrawing blood or giving injections. Arteries usually follow definite pathways. Veins are more difficult to follow because they connect in irregular networks in which many smaller veins merge to form a larger vein. Although only one systemic artery, the aorta, takes oxygenated blood away from the heart (left ventricle), three systemic veins, the ***coronary sinus***, ***superior vena cava***, and ***inferior vena cava***, deliver deoxygenated blood to the right atrium of the heart. The coronary sinus receives blood from the cardiac veins; the superior vena cava receives blood from other veins superior to the diaphragm, except the air sacs (alveoli) of the lungs; the inferior vena cava receives blood from veins inferior to the diaphragm.

■ CHECKPOINT

What are the basic differences between systemic arteries and veins?

Vein	Description and Region Drained
Coronary sinus (KOR-ō-nar-ē; *corona* = crown)	The ***coronary sinus*** is the main vein of the heart; it receives almost all venous blood from the myocardium. It opens into the right atrium between the opening of the inferior vena cava and the tricuspid valve.
Superior vena cava (SVC) (VĒ-na CĀ-va; *vena* = vein; *cava* = cavelike)	The ***SVC*** empties its blood into the superior part of the right atrium. It begins by the union of the right and left brachiocephalic veins and enters the right atrium. The SVC drains the head, neck, chest, and upper limbs.
Inferior vena cava (IVC)	The ***IVC*** is the largest vein in the body. It begins by the union of the common iliac veins, passes through the diaphragm, and enters the inferior part of the right atrium. The IVC drains the abdomen, pelvis, and lower limbs. The inferior vena cava is commonly compressed during the later stages of pregnancy by the enlarging uterus, producing edema of the ankles and feet and temporary varicose veins.

Figure 16.12 Principal veins.

Deoxygenated blood returns to the heart via the superior and inferior venae cavae and the coronary sinus.

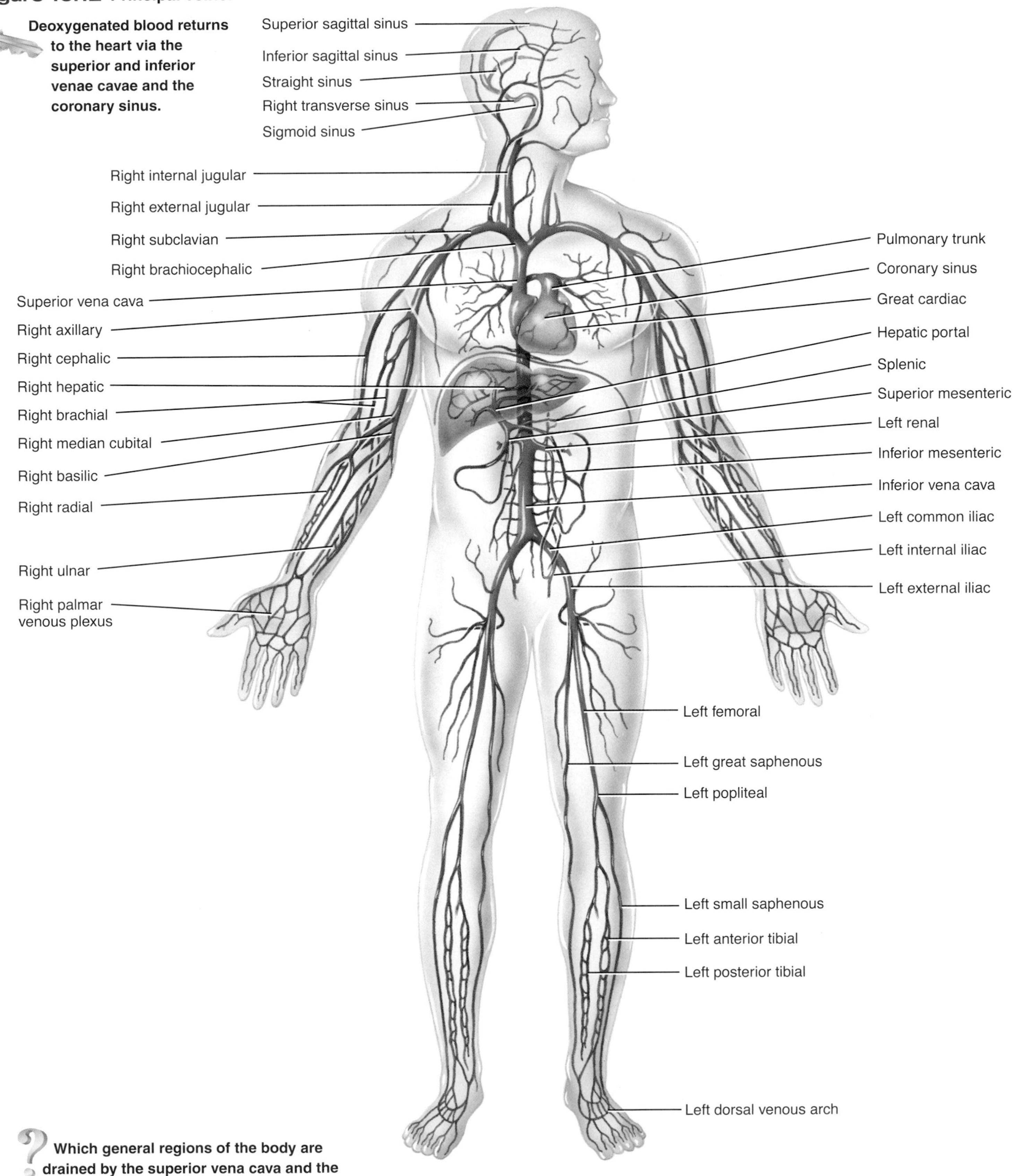

Overall anterior view of the principal veins

Which general regions of the body are drained by the superior vena cava and the inferior vena cava?

Exhibit 16.5 Veins of the Head and Neck *(Figure 16.13)*

OBJECTIVE • Identify the three major veins that drain blood from the head.

• Most blood draining from the head passes into three pairs of veins: the ***internal jugular veins***, ***external jugular veins***, and ***vertebral veins***. Within the brain, all veins drain into dural venous sinuses and then into the internal jugular veins. ***Dural venous sinuses*** are endothelium-lined venous channels between layers of the cranial dura mater.

■ **CHECKPOINT**

Which general areas are drained by the internal jugular, external jugular, and vertebral veins?

Vein	Description and Region Drained
Internal jugular veins (JUG-ū-lar; *jugular* = throat)	The dural venous sinuses (the light blue vessels in Figure 16.13) drain blood from the cranial bones, meninges, and brain. The right and left ***internal jugular veins*** pass inferiorly on either side of the neck lateral to the internal carotid and common carotid arteries. They then unite with the subclavian veins to form the right and left ***brachiocephalic veins*** (brā′-kē-ō-se-FAL-ik; *brachio-* = arm; *-cephalic* = head). From here blood flows into the superior vena cava. The general structures drained by the internal jugular veins are the brain (through the dural venous sinuses), face, and neck.
External jugular veins	The right and left ***external jugular veins*** empty into the subclavian veins. The general structures drained by the external jugular veins are external to the cranium, such as the scalp and superficial and deep regions of the face.
Vertebral veins (VER-te-bral; *vertebra* = vertebrae)	The right and left ***vertebral veins*** empty into the brachiocephalic veins in the neck. They drain deep structures in the neck such as the cervical vertebrae, cervical spinal cord, and some neck muscles.

Figure 16.13 **Principal veins of the head and neck.**

Blood draining from the head passes into the internal jugular, external jugular, and vertebral veins.

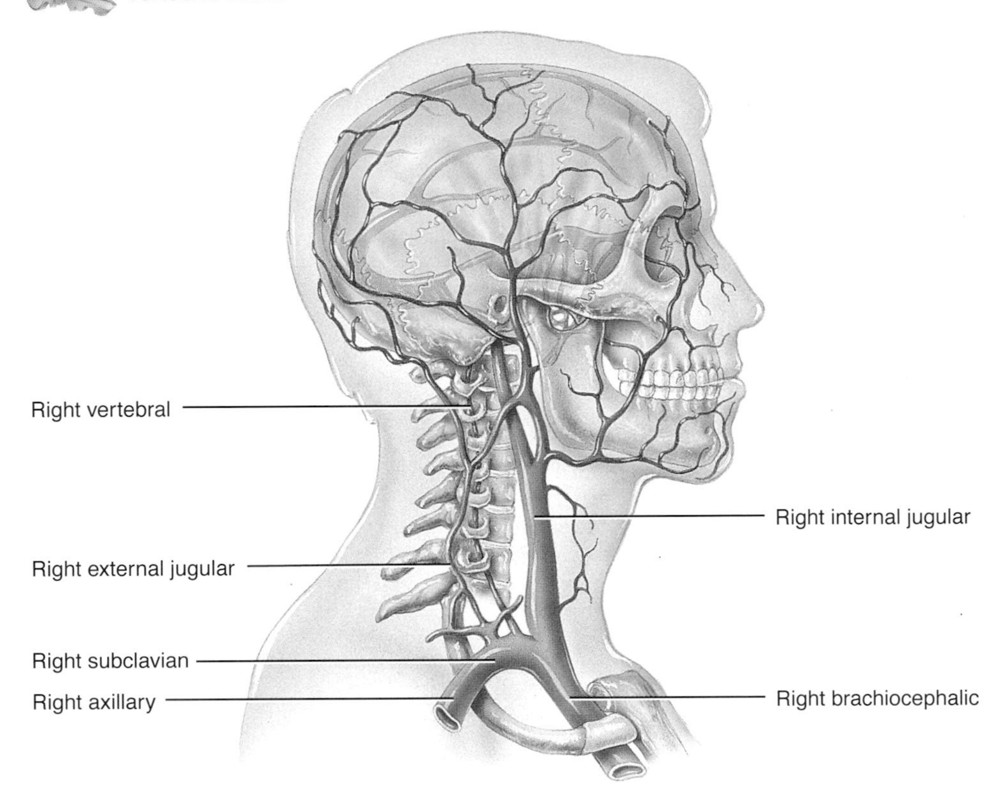

Right lateral view

Into which veins in the neck does all venous blood from the brain drain?

Exhibit 16.6 Veins of the Upper Limbs *(Figure 16.14)*

OBJECTIVE • Identify the principal veins that drain the upper limbs.

• Blood from the upper limbs is returned to the heart by both *superficial* and *deep veins.* Both sets of veins have valves, which are more numerous in the deep veins.

• Superficial veins are larger than deep veins and return most of the blood from the upper limbs.

■ **CHECKPOINT**

Where do the cephalic, basilic, median antebrachial, radial, and ulnar veins originate?

Vein	Description and Region Drained
Superficial Veins	
Cephalic veins (se-FAL-ik = head)	The principal superficial veins that drain the upper limbs originate in the hand and convey blood from the smaller superficial veins into the axillary veins. The ***cephalic veins*** begin on the lateral aspect of the ***dorsal venous networks of the hands (dorsal venous arches)***, networks of veins on the dorsum of the hands (Figure 16.14a) that drain the fingers. The cephalic veins drain blood from the lateral aspect of the upper limbs.
Basilic veins (ba-SIL-ik = royal)	The ***basilic veins*** begin on the medial aspects of the dorsal venous networks of the hands (Figure 16.14b) and drain blood from the medial aspects of the upper limbs. Anterior to the elbow, the basilic veins are connected to the cephalic veins by the ***median cubital veins*** (*cubitus* = elbow), which drain the forearm. If a vein must be punctured for an injection, transfusion, or removal of a blood sample, the median cubital vein is preferred. The basilic veins continue ascending until they join the brachial veins. As the basilic and brachial veins merge in the axillary area, they form the axillary veins.
Median antebrachial veins (an′-tē-BRĀ-kē-al; *ante-* = before, in front of; *brachi-* = arm)	The ***median antebrachial veins (median veins of the forearm)*** begin in the ***palmar venous plexuses***, networks of veins on the palms. The plexuses drain the fingers. The median antebrachial veins ascend in the forearms to join the basilic or median cubital veins, sometimes both. They drain the palms and forearms.
Deep Veins	
Radial veins (RĀ-dē-al = pertaining to the radius)	The paired ***radial veins*** begin at the ***deep palmar venous arches*** (Figure 16.14c). These arches drain the palms. The radial veins drain the lateral aspects of the forearms and pass alongside each radial artery. Just below the elbow joint, the radial veins unite with the ulnar veins to form the brachial veins.
Ulnar veins (UL-nar = pertaining to the ulna)	The paired ***ulnar veins*** begin at the ***superficial palmar venous arches***, which drain the palms and the fingers. The ulnar veins drain the medial aspect of the forearms, pass alongside each ulnar artery, and join with the radial veins to form the brachial veins.
Brachial veins (BRĀ-kē-al)	The paired ***brachial veins*** accompany the brachial arteries. They drain the forearms, elbow joints, and arms. They join with the basilic veins to form the axillary veins.
Axillary veins (AK-si-ler′-ē; *axilla* = armpit)	The ***axillary veins*** ascend to become the subclavian veins. They drain the arms, axillae, and upper part of the chest wall.
Subclavian veins (sub-KLĀ-vē-an; *sub-* = under; *-clavian* = pertaining to the clavicle)	The ***subclavian veins*** are continuations of the axillary veins that unite with the internal jugular veins to form the brachiocephalic veins. The brachiocephalic veins unite to form the superior vena cava. The subclavian veins drain the arms, neck, and thoracic wall.

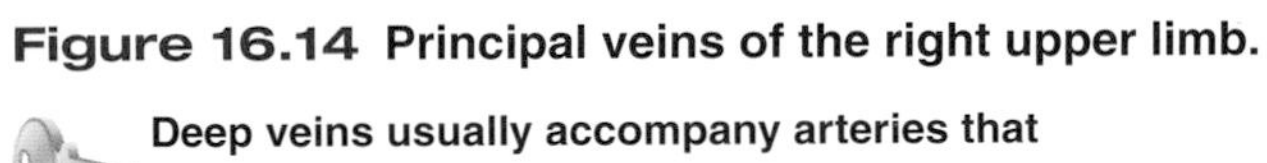

Figure 16.14 Principal veins of the right upper limb.

Deep veins usually accompany arteries that have similar names.

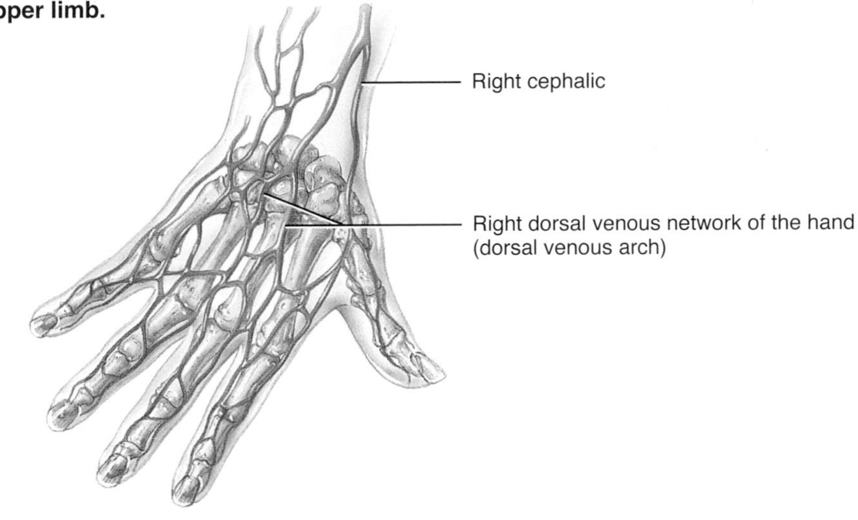

(a) Posterior view of superficial veins of the hand

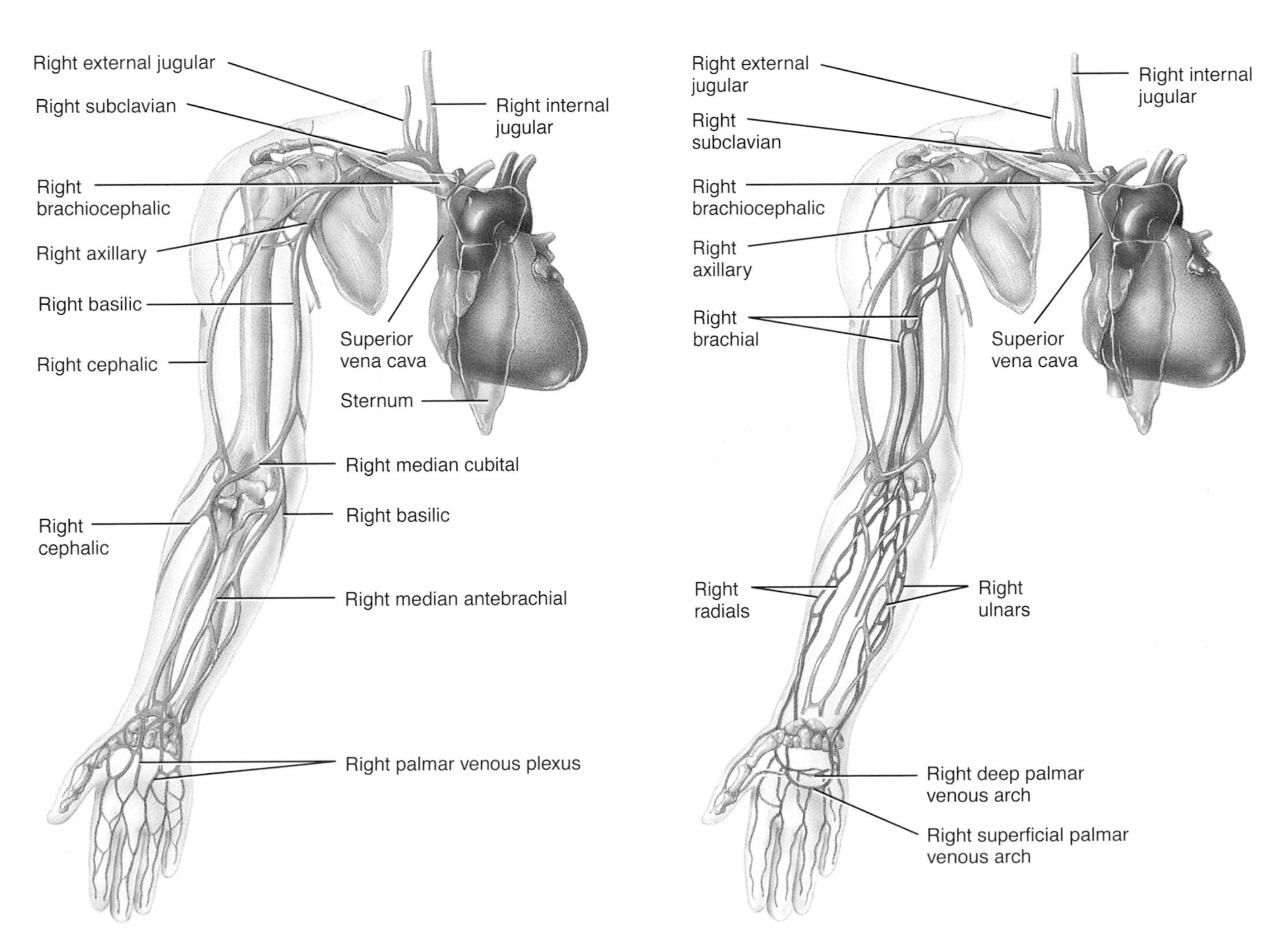

(b) Anterior view of superficial veins

(c) Anterior view of deep veins

From which vein in the upper limb is a blood sample often taken?

Exhibit 16.7 Veins of the Lower Limbs *(Figure 16.15)*

OBJECTIVE • Identify the principal veins that drain the lower limbs.

• As with the upper limbs, blood from the lower limbs is drained by both *superficial* and *deep veins*. The superficial veins often branch with each other and with deep veins along their length. All veins of the lower limbs have valves, which are more numerous than in veins of the upper limbs.

■ **CHECKPOINT**

Why are the great saphenous veins clinically important?

Vein	Description and Region Drained
Superficial Veins	
Great saphenous veins (sa-FĒ-nus = clearly visible)	The ***great saphenous veins***, the longest veins in the body, begin at the medial side of the ***dorsal venous arches*** (VĒ-nus) of the foot, networks of veins on the top of the foot that collect blood from the toes. The great saphenous veins empty into the femoral veins, and they mainly drain the leg and thigh, the groin, external genitals, and abdominal wall. Along their length, the great saphenous veins have from 10 to 20 valves, with more located in the leg than the thigh. The great saphenous veins are often used for prolonged administration of intravenous fluids. This is particularly important in very young children and in patients of any age who are in shock and whose veins are collapsed.The great saphenous veins are also often used as a source of vascular grafts, especially for coronary bypass surgery. In the procedure, the vein is removed and then reversed so that the valves do not obstruct the flow of blood.
Small saphenous veins	The ***small saphenous veins*** begin at the lateral side of the dorsal venous arches of the foot. They empty into the popliteal veins behind the knee. Along their length, the small saphenous veins have from 9 to 12 valves. The small saphenous veins drain the foot and leg.
Deep Veins	
Posterior tibial veins (TIB-ē-al)	The ***deep plantar venous arches*** on the soles drain the toes and ultimately give rise to the paired ***posterior tibial veins***. They accompany the posterior tibial arteries through the leg and drain the foot and posterior leg muscles. About two-thirds the way up the leg, the posterior tibial veins drain blood from the ***fibular (peroneal) veins***, which serve the lateral and posterior leg muscles.
Anterior tibial veins	The paired ***anterior tibial veins*** arise in the dorsal venous arch and accompany each anterior tibial artery. They unite with the posterior tibial veins to form the popliteal vein. The anterior tibial veins drain the ankle joint, knee joint, tibiofibular joint, and anterior portion of the leg.
Popliteal veins (pop′-li-TĒ-al; *popliteus* = hollow behind knee)	The ***popliteal veins*** are formed by the union of the anterior and posterior tibial veins. They drain the skin, muscles, and bones of the knee joint.
Femoral veins (FEM-o-ral)	The ***femoral veins*** accompany each femoral artery and are the continuations of the popliteal veins. They drain the muscles of the thighs, femurs, external genitals, and superficial lymph nodes. The femoral veins enter the pelvic cavity, where they are known as the ***external iliac veins***. The external and internal iliac veins unite to form the common iliac veins, which unite to form the inferior vena cava.

Figure 16.15 Principal veins of the pelvis and lower limbs.

All veins of the lower limbs have valves.

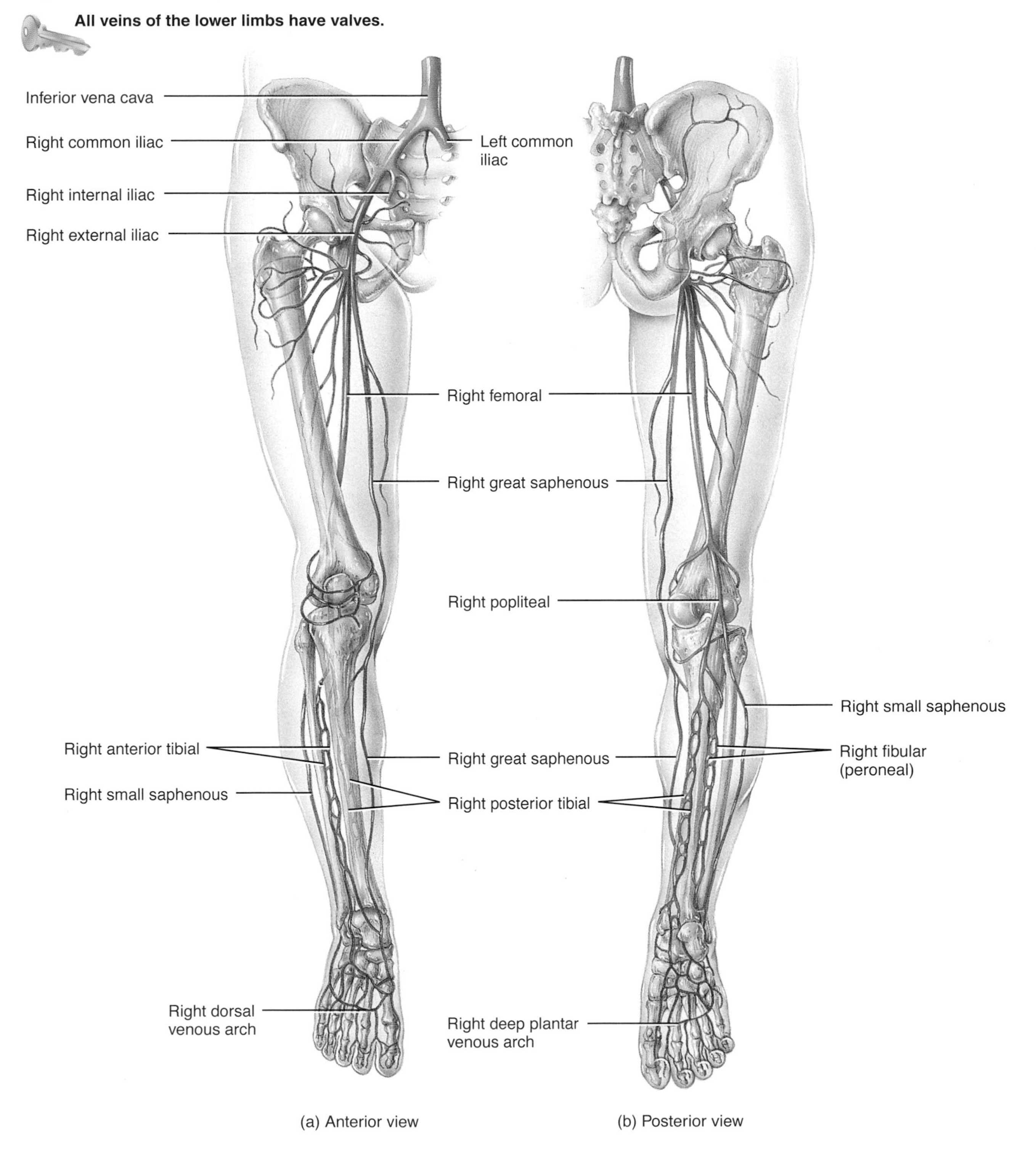

(a) Anterior view (b) Posterior view

Which veins of the lower limb are superficial?

FOCUS ON WELLNESS

Arterial Health—Undoing the Damage of Atherosclerosis

Not so long ago scientists believed that once plaque formed in an artery, it never went away. Medical researchers thought that lifestyle changes and drugs could slow the process of atherosclerosis, but they could not undo damage already done. In recent years, however, researchers have discovered that the body's own healing processes can reverse arterial plaque buildup. Lifestyle changes and drug treatments appear to stabilize the most dangerous atherosclerotic plaques and may even eliminate the need for surgical interventions, such as bypass surgery, in some people.

Stabilizing Dangerous Plaque

The health risk imposed by plaque that accumulates within the artery lining depends upon several factors. Some plaque is fairly stable: It has a low lipid content, is not growing much in size, and has a strong fibrous cap that keeps it from rupturing when blood pressure rises. Unstable plaque is characterized by a large accumulation of lipid in its core and only a thin fibrous cap. In addition, unstable plaques contain a large number of macrophages. In a misguided attempt to heal endothelial damage, macrophages ingest plaque lipids; the net result is increased arterial injury and lipid accumulation. An unstable plaque is apt to rupture, triggering formation of a life-threatening blood clot at the plaque site.

Aggressive Prevention

The first step in preventing, slowing, and possibly reversing artery disease is to control the risk factors associated with its progression. Recommendations for a heart-healthy and artery-healthy lifestyle include no smoking, regular exercise (at least 30 minutes of moderate-intensity exercise per day), stress management, and a heart-healthy diet. Diet recommendations include limiting fat intake and dramatically increasing consumption of plant foods, such as grains, fruits, and vegetables. These recommendations help prevent arterial disease by reducing obesity, blood lipids, platelet stickiness, and blood pressure, and by improving blood glucose control in people at risk for type 2 diabetes.

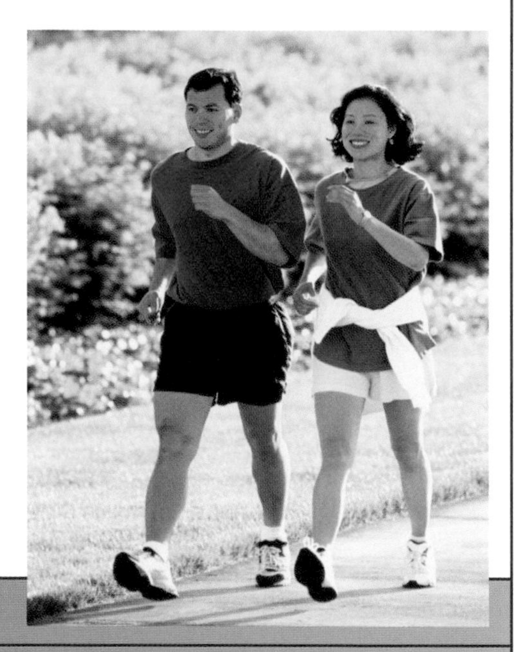

► **THINK IT OVER . . .**

► ***Studies have found that only diets extremely low in fat (10% or fewer calories from fat) lead to plaque regression. Why do you think public health officials generally recommend a diet that supplies up to 30% of its calories from fat? Do you think this recommendation should be lower? Remember, this guideline is supposed to apply to all North Americans, not just those at risk for artery disease.***

Pulmonary Circulation

When deoxygenated blood returns to the heart from the systemic route, it is pumped out of the right ventricle into the lungs. In the lungs, it loses carbon dioxide and picks up oxygen. Now bright red again, the blood returns to the left atrium of the heart and is pumped again into the systemic circulation. The flow of deoxygenated blood from the right ventricle to the air sacs of the lungs and the return of oxygenated blood from the air sacs to the left atrium is called the ***pulmonary circulation*** (see Figure 16.8). The ***pulmonary trunk*** emerges from the right ventricle and then divides into two branches. The ***right pulmonary artery*** runs to the right lung; the ***left pulmonary artery*** goes to the left lung. After birth, the pulmonary arteries are the only arteries that carry deoxygenated blood. On entering the lungs, the branches divide and subdivide until ultimately they form capillaries around the air sacs in the lungs. Carbon dioxide passes from the blood into the air sacs and is exhaled, while inhaled oxygen passes from the air sacs into the blood. The capillaries unite, venules and veins are formed, and, eventually, two ***pulmonary veins*** from each lung transport the oxygenated blood to the left atrium. (After birth, the pulmonary veins are the only veins that carry oxygenated blood.) Contractions of the left ventricle then send the blood into the systemic circulation.

Hepatic Portal Circulation

A vein that carries blood from one capillary network to another is called a ***portal vein***. The hepatic portal vein, formed by the union of the splenic and superior mesenteric veins (Figure 16.16), receives blood from capillaries of digestive organs and delivers it to capillary-like structures in the liver called sinusoids. In the ***hepatic portal circulation*** (*hepat-* = liver), venous blood from the gastrointestinal organs and spleen, rich with substances absorbed from the gastrointestinal tract, is delivered to the hepatic

Figure 16.16 Hepatic portal circulation.

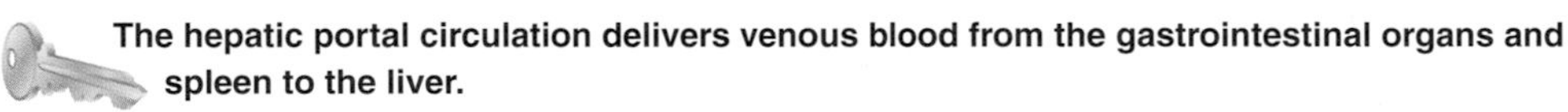

The hepatic portal circulation delivers venous blood from the gastrointestinal organs and spleen to the liver.

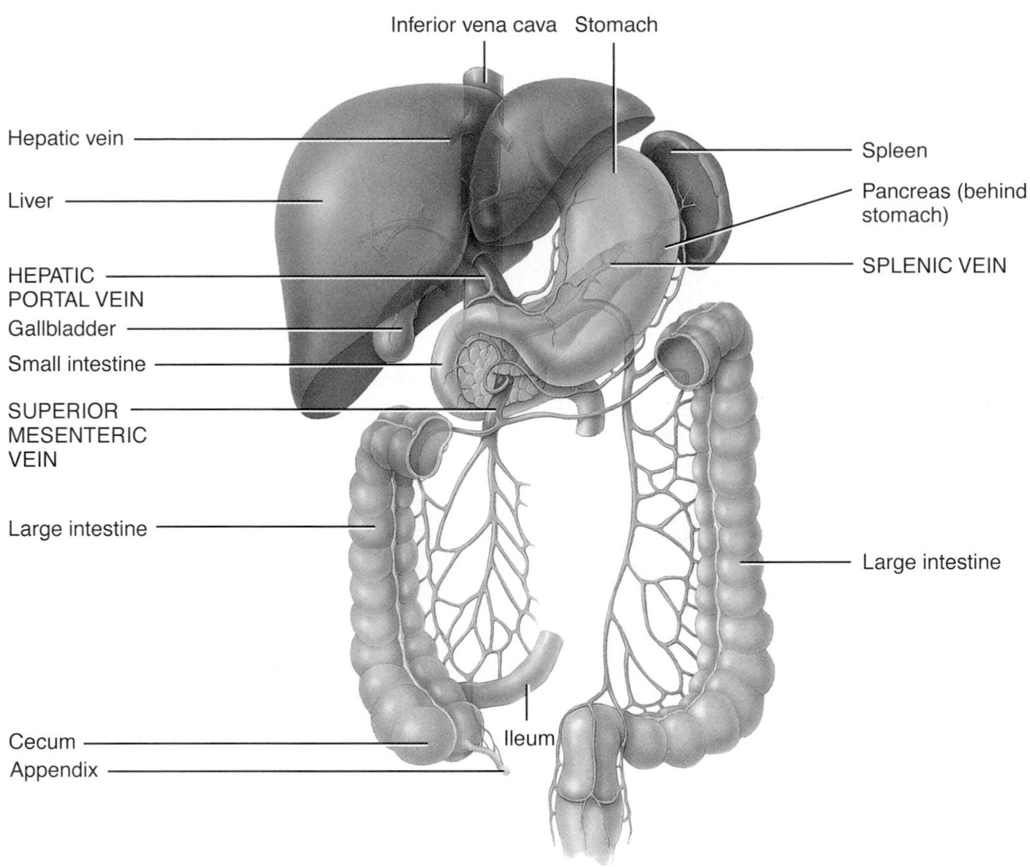

(a) Anterior view of veins draining into the hepatic portal vein

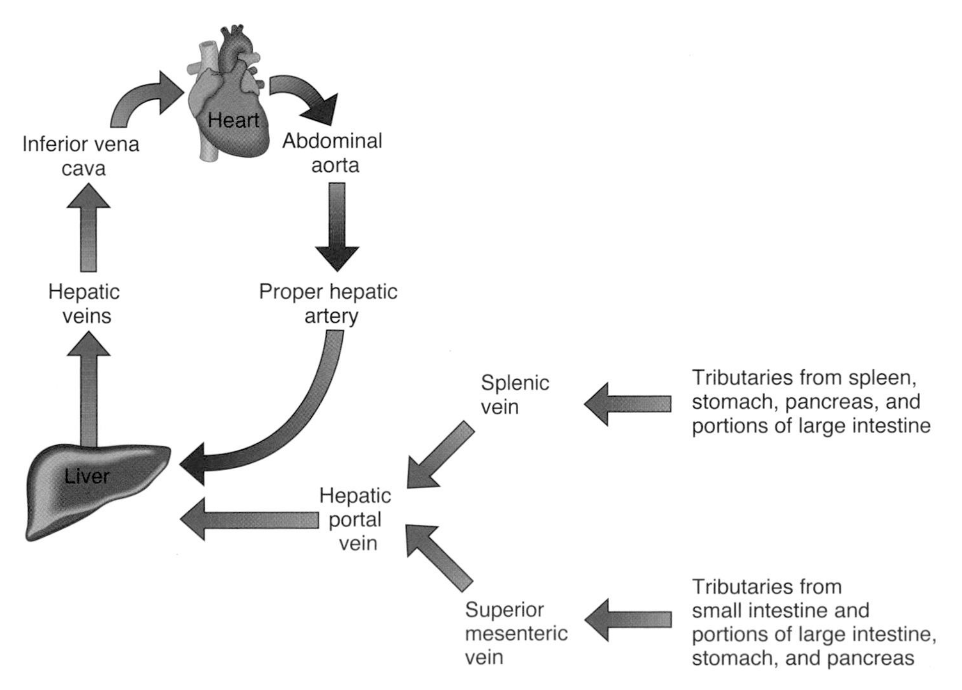

(b) Scheme of principal blood vessels of hepatic portal circulation and arterial supply and venous drainage of liver

Which veins carry blood away from the liver?

portal vein and enters the liver. The liver processes these substances before they pass into the general circulation. At the same time, the liver receives oxygenated blood from the systemic circulation via the hepatic artery. The oxygenated blood mixes with the deoxygenated blood in sinusoids. Ultimately, all blood leaves the sinusoids of the liver through the hepatic veins, which drain into the inferior vena cava.

Fetal Circulation

The circulatory system of a fetus, called ***fetal circulation***, exists only in the fetus and contains special structures that allow the developing fetus to exchange materials with its mother (Figure 16.17). It differs from the postnatal (after birth) circulation because the lungs, kidneys, and gastrointestinal organs do not begin to function until birth. The fetus obtains O_2 and nutrients from and eliminates CO_2 and other wastes into the maternal blood.

The exchange of materials between fetal and maternal circulations occurs through the ***placenta*** (pla-SEN-ta), which forms inside the mother's uterus and attaches to the umbilicus (navel) of the fetus by the ***umbilical cord*** (um-BIL-i-kal). Blood passes from the fetus to the placenta via two ***umbilical arteries*** (Figure 16.17a). These branches of the internal iliac arteries are within the umbilical cord. At the placenta, fetal blood picks up O_2 and nutrients and eliminates CO_2 and wastes. The oxygenated blood returns from the placenta via a single ***umbilical vein***. This vein ascends to the liver of the fetus, where it divides into two branches. Some blood flows through the branch that joins the hepatic portal vein and enters the liver, but most of the blood flows into the second branch, the ***ductus venosus*** (DUK-tus ve-NŌ-sus), which drains into the inferior vena cava.

Deoxygenated blood returning from lower body regions of the fetus mingles with oxygenated blood from the ductus venosus in the inferior vena cava. This mixed blood then enters the right atrium. Deoxygenated blood returning from upper body regions of the fetus enters the superior vena cava and passes into the right atrium.

Most of the fetal blood does not pass from the right ventricle to the lungs, as it does in postnatal circulation, because an opening called the ***foramen ovale*** (fō-RĀ-men ō-VAL-ē) exists in the septum between the right and left atria. About one-third of the blood that enters the right atrium passes through the foramen ovale into the left atrium and joins the systemic circulation. The blood that does pass into the right ventricle is pumped into the pulmonary trunk, but little of this blood reaches the nonfunctioning fetal lungs. Instead, most is sent through the ***ductus arteriosus*** (ar-tē-rē-Ō-sus), a vessel that connects the pulmonary trunk with the aorta, so that most blood bypasses the fetal lungs. The blood in the aorta is carried to all fetal tissues through the systemic circulation. When the common iliac arteries branch into the external and internal iliacs, part of the blood flows into the internal iliacs, into the umbilical arteries, and back to the placenta for another exchange of materials.

After birth, when pulmonary (lung), renal, and digestive functions begin, the following vascular changes occur (Figure 16.17b):

1. When the umbilical cord is tied off, blood no longer flows through the umbilical arteries, they fill with connective tissue, and the distal portions of the umbilical arteries become fibrous cords called ***medial umbilical ligaments***.
2. The umbilical vein collapses but remains as the ***ligamentum teres (round ligament)***, a structure that attaches the umbilicus to the liver.
3. The ductus venosus collapses but remains as the ***ligamentum venosum***, a fibrous cord on the inferior surface of the liver.
4. The placenta is expelled as the "***afterbirth***."
5. The foramen ovale normally closes shortly after birth to become the ***fossa ovalis***, a depression in the interatrial septum. When an infant takes its first breath, the lungs expand and blood flow to the lungs increases. Blood returning from the lungs to the heart increases pressure in the left atrium. This closes the foramen ovale by pushing the valve that guards it against the interatrial septum. Permanent closure occurs in about a year.
6. The ductus arteriosus closes by vasoconstriction almost immediately after birth and becomes the ***ligamentum arteriosum***.

■ CHECKPOINT

10. What are the main functions of the systemic, pulmonary, hepatic portal, and fetal circulations?

AGING AND THE CARDIOVASCULAR SYSTEM

OBJECTIVE • Describe the effects of aging on the cardiovascular system.

General changes in the cardiovascular system associated with aging include increased stiffness of the aorta, reduction in cardiac muscle fiber size, progressive loss of cardiac muscular strength, reduced cardiac output, a decline in maximum heart rate, and an increase in systolic blood pressure. Coronary artery disease (CAD) is the major cause of heart disease and death in older Americans. Congestive heart failure (CHF), a set of symptoms associated with impaired pumping of the heart, is also prevalent in older individuals. Changes in blood vessels that serve brain tissue—for example, atherosclerosis—reduce nourishment to the brain and result in the malfunction or death of brain cells. By age 80, blood flow to the brain is 20% less, and blood flow to the kidneys is 50% less, than it was in the same person at age 30.

■ **CHECKPOINT**

11. What are some of the signs that the cardiovascular system is aging?

• • •

To appreciate the many ways the cardiovascular system contributes to homeostasis of other body systems, examine Focus on Homeostasis: The Cardiovascular System on page 414. Next, in Chapter 17, we will examine the structure and function of the lymphatic system, seeing how it returns excess fluid filtered from capillaries to the cardiovascular system. We will also take a more detailed look at how some white blood cells function as defenders of the body by carrying out immune responses.

Figure 16.17 Fetal circulation and changes at birth. The boxes between parts (a) and (b) describe the fate of certain fetal structures once postnatal circulation is established.

The lungs and gastrointestinal organs do not begin to function until birth.

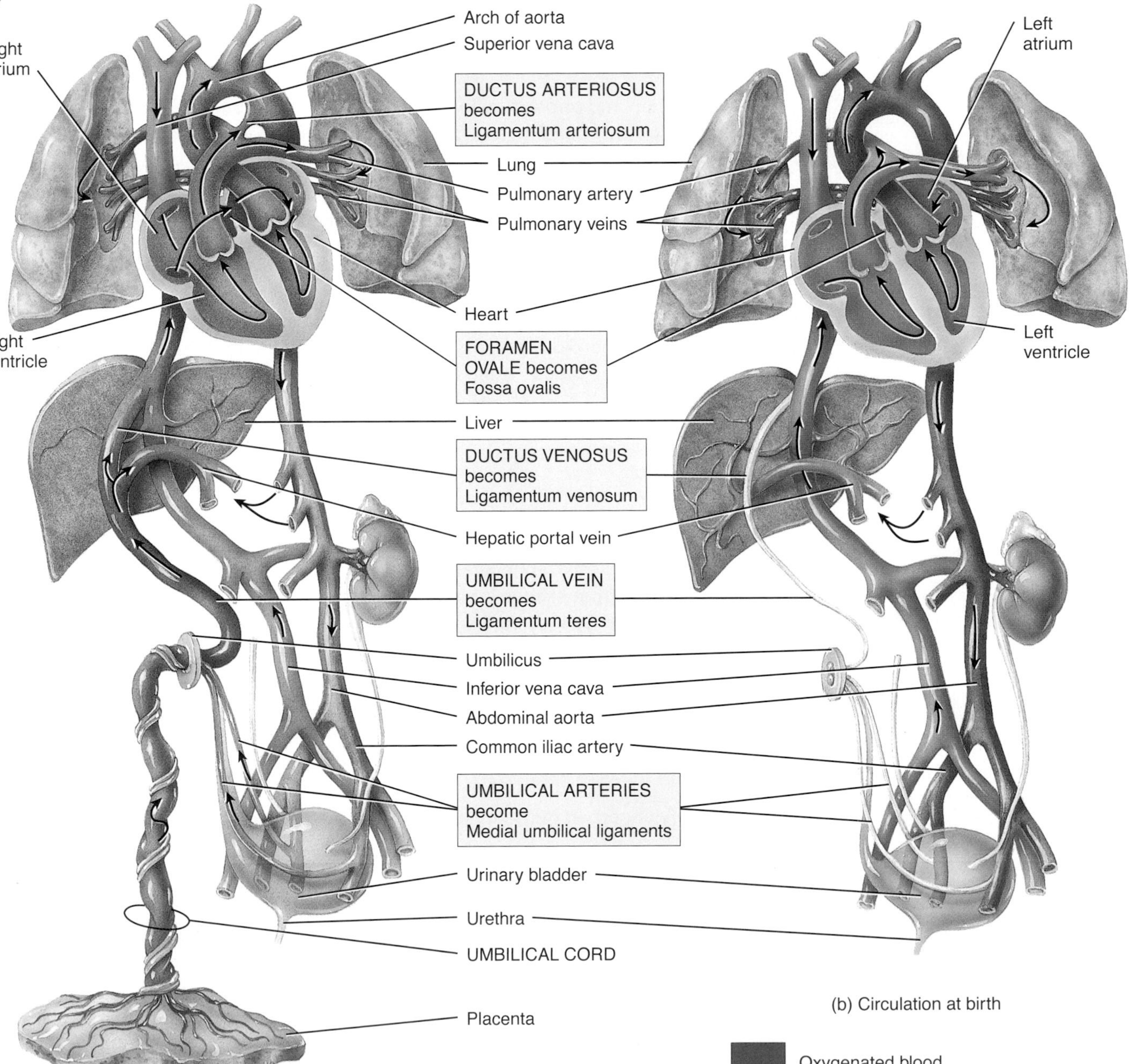

Right atrium
Right ventricle
Arch of aorta
Superior vena cava
DUCTUS ARTERIOSUS becomes Ligamentum arteriosum
Lung
Pulmonary artery
Pulmonary veins
Heart
FORAMEN OVALE becomes Fossa ovalis
Liver
DUCTUS VENOSUS becomes Ligamentum venosum
Hepatic portal vein
UMBILICAL VEIN becomes Ligamentum teres
Umbilicus
Inferior vena cava
Abdominal aorta
Common iliac artery
UMBILICAL ARTERIES become Medial umbilical ligaments
Urinary bladder
Urethra
UMBILICAL CORD
Placenta
Left atrium
Left ventricle

(a) Fetal circulation

(b) Circulation at birth

Oxygenated blood
Mixed oxygenated and deoxygenated blood
Deoxygenated blood

Which structure provides for exchange of materials between mother and fetus?

FOCUS ON HOMEOSTASIS

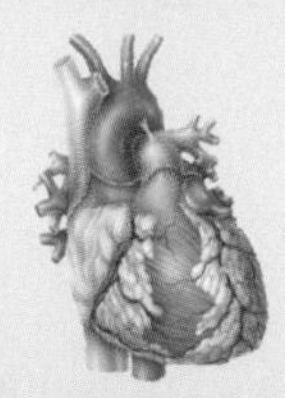

The Cardiovascular System

Body System		Contribution of the Cardiovascular System
For all body systems		The heart pumps blood through blood vessels to body tissues, delivering oxygen and nutrients and removing wastes by means of capillary exchange. Circulating blood keeps body tissues at a proper temperature.
Integumentary system		Blood delivers clotting factors and white blood cells that aid in hemostasis when skin is damaged and contribute to repair of injured skin. Changes in skin blood flow contribute to body temperature regulation by adjusting the amount of heat loss via the skin. Blood flowing in skin may give skin a pink hue.
Skeletal system	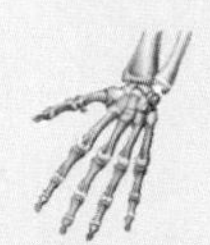	Blood delivers calcium and phosphate ions that are needed for building bone extracellular matrix, hormones that govern building and breakdown of bone extracellular matrix, and erythropoietin that stimulates production of red blood cells by red bone marrow.
Muscular system		Blood circulating through exercising muscles remove heat and lactic acid.
Nervous system	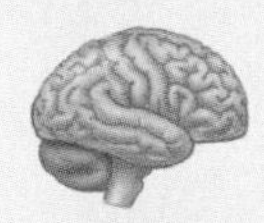	Endothelial cells lining choroid plexuses in brain ventricles help produce cerebrospinal fluid (CSF) and contribute to the blood–brain barrier.
Endocrine system		Circulating blood delivers most hormones to their target tissues. Atrial cells of the heart secrete atrial natriuretic peptide.
Lymphatic system and immunity		Circulating blood distributes lymphocytes, antibodies, and macrophages that carry out immune functions. Lymph forms from excess interstitial fluid, which filters from blood plasma due to blood pressure generated by the heart.
Respiratory system	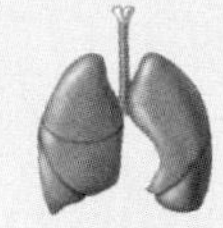	Circulating blood transports oxygen from the lungs to body tissues and carbon dioxide to the lungs for exhalation.
Digestive system	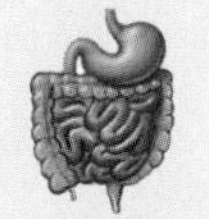	Blood carries newly absorbed nutrients and water to the liver. Blood distributes hormones that aid digestion.
Urinary system	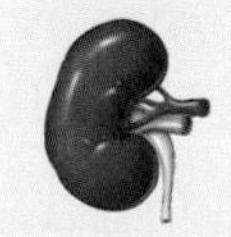	The heart and blood vessels deliver 20% of the resting cardiac output to the kidneys, where blood is filtered, needed substances are reabsorbed, and unneeded substances are eliminated as part of urine, which is excreted.
Reproductive systems	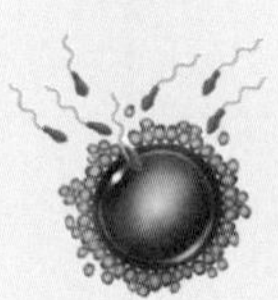	Vasodilation of arterioles in the penis and clitoris causes erection during sexual intercourse. Blood distributes hormones that regulate reproductive functions.

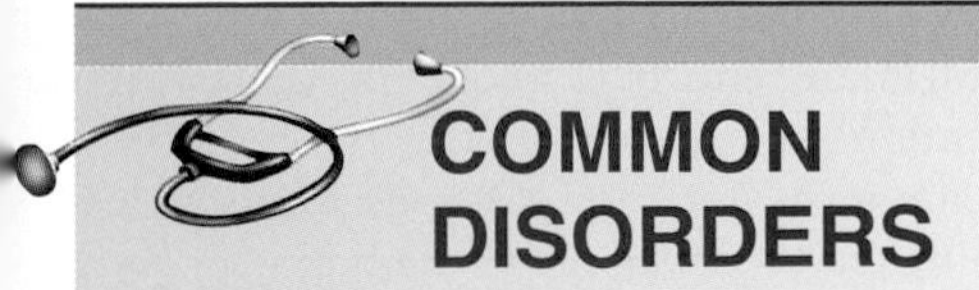

COMMON DISORDERS

Hypertension

About 50 million Americans have ***hypertension***, or persistently high blood pressure. It is the most common disorder affecting the heart and blood vessels and is the major cause of heart failure, kidney disease, and stroke. In May 2003, the Joint National Committee on Prevention, Detection, Evaluation, and Treatment of High Blood Pressure published new guidelines for hypertension because clinical studies have linked what were once considered fairly low pressure readings to an increased risk of cardiovascular disease. The new guidelines are as follows:

Category	*Systolic (mm Hg)*	*Diastolic (mm Hg)*
Normal	Less than 120 *and*	Less than 80
Prehypertension	120–139 *or*	80–89
Stage 1 hypertension	140–159 *or*	90–99
Stage 2 hypertension	Greater than 160 *or*	Greater than 100

Using the new guidelines, the normal classification was previously considered optimal; prehypertension now includes many more individuals previously classified as normal or high-normal; stage 1 hypertension is the same as in previous guidelines; and stage 2 hypertension now combines the previous stage 2 and stage 3 categories since treatment options are the same for the former stages 2 and 3.

Although several categories of drugs can reduce elevated blood pressure, the following lifestyle changes are also effective in managing hypertension:

- *Lose weight*. This is the best treatment for high blood pressure short of using drugs. Loss of even a few pounds helps reduce blood pressure in overweight hypertensive individuals.
- *Limit alcohol intake*. Drinking in moderation may lower the risk of coronary heart disease, mainly among males over 45 and females over 55. Moderation is defined as no more than one 12-oz beer per day for females and no more than two 12-oz beers per day for males.
- *Exercise*. Becoming more physically fit by engaging in moderate activity (such as brisk walking) several times a week for 30 to 45 minutes can lower systolic blood pressure by about 10 mm Hg.
- *Reduce intake of sodium (salt)*. Roughly half the people with hypertension are "salt sensitive." For them, a high-salt diet appears to promote hypertension, and a low-salt diet can lower their blood pressure.
- *Maintain recommended dietary intake of potassium, calcium, and magnesium*. Higher levels of potassium, calcium, and magnesium in the diet are associated with a lower risk of hypertension.
- *Don't smoke*. Smoking has devastating effects on the heart and can augment the damaging effects of high blood pressure by promoting vasoconstriction.
- *Manage stress*. Various meditation and biofeedback techniques help some people reduce high blood pressure. These methods may work by decreasing the daily release of epinephrine and norepinephrine by the adrenal medulla.

Shock

Shock is a failure of the cardiovascular system to deliver enough O_2 and nutrients to meet cellular metabolic needs. The causes of shock are many and varied, but all are characterized by inadequate blood flow to body tissues. Common causes of shock include loss of body fluids, as occurs in hemorrhage, dehydration, burns, excessive vomiting, diarrhea, or sweating. If shock persists, cells and organs become damaged, and cells may die unless proper treatment begins quickly.

Although the symptoms of shock vary with the severity of the condition, the following are commonly observed: systolic blood pressure lower than 90 mm Hg; rapid resting heart rate due to sympathetic stimulation and increased blood levels of epinephrine and norepinephrine; weak, rapid pulse due to reduced cardiac output and fast heart rate; cool, pale skin due to vasoconstriction of skin blood vessels; sweating due to sympathetic stimulation; reduced urine formation and output due to increased levels of aldosterone and antidiuretic hormone (ADH); altered mental state due to reduced oxygen supply to the brain; thirst due to loss of extracellular fluid; and nausea due to impaired circulation to digestive organs.

Aneurysm

An ***aneurysm*** (AN-ū-rizm) is a thin, weakened section of the wall of an artery or a vein that bulges outward, forming a balloonlike sac. Common causes are atherosclerosis, syphilis, congenital blood vessel defects, and trauma. If untreated, the aneurysm enlarges and the blood vessel wall becomes so thin that it bursts. The result is massive hemorrhage along with shock, severe pain, stroke, or death.

MEDICAL TERMINOLOGY AND CONDITIONS

Angiogenesis (an′-jē-ō-JEN-e-sis) Formation of new blood vessels.

Aortography (a′-or-TOG-ra-fē) X-ray examination of the aorta and its main branches after injection of a dye.

Circulation time The time required for a drop of blood to pass from the right atrium, through the pulmonary circulation, back to the left atrium, through the systemic circulation, down to the foot, and back again to the right atrium; normally about 1 minute in a resting person.

Claudication (klaw′-di-KĀ-shun) Pain and lameness or limping caused by defective circulation of the blood in the vessels of the limbs.

Deep vein thrombosis (DVT) The presence of a thrombus (blood clot) in a deep vein of the lower limbs.

Hypotension (hī-pō-TEN-shun) Low blood pressure; most commonly used to describe an acute drop in blood pressure, as occurs during excessive blood loss.

Occlusion (ō-KLOO-zhun) The closure or obstruction of the lumen of a structure such as a blood vessel. An example is an atherosclerotic plaque in an artery.

Orthostatic hypotension (or′-thō-STAT-ik; *ortho-* = straight; *-static* = causing to stand) An excessive lowering of systemic blood pressure when a person stands up; usually a sign of disease. May be caused by excessive fluid loss, certain drugs, and cardiovascular or neurogenic factors. Also called ***postural hypotension***.

Phlebitis (fle-BĪ-tis; *phleb-* = vein) Inflammation of a vein, often in a leg. The condition is often accompanied by pain and redness of the skin over the inflamed vein. It is frequently caused by trauma or bacterial infection.

Syncope (SIN-kō-pē) A temporary cessation of consciousness; a faint. One cause is insufficient blood supply to the brain.

Thrombophlebitis (throm′-bō-fle-BĪ-tis) Inflammation of a vein involving clot formation. Superficial thrombophlebitis occurs in veins under the skin, especially in the calf.

White coat (office) hypertension A stress-induced syndrome found in patients who have elevated blood pressure when being examined by health-care personnel, but otherwise have normal blood pressure.

STUDY OUTLINE

Blood Vessel Structure and Function (p. 386)

1. Arteries carry blood away from the heart. Their walls consist of three layers.
2. The structure of the middle layer gives arteries their two major properties, elasticity and contractility.
3. Arterioles are small arteries that deliver blood to capillaries.
4. Through constriction and dilation, arterioles play a key role in regulating blood flow from arteries into capillaries.
5. Capillaries are microscopic blood vessels through which materials are exchanged between blood and interstitial fluid.
6. Precapillary sphincters regulate blood flow through capillaries.
7. Capillary blood pressure "pushes" fluid out of capillaries into interstitial fluid (filtration).
8. Blood colloid osmotic pressure "pulls" fluid into capillaries from interstitial fluid (reabsorption).
9. Autoregulation refers to local adjustments of blood flow in response to physical and chemical changes in a tissue.
10. Venules are small vessels that emerge from capillaries and merge to form veins. They drain blood from capillaries into veins.
11. Veins consist of the same three layers as arteries but have less elastic tissue and smooth muscle. They contain valves that prevent backflow of blood.
12. Weak venous valves can lead to varicose veins.
13. Venous return, the volume of blood flowing back to the heart through systemic veins, occurs due to the pumping action of the heart, aided by skeletal muscle contractions (the skeletal muscle pump), and breathing (the respiratory pump).

Blood Flow Through Blood Vessels (p. 390)

1. Blood flow is determined by blood pressure and vascular resistance.
2. Blood flows from regions of higher pressure to regions of lower pressure.
3. Blood pressure is highest in the aorta and large systemic arteries; it drops progressively as distance from the left ventricle increases. Blood pressure in the right atrium is close to 0 mm Hg.
4. An increase in blood volume increases blood pressure, and a decrease in blood volume decreases it.
5. Vascular resistance is the opposition to blood flow mainly as a result of friction between blood and the walls of blood vessels.
6. Vascular resistance depends on size of the blood vessel lumen, blood viscosity, and total blood vessel length.
7. Blood pressure and blood flow are regulated by neural and hormonal negative feedback systems and by autoregulation.
8. The cardiovascular center in the medulla oblongata helps regulate heart rate, stroke volume, and size of blood vessel lumen.
9. Vasomotor nerves (sympathetic) control vasoconstriction and vasodilation.
10. Baroreceptors (pressure-sensitive receptors) send impulses to the cardiovascular center to regulate blood pressure.
11. Chemoreceptors (receptors sensitive to concentrations of oxygen, carbon dioxide, and hydrogen ions) also send impulses to the cardiovascular center to regulate blood pressure.
12. Hormones such as angiotensin II, aldosterone, epinephrine, norepinephrine, and antidiuretic hormone raise blood pressure; atrial natriuretic peptide lowers it.

Checking Circulation (p. 394)

1. Pulse is the alternate expansion and elastic recoil of an artery with each heartbeat. It may be felt in any artery that lies near the surface or over a hard tissue.
2. A normal pulse rate is about 70–80 beats per minute.
3. Blood pressure is the pressure exerted by blood on the wall of an artery when the left ventricle undergoes systole and then diastole. It is measured by a sphygmomanometer.
4. Systolic blood pressure (SBP) is the force of blood recorded during ventricular contraction. Diastolic blood pressure (DBP) is the force of blood recorded during ventricular relaxation. The normal blood pressure of a young adult male is less than 120/80 mm Hg.

Circulatory Routes (p. 394)

1. The two major circulatory routes are the systemic circulation and the pulmonary circulation.
2. The systemic circulation takes oxygenated blood from the left ventricle through the aorta to all parts of the body and returns deoxygenated blood to the right atrium.
3. The parts of the aorta include the ascending aorta, the arch of the aorta, the thoracic aorta, and the abdominal aorta. Each part gives off arteries that branch to supply the whole body.
4. Deoxygenated blood is returned to the heart through the systemic veins. All the veins of systemic circulation flow into either the superior or inferior vena cava or the coronary sinus, which empty into the right atrium.
5. The pulmonary circulation takes deoxygenated blood from the right ventricle to the air sacs of the lungs and returns oxygenated blood from the air sacs to the left atrium. It allows blood to be oxygenated for systemic circulation.
6. The hepatic portal circulation collects deoxygenated blood from the veins of the gastrointestinal tract and spleen and directs it into the hepatic portal vein of the liver. This routing allows the liver to extract and modify nutrients and detoxify harmful substances in the blood. The liver also receives oxygenated blood from the hepatic artery.
7. Fetal circulation exists only in the fetus. It involves the exchange of materials between fetus and mother via the placenta. The fetus derives O_2 and nutrients from and eliminates CO_2 and wastes into maternal blood. At birth, when pulmonary (lung), digestive, and liver functions begin, the special structures of fetal circulation are no longer needed.

Aging and the Cardiovascular System (p. 412)

1. General changes associated with aging include reduced elasticity of blood vessels, reduction in cardiac muscle size, reduced cardiac output, and increased systolic blood pressure.
2. The incidence of coronary artery disease (CAD), congestive heart failure (CHF), and atherosclerosis increases with age.

SELF-QUIZ

1. Sensory receptors that monitor changes in the blood pressure to the brain are
   a. chemoreceptors in the aorta
   b. baroreceptors in the carotid arteries
   c. the aortic bodies
   d. precapillary sphincters in the arterioles
   e. proprioceptors in the muscles
2. The blood vessels that allow the exchange of nutrients, wastes, oxygen, and carbon dioxide between the blood and tissues are the
   a. capillaries b. arteries c. venules d. arterioles
   e. veins
3. Substances undergo capillary exchange by means of
   a. simple diffusion and bulk flow
   b. endocytosis, exocytosis, and active transport
   c. simple diffusion and facilitated diffusion
   d. simple diffusion and active transport
   e. filtration, reabsorption, and secretion
4. Blood flows through the blood vessels because of the
   a. establishment of a concentration gradient
   b. elastic recoil of the veins
   c. establishment of a pressure gradient
   d. viscosity (stickiness) of the blood
   e. thinness of the walls of capillaries
5. Which of the following represents pulmonary circulation as the blood flows from the right ventricle?
   a. pulmonary trunk → pulmonary veins → pulmonary capillaries → pulmonary arteries
   b. pulmonary arteries → pulmonary capillaries → pulmonary trunk → pulmonary veins
   c. pulmonary capillaries → pulmonary trunk → pulmonary arteries → pulmonary veins
   d. pulmonary trunk → pulmonary arteries → pulmonary capillaries → pulmonary veins
   e. pulmonary veins → pulmonary capillaries → pulmonary arteries → pulmonary trunk
6. The tissue that allows arteries to stretch is
   a. endothelium b. collagen c. basement membrane
   d. cardiac muscle e. elastic lamina
7. Match the following descriptions to the appropriate blood vessel:

____	a. composed of a single layer of endothelial cells and a basement membrane	A. arteries
____	b. formed by reuniting capillaries	B. arterioles
____	c. carry blood away from heart	C. veins
____	d. regulate blood flow to capillaries	D. venules
____	e. may contain valves	E. capillaries

8. Filtration of substances out of capillaries occurs when the capillary blood pressure
   a. is less than the blood colloid osmotic pressure
   b. and the blood colloid osmotic pressure are equal
   c. is high and the blood colloid osmotic pressure is high
   d. is higher than the blood colloid osmotic pressure
   e. is low and the blood colloid osmotic pressure is low
9. Weakened leg muscles would slow the
   a. blood flow out of the heart
   b. respiratory pump
   c. venous return
   d. ability of arteries to vasodilate
   e. pulse

10. Which of the following statements about blood vessels is true?
- **a.** Capillaries contain valves.
- **b.** Walls of arteries are generally thicker and contain more elastic tissue than walls of veins.
- **c.** Veins carry blood away from the heart.
- **d.** Blood flows most rapidly through veins.
- **e.** Blood pressure in arteries is always lower than in veins.

11. Why is it important that blood flows slowly through the capillaries?
- **a.** It allows time for the materials in the blood to pass through the thick capillary walls.
- **b.** It prevents damage to the capillaries.
- **c.** It permits the efficient exchange of nutrients and wastes between the blood and body cells.
- **d.** It allows the heart time to rest.
- **e.** It allows the blood pressure in capillaries to rise above the blood pressure in the veins.

12. Match the following:

_____	**a.** source of all systemic arteries	**A.** hepatic portal vein
_____	**b.** supplies a lower limb	**B.** pulmonary trunk
_____	**c.** heart's blood system	**C.** pulmonary vein
_____	**d.** returns blood to heart from lower limbs	**D.** common iliac artery
_____	**e.** carries blood to liver	**E.** coronary circulation
_____	**f.** leads to lungs	**F.** inferior vena cava
_____	**g.** returns blood from lungs to heart	**G.** superior vena cava
_____	**h.** supplies blood to brain	**H.** aorta
_____	**i.** returns blood to heart from head and upper body	**I.** cerebral arterial circle

13. For each of the following factors, indicate if it increases **(A)** or decreases **(B)** blood pressure:

_____ **a.** an increase in cardiac output
_____ **b.** hemorrhage
_____ **c.** vasodilation
_____ **d.** vasoconstriction
_____ **e.** stimulation of the heart by the sympathetic nervous system
_____ **f.** hypoxia
_____ **g.** epinephrine
_____ **h.** increase in blood volume
_____ **i.** bradycardia

14. Aldosterone affects blood pressure by
- **a.** increasing heart rate
- **b.** increasing vasoconstriction of arterioles
- **c.** reducing blood volume
- **d.** stimulating release of atrial natriuretic peptide by the heart
- **e.** increasing reabsorption of sodium ions and water by the kidneys

15. In a blood pressure reading of 110/70,
- **a.** 110 represents the diastolic pressure
- **b.** 70 represents the pressure of the blood against the arteries during ventricular relaxation
- **c.** 110 represents the blood pressure and 70 represents the heart rate
- **d.** 70 is the reading taken when the first sound is heard
- **e.** the patient has a severe problem with hypertension

16. Which of the following statements is NOT true?
- **a.** Regulation of blood vessel diameter originates from the vasomotor region of the cerebral cortex.
- **b.** The cerebral cortex may provide input to the CV center.
- **c.** Baroreceptors may stimulate the cardiovascular center.
- **d.** Activation of proprioceptors increases heart rate at the beginning of exercise.
- **e.** Vasomotor tone is due to a moderate level of vasoconstriction.

17. Venous return to the heart is enhanced by all of the following EXCEPT
- **a.** skeletal muscle "milking"
- **b.** valves in veins
- **c.** the pressure difference from venules to the right ventricle
- **d.** vasodilation
- **e.** inhalation during breathing

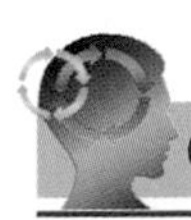

CRITICAL THINKING APPLICATIONS

1. The local anesthetic injected by a dentist often contains a small amount of epinephrine. What effect would epinephrine have on the blood vessels in the vicinity of the dental work? Why might this effect be desired?
2. In this chapter, you've read about varicose veins. Why didn't you read about varicose arteries?
3. Julie was all flustered when she ran in late to her anatomy and physiology lab. She had spilled a cup of coffee on herself while she was weaving in and out of traffic while trying to get around a traffic jam. Then she missed her exit while she was changing the station on the radio, couldn't find a place to park, and missed the lab quiz. The lab today is learning to take blood pressures, and Julie's is high! (It's normally 110 over 70.) What is the physiological explanation for Julie's elevated BP?
4. Peter spent 10 minutes sharpening his favorite knife before carving the roast. Unfortunately, he sliced his finger along with the roast. His wife slapped a towel over the spurting cut and drove him to the emergency room. What type of vessel did Peter cut, and how do you know?

ANSWERS TO FIGURE QUESTIONS

16.1 The femoral artery has the thicker wall; the femoral vein has the wider lumen.

16.2 Metabolically active tissues have more capillaries because they use oxygen and produce wastes more rapidly than inactive tissues.

16.3 Excess filtered fluid and proteins that escape from plasma drain into lymphatic capillaries and are returned by the lymphatic system to the cardiovascular system.

16.4 The skeletal muscle pump and the respiratory pump help boost venous return.

16.5 As blood pressure increases, blood flow increases.

16.6 Vasoconstriction increases vascular resistance, which decreases blood flow through the vasoconstricted blood vessels.

16.7 It happens when you stand up because gravity causes pooling of blood in leg veins as you stand upright, decreasing the blood pressure in your upper body.

16.8 The principal circulatory routes are the systemic and the pulmonary circulations.

16.9 The four parts of the aorta are the ascending aorta, arch of the aorta, thoracic aorta, and abdominal aorta.

16.10 Branches of the arch of the aorta are the brachiocephalic trunk, left common carotid artery, and left subclavian artery.

16.11 The abdominal aorta divides into the common iliac arteries at about the level of the fourth lumbar vertebra.

16.12 The superior vena cava drains regions above the diaphragm (except the cardiac veins and the alveoli of the lungs), and the inferior vena cava drains regions below the diaphragm.

16.13 All venous blood in the brain drains into the internal jugular veins.

16.14 The median cubital vein is often used for withdrawing blood.

16.15 Superficial veins of the lower limbs include the dorsal venous arch and the great saphenous and small saphenous veins.

16.16 The hepatic veins carry blood away from the liver.

16.17 The exchange of materials between mother and fetus occurs across the placenta.

chapter 17

THE LYMPHATIC SYSTEM AND IMMUNITY

did you know?

A good laugh not only feels great, it is good medicine as well. Researchers exploring the health benefits of humor suggest that a good sense of humor enhances your health, partly because it helps to protect the immune system. Laughter increases blood levels of important immune components, such as immunoglobulin A, which helps fight infections in the upper respiratory and gastrointestinal tracts. Some research has shown that laughter may also increase levels of disease-fighting T cells and natural killer cells. Laughter may exert its effects by decreasing feelings of stress, and short-circuiting the stress response and the immunosuppressive effects of stress.

Focus on Wellness, page 431

www.wiley.com/college/apcentral

Maintaining homeostasis in the body requires continual combat against harmful agents in our environment. Despite constant exposure to a variety of ***pathogens*** (PATH-ō-jens), disease-producing microbes such as bacteria and viruses, most people remain healthy. The body surface also endures cuts and bumps, exposure to ultraviolet rays in sunlight, chemical toxins, and minor burns with an array of defenses. In this chapter, we will explore the mechanisms that provide defenses against intruders and promote the repair of damaged body tissues.

Immunity or ***resistance*** is the ability to use our body's defenses to ward off damage or disease. The two types of immunity are (1) innate and (2) adaptive. ***Innate (nonspecific) immunity*** refers to defenses that are present at birth. They are always present and available to provide rapid responses to protect us against disease. Innate immunity does not involve specific recognition of a microbe and acts against all microbes in the same way. However, innate immunity does not have a memory component, that is, it cannot recall a previous contact with a foreign molecule. Among the components of innate immunity are the first line of defense (skin and mucous membranes) and the second line of defense (natural killer cells and phagocytes, inflammation, fever, and antimicrobial substances). Innate immune responses represent immunity's early-warning system and are designed to prevent microbes from gaining access into the body and to help eliminate those that do gain access.

Adaptive (specific) immunity refers to defenses that involve specific recognition of a microbe once it has breached the innate immunity defenses. Adaptive immunity is based on a specific response to a specific microbe; that is, it adapts or adjusts to handle a specific microbe. Unlike innate immunity, adaptive immunity is slower to respond but it does have a memory component. Adaptive immunity involves lymphocytes (a type of white blood cell) called T lymphocytes (T cells) and B lymphocytes (B cells).

looking back to move ahead . . .

- **Veins (page 389)**
- **Cancer (page 66)**
- **Epidermis of Skin (page 99)**
- **Mucous Membranes (page 90)**
- **Phagocytosis (page 51)**

LYMPHATIC SYSTEM

OBJECTIVES • **Describe the components and major functions of the lymphatic system.**

- **Describe the organization of lymphatic vessels and the circulation of lymph.**
- **Compare the structure and functions of the primary and secondary lymphatic organs and tissues.**

The body system responsible for adaptive immunity (and some aspects of innate immunity) is the ***lymphatic*** (lim-FAT-ik) ***system***, which consists of lymph, lymphatic vessels, a number of structures and organs containing lymphatic tissue, and red bone marrow (Figure 17.1). ***Lymphatic tissue*** is a specialized form of reticular connective tissue (see Table 4.2C on page 86) that contains large numbers of lymphocytes.

Most components of blood plasma filter out of blood capillary walls to form *interstitial fluid*, the fluid that surrounds the cells of body tissues. After interstitial fluid passes into lymphatic vessels, it is called ***lymph*** (LIMF = clear fluid). Both fluids are chemically similar to blood plasma. The main difference is that interstitial fluid and lymph contain less protein than blood plasma because most plasma protein molecules are too large to filter through the capillary wall. Each day, about 20 liters of fluid filter from blood into tissue spaces. This fluid must be returned to the cardiovascular system to maintain normal blood volume. About 17 liters of the fluid filtered daily from the arterial end of blood capillaries return to the blood directly by reabsorption at the venous end of the capillaries. The remaining 3 liters per day pass first into lymphatic vessels and are then returned to the blood.

The lymphatic system has three primary functions:

1. **Draining excess interstitial fluid.** Lymphatic vessels drain excess interstitial fluid and leaked proteins from tissue spaces and return them to the blood. This activity helps maintain fluid balance in the body and prevents depletion of vital plasma proteins.
2. **Transporting dietary lipids.** Lymphatic vessels transport the lipids and lipid-soluble vitamins (A, D, E, and K) absorbed by the gastrointestinal tract into the blood.
3. **Carrying out immune responses.** Lymphatic tissue initiates highly specific responses directed against particular microbes or abnormal cells.

Lymphatic Vessels and Lymph Circulation

Lymphatic vessels begin as ***lymphatic capillaries***. These tiny vessels are closed at one end and located in the spaces between cells (Figure 17.2 on page 423). Lymphatic capillaries are slightly larger than blood capillaries and have a unique structure that permits interstitial fluid to flow into them, but not out. The endothelial cells that make up the wall of a lymphatic capillary are not attached end to end, but rather, the ends overlap (Figure 17.2b). When pressure is greater in interstitial fluid than in lymph, the cells separate slightly, like a one-way swinging door, and interstitial fluid enters the lymphatic capillary. When pressure is greater inside the lymphatic capillary, the cells adhere more closely and lymph cannot escape back into interstitial fluid.

Unlike blood capillaries, which link two larger blood vessels that form part of a circuit, lymphatic capillaries begin in the tissues and carry the lymph that forms there toward a larger lymphatic vessel. Just as blood capillaries unite to form venules and veins, lymphatic capillaries unite to form larger and larger ***lymphatic vessels*** (see Figure 17.1). Lymphatic vessels resemble veins in structure but have thinner walls and more valves. Located at intervals along lymphatic vessels are ***lymph nodes***, masses of B cells and T cells that are surrounded by a capsule. Lymph flows through lymph nodes.

From the lymphatic vessels, lymph eventually passes into one of two main channels: the thoracic duct or the right lymphatic duct. The ***thoracic duct***, the main lymph-collecting duct, receives lymph from the left side of the head, neck, and chest; the left upper limb; and the entire body below the ribs. The ***right lymphatic duct*** drains lymph from the upper right side of the body (see Figure 17.1).

Ultimately, the thoracic duct empties its lymph into the junction of the left internal jugular and left subclavian veins, and the right lymphatic duct empties its lymph into the junction of the right internal jugular and right subclavian veins. Thus, lymph drains back into the blood (Figure 17.3 on page 423).

The same two pumps that aid return of venous blood to the heart maintain the flow of lymph:

1. **Skeletal muscle pump**. The "milking action" of skeletal muscle contractions (see Figure 16.4 on page 390) compresses lymphatic vessels (as well as veins) and forces lymph toward the subclavian veins.
2. **Respiratory pump**. Lymph flow is also maintained by pressure changes that occur during inhalation (breathing in). Lymph flows from the abdominal region, where the pressure is higher, toward the thoracic region, where it is lower. When the pressures reverse during exhalation (breathing out), the valves prevent backflow of lymph.

Edema (e-DĒ-ma) is an excessive accumulation of interstitial fluid in tissue spaces. It may be caused by a lymphatic system obstruction, such as an infected lymph node or a blocked lymphatic vessel. Edema may also result from increased capillary blood pressure, which causes excess interstitial fluid to form faster than it can pass into lymphatic vessels or be reabsorbed back into the capillaries. Another cause is lack of skeletal muscle contractions, as in individuals who are paralyzed.

Figure 17.1 Components of the lymphatic system.

The lymphatic system consists of lymph, lymphatic vessels, lymphatic tissues, and red bone marrow.

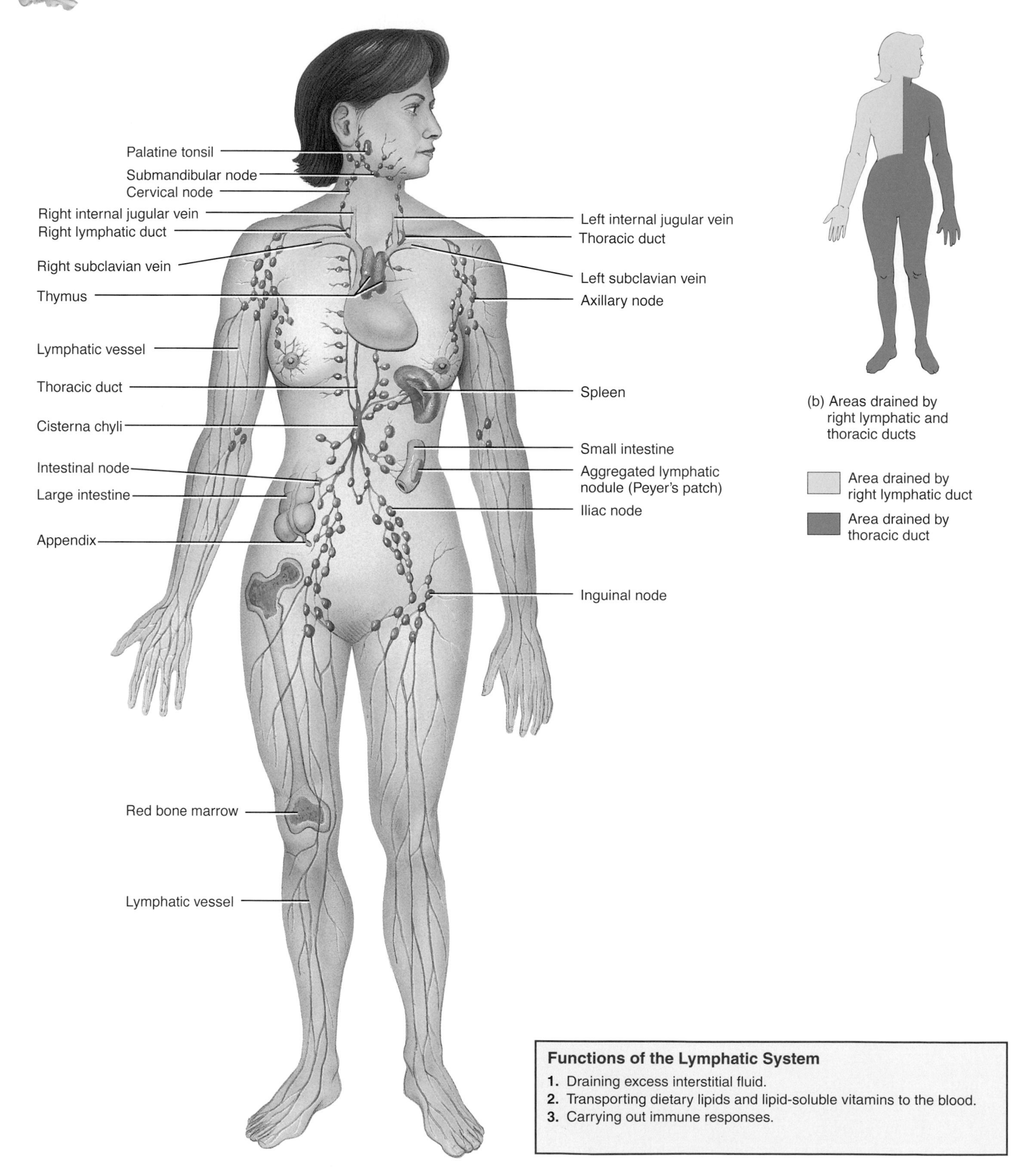

(b) Areas drained by right lymphatic and thoracic ducts

Functions of the Lymphatic System

1. Draining excess interstitial fluid.
2. Transporting dietary lipids and lipid-soluble vitamins to the blood.
3. Carrying out immune responses.

(a) Anterior view of principal components of lymphatic system

What is lymphatic tissue?

Figure 17.2 Lymphatic capillaries.

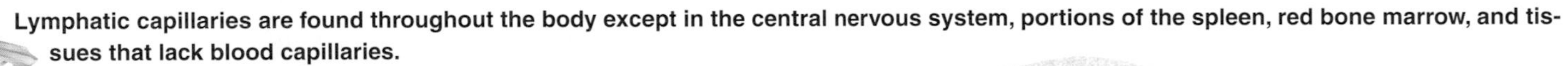

Lymphatic capillaries are found throughout the body except in the central nervous system, portions of the spleen, red bone marrow, and tissues that lack blood capillaries.

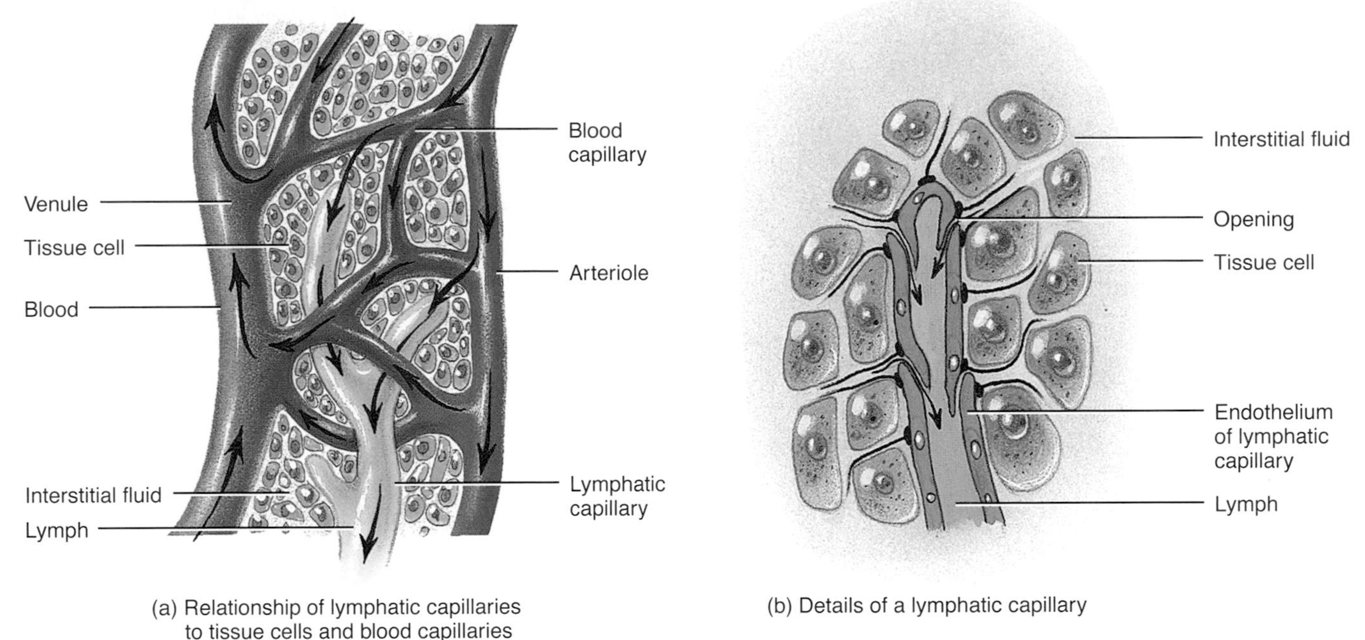

(a) Relationship of lymphatic capillaries to tissue cells and blood capillaries

(b) Details of a lymphatic capillary

Why is lymph more similar to interstitial fluid than it is to blood plasma?

Figure 17.3 Relationship of lymphatic vessels and lymph nodes to the cardiovascular system. Arrows show the direction of flow of lymph and blood.

The sequence of fluid flow is: blood capillaries (blood plasma) → interstitial spaces (interstitial fluid) → lymphatic capillaries (lymph) → lymphatic vessels and lymph nodes (lymph) → lymphatic ducts (lymph) → junction of jugular and subclavian veins (blood plasma).

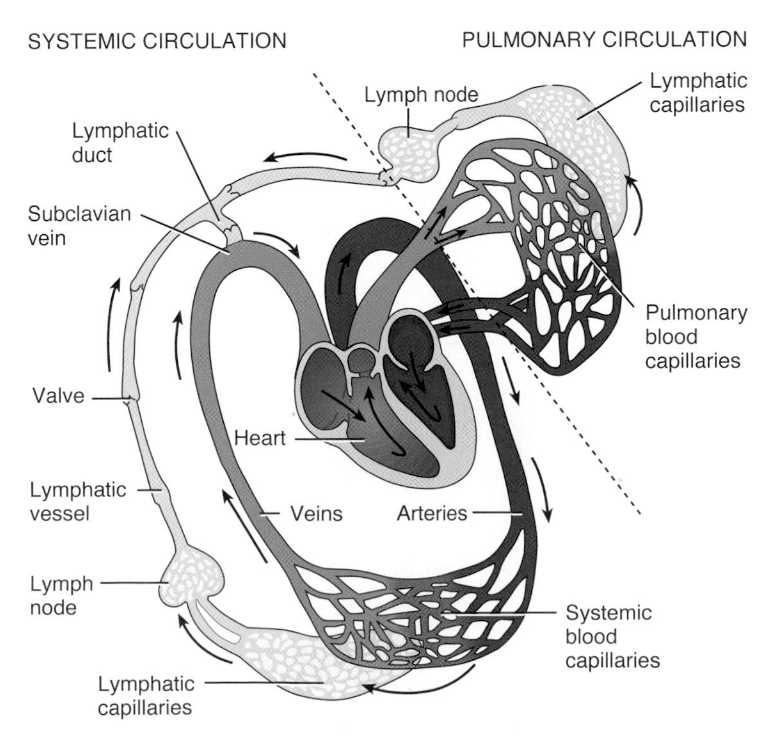

Which vessels of the cardiovascular system (arteries, veins, or capillaries) produce lymph?

Lymphatic Organs and Tissues

Lymphatic organs and tissues, which are widely distributed throughout the body, are classified into two groups based on their functions. ***Primary lymphatic organs*** and ***tissues***, the sites where stem cells divide and develop into mature B cells and T cells, include the ***red bone marrow*** (in flat bones and the ends of the long bones of adults) and the ***thymus***. The ***secondary lymphatic organs*** and ***tissues***, the sites where most immune responses occur, include ***lymph nodes***, the ***spleen***, and ***lymphatic nodules***.

Thymus

The ***thymus*** is a two-lobed organ located posterior to the sternum and medial to the lungs and superior to the heart (see Figure 17.1). It contains large numbers of T cells and scattered dendritic cells (so named for their long, branchlike projections), epithelial cells, and macrophages. Immature T cells migrate from red bone marrow to the thymus, where they multiply and begin to mature. Only about 2% of the immature T cells that arrive in the thymus achieve the proper "education" to "graduate" into mature T cells. The remaining cells die via apoptosis (programmed cell death). Thymic macrophages help clear out the debris of dead and dying cells. Mature T cells leave the thymus via the blood and are carried to lymph nodes, the spleen, and other lymphatic tissues where they populate parts of these organs and tissues.

Lymph Nodes

Located along lymphatic vessels are about 600 bean-shaped ***lymph nodes***. They are scattered throughout the body, both superficially and deep, and usually occur in groups (see Figure 17.1). Lymph nodes are heavily concentrated near the mammary glands and in the axillae and groin. Each node is covered by a capsule of dense connective tissue (Figure 17.4). Internally, different regions of a lymph node may contain B cells that develop into plasma cells, as well as T cells, dendritic cells, and macrophages.

Figure 17.4 Structure of a lymph node (partially sectioned). Green arrows indicate direction of lymph flow into and out of the lymph node.

Lymph nodes are present throughout the body, usually clustered in groups.

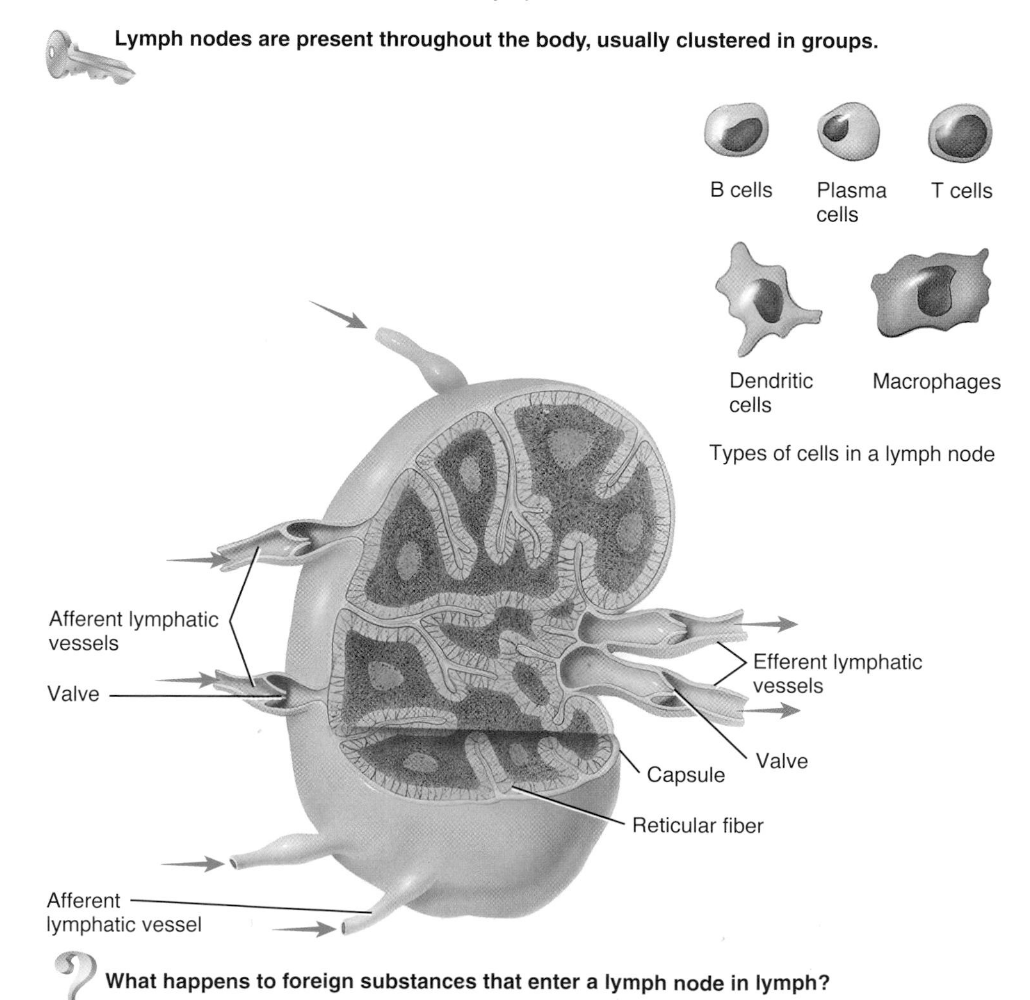

What happens to foreign substances that enter a lymph node in lymph?

Lymph nodes filter lymph, which enters a node through one of several *afferent lymphatic vessels* (*af-* = toward; *-ferrent* = to carry). As lymph flows through the node, foreign substances are trapped by *reticular fibers* within the spaces between cells. Macrophages destroy some foreign substances by phagocytosis, and lymphocytes destroy others by a variety of immune responses. Filtered lymph leaves the other end of the node through one or two *efferent lymphatic vessels* (*ef-* = away). Plasma cells and T cells that have divided many times within a lymph node can also leave the node and circulate to other parts of the body. Valves direct the flow of lymph inward through the afferent lymphatic vessels and outward through the efferent lymphatic vessels (Figure 17.4).

Metastasis (me-TAS-ta-sis; *meta-* = beyond; *stasis* = to stand), the spread of a disease from one part of the body to another, can occur via lymphatic vessels. All malignant tumors eventually metastasize. Cancer cells may travel in the blood or lymph and establish new tumors where they lodge. When metastasis occurs via lymphatic vessels, secondary tumor sites can be predicted according to the direction of lymph flow from the primary tumor site. Cancerous lymph nodes feel enlarged, firm, nontender, and fixed to underlying structures. By contrast, most lymph nodes that are enlarged due to an infection are softer, tender, and movable.

Spleen

The ***spleen*** is the largest single mass of lymphatic tissue in the body (see Figure 17.1). It lies between the stomach and diaphragm and is covered by a capsule of dense connective tissue. The spleen contains two types of tissue called white pulp and red pulp. *White pulp* is lymphatic tissue, consisting mostly of lymphocytes and macrophages. *Red pulp* consists of blood-filled *venous sinuses* and cords of *splenic tissue* consisting of red blood cells, macrophages, lymphocytes, plasma cells, and granular leukocytes.

Blood flowing into the spleen through the splenic artery enters the white pulp. Within the white pulp, B cells and T cells carry out immune responses, while macrophages destroy pathogens by phagocytosis. Within the red pulp, the spleen performs three functions related to blood cells: (1) removal by macrophages of worn out or defective blood cells and platelets; (2) storage of platelets, perhaps up to one-third of the body's supply; and (3) production of blood cells (hemopoiesis) during fetal life.

The spleen is the organ most often damaged in cases of abdominal trauma. A ruptured spleen causes severe internal hemorrhage and shock. Prompt **splenectomy**, removal of the spleen, is needed to prevent bleeding to death. After a splenectomy, other structures, particularly red bone marrow and the liver, can take over functions normally carried out by the spleen.

Lymphatic Nodules

Lymphatic nodules are egg-shaped masses of lymphatic tissue that are not surrounded by a capsule. They are plentiful in the connective tissue of mucous membranes lining the gastrointestinal, urinary, and reproductive tracts and the respiratory airways. Although many lymphatic nodules are small and solitary, some occur as large aggregations in specific parts of the body. Among these are the tonsils in the pharyngeal region and the aggregated lymphatic follicles (Peyer's patches) in the ileum of the small intestine (see Figure 17.1). Aggregations of lymphatic nodules also occur in the appendix. The five ***tonsils***, which form a ring at the junction of the oral cavity, nasal cavity, and throat, are strategically positioned to participate in immune responses against inhaled or ingested foreign substances. The single ***pharyngeal tonsil*** (fa-RIN-jē-al) or ***adenoid*** is embedded in the posterior wall of the upper part of the throat (see Figure 18.2 on page 447). The two ***palatine tonsils*** (PAL-a-tīn) lie at the back of the mouth, one on either side; these are the tonsils commonly removed in a tonsillectomy. The paired ***lingual tonsils*** (LIN-gwal), located at the base of the tongue, may also require removal during a tonsillectomy.

■ **CHECKPOINT**

1. How are interstitial fluid and lymph similar, and how do they differ?
2. What are the roles of the thymus and the lymph nodes in immunity?
3. Describe the functions of the spleen and tonsils.

INNATE IMMUNITY

OBJECTIVE • **Describe the various components of innate immunity.**

Innate immunity includes barriers provided by the skin and mucous membranes. It also includes various internal defenses, such as internal antimicrobial proteins, natural killer cells, phagocytes, inflammation, and fever.

First Line of Defense: Skin and Mucous Membranes

Both ***physical barriers*** and ***chemical barriers*** to pathogens and foreign substances are found in the skin that covers the body and in mucous membranes that line body openings such as the mouth and breathing airways. With its many layers of closely packed, keratinized cells, the *epidermis* (the outer epithelial layer of the skin) provides a formidable physical barrier to the entrance of microbes (see Figure 5.1 on page 98). In addition, continual shedding of the top epidermal cells helps remove microbes at the skin's surface. Bacteria rarely penetrate an intact and healthy epidermis.

The epithelial layer of *mucous membranes* secretes a fluid called *mucus* that lubricates and moistens the surface of a body cavity. Because mucus is sticky, it traps many microbes and foreign substances. The mucous membrane of the nose has mucus-coated *hairs* that trap and filter microbes, dust, and pollutants from inhaled air. The mucous membrane of the upper airways contains *cilia*, microscopic hairlike projections on the surface of the epithelial cells, which propel inhaled dust and microbes that have become trapped in mucus toward the throat.

Other fluids produced by various organs also help protect epithelial surfaces of the skin and mucous membranes. The *lacrimal apparatus* (LAK-ri-mal) of the eyes (see Figure 12.5 on page 294) produces and drains away tears in response to irritants, diluting microbes and keeping them from settling on the surface of the eyes. *Saliva*, produced by the salivary glands, washes microbes from the surfaces of the teeth and from the mucous membrane of the mouth, much like tears wash the eyes. The cleansing of the urethra by the *flow of urine* retards microbial colonization of the urinary system. *Vaginal secretions*, likewise, move microbes out of the body in females. *Defecation* and *vomiting* also expel microbes.

Certain chemicals also contribute to the resistance of the skin and mucous membranes to microbial invasion. Sebaceous (oil) glands of the skin secrete an oily substance called *sebum* that forms a protective film over the surface of the skin. *Perspiration* helps flush microbes from the surface of the skin and contains *lysozyme*, an enzyme capable of breaking down the cell walls of certain bacteria. (Lysozyme is also found in tears, nasal secretions, and tissue fluids). *Gastric juice*, a mixture of hydrochloric acid, enzymes, and mucus in the stomach, destroys many bacteria and most bacterial toxins. *Vaginal secretions* also are slightly acidic, which discourages bacterial growth.

Second Line of Defense: Internal Defenses

Although the skin and mucous membranes are very effective barriers in preventing invasion by pathogens, they may be broken by injuries or everyday activities such as brushing the teeth or shaving. Any pathogens that get past the surface barriers encounter a second line of defense consisting of internal antimicrobial proteins, phagocytes, natural killer cells, inflammation, and fever.

Internal Antimicrobial Proteins

Various body fluids contain four main types of ***antimicrobial proteins*** that discourage microbial growth:

1. Lymphocytes, macrophages, and fibroblasts infected with viruses produce proteins called ***interferons*** (in′-ter-FĒR-ons), or ***IFNs***. After their release by virus-infected cells, IFNs diffuse to uninfected neighboring cells, where they stimulate synthesis of proteins that interfere with viral replication. Viruses can cause disease only if they can replicate within body cells.
2. A group of normally inactive proteins in blood plasma and on plasma membranes makes up the ***complement system***. When activated, these proteins "complement" or enhance certain immune, allergic, and inflammatory reactions. One effect of complement proteins is to create holes in the plasma membrane of the microbe. As a result, extracellular fluid moves into the holes, causing the microbe to burst, a process called *cytolysis*. Another effect of complement is to cause *chemotaxis* (kē′-mō-TAK-sis), the chemical attraction of phagocytes to a site. Some complement proteins cause *opsonization* (op′-son-i-ZĀ-shun), a process in which complement proteins bind to the surface of a microbe and promote phagocytosis.
3. ***Transferrins*** are proteins that bind to iron in blood, milk, saliva, and tears. They inhibit microbial growth by reducing the amount of available iron, a substance needed for bacterial metabolism.
4. ***Antimicrobial peptides*** are newly discovered short-chain proteins that are produced by phagocytes and mucous-membrane epithelial cells. They cause lysis of microbes.

Phagocytes and Natural Killer Cells

When microbes penetrate the skin and mucous membranes or bypass the antimicrobial proteins in blood, the next nonspecific defense consists of phagocytes and natural killer cells.

Phagocytes (*phago-* = eat; *-cytes* = cells) are specialized cells that perform ***phagocytosis*** (*-osis* = process), the ingestion of microbes or other particles such as cellular debris. The two main types of phagocytes are neutrophils and macrophages. When an infection occurs, neutrophils and monocytes migrate to the infected area. During this migration, the monocytes enlarge and develop into actively phagocytic cells called ***macrophages*** (MAK-rō-fā-jez) (see Figure 14.2a on page 348). Some are *wandering macrophages*, which migrate to infected areas. Others are *fixed macrophages*, which remain in certain locations, including the skin and subcutaneous layer, liver, lungs, brain, spleen, lymph nodes, and red bone marrow.

About 5–10% of lymphocytes in the blood are ***natural killer (NK) cells***, which have the ability to kill a wide variety of microbes and certain tumor cells. NK cells also are present in the spleen, lymph nodes, and red bone marrow. Some cancer and AIDS patients have defective or decreased numbers of NK cells. They cause cellular destruction by releasing proteins that destroy the target cell's membrane.

Inflammation

Inflammation is a defensive response of the body to tissue damage. Because inflammation is one of the body's innate defenses, the response of a tissue to a cut is similar to the response to damage caused by burns, radiation, or invasion of bacteria or viruses. The events of inflammation dispose of microbes, toxins, or foreign material at the site of injury, prevent their spread to other tissues, and prepare the site for tissue repair. Thus, inflammation helps restore tissue homeostasis. The four signs and symptoms of inflammation are redness, pain, heat, and swelling. Inflammation can also cause the loss of function in the injured area, depending on the site and extent of the injury.

The stages of inflammation are as follows:

1. In a region of tissue injury, mast cells in connective tissue and basophils and platelets in blood release *histamine*. In response to histamine, two immediate changes occur in the blood vessels: *increased permeability* and *vasodilation*, an increase in the diameter of the blood vessels (Figure 17.5). Increased permeability means that substances normally retained in blood are permitted to pass out of the blood vessels. Vasodilation is an increase in the diameter of the blood vessels and allows more blood to flow to the damaged area and helps remove microbial toxins and dead cells. Increased permeability permits defensive substances such as antibodies and clot-forming chemicals to enter the injured area from the blood.

   From the events that occur during inflammation, it's easy to understand the signs and symptoms. Heat and redness result from the large amount of blood that accumulates in the damaged area. The area swells due to an increased amount of interstitial fluid that has leaked out of the capillaries (edema). Pain results from injury to neurons, from toxic chemicals released by microbes, and from the increased pressure of edema.
2. The increased permeability of capillaries causes leakage of clotting proteins into tissues. Fibrinogen is converted to an insoluble, thick network of fibrin threads, which traps the invading organisms and prevents their spread. The resulting clot isolates the invading microbes and their toxins.
3. Shortly after the inflammatory process starts, phagocytes are attracted to the site of injury by chemotaxis (Figure 17.5). Near the damaged area, neutrophils begin to squeeze through the wall of the blood vessel, a process called *emigration*. Neutrophils predominate in the early stages of infection, but they die off rapidly together with the microbes they have eaten. Within a few hours, monocytes arrive in the infected area. Once in the tissue, they turn into wandering macrophages that engulf damaged tissue, worn-out neutrophils, and invading microbes.
4. Eventually, macrophages also die. Within a few days, a pocket of dead phagocytes and damaged tissue forms; this collection of dead cells and fluid is called ***pus***. At times, pus reaches the surface of the body or drains into an internal cavity and is dispersed; on other occasions the pus remains even after the infection is terminated. In this case, the pus is gradually destroyed over a period of days and is absorbed.

Figure 17.5 Inflammation. Several substances stimulate vasodilation, increased permeability of blood vessels, chemotaxis, emigration, and phagocytosis. Phagocytes migrate from blood to the site of tissue injury.

Inflammation is an innate immune response of the body to tissue damage.

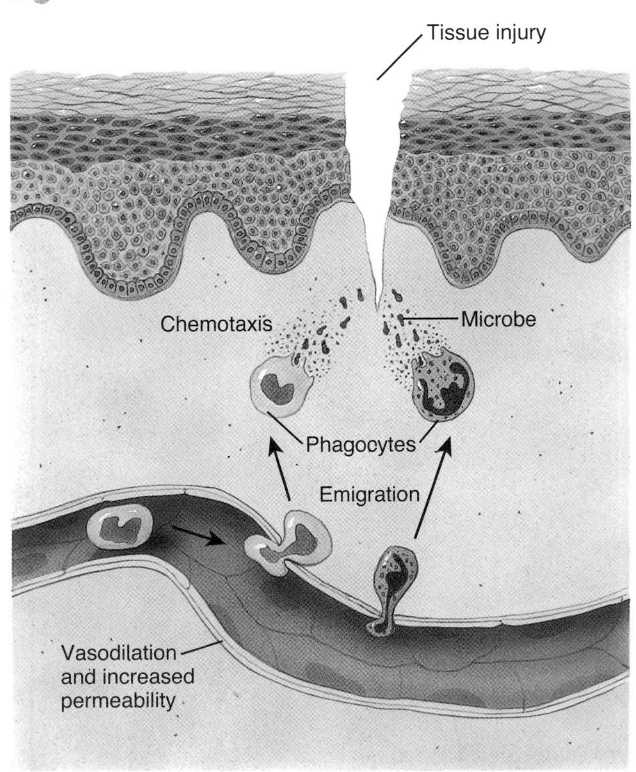

Phagocytes migrate from blood to site of tissue injury

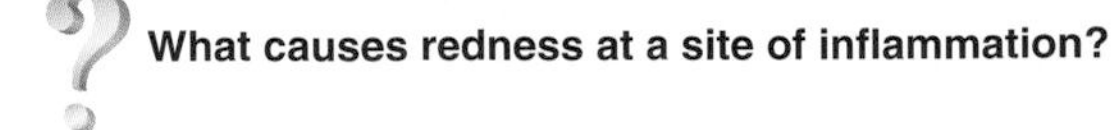
What causes redness at a site of inflammation?

If pus cannot drain out of an inflamed region, the result is an **abscess**—an excessive accumulation of pus in a confined space. Common examples are pimples and boils. When superficial inflamed tissue sloughs off the surface of an organ or tissue, the resulting open sore is called an **ulcer**. People with poor circulation—for instance, diabetics with advanced atherosclerosis—are particularly susceptible to ulcers in the tissues of their legs.

Fever

Fever is an abnormally high body temperature that occurs because the hypothalamic thermostat is reset. It commonly occurs during infection and inflammation. Many bacterial toxins elevate body temperature, sometimes by triggering release of fever-causing substances such as interleukin-1 from macrophages. Elevated body temperature intensifies the effects of interferons, inhibits the growth of some microbes, and speeds up body reactions that aid repair.

Table 17.1 summarizes the components of innate defenses.

■ **CHECKPOINT**

4. What physical and chemical factors provide protection from disease in the skin and mucous membranes?
5. What internal defenses provide protection against microbes that penetrate the skin and mucous membranes?
6. What are the main signs and symptoms of inflammation?

ADAPTIVE IMMUNITY

OBJECTIVES • Define adative immunity and compare it with innate immunity.

- **Explain the relationship between an antigen and an antibody.**
- **Compare the functions of cell-mediated immunity and antibody-mediated immunity.**

The various aspects of innate immunity have one thing in common: They are not specifically directed against a particular type of invader. Adaptive (specific) immunity involves the production of specific types of cells or specific antibodies to destroy a particular antigen. An ***antigen*** is any substance—such as microbes, foods, drugs, pollen, or tissue—that the immune system recognizes as foreign (nonself). The branch of science that deals with the responses of the body to antigens is called ***immunology*** (im′-ū-NOL-ō-jē). The ***immune system*** includes the cells and tissues that carry out immune responses. Normally, a person's adaptive immune system cells recognize and do not attack their own tissues and chemicals. Such lack of reaction against self-tissues is called ***self-tolerance***.

At times, self-tolerance breaks down, which leads to an **autoimmune disease**. Sometimes tissues undergo changes that cause the adaptive immune system to recognize them as foreign antigens and attack them. Among human autoimmune diseases are systemic lupus erythematosus (SLE), Addison's disease, Graves disease, type 1 diabetes mellitus, myasthenia gravis, multiple sclerosis (MS), and ulcerative colitis.

Table 17.1 Summary of Innate Defenses

Component	Functions
First Line of Defense: Skin and Mucous Membranes	
Physical Factors	
Epidermis of skin	Forms a physical barrier to the entrance of microbes.
Mucous membranes	Inhibit the entrance of many microbes, but not as effective as intact skin.
Mucus	Traps microbes in respiratory and gastrointestinal tracts.
Hairs	Filter out microbes and dust in nose.
Cilia	Together with mucus, trap and remove microbes and dust from upper respiratory tract.
Lacrimal apparatus	Tears dilute and wash away irritating substances and microbes.
Saliva	Washes microbes from surfaces of teeth and mucous membranes of mouth.
Flow of urine	Washes microbes from urethra.
Defecation and vomiting	Expel microbes from body.
Chemical Factors	
Sebum	Forms a protective acidic film over the skin surface that inhibits growth of many microbes.
Perspiration	Flushes microbes from skin surface.
Lysozyme	Antimicrobial substance in perspiration, tears, saliva, nasal secretions, and tissue fluids.
Gastric juice	Destroys most bacteria and toxins in stomach.
Vaginal secretions	Slight acidity discourages bacterial growth; flush microbes out of vagina.
Second Line of Defense: Internal Defenses	
Internal Antimicrobial Proteins	
Interferons (IFNs)	Protect uninfected host cells from viral infection.
Complement system	Causes cytolysis of microbes, promotes phagocytosis, and contributes to inflammation.
Transferrins	Inhibit growth of certain bacteria by reducing the amount of available iron.
Antimicrobial peptides	Cause lysis of microbes.
Natural killer (NK) cells	Kill infected target cells by releasing proteins that destroy target cell's membranes.
Phagocytes	Ingest microbes and other particles.
Inflammation	Confines and destroys microbes and initiates tissue repair.
Fever	Intensifies the effects of interferons, inhibits growth of some microbes, and speeds up body reactions that aid repair.

Maturation of T Cells and B Cells

The cells that carry out adaptive immune responses are lymphocytes called B cells and T cells. Both develop from stem cells that originate in red bone marrow (see Figure 14.2 on page 348). B cells complete their development in red bone marrow; immature T cells migrate from red bone marrow to the thymus, where they mature. Before T cells leave the thymus or B cells leave red bone marrow, they begin to make several distinctive proteins that are inserted into their plasma membranes. Some of these proteins function as ***antigen receptors***—molecules capable of recognizing and binding to specific antigens.

Types of Adaptive Immune Responses

Adaptive immunity consists of two types of closely allied responses, both triggered by antigens. In ***cell-mediated immune responses***, some T cells are like an army of soldiers that directly attack the invading antigen. In ***antibody-mediated immune responses***, B cells change into plasma cells, which synthesize and secrete antibodies. A given antibody can bind to and inactivate a specific antigen. Other T cells aid both cell-mediated and antibody-mediated adaptive immune responses. Although each type of response is specialized to combat different aspects of an invasion, a given pathogen can provoke both types of adaptive immune responses.

Antigens and Antibodies

An antigen (meaning *anti*body *gen*erator) causes the body to produce specific antibodies and/or specific T cells that react with it. Entire microbes or parts of microbes may act as antigens. Chemical components of bacterial structures such as flagella, capsules, and cell walls are antigenic, as are bacterial toxins and viral proteins. Other examples of antigens include chemical components of pollen, egg white, incompatible blood cells, and transplanted tissues and organs. The huge variety of antigens in the environment provides myriad opportunities for provoking immune responses.

Located at the plasma membrane surface of most body cells are "self-antigens" known as *major histocompatibility complex (MHC)* proteins. Unless you have an identical twin, your MHC proteins are unique. Thousands to several hundred thousand MHC molecules mark the surface of each of your body cells except red blood cells. MHC proteins are the reason that tissues may be rejected when they are transplanted from one person to another, but their normal function is to help T cells recognize that an antigen is foreign, not self. This recognition is an important first step in any adaptive immune response.

The success of an organ or tissue transplant depends on **histocompatibility** (his′-tō-kom-pat-i-BIL-i-tē), the tissue compatibility between the donor and the recipient. The more similar the MHC antigens, the greater the histocompatibility, and thus the greater the chance that the transplant will not be rejected. In the United States, a nationwide computerized registry helps physicians select the most histocompatible and neediest organ transplant recipients whenever donor organs become available.

Antigens induce plasma cells to secrete proteins known as ***antibodies***. Most antibodies contain four polypeptide chains (Figure 17.6a). At two tips of the chains are ***variable regions***,

Figure 17.6 Structure of an antibody and relationship of an antigen to an antibody.

An antigen stimulates plasma cells to secrete specific antibodies that combine with the antigen.

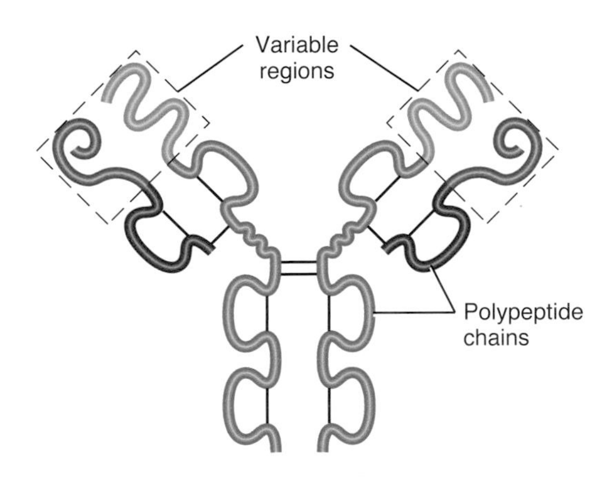

(a) Diagram of an antibody molecule

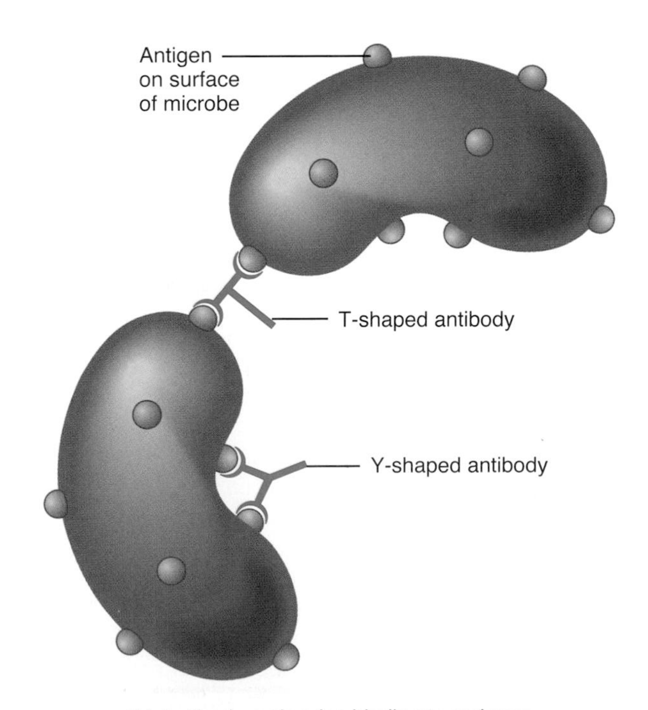

(b) Antibody molecules binding to antigens

What is the function of the variable regions of an antibody?

so named because the sequence of amino acids there varies for each different antibody. The variable regions are the ***antigen-binding sites***, the parts of an antibody that "fit" and bind to a particular antigen, much like your house key fits into its lock. Because the antibody "arms" can move somewhat, an antibody can assume either a T shape or a Y shape. This flexibility enhances the ability of the antibody to bind to two identical antigens at the same time—for example, on the surface of microbes (Figure 17.6b).

Antibodies belong to a group of plasma proteins called globulins, and for this reason they are also known as ***immunoglobulins*** (im′-ū-nō-GLOB-ū-lins). Immunoglobulins are grouped in five different classes, designated IgG, IgA, IgM, IgD, and IgE. Each class has a distinct chemical structure and different functions (Table 17.2). Because they appear first and are relatively short-lived, IgM antibodies indicate a recent invasion. In a sick patient, a high level of IgM against a particular pathogen helps identify the cause of the illness. Resistance of the fetus and newborn to infection stems mainly from maternal IgG antibodies that cross the placenta before birth and IgA antibodies in breast milk after birth.

Processing and Presenting Antigens

For an adaptive immune response to occur, B cells and T cells must recognize that a foreign antigen is present. B cells can recognize and bind to antigens in lymph, interstitial fluid, or blood plasma, but T cells only recognize fragments of antigens that are processed and presented in a certain way.

Table 17.2 Classes of Immunoglobulins

Name and Structure	Characteristics and Functions
IgG	About 80% of all antibodies in the blood. Also found in lymph and the intestines. Protects against bacteria and viruses by enhancing phagocytosis, neutralizing toxins, and triggering the complement system. Only class of antibody to cross the placenta from mother to fetus, thereby conferring considerable immune protection in newborns.
IgA	About 10% to 15% of all antibodies in the blood. Found mainly in sweat, tears, saliva, mucus, breast milk, and gastrointestinal secretions. Levels decrease during stress, lowering resistance to infection. Provides localized protection against bacteria and viruses on mucous membranes.
IgM	About 5% to 10% of all antibodies in the blood. Also found in lymph. First antibody class to be secreted by plasma cells after an initial exposure to any antigen. Activates complement and causes agglutination and lysis of microbes. In blood plasma, the anti-A and anti-B antibodies of the ABO blood group, which bind to A and B antigens during incompatible blood transfusions, are also IgM antibodies (see Figure 14.6 on page 358).
IgD	About 0.2% of all antibodies in the blood. Also found in lymph and on the surfaces of B cells as antigen receptors. Involved in activation of B cells.
IgE	Less than 0.1% of all antibodies in the blood. Also located on mast cells and basophils. Involved in allergic and hypersensitivity reactions and provides protection against parasitic worms.

Focus on Wellness

Lifestyle, Immune Function, and Resistance to Disease

If you want to observe the relationship between lifestyle and immune function, visit a college campus. As the semester progresses and the workload accumulates, an increasing number of students can be found in the waiting rooms of student health services.

Is Stress the Culprit?

Stress has been implicated as hazardous to immune function. Researchers in the field of *psychoneuroimmunology (PNI)* have found many communication pathways that link the nervous, endocrine, and immune systems. Chronic stress affects the immune system in several ways. For example, cortisol, a hormone secreted by the adrenal cortex in association with the stress response, inhibits immune system activity, perhaps one of its energy conservation effects. PNI research supports what many people have observed since the beginning of time: Your thoughts, feelings, moods, and beliefs influence your level of health and the course of disease. Especially toxic to the immune system are feelings of helplessness, hopelessness, fear, and social isolation.

People resistant to the negative health effects of stress are more likely to experience a sense of control over the future, a commitment to their work, expectations of generally positive outcomes for themselves, and feelings of social support. To increase your stress resistance, cultivate an optimistic outlook, get involved in your work, and build good relationships with others.

Or Is Lifestyle at Fault?

When work and stress pile up, health habits can change. Many people smoke or consume more alcohol when stressed, two habits detrimental to optimal immune function. Under stress, people are less likely to eat well or exercise regularly, two habits that enhance immunity.

Adequate sleep and relaxation are especially important for a healthy immune system. But when there aren't enough hours in the day, you may be tempted to steal some from the night. While skipping sleep may give you a few more hours of productive time in the short run, in the long run you end up even farther behind, especially if getting sick keeps you out of commission for several days, blurs your concentration, and blocks your creativity.

Even if you make time to get eight hours of sleep, stress can cause insomnia. If you find yourself tossing and turning at night, it's time to improve your stress management and relaxation skills! Be sure to unwind from the day before going to bed.

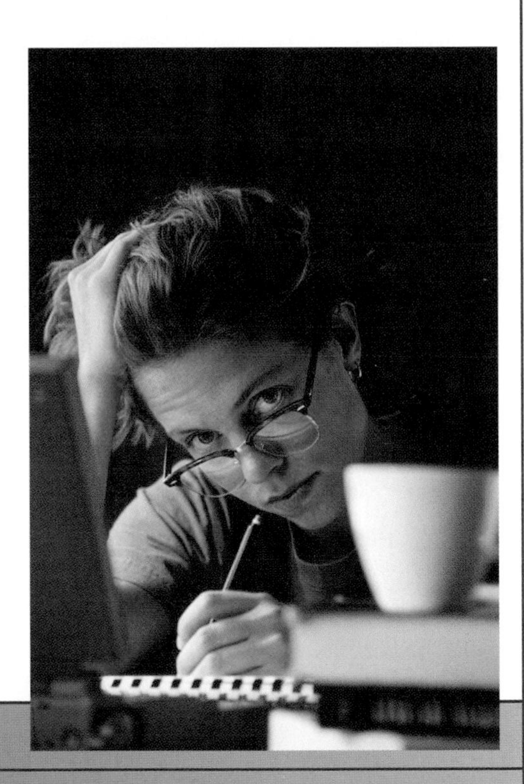

► Think It Over . . .

► *Have you ever observed a connection between stress and illness in your own life?*

In ***antigen processing***, antigenic proteins are broken down into fragments and then combine with MHC molecules. Next the antigen-MHC complex is inserted into the plasma membrane of a body cell. The insertion of the complex into the plasma membrane is called ***antigen presentation***. When an antigenic fragment comes from a *self-protein*, T cells ignore the antigen-MHC complex. However, if the fragment comes from a *foreign protein*, T cells recognize the antigen-MHC as an intruder, and an adaptive immune response takes place.

A special class of cells called ***antigen-presenting cells (APCs)*** process and present antigens. APCs include dendritic cells, macrophages, and B cells. They are strategically located in places where antigens are likely to penetrate innate defenses and enter the body, such as the epidermis and dermis of the skin (Langerhans cells are a type of dendritic cell); mucous membranes that line the respiratory, gastrointestinal, urinary, and reproductive tracts; and lymph nodes. After processing and presenting an antigen, APCs migrate from tissues via lymphatic vessels to lymph nodes.

The steps in the processing and presenting of an antigen by an APC occur as follows (Figure 17.7):

1. ***Ingestion of the antigen.*** Antigen-presenting cells ingest antigens by phagocytosis. Ingestion could occur almost anywhere in the body that invaders, such as microbes, have penetrated the nonspecific defenses.
2. ***Digestion of antigen into fragments.*** Within the APC, protein-digesting enzymes split large antigens into short peptide fragments.
3. ***Synthesis of MHC molecules.*** At the same time, the APC synthesizes MHC molecules and packages them into vesicles.
4. ***Fusion of vesicles.*** The vesicles containing antigen fragments and MHC molecules merge and fuse.
5. ***Binding of fragments to MHC molecules.*** After fusion of the two vesicles, antigen fragments bind to MHC molecules.
6. ***Insertion of antigen-MHC complex into the plasma membrane.*** The combined vesicle that contains antigen-MHC complexes splits open and the antigen-MHC complexes are inserted into the plasma membrane.

After processing an antigen, the APC migrates to lymphatic tissue to present the antigen to T cells. Within lymphatic tissue, a small number of T cells that have the correct antigen receptors recognize and bind to the antigen fragment—MHC complex, triggering either a cell-mediated or an antibody-mediated immune response.

T Cells and Cell-Mediated Immunity

The presentation of an antigen together with MHC molecules by APCs informs T cells that intruders are present in the body and that combative action should begin. But a T cell becomes activated only if its antigen receptor binds to the foreign antigen (antigen recognition) and at the same time it receives a second stimulating signal, a process known as ***costimulation*** (Figure 17.8). A common costimulator is ***interleukin-2 (IL-2)***. The need for two signals is a little like starting and driving a car. When you insert the correct key (antigen) in the ignition (T cell receptor) and turn it, the car starts (recognition of specific antigen), but it cannot move forward until you move the gear shift into drive (costimulation). The need for costimulation probably helps prevent immune responses from occurring accidentally.

Figure 17.7 Processing and presenting of antigen by an antigen-presenting cell (APC).

An APC migrates to a lymphatic tissue where it "presents" a processed antigen to T cells having receptors that fit that particular antigen fragment.

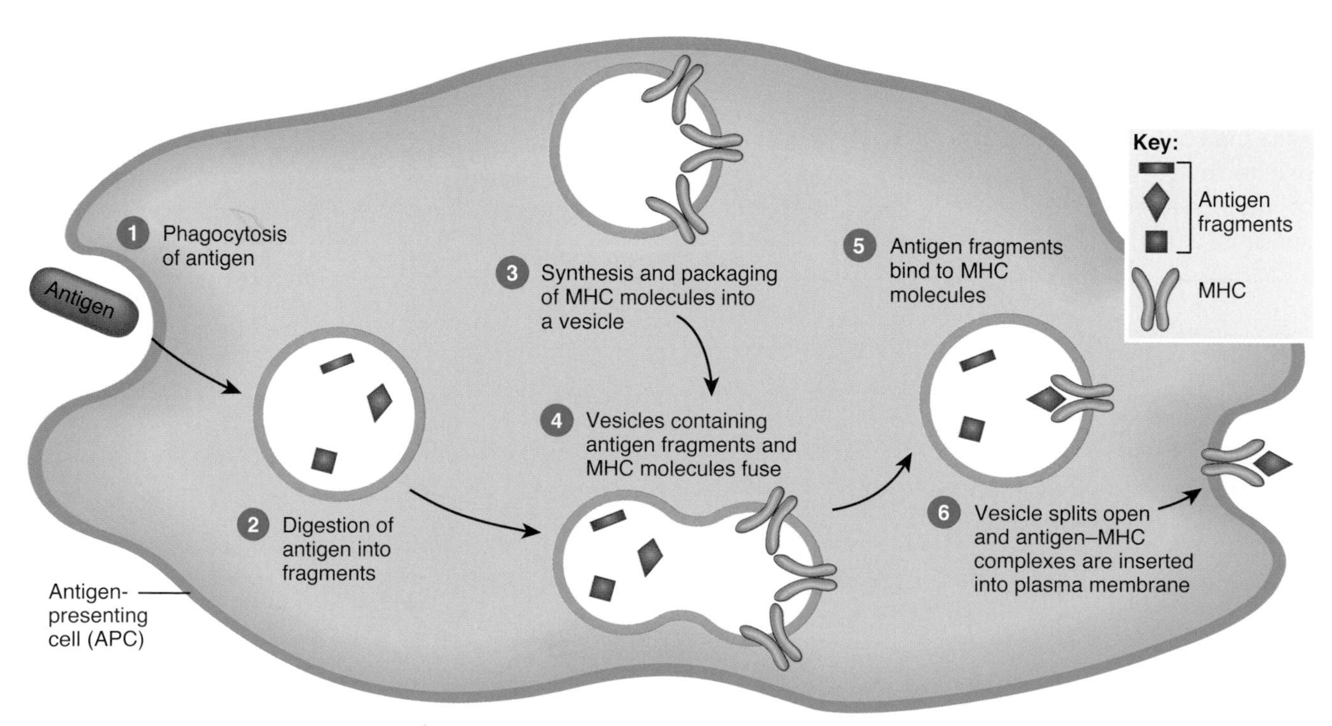

APCs present exogenous antigens in association with MHC-II molecules

Which types of cells can function as APCs?

Figure 17.8 Activation and division of T cells in cell-mediated immunity.

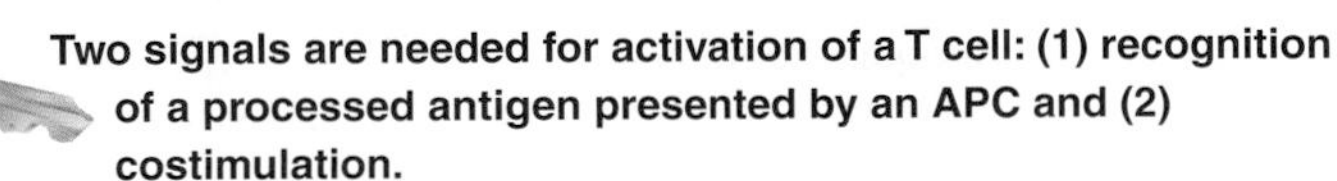

Two signals are needed for activation of a T cell: (1) recognition of a processed antigen presented by an APC and (2) costimulation.

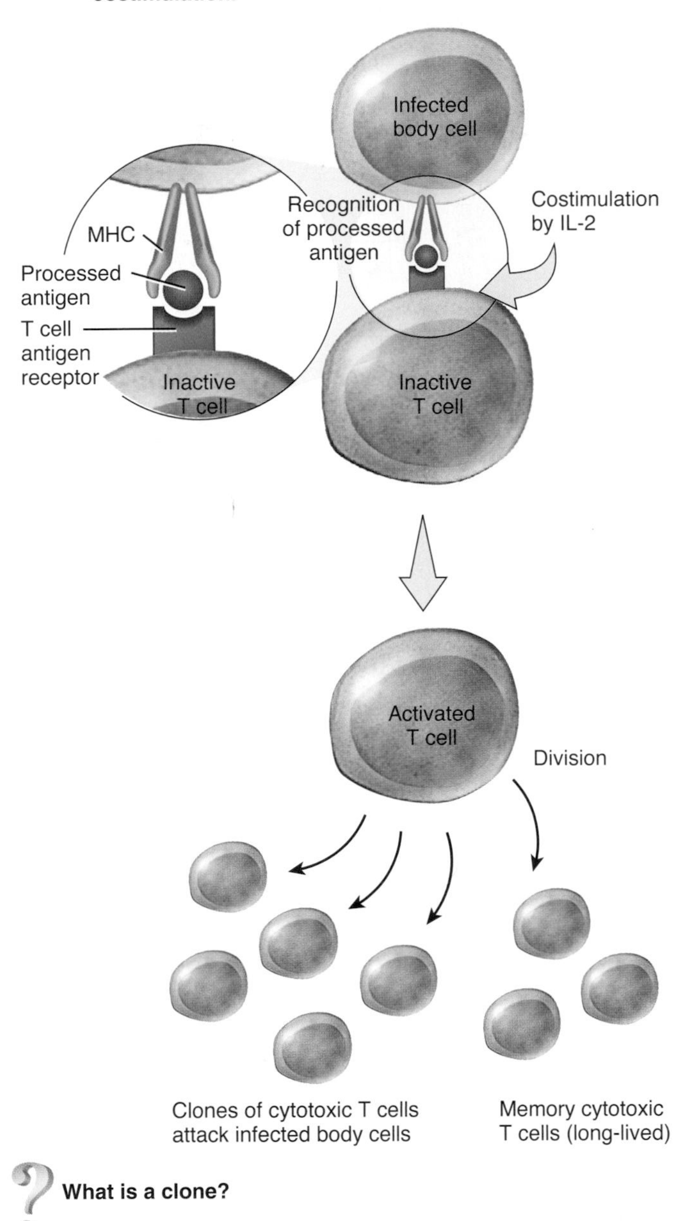

What is a clone?

Once a T cell is activated, it enlarges and divides many times. The resulting population of identical cells, termed a ***clone***, all recognize the same antigen. Before the first exposure to a given antigen, only a few T cells are able to recognize it. Once an adaptive immune response has begun, thousands of T cells can respond. Activation and division of T cells occur in the secondary lymphatic organs and tissues. If you have ever noticed swollen tonsils or lymph nodes in your neck, the continuous division of lymphocytes participating in an adaptive immune response was likely the cause.

The three major types of T cells are helper T cells, cytotoxic T cells, and memory T cells. ***Helper T cells*** help other cells of the adaptive immune system combat intruders. For instance, helper T cells release the costimulator protein interleukin-2, which enhances the activation and division of T cells. Other proteins released by helper T cells attract phagocytes and enhance the phagocytic ability of macrophages. Helper T cells also stimulate the development of B cells into antibody-producing plasma cells and the development of natural killer cells.

To reduce the risk of rejection, recipients of **organ transplants** receive immunosuppressive drugs. One such drug is *cyclosporine*, derived from a fungus, which inhibits secretion of interleukin-2 by helper T cells but has only a minimal effect on B cells. Thus, the risk of rejection is diminished while resistance to some diseases is maintained.

Cytotoxic T cells are the soldiers that march forth to do battle with foreign invaders in cell-mediated adaptive immune responses. The name "cytotoxic" reflects their function—killing cells. Cytotoxic T cells are especially effective against body cells infected by microbes, some tumor cells, and cells of a transplant. After they divide, cytotoxic T cells leave secondary lymphatic organs and tissues and migrate to sites of invasion, infection, and tumor formation. Cytotoxic T cells recognize and attach to cells bearing the antigen that stimulated their activation and division, operating in the following way (Figure 17.9):

1. Cytotoxic T cells, using receptors on their surfaces, recognize and bind to infected target cells that have microbial antigens displayed on their surface. The cytotoxic T cell then releases ***granzymes***, protein-digesting enzymes that trigger apoptosis, the fragmentation of cellular contents (Figure 17.9a). Once the infected cell is destroyed, the released microbes are killed by phagocytes.
2. Cytotoxic T cells can also bind to infected body cells and release two proteins: perforin and granulysin. ***Perforin*** inserts into the plasma membrane of the target cell and creates channels in the membrane (Figure 17.9b). As a result, extracellular fluid flows into the target cell and cytolysis (cell bursting) occurs. ***Granulysin*** enters through the channels and destroys the microbes by creating holes in their plasma membranes. Cytotoxic T cells may also destroy target cells by releasing a toxic molecule called ***lymphotoxin***, which activates enzymes in the target cell. These enzymes cause the target cell's DNA to fragment and the cell dies. In addition, cytotoxic T cells secrete gamma-interferon, which attracts and activates phagocytic cells, and macrophage migration inhibition factor, which prevents migration of phagocytes from the infection site. After detaching from a target cell, a cytotoxic T cell can seek out and destroy another target cell.

Figure 17.9 Action of a cytotoxic T cell. After delivering a "lethal hit," a cytotoxic T cell can detach and attack another target cell displaying the same antigen.

Cytotoxic T cells kill their targets directly by secreting granzymes that trigger apoptosis and perforin that triggers cytolysis of infected target cells.

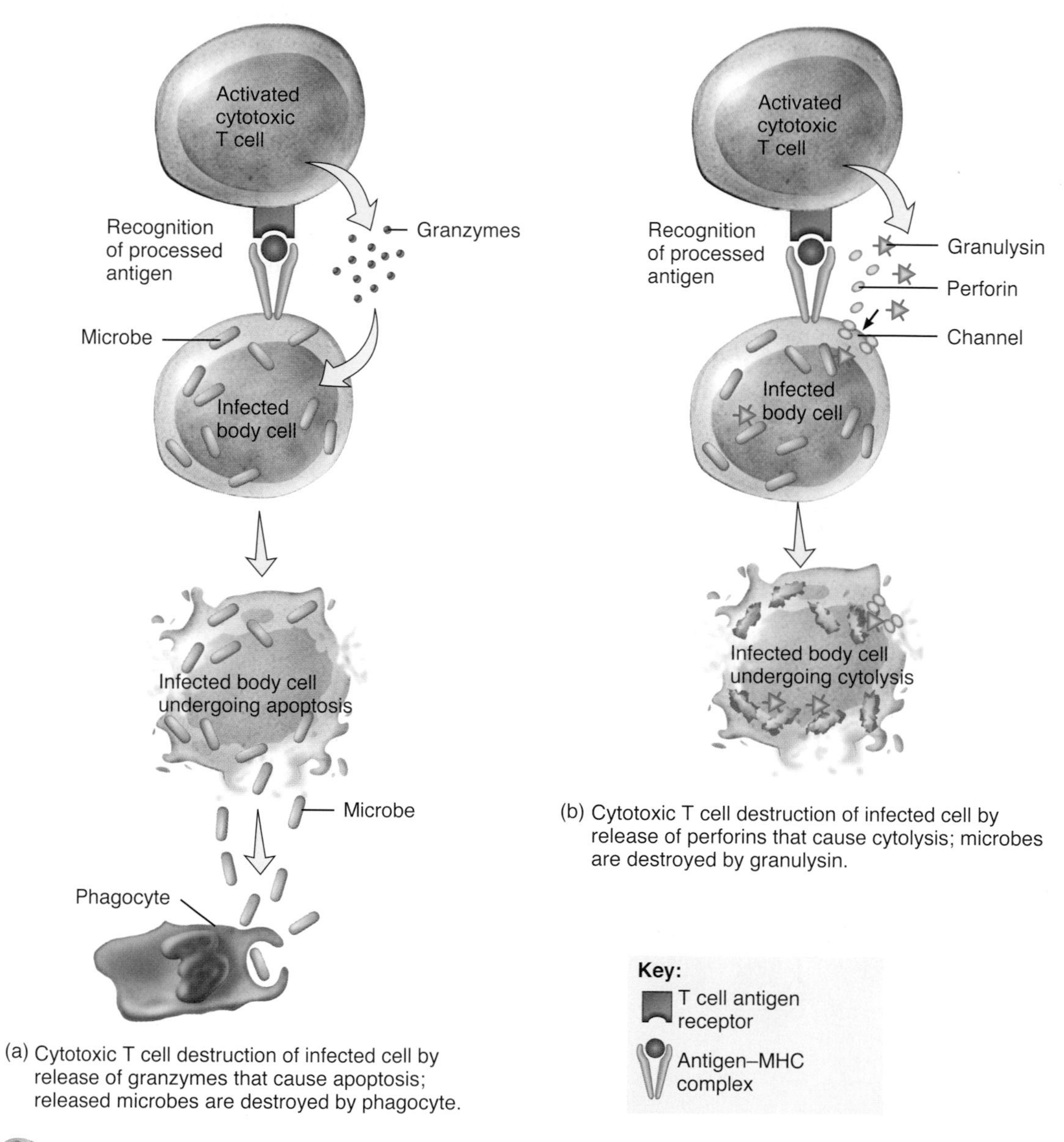

(a) Cytotoxic T cell destruction of infected cell by release of granzymes that cause apoptosis; released microbes are destroyed by phagocyte.

(b) Cytotoxic T cell destruction of infected cell by release of perforins that cause cytolysis; microbes are destroyed by granulysin.

Besides cells infected by microbes, what other types of cells do cytotoxic T cells attack?

Memory T cells remain in lymphatic tissue long after the original infection and are able to recognize the original invading antigen. Should the same antigen invade the body at a later date, the memory T cells initiate a faster reaction than occurred during the first invasion. The second response is so rapid that the pathogens are usually destroyed before any signs or symptoms of the disease occur. Memory T cells may provide immunity to a particular antigen for years. For instance, a person usually has the chickenpox only once because of memory T cells.

■ CHECKPOINT

7. What is the normal function of major histocompatibility complex proteins (self-antigens)?

8. How do antigens arrive at lymphatic tissues?

9. How do antigen-presenting cells process antigens?

10. What are the functions of helper, cytotoxic, and memory T cells?

11. How do cytotoxic T cells kill their targets?

B Cells and Antibody-Mediated Immunity

The body contains not only millions of different T cells, but also millions of different B cells, each capable of responding to a specific antigen. Cytotoxic T cells leave lymphatic tissues to seek out and destroy a foreign antigen, but B cells stay put. In the presence of a foreign antigen, specific B cells in lymph nodes, the spleen, or lymphatic nodules become activated. They then divide and develop into plasma cells that secrete specific antibodies, which in turn circulate in the lymph and blood to reach the sites of invasion.

During activation of a B cell, antigen receptors on the cell surface of a B cell bind to an antigen (Figure 17.10). B-cell antigen receptors are chemically similar to the antibodies that eventually are secreted by the plasma cells. Although B cells can respond to an unprocessed antigen present in lymph or interstitial fluid, their response is much more intense when they process the antigen. Antigen processing in a B cell occurs in the following way: The antigen is taken into the B cell, broken down into fragments and combined with MHC protein, and moved to the B cell surface. Helper T cells recognize the processed antigen-MHC protein complex and deliver the costimulation needed for B cell division and differentiation. The helper T cell releases interleukin-2 and other proteins that function as costimulators to activate B cells.

Figure 17.10 Activation, division, and differentiation of B cells in antibody-mediated immunity.

B cells develop into antibody-secreting plasma cells.

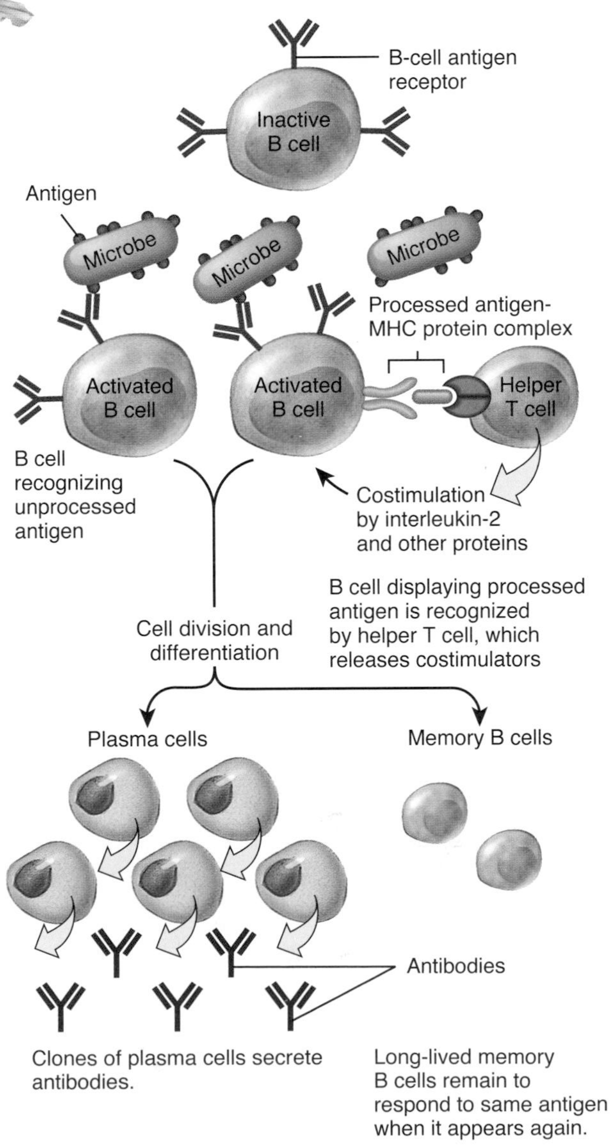

What types of cells respond to a second or subsequent invasion by an antigen?

Some of the activated B cells enlarge, divide, and differentiate into a clone of antibody-secreting ***plasma cells***. A few days or weeks after exposure to an antigen, a plasma cell secretes hundreds of millions of antibodies daily, and secretion occurs for about four or five days, until the plasma cell dies. Most antibodies travel in lymph and blood to the invasion sites. Some activated B cells do not differentiate into plasma cells but rather remain as ***memory B cells*** that are ready to respond more rapidly and forcefully should the same antigen reappear at a future time.

Although the functions of the five classes of antibodies differ somewhat, all attack antigens in several ways:

1. **Neutralizing antigen.** The binding of an antibody to its antigen neutralizes some bacterial toxins and prevents attachment of some viruses to body cells.
2. **Immobilizing bacteria.** Some antibodies cause bacteria to lose their motility, which limits bacterial spread into nearby tissues.
3. **Agglutinating antigen.** Binding of antibodies to antigens may connect pathogens to one another, causing *agglutination*, the clumping together of particles. Phagocytic cells ingest agglutinated microbes more readily.
4. **Activating complement.** Antigen–antibody complexes activate complement proteins, which then work to remove microbes through opsonization and cytolysis.
5. **Enhancing phagocytosis.** Once antigens have bound to an antibody's variable region, the antibody acts as a "flag" that attracts phagocytes. Antibodies enhance the activity of phagocytes by causing agglutination, by activating complement, and by coating microbes so that they are more susceptible to phagocytosis (opsonization).

Table 17.3 summarizes the functions of cells that participate in adaptive immune responses.

An antibody-mediated response typically produces many different antibodies that recognize different parts of an antigen or different antigens of a foreign cell. By contrast, a **monoclonal antibody (MAb)** is a pure antibody produced from a single clone of identical cells grown in the laboratory. Clinical uses of MAbs include the diagnosis of

Table 17.3 Summary of Cell Functions in Adaptive Immune Responses

Cell	Functions
Antigen-presenting Cell (APC)	Processes and presents foreign antigens to T cells. APCs include macrophages, B cells, and dendritic cells.
Helper T Cell	Helps other cells of the immune system combat intruders by releasing the costimulator protein interleukin-2 (IL-2), which enhances the activation and division of T cells. Other proteins attract phagocytes and enhance the phagocytic ability of macrophages. Helper T cells also stimulate the development of B cells into antibody-producing plasma cells and the development of natural killer cells.
Cytotoxic T Cell	Kills host target cells by releasing granzymes that induce apoptosis, perforin that forms channels to cause cytolysis, granulysin that destroys microbes, lymphotoxin that destroys target cell DNA, gamma-interferon that attracts macrophages and increases their phagocytic activity, and macrophage inhibition factor that prevents macrophage migration from site of infection.
Memory T Cell	Remains in lymphatic tissue and recognizes original invading antigen, even years after the first encounter.
B Cell	Differentiates into antibody-producing plasma cell.
Plasma Cell	Descendant of B cell that produces and secretes antibodies.
Memory B Cell	Remains ready to produce a more rapid and forceful secondary response should the same antigen enter the body in the future.

pregnancy, allergies, and diseases such as strep throat, hepatitis, rabies, and some sexually transmitted diseases. MAbs have also been used to detect cancer at an early stage and to determine the extent of metastasis. They may also be useful in preparing vaccines to counteract the rejection associated with transplants, to treat autoimmune diseases, and perhaps to treat AIDS.

Immunological Memory

A hallmark of adaptive immune responses is memory for specific antigens that have triggered immune responses in the past. ***Immunological memory*** is due to the presence of long-lasting antibodies and very long-lived lymphocytes that arise during division and differentiation of antigen-stimulated B cells and T cells.

Primary and Secondary Response

Adaptive immune responses, whether cell-mediated or antibody-mediated, are much quicker and more intense after a second or subsequent exposure to an antigen than after the first exposure. Initially, only a few cells have the correct antigen receptors to respond, and the immune response may take several days to build to maximum intensity. Because thousands of memory cells exist after an initial encounter with an antigen, they can divide and differentiate into plasma cells or cytotoxic T cells within hours the next time the same antigen appears.

One measure of immunological memory is the amount of antibody in blood plasma. After an initial contact with an antigen, no antibodies are present for a few days. Then, the levels of antibodies slowly rise, first IgM and then IgG, followed by a gradual decline (Figure 17.11). This is the ***primary response***. Memory cells may live for decades. Every new encounter with the same antigen causes a rapid division of memory cells. The antibody level after subsequent encounters is far greater than during a primary response and consists mainly of IgG antibodies. This accelerated, more in-

Figure 17.11 Secretion of antibodies. The primary response (after first exposure) is milder than the secondary response (after second or subsequent exposure) to a given antigen.

Immunological memory is the basis for successful immunization by vaccination.

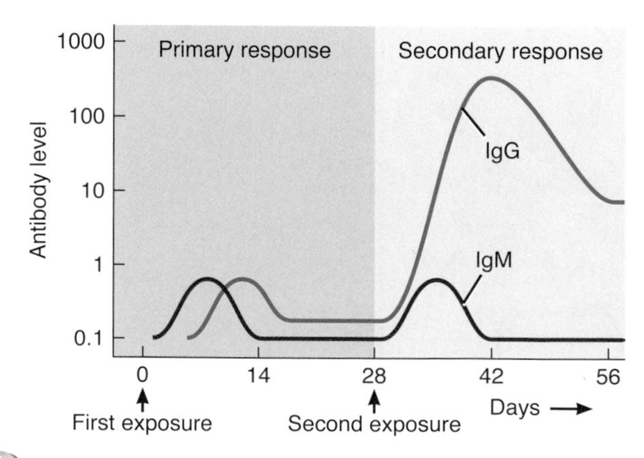

Which type of antibody responds most strongly during the secondary response?

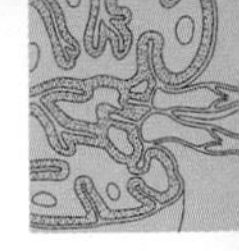

tense response is called the ***secondary response***. Antibodies produced during a secondary response are even more effective than those produced during a primary response. Thus, they are more successful in disposing of the invaders.

Primary and secondary responses occur during microbial infection. When you recover from an infection without taking antimicrobial drugs, it is usually because of the primary response. If the same microbe infects you later, the secondary response could be so swift that the microbes are destroyed before you exhibit any signs or symptoms of infection.

Naturally Acquired and Artificially Acquired Immunity

Immunological memory provides the basis for immunization by vaccination against certain diseases, for instance, polio. When you receive the *vaccine*, which may contain weakened or killed whole microbes or parts of microbes, your B cells and T cells are activated. Should you subsequently encounter the living pathogen as an infecting microbe, your body initiates a secondary response. However, booster doses of some immunizing agents must be given periodically to maintain adequate protection against the pathogen. Table 17.4 summarizes the various types of antigen encounters that provide naturally and artificially acquired immunity.

Table 17.4 Types of Adaptive Immunity

Type	How Acquired
Naturally acquired active immunity	Following exposure to a microbe, antigen recognition by B cells and T cells and costimulation lead to antibody-secreting plasma cells, cytotoxic T cells, and B and T memory cells.
Naturally acquired passive immunity	Transfer of IgG antibodies from mother to fetus across the placenta, or of IgA antibodies from mother to baby in milk during breast-feeding.
Artificially acquired active immunity	Antigens introduced during a vaccination stimulate cell-mediated and antibody-mediated immune responses, leading to production of memory cells. The antigens are pretreated to be immunogenic but not pathogenic; that is, they will trigger an immune response but not cause significant illness.
Artificially acquired passive immunity	Intravenous injection of immunoglobulins (antibodies).

■ **CHECKPOINT**

12. How are cell-mediated and antibody-mediated immune responses similar and different?
13. How is the secondary response to an antigen different from the primary response?

AGING AND THE IMMUNE SYSTEM

OBJECTIVE • Describe the effects of aging on the immune system.

With advancing age, most people become more susceptible to all types of infections and malignancies. Their response to vaccines is decreased, and they tend to produce more autoantibodies (antibodies against their body's own molecules). In addition, the immune system exhibits lowered levels of function. For example, T cells become less responsive to antigens, and fewer T cells respond to infections. This may result from age-related atrophy of the thymus or decreased production of thymic hormones. Because the T cell population decreases with age, B cells are also less responsive. Consequently, antibody levels do not increase as rapidly in response to a challenge by an antigen, resulting in increased susceptibility to various infections. It is for this key reason that elderly individuals are encouraged to get influenza (flu) vaccinations each year.

■ **CHECKPOINT**

14. What are the consequences of decreases in the number of T cells and B cells with advancing age?

• • •

To appreciate the many ways that the lymphatic system and immunity contribute to homeostasis of other body systems, examine Focus on Homeostasis: The Lymphatic System and Immunity on page 438. Next, in Chapter 18, we will explore the structure and function of the respiratory system and see how its operation is regulated by the nervous system. Most importantly, the respiratory system provides for gas exchange—taking in oxygen and blowing off carbon dioxide. The cardiovascular system aids gas exchange by transporting blood containing the gases between the lungs and tissue cells.

FOCUS ON HOMEOSTASIS

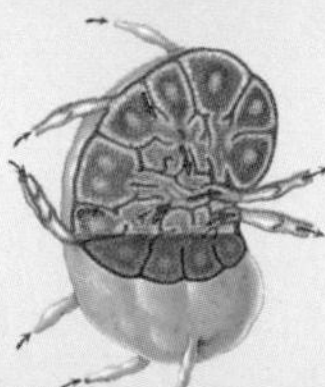

THE LYMPHATIC SYSTEM AND IMMUNITY

BODY SYSTEM		CONTRIBUTION OF THE LYMPHATIC SYSTEM AND IMMUNITY
For all body systems		B cells, T cells, and antibodies protect all body systems from attack by harmful foreign microbes (pathogens), foreign cells, and cancer cells.
Integumentary system		Lymphatic vessels drain excess interstitial fluid and leaked plasma proteins from the dermis of the skin. Immune system cells (Langerhans cells) in the skin help protect the skin. Lymphatic tissue provides IgA antibodies in sweat.
Skeletal system		Lymphatic vessels drain excess interstitial fluid and leaked plasma proteins from connective tissue around bones.
Muscular system		Lymphatic vessels drain excess interstitial fluid and leaked plasma proteins from muscles.
Nervous system	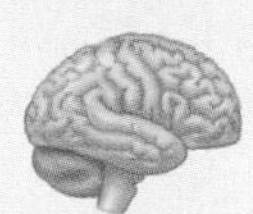	Lymphatic vessels drain excess interstitial fluid and leaked plasma proteins from the peripheral nervous system.
Endocrine system	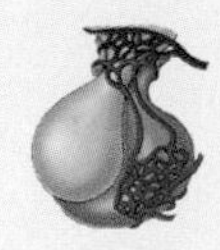	Flow of lymph distributes some hormones. Lymphatic vessels drain excess interstitial fluid and leaked plasma proteins from endocrine glands.
Cardiovascular system	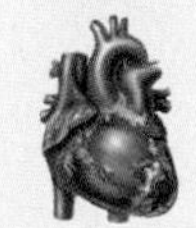	Lymph returns excess fluid filtered from blood capillaries and leaked plasma proteins to venous blood. Macrophages in spleen destroy aged red blood cells and remove debris in blood.
Respiratory system	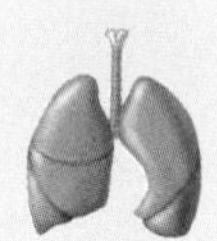	Tonsils, lymphatic nodules in the mucosa, and alveolar macrophages in the lungs help protect airways and lungs from pathogens. Lymphatic vessels drain excess interstitial fluid from the lungs.
Digestive system	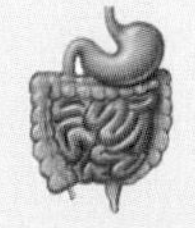	Tonsils and lymphatic nodules in the mucosa help defend against toxins and pathogens that penetrate the body from the gastrointestinal tract. Digestive system provides IgA antibodies in saliva and gastrointestinal secretions. Lymphatic vessels pick up absorbed dietary lipids and fat-soluble vitamins from the small intestine and transport them to the blood. Lymphatic vessels drain excess interstitial fluid and leaked plasma proteins from organs of the digestive system.
Urinary system	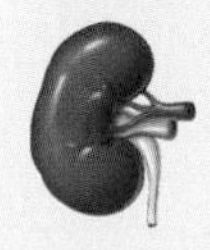	Lymphatic vessels drain excess interstitial fluid and leaked plasma proteins from organs of the urinary system. Lymphatic nodules in the mucosa help defend against toxins and pathogens that penetrate the body via the urethra.
Reproductive systems	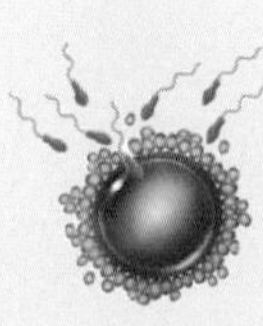	Lymphatic vessels drain excess interstitial fluid and leaked plasma proteins from organs of the reproductive systems. Lymphatic nodules in the mucosa help defend against toxins and pathogens that penetrate the body via the vagina and penis. In females, sperm deposited in the vagina are not attacked as foreign invaders due to components in seminal fluid that inhibit immune responses. IgG antibodies can cross the placenta to provide protection to a developing fetus. Lymphatic tissue provides IgA antibodies in the milk of the nursing mother.

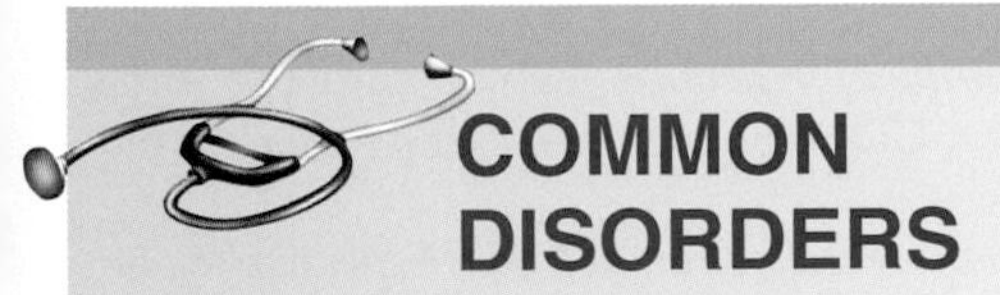

COMMON DISORDERS

AIDS: Acquired Immunodeficiency Syndrome

Acquired immunodeficiency syndrome (AIDS) is a condition in which a person experiences a telltale assortment of infections due to the progressive destruction of immune system cells by the ***human immunodeficiency virus (HIV)***. AIDS represents the end stage of infection by HIV. A person who is infected with HIV may be symptom-free for many years, even while the virus is actively attacking the immune system. In the two decades after the first five cases were reported in 1981, 22 million people died of AIDS. Worldwide, about 40 million people are currently infected with HIV.

HIV Transmission

Because HIV is present in the blood and some body fluids, it is most effectively transmitted (spread from one person to another) by practices that involve the exchange of blood or body fluids. HIV is transmitted in semen or vaginal fluid during unprotected (without a condom) anal, vaginal, or oral sex. HIV also is transmitted by direct blood-to-blood contact, such as occurs in intravenous drug users who share hypodermic needles or health-care professionals who may be accidentally stuck by HIV-contaminated hypodermic needles. In addition, HIV can be transmitted from an HIV-infected mother to her baby at birth or during breast-feeding.

The chances of transmitting or of being infected by HIV during vaginal or anal intercourse can be greatly reduced—although not eliminated—by the use of latex condoms. Public health programs aimed at encouraging drug users not to share needles have proved effective at checking the increase in new HIV infections in this population. Also, giving certain drugs to pregnant HIV-infected women greatly reduces the risk of transmission of the virus to their babies.

HIV is a very fragile virus; it cannot survive for long outside the human body. The virus is not transmitted by insect bites. A person cannot become infected by casual physical contact with an HIV-infected person, such as by hugging or sharing household items. The virus can be eliminated from personal care items and medical equipment by exposing them to heat (135°F for 10 minutes) or by cleaning them with common disinfectants such as hydrogen peroxide, rubbing alcohol, household bleach, or germicidal cleansers such as Betadine® or Hibiclens®. Standard dishwashing and clothes washing also kill HIV.

HIV: Structure and Infection

HIV consists of an inner core of ribonucleic acid (RNA) covered by a protein coat (capsid) surrounded by an outer layer, the envelope, composed of a lipid bilayer penetrated by proteins. Outside a living host cell, a virus is unable to replicate. However, when the virus infects and enters a host cell, its RNA uses the host cell's resources to make thousands of copies of the virus. New viruses eventually leave and then infect other cells.

HIV mainly damages helper T cells. Over 10 billion viral copies may be made each day. The viruses bud so rapidly from an infected cell's plasma membrane that the cell ruptures and dies. In most HIV-infected people, helper T cells are initially replaced as fast as they are destroyed. After several years, however, the body's ability to replace helper T cells is slowly exhausted, and the number of helper T cells in circulation gradually declines.

Signs, Symptoms, and Diagnosis of HIV Infection

Soon after being infected with HIV, most people experience a brief flu-like illness. Common signs and symptoms are fever, fatigue, rash, headache, joint pain, sore throat, and swollen lymph nodes. About 50% of infected people have night sweats. As early as three to four weeks after HIV infection, plasma cells begin secreting antibodies against HIV. These antibodies are detectable in blood plasma and form the basis for some of the screening tests for HIV. When people test "HIV-positive," it usually means they have antibodies to HIV antigens in their bloodstream.

Progression to AIDS

After a period of 2 to 10 years, the virus destroys enough helper T cells that most infected people begin to experience symptoms of immunodeficiency. HIV-infected people commonly have enlarged lymph nodes and experience persistent fatigue, involuntary weight loss, night sweats, skin rashes, diarrhea, and various lesions of the mouth and gums. In addition, the virus may begin to infect neurons in the brain, affecting the person's memory and producing visual disturbances.

As the immune system slowly collapses, an HIV-infected person becomes susceptible to a host of *opportunistic infections*. These are diseases caused by microorganisms that are normally held in check but now proliferate because of the defective immune system. AIDS is diagnosed when the helper T cell count drops below 200 cells per microliter (= cubic millimeter) of blood or when opportunistic infections arise, whichever occurs first. In time, opportunistic infections usually are the cause of death.

Treatment of HIV Infection

At present, infection with HIV cannot be cured. Vaccines designed to block new HIV infections and to reduce the viral load (the number of copies of HIV RNA in a microliter of blood plasma) in those who are already infected are in clinical trials. Meanwhile, two categories of drugs have proved successful in extending the life of many HIV-infected people:

1. ***Reverse transcriptase inhibitors*** interfere with the action of reverse transcriptase, the enzyme that the virus uses to convert its RNA into a DNA copy. Among the drugs in this category are zidovudine (ZDV, previously called AZT), didanosine (ddl), and stavudine (d4T). Trizivir®, approved in 2000 for treatment of HIV infection, combines three reverse transcriptase inhibitors in one pill.
2. ***Protease inhibitors*** interfere with the action of protease, a viral enzyme that cuts proteins into pieces to assemble the coat of

newly produced HIV particles. Drugs in this category include nelfinavir, saquinavir, ritonavir, and indinavir.

In 1996, physicians treating HIV-infected patients widely adopted *highly active antiretroviral therapy (HAART)*—a combination of two differently acting reverse transcriptase inhibitors and one protease inhibitor. Most HIV-infected individuals receiving HAART experience a drastic reduction in viral load and an increase in the number of helper T cells in their blood. Not only does HAART delay the progression of HIV infection to AIDS, but many people with AIDS have seen the remission or disappearance of opportunistic infections and an apparent return to health. Unfortunately, HAART is very costly (exceeding $10,000 per year), the dosing schedule is grueling, and not all people can tolerate the toxic side effects of these drugs. Although HIV may virtually disappear from the blood with drug treatment (and thus a blood test may be "negative" for HIV), the virus typically still lurks in various lymphatic tissues. In such cases, the infected person can still transmit the virus to another person.

Allergic Reactions

A person who is overly reactive to a substance that is tolerated by most other people is said to be ***allergic***. Whenever an allergic reaction takes place, some tissue injury occurs. The antigens that induce an allergic reaction are termed ***allergens***. Common allergens include certain foods (milk, peanuts, shellfish, eggs), antibiotics (penicillin, tetracycline), vaccines (pertussis, typhoid), venoms (honeybee, wasp, snake), cosmetics, chemicals in plants such as poison ivy, pollens, dust, molds, iodine-containing dyes used in certain x-ray procedures, and even microbes.

Type I (anaphylactic) reactions are the most common and typically occur within a few minutes after a person who was previously sensitized to an allergen is reexposed to it. In response to certain allergens, some people produce IgE antibodies that bind to the surface of mast cells and basophils. The next time the same allergen enters the body, it attaches to the IgE antibodies already present. In response, both the mast cells and basophils release histamine, prostaglandins, and other chemicals. Collectively, these chemicals cause vasodilation, increased blood capillary permeability, increased smooth muscle contraction in the airways of the lungs, and increased mucus secretion. As a result, a person may experience inflammatory responses, difficulty in breathing through the narrowed airways, and a runny nose from excess mucus secretion. In ***anaphylactic shock***, which may occur in a susceptible individual who has just received a triggering drug or been stung by a wasp, wheezing and shortness of breath as airways constrict are usually accompanied by shock due to vasodilation and fluid loss from blood. Injecting epinephrine to dilate the airways and strengthen the heartbeat usually is effective in this life-threatening emergency.

Type II (cytotoxic) reactions are caused by antibodies directed against antigens on a person's blood cells or tissue cells. Type II reactions, which may occur in incompatible blood transfusion reactions, damage cells by causing lysis.

Type III (immune-complex) reactions involve antigens, antibodies, and complement. Glomerulonephritis and rheumatoid arthritis (RA) arise in this way.

Type IV (cell-mediated) reactions or delayed hypersensitivity reactions usually appear 12–72 hours after exposure to an allergen. Type IV reactions occur when allergens are taken up by antigen-presenting cells (such as Langerhans cells in the skin) that migrate to lymph nodes and present the allergen to T cells, which then divide. Some of the new T cells return to the site of allergen entry into the body, where they produce gamma-interferon, which activates macrophages, and tumor necrosis factor, which stimulates an inflammatory response. Intracellular bacteria such as *Mycobacterium tuberculosis* trigger this type of cell-mediated immune response, as do certain haptens, such as poison ivy toxin. The skin test for tuberuclosis also is a delayed hypersensitivity reaction.

Infectious Mononucleosis

Infectious mononucleosis or "mono" is a contagious disease caused by the *Epstein-Barr virus (EBV)*. It occurs mainly in children and young adults, and more often in females than in males. The virus commonly enters the body through intimate oral contact such as kissing, which accounts for its being called the "kissing disease." EBV then multiplies in lymphatic tissues and spreads into the blood, where it infects and multiplies in B cells, the primary host cells. Because of this infection, the B cells become enlarged and abnormal in appearance so that they resemble monocytes, the primary reason for the term *mononucleosis*. Besides an elevated white blood cell count, with an abnormally high percentage of lymphocytes, signs and symptoms include fatigue, headache, dizziness, sore throat, enlarged and tender lymph nodes, and fever. There is no cure for infectious mononucleosis, but the disease usually runs its course in a few weeks.

Lymphomas

Lymphomas (lim-FŌ-mas; *lymph-* = clear water; *-oma* = tumor) are cancers of the lymphatic organs, especially the lymph nodes. Most have no known cause. The two main types of lymphomas are Hodgkin disease and non-Hodgkin lymphoma.

Hodgkin disease (HD) is characterized by painless, nontender enlargement of one or more lymph nodes, most commonly in the neck, chest, and axillae (armpits). If the disease has metastasized from these sites, fevers, night sweats, weight loss, and bone pain also occur. HD primarily affects individuals between ages 15 and 35 and those over 60; it is more common in males. If diagnosed early, HD has a 90–95% cure rate.

Non-Hodgkin lymphoma (NHL), which is more common than HD, occurs in all age groups. NHL may start the same way as HD but may also include an enlarged spleen, anemia, and general malaise. Up to half of all individuals with NHL are cured or survive for a lengthy period. Treatment options for both HD and NHL include radiation therapy, chemotherapy, and red bone marrow transplantation.

Systemic Lupus Erythematosus

Systemic lupus erythematosus (er-e-thēm-a-TŌ-sus), ***SLE***, or ***lupus*** (*lupus* = wolf) is a chronic autoimmune disease that affects multiple body systems. Most cases of SLE occur in women between the ages of 15 and 25, more often in blacks than in whites. Although the cause of SLE is not known, both a genetic predisposition to the disease and environmental factors contribute. Females are nine times more likely than males to suffer from SLE. The disorder often occurs in females who exhibit extremely low levels of androgens (male sex hormones).

Signs and symptoms of SLE include joint pain, slight fever, fatigue, oral ulcers, weight loss, enlarged lymph nodes and spleen, photosensitivity, rapid loss of large amounts of scalp hair, and sometimes an eruption across the bridge of the nose and cheeks called a "butterfly rash." The erosive nature of some of the SLE skin lesions was thought to resemble the damage inflicted by the bite of a wolf—thus, the term lupus. Kidney damage occurs as antigen–antibody complexes become trapped in kidney capillaries, thereby obstructing blood filtering. Renal failure is the most common cause of death.

MEDICAL TERMINOLOGY AND CONDITIONS

Allograft (AL-ō-graft; *allo-* = other) A transplant between genetically different individuals of the same species. Skin transplants from other people and blood transfusions are allografts.

Autograft (AW-tō-graft; *auto-* = self) A transplant in which one's own tissue is grafted to another part of the body (such as skin grafts for burn treatment or plastic surgery).

Chronic fatigue syndrome (CFS) A disorder, usually occurring in young female adults, characterized by (1) extreme fatigue that impairs normal activities for at least six months and (2) the absence of other known diseases (cancer, infections, drug abuse, toxicity, or psychiatric disorders) that might produce similar symptoms.

Gamma globulin (GLOB-ū-lin) Suspension of immunoglobulins from blood consisting of antibodies that react with a specific pathogen. It is prepared by injecting the pathogen into animals, removing blood from the animals after antibodies have been produced, isolating the antibodies, and injecting them into a human to provide short-term immunity.

Graft Any tissue or organ used for transplantation or a transplant of such structures.

Lymphadenopathy (lim-fad′-e-NOP-a-thē; *lymph-* = clear fluid; *-pathy* = disease) Enlarged, sometimes tender lymph glands as a response to infection, also called ***swollen glands***.

Splenomegaly (splē′-nō-MEG-a-lē; *mega-* = large) Enlarged spleen.

Tonsillectomy (ton′-si-LEK-tō-mē; *-ectomy* = excision) Removal of a tonsil.

Xenograft (ZEN-ō-graft; *xeno-* = strange or foreign) A transplant between animals of different species. Xenografts from porcine (pig) or bovine (cow) tissue may be used in people as a physiological dressing for severe burns.

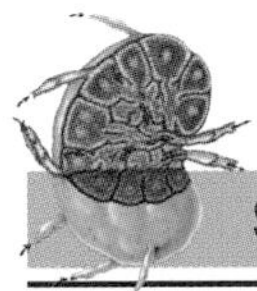

STUDY OUTLINE

Introduction (p. 420)

1. Despite constant exposure to a variety of pathogens (disease-producing microbes such as bacteria and viruses), most people remain healthy.
2. Immunity or resistance is the ability to ward off damage or disease. Innate immunity refers to defenses that are present at birth; they are always present and provide immediate but general protection against invasion by a wide range of pathogens. Adaptive immunity refers to defenses that respond to a particular invader; it involves activation of specific lymphocytes that can combat a specific invader.

Lymphatic System (p. 421)

1. The body system responsible for adaptive immunity (and some aspects of innate immunity) is the lymphatic system, which consists of lymph, lymphatic vessels, structures and organs that contain lymphatic tissue, and red bone marrow.
2. Components of blood plasma filter through blood capillary walls to form interstitial fluid, the fluid that bathes the cells of body tissues. After interstitial fluid passes into lymphatic vessels, it is called lymph. Interstitial fluid and lymph are chemically similar to blood plasma.
3. The lymphatic system drains tissue spaces of excess fluid and returns proteins that have escaped from blood to the cardiovascular system. It also transports lipids and lipid-soluble vitamins from the gastrointestinal tract to the blood, and it protects the body against invasion.
4. Lymphatic vessels begin as lymphatic capillaries in tissue spaces between cells. The lymphatic capillaries merge to form larger lymphatic vessels, which ultimately drain into the thoracic duct or right lymphatic duct. Located at intervals along lymphatic vessels are lymph nodes, masses of B cells and T cells surrounded by a capsule.
5. The passage of lymph is from interstitial fluid, to lymphatic capillaries, to lymphatic vessels and lymph nodes, to the thoracic duct or right lymphatic duct, to the junction of the internal jugular and subclavian veins.
6. Lymph flows due to the "milking action" of skeletal muscle contractions and pressure changes that occur during inhalation. Valves in the lymphatic vessels prevent backflow of lymph.

7. Primary lymphatic organs and tissues are the sites where stem cells divide and develop into mature B cells and T cells. They include the red bone marrow (in flat bones and the ends of the long bones of adults) and the thymus. Stem cells in red bone marrow give rise to mature B cells and to immature T cells that migrate to the thymus, where they mature into functional T cells.
8. The secondary lymphatic organs and tissues are the sites where most immune responses occur. They include lymph nodes, the spleen, and lymphatic nodules.
9. Lymph nodes contain B cells that develop into plasma cells, T cells, dendritic cells, and macrophages. Lymph enters nodes through afferent lymphatic vessels and exits through efferent lymphatic vessels.
10. The spleen is the single largest mass of lymphatic tissue in the body. It is a site where B cells divide into plasma cells and macrophages phagocytize worn-out red blood cells and platelets.
11. Lymphatic nodules are oval-shaped concentrations of lymphatic tissue that are not surrounded by a capsule. They are scattered throughout the mucosa of the gastrointestinal, respiratory, urinary, and reproductive tracts.

Innate Immunity (p. 425)

1. Innate immunity defenses include barriers provided by the skin and mucous membranes (first line of defense). They also include various internal defenses (second line of defense): internal antimicrobial proteins (interferons, complement, transferrin, and antimicrobial peptides), phagocytes (neutrophils and macrophages), natural killer cells (which have the ability to kill a wide variety of infectious microbes and certain tumor cells), inflammation, and fever.
2. Table 17.1 on page 428 summarizes the components of innate immunity.

Adaptive Immunity (p. 428)

1. Adaptive immunity involves the production of specific types of cells or specific antibodies to destroy a particular antigen.
2. An antigen is any substance that the adaptive immune system recognizes as foreign (nonself). Normally, a person's immune system cells exhibit self-tolerance: They recognize and do not attack their own tissues and chemicals.
3. B cells complete their development in red bone marrow, but mature T cells develop in the thymus from immature T cells that migrate from bone marrow.
4. The major histocompatibility complex (MHC) proteins are unique to each person's body cells. All cells except red blood cells display MHC molecules.
5. Antigens induce plasma cells to secrete antibodies, proteins that typically contain four polypeptide chains. The variable regions of an antibody are the antigen-binding sites, where the antibody can bind to a particular antigen.
6. Based on chemistry and structure, antibodies, also known as immunoglobulins (IGs), are grouped in five classes, each with specific functions: IgG, IgA, IgM, IgD, and IgE (see Table 17.2 on page 430). Functionally, antibodies neutralize antigens, immobilize bacteria, agglutinate antigens, activate complement, and enhance phagocytosis.
7. Antigen-presenting cells (APCs) process and present antigens to activate T cells, and they secrete substances that stimulate division of T cells and B cells.
8. There are three main kinds of T cells: helper T cells, which stimulate growth and division of cytotoxic T cells, attract phagocytes, and stimulate development of B cells into antibody-producing plasma cells; cytotoxic T cells, which eliminate invaders by (1) releasing granzymes that cause target cell apoptosis (phagocytes then kill the microbes) and (2) releasing perforin, which causes cytolysis, and granulysin that destroys the microbes; and memory T cells, which recognize previously encountered antigens at a later date.
9. Antibody-mediated immunity refers to destruction of antigens by antibodies, which are produced by descendants of B cells called plasma cells.
10. B cells develop into antibody-producing plasma cells under the influence of chemicals secreted by antigen-presenting cells and helper T cells.
11. Table 17.3 on page 436 summarizes the functions of cells that participate in adaptive immune responses.
12. Immunization against certain microbes is possible because memory B cells and memory T cells remain after a primary response to an antigen. The secondary response provides protection should the same microbe enter the body again. Table 17.4 on page 437 summarizes the various types of antigen encounters that provide naturally and artificially acquired immunity.

Aging and the Immune System (p. 437)

1. With advancing age, individuals become more susceptible to infections and malignancies, respond less well to vaccines, and produce more autoantibodies.
2. T cell responses also diminish with age.

SELF-QUIZ

1. Which of the following is NOT true concerning the lymphatic system?
   a. Lymphatic vessels transport lipids from the gastrointestinal tract to the blood.
   b. Lymph is more similar to interstitial fluid than to blood.
   c. Lymphatic tissue is present in only a few isolated organs in the body.
   d. The unique structure of lymphatic capillaries allows fluid to flow into them but not out of them.
   e. Lymphatic vessels resemble veins in structure.

2. Which of the following are produced by virus-infected cells to protect uninfected cells from viral invasion?
   a. complement molecules **b.** prostaglandins **c.** fibrins **d.** interferons **e.** histamines

3. A blockage in the right lymphatic duct would interfere with lymph drainage from the
   a. left arm **b.** right leg **c.** lower abdomen **d.** left leg **e.** right arm

4. Which of the following best represents lymph flow from the interstitial spaces back to the blood?
   a. lymphatic capillaries → lymphatic ducts → lymphatic vessels → junction of internal jugular and subclavian veins
   b. junction of internal jugular and subclavian veins → lymphatic capillaries → lymphatic vessels → lymphatic ducts
   c. lymphatic capillaries → lymphatic vessels → lymphatic ducts → junction of internal jugular and subclavian veins
   d. lymphatic ducts → lymphatic vessels → lymphatic capillaries → junction of internal jugular and subclavian veins
   e. lymphatic capillaries → lymphatic vessels → junction of internal jugular and subclavian veins → lymphatic ducts

5. Lymph nodes
   a. filter lymph **b.** are another name for tonsils **c.** produce lymph **d.** are a primary storage site for blood **e.** produce a protective mucus

6. Which of the following is NOT true about the role of skin in nonspecific immunity?
   a. Sebum inhibits the growth of certain bacteria.
   b. Epidermal cells produce interferons to destroy viruses.
   c. Shedding of epidermal cells helps remove microbes.
   d. Lysozyme in sweat destroys some bacteria.
   e. The skin forms a physical barrier to prevent entry of microbes.

7. Which of the following statements about B cells is true?
   a. They become functional while in the thymus
   b. Some develop into plasma cells that secrete antibodies.
   c. Some B cells become natural killer cells.
   d. Cytotoxic B cells travel in lymph and blood to react with foreign antigens.
   e. They kill virus-infected cells by secreting perforin.

8. The cells that release granzymes, perforin, granulysin, and lymphotoxin are
   a. cytotoxic T cells **b.** plasma cells **c.** B cells **d.** natural killer cells **e.** helper T cells

9. The secondary response in antibody-mediated immunity
   a. is characterized by a slow rise in antibody levels and then a gradual decline
   b. occurs when you first receive a vaccination against some disease
   c. produces fewer but more responsive antibodies than occur during the primary response
   d. is an intense response by memory cells to produce antibodies when an antigen is contacted again
   e. is rarely seen except in autoimmune disorders

10. The ability of the body's immune system to recognize its own tissues is known as
    a. immunological escape **b.** autoimmunity **c.** nonspecific resistance **d.** hypersensitivity **e.** self-tolerance

11. A disease that causes destruction of helper T cells would result in all of the following effects EXCEPT
    a. inability to produce cytotoxic T cells
    b. alteration of lymph flow
    c. lack of development of plasma cells
    d. decreased production of antibodies
    e. increased risk of developing infections

12. In which lymphatic organ do T cells mature?
    a. thyroid gland **b.** spleen **c.** thymus **d.** red bone marrow **e.** lymph node

13. Place the following steps involved in the process of inflammation in the correct order.
    1. arrival of large numbers of neutrophils
    2. vasodilation and increased permeability of blood vessels
    3. formation of pus
    4. increased migration of monocytes
    5. formation of fibrin network to form a clot
    6. release of histamine
    a. 6, 2, 4, 1, 5, 3 **b.** 3, 6, 1, 4, 2, 5 **c.** 5, 1, 4, 2, 6, 3 **d.** 6, 2, 5, 1, 4, 3 **e.** 4, 6, 1, 3, 2, 5

14. Match the following:
    ____ **a.** destroy antigens by cytolysis
    ____ **b.** stimulate other cells of the adaptive immune response
    ____ **c.** are programmed to recognize the original invading antigen; allow immunity to last for years
    ____ **d.** function in innate immunity
    ____ **e.** develop into plasma cells

    A. natural killer cells
    B. helper T cells
    C. B cells
    D. memory T cells
    E. cytotoxic T cells

15. What happens during opsonization?
    a. engulfment of a microbe by a phagocyte
    b. chemical attraction of a phagocyte
    c. binding of complement to a microbe
    d. attachment of a phagocyte to a microbe
    e. breakdown of a microbe by enzymes

16. All of the following contribute to nonspecific immunity EXCEPT
    a. complement b. immunoglobulins
    c. natural killer cells d. lysozyme e. interferons

17. Inflammation produces
    a. redness due to bleeding
    b. heat due to fever-causing toxins
    c. swelling due to increased permeability of capillaries
    d. pain due to histamine release
    e. mucus due to phagocytosis

18. Place the phases of phagocytosis in the correct order.
    1. adherence to foreign material
    2. chemotaxis of phagocytes
    3. exocytosis of indigestible materials
    4. ingestion of foreign material
    a. 1, 2, 3, 4 b. 2, 1, 4, 3 c. 1, 4, 3, 2 d. 4, 3, 2, 1
    e. 2, 1, 3, 4

19. Antibodies attack antigens by all of the following methods EXCEPT
    a. agglutination of antigens
    b. activation of complement
    c. opsonization to enhance phagocytosis
    d. preventing attachment to body cells
    e. producing acid secretions

20. What is the importance of tonsils in the body's defenses?
    a. They help destroy microbes that are inhaled.
    b. They contain ciliated cells that move trapped pathogens from the breathing passages.
    c. They are needed for T cell maturation.
    d. They are needed for B cell maturation.
    e. They filter lymph.

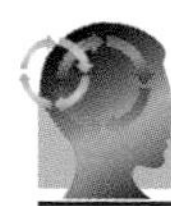

CRITICAL THINKING APPLICATIONS

1. Marcia found a lump in her right breast during her monthly self-examination. The lump was found to be cancerous. The surgeon removed the breast lump, the surrounding tissue, and some lymph nodes. Which nodes were probably removed and why?

2. Years ago, a tonsillectomy was almost considered a "rite of passage" for children in elementary school. It seemed like all children were getting their tonsils removed! Why are tonsils frequently infected in young children?

3. B.J. stepped on a rusty fishhook while walking along the beach. The emergency room nurse removed the fishhook and gave B.J. a tetanus booster. Why?

4. You learned in Chapter 16 that the cornea and lens of the eye are completely lacking in capillaries. How is this fact related to the high success of corneal transplants?

ANSWERS TO FIGURE QUESTIONS

17.1 Lymphatic tissue is reticular connective tissue that contains large numbers of lymphocytes.

17.2 Lymph is more similar to interstitial fluid because its protein content is low.

17.3 Capillaries produce lymph.

17.4 Foreign substances in lymph may be phagocytized by macrophages or destroyed by T cells or antibodies produced by plasma cells.

17.5 Redness is caused by increased blood flow due to vasodilation.

17.6 The variable regions of an antibody can bind specifically to the antigen that triggered its production.

17.7 APCs include macrophages, B cells, and dendritic cells.

17.8 A clone is a population of identical cells.

17.9 Cytotoxic T cells can also attack some tumor cells and transplanted tissue cells.

17.10 Memory T and B cells respond to a second invasion by the same antigen.

17.11 IgG is the antibody secreted in greatest amount during a secondary response.

THE RESPIRATORY SYSTEM chapter 18

did you know?

Cigarette smoking is the single most preventable cause of death and disability worldwide. Smoking disrupts the body's ability to maintain homeostasis and health because it introduces many harmful substances into the body, and wreaks havoc on the fragile tissues of the respiratory system. For example, smoking causes emphysema by progressively destroying the alveoli and bronchioles. The irritation produced by cigarette smoke often leads to chronic bronchitis. Smoking causes many types of cancers, including cancers of the mouth, throat, lungs, esophagus, stomach, kidneys, pancreas, colon, and urinary bladder. The chemicals in cigarette smoke also raise blood pressure and accelerate the process of atherosclerosis.

Focus on Wellness, page 463

www.wiley.com/college/apcentral

Body cells continually use oxygen (O_2) for the metabolic reactions that release energy from nutrient molecules and produce ATP. These same reactions produce carbon dioxide (CO_2). Because an excessive amount of CO_2 produces acidity that can be toxic to cells, excess CO_2 must be eliminated quickly and efficiently. The ***respiratory system***, which includes the nose, pharynx (throat), larynx (voice box), trachea (windpipe), bronchi, and lungs (Figure 18.1), provides for gas exchange, the intake of O_2, and the removal of CO_2. The respiratory system also helps regulate blood pH; contains receptors for the sense of smell; filters, warms, and moistens inspired air; produces sounds; and rids the body of some water and heat in exhaled air.

looking back to move ahead . . .

- **Cartilage (page 89)**
- **Pseudostratified Ciliated Columnar Epithelium (page 75)**
- **Simple Squamous Epithelium (page 75)**
- **Muscles Used in Breathing (page 200)**
- **Diffusion (page 47)**
- **Ions (page 25)**
- **Medulla and Pons (pages 250–253)**

Figure 18.1 Organs of the respiratory system.

The upper respiratory system includes the nose, pharynx, and associated structures. The lower respiratory system includes the larynx, trachea, bronchi, and lungs.

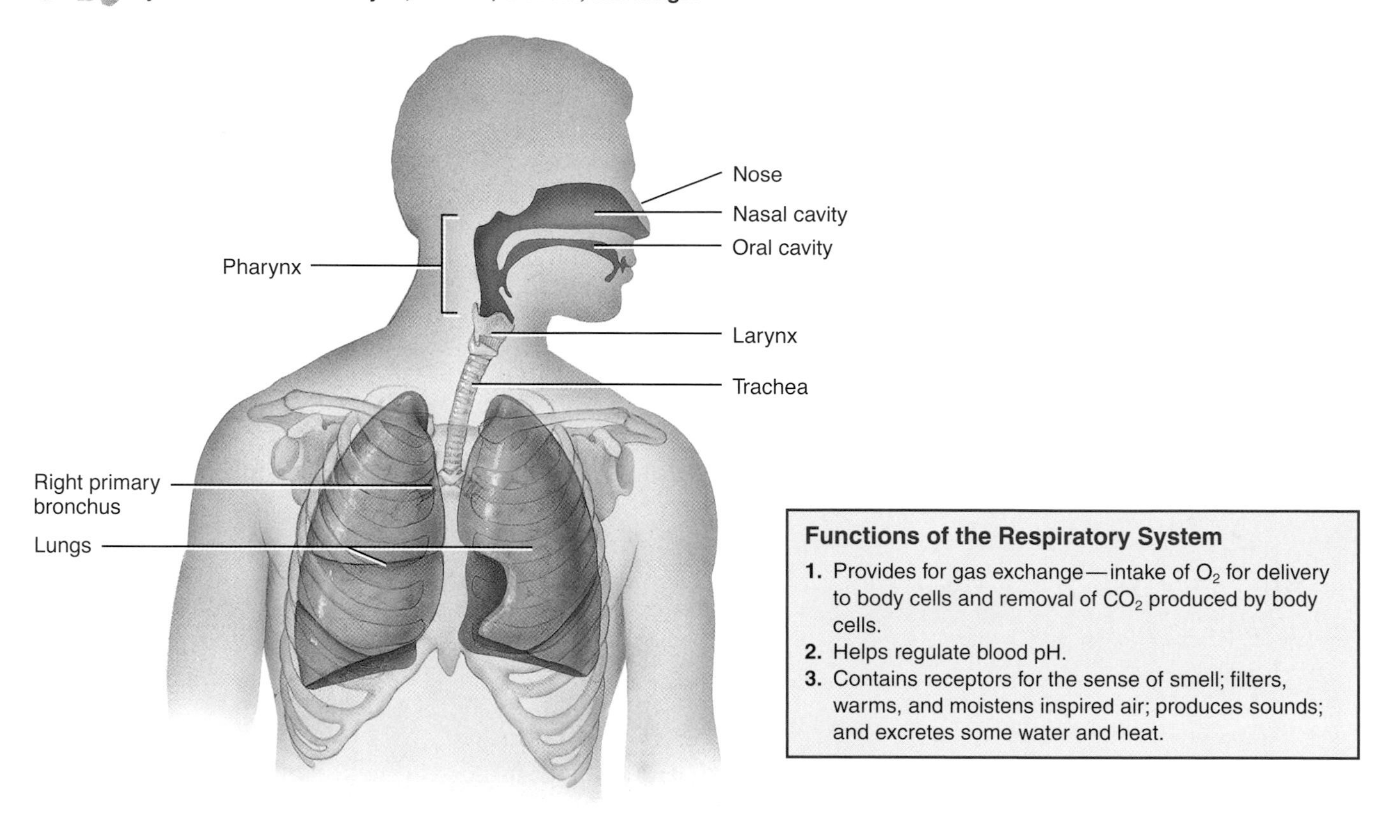

Which structures comprise the conducting zone of the respiratory system?

The branch of medicine that deals with the diagnosis and treatment of diseases of the ears, nose, and throat (ENT) is called ***otorhinolaryngology*** (ō′-tō-rī′-nō-lar′-in-GOL-ō-jē; *oto-* = ear; *rhino-* = nose; *laryngo-* = voice box; *-logy* = study of). A ***pulmonologist*** (*pulmon-* = lung) is a specialist in the diagnosis and treatment of diseases of the lungs.

The entire process of gas exchange in the body, called ***respiration***, occurs in three basic steps:

1. ***Pulmonary ventilation***, or ***breathing***, is the flow of air into and out of the lungs.
2. ***External respiration*** is the exchange of gases between the air spaces (alveoli) of the lungs and the blood in pulmonary capillaries. In this process, pulmonary capillary blood gains O_2 and loses CO_2.
3. ***Internal respiration*** is the exchange of gases between blood in systemic capillaries and tissue cells. The blood loses O_2 and gains CO_2. Within cells, the metabolic reactions that consume O_2 and give off CO_2 during the production of ATP are termed *cellular respiration* (discussed in Chapter 20).

As you can see, two systems are cooperating to supply O_2 and eliminate CO_2—the cardiovascular and respiratory systems. The first two steps are the responsibility of the respiratory system, while the third step is a function of the cardiovascular system.

ORGANS OF THE RESPIRATORY SYSTEM

OBJECTIVE • Describe the structure and functions of the nose, pharynx, larynx, trachea, bronchi, bronchioles, and lungs.

Structurally, the respiratory system consists of two parts: The ***upper respiratory system*** includes the nose, pharynx, and associated structures; the ***lower respiratory system*** consists of the larynx, trachea, bronchi, and lungs. The respiratory system can also be divided into two parts based on function.

The ***conducting zone*** consists of a series of interconnecting cavities and tubes—nose, pharynx, larynx, trachea, bronchi, bronchioles, and terminal bronchioles—that conduct air into the lungs. The ***respiratory zone*** consists of tissues within the lungs where gas exchange occurs—the respiratory bronchioles, alveolar ducts, alveolar sacs, and alveoli.

Nose

The ***nose*** has a visible external portion and an internal portion inside the skull (Figure 18.2). The external nose consists of bone and cartilage covered with skin and lined with mucous membrane. It has two openings called the ***external nares*** (NA-rēz; singular is ***naris***) or ***nostrils***.

The internal nose connects to the throat through two openings called the ***internal nares***. Four paranasal sinuses (frontal, sphenoidal, maxillary, and ethmoidal) and the nasolacrimal ducts also connect to the internal nose. The space inside the internal nose, called the ***nasal cavity***, lies below the cranium and above the mouth. A vertical partition, the ***nasal septum***, divides the nasal cavity into right and left sides. The septum consists of the perpendicular plate of the ethmoid bone, vomer, and cartilage (see Figure 6.7a on page 126).

Rhinoplasty (RĪ-nō-plas′-tē; *-plasty* = to mold or to shape), commonly called a "nose job," is a surgical procedure to alter the shape of the external nose. Although rhinoplasty is often done for cosmetic reasons, it is sometimes performed to repair a fractured nose or a deviated nasal septum. With anesthesia, instruments inserted through the nostrils are used to reshape the nasal cartilage and fracture and reposition the nasal bones, to achieve the desired shape. An internal packing and splint keep the nose in the desired position while it heals.

Figure 18.2 Respiratory organs in the head and neck.

As air passes through the nose, it is warmed, filtered, and moistened.

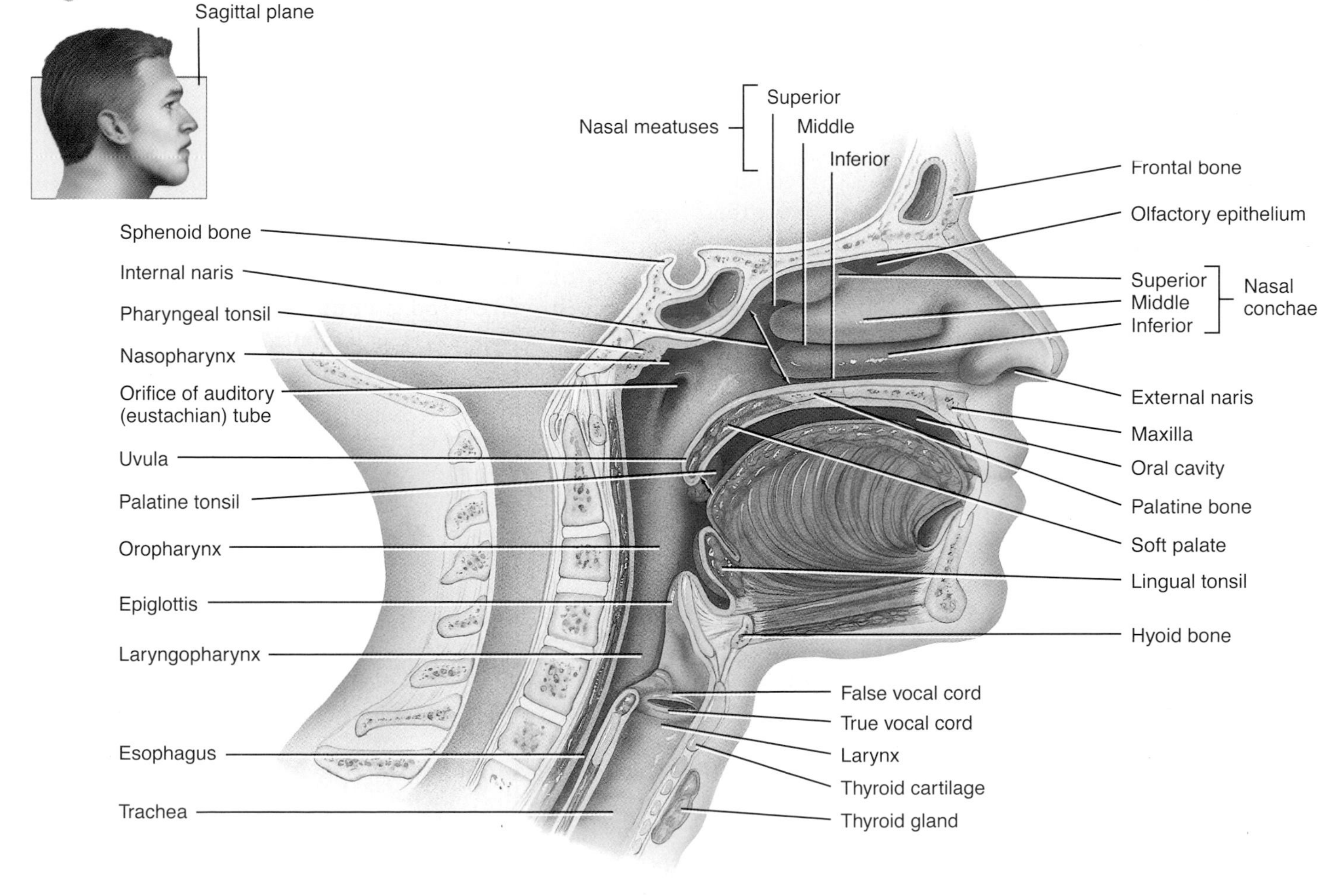

Sagittal section of the left side of the head and neck

What is the path taken by air molecules into and through the nose?

The interior structures of the nose are specialized for three basic functions: (1) filtering, warming, and moistening incoming air; (2) detecting olfactory (smell) stimuli; and (3) modifying the vibrations of speech sounds. When air enters the nostrils, it passes coarse hairs that trap large dust particles. The air then flows over three shelves called the superior, middle, and inferior ***nasal conchae*** (KONG-kē) that extend out of the wall of the cavity. Mucous membrane lines the nasal cavity and the three conchae. As inspired air whirls around the conchae, it is warmed by blood circulating in abundant capillaries. The olfactory receptors lie in the membrane lining the superior nasal conchae and adjacent septum. This region is called the *olfactory epithelium.*

Pseudostratified ciliated columnar epithelial cells and goblet cells line the nasal cavity. Mucus secreted by goblet cells moistens the air and traps dust particles. Cilia move the dust-laden mucus toward the pharynx, at which point it can be swallowed or spit out, thus removing particles from the respiratory tract.

Substances in cigarette smoke **inhibit movement of cilia.** If the cilia are paralyzed, only coughing can remove mucus–dust packages from the airways. This is why smokers cough so much and are more prone to respiratory infections.

Pharynx

The ***pharynx*** (FAIR-inks), or throat, is a funnel-shaped tube that starts at the internal nares and extends partway down the neck (Figure 18.2). It lies just posterior to the nasal and oral cavities and just anterior to the cervical (neck) vertebrae. Its wall is composed of skeletal muscle and lined with mucous membrane. The pharynx functions as a passageway for air and food, provides a resonating chamber for speech sounds, and houses the tonsils, which participate in immunological responses to foreign invaders.

The upper part of the pharynx, called the ***nasopharynx***, connects with the two internal nares and has two openings that lead into the auditory (eustachian) tubes. The posterior wall contains the *pharyngeal tonsil.* The nasopharynx exchanges air with the nasal cavities and receives mucus–dust packages. The cilia of its pseudostratified ciliated columnar epithelium move the mucus–dust packages toward the mouth. The nasopharynx also exchanges small amounts of air with the auditory tubes to equalize air pressure between the pharynx and middle ear. The middle portion of the pharynx, the ***oropharynx***, opens into the mouth and nasopharynx. Two pairs of tonsils, the *palatine tonsils* and *lingual tonsils*, are found in the oropharynx. The lowest portion of the pharynx, the ***laryngopharynx*** (la-rin′-gō-FAIR-inks), connects with both the esophagus (food tube) and the larynx (voice box). Thus, the oropharynx and laryngopharynx both serve as passageways for air as well as for food and drink.

Larynx

The ***larynx*** (LAIR-inks), or voice box, is a short tube of cartilage lined by mucous membrane that connects the pharynx with the trachea (Figure 18.3). It lies in the midline of the neck anterior to the fourth, fifth, and sixth cervical vertebrae (C4 to C6).

The ***thyroid cartilage***, which consists of hyaline cartilage, forms the anterior wall of the larynx. Its common name (Adam's apple) reflects the fact that it is often larger in males than in females due to the influence of male sex hormones during puberty.

The ***epiglottis*** (*epi-* = over; *glottis* = tongue) is a large, leaf-shaped piece of elastic cartilage that is covered with epithelium (see also Figure 18.2). The "stem" of the epiglottis is attached to the anterior rim of the thyroid cartilage and hyoid bone. The broad superior "leaf" portion of the epiglottis is unattached and is free to move up and down like a trap door. During swallowing, the pharynx and larynx rise. Elevation of the pharynx widens it to receive food or drink; elevation of the larynx causes the epiglottis to move down and form a lid over the larynx, closing it off. The closing of the larynx in this way during swallowing routes liquids and foods into the esophagus and keeps them out of the airways below. When anything but air passes into the larynx, a cough reflex attempts to expel the material.

The ***cricoid cartilage*** (KRĪ-koyd) is a ring of hyaline cartilage that forms the inferior wall of the larynx and is attached to the first tracheal cartilage. The paired ***arytenoid cartilages*** (ar′-i-TĒ-noyd), consisting mostly of hyaline cartilage, are located above the cricoid cartilage. They attach to the true vocal cords and pharyngeal muscles and function in voice production. The cricoid cartilage is the landmark for making an emergency airway (a tracheotomy; see page 451).

The Structures of Voice Production

The mucous membrane of the larynx forms two pairs of folds: an upper pair called the ***false vocal cords*** and a lower pair called the ***true vocal cords*** (see Figure 18.2). The false vocal cords hold the breath against pressure in the thoracic cavity when you strain to lift a heavy object, such as a backpack filled with textbooks. They do not produce sound.

The true vocal cords produce sounds during speaking and singing. They contain elastic ligaments stretched between pieces of rigid cartilage like the strings on a guitar. Muscles attach both to the cartilage and to the true vocal cords. When the muscles contract, they pull the elastic ligaments tight, which moves the true vocal cords out into the air passageway. The air pushed against the true vocal cords causes them to vibrate and sets up sound waves in the air in the pharynx, nose, and mouth. The greater the air pressure, the louder the sound.

Figure 18.3 Larynx.

The larynx is composed of cartilage.

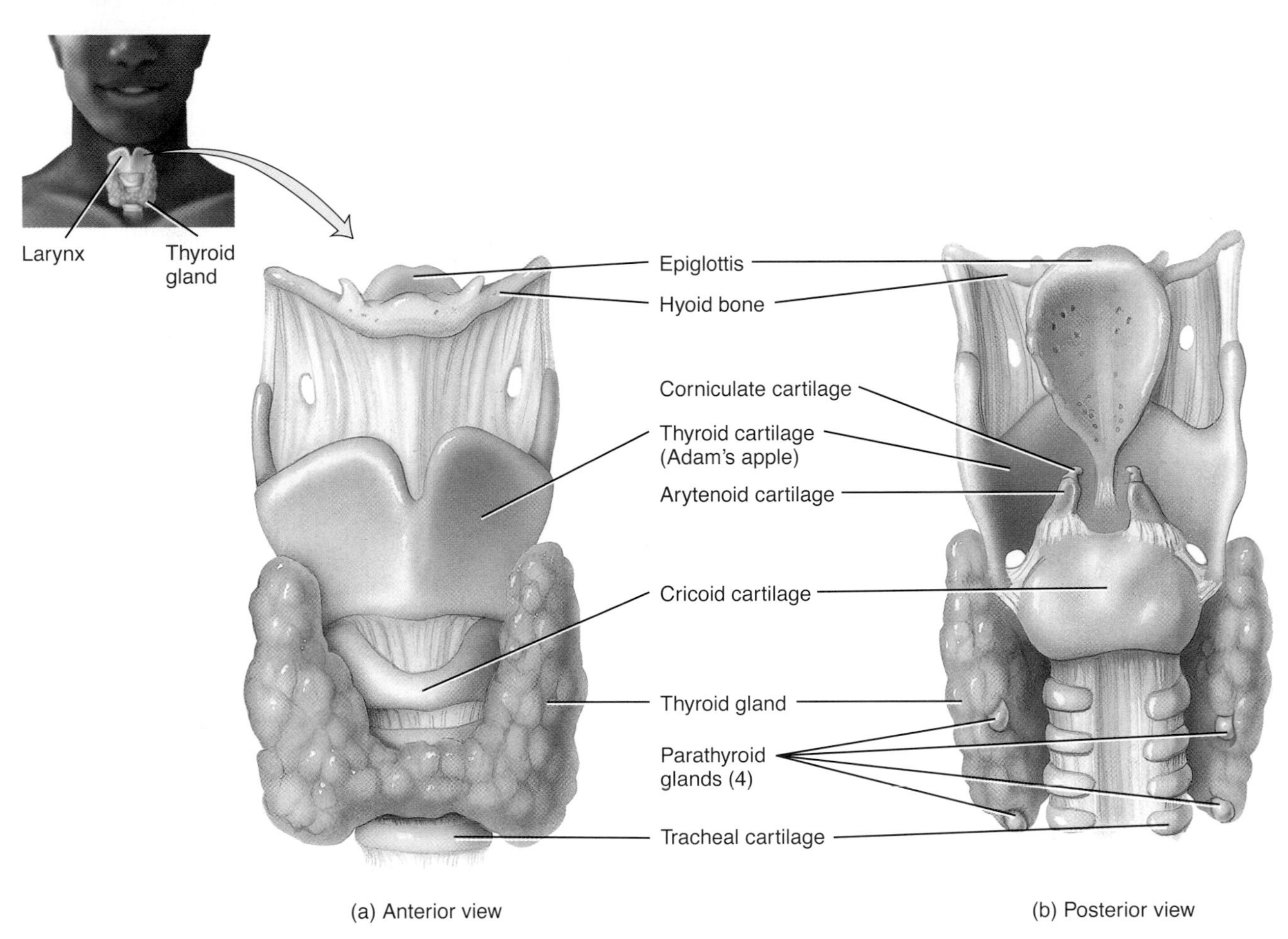

How does the epiglottis prevent foods and liquids from entering the larynx?

Pitch is controlled by the tension of the true vocal cords. If they are pulled taut, they vibrate more rapidly and a higher pitch results. Lower sounds are produced by decreasing the muscular tension. Due to the influence of male sex hormones, vocal cords are usually thicker and longer in males than in females. They therefore vibrate more slowly, giving men a lower range of pitch than women.

Laryngitis is an inflammation of the larynx that is most often caused by a respiratory infection or irritants such as cigarette smoke. Inflammation of the vocal folds causes hoarseness or loss of voice by interfering with the contraction of the folds or by causing them to swell to the point where they cannot vibrate freely. Many long-term smokers acquire a permanent hoarseness from the damage done by chronic inflammation. **Cancer of the larynx** is found almost exclusively in individuals who smoke. The condition is characterized by hoarseness, pain on swallowing, or pain radiating to an ear. Treatment consists of radiation therapy and/or surgery.

■ CHECKPOINT

1. What functions do the respiratory and cardiovascular systems have in common?
2. Compare the structure and functions of the external and internal nose.
3. How does the larynx function in respiration and voice production?

Trachea

The *trachea* (TRĀ-kē-a), or windpipe, is a tubular passageway for air that is located anterior to the esophagus. It extends from the larynx to the upper part of the fifth thoracic vertebra (T5), where it divides into right and left primary bronchi (Figure 18.4).

The wall of the trachea is lined with mucous membrane and is supported by cartilage. The mucous membrane is composed of pseudostratified ciliated columnar epithelium, consisting of ciliated columnar cells, goblet cells, and basal cells (see Table 4.1E on page 78), and provides the same protection against dust as the membrane lining the nasal cavity and larynx. The cilia in the upper respiratory tract move mucus and trapped particles *down* toward the pharynx, but the cilia in the lower respiratory tract move mucus and trapped particles *up* toward the pharynx. The cartilage layer consists of 16 to 20 C-shaped rings of hyaline cartilage stacked one on top of another. The open part of each C-shaped cartilage ring faces the esophagus and permits it to expand slightly into the trachea during swallowing. The solid parts of the C-shaped cartilage rings provide a rigid support so the tracheal wall does not collapse inward and obstruct the air passageway. The rings of cartilage may be felt under the skin below the larynx.

Figure 18.4 Branching of airways from the trachea and lobes of the lungs.

The bronchial tree consists of airways that begin at the trachea and end at the terminal bronchioles.

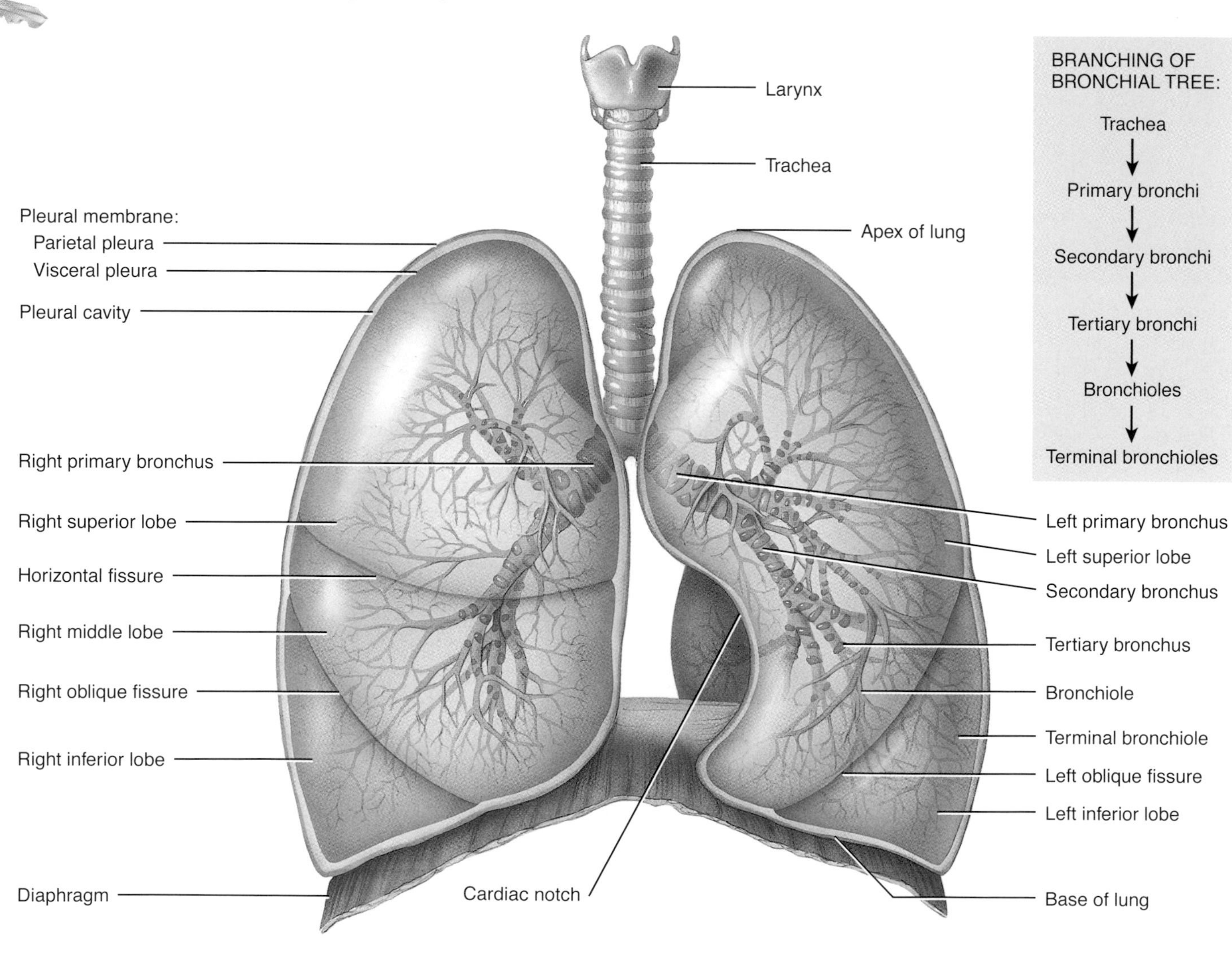

Anterior view

How many lobes and secondary bronchi are present in each lung?

Several conditions may block airflow by obstructing the trachea. The rings of cartilage that support the trachea may be accidentally crushed, the mucous membrane may become inflamed and swell so much that it closes off the passageway, excess mucus secreted by inflamed membranes may clog the lower respiratory passages, or a large object may be aspirated (breathed in). If the obstruction is above the level of the larynx, a **tracheotomy** (trā-kē-O-tō-mē) may be performed. In this procedure, also called a *tracheostomy*, an incision is made in the trachea below the cricoid cartilage and a tracheal tube is inserted to create an emergency air passageway.

Bronchi and Bronchioles

The trachea divides into a ***right primary bronchus*** (BRON-kus = windpipe), which goes to the right lung, and a ***left primary bronchus***, which goes to the left lung (Figure 18.4). Like the trachea, the primary bronchi (BRONG-kē) contain incomplete rings of cartilage and are lined by pseudostratified ciliated columnar epithelium. Pulmonary blood vessels, lymphatic vessels, and nerves enter and exit the lungs with the two bronchi.

On entering the lungs, the primary bronchi divide to form the ***secondary bronchi***, one for each lobe of the lung. (The right lung has three lobes; the left lung has two.) The secondary bronchi continue to branch, forming still smaller bronchi, called ***tertiary bronchi***, that divide several times, ultimately giving rise to smaller ***bronchioles***. Bronchioles, in turn, branch into even smaller tubes called ***terminal bronchioles***. Because the airways resemble an upside-down tree with many branches, their arrangement is known as the ***bronchial tree***.

As the branching becomes more extensive in the bronchial tree, structural changes occur. First, plates of cartilage gradually replace the incomplete rings of cartilage in primary bronchi and finally disappear in the distal bronchioles. Second, as the amount of cartilage decreases, the amount of smooth muscle increases. Smooth muscle encircles the lumen in spiral bands. During exercise, activity in the sympathetic division of the autonomic nervous system (ANS) increases and causes the adrenal medullae to release the hormones epinephrine and norepinephrine. Both chemicals cause relaxation of smooth muscle in the bronchioles, which dilates (widens) the airways. The result is improved airflow, and air reaches the alveoli more quickly.

During an **asthma attack**, bronchiolar smooth muscle goes into spasm. Because there is no supporting cartilage, the spasms can close off the air passageways. Movement of air through constricted bronchioles causes breathing to be more labored. The parasympathetic division of the ANS and mediators of allergic reactions such as histamine also cause narrowing of bronchioles (bronchoconstriction) due to contraction of bronchiolar smooth muscle.

Lungs

The ***lungs*** (= lightweights, because they float) are two spongy, cone-shaped organs in the thoracic cavity. They are separated from each other by the heart and other structures in the mediastinum (see Figure 15.1 on page 365). The ***pleural membrane*** is a double-layered serous membrane that encloses and protects each lung (Figure 18.4). The outer layer is attached to the wall of the thoracic cavity and diaphragm and is called the ***parietal pleura***. The inner layer, the ***visceral pleura***, is attached to the lungs. Between the visceral and parietal pleurae is a narrow space, the ***pleural cavity***, which contains a lubricating fluid secreted by the membranes. This fluid reduces friction between the membranes, allowing them to slide easily over one another during breathing.

The lungs extend from the diaphragm to slightly above the clavicles and lie against the ribs. The broad bottom portion of each lung is its ***base***; the narrow top portion is the ***apex*** (Figure 18.4). The left lung has an indentation, the ***cardiac notch***, in which the heart lies. Due to the space occupied by the heart, the left lung is about 10% smaller than the right lung.

Deep grooves called fissures divide each lung into lobes. The *oblique fissure* divides the left lung into *superior* and *inferior lobes*. The *oblique* and *horizontal fissures* divide the right lung into *superior*, *middle*, and *inferior lobes* (Figure 18.4). Each lobe receives its own secondary bronchus.

Each lung lobe is divided into smaller segments that are supplied by a tertiary bronchus. The segments, in turn, are subdivided into many small compartments called ***lobules*** (Figure 18.5). Each lobule contains a lymphatic vessel, an arteriole, a venule, and a branch from a terminal bronchiole wrapped in elastic connective tissue. Terminal bronchioles subdivide into microscopic branches called ***respiratory bronchioles***, which are lined by nonciliated simple cuboidal epithelium. Respiratory bronchioles, in turn, subdivide into several ***alveolar ducts***. The two or more alveoli that share a common opening to the alveolar duct are called ***alveolar sacs***.

Alveoli

An ***alveolus*** (al-VĒ-ō-lus; plural is ***alveoli***) is a cup-shaped outpouching of an alveolar sac. Many alveoli and alveolar sacs surround each alveolar duct. The walls of alveoli consist mainly of thin *alveolar cells*, which are simple squamous ep-

Figure 18.5 Lobule of the lung.

Alveolar sacs are two or more alveoli that share a common opening into an alveolar duct.

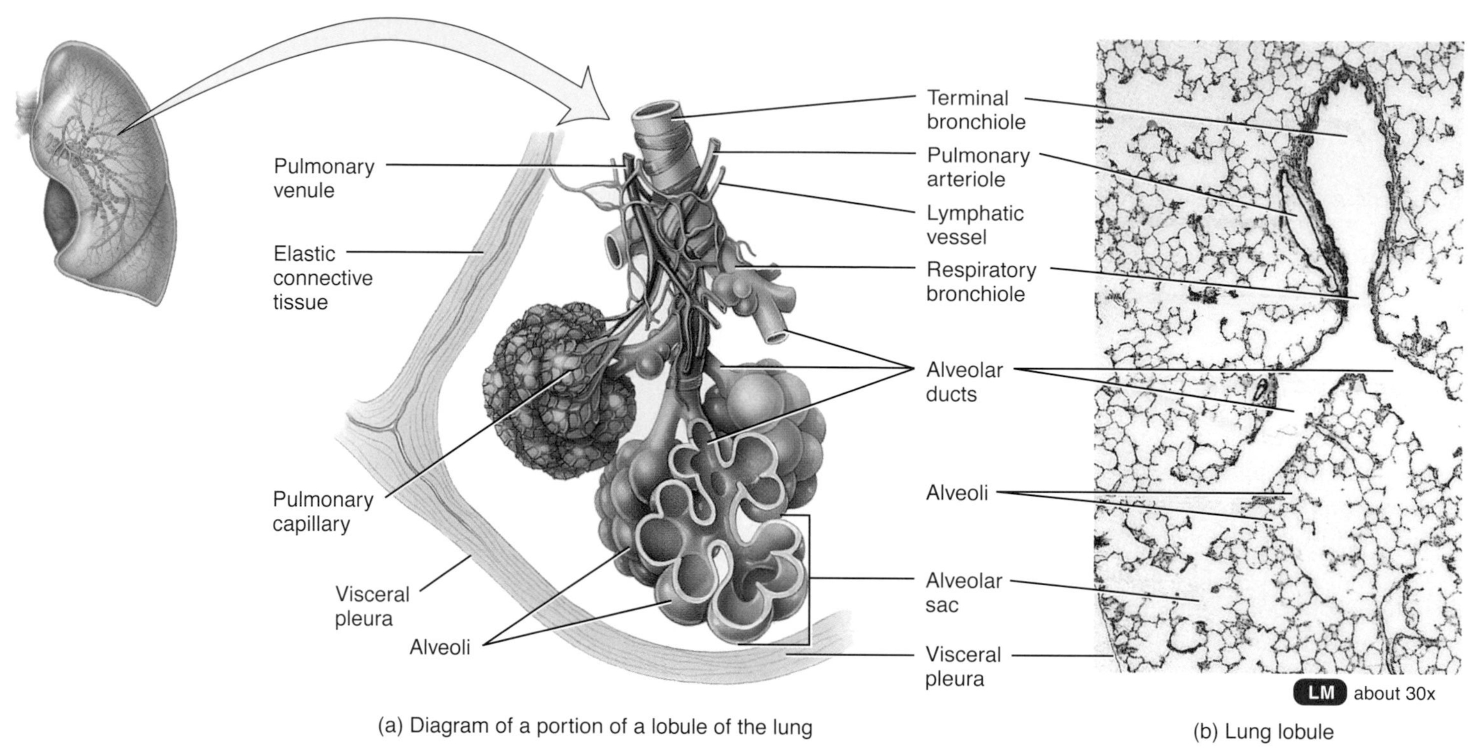

(a) Diagram of a portion of a lobule of the lung

(b) Lung lobule

What are the major parts of a lobule of a lung?

ithelial cells (Figure 18.6). They are the main sites of gas exchange. Scattered among them are *surfactant-secreting cells* that secrete *alveolar fluid*, which keeps the surface between the cells and the air moist. Included in the alveolar fluid is ***surfactant*** (sur-FAK-tant), a mixture of phospholipids and lipoproteins that reduces the tendency of alveoli to collapse. Also present are *alveolar macrophages*, wandering phagocytes that remove fine dust particles and other debris in the alveolar spaces. Underlying the layer of alveolar cells is an elastic basement membrane and a thin layer of connective tissue containing plentiful elastic and reticular fibers (described shortly). Around the alveoli, the pulmonary arteriole and venule form lush networks of blood capillaries (see Figure 18.5a).

The exchange of O_2 and CO_2 between the air spaces in the lungs and the blood takes place by diffusion across the alveolar and capillary walls, which together form the ***respiratory membrane.*** It consists of the following layers (Figure 18.6b):

1. The *alveolar cells* that form the wall of an alveolus.
2. An *epithelial basement membrane* underlying the alveolar cells.
3. A *capillary basement membrane* that is often fused to the epithelial basement membrane.
4. The *endothelial cells* of a capillary.

Despite having several layers, the respiratory membrane is only 0.5 μm* wide. This thin width, far less than the thickness of a sheet of tissue paper, permits O_2 and CO_2 to diffuse efficiently between the blood and alveolar air spaces. Moreover, the lungs contain roughly 300 million alveoli. They provide a huge surface area for the exchange of O_2 and CO_2—about 30 to 40 times greater than the surface area of your skin or half the size of a tennis court!

■ CHECKPOINT

4. What is the bronchial tree? Describe its structure.
5. Where are the lungs located? Distinguish the parietal pleura from the visceral pleura.
6. Where in the lungs does the exchange of O_2 and CO_2 take place?

*1 μm (micrometer) = 1/1,000,000 of a meter or 1/25,000 of an inch.

Figure 18.6 Structure of an alveolus.

The exchange of respiratory gases occurs by diffusion across the respiratory membrane.

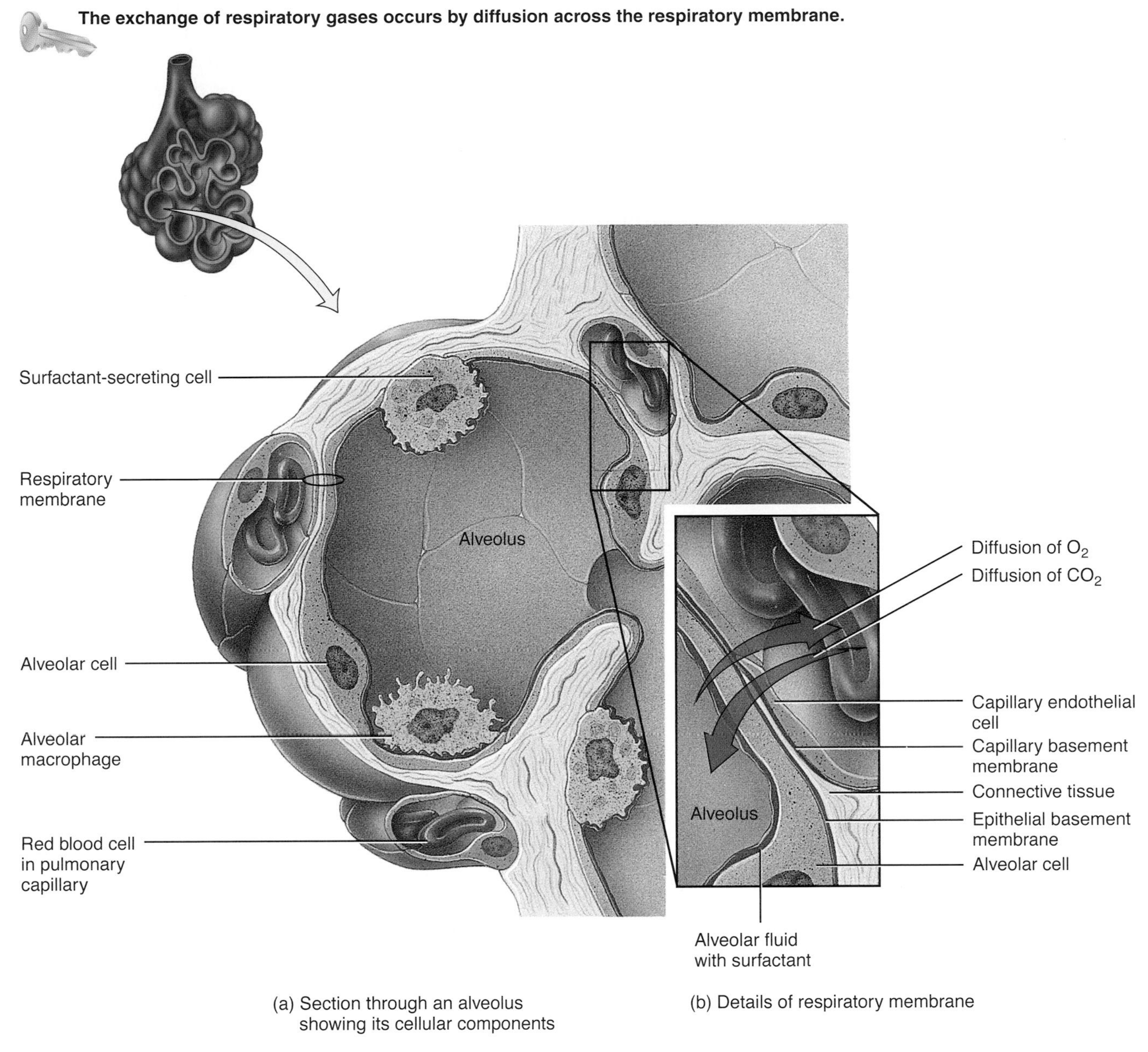

(a) Section through an alveolus showing its cellular components

(b) Details of respiratory membrane

Which cells secrete alveolar fluid?

PULMONARY VENTILATION

OBJECTIVES • Explain how inhalation and exhalation take place.

• Define the various lung volumes and capacities.

Pulmonary ventilation, the flow of air between the atmosphere and the lungs, occurs due to differences in air pressure. We inhale or breathe in when the pressure inside the lungs is less than the atmospheric air pressure. We exhale or breathe out when the pressure inside the lungs is greater than the atmospheric air pressure. Contraction and relaxation of skeletal muscles create the air pressure changes that power breathing.

Muscles of Inhalation and Exhalation

Breathing in is called ***inhalation*** or *inspiration*. The muscles of quiet (unforced) inhalation are the diaphragm, the dome-shaped skeletal muscle that forms the floor of the thoracic cavity, and the external intercostals, which extend between the ribs (Figure 18.7). The diaphragm contracts when it receives nerve impulses from the phrenic nerves. As the diaphragm contracts, it descends and becomes flatter, which causes the volume of the attached lungs to expand. As the external intercostals contract, they pull the ribs upward and outward; the attached lungs follow, further increasing lung volume. Contraction of the diaphragm is responsible for about 75% of the air that enters the lungs during quiet breathing. Advanced pregnancy, obesity, confining clothing, or increased size of the stomach after eating a large meal can impede descent of the diaphragm and may cause shortness of breath.

During deep, labored inhalations, the sternocleidomastoid muscles elevate the sternum, the scalene muscles elevate the two uppermost ribs, and the pectoralis minor muscles elevate the third through fifth ribs. As the ribs and sternum are elevated, the size of the lungs increases (Figure 18.7b). Movements of the pleural membrane aid expansion of the lungs. The parietal and visceral pleurae normally adhere tightly because of the surface tension created by their moist adjoining surfaces. Whenever the thoracic cavity expands, the parietal pleura lining the cavity follows, and the visceral pleura and lungs are pulled along with it.

Breathing out, called ***exhalation*** or *expiration*, begins when the diaphragm and external intercostals relax. Exhalation occurs due to *elastic recoil* of the chest wall and lungs, both of which have a natural tendency to spring back after they have been stretched. Although the alveoli and airways recoil, they don't completely collapse. Because surfactant in alveolar fluid *reduces* elastic recoil, a lack of surfactant causes breathing difficulty by increasing the chance of alveolar collapse.

Figure 18.7 Muscles of inhalation and exhalation and their actions. The pectoralis minor muscle (not shown here) is illustrated in Figures 8.17 and 8.18 on pages 201 and 203.

During quiet inhalation, the diaphragm and external intercostals contract, the lungs expand, and air moves into the lungs. During exhalation, the diaphragm relaxes and the lungs recoil inward, forcing air out of the lungs.

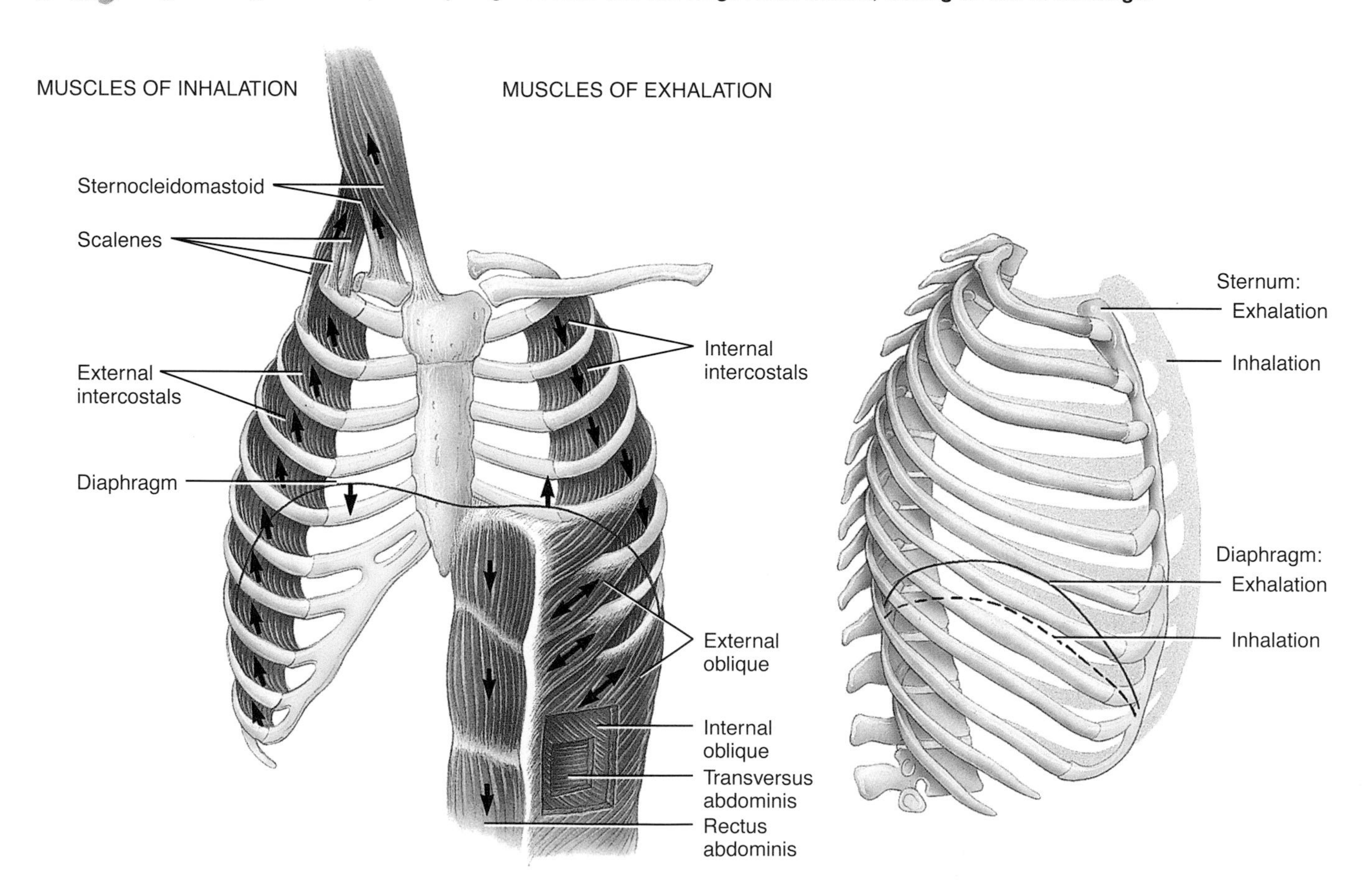

(a) Anterior view of the muscles of inhalation and their actions (left) and the muscles of exhalation and their actions (right)

(b) Lateral view of the changes in size of the thoracic cavity during inhalation and exhalation

What are the main muscles that power your breathing?

Because no muscular contractions are involved, quiet exhalation, unlike quiet inhalation, is a *passive process.* Exhalation becomes *active* only during forceful breathing, such as in playing a wind instrument or during exercise. During these times, muscles of exhalation—the internal intercostals, external oblique, internal oblique, transversus abdominis, and rectus abdominis—contract to move the lower ribs downward and compress the abdominal viscera, thus forcing the diaphragm upward (Figure 18.7a).

Pressure Changes During Ventilation

As the lungs expand, the air molecules inside occupy a larger *volume*, which causes the air *pressure* inside to decrease. (When gas molecules are put into a larger container, they exert a smaller pressure on the walls of the container, in this case the airways and alveoli of the lungs.) Because the atmospheric air pressure is now higher than the ***alveolar pressure***, the air pressure inside the lungs, air moves into the lungs. By contrast, when lung volume decreases, the alveolar pressure increases. (When gas molecules are squeezed into a smaller container, they exert a larger pressure on the walls of the container.) Air then flows from the area of higher pressure in the alveoli to the area of lower pressure in the atmosphere. Figure 18.8 shows the sequence of pressure changes during quiet breathing.

1. At rest just before an inhalation, the air pressure inside the lungs is the same as the pressure of the atmosphere, which is about 760 mm Hg (millimeters of mercury) at sea level.
2. As the diaphragm and external intercostals contract and the overall size of the thoracic cavity increases, the volume of the lungs increases and alveolar pressure de-

Figure 18.8 Pressure changes during pulmonary ventilation.

Air moves into the lungs when alveolar pressure is less than atmospheric pressure, and out of the lungs when alveolar pressure is greater than atmospheric pressure.

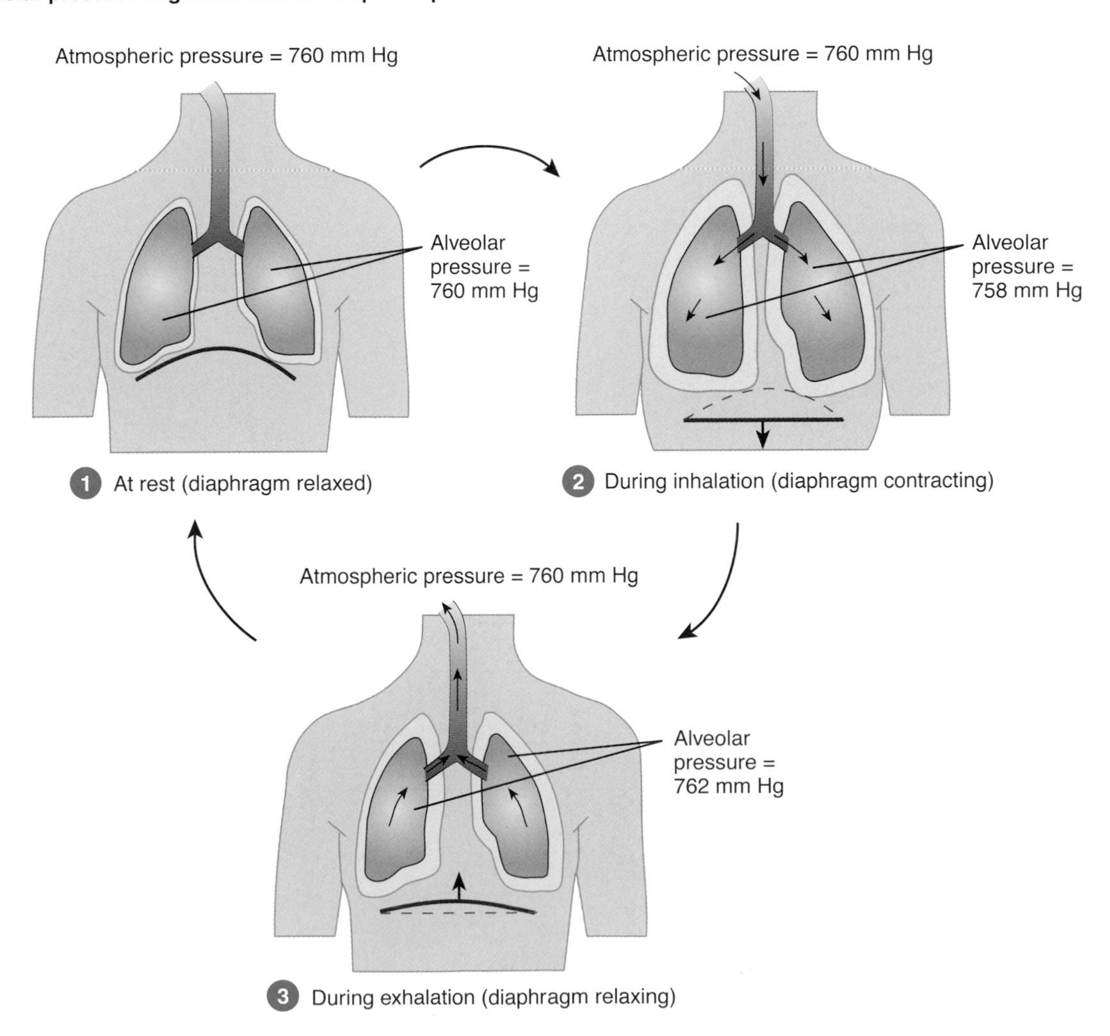

How does the alveolar pressure change during a normal, quiet breath?

creases from 760 to 758 mm Hg. Now there is a pressure difference between the atmosphere and the alveoli, and air flows from the atmosphere (higher pressure) into the lungs (lower pressure).

3. When the diaphragm and external intercostals relax, lung elastic recoil causes the lung volume to decrease, and alveolar pressure rises from 760 to 762 mm Hg. Air then flows from the area of higher pressure in the alveoli to the area of lower pressure in the atmosphere.

Lung Volumes and Capacities

While at rest, a healthy adult breathes about 12 times a minute, with each inhalation and exhalation moving about 500 mL of air into and out of the lungs. The volume of one breath is called the ***tidal volume***. The ***minute ventilation (MV)***—the total volume of air inhaled and exhaled each minute—is equal to breathing rate multiplied by tidal volume:

$$\text{MV} = 12 \text{ breaths/min} \times 500 \text{ mL/breath}$$
$$= 6000 \text{ mL/min or } 6 \text{ liters/min}$$

Tidal volume varies considerably from one person to another and in the same person at different times. About 70% of the tidal volume (350 mL) actually reaches the respiratory bronchioles and alveolar sacs and thus participates in gas exchange. The other 30% (150 mL) does not participate in gas exchange because it remains in the conducting airways of the nose, pharynx, larynx, trachea, bronchi, bronchioles, and terminal bronchioles. Collectively, these conducting airways are known as the ***anatomic dead space***.

The apparatus commonly used to measure respiratory rate and the amount of air inhaled and exhaled during breathing is a ***spirometer*** (*spiro-* = breathe; *meter* = measuring device). The record produced by a spirometer is called a ***spirogram***. Inhalation is recorded as an upward deflection, and exhalation is recorded as a downward deflection (Figure 18.9).

By taking a very deep breath, you can inhale a good deal more than 500 mL. This additional inhaled air, called the ***inspiratory reserve volume***, is about 3100 mL in an average adult male and 1900 mL in an average adult female (Figure 18.9). Even more air can be inhaled if inhalation follows forced exhalation. If you inhale normally and then exhale as forcibly as possible, you should be able to push out consider-

Figure 18.9 Spirogram showing lung volumes and capacities in milliliters (mL). The average values for a healthy adult male and female are indicated, with the values for a female in parentheses. Note that the spirogram is read from right (start of record) to left (end of record).

Lung capacities are combinations of various lung volumes.

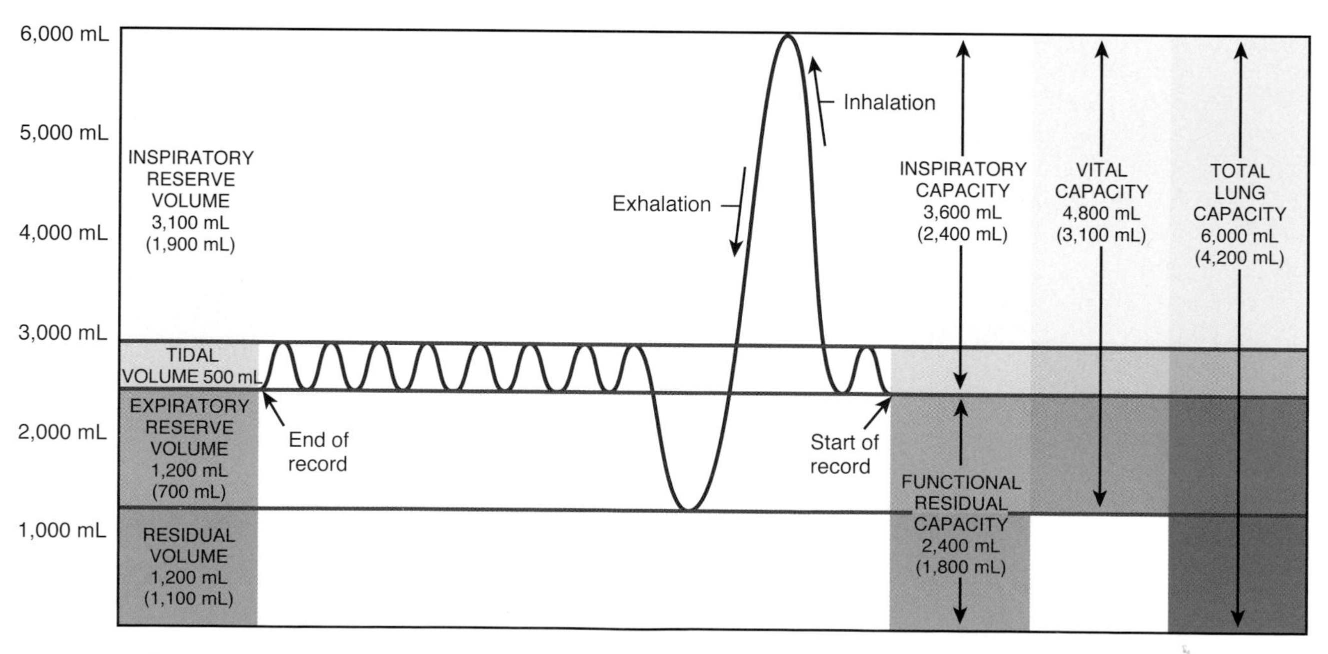

Breathe in as deeply as possible and then exhale as much air as you can. Which lung capacity have you demonstrated?

ably more air in addition to the 500 mL of tidal volume. The extra 1200 mL in males and 700 mL in females is called the ***expiratory reserve volume***. Even after the expiratory reserve volume is expelled, considerable air remains in the lungs and airways. This volume, called the ***residual volume***, amounts to about 1200 mL in males and 1100 mL in females.

Lung *capacities* are combinations of specific lung *volumes* (Figure 18.9). ***Inspiratory capacity*** is the sum of tidal volume and inspiratory reserve volume (500 mL + 3100 mL = 3600 mL in males and 500 mL + 1900 mL = 2400 mL in females). ***Functional residual capacity*** is the sum of residual volume and expiratory reserve volume (1200 mL +1200 mL = 2400 mL in males and 1100 mL + 700 mL = 1800 mL in females). ***Vital capacity*** is the sum of inspiratory reserve volume, tidal volume, and expiratory reserve volume (4800 mL in males and 3100 mL in females). Finally, ***total lung capacity*** is the sum of vital capacity and residual volume (4800 mL + 1200 mL = 6000 mL in males and 3100 mL + 1100 mL = 4200 mL in females). The values given here are typical for young adults. Lung volumes and capacities vary with age (smaller in older people), gender (generally smaller in females), and body size (smaller in shorter people). Lung volumes and capacities provide information about an individual's respiratory status since they are usually abnormal in people with pulmonary disorders.

Breathing Patterns and Modified Respiratory Movements

The term for the normal pattern of quiet breathing is ***eupnea*** (ūp-NĒ-a; *eu-* = good, easy, or normal; *-pnea* = breath). Eupnea can consist of shallow, deep, or combined shallow and deep breathing. A pattern of shallow (chest) breathing, called ***costal breathing***, consists of an upward and outward movement of the chest due to contraction of the external intercostal muscles. A pattern of deep (abdominal) breathing, called ***diaphragmatic breathing***, consists of the outward movement of the abdomen due to the contraction and descent of the diaphragm.

Respirations also provide humans with methods for expressing emotions such as laughing, sighing, and sobbing. Moreover, respiratory air can be used to expel foreign matter from the lower air passages through actions such as sneezing and coughing. Respiratory movements are also modified and controlled during talking and singing. Some of the modified respiratory movements that express emotion or clear the airways are listed in Table 18.1. All these movements are reflexes, but some of them also can be initiated voluntarily.

■ CHECKPOINT

7. Compare what happens during quiet versus labored ventilation.

8. What is the basic difference between a lung volume and a lung capacity?

Table 18.1 Modified Respiratory Movements

Movement	Description
Coughing	A long-drawn and deep inhalation followed by a strong exhalation that suddenly sends a blast of air through the upper respiratory passages. Stimulus for this reflex act may be a foreign body lodged in the larynx, trachea, or epiglottis.
Sneezing	Spasmodic contraction of muscles of exhalation that forcefully expels air through the nose and mouth. Stimulus may be an irritation of the nasal mucosa.
Sighing	A long-drawn and deep inhalation immediately followed by a shorter but forceful exhalation.
Yawning	A deep inhalation through the widely opened mouth producing an exaggerated depression of the mandible. It may be stimulated by drowsiness, fatigue, or someone else's yawning, but precise cause is unknown.
Sobbing	A series of convulsive inhalations followed by a single prolonged exhalation.
Crying	An inhalation followed by many short convulsive exhalations, during which the vocal folds vibrate; accompanied by characteristic facial expressions and tears.
Laughing	The same basic movements as crying, but the rhythm of the movements and the facial expressions usually differ from those of crying.
Hiccupping	Spasmodic contraction of the diaphragm followed by a spasmodic closure of the larynx, which produces a sharp sound on inhalation. Stimulus is usually irritation of the sensory nerve endings of the gastrointestinal tract.

EXCHANGE OF OXYGEN AND CARBON DIOXIDE

OBJECTIVE • Describe the exchange of oxygen and carbon dioxide between alveolar air and blood (external respiration) and between blood and body cells (internal respiration).

Air is a mixture of gases—nitrogen, oxygen, water vapor, carbon dioxide, and others—each of which contributes to the total air pressure. The pressure of a specific gas in a mixture is called its ***partial pressure*** and is denoted as P_x, where the subscript X denotes the chemical formula of the gas. The total pressure of air, the atmospheric pressure, is the sum of all the partial pressures:

$$P_{N_2}\text{ (597.4 mm Hg)} + P_{O_2}\text{ (158.8 mm Hg)}$$
$$+ P_{H_2O}\text{ (3.0 mm Hg)} + P_{CO_2}\text{ (0.3 mm Hg)}$$
$$+ P_{\text{other gases}}\text{ (0.5 mm Hg)}$$
$$= \text{atmospheric pressure (760 mm Hg)}$$

Partial pressures are important because each gas diffuses from areas where its partial pressure is higher to areas where its partial pressure is lower in the body.

External Respiration: Pulmonary Gas Exchange

External respiration, also termed ***pulmonary gas exchange***, is the diffusion of O_2 from air in the alveoli of the lungs to blood in pulmonary capillaries and the diffusion of CO_2 in the opposite direction (Figure 18.10a). External respiration in the lungs converts ***deoxygenated*** (low-oxygen) ***blood*** that comes from the right side of the heart to ***oxygenated*** (high-

Figure 18.10 Changes in partial pressures of oxygen (O_2) and carbon dioxide (CO_2) in mm Hg during external and internal respiration.

Each gas in a mixture of gases diffuses from an area of higher partial pressure of that gas to an area of lower partial pressure of that gas.

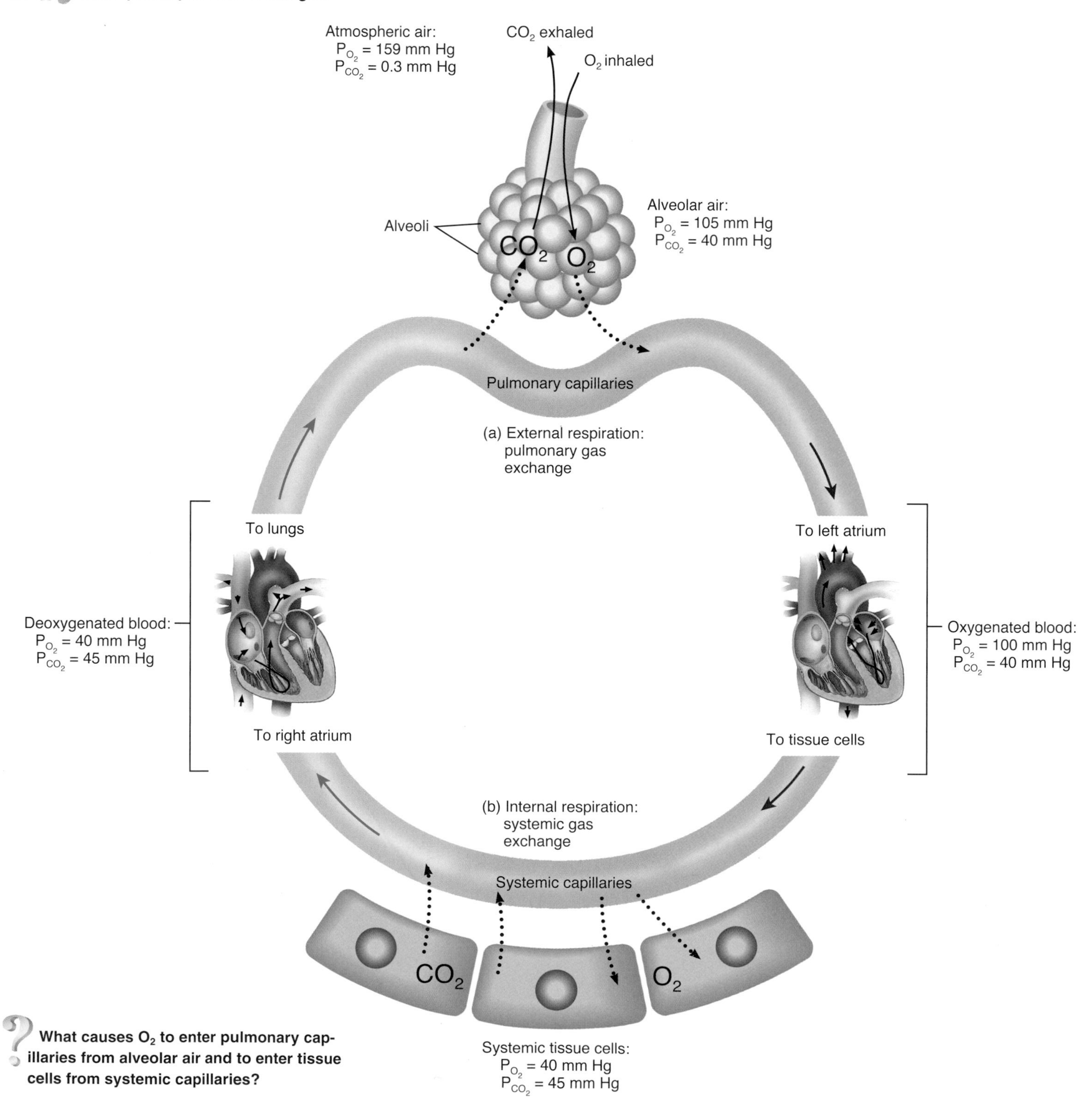

What causes O_2 to enter pulmonary capillaries from alveolar air and to enter tissue cells from systemic capillaries?

oxygen) ***blood*** that returns to the left side of the heart. As blood flows through the pulmonary capillaries, it picks up O_2 from alveolar air and unloads CO_2 into alveolar air. Although this process is commonly called an "exchange" of gases, each gas diffuses *independently* from an area where its partial pressure is higher to an area where its partial pressure is lower. An important factor that affects the rate of external respiration is the total surface area available for gas exchange. Any pulmonary disorder that decreases the functional surface area of the respiratory membrane, for example, emphysema (see page 467), decreases the rate of gas exchange.

O_2 diffuses from alveolar air, where its partial pressure (P_{O_2}) is 105 mm Hg, into the blood in pulmonary capillaries, where P_{O_2} is about 40 mm Hg in a resting person. During exercise, the P_{O_2} of blood entering the pulmonary capillaries is even lower because contracting muscle fibers are using more O_2. Diffusion continues until the P_{O_2} of pulmonary capillary blood increases to 105 mm Hg, matching the P_{O_2} of alveolar air. Blood leaving pulmonary capillaries near alveolar air spaces mixes with a small volume of blood that has flowed through conducting portions of the respiratory system, where gas exchange does not occur. Thus, the P_{O_2} of blood in the pulmonary veins is about 100 mm Hg, slightly less than the P_{O_2} in pulmonary capillaries.

While O_2 is diffusing from alveolar air into deoxygenated blood, CO_2 is diffusing in the opposite direction. The P_{CO_2} of deoxygenated blood is 45 mm Hg in a resting person, compared to the P_{CO_2} of alveolar air, which is 40 mm Hg. Because of this difference in P_{CO_2}, carbon dioxide diffuses from deoxygenated blood into the alveoli until the P_{CO_2} of the blood decreases to 40 mm Hg. Exhalation keeps alveolar P_{CO_2} at 40 mm Hg. Oxygenated blood returning to the left side of the heart in the pulmonary veins thus has a P_{CO_2} of 40 mm Hg.

> As a person ascends in altitude, the total atmospheric pressure decreases, with a parallel decrease in the partial pressure of oxygen. P_{O_2} decreases from 159 mm Hg at sea level to 73 mm Hg at 6000 meters (about 20,000 ft). Alveolar P_{O_2} decreases correspondingly, and less oxygen diffuses into the blood. The common symptoms of **high altitude sickness**—shortness of breath, nausea, and dizziness—are due to a lower level of oxygen in the blood.

Internal Respiration: Systemic Gas Exchange

The left ventricle pumps oxygenated blood into the aorta and through the systemic arteries to systemic capillaries. The exchange of O_2 and CO_2 between systemic capillaries and tissue cells is called ***internal respiration*** or ***systemic gas exchange*** (Figure 18.10b). As O_2 leaves the bloodstream, oxygenated blood is converted into deoxygenated blood. Unlike external respiration, which occurs only in the lungs, internal respiration occurs in tissues throughout the body.

The P_{O_2} of blood pumped into systemic capillaries is higher (100 mm Hg) than the P_{O_2} in tissue cells (about 40 mm Hg at rest) because cells constantly use up O_2 to produce ATP. Due to this pressure difference, oxygen diffuses out of the capillaries into tissue cells, and blood P_{O_2} decreases. While O_2 diffuses from the systemic capillaries into tissue cells, CO_2 diffuses in the opposite direction. Because tissue cells are constantly producing CO_2, the P_{CO_2} of cells (45 mm Hg at rest) is higher than that of systemic capillary blood (40 mm Hg). As a result, CO_2 diffuses from tissue cells through interstitial fluid into systemic capillaries until the P_{CO_2} in the blood increases. The deoxygenated blood then returns to the heart and is pumped to the lungs for another cycle of external respiration.

CHECKPOINT

9. What are the basic differences among pulmonary ventilation, external respiration, and internal respiration?
10. In a person at rest, what is the partial pressure difference that drives diffusion of oxygen into the blood in pulmonary capillaries?

TRANSPORT OF RESPIRATORY GASES

OBJECTIVE • Describe how the blood transports oxygen and carbon dioxide.

The blood transports gases between the lungs and body tissues. When O_2 and CO_2 enter the blood, certain physical and chemical changes occur that aid in gas transport and exchange.

Oxygen Transport

Oxygen does not dissolve easily in water, and therefore only about 1.5% of the O_2 in blood is dissolved in blood plasma, which is mostly water. About 98.5% of blood O_2 is bound to hemoglobin in red blood cells (Figure 18.11).

The heme part of hemoglobin contains four atoms of iron, each capable of binding to a molecule of O_2. Oxygen and ***deoxyhemoglobin*** (Hb) bind in an easily reversible reaction to form ***oxyhemoglobin*** (Hb–O_2):

$$\underset{\text{Deoxyhemoglobin}}{\text{Hb}} + \underset{\text{Oxygen}}{O_2} \underset{\text{Release of } O_2}{\overset{\text{Binding of } O_2}{\rightleftharpoons}} \underset{\text{Oxyhemoglobin}}{\text{Hb–}O_2}$$

When blood P_{O_2} is high, hemoglobin binds with large amounts of O_2 and is *fully saturated;* that is, every available

Figure 18.11 Transport of oxygen and carbon dioxide in the blood.

Most O_2 is transported by hemoglobin as oxyhemoglobin within red blood cells; most CO_2 is transported in blood plasma as bicarbonate ions.

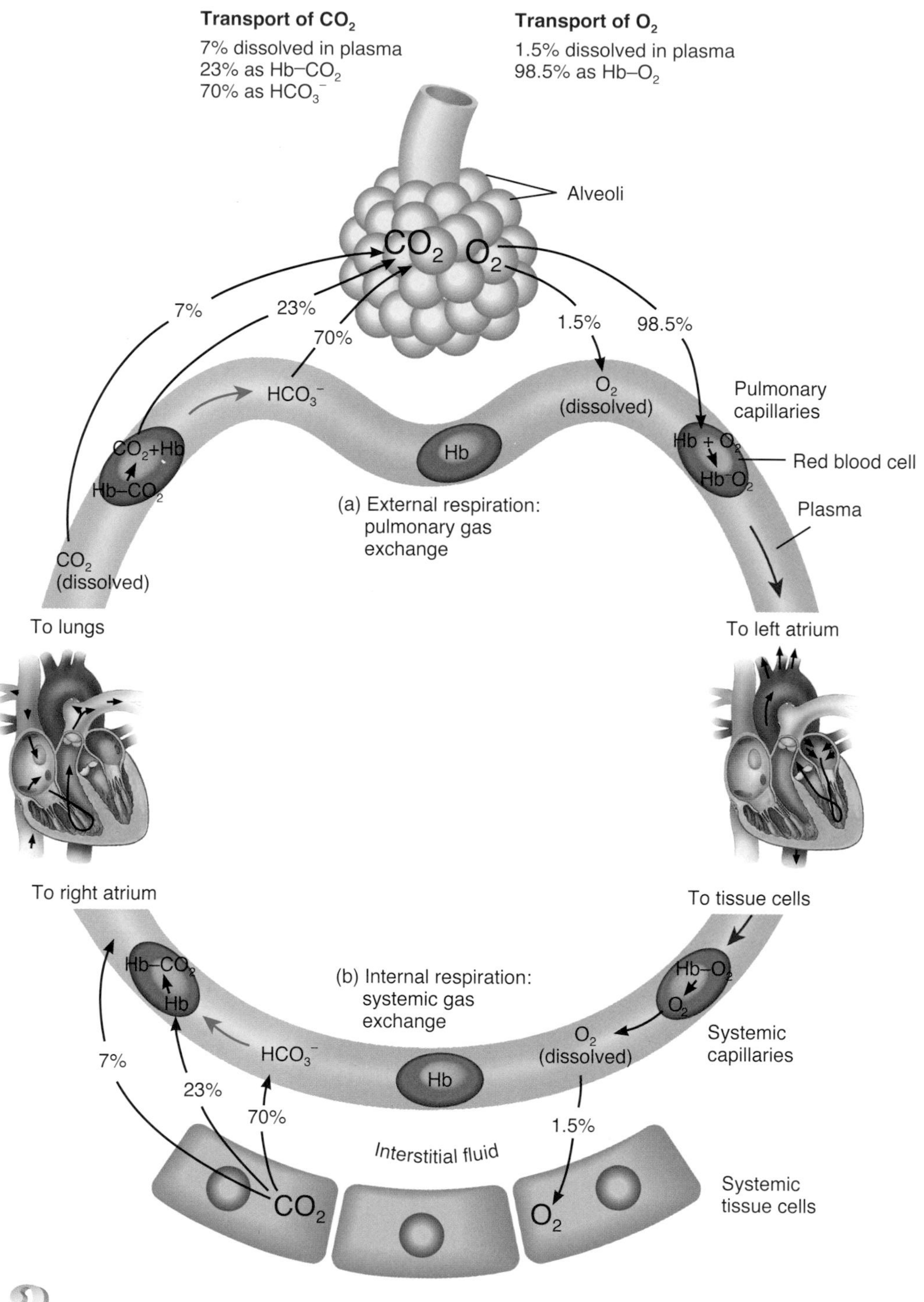

What percentage of oxygen is transported in blood by hemoglobin?

iron atom has combined with a molecule of O_2. When blood P_{O_2} is low, hemoglobin releases O_2. Therefore, in systemic capillaries, where the P_{O_2} is lower, hemoglobin releases O_2, which then can diffuse from blood plasma into interstitial fluid and into tissue cells (Figure 18.11b).

Besides P_{O_2}, several other factors influence the amount of O_2 released by hemoglobin:

1. **Carbon dioxide.** As the P_{CO_2} rises in any tissue, hemoglobin releases O_2 more readily. Thus, hemoglobin releases more O_2 as blood flows through active tissues that are producing more CO_2, such as muscular tissue during exercise.
2. **Acidity.** In an acidic environment, hemoglobin releases O_2 more readily. During exercise, muscles produce lactic acid, which promotes release of O_2 from hemoglobin.
3. **Temperature.** Within limits, as temperature increases, so does the amount of O_2 released from hemoglobin. Active tissues produce more heat, which elevates the local temperature and promotes release of O_2.

Carbon monoxide (CO) is a colorless and odorless gas found in tobacco smoke and in exhaust fumes from automobiles, gas furnaces, and space heaters. CO binds to the heme group of hemoglobin, just as O_2 does, except that CO binds over 200 times more strongly. At a concentration as small as 0.1%, CO combines with half the available hemoglobin molecules and reduces the oxygen-carrying capacity of the blood by 50%. Elevated blood levels of CO cause **carbon monoxide poisoning**, which can cause the lips and oral mucosa to appear bright, cherry-red (the color of hemoglobin with carbon monoxide bound to it). Administering pure oxygen, which speeds up the separation of carbon monoxide from hemoglobin, may rescue the person.

Carbon Dioxide Transport

Carbon dioxide is transported in the blood in three main forms (Figure 18.11):

1. **Dissolved CO_2.** The smallest percentage—about 7%—is dissolved in blood plasma. Upon reaching the lungs, it diffuses into alveolar air and is exhaled.
2. **Bound to amino acids.** A somewhat higher percentage, about 23%, combines with the amino groups of amino acids and proteins in blood. Because the most prevalent protein in blood is hemoglobin (inside red blood cells), most of the CO_2 transported in this manner is bound to hemoglobin. Hemoglobin that has bound CO_2 is termed ***carbaminohemoglobin (Hb–CO₂)***:

$$\underset{\text{Hemoglobin}}{\text{Hb}} + \underset{\text{Carbon dioxide}}{CO_2} \rightleftharpoons \underset{\text{Carbaminohemoglobin}}{\text{Hb–}CO_2}$$

In tissue capillaries P_{CO_2} is relatively high, which promotes formation of carbaminohemoglobin. But in pulmonary capillaries, P_{CO_2} is relatively low, and the CO_2 readily splits apart from hemoglobin and enters the alveoli by diffusion.

3. **Bicarbonate ions.** The greatest percentage of CO_2—about 70%—is transported in blood plasma as ***bicarbonate ions (HCO_3^-)***. As CO_2 diffuses into tissue capillaries and enters the red blood cells, it combines with water to form carbonic acid (H_2CO_3). The enzyme inside red blood cells that drives this reaction is *carbonic anhydrase (CA)*. The carbonic acid then breaks down into hydrogen ions (H^+) and HCO_3^-:

$$\underset{\text{Carbon dioxide}}{CO_2} + \underset{\text{Water}}{H_2O} \overset{CA}{\rightleftharpoons} \underset{\text{Carbonic acid}}{H_2CO_3} \rightleftharpoons \underset{\text{Hydrogen ion}}{H^+} + \underset{\text{Bicarbonate ion}}{HCO_3^-}$$

Thus, as blood picks up CO_2, HCO_3^- accumulates inside RBCs. Some HCO_3^- moves out into the blood plasma, down its concentration gradient. In exchange, chloride ions (Cl^-) move from plasma into the RBCs. This exchange of negative ions, which maintains the electrical balance between blood plasma and RBC cytosol, is known as the *chloride shift*. As a result of these chemical reactions, CO_2 is removed from tissue cells and transported in blood plasma as HCO_3^-.

As blood passes through pulmonary capillaries in the lungs, all these reactions reverse. The CO_2 that was dissolved in plasma diffuses into alveolar air. The CO_2 that was combined with hemoglobin splits and diffuses into the alveoli. The bicarbonate ions (HCO_3^-) reenter the red blood cells from the blood plasma and recombine with H^+ to form H_2CO_3, which splits into CO_2 and H_2O. This CO_2 leaves the red blood cells, diffuses into alveolar air, and is exhaled (Figure 18.11a).

■ **CHECKPOINT**

11. What is the relationship between hemoglobin and P_{O_2}?
12. What factors cause hemoglobin to unload more oxygen as blood flows through capillaries of metabolically active tissues, such as skeletal muscle during exercise?

CONTROL OF RESPIRATION

OBJECTIVE • **Explain how the nervous system controls breathing and list the factors that can alter the rate and depth of breathing.**

At rest, about 200 mL of O_2 are used each minute by body cells. During strenuous exercise, however, O_2 use typically increases 15- to 20-fold in normal healthy adults, and as much as 30-fold in elite endurance-trained athletes. Several mechanisms help match respiratory effort to metabolic demand.

Respiratory Center

The basic rhythm of respiration is controlled by groups of neurons in the brain stem. The area from which nerve impulses are sent to respiratory muscles is called the ***respiratory center*** and consists of groups of neurons in both the medulla oblongata and the pons.

The ***medullary rhythmicity area*** (rith-MIS-i-tē) in the medulla oblongata controls the basic rhythm of respiration. Within the medullary rhythmicity area are both inspiratory and expiratory areas. Figure 18.12 shows the relationships of the inspiratory and expiratory areas during normal quiet breathing and forceful breathing.

During quiet breathing, inhalation lasts for about 2 seconds and exhalation lasts for about 3 seconds. Nerve impulses generated in the ***inspiratory area*** establish the basic rhythm of breathing. While the inspiratory area is active, it generates nerve impulses for about 2 seconds (Figure 18.12a). The impulses propagate to the external intercostal muscles via intercostal nerves and to the diaphragm via the phrenic nerves. When the nerve impulses reach the diaphragm and external intercostal muscles, the muscles contract and inhalation occurs. Even when all incoming nerve connections to the inspiratory area are cut or blocked, neurons in this area still rhythmically discharge impulses that cause inhalation. At the end of 2 seconds, the inspiratory area becomes inactive and nerve impulses cease. With no impulses arriving, the diaphragm and external intercostal muscles relax for about 3 seconds, allowing passive elastic recoil of the lungs and thoracic wall. Then, the cycle repeats.

The neurons of the ***expiratory area*** remain inactive during quiet breathing. However, during forceful breathing, nerve impulses from the inspiratory area activate the expiratory area (Figure 18.12b). Impulses from the expiratory area then cause contraction of the internal intercostal and abdominal muscles, which decreases the size of the thoracic cavity and causes forceful exhalation.

The ***pneumotaxic area*** (noo-mō-TAK-sik; *pneumo-* = air or breath; *-toxic* = arrangement) in the upper pons helps turn off the inspiratory area to shorten the duration of inhalations and to increase breathing rate. The ***apneustic area*** (ap-NOO-stik) in the lower pons sends excitatory impulses to the inspiratory area that activate it and prolong inhalation. The result is a long, deep inhalation.

Regulation of the Respiratory Center

Although the basic rhythm of respiration is set and coordinated by the inspiratory area, the rhythm can be modified in response to inputs from other brain regions, receptors in the peripheral nervous system, and other factors.

Cortical Influences on Respiration

Because the cerebral cortex has connections with the respiratory center, we can voluntarily alter our pattern of breathing. We can even refuse to breathe at all for a short time. Voluntary control is protective because it enables us to prevent water or irritating gases from entering the lungs. The ability to not breathe, however, is limited by the buildup of CO_2 and H^+ in body fluids. When the P_{CO_2} and H^+ concentration

Figure 18.12 Roles of the medullary rhythmicity area in controlling (a) the basic rhythm of quiet respiration and (b) forceful breathing.

During normal, quiet breathing, the expiratory area is inactive. During forceful breathing, the inspiratory area activates the expiratory area.

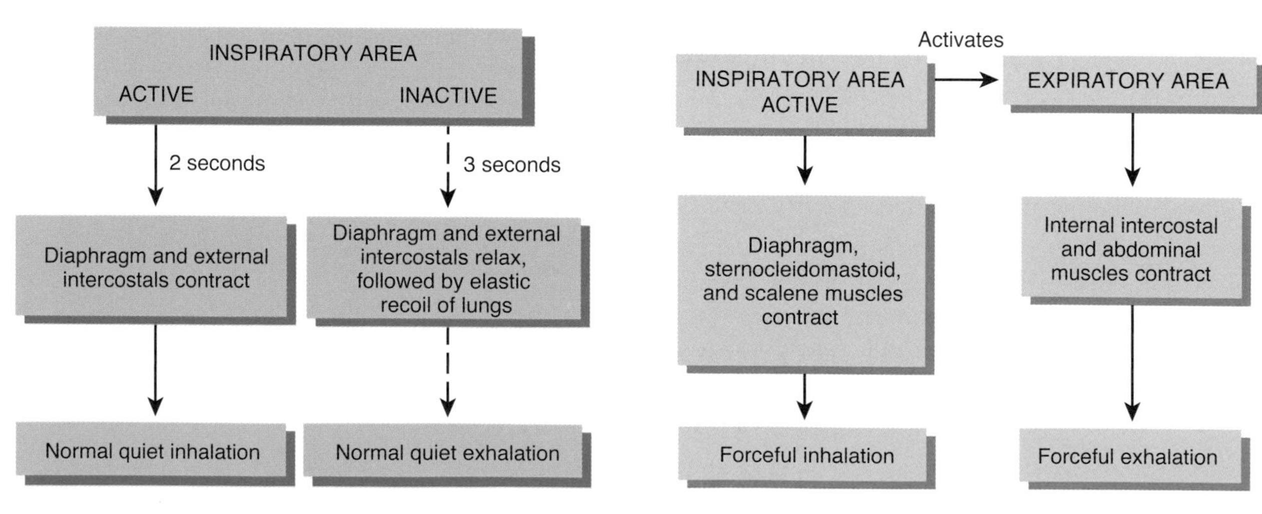

Which nerves convey impulses from the respiratory center to the diaphragm?

Focus on Wellness

Smoking—A Breathtaking Experience

Cigarette smoking is the single most preventable cause of death and disability worldwide. All forms of tobacco use disrupt the body's ability to maintain homeostasis and health. Here are a few of smoking's most deadly effects on the respiratory system.

Where There's Smoke . . .

. . . there is injury. The delicate structure of the alveoli allows you to extract life-giving oxygen and cleanse your body of metabolic wastes. Chronic exposure to smoke gradually destroys lung elasticity. The result is ***emphysema,*** a progressive destruction of the alveoli and collapse of respiratory bronchioles. As a result, oxygen uptake is increasingly more difficult.

Bronchitis is an inflammation of the upper respiratory tract. In smokers, chronic bronchitis may result from the irritation of cigarette smoke. "Smoker's cough" is a symptom of bronchitis. If left untreated, bronchitis increases the risk of more serious infections and permanent airway damage.

Even in the short term, several factors decrease respiratory efficiency in smokers. (1) Nicotine constricts terminal bronchioles, which decreases airflow into and out of the lungs. (2) Carbon monoxide in smoke binds to hemoglobin and reduces its oxygen-carrying capability. (3) Irritants in smoke cause increased mucus secretion by the mucosa of the bronchial tree and swelling of the mucosal lining, both of which impede airflow into and out of the lungs. (4) Irritants in smoke also inhibit the cilia and destroy cilia in the lining of the airways. Thus, excess mucus and foreign debris are not easily removed, which further adds to the difficulty in breathing.

Calling for Cancer

Lung cancer was a rare disease in the early 1900s. Now it is the leading cancer killer for both men and women, thanks primarily to smoking. Cigarette smoke contains a number of known carcinogens, including benzo[*a*]pyrene, *N*-nitrosoamines, and radioactive particles such as radon and polonium, which may initiate and promote the cellular changes leading to lung cancer.

Smoking increases the risk of cancers at other sites as well. Cancers of the oral cavity (mouth, tongue, lip, cheek, and throat) and larynx occur in pipe and cigar smokers as well as cigarette smokers, and in people who use smokeless tobacco. When smoke is inhaled, carcinogens are absorbed from the airways into the bloodstream and can thus contribute to cancers at sites outside the lungs. Cancers of the esophagus, stomach, kidney, pancreas, colon, and urinary bladder are more common in smokers than in nonsmokers. Women who smoke have higher rates of cervical and breast cancer.

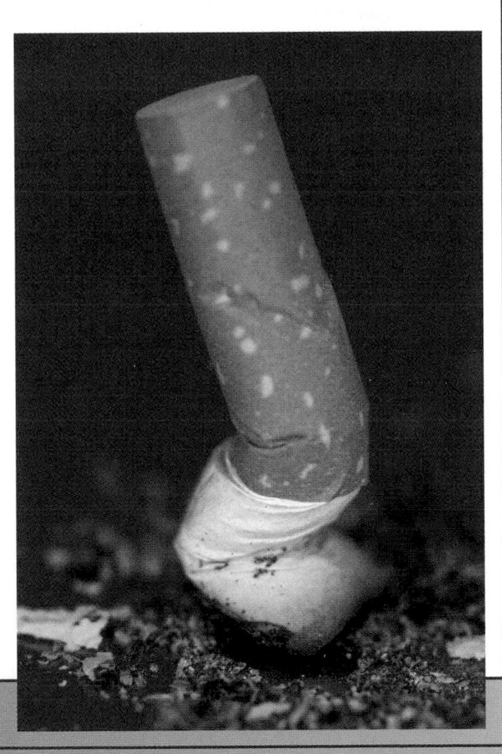

▶ Think It Over . . .

▶ ***Smoking reduces the amount of oxygen reaching body tissues, which interferes with the body's ability to make collagen. How does this help explain why smokers heal more slowly from ulcers, surgery, and fractures?***

reach a certain level, the inspiratory area is strongly stimulated and breathing resumes, whether the person wants it or not. It is impossible for people to kill themselves by voluntarily holding their breath. Even if breath is held long enough to cause fainting, breathing resumes when consciousness is lost. Nerve impulses from the hypothalamus and limbic system also stimulate the respiratory center, allowing emotional stimuli to alter respirations as, for example, in laughing and crying.

Chemoreceptor Regulation of Respiration

Certain chemical stimuli determine how quickly and how deeply we breathe. The respiratory system functions to maintain proper levels of CO_2 and O_2 and is very responsive to changes in the levels of either in body fluids. Sensory neurons that are responsive to chemicals are termed ***chemoreceptors.*** *Central chemoreceptors*, located within the medulla oblongata, respond to changes in H^+ level or P_{CO_2} or both, in cerebrospinal fluid. *Peripheral chemoreceptors*, located within the arch of the aorta and common carotid arteries, are especially sensitive to changes in P_{O_2}, H^+, and P_{CO_2} in the blood.

Because CO_2 is lipid-soluble, it easily diffuses through the plasma membrane into cells, where it combines with water (H_2O) to form carbonic acid (H_2CO_3). Carbonic acid quickly breaks down into H^+ and HCO_3^-. Any increase in

CO_2 in the blood thus causes an increase in H^+ inside cells, and any decrease in CO_2 causes a decrease in H^+.

Normally, the P_{CO_2} in arterial blood is 40 mm Hg. If even a slight increase in P_{CO_2} occurs—a condition called **hypercapnia**—the central chemoreceptors are stimulated and respond vigorously to the resulting increase in H^+ level. The peripheral chemoreceptors also are stimulated by both the high P_{CO_2} and the rise in H^+. In addition, the peripheral chemoreceptors respond to severe **hypoxia**, a deficiency of O_2. If P_{O_2} in arterial blood falls from a normal level of 100 mm Hg to about 50 mm Hg, the peripheral chemoreceptors are strongly stimulated.

The chemoreceptors participate in a negative feedback system that regulates the levels of CO_2, O_2, and H^+ in the blood (Figure 18.13). As a result of increased P_{CO_2}, decreased pH (increased H^+), or decreased P_{O_2}, input from the central and peripheral chemoreceptors causes the inspiratory area to become highly active. Then, the rate and depth of breathing increase. Rapid and deep breathing, called ***hyperventilation***, allows the exhalation of more CO_2 until P_{CO_2} and H^+ are lowered to normal.

If the partial pressure of CO_2 in arterial blood is lower than 40 mm Hg—a condition called **hypocapnia**—the central and peripheral chemoreceptors are not stimulated, and stimulatory impulses are not sent to the inspiratory area. Then, the area sets its own moderate pace until CO_2 accumulates and the P_{CO_2} rises to 40 mm Hg. People who hyperventilate voluntarily and cause hypocapnia can hold their breath for an unusually long period of time. Swimmers were once encouraged to hyperventilate just before a competition. However, this practice is risky because the O_2 level may fall dangerously low and cause fainting before the P_{CO_2} rises high enough to stimulate inhalation. A person who faints on land may suffer bumps and bruises, but one who faints in the water may drown.

Severe deficiency of O_2 depresses activity of the central chemoreceptors and inspiratory area, which then do not respond well to any inputs and send fewer impulses to the muscles of respiration. As the breathing rate decreases or breathing ceases altogether, P_{O_2} falls lower and lower, thereby establishing a positive feedback cycle with a possibly fatal result.

Other Influences on Respiration

Other factors that contribute to regulation of respiration include the following:

- **Limbic system stimulation**. Anticipation of activity or emotional anxiety may stimulate the limbic system, which then sends excitatory input to the inspiratory area, increasing the rate and depth of ventilation.

Figure 18.13 Negative feedback control of breathing in response to changes in blood P_{CO_2}, pH (H^+ level), and P_{O_2}.

An increase in blood P_{CO_2} stimulates the inspiratory center.

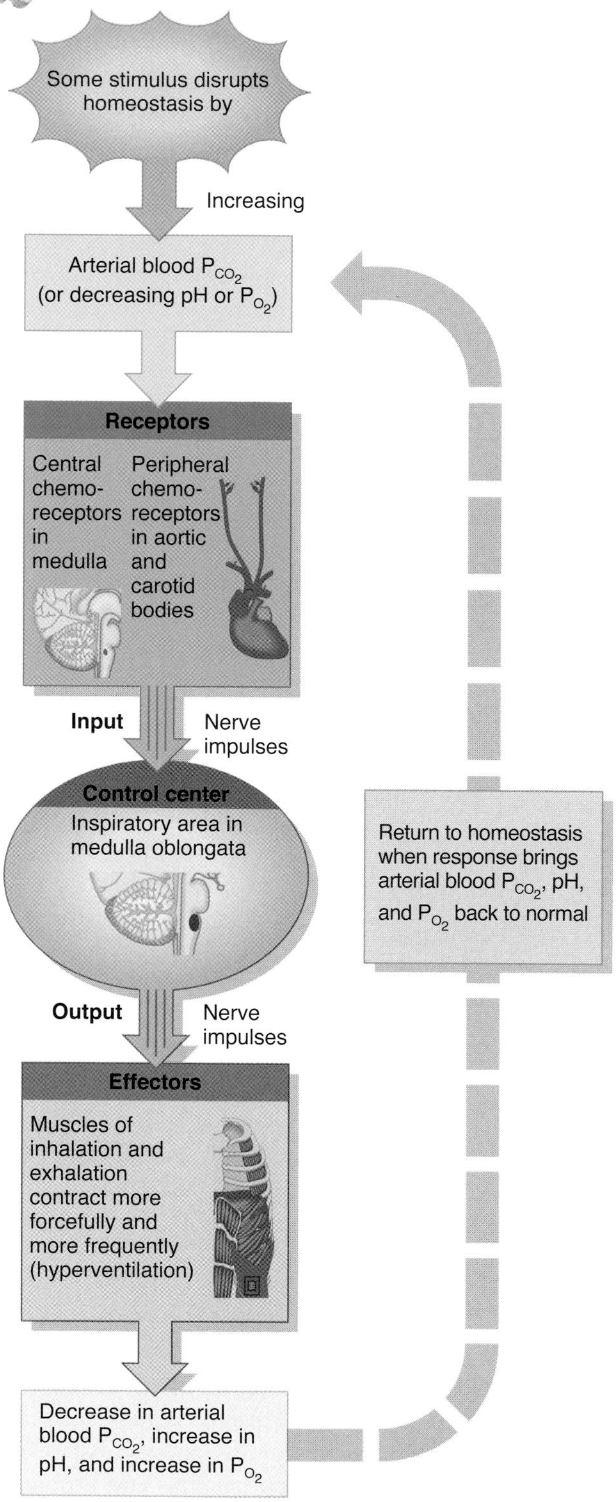

What is the normal arterial blood P_{CO_2}?

- **Proprioceptor stimulation of respiration.** As soon as you start exercising, your rate and depth of breathing increase, even before changes in P_{O_2}, P_{CO_2} or H^+ level occur. The main stimulus for these quick changes in ventilation is input from proprioceptors, which monitor movement of joints and muscles. Nerve impulses from the proprioceptors stimulate the inspiratory area of the medulla oblongata.
- **Temperature.** An increase in body temperature, as occurs during a fever or vigorous muscular exercise, increases the rate of respiration; a decrease in body temperature decreases respiratory rate. A sudden cold stimulus (such as plunging into cold water) causes temporary ***apnea*** (AP-nē-a; *a-* = without; *-pnea* = breath), an absence of breathing.
- **Pain.** A sudden, severe pain brings about brief apnea, but a prolonged somatic pain increases respiratory rate. Visceral pain may slow the respiratory rate.
- **Irritation of airways.** Physical or chemical irritation of the pharynx or larynx brings about an immediate cessation of breathing followed by coughing or sneezing.
- **The inflation reflex.** Located in the walls of bronchi and bronchioles are pressure-sensitive ***stretch receptors.*** When these receptors become stretched during overinflation of the lungs, the inspiratory area is inhibited. As a result, exhalation begins. This reflex is mainly a protective mechanism for preventing excessive inflation of the lungs.

■ **CHECKPOINT**

13. How does the medullary rhythmicity area function in regulating respiration?

14. How do the cerebral cortex, levels of CO_2 and O_2, proprioceptors, inflation reflex, temperature changes, pain, and irritation of the airways modify respiration?

EXERCISE AND THE RESPIRATORY SYSTEM

OBJECTIVE • **Describe the effects of exercise on the respiratory system.**

During exercise, the respiratory and cardiovascular systems make adjustments in response to both the intensity and duration of the exercise. The effects of exercise on the heart were discussed in Chapter 15; here we focus on how exercise affects the respiratory system.

Recall that the heart pumps the same amount of blood to the lungs as to all the rest of the body. Thus, as cardiac output rises, the rate of blood flow through the lungs also increases. If blood flows through the lungs twice as fast as at rest, it picks up twice as much oxygen per minute. In addition, the rate at which O_2 diffuses from alveolar air into the blood increases during maximal exercise because blood flows through a larger percentage of the pulmonary capillaries, providing a greater surface area for diffusion of O_2 into the blood.

When muscles contract during exercise, they consume large amounts of O_2 and produce large amounts of CO_2, forcing the respiratory system to work harder to maintain normal blood gas levels. During vigorous exercise, O_2 consumption and ventilation increase dramatically. At the onset of exercise, an abrupt increase in ventilation, due to activation of proprioceptors, is followed by a more gradual increase. With moderate exercise, the depth of ventilation rather than breathing rate is increased. When exercise is more strenuous, breathing rate also increases.

At the end of an exercise session, an abrupt decrease in ventilation rate is followed by a more gradual decline to the resting level. The initial decrease is due mainly to decreased stimulation of proprioceptors when movement stops or slows. The more gradual decrease reflects the slower return of blood chemistry and blood temperature to resting levels.

AGING AND THE RESPIRATORY SYSTEM

OBJECTIVE • **Describe the effects of aging on the respiratory system.**

With advancing age, the airways and tissues of the respiratory tract, including the alveoli, become less elastic and more rigid; the chest wall becomes more rigid as well. The result is a decrease in lung capacity. In fact, vital capacity (the maximum amount of air that can be expired after maximal inhalation) can decrease as much as 35% by age 70. A decrease in blood level of O_2, decreased activity of alveolar macrophages, and diminished ciliary action of the epithelium lining the respiratory tract occur. Owing to all these age-related factors, elderly people are more susceptible to pneumonia, bronchitis, emphysema, and other pulmonary disorders. Age-related changes in the structure and functions of the lung can also contribute to an older person's reduced ability to perform vigorous exercises, such as running.

■ **CHECKPOINT**

15. How does exercise affect the inspiratory area?

16. What accounts for the decrease in vital capacity with aging?

• • •

To appreciate the many ways that the respiratory system contributes to homeostasis of other body systems, examine Focus on Homeostasis: The Respiratory System on page 466. Next, in Chapter 19, we will see how the digestive system makes nutrients available to body cells so that oxygen provided by the respiratory system can be used for ATP production.

FOCUS ON HOMEOSTASIS

The Respiratory System

Body System		Contribution of the Respiratory System
For all body systems		Provides oxygen and removes carbon dioxide. Helps adjust the pH of body fluids through exhalation of carbon dioxide.
Muscular system		Increased rate and depth of breathing support increased activity of skeletal muscles during exercise.
Nervous system	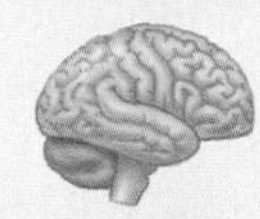	Nose contains receptors for the sense of smell (olfaction). Vibrations of air flowing across the vocal cords produce sounds for speech.
Endocrine system	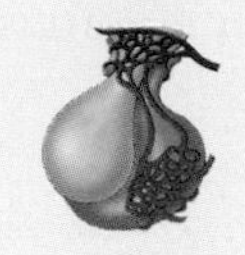	Angiotensin converting enzyme (ACE) in the lungs promotes formation of the hormone angiotensin II, which in turn stimulates the adrenal gland to release the hormone aldosterone.
Cardiovascular system	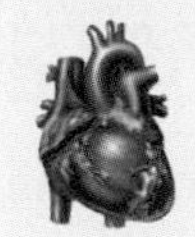	During inhalations, the respiratory pump aids the return of venous blood to the heart.
Lymphatic system and immunity		Hairs in the nose, cilia and mucus in the trachea, bronchi and smaller airways, and alveolar macrophages contribute to nonspecific immunity to disease. The pharynx (throat) contains lymphatic tissue (tonsils). During inhalation, the respiratory pump promotes the flow of lymph.
Digestive system	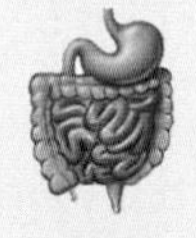	Forceful contraction of the respiratory muscles can assist in defecation.
Urinary system	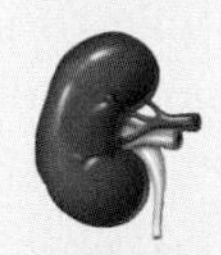	Together, the respiratory and urinary systems regulate the pH of body fluids.
Reproductive systems	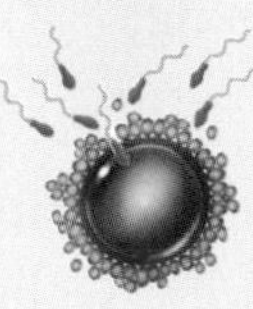	Increased rate and depth of breathing support activity during sexual intercourse. Internal respiration provides oxygen to the developing fetus.

COMMON DISORDERS

Asthma

Asthma (AZ-ma = panting) is a disorder characterized by chronic airway inflammation, airway hypersensitivity to a variety of stimuli, and airway obstruction. The airway obstruction may be due to smooth muscle spasms in the walls of smaller bronchi and bronchioles, swelling of the mucosa of the airways, increased mucus secretion, or damage to the epithelium of the airway. Asthma is at least partially reversible, either spontaneously or with treatment. It affects 3–5% of the U.S. population and is becoming increasingly common in children.

Asthmatics typically react to low concentrations of stimuli that do not normally cause symptoms in people without asthma. Sometimes the trigger is an allergen such as pollen, dust mites, molds, or a particular food. Other common triggers include emotional upset, aspirin, sulfiting agents (used in wine and beer and to keep greens fresh in salad bars), exercise, and breathing cold air or cigarette smoke. Symptoms include difficult breathing, coughing, wheezing, chest tightness, tachycardia, fatigue, moist skin, and anxiety.

Chronic Obstructive Pulmonary Disease

Chronic obstructive pulmonary disease (COPD) is a respiratory disorder characterized by chronic obstruction of airflow. The principal types of COPD are emphysema and chronic bronchitis. In most cases, COPD is preventable because its most common cause is cigarette smoking or breathing secondhand smoke. Other causes include air pollution, pulmonary infection, occupational exposure to dusts and gases, and genetic factors.

Emphysema

Emphysema (em′-fi-SĒ-ma = blown up or full of air) is a disorder characterized by destruction of the walls of the alveoli, which produces abnormally large air spaces that remain filled with air during exhalation. With less surface area for gas exchange, O_2 diffusion across the respiratory membrane is reduced. Blood O_2 level is somewhat lowered, and any mild exercise that raises the O_2 requirements of the cells leaves the patient breathless. As increasing numbers of alveolar walls are damaged, lung elastic recoil decreases due to loss of elastic fibers, and an increasing amount of air becomes trapped in the lungs at the end of exhalation. Over several years, added respiratory exertion increases the size of the chest cage, resulting in a "barrel chest." Emphysema is a common precursor to the development of lung cancer.

Chronic Bronchitis

Chronic bronchitis is a disorder characterized by excessive secretion of bronchial mucus accompanied by a cough. Inhaled irritants lead to chronic inflammation with an increase in the size and number of mucous glands and goblet cells in the airway epithelium. The thickened and excessive mucus produced narrows the airway and impairs the action of cilia. Thus, inhaled pathogens become embedded in airway secretions and multiply rapidly. Besides a cough, symptoms of chronic bronchitis are shortness of breath, wheezing, cyanosis, and pulmonary hypertension.

Lung Cancer

In the United States, ***lung cancer*** is the leading cause of cancer death in both males and females. At the time of diagnosis, lung cancer is usually well advanced. Most people with lung cancer die within a year of the diagnosis, and the overall survival rate is only 10–15%. About 85% of lung cancer cases are due to smoking, and the disease is 10 to 30 times more common in smokers than nonsmokers. Exposure to secondhand smoke also causes lung cancer and heart disease. Other causes of lung cancer are ionizing radiation, such as x-rays, and inhaled irritants, such as asbestos and radon gas.

Symptoms of lung cancer may include a chronic cough, spitting blood from the respiratory tract, wheezing, shortness of breath, chest pain, hoarseness, difficulty swallowing, weight loss, anorexia, fatigue, bone pain, confusion, problems with balance, headache, anemia, low blood platelet count, and jaundice.

Pneumonia

Pneumonia or ***pneumonitis*** is an acute infection or inflammation of the alveoli. It is the most common infectious cause of death in the United States, where an estimated 4 million cases occur annually. When certain microbes enter the lungs of susceptible individuals, they release damaging toxins, stimulating inflammation and immune responses that have damaging side effects. The toxins and immune response damage alveoli and bronchial mucous membranes; inflammation and edema cause the alveoli to fill with debris and fluid, interfering with ventilation and gas exchange. The most common cause is the bacterium *Streptococcus pneumoniae*, but other bacteria, viruses, or fungi may also cause pneumonia.

Tuberculosis

The bacterium *Mycobacterium tuberculosis* produces an infectious, communicable disease called ***tuberculosis (TB)*** that most often affects the lungs and the pleurae but may involve other parts of the body. Once the bacteria are inside the lungs, they multiply and cause inflammation, which stimulates neutrophils and macrophages to migrate to the area and engulf the bacteria to prevent their spread. If the immune system is not impaired, the bacteria may remain dormant for life. Impaired immunity may enable the bacteria to escape into blood and lymph to infect other organs. In many people, symptoms—fatigue, weight loss, lethargy, anorexia, a low-grade fever, night sweats, cough, dyspnea, chest pain, and spitting blood (hemoptysis)—do not develop until the disease is advanced.

Coryza and Influenza

Hundreds of viruses, especially the *rhinoviruses* (*rhin-* = nose), can cause ***coryza*** (ko-RĪ-za) or the ***common cold***. Typical symptoms include sneezing, excessive nasal secretion, dry cough, and congestion. The uncomplicated common cold is not usually accompanied by a fever. Complications include sinusitis, asthma, bronchitis, ear infections, and laryngitis.

Influenza (flu) is also caused by a virus. Its symptoms include chills, fever (usually higher than 101°F, or 38°C), headache, and muscular aches. Coldlike symptoms appear as the fever subsides.

Pulmonary Edema

Pulmonary edema is an abnormal accumulation of interstitial fluid in the interstitial spaces and alveoli of the lungs. The edema may arise from increased pulmonary capillary permeability (pulmonary origin) or increased pulmonary capillary pressure due to congestive heart failure (cardiac origin). The most common symptom is painful or labored breathing. Other symptoms include wheezing, rapid breathing rate, restlessness, a feeling of suffocation, cyanosis, paleness, and excessive perspiration.

MEDICAL TERMINOLOGY AND CONDITIONS

Abdominal thrust maneuver First-aid procedure to clear the airways of obstructing objects. It is performed by applying a quick upward thrust between the navel and lower ribs that causes sudden elevation of the diaphragm and forceful, rapid expulsion of air from the lungs, forcing air out of the trachea to eject the obstructing object. Also used to expel water from the lungs of near-drowning victims before resuscitation is begun. Also known as the ***Heimlich maneuver*** (HĪM-lik ma-NOO-ver).

Asphyxia (as-FIK-sē-a; *sphyxia* = pulse) Oxygen starvation due to low atmospheric oxygen or interference with ventilation, external respiration, or internal respiration.

Aspiration (as′-pi-RĀ-shun) Inhalation into the bronchial tree of a substance other than air, for instance, water, food, or a foreign body.

Bronchoscopy The visual examination of the bronchi through a ***bronchoscope***, an illuminated, tubular instrument that is passed through the mouth (or nose), larynx, and trachea into the bronchi.

Cystic fibrosis (CF) An inherited disease of secretory epithelia that affects the airways, liver, pancreas, small intestine, and sweat glands. Clogging and infection of the airways leads to difficulty in breathing and eventual destruction of lung tissue.

Dyspnea (DISP-nē-a; *dys-* = painful, difficult) Painful or labored breathing.

Epistaxis (ep′-i-STAK-sis) Loss of blood from the nose due to trauma, infection, allergy, malignant growths, or bleeding disorders. It can be arrested by cautery with silver nitrate, electrocautery, or firm packing. Also called ***nosebleed***.

Hypoxia (hī-POK-sē-a; *hypo-* = below or under) A deficiency of O_2 at the tissue level that may be caused by a low P_{O_2} in arterial blood, as from high altitudes; too little functioning hemoglobin in the blood, as in anemia; inability of the blood to carry O_2 to tissues fast enough to sustain their needs, as in heart failure; or inability of tissues to use O_2 properly, as in cyanide poisoning.

Mechanical ventilation The use of an automatically cycling device (ventilator or respirator) to assist breathing. A plastic tube is inserted into the nose or mouth and the tube is attached to a device that forces air into the lungs. Exhalation occurs passively due to the elastic recoil of the lungs.

Pleurisy Inflammation of the pleural membranes, which causes friction during breathing that can be quite painful when the swollen membranes rub against each other. Also known as ***pleuritis***.

Rales (RĀLS) Sounds sometimes heard in the lungs that resemble bubbling or rattling. Different types are due to the presence of an abnormal type or amount of fluid or mucus within the bronchi or alveoli, or to bronchoconstriction that causes turbulent airflow.

Respiratory distress syndrome (RDS) A breathing disorder of premature newborns in which the alveoli do not remain open due to a lack of surfactant. Surfactant reduces surface tension and is necessary to prevent the collapse of alveoli during exhalation.

Respiratory failure A condition in which the respiratory system either cannot supply enough O_2 to maintain metabolism or cannot eliminate enough CO_2 to prevent respiratory acidosis (a higher-than-normal H^+ level in interstitial fluid).

Rhinitis (rī-NĪ-tis; *rhin-* = nose) Chronic or acute inflammation of the mucous membrane of the nose.

Sudden infant death syndrome (SIDS) Death of infants between the ages of 1 week and 12 months thought to be due to hypoxia that occurs while sleeping in a prone position (on the stomach) and rebreathing exhaled air trapped in a depression of the mattress. It is now recommended that normal newborns be placed on their backs for sleeping (remember: “back to sleep”).

Tachypnea (tak′-ip-NĒ-a; *tachy-* = rapid) Rapid breathing rate.

Wheeze (HWĒZ) A whistling, squeaking, or musical high-pitched sound during breathing resulting from a partially obstructed airway.

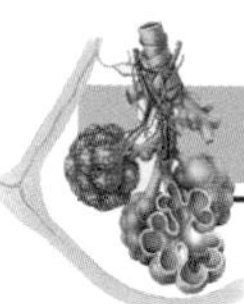

STUDY OUTLINE

Organs of the Respiratory System (p. 446)

1. Respiratory organs include the nose, pharynx, larynx, trachea, bronchi, and lungs, and they act with the cardiovascular system to supply oxygen and remove carbon dioxide from the blood.
2. The external portion of the nose is made of cartilage and skin and is lined with mucous membrane. Openings to the exterior are the external nares.
3. The internal portion of the nose, divided from the external portion by the septum, communicates with the paranasal sinuses and nasopharynx through the internal nares.
4. The nose is adapted for warming, moistening, and filtering air; olfaction; and serving as a resonating chamber for special sounds.
5. The pharynx (throat), a muscular tube lined by a mucous membrane, is divided into the nasopharynx, oropharynx, and laryngopharynx.
6. The nasopharynx functions in respiration. The oropharynx and laryngopharynx function both in digestion and in respiration.
7. The larynx connects the pharynx and the trachea. It contains the thyroid cartilage (Adam’s apple), the epiglottis, the cricoid cartilage, arytenoid cartilages, false vocal cords, and true vocal cords. Taut true vocal cords produce high pitches; relaxed ones produce low pitches.
8. The trachea (windpipe) extends from the larynx to the primary bronchi. It is composed of smooth muscle and C-shaped rings of cartilage and is lined with pseudostratified ciliated columnar epithelium.
9. The bronchial tree consists of the trachea, primary bronchi, secondary bronchi, tertiary bronchi, bronchioles, and terminal bronchioles.
10. Lungs are paired organs in the thoracic cavity enclosed by the pleural membrane. The parietal pleura is the outer layer; the visceral pleura is the inner layer.
11. The right lung has three lobes separated by two fissures; the left lung has two lobes separated by one fissure plus a depression, the cardiac notch.

12. Each lobe consists of lobules, which contain lymphatic vessels, arterioles, venules, terminal bronchioles, respiratory bronchioles, alveolar ducts, alveolar sacs, and alveoli.
13. Exchange of gases (oxygen and carbon dioxide) in the lungs occurs across the respiratory membrane, a thin "sandwich" consisting of alveolar cells, basement membrane, and endothelial cells of a capillary.

Pulmonary Ventilation (p. 453)

1. Pulmonary ventilation (breathing) consists of inhalation and exhalation, the movement of air into and out of the lungs. Air flows from higher to lower pressure.
2. Inhalation occurs when alveolar pressure falls below atmospheric pressure. Contraction of the diaphragm and external intercostals expands the volume of the lungs. Increased volume of the lungs decreases alveolar pressure, and air moves from higher to lower pressure, from the atmosphere into the lungs.
3. Exhalation occurs when alveolar pressure is higher than atmospheric pressure. Relaxation of the diaphragm and external intercostals decreases lung volume, and alveolar pressure increases so that air moves from the lungs to the atmosphere.
4. The sternocleidomastoids, scalenes, and pectoralis minors contribute to forced inhalation. Forced exhalation involves contraction of the internal intercostals, external oblique, internal oblique, transversus abdominis, and rectus abdominis.
5. The minute ventilation is the total air taken in during 1 minute (breathing rate per minute multiplied by tidal volume).
6. The lung volumes are tidal volume, inspiratory reserve volume, expiratory reserve volume, and residual volume.
7. Lung capacities, the sum of two or more lung volumes, include inspiratory, functional residual, vital, and total.

Exchange of Oxygen and Carbon Dioxide (p. 457)

1. The partial pressure of a gas (P) is the pressure exerted by that gas in a mixture of gases.
2. Each gas in a mixture of gases exerts its own pressure and behaves as if no other gases are present.
3. In external and internal respiration, O_2 and CO_2 move from areas of higher partial pressure to areas of lower partial pressure.
4. External respiration is the exchange of gases between alveolar air and pulmonary blood capillaries. It is aided by a thin respiratory membrane, a large alveolar surface area, and a rich blood supply.
5. Internal respiration is the exchange of gases between systemic tissue capillaries and systemic tissue cells.

Transport of Respiratory Gases (p. 459)

1. Most oxygen, 98.5%, is carried by the iron atoms of the heme in hemoglobin; 1.5% is dissolved in plasma.
2. The association of O_2 and hemoglobin is affected by P_{O_2}, pH, temperature, and P_{CO_2}.
3. Hypoxia refers to O_2 deficiency at the tissue level.
4. Carbon dioxide is transported in three ways. About 7% is dissolved in plasma, 23% combines with the globin of hemoglobin, and 70% is converted to bicarbonate ions (HCO_3^-).

Control of Respiration (p. 461)

1. The respiratory center consists of a medullary rhythmicity area (inspiratory and expiratory areas) in the medulla oblongata and groups of neurons in the pons.
2. The inspiratory area sets the basic rhythm of respiration.
3. Respirations may be modified by several factors, including cortical influences; chemical stimuli, such as levels of O_2, CO_2, and H^+; limbic system stimulation; proprioceptor input; temperature; pain; the inflation reflex; and irritation to the airways.

Exercise and the Respiratory System (p. 465)

1. The rate and depth of ventilation change in response to both the intensity and duration of exercise.
2. The abrupt increase in ventilation at the start of exercise is due to neural changes that send excitatory impulses to the inspiratory area in the medulla oblongata. The more gradual increase in ventilation during moderate exercise is due to chemical and physical changes in the bloodstream.

Aging and the Respiratory System (p. 465)

1. Aging results in decreased vital capacity, decreased blood level of O_2, and diminished alveolar macrophage activity.
2. Elderly people are more susceptible to pneumonia, emphysema, bronchitis, and other pulmonary disorders.

SELF-QUIZ

1. Which of the following is NOT true concerning the pharynx?
   a. Food, drink, and air pass through the oropharynx and laryngopharynx.
   b. The auditory (eustachian) tubes have openings in the nasopharynx.
   c. The pseudostratified ciliated epithelium of the nasopharynx helps move dust-laden mucus toward the mouth.
   d. The palatine and lingual tonsils are located in the laryngopharynx.
   e. The wall of the pharynx is composed of skeletal muscle lined with mucous membranes.
2. During speaking, you raise your voice's pitch. This is possible because
   a. the epiglottis vibrates rapidly
   b. you have increased the air pressure pushing against the vocal cords
   c. you have increased the tension on the true vocal cords
   d. your true vocal cords have become thicker and longer
   e. the true vocal cords begin to vibrate more slowly
3. Johnny is having an asthma attack and feels as if he cannot breathe. Why?
   a. His diaphragm is not contracting.

b. Spasms in the bronchiole smooth muscle have blocked airflow to the alveoli.

c. Excess mucus production is interfering with airflow into the lungs.

d. The epiglottis has closed and air is not entering the lungs.

e. Insufficient surfactant is being produced.

4. Which sequence of events best describes inhalation?

a. contraction of diaphragm → increase in size of thoracic cavity → decrease in alveolar pressure **b.** relaxation of diaphragm → decrease in size of thoracic cavity → increase in alveolar pressure **c.** contraction of diaphragm → decrease in size of thoracic cavity → decrease in alveolar pressure **d.** relaxation of diaphragm → increase in size of thoracic cavity → increase in alveolar pressure **e.** contraction of diaphragm → decrease in size of thoracic cavity → increase in alveolar pressure

5. Which of the following does NOT help keep air passages clean?

a. nostril hairs **b.** alveolar macrophages
c. capillaries in the nasal cavities
d. cilia in the upper and lower respiratory tracts **e.** mucus

6. If the total pressure of a mixture of gases is 760 mm Hg and gas Z makes up 20% of the total mixture, then the partial pressure of gas Z would be:

a. 152 mm Hg **b.** 175 mm Hg **c.** 225 mm Hg
d. 608 mm Hg **e.** 760 mm Hg

7. How does hypercapnia affect respiration?

a. It increases the rate of respiration.
b. It decreases the rate of respiration.
c. It causes hypoventilation.
d. It does not change the rate of respiration.
e. It activates stretch receptors in the lungs.

8. Air would flow into the lungs along the following route:

1. bronchioles **2.** primary bronchi
3. secondary bronchi **4.** terminal bronchioles
5. tertiary bronchi **6.** trachea

a. 6, 1, 2, 3, 5, 4 **b.** 6, 5, 3, 4, 2, 1 **c.** 6, 2, 3, 5, 4, 1
d. 6, 2, 3, 5, 1, 4 **e.** 6, 1, 4, 5, 3, 2

9. Match the following:

____ **a.** normally inactive; when activated, causes contraction of internal intercostals and abdominal muscles and forced exhalation
____ **b.** located in pons; stimulates inspiratory area to prolong inhalation
____ **c.** sets basic rhythm of respiration; located in medulla
____ **d.** transmits inhibitory impulses to inspiratory area; located in pons
____ **e.** allows voluntary alteration of breathing patterns

A. inspiratory area
B. expiratory area
C. pneumotaxic area
D. apneustic area
E. cerebral cortex

10. Under normal body conditions, hemoglobin releases oxygen more readily when

a. body temperature increases
b. blood acidity decreases
c. blood pH increases
d. blood oxygen partial pressure is high
e. blood CO_2 is low

11. Match the following:

____ **a.** decreased carbon dioxide levels
____ **b.** normal, quiet breathing
____ **c.** rapid breathing
____ **d.** exchange of gases between the blood and lungs
____ **e.** inhalation and exhalation
____ **f.** increased carbon dioxide levels
____ **g.** exchange of gases between blood and tissue cells
____ **h.** absence of breathing

A. external respiration
B. apnea
C. hypercapnia
D. eupnea
E. internal respiration
F. hypocapnia
G. pulmonary ventilation
H. hyperventilation

12. Which of the following statements is NOT true concerning the lungs?

a. The lungs contain about 300 million alveoli.
b. The left lung is thicker and broader because the liver lies below it.
c. The right lung is composed of three lobes.
d. The top portion of the lung is the apex.
e. The lungs are surrounded by a serous membrane.

13. Exhalation

a. occurs when alveolar pressure reaches 758 mm Hg
b. is normally considered an active process requiring muscle contraction
c. occurs when alveolar pressure is greater than atmospheric pressure
d. involves the expansion of the pleural membranes
e. occurs when the atmospheric pressure is equal to the pressure in the lungs

14. In which structures would you find simple squamous epithelium?

a. secondary bronchi **b.** larynx and pharynx
c. tertiary bronchi **d.** primary bronchi
e. alveoli

15. Overinflation of the lungs is prevented by

a. the inflation reflex **b.** pain in the pleural membranes **c.** nerve impulses from proprioceptors **d.** control from the cerebral cortex **e.** controlling blood pressure

16. The function of goblet cells in the nasal cavities is to

a. warm the air entering the nose
b. produce mucus to trap inhaled dust
c. increase the surface area inside the nose
d. help produce speech
e. exchange O_2 and CO_2 within the nasal cavities

17. Decreasing the surface area of the respiratory membrane would affect
   a. internal respiration b. inhalation c. speech
   d. external respiration e. mucus production
18. In which form is carbon dioxide NOT carried in the blood?
   a. bicarbonate ion
   b. bound to globin
   c. oxyhemoglobin
   d. carbaminohemoglobin
   e. dissolved in plasma
19. Fill in the blanks in the following chemical reactions.

   CO_2 + ______ ⇌ H_2CO_3 ⇌ H^+ + ______

   a. HCO_3^-, O_2 b. HCO_3^-, H_2O c. H^+, H_2O
   d. O_2, HCO_3^- e. H_2O, HCO_3^-
20. Of the following, which would have the highest partial pressure of oxygen?
   a. alveolar air at the end of exhalation
   b. rapidly contracting skeletal muscle fibers
   c. alveolar air immediately after inhalation
   d. blood flowing into the lungs from the right side of the heart
   e. blood returning to the heart from the tissue cells
21. Match the following:

____	a. forceful exhalation of air	A. vital capacity
____	b. inspiratory reserve volume + tidal volume + expiratory reserve volume	B. inspiratory reserve volume
____	c. volume of air moved during normal quiet breathing	C. residual volume
____	d. air remaining after forced exhalation	D. expiratory reserve volume
____	e. forceful inhalation of air	E. tidal volume

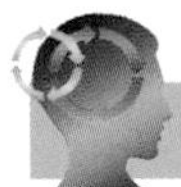

CRITICAL THINKING APPLICATIONS

1. Your three-year-old nephew Levi likes to get his own way all the time! Right now, Levi wants to eat 20 chocolate kisses (1 for each finger and toe), but you'll only give him one for each year of his age. He is at this moment "holding my breath until I turn blue and won't you be sorry!" Is he in danger of death?
2. Katie was diagnosed with exercise-induced asthma after she reported trouble catching her breath during a swim meet. Exercise-induced asthma is a particularly annoying condition for an athlete because the body's response to exercise is the exact opposite of the body's need. Explain this statement.
3. Brianna has a flare for being dramatic. "I can't come to work today," she whispered, "I've got laryngitis and a horrible case of coryza." What is wrong with Brianna?
4. The entire tour group was in fine health when they left the coast of China for their next stop—Tibet! After touring the mountainous area for a day, many of the group felt dizzy, nauseous and exhausted. They were hyperventilating and could not catch their breath. The local physician had seen this condition many times before in people that did not take the time to acclimate to the mountains. What caused the tour group's symptoms?

ANSWERS TO FIGURE QUESTIONS

18.1 The conducting zone of the respiratory system includes the nose, pharynx, larynx, trachea, bronchi, and bronchioles (except the respiratory bronchioles).

18.2 Air molecules flow through the external nares, the nasal cavity, and then the internal nares.

18.3 During swallowing, the epiglottis closes over the larynx to block food and liquids from entering.

18.4 There are two lobes and two secondary bronchi in the left lung and three lobes and three secondary bronchi in the right lung.

18.5 A lung lobule includes a lymphatic vessel, arteriole, venule, and branch of a terminal bronchiole wrapped in elastic connective tissue.

18.6 Surfactant-secreting cells secrete alveolar fluid, which includes surfactant.

18.7 The main muscles that cause quiet breathing are the diaphragm and external intercostals.

18.8 Alveolar pressure is 758 mm Hg during inhalation; alveolar pressure during exhalation is 762 mm Hg.

18.9 You demonstrate vital capacity when you breathe in as deeply as possible and then exhale as much air as you can.

18.10 Oxygen enters pulmonary capillaries from alveolar air and enters tissue cells from systemic capillaries due to differences in P_{O_2}.

18.11 Hemoglobin transports about 98.5% of the oxygen carried in blood.

18.12 The phrenic nerves stimulate the diaphragm to contract.

18.13 Normal arterial blood P_{CO_2} is 40 mm Hg.

chapter 19 THE DIGESTIVE SYSTEM

did you know? ***While a little alcohol may reduce the risk of heart disease, the overconsumption of alcohol leads to many health problems. The organ that sustains the most damage from alcohol abuse is the liver. The liver converts alcohol into acetaldehyde, which is even more toxic to the body than alcohol. This process is associated with the deposition of fatty compounds in the liver. If drinking continues, this condition progresses to an inflammatory condition called alcohol hepatitis, and then to cirrhosis, a condition in which scar tissue replaces functional liver tissue. Risk of cancer of the liver is 30% higher in people with cirrhosis.***

Focus on Wellness, page 494

www.wiley.com/college/apcentral

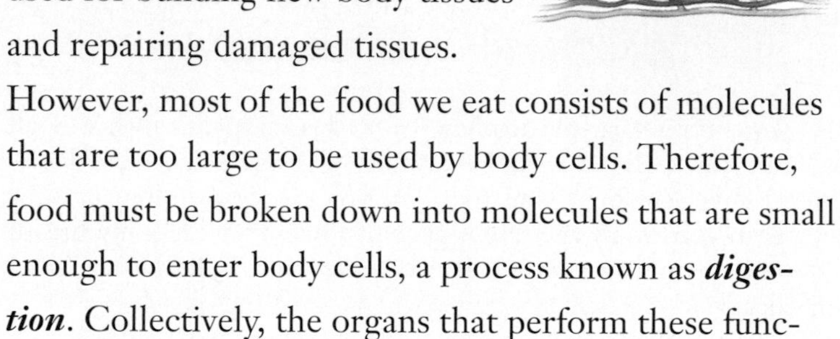

The food we eat contains a variety of nutrients, which are used for building new body tissues and repairing damaged tissues. However, most of the food we eat consists of molecules that are too large to be used by body cells. Therefore, food must be broken down into molecules that are small enough to enter body cells, a process known as ***digestion***. Collectively, the organs that perform these functions are known as the ***digestive system***.

The medical specialty that deals with the structure, function, diagnosis, and treatment of diseases of the stomach and intestines is ***gastroenterology*** (gas′-trō-en′-ter-OL-ō-jē; *gastro-* = stomach; *entero-* = intestines; *-logy* = study of). The medical specialty that deals with the diagnosis and treatment of disorders of the rectum and anus is ***proctology*** (prok-TOL-ō-jē; *proct-* = rectum).

looking back to move ahead . . .

- Mucous Membranes (page 90)
- Serous Membranes (page 91)
- Smooth Muscle Tissue (page 186)
- Muscles that Move the Mandible (page 195)
- Negative Feedback System (page 8)
- Simple Columnar Epithelium (page 75)
- Carbohydrates, Lipids, Proteins (pages 31–37)
- Enzymes (page 37)

OVERVIEW OF THE DIGESTIVE SYSTEM

OBJECTIVE • Identify the organs of the digestive system and their basic functions.

Two groups of organs compose the digestive system (Figure 19.1): the gastrointestinal tract and the accessory digestive organs. The ***gastrointestinal (GI) tract*** is a continuous tube that extends from the mouth to the anus. The GI tract contains food from the time it is eaten until it is digested and absorbed or eliminated from the body. Organs of the gastrointestinal tract include the mouth, pharynx, esophagus, stomach, small intestine, and large intestine. The teeth, tongue, salivary glands, liver, gallbladder, and pancreas serve as ***accessory digestive organs***. Teeth aid in the physical breakdown of food, and the tongue assists in chewing and swallowing. The other accessory digestive organs never come into direct contact with food. The secretions that they produce or store flow into the GI tract through ducts and aid in the chemical breakdown of food.

Overall, the digestive system performs six basic processes:

1. **Ingestion**. This process involves taking foods and liquids into the mouth (eating).
2. **Secretion**. Each day, cells within the walls of the GI tract and accessory organs secrete a total of about 7 liters of water, acid, buffers, and enzymes into the lumen of the tract.
3. **Mixing and propulsion**. Alternating contraction and relaxation of smooth muscle in the walls of the GI tract mix

Figure 19.1 Organs of the digestive system and related structures.

Organs of the gastrointestinal (GI) tract are the mouth, pharynx, esophagus, stomach, small intestine, and large intestine. Accessory digestive organs are the teeth, tongue, salivary glands, liver, gallbladder, and pancreas.

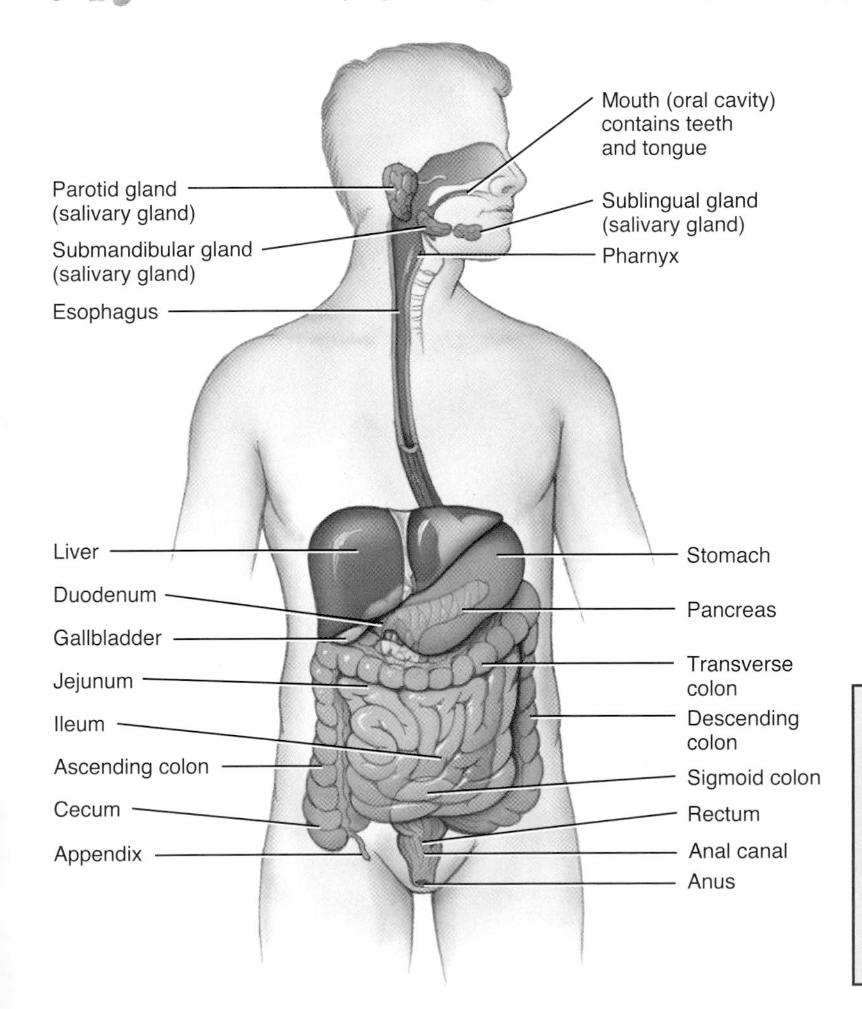

Functions of the Digestive System

1. Ingestion: taking food into the mouth.
2. Secretion: release of water, acid, buffers, and enzymes into the lumen of the GI tract.
3. Mixing and propulsion: churning and pushing food through the GI tract.
4. Digestion: mechanical and chemical breakdown of food.
5. Absorption: passage of digested products from the GI tract into the blood and lymph.
6. Defecation: elimination of feces from the GI tract.

Right lateral view of head and neck and anterior view of trunk

Which accessory digestive organs assist in the physical breakdown of food?

food and secretions and propel them toward the anus. The ability of the GI tract to mix and move material along its length is termed ***motility***.

4. **Digestion**. Mechanical and chemical processes break down ingested food into small molecules. In ***mechanical digestion*** the teeth cut and grind food before it is swallowed, and then smooth muscles of the stomach and small intestine churn the food. As a result, food molecules become dissolved and thoroughly mixed with digestive enzymes. In ***chemical digestion*** the large carbohydrate, lipid, protein, and nucleic acid molecules in food are broken down into smaller molecules by digestive enzymes.
5. **Absorption**. The entrance of ingested and secreted fluids, ions, and the small molecules that are products of digestion into the epithelial cells lining the lumen of the GI tract is called ***absorption***. The absorbed substances pass into interstitial fluid and then into blood or lymph and circulate to cells throughout the body.
6. **Defecation**. Wastes, indigestible substances, bacteria, cells shed from the lining of the GI tract, and digested materials that were not absorbed leave the body through the anus in a process called ***defecation***. The eliminated material is termed ***feces***.

CHECKPOINT

1. Which components of the digestive system are GI tract organs and which are accessory digestive organs?
2. Which organs of the digestive system come in contact with food, and what are some of their digestive functions?

LAYERS OF THE GI TRACT AND THE OMENTUM

OBJECTIVE • Describe the four layers that form the wall of the gastrointestinal tract.

The wall of the GI tract, from the lower esophagus to the anal canal, has the same basic, four-layered arrangement of tissues. The four layers of the tract, from the inside out, are the mucosa, submucosa, muscularis, and serosa (Figure 19.2).

Figure 19.2 Layers of the gastrointestinal tract. Variations in this basic plan may be seen in the stomach (Figure 19.8), small intestine (Figure 19.13), and large intestine (Figure 19.16).

The four layers of the GI tract from inside to outside are the mucosa, submucosa, muscularis, and serosa.

Mesentery
Enteric neurons in submucosa
Gland in mucosa
Vein
Duct of gland outside tract (such as pancreas)
Glands in submucosa
Lymphatic tissue
Artery
Nerve
Lumen
MUCOSA:
Epithelium
Lamina propria
Muscularis mucosae
Enteric neurons in muscularis
SUBMUCOSA
MUSCULARIS:
Circular muscle
Longitudinal muscle
SEROSA:
Areolar connective tissue
Epithelium

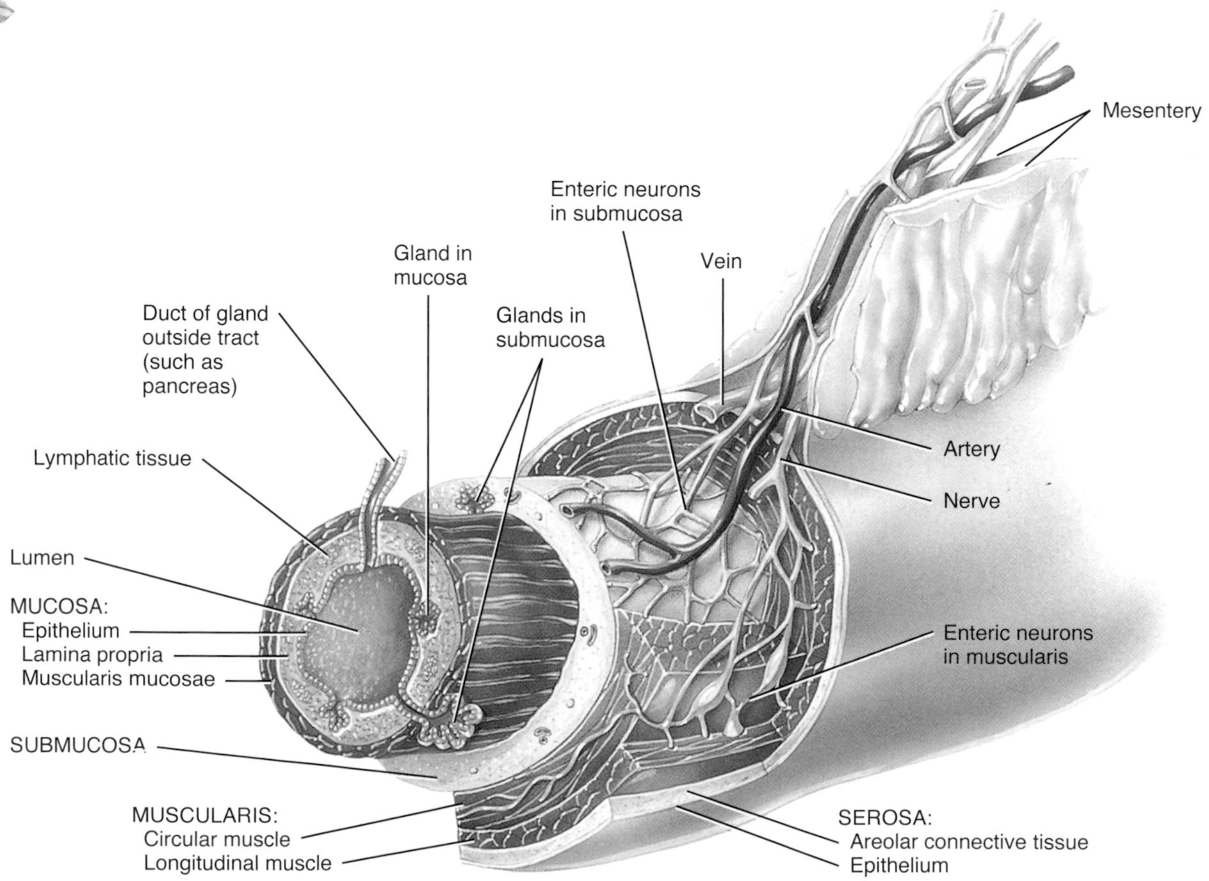

What is the function of the nerves in the wall of the gastrointestinal tract?

1. **Mucosa**. The ***mucosa***, or inner lining of the tract, is a mucous membrane. It is composed of a layer of epithelium in direct contact with the contents of the GI tract, a layer of areolar connective tissue called the ***lamina propria***, and a thin layer of smooth muscle called the ***muscularis mucosae***. Contractions of the muscularis mucosae create folds in the mucosa that increase the surface area for digestion and absorption. The mucosa also contains prominent lymphatic nodules that protect against the entry of pathogens through the GI tract.
2. **Submucosa**. The ***submucosa*** consists of areolar connective tissue that binds the mucosa to the muscularis. It contains many blood and lymphatic vessels that receive absorbed food molecules. Also located in the submucosa are networks of neurons that are a part of the ***enteric nervous system (ENS)***, the "brain of the gut." ENS neurons within the submucosa control the secretions of the organs of the GI tract.
3. **Muscularis**. As its name implies, the ***muscularis*** of the GI tract is a thick layer of muscle. In the mouth, pharynx, and upper esophagus, it consists in part of *skeletal muscle* that produces voluntary swallowing. Skeletal muscle also forms the external anal sphincter, which permits voluntary control of defecation. Recall that a sphincter is a thick circle of muscle around an opening. In the rest of the tract, the muscularis consists of *smooth muscle*, usually arranged as an inner sheet of circular fibers and an outer sheet of longitudinal fibers. Involuntary contractions of these smooth muscles help break down food physically, mix it with digestive secretions, and propel it along the tract. ENS neurons within the muscularis control the frequency and strength of its contractions.
4. **Serosa and peritoneum**. The ***serosa***, the outermost layer around organs of the GI tract below the diaphragm, is a membrane composed of simple squamous epithelium and areolar connective tissue. The serosa secretes a slippery, watery fluid that allows the tract to glide easily against other organs. The serosa is also called the *visceral peritoneum* (per′-i-tō-NE-um = to stretch over). Recall from Chapter 4 that the *peritoneum* is the largest serous membrane of the body. The *parietal peritoneum* lines the wall of the abdominal cavity; the visceral peritoneum covers organs in the cavity.

In addition to binding the organs to each other and to the walls of the abdominal cavity, the peritoneal folds contain blood vessels, lymphatic vessels, and nerves that supply the abdominal organs. The ***greater omentum*** (ō-MENT-um = fat skin) drapes over the transverse colon and small intestine like a "fatty apron" (Figure 19.3a, b). The many lymph nodes of the greater omentum contribute macrophages and anti-

Figure 19.3 Views of the abdomen and pelvis. The relationship of the parts of the peritoneum (greater omentum and mesentery) to each other and to organs of the digestive system is shown.

The peritoneum is the largest serous membrane in the body.

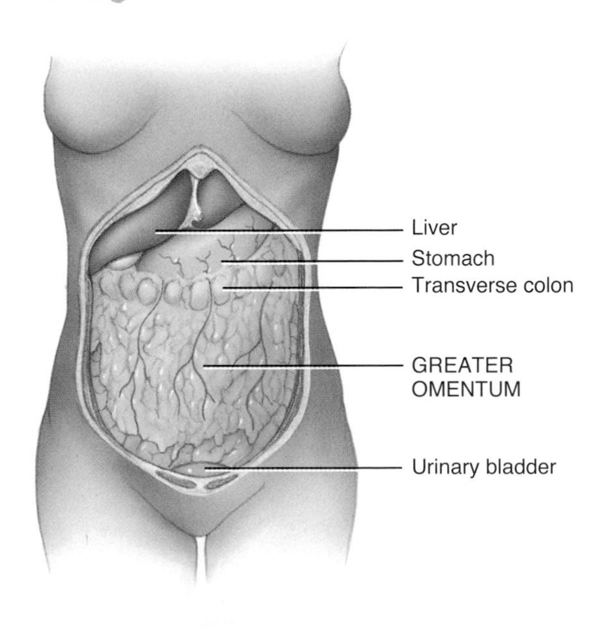

(a) Anterior view

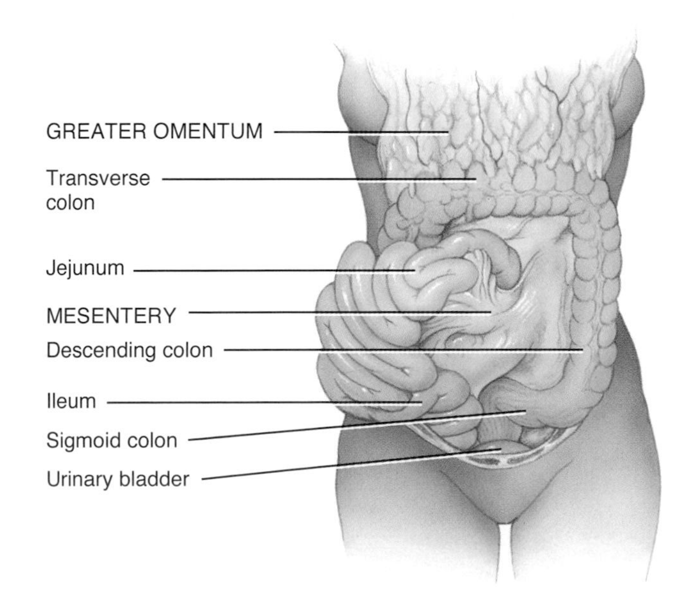

(b) Anterior view (greater omentum lifted and small intestine reflected to right side)

Which part of the peritoneum binds the small intestine to the posterior abdominal wall?

body-producing plasma cells that help combat and contain infections of the GI tract. The greater omentum normally contains considerable adipose tissue. Its adipose tissue content can greatly expand with weight gain, giving rise to the characteristic "beer belly" seen in some overweight individuals. A part of the peritoneum, the ***mesentery*** (MEZ-en-ter′-ē; *mes-* = middle), binds the small intestine to the posterior abdominal wall (Figure 19.3b).

A common cause of **peritonitis**, an acute inflammation of the peritoneum, is contamination of the peritoneum by infectious microbes, which can result from accidental or surgical wounds in the abdominal wall, or from perforation or rupture of abdominal organs.

■ **CHECKPOINT**

3. Where along the GI tract is the muscularis composed of skeletal muscle? Is control of this skeletal muscle voluntary or involuntary?
4. Where are the visceral peritoneum and parietal peritoneum located?

MOUTH

OBJECTIVES • Identify the locations of the salivary glands, and describe the functions of their secretions.
- **Describe the structure and functions of the tongue.**
- **Identify the parts of a typical tooth, and compare deciduous and permanent dentitions.**

The ***mouth*** or ***oral cavity*** is formed by the cheeks, hard and soft palates, and tongue (Figure 19.4). The cheeks form the lateral walls of the oral cavity. The *lips* are fleshy folds around the opening of the mouth. Both the cheeks and lips are covered on the outside by skin and on the inside by a mucous membrane. During chewing, the lips and cheeks help keep food between the upper and lower teeth. They also assist in speech.

The *hard palate*, consisting of the maxillae and palatine bones, forms most of the roof of the mouth. The rest is formed by the muscular *soft palate*. Hanging from the soft palate is a projection called the *uvula* (Ū-vū-la). During swallowing, the uvula moves upward with the soft palate, which prevents entry of swallowed foods and liquids into the nasal

Figure 19.4 Structures of the mouth (oral cavity).

The cheeks, hard and soft palates, and tongue form the mouth.

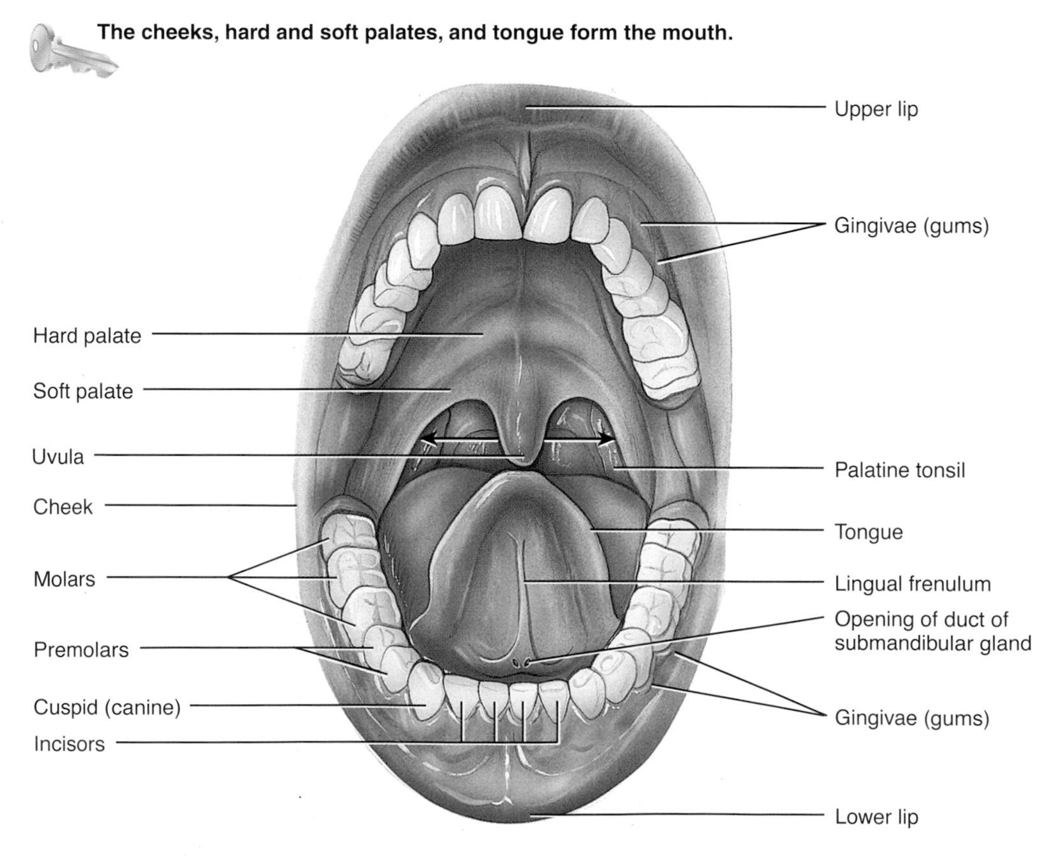

What are the functions of the muscles of the tongue?

cavity. At the back of the soft palate, the mouth opens into the oropharynx. The *palatine tonsils* are just posterior to the opening.

Tongue

The ***tongue*** forms the floor of the oral cavity. It is an accessory digestive organ composed of skeletal muscle covered with mucous membrane (see Figure 12.4 on page 292).

The muscles of the tongue maneuver food for chewing, shape the food into a rounded mass, force the food to the back of the mouth for swallowing, and alter the shape and size of the tongue for swallowing and speech. The ***lingual frenulum*** (LING-gwal FREN-ū-lum; *lingua* = tongue; *frenum* = bridle), a fold of mucous membrane in the midline of the undersurface of the tongue, limits the movement of the tongue posteriorly (Figure 19.4). The lingual tonsils lie at the base of the tongue (see Figure 12.4a). The upper surface and sides of the tongue are covered with projections called ***papillae*** (pa-PIL-ē), some of which contain taste buds.

Salivary Glands

The three pairs of ***salivary glands*** are accessory organs of digestion that lie outside the mouth and release their secretions into ducts emptying into the oral cavity (see Figure 19.1). The ***parotid glands*** are located inferior and anterior to the ears between the skin and the masseter muscle. The ***submandibular glands*** are found in the floor of the mouth; they are medial and partly inferior to the mandible. The ***sublingual glands*** are beneath the tongue and superior to the submandibular glands.

The fluid secreted by the salivary glands, called ***saliva***, is composed of 99.5% water and 0.5% solutes. The water in saliva helps dissolve foods so they can be tasted and digestive reactions can begin. One of the solutes, the digestive enzyme ***salivary amylase***, begins the digestion of starches in the mouth. Mucus in saliva lubricates food so it can easily be swallowed. The enzyme lysozyme kills bacteria, thereby protecting the mouth's mucous membrane from infection and the teeth from decay.

Secretion of saliva, called ***salivation*** (sal-i-VĀ-shun), is controlled by the autonomic nervous system. Normally, parasympathetic stimulation promotes continuous secretion of a moderate amount of saliva, which keeps the mucous membranes moist and lubricates the movements of the tongue and lips during speech. Sympathetic stimulation dominates during stress, resulting in dryness of the mouth.

Teeth

The ***teeth (dentes)*** are accessory digestive organs located in bony sockets of the mandible and maxillae. The sockets are covered by the *gingivae* (JIN-ji-vē; singular is *gingiva*) or *gums* and are lined with the *periodontal ligament* (*peri-* = around; *odont-* = tooth). This dense fibrous connective tissue anchors the teeth to bone (Figure 19.5).

A typical tooth has three major external regions: the crown, root, and neck. The ***crown*** is the visible portion above the level of the gums. The ***root*** consists of one to three projections embedded in the socket. The ***neck*** is the junction line of the crown and root, near the gum line.

Internally, ***dentin*** forms the majority of the tooth. Dentin consists of a calcified connective tissue that gives the tooth its basic shape and rigidity. The dentin of the crown is covered by ***enamel*** that consists primarily of calcium phosphate and calcium carbonate. Enamel, the hardest substance in the body and the richest in calcium salts (about 95% of its dry weight), protects the tooth from the wear and tear of chewing. It is also a barrier against acids that easily dissolve the dentin. The dentin of the root is covered by ***cementum***, a bonelike substance that attaches to the root to the periodontal ligament. The dentin of a tooth encloses the ***pulp cavity***, a

Figure 19.5 Parts of a typical tooth.

There are 20 teeth in a complete deciduous set and 32 teeth in a complete permanent set.

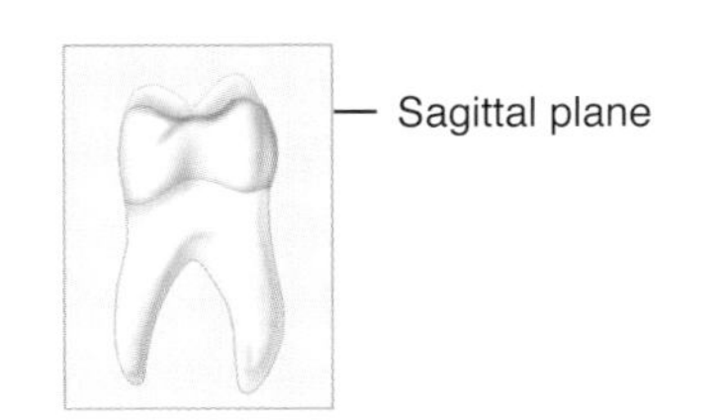

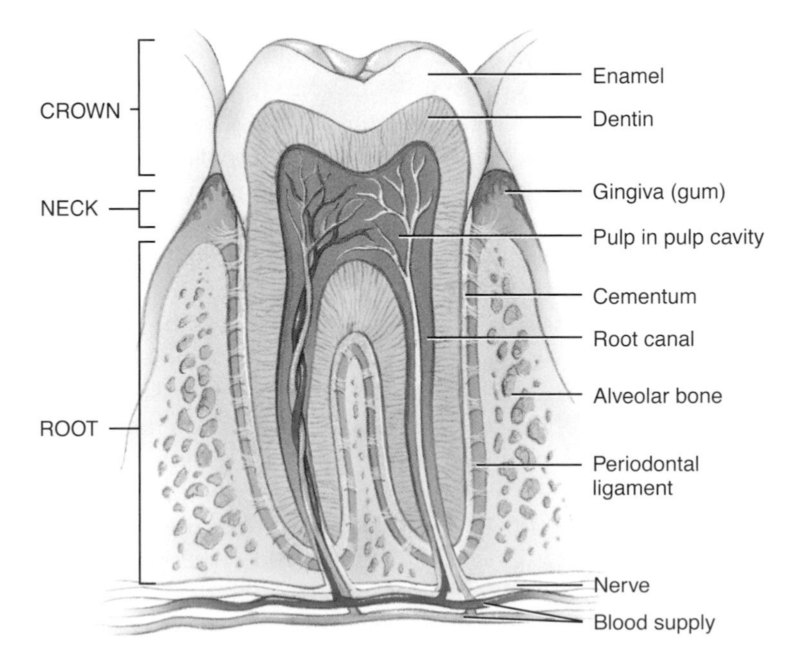

Sagittal section of a mandibular (lower) molar

What type of tissue is the main component of teeth?

space in the crown filled with ***pulp***, a connective tissue containing blood vessels, nerves, and lymphatic vessels. Narrow extensions of the pulp cavity run through the root of the tooth and are called ***root canals***. Each root canal has an opening at its base through which blood vessels bring nourishment, lymphatic vessels offer protection, and nerves provide sensation.

Humans have two sets of teeth. The ***deciduous teeth*** begin to erupt at about 6 months of age, and one pair appears about each month thereafter until all 20 are present. They are generally lost in the same sequence between 6 and 12 years of age. The ***permanent teeth*** appear between age 6 and adulthood. There are 32 teeth in a complete permanent set.

Humans have different teeth for different functions (see Figure 19.4). ***Incisors*** are closest to the midline, are chisel-shaped, and are adapted for cutting into food; ***cuspids*** (canines) are next to the incisors and have one pointed surface (cusp) to tear and shred food; ***premolars*** have two cusps to crush and grind food; and ***molars*** have three or more blunt cusps to crush and grind food.

Root canal therapy is a multistep procedure in which all traces of pulp tissue are removed from the pulp cavity and root canals of a badly diseased tooth. After a hole is made in the tooth, the root canals are filed out and irrigated to remove bacteria. Then, the canals are treated with medication and sealed tightly. The damaged crown is then repaired.

Digestion in the Mouth

Mechanical digestion in the mouth results from chewing, or ***mastication*** (mas′-ti-KĀ-shun = to chew), in which food is manipulated by the tongue, ground by the teeth, and mixed with saliva. As a result, the food is reduced to a soft, flexible, easily swallowed mass called a ***bolus*** (= lump).

Dietary carbohydrates are either monosaccharide and disaccharide sugars or complex polysaccharides such as glycogen and starches (see page 31). Most of the carbohydrates we eat are starches from plant sources, but only monosaccharides (glucose, fructose, and galactose) can be absorbed into the bloodstream. Thus, ingested starches must be broken down into monosaccharides. Salivary amylase begins the breakdown of starch by breaking particular chemical bonds between the glucose subunits. The resulting products include the disaccharide maltose (2 glucose subunits), the trisaccharide maltotriose (3 glucose subunits), and larger fragments called dextrins (5 to 10 glucose subunits). Salivary amylase in the swallowed food continues to act for about an hour until it is inactivated by stomach acids.

■ **CHECKPOINT**

5. What structures form the mouth (oral cavity)?

6. How is saliva secretion regulated?

7. What is a bolus? How is it formed?

PHARYNX AND ESOPHAGUS

OBJECTIVE • Describe the location, structure, and functions of the pharynx and esophagus.

When food is swallowed, it passes from the mouth into the ***pharynx*** (FAIR-inks), a funnel-shaped tube that is composed of skeletal muscle and lined by mucous membrane. It extends from the internal nares to the esophagus posteriorly and the larynx anteriorly (Figure 19.6a). The nasopharynx is involved in respiration (see Figure 18.2 on page 447); food that is swallowed passes from the mouth into the oropharynx and laryngopharynx before passing into the esophagus. Muscular contractions of the oropharynx and laryngopharynx help propel food into the esophagus.

The ***esophagus*** (e-SOF-a-gus = eating gullet) is a muscular tube lined with stratified squamous epithelium that lies posterior to the trachea. It begins at the end of the laryngopharynx, passes through the mediastinum and diaphragm, and connects to the superior aspect of the stomach. It transports food to the stomach and secretes mucus. At each end of the esophagus, the muscularis forms two sphincters—the ***upper esophageal sphincter (UES)*** (e-sof′-a-JĒ-al), which consists of skeletal muscle, and the ***lower esophageal sphincter (LES)***, which consists of smooth muscle. The upper esophageal sphincter regulates the movement of food from the pharynx into the esophagus; the lower esophageal sphincter regulates the movement of food from the esophagus into the stomach.

Swallowing, the movement of food from the mouth to the stomach, involves the mouth, pharynx, and esophagus and is helped by saliva and mucus. Swallowing is divided into three stages: the voluntary, pharyngeal, and esophageal stages.

In the ***voluntary stage*** of swallowing, the bolus is forced to the back of the mouth cavity and into the oropharynx by the movement of the tongue upward and backward against the palate. With the passage of the bolus into the oropharynx, the involuntary ***pharyngeal stage*** of swallowing begins (Figure 19.6b). Breathing is temporarily interrupted when the soft palate and uvula move upward to close off the nasopharynx, the epiglottis seals off the larynx, and the vocal cords come together. After the bolus passes through the oropharynx, the respiratory passageways reopen and breath-

Figure 19.6 Swallowing. During the pharyngeal stage of swallowing (b), the tongue rises against the palate, the nasopharynx is closed off, the larynx rises, the epiglottis seals off the larynx, and the bolus passes into the esophagus. During the esophageal stage of swallowing (c), food moves through the esophagus into the stomach via peristalsis.

Swallowing moves food from the mouth into the stomach.

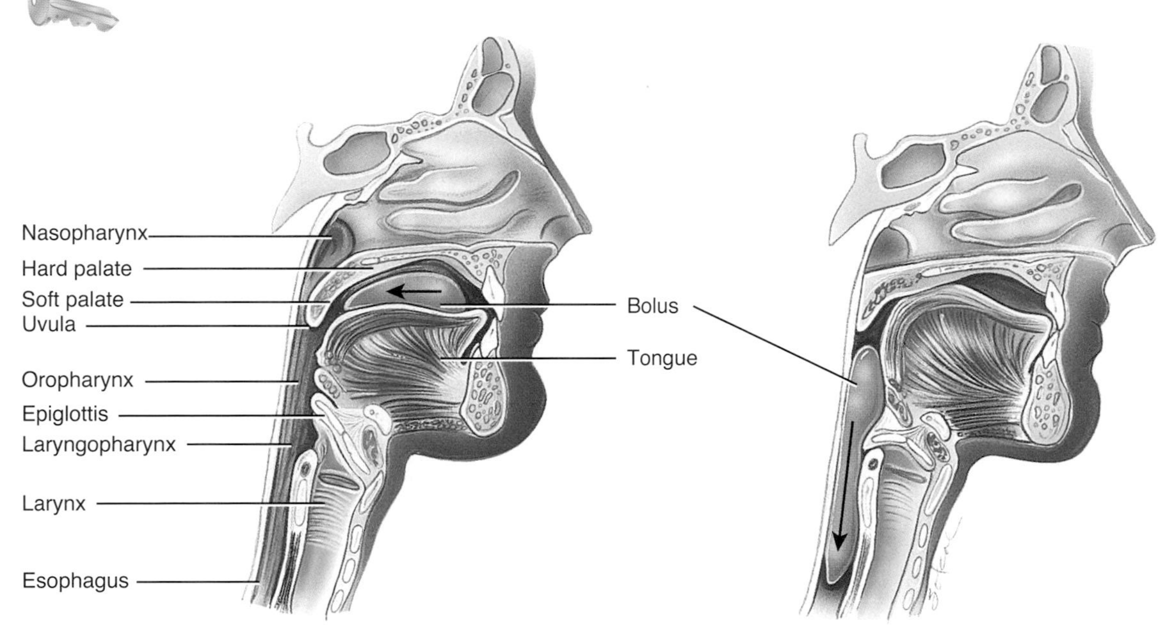

(a) Position of structures before swallowing

(b) During the pharyngeal stage of swallowing

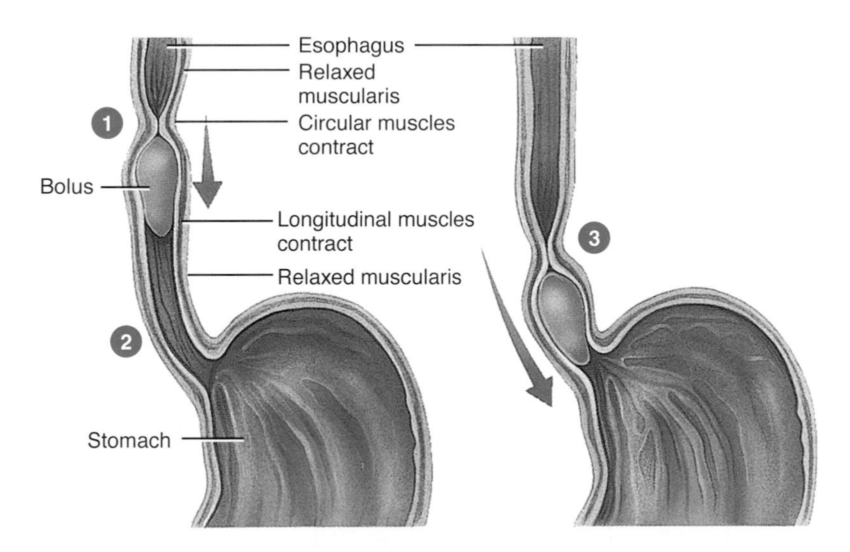

(c) Esophageal stage of swallowing

Is swallowing a voluntary or an involuntary action?

ing resumes. Once the upper esophageal sphincter relaxes, the bolus moves into the esophagus.

In the ***esophageal stage***, food is pushed through the esophagus by a process called ***peristalsis*** (Figure 19.6c):

1. The circular muscle fibers in the section of esophagus above the bolus contract, constricting the wall of the esophagus and squeezing the bolus downward.
2. Longitudinal muscle fibers around the bottom of the bolus contract, shortening the section of the esophagus below the bolus and pushing its walls outward.
3. After the bolus moves into the new section of the esophagus, the circular muscles above it contract, and the cycle repeats. The contractions move the bolus down the esophagus toward the stomach. As the bolus approaches the end of the esophagus, the lower esophageal sphincter relaxes and the bolus moves into the stomach.

> Sometimes, after food has entered the stomach, the lower esophageal sphincter fails to close adequately and the stomach contents can back up (reflux) into the lower esophagus, a condition known as **gastroesophageal reflux disease (GERD)**. Reflux of acid from the stomach can irritate the esophageal wall, causing a burning sensation known as **heartburn**. Although it is experienced in a region very near the heart, heartburn is unrelated to any cardiac problem. GERD also may increase the risk of esophageal cancer.

■ **CHECKPOINT**

8. How does a bolus pass from the mouth into the stomach?

STOMACH

OBJECTIVE • Describe the location, structure, and functions of the stomach.

The ***stomach*** is a J-shaped enlargement of the GI tract directly below the diaphragm. The stomach connects the esophagus to the duodenum, the first part of the small intestine (Figure 19.7). Because a meal can be eaten much more quickly than the intestines can digest and absorb it, one of the functions of the stomach is to serve as a mixing chamber and holding reservoir. At appropriate intervals after food is ingested, the stomach forces a small quantity of material into the duodenum. The position and size of the stomach vary continually; the diaphragm pushes it inferiorly with each in-

Figure 19.7 External and internal anatomy of the stomach. The dashed lines indicate the approximate borders of the regions of the stomach.

The four regions of the stomach are the cardia, fundus, body, and pylorus.

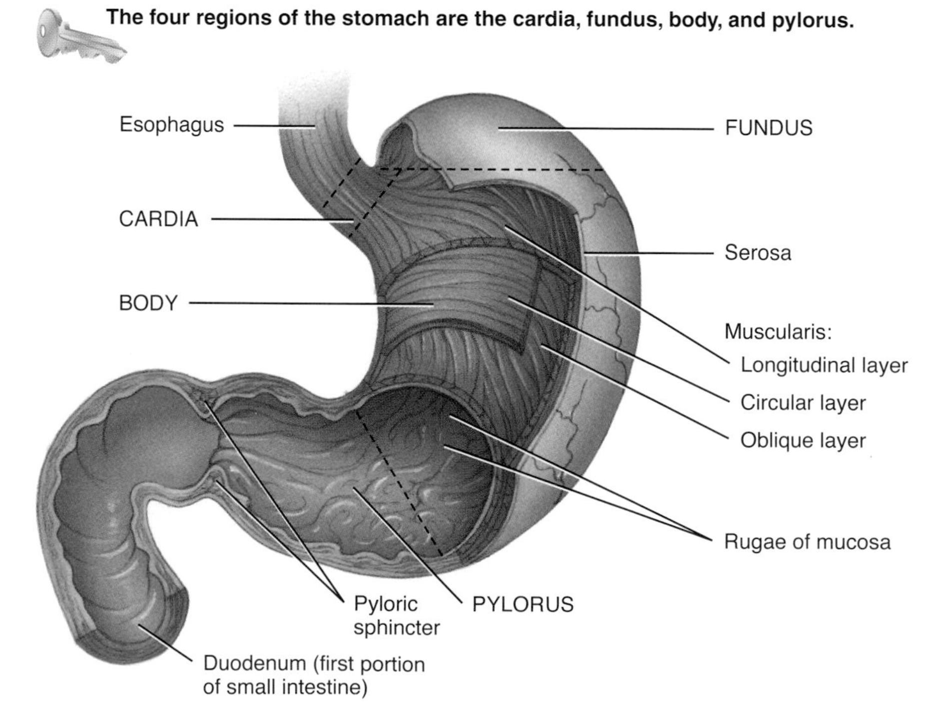

Does your stomach still have rugae after a very big meal?

halation and pulls it superiorly with each exhalation. Empty, it is about the size of a large sausage, but it is the most elastic part of the GI tract and can accommodate a large quantity of food.

Structure of the Stomach

The stomach has four main regions: cardia, fundus, body, and pylorus (Figure 19.7). The ***cardia*** (CAR-dē-a) surrounds the superior opening of the stomach. The stomach then curves upward. The portion superior and to the left of the cardia is the ***fundus*** (FUN-dus). Inferior to the fundus is the large central portion of the stomach, called the ***body***. The narrow, most inferior region is the ***pylorus*** (pī-LOR-us; *pyl-* = gate; *-orus* = guard). Between the pylorus and duodenum is the ***pyloric sphincter***.

The stomach wall is composed of the same four basic layers as the rest of the GI tract (mucosa, submucosa, muscularis, serosa), with certain differences. When the stomach is empty, the mucosa lies in large folds, called ***rugae*** (ROO-jē = wrinkles). The surface of the mucosa is a layer of nonciliated simple columnar epithelial cells called ***surface mucous cells*** (Figure 19.8). Epithelial cells also extend downward and form columns of secretory cells called ***gastric glands*** that line narrow channels called ***gastric pits***. Secretions from the gastric glands flow into the gastric pits and then into the lumen of the stomach.

The gastric glands contain three types of *exocrine gland cells* that secrete their products into the stomach lumen: mucous neck cells, chief cells, and parietal cells. Both surface mucous cells and ***mucous neck cells*** secrete mucus (Figure 19.9). The ***chief cells*** secrete an inactive gastric enzyme called ***pepsinogen***. ***Parietal cells*** produce hydrochloric acid, which kills many microbes in food and helps convert pepsinogen to the active digestive enzyme ***pepsin***. Parietal cells also secrete *intrinsic factor*, which is involved in the absorption of vitamin B_{12}. Inadequate production of intrinsic factor can result in pernicious anemia because vitamin B_{12} is needed for red blood cell production. The secretions of the mucous, chief, and parietal cells are collectively called ***gastric juice***. The ***G cells***, a fourth type of cell in the gastric glands, secrete the hormone ***gastrin*** into the bloodstream.

Figure 19.8 Layers of the stomach.

Secretions from the gastric glands flow into the gastric pits and then into the lumen of the stomach.

Lumen of stomach
Gastric pits
Simple columnar epithelium
Lamina propria
Gastric gland
Lymphatic nodule
Muscularis mucosae
Lymphatic vessel
Venule
Arteriole
Oblique layer of muscle
Circular layer of muscle
Enteric neurons in muscularis
Longitudinal layer of muscle
MUCOSA
SUBMUCOSA
MUSCULARIS
SEROSA

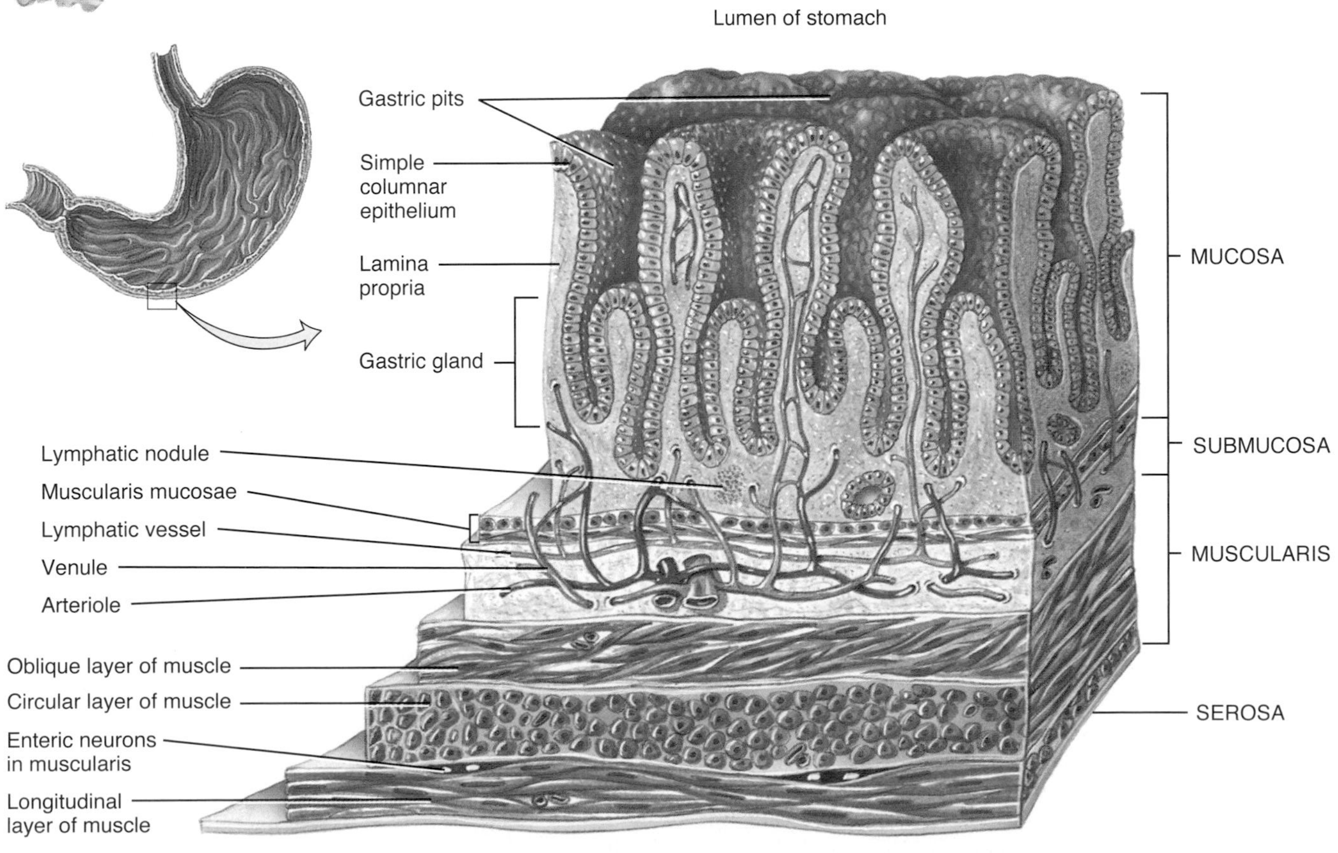

Three-dimensional view of layers of the stomach

Which stomach layer is in contact with swallowed food?

Figure 19.9 Sectional view of gastric glands and types of cells in the stomach mucosa.

Gastric juice is the combined secretions of mucous cells, chief cells, and parietal cells.

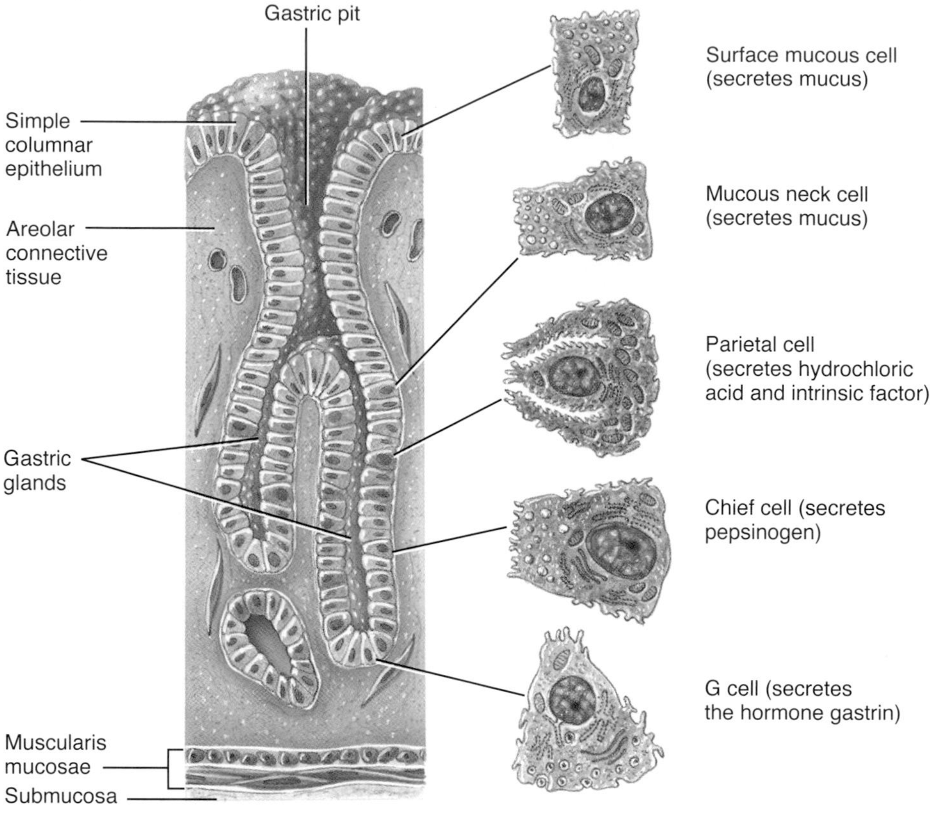

Gastric pit
Simple columnar epithelium
Areolar connective tissue
Gastric glands
Muscularis mucosae
Submucosa
Surface mucous cell (secretes mucus)
Mucous neck cell (secretes mucus)
Parietal cell (secretes hydrochloric acid and intrinsic factor)
Chief cell (secretes pepsinogen)
G cell (secretes the hormone gastrin)

Which type of cell shown here is part of the endocrine system (secretes a hormone)?

The submucosa of the stomach is composed of areolar connective tissue that connects the mucosa to the muscularis. The muscularis has three rather than two layers of smooth muscle: an outer longitudinal layer, a middle circular layer, and an inner oblique layer (see Figure 19.7). The serosa covering the stomach, composed of simple squamous epithelium and areolar connective tissue, is part of the visceral peritoneum.

Digestion and Absorption in the Stomach

Once food reaches the stomach, the stomach wall is stretched and the pH of the stomach contents increases because proteins in food have buffered some of the stomach acid. These changes in the stomach trigger nerve impulses that stimulate the flow of gastric juice and initiate ***mixing waves***, gentle, rippling peristaltic movements of the muscularis. These waves macerate food and mix it with the secretions of the gastric glands, producing ***chyme*** (KĪM = juice), a thick liquid with the consistency of pea soup. Each mixing wave forces a small amount of chyme through the partially closed pyloric sphincter into the duodenum, a process called ***gastric emptying***. Most of the chyme is forced back into the body of the stomach. The next mixing wave pushes chyme forward again and forces a little more into the duodenum. After the stomach has emptied some of its contents into the duodenum, reflexes begin to slow the exit of chyme from the stomach. This prevents overloading of the duodenum with more chyme than it can handle. Foods rich in carbohydrate spend the least time in the stomach; high-protein foods remain somewhat longer, and gastric emptying is slowest after a meal containing large amounts of fat.

Vomiting is the forcible expulsion of the contents of the upper GI tract (stomach and sometimes duodenum) through the mouth. The strongest stimuli for vomiting are irritation and excessive distension of the stomach. Other stimuli include unpleasant sights, general anesthesia, dizziness, and certain drugs such as morphine. Prolonged vomiting, especially in infants and elderly people, can be serious because the loss of acidic gastric juice can lead to alkalosis (higher than normal blood pH), dehydration, and damage to the esophagus and teeth.

The main event of chemical digestion in the stomach is the beginning of protein digestion by the enzyme pepsin, which breaks peptide bonds between the amino acids of proteins. As a result, the proteins become fragmented into *peptides*, smaller strings of amino acids. Pepsin is most effective in the very acidic environment of the stomach, which has a pH of 2. What keeps pepsin from digesting the protein in stomach cells along with the food? First, recall that chief cells secrete pepsin in an inactive form (pepsinogen). It is not converted into active pepsin until it contacts hydrochloric acid in gastric juice. Second, mucus secreted by mucous cells coats the mucosa, forming a thick barrier between the cells of the stomach lining and the gastric juice.

The epithelial cells of the stomach are impermeable to most materials, so little absorption occurs. However, mucous cells of the stomach absorb some water, ions, and short-chain fatty acids, as well as certain drugs (especially aspirin) and alcohol.

■ **CHECKPOINT**

9. What are the components of gastric juice?
10. What is the role of pepsin? Why is it secreted in an inactive form?
11. What substances are absorbed in the stomach?

PANCREAS

OBJECTIVE • Describe the location, structure, and functions of the pancreas.

From the stomach, chyme passes into the small intestine. Because chemical digestion in the small intestine depends on activities of the pancreas, liver, and gallbladder, we first consider these accessory digestive organs and their contributions to digestion in the small intestine.

Structure of the Pancreas

The ***pancreas*** (*pan-* = all; *-creas* = flesh) lies behind the stomach (see Figure 19.1). Secretions pass from the pancreas to the duodenum via the ***pancreatic duct***, which unites with the common bile duct from the liver and gallbladder, forming a common duct to the duodenum (Figure 19.10).

The pancreas is made up of small clusters of glandular epithelial cells, most of which are arranged in clusters called ***acini*** (AS-i-nī). The acini constitute the *exocrine* portion of the organ (see Figure 13.11 on page 328). The cells within acini secrete a mixture of fluid and digestive enzymes called ***pancreatic juice***. The remaining 1% of the cells are orga-

Figure 19.10 Relation of the pancreas to the liver, gallbladder, and duodenum. The inset shows details of the common bile duct and pancreatic duct forming the common duct.

Pancreatic juice in the pancreatic duct and bile in the common bile duct both flow into the common duct to the duodenum.

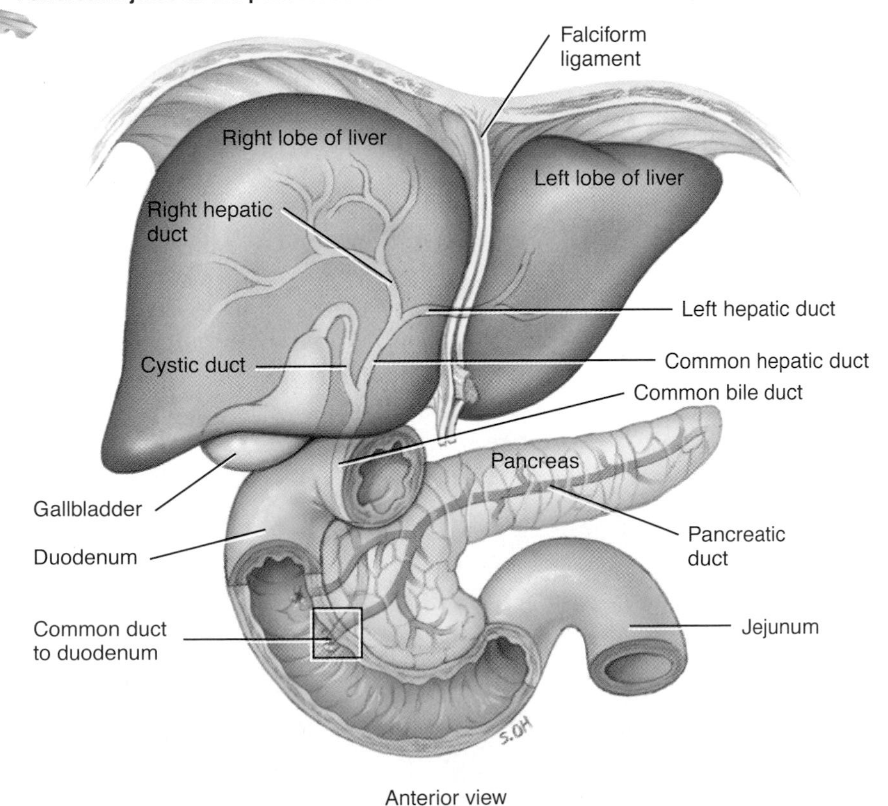

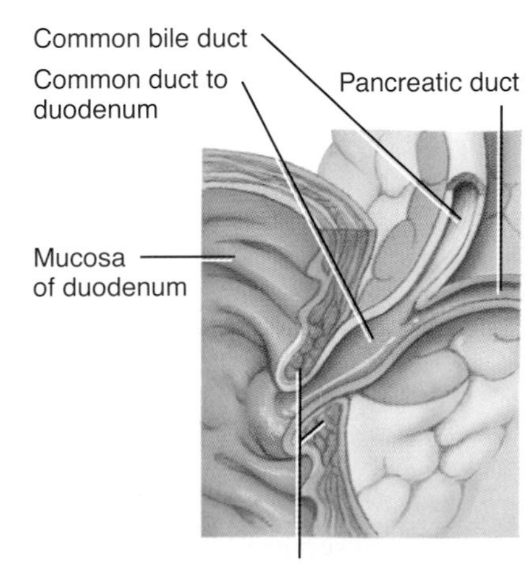

What substances are present in pancreatic juice?

nized into clusters called ***pancreatic islets (islets of Langerhans)***, the *endocrine* portion of the pancreas. These cells secrete the hormones glucagon, insulin, somatostatin, and pancreatic polypeptide, which are discussed in Chapter 13.

Pancreatic Juice

Pancreatic juice is a clear, colorless liquid that consists mostly of water, some salts, sodium bicarbonate, and enzymes. The bicarbonate ions give pancreatic juice a slightly alkaline pH (7.1 to 8.2), which inactivates pepsin from the stomach and creates the optimal environment for activity of enzymes in the small intestine. The enzymes in pancreatic juice include a starch-digesting enzyme called ***pancreatic amylase***; several protein-digesting enzymes including ***trypsin*** (TRIP-sin), ***chymotrypsin*** (kī′-mō-TRIP-sin), and ***carboxypeptidase*** (kar-bok′-sē-PEP-ti-dās); the main triglyceride-digesting enzyme in adults, called ***pancreatic lipase***; and nucleic acid–digesting enzymes called ***ribonuclease*** and ***deoxyribonuclease***. The protein-digesting enzymes are produced in an inactive form, which prevents them from digesting the pancreas itself. Upon reaching the small intestine, the inactive form of trypsin is activated by an enzyme called ***enterokinase***. In turn, trypsin activates the other protein-digesting pancreatic enzymes.

Pancreatic cancer usually affects people over 50 years of age and occurs more frequently in males. Typically, there are few symptoms until the disorder reaches an advanced stage and often not until it has metastasized to other parts of the body such as the lymph nodes, liver, or lungs. The disease is nearly always fatal and is the fourth most common cause of death from cancer in the United States. Pancreatic cancer has been linked to fatty foods, high alcohol consumption, genetic factors, smoking, and chronic pancreatitis.

■ **CHECKPOINT**

12. What are the pancreatic acini? How do their functions differ from those of the pancreatic islets?

13. What is the role of enterokinase?

LIVER AND GALLBLADDER

OBJECTIVE • Describe the location, structure, and functions of the liver and gallbladder.

In an average adult, the ***liver*** weighs 1.4 kg (about 3 lb) and, after the skin, is the second largest organ of the body. It is located below the diaphragm, mostly on the right side of the body. A connective tissue capsule covers the liver, which in turn is covered by peritoneum, the serous membrane that covers all the viscera. The ***gallbladder*** (*gall-* = bile) is a pear-shaped sac that hangs from the lower front margin of the liver (Figure 19.10).

Structure of the Liver and Gallbladder

The lobes of the liver are made up of many functional units called ***lobules*** (Figure 19.11). A lobule consists of specialized epithelial cells, called ***hepatocytes*** (*hepat-* = liver; *-cytes* = cells), arranged around a ***central vein***. In addition, the liver has highly-permeable capillaries called ***sinusoids***. Also present in the sinusoids are fixed phagocytes called *Kupffer cells*, which destroy worn-out blood cells, bacteria, and other foreign matter in the venous blood draining from the gastrointestinal tract.

Bile, which is secreted by hepatocytes, enters *bile canaliculi* (kan′-a-LIK-ū-lī = small canals), which are narrow intercellular canals that empty into *bile ducts* at the periphery of the lobules. The bile ducts merge and eventually form the ***right*** and ***left hepatic ducts***, which unite and exit the liver as the ***common hepatic duct*** (see Figure 19.10). Farther on, the common hepatic duct joins the ***cystic duct*** (*cystic* = bladder) from the gallbladder to form the ***common bile duct***. When the small intestine is empty, the sphincter around the common duct at the entrance to the duodenum closes, and bile backs up into the cystic duct to the gallbladder for storage.

Bile

Bile salts in bile aid in ***emulsification***, the breakdown of large lipid globules into a suspension of small lipid globules, and in absorption of lipids following their digestion. The small lipid globules formed as a result of emulsification present a very large surface area so that pancreatic lipase can digest them rapidly. The principal bile pigment is ***bilirubin***, which is derived from heme. When worn-out red blood cells are broken down, iron, globin, and bilirubin are released. The iron and globin are recycled, but some of the bilirubin is excreted in bile. Bilirubin eventually is broken down in the intestine, and one of its breakdown products (stercobilin) gives feces their normal brown color (see Figure 14.3 on page 350). After they have served as emulsifying agents, most bile salts are reabsorbed by active transport in the final portion of the small intestine (ileum) and enter portal blood flowing toward the liver.

The components of bile sometimes crystallize and form **gallstones**. As they grow in size and number, gallstones may cause intermittent or complete obstruction to the flow of bile from the gallbladder into the duodenum. Treatment consists of using gallstone-dissolving drugs or *lithotripsy*, a shock-wave therapy that smashes the gallstones into particles small enough to pass through the ducts. For people with recurrent gallstones or for whom drugs or lithotripsy is not indicated, *cholecystectomy*—the removal of the gallbladder and its contents—is necessary.

Figure 19.11 Structure of the liver.

A liver lobule consists of hepatocytes arranged around a central vein.

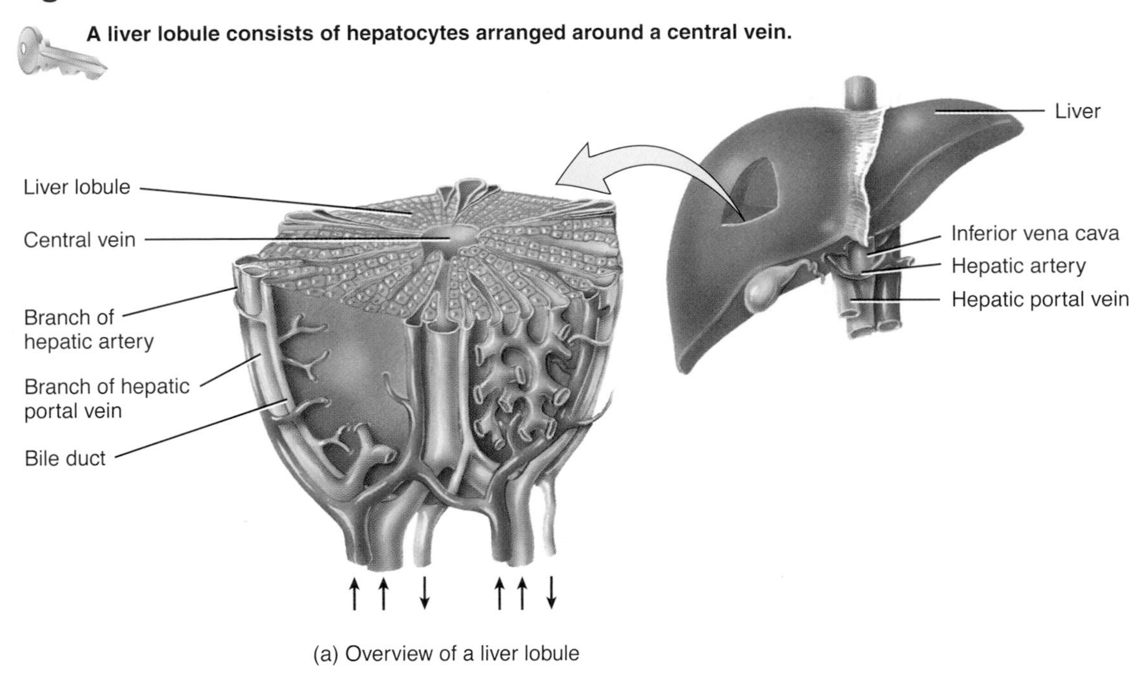

(a) Overview of a liver lobule

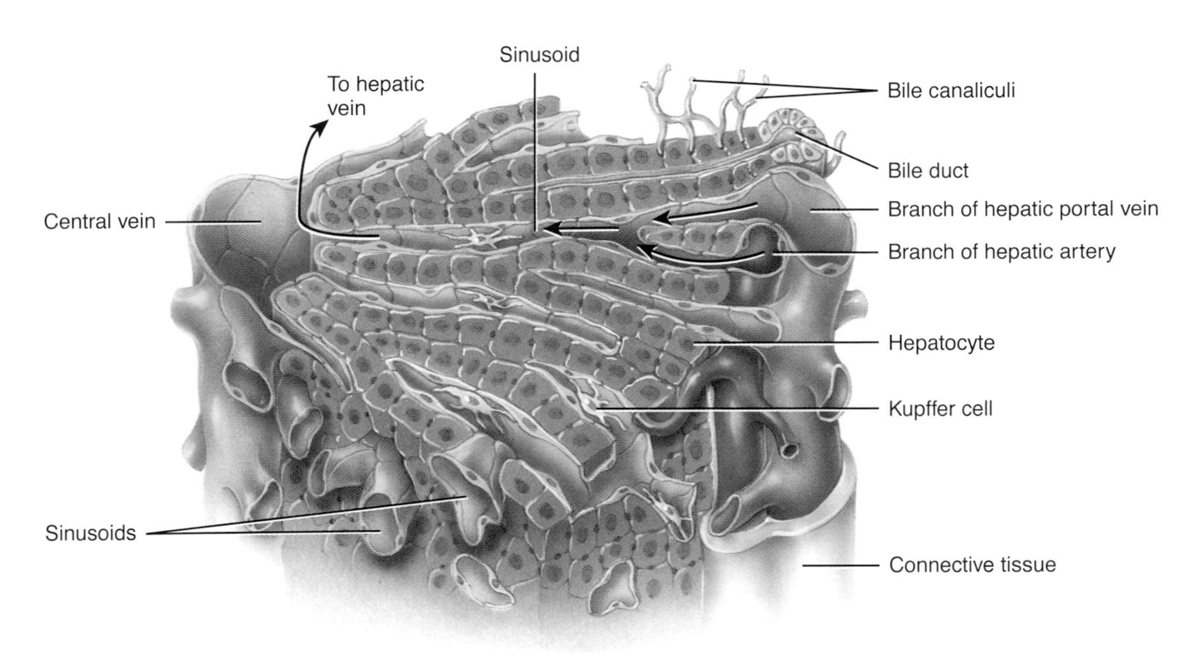

(b) Details of a portion of a liver lobule

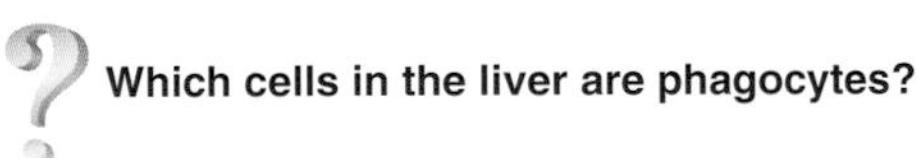
Which cells in the liver are phagocytes?

Functions of the Liver

The liver performs many other vital functions in addition to the secretion of bile and bile salts and the phagocytosis of bacteria and dead or foreign material by the Kupffer cells. Many of these are related to metabolism and are discussed in Chapter 20. Briefly, however, other functions of the liver include the following:

1. **Carbohydrate metabolism**. The liver is especially important in maintaining a normal blood glucose level. When blood glucose is low, the liver can break down glycogen to glucose and release glucose into the bloodstream. The liver can also convert certain amino acids and lactic acid to glucose, and it can convert other sugars, such as fructose and galactose, into glucose. When blood

glucose is high, as occurs just after eating a meal, the liver converts glucose to glycogen and triglycerides for storage.

2. **Lipid metabolism**. Hepatocytes store some triglycerides; break down fatty acids to generate ATP; synthesize lipoproteins, which transport fatty acids, triglycerides, and cholesterol to and from body cells; synthesize cholesterol; and use cholesterol to make bile salts.
3. **Protein metabolism**. Hepatocytes remove the amino group ($-NH_2$) from amino acids so that the amino acids can be used for ATP production or converted to carbohydrates or fats. They also convert the resulting toxic ammonia (NH_3) into the much less toxic urea, which is excreted in urine. Hepatocytes also synthesize most plasma proteins, such as globulins, albumin, prothrombin, and fibrinogen.
4. **Processing of drugs and hormones**. The liver can detoxify substances such as alcohol or secrete drugs such as penicillin, erythromycin, and sulfonamides into bile. It can also inactivate thyroid hormones and steroid hormones such as estrogens and aldosterone.
5. **Excretion of bilirubin**. Bilirubin, derived from the heme of aged red blood cells, is absorbed by the liver from the blood and secreted into bile. Most of the bilirubin in bile is metabolized in the small intestine by bacteria and eliminated in feces.
6. **Storage of vitamins and minerals**. In addition to storing glycogen, the liver stores certain vitamins (A, D, E, and K) and minerals (iron and copper), which are released from the liver when needed elsewhere in the body.
7. **Activation of vitamin D**. The skin, liver, and kidneys participate in synthesizing the active form of vitamin D.

■ **CHECKPOINT**

14. How are the liver and gallbladder connected to the duodenum?

15. What is the function of bile?

16. List all of the functions of the liver.

SMALL INTESTINE

OBJECTIVE • **Describe the location, structure, and functions of the small intestine.**

Within 2 to 4 hours after eating a meal, the stomach has emptied its contents into the small intestine, where the major events of digestion and absorption occur. The ***small intestine*** averages 2.5 cm (1 in.) in diameter; its length is about 3 m (10 ft) in a living person and about 6.5 m (21 ft) in a cadaver due to the loss of smooth muscle tone after death.

Structure of the Small Intestine

The small intestine has three portions (Figure 19.12a): the duodenum, the jejunum, and the ileum. The ***duodenum*** (doo′-ō-DĒ-num), the shortest part (about 25 cm or 10 in.), attaches to the pylorus of the stomach. *Duodenum* means "twelve"; the structure is so named because it is about as long as the width of 12 fingers. The ***jejunum*** (jē-JOO-num = empty) is about 1 m (3 ft) long and is so named because it is empty at death. The final portion of the small intestine, the ***ileum*** (IL-ē-um = twisted), measures about 2 m (6 ft) and joins the large intestine at the ***ileocecal sphincter*** (il′-ē-ō-SĒ-kal).

Figure 19.12 External and internal anatomy of the small intestine. (a) Portions of the small intestine are the duodenum, jejunum, and ileum. (b) Circular folds increase the surface area for digestion and absorption in the small intestine.

Most digestion and absorption occur in the small intestine.

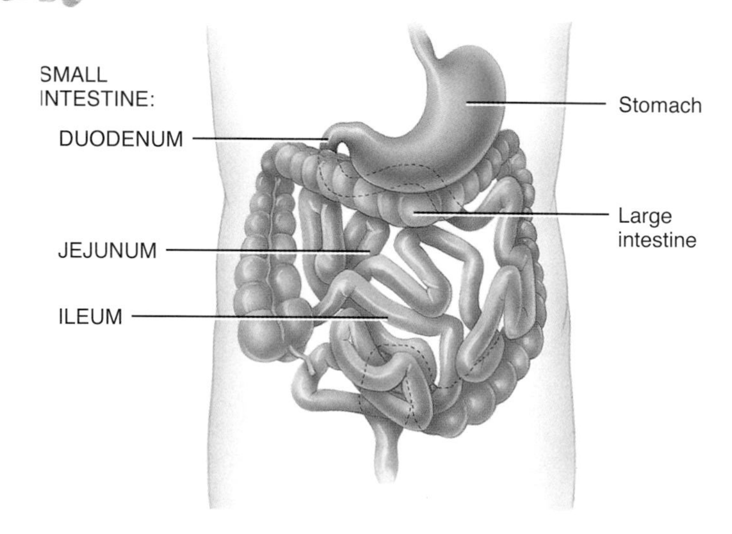

(a) Anterior view of external anatomy

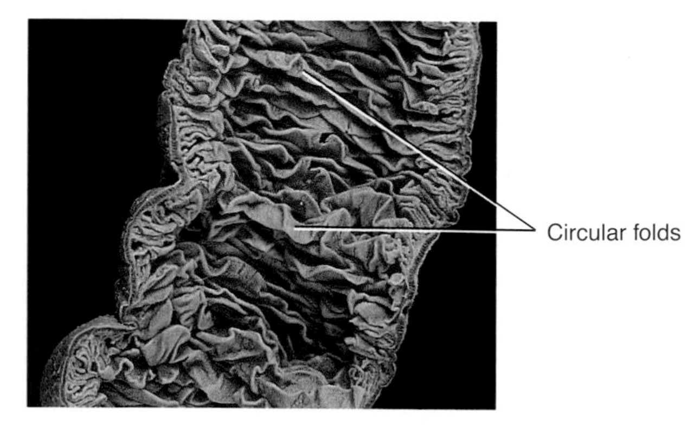

(b) Internal anatomy of the jejunum

Which segment of the small intestine is the longest?

The wall of the small intestine is composed of the same four layers that make up most of the GI tract: mucosa, submucosa, muscularis, and serosa (Figure 19.13). The epithelial layer of the small intestinal mucosa consists of simple columnar epithelium that contains many types of cells. ***Absorptive cells*** of the epithelium contain microvilli and digest and absorb nutrients in small intestinal chyme. Also present in the epithelium are ***goblet cells***, which secrete mucus. The small intestinal mucosa contains ***intestinal glands***, which are deep crevices lined by epithelial cells that secrete intestinal juice. Besides absorptive cells and goblet cells, intestinal glands also contain three types of endocrine cells that secrete hormones into the bloodstream: ***S cells***, ***CCK cells***, and ***K cells***, which secrete ***secretin*** (se-KRĒ-tin), ***cholecystokinin*** (kō-le-sis′-tō-KĪN-in) or ***CCK***, and ***glucose-dependent insulinotropic peptide***, or ***GIP***, respectively (see Table 19.2 on page 495 for secretin and CCK and Table 13.2 on page 334 for GIP). The lamina propria of the small intestinal mucosa contains areolar connective tissue that has an abundance of lymphatic tissue, which helps defend against pathogens in food. The submucosa of the duodenum contains ***duodenal glands*** that secrete an alkaline mucus. It helps neutralize gastric acid in the chyme. The muscularis of the small intestine consists of two layers of smooth muscle—an outer longitudinal layer and an inner circular layer. The serosa is composed of simple squamous epithelium and areolar connective tissue.

Even though the wall of the small intestine is composed of the same four basic layers as the rest of the GI tract, special structural features of the small intestine facilitate the process of digestion and absorption. These structural features include circular folds, villi, and microvilli. ***Circular folds*** are permanent ridges of the mucosa and submucosa that enhance absorption by increasing surface area and causing the chyme to spiral, rather than move in a straight line, as it passes through the small intestine (see Figure 19.12b). Also present in the small intestine are numerous ***villi*** (= tufts of hair; singular is *villus*), fingerlike projections of the mucosa that increase the surface area of the intestinal epithelium. Each villus consists of a layer of simple columnar epithelium surrounding a core of lamina propria. Within the core are an arteriole, a venule, a blood capillary network, and a ***lacteal*** (LAK-tē-al = milky), which is a lymphatic capillary. Nutrients absorbed by the epithelial cells covering the villus pass through the wall of a capillary or a lacteal to enter blood or lymph, respectively. Besides circular folds and villi, the small intestine also has ***microvilli*** (mī-krō-VIL-ī; *micro-* = small), tiny

Figure 19.13 Structure of the small intestine.

Circular folds, villi, and microvilli increase the surface area for digestion and absorption in the small intestine.

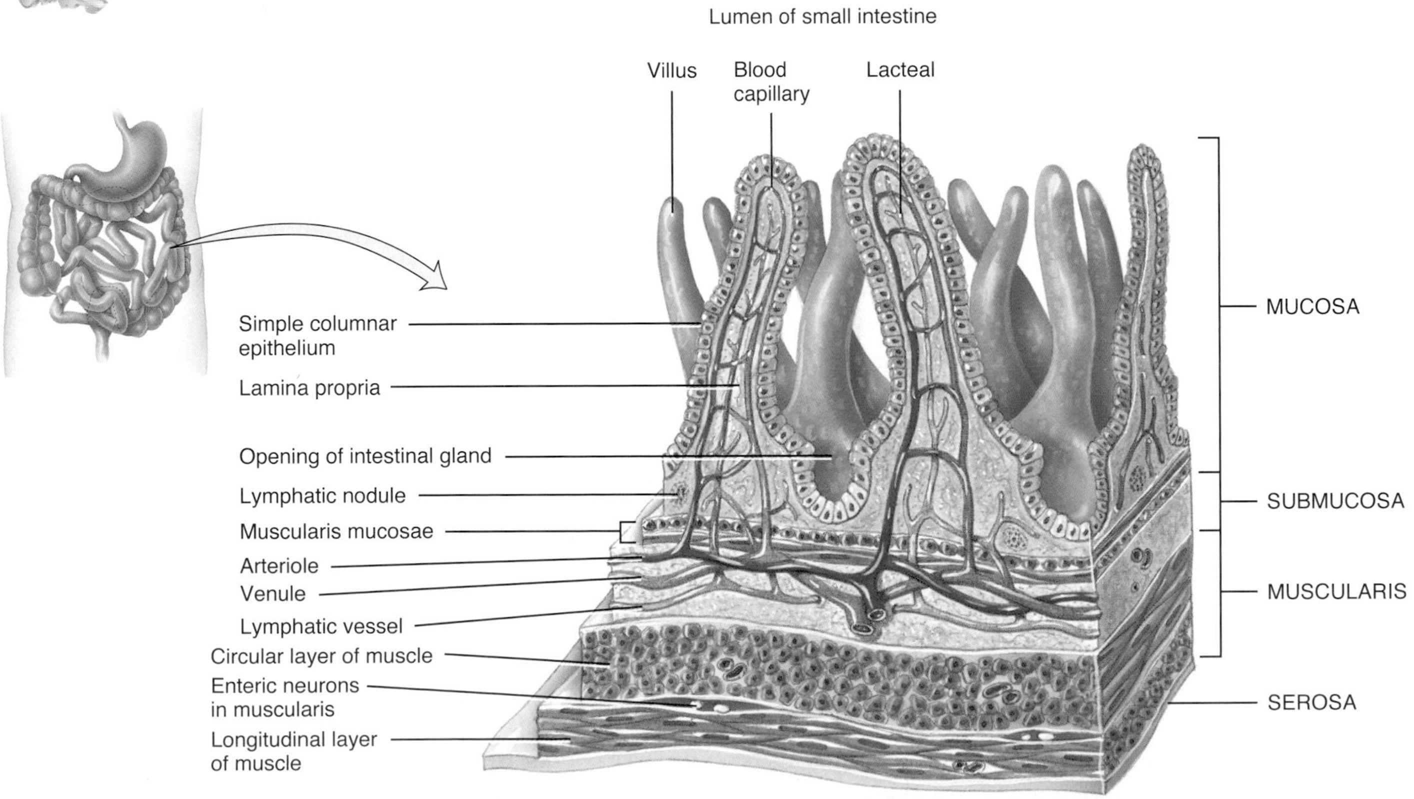

Three-dimensional view of layers of the small intestine showing villi

Where are the cells located that absorb dietary nutrients?

projections of the plasma membrane of absorptive cells that increase the surface area of these cells (see Figure 19.14). Thus, digested nutrients can move rapidly into absorptive cells.

Intestinal Juice

Intestinal juice, secreted by the intestinal glands, is a watery clear yellow fluid with a slightly alkaline pH of 7.6 that contains some mucus. Together, pancreatic and intestinal juices provide a liquid medium that aids absorption of substances from chyme as they come in contact with the microvilli. Intestinal enzymes are synthesized in the absorptive cells that line the villi. Most digestion by enzymes of the small intestine occurs in or on the surface of these absorptive cells.

Mechanical Digestion in the Small Intestine

Two types of movements contribute to intestinal motility in the small intestine: segmentations and peristalsis. ***Segmentations*** are localized contractions that slosh chyme back and forth, mixing it with digestive juices and bringing food particles into contact with the mucosa for absorption. The movements are similar to alternately squeezing the middle and the ends of a capped tube of toothpaste. They do not push the intestinal contents along the tract.

After most of a meal has been absorbed, segmentation stops; peristalsis begins in the lower portion of the stomach and pushes chyme forward along a short stretch of small intestine. The peristaltic wave slowly migrates down the small intestine, reaching the end of the ileum in 90 to 120 minutes. Then another wave of peristalsis begins in the stomach. Altogether, chyme remains in the small intestine for 3 to 5 hours.

Chemical Digestion in the Small Intestine

The chyme entering the small intestine contains partially digested carbohydrates and proteins. The completion of digestion in the small intestine is a collective effort of pancreatic juice, bile, and intestinal juice. Once digestion is completed, the final products of digestion are ready for absorption.

Starches and dextrins not reduced to maltose by the time chyme leaves the stomach are broken down by ***pancreatic amylase***, an enzyme in pancreatic juice that acts in the small intestine. Three enzymes located at the surface of small intestinal absorptive cells complete the digestion of disaccharides, breaking them down into monosaccharides, which are small enough to be absorbed. ***Maltase*** splits maltose into two molecules of glucose. ***Sucrase*** breaks sucrose into a molecule of glucose and a molecule of fructose. ***Lactase*** digests lactose into a molecule of glucose and a molecule of galactose.

In some people the absorptive cells of the small intestine fail to produce enough lactase. This results in a condition called **lactose intolerance**, in which undigested lactose in chyme retains fluid in the feces, and bacterial fermentation of lactose produces gases. Symptoms of lactose intolerance include diarrhea, gas, bloating, and abdominal cramps after consumption of milk and other dairy products. The severity of symptoms varies from relatively minor to sufficiently serious to require medical attention.

Enzymes in pancreatic juice (trypsin, chymotrypsin, elastase, and carboxypeptidase) continue the digestion of proteins begun in the stomach, though their actions differ somewhat because each splits the peptide bond between different amino acids. Protein digestion is completed by ***peptidases***, enzymes produced by absorptive cells that line the villi. The final products of protein digestion are amino acids, dipeptides, and tripeptides.

In an adult, most lipid digestion occurs in the small intestine. In the first step of lipid digestion, bile salts emulsify large globules of triglycerides and lipids into small lipid globules, giving pancreatic lipase easy access. Recall that triglycerides consist of a molecule of glycerol with three attached fatty acids (see Figure 2.10 on page 33). In the second step, ***pancreatic lipase***, found in pancreatic juice, breaks down each triglyceride molecule by removing two of the three fatty acids from glycerol; the third remains attached to the glycerol. Thus, fatty acids and monoglycerides are the end products of triglyceride digestion.

Pancreatic juice contains two nucleases: ***ribonuclease***, which digests RNA, and ***deoxyribonuclease***, which digests DNA. The nucleotides that result from the action of the two nucleases are further digested by small intestinal enzymes into pentoses, phosphates, and nitrogenous bases.

Table 19.1 summarizes the enzymes that contribute to digestion.

Absorption in the Small Intestine

All the mechanical and chemical phases of digestion from the mouth down through the small intestine are directed toward changing food into molecules that can undergo ***absorption***. Recall that absorption refers to the movement of small molecules through the absorptive epithelial cells of the mucosa into the underlying blood and lymphatic vessels. About 90% of all absorption takes place in the small intestine. The other 10% occurs in the stomach and large intestine. Absorption in the small intestine occurs by simple diffusion, facilitated diffusion, osmosis, and active transport. Any undigested or unabsorbed material left in the small intestine is passed on to the large intestine.

Absorption of Monosaccharides

All carbohydrates are absorbed as monosaccharides. Glucose and galactose are transported into absorptive cells of the villi by active transport. Fructose is transported by facilitated diffusion (Figure 19.14a on page 490). After absorption, monosaccharides are transported out of the epithelial cells by facil-

Table 19.1 Summary of Digestive Enzymes

Enzyme	Source	Substrate	Product
Carbohydrate Digesting			
Salivary amylase	Salivary glands.	Starches.	Maltose (disaccharide), maltotriose (trisaccharide), and dextrins.
Pancreatic amylase	Pancreas.	Starches.	Maltose, maltotriose, and dextrins.
Maltase	Small intestine.	Maltose.	Glucose.
Sucrase	Small intestine.	Sucrose.	Glucose and fructose.
Lactase	Small intestine.	Lactose.	Glucose and galactose.
Protein Digesting			
Pepsin	Stomach (chief cells).	Proteins.	Peptides.
Trypsin	Pancreas.	Proteins.	Peptides.
Chymotrypsin	Pancreas.	Proteins.	Peptides.
Carboxypeptidase	Pancreas.	Amino acid at carboxyl (acid) end of peptides.	Peptides and amino acids.
Peptidases	Small intestine.	Amino acid at amino end of peptides and dipeptides.	Peptides and amino acids.
Lipid Digesting			
Pancreatic lipase	Pancreas.	Triglycerides (fats) that have been emulsified by bile salts.	Fatty acids and monoglycerides.
Nucleases			
Ribonuclease	Pancreas.	Ribonucleic acid.	Nucleotides.
Deoxyribonuclease	Pancreas.	Deoxyribonucleic acid.	Nucleotides.

itated diffusion into the blood capillaries, which drain into venules of the villi. From here, monosaccharides are carried to the liver via the hepatic portal vein, then through the heart and to the general circulation (Figure 19.14b).

Absorption of Amino Acids

Enzymes break down dietary proteins into amino acids, dipeptides, and tripeptides, which are absorbed mainly in the duodenum and jejunum. About half of the absorbed amino acids are present in food, but half come from proteins in digestive juices and dead cells that slough off the mucosa! Amino acids, dipeptides, and tripeptides enter absorptive cells of the villi via active transport (Figure 19.14a). Inside the epithelial cells, peptides are digested into amino acids, which leave via diffusion and enter blood capillaries. Like monosaccharides, amino acids are carried in hepatic portal blood to the liver (Figure 19.14b). If not removed by liver cells, amino acids enter the general circulation. From there, body cells take up amino acids for use in protein synthesis and ATP production.

Absorption of Ions and Water

Absorptive cells lining the small intestine also absorb most of the ions and water that enter the GI tract in food, drink, and digestive secretions. Major ions absorbed in the small intestine include sodium, potassium, calcium, iron, magnesium, chloride, phosphate, nitrate, and iodide. All water absorption in the GI tract, about 9 liters (a little more than 2 gallons) daily, occurs via osmosis. When monosaccharides, amino acids, peptides, and ions are absorbed, they "pull" water along by osmosis.

Absorption of Lipids and Bile Salts

Lipases break down triglycerides into monoglycerides and fatty acids. The fatty acids can be either short-chain fatty acids (with fewer than 10–12 carbons) or long-chain fatty acids. The short-chain fatty acids are absorbed via simple diffusion into absorptive cells of the villi and then pass into blood capillaries along with monosaccharides and amino acids (Figure 19.14a). Bile salts emulsify the larger lipids, forming many ***micelles*** (mī-SELZ = small morsels), tiny droplets that include some bile salt molecules along with the long-chain fatty acids, monoglycerides, cholesterol, and other dietary lipids (Figure 19.14a). From micelles, these lipids diffuse into absorptive cells of the villi where they are packaged into ***chylomicrons***, large spherical particles that are coated with proteins. Chylomicrons leave the epithelial cells via exocytosis and enter lymphatic fluid within a lacteal. Thus, most absorbed dietary lipids bypass the hepatic portal

Figure 19.14 Absorption of digested nutrients in the small intestine. For simplicity, all digested foods are shown in the lumen of the small intestine, even though some nutrients are digested at the surface of or in absorptive epithelial cells of the villi.

Long-chain fatty acids and monoglycerides are absorbed into lacteals; other products of digestion enter blood capillaries.

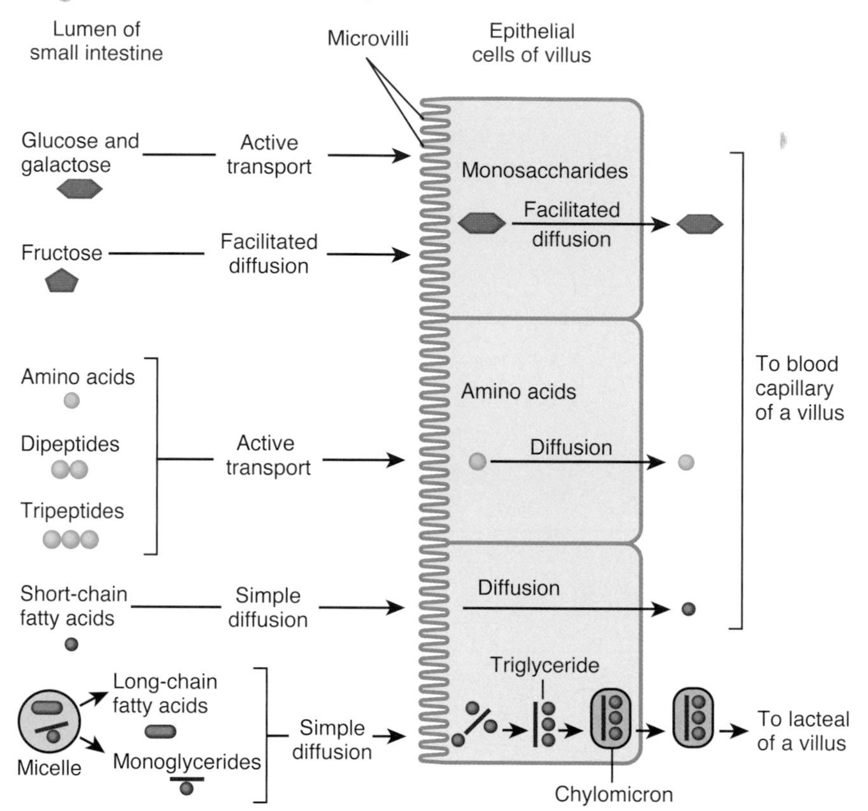

(a) Mechanisms for movement of nutrients through absorptive epithelial cells of the villi

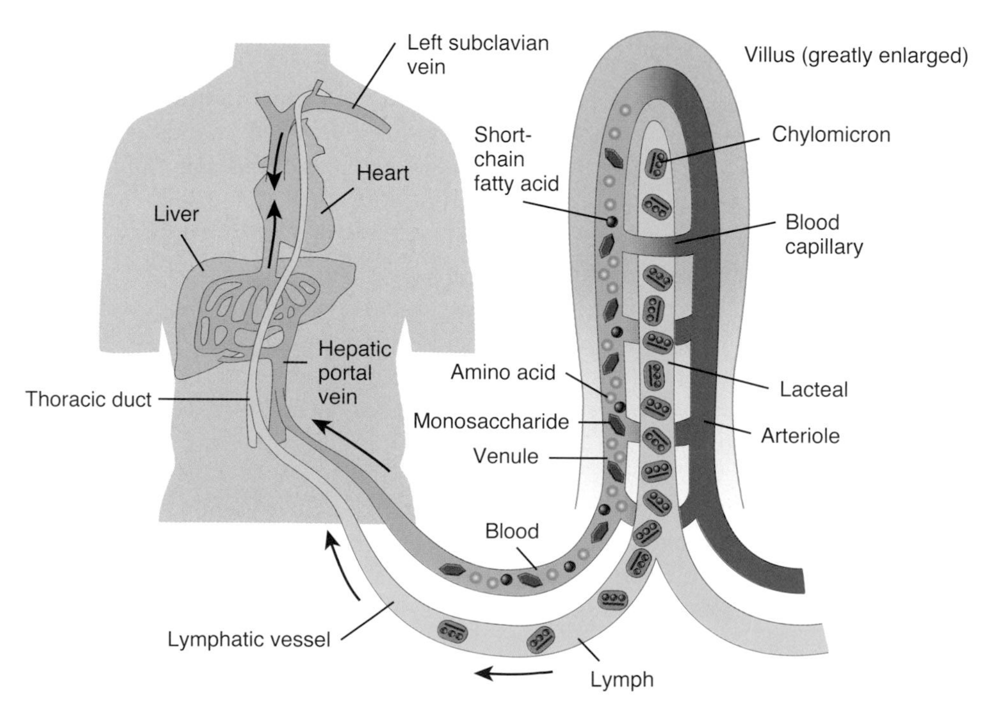

(b) Movement of absorbed nutrients into the blood and lymph

How are fat-soluble vitamins (A, D, E, and K) absorbed?

circulation because they enter lymphatic vessels instead of blood capillaries. Lymphatic fluid carrying chylomicrons from the small intestine passes into the thoracic duct and in due course empties into the left subclavian vein (Figure 19.14b). As blood passes through capillaries in adipose tissue and the liver, chylomicrons are removed and their lipids are stored for future use.

When chyme reaches the ileum, most of the bile salts are reabsorbed and returned by the blood to the liver for recycling. Insufficient bile salts, due to either obstruction of the bile ducts or liver disease, can result in the loss of up to 40% of dietary lipids in feces due to diminished lipid absorption.

Absorption of Vitamins

Fat-soluble vitamins (A, D, E, and K) are included along with ingested dietary lipids in micelles and are absorbed via simple diffusion. Most water-soluble vitamins, such as the B vitamins and vitamin C, are absorbed by simple diffusion. Vitamin B_{12} must be combined with intrinsic factor (produced by the stomach) for its absorption via active transport in the ileum.

■ **CHECKPOINT**

17. In what ways are the mucosa and submucosa of the small intestine adapted for digestion and absorption?

18. Explain the function of each digestive enzyme.

19. Define absorption. How are the end products of carbohydrate and protein digestion absorbed? How are the end products of lipid digestion absorbed?

20. By what routes do absorbed nutrients reach the liver?

LARGE INTESTINE

OBJECTIVE • Describe the location, structure, and functions of the large intestine.

The large intestine is the last part of the GI tract. Its overall functions are the completion of absorption, the production of certain vitamins, the formation of feces, and the expulsion of feces from the body.

Structure of the Large Intestine

The ***large intestine*** averages about 6.5 cm (2.5 in.) in diameter and about 1.5 m (5 ft) in length. It extends from the ileum to the anus and is attached to the posterior abdominal wall by its mesentery (see Figure 19.3b). The large intestine has four principal regions: cecum, colon, rectum, and anal canal (Figure 19.15).

Figure 19.15 Anatomy of the large intestine.

The regions of the large intestine are the cecum, colon, rectum, and anal canal.

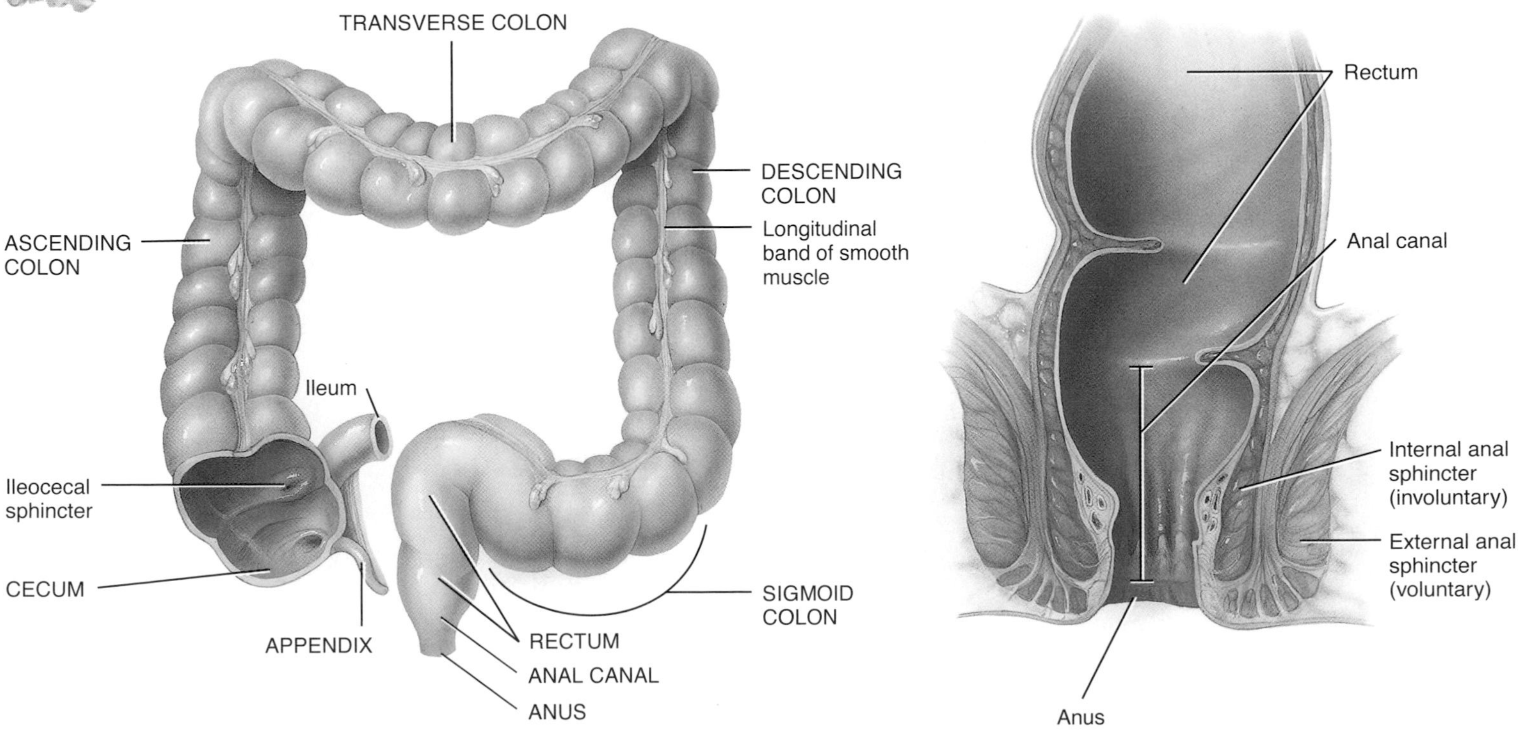

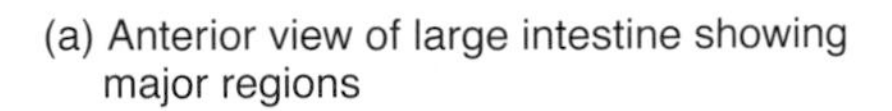

(a) Anterior view of large intestine showing major regions

(b) Frontal section of anal canal

What are the functions of the large intestine?

At the opening of the ileum into the large intestine is a valve called the ***ileocecal sphincter***. It allows materials from the small intestine to pass into the large intestine. Inferior to the ileocecal sphincter is the first segment of large intestine, called the ***cecum***. Attached to the cecum is a twisted coiled tube called the ***appendix***.

The open end of the cecum merges with the longest portion of the large intestine, called the ***colon*** (= food passage). The colon is divided into ascending, transverse, descending, and sigmoid portions. The ***ascending colon*** ascends on the right side of the abdomen, reaches the undersurface of the liver, and turns to the left. The colon continues across the abdomen to the left side as the ***transverse colon***. It curves beneath the lower border of the spleen on the left side and passes downward as the ***descending colon***. The S-shaped ***sigmoid colon*** begins near the iliac crest of the left hip bone and ends as the ***rectum***.

The last 2 to 3 cm (1 in.) of the rectum is called the ***anal canal***. The opening of the anal canal to the exterior is called the ***anus***. It has an internal sphincter of smooth (involuntary) muscle and an external sphincter of skeletal (voluntary) muscle. Normally, the anal sphincters are closed except during the elimination of feces.

The wall of the large intestine contains the typical four layers found in the rest of the GI tract: mucosa, submucosa, muscularis, and serosa. The epithelium of the mucosa is simple columnar epithelium that contains mostly absorptive cells and goblet cells (Figure 19.16). The cells form long tubes called *intestinal glands*. The absorptive cells function primarily in ion and water absorption. The goblet cells secrete mucus that lubricates the contents of the colon. Lymphatic nodules also are found in the mucosa. Compared to the small intestine, the mucosa of the large intestine does not have as many structural adaptations that increase surface area. There are no circular folds or villi; however, microvilli of the absorptive cells are present. Consequently, much more absorption occurs in the small intestine than in the large intestine. The muscularis consists of an external layer of longitudinal muscles and an internal layer of circular muscles. Unlike other parts of the gastrointestinal tract, the outer longitudinal layer of the muscularis is bundled into three longitudinal bands that run the length of most of the large intestine (see Figure 19.15a). Contractions of the muscularis gather the colon into a series of pouches, which give the colon a puckered appearance.

Polyps in the colon are generally slow-developing benign growths that arise from the mucosa of the large intestine. Often, they do not cause symptoms. If symptoms do occur, they include diarrhea, blood in the feces, and mucus discharged from the anus. The polyps are removed by colonoscopy or surgery because some of them may become cancerous.

Figure 19.16 Structure of the large intestine.

Intestinal glands formed by absorptive cells and goblet cells extend the full thickness of the mucosa.

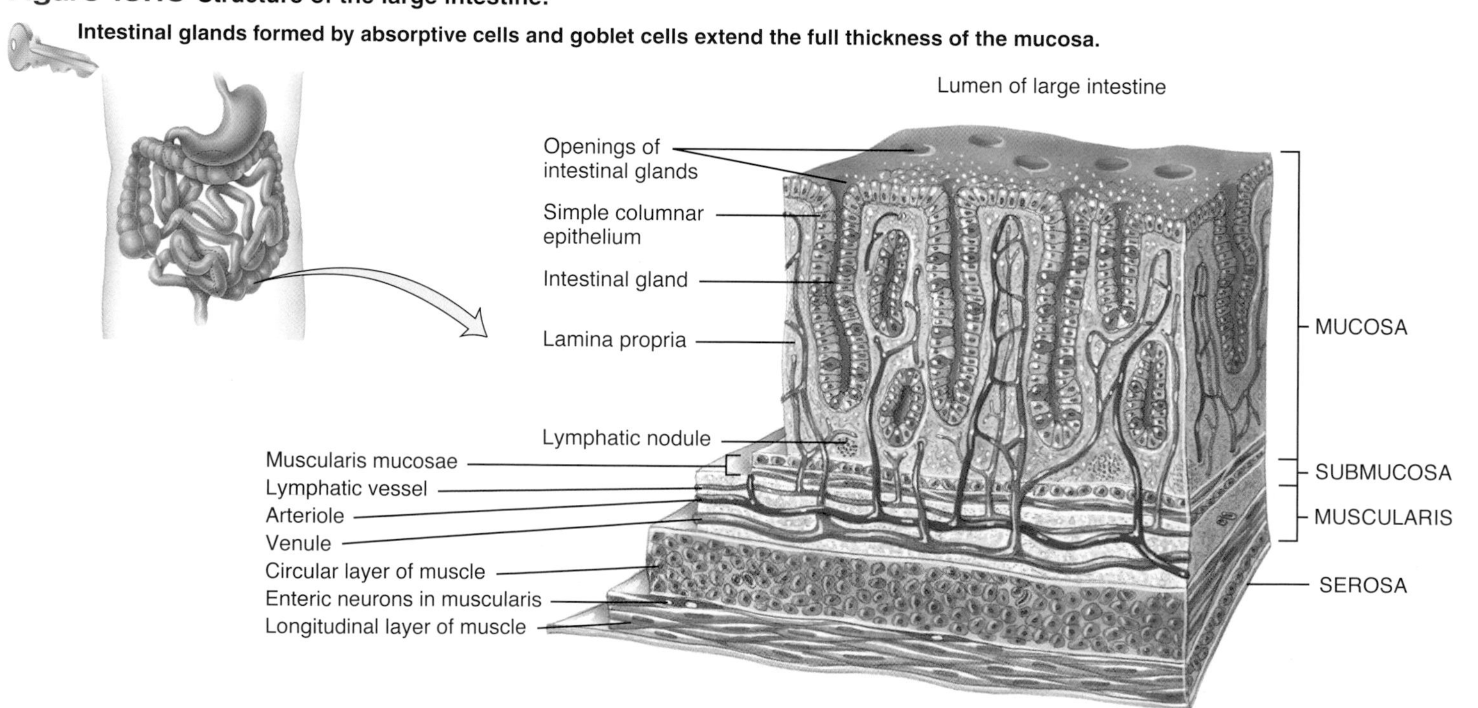

Three-dimensional view of layers of the large intestine

How does the muscularis of the large intestine differ from that of other parts of the GI tract?

Digestion and Absorption in the Large Intestine

The passage of chyme from the ileum into the cecum is regulated by the ileocecal sphincter. The sphincter normally remains slightly contracted so that the passage of chyme is usually a slow process. Immediately after a meal, a reflex intensifies peristalsis, forcing any chyme in the ileum into the cecum. ***Peristalsis*** occurs in the large intestine at a slower rate than in other portions of the GI tract. Characteristic of the large intestine is ***mass peristalsis***, a strong peristaltic wave that begins in the middle of the colon and drives the colonic contents into the rectum. Food in the stomach initiates mass peristalsis, which usually takes place three or four times a day, during or immediately after a meal.

The final stage of digestion occurs in the colon through the activity of bacteria that normally inhabit the lumen. The glands of the large intestine secrete mucus but no enzymes. Bacteria ferment any remaining carbohydrates and release hydrogen, carbon dioxide, and methane gases. These gases contribute to flatus (gas) in the colon, termed *flatulence* when it is excessive. Bacteria also break down the remaining proteins to amino acids and decompose bilirubin to simpler pigments, including stercobilin, which give feces their brown color. Several vitamins needed for normal metabolism, including some B vitamins and vitamin K, are bacterial products that are absorbed in the colon.

Although most water absorption occurs in the small intestine, the large intestine also absorbs a significant amount. The large intestine also absorbs ions, including sodium and chloride, and some dietary vitamins.

By the time chyme has remained in the large intestine 3 to 10 hours, it has become solid or semisolid as a result of water absorption and is now called ***feces***. Chemically, feces consist of water, inorganic salts, sloughed-off epithelial cells from the mucosa of the gastrointestinal tract, bacteria, products of bacterial decomposition, unabsorbed digested materials, and indigestible parts of food.

The Defecation Reflex

Mass peristaltic movements push fecal material from the sigmoid colon into the rectum. The resulting distension of the rectal wall stimulates stretch receptors, which initiates a ***defecation reflex*** that empties the rectum. Impulses from the spinal cord travel along parasympathetic nerves to the descending colon, sigmoid colon, rectum, and anus. The resulting contraction of the longitudinal rectal muscles shortens the rectum, thereby increasing the pressure within it. This pressure plus parasympathetic stimulation opens the internal sphincter. The external sphincter is voluntarily controlled. If it is voluntarily relaxed, defecation occurs and the feces are expelled through the anus; if it is voluntarily constricted, defecation can be postponed. Voluntary contractions of the diaphragm and abdominal muscles aid defecation by increasing the pressure within the abdomen, which pushes the walls of the sigmoid colon and rectum inward. If defecation does not occur, the feces back up into the sigmoid colon until the next wave of mass peristalsis stimulates the stretch receptors. In infants, the defecation reflex causes automatic emptying of the rectum because voluntary control of the external anal sphincter has not yet developed.

Diarrhea (dī-a-RĒ-a; *dia-* = through; *rrhea* = flow) is an increase in the frequency, volume, and fluid content of the feces caused by increased motility of and decreased absorption by the intestines. When chyme passes too quickly through the small intestine and feces pass too quickly through the large intestine, there is not enough time for absorption. Frequent diarrhea can result in dehydration and electrolyte imbalances. Excessive motility may be caused by lactose intolerance, stress, and microbes that irritate the gastrointestinal mucosa.

Constipation (kon′-sti-PĀ-shun; *con-* = together; *stip-* = to press) refers to infrequent or difficult defecation caused by decreased motility of the intestines. Because the feces remain in the colon for prolonged periods, excessive water absorption occurs, and the feces become dry and hard. Constipation may be caused by poor habits (delaying defecation), spasms of the colon, insufficient fiber in the diet, inadequate fluid intake, lack of exercise, emotional stress, or certain drugs.

■ **CHECKPOINT**

21. What activities occur in the large intestine to change its contents into feces?

22. What is defecation and how does it occur?

PHASES OF DIGESTION

OBJECTIVES • **Describe the three phases of digestion.**
• **Describe the major hormones that regulate digestive activities.**

Digestive activities occur in three overlapping phases: the cephalic phase, the gastric phase, and the intestinal phase.

Cephalic Phase

During the ***cephalic phase*** of digestion, the smell, sight, sound, or thought of food activates neural centers in the brain. The brain then activates the facial (VII), glossopharyn-

FOCUS ON WELLNESS

Emotional Eating—Consumed by Food

In addition to keeping us alive, eating serves countless psychological, social, and cultural purposes. We eat to celebrate, punish, comfort, defy, and deny. Eating in response to emotional drives, such as feeling stressed, bored, or tired, rather than in response to true physical hunger, is called *emotional eating*.

Food as Emotional Rescue

Emotional eating is so common that, within limits, it is considered well within the range of normal behavior. Who hasn't at one time or another headed for the refrigerator after a bad day? Problems arise when emotional eating becomes so excessive that it interferes with health. Physical health problems include obesity and associated disorders such as hypertension and heart disease. Psychological health problems include poor self-esteem; an inability to cope effectively with feelings of stress; and in extreme cases, eating disorders.

For emotional eaters, the drive to eat often masks unpleasant feelings such as boredom, loneliness, depression, anxiety, anger, or fatigue. Eating provides comfort and solace, numbing pain and "feeding the hungry heart." Some emotional overeaters say that stuffing themselves with food becomes a metaphor for suppressing undesirable feelings.

Eating may provide a biochemical "fix" as well. Emotional eaters typically overeat carbohydrate foods (sweets and starches), which may raise brain serotonin levels and lead to feelings of relaxation. Food becomes a way to self-medicate when negative emotions arise.

Consumed by Food

In extreme cases, eating becomes an addiction, and the drive to consume excessive amounts of food begins to take over a person's life. People with bulimia or binge-eating disorder have an overwhelmingly urgent and totally uncontrollable drive to eat, causing them to consume huge volumes of food several times a week, sometimes several times a day. People with bulimia try to purge the calories they have consumed by vomiting, exercising excessively, or using laxatives and diuretics, but people with binge-eating disorder usually do not.

Eating disorders can be very dangerous and even lethal, requiring prompt, comprehensive, and in-depth professional treatment that helps people cope with the underlying psychological issues. Therapy for emotional eaters requires addressing the emotions that trigger overeating and devising effective coping strategies that eliminate the need to deal with stress by overeating.

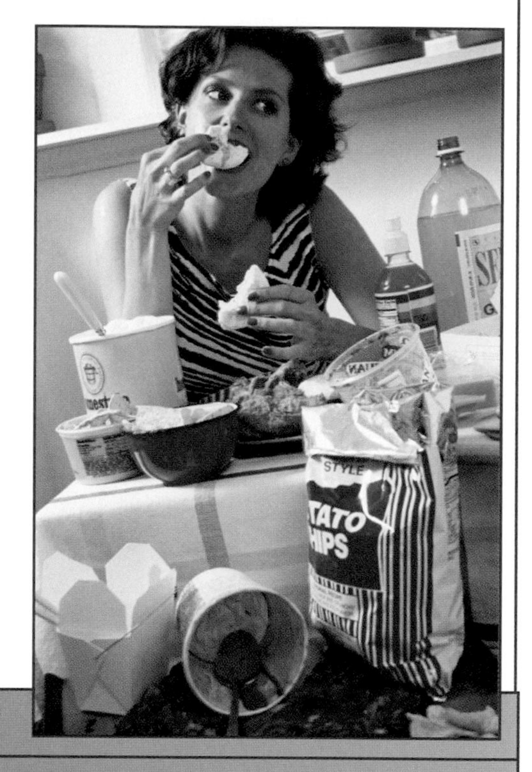

► **THINK IT OVER . . .**

► ***Why might repeated attempts to lose weight with very restrictive diets lead to emotional overeating?***

geal (IX), and vagus (X) nerves. The facial and glossopharyngeal nerves stimulate the salivary glands to secrete saliva, while the vagus nerves stimulate the gastric glands to secrete gastric juice. The purpose of the cephalic phase of digestion is to prepare the mouth and stomach for food that is about to be eaten.

Gastric Phase

Once food reaches the stomach, the ***gastric phase*** of digestion begins. The purpose of this phase of digestion is to continue gastric secretion and to promote gastric motility. Gastric secretion during the gastric phase is regulated by the hormone ***gastrin***. Gastrin is released from the G cells of the gastric glands in response to several stimuli: stretching of the stomach by chyme, partially digested proteins in chyme, caffeine in chyme, and the high pH of chyme due to the presence of food in the stomach. Gastrin stimulates gastric glands to secrete large amounts of gastric juice. It also strengthens the contraction of the lower esophageal sphincter to prevent reflux of acid chyme into the esophagus, increases motility of the stomach, and relaxes the pyloric sphincter, which promotes gastric emptying.

Intestinal Phase

The ***intestinal phase*** of digestion begins once food enters the small intestine. In contrast to the activities initiated during the cephalic and gastric phases, which stimulate stomach secretory activity and motility, those occurring during the intestinal phase have inhibitory effects that slow the exit of chyme from the stomach and prevent overloading of the duodenum with more chyme than it can handle. In addition, responses occurring during the intestinal phase promote the continued digestion of foods that have reached the small intestine.

The activities of the intestinal phase are mediated by two major hormones secreted by the small intestine: cholecystokinin and secretin. ***Cholecystokinin (CCK)*** is secreted by CCK cells in intestinal glands of the small intestine in response to chyme containing amino acids from partially digested proteins and fatty acids from partially digested triglycerides. CCK stimulates secretion of pancreatic juice that is rich in digestive enzymes. It also causes contraction of the wall of the gallbladder, which squeezes stored bile out of the gallbladder into the cystic duct and through the common bile duct. In addition, CCK slows gastric emptying by promoting contraction of the pyloric sphincter, and it produces satiety (feeling full to satisfaction) by acting on the hypothalamus in the brain.

Acidic chyme entering the duodenum stimulates the release of ***secretin*** from S cells in intestinal glands of the small intestine. In turn, secretin stimulates the flow of pancreatic juice that is rich in bicarbonate (HCO_3^-) ions to buffer the acidic chyme that enters the duodenum from the stomach.

Table 19.2 summarizes the major hormones that control digestion.

■ **CHECKPOINT**

23. What are the stimuli that cause the cephalic phase of digestion?

24. Compare and contrast the activities that occur during the gastric phase of digestion with those that occur during the intestinal phase of digestion.

AGING AND THE DIGESTIVE SYSTEM

OBJECTIVE • Describe the effects of aging on the digestive system.

Changes in the digestive system associated with aging include decreased secretory mechanisms, decreased motility of the digestive organs, loss of strength and tone of the muscular tissue and its supporting structures, changes in sensory feedback regarding enzyme and hormone release, and diminished response to pain and internal sensations. In the upper portion of the GI tract, common changes include reduced sensitivity to mouth irritations and sores, loss of taste, periodontal disease, difficulty in swallowing, hiatal hernia, gastritis, and peptic ulcer disease. Changes that may appear in the small intestine include duodenal ulcers, maldigestion, and malabsorption. Other pathologies that increase in incidence with age are appendicitis, gallbladder problems, jaundice, cirrhosis of the liver, and acute pancreatitis. Changes in the large intestine such as constipation, hemorrhoids, and diverticular disease may also occur. The incidence of cancer of the colon or rectum increases with age.

■ **CHECKPOINT**

25. List several changes in the upper and lower portions of the GI tract associated with aging.

• • •

Now that our exploration of the digestive system is completed, you can appreciate the many ways that this system contributes to homeostasis of other body systems by examining Focus on Homeostasis: The Digestive System on page 496. Next, in Chapter 20, you will discover how the nutrients absorbed by the GI tract are utilized in metabolic reactions by the body tissues.

Table 19.2 Major Hormones That Control Digestion

Hormone	Where Produced	Stimulant	Action
Gastrin	Stomach mucosa (pyloric region).	Stretching of stomach, partially digested proteins and caffeine in stomach, and high pH of stomach chyme.	Stimulates secretion of gastric juice, increases motility of GI tract, and relaxes pyloric sphincter.
Secretin	Intestinal mucosa.	Acidic chyme that enters the small intestine.	Stimulates secretion of pancreatic juice rich in bicarbonate ions.
Cholecystokinin (CCK)	Intestinal mucosa.	Amino acids and fatty acids in chyme in small intestine.	Inhibits gastric emptying, stimulates secretion of pancreatic juice rich in digestive enzymes, causes ejection of bile from the gallbladder, and induces a feeling of satiety (feeling full to satisfaction).

FOCUS ON HOMEOSTASIS

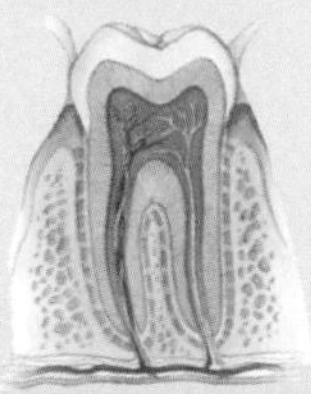

The Digestive System

Body System		Contribution of the Digestive System
For all body systems		The digestive system breaks down dietary nutrients into forms that can be absorbed and used by body cells for producing ATP and building body tissues; absorbs water, minerals, and vitamins needed for the growth and functions of body tissues; and eliminates wastes from body tissues in feces.
Integumentary system	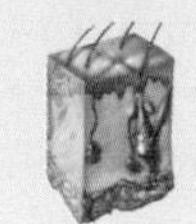	The small intestine absorbs vitamin D, which the skin and kidneys modify to produce the hormone calcitriol. Excess dietary calories are stored as triglycerides in adipose cells in the dermis and subcutaneous layer.
Skeletal system	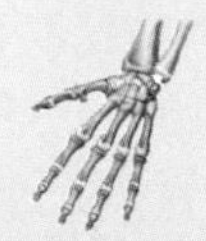	The small intestine absorbs dietary calcium and phosphorus salts needed to build bone extracellular matrix.
Muscular system		The liver can convert lactic acid produced by muscles during exercise to glucose.
Nervous system	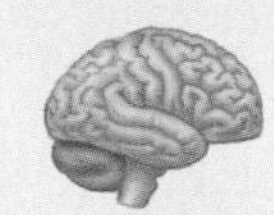	Gluconeogenesis (synthesis of new glucose molecules) in the liver plus digestion and absorption of dietary carbohydrates provide glucose, needed for ATP production by neurons.
Endocrine system	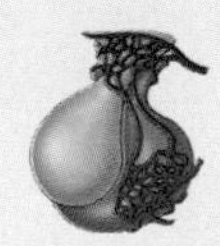	The liver inactivates some hormones, ending their activity. Pancreatic islets release insulin and glucagon. Cells in the mucosa of the stomach and small intestine release hormones that regulate digestive activities.
Cardiovascular system	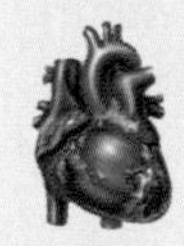	The GI tract absorbs water that helps maintain blood volume and iron that is needed for the synthesis of hemoglobin in red blood cells. Bilirubin from hemoglobin breakdown is partially excreted in feces. The liver synthesizes most plasma proteins.
Lymphatic system and immunity		The acidity of gastric juice destroys bacteria and most toxins in the stomach.
Respiratory system	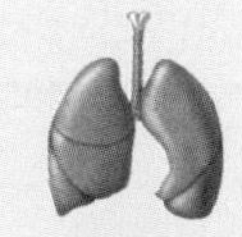	The pressure of abdominal organs against the diaphragm helps expel air quickly during a forced exhalation.
Urinary system	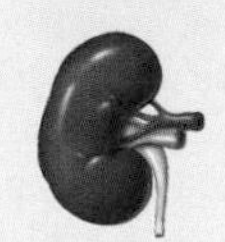	Absorption of water by the GI tract provides water needed to excrete waste products in urine.
Reproductive systems	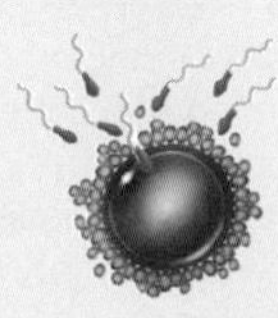	Digestion and absorption provides adequate nutrients, including fats, for normal development of reproductive structures, for the production of gametes (oocytes and sperm), and for fetal growth and development during pregnancy.

COMMON DISORDERS

Dietary Fiber and the Digestive System

Dietary fiber consists of indigestible plant substances, such as cellulose, lignin, and pectin, found in fruits, vegetables, grains, and beans. ***Insoluble fiber***, which does not dissolve in water, includes the structural parts of plants such as fruit and vegetable skins and the bran coating around wheat and corn kernels. Insoluble fiber passes through the GI tract largely unchanged and speeds up the passage of material through the tract. ***Soluble fiber***, which does dissolve in water, forms a gel that slows the passage of materials through the tract. It is found in abundance in beans, oats, barley, broccoli, prunes, apples, and citrus fruits. It tends to slow the passage of material through the tract.

People who choose a fiber-rich diet may reduce their risk of developing obesity, diabetes, atherosclerosis, gallstones, hemorrhoids, diverticulitis, appendicitis, and colon cancer. Insoluble fiber may help protect against colon cancer, and soluble fiber may help lower blood cholesterol level.

Dental Caries

Dental caries, or tooth decay, involves a gradual demineralization (softening) of the enamel and dentin by bacterial acids. If untreated, various microorganisms may invade the pulp, causing inflammation and infection with subsequent death of the pulp. Such teeth are treated by root canal therapy.

Periodontal Disease

Periodontal disease refers to a variety of conditions characterized by inflammation and degeneration of the gums, bone, periodontal ligament, and cementum. Periodontal diseases are often caused by poor oral hygiene; by local irritants, such as bacteria, impacted food, and cigarette smoke; or by a poor "bite."

Peptic Ulcer Disease

Five to ten percent of the U.S. population develops ***peptic ulcer disease (PUD)*** each year. An ***ulcer*** is a craterlike lesion in a membrane; ulcers that develop in areas of the GI tract exposed to acidic gastric juice are called ***peptic ulcers***. The most common complication of peptic ulcers is bleeding, which can lead to anemia. In acute cases, peptic ulcers can lead to shock and death. Three distinct causes of PUD are recognized: (1) the bacterium *Helicobacter pylori*, (2) nonsteroidal anti-inflammatory drugs (NSAIDs) such as aspirin, and (3) hypersecretion of HCl.

Appendicitis

Appendicitis is an inflammation of the appendix. Appendectomy (surgical removal of the appendix) is recommended in all suspected cases because it is safer to operate than to risk gangrene, rupture, and peritonitis.

Colorectal Cancer

Colorectal cancer is among the deadliest of malignancies. An inherited predisposition contributes to more than half of all cases of colorectal cancer. Intake of alcohol and diets high in animal fat and protein are associated with increased risk of colorectal cancer; dietary fiber, retinoids, calcium, and selenium may be protective. Signs and symptoms of colorectal cancer include diarrhea, constipation, cramping, abdominal pain, and rectal bleeding. Screening for colorectal cancer includes testing for blood in the feces, digital rectal examination, sigmoidoscopy, colonoscopy, and barium enema.

Diverticular Disease

Diverticulosis is the development of diverticula, saclike outpouchings of the wall of the colon in places where the muscularis has become weak. Many people who develop diverticulosis have no symptoms and experience no complications. About 15% of people with diverticulosis eventually develop an inflammation known as ***diverticulitis***, characterized by pain, either constipation or increased frequency of defecation, nausea, vomiting, and low-grade fever. Patients who change to high-fiber diets often show marked relief of symptoms.

Hepatitis

Hepatitis is an inflammation of the liver caused by viruses, drugs, and chemicals, including alcohol.

Hepatitis A (infectious hepatitis), caused by the hepatitis A virus, is spread by fecal contamination of food, clothing, toys, eating utensils, and so forth (fecal–oral route). It does not cause lasting liver damage.

Hepatitis B, caused by the hepatitis B virus, is spread primarily by sexual contact and contaminated syringes and transfusion equipment. It can also be spread by any secretion via saliva and tears. Hepatitis B can produce chronic liver inflammation. Vaccines are available for hepatitis B and are required for certain individuals, such as health-care providers.

Hepatitis C, caused by the hepatitis C virus, is clinically similar to hepatitis B. It is often spread by blood transfusions and can cause cirrhosis and liver cancer.

Hepatitis D is caused by the hepatitis D virus. It is transmitted like hepatitis B. A person must be infected with hepatitis B to contract hepatitis D. Hepatitis D results in severe liver damage and has a fatality rate higher than that due to infection with hepatitis B virus alone.

Hepatitis E is caused by the hepatitis E virus and is spread like hepatitis A. Although it does not cause chronic liver disease, the hepatitis E virus is responsible for a very high death rate in pregnant women.

MEDICAL TERMINOLOGY AND CONDITIONS

Anorexia nervosa A chronic disorder characterized by self-induced weight loss, negative perception of body image, and physiological changes that result from nutritional depletion. Patients have a fixation on weight control and often abuse laxatives, which worsens their fluid and electrolyte imbalances and nutrient deficiencies. The disorder is found predominantly in young, single females, and it may be inherited. Individuals may become emaciated and may ultimately die of starvation or one of its complications.

Bulimia (*bu-* = ox; *-limia* = hunger) or ***binge-purge syndrome*** A disorder characterized by overeating at least twice a week followed by purging by self-induced vomiting, strict dieting or fasting, vigorous exercise, or use of laxatives or diuretics; it occurs in response to fears of being overweight, stress, depression, and physiological disorders such as hypothalamic tumors.

Canker sore (KANG-ker) Painful ulcer on the mucous membrane of the mouth that affects females more often than males, usually between ages 10 and 40; it may be an autoimmune reaction or result from a food allergy.

Cholecystitis (kō′-lē-sis-TĪ-tis; *chole-* = bile; *cyst-* = bladder; *-itis* = inflammation of) In some cases, an autoimmune inflammation of the gallbladder; other cases are caused by obstruction of the cystic duct by bile stones.

Cirrhosis Distorted or scarred liver as a result of chronic inflammation due to hepatitis, chemicals that destroy hepatocytes, parasites that infect the liver, or alcoholism; the hepatocytes are replaced by fibrous or adipose connective tissue. Symptoms include jaundice, edema in the legs, uncontrolled bleeding, and increased sensitivity to drugs.

Colostomy (kō-LOS-tō-mē; *-stomy* = provide an opening) The diversion of the fecal stream through an opening in the colon, creating a surgical "stoma" (artificial opening) that is affixed to the exterior of the abdominal wall. This opening serves as a substitute anus through which feces are eliminated into a bag worn on the abdomen.

Inflammatory bowel disease (in-FLAM-a-tō′-rē BOW-el) Disorder that exists in two forms: (1) Crohn's disease, an inflammation of the gastrointestinal tract, especially the distal ileum and proximal colon, in which the inflammation may extend from the mucosa through the serosa, and (2) ulcerative colitis, an inflammation of the mucosa of the gastrointestinal tract, usually limited to the large intestine and usually accompanied by rectal bleeding.

Irritable bowel syndrome (IBS) Disease of the entire gastrointestinal tract in which a person reacts to stress by developing symptoms (such as cramping and abdominal pain) associated with alternating patterns of diarrhea and constipation. Excessive amounts of mucus may appear in feces; other symptoms include flatulence, nausea, and loss of appetite.

Malocclusion (mal′-ō-KLOO-zhun; *mal-* = bad; *occlusion* = to fit together) Condition in which the surfaces of the maxillary (upper) and mandibular (lower) teeth fit together poorly.

Nausea (NAW-sē-a = seasickness) Discomfort characterized by a loss of appetite and the sensation of impending vomiting. Its causes include local irritation of the gastrointestinal tract, a systemic disease, brain disease or injury, overexertion, or the effects of medication or drug overdose.

Traveler's diarrhea Infectious disease of the gastrointestinal tract that results in loose, urgent bowel movements; cramping; abdominal pain; malaise; nausea; and occasionally fever and dehydration. It is acquired through ingestion of food or water contaminated with fecal material typically containing bacteria (especially *Escherichia coli*); viruses or protozoan parasites are a less common cause.

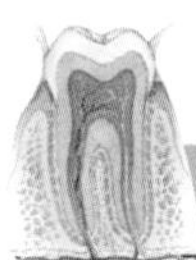

STUDY OUTLINE

Introduction (p. 472)

1. The breakdown of larger food molecules into smaller molecules is called digestion; the passage of these smaller molecules into blood and lymph is termed absorption.
2. The organs that collectively perform digestion and absorption constitute the digestive system.

Overview of the Digestive System (p. 473)

1. The GI tract is a continuous tube extending from the mouth to the anus.
2. The accessory digestive organs include the teeth, tongue, salivary glands, liver, gallbladder, and pancreas.
3. Digestion includes six basic processes: ingestion, secretion, mixing and propulsion, mechanical and chemical digestion, absorption, and defecation.

Layers of the GI Tract and Omentum (p. 474)

1. The basic arrangement of layers in most of the gastrointestinal tract, from the inside to the outside, is the mucosa, submucosa, muscularis, and serosa.
2. Parts of the peritoneum include the mesentery and greater omentum.

Mouth (p. 476)

1. The mouth is formed by the cheeks, hard and soft palates, lips, and tongue, which aid mechanical digestion.
2. The tongue forms the floor of the oral cavity. It is composed of skeletal muscle covered with mucous membrane. The superior surface and lateral areas of the tongue are covered with papillae. Some papillae contain taste buds.
3. Most saliva is secreted by the salivary glands, which lie outside the mouth and release their secretions into ducts that empty into the oral cavity. There are three pairs of salivary glands: parotid, submandibular, and sublingual. Saliva lubricates food and starts the chemical digestion of carbohydrates. Salivation is controlled by the autonomic nervous system.

4. The teeth, or dentes, project into the mouth and are adapted for mechanical digestion. A typical tooth consists of three principal portions: crown, root, and neck. Teeth are composed primarily of dentin and are covered by enamel, the hardest substance in the body. Humans have two sets of teeth: deciduous and permanent.
5. Through mastication, food is mixed with saliva and shaped into a bolus.
6. Salivary amylase begins the digestion of starches in the mouth.

Pharynx and Esophagus (p. 478)

1. Food that is swallowed passes from the mouth into the oropharynx.
2. From the oropharynx, food passes into the laryngopharynx.
3. The esophagus is a muscular tube that connects the pharynx to the stomach.
4. Swallowing moves a bolus from the mouth to the stomach by peristalsis. It consists of a voluntary stage, pharyngeal stage (involuntary), and esophageal stage (involuntary).

Stomach (p. 480)

1. The stomach connects the esophagus to the duodenum.
2. The main regions of the stomach are the cardia, fundus, body, and pylorus.
3. Adaptations of the stomach for digestion include rugae; glands that produce mucus, hydrochloric acid, a protein-digesting enzyme (pepsin), intrinsic factor, and gastrin; and a three-layered muscularis for efficient mechanical movement.
4. Mechanical digestion consists of mixing waves that macerate food and mix it with gastric juice, forming chyme.
5. Chemical digestion consists of the conversion of proteins into peptides by pepsin.
6. The stomach wall is impermeable to most substances. Among the substances the stomach can absorb are water, ions, short-chain fatty acids, some drugs, and alcohol.

Pancreas (p. 483)

1. Secretions pass from the pancreas to the duodenum via the pancreatic duct.
2. Pancreatic islets (islets of Langerhans) secrete hormones and constitute the endocrine portion of the pancreas.
3. Acinar cells, which secrete pancreatic juice, constitute the exocrine portion of the pancreas.
4. Pancreatic juice contains enzymes that digest starch (pancreatic amylase); proteins (trypsin, chymotrypsin, and carboxypeptidase); triglycerides (pancreatic lipase); and nucleic acids (nucleases).

Liver and Gallbladder (p. 484)

1. The liver has left and right lobes. The gallbladder is a sac located in a depression under the liver that stores and concentrates bile produced by the liver.
2. The lobes of the liver are made up of lobules that contain hepatocytes (liver cells), sinusoids, Kupffer cells, and a central vein.
3. Hepatocytes produce bile that is carried by a duct system to the gallbladder for concentration and temporary storage.
4. Bile's contribution to digestion is the emulsification of dietary lipids.
5. The liver also functions in carbohydrate, lipid, and protein metabolism; processing of drugs and hormones; excretion of bilirubin; synthesis of bile salts; storage of vitamins and minerals; phagocytosis; and activation of vitamin D.

Small Intestine (p. 486)

1. The small intestine extends from the pyloric sphincter to the ileocecal sphincter. It is divided into the duodenum, the jejunum, and the ileum.
2. The small intestine is highly adapted for digestion and absorption. Its glands produce enzymes and mucus, and the microvilli, villi, and circular folds of its wall provide a large surface area for digestion and absorption.
3. Mechanical digestion in the small intestine involves segmentations and migrating waves of peristalsis.
4. Enzymes in pancreatic juice, bile, and the microvilli of the absorptive cells of the small intestine bile break down disaccharides to monosaccharides; protein digestion is completed by peptidase enzymes; triglycerides are broken down into fatty acids and monoglycerides by pancreatic lipase; and nucleases break down nucleic acids to pentoses and nitrogenous bases.
5. Absorption is the passage of nutrients from digested food in the gastrointestinal tract into the blood or lymph. Absorption, which occurs mostly in the small intestine, occurs by means of simple diffusion, facilitated diffusion, osmosis, and active transport.
6. Monosaccharides, amino acids, and short-chain fatty acids pass into the blood capillaries.
7. Long-chain fatty acids and monoglycerides are absorbed as part of micelles, resynthesized to triglycerides, and transported in chylomicrons to the lacteal of a villus.
8. The small intestine also absorbs water, electrolytes, and vitamins.

Large Intestine (p. 491)

1. The large intestine extends from the ileocecal sphincter to the anus. Its regions include the cecum, colon, rectum, and anal canal.
2. The mucosa contains numerous absorptive cells that absorb water and goblet cells that secrete mucus.
3. Mass peristalsis is a strong peristaltic wave that drives the contents of the colon into the rectum.
4. In the large intestine, substances are further broken down, and some vitamins are synthesized through bacterial action.
5. The large intestine absorbs water, electrolytes, and vitamins.
6. Feces consist of water, inorganic salts, epithelial cells, bacteria, and undigested foods.
7. The elimination of feces from the rectum is called defecation. Defecation is a reflex action aided by voluntary contractions of the diaphragm and abdominal muscles and relaxation of the external anal sphincter.

Phases of Digestion (p. 493)

1. Digestive activities occur in three overlapping phases: cephalic phase, gastric phase, and intestinal phase.

2. During the cephalic phase of digestion, salivary glands secrete saliva and gastric glands secrete gastric juice in order to prepare the mouth and stomach for food that is about to be eaten.
3. The presence of food in the stomach causes the gastric phase of digestion, which promotes gastric juice secretion and gastric motility.
4. During the intestinal phase of digestion, food is digested in the small intestine. In addition, gastric motility and gastric secretion decrease in order to slow the exit of chyme from the stomach, which prevents the small intestine from being overloaded with more chyme than it can handle.
5. The activities that occur during the various phases of digestion are coordinated by hormones. Table 19.2 on page 495 summarizes the major hormones that control digestion.

Aging and the Digestive System (p. 495)

1. General changes with age include decreased secretory mechanisms, decreased motility, and loss of tone.
2. Specific changes may include loss of taste, hernias, peptic ulcer disease, constipation, hemorrhoids, and diverticular diseases.

SELF-QUIZ

1. Which of the following is NOT an accessory digestive organ?
   a. teeth b. salivary glands c. liver d. pancreas e. esophagus
2. Chewing food is an example of
   a. absorption b. mechanical digestion c. secretion d. chemical digestion e. ingestion
3. Which of the following is mismatched?
   a. submucosa, enteric nervous system (ENS)
   b. muscularis, lacteal c. serosa, greater omentum
   d. mucosa, villi e. serosa, visceral peritoneum
4. Most chemical digestion occurs in the
   a. liver b. stomach c. duodenum d. colon e. pancreas
5. Absorption is defined as
   a. the elimination of solid wastes from the digestive system
   b. a reflex action controlled by the autonomic nervous system
   c. the breakdown of foods by enzymes
   d. the passage of nutrients from the gastrointestinal tract into the bloodstream
   e. the mechanical breakdown of triglycerides
6. The exposed portions of the teeth that you clean with a toothbrush are the
   a. crowns b. periodontal ligaments c. roots d. pulp cavities e. gingivae
7. The smell of your favorite food cooking makes "your mouth water"; this is due to
   a. sympathetic stimulation of the salivary glands
   b. mastication
   c. parasympathetic stimulation of the salivary glands
   d. increased mucus secretion by the pharynx
   e. the enteric nervous system
8. Match the following:

____ a. carries bile		A. pyloric sphincter
____ b. proteins combined with triglycerides and cholesterol		B. circular folds
____ c. surrounds the opening between the stomach and duodenum		C. micelles
____ d. secrete pancreatic juice		D. cystic duct
____ e. increase surface area in small intestine		E. ileocecal sphincter
____ f. bile salts combined with partially digested lipids		F. rugae
____ g. located between the opening of the small and large intestine		G. chylomicrons
____ h. large mucosal folds in stomach		H. acini

9. Which of the following correctly describes the esophagus?
   a. Food enters the esophagus from the pyloric region of the stomach.
   b. The movement of food through the entire esophagus is under voluntary control.
   c. It allows the passage of chyme.
   d. It produces several enzymes that aid in the digestion of food.
   e. It is a muscular tube extending from the pharynx to the stomach.
10. If an incision were made into the stomach, the tissue layers would be cut in what order?
    a. mucosa, muscularis, serosa, submucosa
    b. mucosa, muscularis, submucosa, serosa
    c. serosa, muscularis, mucosa, submucosa
    d. muscularis, submucosa, mucosa, serosa
    e. serosa, muscularis, submucosa, mucosa
11. Most water absorption in the digestive tract occurs in the
    a. small intestine b. stomach c. mouth d. liver e. large intestine

12. Which of the following would NOT result in secretion of gastric juices in the stomach?
    a. secretion of gastrin
    b. stimulation by the vagus nerves
    c. the presence of partially digested proteins
    d. stretching of the stomach
    e. stimulation by the sympathetic nervous system
13. Bile
    a. is produced in the gallbladder
    b. is an enzyme that breaks down carbohydrates
    c. emulsifies triglycerides
    d. is required for the absorption of amino acids
    e. enters the small intestine through the right hepatic duct
14. Which of the following is NOT a function of the liver?
    a. processing newly absorbed nutrients
    b. producing enzymes that digest proteins
    c. breaking down old red blood cells
    d. detoxifying certain poisons
    e. producing bile
15. The purpose of villi in the small intestine is to
    a. aid in the movement of food through the small intestines
    b. phagocytize microbes
    c. produce digestive enzymes
    d. increase the surface area for absorption of digested nutrients
    e. produce acidic secretions
16. Which of the following is NOT produced in the stomach?
    a. sodium bicarbonate ($NaHCO_3$)
    b. gastrin
    c. pepsinogen
    d. mucus
    e. hydrochloric acid (HCl)
17. Which of the following is NOT correctly paired?
    a. esophagus, peristalsis
    b. mouth, mastication
    c. large intestine, mass peristalsis
    d. small intestine, segmentations
    e. stomach, emulsification
18. The enzyme pancreatic lipase digests triglycerides into
    a. glucose
    b. amino acids
    c. fatty acids and monoglycerides
    d. nucleic acids
    e. amylase
19. Place the following in the correct order as food passes from the small intestine:
    1. sigmoid colon
    2. transverse colon
    3. ascending colon
    4. rectum
    5. cecum
    6. descending colon

    a. 1, 3, 2, 6, 5, 4 b. 5, 1, 6, 2, 3, 4 c. 4, 1, 6, 2, 3, 5
    d. 2, 3, 5, 6, 4, 1 e. 5, 3, 2, 6, 1, 4
20. Lacteals function
    a. in the absorption of lipids in chylomicrons
    b. to produce bile in the liver
    c. in the absorption of electrolytes
    d. in the fermentation of carbohydrates in the large intestine
    e. to produce salivary amylas

CRITICAL THINKING APPLICATIONS

1. Four out of five dentists think that you should chew sugarless gum, but all five think that you should brush your teeth. Why?
2. The discussion around the bridge table was growing heated. Edna is convinced that lactose intolerance is the cause of her constipation. Gertrude insists that lactose intolerance has nothing to do with bowel problems but is the cause of her heartburn. Of course, neither of these ladies has eaten dairy products for years (which may help explain their osteoporosis). Please settle the argument.
3. Jared put a plastic spider in his sister's drink as a joke. Unfortunately, his mother doesn't think the joke was very funny because his sister swallowed it and now they're all at the ER (emergency room). The doctor suspects that the spider may have lodged at the junction of the stomach and the duodenum. Name the sphincter at this junction. Trace the path taken by the plastic spider on its journey to its new home. What procedure could the doctor use to view the interior of the stomach? What structures may be viewed in the stomach (besides the spider)?
4. Jerry hadn't eaten all day when he bought a dried out, lukewarm hot dog from a street vendor for lunch. A few hours later, he was a victim of food poisoning and was desperately seeking a bathroom. After vomiting several times, Jerry noticed that he was expelling a greenish-yellow liquid. The hot dog may have been a bit shriveled, but it wasn't green! What is the source of this colored fluid?

ANSWERS TO FIGURE QUESTIONS

19.1 The teeth cut and grind food.

19.2 Nerves in its wall help regulate secretions and contractions of the gastrointestinal tract.

19.3 Mesentery binds the small intestine to the posterior abdominal wall.

19.4 Muscles of the tongue maneuver food for chewing, shape food into a bolus, force food to the back of the mouth for swallowing, and alter the shape of the tongue for swallowing and speech production.

19.5 The main component of teeth is a connective tissue called dentin.

19.6 Swallowing is both voluntary and involuntary. Initiation of swallowing, carried out by skeletal muscles, is voluntary. Completion of swallowing—moving a bolus along the esophagus and into the stomach—involves peristalsis of smooth muscle and is involuntary.

19.7 After a very large meal, the stomach probably does not have rugae because as the stomach fills, the rugae stretch out.

19.8 The simple columnar epithelial cells of the mucosa are in contact with food in the stomach.

19.9 G cells, which secrete the hormone gastrin, are part of the endocrine system.

19.10 Pancreatic juice is a mixture of water, salts, bicarbonate ions, and digestive enzymes.

19.11 Kupffer cells in the liver are phagocytes.

19.12 The ileum is the longest portion of the small intestine.

19.13 The absorptive cells cover the surface of the villi.

19.14 Fat-soluble vitamins are absorbed by diffusion from micelles.

19.15 Functions of the large intestine include completion of absorption, synthesis of certain vitamins, formation of feces, and elimination of feces.

19.16 The muscularis of the large intestine forms three longitudinal bands that gather the colon into a series of pouches.

NUTRITION AND METABOLISM chapter 20

did you know?

When energy intake (how many Calories you consume) exceeds energy expenditure (how many Calories you burn), the extra energy is saved in the form of body fat. Over the years excess fat storage can lead to obesity. The remedy? Eat less and exercise more, in order to create an energy imbalance that coaxes the body into using stored fat for fuel. Sounds simple, but many people have a very difficult time losing weight and keeping it off. Studies of people who win at losing weight have found that significant, lifelong changes in eating and exercise behaviors are the key to long-term weight control success.

Focus on Wellness, page 518

www.wiley.com/college/apcentral

The food we eat is our only source of energy for performing biological work. Many molecules needed to maintain cells and tissues can be made from building blocks within the body; others must be obtained in food because we cannot make them. Food molecules absorbed by the gastrointestinal (GI) tract have three main fates:

1. *To supply energy* for sustaining life processes, such as active transport, DNA replication, protein synthesis, muscle contraction, maintenance of body temperature, and cell division.
2. *To serve as building blocks* for the synthesis of more complex molecules, such as muscle proteins, hormones, and enzymes.
3. *Storage for future use.* For example, glycogen is stored in liver cells, and triglycerides are stored in adipose cells.

In this chapter we will discuss the major groups of nutrients; guidelines for healthy eating; how each group of nutrients is used for ATP production, growth, and repair of the body; and how various factors affect the body's metabolic rate.

looking back to move ahead . . .

- **Main Chemical Elements in the Body (page 23)**
- **Enzymes (page 37)**
- **Carbohydrates, Lipids, Proteins (pages 31–37)**
- **Negative Feedback System (page 8)**
- **Functions of the Liver (page 485)**
- **Hypothalamus and Body Temperature Regulation (page 254)**

NUTRIENTS

OBJECTIVES • Define a nutrient and identify the six main types of nutrients.

• List the guidelines for healthy eating.

Nutrients are chemical substances in food that body cells use for growth, maintenance, and repair. The six main types of nutrients are carbohydrates, lipids, proteins, water, minerals, and vitamins. ***Essential nutrients*** are specific nutrient molecules that the body cannot make in sufficient quantity to meet its needs and thus must be obtained from the diet. Some amino acids, some fatty acids, vitamins, and minerals are essential nutrients. The structures and functions of carbohydrates, proteins, lipids, and water were discussed in Chapter 2. In this chapter, we discuss some guidelines for healthy eating and the roles of minerals and vitamins in metabolism.

Guidelines for Healthy Eating

Each gram of protein or carbohydrate in food provides about 4 Calories; 1 gram of fat (lipids) provides about 9 Calories. We do not know with certainty what levels and types of carbohydrate, fat, and protein are optimal in the diet. Different populations around the world eat radically different diets that are adapted to their particular lifestyles. However, many experts recommend the following distribution of calories: 50–60% from carbohydrates, with less than 15% from simple sugars; less than 30% from fats (triglycerides are the main type of dietary fat), with no more than 10% as saturated fats; and about 12–15% from proteins.

The guidelines for healthy eating are to:

- Eat a variety of foods.
- Maintain a healthy weight.
- Choose foods low in fat, saturated fat, and cholesterol.
- Eat plenty of vegetables, fruits, and grain products.
- Use sugars in moderation only.

In 2005, the United States Department of Agriculture (USDA) introduced a new food pyramid called ***My Pyramid***, which represents a *personalized* approach to making healthy food choices and maintaining regular physical activity. By consulting a chart, it is possible to determine your calorie level based on your gender, age, and activity level. Once this is determined, you can choose the type and amount of food to be consumed.

If you carefully examine the My Pyramid in Figure 20.1, you will note that the six color bands represent the five basic food groups plus oils. Foods from all bands are needed each day. Also note that the overall size of the bands suggests the proportion of food a person should choose on a daily basis. The wider base of each band represents foods with little or no solid fats or added sugars and these foods should be selected more often. The narrower top of each band represents foods with more added sugars and solid fats, which should be selected less frequently. The person climbing the steps is a reminder of the need for daily physical activity.

As an example of how the My Pyramid works, let's assume based upon consulting a chart that the calorie level of an 18-year-old moderately active female is 2000 Calories and that of an 18-year-old moderately active male is 2800 Calories. Accordingly, it is suggested that the following foods should be chosen in the following amounts:

Calorie level	2000	2800
Fruits (includes all fresh, frozen, canned, and dried fruits and fruit juices)	2 cups	2.5 cups
Vegetables (includes all fresh, frozen, canned, and dried vegetables and vegetable juices)	2.5 cups	3.5 cups
Grains (includes all foods made from wheat, rice, oats, cornmeal, and barley such as bread, cereals, oatmeal, rice, pasta, crackers, tortillas, and grits)	6 oz	10 oz
Meats and beans (includes lean meat, poultry, fish, eggs, peanut butter, beans, nuts, and seeds)	5.5 oz	7 oz
Milk group (includes milk products and foods made from milk that retain their calcium content such as cheeses and yogurt)	3 cups	3 cups
Oils (Choose mostly fats that contain monounsaturated and polyunsaturated fatty acids such as fish, nuts, seeds, and vegetable oils)	6 tsp	8 tsp

In addition, you should choose and prepare foods with little salt. In fact, sodium intake should be less than 2300 mg per day. If you choose to drink alcohol, it should be consumed in moderation (no more than 1 drink per day for women and 2 drinks per day for men). A drink is defined as 12 oz of regular beer, 5 oz of wine, or $1^1/_2$ oz of 80 proof distilled spirits.

Minerals

Minerals are inorganic elements that constitute about 4% of the total body weight and are concentrated most heavily in the skeleton. Minerals with known functions in the body include calcium, phosphorus, potassium, sulfur, sodium, chloride, fluoride, magnesium, iron, iodide, manganese, cobalt, copper, zinc, selenium, and chromium. Others—aluminum, boron, silicon, and molybdenum—are present but may have no functions. Typical diets supply adequate amounts of potassium, sodium, chloride, and magnesium. Some attention must be paid to eating foods that provide

Figure 20.1 My Pyramid.

My Pyramid is a new personalized approach to making healthy food choices and maintaining regular physical activity.

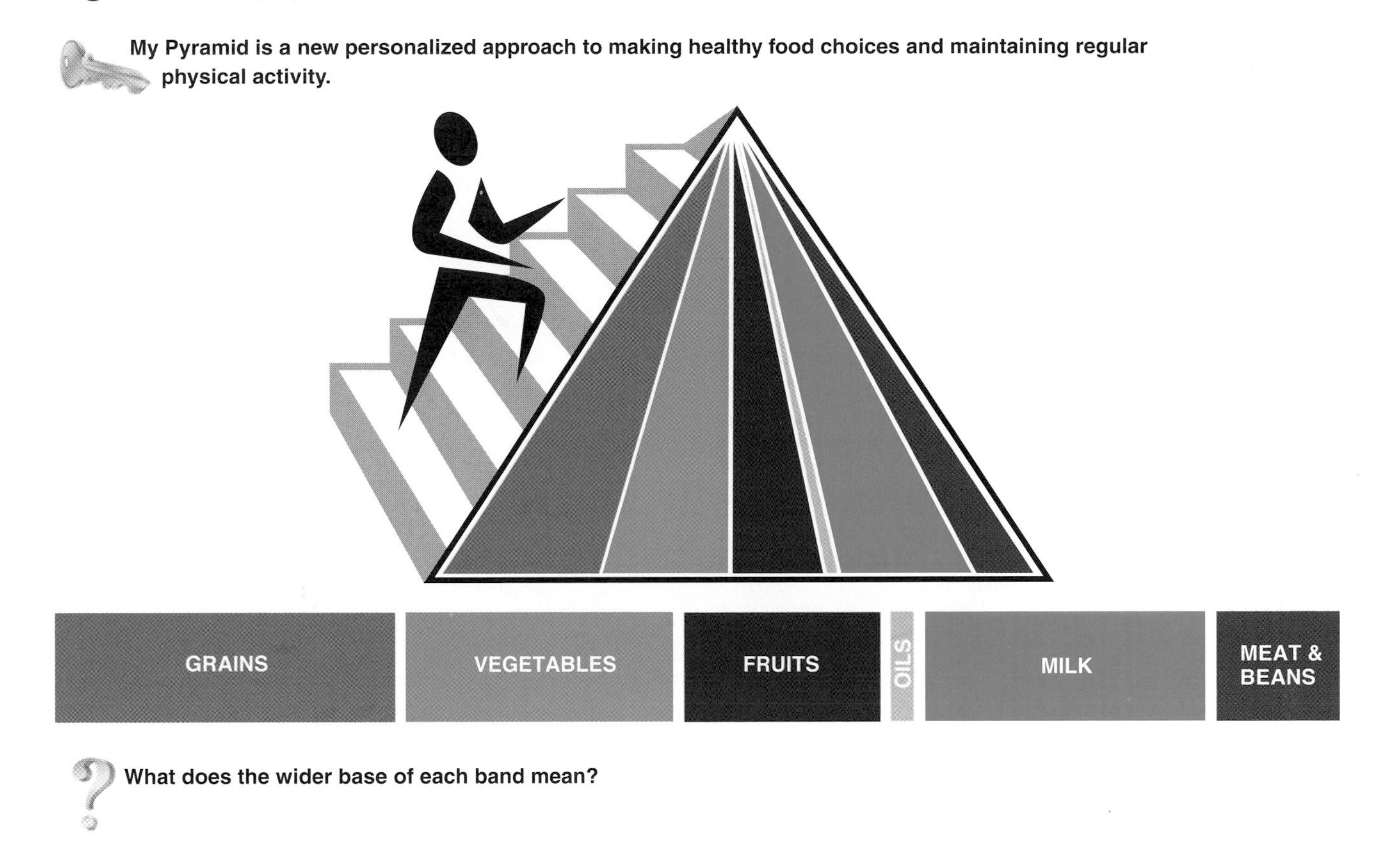

What does the wider base of each band mean?

enough calcium, phosphorus, iron, and iodine. Excess amounts of most minerals are excreted in the urine and feces.

A major role of minerals is to help regulate enzymatic reactions. Calcium, iron, magnesium, and manganese are part of some coenzymes. Magnesium also serves as a catalyst for the conversion of ADP to ATP. Minerals such as sodium and phosphorus work in buffer systems, which help control the pH of body fluids. Sodium also helps regulate the osmosis of water and, with other ions, is involved in the generation of nerve impulses. Table 20.1 describes the roles of several minerals in various body functions.

Vitamins

Organic nutrients required in small amounts to maintain growth and normal metabolism are called ***vitamins***. Unlike carbohydrates, lipids, or proteins, vitamins do not provide energy or serve as the body's building materials. Most vitamins with known functions serve as coenzymes.

Most vitamins cannot be synthesized by the body and must be ingested. Other vitamins, such as vitamin K, are produced by bacteria in the GI tract and then absorbed. The body can assemble some vitamins if the raw materials, called ***provitamins***, are provided. For example, vitamin A is produced by the body from the provitamin beta-carotene, a chemical present in orange and yellow vegetables such as carrots and in dark green vegetables such as spinach. No single food contains all the vitamins required by the body—one of the best reasons to eat a varied diet.

Vitamins are divided into two main groups: fat-soluble and water-soluble. The ***fat-soluble vitamins*** are vitamins A, D, E, and K. They are absorbed along with dietary lipids in the small intestine and packaged into chylomicrons. They cannot be absorbed in adequate quantity unless they are ingested with other lipids. Fat-soluble vitamins may be stored in cells, particularly in the liver. The ***water-soluble vitamins*** include the B vitamins and vitamin C. They are dissolved in body fluids. Excess quantities of these vitamins are not stored but instead are excreted in the urine.

Besides their other functions, three vitamins—C, E, and beta-carotene (a provitamin)—are termed ***antioxidant vitamins*** because they inactivate oxygen free radicals. Recall that free radicals are highly reactive ions or molecules that carry an unpaired electron in their outermost electron shell. Free radicals damage cell membranes, DNA, and other cellular structures and contribute to the formation of atherosclerotic plaques. Some free radicals arise naturally in the body, and others come from environmental hazards such as tobacco smoke and radiation. Antioxidant vitamins are thought to play a role in protecting against some kinds of cancer, reduc-

Table 20.1 Minerals Vital to the Body

Mineral	Comments	Importance
Calcium	Most abundant mineral in body. Appears in combination with phosphates. About 99% is stored in bone and teeth. Blood Ca^{2+} level is controlled by parathyroid hormone (PTH). Calcitriol promotes absorption of dietary calcium. Sources are milk, egg yolk, shellfish, and leafy green vegetables.	Formation of bones and teeth, blood clotting, normal muscle and nerve activity, endocytosis and exocytosis, cellular motility, chromosome movement during cell division, glycogen metabolism, and release of neurotransmitters and hormones.
Phosphorus	About 80% is found in bones and teeth as phosphate salts. Blood phosphate level is controlled by parathyroid hormone (PTH). Sources are dairy products, meat, fish, poultry, and nuts.	Formation of bones and teeth. Phosphates constitute a major buffer system of blood. Plays important role in muscle contraction and nerve activity. Component of many enzymes. Involved in energy transfer (ATP). Component of DNA and RNA.
Potassium	Major cation (K^+) in intracellular fluid. Excess excreted in urine. Present in most foods (meats, fish, poultry, fruits, and nuts).	Needed for generation and conduction of action potentials in neurons and muscle fibers.
Sulfur	Component of many proteins (such as insulin and chondroitin sulfate), electron carriers in electron transport chain, and some vitamins (thiamine and biotin). Sources include beef, liver, lamb, fish, poultry, eggs, cheese, and beans.	As component of hormones and vitamins, regulates various body activities. Needed for ATP production by electron transport chain.
Sodium	Most abundant cation (Na^+) in extracellular fluids; some found in bones. Normal intake of NaCl (table salt) supplies more than the required amounts.	Strongly affects distribution of water through osmosis. Part of bicarbonate buffer system. Functions in nerve and muscle action potential conduction.
Chloride	Major anion (Cl^-) in extracellular fluid. Sources include table salt (NaCl), soy sauce, and processed foods.	Plays role in acid–base balance of blood, water balance, and formation of HCl in stomach.
Magnesium	Important cation (Mg^{2+}) in intracellular fluid. Excreted in urine and feces. Widespread in various foods, such as green leafy vegetables, seafood, and whole-grain cereals.	Required for normal functioning of muscle and nervous tissue. Participates in bone formation. Constituent of many coenzymes.
Iron	About 66% found in hemoglobin of blood. Normal losses of iron occur by shedding of hair, epithelial cells, and mucosal cells, and in sweat, urine, feces, bile, and blood lost during menstruation. Sources are meat, liver, shellfish, egg yolk, beans, legumes, dried fruits, nuts, and cereals.	As component of hemoglobin, reversibly binds O_2. Component of cytochromes in electron transport chain.
Iodide	Essential component of thyroid hormones. Sources are seafood, iodized salt, and vegetables grown in iodine-rich soils.	Required by thyroid gland to synthesize thyroid hormones, which regulate metabolic rate.
Manganese	Some stored in liver and spleen.	Activates several enzymes. Needed for hemoglobin synthesis, urea formation, growth, reproduction, lactation, and bone formation.
Copper	Some stored in liver and spleen. Sources include eggs, whole-wheat flour, beans, beets, liver, fish, spinach, and asparagus.	Required with iron for synthesis of hemoglobin. Component of coenzymes in electron transport chain and enzyme necessary for melanin formation.
Cobalt	Constituent of vitamin B_{12}.	As part of vitamin B_{12}, required for erythropoiesis.
Zinc	Important component of certain enzymes. Widespread in many foods, especially meats.	As a component of carbonic anhydrase, important in carbon dioxide metabolism. Necessary for normal growth and wound healing, normal taste sensations and appetite, and normal sperm counts in males. As a component of peptidases, it is involved in protein digestion.
Fluoride	Components of bones, teeth, other tissues.	Appears to improve tooth structure and inhibit tooth decay.
Selenium	Important component of certain enzymes. Found in seafood, meat, chicken, tomatoes, egg yolk, milk, mushrooms, and garlic, and cereal grains grown in selenium-rich soil.	Needed for synthesis of thyroid hormones, sperm motility, and proper functioning of the immune system. Also functions as an antioxidant. Prevents chromosome breakage and may play a role in preventing certain birth defects, miscarriage, prostate cancer, and coronary artery disease.
Chromium	Found in high concentrations in brewer's yeast. Also found in wine and some brands of beer.	Needed for normal activity of insulin in carbohydrate and lipid metabolism.

ing the buildup of atherosclerotic plaque, delaying some effects of aging, and decreasing the chance of cataract formation in the lenses of the eyes. Table 20.2 lists the principal vitamins, their sources, their functions, and related deficiency disorders.

Most nutritionists recommend eating a balanced diet that includes a variety of foods rather than taking **vitamin supplements** or **mineral supplements**, except in special circumstances. Common examples of necessary supplementations include iron for women who have excessive menstrual bleeding; iron and calcium for women who are pregnant or breast-feeding; folic acid (folate) for all women who may become pregnant, to reduce the risk of fetal neural tube defects; calcium for most adults, because they do not receive the recommended amount in their diets; and vitamin B_{12} for strict vegetarians, who eat no meat. Because most North Americans do not ingest in their food the high levels of antioxidant vitamins thought to have beneficial effects, some experts recommend supplementing vitamins C and E. More is not always better; larger doses of vitamins or minerals can be very harmful.

■ **CHECKPOINT**

1. Describe the My Pyramid and give examples of foods from each food group.
2. Briefly describe the functions of the minerals calcium and sodium in the body.
3. Explain how vitamins are different from minerals, and distinguish between a fat-soluble vitamin and a water-soluble vitamin.

METABOLISM

OBJECTIVES • **Define metabolism and describe its importance in homeostasis.**

• **Explain how the body uses carbohydrates, lipids, and proteins.**

Metabolism (me-TAB-ō-lizm; *metabol-* = change) refers to all the chemical reactions of the body. Recall from Chapter 2 that chemical reactions occur when chemical bonds between substances are formed or broken, and that ***enzymes*** serve as catalysts to speed up chemical reactions. Some enzymes require the presence of an ion such as calcium, iron, or zinc. Other enzymes work together with ***coenzymes***, which function as temporary carriers of atoms being removed from or added to a substrate during a reaction. Many coenzymes are derived from vitamins. Examples include the coenzyme ***NAD^+***, derived from the B vitamin niacin, and the coenzyme ***FAD***, derived from vitamin B_2 (riboflavin).

The body's metabolism may be thought of as an energy-balancing act between anabolic (synthesis) and catabolic (decomposition) reactions. Chemical reactions that combine simple substances into more complex molecules are collectively known as ***anabolism*** (a-NAB-ō-lizm; *ana-* = upward). Overall, anabolic reactions use more energy than they produce. The energy they use is supplied by catabolic reactions (Figure 20.2). One example of an anabolic process is the formation of peptide bonds between amino acids, combining them into proteins.

The chemical reactions that break down complex organic compounds into simple ones are collectively known as ***catabolism*** (ka-TAB-ō-lizm; *cata-* = downward). Catabolic reactions release the energy stored in organic molecules. This energy is transferred to molecules of ATP and then used to power anabolic reactions. Important sets of catabolic reactions occur during glycolysis, the Krebs cycle, and the electron transport chain, which are discussed shortly.

About 40% of the energy released in catabolism is used for cellular functions; the rest is converted to heat, some of which helps maintain normal body temperature. Excess heat is lost to the environment. Compared with machines, which typically convert only 10–20% of energy into work, the 40% efficiency of the body's metabolism is impressive. Still, the body has a continuous need to take in and process external sources of energy so that cells can synthesize enough ATP to sustain life.

Figure 20.2 Role of ATP in linking anabolic and catabolic reactions. When complex molecules are split apart (catabolism, at left), some of the energy is transferred to form ATP and the rest is given off as heat. When simple molecules are combined to form complex molecules (anabolism, at right), ATP provides the energy for synthesis, and again some energy is given off as heat.

The coupling of energy-releasing and energy-requiring reactions is achieved through ATP.

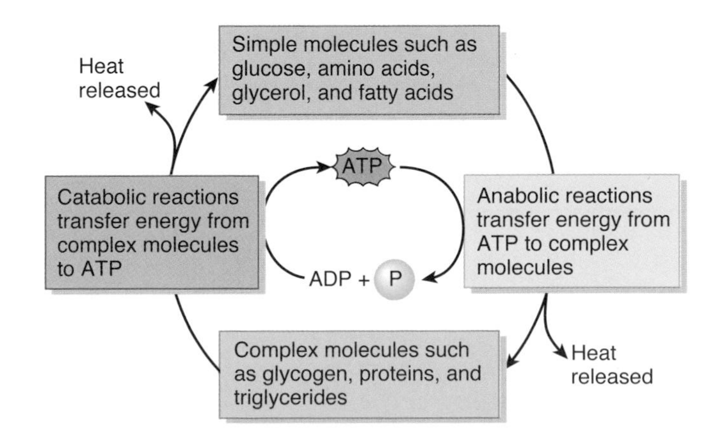

In a pancreatic cell that produces digestive enzymes, does anabolism or catabolism predominate?

Table 20.2 The Principal Vitamins

Vitamin	Comment and Source	Functions	Deficiency Symptoms and Disorders
Fat-soluble Vitamins	**All require bile salts and some dietary lipids for adequate absorption.**		
A	Formed from provitamin beta-carotene (and other provitamins) in GI tract. Stored in liver. Sources of carotene and other provitamins include orange, yellow, and green vegetables; sources of vitamin A include liver and milk.	Maintains general health and vigor of epithelial cells. Beta-carotene acts as an antioxidant to inactivate free radicals.	Atrophy and keratinization of epithelium, leading to dry skin and hair; increased incidence of ear, sinus, respiratory, urinary, and digestive system infections; inability to gain weight; drying of cornea; and skin sores.
		Essential for formation of light-sensitive pigments in photoreceptors of retina.	**Night blindness** or decreased ability for dark adaptation.
		Aids in growth of bones and teeth by helping to regulate activity of osteoblasts and osteoclasts.	Slow and faulty development of bones and teeth.
D	In the presence of sunlight, the skin, liver, and kidneys produce the active form of vitamin D (calcitriol). Stored in tissues to slight extent. Most excreted in bile. Dietary sources include fish-liver oils, egg yolk, and fortified milk.	Essential for absorption of calcium and phosphorus from GI tract. Works with parathyroid hormone (PTH) to maintain Ca^{2+} homeostasis.	Defective utilization of calcium by bones leads to ***rickets*** in children and ***osteomalacia*** in adults. Possible loss of muscle tone.
E (tocopherols)	Stored in liver, adipose tissue, and muscles. Sources include fresh nuts and wheat germ, seed oils, and green leafy vegetables.	Inhibits catabolism of certain fatty acids that help form cell structures, especially membranes. Involved in formation of DNA, RNA, and red blood cells. May promote wound healing, contribute to the normal structure and functioning of the nervous system, and prevent scarring. Acts as an antioxidant to inactivate free radicals.	Abnormal structure and function of mitochondria, lysosomes, and plasma membranes. A possible consequence is ***hemolytic anemia***.
K	Produced by intestinal bacteria. Stored in liver and spleen. Dietary sources include spinach, cauliflower, cabbage, and liver.	Coenzyme essential for synthesis of several clotting factors by liver, including prothrombin.	Delayed clotting time results in excessive bleeding.
Water-soluble Vitamins	**Dissolved in body fluids. Most are not stored in body. Excess intake is eliminated in urine.**		
B_1 (thiamine)	Rapidly destroyed by heat. Sources include whole-grain products, eggs, pork, nuts, liver, and yeast.	Acts as a coenzyme for many different enzymes that break carbon-to-carbon bonds and are involved in carbohydrate metabolism of pyruvic acid to CO_2 and H_2O. Essential for synthesis of the neurotransmitter acetylcholine.	Buildup of pyruvic and lactic acids and insufficient production of ATP for muscle and nerve cells leads to: (1) ***beriberi***, partial paralysis of smooth muscle of GI tract, causing digestive disturbances, skeletal muscle paralysis, and atrophy of limbs; (2) ***polyneuritis***, due to degeneration of myelin sheaths: impaired reflexes, impaired sense of touch, stunted growth in children, and poor appetite.

Vitamin	Comment and Source	Functions	Deficiency Symptoms and Disorders
Water-soluble *(continued)*			
B_2 (riboflavin)	Small amounts supplied by bacteria of GI tract. Dietary sources include yeast, liver, beef, veal, lamb, eggs, whole-grain products, asparagus, peas, beets, and peanuts.	Component of certain coenzymes (for example, FAD and FMN) in carbohydrate and protein metabolism, especially in cells of the eyes, skin, mucosa of the intestine, and blood.	Faulty use of oxygen resulting in blurred vision, cataracts, and corneal ulcerations. Also dermatitis and cracking of skin, lesions of intestinal mucosa, and one type of anemia.
Niacin (nicotinamide)	Derived from amino acid tryptophan. Sources include yeast, meats, liver, fish, whole-grain products, peas, beans, and nuts.	Essential component of NAD and NADP, coenzymes in oxidation–reduction reactions. In lipid metabolism, inhibits production of cholesterol and assists in triglyceride breakdown.	***Pellagra***, characterized by dermatitis, diarrhea, and psychological disturbances.
B_6 (pyridoxine)	Synthesized by bacteria of GI tract. Stored in liver, muscles, and brain. Other sources include salmon, yeast, tomatoes, yellow corn, spinach, whole grain products, liver, and yogurt.	Essential coenzyme for normal amino acid metabolism. Assists production of circulating antibodies. May function as coenzyme in triglyceride metabolism.	Dermatitis of eyes, nose, and mouth, retarded growth, and nausea.
B_{12} (cyanocobalamin)	Only B vitamin not found in vegetables; only vitamin containing cobalt. Absorption from GI tract depends on intrinsic factor secreted by the stomach mucosa. Sources include liver, kidney, milk, eggs, cheese, and meat.	Coenzyme necessary for red blood cell formation, formation of the amino acid methionine, entrance of some amino acids into Krebs cycle, and synthesis of choline (used to make acetylcholine).	Pernicious anemia, neuropsychiatric abnormalities (ataxia, memory loss, weakness, personality and mood changes, and abnormal sensations), and impaired activity of osteoblasts.
Pantothenic acid	Some produced by bacteria of GI tract. Stored primarily in liver and kidneys. Other sources include liver, kidneys, yeast, green vegetables, and cereal.	Constituent of coenzyme A, which is used to transfer acetyl groups into Krebs cycle; conversion of lipids and amino acids into glucose; and synthesis of cholesterol and steroid hormones.	Fatigue, muscle spasms, insufficient production of adrenal steroid hormones, vomiting, and insomnia.
Folic acid (folate, folacin)	Synthesized by bacteria of GI tract. Dietary sources include green leafy vegetables, broccoli, asparagus, breads, dried beans, and citrus fruits.	Component of enzyme systems synthesizing nitrogenous bases of DNA and RNA. Essential for normal production of red and white blood cells.	Production of abnormally large red blood cells. Higher risk of neural tube defects in babies born to folic acid-deficient mothers.
Biotin	Synthesized by bacteria of GI tract. Dietary sources include yeast, liver, egg yolk, and kidneys.	Essential coenzyme for conversion of pyruvic acid to oxaloacetic acid and synthesis of fatty acids and purines.	Mental depression, muscular pain, dermatitis, fatigue, and nausea.
C (ascorbic acid)	Rapidly destroyed by heat. Some stored in glandular tissue and plasma. Sources include citrus fruits, strawberries, melons, tomatoes, and green vegetables.	Promotes protein synthesis including synthesis of collagen in connective tissue. As coenzyme, may combine with poisons, rendering them harmless until excreted. Works with antibodies, promotes wound healing, and functions as an antioxidant.	Scurvy; anemia; many symptoms related to poor collagen formation, including tender swollen gums, loosening of teeth, poor wound healing, bleeding, impaired immune responses, and retardation of growth.

Carbohydrate Metabolism

During digestion, polysaccharide and disaccharide carbohydrates are catabolized to monosaccharides—glucose, fructose, and galactose—which are absorbed in the small intestine. Shortly after their absorption, however, fructose and galactose are converted to glucose. Thus, the story of carbohydrate metabolism is really the story of glucose metabolism.

Because glucose is the body's preferred source for synthesizing ATP, the fate of glucose absorbed from the diet depends on the needs of body cells. If the cells require ATP immediately, they oxidize the glucose. Glucose not needed for immediate ATP production may be converted to glycogen for storage in liver cells and skeletal muscle fibers. If these glycogen stores are full, the liver cells can transform the glucose to triglycerides for storage in adipose tissue. At a later time, when the cells need more ATP, the glycogen and triglycerides can be converted back to glucose. Cells throughout the body also can use glucose to make certain amino acids, the building blocks of proteins.

Before glucose can be used by body cells, it must pass through the plasma membrane by facilitated diffusion and enter the cytosol. Insulin increases the rate of facilitated diffusion of glucose.

Glucose Catabolism

The catabolism of glucose to produce ATP is known as ***cellular respiration***. Overall, its many reactions can be summarized as follows.

1 glucose + 6 oxygen ⟶ 36–38 ATP
+ 6 carbon dioxide + 6 water

Four interconnecting sets of chemical reactions contribute to cellular respiration (Figure 20.3):

Figure 20.3 Cellular respiration.

The catabolism of glucose to produce ATP involves glycolysis, the formation of acetyl coenzyme A, the Krebs cycle, and the electron transport chain.

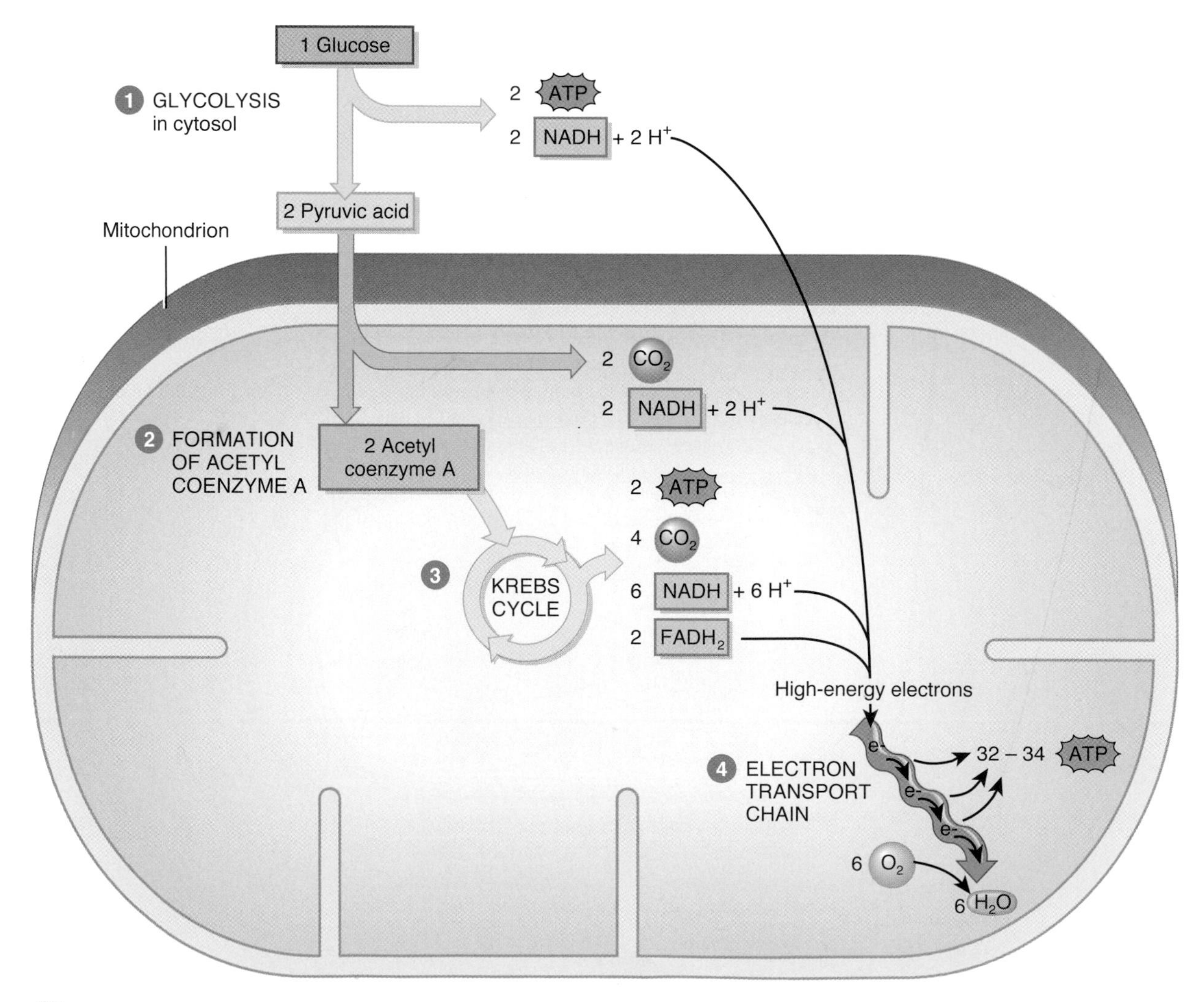

How many molecules of ATP are produced during the complete catabolism of one molecule of glucose?

1. During ***glycolysis*** (glī-KOL-i-sis; *glyco-* = sugar; *-lysis* = breakdown), reactions that take place in the cytosol convert one six-carbon glucose molecule into two three-carbon pyruvic acid molecules. The reactions of glycolysis directly produce two ATPs. They also transfer some chemical energy, in the form of high-energy electrons, from glucose to the coenzyme NAD^+, forming two $NADH + H^+$. Because glycolysis does not require oxygen, it is a way to produce ATP anaerobically (without oxygen) and is known as ***anaerobic cellular respiration***. If oxygen is available, however, most cells next convert pyruvic acid to acetyl coenzyme A.
2. The formation of ***acetyl coenzyme A*** is a transition step that prepares pyruvic acid for entrance into the Krebs cycle. First, pyruvic acid enters a mitochondrion and is converted to a two-carbon fragment by removing a molecule of carbon dioxide (CO_2). Molecules of CO_2 produced during glucose catabolism diffuse into the blood and are eventually exhaled. Then, the coenzyme NAD^+ takes away a hydrogen atom (H_2), in the process becoming $NADH + H^+$. Finally, the remaining atoms, called an ***acetyl group***, are attached to coenzyme A, to form acetyl coenzyme A.
3. The ***Krebs cycle*** is a series of reactions that transfer the chemical energy from acetyl coenzyme A to two other coenzymes—NAD^+ and FAD—thereby forming $NADH + H^+$ and $FADH_2$. Krebs cycle reactions also produce CO_2 and one ATP for each acetyl coenzyme A that enters the Krebs cycle. To harvest the energy in NADH and $FADH_2$, their high-energy electrons must first go through the electron transport chain.
4. Through the reactions of the ***electron transport chain***, the energy in $NADH + H^+$ and $FADH_2$ is used to synthesize ATP. As the coenzymes pass their high-energy electrons through a series of "electron carriers," ATP is synthesized. Finally, lower-energy electrons are passed to oxygen in a reaction that produces water. Because the Krebs cycle and the electron transport chain together require oxygen to produce ATP, they are known as ***aerobic cellular respiration***.

Glucose Anabolism

Even though most of the glucose in the body is catabolized to generate ATP, glucose may take part in or be formed via several anabolic reactions. One is the synthesis of glycogen; another is the synthesis of new glucose molecules from some of the products of protein and lipid breakdown.

If glucose is not needed immediately for ATP production, it combines with many other molecules of glucose to form a long-chain molecule called ***glycogen*** (Figure 20.4). Synthesis of glycogen is stimulated by insulin. The body can store about 500 grams (about 1.1 lb) of glycogen, roughly 75% in skeletal muscle fibers and the rest in liver cells.

If blood glucose level falls below normal, glucagon is released from the pancreas and epinephrine is released from the adrenal medullae. These hormones stimulate breakdown of glycogen into its glucose subunits (Figure 20.4). Liver cells release this glucose into the blood, and body cells pick it up to use for ATP production. Glycogen breakdown usually occurs between meals.

Figure 20.4 Reactions of glucose anabolism: synthesis of glycogen, breakdown of glycogen, and synthesis of glucose from amino acids, lactic acid, or glycerol.

About 500 grams (1.1 lb) of glycogen are stored in skeletal muscles and the liver.

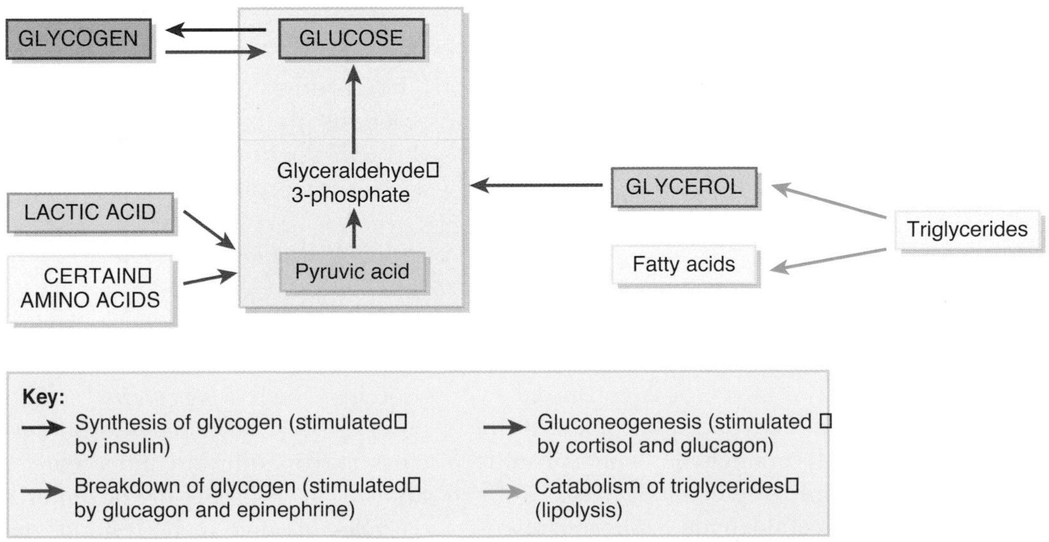

GLYCOGEN
GLUCOSE
Glyceraldehyde 3-phosphate
GLYCEROL
Triglycerides
LACTIC ACID
Pyruvic acid
Fatty acids
CERTAIN AMINO ACIDS

Key:
Synthesis of glycogen (stimulated by insulin)
Breakdown of glycogen (stimulated by glucagon and epinephrine)
Gluconeogenesis (stimulated by cortisol and glucagon)
Catabolism of triglycerides (lipolysis)

Which body cells can synthesize glucose from amino acids?

The amount of glycogen stored in the liver and skeletal muscles varies and can be completely used up during long-term athletic endeavors. Thus, many marathon runners and other endurance athletes follow a precise exercise and dietary regimen that includes eating large amounts of complex carbohydrates, such as pasta and potatoes, in the three days before an event. This practice, called **carbohydrate loading**, helps maximize the amount of glycogen available for ATP production in muscles. For athletic events lasting more than an hour, carbohydrate loading has been shown to increase an athlete's endurance.

When your liver runs low on glycogen, it is time to eat. If you don't, your body starts catabolizing triglycerides (fats) and proteins. Actually, the body normally catabolizes some of its triglycerides and proteins, but large-scale triglyceride and protein catabolism does not happen unless you are starving, eating very few carbohydrates, or suffering from an endocrine disorder.

Liver cells can convert the glycerol part of triglycerides, lactic acid, and certain amino acids to glucose (Figure 20.4). The series of reactions that form glucose from these noncarbohydrate sources is called ***gluconeogenesis*** (gloo′-kō-nē′-ō-JEN-e-sis; *neo-* = new). This process releases glucose into the blood, thereby keeping blood glucose level normal during the hours between meals when glucose is not being absorbed. Gluconeogenesis occurs when the liver is stimulated by cortisol from the adrenal cortex and glucagon from the pancreas.

Lipid Metabolism

Lipids, like carbohydrates, may be catabolized to produce ATP. If the body has no immediate need to use lipids in this way, they are stored as triglycerides in adipose tissue throughout the body, and in the liver. A few lipids are used as structural molecules or to synthesize other substances. Two essential fatty acids that the body cannot synthesize are linoleic acid and linolenic acid. Dietary sources of these lipids include vegetable oils and leafy vegetables.

Lipid Catabolism

Muscle, liver, and adipose cells routinely catabolize fatty acids from triglycerides to produce ATP. First, the triglycerides are split into glycerol and fatty acids—a process called ***lipolysis*** (li-POL-i-sis) (Figure 20.5). The hormones epinephrine, norepinephrine, and cortisol enhance lipolysis.

The glycerol and fatty acids that result from lipolysis are catabolized via different pathways. Glycerol is converted by many cells of the body to glyceraldehyde 3-phosphate. If the ATP supply in a cell is high, glyceraldehyde 3-phosphate is converted into glucose, an example of gluconeogenesis. If the ATP supply in a cell is low, glyceraldehyde 3-phosphate enters the catabolic pathway to pyruvic acid.

Fatty acid catabolism begins as enzymes remove two carbon atoms at a time from the fatty acid and attach them to molecules of coenzyme A, forming acetyl coenzyme A (acetyl CoA). Then the acetyl CoA enters the Krebs cycle (Figure 20.5). A 16-carbon fatty acid such as palmitic acid can yield as many as 129 ATPs via the Krebs cycle and the electron transport chain.

As part of normal fatty acid catabolism, the liver converts some acetyl CoA molecules into substances known as ***ketone bodies*** (Figure 20.5). Ketone bodies then leave the liver to enter body cells, where they are broken down into acetyl CoA, which enters the Krebs cycle.

The level of ketone bodies in the blood normally is very low because other tissues use them for ATP production as fast as they are formed. When the concentration of ketone bodies in the blood rises above normal—a condition called **ketosis**—the ketone bodies, most of which are acids, must be buffered. If too many accumulate, blood pH falls. When a diabetic becomes seriously insulin deficient, one of the telltale signs is the sweet smell on the breath from the ketone body acetone. Prolonged ketosis can lead to **acidosis**, an abnormally low blood pH that can result in death.

Lipid Anabolism

Insulin stimulates liver cells and adipose cells to synthesize triglycerides when more calories are consumed than are needed to satisfy ATP needs (Figure 20.5). Excess dietary carbohydrates, proteins, and fats all have the same fate—they are converted into triglycerides. Certain amino acids can undergo the following reactions: amino acids → acetyl CoA → fatty acids → triglycerides. The use of glucose to form lipids takes place via two pathways:

1. glucose → glyceraldehyde 3-phosphate → glycerol; and
2. glucose → glyceraldehyde 3-phosphate → acetyl CoA → fatty acids.

The resulting glycerol and fatty acids can undergo anabolic reactions to become stored triglycerides, or they can go through a series of anabolic reactions to produce other lipids such as lipoproteins, phospholipids, and cholesterol.

Lipid Transport in Blood

Most lipids, such as triglycerides and cholesterol, are not water-soluble. For transport in watery blood, such molecules first are made more water soluble by combining them with proteins. Such ***lipoproteins*** are spherical particles with an outer shell of proteins, phospholipids, and cholesterol molecules surrounding an inner core of triglycerides and other lipids. The proteins in the outer shell help the lipoprotein particles dissolve in body fluids and also have specific functions.

Figure 20.5 Metabolism of lipids. Lipolysis is the breakdown of triglycerides into glycerol and fatty acids. Glycerol may be converted to glyceraldehyde 3-phosphate, which can then be converted to glucose or enter the Krebs cycle. Fatty acid fragments enter the Krebs cycle as acetyl coenzyme A. Fatty acids also can be converted into ketone bodies.

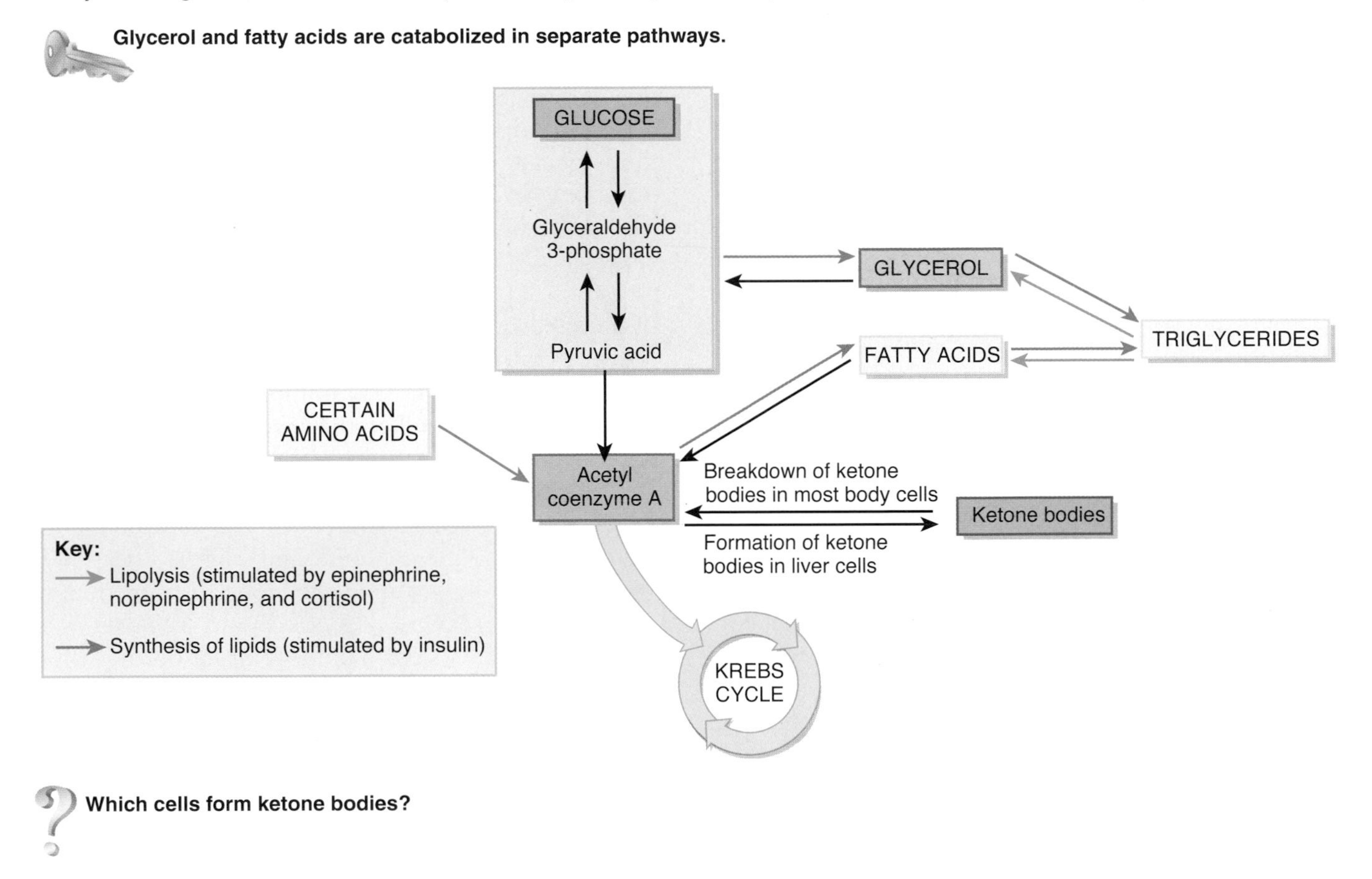

Which cells form ketone bodies?

Lipoproteins are transport vehicles: They provide delivery and pickup services so that lipids can be available when cells need them or removed when they are not needed. Lipoproteins are categorized and named mainly according to their size and density. From largest and lightest to smallest and heaviest, the four major types of lipoproteins are chylomicrons, very low-density lipoproteins, low-density lipoproteins, and high-density lipoproteins.

1. ***Chylomicrons*** form in absorptive epithelial cells of the small intestine and transport dietary lipids to adipose tissue for storage.
2. ***Very low-density lipoproteins (VLDLs)*** transport triglycerides made in liver cells to adipose cells for storage. After depositing some of their triglycerides in adipose cells, VLDLs are converted to LDLs.
3. ***Low-density lipoproteins (LDLs)*** carry about 75% of the total cholesterol in blood and deliver it to cells throughout the body for use in repair of cell membranes and synthesis of steroid hormones and bile salts.
4. ***High-density lipoproteins (HDLs)*** remove excess cholesterol from body cells and transport it to the liver for elimination.

When present in excessive numbers, LDLs deposit cholesterol in and around smooth muscle fibers in arteries, forming fatty plaques that increase the risk of coronary artery disease (see page 379). For this reason, the cholesterol in LDLs, called LDL-cholesterol, is known as **"bad" cholesterol**. Eating a high-fat diet increases the production of VLDLs, which elevates the LDL level and increases the formation of fatty plaques. Because HDLs prevent accumulation of cholesterol in the blood, a high HDL level is associated with decreased risk of coronary artery disease. For this reason, HDL-cholesterol is known as **"good" cholesterol**.

Desirable levels of blood cholesterol in adults are total cholesterol under 200 mg/dL, LDL under 130 mg/dL, and HDL over 40 mg/dL. The ratio of total cholesterol to HDL cholesterol predicts the risk of developing coronary artery disease. A person with a total cholesterol of 180 mg/dL and HDL of 60 mg/dL has a risk ratio of 3. Ratios above 4 are considered undesirable; the higher the ratio, the greater the risk of developing coronary artery disease.

Protein Metabolism

During digestion, proteins are broken down into amino acids. Unlike carbohydrates and triglycerides, proteins are not warehoused for future use. Instead, their amino acids are either oxidized to produce ATP or used to synthesize new proteins for growth and repair of body tissues. Excess dietary amino acids are converted into glucose (gluconeogenesis) or triglycerides.

The active transport of amino acids into body cells is stimulated by insulinlike growth factors (IGFs) and insulin. Almost immediately after digestion, amino acids are reassembled into proteins. Many proteins function as enzymes; other proteins are involved in transportation (hemoglobin) or serve as antibodies, clotting factors (fibrinogen), hormones (insulin), or contractile elements in muscle fibers (actin and myosin). Several proteins serve as structural components of the body (collagen, elastin, and keratin).

Protein Catabolism

A certain amount of protein catabolism occurs in the body each day, stimulated mainly by cortisol from the adrenal cortex. Proteins from worn-out cells (such as red blood cells) are broken down into amino acids. Some amino acids are converted into other amino acids, peptide bonds are reformed, and new proteins are made as part of the recycling process. Liver cells convert some amino acids to fatty acids, ketone bodies, or glucose. Figure 20.4 shows the conversion of amino acids into glucose (gluconeogenesis). Figure 20.5 shows the conversion of amino acids into fatty acids or ketone bodies.

Amino acids also are oxidized to generate ATP. Before amino acids can enter the Krebs cycle, however, their amino group ($-NH_2$) must first be removed, a process called ***deamination*** (dē-am′-i-NĀ-shun). Deamination occurs in liver cells and produces ammonia (NH_3). Liver cells then convert the highly toxic ammonia to urea, a relatively harmless substance that is excreted in the urine.

Protein Anabolism

Protein anabolism, the formation of peptide bonds between amino acids to produce new proteins, is carried out on the ribosomes of almost every cell in the body, directed by the cells' DNA and RNA. Insulinlike growth factors, thyroid hormones, insulin, estrogens, and testosterone stimulate protein synthesis. Because proteins are a main component of most cell structures, adequate dietary protein is especially essential during the growth years, during pregnancy, and when tissue has been damaged by disease or injury. Once dietary intake of protein is adequate, eating more protein does not increase bone or muscle mass; only a regular program of forceful, weight-bearing muscular activity accomplishes that goal.

Of the 20 amino acids in the human body, 10 are ***essential amino acids***: They must be present in the diet because they cannot be synthesized in the body in adequate amounts. ***Nonessential amino acids*** are those synthesized by the body. They are formed by the transfer of an amino group from an amino acid to pyruvic acid or to an acid in the Krebs cycle. Once the appropriate essential and nonessential amino acids are present in cells, protein synthesis occurs rapidly. Table 20.3 summarizes the processes occurring in both catabolism and anabolism of carbohydrates, lipids, and proteins.

Phenylketonuria (fen′-il-kē′-tō-NOO-rē-a) or **PKU** is a genetic error of protein metabolism characterized by elevated blood levels of the amino acid phenylalanine. Most children with phenylketonuria have a mutation in the gene that codes for the enzyme needed to convert phenylalanine into the amino acid tyrosine, which can enter the Krebs cycle. Because the enzyme is deficient, phenylalanine cannot be metabolized, and what is not used in protein synthesis builds up in the blood. If untreated, the disorder causes vomiting, rashes, seizures, growth deficiency, and severe mental retardation. Newborns are screened for PKU, and mental retardation can be prevented by restricting the child to a diet that supplies only the amount of phenylalanine needed for growth, although learning disabilities may still ensue. Because the artificial sweetener aspartame (NutraSweet®) contains phenylalanine, its consumption must be restricted in children with PKU.

■ CHECKPOINT

4. What happens during glycolysis?
5. What happens in the electron transport chain?
6. Which reactions produce ATP during the complete oxidation of a molecule of glucose?
7. What is gluconeogenesis, and why is it important?
8. What is the difference between anabolism and catabolism?
9. How does ATP provide a link between anabolism and catabolism?
10. What are the functions of the proteins in lipoproteins?
11. Which lipoprotein particles contain "good" and "bad" cholesterol, and why are these terms used?
12. Where are triglycerides stored in the body?
13. What are ketone bodies? What is ketosis?
14. What are the possible fates of the amino acids from protein catabolism?

Table 20.3 Summary of Metabolism

Process	Comment
Carbohydrate Metabolism	
Glucose catabolism	Complete catabolism of glucose (cellular respiration) is the chief source of ATP in most cells. It consists of glycolysis, the Krebs cycle, and the electron transport chain. One molecule of glucose yields 36–38 molecules of ATP.
Glycolysis	Conversion of glucose into pyruvic acid, with net production of two ATP per glucose molecule; reactions do not require oxygen (anaerobic cellular respiration).
Krebs cycle	Series of reactions in which coenzymes (NAD^+ and FAD) pick up high-energy electrons. Some ATP is produced. CO_2, H_2O, and heat are byproducts. Reactions are aerobic (aerobic cellular respiration).
Electron transport chain	Third set of reactions in glucose catabolism in which electrons are passed from one carrier to the next and most of the ATP is produced. Reactions are aerobic (aerobic cellular respiration).
Glucose anabolism	Some glucose is converted into glycogen for storage if not needed immediately for ATP production. Glycogen can be converted back to glucose for use in ATP production. Gluconeogenesis is the synthesis of glucose from amino acids, glycerol, or lactic acid.
Lipid Metabolism	
Triglyceride catabolism	Triglycerides are broken down into glycerol and fatty acids. Glycerol may be converted into glucose (gluconeogenesis) or catabolized via glycolysis. Fatty acids are converted into acetyl CoA that can enter the Krebs cycle for ATP production or be used to form ketone bodies.
Triglyceride anabolism	Synthesis of triglycerides from glucose and amino acids. Triglycerides are stored in adipose tissue.
Protein Metabolism	
Catabolism	Amino acids are deaminated to enter the Krebs cycle. Ammonia formed during deamination is converted to urea in the liver and excreted in the urine. Amino acids may be converted into glucose (gluconeogenesis), fatty acids, or ketone bodies.
Anabolism	Protein synthesis is directed by DNA and uses the cell's RNA and ribosomes.

METABOLISM AND BODY HEAT

OBJECTIVES • **Explain how body heat is produced and lost.**

• **Describe how body temperature is regulated.**

We now consider the relationship of foods to body heat, heat production and loss, and the regulation of body temperature.

Measuring Heat

Heat is a form of energy that can be measured as ***temperature*** and expressed in units called calories. A ***calorie (cal)*** is defined as the amount of heat required to raise the temperature of 1 gram of water 1°C. Because the calorie is a relatively small unit, the ***kilocalorie (kcal)*** or ***Calorie (Cal)*** (always spelled with an uppercase C) is often used to measure the body's metabolic rate and to express the energy content of foods. A kilocalorie equals 1000 calories. Thus, when we say that a particular food item contains 500 Calories, we are actually referring to kilocalories. Knowing the caloric value of foods is important. If we know the amount of energy the body uses for various activities, we can adjust our food intake by taking in only enough kilocalories to sustain our activities.

Body Temperature Homeostasis

The body produces more or less heat depending on the rates of metabolic reactions. Homeostasis of body temperature can be maintained only if the rate of heat production by metabolism equals the rate of heat loss from the body. Thus, it is important to understand the ways in which heat can be produced and lost.

Body Heat Production

Most of the heat produced by the body comes from the catabolism of the food we eat. The rate at which this heat is produced, the ***metabolic rate***, is measured in kilocalories. Because many factors affect metabolic rate, it is measured under standard conditions, with the body in a quiet, resting, and fasting condition called the ***basal state***. The measurement obtained is the ***basal metabolic rate (BMR)***. BMR is 1200 to 1800 Calories per day in adults, which amounts to about 24 Calories per kilogram of body mass in adult males and 22 Calories per kilogram in adult females.

The added Calories needed to support daily activities, such as digestion and walking, range from 500 Calories for a small, relatively sedentary person to over 3000 Calories for a person in training for Olympic-level competitions. The following factors affect metabolic rate:

1. **Exercise.** During strenuous exercise the metabolic rate increases by as much as 15 to 20 times the BMR.
2. **Hormones.** Thyroid hormones are the main regulators of BMR, which increases as the blood levels of thyroid hormones rise. Testosterone, insulin, and human growth hormone can increase the metabolic rate by 5–15%.
3. **Nervous system.** During exercise or in a stressful situation, the sympathetic division of the autonomic nervous system releases norepinephrine, and it stimulates release of the hormones epinephrine and norepinephrine by the adrenal medulla. Both epinephrine and norepinephrine increase the metabolic rate of body cells.
4. **Body temperature.** The higher the body temperature, the higher the metabolic rate. As a result, metabolic rate is substantially increased during a fever.
5. **Ingestion of food.** The ingestion of food, especially proteins, can raise metabolic rate by 10–20%.
6. **Age.** The metabolic rate of a child, in relation to its size, is about double that of an elderly person due to the high rates of growth-related reactions in children.
7. **Other factors.** Other factors that affect metabolic rate are gender (lower in females, except during pregnancy and lactation), climate (lower in tropical regions), sleep (lower), and malnutrition (lower).

Body Heat Loss

Because body heat is continuously produced by metabolic reactions, heat must also be removed continuously or body temperature would rise steadily. The principal routes of heat loss from the body to the environment are radiation, conduction, convection, and evaporation.

1. **Radiation** is the transfer of heat in the form of infrared rays between a warmer object and a cooler one without physical contact. Your body loses heat by radiating more infrared waves than it absorbs from cooler objects. If surrounding objects are warmer than you are, you absorb more heat by radiation than you lose.
2. **Conduction** is the heat exchange that occurs between two materials that are in direct contact. Body heat is lost by conduction to solid materials in contact with the body, such as your chair, clothing, and jewelry. Heat can also be gained by conduction, for example, while soaking in a hot tub.
3. **Convection** is the transfer of heat by the movement of a gas or a liquid between areas of different temperatures. The contact of air or water with your body results in heat transfer by both conduction and convection. When cool air makes contact with the body, it becomes warmed and is carried away by convection currents. The faster the air moves—for example, by a breeze or a fan—the faster the rate of convection.
4. **Evaporation** is the conversion of a liquid to a vapor. Under typical resting conditions, about 22% of heat loss occurs through evaporation of water—a daily loss of about 300 mL in exhaled air and 400 mL from the skin surface. Evaporation provides the main defense against overheating during exercise. Under extreme conditions, a maximum of about 3 liters of sweat can be produced each hour, removing more than 1700 kcal of heat if all of it evaporates. Sweat that drips off the body rather than evaporating removes very little heat.

Regulation of Body Temperature

If the amount of heat production equals the amount of heat loss, you maintain a nearly constant body temperature near 37°C (98.6°F). If your heat-producing mechanisms generate more heat than is lost by your heat-losing mechanisms, your body temperature rises. For example, strenuous exercise and some infections elevate body temperature. If you lose heat faster than you produce it, your body temperature falls. Immersion in cold water, certain diseases such as hypothyroidism, and some drugs such as alcohol and antidepressants can cause body temperature to fall. An elevated temperature may destroy body proteins, and a depressed temperature may cause cardiac arrhythmias; both can lead to death.

The balance between heat production and heat loss is controlled by neurons in the hypothalamus. These neurons generate more nerve impulses when blood temperature increases and fewer impulses when blood temperature decreases. If body temperature falls, mechanisms that help conserve heat and increase heat production act by means of several negative feedback loops to raise the body temperature to normal (Figure 20.6). Thermoreceptors send nerve impulses to the hypothalamus, which produces a releasing hormone called thyrotropin-releasing hormone (TRH). TRH in turn stimulates the anterior pituitary to release thyroid-stimulating hormone (TSH). Nerve impulses from the hypothalamus and TSH then activate several effectors:

- Sympathetic nerves cause blood vessels of the skin to constrict (vasoconstriction). The decrease of blood flow slows the rate of heat loss from the skin. Because less heat is lost, body temperature increases even if the metabolic rate remains the same.
- Sympathetic nerves stimulate the adrenal medulla to release epinephrine and norepinephrine into the blood. These hormones increase cellular metabolism, which increase heat production.
- The hypothalamus stimulates parts of the brain that increase muscle tone. As muscle tone increases in one mus-

cle (the agonist), the small contractions stretch muscle spindles in its antagonist muscle, initiating a stretch reflex. The resulting contraction in the antagonist stretches muscle spindles in the agonist, and it too develops a stretch reflex. This repetitive cycle—called ***shivering***—greatly increases the rate of heat production. During maximal shivering, body heat production can rise to about four times the basal rate in just a few minutes.

- The thyroid gland responds to TSH by releasing more thyroid hormones into the blood, increasing the metabolic rate.

If body temperature rises above normal, a negative feedback system opposite to the one depicted in Figure 20.6 goes into action. The higher temperature of the blood stimulates the hypothalamus. Nerve impulses cause dilation of blood

Figure 20.6 Negative feedback mechanisms that increase heat production.

When stimulated, the heat-promoting center in the hypothalamus raises body temperature.

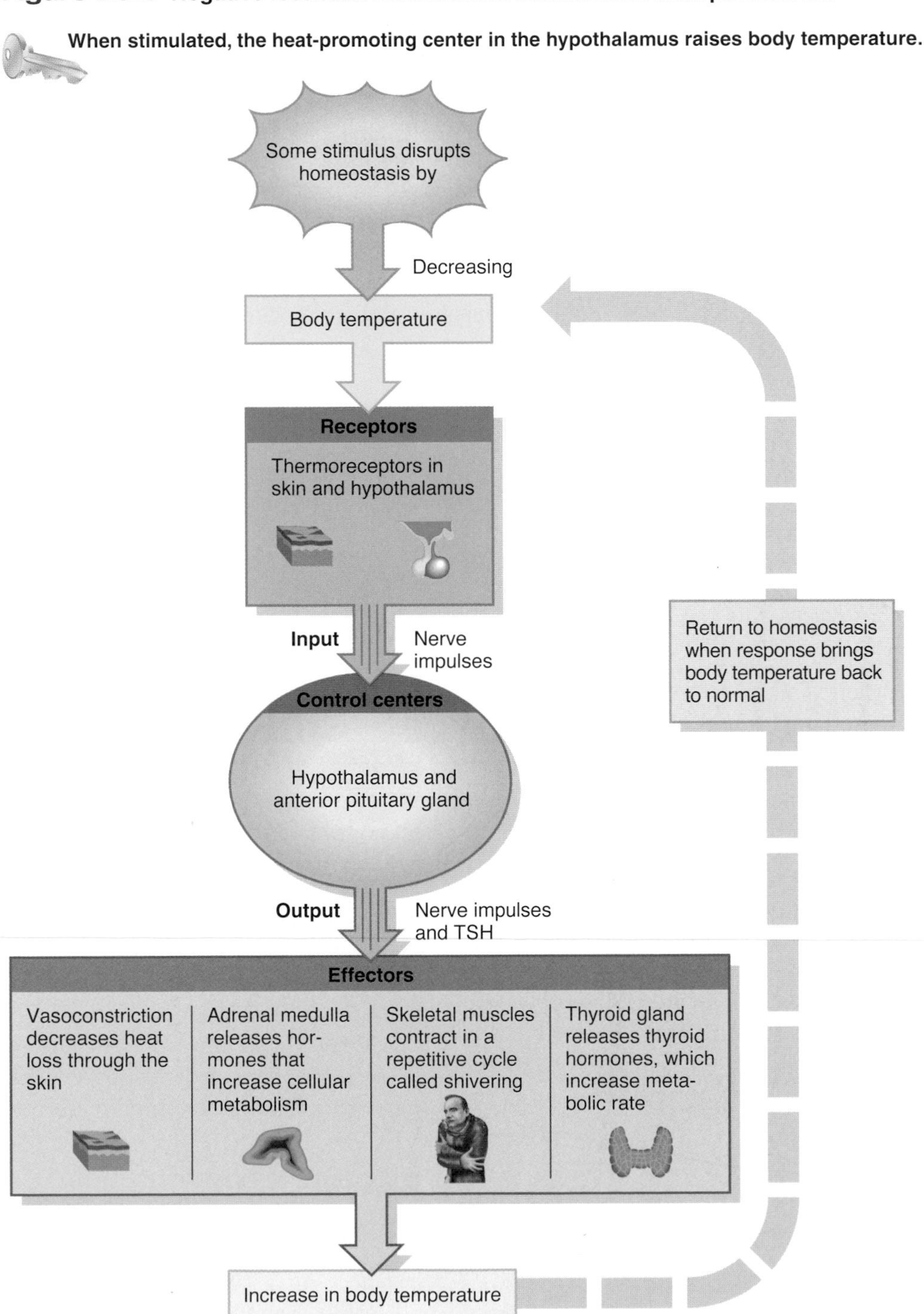

What factors can increase metabolic rate and thus increase heat production?

Focus on Wellness

Exercise Training—Metabolic Workout

Athletes spend hours a day training for their sports. Many physiological changes occur as a result of all this training, including an increased ability to produce ATP for muscle contraction. These improvements are specific to the metabolic pathways that are used during the training. Athletes design their training programs to challenge the ATP production system or systems most vital to their sports.

Metabolic Power

Some sports require a short burst of power, high energy output that lasts only a few seconds. Such events include the 100-meter sprint, the shot put, the discus throw, and the 25-meter swim. Muscle contraction for these events is supplied primarily by existing ATP and creatine phosphate. (Recall from Chapter 8 that creatine phosphate can donate a phosphate group to ADP to restore ATP.)

Athletes improve their ability to generate power by practicing their events, over and over. Power lifting—forcefully lifting very heavy weights—also challenges the muscles to produce more ATP faster. In response to such training, the concentration of enzymes required for these ATP-production pathways as well as the levels of ATP and creatine phosphate increases in trained muscles.

Going for the Burn

Many athletic events require well-trained glycolytic pathways. Anaerobic glycolysis, combined with ATP and creatine phosphate, provides most of the energy for high-intensity exercise lasting up to 90 seconds, such as a 400-meter run or a 100-meter swim. Many sports, such as basketball, soccer, and tennis, require bursts of high-ATP production by anaerobic glycolysis interspersed with somewhat lower energy output.

Athletes train the anaerobic glycolytic pathway by exercising at high intensities for periods of a minute or longer. Interval training includes both high-intensity work and periods of lower-intensity work or rest. In response to high-intensity exercise, the concentration of enzymes required for anaerobic glycolysis increases.

Going the Distance

Most athletic events, including all events lasting longer than a few minutes, require aerobic cellular respiration. Athletes improve these ATP-production pathways by exercising at moderate to vigorous intensities for extended periods of time, with or without high-intensity intervals. Aerobic training increases the size and number of mitochondria as well as the concentration of enzymes required for ATP production by aerobic pathways.

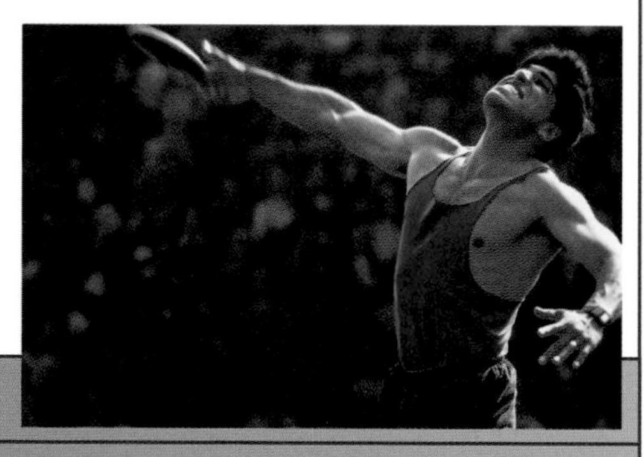

► **Think It Over . . .**

► ***Do you think that the burning sensation that can develop with high-intensity exercise, producing ATP via anaerobic glycolysis, represents the breakdown of adipose tissue?***

vessels in the skin. The skin becomes warm, and the excess heat is lost to the environment by radiation and conduction as an increased volume of blood flows from the warmer interior of the body into the cooler skin. At the same time, metabolic rate decreases, and the high temperature of the blood stimulates sweat glands of the skin by means of hypothalamic activation of sympathetic nerves. As the water in sweat evaporates from the surface of the skin, the skin is cooled. All these responses counteract heat-promoting effects and help return body temperature to normal.

Hypothermia is a lowering of core body temperature to 35°C (95°F) or below. Causes of hypothermia include an overwhelming cold stress (immersion in icy water), metabolic diseases (hypoglycemia, adrenal insufficiency, or hypothyroidism), drugs (alcohol, antidepressants, sedatives, or tranquilizers), burns, and malnutrition. Symptoms of hypothermia include sensation of cold, shivering, confusion, vasoconstriction, muscle rigidity, slow heart rate, loss of spontaneous movement, and coma. Death is usually caused by cardiac arrhythmias. Because the elderly have reduced metabolic protection against a cold environment coupled with a reduced perception of cold, they are at greater risk for developing hypothermia.

■ **CHECKPOINT**

15. In what ways can a person lose heat to or gain heat from the surroundings? How is it possible for a person to lose heat on a sunny beach when the temperature is 40°C (104°F) and the humidity is 85 percent?

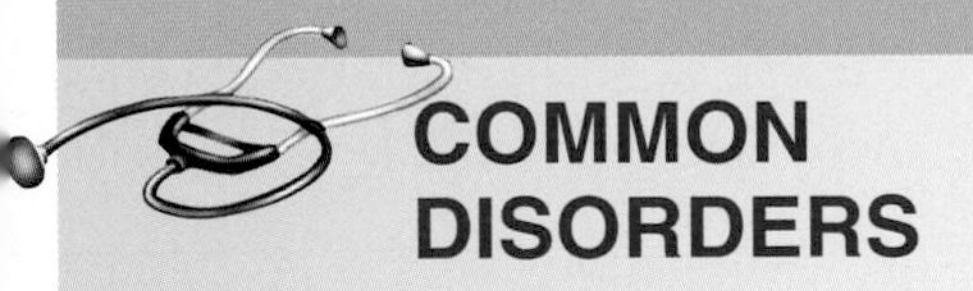

COMMON DISORDERS

Fever

A *fever* is an elevation of body temperature that results from a resetting of the hypothalamic thermostat. The most common causes of fever are viral or bacterial infections and bacterial toxins; other causes are ovulation, excessive secretion of thyroid hormones, tumors, and reactions to vaccines. When phagocytes ingest certain bacteria, they are stimulated to secrete a ***pyrogen*** (PĪ-rō-gen; *pyro-* = fire; *-gen* = produce), a fever-producing substance. The pyrogen circulates to the hypothalamus and induces secretion of prostaglandins. Some prostaglandins can reset the hypothalamic thermostat at a higher temperature, and temperature-regulating reflex mechanisms then act to bring body temperature up to this new setting. *Antipyretics* are agents that relieve or reduce fever. Examples include aspirin, acetaminophen (Tylenol®), and ibuprofen (Advil®), all of which reduce fever by inhibiting synthesis of certain prostaglandins.

Although death results if core temperature rises above 44–46°C (112–114°F), up to a point, fever is beneficial. For example, a higher temperature intensifies the effect of interferon and the phagocytic activities of macrophages while hindering replication of some pathogens. Because fever increases heart rate, infection-fighting white blood cells are delivered to sites of infection more rapidly. In addition, antibody production and T cell proliferation increase.

Obesity

Obesity is body weight more than 20% above a desirable standard due to an excessive accumulation of adipose tissue; it affects one-third of the adult population in the United States. (An athlete may be *overweight* due to a higher-than-normal amount of muscle tissue without being obese.) Even moderate obesity is hazardous to health; it is implicated as a risk factor in cardiovascular disease, hypertension, pulmonary disease, non-insulin-dependent diabetes mellitus, arthritis, certain cancers (breast, uterus, and colon), varicose veins, and gallbladder disease.

In a few cases, obesity may result from trauma to or tumors in the food-regulating centers in the hypothalamus. In most cases of obesity, no specific cause can be identified. Contributing factors include genetic factors, eating habits taught early in life, overeating to relieve tension, and social customs.

MEDICAL TERMINOLOGY AND CONDITIONS

Heat cramps Cramps that result from profuse sweating. The salt lost in sweat causes painful contractions of muscles; such cramps tend to occur in muscles used while working but do not appear until the person relaxes once the work is done. Drinking salted liquids usually leads to rapid improvement.

Heatstroke (sunstroke) A severe and often fatal disorder caused by exposure to high temperatures. Blood flow to the skin is decreased, perspiration is greatly reduced, and body temperature rises sharply because of failure of the hypothalamic thermostat. Body temperature may reach 43°C (110°F). Treatment, which must be undertaken immediately, consists of cooling the body by immersing the victim in cool water and by administering fluids and electrolytes.

Kwashiorkor (kwash′-ē-OR-kor) A disorder in which protein intake is deficient despite normal or nearly normal caloric intake, characterized by edema of the abdomen, enlarged liver, decreased blood pressure, low pulse rate, lower than normal body temperature, and sometimes mental retardation. Because the main protein in corn lacks two essential amino acids, which are needed for growth and tissue repair, many African children whose diet consists largely of cornmeal develop kwashiorkor.

Malnutrition (*mal-* = bad) An imbalance of total caloric intake or intake of specific nutrients, which can be either inadequate or excessive.

Marasmus (mar-AZ-mus) A type of undernutrition that results from inadequate intake of both protein and calories. Its characteristics include retarded growth, low weight, muscle wasting, emaciation, dry skin, and thin, dry, dull hair.

STUDY OUTLINE

Introduction (p. 503)

1. The food we eat is our only source of energy for performing biological work; it also provides essential substances that we cannot synthesize.
2. Food molecules absorbed by the gastrointestinal tract are used to supply energy for life processes, serve as building blocks during synthesis of complex molecules, or are stored for future use.

Nutrients (p. 504)

1. Nutrients include carbohydrates, lipids, proteins, water, minerals, and vitamins.

2. Nutrition experts suggest dietary calories be 50–60% from carbohydrates, 30% or less from fats, and 12–15% from proteins.
3. The My Pyramid guide represents a personalized approach to making healthy food choices and maintaining regular physical activity.
4. Some minerals known to perform essential functions include calcium, phosphorus, potassium, sodium, chloride, magnesium, iron, manganese, copper, and zinc. Their functions are summarized in Table 20.1 on page 506.
5. Vitamins are organic nutrients that maintain growth and normal metabolism. Many function as coenzymes.
6. Fat-soluble vitamins are absorbed with fats and include vitamins A, D, E, and K; water-soluble vitamins are absorbed with water and include the B vitamins and vitamin C.
7. The functions of the principal vitamins and their deficiency disorders are summarized in Table 20.2 on pages 508–509.

Metabolism (p. 507)

1. Metabolism refers to all chemical reactions of the body and has two phases: catabolism and anabolism. Anabolism consists of reactions that combine simple substances into more complex molecules. Catabolism consists of reactions that break down complex organic compounds into simple ones.
2. Metabolic reactions are catalyzed by enzymes, proteins that speed up chemical reactions without being changed.
3. Anabolic reactions require energy, which is supplied by catabolic reactions.
4. During digestion, polysaccharides and disaccharides are converted to glucose.
5. Glucose moves into cells by facilitated diffusion, which is stimulated by insulin. Some glucose is catabolized by cells to produce ATP. Excess glucose can be stored by the liver and skeletal muscles as glycogen or converted to fat.
6. Glucose catabolism is also called cellular respiration. The complete catabolism of glucose to produce ATP involves glycolysis, the Krebs cycle, and the electron transport chain. It can be represented as follows: 1 glucose + 6 oxygen $\rightarrow$ 36 or 38 ATP + 6 carbon dioxide + 6 water.
7. Glycolysis is also called anaerobic respiration because it occurs without oxygen. During glycolysis, which occurs in the cytosol, one glucose molecule is broken down into two molecules of pyruvic acid. Glycolysis yields a net of two ATP and two $NADH + H^+$.
8. When oxygen is plentiful, most cells convert pyruvic acid to acetyl coenzyme A, which enters the Krebs cycle.
9. The Krebs cycle occurs in mitochondria. The chemical energy originally contained in glucose, pyruvic acid, and acetyl coenzyme A is transferred to the coenzymes NADH and $FADH_2$.
10. The electron transport chain is a series of reactions that occur in mitochondria in which the energy in the reduced coenzymes is transferred to ATP.
11. The conversion of glucose to glycogen for storage occurs extensively in liver and skeletal muscle fibers and is stimulated by insulin. The body can store about 500 g of glycogen.
12. The breakdown of glycogen to glucose occurs mainly between meals.
13. Gluconeogenesis is the conversion of glycerol, lactic acid, or amino acids to glucose.
14. Some triglycerides may be catabolized to produce ATP; others are stored in adipose tissue. Other lipids are used as structural molecules or to synthesize other substances.
15. Triglycerides must be split into fatty acids and glycerol before they can be catabolized. Glycerol can be transformed into glucose by conversion into glyceraldehyde 3-phosphate. Fatty acids are catabolized through formation of acetyl coenzyme A, which can enter the Krebs cycle.
16. The formation of ketone bodies by the liver is a normal phase of fatty acid catabolism, but an excess of ketone bodies, called ketosis, may cause acidosis.
17. The conversion of glucose or amino acids into lipids is stimulated by insulin.
18. Lipoproteins transport lipids in the bloodstream. Types of lipoproteins include chylomicrons, which carry dietary lipids to adipose tissue; very-low-density lipoproteins (VLDLs), which carry triglycerides from the liver to adipose tissue; low-density lipoproteins (LDLs), which deliver cholesterol to body cells; and high-density lipoproteins (HDLs), which remove excess cholesterol from body cells and transport it to the liver for elimination.
19. Amino acids, under the influence of insulinlike growth factors and insulin, enter body cells by means of active transport. Inside cells, amino acids are reassembled into proteins that function as enzymes, hormones, structural elements, and so forth; stored as fat or glycogen; or used for ATP production.
20. Before amino acids can be catabolized, they must deaminated. Liver cells convert the resulting ammonia to urea, which is excreted in urine.
21. Amino acids may also be converted into glucose, fatty acids, and ketone bodies.
22. Protein synthesis is stimulated by insulinlike growth factors, thyroid hormones, insulin, estrogen, and testosterone. It is directed by DNA and RNA and carried out on ribosomes.
23. Table 20.3 on page 515 summarizes carbohydrate, lipid, and protein metabolism.

Metabolism and Body Heat (p. 515)

1. A calorie is the amount of energy required to raise the temperature of 1 gram of water 1°C.
2. The Calorie is the unit of heat used to express the caloric value of foods and to measure the body's metabolic rate. One Calorie equals 1000 calories, or 1 kilocalorie.
3. Most body heat is a result of catabolism of the food we eat. The rate at which this heat is produced is known as the metabolic rate and is affected by exercise, hormones, the nervous system, body temperature, ingestion of food, age, gender, climate, sleep, and nutrition.
4. Measurement of the metabolic rate under basal conditions is called the basal metabolic rate (BMR).
5. Mechanisms of heat loss are radiation, conduction, convection, and evaporation.

6. Radiation is the transfer of heat from a warmer object to a cooler object without physical contact.
7. Conduction is the transfer of heat between two objects in contact with each other.
8. Convection is the transfer of heat by the movement of a liquid or gas between areas of different temperatures.
9. Evaporation is the conversion of a liquid to a vapor; in the process, heat is lost.
10. A normal body temperature is maintained by negative feedback loops that regulate heat-producing and heat-losing mechanisms.
11. Responses that produce or retain heat when body temperature falls are vasoconstriction; release of epinephrine, norepinephrine, and thyroid hormones; and shivering.
12. Responses that increase heat loss when body temperature rises include vasodilation, decreased metabolic rate, and evaporation of sweat.

SELF-QUIZ

1. Creating a protein from amino acids is an example of
   a. deamination b. anabolism
   c. gluconeogenesis d. catabolism
   e. cellular respiration
2. Free radicals
   a. are a type of provitamin
   b. are essential amino acids
   c. can cause damage to cellular structures
   d. help regulate enzymatic reactions
   e. are a form of energy
3. Which of the following statements about vitamins is NOT true?
   a. Most vitamins are synthesized by the body cells.
   b. Vitamins can act as coenzymes.
   c. Vitamin K is produced by bacteria in the GI tract.
   d. Lipid-soluble vitamins may be stored in the liver.
   e. Excess water-soluble vitamins are excreted in urine.
4. Match the following:
   ____ a. precursor for vitamin A
   ____ b. form in which lipids are transported in the blood plasma
   ____ c. needed to convert ADP to ATP
   ____ d. derived from vitamin B_2 (riboflavin)

   A. lipoproteins
   B. FAD
   C. beta-carotene
   D. magnesium
5. Body temperature is controlled by the
   a. pons b. thyroid gland
   c. hypothalamus d. adrenal medulla
   e. autonomic nervous system
6. The removal of an amino group ($—NH_2$) from amino acids entering the Krebs cycle is known as
   a. deamination b. convection c. ketogenesis
   d. lipolysis e. aerobic respiration
7. If your diet is low in carbohydrates, which compound(s) does your body begin to catabolize next for ATP production?
   a. vitamins b. lipids c. minerals d. cholesterol
   e. amino acids
8. Cellular respiration includes the following steps in order:
   a. Krebs cycle, glycolysis, electron transport chain
   b. Krebs cycle, electron transport chain, glycolysis
   c. glycolysis, electron transport chain, Krebs cycle
   d. electron transport chain, Krebs cycle, glycolysis
   e. glycolysis, Krebs cycle, electron transport chain
9. Which of the following is most often used to synthesize ATP?
   a. galactose b. triglycerides c. amino acids
   d. glucose e. glycerol
10. How does glucose enter the cytosol of cells?
    a. facilitated diffusion b. simple diffusion
    c. active transport d. osmosis e. electron transport
11. Sweat drying from a person's skin surface causes loss of body heat by
    a. radiation b. conduction c. convection
    d. evaporation e. conversion
12. Glycolysis
    a. requires the presence of oxygen
    b. produces two ATP molecules per glucose molecule
    c. takes place in mitochondria
    d. is also known as the Krebs cycle
    e. is the conversion of glucose to glycogen
13. Which of the following statements is NOT true?
    a. Triglycerides are stored in adipose tissue.
    b. Chylomicrons enter the blood by way of lacteals in the intestines.
    c. Most of the body's cholesterol is carried in low-density lipoproteins.
    d. Lipids can be stored in the liver.
    e. High-density lipoproteins contribute to the formation of fatty plaques.
14. Which of the following equations summarizes the complete catabolism of a molecule of glucose?
    a. glucose + 6 water $\rightarrow$ 36 or 38 ATP + 6 CO_2 + 6 O_2
    b. glucose + 6 O_2 $\rightarrow$ 36 or 38 ATP + 6 CO_2 + 6 water
    c. glucose + ATP $\rightarrow$ 31 or 38 CO_2 + 6 water
    d. glucose + pyruvic acid $\rightarrow$ 36 or 38 ATP + 6 O_2
    e. glucose + citric acid $\rightarrow$ 31 or 38 ATP + 6 CO_2

15. Those amino acids that cannot by synthesized by the body and must be obtained from the diet are known as

a. coenzymes **b.** ketones **c.** essential amino acids **d.** nonessential amino acids **e.** polypeptides

16. Which of the following would NOT increase the metabolic rate?

a. increased levels of thyroid hormones
b. epinephrine
c. old age
d. fever
e. exercise

17. FAD and NAD^+ are examples of

a. nutrients **b.** antioxidants **c.** pyrogens **d.** coenzymes **e.** minerals

18. The process by which glucose is formed from amino acids is

a. gluconeogenesis **b.** deamination **c.** anaerobic respiration **d.** ketogenesis **e.** glycolysis

19. All of the following can contribute to an increase in body temperature EXCEPT

a. shivering
b. release of thyroid hormones
c. sympathetic stimulation of the adrenal medulla
d. vasodilation of blood vessels in the skin
e. activation of the hypothalamus

20. Match the following:

____ **a.** conversion of glucose to pyruvic acid	**A.** catabolism
____ **b.** the complete breakdown of glucose	**B.** anabolism
____ **c.** building simple molecules into more complex ones	**C.** glycolysis
____ **d.** NAD^+ and FAD pick up high-energy electrons	**D.** cellular respiration
____ **e.** the breakdown of organic compounds	**E.** Krebs cycle

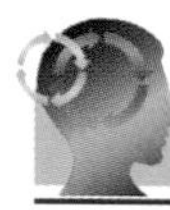

CRITICAL THINKING APPLICATIONS

1. Carla and Ashley, members of their college's tennis team, ate lunch at McDonald's before their afternoon practice. Carla had a Quarter Pounder with cheese, small french fries, and a small chocolate shake; Ashley had a Chicken McGrill (plain, without mayonnaise), a garden salad with fat-free vinaigrette, and a glass of 1% low-fat milk. Critique their choices based on the recommended distributions of calories. How many calories are left for breakfast, dinner, and snacks?

Quarter Pounder Meal (percentages of total calories):

total fats = 41%
saturated fats = 17%
total carbohydrates = 46%
simple sugars = 23%
proteins = 13%
total Calories = 1085

Grilled Chicken Sandwich Meal (percentages of total calories):

total fats = 24%
saturated fats = 9%
total carbohydrates = 47%
simple sugars = 15%
proteins = 29%
total Calories = 592

2. It's noon on a hot summer day, the sun is directly overhead, and a group of sunbathers roasts on the beach. What mechanism causes their body temperature to increase? Several of the sunbathers jump into the cool water. What mechanisms decrease their body temperature?

3. Shannon is a morning person but her roommate Darla is not. In fact, Shannon teases Darla for being a classic example of "BMR" (barely mentally responsive) during her 8 A.M. class. What does BMR really mean? How is metabolism measured?

4. Rob swallows a multivitamin tablet every morning and an antioxidant tablet containing beta-carotene, vitamin C, and vitamin E with his dinner every night. What are the functions of antioxidants in the body? What happens to the antioxidants if any exceed his daily requirements?

ANSWERS TO FIGURE QUESTIONS

20.1 The wider base of each band represents foods with little or no solid fats or added sugars.

20.2 The formation of digestive enzymes in the pancreas is part of anabolism.

20.3 Complete catabolism of glucose yields 36–38 ATP.

20.4 Liver cells can carry out gluconeogenesis.

20.5 Liver cells form ketone bodies.

20.6 Exercise, the sympathetic nervous system, hormones (epinephrine, norepinephrine, thyroid hormones, testosterone, human growth hormone), elevated body temperature, and ingestion of food increase metabolic rate.

THE URINARY SYSTEM

did you know? ***A kidney stone is a hard mass, usually composed of calcium oxalate, uric acid, or calcium phosphate crystals. The medical term for a kidney stone is renal calculus. Researchers do not yet know why some people are predisposed to developing kidney stones. About 90% of stones will pass on their own within three to six weeks, so patients are usually advised to try lifestyle changes before going on to medical treatment. Increased fluid intake (3 to 4 quarts of fluid, preferably water, per day), and changes in diet and medications are often sufficient treatment.***

Focus on Wellness, page 535

www.wiley.com/college/apcentral

As body cells carry out their metabolic functions, they consume oxygen and nutrients and produce substances, such as carbon dioxide, that have no useful functions and need to be eliminated from the body. While the respiratory system rids the body of carbon dioxide, the urinary system disposes of most other unneeded substances. As you will learn in this chapter, however, the urinary system is not merely concerned with waste disposal; it carries out a number of other important functions as well.

looking back to move ahead . . .

- Transport Across the Plasma Membrane (page 47)
- Simple Cuboidal Epithelium (page 75)
- Transitional Epithelium (page 82)
- Actions of Antidiuretic Hormone (ADH) (page 322)
- Vitamin D, Calcitriol, and Calcium Homeostasis (page 326)
- Renin–Angiotensin–Aldosterone Pathway (page 331)
- Filtration and Reabsorption in Capillaries (pages 388-389)
- Blood Colloid Osmotic Pressure (page 388)

OVERVIEW OF THE URINARY SYSTEM

OBJECTIVE • **List the components of the urinary system and their general functions.**

The ***urinary system*** consists of two kidneys, two ureters, one urinary bladder, and one urethra (Figure 21.1). After the kidneys filter blood, they return most of the water and many of the solutes to the bloodstream. The remaining water and solutes constitute ***urine***, which passes through the ureters and is stored in the urinary bladder until it is expelled from the body through the urethra. ***Nephrology*** (nef-ROL-ō-jē; *nephro-* = kidney; *-logy* = study of) is the scientific study of the anatomy, physiology, and disorders of the kidneys. The branch of medicine that deals with the male and female urinary systems and the male reproductive system is ***urology*** (ū-ROL-ō-jē; *uro-* = urine). A physician who specializes in this branch of medicine is called a ***urologist*** (ū-ROL-ō-jist).

The kidneys do the major work of the urinary system. The other parts of the system are primarily passageways and temporary storage areas. Functions of the kidneys include the following:

- **Regulation of ion levels in the blood.** The kidneys help regulate the blood levels of several ions, most importantly sodium ions (Na^+), potassium ions (K^+), calcium ions (Ca^{2+}), chloride ions (Cl^-), and phosphate ions (HPO_4^{2-}).
- **Regulation of blood volume and blood pressure.** The kidneys adjust the volume of blood in the body by returning water to the blood or eliminating it in the urine. They help regulate blood pressure by secreting the enzyme renin, which activates the renin–angiotensin–aldosterone pathway (see Figure 13.14 on page 331), by adjusting blood flow into and out of the kidneys, and by adjusting blood volume.
- **Regulation of blood pH.** The kidneys regulate the concentration of H^+ in the blood by excreting a variable amount of H^+ in the urine. They also conserve blood bicarbonate ions (HCO_3^-), an important buffer of H^+. Both activities help regulate blood pH.

Figure 21.1 Organs of the female urinary system in relation to surrounding structures.

Urine formed by the kidneys passes first into the ureters, then to the urinary bladder for storage, and finally through the urethra for elimination from the body.

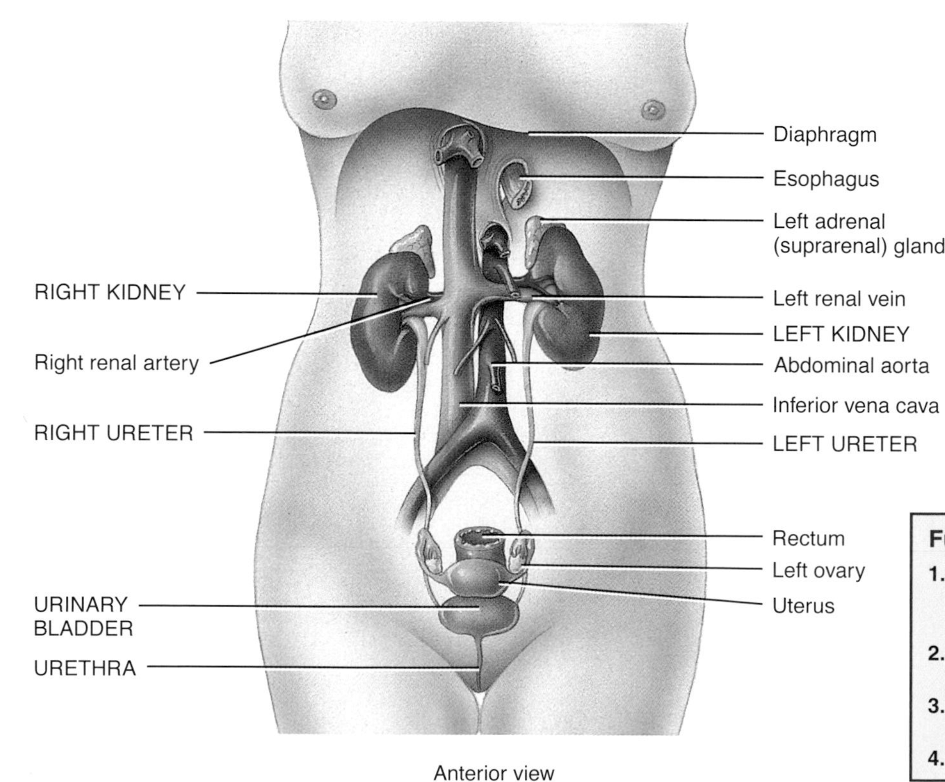

Functions of the Urinary System

1. The kidneys regulate blood volume and composition, help regulate blood pressure and pH, produce two hormones, and excrete wastes.
2. The ureters transport urine from the kidneys to the urinary bladder.
3. The urinary bladder stores urine and expels it into the urethra.
4. The urethra discharges urine from the body.

Which organ of the urinary system does most of the work to form urine?

- **Production of hormones.** The kidneys produce two hormones. *Calcitriol*, the active form of vitamin D, helps regulate calcium homeostasis (see Figure 13.10 on page 327), and *erythropoietin* stimulates production of red blood cells (see Figure 14.4 on page 351).
- **Excretion of wastes.** By forming urine, the kidneys help excrete ***wastes***—substances that have no useful function in the body. Some wastes excreted in urine result from metabolic reactions in the body. These include ammonia and urea from the breakdown of amino acids; bilirubin from the breakdown of hemoglobin; creatinine from the breakdown of creatine phosphate in muscle fibers; and uric acid from the breakdown of nucleic acids. Other wastes excreted in urine are foreign substances from the diet, such as drugs and environmental toxins.

CHECKPOINT

1. What are wastes, and how do the kidneys take part in their removal from the body?

STRUCTURE OF THE KIDNEYS

OBJECTIVE • Describe the structure and blood supply of the kidneys.

The ***kidneys*** (KID-nēz) are a pair of reddish organs shaped like kidney beans (Figure 21.2). They lie on either side of the vertebral column between the peritoneum and the back wall of the abdominal cavity at the level of the 12th thoracic and first three lumbar vertebrae. The 11th and 12th pairs of ribs provide some protection for the superior parts of the kidneys. The right kidney is slightly lower than the left because the liver occupies a large area above the kidney on the right side.

External Anatomy of the Kidneys

An adult kidney is about the size of a bar of bath soap. Near the center of the medial border is an indentation called the ***renal hilum*** (HĪ-lum), through which the ureter leaves the

Figure 21.2 **Structure of the kidney.**

A renal capsule covers the kidney. Internally, the two main regions are the renal cortex and the renal medulla.

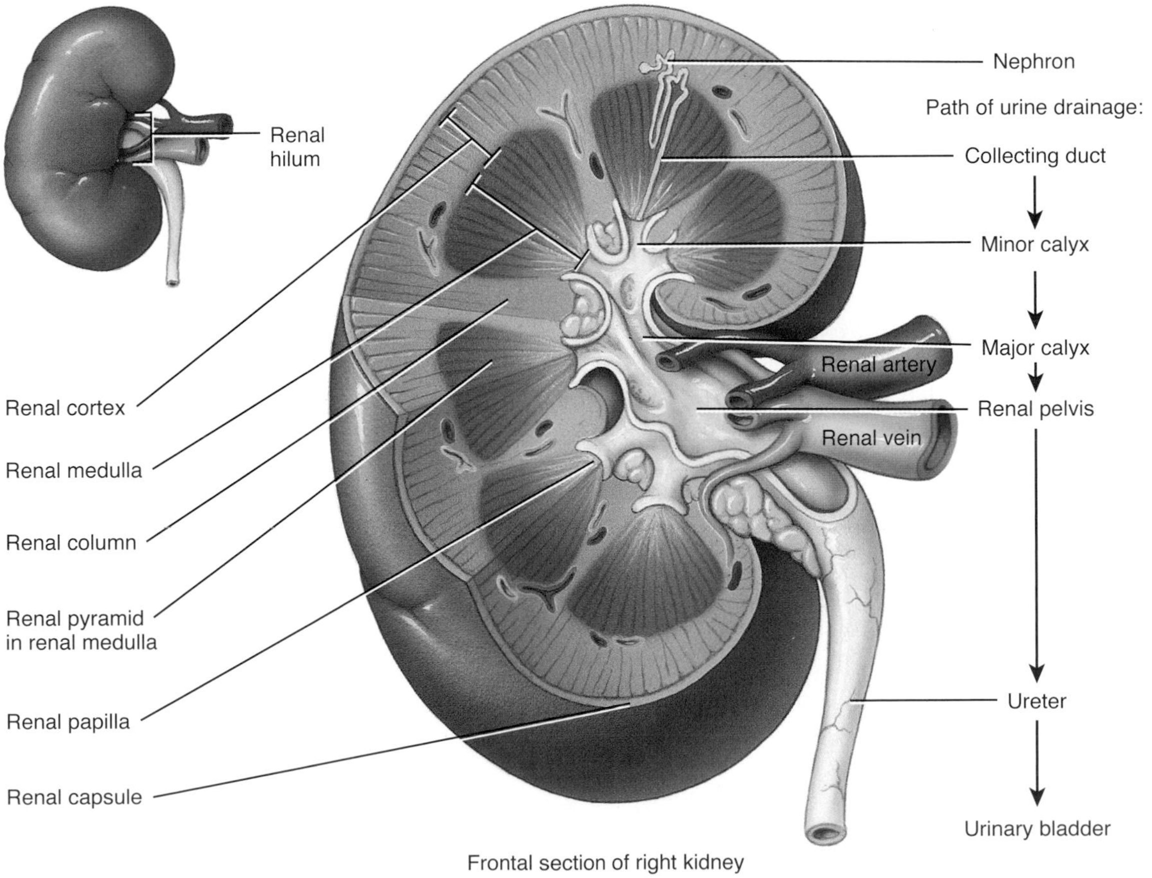

Frontal section of right kidney

Where in the kidney are the renal pyramids located?

kidney and blood vessels, lymphatic vessels, and nerves enter and exit. Surrounding each kidney is the smooth, transparent *renal capsule*, a connective tissue sheath that helps maintain the shape of the kidney and serves as a barrier against trauma (Figure 21.2). Adipose (fatty) tissue surrounds the renal capsule and cushions the kidney. Along with a thin layer of dense irregular connective tissue, the adipose tissue anchors the kidney to the posterior abdominal wall.

Internal Anatomy of the Kidneys

Internally, the kidneys have two main regions: an outer light-red region called the ***renal cortex*** (*cortex* = rind or back) and an inner, darker red-brown region called the ***renal medulla*** (*medulla* = inner portion) (Figure 21.2). Within the renal medulla are several cone-shaped ***renal pyramids***. Extensions of the renal cortex, called ***renal columns***, fill the spaces between renal pyramids.

Urine formed in the kidney drains into a large, funnel-shaped cavity called the ***renal pelvis*** (*pelv-* = basin). The rim of the renal pelvis contains cuplike structures called ***major*** and ***minor calyces*** (KAL-i-sēz = cups; singular is *calyx*). Urine flows from several ducts within the kidney into a minor calyx and from there through a major calyx into the renal pelvis, which connects to a ureter. Water and solutes in the fluid that drains into the renal pelvis remain in the urine and are *excreted* (eliminated from the body).

Renal Blood Supply

About 20–25% of the resting cardiac output—1200 milliliters of blood per minute—flows into the kidneys through the right and left ***renal arteries*** (Figure 21.3). Within each kidney, the renal artery divides into smaller and smaller vessels (*segmental, interlobar, arcuate, interlobular arteries*) that eventually deliver blood to the ***afferent arterioles*** (*af-* = toward; *-ferre* = to carry). Each afferent arteriole divides into a tangled capillary network called a ***glomerulus*** (glō-MER-ū-lus = little ball; plural is *glomeruli*).

The capillaries of the glomerulus reunite to form an ***efferent arteriole*** (*ef-* = out). Upon leaving the glomerulus, each efferent arteriole divides to form a network of capillaries around the kidney tubules (described shortly). These ***peritubular capillaries*** (*peri-* = around) eventually reunite to form *peritubular veins*, which merge into *interlobular, arcuate,* and *interlobar veins*. Ultimately, all these smaller veins drain into the ***renal vein***.

Nephrons

The functional units of the kidney are the ***nephrons*** (NEF-ronz), numbering about a million in each kidney (Figure 21.4 on page 528). A nephron consists of two parts: a ***renal corpuscle*** (KOR-pus-ul = tiny body), where blood plasma is filtered, and a ***renal tubule*** into which the filtered fluid, called ***glomerular filtrate***, passes. As the fluid moves through the renal tubules, wastes and excess substances are added, and useful materials are returned to the blood in the peritubular capillaries.

The two parts that make up a renal corpuscle are the ***glomerulus*** and the ***glomerular (Bowman's) capsule***, a double-walled cup of epithelial cells that surrounds the glomerular capillaries. Glomerular filtrate first enters the glomerular capsule and then passes into the renal tubule. In the order that fluid passes through them, the three main sections of the renal tubule are the ***proximal convoluted tubule***, the ***loop of Henle***, and the ***distal convoluted tubule***. *Proximal* denotes the part of the tubule attached to the glomerular capsule, and *distal* denotes the part that is farther away. *Convoluted* means the tubule is tightly coiled rather than straight. The renal corpuscle and both convoluted tubules lie within the renal cortex; the loop of Henle extends into the renal medulla. The first part of the loop of Henle begins in the renal cortex and extends downward into the renal medulla, where it is called the ***descending limb of the loop of Henle*** (Figure 21.4). It then makes a hairpin turn and returns to the renal cortex as the ***ascending limb of the loop of Henle***. The distal convoluted tubules of several nephrons empty into a common ***collecting duct***.

The **number of nephrons** is constant from birth. New nephrons do not form to replace those that are injured or diseased. Signs of kidney damage often are not apparent, however, until the majority of nephrons are damaged because the remaining functional nephrons adapt to handle a larger-than-normal load. Surgical removal of one kidney, for example, stimulates enlargement of the remaining kidney, which eventually is able to filter blood at 80% of the rate of two normal kidneys.

■ **CHECKPOINT**

2. Which structures help protect and cushion the kidneys?
3. What is the functional unit of the kidney? Describe its structure.

Figure 21.3 Blood supply of the right kidney.

The renal arteries deliver about 25% of the resting cardiac output to the kidneys.

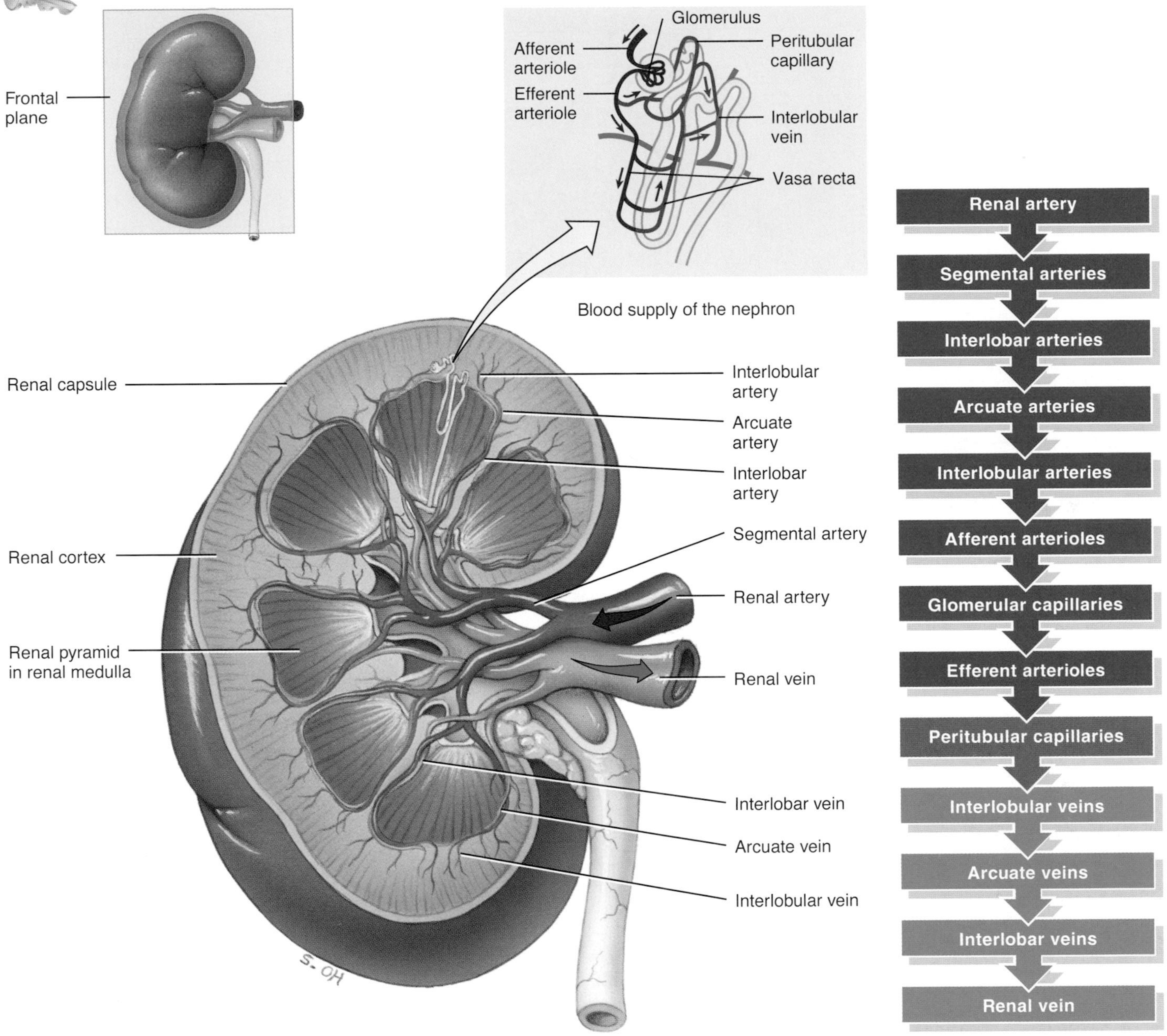

Frontal plane

Glomerulus
Afferent arteriole
Efferent arteriole
Peritubular capillary
Interlobular vein
Vasa recta

Blood supply of the nephron

Renal capsule
Renal cortex
Renal pyramid in renal medulla
Interlobular artery
Arcuate artery
Interlobar artery
Segmental artery
Renal artery
Renal vein
Interlobar vein
Arcuate vein
Interlobular vein

(a) Anterior view of frontal section of right kidney

Renal artery
Segmental arteries
Interlobar arteries
Arcuate arteries
Interlobular arteries
Afferent arterioles
Glomerular capillaries
Efferent arterioles
Peritubular capillaries
Interlobular veins
Arcuate veins
Interlobar veins
Renal vein

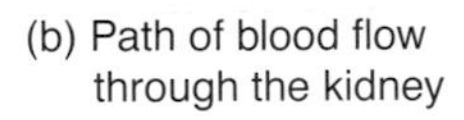

(b) Path of blood flow through the kidney

How much blood enters the renal arteries each minute?

Figure 21.4 The parts of a nephron, collecting duct, and associated blood vessels.

Nephrons are the functional units of the kidneys.

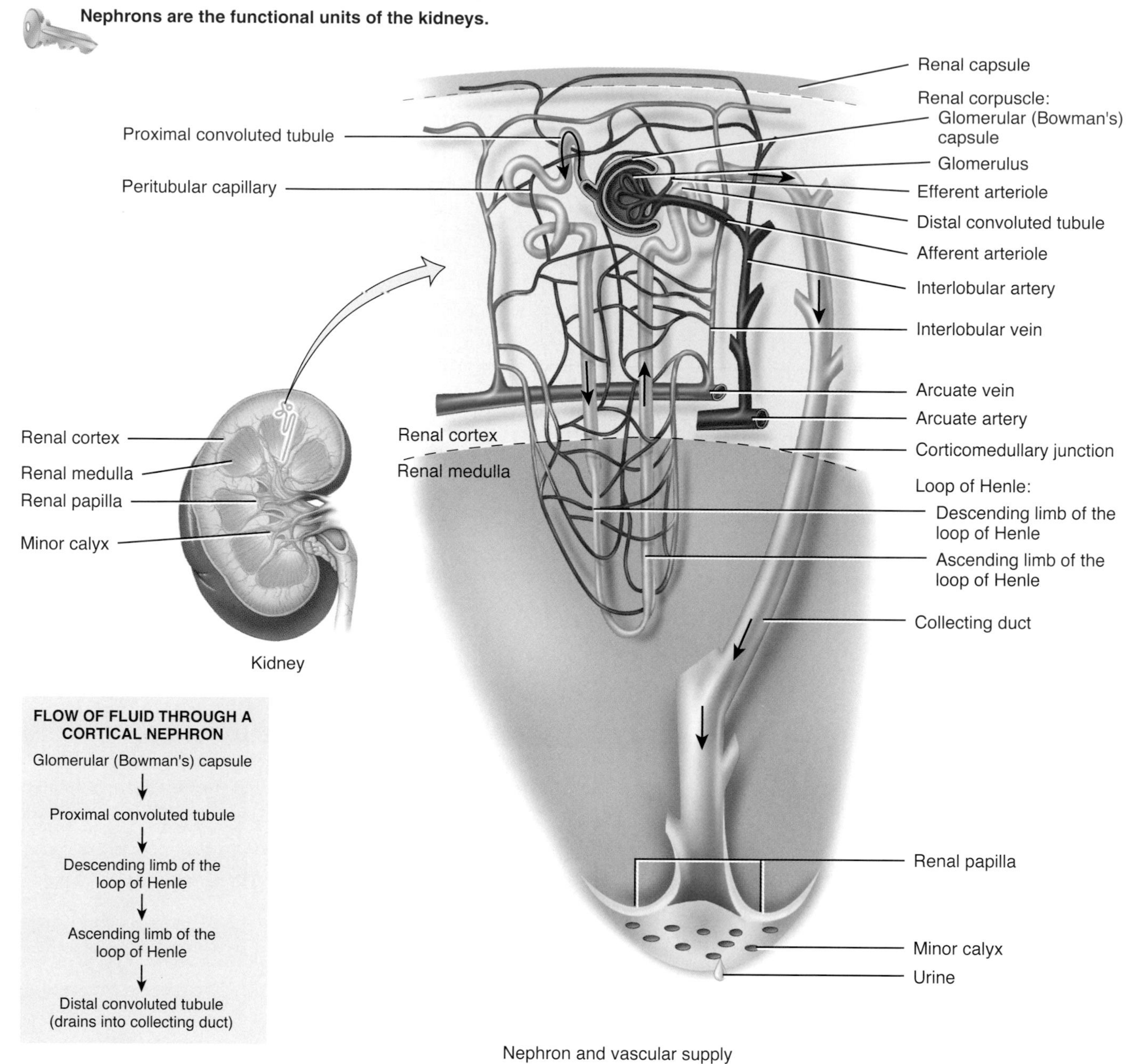

Nephron and vascular supply

A water molecule has just entered the proximal convoluted tubule of a nephron. Which parts of the nephron will it travel through (in order) to reach the renal pelvis in a drop of urine?

FUNCTIONS OF THE NEPHRON

OBJECTIVE • Identify the three basic functions performed by nephrons and collecting ducts and indicate where each occurs.

To produce urine, nephrons and collecting ducts perform three basic processes—glomerular filtration, tubular reabsorption, and tubular secretion (Figure 21.5):

1. Filtration is the forcing of fluids and dissolved substances smaller than a certain size through a membrane by pressure. ***Glomerular filtration*** is the first step of urine production: Blood pressure forces water and most solutes in blood plasma across the wall of glomerular capillaries, forming glomerular filtrate. Filtration occurs in glomeruli just as it occurs in other capillaries (see Figure 16.3 on page 389).
2. ***Tubular reabsorption*** occurs as filtered fluid flows along the renal tubule and through the collecting duct: Tubule

Figure 21.5 Overview of functions of a nephron. Excreted substances remain in the urine and eventually leave the body.

Glomerular filtration occurs in the renal corpuscle; tubular reabsorption and tubular secretion occur all along the renal tubule and collecting duct.

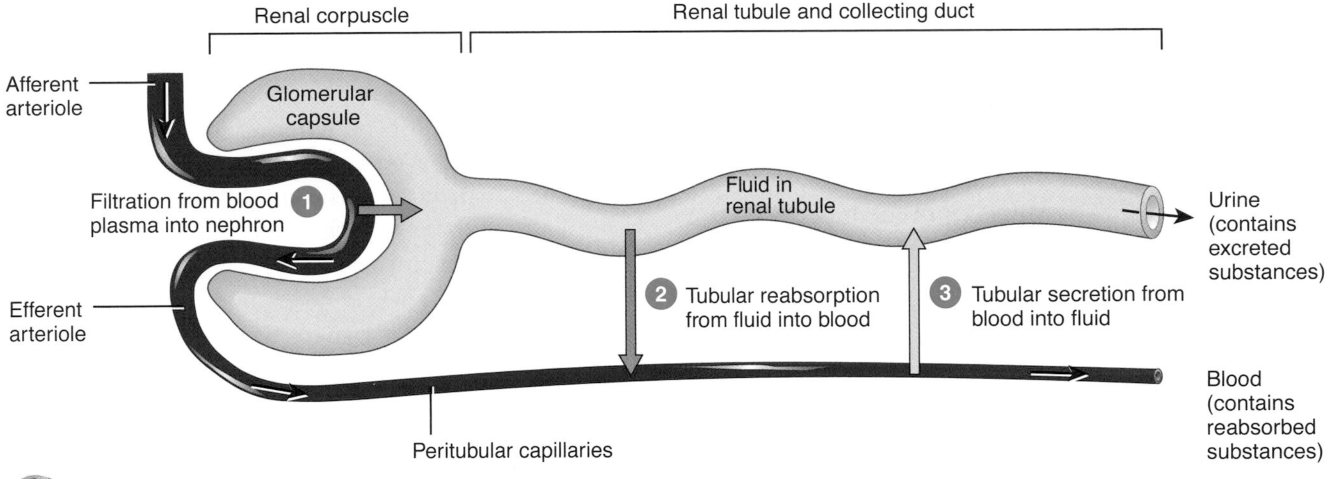

When the renal tubules secrete the drug penicillin, is the drug being added to or removed from the blood?

and duct cells return about 99% of the filtered water and many useful solutes to the blood flowing through peritubular capillaries.

 ***Tubular secretion*** also takes place as fluid flows along the tubule and through the collecting duct: The tubule and duct cells remove substances, such as wastes, drugs, and excess ions, from blood in the peritubular capillaries and transport them into the fluid in the renal tubules.

As nephrons perform their functions, they help maintain homeostasis of the blood's volume and composition. The situation is somewhat similar to a recycling center: Garbage trucks dump refuse into an input hopper, where the smaller refuse passes onto a conveyor belt (glomerular filtration of blood plasma). As the conveyor belt carries the garbage along, workers remove useful items, such as aluminum cans, plastics, and glass containers (reabsorption). Other workers place additional garbage and larger items onto the conveyor belt (secretion). At the end of the belt, all remaining garbage falls into a truck for transport to the landfill (excretion of wastes in urine).

Table 21.1 compares the substances that are filtered, reabsorbed, and excreted in urine per day in an adult male. Although the values shown are typical, they vary considerably according to diet. The following sections describe each of the three steps that contribute to urine formation in more detail.

Table 21.1 Substances Filtered, Reabsorbed, and Excreted in Urine per Day

Substance	Filtered* (enters renal tubule)	Reabsorbed (returned to blood)	Excreted in Urine
Water	180 liters	178–179 liters	1–2 liters
Chloride ions (Cl^-)	640 g	633.7 g	6.3 g
Sodium ions (Na^+)	579 g	575 g	4 g
Bicarbonate ions (HCO_3^-)	275 g	274.97 g	0.03 g
Glucose	162 g	162 g	0
Urea	54 g	24 g	30 g†
Potassium ions (K^+)	29.6 g	29.6 g	2.0 g‡
Uric acid	8.5 g	7.7 g	0.8 g
Creatinine	1.6 g	0	1.6 g

*Assuming glomerular filtration is 180 liters per day.
†In addition to being filtered and reabsorbed, urea is secreted.
‡After virtually all filtered K^+ is reabsorbed in the convoluted tubules and loop of Henle, a variable amount of K^+ is secreted in the collecting duct.

Glomerular Filtration

Two layers of cells compose the capsule that surrounds the glomerular capillaries (Figure 21.6). Think of the renal corpuscle as a fist (the glomerular capillaries) pushed into a limp balloon (the glomerular capsule) until the fist is covered by two layers of the balloon with a space, the ***capsular space***, in between. The cells that make up the inner wall of the glomerular capsule, called *podocytes*, adhere closely to the endothelial cells of the glomerulus. Together, the podocytes and glomerular endothelium form a *filtration membrane* that permits the passage of water and solutes from the blood into the capsular space. Blood cells and most plasma proteins remain in the blood because they are too large to pass through the filtration membrane. Simple squamous epithelial cells form the outer layer of the glomerular capsule.

Net Filtration Pressure

The pressure that causes filtration is the blood pressure in the glomerular capillaries. Two other pressures oppose glomerular filtration: (1) blood colloid osmotic pressure (see page 388) and (2) glomerular capsule pressure (due to fluid already in the capsular space and renal tubule). When either of these pressures increases, glomerular filtration decreases. Normally, blood pressure is greater than the two opposing pressures, producing a ***net filtration pressure*** of about 10 mm Hg. Net filtration pressure forces a large volume of fluid into the capsular space, about 150 liters daily in females and 180 liters daily in males.

Because the efferent arteriole is smaller in diameter than the afferent arteriole, it helps raise the blood pressure in the glomerular capillaries. When blood pressure increases or decreases slightly, changes in the diameters of the afferent and efferent arterioles can actually keep net filtration pressure steady to maintain normal glomerular filtration. Constriction of the afferent arteriole decreases blood flow into the glomerulus, which decreases net filtration pressure. Constriction of the efferent arteriole slows outflow of blood and increases net filtration pressure.

Conditions that greatly reduce blood pressure, for instance severe hemorrhage, may cause glomerular blood pressure to fall so low that net filtration pressure drops despite constriction of efferent arterioles. Then, glomerular filtration slows, or even stops entirely. The result is **oliguria** (*olig-* = scanty; *-uria* = urine production), a daily urine output between 50 and 250 mL, or **anuria**, a daily urine output of less than 50 mL. Obstructions, such as a kidney stone that blocks a ureter or an enlarged prostate that blocks the urethra in a male, can also decrease net filtration pressure and thereby reduce urine output.

Figure 21.6 Glomerular filtration, the first step in urine formation.

Glomerular filtrate (red arrows) passes into the capsular space and then into the proximal convoluted tubule.

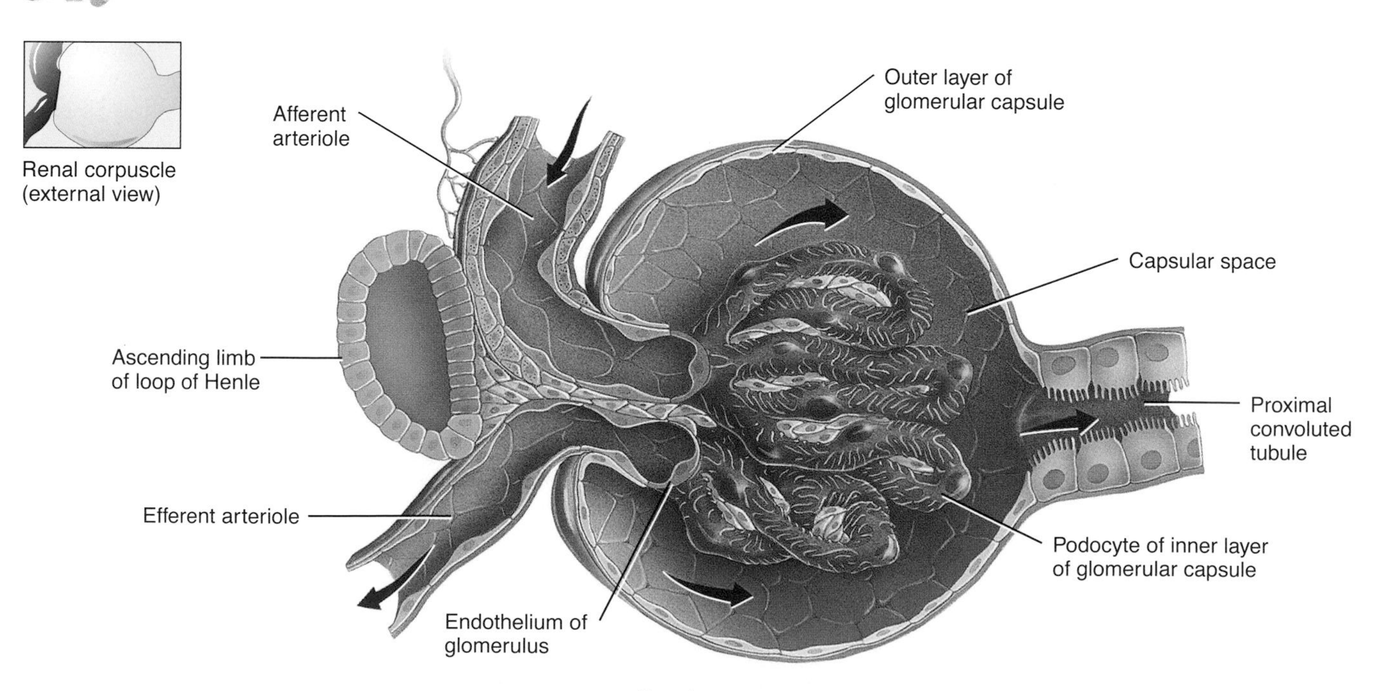

Which cells make up the filtration membrane in the renal corpuscle?

Glomerular Filtration Rate

The amount of filtrate that forms in both kidneys every minute is called the ***glomerular filtration rate (GFR)***. In adults, the GFR is about 105 mL/min in females and 125 mL/min in males. It is very important for the kidneys to maintain a constant GFR. If the GFR is too high, needed substances pass so quickly through the renal tubules that they are unable to be reabsorbed and pass out of the body as part of urine. On the other hand, if the GFR is too low, nearly all the filtrate is reabsorbed, and waste products are not adequately excreted.

Atrial natriuretic peptide (ANP) is a hormone that promotes loss of sodium ions and water in the urine in part because it increases glomerular filtration rate. Cells in the atria of the heart secrete more ANP if the heart is stretched more, as occurs when blood volume increases. ANP then acts on the kidneys to increase loss of sodium ions and water in urine, which reduces the blood volume back to normal.

Like most blood vessels of the body, those of the kidneys are supplied by sympathetic neurons of the autonomic nervous system. When these neurons are active, they cause vasoconstriction. At rest, sympathetic stimulation is low and the afferent and efferent arterioles are relatively dilated. With greater sympathetic stimulation, as occurs during exercise or hemorrhage, the afferent arterioles are constricted more than the efferent arterioles. As a result, blood flow into glomerular capillaries is greatly decreased, net filtration pressure decreases, and GFR drops. These changes reduce urine output, which helps conserve blood volume and permits greater blood flow to other body tissues.

Tubular Reabsorption and Secretion

Tubular reabsorption—returning most of the filtered water and many of the filtered solutes to the blood—is the second basic function of the nephrons and collecting ducts. The filtered fluid becomes *tubular fluid* once it enters the proximal convoluted tubule. Due to reabsorption and secretion, the composition of tubular fluid changes as it flows along the nephron tubule and through a collecting duct. Typically, about 99% of the filtered water is reabsorbed. Only 1% of the water in glomerular filtrate actually leaves the body in *urine*, the fluid that drains into the renal pelvis.

Epithelial cells all along the renal tubules and collecting ducts carry out tubular reabsorption (Figure 21.7). Some solutes are passively reabsorbed by diffusion; others are reabsorbed by active transport. Proximal convoluted tubule cells make the largest contribution, reabsorbing 65% of the filtered water, 100% of the filtered glucose and amino acids, and large quantities of various ions such as sodium (Na^+), potassium (K^+), chloride (Cl^-), bicarbonate (HCO_3^-), calcium (Ca^{2+}), and magnesium (Mg^{2+}). Reabsorption of solutes also promotes reabsorption of water in the following way. The movement of solutes into peritubular capillaries decreases the solute concentration of the tubular fluid but increases the solute concentration in the peritubular capillaries. As a result, water moves by osmosis into peritubular capillaries. Cells located distal to the proximal convoluted tubule carry out fine-tune reabsorption to maintain homeostatic balances of water and selected ions. To appreciate the huge extent of tubular reabsorption, look back at Table 21.1 on page 529 and compare the amounts of substances that are filtered, reabsorbed, and excreted in urine.

When the blood concentration of glucose rises above normal, transporters in the proximal convoluted tubules may not be able to work fast enough to reabsorb all of the filtered glucose. As a result, some glucose remains in the urine, a condition called **glucosuria** (gloo′-kō-SOO-rē-a). The most common cause of glucosuria is diabetes mellitus, in which the blood glucose level may rise far above normal because insulin activity is deficient. Because "water follows solutes" as tubular reabsorption takes place, any condition that reduces reabsorption of filtered solutes also increases the amount of water lost in urine. **Polyuria** (pol′-ē-Ū-rē-a; *poly-* = too much), excessive excretion of urine, usually accompanies glucosuria and is a common symptom of diabetes.

The third function of the nephrons and collecting ducts is ***tubular secretion***, the transfer of materials from the blood through tubule cells and into tubular fluid. As is the case for tubular reabsorption, tubular secretion takes place all along the renal tubules and collecting ducts and occurs via both passive diffusion and active transport processes. Secreted substances include hydrogen ions (H^+), K^+, ammonia (NH_3), urea, creatinine (a waste from creatine in muscle cells), and certain drugs such as penicillin. Tubular secretion helps eliminate these substances from the body.

Ammonia is a poisonous waste product that is produced when amino groups are removed from amino acids. Liver cells convert most ammonia to urea, which is a less-toxic compound. Although tiny amounts of urea and ammonia are present in sweat, most excretion of these nitrogen-containing waste products occurs in the urine. Urea and ammonia in blood are both filtered at the glomerulus and secreted by proximal convoluted tubule cells into the tubular fluid. Secretion of excess K^+ for elimination in the urine also is very important. Tubule cell secretion of K^+ varies with dietary intake of potassium to maintain a stable level of K^+ in body fluids.

Tubular secretion also helps control blood pH. A normal blood pH of 7.35 to 7.45 is maintained, even though the typical high-protein diet in North America provides more acid-producing foods than alkali-producing foods. To eliminate acids, the cells of the renal tubules secrete H^+ into the tubular fluid, which helps maintain the pH of blood in the normal range. Due to H^+ secretion, urine is typically acidic (has a pH below 7).

Figure 21.7 Filtration, reabsorption, and secretion in the nephrons and collecting ducts. Percentages refer to the amounts initially filtered at the glomerulus.

Filtration occurs in the renal corpuscle; reabsorption occurs all along the renal tubule and collecting ducts.

RENAL CORPUSCLE

Glomerular filtration rate: 105–125 mL/min

Filtered substances: water and all solutes present in blood (except proteins) including ions, glucose, amino acids, creatinine

DISTAL CONVOLUTED TUBULE

Reabsorption (into blood) of:

Water	10–15% (osmosis)
Na^+	5%
Cl^-	5%
Ca^{2+}	variable

PROXIMAL CONVOLUTED TUBULE

Reabsorption (into blood) of filtered:

Water	65% (osmosis)
Glucose	100%
Amino acids	100%
Na^+	65%
K^+	65%
Cl^-	50%
HCO_3^-	80–90%
Ca^{2+}, Mg^{2+}	variable
Urea	50%

Secretion (into urine) of:

H^+	variable
Ammonia	variable
Urea	variable
Creatinine	small amount

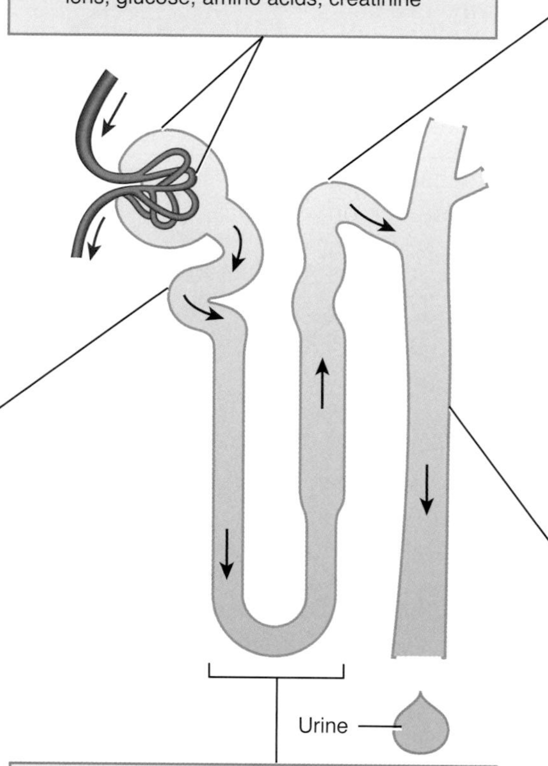

LAST PART OF DISTAL TUBULE AND COLLECTING DUCT

Reabsorption (into blood) of:

Water	5–9% (insertion of water channels stimulated by ADH)
Na^+	1–4%
HCO_3^-	variable amount
Urea	variable

Secretion (into urine) of:

K^+	variable amount to adjust for dietary intake (leakage channels)
H^+	variable amounts to maintain acid–base homeostasis (H^+ pumps)

Tubular fluid leaving the collecting duct is dilute when ADH level is low and concentrated when ADH level is high.

LOOP OF HENLE

Reabsorption (into blood) of:

Water	15% (osmosis in descending limb)
Na^+	20–30% (ascending limb)
K^+	20–30% (ascending limb)
Cl^-	35% (ascending limb)
HCO_3^-	10–20%
Ca^{2+}, Mg^{2+}	variable

Secretion (into urine) of:

Urea	variable

In which segments of the nephrons and collecting ducts does secretion occur?

Hormonal Regulation of Nephron Functions

Hormones affect the extent of Na^+, Cl^-, and water reabsorption as well as K^+ secretion by the renal tubules. The most important hormonal regulators of ion reabsorption and secretion are ***angiotensin II*** and ***aldosterone***. In the proximal convoluted tubules, angiotensin II enhances reabsorption of Na^+ and Cl^-. Angiotensin II also stimulates the adrenal cortex to release aldosterone, a hormone that in turn stimulates the collecting ducts to reabsorb more Na^+ and Cl^- and secrete more K^+. When more Na^+ and Cl^- are reabsorbed, then more water is also reabsorbed by osmosis. Aldosterone-stimulated secretion of K^+ is the major regulator of blood K^+ level. An elevated level of K^+ in plasma causes serious disturbances in cardiac rhythm or even cardiac arrest. Besides increasing glomerular filtration rate, the hormone ***atrial natriuretic peptide (ANP)*** plays a minor role in inhibiting the reabsorption of Na^+ (and Cl^- and water) by the renal tubules. As GFR increases and Na^+, Cl^-, and water reabsorption decrease, more water and salt are lost in the urine. The final effect is to lower blood volume.

The major hormone that regulates water reabsorption is ***antidiuretic hormone (ADH)***, which operates via negative feedback (Figure 21.8). When the concentration of water in the blood decreases by as little as 1%, osmoreceptors in the hypothalamus stimulate release of ADH from the posterior pituitary. A second powerful stimulus for ADH secretion is a decrease in blood volume, as occurs in hemorrhaging or severe dehydration. ADH acts on tubule cells in the last part of the distal convoluted tubules and throughout the collecting ducts. In the absence of ADH, these parts of the renal tubule have a very low permeability to water. ADH increases the water permeability of these tubule cells by causing insertion of proteins that function as water channels into their plasma membranes. When the water permeability of the tubule cells increases, water molecules move from the tubular fluid into the cells and then into the blood. The kidneys can produce as little as 400–500 mL of very concentrated urine each day when ADH concentration is maximal, for instance during severe dehydration. When ADH level declines, the water channels are removed from the membranes. The kidneys produce a large volume of dilute urine when ADH level is low.

Diuretics are substances that slow reabsorption of water by the kidneys and thereby cause *diuresis*, an elevated urine flow rate. Naturally occurring diuretics include *caffeine* in coffee, tea, and cola sodas, which inhibits Na^+ reabsorption, and *alcohol* in beer, wine, and mixed drinks, which inhibits secretion of ADH. In a condition known as *diabetes insipidus*, ADH secretion is inadequate or the ADH receptors are faulty, and a person may excrete up to 20 liters of very dilute urine daily.

Figure 21.8 Negative feedback regulation of water reabsorption by ADH.

When ADH level is high, the kidneys reabsorb more water.

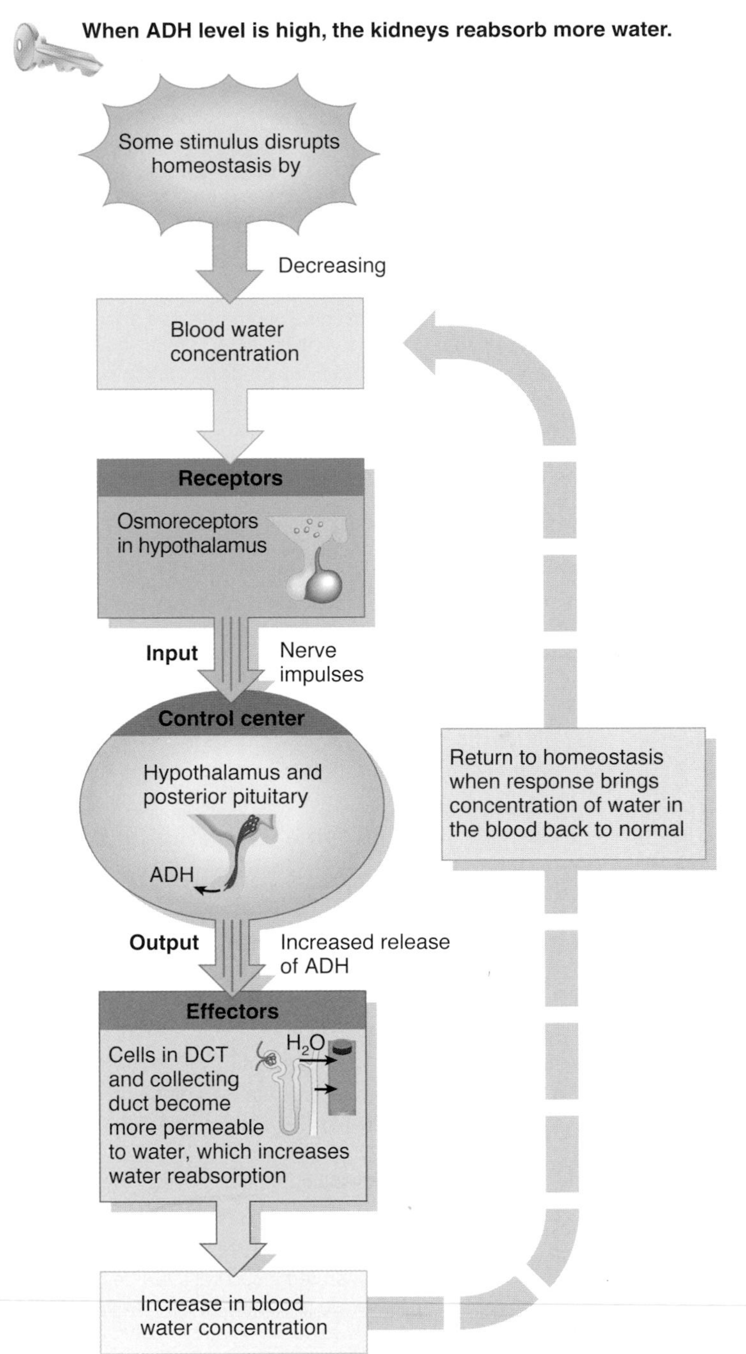

Would blood ADH level be higher or lower than normal in a person who has just completed a 5-km run without drinking any water?

Components of Urine

An analysis of the volume and physical, chemical, and microscopic properties of urine, called a ***urinalysis***, tells us much about the state of the body. Table 21.2 summarizes the principal physical characteristics of urine.

Table 21.2 Physical Characteristics of Normal Urine

Characteristic	Description
Volume	One to two liters (about 1 to 2 quarts) in 24 hours but varies considerably.
Color	Yellow or amber, but varies with urine concentration and diet. Color is due to urochrome (pigment produced from breakdown of bile) and urobilin (from breakdown of hemoglobin). Concentrated urine is darker in color. Diet, medications, and certain diseases affect color.
Turbidity	Transparent when freshly voided, but becomes turbid (cloudy) after a while.
Odor	Mildly aromatic but becomes ammonia-like after a time. Some people inherit the ability to form methylmercaptan from digested asparagus, which gives urine a characteristic odor.
pH	Ranges between 4.6 and 8.0; average 6.0; varies considerably with diet. High-protein diets increase acidity; vegetarian diets increase alkalinity.
Specific gravity	Specific gravity (density) is the ratio of the weight of a volume of a substance to the weight of an equal volume of distilled water. Urine specific gravity ranges from 1.001 to 1.035. The higher the concentration of solutes, the higher the specific gravity.

The volume of urine eliminated per day in a normal adult is 1 to 2 liters (about 1 to 2 quarts). Water accounts for about 95% of the total volume of urine. In addition to urea, creatinine, potassium, and ammonia, typical solutes normally present in urine include uric acid as well as sodium, chloride, magnesium, sulfate, phosphate, and calcium ions.

If disease alters body metabolism or kidney function, traces of substances not normally present may appear in the urine, or normal constituents may appear in abnormal amounts. Table 21.3 lists several abnormal constituents in urine that may be detected as part of a urinalysis.

■ CHECKPOINT

4. How does blood pressure promote filtration of blood in the kidneys?
5. What solutes are reabsorbed and secreted as fluid moves along the renal tubules?
6. How do angiotensin II, aldosterone, and antidiuretic hormone regulate tubular reabsorption and secretion?
7. What are the characteristics of normal urine?

Table 21.3 Summary of Abnormal Constituents in Urine

Abnormal Constituent	Comments
Albumin	A normal constituent of blood plasma that usually appears in only very small amounts in urine because it is too large to be filtered. The presence of excessive albumin in the urine, *albuminuria* (al′-bū-mi-NOO-rē-a), indicates an increase in the permeability of filtering membranes due to injury or disease, increased blood pressure, or damage to kidney cells.
Glucose	*Glucosuria*, the presence of glucose in the urine, usually indicates diabetes mellitus.
Red blood cells (erythrocytes)	*Hematuria* (hēm-a-TOO-rē-a), the presence of hemoglobin from ruptured red blood cells in the urine, can occur with acute inflammation of the urinary organs as a result of disease or irritation from kidney stones, tumors, trauma, and kidney disease.
White blood cells (leukocytes)	The presence of white blood cells and other components of pus in the urine, referred to as *pyuria* (pī-Ū-rē-a), indicates infection in the kidneys or other urinary organs.
Ketone bodies	High levels of ketone bodies in the urine, called *ketonuria* (kē-tō-NOO-rē-a), may indicate diabetes mellitus, anorexia, starvation, or too little carbohydrate in the diet.
Bilirubin	When red blood cells are destroyed by macrophages, the globin portion of hemoglobin is split off and the heme is converted to biliverdin. Most of the biliverdin is converted to bilirubin. An above-normal level of bilirubin in urine is called *bilirubinuria* (bil′-ē-roo-bi-NOO-rē-a).
Urobilinogen	The presence of urobilinogen (breakdown product of hemoglobin) in urine is called *urobilinogenuria* (ū′-rō-bi-lin′-ō-jē-NOO-rē-a). Trace amounts are normal, but elevated urobilinogen may be due to hemolytic or pernicious anemia, infectious hepatitis, obstruction of bile ducts, jaundice, cirrhosis, congestive heart failure, or infectious mononucleosis.
Casts	*Casts* are tiny masses of material that have hardened and assumed the shape of the lumen of a tubule in which they formed. They are flushed out of the tubule when glomerular filtrate builds up behind them. Casts are named after the cells or substances that compose them or based on their appearance. For example, there are white blood cell casts, red blood cell casts, and epithelial cell casts (cells from renal tubules).
Microbes	The number and type of bacteria vary with specific infections in the urinary tract. One of the most common is *E. coli.* The most common fungus to appear in urine is *Candida albicans,* a cause of vaginitis. The most frequent protozoan seen is *Trichomonas vaginalis*, a cause of vaginitis in females and urethritis in males.

FOCUS ON WELLNESS

Infection Prevention for Recurrent UTIs

Urinary tract infections (UTIs) are the most common bacterial infections and the second most common illness (after colds) among women. About 10–15% of women develop UTIs several times a month. Men get UTIs, too, but much less frequently. The female's shorter urethra allows bacteria to enter the urinary bladder more easily. In addition, the urethral and anal openings are closer in females. Most first-time UTIs are caused by *Escherichia coli (E. coli)* bacteria that have migrated to the urethra from the anal area. *E. coli* bacteria are necessary for proper digestion and are welcome in the intestinal tract, but they cause much pain and suffering if they infect the urinary system.

Infection Prevention

Personal hygiene is the first line of prevention. Care must be taken to avoid transporting bacteria from the anal area to the urethra. Girls should be taught to wipe from front to back and to wash hands thoroughly after using the toilet. When bathing, women and girls should wash from front to back as well.

Menstrual blood provides an excellent growth medium for bacteria. Sanitary napkins and tampons should be changed often. Some women find that switching from tampons to napkins or from napkins to tampons reduces the frequency of UTIs. Deodorant tampons and napkins and superabsorbent tampons can increase irritation.

People who are prone to UTIs should drink at least 2 to 2.5 liters of fluid daily. Drinking cranberry and blueberry juice may help to decrease bacterial growth in the urinary bladder. Voiding frequently, every 2 to 3 hours, helps prevent recurrent UTIs because it expels bacteria and eliminates the urine needed for their growth.

Partners in Health

Sexual intercourse is frequently associated with the onset of UTIs in women. Women who find that sex brings on UTIs learn to develop and teach their partners stringent personal hygiene. Women should drink plenty of water before and after sex and urinate as soon afterward as possible. This flushes out bacteria that may have entered the urethra.

At times a woman's partner may be the source of bacterial transmission. When UTIs continue to recur, he should be tested for asymptomatic urethritis, which is the term for any bacterial infection of the urethra other than gonorrhea. Sometimes treating the partner with antibiotics cures both parties.

► THINK IT OVER . . .

► ***One of the basic tenets of the wellness philosophy is that the health-care system works best when patients work as partners with their providers to understand, treat, and prevent illness. Explain why treatment of recurrent UTIs is a good illustration of this belief.***

TRANSPORTATION, STORAGE, AND ELIMINATION OF URINE

OBJECTIVE • Describe the structure and functions of the ureters, urinary bladder, and urethra.

As you learned earlier in the chapter, urine produced by the nephrons drains into the minor calyces, which join to become major calyces that unite to form the renal pelvis (see Figure 21.2). From the renal pelvis, urine drains first into the ureters and then into the urinary bladder; urine is then discharged from the body through the urethra (see Figure 21.1).

Ureters

Each of the two ***ureters*** (Ū-re-ters or ū-RĒ-ters) transports urine from the renal pelvis of one of the kidneys to the urinary bladder (see Figure 21.1). The ureters pass under the urinary bladder for several centimeters, causing the bladder to compress the ureters and thus prevent backflow of urine when pressure builds up in the bladder during urination. If this physiological valve is not operating, cystitis (urinary bladder inflammation) may develop into a kidney infection.

The wall of the ureter consists of three layers. The inner layer is the mucosa, containing *transitional epithelium* (see Table 4.1I on page 80) with an underlying layer of areolar connective tissue. Transitional epithelium is able to stretch—a marked advantage for any organ that must accommodate a variable volume of fluid. Mucus secreted by the goblet cells of the mucosa prevents the cells from coming in contact with urine, the solute concentration and pH of which may differ drastically from the cytosol of cells that form the wall of the ureters. The middle layer consists of smooth muscle. Urine is transported from the renal pelvis to the urinary bladder primarily by peristaltic contractions of this smooth muscle, but the fluid pressure of the urine and gravity may also contribute. The outer layer consists of areolar connective tissue containing blood vessels, lymphatic vessels, and nerves.

Urinary Bladder

The ***urinary bladder*** is a hollow muscular organ situated in the pelvic cavity behind the pubic symphysis (Figure 21.9). In males, it is directly in front of the rectum (see Figure 23.1 on page 557). In females, it is in front of the vagina and below the uterus. Folds of the peritoneum hold the urinary bladder in position. The shape of the urinary bladder depends on how much urine it contains. When empty, it looks like a deflated balloon. It becomes spherical when slightly stretched and, as urine volume increases, becomes pear-shaped and rises into the abdominal cavity. Urinary bladder capacity averages 700–800 mL. It is smaller in females because the uterus occupies the space just superior to the urinary bladder. Toward the base of the bladder, the ureters drain into the urinary bladder via the *ureteral openings*. Like the ureters, the mucosa of the urinary bladder contains transitional epithelium. The muscular layer of the urinary bladder wall consists of three layers of smooth muscle called the ***detrusor muscle*** (de-TROO-ser = to push down). The peritoneum, which covers the superior surface of the urinary bladder, forms a serous outer coat; the rest of the urinary bladder has a fibrous outer covering.

Urethra

The ***urethra*** (ū-RĒ-thra), the terminal portion of the urinary system, is a small tube leading from the floor of the urinary bladder to the exterior of the body (Figure 21.9). In females,

Figure 21.9 Ureters, urinary bladder, and urethra (female).

Urine is stored in the urinary bladder until it is expelled by micturition.

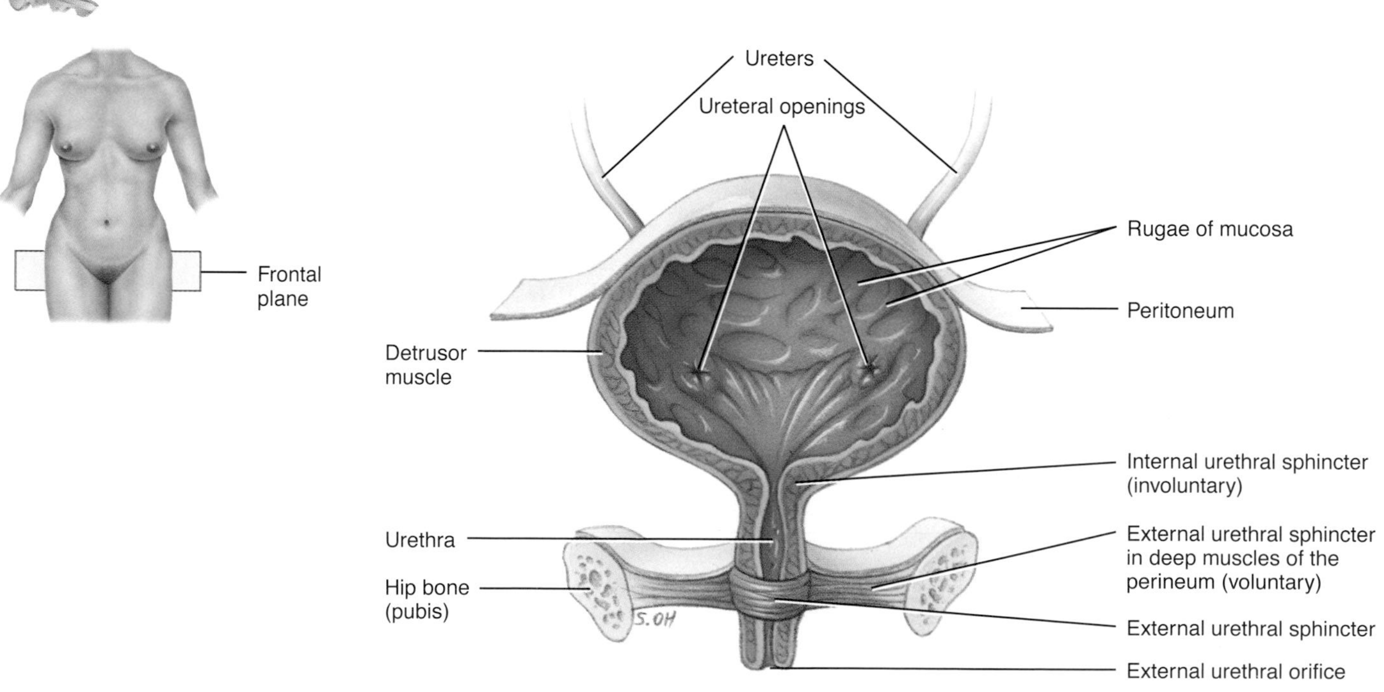

What is a lack of voluntary control over micturition called?

it lies directly behind the pubic symphysis and is embedded in the front wall of the vagina. The opening of the urethra to the exterior, the ***external urethral orifice***, lies between the clitoris and vaginal opening. In males, the urethra passes vertically through the prostate, the deep perineal muscles, and finally the penis (see Figures 23.1 and 23.6 on pages 557 and 564).

Around the opening to the urethra is an ***internal urethral sphincter*** composed of smooth muscle. The opening and closing of the internal urethral sphincter is involuntary. Below the internal sphincter is the ***external urethral sphincter***, which is composed of skeletal muscle and is under voluntary control. In both males and females, the urethra is the passageway for discharging urine from the body. The male urethra also serves as the duct through which semen is ejaculated.

Micturition

The urinary bladder stores urine prior to its elimination and then expels urine into the urethra by an act called ***micturition*** (mik′-too-RI-shun = to urinate), commonly known as *urination*. Micturition requires a combination of involuntary and voluntary muscle contractions. When the volume of urine in the urinary bladder exceeds 200 to 400 mL, pressure within the bladder increases considerably, and stretch receptors in its wall transmit nerve impulses into the spinal cord. These impulses propagate to the lower part of the spinal cord and trigger a reflex called the ***micturition reflex***. In this reflex, parasympathetic impulses from the spinal cord cause *contraction* of the detrusor muscle and *relaxation* of the internal urethral sphincter muscle. Simultaneously, the spinal cord inhibits somatic motor neurons, causing relaxation of skeletal muscle in the external urethral sphincter. Upon contraction of the urinary bladder wall and relaxation of the sphincters, urination takes place. Urinary bladder filling causes a sensation of fullness that initiates a conscious desire to urinate before the micturition reflex actually occurs. Although emptying of the urinary bladder is a reflex, in early childhood we learn to initiate it and stop it voluntarily. Through learned control of the external urethral sphincter muscle and certain muscles of the pelvic floor, the cerebral cortex can initiate micturition or delay it for a limited period of time.

An inability to prevent micturition is termed **incontinence**. Under about two years of age, incontinence is normal because neurons to the external urethral sphincter muscle are not completely developed. Infants void whenever the urinary bladder is sufficiently distended to trigger the reflex. In *stress incontinence*, physical stresses that increase abdominal pressure, such as coughing, sneezing, laughing, exercising, pregnancy, or simply walking, cause leakage of urine from the urinary bladder. Smokers have twice the risk of developing incontinence as nonsmokers.

■ **CHECKPOINT**

8. What forces help propel urine from the renal pelvis to the urinary bladder?
9. What is micturition? How does the micturition reflex occur?
10. How does the location of the urethra compare in males and females?

AGING AND THE URINARY SYSTEM

OBJECTIVE • **Describe the effects of aging on the urinary system.**

With aging, the kidneys shrink in size, have a decreased blood flow, and filter less blood. The mass of the two kidneys decreases from an average of 260 g in 20-year-olds to less than 200 g by age 80. Likewise, renal blood flow and filtration rate decline by 50% between ages 40 and 70. Kidney diseases that become more common with age include acute and chronic kidney inflammations and renal calculi (kidney stones). Because the sensation of thirst diminishes with age, older individuals also are susceptible to dehydration. Urinary tract infections are more common among the elderly, as are polyuria, nocturia (excessive urination at night), increased frequency of urination, dysuria (painful urination), urinary retention or incontinence, and hematuria (blood in the urine).

■ **CHECKPOINT**

11. Why are older individuals more susceptible to dehydration?

• • •

To appreciate the many ways that the urinary system contributes to homeostasis of other body systems, examine Focus on Homeostasis: The Urinary System on page 538. Next, in Chapter 22, we will see how the kidneys and lungs contribute to maintenance of homeostasis of body fluid volume, ion levels in body fluids, and acid–base balance.

FOCUS ON HOMEOSTASIS

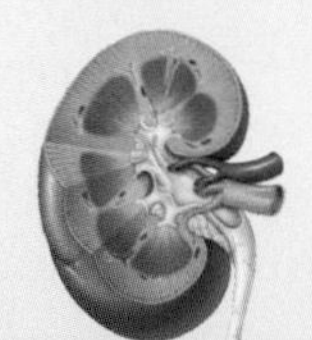

THE URINARY SYSTEM

BODY SYSTEM		CONTRIBUTION OF THE URINARY SYSTEM
For all body systems		The kidneys regulate the volume, composition, and pH of body fluids by removing wastes and excess substances from blood and excreting them in the urine. The ureters transport urine from the kidneys to the urinary bladder, which stores urine until it is eliminated through the urethra.
Integumentary system	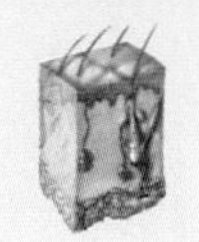	The kidneys and skin both contribute to the synthesis of calcitriol, the active form of vitamin D.
Skeletal system	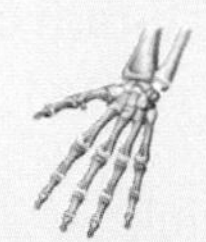	The kidneys help adjust the levels of blood calcium and phosphates needed for building bone extracellular matrix.
Muscular system		The kidneys help adjust the level of blood calcium, needed for contraction of muscles.
Nervous system	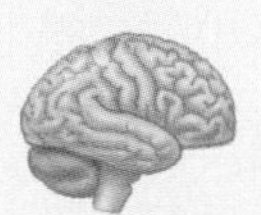	The kidneys perform gluconeogenesis (synthesis of glucose from certain amino acids and lactic acid), thereby providing glucose for ATP production in neurons, especially during fasting or starvation.
Endocrine system	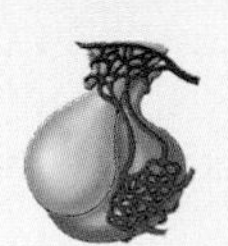	The kidneys participate in the synthesis of calcitriol, the active form of vitamin D, and release erythropoietin, the hormone that stimulates the production of red blood cells.
Cardiovascular system	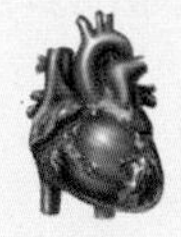	By increasing or decreasing reabsorption of water filtered from the blood, the kidneys help adjust the blood volume and blood pressure. Renin released by the kidneys raises blood pressure. Some bilirubin (from hemoglobin breakdown) is converted to a yellow pigment (urobilin), which is excreted in the urine.
Lymphatic system and immunity	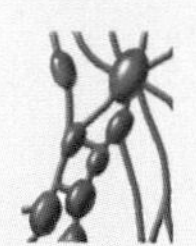	By increasing or decreasing the reabsorption of water filtered from blood, the kidneys help adjust the volume of interstitial fluid and lymph. Urination flushes microbes out of the urethra.
Respiratory system	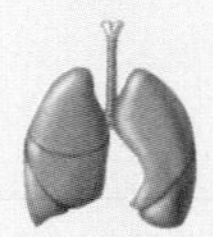	The kidneys and lungs cooperate in adjusting the pH of body fluids.
Digestive system	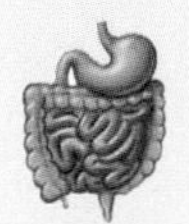	The kidneys help synthesize calcitriol, the active form of vitamin D, which is needed for absorption of dietary calcium.
Reproductive systems	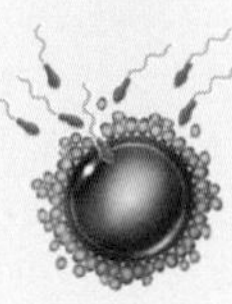	In males, the portion of the urethra that extends through the prostate and penis is a passageway for semen as well as urine.

COMMON DISORDERS

Glomerulonephritis

Glomerulonephritis is an inflammation of the glomeruli of the kidney. One of the most common causes is an allergic reaction to the toxins produced by streptococcal bacteria that have recently infected another part of the body, especially the throat. Because inflamed and swollen glomeruli allow blood cells and plasma proteins to enter the filtrate, the urine contains many red blood cells (hematuria) and large amounts of protein.

Renal Failure

Renal failure is a decrease or cessation of glomerular filtration. In ***acute renal failure (ARF)*** the kidneys abruptly stop working entirely (or almost entirely). The main feature of ARF is the suppression of urine flow, leading to oliguria or anuria. Causes include low blood volume (for example, due to hemorrhage); decreased cardiac output; damaged renal tubules; kidney stones; or reactions to the dyes used to visualize blood vessels in angiograms, nonsteroidal anti-inflammatory drugs, and some antibiotic drugs.

Chronic renal failure (CRF) refers to a progressive and usually irreversible decline in glomerular filtration rate (GFR). CRF may result from chronic glomerulonephritis, pyelonephritis, polycystic kidney disease, or traumatic loss of kidney tissue. The final stage of CRF is called ***end-stage renal failure*** and occurs when about 90% of the nephrons have been lost. At this stage, GFR diminishes to 10–15% of normal, oliguria is present, and blood levels of nitrogen-containing wastes and creatinine are high. People with end-stage renal failure require dialysis therapy and are possible candidates for a kidney transplant operation.

Polycystic Kidney Disease

Polycystic kidney disease (PKD) is one of the most common inherited disorders. In PKD, the kidney tubules become riddled with hundreds or thousands of cysts (fluid-filled cavities). In addition, inappropriate apoptosis (programmed cell death) of cells in noncystic tubules leads to progressive impairment of renal function and eventually to end-stage renal failure.

People with PKD also may have cysts and apoptosis in the liver, pancreas, spleen, and gonads; increased risk of cerebral aneurysms; heart valve defects; and diverticuli in the colon. Typically, symptoms are not noticed until adulthood, when patients may have back pain, urinary tract infections, blood in the urine, hypertension, and large abdominal masses. Using drugs to restore normal blood pressure, restricting protein and salt in the diet, and controlling urinary tract infections may slow progression to renal failure.

MEDICAL TERMINOLOGY AND CONDITIONS

Dialysis (dī-AL-i-sis; *dialyo* = to separate) is the separation of large solutes from smaller ones by diffusion through a selectively permeable membrane. It is used to cleanse a person's blood artificially when the kidneys are so impaired by disease or injury that they are unable to function adequately. One method of dialysis is ***hemodialysis*** (hē-mō-dī-AL-i-sis; *hemo-* = blood) which filters the patient's blood directly by removing wastes and excess electrolytes and fluid and then returning the cleansed blood to the patient. Blood removed from the body is delivered to a *hemodialyzer* (artificial kidney). Inside the hemodialyzer, blood flows through a *dialysis membrane*, which contains pores large enough to permit the diffusion of small solutes. A special solution, called the *dialysate* (dī-AL-i-sāt) is pumped into the hemodialyzer so that it surrounds the dialysis membrane. The dialysate is specially formulated to maintain diffusion gradients that remove wastes from the blood (for example, urea, creatinine, uric acid, excess phosphate, potassium, and sulfate ions) and add needed substances (for example, glucose and bicarbonate ions) to it. As a rule, most people on hemodialysis require about 6–12 hours a week, typically divided into three sessions.

Dysuria (dis-Ū-rē-a; *dys* = painful; *uria* = urine) Painful urination.

Enuresis (en′-ū-RĒ-sis; = to void urine) Involuntary voiding of urine after the age at which voluntary control has typically been attained.

Intravenous pyelogram (in′-tra-VĒ-nus PĪ-el-ō-gram′; *intra-* = within; *veno-* = vein; *pyelo-* = pelvis of kidney; *-gram* = record) or ***IVP*** Radiograph (x-ray film) of the kidneys after venous injection of a dye.

Kidney stones Insoluble stones occasionally formed from solidification of the crystals of urine salts. Can be caused by ingestion of excessive mineral salts, insufficient water intake, abnormally alkaline or acidic urine, or overactive parathyroid glands. Usually form in the renal pelvis. Often cause intense pain. Also termed ***renal calculi***.

Nocturnal enuresis (nok-TUR-nal en′-ū-RĒ-sis) Discharge of urine during sleep, resulting in bed-wetting; occurs in about 15% of 5-year-old children and generally resolves spontaneously, afflicting only about 1% of adults. Possible causes include smaller-than-normal urinary bladder capacity, failure to awaken in response to a full urinary bladder, and above-normal production of urine at night. Also termed ***nocturia***.

Urinary retention A failure to completely or normally void urine; may be due to an obstruction in the urethra or neck of the urinary bladder, to nervous contraction of the urethra, or to lack of urge to urinate. In men, an enlarged prostate may constrict the urethra and cause urinary retention. If urinary retention is prolonged, a catheter (slender rubber drainage tube) must be placed into the urethra to drain the urine.

STUDY OUTLINE

Overview of the Urinary System (p. 524)

1. The organs of the urinary system include the kidneys, ureters, urinary bladder, and urethra.
2. After the kidneys filter blood and return most of the water and many solutes to the blood, the remaining water and solutes constitute urine.
3. The kidneys regulate blood ionic composition, blood volume, blood pressure, and blood pH.
4. The kidneys also release calcitriol and erythropoietin and excrete wastes and foreign substances.

Structure of the Kidneys (p. 525)

1. The kidneys lie on either side of the vertebral column between the peritoneum and the back wall of the abdominal cavity.
2. Each kidney is enclosed in a renal capsule, which is surrounded by adipose tissue.
3. Internally, the kidneys consist of a renal cortex, renal medulla, renal pyramids, renal columns, calyces, and a renal pelvis.
4. Blood enters the kidney through the renal artery and leaves through the renal vein.
5. The nephron is the functional unit of the kidney. A nephron consists of a renal corpuscle (glomerulus and glomerular or Bowman's capsule) and a renal tubule (proximal convoluted tubule, descending limb of the loop of Henle, ascending limb of the loop of Henle, and distal convoluted tubule). The distal convoluted tubules of several nephrons empty into a common collecting duct.

Functions of the Nephron (p. 528)

1. Nephrons perform three basic tasks: glomerular filtration, tubular reabsorption, and tubular secretion.
2. Together, the podocytes and glomerular endothelium form a leaky filtration membrane that permits the passage of water and solutes from the blood into the capsular space. Blood cells and most plasma proteins remain in the blood because they are too large to pass through the filtration membrane. The pressure that causes filtration is the blood pressure in the glomerular capillaries.
3. Table 21.1 on page 529 describes the substances that are filtered, reabsorbed, and excreted in urine on a daily basis.
4. The amount of filtrate that forms in both kidneys every minute is the glomerular filtration rate (GFR). Atrial natriuretic peptide (ANP) increases GFR; sympathetic stimulation decreases GFR.
5. Epithelial cells all along the renal tubules and collecting ducts carry out tubular reabsorption and tubular secretion. Tubular reabsorption retains substances needed by the body, including water, glucose, amino acids, and ions such as sodium (Na^+), potassium (K^+), chloride (Cl^-), bicarbonate (HCO_3^-), calcium (Ca^{2+}), and magnesium (Mg^{2+}).
6. Angiotensin II enhances reabsorption of Na^+ and Cl^-. Angiotensin II also stimulates the adrenal cortex to release aldosterone, which stimulates the collecting ducts to reabsorb more Na^+ and Cl^- and secrete more K^+. Atrial natriuretic peptide inhibits reabsorption of Na^+ (and Cl^- and water) by the renal tubules, which reduces blood volume.
7. Most water is reabsorbed by osmosis together with reabsorbed solutes, mainly in the proximal convoluted tubule. Reabsorption of the remaining water is regulated by antidiuretic hormone (ADH) in the last part of the distal convoluted tubule and collecting duct.
8. Tubular secretion discharges chemicals not needed by the body into the urine. Included are excess ions, nitrogenous wastes, hormones, and certain drugs. The kidneys help maintain blood pH by secreting H^+. Tubular secretion also helps maintain proper levels of K^+ in the blood.
9. Table 21.2 on page 534 describes the physical characteristics of urine that are evaluated in a urinalysis: color, odor, turbidity, pH, and specific gravity.
10. Chemically, normal urine contains about 95% water and 5% solutes.
11. Table 21.3 on page 534 lists the abnormal constituents that can be diagnosed through urinalysis, including albumin, glucose, red blood cells, white blood cells, ketone bodies, bilirubin, urobilinogen, casts, and microbes.

Transportation, Storage, and Elimination of Urine (p. 535)

1. The ureters transport urine from the renal pelves of the right and left kidneys to the urinary bladder and consist of a mucosa, muscularis, and adventitia.
2. The urinary bladder is posterior to the pubic symphysis. Its function is to store urine prior to micturition.
3. The mucosa of the urinary bladder contains stretchy transitional epithelium. The muscular layer of the wall consists of three layers of smooth muscle together referred to as the detrusor muscle.
4. The urethra is a tube leading from the floor of the urinary bladder to the exterior. Its function is to discharge urine from the body.
5. The micturition reflex discharges urine from the urinary bladder by means of parasympathetic impulses that cause contraction of the detrusor muscle and relaxation of the internal urethral sphincter muscle, and by inhibition of somatic motor neurons to the external urethral sphincter.

Aging and the Urinary System (p. 537)

1. With aging, the kidneys shrink in size, have lowered blood flow, and filter less blood.
2. Common problems related to aging include urinary tract infections, increased frequency of urination, urinary retention or incontinence, and renal calculi (kidney stones).

SELF-QUIZ

1. Which of the following is NOT a function of the urinary system?
   a. regulation of blood volume and composition
   b. stimulation of red blood cell production
   c. regulation of body temperature
   d. regulation of blood pressure
   e. regulation of blood pH

2. Which of the following structures is located in the renal cortex?
   a. the renal pyramid b. the renal column
   c. the major calyx d. the minor calyx
   e. the renal corpuscle

3. Which of the following increases water reabsorption in the distal convoluted tubules and collecting ducts?
   a. antidiuretic hormone (ADH) b. angiotensin II
   c. atrial natriuretic peptide (ANP) d. diuretics
   e. glucosuria

4. The major openings located in the base of the bladder are the
   a. renal artery, renal vein, urethra
   b. renal artery, renal vein, ureter
   c. ureter, urethra, collecting tubes
   d. urethra and two ureters
   e. external urethral sphincter and papillary ducts

5. Which statement does NOT describe the kidneys?
   a. They are protected by the 11th and 12th pairs of ribs.
   b. The average adult kidney is 11 cm (4 inches) long and 6 cm (2 inches) wide.
   c. The left kidney is lower than the right to accommodate the large size of the liver.
   d. Each kidney is surrounded by adipose and connective tissue
   e. The kidneys are surrounded by a renal capsule.

6. Place the following structures in the correct order for the flow of urine:
   1. renal tubules 2. minor calyx 3. renal pelvis
   4. major calyx 5. collecting ducts 6. ureters
   a. 1, 2, 4, 3, 6, 5 b. 5, 1, 4, 2, 3, 6 c. 5, 1, 2, 4, 3, 6
   d. 3, 5, 1, 2, 4, 6 e. 1, 5, 2, 4, 3, 6

7. The functional unit of the kidney where urine is produced is the
   a. nephron b. pyramid c. pelvis d. glomerulus
   e. calyx

8. What causes filtration of plasma across the filtration membrane?
   a. a full urinary bladder
   b. control by the nervous system
   c. water retention
   d. the pressure of the blood
   e. the pressure of urine in the glomerulus

9. Glomerular filtration rate (GFR) is the
   a. rate of urinary bladder filling
   b. amount of filtrate formed in both kidneys each minute
   c. amount of filtrate reabsorbed at the collecting ducts
   d. amount of blood delivered to the kidneys each minute
   e. amount of urine formed per hour

10. Which of the following is secreted into the urine from the blood?
    a. hydrogen ions (H^+) b. amino acids c. glucose
    d. water e. white blood cells

11. In the nephron, tubular fluid that is reabsorbed from the renal tubules enters the
    a. glomerulus b. peritubular capillaries
    c. efferent arteriole d. afferent arteriole e. renal artery

12. Place the following structures in the correct order as they are involved in the formation of urine in the nephrons.
    1. distal convoluted tubule
    2. renal corpuscle
    3. descending limb of loop of Henle
    4. proximal convoluted tubule
    5. collecting duct
    6. ascending limb of loop of Henle
    a. 4, 1, 6, 3, 2, 5 b. 2, 6, 3, 1, 5, 4 c. 2, 4, 3, 6, 5, 1
    d. 5, 1, 4, 3, 6, 2 e. 2, 4, 3, 6, 1, 5

13. Blood is carried out of the glomerulus by the
    a. renal artery
    b. afferent arteriole
    c. peritubular venule
    d. segmental artery
    e. efferent arteriole

14. Which of the following increases glomerular filtration rate (GFR)?
    a. atrial natriuretic peptide (ANP)
    b. constriction of the afferent arterioles
    c. increased sympathetic stimulation to the afferent arterioles
    d. ADH
    e. angiotensin II

15. Which of the following statements concerning tubular reabsorption is NOT true?
    a. Most reabsorption occurs in the proximal convoluted tubules.
    b. Tubular reabsorption is a selective process.
    c. Tubular reabsorption of excess potassium ions (K^+) maintains the correct blood level of K^+.
    d. The reabsorption of water in the proximal convoluted tubules depends upon sodium ion (Na^+) reabsorption.
    e. Tubular reabsorption allows the body to retain most filtered nutrients.

16. The micturition reflex
    a. is under the control of hormones
    b. is activated by low pressure in the urinary bladder
    c. depends upon contraction of the internal urethral sphincter muscle
    d. is an involuntary reflex over which normal adults have voluntary control
    e. is also known as incontinence
17. Which of the following is NOT normally present in glomerular filtrate?
    a. blood cells b. glucose
    c. nitrogenous wastes such as urea d. amino acids
    e. water
18. Urine formation requires which of the following?
    a. glomerular filtration and tubular secretion only
    b. glomerular filtration and tubular reabsorption only
    c. glomerular filtration, tubular reabsorption, and tubular secretion
    d. tubular reabsorption, tubular filtration, and tubular secretion
    e. tubular secretion and tubular reabsorption only
19. The transport of urine from the renal pelvis into the urinary bladder is the function of the
    a. urethra b. efferent arteriole c. afferent arteriole
    d. renal pyramids e. ureters
20. Incontinence is
    a. failure of the urinary bladder to expel urine
    b. a lack of voluntary control over the micturition reflex
    c. an inability of the kidneys to produce urine
    d. an ability to consciously control micturition
    e. a form of kidney dialysis

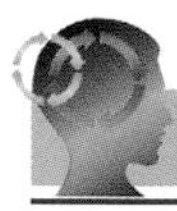

CRITICAL THINKING APPLICATIONS

1. Yesterday, you attended a large, outdoor party where beer was the only beverage available. You remember having to urinate many, many times yesterday, and today you're very thirsty. What hormone is affected by alcohol, and how does this affect your kidney function?
2. Sarah is an "above average" 1-year-old whose parents would like her to be the first toilet-trained child in preschool. However, in this case at least, Sarah is average for her age and remains incontinent. Should her parents be concerned by this lack of success?
3. Kayla is a healthy, VERY active 4-year old. She doesn't like to take the time to go to the bathroom because, as she says, "I might miss somethin'." Her mother is worried that Kayla's kidneys may stop working when her urinary bladder is full. Should her mother be concerned?
4. Assume that the length of a nephron's twisted and convoluted renal tubules is about the same as the width of the kidney. How many meters of tubules would one kidney contain?

ANSWERS TO FIGURE QUESTIONS

21.1 By forming urine, the kidneys do the major work of the urinary system.

21.2 The renal pyramids are located in the renal medulla.

21.3 About 1200 mL of blood enters the kidneys each minute.

21.4 The water molecule will travel from the proximal convoluted tubule → descending limb of the loop of Henle → ascending limb of the loop of Henle → distal convoluted tubule → collecting duct → minor calyx → major calyx → renal pelvis.

21.5 Secreted penicillin is being removed from the blood.

21.6 Podocytes and the glomerular endothelium make up the filtration membrane.

21.7 Secretion occurs in the proximal convoluted tubule, the loop of Henle, the last part of the distal convoluted tubule, and the collecting duct.

21.8 The blood level of ADH would be higher than normal after a 5-km run, due to loss of body water in sweat.

21.9 A lack of voluntary control over micturition is termed incontinence.

FLUID, ELECTROLYTE, AND ACID–BASE BALANCE

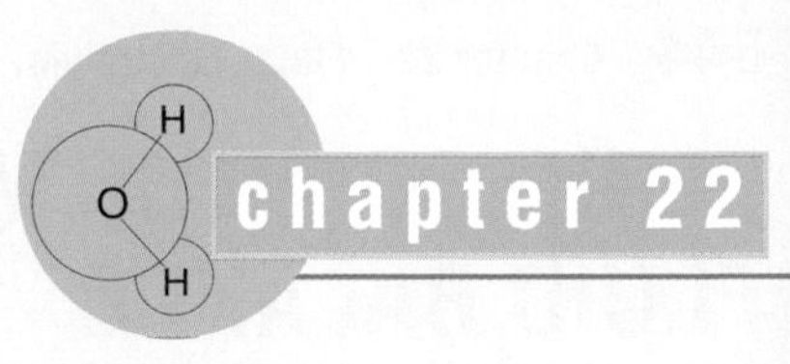

did you know? ***High blood pressure often responds to lifestyle treatments, such as increasing physical activity and developing better eating habits. The DASH (Dietary Approaches to Stopping Hypertension) diet is the most successful dietary treatment for hypertension. It may work partly by improving fluid balance through changes in the concentrations of many of the body's electrolytes. The DASH diet is low in sodium, and high in calcium, potassium, and magnesium. The DASH diet also encourages the consumption of low-fat dairy products, and plenty of fruits, vegetables, and whole grains. Fish, poultry, dried beans, and nuts are on the menu, but only small amounts of red meat, sweets, and salty foods are permitted.***

Focus on Wellness, page 551

www.wiley.com/college/apcentral

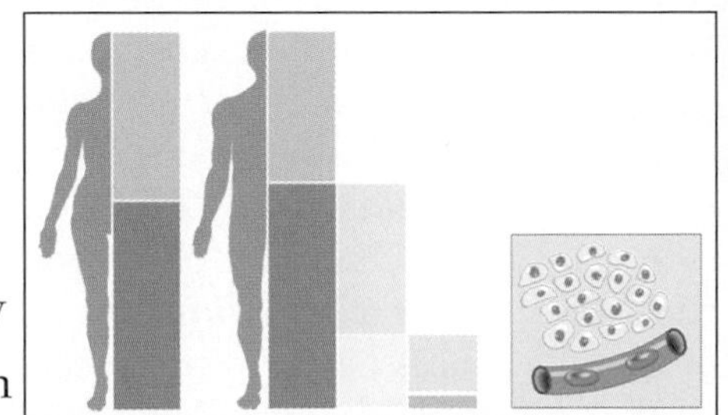

In Chapter 21 you learned how the kidneys form urine. One important function of the kidneys is to help maintain fluid balance in the body. The water and dissolved solutes in the body constitute the ***body fluids***. Regulatory mechanisms involving the kidneys and other organs normally maintain homeostasis of the body fluids. Malfunction in any or all of them may seriously endanger the functioning of organs throughout the body. In this chapter, we will explore the mechanisms that regulate the volume and distribution of body fluids and examine the factors that determine the concentrations of solutes and the pH of body fluids.

looking back to move ahead . . .

- **Acids, Bases, and pH (page 30)**
- **Intracellular and Extracellular Fluid (page 47)**
- **Osmosis (page 49)**
- **Antidiuretic Hormone (ADH) (page 322)**
- **Hormonal Regulation of Calcium in Body Fluids (pages 325–327)**
- **Renin–Angiotensin–Aldosterone Pathway (page 331)**
- **Control of Breathing Rate and Depth (page 461)**
- **Ions Reabsorbed and Secreted in the Kidneys (page 531)**
- **Negative Feedback Regulation of ADH Secretion (page 533)**

FLUID COMPARTMENTS AND FLUID BALANCE

OBJECTIVES • Compare the locations of intracellular fluid (ICF) and extracellular fluid (ECF), and describe the various fluid compartments of the body.

• Describe the sources of water and solute gain and loss, and explain how each is regulated.

In lean adults, body fluids make up between 55% and 60% of total body mass (Figure 22.1). Fluids are present in two main "compartments"—inside cells and outside cells. About two-thirds of body fluid is ***intracellular fluid (ICF)*** (*intra-* = within) or ***cytosol***, the fluid within cells. The other third, called ***extracellular fluid (ECF)*** (*extra-* = outside), is outside cells and includes all other body fluids. About 80% of the ECF is ***interstitial fluid*** (*inter-* = between), which occupies the spaces between tissue cells, and about 20% of the ECF is ***blood plasma***, the liquid portion of the blood. Other extracellular fluids that are grouped with interstitial fluid include lymph in lymphatic vessels; cerebrospinal fluid in the nervous system; synovial fluid in joints; aqueous humor and vitreous body in the eyes; endolymph and perilymph in the ears; and pleural, pericardial, and peritoneal fluids between serous membranes of the lungs, heart, and abdominal organs.

Two "barriers" separate intracellular fluid, interstitial fluid, and blood plasma.

1. The *plasma membrane* of each cell separates intracellular fluid from the surrounding interstitial fluid. You learned in Chapter 3 that the plasma membrane is a selectively permeable barrier: It allows some substances to cross but blocks the movement of other substances. In addition, active transport pumps work continuously to maintain different concentrations of certain ions in the cytosol and interstitial fluid.
2. *Blood vessel walls* separate the interstitial fluid from blood plasma. Only in capillaries, the smallest blood vessels, are the walls thin enough and leaky enough to permit the exchange of water and solutes between blood plasma and interstitial fluid.

The body is in ***fluid balance*** when the required amounts of water and solutes are present and are correctly proportioned among the various compartments. ***Water*** is by far the

Figure 22.1 Body fluid compartments.

In lean adults, fluids make up 55–60% of body mass.

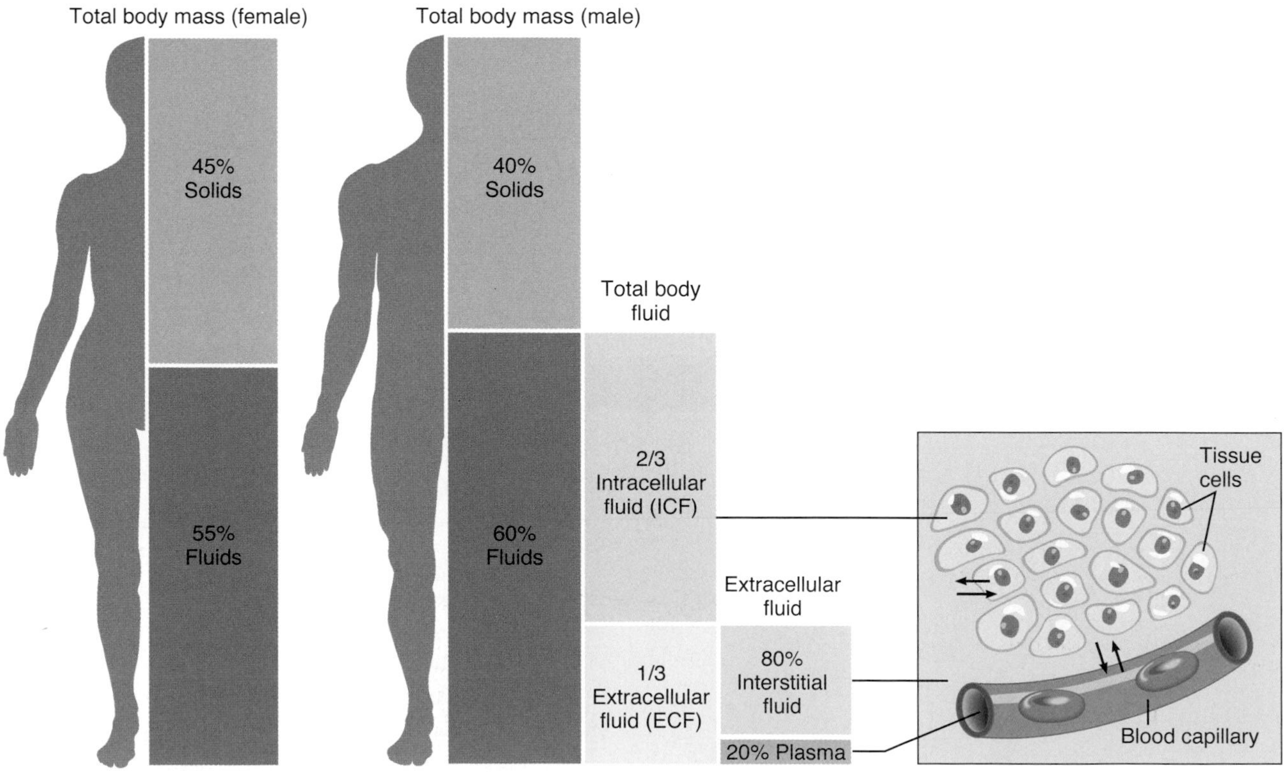

(a) Distribution of body solids and fluids in an average lean, adult female and male

(b) Exchange of water among body fluid compartments

What is body fluid?

largest single component of the body, making up 45–75% of total body mass, depending on age and gender.

The processes of filtration, reabsorption, diffusion, and osmosis provide for the continual exchange of water and solutes among body fluid compartments (Figure 22.1b). Yet, the volume of fluid in each compartment remains remarkably stable. Because osmosis is the primary means of water movement between intracellular fluid and interstitial fluid, the concentration of solutes in these fluids determines the *direction* of water movement. Most solutes in body fluids are ***electrolytes***, inorganic compounds that break apart into ions when dissolved in water. They are the main contributors to the osmotic movement of water. Fluid balance depends primarily on electrolyte balance so the two are closely interrelated. Because intake of water and electrolytes rarely occurs in exactly the same proportions as their presence in body fluids, the ability of the kidneys to excrete excess water by producing dilute urine, or to excrete excess electrolytes by producing concentrated urine, is of utmost importance in the maintenance of homeostasis.

Sources of Body Water Gain and Loss

The body can gain water by ingestion and by metabolic reactions (Figure 22.2). The main sources of body water are ingested liquids (about 1600 mL) and moist foods (about 700 mL) absorbed from the gastrointestinal (GI) tract, which total about 2300 mL/day. The other source of water is ***metabolic water*** that is produced in the body during chemical reactions. Most of it is produced during aerobic cellular respiration (see Figure 20.3 on page 510) and to a smaller extent during dehydration synthesis reactions (see Figure 2.8 on page 32). Metabolic water gain accounts for about 200 mL/day. Thus, daily water gain totals about 2500 mL.

Normally, body fluid volume remains constant because water loss equals water gain. Water loss occurs in four ways (Figure 22.2). Each day the kidneys excrete about 1500 mL in urine, about 600 mL evaporates from the skin surface, the lungs exhale about 300 mL as water vapor, and the gastrointestinal tract eliminates about 100 mL in feces. In women of reproductive age, additional water is lost in menstrual flow. On average, daily water loss totals about 2500 mL. The amount of water lost by a given route can vary considerably over time. For example, water may literally pour from the skin in the form of sweat during strenuous exertion. In other cases, water may be lost in vomit or diarrhea during a GI tract infection.

Regulation of Body Water Gain

An area in the hypothalamus known as the ***thirst center*** governs the urge to drink. When water loss is greater than water gain, ***dehydration***—a decrease in volume and an increase in osmotic pressure of body fluids—stimulates thirst (Figure 22.3). When body mass decreases by 2% due to fluid loss, mild dehydration exists. A decrease in blood volume causes blood pressure to fall.

Figure 22.2 Water balance: Sources of daily water gain and loss under normal conditions. Numbers are average volumes for adults.

Normally, daily water loss and water gain are both equal to 2500 mL.

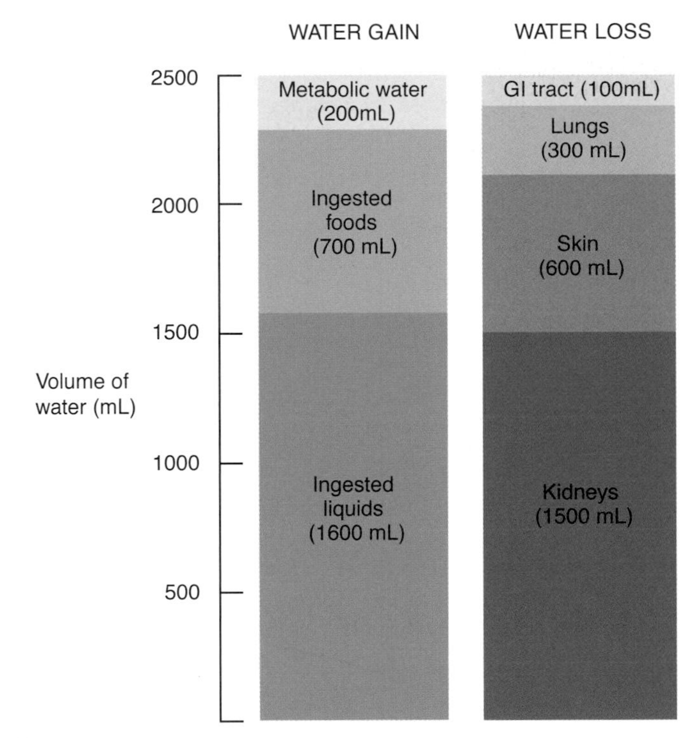

How would a diuretic drug affect a person's water balance?

This change stimulates the kidneys to release renin, which promotes the formation of angiotensin II. Osmoreceptors in the hypothalamus and increased angiotensin II in the blood both stimulate the thirst center in the hypothalamus. Other signals that stimulate thirst come from neurons in the mouth that detect dryness due to a decreased flow of saliva. As a result, the sensation of thirst increases, which usually leads to increased fluid intake (if fluids are available) and restoration of normal fluid volume. Overall, fluid gain balances fluid loss.

Sometimes the sensation of thirst does not occur quickly enough or access to fluids is restricted, and significant dehydration ensues. This happens most often in elderly people, in infants, and in those who are in a confused mental state. In situations where heavy sweating or fluid loss from diarrhea or vomiting occurs, it is wise to start replacing body fluids by drinking fluids even before the sensation of thirst appears.

Regulation of Water and Solute Loss

Elimination of *excess* body water or solutes occurs mainly by controlling the amount lost in urine. The extent of *urinary salt (NaCl) loss is the main factor that determines body fluid volume*. The reason is that in osmosis "water follows solutes,"

Figure 22.3 **Pathways through which dehydration stimulates thirst.**

Dehydration occurs when water loss is greater than water gain.

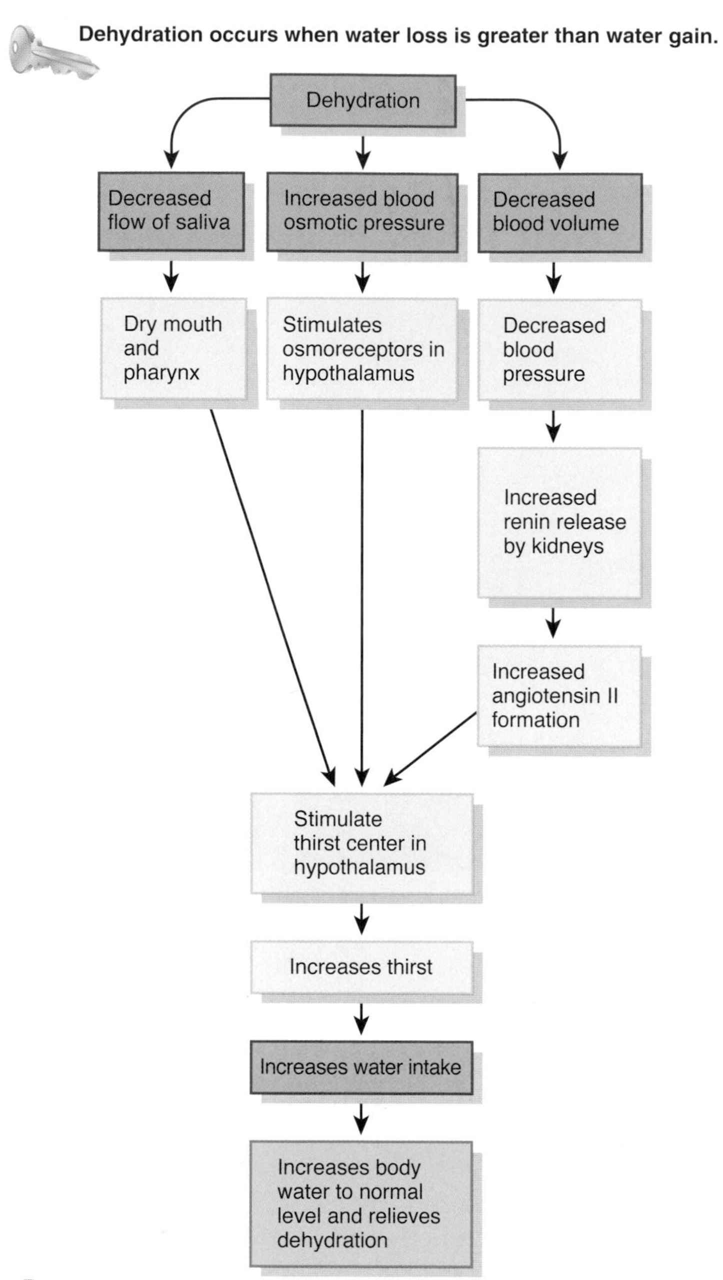

Does regulation of these pathways occur via negative or positive feedback? Why?

and the two main solutes in extracellular fluid (and in urine) are sodium ions (Na^+) and chloride ions (Cl^-). Because our daily diet contains a highly variable amount of NaCl, urinary excretion of Na^+ and Cl^- must also vary to maintain homeostasis. Three hormones regulate the extent of renal Na^+ and Cl^- reabsorption (and thus how much is lost in the urine): ***atrial natriuretic peptide (ANP), angiotensin II***, and ***aldosterone***.

Figure 22.4 depicts the sequence of changes that occur after a salty meal. The resulting increase in blood volume stretches the atria of the heart and promotes the release of atrial natriuretic peptide. ANP promotes ***natriuresis***, ele-

Figure 22.4 **Hormonal regulation of renal Na^+ and Cl^- reabsorption.**

The three main hormones that regulate renal Na^+ and Cl^- reabsorption (and thus the amount lost in the urine) are angiotensin II, aldosterone, and atrial natriuretic peptide.

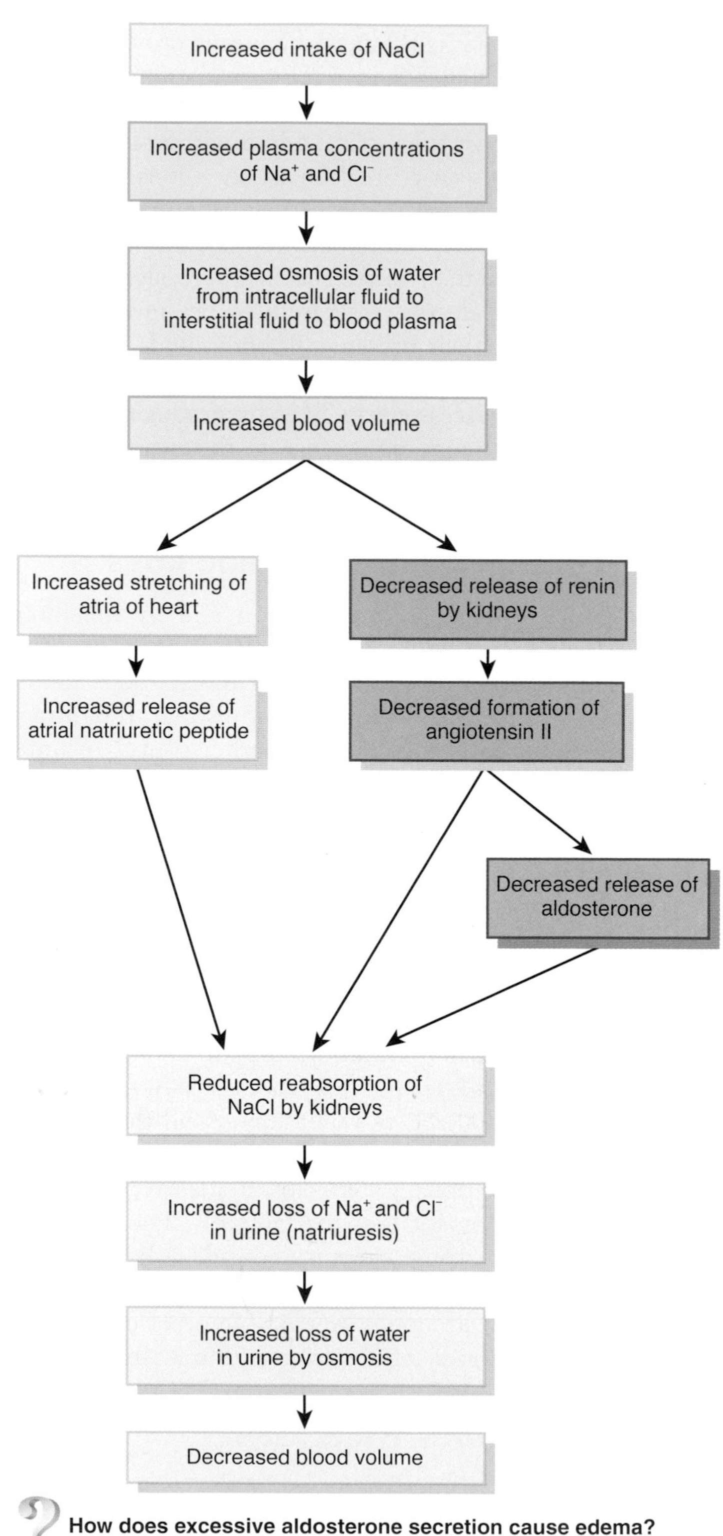

How does excessive aldosterone secretion cause edema?

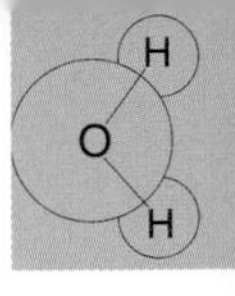

vated urinary loss of Na^+ (and Cl^-) and water, which decreases blood volume. The initial increase in blood volume also slows release of renin from the kidneys. As renin level declines, less angiotensin II is formed. With less angiotensin II, the kidney tubules reabsorb less Na^+, Cl^-, and water. In addition, less angiotensin II leads to less aldosterone and further slowing of Na^+ and Cl^- reabsorption in the renal tubules. More filtered Na^+ and Cl^- thus remain in the tubular fluid to be excreted in the urine. The osmotic consequence of excreting more Na^+ and Cl^- is loss of more water in urine, which decreases blood volume and blood pressure. By contrast, when someone becomes dehydrated, higher levels of angiotensin II and aldosterone promote urinary reabsorption of Na^+ and Cl^- (and water by osmosis with the solutes) and thereby conserve the volume of body fluids by reducing urinary loss.

The major hormone that regulates water loss is ***antidiuretic hormone (ADH)***. An increase in the osmotic pressure of body fluids (a decrease in the water concentration of the fluids) stimulates release of ADH (see Figure 21.8 on page 533). ADH promotes the insertion of water channels into the plasma membranes of cells in the collecting ducts of the kidneys. As a result, the permeability of these cells to water increases, and water moves from the tubular fluid into the cells and then into the bloodstream. By contrast, intake of plain water decreases the osmotic pressure of blood and interstitial fluid. Within minutes, ADH secretion shuts down, and soon its blood level is close to zero. Then, the water channels are removed from the membranes. As the number of water channels decreases, more water is lost in the urine.

Table 22.1 summarizes the factors that maintain body water balance.

If excess sodium ions remain in the body because the kidneys fail to excrete enough of them, water is also osmotically retained. The result is increased blood volume, increased blood pressure, and **edema**, an abnormal accumulation of interstitial fluid. Renal failure and excessive aldosterone secretion are two causes of Na^+ retention. Excessive urinary loss of Na^+, by contrast, has the osmotic effect of causing excessive water loss, which results in **hypovolemia**, an abnormally low blood volume. Hypovolemia related to Na^+ loss is most often due to inadequate secretion of aldosterone.

Movement of Water Between Fluid Compartments

Intracellular and interstitial fluids normally have the same osmotic pressure, so cells neither shrink nor swell. An increase in the osmotic pressure of interstitial fluid draws water out of cells, so they shrink slightly. A decrease in the osmotic pressure of interstitial fluid causes cells to swell. Changes in osmotic pressure most often result from changes in the concentration of Na^+. A decrease in the osmotic pressure of interstitial fluid inhibits secretion of ADH. Normally functioning kidneys then excrete excess water in the urine, which raises the osmotic pressure of body fluids to normal. As a result, body cells swell only slightly, and only for a brief period of time.

When a person steadily consumes water faster than the kidneys can excrete it (the maximum urine flow rate is about 15 mL/min) or when kidney function is poor, the decreased Na^+ concentration of interstitial fluid causes water to move by osmosis from interstitial fluid into intracellular fluid. The result may be **water intoxication**, a state in which excessive body water causes cells to swell dangerously, producing convulsions, coma, and possibly death. To prevent this dire sequence of events, solutions given for intravenous or **oral rehydration therapy (ORT)** include a small amount of table salt (NaCl).

■ CHECKPOINT

1. What is the approximate volume of each of your body fluid compartments?
2. Which routes of water gain and loss from the body are regulated?
3. How do angiotensin II, aldosterone, atrial natriuretic peptide, and antidiuretic hormone regulate the volume and osmotic pressure of body fluids?

Table 22.1 Summary of Factors that Maintain Body Water Balance

Factor	Mechanism	Effect
Thirst center in hypothalamus	Stimulates desire to drink fluids.	Water gain if thirst is quenched.
Angiotensin II	Stimulates secretion of aldosterone.	Reduces loss of water in urine.
Aldosterone	By promoting urinary reabsorption of Na^+ and Cl^-, increases water reabsorption via osmosis.	Reduces loss of water in urine.
Atrial natriuretic peptide (ANP)	Promotes natriuresis, elevated urinary excretion of Na^+ (and Cl^-), accompanied by water.	Increases loss of water in urine.
Antidiuretic hormone (ADH)	Promotes insertion of water-channel proteins into the plasma membranes of cells in the collecting ducts of the kidneys. As a result, the water permeability of these cells increases and more water is reabsorbed.	Reduces loss of water in urine.

ELECTROLYTES IN BODY FLUIDS

OBJECTIVES • **Compare the electrolyte composition of the three major fluid compartments: plasma, interstitial fluid, and intracellular fluid.**

- **Discuss the functions of sodium, chloride, potassium, and calcium ions, and explain how their concentrations are regulated.**

The ions formed when electrolytes break apart serve four general functions in the body:

1. Because they are largely confined to particular fluid compartments and are more numerous than nonelectrolytes, certain ions *control the osmosis of water between fluid compartments.*
2. Ions *help maintain the acid–base balance* required for normal cellular activities.
3. Ions *carry electrical current,* which allows production of action potentials.
4. Several ions *serve as cofactors* needed for optimal activity of enzymes.

Figure 22.5 compares the concentrations of the main electrolytes and protein anions in extracellular fluid (blood plasma and interstitial fluid) and intracellular fluid. The chief difference between the two extracellular fluids is that blood plasma contains many protein anions, but interstitial fluid has very few. Because normal capillary membranes are virtually impermeable to proteins, only a few plasma proteins leak out of blood vessels into the interstitial fluid. This difference in protein concentration is largely responsible for the blood colloid osmotic pressure, the difference in osmotic pressure between blood plasma and interstitial fluid. The other components of the two extracellular fluids are similar.

The electrolyte content of intracellular fluid differs considerably from that of extracellular fluid. Sodium ions (Na^+) are the most abundant extracellular ions, representing about 90% of extracellular cations. Na^+ plays a pivotal role in fluid and electrolyte balance because it accounts for almost half of the osmotic pressure of extracellular fluid. Na^+ is necessary for the generation and conduction of action potentials in neurons and muscle fibers. As you learned earlier in this chapter, the Na^+ level in the blood is controlled by aldosterone, antidiuretic hormone, and atrial natriuretic peptide.

Chloride ions (Cl^-) are the most prevalent anions in extracellular fluid. Because most plasma membranes contain many Cl^- leakage channels, Cl^- moves easily between the extracellular and intracellular compartments. For this reason, Cl^- can help balance the level of anions in different fluid

Figure 22.5 Electrolyte and protein anion concentrations in blood plasma, interstitial fluid, and intracellular fluid. The height of each column represents the milliequivalents per liter (mEq/liter), the total number of cations or anions (positive or negative electrical charges) in a given volume of solution.

The electrolytes present in extracellular fluids are different from those present in intracellular fluid.

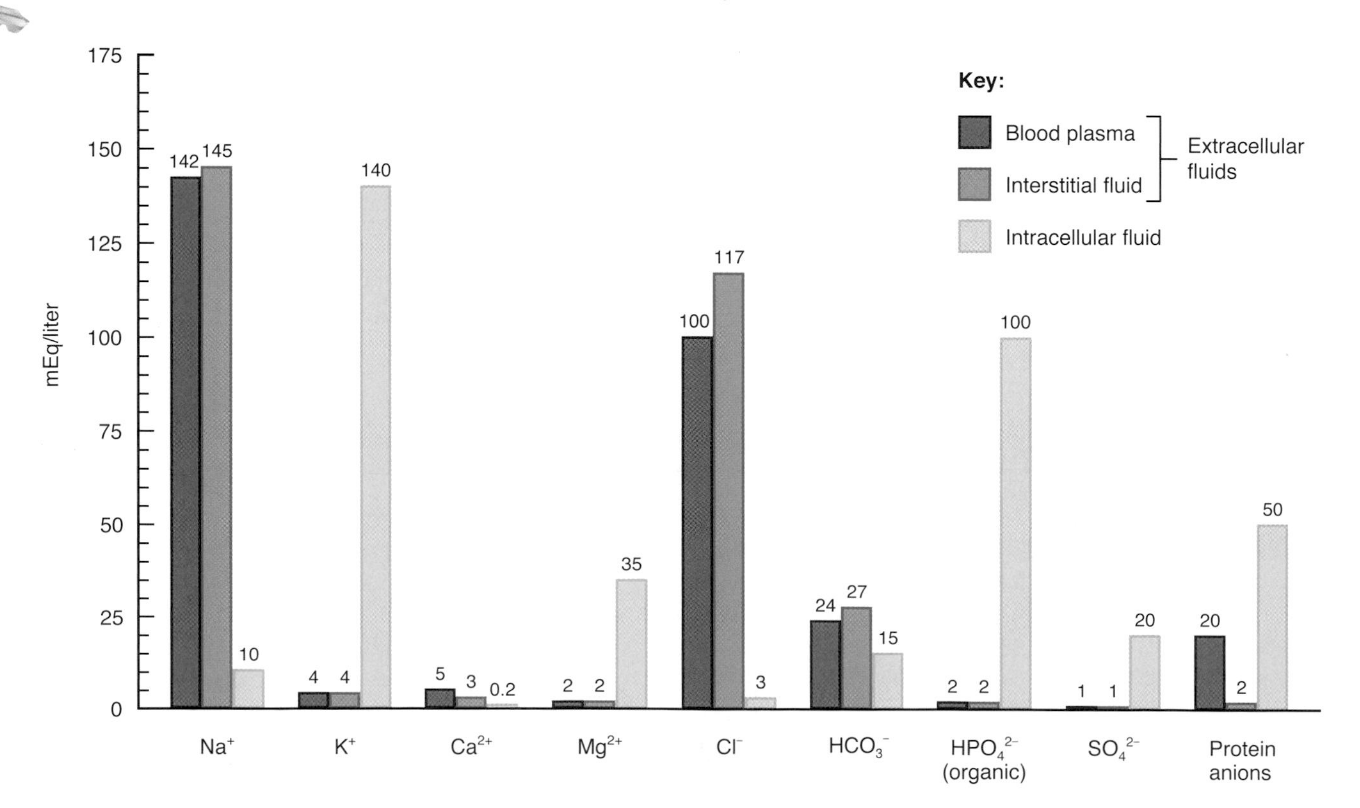

What is the major cation in ECF?

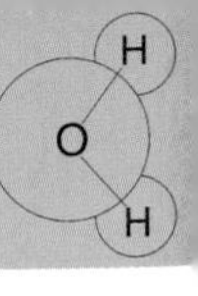

compartments. As you also learned earlier, ADH helps regulate Cl^- balance in body fluids by regulating the extent of water loss in urine. Processes that increase or decrease renal reabsorption of sodium ions also affect reabsorption of chloride ions. The negatively charged Cl^- follows the positively charged Na^+ due to the electrical attraction of oppositely charged particles.

Potassium ions (K^+), the most abundant cations in intracellular fluid, play a key role in establishing the resting membrane potential and in the repolarization phase of action potentials in neurons and muscle fibers. When K^+ moves into or out of cells, it often is exchanged for H^+ and thereby helps regulate the pH of body fluids. The level of K^+ in blood plasma is controlled mainly by aldosterone. When blood plasma K^+ is high, more aldosterone is secreted into the blood. Aldosterone then stimulates the renal collecting ducts to secrete more K^+ and excess K^+ is lost in the urine. Conversely, when blood plasma K^+ is low, aldosterone secretion decreases and less K^+ is excreted in urine.

About 98% of the calcium in adults is in the skeleton and teeth, where it is combined with phosphates to form mineral salts. In body fluids, calcium is mainly an extracellular cation (Ca^{2+}). Besides contributing to the hardness of bones and teeth, Ca^{2+} plays important roles in blood clotting, neurotransmitter release, maintenance of muscle tone, and excitability of nervous and muscle tissue.

The two main regulators of Ca^{2+} level in blood plasma are parathyroid hormone (PTH) and calcitriol, the form of vitamin D that acts as a hormone (see Figure 13.10 on page 327). A low plasma Ca^{2+} level promotes release of more PTH, which increases bone *resorption* by stimulating osteoclasts in bone tissue to release Ca^{2+} (and phosphate) from mineral salts of bone matrix. PTH also enhances *reabsorption* of Ca^{2+} from glomerular filtrate back into blood and increases production of calcitriol (which in turn increases Ca^{2+} *absorption* from the gastrointestinal tract).

Table 22.2 describes the imbalances that result from the deficiency or excess of several electrolytes.

Table 22.2 Blood Electrolyte Imbalances

	Deficiency		Excess	
Electrolyte*	**Name and Causes**	**Signs and Symptoms**	**Name and Causes**	**Signs and Symptoms**
Sodium (Na^+) 136–148 mEq/liter	**Hyponatremia** (hī-pō-na-TRĒ-mē-a) may be due to decreased sodium intake; increased sodium loss through vomiting, diarrhea, aldosterone deficiency, or taking certain diuretics; and excessive water intake.	Muscular weakness; dizziness, headache, and hypotension; tachycardia and shock; mental confusion, stupor, and coma.	**Hypernatremia** may occur with dehydration, water deprivation, or excessive sodium in the diet or in intravenous fluids.	Intense thirst, hypertension, edema, agitation, and convulsions.
Chloride (Cl^-) 95–105 mEq/liter	**Hypochloremia** (hī-pō-klō-RĒ-mē-a) may be due to excessive vomiting, water intoxication, aldosterone deficiency, congestive heart failure, and therapy with certain diuretics such as furosemide (Lasix®).	Muscle spasms, metabolic alkalosis, shallow ventilations, hypotension, and tetany.	**Hyperchloremia** may result from dehydration due to water loss or water deprivation, excessive chloride intake, or severe renal failure, aldosterone excess, certain types of acidosis, or some drugs.	Lethargy, weakness, metabolic acidosis, and rapid, deep breathing.
Potassium (K^+) 3.5–5.0 mEq/liter	**Hypokalemia** (hī-pō-ka-LĒ-mē-a) may result from excessive fluid loss due to vomiting or diarrhea, decreased potassium intake, aldosterone excess, kidney disease, or therapy with some diuretics.	Muscle fatigue, flaccid paralysis, mental confusion, increased urine output, shallow ventilations, and changes in the electrocardiogram.	**Hyperkalemia** may be due to excessive potassium intake, renal failure, aldosterone deficiency, or crushing injuries to body tissues.	Irritability, nausea, vomiting, diarrhea, muscular weakness; can cause death by inducing ventricular fibrillation.
Calcium (Ca^{2+}) Total = 9–10.5 mg/dL; ionized = 4.5–5.5 mEq/liter	**Hypocalcemia** (hī-pō-kal-SĒ-mē-a) may be due to increased calcium loss, reduced calcium intake, elevated levels of phosphate, or parathyroid hormone deficiency.	Numbness and tingling of the fingers; hyperactive reflexes, muscle cramps, tetany, and convulsions; bone fractures; spasms of laryngeal muscles that can cause death by asphyxiation.	**Hypercalcemia** may result from hyperparathyroidism, some cancers, excessive intake of vitamin D, and Paget's disease of bone.	Lethargy, weakness, anorexia, nausea, vomiting, polyuria, itching, bone pain, depression, confusion, paresthesia, stupor, and coma.

*Values are normal ranges of blood plasma levels in adults.

People who are at risk for **fluid and electrolyte imbalances** include those who depend on others for fluid and food, such as infants, the elderly, and the hospitalized. Also at risk are individuals undergoing medical treatment that involves intravenous infusions, drainages or suctions, and urinary catheters. People who receive diuretics experience excessive fluid losses and require increased fluid intake; those who experience fluid retention and have fluid restrictions are also at risk. Finally at risk are postoperative individuals, severe burn or trauma cases, individuals with chronic diseases (congestive heart failure, diabetes, chronic obstructive lung disease, and cancer), people in confinement, and individuals with altered levels of consciousness who may be unable to communicate needs or respond to thirst.

■ **CHECKPOINT**

4. What are the functions of electrolytes in the body?

ACID–BASE BALANCE

OBJECTIVES • Compare the roles of buffers, exhalation of carbon dioxide, and kidney excretion of H^+ in maintaining the pH of body fluids.

- **Define acid–base imbalances, describe their effects on the body, and explain how they are treated.**

From our discussion thus far, it should be clear that various ions play different roles in helping to maintain homeostasis. A major homeostatic challenge is keeping the H^+ level (pH) of body fluids in the appropriate range. This task—the maintenance of acid–base balance—is of critical importance because the three-dimensional shape of all body proteins, which enables them to perform specific functions, is very sensitive to the most minor changes in pH. When the diet contains a large amount of protein, as is typical in North America, cellular metabolism produces more acids than bases and thus tends to acidify the blood.

In a healthy person, the pH of systemic arterial blood remains between 7.35 and 7.45. The removal of H^+ from body fluids and its subsequent elimination from the body depend on three major mechanisms: buffer systems, exhalation of carbon dioxide, and kidney excretion of H^+ into the urine.

The Actions of Buffer Systems

Buffers are substances that act quickly to temporarily bind H^+, removing the highly reactive, excess H^+ from solution but not from the body. Buffers prevent rapid, drastic changes in the pH of a body fluid by converting strong acids and bases into weak acids and bases. Strong acids release H^+ more readily than weak acids and thus contribute more free hydrogen ions. Similarly, strong bases raise pH more than weak ones. The principal buffer systems of the body fluids are the protein buffer system, the carbonic acid–bicarbonate buffer system, and the phosphate buffer system.

Protein Buffer System

Many proteins can act as buffers. Altogether, proteins in body fluids comprise the ***protein buffer system***, which is the most abundant buffer in intracellular fluid and plasma. Hemoglobin is an especially good buffer within red blood cells, and albumin is the main protein buffer in blood plasma. Recall that proteins are composed of amino acids, organic molecules that contain at least one carboxyl group ($-COOH$) and at least one amino group ($-NH_2$); these groups are the functional components of the protein buffer system. The carboxyl group releases H^+ when pH rises. The H^+ is then able to react with any excess OH^- in the solution to form water. The amino group combines with H^+, forming an $-NH_3^+$ group, when pH falls. Thus, proteins can buffer both acids and bases.

Carbonic Acid–Bicarbonate Buffer System

The ***carbonic acid–bicarbonate buffer system*** is based on the *bicarbonate ion* (HCO_3^-), which can act as a weak base, and *carbonic acid* (H_2CO_3), which can act as a weak acid. HCO_3^- is a significant anion in both intracellular and extracellular fluids (Figure 22.5). Because the kidneys reabsorb filtered HCO_3^-, this important buffer is not lost in the urine. If there is an excess of H^+, the HCO_3^- can function as a weak base and remove the excess H^+ as follows:

$$\underset{\text{Hydrogen ion}}{H^+} + \underset{\substack{\text{Bicarbonate ion}\\\text{(weak base)}}}{HCO_3^-} \longrightarrow \underset{\text{Carbonic acid}}{H_2CO_3}$$

Conversely, if there is a shortage of H^+, the H_2CO_3 can function as a weak acid and provide H^+ as follows:

$$\underset{\substack{\text{Carbonic acid}\\\text{(weak acid)}}}{H_2CO_3} \longrightarrow \underset{\text{Hydrogen ion}}{H^+} + \underset{\text{Bicarbonate ion}}{HCO_3^-}$$

Phosphate Buffer System

The ***phosphate buffer system*** acts via a mechanism similar to the carbonic acid–bicarbonate buffer system. The components of the phosphate buffer system are the ions *dihydrogen phosphate* ($H_2PO_4^-$) and *monohydrogen phosphate* (HPO_4^{2-}). Recall that phosphates are major anions in intracellular fluid and minor ones in extracellular fluids (Figure 22.5). The dihydrogen phosphate ion acts as a weak acid and is capable of buffering strong bases such as OH^-, as follows:

$$\underset{\substack{\text{Hydroxide ion}\\\text{(strong base)}}}{OH^-} + \underset{\substack{\text{Dihydrogen}\\\text{phosphate}\\\text{(weak acid)}}}{H_2PO_4^-} \longrightarrow \underset{\text{Water}}{H_2O} + \underset{\substack{\text{Monohydrogen}\\\text{phosphate}\\\text{(weak base)}}}{HPO_4^{2-}}$$

FOCUS ON WELLNESS

Prolonged Physical Activity—A Challenge to Fluid and Electrolyte Balance

Heavy or prolonged physical activity can lead to dehydration and disrupt fluid and electrolyte balance. Strenuous exercise in hot weather may cause the loss of over 2 liters (about 2 qt) of water per hour from the skin and lungs. Such losses can lead to dehydration and elevated body temperature if fluids are not replaced.

Don't Sweat it

Dehydration is a loss of body fluid that amounts to 1% or more of total body weight. It is most common during physical activity at a high temperature but can also occur during strenuous exercise at lower temperature. Fluid deficits of 5% are common in athletic events such as football, soccer, tennis, and long-distance running. Symptoms include irritability, fatigue, and loss of appetite.

With dehydration, water is lost from all body fluid compartments. The decrease in blood volume impairs physical performance because it decreases the amount of blood the heart can pump per beat. Muscles need oxygen to work; as cardiac output is reduced, muscle performance declines. The body tries to maintain blood volume to the muscles by constricting blood vessels in the skin, so less heat is lost and body temperature rises. Intracellular electrolyte changes may also occur.

Thirst is the body's signal that its water level is getting too low. Unfortunately, thirst is not a reliable indicator of fluid needs. People tend to drink just enough to relieve their parched throats. The thirst mechanism is especially unreliable in children and older adults. Aging decreases the kidneys' ability to retain water when the body needs fluids, which increases the susceptibility to dehydration.

► THINK IT OVER . . .

► ***Sports drinks contain electrolytes such as sodium and potassium. Why might such drinks help a dehydrated person regain normal hydration levels better than plain water?***

The monohydrogen phosphate ion, in contrast, acts as a weak base and is capable of buffering the H^+ released by a strong acid such as hydrochloric acid (HCl):

$$\underset{\text{Hydrogen ion (strong acid)}}{H^+} + \underset{\text{Monohydrogen phosphate (weak base)}}{HPO_4^{2-}} \longrightarrow \underset{\text{Dihydrogen phosphate (weak acid)}}{H_2PO_4^-}$$

Because the concentration of phosphates is highest in intracellular fluid, the phosphate buffer system is an important regulator of pH in the cytosol. It also acts to a smaller degree in extracellular fluids, and it buffers acids in urine.

Exhalation of Carbon Dioxide

Breathing plays an important role in maintaining the pH of body fluids. An increase in the carbon dioxide (CO_2) concentration in body fluids increases H^+ concentration and thus lowers the pH (makes body fluids more acidic). Conversely, a decrease in the CO_2 concentration of body fluids raises the pH (makes body fluids more alkaline). These chemical interactions are illustrated by the following reversible reactions.

$$\underset{\text{Carbon dioxide}}{CO_2} + \underset{\text{Water}}{H_2O} \rightleftharpoons \underset{\text{Carbonic acid}}{H_2CO_3} \rightleftharpoons \underset{\text{Hydrogen ion}}{H^+} + \underset{\text{Bicarbonate ion}}{HCO_3^-}$$

Changes in the rate and depth of breathing can alter the pH of body fluids within a couple of minutes. With increased ventilation, more CO_2 is exhaled, the reaction goes from right to left, H^+ concentration falls, and blood pH rises. If ventilation is slower than normal, less carbon dioxide is exhaled, and the blood pH falls.

The pH of body fluids and the rate and depth of breathing interact via a negative feedback loop (Figure 22.6). When the blood acidity increases, the decrease in pH (increase in concentration of H^+) is detected by chemoreceptors in the medulla oblongata and in the aortic and carotid bodies, both of which stimulate the inspiratory area in the medulla oblongata. As a result, the diaphragm and other respiratory muscles contract more forcefully and frequently, so more CO_2 is exhaled, driving the reaction to the left. As less H_2CO_3 forms and fewer H^+ are present, blood pH increases. When the response brings blood pH (H^+ concentration) back to normal, there is a return to acid–base homeostasis.

By contrast, if the pH of the blood increases, the respiratory center is inhibited and the rate and depth of breathing decreases. Then, CO_2 accumulates in the blood and its H^+ concentration increases. This respiratory mechanism is powerful, but it can only regulate the concentration of one acid—carbonic acid.

Kidney Excretion of H^+

The slowest mechanism for removal of acids is also the only way to eliminate most acids that form in the body: Cells of the renal tubules secrete H^+, which then is excreted in urine. Also, because the kidneys synthesize new HCO_3^- and reabsorb filtered HCO_3^-, this important buffer is not lost in the urine. Because of the contributions of the kidneys to acid–base balance, it's not surprising that renal failure can quickly cause death.

Table 22.3 summarizes the mechanisms that maintain pH of body fluids.

Acid–Base Imbalances

Acidosis is a condition in which arterial blood pH is below 7.35. The principal physiological effect of acidosis is depression of the central nervous system through depression of synaptic transmission. If the systemic arterial blood pH falls below 7, depression of the nervous system is so severe that the individual becomes disoriented, then becomes comatose, and may die.

In ***alkalosis***, arterial blood pH is higher than 7.45. A major physiological effect of alkalosis is overexcitability in both the central nervous system and peripheral nerves. Neurons conduct impulses repetitively, even when not stimulated; the results are nervousness, muscle spasms, and even convulsions and death.

A change in blood pH that leads to acidosis or alkalosis may be countered by ***compensation***, the physiological response to an acid–base imbalance that acts to normalize arterial blood pH. Compensation may be either *complete*, if pH indeed is brought within the normal range, or *partial*, if systemic arterial blood pH is still lower than 7.35 or higher than 7.45. If a person has altered blood pH due to metabolic causes, hyperventilation or hypoventilation can help bring blood pH back toward the normal range; this form of compensation, termed ***respiratory compensation***, occurs within minutes and reaches its maximum within hours. If, however, a person has altered blood pH due to respiratory causes, then ***renal compensation***—changes in secretion of H^+ and reabsorption of HCO_3^- by the kidney tubules—can help reverse the change. Renal compensation may begin in minutes, but it takes days to reach maximum effectiveness.

■ **CHECKPOINT**

5. How do proteins, bicarbonate ions, and phosphate ions help maintain the pH of body fluids?

6. What are the major physiological effects of acidosis and alkalosis?

Figure 22.6 Negative feedback regulation of blood pH by the respiratory system.

Exhalation of carbon dioxide lowers the H^+ concentration of blood.

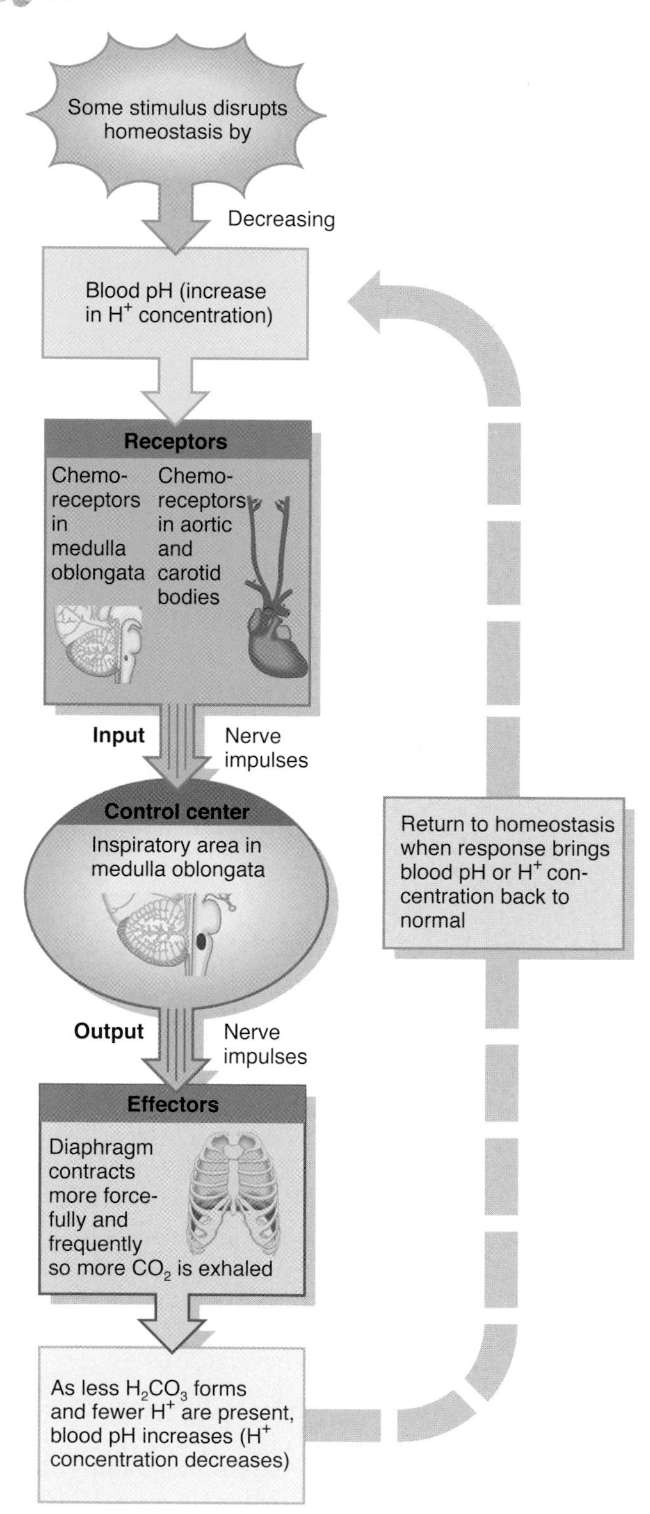

If you hold your breath for 30 seconds, what is likely to happen to your blood pH?

Table 22.3 Mechanisms That Maintain pH of Body Fluids

Mechanism	Comments
Buffer Systems	Convert strong acids and bases into weak acids and bases, preventing drastic changes in body fluid pH.
Proteins	The most abundant buffers in body cells and blood. Hemoglobin is a buffer in the cytosol of red blood cells; albumin is a buffer in blood plasma.
Carbonic acid–bicarbonate	Important regulators of blood pH. The most abundant buffers in extracellular fluid.
Phosphates	Important buffers in intracellular fluid and in urine.
Exhalation of CO_2	With increased exhalation of CO_2, pH rises (fewer H^+). With decreased exhalation of CO_2, pH falls (more H^+).
Kidneys	Kidney tubules secrete H^+ into the urine and reabsorb HCO_3^- so it is not lost in the urine.

AGING AND FLUID, ELECTROLYTE, AND ACID–BASE BALANCE

OBJECTIVE • Describe the changes in fluid, electrolyte, and acid–base balance that may occur with aging.

By comparison with children and younger adults, older adults often have an impaired ability to maintain fluid, electrolyte, and acid–base balance. With increasing age, many people have a decreased volume of intracellular fluid and decreased total body potassium due to declining skeletal muscle mass and increasing mass of adipose tissue (which contains very little water). Age-related decreases in respiratory and renal functioning may compromise acid–base balance by slowing the exhalation of CO_2 and the excretion of excess acids in urine. Other kidney changes, such as decreased blood flow, decreased glomerular filtration rate, and reduced sensitivity to antidiuretic hormone, have an adverse effect on the ability to maintain fluid and electrolyte balance. Due to a decrease in the number and efficiency of sweat glands, water loss from the skin declines with age. Because of these age-related changes, older adults are susceptible to several fluid and electrolyte disorders.

■ **CHECKPOINT**

7. How are skeletal muscle mass and adipose tissue related to fluid and electrolyte imbalance?

STUDY OUTLINE

Fluid Compartments and Fluid Balance (p. 544)

1. The water and dissolved solutes in the body constitute the body fluids.
2. About two-thirds of the body's fluid is located within cells and is called intracellular fluid (ICF). The other one-third, called extracellular fluid (ECF), includes all other body fluids. About 80% of the ECF is interstitial fluid, which occupies the microscopic spaces between tissue cells, and about 20% of the ECF is blood plasma, the liquid portion of the blood.
3. Fluid balance means that the various body compartments contain the normal amount of water and solutes.
4. Water is the largest single component in the body, about 55–60% of total body mass in lean adults.
5. An electrolyte is an inorganic substance that dissociates into ions in solution. Fluid balance and electrolyte balance are interrelated.
6. Daily water gain and loss are each about 2500 mL. Sources of water gain are ingested liquids and foods and water produced by metabolic reactions (metabolic water). Water is lost from the body through urination, evaporation from the skin surface, exhalation of water vapor, and defecation. In women, menstrual flow is an additional route for loss of body water.
7. The main way to regulate body water gain is by adjusting the volume of water intake. The thirst center in the hypothalamus governs the urge to drink.
8. Angiotensin II and aldosterone reduce urinary loss of Na^+ and Cl^- and thereby increase the volume of body fluids. Atrial natriuretic peptide promotes natriuresis, elevated excretion of Na^+ (and Cl^-) and water, which decreases blood volume.
9. Table 22.1 on page 547 summarizes the factors that maintain water balance.

Electrolytes in Body Fluids (p. 548)

1. Electrolytes control the osmosis of water between fluid compartments, help maintain acid–base balance, carry electrical current, and act as enzyme cofactors.

2. Sodium ions (Na^+) are the most abundant extracellular ions. They are involved in action potentials, muscle contraction, and fluid and electrolyte balance. Na^+ level is controlled by aldosterone, antidiuretic hormone, and atrial natriuretic peptide.
3. Chloride ions (Cl^-) are the major extracellular anions. They play a role in regulating osmotic pressure and forming HCl in gastric juice. Cl^- level is controlled by processes that increase or decrease kidney reabsorption of Na^+.
4. Potassium ions (K^+) are the most abundant cations in intracellular fluid. They play a key role in establishing the resting membrane potential in neurons and muscle fibers, and contribute to regulation of pH. K^+ level is controlled by aldosterone.
5. Calcium is the most abundant mineral in the body. Calcium salts are structural components of bones and teeth. Ca^{2+}, which are principally extracellular cations, function in blood clotting, neurotransmitter release, and contraction of muscle. Ca^{2+} level is controlled mainly by parathyroid hormone and calcitriol.
6. Table 22.2 on page 549 describes the imbalances that result from deficiency or excess of important body electrolytes.

Acid–Base Balance (p. 550)

1. The normal pH of systemic arterial blood is 7.35 to 7.45.
2. Homeostasis of pH is maintained by buffer systems, by exhalation of carbon dioxide, and by kidney excretion of H^+ and reabsorption of HCO_3^-. Table 22.3 on page 553 summarizes the mechanisms that maintain pH of body fluids.
3. Acidosis is a systemic arterial blood pH below 7.35; its principal effect is depression of the central nervous system (CNS). Alkalosis is a systemic arterial blood pH above 7.45; its principal effect is overexcitability of the CNS.

Aging and Fluid, Electrolyte, and Acid–Base Balance (p. 553)

1. With increasing age, there is decreased intracellular fluid volume and decreased potassium due to declining skeletal muscle mass.
2. Decreased kidney function adversely affects fluid and electrolyte balance.

SELF-QUIZ

1. Normally, most of the body's water is lost through
   a. the gastrointestinal tract b. cellular respiration
   c. exhalation by the lungs d. excretion of urine
   e. evaporation from the skin
2. Substances that dissociate into ions when dissolved in body fluids are
   a. neurotransmitters b. enzymes c. nonelectrolytes
   d. hormones e. electrolytes
3. Which of the following statements about sodium is NOT true?
   a. Sodium ions are the most abundant intracellular ions
   b. Sodium is necessary for generating action potentials in neurons.
   c. Excess sodium ions can cause edema.
   d. Sodium levels are regulated by the kidneys.
   e. Aldosterone helps regulate the concentration of sodium in the blood.
4. Parathyroid hormone (PTH) controls blood level of
   a. magnesium b. sodium c. calcium d. potassium
   e. chloride
5. Fluid movement between intracellular fluid and extracellular fluid depends primarily on the concentration of which ion in extracellular fluid?
   a. sodium b. potassium c. calcium d. phosphate
   e. magnesium
6. Which of the following are mismatched?
   a. the most abundant extracellular anion, Cl^-
   b. the most abundant mineral in the body, Ca^{2+}
   c. the most abundant extracellular cation, Na^+
   d. the most abundant intracellular cation, K^+
   e. the most abundant intracellular anion, HCO_3^-
7. Which of the following statements concerning acid–base balance in the body is NOT true?
   a. An increase in respiration rate increases pH of body fluids.
   b. Normal pH of extracellular fluid is 7.35 to 7.45.
   c. Buffers are an important mechanism in the maintenance of pH balance.
   d. A blood pH of 7.2 is called alkalosis.
   e. Respiratory acidosis is characterized by a high level of CO_2 in body fluids.
8. Most human buffer systems consist of
   a. a weak acid and a weak base
   b. a strong acid and a strong base
   c. a strong acid such as HCl
   d. an electrolyte and nonelectrolyte
   e. a weak base and a gas
9. The most abundant buffer in body cells and plasma is the ________ buffer system.
   a. hemoglobin b. carbonic acid c. protein
   d. bicarbonate e. phosphate
10. Most (80%) of the extracellular fluid is part of the body's
    a. interstitial fluid b. lymph c. cerebrospinal fluid
    d. plasma e. synovial fluid
11. Which hormone stimulates the kidneys to secrete more K^+?
    a. atrial natriuretic peptide b. angiotensin
    c. aldosterone d. antidiuretic hormone
    e. parathyroid hormone
12. Most of the body's water comes from
    a. cellular respiration b. adipose tissue
    c. urine production d. water intoxication
    e. ingested liquids and foods

13. The thirst center can be activated by all of the following EXCEPT
    a. angiotensin II
    b. an increase in blood volume
    c. a decrease in flow from salivary glands
    d. a decrease in blood pressure
    e. an increase in blood osmotic pressure

14. The center for thirst is located in the
    a. kidneys b. adrenal cortex c. hypothalamus
    d. cerebral cortex e. liver

15. Which of the following is NOT one of the functions of electrolytes in the body?
    a. control of fluid movement between the extracellular and intracellular compartments
    b. regulation of pH
    c. enzyme cofactor
    d. energy source
    e. carrier of electric current

16. Aldosterone is secreted in response to
    a. increased blood pressure b. decreased blood volume
    c. increased calcium levels d. increased sodium levels
    e. increased water levels

17. What is the importance of buffer systems in the body?
    a. They help maintain the calcium and phosphate balances of bone.
    b. They control the body's water balance.
    c. They prevent drastic changes in the body's pH.
    d. They help regulate blood volume.
    e. They are responsible for the operation of the body's sodium pump.

18. Match the following:

____	a. extracellular cation; structural component of bones and teeth	A. calcium
____	b. most abundant anion in extracellular fluid	B. chloride
____	c. most abundant extracellular cation; needed for generation and conduction of action potentials	C. potassium
____	d. most abundant cation in intracellular fluid; involved in nerve and muscle homeostasis	D. sodium

CRITICAL THINKING APPLICATIONS

1. José was grazing his way through lunch by eating at street vendors. He had a large order of fries with extra salt, then a foot-long hot dog with ketchup (a very high sodium content lunch). Next, José bought a large bottled-water and drank the entire bottle. How will his body respond to this lunch?

2. One-year-old Timon had a busy morning at the "mom and tot" swim program. Today's lesson included lots of underwater exercises in blowing bubbles. After the lesson, Timon seemed disoriented and then suffered a convulsion. The emergency room nurse thinks the swim class has something to do with Timon's problem. What is wrong with Timon?

3. Mike and Jennie are the same height and both weigh 150 lb., but when Mike and Jennie measured their blood alcohol after drinking identical alcoholic beverages, Jennie's blood alcohol level was higher than Mike's. In the body, alcohol is transported in the body fluids. Use your knowledge about the differences in body water between males and females to explain the difference in alcohol level.

4. Alex was 15 minutes late for A&P class. While searching for his pen, he thought he heard the instructor say something about the heart affecting water balance but he had thought it was the other way around. Alex just decided to ignore the whole thing. Bad move, Alex! Explain the relationship of the heart to fluid balance.

ANSWERS TO FIGURE QUESTIONS

22.1 The term body fluid refers to body water and its dissolved substances.

22.2 A diuretic drug increases urine flow rate, thereby increasing loss of fluid from the body and decreasing the volume of body fluids.

22.3 Negative feedback is in operation because the result (an increase in fluid intake) is opposite to the initiating stimulus (dehydration).

22.4 Elevated aldosterone promotes abnormally high renal reabsorption of NaCl and water, which expands blood volume and increases blood pressure. Increased blood pressure causes more fluid to filter out of capillaries and accumulate in the interstitial fluid, a condition called edema.

22.5 The major cation in ECF is Na^+.

22.6 Breath holding causes blood pH to decrease slightly as CO_2 and H^+ accumulate.

chapter 23 THE REPRODUCTIVE SYSTEMS

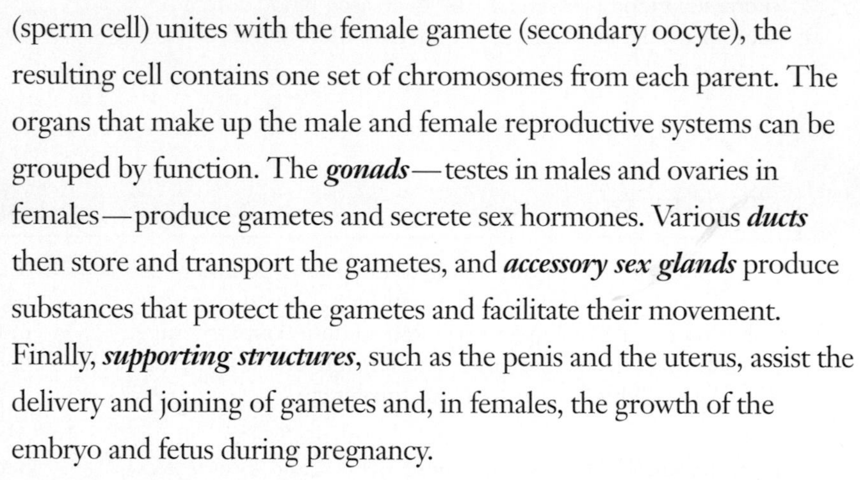

did you know?

Abstaining from sexual contact prevents the risk of sexually transmitted infections. For couples who engage in sexual contact, safer-sex practices at least reduce the risk of acquiring an infection. Safer sex practices mean avoiding contact with oral or genital sores and avoiding the exchange of body fluids, such as semen, blood, and vaginal secretions. Latex condoms provide the best protection against sexually transmitted infections for couples having sexual intercourse. Studies of couples in which one partner was infected with HIV have found that when latex condoms were used consistently, HIV transmission to partners was very low (zero in one study, 2 out of 171 in another).

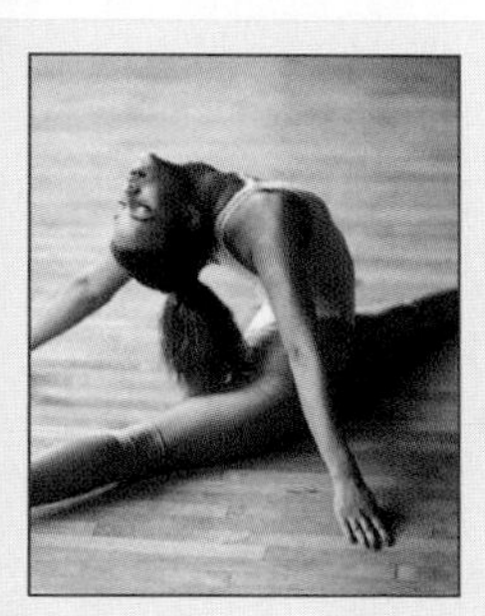

Focus on Wellness, page 576

www.wiley.com/college/apcentral

Sexual reproduction is the process by which organisms produce offspring by making germ cells called ***gametes*** (GAM-ēts = spouses). After ***fertilization***, when the male gamete (sperm cell) unites with the female gamete (secondary oocyte), the resulting cell contains one set of chromosomes from each parent. The organs that make up the male and female reproductive systems can be grouped by function. The ***gonads***—testes in males and ovaries in females—produce gametes and secrete sex hormones. Various ***ducts*** then store and transport the gametes, and ***accessory sex glands*** produce substances that protect the gametes and facilitate their movement. Finally, ***supporting structures***, such as the penis and the uterus, assist the delivery and joining of gametes and, in females, the growth of the embryo and fetus during pregnancy.

Gynecology (gī′-ne-KOL-ō-jē; *gyneco-* = woman; *-logy* = study of) is the specialized branch of medicine concerned with the diagnosis and treatment of diseases of the female reproductive system. As noted in Chapter 21, ***urology*** (ū-ROL-ō-jē) is the study of the urinary system. Urologists also diagnose and treat diseases and disorders of the male reproductive system. The branch of medicine that deals with male disorders, especially infertility and sexual dysfunction, is called ***andrology*** (an-DROL-ō-jē; *andro-* = masculine).

looking back to move ahead . . .

- Somatic Cell Division (page 62)
- Sympathetic and Parasympathetic Divisions of the Autonomic Nervous System (page 274)
- Hormones of the Hypothalamus and Pituitary Gland (page 319)

MALE REPRODUCTIVE SYSTEM

OBJECTIVES • Describe the location, structure, and functions of the organs of the male reproductive system.

- **Describe how sperm cells are produced.**
- **Explain the roles of hormones in regulating male reproductive functions.**

The organs of the ***male reproductive system*** are the testes; a system of ducts (epididymis, ductus deferens, ejaculatory ducts, and urethra); accessory sex glands (seminal vesicles, prostate, and bulbourethral glands); and several supporting structures, including the scrotum and the penis (Figure 23.1). The testes produce sperm and secrete hormones. Sperm are transported and stored, helped to mature, and conveyed to the exterior by a system of ducts. Semen contains sperm plus the secretions provided by the accessory sex glands.

Scrotum

The ***scrotum*** (SKRŌ-tum = bag) is a pouch that supports the testes; it consists of loose skin, superficial fascia, and smooth muscle (Figure 23.1). Internally, a septum divides the scrotum into two sacs, each containing a single testis.

The production and survival of sperm is optimal at a temperature that is about 2-3°C below normal body temperature. This lowered body temperature is maintained within

Figure 23.1 Male organs of reproduction and surrounding structures.

Reproductive organs are adapted to produce new individuals and pass genetic material from one generation to the next.

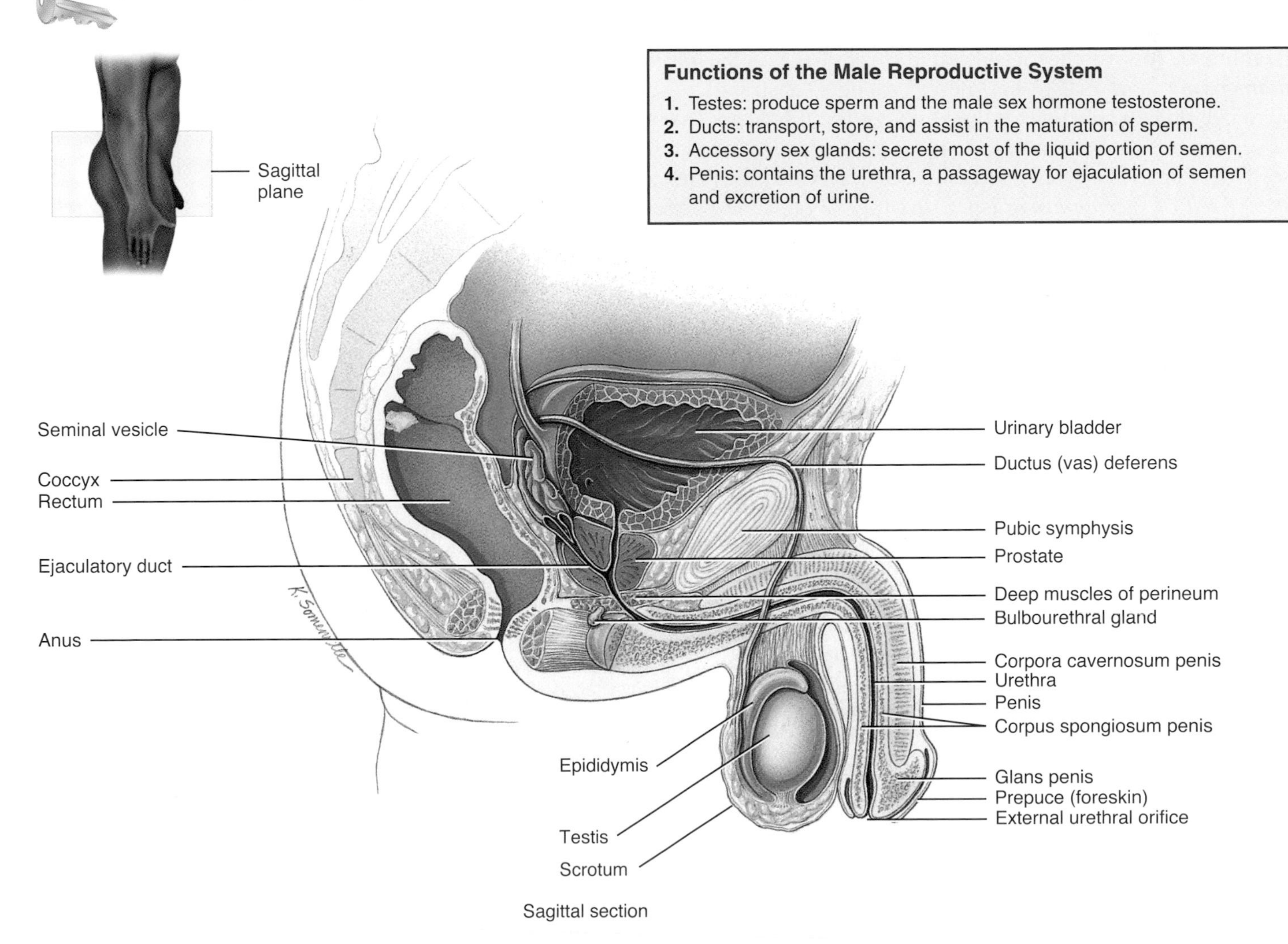

Among the male organs of reproduction, how is the penis classified functionally?

the scrotum because it is outside the pelvic cavity. On exposure to cold, skeletal muscles contract to elevate the testes, moving them closer to the pelvic cavity, where they can absorb body heat. Exposure to warmth causes relaxation of the skeletal muscles and descent of the testes, increasing the surface area exposed to the air, so that the testes can give off excess heat to their surroundings.

Testes

The ***testes*** (TES-tēz; singular is *testis*), or *testicles*, are paired oval glands that develop on the embryo's posterior abdominal wall and usually begin their descent into the scrotum in the seventh month of fetal development.

The testes are covered by a dense ***white fibrous capsule*** that extends inward and divides each testis into internal compartments called ***lobules*** (Figure 23.2a). Each of the 200 to 300 lobules contains one to three tightly coiled ***seminiferous tubules*** (*semin-* = seed; *fer-* = to carry) that produce sperm by a process called spermatogenesis (described shortly).

Seminiferous tubules are lined with ***spermatogenic*** (sperm-forming) ***cells*** (Figure 23.2b). Positioned against the basement membrane, toward the outside of the tubules, are the ***spermatogonia*** (sper-ma′-tō-GŌ-nē-a; *-gonia* = offspring), the stem cell precursors. Toward the lumen of the tubule are layers of cells in order of advancing maturity: primary spermatocytes, secondary spermatocytes, spermatids, and sperm cells. After a ***sperm cell*** or ***spermatozoon*** (sper′-ma-tō-ZŌ-on; *-zoon* = life) has formed, it is released into the lumen of the seminiferous tubule.

Large ***Sertoli cells***, located between the developing sperm cells in the seminiferous tubules, support, protect, and nourish spermatogenic cells; phagocytize degenerating spermatogenic cells; secrete fluid for sperm transport; and release the hormone inhibin, which helps regulate sperm production. Between the seminiferous tubules are clusters of ***Leydig cells***. These cells secrete the hormone ***testosterone***, the most important androgen. An ***androgen*** (AN-drō-jen) is a hormone that promotes the development of masculine characteristics. Testosterone also promotes a man's libido (sex drive).

> The condition in which the testes do not descend into the scrotum is called **cryptorchidism** (krip-TOR-ki-dizm; *crypt-* = hidden; *orchid* = testis). It occurs in about 3% of full-term infants and about 30% of premature infants. Untreated bilateral cryptorchidism causes sterility due to the higher temperature of the pelvic cavity. The chance of testicular cancer is 30 to 50 times greater in cryptorchid testes, possibly due to abnormal division of germ cells caused by the higher temperature of the pelvic cavity. The testes of about 80% of boys with cryptorchidism will descend spontaneously during the first year of life. When the testes remain undescended, the condition can be corrected surgically, ideally before 18 months of age.

Spermatogenesis

The process by which the seminiferous tubules of the testes produce sperm is called ***spermatogenesis*** (sper-ma′-tō-JEN-e-sis). It consists of three stages: meiosis I, meiosis II, and spermiogenesis. We begin with meiosis.

OVERVIEW OF MEIOSIS As you learned in Chapter 3, most body cells (somatic cells), such as brain cells, stomach cells, kidney cells, and so forth, contain 23 pairs of chromosomes, or a total of 46 chromosomes. One member of each pair is inherited from each parent. The two chromosomes that make up each pair are called ***homologous chromosomes*** (hō-MOL-ō-gus; *homo-* = same); they contain similar genes arranged in the same (or almost the same) order. Because somatic cells contain two sets of chromosomes, they are termed ***diploid cells*** (DIP-loyd; *dipl-* = double; *-oid* = form), symbolized as ***2n***. Gametes differ from somatic cells because they contain a single set of 23 chromosomes, symbolized as ***n***; they are thus said to be ***haploid*** (HAP-loyd; *hapl-* = single).

In sexual reproduction, an organism results from the fusion of two different gametes, one produced by each parent. If each gamete had the same number of chromosomes as somatic cells, then the number of chromosomes would double each time fertilization occurred. Instead, gametes receive a single set of chromosomes by means of a special type of reproductive cell division called ***meiosis*** (mī-Ō-sis; *mei-* = lessening; *-osis* = condition of). Meiosis occurs in two successive stages: ***meiosis I*** and ***meiosis II***. First, we will examine how meiosis occurs during spermatogenesis. Later in the chapter, we will follow the steps of meiosis during oogenesis, the production of female gametes.

STAGES OF SPERMATOGENESIS Spermatogenesis begins during puberty and continues throughout life. The time from onset of cell division in a ***spermatogonium*** until sperm are released into the lumen of a seminiferous tubule is 65 to 75 days. The spermatogonia contain the diploid number of chromosomes (46). After a spermatogonium undergoes mitosis, one cell stays near the basement membrane as a spermatogonium, so stem cells remain for future mitosis (Figure 23.3 on page 560). The other cell differentiates into a ***primary spermatocyte*** (sper-MA-tō-sīt′). Like spermatogonia, primary spermatocytes are diploid. Spermatogenesis proceeds as follows:

1. **Meiosis I.** During the interphase that precedes meiosis I, the chromosomes replicate, as also occurs in the interphase before mitosis in somatic cell division (see page 62). The 46 chromosomes, now each made up of two identical "sister" chromatids, line up as 23 pairs of homologous chromosomes. (By contrast, pairing of homologous chromosomes does not occur during mitosis.) The four chromatids of each homologous pair then twist around one another. At this time, portions of one chromatid may be exchanged with portions of another; such an exchange is termed ***crossing-over***. Crossing-over results in ***genetic recombination***, the formation of new

Figure 23.2 Anatomy and histology of the testes. (a) Spermatogenesis occurs in the seminiferous tubules. (b) The stages of spermatogenesis. Arrows in (b) indicate the progression from least mature to most mature spermatogenic cells. The (*n*) and (2*n*) refer to haploid and diploid chromosome number, to be described shortly.

The male gonads are the testes, which produce haploid sperm.

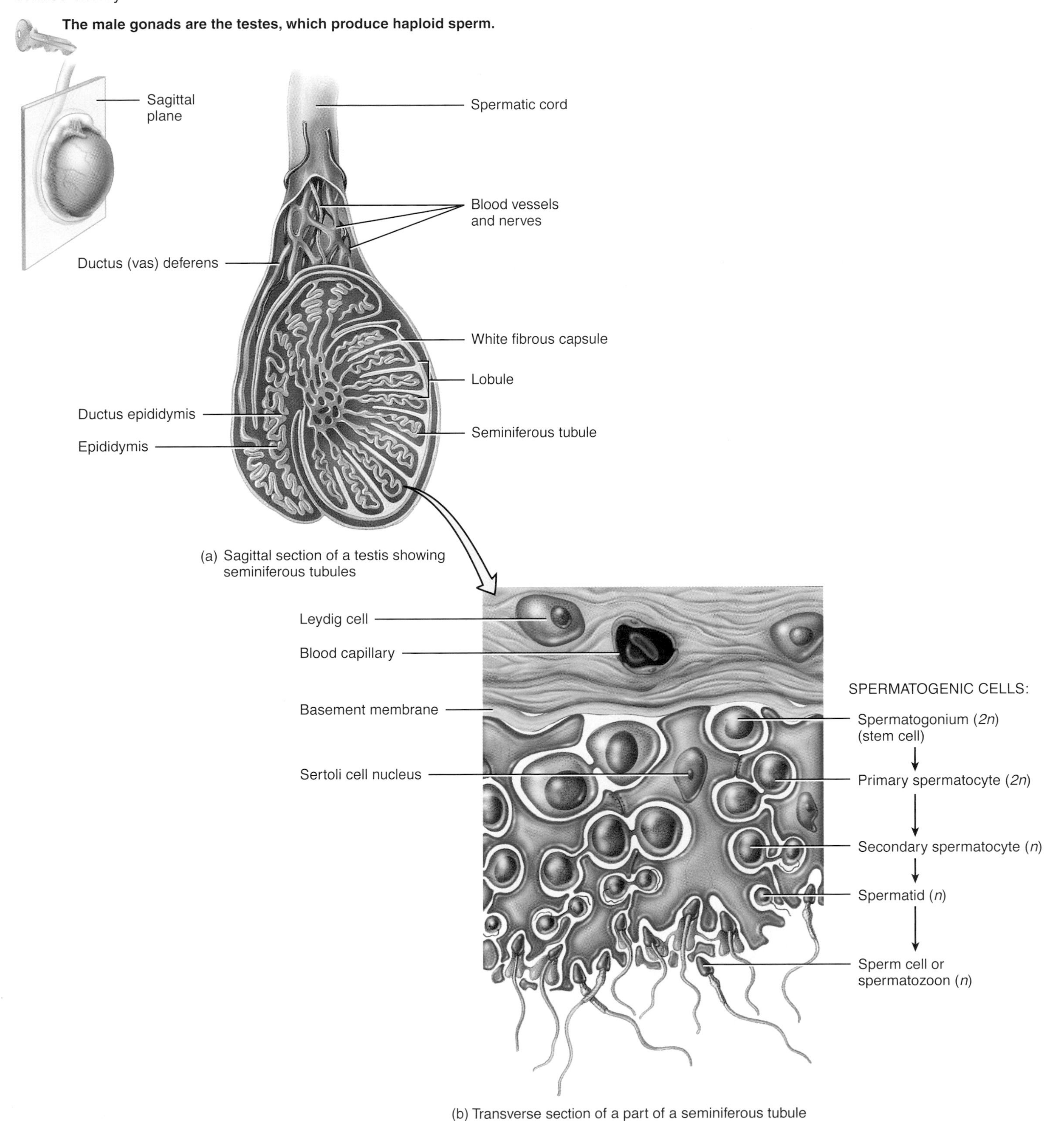

(a) Sagittal section of a testis showing seminiferous tubules

(b) Transverse section of a part of a seminiferous tubule

Which spermatogenic cells in a seminiferous tubule are least mature?

Figure 23.3 Spermatogenesis. The designation $2n$ means diploid (46 chromosomes); n means haploid (23 chromosomes).

Spermiogenesis is the process whereby spermatids mature into sperm.

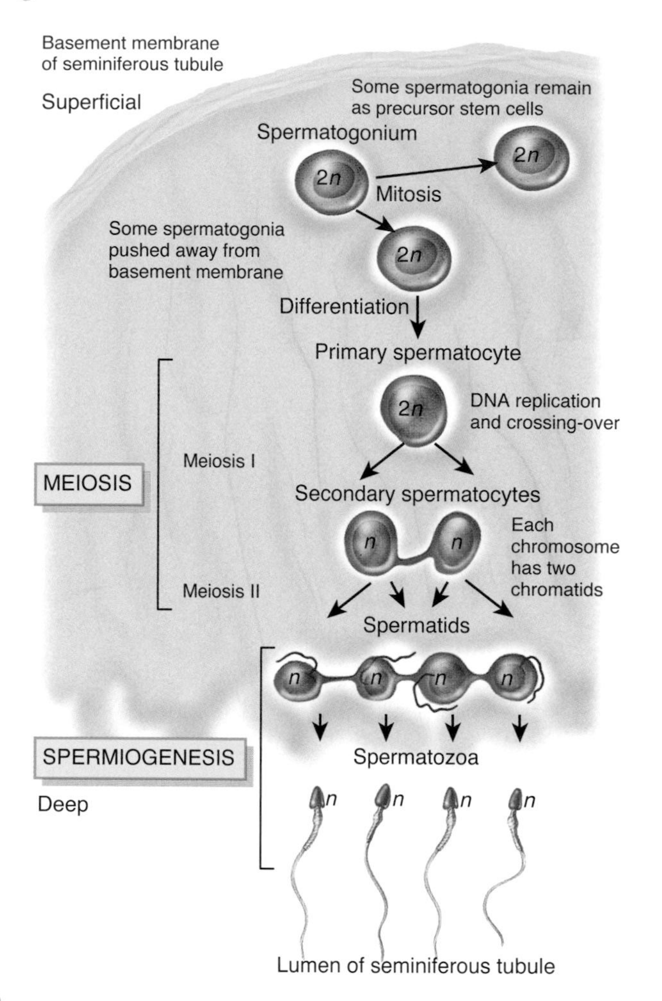

What is the significance of crossing-over?

combinations of genes. As a result, the sperm eventually produced are genetically unlike one another and unlike the parent cell that produced them.

Next, the members of each homologous pair separate, with one member of each pair moving to opposite ends of the cell. The sister chromatids, held by a centromere, remain together. (During mitosis, the sister chromatids move to opposite ends of the cell.) The net effect of meiosis I is that each resulting cell contains a haploid set of chromosomes.

The cells formed by meiosis I are haploid ***secondary spermatocytes***, having 23 (replicated) chromosomes. Each chromosome within a secondary spermatocyte is made up of two chromatids (two copies of the DNA) still attached by a centromere. The genes on each chromatid may be rearranged as a result of crossing-over.

2. **Meiosis II.** In meiosis II there is no further replication of DNA. The chromosomes of the secondary spermatocytes line up in single file near the center of the nucleus, and this time the chromatids separate, as also occurs in mitosis. The cells formed from meiosis II, termed ***spermatids***, contain 23 chromosomes, each of which is composed of a single chromatid.
3. **Spermiogenesis.** In the final stage of spermatogenesis, called ***spermiogenesis*** (sper′-mē-ō-JEN-e-sis), each haploid spermatid develops into a single ***sperm cell*** (see Figure 23.2b).

Sperm

Sperm are produced at the rate of about 300 million per day. Once ejaculated, most do not survive more than 48 hours in the female reproductive tract. The major parts of a sperm cell are the head and the tail (Figure 23.4). The ***head*** contains the nuclear material (DNA) and an ***acrosome*** (*acro-* = atop), a vesicle containing enzymes that aid penetration by the sperm

Figure 23.4 Parts of a sperm cell.

About 300 million sperm mature each day.

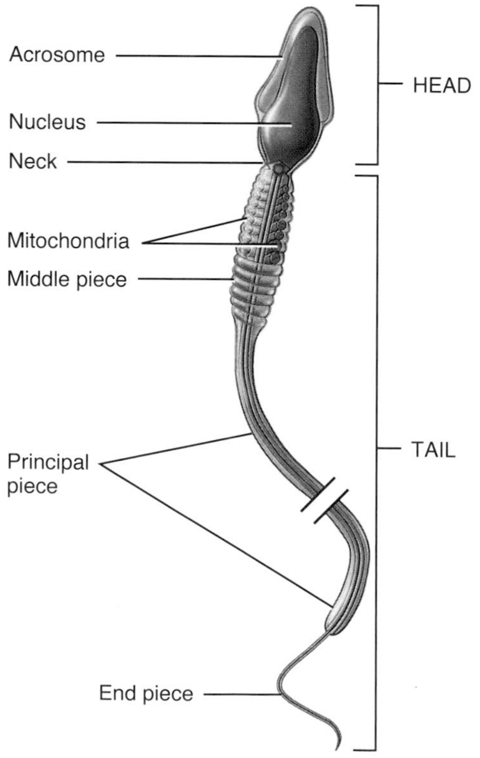

Acrosome
Nucleus
Neck
HEAD
Mitochondria
Middle piece
Principal piece
TAIL
End piece

What is the function of the sperm middle piece?

cell into a secondary oocyte. The ***tail*** of a sperm cell is subdivided into four parts: neck, middle piece, principal piece, and end piece. The *neck* is the constricted region just behind the head. The *middle piece* contains mitochondria that provide ATP for locomotion. The *principal piece* is the longest portion of the tail and the *end piece* is the terminal, tapering portion of the tail.

Hormonal Control of the Testes

At the onset of puberty, neurosecretory cells in the hypothalamus increase their secretion of ***gonadotropin-releasing hormone*** (***GnRH***). This hormone, in turn, stimulates the anterior pituitary to increase its secretion of ***luteinizing hormone (LH)*** and ***follicle-stimulating hormone (FSH)***. Figure 23.5 shows the hormones and negative feedback cycles that control the Leydig and Sertoli cells of the testes and stimulate spermatogenesis.

LH stimulates Leydig cells, which are located between seminiferous tubules, to secrete the hormone ***testosterone*** (tes-TOS-te-rōn). This steroid hormone is synthesized from cholesterol in the testes and is the principal androgen. Testosterone acts in a negative feedback manner to suppress secretion of LH by the anterior pituitary and to suppress secretion of GnRH by hypothalamic neurosecretory cells. In some target cells, such as those in the external genitals and prostate, an enzyme converts testosterone to another androgen called ***dihydrotestosterone (DHT)***.

FSH and testosterone act together to stimulate spermatogenesis. Once the degree of spermatogenesis required for male reproductive functions has been achieved, Sertoli cells release ***inhibin***, a hormone named for its inhibition of FSH secretion by the anterior pituitary (Figure 23.5). Inhibin thus inhibits the secretion of hormones needed for spermatogenesis. If spermatogenesis is proceeding too slowly, less inhibin is released, which permits more FSH secretion and an increased rate of spermatogenesis.

Testosterone and dihydrotestosterone both bind to the same androgen receptors, producing several effects:

- ***Prenatal development.*** Before birth, testosterone stimulates the male pattern of development of reproductive system ducts and the descent of the testes. DHT, by contrast, stimulates development of the external genitals. Testosterone also is converted in the brain to estrogens (feminizing hormones), which may play a role in the development of certain regions of the brain in males.
- ***Development of male sexual characteristics.*** At puberty, testosterone and DHT bring about development and enlargement of the male sex organs and the development of masculine secondary sexual characteristics. These include muscular and skeletal growth that results in wide shoulders and narrow hips; pubic, axillary, facial, and chest hair

Figure 23.5 Hormonal control of spermatogenesis and actions of testosterone and dihydrotestosterone (DHT). Dashed red lines indicate negative feedback inhibition.

Release of FSH is stimulated by GnRH and inhibited by inhibin; release of LH is stimulated by GnRH and inhibited by testosterone.

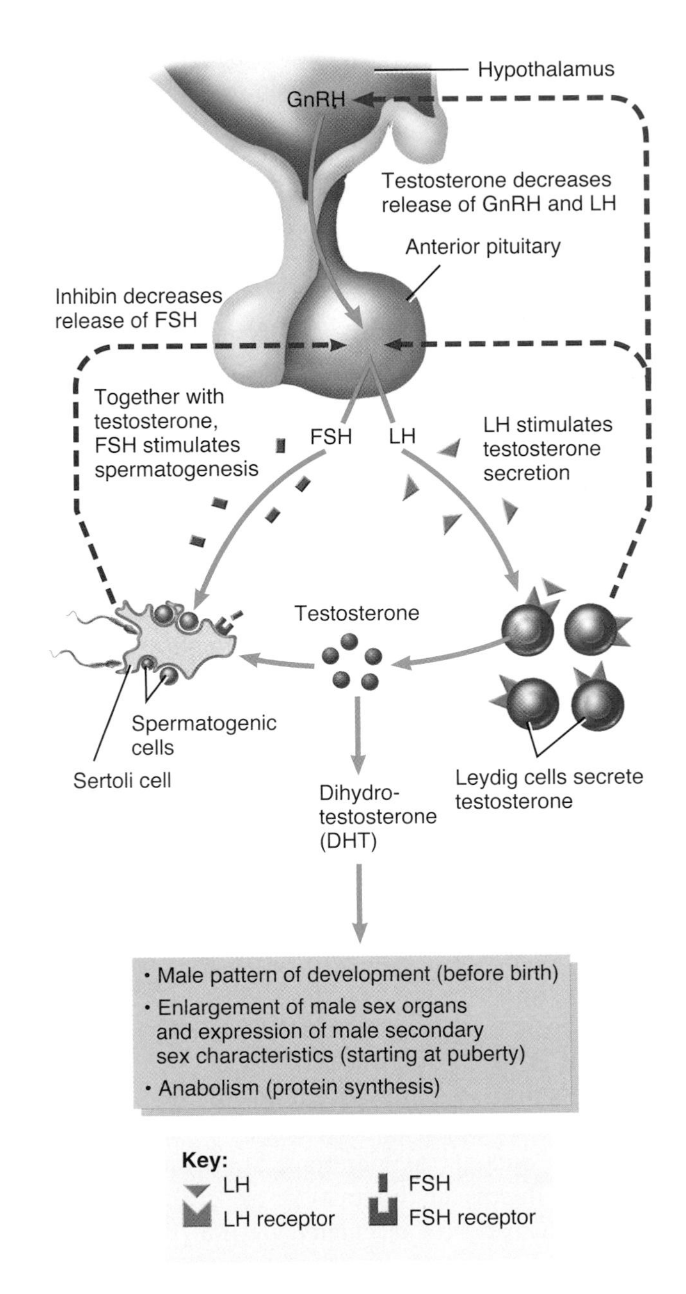

Which cells secrete inhibin?

(within hereditary limits); thickening of the skin; increased sebaceous (oil) gland secretion; and enlargement of the larynx and consequent deepening of the voice.

- ***Development of sexual function.*** Androgens contribute to male sexual behavior and spermatogenesis and to sex drive (libido) in both males and females. Recall that the adrenal cortex is the main source of androgens in females.
- ***Stimulation of anabolism.*** Androgens are anabolic hormones; that is, they stimulate protein synthesis. This effect is obvious in the heavier muscle and bone mass of most men as compared to women.

■ **CHECKPOINT**

1. How does the scrotum protect the testes?
2. What are the principal events of spermatogenesis and where do they occur?
3. What are the roles of FSH, LH, testosterone, and inhibin in the male reproductive system? How is secretion of these hormones controlled?

Ducts

Following spermatogenesis, pressure generated by the continual release of sperm and fluid secreted by Sertoli cells propels sperm and fluid through the seminiferous tubules and into the epididymis (see Figure 23.2a).

Epididymis

The ***epididymis*** (ep′-i-DID-i-mis; *epi-* = above or over; *-didymis* = testis; plural is *epididymides*) is a comma-shaped organ that lies along the posterior border of the testis (see Figures 23.1 and 23.2a). Each epididymis consists mostly of the tightly coiled ***ductus epididymis***. Functionally, the ductus epididymis is the site of *sperm maturation*, the process by which sperm acquire motility and the ability to fertilize a secondary oocyte. This occurs over a 10- to 14-day period. The ductus epididymis also stores sperm and helps propel them during sexual arousal by peristaltic contraction of its smooth muscle into the ductus (vas) deferens. Sperm may remain in storage in the ductus epididymis for several months. Any stored sperm that are not ejaculated by that time are eventually reabsorbed.

Ductus Deferens

At the end of the epididymis, the ductus epididymis becomes less convoluted, and its diameter increases. Beyond the epididymis, the duct is termed the ***ductus deferens*** or ***vas deferens*** (VAS DEF-er-enz; *vas* = vessel; *de-* = away). See Figure 23.2a. The ductus deferens ascends along the posterior border of the epididymis and penetrates the inguinal canal, a passageway in the front abdominal wall. Then, it enters the pelvic cavity, where it loops over the side and down the posterior surface of the urinary bladder (see Figure 23.1). The ductus deferens has a heavy coat of three layers of muscle. Functionally, the ductus deferens stores sperm, which can remain viable here for up to several months. The ductus deferens also conveys sperm from the epididymis toward the urethra during sexual arousal by peristaltic contractions of the muscular coat.

Accompanying the ductus deferens as it ascends in the scrotum are blood vessels, autonomic nerves, and lymphatic vessels that together make up the ***spermatic cord***, a supporting structure of the male reproductive system (see Figure 23.2a).

Ejaculatory Ducts

The ***ejaculatory ducts*** (e-JAK-yū-la-tō′-rē; *ejacul-* = to expel) (see Figure 23.1) are formed by the union of the duct from the ductus deferens and the seminal vesicles (to be described shortly). The ejaculatory ducts eject sperm into the urethra.

Urethra

The ***urethra*** is the terminal duct of the male reproductive system, serving as a passageway for both sperm and urine. In the male, the urethra passes through the prostate, deep perineal muscles, and penis (see Figure 23.1). The opening of the urethra to the exterior is called the ***external urethral orifice***.

Accessory Sex Glands

The ducts of the male reproductive system store and transport sperm cells, but the ***accessory sex glands*** secrete most of the liquid portion of semen.

The paired ***seminal vesicles*** (VES-i-kuls) are pouchlike structures, lying posterior to the base of the urinary bladder and anterior to the rectum (see Figure 23.1). They secrete an alkaline, viscous fluid that contains fructose, prostaglandins, and clotting proteins (unlike those found in blood). The alkaline nature of the fluid helps to neutralize the acidic environment of the male urethra and female reproductive tract that otherwise would inactivate and kill sperm. The fructose is used for ATP production by sperm. Prostaglandins contribute to sperm motility and viability and may also stimulate muscular contraction within the female reproductive tract. Clotting proteins help semen coagulate after ejaculation. Fluid secreted by the seminal vesicles normally constitutes about 60% of the volume of semen.

The ***prostate*** (PROS-tāt) is a single, doughnut-shaped gland about the size of a golf ball (see Figure 23.1). It is inferior to the urinary bladder and surrounds the upper portion of the urethra. The prostate slowly increases in size from birth to puberty, and then it expands rapidly. The size attained by age 30 remains stable until about age 45, when further enlargement may occur. The prostate secretes a milky, slightly acidic fluid (pH about 6.5) that contains (1) *citric acid*,

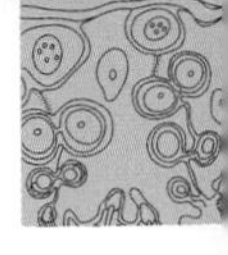

which can be used by sperm for ATP production via the Krebs cycle (see page 511); (2) acid phosphatase (the function of which is unknown); and (3) several protein-digesting enzymes, such as *prostate-specific antigen (PSA)*. Prostatic secretions make up about 25% of the volume of semen.

The paired ***bulbourethral*** (bul′-bō-ū-RĒ-thral) ***glands*** are about the size of peas. They are located inferior to the prostate on either side of the urethra (see Figure 23.1). During sexual arousal, the bulbourethral glands secrete an alkaline substance into the urethra that protects the passing sperm by neutralizing acids from urine in the urethra. At the same time, they secrete mucus that lubricates the end of the penis and the lining of the urethra, thereby decreasing the number of sperm damaged during ejaculation.

Semen

Semen (= seed) is a mixture of sperm and the secretions of the seminal vesicles, prostate, and bulbourethral glands. The volume of semen in a typical ejaculation is 2.5 to 5 milliliters, with 50 to 150 million sperm per milliliter. When the number falls below 20 million per milliliter, the male is likely to be infertile. A very large number of sperm is required for fertilization because only a tiny fraction ever reaches the secondary oocyte.

Despite the slight acidity of prostatic fluid, semen has a slightly alkaline pH of 7.2 to 7.7 due to the higher pH and larger volume of fluid from the seminal vesicles. The prostatic secretion gives semen a milky appearance, and fluids from the seminal vesicles and bulbourethral glands give it a sticky consistency. Semen also contains an antibiotic that can destroy certain bacteria. The antibiotic may help control the abundance of naturally occurring bacteria in the semen and in the lower female reproductive tract. The presence of blood in semen is called ***hemospermia*** (hē-mō-SPER-mē-a; *hemo-* = blood; *-sperma* = seed). In most cases, it is caused by inflammation of the blood vessels lining the seminal vesicles; it is usually treated with antibiotics.

Penis

The ***penis*** contains the urethra and is a passageway for the ejaculation of semen and the excretion of urine (see Figure 23.1). It is cylindrical in shape and consists of a root, a body, and the glans penis. The ***root of the penis*** is the attached portion (proximal portion). The ***body of the penis*** is composed of three cylindrical masses of tissue. The two dorsolateral masses are called the ***corpora cavernosa penis*** (*corpora* = main bodies; *cavernosa* = hollow). The smaller midventral mass, the ***corpus spongiosum penis***, contains the urethra. All three masses are enclosed by fascia (a sheet of fibrous connective tissue) and skin and consist of erectile tissue permeated by blood sinuses.

The distal end of the corpus spongiosum penis is a slightly enlarged region called the ***glans penis***. In the glans penis is the opening of the urethra (the ***external urethral orifice***) to the exterior. Covering the glans in an uncircumcised penis is the loosely fitting ***prepuce*** (PRĒ-poos), or ***foreskin***.

Circumcision (= to cut around) is a surgical procedure in which part or the entire prepuce is removed. It is usually performed just after delivery, 3 to 4 days after birth, or on the eighth day as part of a Jewish religious rite. Although most health-care professionals find no medical justification for circumcision, some feel that it has benefits, such as a lower risk of urinary tract infections, protection against penile cancer, and possibly a lower risk for sexually transmitted diseases. Indeed, studies in several African villages have found lower rates of HIV infection among circumcised men.

Most of the time, the penis is flaccid (limp) because its arteries are vasoconstricted, which limits blood flow. The first visible sign of sexual excitement is ***erection***, the enlargement and stiffening of the penis. Parasympathetic impulses cause release of neurotransmitters and local hormones, including the gas nitric oxide, which relaxes vascular smooth muscle in the penile arteries. The arteries supplying the penis dilate, and large quantities of blood enter the blood sinuses. Expansion of these spaces compresses the veins draining the penis, so blood outflow is slowed.

Ejaculation (ē-jak-ū-LĀ-shun; *ejectus-* = to throw out), the powerful release of semen from the urethra to the exterior, is a sympathetic reflex coordinated by the lumbar portion of the spinal cord. As part of the reflex, the smooth muscle sphincter at the base of the urinary bladder closes. Thus, urine is not expelled during ejaculation, and semen does not enter the urinary bladder. Even before ejaculation occurs, peristaltic contractions in the ductus deferens, seminal vesicles, ejaculatory ducts, and prostate propel semen into the penile portion of the urethra. Typically, this leads to ***emission*** (ē-MISH-un), the discharge of a small volume of semen before ejaculation. Emission may also occur during sleep (nocturnal emission). The penis returns to its flaccid state when the arteries constrict, and pressure on the veins is relieved.

Erectile dysfunction (ED), previously termed *impotence*, is the consistent inability of an adult male to ejaculate or to attain or hold an erection long enough for sexual intercourse. Many cases of impotence are caused by insufficient release of nitric oxide. The drug sildenafil (Viagra®) enhances the effect of nitric oxide.

■ CHECKPOINT

4. Trace the course of sperm through the system of ducts from the seminiferous tubules through the urethra.
5. What is semen? What is its function?

FEMALE REPRODUCTIVE SYSTEM

OBJECTIVES • Describe the location, structure, and functions of the organs of the female reproductive system.
• Describe how oocytes are produced.

The organs of the ***female reproductive system*** (Figure 23.6) include the ovaries; the uterine (fallopian) tubes, or oviducts; the uterus; the vagina; and external organs, which are collectively called the vulva, or pudendum. The mammary glands also are considered part of the female reproductive system.

Ovaries

The ***ovaries*** (= egg receptacles) are paired organs that produce secondary oocytes (cells that develop into mature ova, or eggs, following fertilization) and hormones, such as progesterone and estrogens (the female sex hormones), inhibin, and relaxin. The ovaries arise from the same embryonic tissue as the testes, and they are the size and shape of unshelled almonds. One ovary lies on each side of the pelvic cavity, held in place by ligaments. Figure 23.7 shows the histology of an ovary.

The ***germinal epithelium*** is a layer of simple epithelium (low cuboidal or squamous) that covers the surface of the ovary. Deep to the germinal epithelium is the ***ovarian cortex***,

Figure 23.6 Female organs of reproduction and surrounding structures.

The female organs of reproduction include the ovaries, uterine (fallopian) tubes, uterus, vagina, vulva, and mammary glands.

Functions of the Female Reproductive System

1. Ovaries: produce secondary oocytes and hormones, including estrogens, progesterone, inhibin, and relaxin.
2. Uterine tubes: transport a secondary oocyte to the uterus, and normally are the sites where fertilization occurs.
3. Uterus: site of implantation of a fertilized ovum, development of the fetus during pregnancy, and labor.
4. Vagina: receives the penis during sexual intercourse and is a passageway for childbirth.
5. Mammary glands: synthesize, secrete, and eject milk for nourishment of the newborn.

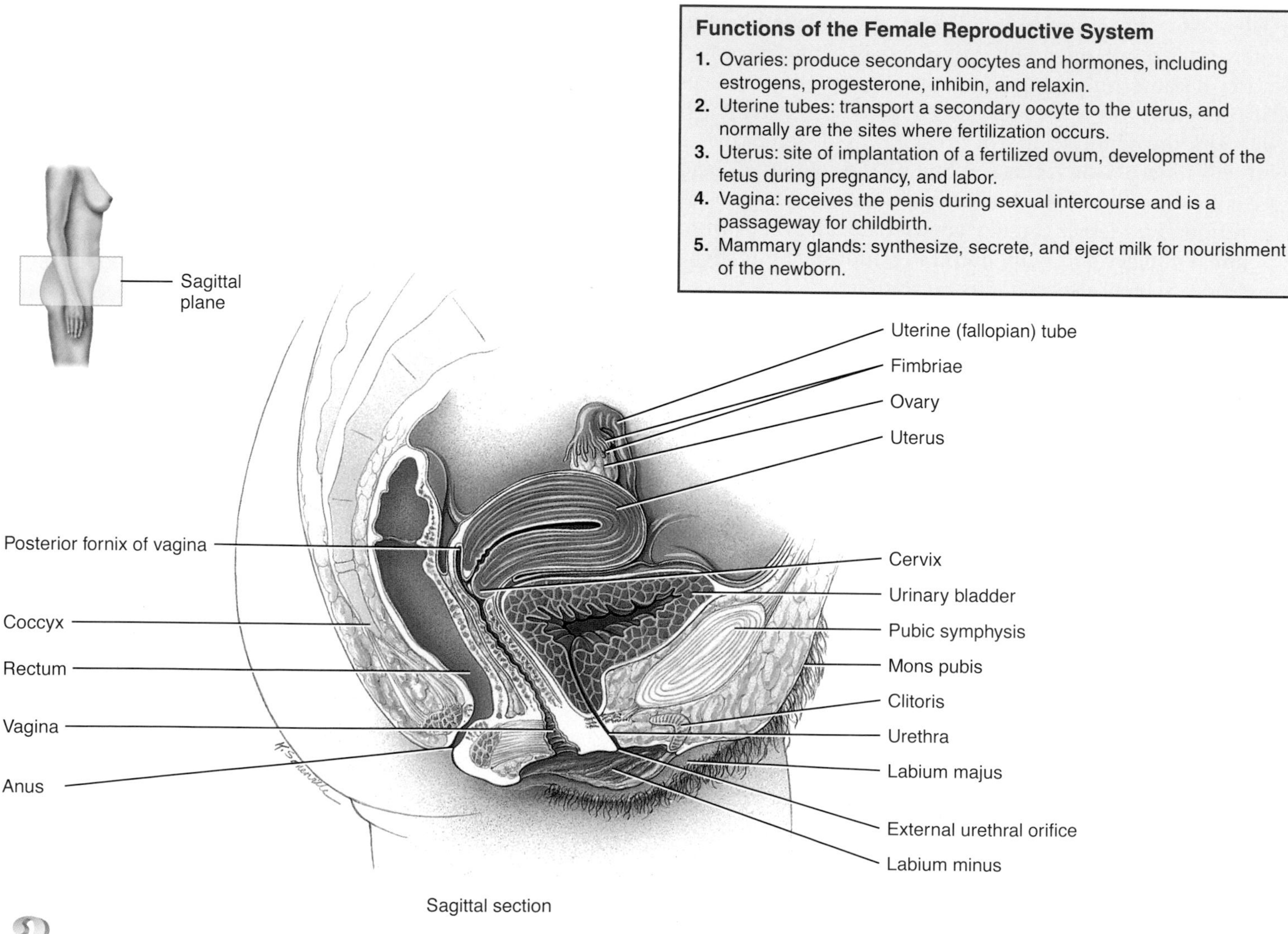

What term refers to the external genitals of the female?

Figure 23.7 Histology of the ovary. The arrows indicate the sequence of developmental stages that occur as part of the maturation of an ovum during the ovarian cycle.

The ovaries are the female gonads; they produce haploid oocytes.

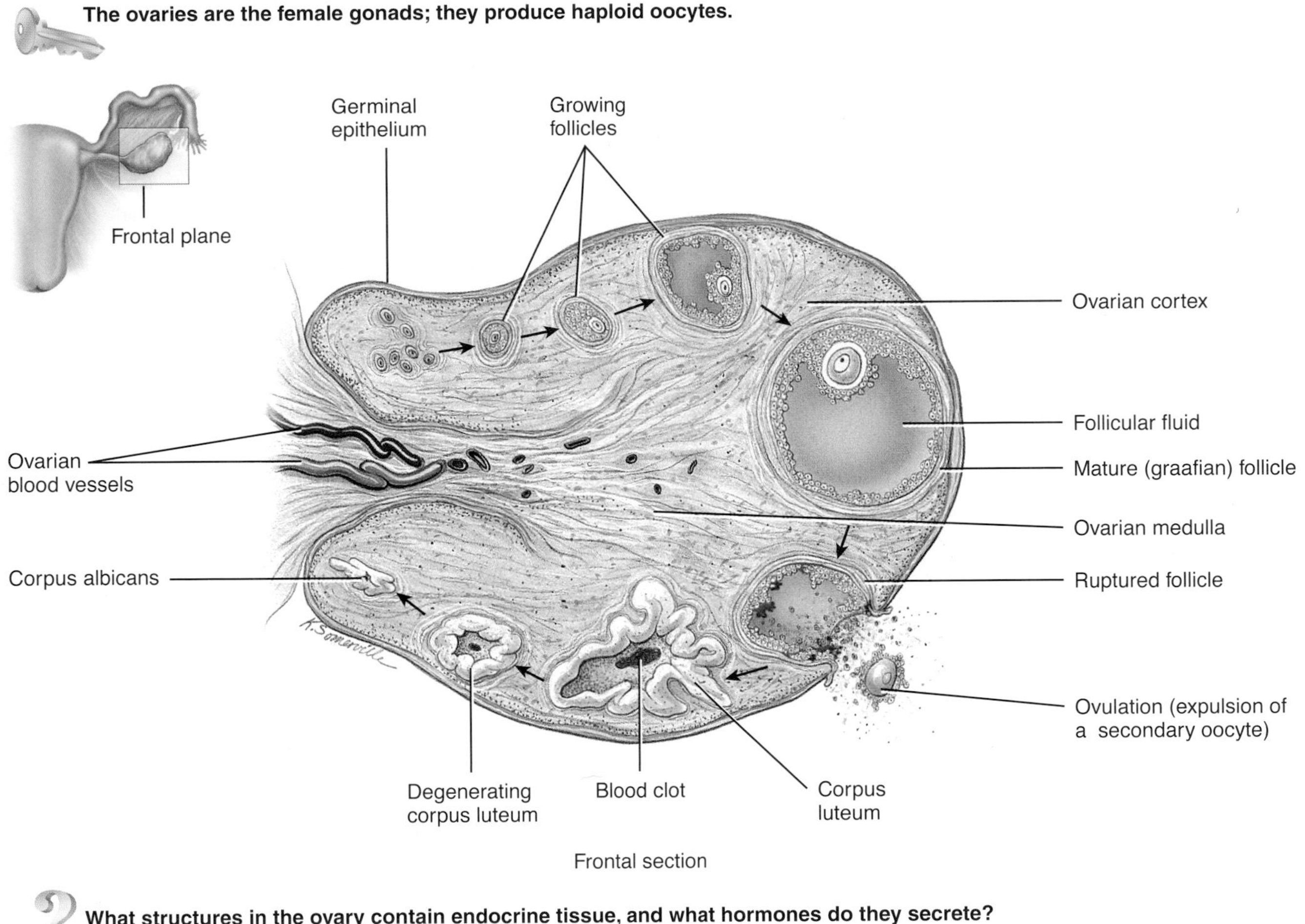

What structures in the ovary contain endocrine tissue, and what hormones do they secrete?

a region of dense connective tissue that contains ovarian follicles. Each ***ovarian follicle*** (*folliculus* = little bag) consists of an ***oocyte*** and a variable number of surrounding cells that nourish the developing oocyte and begin to secrete estrogens as the follicle grows larger. The follicle enlarges until it is a ***mature (graafian) follicle***, a large, fluid-filled follicle that is preparing to rupture and expel a secondary oocyte. The remnants of an ovulated follicle develop into a ***corpus luteum*** (= yellow body). The corpus luteum produces progesterone, estrogens, relaxin, and inhibin until it degenerates and turns into fibrous tissue called a ***corpus albicans*** (= white body). The ***ovarian medulla*** is a region deep to the ovarian cortex that consists of loose connective tissue and contains blood vessels, lymphatic vessels, and nerves.

An **ovarian cyst** is a fluid-filled sac in or on an ovary. Such cysts are relatively common, are usually noncancerous, and frequently disappear on their own. Cancerous cysts are more likely to occur in women over 40. Ovarian cysts may cause pain, pressure, a dull ache, or fullness in the abdomen; pain during sexual intercourse; delayed, painful, or irregular menstrual periods; abrupt onset of sharp pain in the lower abdomen; and/or vaginal bleeding. Most ovarian cysts require no treatment, but larger ones (more than 5 cm or 2 in.) may be removed surgically.

Oogenesis

Formation of gametes in the ovaries is termed ***oogenesis*** (ō′-ō-JEN-e-sis; *oo-* = egg). Unlike spermatogenesis, which begins in males at puberty, oogenesis begins in females before they are even born. Also, males produce new sperm throughout life, while females have all the eggs they will ever have by birth. Oogenesis occurs in essentially the same manner as spermatogenesis. It involves meiosis and maturation.

MEIOSIS I During early fetal development, cells in the ovaries differentiate into ***oogonia*** (ō′-ō-GŌ-nē-a), which can give rise to cells that develop into secondary oocytes

(Figure 23.8). Before birth, most of these cells degenerate, but a few develop into larger cells called ***primary oocytes*** (Ō-ō-sīts). These cells begin meiosis I during fetal development but do not complete it until after puberty. At birth, 200,000 to 2,000,000 primary oocytes remain in each ovary. Of these, about 40,000 remain at puberty, but only 400 go on to mature and ovulate during a woman's reproductive lifetime. The remainder degenerate.

After puberty, hormones secreted by the anterior pituitary stimulate the resumption of oogenesis each month. Meiosis I resumes in several primary oocytes, although in each cycle only one follicle typically reaches the maturity needed for ovulation. The diploid primary oocyte completes meiosis I, resulting in two haploid cells of unequal size, both with 23 chromosomes (n) of two chromatids each. The smaller cell produced by meiosis I, called the *first polar body*, is essentially a packet of discarded nuclear material; the larger cell, known as the ***secondary oocyte***, receives most of the cytoplasm. Once a secondary oocyte is formed, it begins meiosis II and then stops. The follicle in which these events are taking place—the mature (graafian) follicle—soon ruptures and releases its secondary oocyte, a process known as ***ovulation***.

Figure 23.8 Oogenesis. Diploid cells ($2n$) have 46 chromosomes; haploid cells (n) have 23 chromosomes.

In an oocyte, meiosis II is completed only if fertilization occurs.

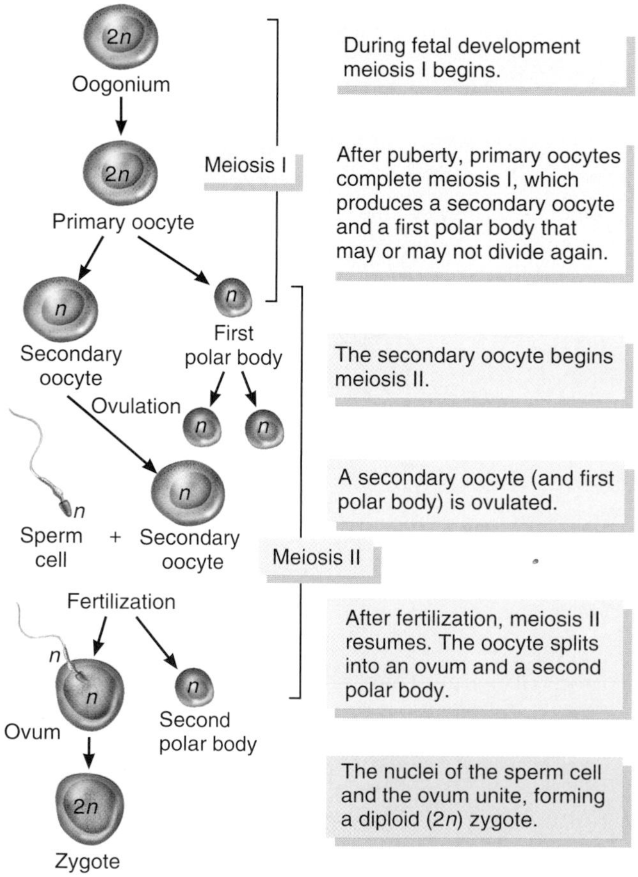

During fetal development meiosis I begins.

After puberty, primary oocytes complete meiosis I, which produces a secondary oocyte and a first polar body that may or may not divide again.

The secondary oocyte begins meiosis II.

A secondary oocyte (and first polar body) is ovulated.

After fertilization, meiosis II resumes. The oocyte splits into an ovum and a second polar body.

The nuclei of the sperm cell and the ovum unite, forming a diploid ($2n$) zygote.

How does the age of a primary oocyte in a female compare with the age of a primary spermatocyte in a male?

MEIOSIS II At ovulation, usually a single secondary oocyte (with the first polar body) is expelled into the pelvic cavity and swept into the uterine (fallopian) tube. If a sperm penetrates the secondary oocyte (fertilization), meiosis II resumes. The secondary oocyte splits into two haploid (n) cells of unequal size. The larger cell is the ***ovum***, or mature egg; the smaller one is the *second polar body*. The nuclei of the sperm cell and the ovum then unite, forming a diploid ($2n$) ***zygote***. The first polar body may also undergo another division to produce two polar bodies. If it does, the primary oocyte ultimately gives rise to a single haploid (n) ovum and three haploid (n) polar bodies. Thus, each primary oocyte gives rise to a single gamete (secondary oocyte, which becomes an ovum after fertilization); in contrast, each primary spermatocyte produces four gametes (sperm).

Uterine Tubes

Females have two ***uterine (fallopian) tubes*** that extend laterally from the uterus and transport the secondary oocytes from the ovaries to the uterus (Figure 23.9). The open, funnel-shaped end of each tube, the ***infundibulum***, lies close to the ovary but is open to the pelvic cavity. It ends in a fringe of fingerlike projections called ***fimbriae*** (FIM-brē-ē = fringe). From the infundibulum, the uterine tubes extend medially, attaching to the upper and outer corners of the uterus.

After ovulation, local currents produced by movements of the fimbriae, which surround the surface of the mature follicle just before ovulation occurs, sweep the secondary oocyte into the uterine tube. The oocyte is then moved along the tube by cilia in the tube's mucous lining and peristaltic contractions of its smooth muscle layer.

The usual site for fertilization of a secondary oocyte by a sperm cell is in the uterine tube. Fertilization may occur any time up to about 24 hours after ovulation. The fertilized ovum (zygote) descends into the uterus within seven days. Unfertilized secondary oocytes disintegrate.

Uterus

The ***uterus*** (*womb*) serves as part of the pathway for sperm deposited in the vagina to reach the uterine tubes. It is also the site of implantation of a fertilized ovum, development of the fetus during pregnancy, and labor. During reproductive cycles when implantation does not occur, the uterus is the source of menstrual flow. The uterus is situated between the

Figure 23.9 Uterus and associated structures. In the left side of the drawing, the uterine tube and uterus have been sectioned to show internal structures.

The uterus is the site of menstruation, implantation of a fertilized ovum, development of a fetus, and labor.

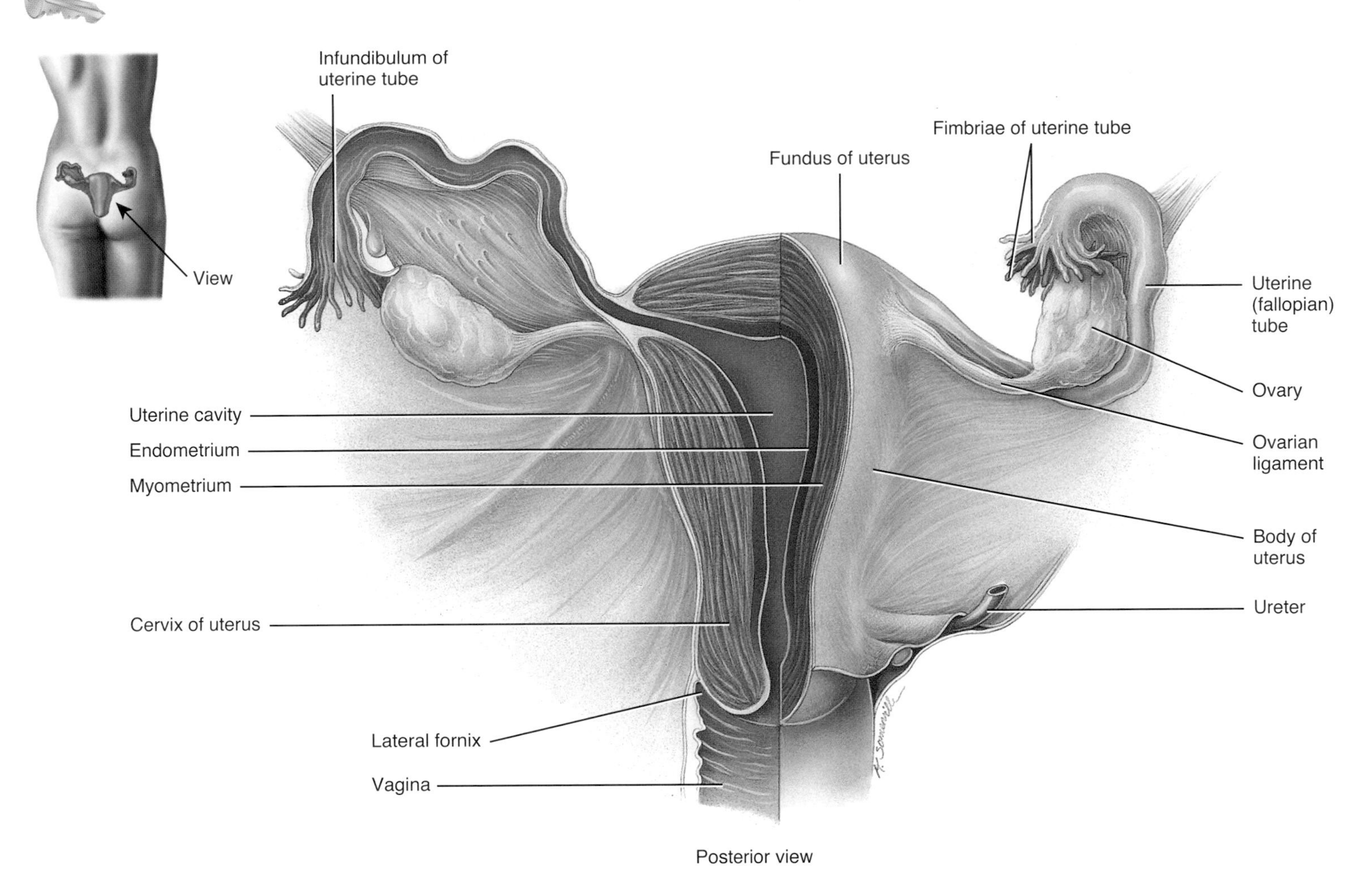

Which part of the uterine lining rebuilds after each menstruation?

urinary bladder and the rectum and is shaped like an inverted pear.

Parts of the uterus include the dome-shaped portion superior to the uterine tubes called the ***fundus***, the tapering central portion called the ***body***, and the narrow portion opening into the vagina called the ***cervix***. The interior of the body of the uterus is called the ***uterine cavity*** (Figure 23.9).

The middle muscular layer of the uterus, the ***myometrium*** (*myo-* = muscle), consists of smooth muscle and forms the bulk of the uterine wall. During childbirth, coordinated contractions of uterine muscles help expel the fetus.

The innermost part of the uterine wall, the ***endometrium*** (*endo* = within), is a mucous membrane. It nourishes a growing fetus or is shed each month during menstruation if fertilization does not occur. The endometrium contains many *endometrial glands* whose secretions nourish sperm and the zygote.

Hysterectomy (hiss-ter-EK-tō-mē; *hyster-* = uterus), the surgical removal of the uterus, is the most common gynecological operation. It may be indicated in conditions such as fibroids, endometriosis, pelvic inflammatory disease, recurrent ovarian cysts, excessive uterine bleeding, and cancer of the cervix, uterus, or ovaries. In a *partial hysterectomy*, the body of the uterus is removed but the cervix is left in place. A *complete hysterectomy* is the removal of both the body and cervix of the uterus.

Vagina

The ***vagina*** (va-JĪ-na = sheath) is a tubular canal that extends from the exterior of the body to the uterine cervix (Figure 23.9). It is the receptacle for the penis during sexual intercourse, the outlet for menstrual flow, and the passageway for childbirth. The vagina is situated between the urinary bladder and the rectum. A recess, called the ***fornix*** (= arch or vault), surrounds the cervix. When properly inserted, a contraceptive diaphragm rests on the fornix, covering the cervix.

The mucosa of the vagina contains large stores of glycogen, the decomposition of which produces organic acids. The resulting acidic environment retards microbial growth, but it also is harmful to sperm. Alkaline components of semen, mainly from the seminal vesicles, neutralize the acidity of the vagina and increase viability of sperm. The muscular layer is composed of smooth muscle that can stretch to receive the penis during intercourse and allow for childbirth. There may be a thin fold of mucous membrane called the ***hymen*** (= membrane) partially covering the ***vaginal orifice***, the vaginal opening (see Figure 23.10).

Perineum and Vulva

The ***perineum*** (per′-i-NĒ-um) is the diamond-shaped area between the thighs and buttocks of both males and females that contains the external genitals and anus (Figure 23.10).

The term ***vulva*** (VUL-va = to wrap around), or ***pudendum*** (pū-DEN-dum), refers to the external genitals of the female (Figure 23.10). The ***mons pubis*** (MONZ PŪ-bis; *mons* = mountain) is an elevation of adipose tissue covered by coarse pubic hair, which cushions the pubic symphysis. From the mons pubis, two longitudinal folds of skin, the ***labia majora*** (LĀ-bē-a ma-JŌ-ra; *labia* = lips; *majora* = larger), extend down and back (singular is *labium majus*). In females the labia majora develop from the same embryonic tissue that the scrotum develops from in males. The labia majora contain adipose tissue and sebaceous (oil) and sudoriferous (sweat) glands. Like the mons pubis, they are covered by pubic hair. Medial to the labia majora are two folds of skin called the ***labia minora*** (mī-NŌ-ra = smaller; singular is *labium minus*). The labia minora do not contain pubic hair or fat and have few sudoriferous (sweat) glands; they do, however, contain numerous sebaceous (oil) glands.

Figure 23.10 Components of the vulva.

Like the penis, the clitoris is capable of erection upon sexual stimulation.

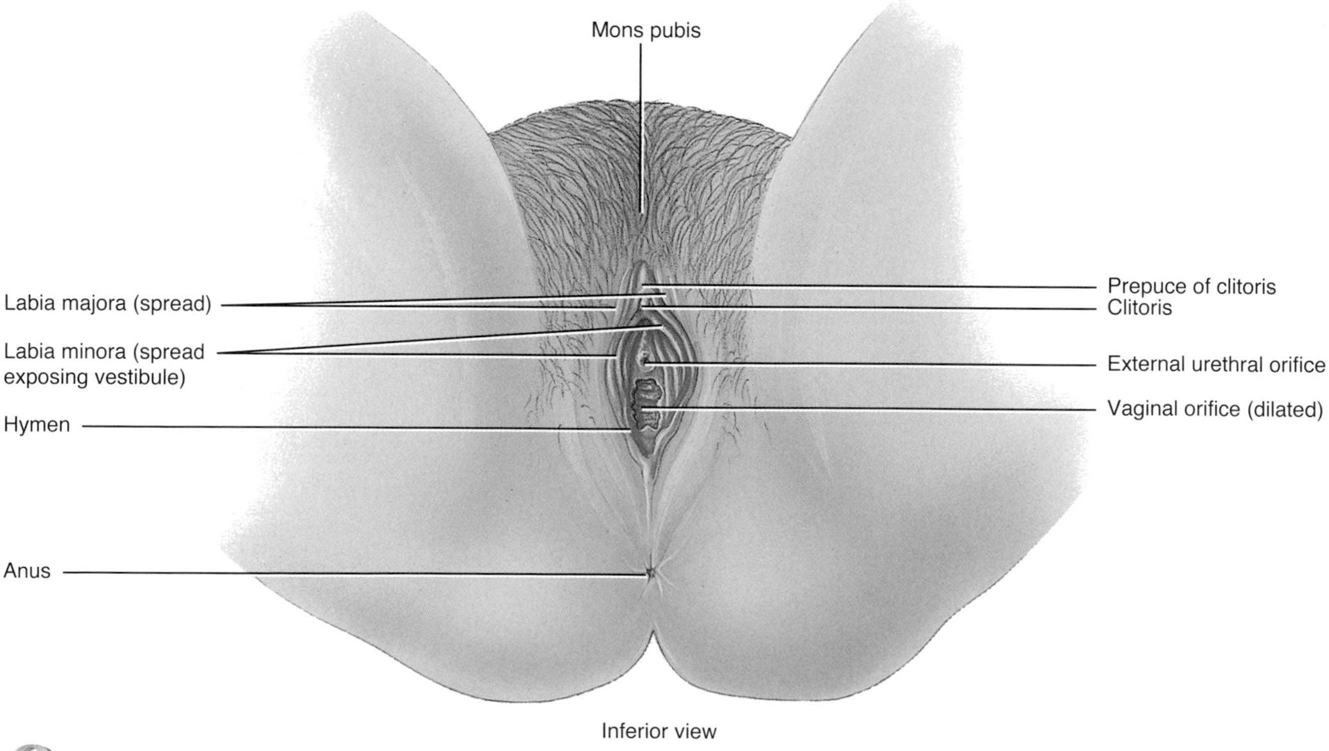

What surface structures are anterior to the vaginal opening?

The ***clitoris*** (KLIT-o-ris) is a small, cylindrical mass of erectile tissue and nerves. It is located at the anterior junction of the labia minora. A layer of skin called the ***prepuce*** (PRĒ-poos), also known as the *foreskin*, is formed at a point where the labia minora unite and cover the body of the clitoris. The exposed portion of the clitoris is the ***glans***. Like the penis, the clitoris is capable of enlargement upon sexual stimulation.

The region between the labia minora is called the ***vestibule***. In the vestibule are the hymen (if present); ***vaginal orifice***, the opening of the vagina to the exterior; ***external urethral orifice***, the opening of the urethra to the exterior; and on either side of the external urethral orifice, the openings of the ducts of the ***paraurethral glands***. These glands in the wall of the urethra secrete mucus. The male's prostate develops from the same embryonic tissue as the female's paraurethral glands. On either side of the vaginal orifice itself are the ***greater vestibular glands***, which produce a small quantity of mucus during sexual arousal and intercourse that adds to cervical mucus and provides lubrication. In males, the bulbourethral glands are equivalent structures.

During childbirth, if the vagina is too small to accommodate the head of an emerging fetus, the skin, vaginal epithelium, subcutaneous fat, and muscle of the perineum may tear. Moreover, the tissues of the rectum may be damaged. To avoid such damage, a small incision called an **episiotomy** (e-piz-ē-OT-ō-mē; *episi-* = vulva or pubic region; *-otomy* = incision) is made in the perineal skin and underlying tissues just prior to delivery. After delivery, the episiotomy is sutured in layers.

Mammary Glands

The ***mammary glands*** (*mamma* = breast), located in the breasts, are modified sudoriferous (sweat) glands that produce milk. The breasts lie over the pectoralis major and serratus anterior muscles and are attached to them by a layer of connective tissue (Figure 23.11). Each breast has one pigmented projection, the ***nipple***, with a series of closely spaced openings of ducts where milk emerges. The circular pigmented area of skin surrounding the nipple is called the

Figure 23.11 Mammary glands.

The mammary glands function in the synthesis, secretion, and ejection of milk (lactation).

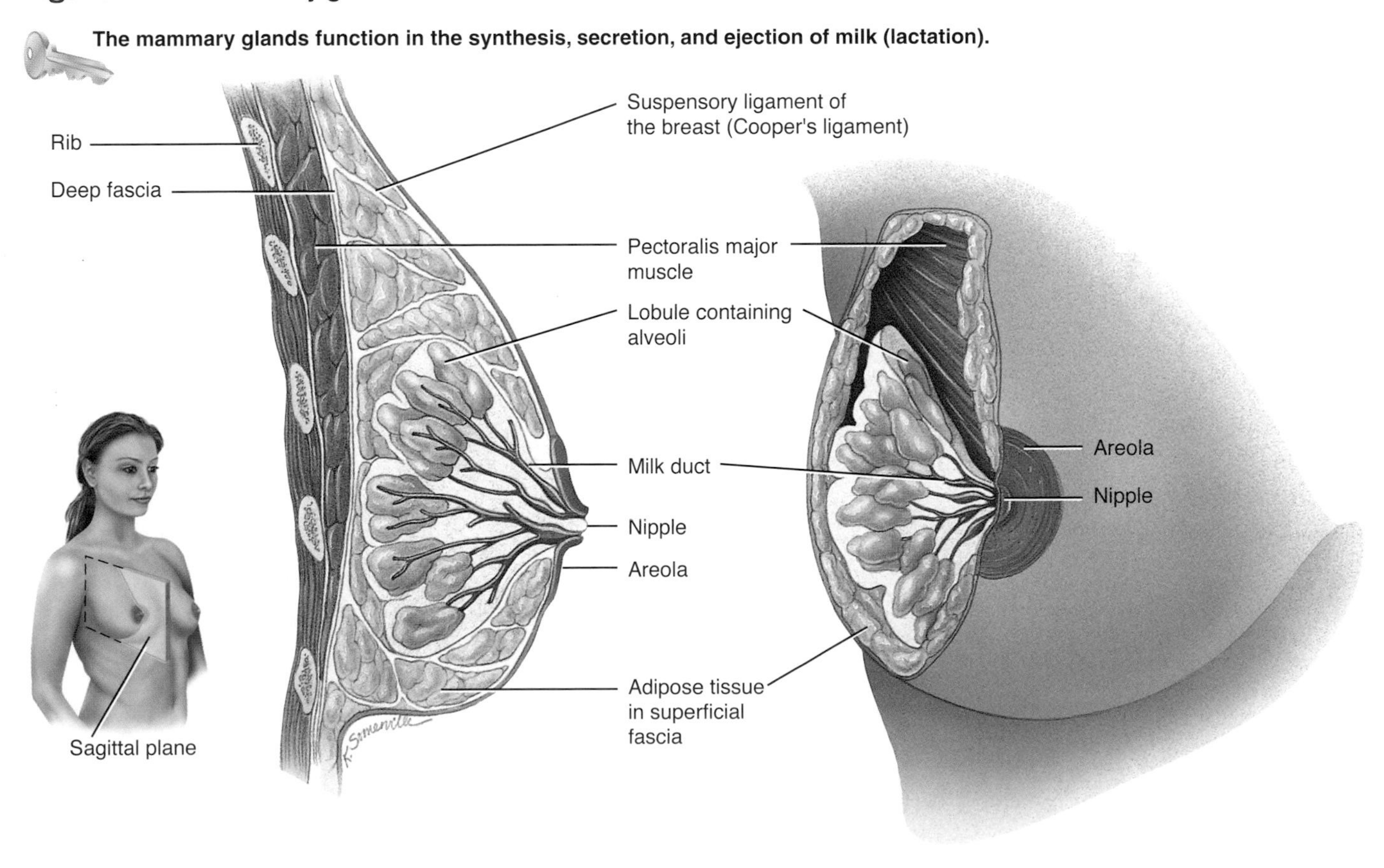

What hormone regulates the ejection of milk from the mammary glands?

areola (a-RĒ-ō-la = small space). This region appears rough because it contains modified sebaceous (oil) glands. Internally, each mammary gland consists of 15 to 20 ***lobes*** arranged radially and separated by adipose tissue and strands of connective tissue called ***suspensory ligaments of the breast (Cooper's ligaments)***, which support the breast. In each lobe are smaller ***lobules***, in which milk-secreting glands called ***alveoli*** (= small cavities) are found. When milk is being produced, it passes from the alveoli into a series of tubules that drain toward the nipple.

At birth, the mammary glands are undeveloped and appear as slight elevations on the chest. With the onset of puberty, under the influence of estrogens and progesterone, the female breasts begin to develop. The duct system matures and fat is deposited, which increases breast size. The areola and nipple also enlarge and become more darkly pigmented.

The functions of the mammary glands are the synthesis, secretion, and ejection of milk; these functions, called ***lactation***, are associated with pregnancy and childbirth. Milk production is stimulated largely by the hormone prolactin from the anterior pituitary, with contributions from progesterone and estrogens. The ejection of milk is stimulated by oxytocin, which is released from the posterior pituitary in response to the sucking of an infant on the mother's nipple (suckling).

> The breasts of females are highly susceptible to cysts and tumors. In **fibrocystic disease**, the most common cause of breast lumps in females, one or more cysts (fluid-filled sacs) and thickening of alveoli develop. The condition, which occurs mainly in females between the ages of 30 and 50, is probably due to a relative excess of estrogens or a deficiency of progesterone in the postovulatory phase of the reproductive cycle (discussed shortly). Fibrocystic disease usually causes one or both breasts to become lumpy, swollen, and tender a week or so before menstruation begins.

■ **CHECKPOINT**

6. Describe the principal events of oogenesis.
7. Where are the uterine tubes located? What is their function?
8. Describe the histology of the uterus.
9. What is the function of the vagina? Describe its histology.
10. Describe the structure of the mammary glands. How are they supported?

THE FEMALE REPRODUCTIVE CYCLE

OBJECTIVE • **Describe the major events of the ovarian and uterine cycles.**

During their reproductive years, nonpregnant females normally exhibit cyclical changes in the ovaries and uterus. Each cycle takes about a month and involves both oogenesis and preparation of the uterus to receive a fertilized ovum. Hormones secreted by the hypothalamus, anterior pituitary, and ovaries control the main events. You have already learned about the ***ovarian cycle***, the series of events in the ovaries that occur during and after the maturation of an oocyte. Steroid hormones released by the ovaries control the ***uterine (menstrual) cycle***, a concurrent series of changes in the endometrium of the uterus to prepare it for the arrival of a fertilized ovum that will develop there until birth. If fertilization does not occur, the levels of ovarian hormones decrease, which causes part of the endometrium to slough off. The general term ***female reproductive cycle*** encompasses the ovarian and uterine cycles, the hormonal changes that regulate them, and the related cyclical changes in the breasts and cervix.

Hormonal Regulation of the Female Reproductive Cycle

Gonadotropin-releasing hormone (GnRH) secreted by the hypothalamus controls the ovarian and uterine cycles (Figure 23.12). GnRH stimulates the release of ***follicle-stimulating hormone (FSH)*** and ***luteinizing hormone (LH)*** from the anterior pituitary. FSH, in turn, initiates follicular growth and the secretion of estrogens by the growing follicles. LH stimulates the further development of ovarian follicles and their full secretion of estrogens. At midcycle, LH triggers ovulation and then promotes formation of the corpus luteum, the reason for the name luteinizing hormone. Stimulated by LH, the corpus luteum produces and secretes estrogens, progesterone, relaxin, and inhibin.

Estrogens secreted by ovarian follicles have several important functions throughout the body:

- Estrogens promote the development and maintenance of female reproductive structures, feminine secondary sex characteristics, and the mammary glands. The secondary sex characteristics include distribution of adipose tissue in the breasts, abdomen, mons pubis, and hips; a broad pelvis; and the pattern of hair growth on the head and body.
- Estrogens stimulate protein synthesis, acting together with insulinlike growth factors, insulin, and thyroid hormones.
- Estrogens lower blood cholesterol level, which is probably the reason that women under age 50 have a much lower risk of coronary artery disease than do men of comparable age.

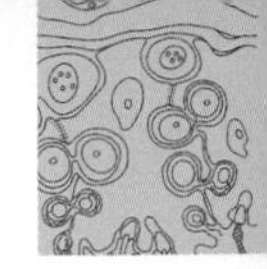

Figure 23.12 The female reproductive cycle. The length of the female reproductive cycle typically is 24 to 36 days; the preovulatory phase is more variable in length than the other phases. (a) Events in the ovarian and uterine cycles and the release of anterior pituitary hormones are correlated with the sequence of the cycle's four phases. In the cycle shown, fertilization and implantation have not occurred. (b) Relative concentrations of anterior pituitary hormones (FSH and LH) and ovarian hormones (estrogens and progesterone) during the phases of a normal female reproductive cycle.

Estrogens are secreted by the dominant follicle before ovulation; after ovulation, both progesterone and estrogens are secreted by the corpus luteum.

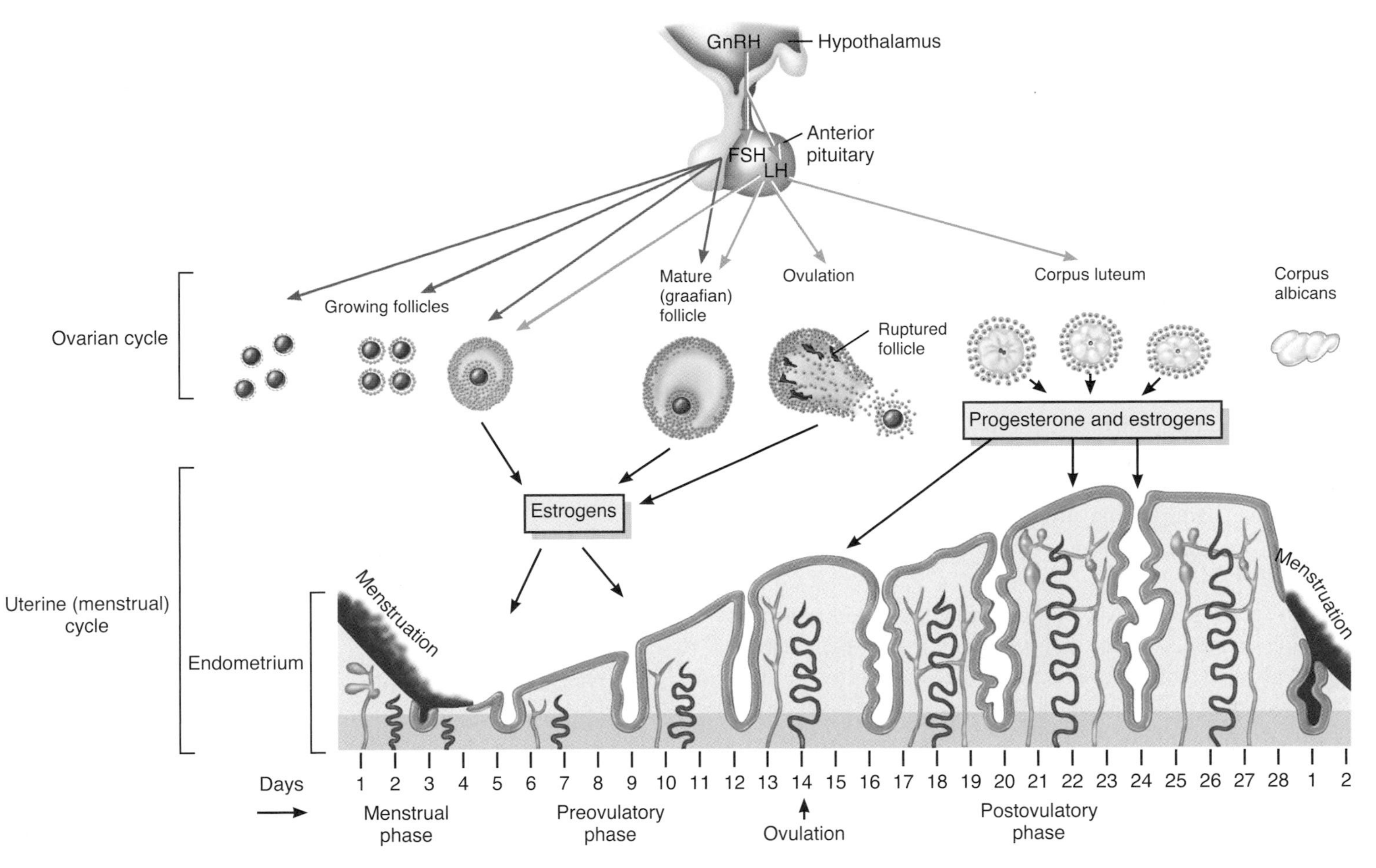

(a) Hormonal regulation of changes in the ovary and uterus

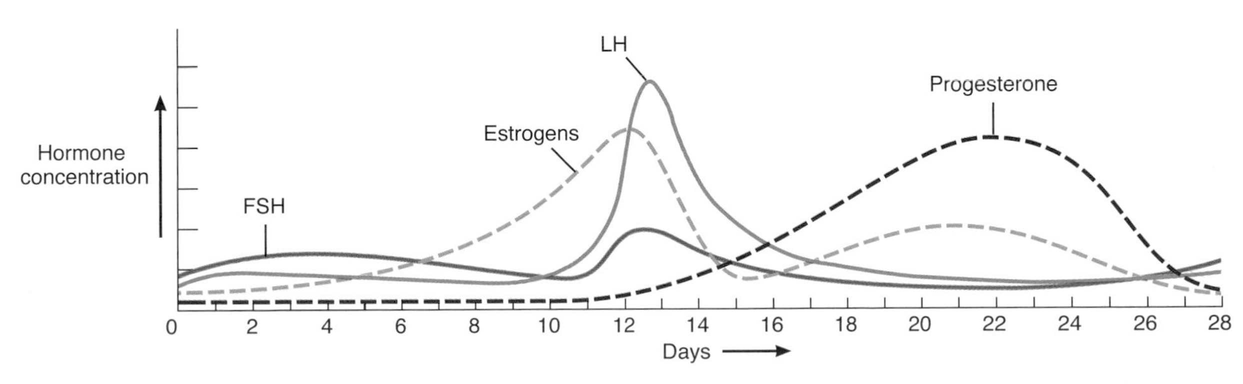

(b) Changes in concentration of anterior pituitary and ovarian hormones

Which hormones are responsible for the proliferative phase of endometrial growth, for ovulation, for growth of the corpus luteum, and for the surge of LH at midcycle?

Progesterone, secreted mainly by cells of the corpus luteum, acts together with estrogens to prepare and then maintain the endometrium for implantation of a fertilized ovum and to prepare the mammary glands for milk secretion.

A small quantity of ***relaxin***, produced by the corpus luteum during each monthly cycle, relaxes the uterus by inhibiting contractions of the myometrium. Presumably, implantation of a fertilized ovum occurs more readily in a "quiet" uterus. During pregnancy, the placenta produces much more relaxin, and it continues to relax uterine smooth muscle. At the end of pregnancy, relaxin also increases the flexibility of the pubic symphysis and helps dilate the uterine cervix, both of which ease delivery of the baby.

Inhibin is secreted by growing follicles and by the corpus luteum after ovulation. It inhibits secretion of FSH and, to a lesser extent, LH.

Phases of the Female Reproductive Cycle

The duration of the female reproductive cycle varies from 24 to 35 days. For this discussion we assume a duration of 28 days and divide it into four phases: the menstrual phase, the preovulatory phase, ovulation, and the postovulatory phase (Figure 23.12). Because they occur at the same time, the events of the ovarian cycle (events in the ovaries) and menstrual cycle (events in the uterus) will be discussed together.

Menstrual Phase

The ***menstrual phase*** (MEN-stroo-al), also called ***menstruation*** (men′-stroo-Ā-shun) or ***menses*** (= month), lasts for roughly the first five days of the cycle. (By convention, the first day of menstruation marks the first day of a new cycle).

EVENTS IN THE OVARIES During the menstrual phase, several ovarian follicles grow and enlarge.

EVENTS IN THE UTERUS Menstrual flow from the uterus consists of 50 to 150 mL of blood and tissue cells from the endometrium. This discharge occurs because the declining level of ovarian hormones (progesterone and estrogens) causes the uterine arteries to constrict. As a result, the cells they supply become oxygen-deprived and start to die. Eventually, part of the endometrium sloughs off. The menstrual flow passes from the uterine cavity to the cervix and through the vagina to the exterior.

Preovulatory Phase

The ***preovulatory phase*** is the time between the end of menstruation and ovulation. The preovulatory phase of the cycle accounts for most of the variation in cycle length. In a 28-day cycle, it lasts from days 6 to 13.

EVENTS IN THE OVARIES Under the influence of FSH, several follicles continue to grow and begin to secrete estrogens and inhibin. By about day 6, a single follicle in one of the two ovaries has outgrown all the others to become the *dominant follicle*. Estrogens and inhibin secreted by the dominant follicle decrease the secretion of FSH (Figure 23.12b, see days 8 to 11), which causes other, less well-developed follicles to stop growing and die.

The one dominant follicle becomes the ***mature (graafian) follicle***. The mature follicle continues to enlarge until it is ready for ovulation, forming a blisterlike bulge on the surface of the ovary. During maturation, the follicle continues to increase its production of estrogens under the influence of an increasing level of LH.

With reference to the ovarian cycle, the menstrual phase and preovulatory phase together are termed the ***follicular phase*** (fō-LIK-ū-lar) because ovarian follicles are growing and developing.

EVENTS IN THE UTERUS Estrogens liberated into the blood by growing ovarian follicles stimulate the repair of the endometrium. As the endometrium thickens, the short, straight endometrial glands develop, and the arterioles coil and lengthen.

Ovulation

Ovulation, the rupture of the mature (graafian) follicle and the release of the secondary oocyte into the pelvic cavity, usually occurs on day 14 in a 28-day cycle.

The high levels of estrogens during the last part of the preovulatory phase exert a *positive feedback* effect on both LH and GnRH. A high level of estrogens stimulates the hypothalamus to release more gonadotropin-releasing hormone (GnRH) and the anterior pituitary to produce more LH. GnRH then promotes the release of even more LH. The resulting surge of LH (Figure 23.12b) brings about rupture of the mature (graafian) follicle and expulsion of a secondary oocyte. An over-the-counter home test that detects the LH surge associated with ovulation can be used to predict ovulation a day in advance.

Postovulatory Phase

The ***postovulatory phase*** of the female reproductive cycle is the time between ovulation and onset of the next menstruation. This phase is the most constant in duration and lasts for 14 days, from days 15 to 28 in a 28-day cycle.

EVENTS IN ONE OVARY After ovulation, the mature follicle collapses. Stimulated by LH, the remaining follicular cells enlarge and form the corpus luteum, which secretes progesterone, estrogens, relaxin, and inhibin. With reference to the ovarian cycle, this phase is also called the ***luteal phase***.

Subsequent events depend on whether or not the oocyte is fertilized. If the oocyte is not fertilized, the corpus luteum lasts for only two weeks, after which its secretory activity declines, and it degenerates into a corpus albicans (Figure 23.12). As the levels of progesterone, estrogens, and inhibin decrease, release of GnRH, FSH, and LH rises due to loss of negative feedback suppression by the ovarian hormones. Then, follicular growth resumes and a new ovarian cycle begins.

If the secondary oocyte is fertilized and begins to divide, the corpus luteum persists past its normal two-week lifespan. It is "rescued" from degeneration by ***human chorionic gonadotropin*** (kōr′-ē-ON-ik) ***(hCG)***, a hormone produced by the embryo beginning about eight days after fertilization. Like LH, hCG stimulates the secretory activity of the corpus luteum. The presence of hCG in maternal blood or urine is an indicator of pregnancy, and hCG is the hormone detected by home pregnancy tests.

EVENTS IN THE UTERUS Progesterone and estrogens produced by the corpus luteum promote growth of the endometrial glands, which begin to secrete glycogen, and vascularization and thickening of the endometrium. These preparatory changes peak about one week after ovulation, at the time a fertilized ovum might arrive at the uterus.

Figure 23.13 summarizes the hormonal interactions and cyclical changes in the ovaries and uterus during the ovarian and menstrual cycles.

Figure 23.13 Summary of hormonal interactions in the ovarian and menstrual cycles.

Hormones from the anterior pituitary regulate ovarian function, and hormones from the ovaries regulate the changes in the endometrial lining of the uterus.

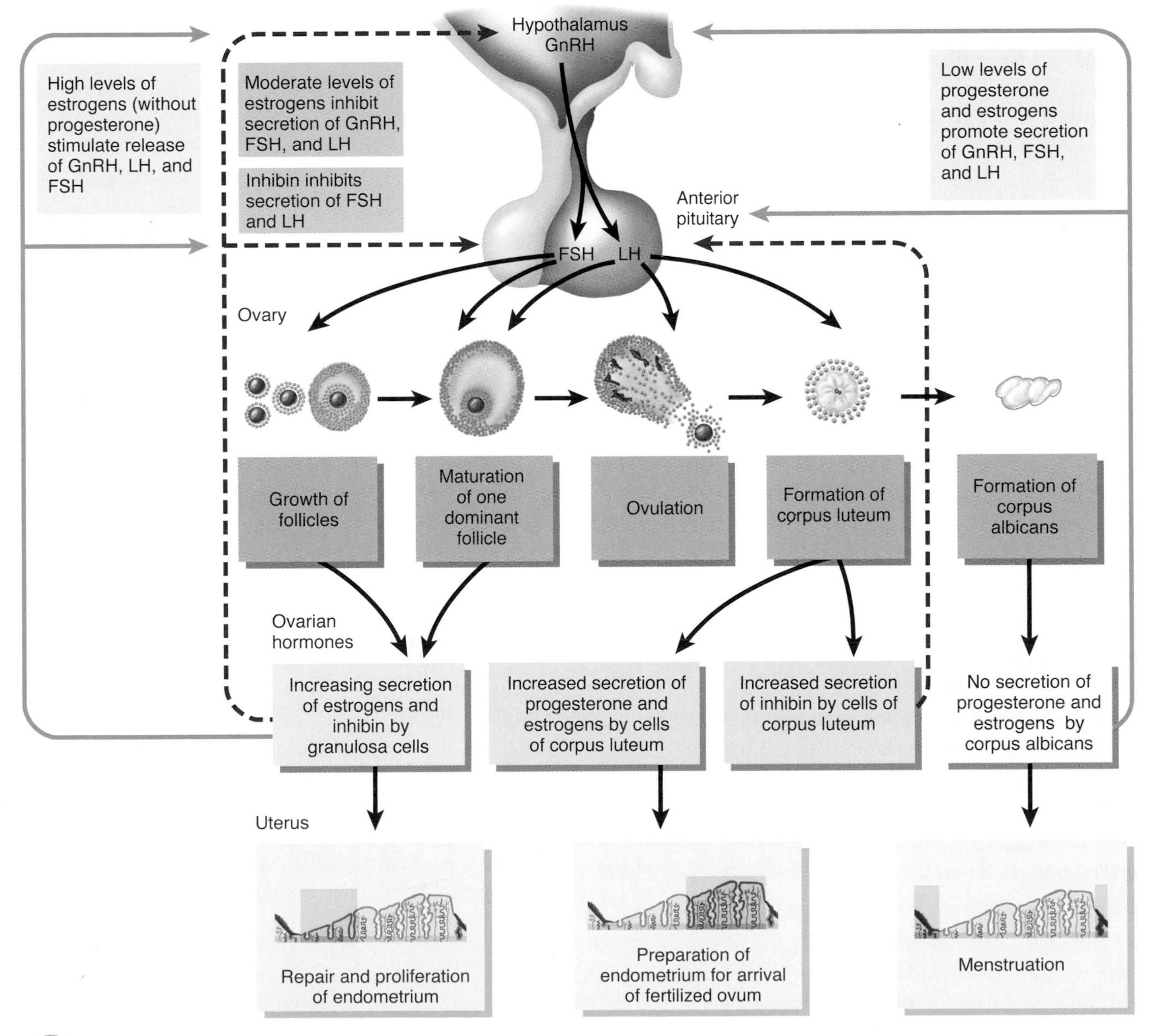

When declining levels of estrogens and progesterone stimulate secretion of GnRH, is this a positive or negative feedback effect? Why?

■ **CHECKPOINT**

11. Describe the function of each of the following hormones in the uterine and ovarian cycles: GnRH, FSH, LH, estrogens, progesterone, and inhibin.

12. Briefly outline the major events and hormonal changes of each phase of the uterine cycle, and correlate them with the events of the ovarian cycle.

13. Prepare a labeled diagram of the major hormonal changes that occur during the uterine and ovarian cycles.

Table 23.1 Failure Rates of Several Birth Control Methods

Method	Failure Rates*	
	Perfect Use†	Typical Use
None	85%	85%
Complete abstinence	0%	0%
Surgical sterilization		
Vasectomy	0.10%	0.15%
Tubal ligation	0.5%	0.5%
Hormonal methods		
Oral contraceptives	0.1%	3%‡
Depo-provera®	0.05%	0.05%
Intrauterine device		
Copper T 380A	0.6%	0.8%
Spermicides	6%	26%‡
Barrier methods		
Male condom	3%	14%‡
Vaginal pouch	5%	21%‡
Diaphragm	6%	20%‡
Periodic abstinence		
Rhythm	9%	25%‡
Sympto-thermal	2%	20%‡

*Defined as percentage of women having an unintended pregnancy during the first year of use.

†Failure rate when the method is used correctly and consistently.

‡Includes couples who forgot to use the method.

BIRTH CONTROL METHODS AND ABORTION

OBJECTIVE • Compare the various types of birth control methods and outline the effectiveness of each.

No single, ideal method of ***birth control*** exists. The only method of preventing pregnancy that is 100% reliable is total ***abstinence***, the avoidance of sexual intercourse. Several other methods are available, including surgical sterilization, hormonal methods, intrauterine devices, spermicides, barrier methods, and periodic abstinence. Table 23.1 provides the failure rates for each method of birth control. We will also discuss induced abortion, the intentional termination of pregnancy.

Surgical Sterilization

Sterilization is a procedure that renders an individual incapable of reproduction. The most common means of sterilization of males is ***vasectomy*** (va-SEK-tō-mē; *-ectomy* = cut out), in which a portion of each ductus deferens is removed. Even though sperm production continues in the testes after a vasectomy, sperm can no longer reach the exterior. Instead, they degenerate and are destroyed by phagocytosis. Blood testosterone level is normal, so a vasectomy has no effect on sexual desire or performance. Sterilization in females most often is achieved by performing a ***tubal ligation*** (lī-GĀ-shun), in which both uterine tubes are tied closed and then cut. As a result, the secondary oocyte cannot pass through the uterine tubes, and sperm cannot reach the oocyte.

Hormonal Methods

Aside from total abstinence or surgical sterilization, hormonal methods are the most effective means of birth control. Used by 50 million women worldwide, ***oral contraceptives*** ("the pill") contain various mixtures of synthetic estrogens and progestins (chemicals with actions similar to those of progesterone). They prevent pregnancy mainly by negative feedback inhibition of anterior pituitary secretion of FSH and LH. The low levels of FSH and LH usually prevent development of a dominant follicle. As a result, the level of estrogens does not rise, the midcycle LH surge does not occur, and ovulation is not triggered. Thus, there is no secondary oocyte available for fertilization. If taken properly, the pill is close to 100% effective.

Among the noncontraceptive benefits of oral contraceptives are regulation of the length of menstrual cycles and decreased menstrual flow (and therefore decreased risk of anemia). The pill also provides protection against endometrial and ovarian cancers and reduces the risk of endometriosis. However, oral contraceptives may not be advised for women with a history of blood clotting disorders, cerebral blood vessel damage, migraine headaches, hypertension, liver malfunction, or heart disease. Women who take the pill and smoke face far higher odds of having a heart attack or stroke than do nonsmoking pill users. Smokers should quit smoking or use an alternative method of birth control.

The same hormones found in oral contraceptives are used for ***emergency contraception (EC)***, the so-called "morn-

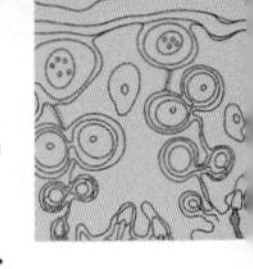

ing-after pill." The relatively high levels of estrogens and progestin in EC pills provide negative feedback inhibition of FSH and LH secretion. Loss of the stimulating effects of these gonadotropic hormones causes the ovaries to cease secretion of their own estrogens and progesterone. In turn, declining levels of estrogens and progesterone induce shedding of the uterine lining, thereby blocking implantation.

Other hormonal methods of contraception are also available:

- **Norplant®** consists of six slender hormone-containing capsules that are surgically implanted under the skin of the arm using local anesthesia. They slowly and continually release a progestin, which inhibits ovulation and thickens the cervical mucus. The effects last for 5 years, and Norplant® is about as reliable as sterilization. Removing the Norplant® capsules restores fertility.
- **Depo-provera®**, which is given as an intramuscular injection once every 3 months, contains progestin that prevents maturation of the ovum and causes changes in the uterine lining that make pregnancy less likely.
- **Lunelle®** is a once-a-month intramuscular injection. It contains estrogens and progestin and acts like an oral contraceptive.
- **Birth control skin patches** contain estrogens and progestin and are placed on the skin once a week for three weeks. Each week the patch is removed and a new one is placed on a different area of the skin. During the fourth week no patch is used so that menstruation can occur.
- The **vaginal ring** is a doughnut-shaped ring that fits in the vagina and releases either a progestin alone or a progestin and an estrogen. It is worn for 3 weeks and removed for 1 week to allow menstruation to occur.

Intrauterine Devices

An ***intrauterine device (IUD)*** is a small object made of plastic, copper, or stainless steel that is inserted into the cavity of the uterus. IUDs cause changes in the uterine lining that prevent implantation of a fertilized ovum. The IUD most commonly used in the United States today is the Copper T 380A, which is approved for up to 10 years of use and has long-term effectiveness comparable to that of tubal ligation. Some women cannot use IUDs because of expulsion, bleeding, or discomfort.

Spermicides

Various foams, creams, jellies, suppositories, and douches that contain sperm-killing agents, or ***spermicides***, make the vagina and cervix unfavorable for sperm survival and are available without prescription. The most widely used spermicide is nonoxynol-9, which kills sperm by disrupting their plasma membrane. A spermicide is more effective when used together with a barrier method such as a diaphragm or a condom.

Barrier Methods

Barrier methods are designed to prevent sperm from gaining access to the uterine cavity and uterine tubes. In addition to preventing pregnancy, barrier methods may also provide some protection against sexually transmitted diseases (STDs) such as AIDS. In contrast, oral contraceptives and IUDs confer no such protection. Among the barrier methods are use of a condom, a vaginal pouch, or a diaphragm.

A ***condom*** is a nonporous, latex covering placed over the penis that prevents deposition of sperm in the female reproductive tract. A ***vaginal pouch***, sometimes called a female condom, is made of two flexible rings connected by a polyurethane sheath. One ring lies inside the sheath and is inserted to fit over the cervix; the other ring remains outside the vagina and covers the female external genitals.

A ***diaphragm*** is a rubber, dome-shaped structure that fits over the cervix and is used in conjunction with a spermicide. It can be inserted up to 6 hours before intercourse. The diaphragm stops most sperm from passing into the cervix, and the spermicide kills most sperm that do get by. Although diaphragm use does decrease the risk of some STDs, it does not fully protect against HIV infection.

Periodic Abstinence

A couple can use their knowledge of the physiological changes that occur during the female reproductive cycle to decide either to abstain from intercourse on those days when pregnancy is a likely result, or to plan intercourse on those days if they wish to conceive a child. In females with normal and regular menstrual cycles, these physiological events help to predict the day on which ovulation is likely to occur.

The first physiologically based method, developed in the 1930s, is known as the ***rhythm method (natural family planning)***. It takes advantage of the fact that a secondary oocyte is fertilizable for only 24 hours and is available for only 3 to 5 days in each reproductive cycle. During this time (3 days before ovulation, the day of ovulation, and 3 days after ovulation) the couple abstains from intercourse. The effectiveness of the rhythm method for birth control is poor in many women due to the irregularity of their cycles.

Another system is the ***sympto-thermal method***, in which couples are instructed to know and understand certain signs of fertility. The signs of ovulation include increased basal body temperature; the production of abundant clear, stretchy cervical mucus; and pain associated with ovulation (mittelschmerz). If a couple abstains from sexual intercourse when the signs of ovulation are present and for 3 days afterward, the chance of pregnancy is decreased. A big problem with this method is that fertilization is very likely if intercourse occurs one or two days *before* ovulation.

Focus on Wellness

The Female Athlete Triad—Disordered Eating, Amenorrhea, and Premature Osteoporosis

The female reproductive cycle can be disrupted by many factors, including weight loss, low body weight, disordered eating, and vigorous physical activity. Many athletes experience intense pressure from coaches, parents, peers, and themselves to lose weight to improve performance. Consequently, many develop disordered eating behaviors and engage in other harmful weight-loss practices in a struggle to maintain a very low body weight. The athletes with the highest rates of menstrual irregularity include runners, gymnasts, dancers, figure skaters, and divers.

Sticks and Stones and . . . the Female Athlete Triad?

Menstrual irregularity should never be ignored, because it may be caused by a serious underlying disorder for which the athlete should receive prompt medical treatment. Even when menstrual irregularity is apparently caused by disordered eating and physical training, and not associated with another physical disorder, it is still a cause for concern. One reason is that women with amenorrhea, the absence of menstrual cycles, are at increased risk for premature osteoporosis. The observation that three conditions—disordered eating, amenorrhea, and osteoporosis—tend to occur together in female athletes led researchers to coin the term "female athlete triad."

Why osteoporosis? Remember that the ovarian follicles produce estrogens when stimulated by FSH and LH. If ovulation is not occurring, then the ovarian follicles, and later the corpus luteum, are not producing estrogens. Chronically low levels of estrogens are associated with loss of bone minerals, as estrogens help bones retain calcium. The loss of the protective effect of estrogens explains why many women experience a decline in bone density after menopause, when levels of estrogens drop. Amenorrheic runners have been shown to experience a similar effect. In one study, amenorrheic runners in their 20s had bone densities similar to those of postmenopausal women 50 to 70 years old. Short periods of menstrual irregularity in young athletes may cause no lasting harm, but long-term cessation of the menstrual cycle may be accompanied by a loss of bone mass or, in adolescent athletes, a failure to achieve an adequate bone mass, both of which can lead to premature osteoporosis and irreversible bone damage.

It is ironic that dedicated athletes should experience premature osteoporosis, because physical activity in general is associated with a *reduced* risk of osteoporosis. Exercise has been shown to increase bone density, especially if the exercise involves bone stress, such as running and aerobic dancing. However, in the presence of disordered eating and overtraining, exercise may simply add insult to injury.

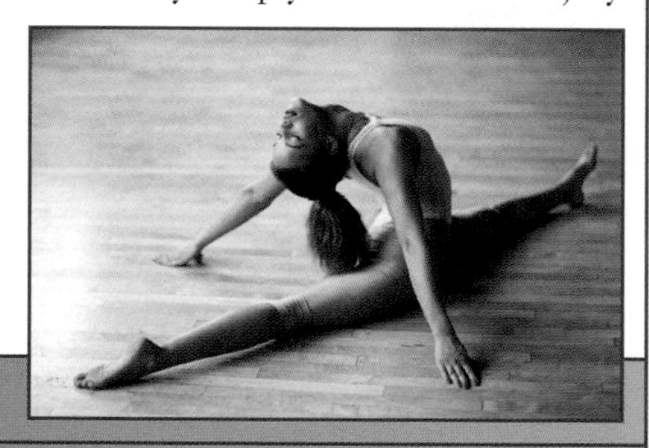

► **Think It Over . . .**

► ***Do you think that girls and women should be discouraged from participating in athletics and other forms of vigorous physical activities because of the female athlete triad?***

Abortion

Abortion refers to the premature expulsion of the products of conception from the uterus, usually before the 20th week of pregnancy. An abortion may be spontaneous (naturally occurring; also called a *miscarriage*) or induced (intentionally performed). Induced abortions may be performed by vacuum aspiration (suction), infusion of a saline solution, or surgical evacuation (scraping).

Certain drugs, most notably RU 486, can induce a so-called nonsurgical abortion. ***RU 486 (mifepristone)*** blocks the action of progesterone by binding to and blocking progesterone receptors. Within 12 hours after taking RU 486, the endometrium starts to degenerate, and within 72 hours, it begins to slough off. A form of prostaglandin E (misoprostol), which stimulates uterine contractions, is given after RU 486 to aid in expulsion of the endometrium. RU 486 can be taken up to 5 weeks after conception. One side effect of the drug is uterine bleeding.

■ CHECKPOINT

14. How do oral contraceptives reduce the likelihood of pregnancy?

15. Why do some methods of birth control protect against sexually transmitted diseases, but others do not?

AGING AND THE REPRODUCTIVE SYSTEMS

OBJECTIVE • Describe the effects of aging on the reproductive systems.

During the first decade of life, the reproductive system is in a juvenile state. At about age 10, hormone-directed changes start to occur in both sexes. ***Puberty*** (PŪ-ber-tē = a ripe age) is the period when secondary sexual characteristics begin to develop and the potential for sexual reproduction is reached. Onset of puberty is marked by bursts of LH and FSH secretion, each triggered by a burst of GnRH. The stimuli that cause the GnRH bursts are still unclear, but a role for the hormone leptin is starting to unfold. Just before puberty, leptin levels rise in proportion to adipose tissue mass. Leptin may signal the hypothalamus that long-term energy stores (triglycerides in adipose tissue) are adequate for reproductive functions to begin.

In females, the reproductive cycle normally occurs once each month from ***menarche*** (me-NAR-kē), the first menses, to ***menopause***, the permanent cessation of menses. Thus, the female reproductive system has a time-limited span of fertility between menarche and menopause. Between the ages of 40 and 50 the pool of remaining ovarian follicles becomes exhausted. As a result, the ovaries become less responsive to hormonal stimulation. The production of estrogens declines, despite copious secretion of FSH and LH by the anterior pituitary. Many women experience hot flashes and heavy sweating, which coincide with bursts of GnRH release. Other symptoms of menopause are headache, hair loss, muscular pains, vaginal dryness, insomnia, depression, weight gain, and mood swings. Some atrophy of the ovaries, uterine tubes, uterus, vagina, external genitalia, and breasts occurs in postmenopausal women. Due to loss of estrogens, most women also experience a decline in bone mineral density after menopause. Sexual desire (libido) does not show a parallel decline; it may be maintained by adrenal androgens. The risk of having uterine cancer peaks at about 65 years of age, but cervical cancer is more common in younger women.

In males, declining reproductive function is much more subtle than in females. Healthy men often retain reproductive capacity into their eighties or nineties. At about age 55 a decline in testosterone synthesis leads to reduced muscle strength, fewer viable sperm, and decreased sexual desire. However, abundant sperm may be present even in old age.

Enlargement of the prostate to two to four times its normal size occurs in approximately one-third of all males over age 60. This condition, called ***benign prostatic hyperplasia (BPH)***, is characterized by frequent urination, nocturia (bedwetting), hesitancy in urination, decreased force of urinary stream, postvoiding dribbling, and a sensation of incomplete emptying.

■ CHECKPOINT

16. What changes occur in males and females at puberty?

17. What do the terms menarche and menopause mean?

• • •

To appreciate the many ways that the reproductive systems contribute to homeostasis of other body systems, examine Focus on Homeostasis: The Reproductive Systems on page 578. Next, in Chapter 24, you will explore the major events that occur during pregnancy and you will discover how genetics (inheritance) plays a role in the development of a child.

FOCUS ON HOMEOSTASIS

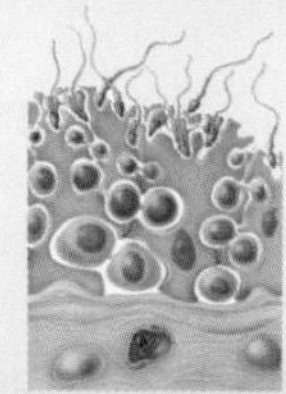

THE REPRODUCTIVE SYSTEMS

BODY SYSTEM		CONTRIBUTION OF THE REPRODUCTIVE SYSTEMS
For all body systems		**The male and female reproductive systems produce gametes (oocytes and sperm) that unite to form embryos and fetuses, which contain cells that divide and differentiate to form all of the organ systems of the body.**
Integumentary system	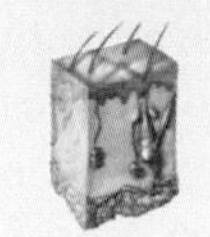	**Androgens promote the growth of body hair. Estrogens stimulate the deposition of fat in the breasts, abdomen, and hips. Mammary glands produce milk. Skin stretches during pregnancy as the fetus enlarges.**
Skeletal system	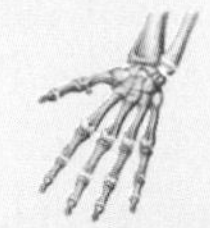	**Androgens and estrogens stimulate the growth and maintenance of bones of the skeletal system.**
Muscular system		**Androgens stimulate the growth of skeletal muscles.**
Nervous system	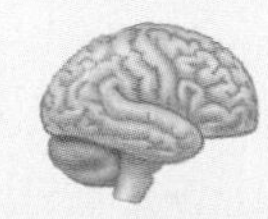	**Androgens influence libido (sex drive). Estrogens may play a role in the development of certain regions of the brain in males.**
Endocrine system		**Testosterone and estrogens exert feedback effects on the hypothalamus and anterior pituitary gland.**
Cardiovascular system		**Estrogens lower blood cholesterol level and may reduce the risk of coronary artery disease in women under age 50.**
Lymphatic system and immunity	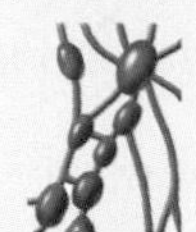	**The presence of an antibiotic-like chemical in semen and the acidic pH of vaginal fluid provide innate immunity against microbes in the reproductive tract.**
Respiratory system	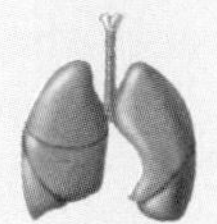	**Sexual arousal increases the rate and depth of breathing.**
Digestive system	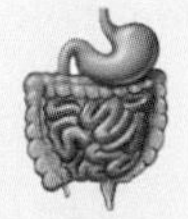	**The presence of the fetus during pregnancy crowds the digestive organs, which leads to heartburn and constipation.**
Urinary system	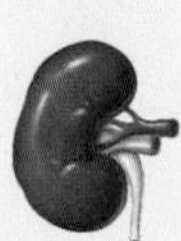	**In males, the portion of the urethra that extends through the prostate and penis is a passageway for urine as well as semen.**

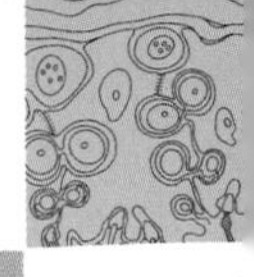

COMMON DISORDERS

Reproductive System Disorders in Males

Testicular Cancer

Testicular cancer is the most common cancer, and also one of the most curable, in males between the ages of 20 and 35. An early sign of testicular cancer is a mass in the testis, often associated with a sensation of testicular heaviness or a dull ache in the lower abdomen; pain usually does not occur. All males should perform regular testicular self-examinations.

Prostate Disorders

Because the prostate surrounds part of the urethra, any prostatic infection, enlargement, or tumor can obstruct the flow of urine. Acute and chronic infections of the prostate are common in adult males, often in association with inflammation of the urethra. In ***acute prostatitis***, the prostate becomes swollen and tender. ***Chronic prostatitis*** is one of the most common chronic infections in men of the middle and later years; on examination, the prostate feels enlarged, soft, and very tender, and its surface outline is irregular.

Prostate cancer is the leading cause of death from cancer in men in the United States. A blood test can measure the level of prostate-specific antigen (PSA) in the blood. The amount of PSA, which is produced only by prostate epithelial cells, increases with enlargement of the prostate and may indicate infection, benign enlargement, or prostate cancer. Males over the age of 40 should have an annual examination of the prostate. In a ***digital rectal exam***, a physician palpates the prostate through the rectum with the fingers (digits). Many physicians also recommend an annual PSA test for males over 50. Treatment for prostate cancer may involve surgery, radiation, hormonal therapy, and chemotherapy. Because many prostate cancers grow very slowly, some urologists recommend "watchful waiting" before treating small tumors in men over age 70.

Reproductive System Disorders in Females

Premenstrual Syndrome

Premenstrual syndrome (PMS) is a cyclical disorder of severe physical and emotional distress. It appears during the postovulatory phase of the female reproductive cycle and dramatically disappears when menstruation begins. The signs and symptoms are highly variable from one woman to another. They may include edema, weight gain, breast swelling and tenderness, abdominal distension, backache, joint pain, constipation, skin eruptions, fatigue and lethargy, greater need for sleep, depression or anxiety, irritability, mood swings, headache, poor coordination and clumsiness, and cravings for sweet or salty foods. The cause of PMS is unknown. For some women, getting regular exercise; avoiding caffeine, salt, and alcohol; and eating a diet that is high in complex carbohydrates and lean proteins can bring considerable relief.

Endometriosis

Endometriosis (en′-dō-mē′-trē-Ō-sis; *endo-* = within; *metri-* = uterus; *-osis* = condition of or disease) is characterized by the growth of endometrial tissue outside the uterus. The tissue enters the pelvic cavity via the open uterine tubes and may be found in any of several sites—on the ovaries, the outer surface of the uterus, the sigmoid colon, pelvic and abdominal lymph nodes, the cervix, the abdominal wall, the kidneys, and the urinary bladder. Endometrial tissue responds to hormonal fluctuations, whether it is inside or outside the uterus, by first proliferating and then breaking down and bleeding. When this occurs outside the uterus, it can cause inflammation, pain, scarring, and infertility. Symptoms include premenstrual pain or unusually severe menstrual pain.

Breast Cancer

One in eight women in the United States faces the prospect of ***breast cancer***, the second-leading cause of female deaths from cancer. Early detection by breast self-examination and mammograms is the best way to increase the chance of survival.

The most effective technique for detecting tumors less than 1 cm (0.4 in.) in diameter is ***mammography*** (mam-OG-ra-fē; *-graphy* = to record), a type of radiography using very sensitive x-ray film. The image of the breast, called a ***mammogram***, is best obtained by compressing the breasts, one at a time, using flat plates. A supplementary procedure for evaluating breast abnormalities is ***ultrasonography***. Although ultrasonography cannot detect tumors smaller than 1 cm in diameter, it can be used to determine whether a lump is a benign, fluid-filled cyst or a solid (and therefore possibly malignant) tumor.

Among the factors that increase the risk of developing breast cancer are (1) a family history of breast cancer, especially in a mother or sister; (2) never having borne a child or having a first child after age 35; (3) previous cancer in one breast; (4) exposure to ionizing radiation, such as x-rays; (5) excessive alcohol intake; and (6) cigarette smoking.

The American Cancer Society recommends the following steps to help diagnose breast cancer as early as possible:

- All women over 20 should develop the habit of monthly breast self-examination.
- A physician should examine the breasts every three years when a woman is between the ages of 20 and 39, and every year after age 40.
- A mammogram should be taken in women between the ages of 35 and 39, to be used later for comparison (baseline mammogram).
- Women with no symptoms should have a mammogram every year after age 40.
- Women of any age with a history of breast cancer, a strong family history of the disease, or other risk factors should consult a physician to determine a schedule for mammography.

Treatment for breast cancer may involve hormone therapy, chemotherapy, radiation therapy, ***lumpectomy*** (removal of the tumor and the immediate surrounding tissue), a modified or radical mastectomy, or a combination of these approaches. A ***radical mastectomy*** (*mast-* = breast) involves removal of the affected breast along with the underlying pectoral muscles and the axillary lymph nodes. (Lymph nodes are removed because the spread of cancerous cells usually occurs through lymphatic or blood vessels.) Radiation treatment and chemotherapy may follow the surgery to ensure the destruction of any stray cancer cells. In some cases of metastatic (spreading) breast cancer, Herceptin®, a monoclonal antibody drug that targets an antigen on the surface of breast cancer cells, can cause regression of the tumors and retard progression of the disease. Finally, two promising drugs for breast cancer *prevention* are now on the market—tamoxifen (Nolvadex®) and raloxifene (Evista®).

Ovarian Cancer

Ovarian cancer is the sixth most common form of cancer in females, but the leading cause of death from all gynecological malignancies (excluding breast cancer) because it is difficult to detect before it metastasizes (spreads) beyond the ovaries. Risk factors associated with ovarian cancer include age (usually over age 50); race (whites are at highest risk); family history of ovarian cancer; more than 40 years of active ovulation; *nulliparity* (no pregnancies) or first pregnancy after age 30; a high-fat, low-fiber, vitamin A–deficient diet; and prolonged exposure to asbestos and talc. Early ovarian cancer may have no symptoms or mild ones such as abdominal discomfort, heartburn, nausea, loss of appetite, bloating, and flatulence. Later-stage signs and symptoms include an enlarged abdomen, abdominal and/or pelvic pain, persistent gastrointestinal disturbances, urinary complications, menstrual irregularities, and heavy menstrual bleeding.

Cervical Cancer

Cervical cancer, cancer of the uterine cervix, starts with ***cervical dysplasia*** (dis-PLĀ-zha), a change in the shape, growth, and number of cervical cells. The cells may either return to normal or progress to cancer. In most cases cervical cancer may be detected in its earliest stages by a Pap smear. Some evidence links cervical cancer to the virus that causes genital warts (human papilloma virus). Increased risk is associated with a large number of sexual partners, first intercourse at a young age, and smoking cigarettes.

Vulvovaginal Candidiasis

Candida albicans is a yeastlike fungus that commonly grows on mucous membranes of the gastrointestinal and genitourinary tracts. The organism is responsible for ***vulvovaginal candidiasis*** (vul′-vō-VAJ-i-nal can′-di-DĪ-a-sis), the most common form of ***vaginitis*** (vaj′-i-NĪ-tis), inflammation of the vagina. Candidiasis, commonly referred to as a yeast infection, is characterized by severe itching; a thick, yellow, cheesy discharge; a yeasty odor; and pain. The disorder, experienced at least once by about 75% of females, is usually a result of proliferation of the fungus following antibiotic therapy for another condition. Predisposing conditions include the use of oral contraceptives or cortisone-like medications, pregnancy, and diabetes.

Sexually Transmitted Diseases

A ***sexually transmitted disease (STD)*** is one that is spread by sexual contact. AIDS and hepatitis B, which are sexually transmitted diseases that also may be contracted in other ways, are discussed in Chapters 17 and 19, respectively.

Chlamydia

Chlamydia (kla-MID-ē-a) is a sexually transmitted disease caused by the bacterium *Chlamydia trachomatis* (*chlamy-* = cloak). This unusual bacterium cannot reproduce outside body cells; it "cloaks" itself inside cells, where it divides. At present, chlamydia is the most prevalent sexually transmitted disease in the United States. In most cases the initial infection is asymptomatic and thus difficult to recognize clinically. In males, urethritis is the principal result, causing a clear discharge and burning, frequent, and painful urination. Without treatment, the epididymides may also become inflamed, leading to male sterility. In 70% of females with chlamydia, symptoms are absent, but chlamydia is the leading cause of pelvic inflammatory disease. The uterine tubes may also become inflamed, which increases the risk of female infertility due to the formation of scar tissue in the tubes.

Gonorrhea

Gonorrhea (gon′-ō-RĒ-a) is caused by the bacterium *Neisseria gonorrhoeae*. Discharges from infected mucus membranes are the source of transmission of the bacteria either during sexual contact or during the passage of a newborn through the birth canal. Males usually experience urethritis with profuse pus drainage and painful urination. In females, infection typically occurs in the vagina, often with a discharge of pus. In females, the infection and consequent inflammation can proceed from the vagina into the uterus, uterine tubes, and pelvic cavity. Thousands of women are made infertile by gonorrhea every year as a result of scar tissue formation that closes the uterine tubes. Transmission of bacteria in the birth canal to the eyes of a newborn can result in blindness.

Syphilis

Syphilis, caused by the bacterium *Treponema pallidum*, is transmitted through sexual contact or exchange of blood, or through the placenta to a fetus. The disease progresses through several stages. During the *primary stage*, the chief sign is a painless open sore, called a ***chancre*** (SHANG-ker), at the point of contact. The chancre heals within 1 to 5 weeks. From 6 to 24 weeks later, signs and symptoms such as a skin rash, fever, and aches in the joints and muscles usher in the *secondary stage*, which is systemic—the infection spreads to all major body systems. When signs of organ degeneration appear, the disease is said to be in the *tertiary stage*. If the nervous system is involved, the tertiary stage is called ***neurosyphilis***. As motor areas become extensively damaged, victims may be unable to control urine and bowel movements; eventually they may become bedridden, unable even to feed themselves. Damage to the cerebral cortex produces memory loss and personality changes that range from irritability to hallucinations.

Genital Herpes

Genital herpes is caused by type 2 herpes simplex virus (HSV-2), producing painful blisters on the prepuce, glans penis, and penile shaft in males, and on the vulva or sometimes high up in the vagina in females. The blisters disappear and reappear in most patients, but the virus itself remains in the body; there is no cure. A related virus, type 1 herpes simplex virus (HSV-1), causes cold sores on the mouth and lips. Infected individuals typically experience recurrences of symptoms several times a year.

MEDICAL TERMINOLOGY AND CONDITIONS

Amenorrhea (ā-men′-ō-RĒ-a; *a-* = without; *men-* = month; *-rrhea* = a flow) The absence of menstruation; it may be caused by a hormone imbalance, obesity, extreme weight loss, or very low body fat as may occur during rigorous athletic training.

Dysmenorrhea (dis-men′-ō-RĒ-a; *dys-* = difficult or painful) Painful menstruation; the term is usually reserved to describe menstrual symptoms that are severe enough to prevent a woman from functioning normally for one or more days each month. Some cases are caused by uterine tumors, ovarian cysts, pelvic inflammatory disease, or intrauterine devices.

Endocervical curettage (kū′-re-TAHZH; *curette* = scraper) A procedure in which the cervix is dilated and the endometrium of the uterus is scraped with a spoon-shaped instrument called a curette; commonly called a D and C (dilation and curettage).

Fibroids (FĪ-broyds; *fibro-* = fiber; *-eidos* = resemblance) Noncancerous tumors in the myometrium of the uterus composed of muscular and fibrous tissue. Their growth appears to be related to high levels of estrogens. They do not occur before puberty and usually stop growing after menopause. Symptoms include abnormal menstrual bleeding, and pain or pressure in the pelvic area.

Menorrhagia (men-ō-RA-jē-a; *meno-* = menstruation; *-rhage* = to burst forth). Excessively prolonged or profuse menstrual period. May be due to a disturbance in hormonal regulation of the menstrual cycle, pelvic infection, medications (anticoagulants), fibroids, endometriosis, or intrauterine devices.

Oophorectomy (ō′-of-ō-REK-tō-mē; *oophor-* = bearing eggs) Removal of the ovaries.

Ovarian cyst The most common form of ovarian tumor, in which a fluid-filled follicle or corpus luteum persists and continues growing.

Papanicolaou test (pa′-pa-ni′-kō-LĀ-oo), or ***Pap smear*** A test to detect uterine cancer in which a few cells from the cervix and the part of the vagina surrounding the cervix are removed with a swab and examined microscopically. Malignant cells have a characteristic appearance that allows diagnosis even before symptoms occur.

Pelvic inflammatory disease (PID) A collective term for any extensive bacterial infection of the pelvic organs, especially the uterus, uterine tubes, or ovaries, which is characterized by pelvic soreness, lower back pain, abdominal pain, and urethritis. Often the early symptoms of PID occur just after menstruation. As infection spreads and cases advance, fever may develop, along with painful abscesses of the reproductive organs.

Salpingectomy (sal′-pin-JEK-tō-mē; *salpingo* = tube) Removal of a uterine (fallopian) tube.

Smegma (SMEG-ma) The secretion, consisting principally of sloughed off epithelial cells, found chiefly around the external genitals and especially under the foreskin of the male.

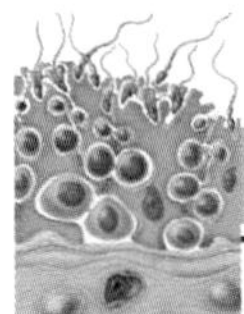

STUDY OUTLINE

Introduction (p. 556)

1. Sexual reproduction is the process of producing offspring by the union of gametes (oocytes and sperm).
2. The organs of reproduction are grouped as gonads (produce gametes), ducts (transport and store gametes), accessory sex glands (produce materials that support gametes), and supporting structures.

Male Reproductive System (p. 557)

1. The male reproductive system includes the testes, epididymis, ductus (vas) deferens, ejaculatory ducts, urethra, seminal vesicles, prostate, bulbourethral (Cowper's) glands, scrotum and penis.
2. The scrotum is a sac that supports and regulates the temperature of the testes.
3. The male gonads include the testes, oval-shaped organs in the scrotum that contain the seminiferous tubules, in which sperm cells develop; Sertoli cells, which nourish sperm cells and produce inhibin; and Leydig cells, which produce the male sex hormone testosterone.
4. Spermatogenesis occurs in the testes and consists of meiosis I, meiosis II, and spermiogenesis. It results in the formation of four haploid sperm cells from a primary spermatocyte.
5. Mature sperm consist of a head and a tail. Their function is to fertilize a secondary oocyte.
6. At puberty, gonadotropin-releasing hormone (GnRH) stimulates anterior pituitary secretion of LH and FSH. LH stimulates Leydig cells to produce testosterone. FSH and testosterone initiate spermatogenesis.

7. Testosterone controls the growth, development, and maintenance of sex organs; stimulates bone growth, protein anabolism, and sperm maturation; and stimulates development of male secondary sex characteristics.
8. Inhibin is produced by Sertoli cells; its inhibition of FSH helps regulate the rate of spermatogenesis.
9. Sperm are transported out of the testes into an adjacent organ, the epididymis, where their motility increases.
10. The ductus (vas) deferens stores sperm and propels them toward the urethra during ejaculation. Removing part of the vas deferens to prevent fertilization is called vasectomy.
11. The ejaculatory ducts are formed by the union of the ducts from the seminal vesicles and vas deferens, and they eject sperm into the urethra.
12. The male urethra passes through the prostate, deep perineal muscles, and penis.
13. The seminal vesicles secrete an alkaline, viscous fluid that constitutes about 60% of the volume of semen and contributes to sperm viability.
14. The prostate secretes a slightly acidic fluid that constitutes about 25% of the volume of semen and contributes to sperm motility.
15. The bulbourethral glands secrete mucus for lubrication and an alkaline substance that neutralizes acid.
16. Semen is a mixture of sperm and seminal fluid; it provides the fluid in which sperm are transported, supplies nutrients, and neutralizes the acidity of the male urethra and the vagina.
17. The penis consists of a root, a body, and a glans penis. It functions to introduce sperm into the vagina. Expansion of its blood sinuses under the influence of sexual excitation is called erection.

Female Reproductive System (p. 564)

1. The female organs of reproduction include the ovaries (gonads), uterine (fallopian) tubes, uterus, vagina, and vulva.
2. The mammary glands are also considered part of the reproductive system.
3. The female gonads are the ovaries, located in the upper pelvic cavity on either side of the uterus.
4. Ovaries produce secondary oocytes; discharge secondary oocytes (the process of ovulation); and secrete estrogens, progesterone, relaxin, and inhibin.
5. Oogenesis (production of haploid secondary oocytes) begins in the ovaries. The oogenesis sequence includes meiosis I and meiosis II. Meiosis II is completed only after an ovulated secondary oocyte is fertilized by a sperm cell.
6. The uterine (fallopian) tube, which transports a secondary oocyte from an ovary to the uterus, is the normal site of fertilization.
7. The uterus is an organ the size and shape of an inverted pear that functions in menstruation, implantation of a fertilized ovum, development of a fetus during pregnancy, and labor. It also is part of the pathway for sperm to reach a uterine tube to fertilize a secondary oocyte.
8. The innermost layer of the uterine wall is the endometrium, which undergoes marked changes during the menstrual cycle.
9. The vagina is a passageway for the menstrual flow, the receptacle for the penis during sexual intercourse, and the lower portion of the birth canal. The smooth muscle of the vaginal wall makes it capable of considerable stretching.
10. The vulva, a collective term for the external genitals of the female, consists of the mons pubis, labia majora, labia minora, clitoris, vestibule, vaginal and urethral orifices, paraurethral glands, and greater vestibular glands.
11. The mammary glands of the female breasts are modified sweat glands located over the pectoralis major muscles. Their function is to secrete and eject milk (lactation).
12. Mammary gland development depends on estrogens and progesterone.
13. Milk production is stimulated by prolactin, estrogens, and progesterone; milk ejection is stimulated by oxytocin.

Female Reproductive Cycle (p. 570)

1. The female reproductive cycle includes the ovarian and menstrual cycles. The function of the ovarian cycle is development of a secondary oocyte; that of the menstrual cycle is preparation of the endometrium each month to receive a fertilized egg.
2. The ovarian and menstrual cycles are controlled by GnRH from the hypothalamus, which stimulates the release of FSH and LH by the anterior pituitary.
3. FSH stimulates development of follicles and initiates secretion of estrogens by the follicles. LH stimulates further development of the follicles, secretion of estrogens by follicular cells, ovulation, formation of the corpus luteum, and the secretion of progesterone and estrogens by the corpus luteum.
4. Estrogens stimulate the growth, development, and maintenance of female reproductive structures; the development of secondary sex characteristics; and protein synthesis.
5. Progesterone works together with estrogens to prepare the endometrium for implantation and the mammary glands for milk synthesis.
6. Relaxin increases the flexibility of the pubic symphysis and helps dilate the uterine cervix to ease delivery of a baby.
7. During the menstrual phase, part of the endometrium is shed, discharging blood and tissue cells.
8. During the preovulatory phase, a group of follicles in the ovaries begins to undergo maturation. One follicle outgrows the others and becomes dominant while the others die. At the same time, endometrial repair occurs in the uterus. Estrogens are the dominant ovarian hormones during the preovulatory phase.
9. Ovulation is the rupture of the dominant mature (graafian) follicle and the release of a secondary oocyte into the pelvic cavity. It is brought about by a surge of LH.
10. During the postovulatory phase, both progesterone and estrogens are secreted in large quantity by the corpus luteum of the ovary, and the uterine endometrium thickens in readiness for implantation.

11. If fertilization and implantation do not occur, the corpus luteum degenerates, and the resulting low level of progesterone and estrogens allows discharge of the endometrium (menstruation) followed by the initiation of another reproductive cycle.
12. If fertilization and implantation occur, the corpus luteum is maintained by hCG.

Birth Control Methods and Abortion (p. 574)

1. Birth control methods include surgical sterilization (vasectomy, tubal ligation), hormonal methods, intrauterine devices, spermicides, barrier methods (condom, vaginal pouch, diaphragm), and periodic abstinence. Table 23.1 on page 574 provides failure rates of the various methods of birth control. Abstinence is the only foolproof method of birth control.
2. Contraceptive pills of the combination type contain estrogens and progestins in concentrations that decrease the secretion of FSH and LH and thereby inhibit development of ovarian follicles and ovulation.
3. An abortion is the premature expulsion from the uterus of the products of conception; it may be spontaneous or induced. RU 486 can induce abortion by blocking the action of progesterone.

Aging and the Reproductive Systems (p. 577)

1. Puberty is the period of time when secondary sex characteristics begin to develop and the potential for sexual reproduction arises. In older females, levels of progesterone and estrogens decrease, resulting in changes in menstruation and then menopause.
2. In older males, decreased levels of testosterone are associated with decreased muscle strength, waning sexual desire, and fewer viable sperm; prostate disorders are common.

SELF-QUIZ

1. The testes are located in the scrotum because
 a. they must be separated from all other organs or sterility can occur
 b. sperm and hormone production and survival require a temperature lower than the normal body temperature
 c. the scrotum supplies the necessary hormones for sperm maturation
 d. sperm in the testes cannot survive without the nutrients supplied by the scrotum
 e. the scrotum produces alkaline fluids that neutralize the acids in the male urethra

2. Match the following:

____ **a.** cells that support, protect and nourish developing spermatogonia	**A.** corpus luteum
____ **b.** contain developing oocytes	**B.** Leydig cells
____ **c.** immature sperm cells	**C.** Sertoli cells
____ **d.** cells that secrete testosterone	**D.** follicles
____ **e.** produce progesterone and estrogens	**E.** spermatogonia

3. Removal of the prostate would
 a. interfere with sperm production
 b. inhibit testosterone release
 c. decrease the volume of semen by about 75%
 d. cause semen to become more acidic
 e. affect semen clotting

4. Which of the following is true?
 a. Meiosis is the process by which somatic (body) cells divide
 b. The haploid chromosome number is symbolized by $2n$
 c. Meiosis I results in diploid spermatocytes.
 d. Gametes contain the haploid chromosome number.
 e. Gametes contain 46 chromosomes in their nuclei.

5. The uterus is the site of all of the following except
 a. menstruation
 b. implantation of a fertilized ovum
 c. ovulation
 d. labor
 e. development of the fetus

6. Menstruation is triggered by a
 a. rapid rise in luteinizing hormone (LH)
 b. rapid fall in luteinizing hormone (LH)
 c. drop in estrogens and progesterone
 d. rise in estrogens and progesterone
 e. rise in inhibin

7. An inflammation of the seminiferous tubules would interfere with the ability to
 a. secrete testosterone **b.** produce sperm
 c. void urine **d.** make semen alkaline
 e. regulate the temperature in the scrotum

8. Erection of the penis involves the release of which neurotransmitter?
 a. norepinephrine **b.** serotonin
 c. glycine **d.** dopamine
 e. nitric oxide

9. Which of the following is NOT a function of semen?
 a. transport sperm
 b. lubricate the reproductive tract
 c. provide an acidic environment needed for fertilization
 d. provide nourishment for sperm
 e. produce antibiotics to destroy some bacteria

10. Prior to ejaculation, sperm are stored in the
 a. Leydig cells **b.** scrotum **c.** Sertoli cells
 d. prostate **e.** ductus (vas) deferens

11. Place the following in the correct order for the passage of sperm from the testes to the outside of the body.
    1. urethra
    2. ductus (vas) deferens
    3. seminiferous tubules
    4. ejaculatory duct
    5. external urethral orifice
    6. epididymis

    a. 6, 3, 2, 4, 1, 5 **b.** 3, 2, 6, 4, 1, 5 **c.** 3, 6, 2, 4, 1, 5
    d. 3, 6, 2, 4, 5, 1 **e.** 2, 4, 6, 1, 3, 5

12. In males, the gland that surrounds the urethra at the base of the urinary bladder is the
    a. glans penis **b.** prostate
    c. seminal vesicle **d.** bulbourethral gland
    e. greater vestibular gland

13. An oocyte is moved towards the uterus by
    a. peristaltic contractions of the uterine (Fallopian) tubes
    b. contraction of the uterus
    c. gravity
    d. swimming
    e. flagella

14. Fertilization normally occurs in the
    a. vagina **b.** cervix **c.** uterus **d.** ovary
    e. uterine tube

15. In the female reproductive system, lubricating mucus is produced by the
    a. vulva
    b. clitoris
    c. mons pubis
    d. greater vestibular glands
    e. sudoriferous glands

16. Ovarian follicles mature during
    a. menstruation
    b. ovulation
    c. the postovulatory phase
    d. the preovulatory phase
    e. the secretory phase

17. Match the following:
    ____ **a.** The enlargement and stiffening of the penis
    ____ **b.** The discharge of a small volume of semen before ejaculation
    ____ **c.** The maturation of spermatids into sperm
    ____ **d.** The powerful release of semen from the urethra to the exterior

    A. emission
    B. ejaculation
    C. erection
    D. spermiogenesis

18. The portion of the uterus responsible for its contraction is the
    a. fundus **b.** infundibulum **c.** endometrium
    d. myometrium **e.** perineum

19. Match the following:
    ____ **a.** released by the hypothalamus to regulate the ovarian cycle
    ____ **b.** stimulates the initial secretion of estrogens by growing follicles
    ____ **c.** stimulates ovulation
    ____ **d.** stimulate growth, development, and maintenance of the female reproductive system
    ____ **e.** works with estrogens to prepare the uterus for implantation of a fertilized ovum
    ____ **f.** assists with labor by helping to dilate the cervix and increase flexibility of the pubic symphysis
    ____ **g.** inhibits release of FSH by the anterior pituitary

    A. luteinizing hormone (LH)
    B. gonadotropin-releasing hormone (GnRH)
    C. relaxin
    D. progesterone
    E. inhibin
    F. follicle-stimulating hormone (FSH)
    G. estrogens

20. Birth control pills are a combination of ovarian hormones that prevent pregnancy by
    a. neutralizing the pH of the vagina
    b. inhibiting motility of the sperm
    c. causing early ovulation, before the follicle is mature
    d. preventing sperm from entering the uterus
    e. inhibiting the secretion of LH and FSH from the pituitary glan

CRITICAL THINKING APPLICATIONS

1. Thirty-five year-old Janelle has been advised to have a complete hysterectomy due to medical problems. She is worried that the procedure will cause menopause. Explain what is involved in the procedure and the likelihood that the procedure will result in menopause.

2. Phil has promised his wife that he will get a vasectomy after the birth of their next child. He is a little concerned, however, about the possible effects on his virility. What would you tell Phil about the procedure?

3. Julio and his wife have been trying unsuccessfully to become pregnant. The fertility clinic suggested that the problem may have something to do with Julio's habits of wearing very close-fitting briefs during the day and taking a long nightly soak in his hot tub. What effect could this have on fertility?

4. Your uncle Mike has just been diagnosed with an enlarged prostate (benign prostatic hyperplasia). What are the symptoms of this condition? What is the effect on the semen of removal of the prostate?

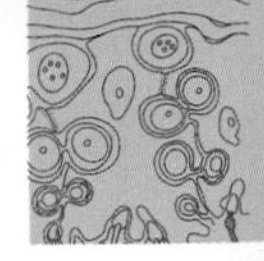

ANSWERS TO FIGURE QUESTIONS

23.1 Functionally, the penis is considered a supporting structure.

23.2 Spermatogonia (stem cells) are the least mature.

23.3 Crossing-over permits the formation of new combinations of genes from maternal and paternal chromosomes.

23.4 The middle piece contains mitochondria, which produce ATP that provides energy for locomotion of sperm.

23.5 The Sertoli cells secrete inhibin.

23.6 The female external genitals are collectively referred to as the vulva or pudendum.

23.7 Ovarian follicles secrete estrogens, and the corpus luteum secretes estrogens, progesterone, relaxin, and inhibin.

23.8 Primary oocytes are present in the ovary at birth, so they are as old as the woman is. In males, primary spermatocytes are continually being formed from spermatogonia and thus are only a few days old.

23.9 The endometrium is rebuilt after each menstruation.

23.10 The mons pubis, clitoris, prepuce, and external urethral orifice are anterior to the vaginal opening.

23.11 Oxytocin regulates milk ejection from the mammary glands.

23.12 The hormones responsible for the proliferative phase of endometrial growth are estrogens; for ovulation, LH; for growth of the corpus luteum, LH; and for the midcycle surge of LH, estrogens.

23.13 This is negative feedback because the response is opposite to the stimulus. Decreasing levels of estrogens and progesterone stimulate release of GnRH, which, in turn, increases production and release of estrogens.

chapter 24 DEVELOPMENT AND INHERITANCE

did you know? ***Human milk provides perfect nutrition for human infants. It also supplies important digestive enzymes and hormones that promote healthy development. Breast milk contains important substances that help babies fight infections. These include secretory immunoglobulin A (IgA) antibodies, formed by the mother in response to infectious agents in the mother's (and baby's) environment. These antibodies summon an immune response without harming helpful flora in the gastrointestinal (GI) tract or causing inflammation, a process that can harm the baby more than the infection. Human milk also contains mucins, oligosaccharides (sugar chains), and glycoproteins (carbohydrate-protein compounds) that bind to microbes and prevent them from infecting the baby's GI tract.***

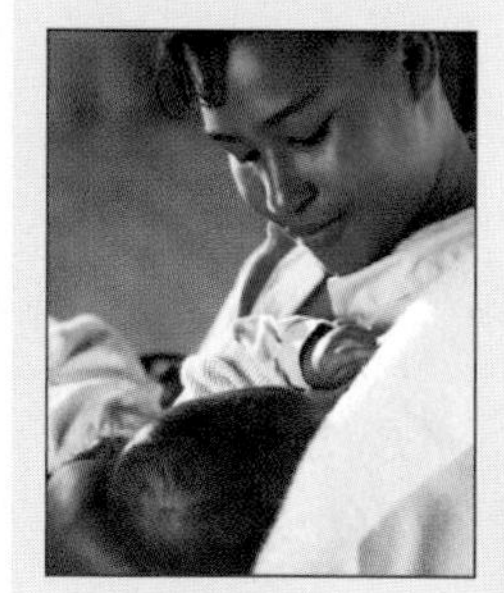

Focus on Wellness, page 603

www.wiley.com/college/apcentral

Once sperm and a secondary oocyte have developed through meiosis and maturation, and the sperm have been deposited in the vagina, pregnancy can occur. ***Pregnancy*** is a sequence of events that begins with fertilization, proceeds to implantation, embryonic development, and fetal development, and normally ends with birth about 38 weeks later, or 40 weeks after the last menstrual period.

Developmental biology is the study of the extraordinary sequence of events from the fertilization of a secondary oocyte to the formation of an adult organism. From fertilization through the eighth week of development, the developing human is called an ***embryo*** (*em-* = into; *-bryo* = grow), and this is the ***embryonic period***. ***Embryology*** (em-brē-OL-ō-jē) is the study of development from the fertilized egg through the eighth week. The ***fetal period*** begins at week nine and continues until birth. During this time, the developing human is called a ***fetus*** (FĒ-tus = offspring).

Obstetrics (ob-STET-riks; *obstetrix* = midwife) is the branch of medicine that deals with the management of pregnancy, labor, and the ***neonatal period***, the first 28 days after birth. ***Prenatal development*** (prē-NĀ-tal; *pre-* = before; *natal* = birth) is the time from fertilization to birth and includes both the embryonic and fetal periods.

In this chapter, we focus on the developmental sequence from fertilization through implantation, embryonic and fetal development, labor, and birth. We will also consider the concept of inheritance.

looking back to move ahead . . .

- Somatic Cell Division (page 62)
- Testes and Ovaries (pages 558, 564)
- Uterine Tubes and Uterus (pages 566–567)
- Estrogens and Progesterone (page 570)
- Positive Feedback System (page 8)
- Mammary Glands (page 569)
- Oxytocin (page 322)
- Prolactin (page 320)

EMBRYONIC PERIOD

OBJECTIVE • Explain the major developmental events that occur during the embryonic period.

First Week of Development

The first week of development is characterized by several significant events including fertilization, cleavage of the zygote, blastocyst formation, and implantation.

Fertilization

During ***fertilization*** (fer-til-i-ZĀ-shun; *fertil-* = fruitful), the genetic material from a haploid sperm cell and a haploid secondary oocyte merges into a single diploid nucleus (Figure 24.1). Of approximately 300 million sperm introduced into the vagina, fewer than 2 million reach the cervix of the uterus and only about 200 reach the secondary oocyte. Fertilization normally occurs in the uterine (fallopian) tube within 12 to 24 hours after ovulation. Sperm can remain viable for about 48 hours after deposition in the vagina, although a secondary oocyte is viable for only about 24 hours after ovulation. Thus, pregnancy is *most likely* to occur if intercourse takes place during a 3-day "window"—from 2 days before ovulation to 1 day after ovulation.

Sperm swim from the vagina into the cervical canal propelled by the whiplike movements of their tails (flagella). The passage of sperm through the rest of the uterus and then into the uterine tube results mainly from contractions of the walls of these organs. Prostaglandins in semen are believed to stimulate uterine motility at the time of intercourse and to aid in the movement of sperm through the uterus and into the uterine tube. Sperm that reach the vicinity of the oocyte within minutes after ejaculation *are not capable* of fertilizing it until about seven hours later. During this time in the female reproductive tract, mostly in the uterine tube, sperm undergo ***capacitation*** (ka-pas′-i-TĀ-shun; *capacit-* = capable of), a series of functional changes that cause the sperm's tail to beat even more vigorously and prepare its plasma membrane to fuse with the oocyte's plasma membrane.

For fertilization to occur, a sperm cell first must penetrate the ***corona radiata*** (kō-RŌ-na = crown; rā-dē-A-ta = to shine), the cells that surround the secondary oocyte, and then the ***zona pellucida*** (ZŌ-na = zone; pe-LOO-si-da = allowing passage of light), the clear glycoprotein layer between the corona radiata and the oocyte's plasma membrane (Figure 24.1). One of the glycoproteins in the zona pellucida acts as a sperm receptor. Its binding to specific membrane proteins in the sperm head triggers the release of enzymes from the acrosome. The acrosomal enzymes digest a path through the zona pellucida as the lashing sperm tail pushes the sperm cell onward. Although many sperm bind to the zona pellucida and release their enzymes, only the first sperm cell to penetrate the entire zona pellucida and reach the oocyte's plasma membrane fuses with the oocyte. The fusion of a sperm with a secondary oocyte sets in motion events that block fertilization by more than one sperm cell.

Once a sperm cell enters a secondary oocyte, the oocyte first must complete meiosis II. It divides into a larger ovum (mature egg) and a smaller second polar body that fragments and disintegrates (see Figure 23.8 on page 566). The nucleus in the head of the sperm and the nucleus of the fertilized ovum fuse, producing a single diploid nucleus that contains 23 chromosomes from each cell. Thus, the fusion of the haploid (n) cells restores the diploid number ($2n$) of 46 chromosomes. The fertilized ovum now is called a ***zygote*** (ZĪ-gōt; *zygon* = yolk).

Figure 24.1 Fertilization. A sperm cell penetrating the corona radiata and zona pellucida around a secondary oocyte.

During fertilization, genetic material from a sperm cell and a secondary oocyte merge to form a single diploid nucleus.

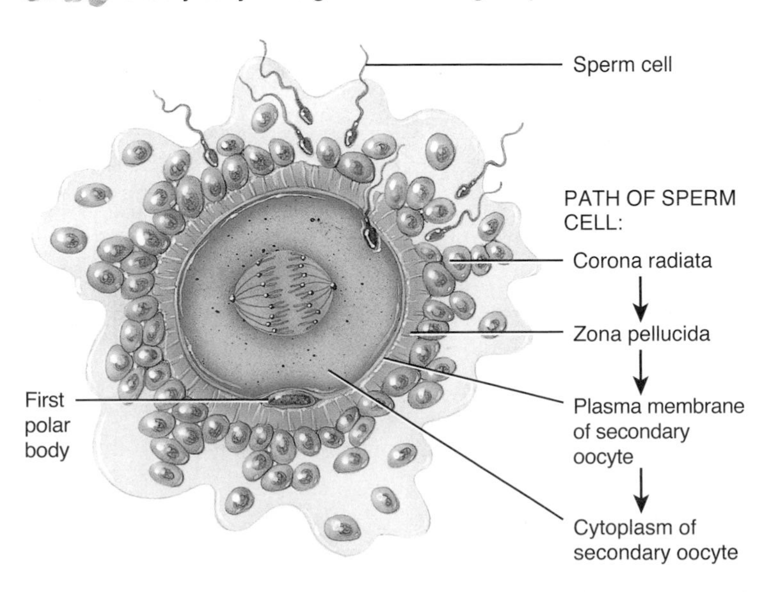

What is capacitation?

> **Dizygotic (fraternal) twins** are produced from the independent release of two secondary oocytes and the subsequent fertilization of each by different sperm. They are the same age and in the uterus at the same time, but they are genetically as dissimilar as are any other siblings. Dizygotic twins may or may not be the same sex. Because **monozygotic (identical) twins** develop from a single fertilized ovum, they contain exactly the same genetic material and are always the same sex. Monozygotic twins arise from separation of the developing zygote into two embryos, which occurs within 8 days after fertilization 99% of the time. Separations that occur later than 8 days are likely to produce **conjoined twins**, a situation in which the twins are joined together and share some body structures.

Early Embryonic Development

After fertilization, rapid mitotic cell divisions of the zygote called ***cleavage*** (KLĒV-ij) take place (Figure 24.2). The first division of the zygote begins about 24 hours after fertilization and is completed about 6 hours later. Each succeeding division takes slightly less time. By the second day after fertilization, the second cleavage is completed and there are four cells (Figure 24.2b). By the end of the third day, there are 16 cells. The progressively smaller cells produced by cleavage are called ***blastomeres*** (BLAS-tō-mērz; *blasto-* = germ or sprout; *-meres* = parts). Successive cleavages eventually produce a solid sphere of cells called the ***morula*** (MOR-ū-la = mulberry). The morula is still surrounded by the zona pellucida and is about the same size as the original zygote (Figure 24.2c).

By the end of the fourth day, the number of cells in the morula increases as it continues to move through the uterine tube toward the uterine cavity. When the morula enters the uterine cavity on day 4 or 5, a glycogen-rich secretion from the glands of the endometrium of the uterus penetrates the morula, collects between the blastomeres, and reorganizes them around a large fluid-filled cavity called the ***blastocyst cavity*** (BLAS-tō-sist; *blasto-* = germ or sprout; *-cyst* = bag) (Figure 24.2e). With the formation of this cavity, the developing mass is then called the ***blastocyst***. Though it now has hundreds of cells, the blastocyst is still about the same size as the original zygote. Further rearrangement of the blastomeres results in the formation of two distinct structures: the inner cell mass and trophoblast (Figure 24.2e). The ***inner cell mass*** is located internally and eventually develops into the embryo. The ***trophoblast*** (TRŌF-ō-blast; *tropho-* = develop or nourish) is an outer superficial layer of cells that forms the wall of the blastocyst. It will ultimately develop into the fetal portion of the placenta, the site of exchange of nutrients and wastes between the mother and fetus.

Stem cells are unspecialized cells (cells without a particular function) that have the ability to divide for long periods and develop into specialized cells. Based on their potential, stem cells are classified into three types:

(1) *Totipotent stem cells* (tō-TIP-ō-tent; *totus-* = whole; *-potentia* = power) have the potential to form all cells of an entire organism. An example is a zygote (fertilized ovum).

(2) *Pluripotent stem cells* (ploo-RIP-ō-tent; *plur-* = several) have the potential to develop into many (but not all) different types of cells of an organism. Examples are inner cell mass cells.

(3) *Multipotent stem cells* (mul-TIP-ō-tent) have the potential to develop into a few different types of cells of an organism. Examples are myeloid and lymphoid stem cells that develop into blood cells.

Pluripotent stem cells currently used in research are derived from (1) extra embryos that were destined to be used for infertility treatments but were not needed and from (2) nonliving fetuses terminated during the first trimester of pregnancy. Because pluripotent stem cells give rise to almost all cell types in the body, they are extremely important in research and health care. For example, they might be used to generate cells and tissues for transplantation to treat conditions such as cancer, Parkinson and Alzheimer disease, spinal cord injury, diabetes, heart disease, stroke, burns, birth defects, osteoarthritis, and rheumatoid arthritis.

On October 13, 2001, researchers reported cloning of the first human embryo to grow cells to treat human diseases. **Therapeutic cloning** is envisioned as a procedure in which the genetic material of a patient with a particular disease is used to create pluripotent stem cells to treat the disease. Using the principles of therapeutic cloning, scientists hope to make an embryo clone of a patient, remove the pluripotent stem cells from the embryo, and then use them to grow tissues to treat particular diseases and disorders.

Scientists are also investigating the potential clinical applications of using *adult stem cells*, stem cells that remain in the body throughout adulthood. Studies have suggested that stem cells in human adult red bone marrow have the ability to differentiate into cells of the liver, kidney, heart, lung, skeletal muscle, skin, and organs of the gastrointestinal tract. In theory, adult stem cells from red bone marrow could be harvested from a patient and then used to repair other tissues and organs in that patient's body without having to use stem cells from embryos.

The blastocyst remains free within the uterine cavity for about 2 days before it attaches to the uterine wall. About 6 days after fertilization, the blastocyst loosely attaches to the endometrium, a process called ***implantation*** (Figure 24.3). As the blastocyst implants, it orients with the inner cell mass toward the endometrium (Figure 24.3).

The major events associated with the first week of development are summarized in Figure 24.4 on page 590.

Figure 24.2 Cleavage and the formation of the morula and blastocyst.

Cleavage refers to the early, rapid mitotic divisions of a zygote.

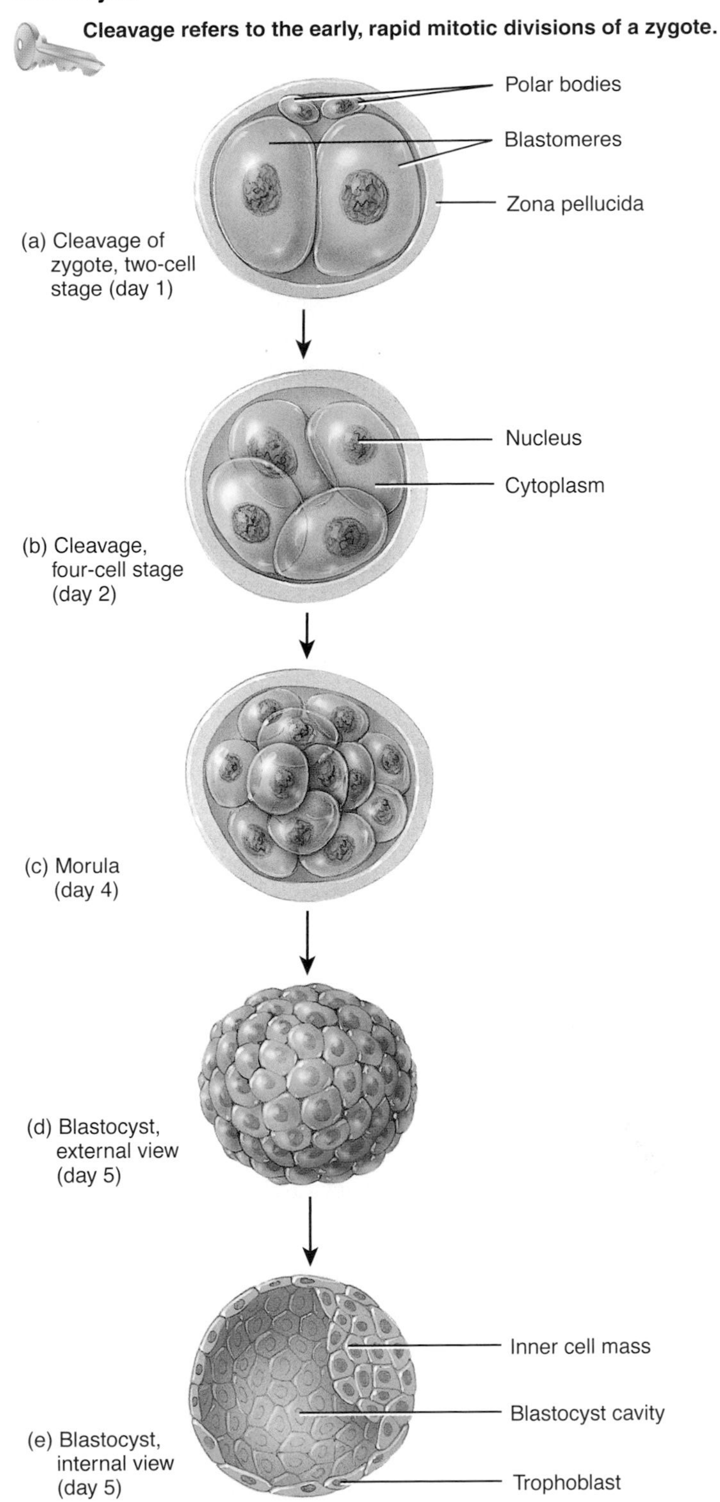

What is the histological difference between a morula and a blastocyst?

Figure 24.3 Relation of a blastocyst to the endometrium of the uterus at the time of implantation.

Implantation, the attachment of a blastocyst to the endometrium, occurs about 6 days after fertilization.

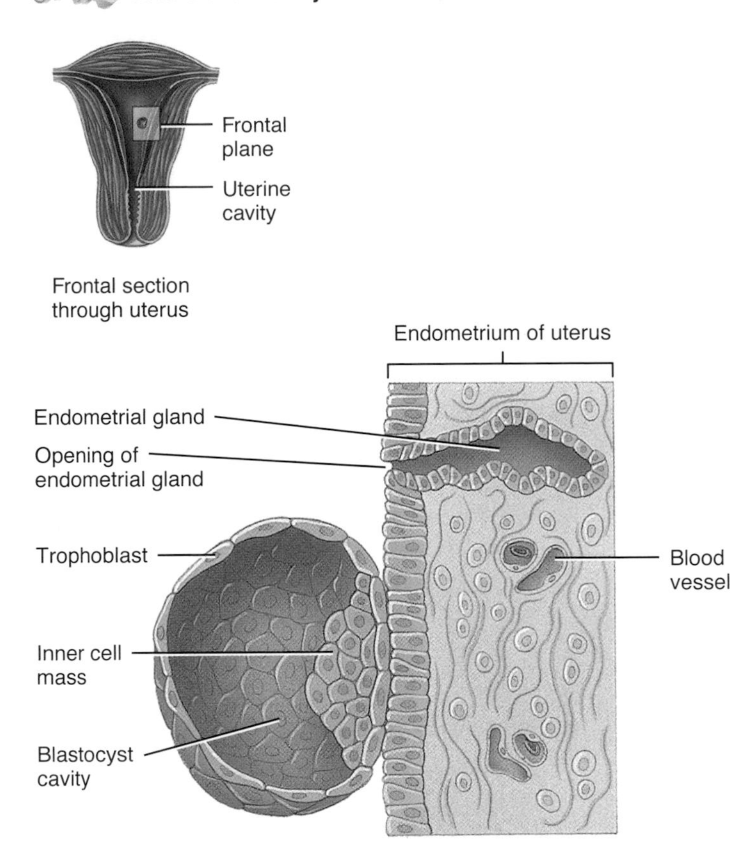

Frontal section through endometrium of uterus and blastocyst, about 6 days after fertilization

How does the blastocyst merge with and burrow into the endometrium?

Ectopic pregnancy (ek-TOP-ik; *ec-* = out of; *-topic* = place) is the development of an embryo or fetus outside the uterine cavity. An ectopic pregnancy usually occurs when movement of the fertilized ovum through the uterine tube is impaired. Situations that impair movement include scarring due to a prior tubal infection, decreased motility of the uterine tube smooth muscle, or abnormal tubal anatomy. Although the most common site of ectopic pregnancies is the uterine tube, ectopic pregnancies may also occur in the ovary, abdominal cavity, or uterine cervix. The signs and symptoms of ectopic pregnancy include one or two missed menstrual cycles followed by bleeding and acute abdominal and pelvic pain. Unless removed, the developing embryo can rupture the uterine tube, often resulting in death of the mother.

Figure 24.4 Summary of events associated with the first week of development.

Fertilization usually occurs in the uterine tube.

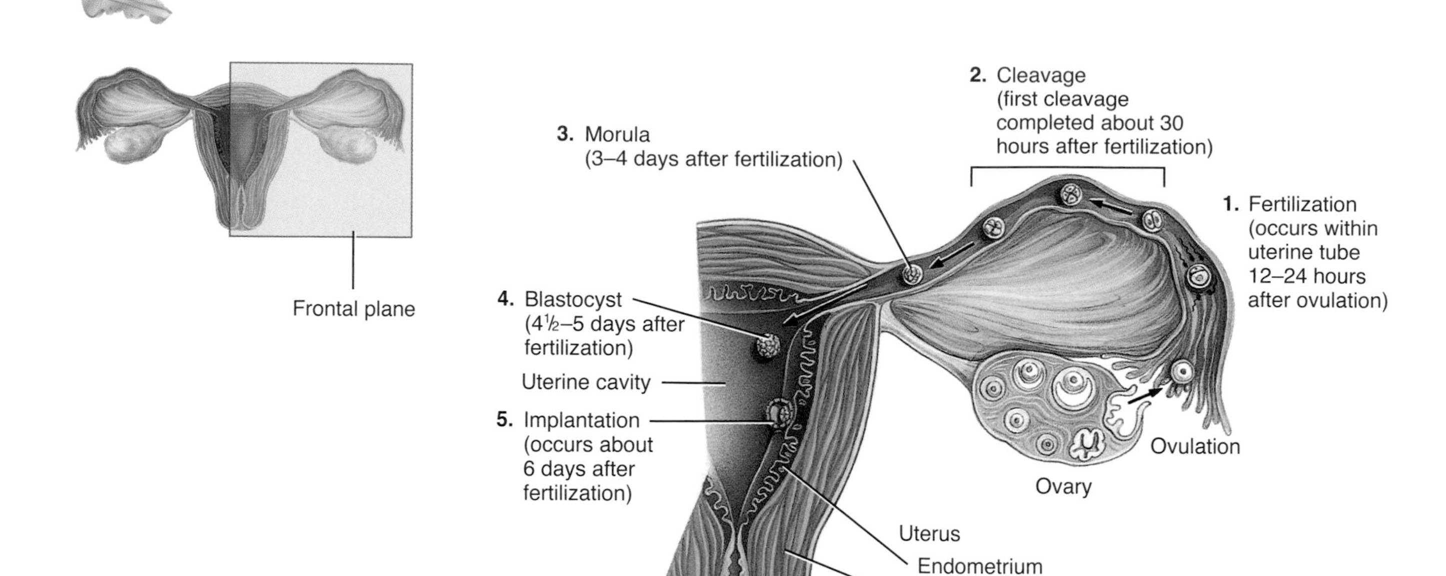

Frontal section through uterus, uterine tube, and ovary

At implantation, how is the blastocyst oriented?

■ **CHECKPOINT**

1. Where does fertilization normally occur?
2. Describe the layers of a blastocyst and their eventual fates.
3. When, where, and how does implantation occur?

Second Week of Development

About 8 days after fertilization, the trophoblast develops into two layers: a ***syncytiotrophoblast*** (sin-sīt′-ē-ō-TRŌF-ō-blast) and a ***cytotrophoblast*** (sī-tō-TRŌF-ō-blast) (Figure 24.5a). The two layers of trophoblast become part of the chorion (one of the fetal membranes) as they undergo further growth (see Figure 24.8 inset). During implantation, the syncytiotrophoblast secretes enzymes that enable the blastocyst to penetrate the uterine lining. Another secretion of the trophoblast is human chorionic gonadotropin (hCG), a hormone which sustains secretion of progesterone and estrogens by the corpus luteum. These hormones maintain the uterine lining in a secretory state and thereby prevent menstruation. About the ninth week of pregnancy the placenta is fully developed and produces the progesterone and estrogens that continue to sustain the pregnancy.

Cells of the inner cell mass also differentiate into two layers around 8 days after fertilization: a ***hypoblast (primitive endoderm)*** and ***epiblast (primitive ectoderm)*** (Figure 24.5b). Cells of the hypoblast and epiblast together form a flat disc referred to as the ***bilaminar embryonic disc*** (bī-LAM-in-ar = two-layered). In addition, a small cavity appears within the epiblast and eventually enlarges to form the ***amniotic cavity*** (am-nē-OT-ik; *amnio-* = lamb).

As the amniotic cavity enlarges, a thin protective membrane called the ***amnion*** (AM-nē-on) develops from the epiblast (Figure 25.5a). With growth of the embryo, the amnion eventually surrounds the entire embryo (see Figure 24.8 inset), creating the amniotic cavity that becomes filled with ***amniotic fluid***. Amniotic fluid serves as a shock absorber for the fetus, helps regulate fetal body temperature, helps prevent drying out, and prevents adhesions between the skin of the fetus and surrounding tissues.

> Embryonic cells are normally sloughed off into amniotic fluid. They can be examined in a procedure called **amniocentesis** (am′-nē-ō-sen-TĒ-sis; *amnio-* = amnion; *-centesis* = puncture to remove fluid), which involves withdrawing some of the amniotic fluid that bathes the developing fetus and analyzing the fetal cells and dissolved substances (see page 607).

Figure 24.5 Principal events of the second week of development.

About 8 days after fertilization, the trophoblast develops into a syncytiotrophoblast and a cytotrophoblast; the inner cell mass develops into a hypoblast and epiblast (bilaminar embryonic disc).

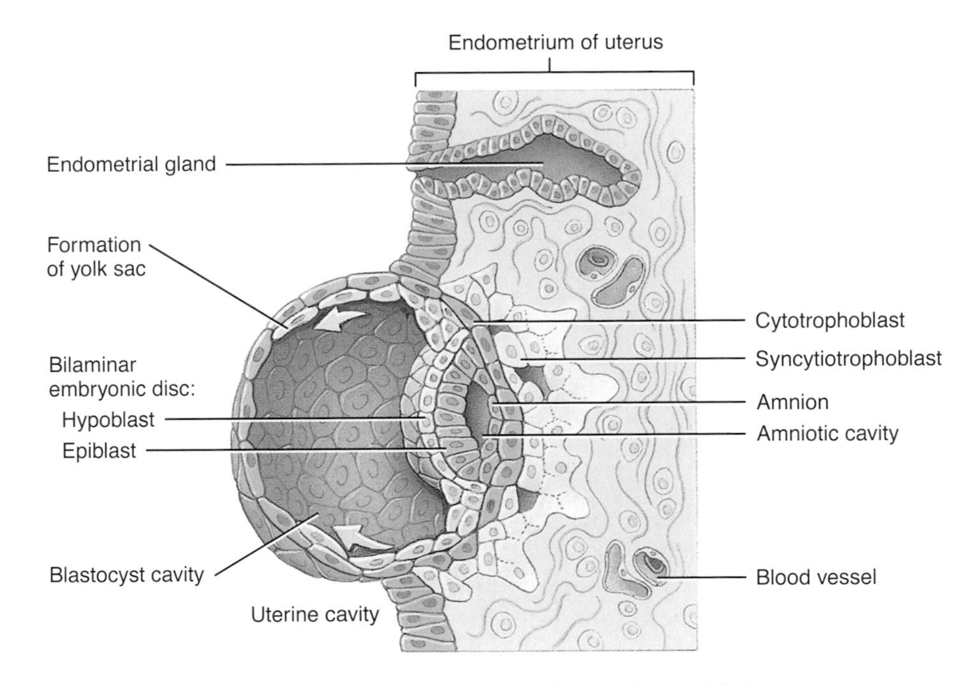

(a) Frontal section through endometrium of uterus showing blastocyst, about 8 days after fertilization

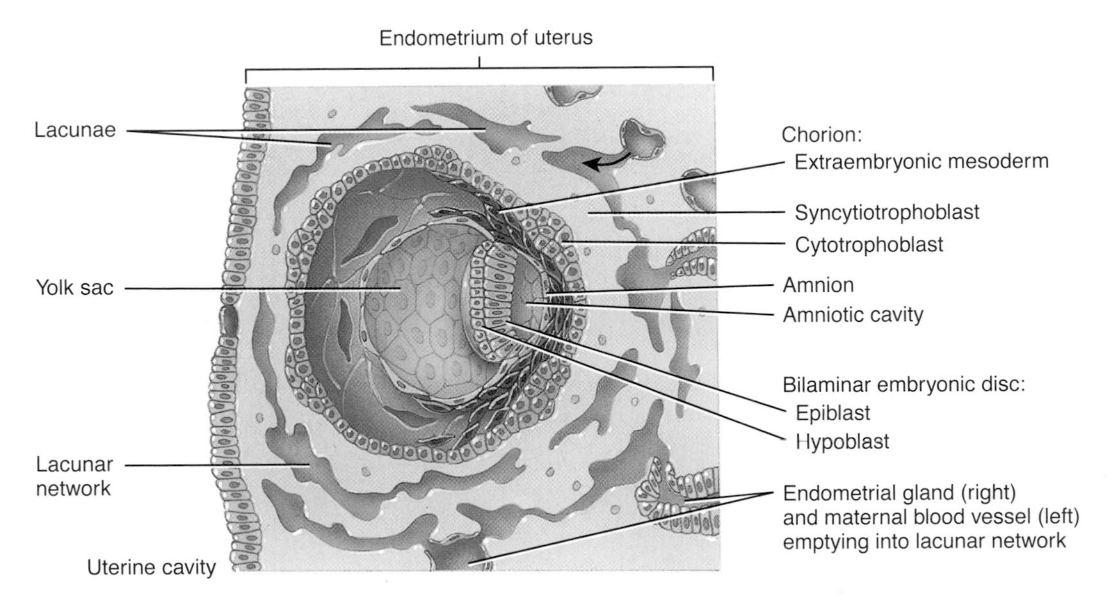

(b) Frontal section through endometrium of uterus showing blastocyst, about 12 days after fertilization

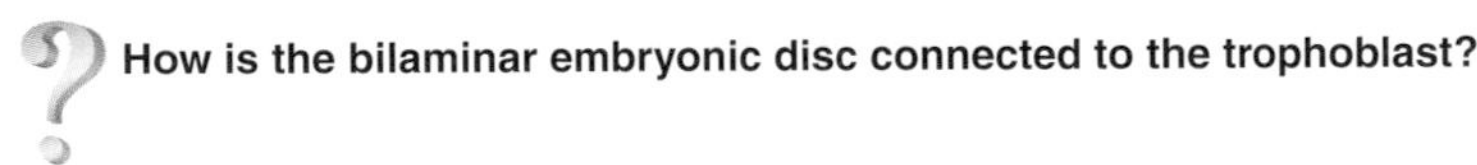

How is the bilaminar embryonic disc connected to the trophoblast?

Also on the eighth day after fertilization, cells of the hypoblast migrate and cover the inner surface of the blastocyst wall (Figure 24.5a), forming the wall of the ***yolk sac***, formerly called the blastocyst cavity (Figure 24.5b). The yolk sac has several important functions in humans. It supplies nutrients to the embryo during the second and third weeks of development, is the source of blood cells from the third through sixth weeks, contains the first cells (primordial germ cells) that will eventually migrate into the developing gonads, and forms part of the gut (gastrointestinal tract). Finally, the yolk sac functions as a shock absorber and helps prevent drying out of the embryo.

On the ninth day after fertilization, the blastocyst becomes completely embedded in the endometrium and small spaces called ***lacunae*** (la-KOO-nē = little lakes) develop within the trophoblast (Figure 24.5b). By the twelfth day of development, the lacunae fuse to form larger, interconnecting spaces called ***lacunar networks***. Maternal blood and glandular secretions enter the lacunar networks, serving as both a rich source of materials for embryonic nutrition and a disposal site for the embryo's wastes.

About the twelfth day after fertilization, mesodermal cells derived from the yolk sac form a connective tissue (mesenchyme) around the amnion and yolk sac called the ***extraembryonic mesoderm*** (Figure 24.5b). The extraembryonic mesoderm and the two layers of the trophoblast together form the ***chorion*** (KOR-ē-on = membrane) (Figure 24.5b). It surrounds the embryo and, later, the fetus (see Figure 24.8 inset). Eventually the chorion becomes the principal embryonic part of the placenta, the structure for exchange of materials between mother and fetus. The chorion protects the embryo and fetus from the immune responses of the mother and also produces human chorionic gonadotropin (hCG), an important hormone of pregnancy.

By the end of the second week of development, the bilaminar embryonic disc becomes connected to the trophoblast by a band of extraembryonic mesoderm called the ***connecting (body) stalk*** (see Figure 24.6 inset), the future umbilical cord.

■ **CHECKPOINT**

4. What are the functions of the trophoblast?

5. Describe the formation of the amnion, yolk sac, and chorion and explain their functions.

Third Week of Development

The third week of development begins a six-week period of rapid embryonic development and differentiation. During the third week, the three primary germ layers are established and lay the groundwork for organ development in weeks four through eight.

Gastrulation

The first major event of the third week of development is called ***gastrulation*** (gas′-troo-LĀ-shun). In this process, the bilaminar (two-layered) embryonic disc transforms into a trilaminar (three-layered) embryonic disc consisting of three primary germ layers, the ectoderm, mesoderm, and endoderm. The ***primary germ layers*** are the major embryonic tissues from which the various tissues and organs of the body develop.

As part of gastrulation, cells of the epiblast move inward and detach from it (Figure 24.6b). Some of the cells push out other cells of the hypoblast, forming the ***endoderm*** (*endo-* = inside; *-derm* = skin). Other cells remain between the epiblast and newly formed endoderm to form the ***mesoderm*** (*meso-* = middle). Cells remaining in the epiblast then form the ***ectoderm*** (*ecto-* = outside). As the embryo develops, the endoderm ultimately becomes the epithelial lining of the gastrointestinal tract, respiratory tract, and several other organs. The mesoderm gives rise to muscle, bone, and other connective tissues. The ectoderm develops into the epidermis of the skin and the nervous system.

About 22 to 24 days after fertilization, mesodermal cells form a solid cylinder of cells called the ***notochord*** (nō-tō-KORD; *noto-* = back; *-chord* = cord). It stimulates mesodermal cells to form parts of the backbone and intervertebral discs. The notochord also stimulates ectodermal cells over it to form the ***neural plate*** (see Figure 24.9a). By the end of the third week, the lateral edges of the neural plate become more elevated and form the ***neural fold***. The depressed midregion is called the ***neural groove***. Generally, the neural folds approach each other and fuse, thus converting the neural plate into a ***neural tube***. Neural tube cells then develop into the brain and spinal cord. The process by which the neural plate, neural folds, and neural tube form is called ***neurulation*** (noor-oo-LĀ-shun).

Neural tube defects (NTDs) are caused by problems with the normal development and closure of the neural tube. These include spina bifida (discussed on page 150) and **anencephaly** (an′-en-SEPH-a-lē; *an-* = without; *encephal* = brain). In anencephaly, the cranial bones fail to develop and certain parts of the brain remain in contact with amniotic fluid and degenerate. Usually, the part of the brain that controls vital functions such as breathing and regulation of the heart is also affected. Infants with anencephaly are stillborn or die within a few days after birth. The condition occurs about once in every 1000 births and is 2 to 4 times more common in female infants than males. Neural tube defects are associated with low levels of folic acid, one of the B vitamins.

Development of the Allantois, Chorionic Villi, and Placenta

The wall of the yolk sac forms a small vascularized outpouching called the ***allantois*** (a-LAN-tō-is; *allant-* = sausage) (see Figure 24.8 inset). In most other mammals, the allantois is used for gas exchange and waste removal. Because of the role of the human placenta in these activities, the allantois is not a prominent structure in humans. Nevertheless, it does function in early formation of blood and blood vessels and it is associated with the development of the urinary bladder.

By the end of the second week of development, ***chorionic villi*** (ko-rē-ON-ik VIL-ī) begin to develop. These fingerlike projections consist of chorion (syncytiotrophoblast surrounded by cytotrophoblast) and contain fetal blood vessels (Figure 24.7 on page 594). By the end of the third week, blood capillaries that

Figure 24.6 Gastrulation.

Gastrulation involves the rearrangement and migration of cells from the epiblast.

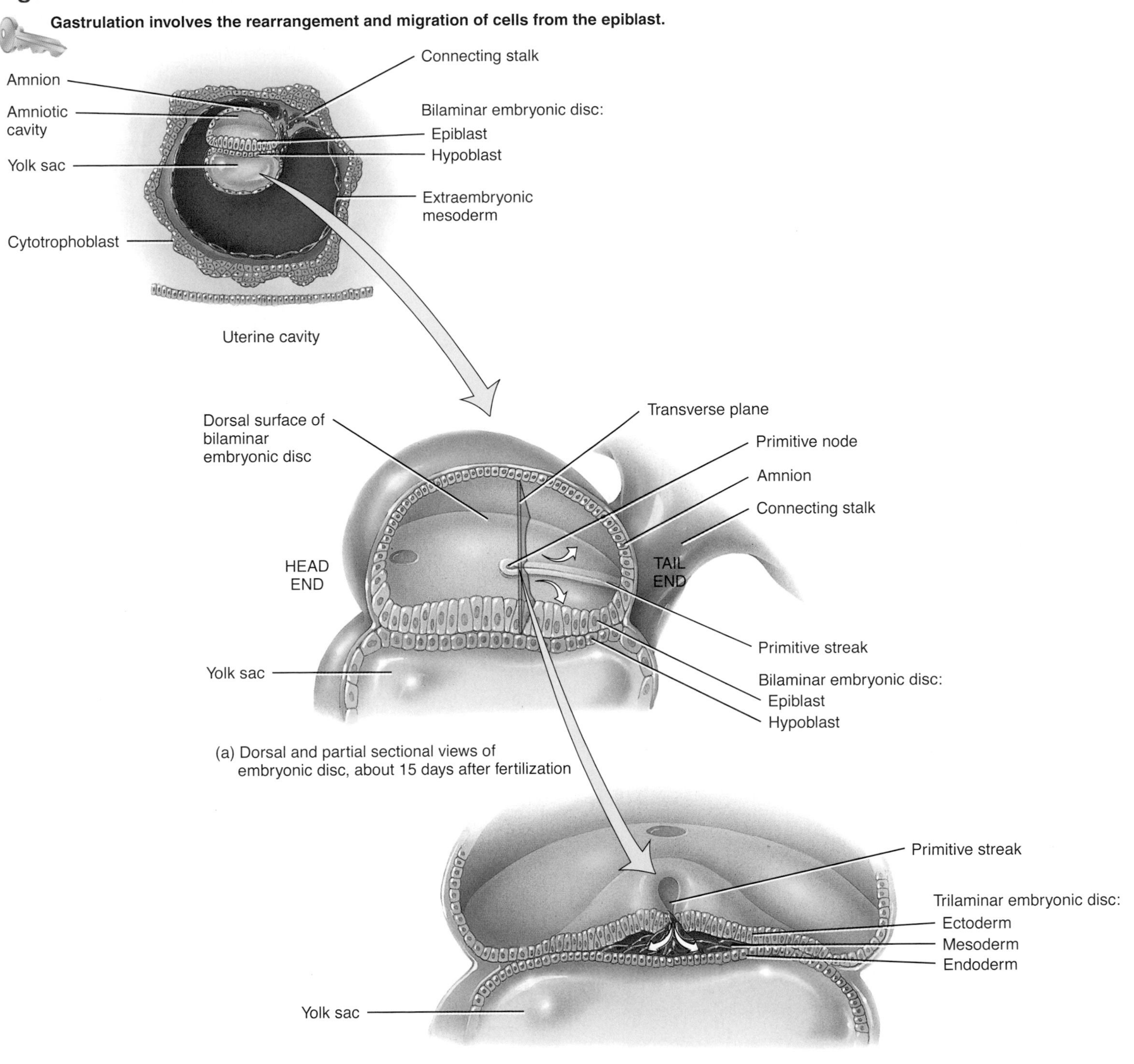

(a) Dorsal and partial sectional views of embryonic disc, about 15 days after fertilization

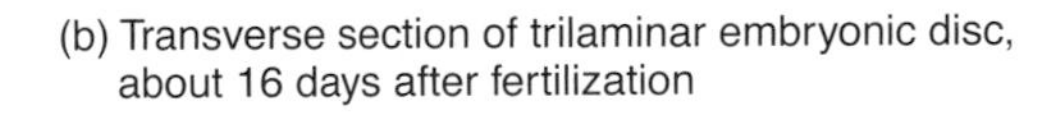

(b) Transverse section of trilaminar embryonic disc, about 16 days after fertilization

What is the significance of gastrulation?

develop in the chorionic villi connect to the embryonic heart by way of the umbilical arteries and umbilical vein. As a result, maternal and fetal blood vessels are in close proximity. Note, however, that maternal and fetal blood vessels do not join, and the blood they carry *does not normally mix*. Instead, oxygen and nutrients in the mother's blood diffuse across the cell membranes into the capillaries of the chorionic villi. Waste products such as carbon dioxide diffuse in the opposite direction.

The ***placenta*** (pla-SEN-ta = flat cake) is the site of the exchange of nutrients and wastes between the mother and fetus. The placenta is unique because it develops from two separate individuals, the mother and the fetus. By the beginning of the twelfth week, the placenta has two distinct parts: (1) the fetal portion formed by the chorionic villi and (2) the maternal portion formed by part of the endometrium of the uterus (Figure 24.8a). When fully developed, the placenta is shaped like a pancake (Figure 24.8b). Most microorganisms cannot pass through it, but certain viruses, such as those that cause AIDS, German measles, chickenpox, measles, encephalitis, and poliomyelitis, can cross the placenta as well as many drugs, alcohol, and some other substances that can cause birth defects. The placenta also stores nutrients such as carbohydrates, proteins, calcium, and iron, which are released into fetal circulation as required, and it produces several hormones that are necessary to maintain pregnancy (discussed later).

The actual connection between the placenta and embryo, and later the fetus, is through the ***umbilical cord*** (um-BIL-i-kul = navel), which develops from the connecting stalk. The umbilical cord consists of two umbilical arteries that carry deoxygenated fetal blood to the placenta, one umbilical vein that carries oxygenated maternal blood into the fetus, and supporting mucous connective tissue. A layer of amnion surrounds the entire umbilical cord and gives it a shiny appearance (Figure 24.8a).

After the birth of the baby, the placenta detaches from the uterus and is therefore termed the ***afterbirth***. At this time, the umbilical cord is tied off and then severed, leaving the baby on its own. The small portion (about an inch) of the cord that remains attached to the infant begins to wither and falls off, usually within 12 to 15 days after birth. The area where the cord was attached becomes covered by a thin layer of skin, and scar tissue forms. The scar is the ***umbilicus*** (navel).

Pharmaceutical companies use human placentas as a source of hormones, drugs, and blood; portions of placentas are also used for burn coverage. The placental and umbilical cord veins can also be used in blood vessel grafts, and cord blood can be frozen to provide a future source of pluripotent stem cells, for example, to repopulate red bone marrow following radiotherapy for cancer.

> In some cases, the entire placenta or part of it may become implanted in the inferior portion of the uterus, near or covering the cervix. This condition is called **placenta previa** (PRĒ-vē-a = before or in front of). Although placenta previa may lead to spontaneous abortion, it also occurs in approximately 1 in 250 live births. It is dangerous to the fetus because it may cause premature birth and intrauterine hypoxia due to maternal bleeding. Maternal mortality is increased due to hemorrhage and infection. The most important symptom is sudden, painless, bright-red vaginal bleeding in the third trimester. Cesarean section is the preferred method of delivery in placenta previa.

Figure 24.7 Development of the chorionic villi.

Blood vessels in the chorionic villi connect to the embryonic heart via the umbilical arteries and umbilical vein.

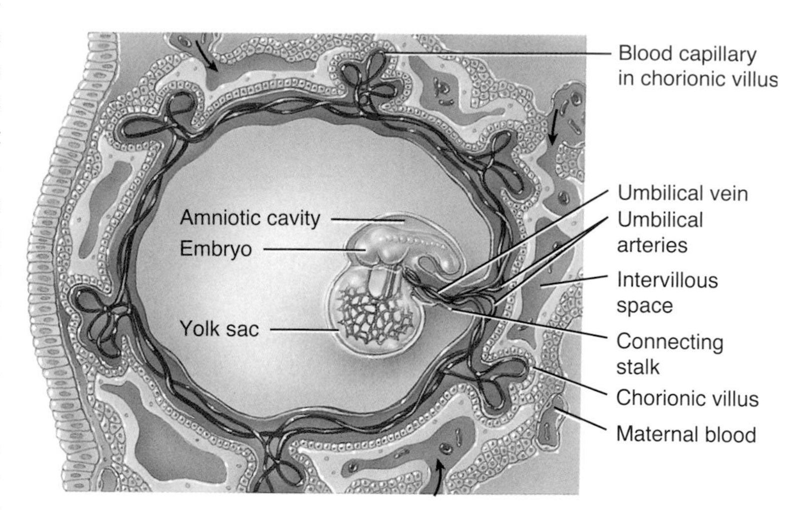

Frontal section through uterus showing an embryo and its vascular supply, about 21 days after fertilization

Why is development of the chorionic villi important?

Fourth Through Eighth Weeks of Development

The fourth through eighth weeks of development are very significant in embryonic development because all major organs appear during this time. By the end of the eighth week, all the major body systems have begun to develop, although their functions for the most part are minimal.

During the fourth week after fertilization, the embryo undergoes dramatic changes in shape and size, nearly tripling its size. It is essentially converted from a flat, two-dimensional trilaminar embryonic disc to a three-dimensional cylinder, a process called ***embryonic folding***.

The first distinguishable structures are those in the head area. The first sign of a developing ear is a thickened area of ectoderm, the ***otic placode*** (future internal ear), which can be distinguished about 22 days after fertilization (see Figure 24.9d). The eyes also begin their development about 22 days after fertilization. This is evidenced by a thickened area of ectoderm called the ***lens placode*** (see Figure 24.9c).

By the middle of the fourth week, the upper limbs begin their development as outgrowths of mesoderm covered by ectoderm called ***upper limb buds*** (see Figure 24.9c, d). By the end of the fourth week, the ***lower limb buds*** develop. The heart also forms a distinct projection on the ventral surface of the embryo called the ***heart prominence*** (see Figure 24.9c). A ***tail*** is also a distinguishing feature of an embryo at the end of the fourth week (see Figure 24.9c).

During the fifth week of development, there is a very rapid development of the brain, so growth of the head is con-

Figure 24.8 Placenta and umbilical cord.

The placenta is formed by the chorionic villi of the embryo and part of the endometrium of the mother.

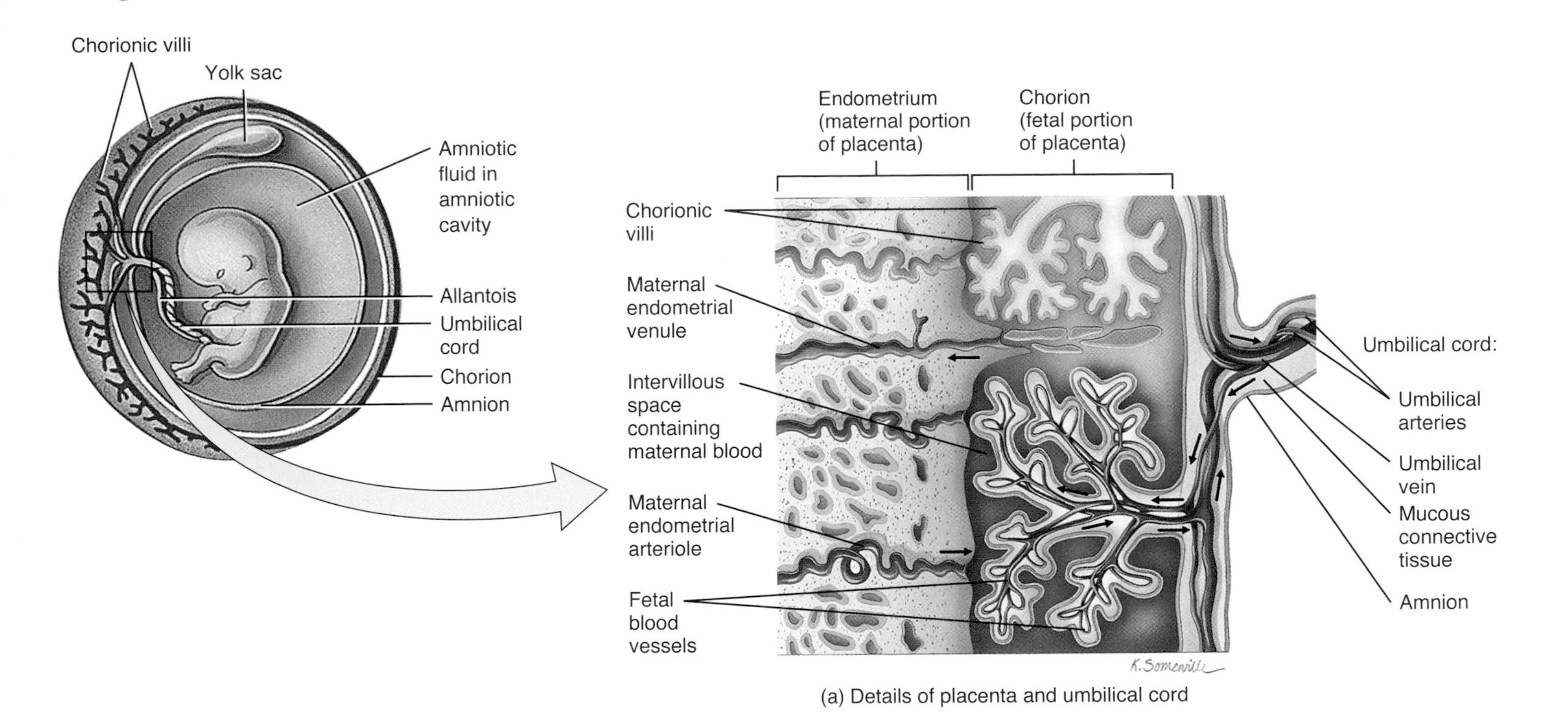

(a) Details of placenta and umbilical cord

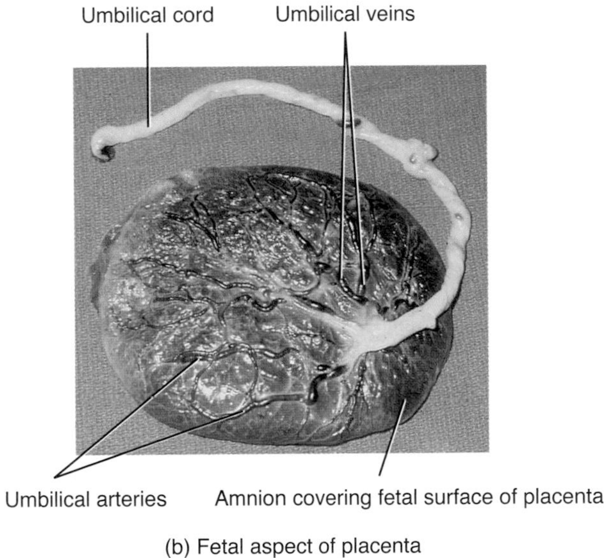

(b) Fetal aspect of placenta

What is the function of the placenta?

siderable. By the end of the sixth week, the head grows even larger relative to the trunk, and the limbs show substantial development. In addition, the neck and trunk begin to straighten, and the heart is now four-chambered. By the seventh week, the various regions of the limbs become distinct and the beginnings of digits appear (see Figure 24.9e). At the start of the eighth week, the final week of the embryonic period, the digits of the hands are short and webbed and the tail is still visible, but shorter. In addition, the eyes are open and the auricles of the ears are visible. By the end of the eighth week, all regions of the limbs are apparent and the digits are distinct and no longer webbed. Also, the eyelids come together and may fuse, the tail disappears, and the external genitals begin to differentiate. The embryo now has clearly human characteristics.

■ **CHECKPOINT**

6. How do the three primary germ layers form? Why are they important?
7. Describe how neurulation occurs. Why is it significant?
8. How does the placenta form and what is its function?
9. Why are the second through fourth weeks of development so crucial?
10. What changes occur in the limbs during the second half of the embryonic period?

FETAL PERIOD

OBJECTIVE • **Define the fetal period and outline its major events.**

During the fetal period, tissues and organs that developed during the embryonic period grow and differentiate. Very few new structures appear during the fetal period, but the rate of body growth is remarkable, especially during the second half of intrauterine life. For example, during the

Table 24.1 Summary of Changes During Embryonic and Fetal Development

Time	Approximate Size and Weight	Representative Changes
Embryonic Period		
1–4 weeks	0.6 cm (3/16 in.)	Primary germ layers and notochord develop. Neurulation occurs. Brain development begins. Blood vessel formation begins and blood forms in yolk sac, allantois, and chorion. Heart forms and begins to beat. Chorionic villi develop and placental formation begins. The embryo folds. The primitive gut and limb buds develop. Eyes and ears begin to develop, tail forms, and body systems begin to form.
5–8 weeks	3 cm (1.25 in.) 1 g (1/30 oz)	Brain development continues. Limbs become more distinct and digits appear. Heart becomes four-chambered. Eyes are far apart and eyelids are fused. Nose develops and is flat. Face is more human-like. Ossification begins. Blood cells start to form in liver. External genitals begin to differentiate. Tail disappears. Major blood vessels form. Many internal organs continue to develop.
Fetal Period		
9–12 weeks	7.5 cm (3 in.) 30 g (1 oz)	Head constitutes about half the length of the fetal body, and fetal length nearly doubles. Brain continues to enlarge. Face is broad, with eyes fully developed, closed, and widely separated. Nose develops a bridge. External ears develop and are low set. Ossification continues. Upper limbs almost reach final relative length but lower limbs are not quite as well developed. Heartbeat can be detected. Gender is distinguishable from external genitals. Urine secreted by fetus is added to amniotic fluid. Red bone marrow, thymus, and spleen participate in blood cell formation. Fetus begins to move, but its movements cannot be felt yet by the mother. Body systems continue to develop.
13–16 weeks	18 cm (6.5–7 in.) 100 g (4 oz)	Head is relatively smaller than rest of body. Eyes move medially to their final positions, and ears move to their final positions on the sides of the head. Lower limbs lengthen. Fetus appears even more humanlike. Rapid development of body systems occurs.
17–20 weeks	25–30 cm (10–12 in.) 200–450 g (0.5–1 lb)	Head is more proportionate to rest of body. Eyebrows and head hair are visible. Growth slows but lower limbs continue to lengthen. Vernix caseosa (fatty secretions of sebaceous glands and dead epithelial cells) and lanugo (delicate fetal hair) cover fetus. Brown fat forms and is the site of heat production. Fetal movements are commonly felt by mother (quickening).
21–25 weeks	27–35 cm (11–14 in.) 550–800 g (1.25–1.5 lb)	Head becomes even more proportionate to rest of body. Weight gain is substantial, and skin is pink and wrinkled. By 24 weeks, lung cells begin to produce surfactant.
26–29 weeks	32–42 cm (13–17 in.) 1110–1350 g (2.5–3 lb)	Head and body are more proportionate and eyes are open. Toenails are visible. Body fat is 3.5% of total body mass and additional subcutaneous fat smoothes out some wrinkles. Testes begin to descend toward scrotum at 28 to 32 weeks. Red bone marrow is major site of blood cell production. Many fetuses born prematurely during this period survive if given intensive care because lungs can provide adequate ventilation and central nervous system is developed sufficiently to control breathing and body temperature.
30–34 weeks	41–45 cm (16.5–18 in.) 2000–2300 g (4.5–5 lb)	Skin is pink and smooth. Fetus assumes upside down position. Pupillary reflex is present by 30 weeks. Body fat is 8% of total body mass. Fetuses 33 weeks and older usually survive if born prematurely.
35–38 weeks	50 cm (20 in.) 3200–3400 g (7–7.5 lb)	By 38 weeks, circumference of fetal abdomen is greater than that of head. Skin is usually bluish-pink, and growth slows as birth approaches. Body fat is 16% of total body mass. Testes are usually in scrotum in full-term male infants. Even after birth, an infant is not completely developed; an additional year is required, especially for complete development of the nervous system.

last two and one half months of intrauterine life, half of the full-term weight is added. At the beginning of the fetal period, the head is half the length of the body. By the end of the fetal period, the head size is only one-quarter the length of the body. During the same period, the fetal limbs also increase in size from one-eighth to one-half the fetal length. The fetus is also less vulnerable to the damaging effects of drugs, radiation, and microbes than it was as an embryo.

A summary of the major developmental events of the embryonic and fetal period is presented in Table 24.1 and illustrated in Figure 24.9.

■ CHECKPOINT

11. What are the general developmental trends during the fetal period?

12. Using Table 24.1 as a guide, select any one body structure in weeks 9 through 12 and trace its development through the remainder of the fetal period.

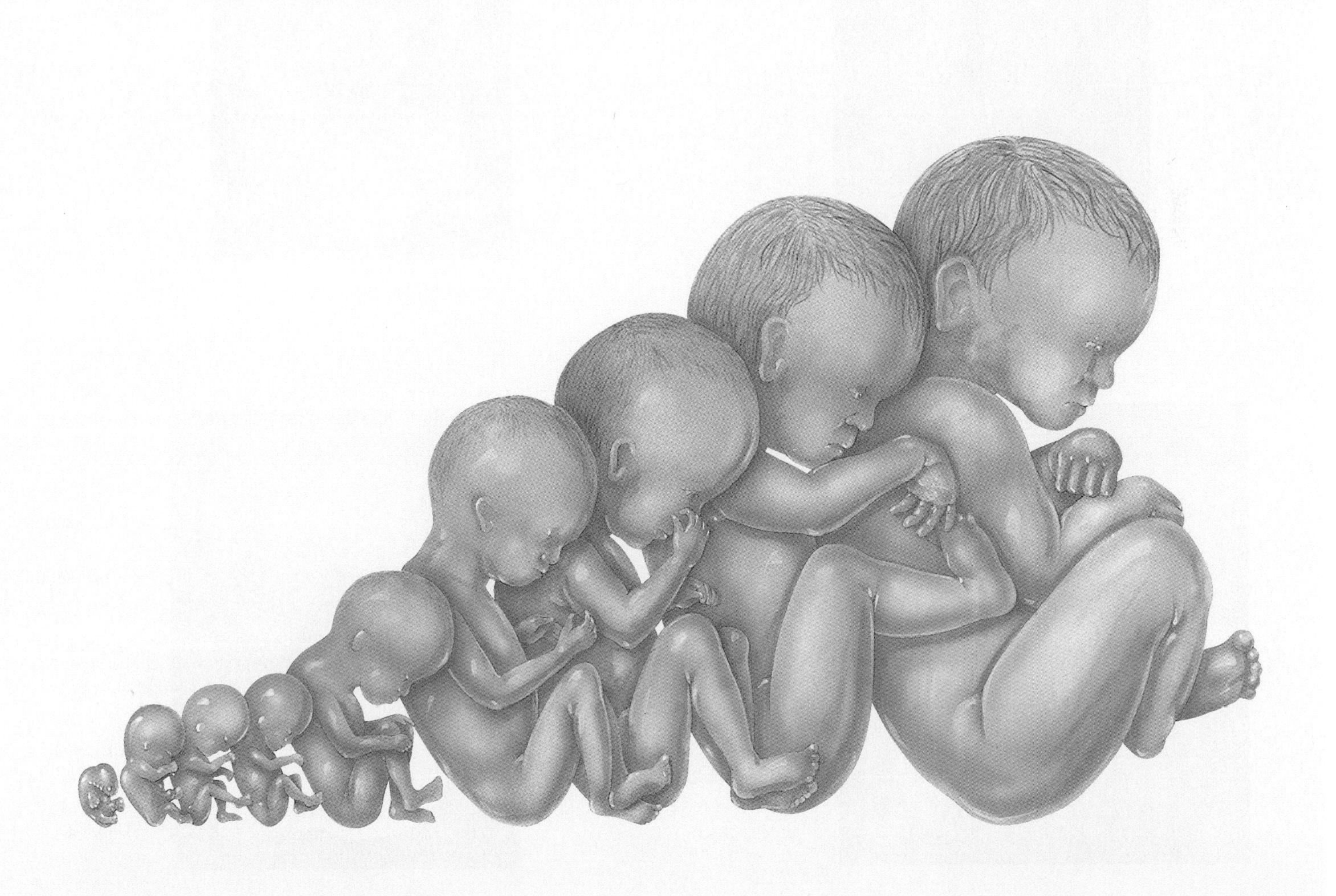

Figure 24.9 Summary of representative developmental events of the embryonic and fetal periods. The embryos and fetuses are not shown at their actual sizes.

Development during the fetal period is mostly concerned with the growth and differentiation of tissues and organs formed during the embryonic period.

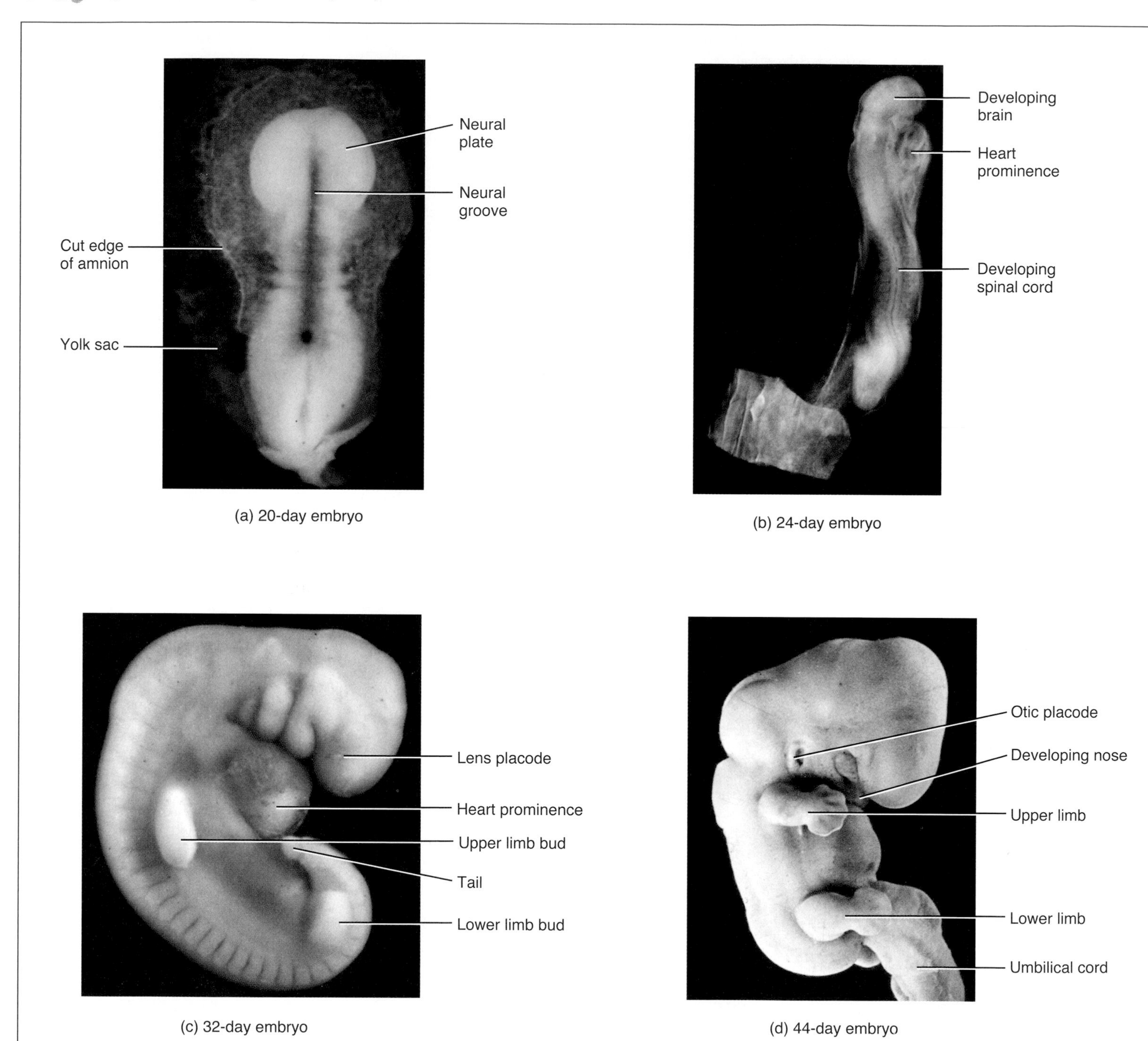

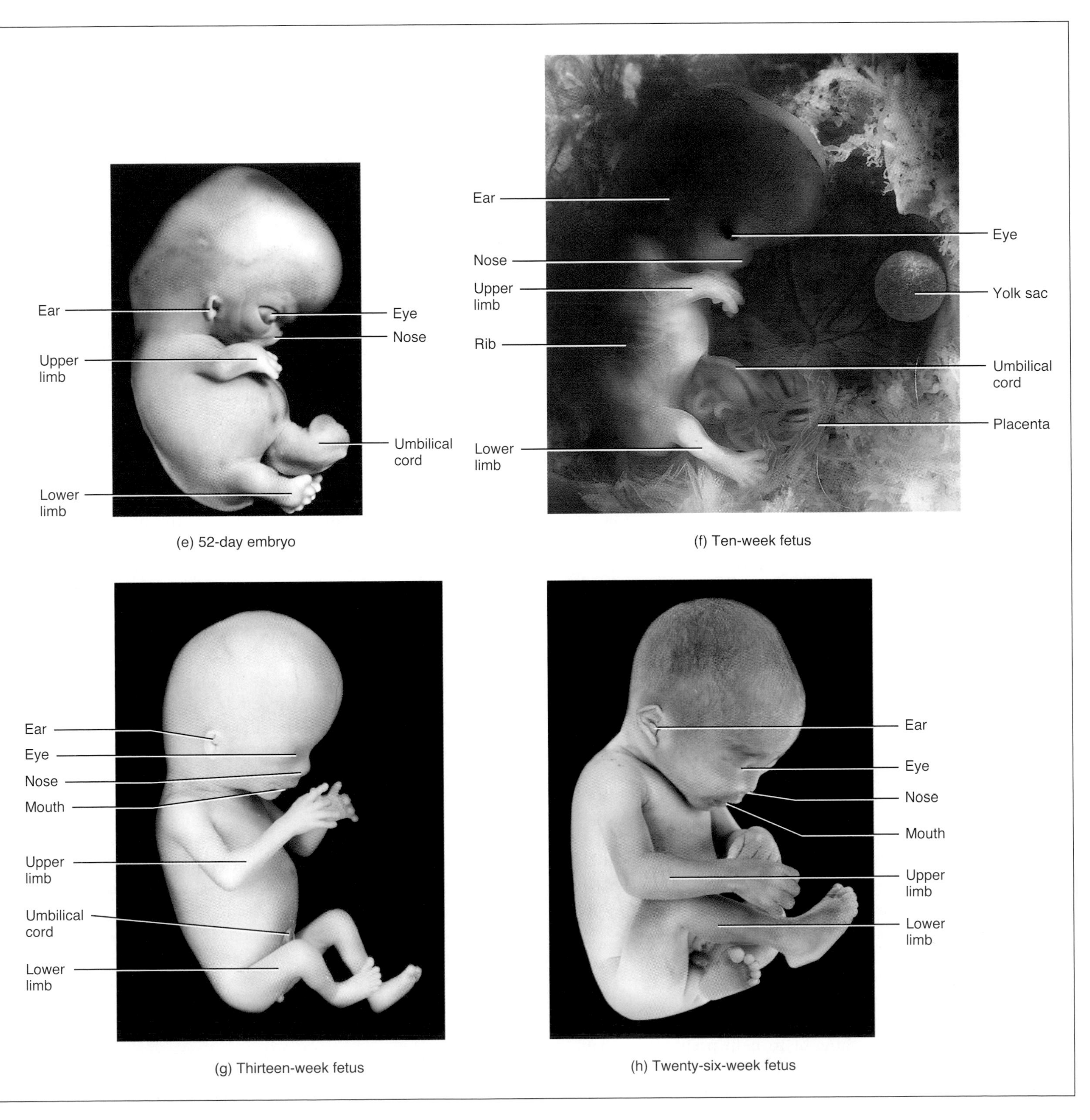

(e) 52-day embryo

(f) Ten-week fetus

(g) Thirteen-week fetus

(h) Twenty-six-week fetus

How does mid-fetal weight compare to end-fetal weight?

MATERNAL CHANGES DURING PREGNANCY

OBJECTIVES • Describe the sources and functions of the hormones secreted during pregnancy.

• Describe the hormonal, anatomical, and physiological changes in the mother during pregnancy.

Hormones of Pregnancy

During the first 3 to 4 months of pregnancy, the corpus luteum in the ovary continues to secrete progesterone and estrogens, which maintain the lining of the uterus during pregnancy and prepare the mammary glands to secrete milk. The amounts secreted by the corpus luteum, however, are only slightly more than those produced after ovulation in a normal menstrual cycle. From the third month through the remainder of the pregnancy, the placenta itself provides the high levels of progesterone and estrogens required. The chorion secretes ***human chorionic gonadotropin (hCG)*** into the blood. In turn, hCG stimulates the corpus luteum to continue production of progesterone and estrogens—an activity required to prevent menstruation and for the continued attachment of the embryo and fetus to the lining of the uterus. By the eighth day after fertilization, hCG can be detected in the blood and urine of a pregnant woman. Peak secretion of hCG occurs at about the ninth week of pregnancy. During the fourth and fifth months the hCG level decreases sharply and then levels off until childbirth.

The chorion begins to secrete estrogens after the first 3 to 4 weeks of pregnancy and progesterone by the sixth week. These hormones are secreted in increasing quantities until the time of birth. From the third month to the ninth month, the placenta supplies the levels of progesterone and estrogens and progesterone needed to maintain the pregnancy. A high level of progesterone ensures that the uterine myometrium is relaxed and that the cervix is tightly closed. After delivery, estrogens and progesterone in the blood decrease to normal levels.

Relaxin, a hormone produced first by the corpus luteum of the ovary and later by the placenta, increases the flexibility of the pubic symphysis and ligaments of the sacroiliac and sacrococcygeal joints and helps dilate the uterine cervix during labor. Both of these actions ease delivery of the baby.

A third hormone produced by the chorion of the placenta is ***human placental lactogen (hPL)***. The rate of secretion of hPL increases in proportion to placental mass, reaching maximum levels after 32 weeks and remaining relatively constant after that. It is thought to help prepare the mammary glands for lactation, enhance maternal growth by increasing protein synthesis, and regulate certain aspects of metabolism in the mother and fetus.

The hormone most recently found to be produced by the placenta is ***corticotropin-releasing hormone (CRH)***, which in nonpregnant people is secreted only by the hypothalamus. CRH is now thought to be part of the "clock" that establishes the timing of birth. Women who have higher levels of CRH earlier in pregnancy are more likely to deliver prematurely; those who have low levels are more likely to deliver after the due date. CRH from the placenta has a second important effect: It increases secretion of cortisol, which is needed for maturation of the fetal lungs and the production of surfactant.

Early pregnancy tests detect the tiny amounts of human chorionic gonadotropin (hCG) in the urine that begin to be excreted about 8 days after fertilization. The test kits can detect pregnancy as early as the first day of a missed menstrual period—that is, at about 14 days after fertilization. Chemicals in the kits produce a color change if a reaction occurs between hCG in the urine and hCG antibodies included in the kit.

Changes During Pregnancy

By about the end of the third month of pregnancy, the uterus occupies most of the pelvic cavity. As the fetus continues to grow, the uterus extends higher into the abdominal cavity. Toward the end of a full-term pregnancy, the uterus fills almost the entire abdominal cavity, reaching almost to the xiphoid process of the sternum. It pushes the maternal intestines, liver, and stomach superiorly, elevates the diaphragm, and widens the thoracic cavity.

Changes in the skin during pregnancy are more apparent in some women than in others. Included are increased pigmentation around the eyes and cheekbones in a masklike pattern, in the areolae of the breasts, and in the lower abdomen. Striae (stretch marks) over the abdomen can occur as the uterus enlarges, and hair loss increases. Pregnancy-induced physiological changes include weight gain due to the fetus, amniotic fluid, the placenta, uterine enlargement, and increased total body water; increased storage of proteins, triglycerides, and minerals; marked breast enlargement in preparation for lactation; and lower back pain due to lordosis.

Several changes occur in the maternal cardiovascular system. Stroke volume increases by about 30% and cardiac output rises by 20–30% due to increased maternal blood flow to the placenta and increased metabolism. Heart rate increases 10–15% and blood volume increases 30–50%, mostly during the second half of pregnancy. These increases are necessary to meet the additional demands of the fetus for nutrients and oxygen.

Pulmonary function is also altered during pregnancy to meet the added oxygen demands of the fetus. Tidal volume can increase by 30–40%, expiratory reserve volume can be

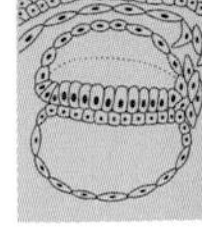

reduced by up to 40%, minute ventilation (the total volume of air inhaled and exhaled each minute) can increase by up to 40%, and total body oxygen consumption can increase by about 10–20%. Dyspnea (difficult breathing) also occurs as the expanding uterus pushes on the diaphragm.

With regard to the gastrointestinal tract, pregnant women experience an increase in appetite. Pressure on the stomach may force the stomach contents superiorly into the esophagus, resulting in heartburn. A general decrease in GI tract motility can cause constipation, delay gastric emptying time, and produce nausea, vomiting, and heartburn. Pressure on the urinary bladder by the enlarging uterus can produce urinary symptoms, such as increased frequency and urgency of urination, and stress incontinence.

Changes in the reproductive system include edema and increased blood flow to the vagina. The uterus increases from its nonpregnant mass of 60–80 g to 900–1200 g at term because of increased numbers of muscle fibers in the myometrium in early pregnancy and enlargement of muscle fibers during the second and third trimesters.

■ **CHECKPOINT**

13. List the hormones involved in pregnancy, and describe the functions of each.

14. What structural and functional changes occur in the mother during pregnancy?

EXERCISE AND PREGNANCY

OBJECTIVE • Explain the effects of pregnancy on exercise and of exercise on pregnancy.

Only a few changes in early pregnancy affect exercise. A pregnant woman may tire more easily than usual, or morning sickness (nausea and sometimes vomiting) may interfere with regular exercise. As the pregnancy progresses, weight is gained and posture changes, so more energy is needed to perform activities, and certain maneuvers (sudden stopping, changes in direction, rapid movements) are more difficult to execute. In addition, certain joints, especially the pubic symphysis, become less stable in response to the increased level of the hormone relaxin. As compensation, many mothers-to-be walk with widely spread legs and a shuffling motion.

Although blood shifts from viscera (including the uterus) to the muscles and skin during exercise, there is no evidence of inadequate blood flow to the placenta. The heat generated during exercise may cause dehydration and further increase body temperature. During early pregnancy especially, excessive exercise and heat buildup should be avoided because elevated body temperature has been implicated in neural tube defects. Exercise has no known effect on lactation, provided a woman remains hydrated and wears a bra that provides good support. Overall, moderate physical activity does not endanger the fetus of a healthy woman who has a normal pregnancy.

Among the benefits of exercise to the mother during pregnancy are a greater sense of well-being and fewer minor complaints.

■ **CHECKPOINT**

15. How do changes during early and late pregnancy affect the ability to exercise?

LABOR AND DELIVERY

OBJECTIVE • Explain the events associated with the three stages of labor.

Labor is the process by which the fetus is expelled from the uterus through the vagina. ***Parturition*** (par′-toor-ISH-un; *parturit-* = childbirth) also means giving birth.

Progesterone inhibits uterine contractions. Toward the end of pregnancy, the levels of estrogens in the mother's blood rise sharply, producing changes that overcome the inhibiting effects of progesterone. Estrogens also stimulate the placenta to release prostaglandins. Prostaglandins induce production of enzymes that digest collagen fibers in the cervix, causing it to soften. High levels of estrogens cause uterine muscle fibers to display receptors for oxytocin, the hormone that stimulates uterine contractions. Relaxin assists by increasing the flexibility of the pubic symphysis and helping dilate the uterine cervix.

The control of labor contractions occurs via a positive feedback cycle. Uterine contractions force the baby's head or body into the uterine cervix, which stretches the cervix. This stimulates stretch receptors in the cervix to send nerve impulses to the hypothalamus, causing it to release oxytocin. Oxytocin stimulates more forceful uterine contractions, which stretches the cervix more, and promotes secretion of more oxytocin. The positive feedback system is broken with the birth of the infant, which decreases stretching of the cervix.

Uterine contractions occur in waves (quite similar to peristaltic waves) that start at the top of the uterus and move downward, eventually expelling the fetus. ***True labor*** begins when uterine contractions occur at regular intervals, usually producing pain. As the interval between contractions shortens, the contractions intensify. Another symptom of true labor in some women is localization of pain in the back that is intensified by walking. The reliable indicator of true labor is dilation of the cervix and the "show," a discharge of a blood-containing mucus that appears in the cervical canal during labor. In ***false labor***, pain is felt in the abdomen at irregular intervals, but it does not intensify and walking does not alter it significantly. There is no "show" and no cervical dilation.

True labor can be divided into three stages:

1. **Stage of dilation.** The time from the onset of labor to the complete dilation of the cervix is the ***stage of dilation***. This stage, which typically lasts 6–12 hours, features regular contractions of the uterus, usually a rupturing of the amniotic sac, and complete dilation (to 10 cm) of the cervix. If the amniotic sac does not rupture spontaneously, it is ruptured intentionally.
2. **Stage of expulsion.** The time (10 minutes to several hours) from complete cervical dilation to delivery of the baby is the ***stage of expulsion***.
3. **Placental stage.** The time (5–30 minutes or more) after delivery until the placenta or "afterbirth" is expelled by powerful uterine contractions is the ***placental stage***. These contractions also constrict blood vessels that were torn during delivery, thereby reducing the likelihood of hemorrhage.

As a rule, labor lasts longer with first babies, typically about 14 hours. For women who have previously given birth, the average duration of labor is about 8 hours—although the time varies enormously among births.

Delivery of a physiologically immature baby carries certain risks. A ***premature infant*** or "preemie" is generally considered a baby who weighs less than 2500 g (5.5 lb) at birth. Poor prenatal care, drug abuse, history of a previous premature delivery, and mother's age below 16 or above 35 increase the chance of premature delivery. The body of a premature infant is not yet ready to sustain some critical functions, and thus its survival is uncertain without medical intervention. The major problem after delivery of an infant under 36 weeks of gestation is respiratory distress syndrome (RDS) of the newborn due to insufficient surfactant. RDS can be eased by use of artificial surfactant and a ventilator that delivers oxygen until the lungs can operate on their own.

About 7% of pregnant women do not deliver by 2 weeks after their due date. Such infants are called *post term babies* or *post date babies*. They carry an increased risk of brain damage to the fetus, and even fetal death, due to inadequate supplies of oxygen and nutrients from an aging placenta. Post-term deliveries may be facilitated by inducing labor, initiated by administration of oxytocin (Pitocin®), or by surgical delivery (cesarean section).

Following the delivery of the baby and placenta is a 6-week period during which the maternal reproductive organs and physiology return to the prepregnancy state. This period is called the ***puerperium*** (pū′-er-PER-ē-um).

Dystocia (dis-TŌ-sē-a; *dys-* = painful or difficult; *toc-* = birth), or difficult labor, may result either from an abnormal position (presentation) of the fetus or a birth canal of inadequate size to permit vaginal delivery. In a **breech presentation**, for example, the fetal buttocks or lower limbs, rather than the head, enter the birth canal first; this occurs most often in premature births. If fetal or maternal distress prevents a vaginal birth, the baby may be delivered surgically through an abdominal incision. A low, horizontal cut is made through the abdominal wall and lower portion of the uterus, through which the baby and placenta are removed. Even though it is popularly associated with the birth of Julius Caesar, the true reason this procedure is termed a **cesarean section (C-section)** is because it was described in Roman Law, *lex cesarea*, about 600 years before Julius Caesar was born. Even a history of multiple C-sections need not exclude a pregnant woman from attempting a vaginal delivery.

■ **CHECKPOINT**

16. What hormonal changes induce labor?
17. What happens during the stage of dilation, the stage of expulsion, and the placental stage of true labor?

LACTATION

OBJECTIVE • Discuss the hormonal control of lactation.

Lactation (lak′-TĀ-shun; *lact-* = milk) is the production and ejection of milk from the mammary glands. A principal hormone in promoting milk production is ***prolactin (PRL)***, which is secreted from the anterior pituitary gland. Even though prolactin levels increase as the pregnancy progresses, no milk production occurs because progesterone inhibits the effects of prolactin. After delivery, the levels of progesterone and estrogens in the mother's blood decrease, and the inhibition is removed. The principal stimulus in maintaining prolactin production during lactation is the sucking action of the infant. Suckling initiates nerve impulses from stretch receptors in the nipples to the hypothalamus, and more prolactin is released by the anterior pituitary.

Oxytocin causes the release of milk into the mammary ducts. Milk formed by the glandular cells of the breasts is stored until the baby begins active suckling. Stimulation of touch receptors in the nipple initiates sensory nerve impulses that are relayed to the hypothalamus. In response, secretion of oxytocin from the posterior pituitary increases.

FOCUS ON WELLNESS

Breast Milk—Mother Nature's Approach to Infection Prevention

Physicians in both industrialized and developing countries have long observed that breast-fed babies contract fewer infections than do babies fed formula. This difference is due, in part, to a number of ingredients in breast milk that enhance an infant's ability to fight disease, including antibodies and some immune cells.

Keeping the Tract Intact When Under Attack

Several substances in breast milk enhance immunity in the baby's gastrointestinal (GI) tract. One family of these is the *secretory immunoglobulin A (IgA)* antibodies. When the baby's mother encounters pathogens, she manufactures antibodies specific to each one. The antibodies pass into her breast milk and escape breakdown in the baby's GI tract because they are protected by the so-called secretory component. Once in the baby's GI tract, the antibodies bind with the targeted infectious agents and prevent them from passing through the lining of the GI tract. This protection is especially important in the earliest days of life, because the infant does not begin to make his or her own secretory IgA until several weeks or months after birth.

The secretory IgA antibodies disable pathogens without harming helpful GI tract flora or causing inflammation. This is important because although inflammation helps fight infection, sometimes the process overwhelms the GI tract. An infant may suffer more from the inflammatory process than the infection itself when inflammation destroys healthy tissue.

The large quantities of the immune system molecule *interleukin-10* found in breast milk also help inhibit inflammation. And a substance called *fibronectin* enhances the phagocytic activity of macrophages, inhibits inflammation, and helps repair tissues damaged by inflammation.

Several other breast-milk molecules help disable harmful microbes. Mucins, certain oligosaccharides (sugar chains), and glycoproteins (carbohydrate–protein compounds) bind to microbes and prevent them from gaining a foothold on the lining of the GI tract. Many of breast milk's immune cells, including T lymphocytes and macrophages, attack invading microbes directly.

Breast-milk compounds help in other ways as well. Some decrease the supply of nutrients such as iron and vitamin B_{12} needed by harmful bacteria to survive. A substance called *bifidus factor* promotes the growth of helpful gut flora, which help crowd out pathogens. *Retinoic acids*, a group of vitamin A precursors, reduce the ability of viruses to replicate. And some of the hormones and growth factors present in breast milk stimulate the baby's GI tract to mature more quickly, making it less vulnerable to dangerous invaders.

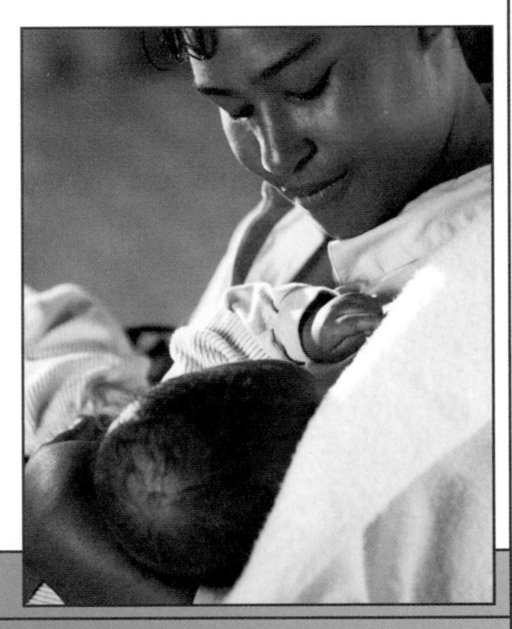

▶ THINK IT OVER . . .

▶ ***Why do you think that preventing infection through breast-feeding is preferable to giving babies antibiotics?***

Oxytocin stimulates contraction of smooth-muscle-like cells surrounding the glandular cells and ducts. The resulting compression moves the milk from the alveoli of the mammary glands into the mammary ducts, where it can be suckled.

During late pregnancy and the first few days after birth, the mammary glands secrete a cloudy fluid called ***colostrum*** (kō-LOS-trum). Although it is not as nutritious as milk—it contains less lactose and virtually no fat—colostrum serves adequately until the appearance of true milk on about the fourth day. Colostrum and maternal milk contain important antibodies that protect the infant during the first few months of life.

Lactation often blocks ovarian cycles for the first few months following delivery, if the frequency of sucking is about 8–10 times a day. This effect is inconsistent, however, and ovulation commonly precedes the first menstrual period after delivery of a baby. As a result, the mother can never be certain she is not fertile. Breast-feeding is therefore not a very reliable birth control measure.

A primary benefit of ***breast-feeding*** is nutritional: Human milk is a sterile solution that contains amounts of fatty acids, lactose, amino acids, minerals, vitamins, and water that are ideal for the baby's digestion, brain development, and growth. Breast-feeding also benefits infants in other ways, as indicated in the Focus on Wellness on page 603.

Years before oxytocin was discovered, it was common practice in midwifery to let a first-born twin nurse at the mother's breast to speed the birth of the second child. Now we know why this practice is helpful—it stimulates the release of oxytocin. Even after a single birth, nursing promotes expulsion of the placenta (afterbirth) and helps the uterus return to its normal size. Synthetic oxytocin (Pitocin®) is often given to induce labor or to increase uterine tone and control hemorrhage just after parturition.

■ **CHECKPOINT**

18. Which hormones contribute to lactation? What is the function of each?

INHERITANCE

OBJECTIVE • Define inheritance, and explain the inheritance of dominant, recessive, and sex-linked traits.

As previously indicated, the genetic material of a father and a mother unite when a sperm cell fuses with a secondary oocyte to form a zygote. Children resemble their parents because they inherit traits passed down from both parents. We now examine some of the principles involved in that process, called inheritance.

Inheritance is the passage of hereditary traits from one generation to the next. It is the process by which you acquired your characteristics from your parents and may transmit some of your traits to your children. The branch of biology that deals with inheritance is called ***genetics*** (je-NET-iks). The area of health care that offers advice on genetic problems (or potential problems) is called ***genetic counseling***.

Genotype and Phenotype

The nuclei of all human cells except gametes contain 23 pairs of chromosomes—the diploid number ($2n$). One chromosome in each pair came from the mother, and the other came from the father. Each ***homolog***—one of the two chromosomes that make up a pair—contains genes that control the same traits. If a chromosome contains a gene for body hair, for example, its homolog will also contain a gene for body hair in the same position on the chromosome. Such alternative forms of a gene that code for the same trait and are at the same location on homologous chromosomes are called ***alleles*** (ah-LĒLZ). For example, one allele of a body hair gene might code for coarse hair, and another allele for fine hair. A ***mutation*** (mū-TĀ-shun; *muta-* = change) is a permanent heritable change in an allele that produces a different variant of the same trait.

The relationship of genes to heredity is illustrated by examining the alleles involved in a disorder called ***phenylketonuria*** or ***PKU***. People with PKU lack phenylalanine hydroxylase, an enzyme that converts the amino acid phenylalanine into tyrosine, another amino acid. If infants with PKU eat foods containing phenylalanine, high levels of phenylalanine build up in the blood. The result is severe brain damage and mental retardation. The allele that codes for phenylalanine hydroxylase is symbolized as *P;* the mutated allele that fails to produce a functional enzyme is symbolized as *p*. The chart in Figure 24.10, which shows the possible combinations of gametes from two parents who each have one *P* and one *p* allele, is called a ***Punnett square***. In constructing a Punnett square, the possible paternal alleles in sperm are written at the left side and the possible maternal alleles in ova (or secondary oocytes) are written at the top. The four spaces on the chart show how the alleles can combine in zygotes formed by the union of these sperm and ova to produce the three different genetic makeups, or ***genotypes*** (JĒ-nō-tīps): *PP*, *Pp*, or *pp*. Notice from the Punnett square that 25% of the offspring will have the *PP* genotype, 50% will have the *Pp* genotype, and 25% will have the *pp* genotype. People who inherit *PP* or *Pp* genotypes do not have PKU; those with a *pp* genotype suffer from the disorder. Although people with a *Pp* genotype have one PKU allele (*p*), the allele that codes for the normal trait (*P*) is more dominant. An allele that dominates or masks the presence of another allele and is fully expressed (*P* in this example) is said to be a ***dominant allele***, and the trait expressed is called a dominant trait. The allele whose presence is completely masked (*p* in this example) is said to be a ***recessive allele***, and the trait it controls is called a recessive trait.

By tradition, the symbols for genes are written in italics, with dominant alleles written in capital letters and recessive alleles in lowercase letters. A person with the same alleles on homologous chromosomes (for example, *PP* or *pp*) is said to be ***homozygous*** for the trait. *PP* is homozygous dominant, and *pp* is homozygous recessive. An individual with different alleles on homologous chromosomes (for example, *Pp*) is said to be ***heterozygous*** for the trait.

Phenotype (FĒ-nō-tīp; *pheno-* = showing) refers to how the genetic makeup is expressed in the body; it is the physical or outward expression of a gene. A person with *Pp* (a heterozygote) has a different *genotype* from a person with *PP* (a homozygote), but both have the same *phenotype*—normal production of phenylalanine hydroxylase. Heterozygous individuals who carry a recessive gene but do not express it (*Pp*) can pass the gene on to their offspring. Such individuals are called *carriers* of the recessive gene.

Alleles that code for normal traits do not always dominate over those that code for abnormal ones, but dominant alleles for severe disorders usually are lethal and cause death of the embryo or fetus. One exception is Huntington disease (HD), which is caused by a dominant allele that does not express itself until adulthood. Both homozygous dominant and

Figure 24.10 Inheritance of phenylketonuria (PKU).

Genotype refers to genetic makeup; phenotype refers to the physical or outward expression of a gene.

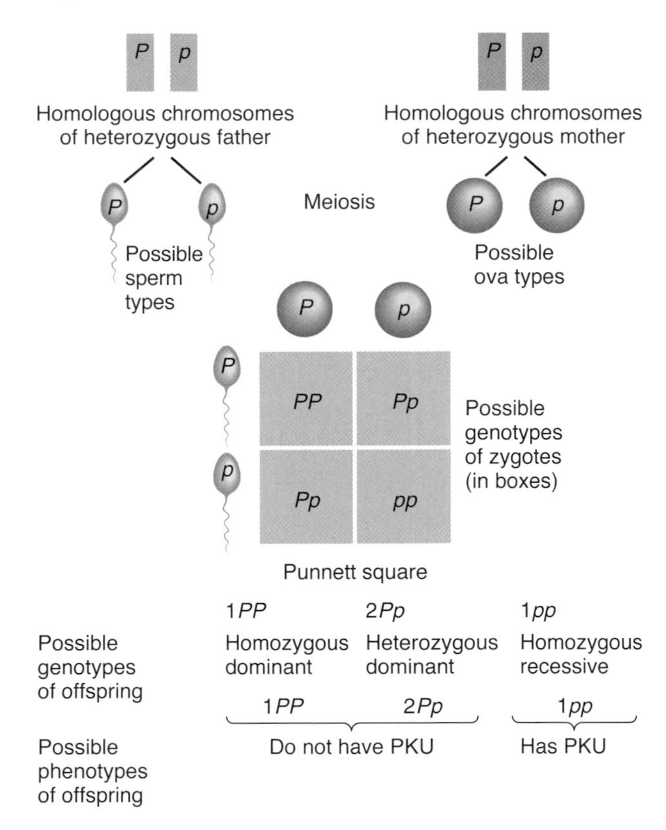

If parents have the genotypes shown here, what is the percent chance that their first child will have PKU? What is the chance of PKU occurring in their second child?

heterozygous people exhibit the disease; homozygous recessive people are normal. HD causes progressive degeneration of the nervous system and eventual death, but because symptoms typically do not appear until after age 30 or 40, many afflicted individuals have already passed the allele for the condition on to their children.

In ***incomplete dominance***, neither member of an allelic pair is dominant over the other, and the heterozygote has a phenotype intermediate between the homozygous dominant and the homozygous recessive phenotypes. An example of incomplete dominance in humans is the inheritance of ***sickle-cell disease (SCD)***. People with the homozygous dominant genotype Hb^AHb^A form normal hemoglobin; those with the homozygous recessive genotype Hb^SHb^S have sickle-cell disease and severe anemia. Although they are usually healthy, those with the heterozygous genotype Hb^AHb^S have minor problems with anemia because half their hemoglobin is normal and half is not. Heterozygotes are carriers, and they are said to have *sickle-cell trait*.

Although a single individual inherits only two alleles for each gene, some genes may have more than two alternate forms, and this is the basis for ***multiple-allele inheritance***. One example of multiple-allele inheritance is the inheritance of the ABO blood group. The four blood types (phenotypes) of the ABO group—A, B, AB, and O—result from the inheritance of six combinations of three different alleles of a single gene called the *I* gene: (1) allele I^A produces the A antigen, (2) allele I^B produces the B antigen, and (3) allele *i* produces neither A nor B antigen. Each person inherits two *I*-gene alleles, one from each parent, that give rise to the various phenotypes. The six possible genotypes produce four blood types, as follows:

Genotype	Blood Type (Phenotype)
I^AI^A or I^Ai	A
I^BI^B or I^Bi	B
I^AI^B	AB
ii	O

Notice that both I^A and I^B are inherited as dominant traits, and *i* is inherited as a recessive trait. An individual with type AB blood has characteristics of both type A and type B red blood cells.

Autosomes and Sex Chromosomes

When viewed under a microscope, the 46 human chromosomes in a normal somatic cell can be identified by their size, shape, and staining pattern to be members of 23 different pairs of chromosomes. In 22 of the pairs, the homologous chromosomes look alike and have the same appearance in both males and females; these 22 pairs are called ***autosomes***. The two members of the 23rd pair are termed the ***sex chromosomes***; they look different in males and females (Figure 24.11a). In females, the pair consists of two chromosomes called X chromosomes. One X chromosome is also present in males, but its mate is a much smaller chromosome called a Y chromosome.

When a spermatocyte undergoes meiosis to reduce its chromosome number, it gives rise to two sperm that contain an X chromosome and two sperm that contain a Y chromosome. Oocytes have no Y chromosomes and produce only X-containing gametes. If the secondary oocyte is fertilized by an X-bearing sperm, the offspring normally is female (XX). Fertilization by a Y-bearing sperm produces a male (XY). Thus, an individual's sex is determined by the father's chromosomes (Figure 24.11b). The prime male-determining gene is one called ***SRY (sex-determining region of the Y chromosome)***. *SRY* acts as a switch to turn on the male pattern of development. Only if the *SRY* gene is present and functional in a fertilized ovum will the fetus develop testes and differentiate into a male; in the absence of *SRY*, the fetus will develop ovaries and differentiate into a female.

The sex chromosomes also are responsible for the transmission of several nonsexual traits. Many of the genes for these traits are present on X chromosomes but are absent

Figure 24.11 Inheritance of gender (sex). In (a) the sex chromosomes, pair 23, are indicated in the colored box.

Gender is determined at the time of fertilization by the sex chromosome of the sperm cell.

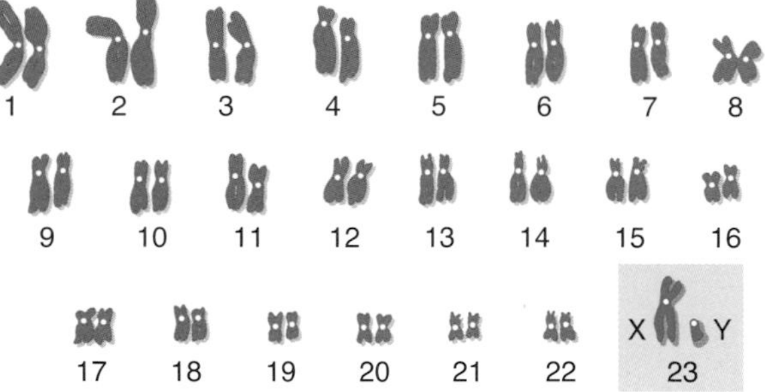

(a) Normal human male chromosomes

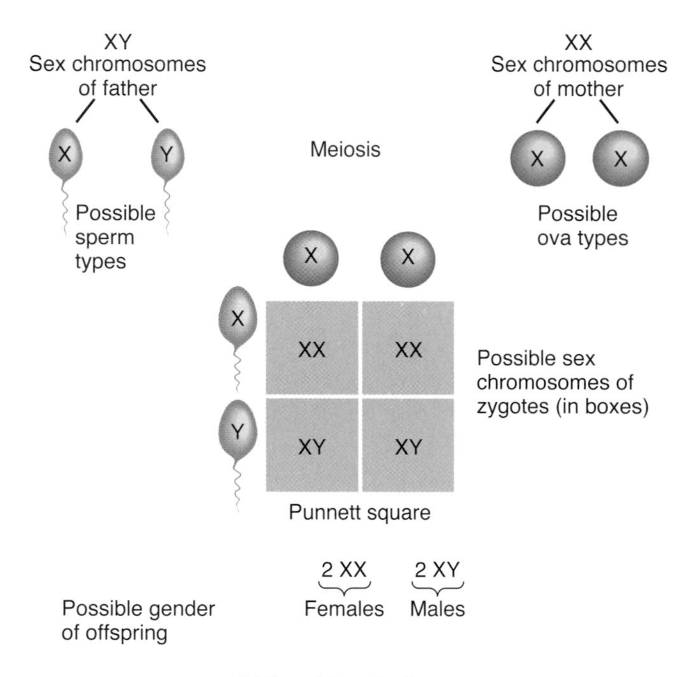

(b) Sex determination

What are chromosomes other than sex chromosomes called?

from Y chromosomes. This feature produces a pattern of heredity, termed ***sex-linked inheritance***, that is different from the patterns already described.

One example of sex-linked inheritance is ***red–green color blindness***, the most common type of color blindness. This condition is characterized by a deficiency in either red- or green-sensitive cones, so red and green are seen as the same color (either red or green, depending on which cone is present). The gene for red–green color blindness is a recessive one designated *c*. Normal color vision, designated *C*, dominates. The *C/c* genes are located only on the X chromosome, and thus the ability to see colors depends entirely on the X chromosomes. The possible combinations are as follows:

Genotype	Phenotype
X^CX^C	Normal female
X^CX^c	Normal female (but a carrier of the recessive gene)
X^cX^c	Red–green color-blind female
X^CY	Normal male
X^cY	Red–green color-blind male

Only females who have two X^c genes are red–green color blind. This rare situation can result only from the mating of a color-blind male and a color-blind or carrier female. (In X^CX^c females the trait is masked by the normal, dominant gene.) Because males do not have a second X chromosome that could mask the trait, all males with an X^c gene will be red–green color blind. Figure 24.12 illustrates the inheritance of red–green color blindness in the offspring of a normal male and a carrier female. Traits inherited in the manner just described are called ***sex-linked traits***. The most common type of ***hemophilia***—a condition in which the blood fails to clot or clots very slowly after an injury—is also a sex-linked trait.

■ CHECKPOINT

19. What do the terms genotype, phenotype, dominant, recessive, homozygous, and heterozygous mean?

20. Define incomplete dominance and give an example.

21. What is multiple-allele inheritance? Give an example.

22. How is the development of gender determined?

23. Define and provide an example of sex-linked inheritance.

Figure 24.12 An example of the inheritance of red–green color blindness.

Red–green color blindness and hemophilia are examples of sex-linked traits.

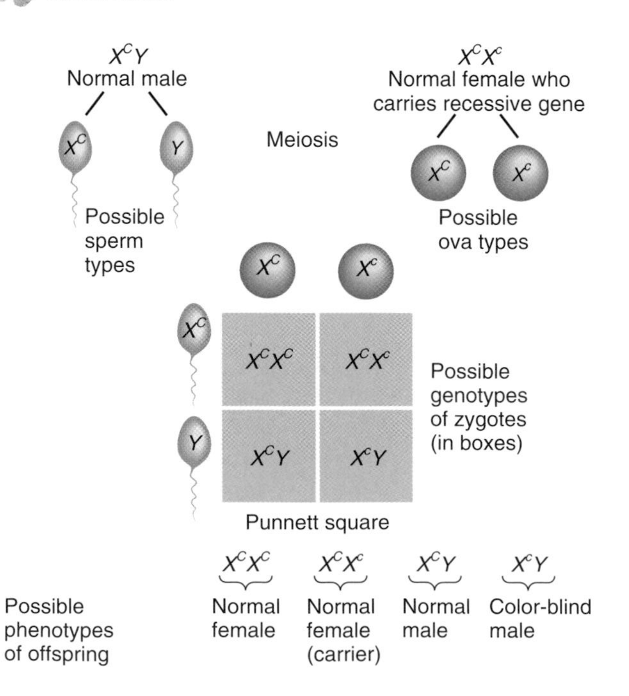

What is the genotype of a red–green color-blind female?

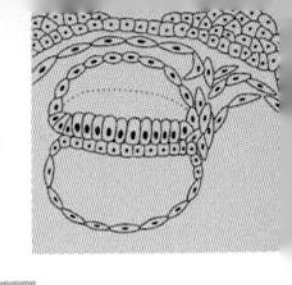

COMMON DISORDERS

Infertility

Female infertility, or the inability to conceive, occurs in about 10% of all women of reproductive age in the United States. Female infertility may be caused by ovarian disease, obstruction of the uterine tubes, or conditions in which the uterus is not adequately prepared to receive a fertilized ovum. ***Male infertility (sterility)*** is an inability to fertilize a secondary oocyte; it does not imply erectile dysfunction (impotence). Male fertility requires production of adequate quantities of viable, normal sperm by the testes, unobstructed transport of sperm through the ducts, and satisfactory deposition of sperm in the vagina. The seminiferous tubules of the testes are sensitive to many factors—x rays, infections, toxins, malnutrition, and higher-than-normal scrotal temperatures—that may cause degenerative changes and produce male sterility.

To begin and maintain a normal reproductive cycle, a female must have a minimum amount of body fat. Even a moderate deficiency of fat—10% to 15% below normal weight for height—may delay the onset of menstruation, inhibit ovulation during the reproductive cycle, or cause amenorrhea (cessation of menstruation). Both dieting and intensive exercise may reduce body fat below the minimum amount and lead to infertility that is reversible, if weight gain or reduction of intensive exercise or both occurs. Studies of very obese women indicate that they, like very lean ones, experience problems with amenorrhea and infertility. Males also experience reproductive problems in response to undernutrition and weight loss. For example, they produce less prostatic fluid and reduced numbers of sperm having decreased motility.

Many fertility-expanding techniques now exist for assisting infertile couples to have a baby.

- To achieve ***in vitro fertilization (IVF)***—fertilization in a laboratory dish—the mother-to-be is given follicle-stimulating hormone (FSH) soon after menstruation, so that several secondary oocytes, rather than the typical single oocyte, will be produced (superovulation). When several follicles have reached the appropriate size, a small incision is made near the umbilicus, and the secondary oocytes are aspirated from the stimulated follicles. They are then transferred to a solution containing sperm, where fertilization takes place.
- ***Intracytoplasmic sperm injection (ICSI)***, the injection of a sperm or spermatid into an oocyte's cytoplasm, has been used when infertility is due to impairments in sperm motility or to the failure of spermatids to develop into spermatozoa. When the zygote achieved by IVF or ICSI reaches the 8-cell or 16-cell stage, it is introduced into the uterus for implantation and subsequent growth.
- In ***embryo transfer***, a man's semen is used to artificially inseminate a fertile secondary oocyte donor. After fertilization in the donor's uterine tube, the morula or blastocyst is transferred from the donor to the infertile woman, who then carries it (and subsequently the fetus) to term.
- In ***gamete intrafallopian transfer (GIFT)*** the goal is to mimic the normal process of conception by uniting sperm and secondary oocyte in the prospective mother's uterine tubes. It is an attempt to bypass conditions in the female reproductive tract that might prevent fertilization, such as high acidity or inappropriate mucus. In this procedure, a woman is given FSH and LH to stimulate the production of several secondary oocytes, which are aspirated from the mature follicles, mixed outside the body with a solution containing sperm, and then immediately inserted into the uterine tubes.

Down Syndrome

Down syndrome is a disorder that most often results during meiosis when an extra chromosome 21 passes to one of the gametes. Most of the time the extra chromosome comes from the mother, a not-too-surprising finding given that all her oocytes began meiosis when she herself was a fetus. They may have been exposed to chromosome-damaging chemicals and radiation for years. (Sperm, by contrast, usually are less than 10 weeks old at the time they fertilize a secondary oocyte.) The chance of conceiving a baby with this syndrome, which is less than 1 in 3000 for women under age 30, increases to 1 in 300 in the 35 to 39 age group, and to 1 in 9 at age 48.

Down syndrome is characterized by mental retardation; retarded physical development (short stature and stubby fingers); distinctive facial structures (large tongue, flat profile, broad skull, slanting eyes, and round head); and malformations of the heart, ears, hands, and feet. Sexual maturity is rarely attained.

MEDICAL TERMINOLOGY AND CONDITIONS

Amniocentesis (am′-nē-ō-sen-TĒ-sis; *amnio-* = amnion; *-centesis* = puncture to remove fluid) A prenatal diagnostic procedure that involves withdrawal of amniotic fluid and analysis of fetal cells and dissolved substances to test for the presence of genetic disorders such as Down syndrome, hemophilia, Tay-Sachs disease, sickle-cell disease, and certain muscular dystrophies. Usually performed at 14–18 weeks of gestation and carries about a 0.5% chance of spontaneous abortion after the procedure.

Breech presentation A malpresentation in which the fetal buttocks or lower limbs present into the maternal pelvis; the most common cause is prematurity.

Chorionic villi sampling (kō-rē-ON-ik VIL-i) or ***CVS*** A prenatal diagnostic procedure that involves removal of chorionic villi tissue to examine it for the same genetic disorder as amniocentesis. May be performed as early as eight weeks of gestation; results are available in a few days. Causes about a 1–2% chance of spontaneous abortion after the procedure.

Conceptus (kon-SEP-tus) Includes all structures that develop from a zygote and includes an embryo plus the embryonic part of the placenta and associated membranes (chorion, amnion, yolk sac, and allantois).

Emesis gravidarum (EM-e-sis gra-VID-ar-um; *emeo* = to vomit; *gravida* = a pregnant woman) Episodes of nausea and possibly vomiting that are most likely to occur in the morning during the early stages of pregnancy; also called ***morning sickness***. Its cause is unknown, but the high levels of human chorionic gonadotropin (hCG) secreted by the placenta, and of progesterone secreted by the ovaries, have been implicated. In some women the severity of these symptoms requires hospitalization for intravenous feeding.

Fertilization age Two weeks less than the gestational age, since a secondary oocyte is not fertilized until about two weeks after the last normal menstrual period (LNMP).

Fetal alcohol syndrome (FAS) A specific pattern of fetal malformation due to intrauterine exposure to alcohol. FAS is one of the most common causes of mental retardation and the most common preventable cause of birth defects in the United States. The symptoms of FAS may include slow growth before and after birth, characteristic facial features (short palpebral fissures, a thin upper lip, and sunken nasal bridge), defective heart and other organs, malformed limbs, genital abnormalities, and central nervous system damage. Behavioral problems, such as hyperactivity, extreme nervousness, reduced ability to concentrate, and an inability to appreciate cause-and-effect relationships, are common.

Fetal surgery A surgical procedure performed on a fetus; in some cases the uterus is opened and the fetus is operated on directly. Fetal surgery has been used to repair diaphragmatic hernias and remove lesions in the lungs.

Fetal ultrasonography (ul′-tra-son-OG-ra-fē) A prenatal diagnostic procedure that uses ultrasound to confirm pregnancy, identify multiple pregnancies, determine fetal age, evaluate fetal viability and growth, determine fetal position, identify fetal-maternal abnormalities, and assist in procedures such as amniocentesis.

Gestational age (jes-TĀ-shun-al; *gestatus* = to bear) The age of an embryo or fetus calculated from the presumed first day of the last normal menstrual period (LNMP).

Lethal gene (LĒ-thal jēn; *lethum* = death) A gene that, when expressed, results in death either in the embryonic state or shortly after birth.

Metafemale syndrome A sex chromosome disorder characterized by at least three X chromosomes (XXX) that occurs about once in every 700 births. These females have underdeveloped genital organs and limited fertility. Generally, they are mentally retarded.

Preeclampsia (prē′-e-KLAMP-sē-a) A syndrome of pregnancy characterized by sudden hypertension, large amounts of protein in urine, and generalized edema; possibly related to an autoimmune or allergic reaction to the presence of a fetus. When the condition is also associated with convulsions and coma, it is referred to as ***eclampsia***.

Puerperal fever (pū-ER-per-al; *puer* = child) A maternal infectious disease of childbirth, also called puerperal sepsis and childbed fever. The disease, which results from an infection originating in the birth canal, affects the endometrium. It may spread to other pelvic structures and lead to septicemia.

Teratogen (TER-a-tō-jen; *terato-* = monster; *-gen* = creating) Any agent or influence that causes developmental defects in the embryo. Examples include alcohol, pesticides, industrial chemicals, antibiotics, thalidomide, LSD, and cocaine.

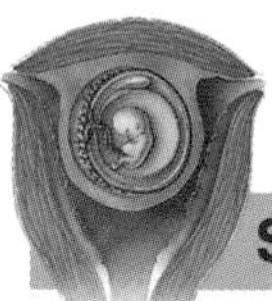

STUDY OUTLINE

Embryonic Period (p. 587)

1. Pregnancy is a sequence of events that begins with fertilization and proceeds to implantation, embryonic development, and fetal development. It normally ends in birth.
2. During fertilization a sperm cell penetrates a secondary oocyte and their nuclei unite. Penetration of the zona pellucida is facilitated by enzymes in the sperm's acrosome. The resulting cell is a zygote.
3. Normally, only one sperm cell fertilizes a secondary oocyte.
4. Early rapid cell division of a zygote is called cleavage, and the cells produced by cleavage are called blastomeres. The solid sphere of cells produced by cleavage is a morula.
5. The morula develops into a blastocyst, a hollow ball of cells differentiated into a trophoblast and an inner cell mass.
6. The attachment of a blastocyst to the endometrium is termed implantation.
7. The trophoblast develops into the syncytiotrophoblast and cytotrophoblast, both of which become part of the chorion.
8. The inner cell mass differentiates into hypoblast and epiblast, the bilaminar (two-layered) embryonic disc.
9. The amnion is a thin protective membrane that develops from the cytotrophoblast.
10. The hypoblast forms the yolk sac, which transfers nutrients to the embryo, forms blood cells, produces primordial germ cells, and forms part of the gut.
11. Blood and secretions enter lacunar networks to supply nutrition to and remove wastes from the embryo.

12. The extraembryonic mesoderm and trophoblast form the chorion, the principal embryonic part of the placenta.
13. The third week of development is characterized by gastrulation, the conversion of the bilaminar disc into a trilaminar (three-layered) embryo consisting of ectoderm, mesoderm, and endoderm.
14. The three primary germ layers form all tissues and organs of the developing organism.
15. The process by which the neural plate, neural folds, and neural tube form is called neurulation. The brain and spinal cord develop from the neural tube.
16. Chorionic villi, projections of the chorion, connect to the embryonic heart so that maternal and fetal blood vessels are brought into close proximity. Thus, nutrients and wastes are exchanged between maternal and fetal blood.
17. Placentation refers to formation of the placenta, the site of exchange of nutrients and wastes between the mother and fetus. The placenta also functions as a protective barrier, stores nutrients, and produces several hormones to maintain pregnancy.
18. The actual connection between the placenta and embryo (and later the fetus) is the umbilical cord.
19. The formation of body organs and systems occurs during the fourth week of development.
20. By the end of the fourth week, upper and lower limb buds develop and by the end of the eighth week the embryo has clearly human features.

Fetal Period (p. 596)

1. The fetal period is primarily concerned with the growth and differentiation of tissues and organs that developed during the embryonic period.
2. The rate of body growth is remarkable, especially during the ninth and sixteenth weeks.
3. The principal changes associated with embryonic and fetal growth are summarized in Table 24.1 on pages 596–597.

Maternal Changes During Pregnancy (p. 600)

1. Pregnancy is maintained by human chorionic gonadotropin (hCG), estrogens, and progesterone.
2. Relaxin increases flexibility of the pubic symphysis and helps dilate the uterine cervix near the end of pregnancy.
3. Human placental lactogen (hPL) contributes to breast development, protein anabolism, and catabolism of glucose and fatty acids.
4. Corticotropin-releasing hormone, produced by the placenta, is thought to establish the timing of birth, and stimulates the secretion of cortisol by the fetal adrenal gland.
5. During pregnancy, several anatomical and physiological changes occur in the mother.

Exercise and Pregnancy (p. 601)

1. During pregnancy, some joints become less stable, and certain maneuvers are more difficult to execute.
2. Moderate physical activity does not endanger the fetus in a normal pregnancy.

Labor and Delivery (p. 601)

1. Labor is the process by which the fetus is expelled from the uterus through the vagina to the outside. True labor involves dilation of the cervix, expulsion of the fetus, and delivery of the placenta.
2. Oxytocin stimulates uterine contractions.

Lactation (p. 602)

1. Lactation refers to the production and ejection of milk by the mammary glands.
2. Milk production is influenced by prolactin (PRL), estrogens, and progesterone.
3. Milk ejection is stimulated by oxytocin.
4. A few of the many benefits of breast-feeding include ideal nutrition for the infant, protection from disease, and decreased likelihood of developing allergies.

Inheritance (p. 604)

1. Inheritance is the passage of hereditary traits from one generation to the next.
2. The genetic makeup of an organism is called its genotype; the traits expressed are called its phenotype.
3. Dominant genes control a particular trait; expression of recessive genes is masked by dominant genes.
4. In incomplete dominance, neither member of an allelic pair dominates; phenotypically, the heterozygote is intermediate between the homozygous dominant and the homozygous recessive. An example is sickle-cell disease.
5. In multiple-allele inheritance, genes have more than two alternative forms. An example is the inheritance of ABO blood groups.
6. Each somatic cell has 46 chromosomes—22 pairs of autosomes and 1 pair of sex chromosomes.
7. In females, the sex chromosomes are two X chromosomes; in males, they are one X chromosome and a much smaller Y chromosome, which normally includes the prime male-determining gene, called *SRY*.
8. If the *SRY* gene is present and functional in a fertilized ovum, the fetus will develop testes and differentiate into a male. In the absence of *SRY*, the fetus will develop ovaries and differentiate into a female.
9. Red–green color blindness and hemophilia result from recessive genes located on the X chromosome. They are sex-linked traits that occur primarily in males because of the absence of any counterbalancing dominant genes on the Y chromosome.

SELF-QUIZ

1. The change in a sperm that allows it to fertilize an egg is known as
   a. vasocongestion **b.** capacitation **c.** cleavage
   d. gestation **e.** meiosis

2. Match the following:

____	**a.** early division of the zygote that increases the cell number but not size	**A.** blastocyst
____	**b.** solid mass of cells 3 to 4 days following fertilization	**B.** trophoblast
____	**c.** a hollow ball of cells found in the uterine cavity about 5 days after fertilization	**C.** morula
____	**d.** portion of the blastocyst that develops into the embryo	**D.** inner cell mass
____	**e.** portion of the blastocyst that forms the fetal portion of the placenta	**E.** zygote
____	**f.** results from fertilization of the sperm and egg	**F.** cleavage

3. The principal hormone in promoting milk production is
   a. relaxin **b.** prolactin (PRL) **c.** oxytocin
   d. cortisol **e.** human chorionic gonadotropin (hCG)

4. The placental hormone that appears to affect the timing of birth is
   a. cortisol **b.** human placental lactogen (hPL)
   c. relaxin **d.** corticotropin-releasing hormone (CRH)
   e. human chorionic gonadotropin (hCG)

5. Which is not true concerning the chorion?
   a. It produces hCG.
   b. It becomes the principal embryonic part of the placenta.
   c. It protects the embryo and fetus from immune reponses of the mother.
   d. It develops into the brain and spinal cord.
   e. It surrounds the embryo and later the fetus.

6. Which of the following is the embryonic membrane that is closest to the developing fetus?
   a. amnion **b.** umbilicus **c.** chorion **d.** placenta
   e. zona pellucida

7. The hormone that causes a home pregnancy test to show a positive result for pregnancy is
   a. follicle-stimulating hormone (FSH)
   b. progesterone
   c. human placental lactogen (hPL)
   d. luteinizing hormone (LH)
   e. human chorionic gonadotropin (hCG)

8. The transformation of a two-layered embryonic disc into a three-layered embryonic disc is called
   a. gastrulation **b.** neurulation **c.** fertilization
   d. implantation **e.** cleavage

9. Match the following:

____	**a.** becomes part of the placenta	**A.** yolk sac
____	**b.** becomes muscle and bone	**B.** chorion
____	**c.** develops into the epidermis and nervous system	**C.** endoderm
____	**d.** is the source of blood cells from the third through the sixth week of development	**D.** ectoderm
____	**e.** becomes the epithelial lining of the respiratory and gastrointestinal tracts	**E.** mesoderm

10. The period of time from conception of the zygote to delivery of the fetus is called
    a. fertilization **b.** placentation **c.** gestation
    d. implantation **e.** gastrulation

11. Homologous chromosomes
    a. contain genes that control the same trait
    b. contain genes that control different traits
    c. are inherited only from the mother
    d. are inherited only from the father
    e. contain all identical alleles

12. Sex-linked traits are carried on the
    a. autosomes
    b. X and Y chromosomes
    c. X chromosomes only
    d. Y chromosomes only
    e. Y chromosomes in males and X chromosomes in females

13. A person who is homozygous for a dominant trait on an autosome would have the genotype
    a. *Aa* **b.** *AA* **c.** *aa* **d.** X^AX^A **e.** X^aX^a

14. The genotype of a normal female is
    a. XY and 44 autosomes **b.** 46 autosomes **c.** 46 X chromosomes **d.** XX and 44 autosomes **e.** XX and 46 autosomes

15. Which of the following is NOT a change that occurs in a female during pregnancy?
    a. increased cardiac output
    b. decreased pulmonary expiratory reserve volume
    c. decreased gastrointestinal tract motility
    d. increased frequency and urgency of urination
    e. decreased production of estrogens

16. The afterbirth is expelled from the uterus during the ________ stage of labor.
    a. parturition **b.** placental **c.** dilation
    d. expulsion **e.** puerperium

CRITICAL THINKING APPLICATIONS

1. Your neighbor has put up a sign to announce the birth of their twins, a girl and a boy. Another neighbor said "Oh, how sweet! I wonder if they're identical twins." Without even seeing the twins, what can you tell her?

2. The science class was studying genetics at school and Kendra came home in tears. She told her older sister, "We were doing our family tree and when I filled in our traits, I discovered that Mom and Dad can't be my real parents' cause the traits don't match!" It turns out that Mom and Dad can roll their tongues but Kendra can't. Could Mom and Dad still be Kendra's parents?

3. Felicity is concerned about the health of her unborn baby. Due to Felicity's medical history, her physician wants to test for the presence of a genetic disorder. Felicity is afraid that this will hurt the baby. The physician reassures Felicity that the procedure will not touch the baby even though it will get a sample of fetal tissue. How is this possible?

4. Baby Peterson was brought into the health clinic by a concerned grandparent. The alcoholic mother had abandoned the baby, and the grandparent was worried that the baby was growing slowly and had not developed normally. One of the neighbors, a nurse, had told the grandparent that the baby's face was characteristic of a syndrome that she'd seen before. What is Baby Peterson's problem?

ANSWERS TO FIGURE QUESTIONS

24.1 Capacitation refers to the functional changes in sperm after they have been deposited in the female reproductive tract that enable them to fertilize a secondary oocyte.

24.2 A morula is a solid ball of cells; a blastocyst consists of a rim of cells (trophoblast) surrounding a cavity (blastocyst cavity) and an inner cell mass.

24.3 The blastocyst secretes digestive enzymes that eat away the endometrial lining at the site of implantation.

24.4 At implantation, the blastocyst is oriented so that the inner cell mass is closest to the endometrium.

24.5 The bilaminar embryonic disc is attached to the trophoblast by the connecting stalk.

24.6 Gastrulation converts a bilaminar embryonic disc into a trilaminar embryonic disc.

24.7 Chorionic villi help to bring the fetal and maternal blood vessels close to each other.

24.8 The placenta functions in the exchange of materials between the fetus and mother, serves as a protective barrier against many microbes, and stores nutrients.

24.9 During this time, fetal weight doubles.

24.10 The odds that a child will have PKU are the same for each child—25%.

24.11 The chromosomes that are not sex chromosomes are called autosomes.

24.12 A red–green color-blind female has an X^cX^c genotype.

SELF-QUIZ ANSWERS

Chapter 1

1. d **2.** b **3a.** G **b.** B **c.** E **d.** C **e.** F **f.** A **g.** D **4a.** nervous system **b.** brain, spinal cord, nerves, sense organs **c.** lymphatic system **d.** Returns proteins and fluid to blood; sites of lymphocyte maturation and proliferation to protect against disease; carries lipids from digestive system to blood **e.** Respiratory System **f.** lungs, pharynx, larynx, trachea, bronchial tubes **g.** testes, ovaries, vagina, uterine tubes, uterus, penis **h.** reproduces the organism and releases hormones **5.** d **6.** a **7.** c **8.** e **9a.** C **b.** D **c.** A **d.** B **10.** a **11.** a **12.** d **13.** c **14.** e **15.** d **16.** c **17.** b **18.** b **19.** e **20a.** D **b.** A **c.** H **d.** F **e.** G **f.** E **g.** C **h.** B

Chapter 2

1. d **2.** c **3.** a **4.** e **5.** d **6.** b **7.** b **8.** c **9.** a **10.** d **11.** e **12.** d **13.** c **14a.** R, D **b.** D **c.** D **d.** R **e.** R **f.** D **g.** D **h.** R **i.** R, D **j.** R, D **15.** a **16.** b **17.** e **18.** a **19.** carbon, hydrogen, oxygen, nitrogen **20a.** D **b.** C **c.** A **d.** E **e.** B

Chapter 3

1. b **2.** e **3.** a **4.** a **5.** c **6.** d **7.** b **8.** d **9.** a **10a.** B **b.** F **c.** G **d.** H **e.** C **f.** E **g.** D **h.** A **11.** e **12.** c **13.** e **14.** d **15.** b **16.** b **17a.** C **b.** E **c.** F **d.** B **e.** D **f.** A **18.** e **19.** e **20.** c

Chapter 4

1. c **2.** d **3.** b **4.** d **5.** a **6a.** C **b.** G **c.** E **d.** D **e.** H **f.** F **g.** B **h.** A **7.** a **8.** e **9.** e **10.** b **11.** a **12.** c **13.** c **14.** d **15.** e **16.** d **17.** a **18.** c **19.** a **20.** b

Chapter 5

1. d **2.** d **3.** a **4.** e **5.** e **6.** a **7.** c **8.** c **9.** a **10.** a **11.** b **12.** a **13.** c **14.** b **15.** d **16.** e **17.** b **18a.** D **b.** F **c.** E **d.** G **e.** H **f.** A **g.** C **h.** B **i.** I **19.** a **20.** b

Chapter 6

1a. C **b.** E **c.** D **d.** A **e.** B **2.** e **3.** a **4a.** E **b.** D **c.** A **d.** C **e.** B **5.** b **6.** e **7.** a **8.** a **9.** c **10.** d **11.** b **12a.** C **b.** D **c.** B **d.** A **13.** e **14.** b **15.** e **16a.** AX **b.** AP **c.** AP **d.** AX **e.** AP **f.** AP **g.** AP **h.** AX **i.** AP **j.** AP **k.** AX **l.** AP **m.** AX **n.** AP **o.** AX **p.** AP **q.** AX **r.** AX **s.** AP **t.** AX **u.** AX **v.** AP **w.** AX **x.** AX **y.** AX **z.** AP **aa.** AP **bb.** AP **cc.** AX **dd.** AX

Chapter 7

1. d **2.** a **3.** c **4.** b **5.** d **6.** e **7.** b **8a.** D **b.** A **c.** E **d.** C **e.** B **9.** a **10.** c **11.** e **12.** a **13.** b **14.** e **15a.** A **b.** G **c.** B **d.** C **e.** D **f.** E **g.** F **h.** J **i.** I **j.** H

Chapter 8

1. e **2a.** D **b.** E **c.** B **d.** A **e.** C **3.** d **4a.** C **b.** D **c.** A **d.** E **e.** B **5.** c **6.** a **7.** a **8a.** SM, CA **b.** SK **c.** SK, CA **d.** CA **e.** SK **f.** SK **g.** SM **h.** SM **i.** SK **j.** CA **9.** b **10.** b **11.** e **12.** b **13.** d **14.** a **15.** c **16.** a **17.** d **18a.** E **b.** D **c.** G **d.** H **e.** B **f.** M **g.** O **h.** F **i.** J **j.** C **k.** K **l.** A **m.** L **n.** I **o.** N a **19a.** C **b.** F **c.** D **d.** B **e.** G **f.** H **g.** A **h.** E **20a.** C **b.** E **c.** B **d.** D **e.** F **f.** A

Chapter 9

1. c **2.** e **3.** a **4.** d **5.** a **6.** e **7.** c **8.** a **9.** c **10.** e **11.** b **12.** d **13.** b **14.** d **15a.** D **b.** E **c.** F **d.** C **e.** A **f.** B **16a.** I **b.** A **c.** D **d.** O **e.** E **f.** P **g.** H **h.** J **i.** G **j.** C **k.** B **l.** L **m.** N **n.** K **o.** F **p.** M

Chapter 10

1. e **2.** a **3.** c **4.** c **5.** d **6.** e **7.** c **8.** a **9.** c **10.** a **11.** b **12.** e **13.** e **14.** a **15.** b **16a.** C **b.** D **c.** A **d.** B **17.** b **18.** d **19.** c, e **20a.** G **b.** H **c.** K **d.** J **e.** F **f.** I **g.** A **h.** D **i.** C **j.** B **k.** E

Chapter 11

1. c **2.** a **3.** c **4.** e **5.** d **6.** a **7.** d **8.** b **9.** d **10.** e **11.** b **12.** d **13.** b **14a.** C **b.** F **c.** E **d.** B **e.** A **f.** D **15a.** S **b.** P **c.** P **d.** S **e.** S **f.** P **g.** S **h.** S **i.** S

Chapter 12

1. b **2.** b **3.** a **4.** d **5.** e **6.** c **7a.** J **b.** E **c.** F **d.** I **e.** C **f.** G **g.** H **h.** A **i.** B **j.** D **8.** e **9.** d

10. c **11.** a **12.** e **13.** e **14.** c **15a.** D **b.** F **c.** G **d.** B **e.** C **f.** A **g.** E **16.** a **17.** e **18.** c **19.** d **20.** a

Chapter 13

1. d **2.** e **3.** c **4.** a **5.** d **6.** a **7.** e **8.** a **9.** b **10.** c **11.** c **12.** a **13.** b **14.** d **15.** b **16.** c **17a.** C **b.** F **c.** G **d.** A **e.** D **f.** E **g.** B **18.** b **19a.** B **b.** C **c.** E **d.** A **e.** F **f.** D **20a.** R **b.** F **c.** F **d.** R **e.** F **f.** E **g.** R **h.** E **i.** F

Chapter 14

1. c **2a.** D **b.** G **c.** C **d.** A **e.** E **f.** F **g.** B **3.** c **4.** d **5.** e **6.** e **7.** b **8.** b **9.** d **10.** a **11.** c **12.** a **13a.** E **b.** D **c.** C **d.** A **e.** B **14.** e **15.** b **16.** a **17.** e **18.** a **19.** b **20.** c

Chapter 15

1a. D **b.** H **c.** F **d.** G **e.** C **f.** B **g.** A **h.** E **2.** e **3.** e **4.** a **5.** c **6.** c **7.** a **8.** c **9.** b **10.** d **11.** d **12.** a **13.** c **14.** a **15.** d **16.** d **17.** b **18.** c **19a.** A **b.** A **c.** A **d.** B **e.** A **f.** B **g.** B **h.** A **i.** B **j.** A **k.** B **l.** B **m.** A **20a.** B **b.** G **c.** A **d.** F **e.** D **f.** C **g.** E

Chapter 16

1. b **2.** a **3.** a **4.** c **5.** d **6.** e **7a.** E **b.** D **c.** A **d.** B **e.** C **8.** d **9.** c **10.** b **11.** c **12a.** H **b.** D **c.** E **d.** F **e.** A **f.** B **g.** C **h.** I **i.** G **13a.** A **b.** B **c.** B **d.** A **e.** A **f.** A **g.** A **h.** A **i.** B **14.** e **15.** b **16.** a **17.** d

Chapter 17

1. c **2.** d **3.** e **4.** c **5.** a **6.** b **7.** b **8.** a **9.** d **10.** e **11.** b **12.** c **13.** d **14a.** E **b.** B **c.** D **d.** A **e.** C **15.** c **16.** b **17.** c **18.** b **19.** e **20.** a

Chapter 18

1. d **2.** c **3.** b **4.** a **5.** c **6.** a **7.** a **8.** d **9a.** B **b.** D **c.** A **d.** C **e.** E **10.** a **11a.** F **b.** D **c.** H **d.** A **e.** G **f.** C **g.** E **h.** B **12.** b **13.** c **14.** e **15.** a **16.** b **17.** d **18.** c **19.** e **20.** c **21a.** D **b.** A **c.** E **d.** C **e.** B

Chapter 19

1. e **2.** b **3.** b **4.** c **5.** d **6.** a **7.** c **8a.** D **b.** G **c.** A **d.** H **e.** B **f.** C **g.** E **h.** F **9.** e **10.** e **11.** a **12.** e **13.** c **14.** b **15.** d **16.** a **17.** e **18.** c **19.** e **20.** a

Chapter 20

1. b **2.** c **3.** a **4a.** C **b.** A **c.** D **d.** B **5.** c **6.** a **7.** b **8.** e **9.** d **10.** a **11.** d **12.** b **13.** e **14.** b **15.** c **16.** c **17.** d **18.** a **19.** d **20a.** C **b.** D **c.** B **d.** E **e.** A

Chapter 21

1. c **2.** e **3.** a **4.** d **5.** c **6.** e **7.** a **8.** d **9.** b **10.** a **11.** b **12.** e **13.** e **14.** a **15.** c **16.** d **17.** a **18.** c **19.** e **20.** b

Chapter 22

1. d **2.** e **3.** a **4.** c **5.** a **6.** e **7.** d **8.** a **9.** c **10.** a **11.** c **12.** e **13.** b **14.** c **15.** d **16.** b **17.** c **18a.** A **b.** B **c.** D **d.** C

Chapter 23

1. b **2a.** C **b.** D **c.** E **d.** B **e.** A **3.** e **4.** d **5.** c **6.** c **7.** b **8.** e **9.** c **10.** e **11.** c **12.** b **13.** a **14.** e **15.** d **16.** d **17a.** C **b.** A **c.** D **d.** B **18.** d **19a.** B **b.** F **c.** A **d.** G **e.** D **f.** C **g.** E **20.** e

Chapter 24

1. b **2a.** F **b.** C **c.** A **d.** D **e.** B **f.** E **3.** b **4.** d **5.** d **6.** a **7.** e **8.** a **9a.** B **b.** E **c.** D **d.** A **e.** C **10.** c **11.** a **12.** c **13.** b **14.** d **15.** e **16.** b

ANSWERS TO CRITICAL THINKING APPLICATIONS

Chapter 1

1. In the anatomical position, the arm would hang along the lateral side of the trunk with the palm facing forward.

2. The astronauts should observe the lumps to see if some of the following processes that may indicate life are present: metabolism, responsiveness, movement, growth, differentiation, and reproduction.

3. Anatomically speaking, Guy's answer does not make sense. Caudal means inferior or away from the head, dorsal means the back, and sural is the calf of the leg. The groin area is located on the anterior of the trunk, near the top of the leg.

4. The midsagittal plane divides the body into equal right and left halves. The transverse (cross-sectional or horizontal) plane divides the top from the bottom of the body.

Chapter 2

1. The protein in the milk is denatured by the acid in the lemon juice. Denaturation changes the characteristic shape of the milk protein so that it is no longer soluble.

2. The human body is composed of 26 elements including C, H, O, and N which make up about 96% of the body's mass. Over half the body's weight is H_2O (water).

3. Albert, Jr. doesn't understand pH. Each increase of one whole number on the pH scale represents a 10-fold decrease in H^+ concentration. The pH 3.5 mixture is 10 times less acidic (or 10 times more alkaline) than the pH 2.5 mixture.

4. Simply adding water to table sugar does not cause it to break apart into monosaccharides. The water acts as a solvent, dissolving the sucrose, and forming a sucrose-water solution. To complete the breakdown of table sugar to glucose and fructose would require the presence of the enzyme sucrase.

Chapter 3

1. Lysosomes contain digestive enzymes that digest bone tissue and release the stored calcium.

2. Sea water is hypertonic to the body. It contains a higher concentration of solutes (NaCl) than the body's cells. Drinking sea water would cause crenation of cells.

3. Mucin (protein) is synthesized on the ribosomes attached to the rough ER. The mucin will travel from the rough ER to a transport vesicle to the Golgi complex. The protein will be modified to a glycoprotein while being transported through the Golgi's cisterns (again using transfer vesicles). The mucin will be packaged in a secretory vesicle and move to the plasma membrane where it will be released by exocytosis.

4. The sodium-potassium pump transports sodium out of the cell and potassium into the cell by active transport. ATP powers the pump.

Chapter 4

1. The pins are stuck through the keratinized, stratified squamous layer of the skin. There is no bleeding because epithelium is avascular.

2. Collagen is the most abundant protein in the body. Bone, cartilage and dense connective tissue contain abundant collagen. Collagen is found in tendons, ligaments, and aponeuroses.

3. The uterine tube is lined with ciliated simple columnar epithelium. The cilia (hair) help move the ovum.

4. Along with the adipose tissue (fat), the chicken leg contains skeletal muscle (meat), bone tissue (leg bone), dense regular (tendons) and dense irregular (periosteum) connective tissue, epithelial tissue, areolar tissue (skin), and blood tissue.

Chapter 5

1. The hair shaft is fused, dead keratinized cells. At the base of the hair, the hair bulb contains the living matrix where cell divison occurs. Hair root plexuses (nervous tissue) surround each hair follicle.

2. The epidermal layer is repaired by cell division of keratinocytes strating at the deepest layer of the epidermis, the stratum basale. The keratinocytes move toward the surface through the stratum spinosum, stratum granulosum, and stratum corneum. The cells become fully kertatinized, flat and dead as they travel to the surface. The process takes about 2–4 weeks.

3. A callus is an abnormal thickening of the stratum corneum due to constant friction/rubbing. Warts are caused by papilloma virus infection resulting in uncontrolled epithelial cell growth. Athlete's foot is a fungal infection.

4. Jeremy's blackheads are caused by oil (sebum) accumulation in the sebaceous glands. Sebum secretion often increases after puberty. The color is due to melanin and oxidized oil.

Chapter 6

1. J.R. fractured his tibia and fibula, the styloid process of the radius, and the scaphoid.

2. The soft spots are fibrous connective tissue membranes that fill the spaces between the skull bones. The fontanels ossify by intramembranous ossification.

3. Due to her age and gender, Olga probably has osteoporosis. Bone loss is due to increased calcium loss and decreased production of growth hormone and estrogen. Shrinkage of the vertebrae results in hunched back and loss of height.

4. Kate fractured the top (under the kneecap) of the larger of the two leg bones. Rapid swelling is caused by blood from torn vessels and the damaged vascular bone tissue. Synovial fluid may also contribute to swelling.The body requires calcium, phosphorous, magnesium, vitamins A, C, and D, hGH and other hormones, and protein for bone matrix in order for Kate's injury to heal.

Chapter 7

1. Flex the knees, flex the hip (on the side with knee up), hyperextend the neck, flex fingers, flex and extend at elbow and shoulder, and depress mandible.

2. The hip joint is a diarthrotic, synovial joint of the ball-and-socket type formed by the head of the femur fitting into the acetabulum of the hip bone. The movements are extension/flexion, abduction/adduction, rotation, and circumduction.

3. The ACL is the anterior cruciate ligament. It connects the tibia to the femur, running posteriorly and laterally. The ACL works along with other internal and external ligaments to stabilize the knee joint.

4. Drew has a dislocation or luxation of his shoulder joint. The clavicle has seperated from the scapula resulting in the bump in the shoulder area, the lack of arm function, and the pain.

Chapter 8

1. The skeletal muscles will not contract without receiving a signal from the neurotransmitter acetylcholine. Since release of ACh is blocked, skeletal muscles will not function.

2. Ali used the orbicularis oris (puckering), frontalis (eyebrows), zygomaticus (cheeks), and buccinator (cheeks).

3. Kate's leg muscles atrophied from loss of myofibrils due to lack of use of the muscles. Kate will need to exercise in order to increase muscle size by building up myofibrils, mitochondria, and sacroplasmic reticulum.

4. Running 10 miles is an aerobic activity which, when repeated over time, will increase blood supply to the muscles and increased endurance. Weight lifting is an anaerobic activity which will result in muscle hypertrophy over time and increased strength.

Chapter 9

1. Smelling coffee and hearing alarm are somatic sensory, stretching and yawning are somatic motor, salivating is autonomic (parasympathetic) motor.

2. This is not a compliment. Gray matter, located on the outer surface of the brain, is composed of neuroglia and neuronal dendrites, cell bodies, and unmyelinated axons. It is the site of all communications among neurons and is crucial for carrying out the functions of the nervous system.

3. Neuropeptides such as endorphins are found in the brain. They are related to feelings of pleasure and are natural painkillers.

4. A myelin sheath increases the speed of nerve impulse conduction (propagation). Because myelination is not complete in infants, their responses are slower and less coordinated than in older children.

Chapter 10

1. The brachial plexus runs through the axillary region and supplies the arm and hands. Kate's weight has put pressure on the brachial plexus and interrupted nerve impulse transmission.

2. The "numbing" anesthetic caused anesthesia (loss of feeling) in Dennis' lower lip due to blocking the left and right mandibular branches of cranial nerve V. The numb right upper lip is due to a nerve block of the maxillary branch of cranial nerve V.

3. The primary motor area in the left precentral gyrus and Broca's area in the left frontal lobe were damaged by the stroke. Both of these areas are in the cerebral cortex.

4. A receptor in Kyle's foot detected the pain from the pin, the impulse traveled along a sensory nerve to the spinal cord which passed the signal along to a motor neuron that stimulated the muscles in Kyle's leg muscles to contract resulting in a withdrawal reflex.

Chapter 11

1. The parasympathetic division of the ANS directs rest-and-digestive activities. The organs of the digestive system will have increased activity to digest food, absorb nutrients, and defecate wastes. In the relaxed condition, the body will also exhibit slower heart rate, and bronchoconstriction.

2. The supermom effect was due to activation of the sympathetic nervous system resulting in a fight-or-flight response. The heart rate, force of contraction, and blood pressure increased, blood flow to muscles increased, and glucose and ATP production increased. The release of the hormones epinephrine and norepinephrine also increased.

3. The goose bumps are a sympathetic nervous system response. The cell bodies of sympathetic preganglionic neurons are in the thoracolumbar (T1–L2) segments of the spinal cord, their axons exit in the anterior roots of spinal nerves, and extend out to a sympathetic ganglion. From there, postganglionic neurons extend to hair follicle smooth muscles (arrector pili), which produce goose bumps when they contract.

4. Dual innervation refers to the innervation of most organs by both the sympathetic and parasympathetic divisions of the autonomic nervous system, not to the presence of more than one head.

Chapter 12

1. Olfactory receptors for smell are rapidly adapting. Olfactory receptors adapt by about 50% in the first second. The aroma of coffee is still present but the responsiveness of the receptors to the stimulus has decreased.

2. The saccule and utricle in the vestibular apparatus of the internal ear respond to head position and maintain static equilibrium.

The otolithic membrane of the maculae moves in response to head movement, stimulating hair cells and triggering impulses that travel along cranial nerve VIII.

3. The cornea is responsible for about 75% of the total refraction of light rays entering the eye. Changing the shape of the cornea by shaving off surface layers will alter refraction of light and change the focus of the image on the retina, hopefully improving visual acuity.

4. The optometrist used eye drops to temporarily paralyze the muscles of the iris during the examination. The radial muscles are contracted in the paralyzed state resulting in dilation of Kate's pupils. Light sensitivity occurs because the circular muscles are also paralyzed and cannot contract to constrict the pupil in response to the bright light.

Chapter 13

1. Patrick has Type 1 diabetes mellitus due to the destruction of the pancreatic beta cells. He must have injections of insulin to metabolize glucose. His aunt has Type 2 diabetes mellitus. She still produces insulin but her body cells have decreased sensitivity to the hormone.

2. Eddie had giantism, an abnormal increase in long bone length caused by oversecretion of human growth hormone (hGH) during childhood. In adulthood, the excess hGH caused enlargement of jaws, hands, feet and other tissues resulting in acromegaly.

3. Melatonin is released by the pineal gland during darkness and sleep. Melatonin helps set the biological clock, controlled by the hypothalamus, which sets sleep patterns. SAD may be caused by excess melatonin. Bright light inhibits melatonin secretion and is a treatment for SAD.

4. Dehydration will stimulate the release of ADH from the posterior pituitary. ADH will increase water retention by the kidneys, decrease sweating and constrict arterioles which will raise BP. Epinephrine and norepinephrine will be released by the adrenal medullae in response to stress.

Chapter 14

1. Bilirubin is a pigment formed from the breakdown of heme from the hemoglobin of old RBC's phagocytized by the liver. If the bile ducts do not transport the bilirubin in bile away from the liver, the bilirubin will build up in blood and other tissues causing a yellow color in the skin and eyes called jaundice.

2. Since the mother is Rh^+, she will not produce antibodies against the Rh antigen. HDN should not be a problem with this baby.

3. A bluish-purple color in the nail beds is seen in cyanosis which is caused by prolonged lack of oxygen (hypoxia).

4. Pluripotent stem cells can transform into myeloid stem cells and lymphoid stem cells and, from these, can produce all of the formed elements of blood: RBCs, platelets, and WBCs (monocytes, eosinophils, neutrophils, basophils, and lymphocytes).

Chapter 15

1. A pacemaker sends electrical impulses into the right side of the heart that stimulate contraction of the heart muscle. A pacemaker is used in conditions in which the heart rhythm is irregular to take over the function of the SA node.

2. The sudden appearance of the car activated the sympathetic division of the nervous system. The sympathetic signals from the cardiovascular (CV) center in the medulla oblongata travel down the spinal cord to the cardiac accelerator nerves which release norepinephrine that increases heart rate and forcefulness of contraction.

3. SV = CO $\div$ HR (beats/min). Assuming an average CO of 5250 mL/min at rest, SV = 5250 mL/min $\div$ 40 beats/min = 131.25 mL. Rearranging the equation, CO = SV – HR (beats/min). With exercise, Jean-Claude's CO = 131.25 mL $\times$ 60 beats/min = 7875 mL/min.

4. Aunt Frieda's heart murmur may indicate a faulty heart valve. The peripheral edema in her feet may be related to the valve disorder or may be a sign of congestive heart failure of the right ventricle.

Chapter 16

1. One effect of epinephrine is to cause vasoconstriction of arterioles. If these vessels were constricted temporarily, this would reduce the blood flow locally at the site of dental work and reduce bleeding.

2. Varicose veins are caused by weak venous valves that allow the backflow of blood. With aging, venous walls may lose their elasticity and become stretched and distended with blood. Arteries rarely become distended because they have thicker inner and middle layers than veins and do not contain valves.

3. Julie's sympathetic nervous system is responding to the stress of her bad day. Impulses travel from the cardiovascular (CV) center to the heart which increases its rate and force of contraction. Impulses from the CV center also travel to the blood vessel walls resulting in vasoconstriction, which increases BP. Increased levels epinephrine and NE due to stress amplify these effects.

4. Peter cut an artery. Blood flows from arteries in rapid spurts due to the high pressure generated by ventricular contraction.

Chapter 17

1. The right axillary lymph nodes were probably removed because lymph flowing in lymphatics away from the tumor (breast lump) was filtered by the axillary lymph nodes. Cancerous cells from the tumor may be carried in the lymph to the axillary node and spread the cancer by metastasis.

2. The five tonsils are positioned near the oral cavity, nasal cavity and throat—the ideal location to intercept microbes and other foreign invaders that enter the body through the mouth or nose. Young children will be exposed to many infectious agents and need to develop immunity to these invaders.

3. Initial tetanus immunization provided artificially acquired active immunity. The booster dose is needed to maintain immunity against the tetanus bacteria and toxin.

4. Without a blood supply, antibodies and T cells do not have easy access to the cornea. Therefore no immune response occurs to reject the transplanted foreign cornea.

Chapter 18

1. Holding the breath will cause blood levels of CO_2 and H^+ to increase and O_2 to decrease. These changes will strongly stimulate the inspiratory area, which will send impulses to resume breathing, whether Levi is still conscious or not.

2. Exercise normally induces the sympathetic nervous system to send signals to dilate the bronchioles, which increases air flow and oxygen supply. Asthma causes constriction of the bronchioles, making inhalation more difficult and reducing the airflow.

3. Brianna has a viral infection—the common cold (coryza). Laryngitis, an inflammation of the larynx, is a common complication from a cold. Inflammation of the true vocal cords can result in loss of voice.

4. The tour group members have altitude sickness. At high elevation, the partial pressure of oxygen is not sufficient for unadapted individuals to maintain an adequate level of oxygen in their blood.

Chapter 19

1. You should brush your teeth to remove food residue and bacteria. When bacteria metabolize the sugars left on teeth after a meal, they produce acid that can demineralize enamel. Dental caries (cavities) can result.

2. Edna and Gertrude are both wrong. Lactose intolerance causes symptoms of cramping, diarrhea, and flatulence from excess gas in the large intestine.

3. The pyloric sphincter is located at this junction. The spider traveled from the mouth through the oro- and laryngopharynx, then the esophagus and finally entered the stomach. Gastroscopy of the stomach with an endoscope would reveal mucus-coated epithelium, gastric pits, and rugae.

4. The repetitive vomiting had emptied Jerry's stomach of hot dog and gastric juice and now he was expelling fluid containing bile. Bile pigments such as bilirubin give bile its color. Bile is produced by the liver.

Chapter 20

1. The recommended percentages of Calories are: total fats = less than 30%, saturated fats = 10%, total carbohydrates = 50–60%, simple sugars = less than 15%, proteins = 12–15%. Total Calories for active women = 2200. The quarter pounder meal leaves 1115 Calories for the day. The grilled chicken leaves 1608 Calories.

2. Body temperature will increase due to radiation from the sun and hot surrounding sand, and possibly conduction from lying on the hot sand. Heat will be lost to the water by conduction and convection.

3. BMR means **b**asal **m**etabolic **r**ate. BMR is the measurement of the rate at which heat is produced, i.e., the metabolic rate, under standard conditions designed to be as close as possible to the basal state. Factors that would affect metabolic rate and should be avoided while measuring the BMR include exercise, stress, increased temperature, eating (all increase), sleep (lower).

4. Antioxidants protect against free radical damage to cell membranes, DNA, and blood vessel walls. Vitamin C is water-soluble so excess quantities will be excreted in the urine. Vitamins A (from the beta-carotene) and E are lipid-soluble and might accumulate to toxic levels in tissues such as the liver.

Chapter 21

1. Alcohol inhibits the secretion of antidiuretic hormone (ADH). ADH is secreted when the hypothalamus detects a decrease in the amount of water in the blood. ADH makes the collecting ducts and distal portion of the distal convoluted tubules more permeable to water so that it can be reabsorbed.

2. Incontinence (lack of involuntary control over micturition) is normal in children of Sarah's age. Neurons to the external urethral sphincter are not completely developed until after about 2 years of age. The desire to control micturition voluntarily must also be present and initiated by the cerebral cortex.

3. No. Glomerular filtration is mainly driven by blood pressure and opposed by glomerular capsular pressure, not pressure from urine in the bladder. Under normal physiological conditions, urine remains in the bladder and does not back up into the kidneys.

4. 5 cm (width of kidney) × 1 million nephrons per kidney = 5×10^6 cm = 5×10^4 m = 50 kilometers.

Chapter 22

1. The high sodium concentration will stimulate secretion of ANP to reduce Na^+ concentration in the blood and to reduce blood volume.

2. Timon has water intoxication. He had excessive intake of water during his swim class causing his body fluids to become too dilute or hypotonic. Water moved into his cells by osmosis, swelling the cells and resulting in convulsions.

3. A lean male's body mass is about 60% water while a lean female's is about 55% water due to a greater amount of subcutaneous fat in the female. Mike will have more fluid volume (60% × 150 lb.) for the alcohol to dissolve in than Jennie (55% × 150 lb.), so Jennie's blood alochol level will be higher than Mike's.

4. An increase in blood volume stretches the atria causing the release of atrial natriuretic peptide (ANP). ANP promotes excretion of Na^+ (natriuresis) in the urine. Increased loss of water goes along with loss of Na^+ and reduces blood volume.

Chapter 23

1. A complete hysterectomy is the surgical removal of the body and cervix of the uterus. The ovaries produce estrogens and progesterone, not the uterus, so menopause will not result since these organs will still be left in place.

2. Vasectomy cuts the ductus deferens so that sperm can not be transported out of the body. The function of the testes is not effected. The Leydig cells secrete the hormone testosterone which maintains male sex characteristics and sex drive. Vesectomy will not affect production of the hormone or its transport to the rest of the body via the blood.

3. Sperm production is optimal at a temperature slightly below normal body temperature. The higher temperature that results from Julio wearing very close-fitting briefs and soaking in the hot tub inhibits sperm production and survival and therefore fertility.

4. Symptoms of BPH include frequent urination, nocturia, hesitancy, decreased force, and incomplete emptying. The prostatic secretions supply a milky appearance to the semen, nutrients for the sperm, lysozyme, and clotting enzymes and PSA (prostate specific antigen) which liquefy semen. Without the prostatic contribution, the semen volume would decrease about 25%.

Chapter 24

1. The neighbor's twins are fraternal or dizygotic. If the cells resulting from cleavage of a single fertilized ovum split into 2 separate groups and continue developing into 2 babies, then identical or monozygotic twins result. Since identical twins came from the same original fertilized ovum, they must contain the same genetic information and be the same gender.

2. Yes. Tongue rolling is a dominant trait. Kendra's parents are both heterozygous for the tongue rolling gene (Tt). The dominant (T) gene determines their tongue rolling ability but they each passed on the recessive (t) gene to Kendra. Kendra is homozygous recessive (tt) and can't roll her tongue.

3. The physician may obtain a sample of fetal tissue from the amniotic fluid, which contains sloughed off fetal cells, or from the chorionic villi, which is fetal placental tissue. Amniocentesis and chorionic villi sampling do not take samples from the actual baby (fetus).

4. The baby has fetal alcohol syndrome (FAS) caused by intrauterine exposure to alcohol. The characteristic facial features are thin upper lip, sunken nasal bridge and short palpebral fissures. Other medical problems include damage to the central nervous system, heart, limbs, and genitals as well as behavioral and learning problems.

Pronunciation Key

1. The most strongly accented syllable appears in capital letters, for example, bilateral (bī-LAT-er-al) and diagnosis (dī-ag-NŌ-sis).
2. If there is a secondary accent, it is noted by a prime ('), for example, constitution (kon'-sti-TOO-shun) and physiology (fiz'-ē-OL-ō-jē). Any additional secondary accents are also noted by a prime, for example, decarboxylation (dē'-kar-bok'-si-LĀ-shun).
3. Vowels marked by a line above the letter are pronounced with the long sound, as in the following common words:
   ā as in *māke* — *ō* as in *pōle*
   ē as in *bē* — *ū* as in *cute*
   ī as in *īvy*
4. Vowels not marked by a line above the letter are pronounced with the short sound, as in the following words:
   a as in *above* or *at* — *o* as in *not*
   e as in *bet* — *u* as in *bud*
   i as in *sip*
5. Other vowel sounds are indicated as follows:
   oy as in *oiloo* as in *root*
6. Consonant sounds are pronounced as in the following words:
   b as in *bat* — *m* as in *mother*
   ch as in *chair* — *n* as in *no*
   d as in *dog* — *p* as in *pick*
   f as in *father* — *r* as in *rib*
   g as in *get* — *s* as in *so*
   h as in *hat* — *t* as in *tea*
   j as in *jump* — *v* as in *very*
   k as in *can* — *w* as in *welcome*
   ks as in *tax* — *z* as in *zero*
   kw as in *quit* — *zh* as in *lesion*
   l as in *let*

A

Abdomen (ab-DŌ-men *or* AB-dō-men) The area between the diaphragm and pelvis.

Abdominal cavity (ab-DŌM-i-nal) Superior portion of the abdominopelvic cavity that contains the stomach, spleen, liver, gallbladder, most of the small intestine, and part of the large intestine.

Abdominal thrust maneuver A first-aid procedure for choking. Employs a quick, upward thrust against the diaphragm that forces air out of the lungs with sufficient force to eject any lodged material. Also called the **Heimlich maneuver** (HĪM-lik).

Abdominopelvic cavity (ab-dom'-i-nō-PEL-vic) Inferior component of the ventral body cavity that is subdivided into a superior abdominal cavity and an inferior pelvic cavity.

Abduction (ab-DUK-shun) Movement away from the midline of the body.

Abortion (a-BOR-shun) The premature loss (spontaneous) or removal (induced) of the embryo or nonviable fetus; miscarriage due to a failure in the normal process of developing or maturing.

Abscess (AB-ses) A localized collection of pus and liquefied tissue in a cavity.

Absorption (ab-SORP-shun) Intake of fluids or other substances by cells of the skin or mucous membranes; the passage of digested foods from the gastrointestinal tract into blood or lymph.

Accommodation (a-kom-ō-DĀ-shun) An increase in the curvature of the lens of the eye to adjust for near vision.

Acetabulum (as'-e-TAB-ū-lum) The rounded cavity on the external surface of the hip bone that receives the head of the femur.

Acetylcholine (ACh) (as'-e-til-KŌ-lēn) A neurotransmitter liberated by many peripheral nervous system neurons and some central nervous system neurons. It is excitatory at neuromuscular junctions but inhibitory at some other synapses (for example, it slows heart rate).

Achalasia (ak'-a-LĀ-zē-a) A condition, in which the lower esophageal sphincter fails to relax normally as food approaches. A whole meal may become lodged in the esophagus and enter the stomach very slowly. Distension of the esophagus results in chest pain that is often confused with pain originating from the heart.

Acid (AS-id) A proton donor, or a substance that dissociates into hydrogen ions (H^+); characterized by an excess of hydrogen ions and a pH less than 7.

Acidosis (as-i-DŌ-sis) A condition in which blood pH is below 7.35. Also known as **acidemia.**

Acini (AS-i-nē) Groups of cells in the pancreas that secrete digestive enzymes.

Acoustic (a-KOOS-tik) Pertaining to sound or the sense of hearing.

Acquired immunodeficiency syndrome (AIDS) A disease caused by the human immunodeficiency virus (HIV). Characterized by a positive HIV-antibody test, low helper T cell count, and certain indicator diseases (for example Kaposi's sarcoma, *Pneumocystis carinii* pneumonia, tuberculosis, fungal diseases). Other symptoms include fever or night sweats, coughing, sore throat, fatigue, body aches, weight loss, and enlarged lymph nodes.

Acrosome (AK-rō-sōm) A lysosomelike organelle in the head of a sperm cell containing enzymes that facilitate the penetration of a sperm cell into a secondary oocyte.

Actin (AK-tin) A contractile protein that is part of the thin filaments in muscle fibers.

Action potential An electrical signal that propagates along the membrane of a neuron or muscle fiber (cell); a rapid change in membrane potential that involves a depolarization followed by a repolarization. Also called a **nerve action potential** or **nerve impulse** as it relates to a neuron, and a **muscle action potential** as it relates to a muscle fiber.

Active transport The movement of substances across cell membranes against a concentration gradient, requiring the expenditure of cellular energy (ATP).

Acute (a-KŪT) Having rapid onset, severe symptoms, and a short course; not chronic.

Adaptation (ad′-ap-TĀ-shun) The adjustment of the pupil of the eye to changes in light intensity. The property by which a sensory neuron relays a decreased frequency of action potentials from a receptor, even though the strength of the stimulus remains constant; the decrease in perception of a sensation over time while the stimulus is still present.

Adduction (ad-DUK-shun) Movement toward the midline of the body.

Adenoids (AD-e-noyds) The pharyngeal tonsils.

Adenosine triphosphate (ATP) (a-DEN-ō-sēn trī-FOS-fāt) The main energy currency in living cells; used to transfer the chemical energy needed for metabolic reactions. ATP consists of the purine base *adenine* and the five-carbon sugar *ribose*, to which are added, in linear array, three *phosphate* groups.

Adenylate cyclase (a-DEN-i-lāt SĪ-klās) An enzyme that is activated when certain neurotransmitters or hormones bind to their receptors; the enzyme that converts ATP into cyclic AMP, an important second messenger.

Adipocyte (AD-i-pō-sīt) Fat cell, derived from a fibroblast.

Adipose tissue (AD-i-pōz) Tissue composed of adipocytes specialized for triglyceride storage and present in the form of soft pads between various organs for support, protection, and insulation.

Adrenal cortex (a-DRĒ-nal KOR-teks) The outer portion of an adrenal gland, divided into three zones; the zona glomerulosa secretes mineralocorticoids, the zona fasciculata secretes glucocorticoids, and the zona reticularis secretes androgens.

Adrenal glands Two glands located superior to each kidney. Also called the **suprarenal glands** (soo′-pra-RĒ-nal).

Adrenal medulla (me-DUL-a) The inner part of an adrenal gland, consisting of cells that secrete epinephrine, norepinephrine, and a small amount of dopamine in response to stimulation by sympathetic preganglionic neurons.

Adrenergic neuron (ad′-ren-ER-jik) A neuron that releases epinephrine (adrenaline) or norepinephrine (noradrenaline) as its neurotransmitter.

Adrenocorticotropic hormone (ACTH) (ad-rē′-nō-kor-ti-kō-TRŌP-ik) A hormone produced by the anterior pituitary that influences the production and secretion of certain hormones of the adrenal cortex.

Adventitia (ad-ven-TISH-a) The outermost covering of a structure or organ.

Aerobic (air-Ō-bik) Requiring molecular oxygen.

Afferent arteriole (AF-er-ent ar-TĒ-rē-ōl) A blood vessel of a kidney that divides into the capillary network called a glomerulus; there is one afferent arteriole for each glomerulus.

Agglutination (a-gloo′-ti-NĀ-shun) Clumping of microorganisms or blood cells, typically due to an antigen–antibody reaction.

Aggregated lymphatic follicles Clusters of lymph nodules that are most numerous in the ileum. Also called **Peyer's patches** (PĪ-erz).

Albinism (AL-bin-izm) Abnormal, nonpathological, partial, or total absence of pigment in skin, hair, and eyes.

Albumin (al-BŪ-min) The most abundant (60%) and smallest of the plasma proteins; it is the main contributor to blood colloid osmotic pressure (BCOP).

Aldosterone (al-DOS-ter-ōn) A mineralocorticoid produced by the adrenal cortex that promotes sodium and water reabsorption by the kidneys and potassium excretion in urine.

Alkaline (AL-ka-līn) Containing more hydroxide ions (OH^-) than hydrogen ions (H^+); a pH higher than 7.

Alkalosis (al-ka-LŌ-sis) A condition in which blood pH is higher than 7.45. Also known as **alkalemia.**

Allantois (a-LAN-tō-is) A small, vascularized outpouching of the yolk sac that serves as an early site for blood formation and development of the urinary bladder.

Alleles (a-LĒLZ) Alternate forms of a single gene that control the same inherited trait (such as type A blood) and are located at the same position on homologous chromosomes.

Allergen (AL-er-jen) An antigen that evokes a hypersensitivity reaction.

Alpha cell (AL-fa) A type of cell in the pancreatic islets (islets of Langerhans) in the pancreas that secretes the hormone glucagon.

Alveolar duct (al-VĒ-ō-lar) Branch of a respiratory bronchiole around which alveoli and alveolar sacs are arranged.

Alveolar macrophage (MAK-rō-fāj) Highly phagocytic cell found in the alveolar walls of the lungs. Also called a **dust cell.**

Alveolar pressure Air pressure within the lungs.

Alveolar sac A cluster of alveoli that share a common opening.

Alveolus (al-VĒ-ō-lus) A small hollow or cavity; an air sac in the lungs; milk-secreting portion of a mammary gland. *Plural is* **alveoli** (al-VĒ-ol-ī).

Alzheimer disease (AD) (ALTZ-hī-mer) Disabling neurological disorder characterized by dysfunction and death of specific cerebral neurons, resulting in widespread intellectual impairment, personality changes, and fluctuations in alertness.

Amnesia (am-NE-zē-a) A lack or loss of memory.

Amenorrhea (ā-men-ō-RĒ-a) Absence of menstruation.

Amino acid (a-MĒ-nō) An organic acid, containing an acidic carboxyl group (—COOH) and a basic amino group (—NH_2); the monomer used to synthesize polypeptides and proteins.

Amnion (AM-nē-on) A thin, protective fetal membrane that develops from the epiblast; holds the fetus suspended in amniotic fluid. Also called the **"bag of waters."**

Amniotic fluid (am′-nē-OT-ik) Fluid in the amniotic cavity, the space between the developing embryo (or fetus) and amnion; the fluid is initially produced as a filtrate from maternal blood and later includes fetal urine. It functions as a shock absorber, helps regulate fetal body temperature, and helps prevent desiccation.

Amphiarthrosis (am′-fē-ar-THRŌ-sis) A slightly movable joint, in which the articulating bony surfaces are separated by fibrous connective tissue or fibrocartilage to which both are attached; types are syndesmosis and symphysis.

Ampulla (am-PUL-la) A saclike dilation of a canal or duct.

Anabolism (a-NAB-ō-lizm) Synthetic, energy-requiring reactions whereby small molecules are built up into larger ones.

Anaerobic (an-ar-Ō-bik) Not requiring oxygen.

Anal canal (Ā-nal) The last 2 or 3 cm (1 in.) of the rectum; opens to the exterior through the anus.

Anal column A longitudinal fold in the mucous membrane of the anal canal that contains a network of arteries and veins.

Anal triangle The subdivision of the female or male perineum that contains the anus.

Analgesia (an-al-JĒ-zē-a) Pain relief; absence of the sensation of pain.

Anaphase (AN-a-fāz) The third stage of mitosis in which the chromatids that have separated at the centromeres move to opposite poles of the cell.

Anaphylaxis (an′-a-fi-LAK-sis) A hypersensitivity (allergic) reaction in which IgE antibodies attach to mast cells and basophils, causing them to produce mediators of anaphylaxis (histamine, leukotrienes, kinins, and prostaglandins) that bring about increased blood permeability, increased smooth muscle contraction, and increased mucus production. Examples are hay fever, hives, and anaphylactic shock.

Anastomosis (a-nas-tō-MŌ-sis) An anatomical connection between tubular structure, especially arteries.

Anatomical position (an′-a-TOM-i-kal) A position of the body universally used in anatomical descriptions in which the body is erect, the head is level, the eyes face forward, the upper limbs are at the sides, the palms face forward, and the feet are flat on the floor.

Anatomic dead space Spaces of the nose, pharynx, larynx, trachea, bronchi, and bronchioles totaling about 150 mL; air in the anatomic dead space does not reach the alveoli to participate in gas exchange.

Anatomy (a-NAT-ō-mē) The structure or study of structure of the body and the relation of its parts to each other.

Androgens (AN-drō-jenz) Masculinizing sex hormones produced by the testes in males and the adrenal cortex in both genders; responsible for libido (sexual desire); the two main androgens are testosterone and dihydrotestosterone.

Anemia (a-NĒ-mē-a) Condition of the blood in which the number of functional red blood cells or their hemoglobin content is below normal.

Anesthesia (an′-es-THĒ-zē-a) A total or partial loss of feeling or sensation; may be general or local.

Aneurysm (AN-ū-rizm) A saclike enlargement of a blood vessel caused by a weakening of its wall.

Angina pectoris (an-JI-na *or* AN-ji-na PEK-tō-ris) A pain in the chest related to reduced coronary circulation due to coronary artery disease (CAD) or spasms of vascular smooth muscle in coronary arteries.

Angiotensin (an-jē-ō-TEN-sin) Either of two forms of a protein associated with regulation of blood pressure. Angiotensin I is produced by the action of renin on angiotensinogen and is converted by the action of ACE (angiotensin-converting enzyme) into angiotensin II, which stimulates aldosterone secretion by the adrenal cortex, stimulates the sensation of thirst, and causes vasoconstriction with resulting increase in systemic vascular resistance.

Anion (AN-ī-on) A negatively charged ion. Examples are the chloride ion (Cl^-) and bicarbonate ion (HCO_3^-).

Anoxia (an-OK-sē-a) Deficiency of oxygen.

Antagonist (an-TAG-ō-nist) A muscle that has an action opposite that of the prime mover (agonist) and yields to the movement of the prime mover.

Anterior (an-TĒR-ē-or) Nearer to or at the front of the body. Equivalent to **ventral** in bipeds.

Anterior pituitary Anterior lobe of the pituitary gland. Also called the **adenohypophysis** (ad′-e-nō-hī-POF-i-sis).

Anterior root The structure composed of axons of motor (efferent) neurons that emerges from the anterior aspect of the spinal cord and extends laterally to join a posterior root, forming a spinal nerve. Also called a **ventral root.**

Anterolateral pathway (an′-ter-ō-LAT-er-al) Sensory pathway that conveys information related to pain, temperature, crude touch, pressure, tickle, and itch.

Antibody (AN-ti-bod′-ē) A protein produced by plasma cells in response to a specific antigen; the antibody combines with that antigen to neutralize, inhibit, or destroy it. Also called an **immunoglobulin** (im-ū-nō-GLOB-ū-lin) or **Ig.**

Antibody-mediated immunity That component of immunity in which B lymphocytes (B cells) develop into plasma cells that produce antibodies that destroy antigens. Also called **humoral** (HŪ-mor-al) **immunity**.

Anticoagulant (an-tī-cō-AG-ū-lant) A substance that can delay, suppress, or prevent the clotting of blood.

Antidiuretic (an′-ti-dī-ū-RET-ik) Substance that inhibits urine formation.

Antidiuretic hormone (ADH) Hormone produced by neurosecretory cells in the hypothalamus that stimulates water reabsorption from kidney tubule cells into the blood and vasoconstriction of arterioles. Also called **vasopressin** (vāz-ō-PRES-in).

Antigen (AN-ti-jen) A substance that has the ability to provoke an immune response and the ability to react with the antibodies or cells that result from the immune response; contraction of *anti*body *gen*erator.

Antigen-presenting cell (APC) Special class of migratory cell that processes and presents antigens to T cells during an immune response; APCs include macrophages, B cells, and dendritic cells, which are present in the skin, mucous membranes, and lymph nodes.

Anuria (an-Ū-rē-a) Daily urine output of less than 50 mL.

Anus (Ā-nus) The distal end and outlet of the rectum.

Aorta (ā-OR-ta) The main systemic trunk of the arterial system of the body that emerges from the left ventricle.

Aortic body (ā-OR-tik) Cluster of chemoreceptors on or near the arch of the aorta that respond to changes in blood levels of oxygen, carbon dioxide, and hydrogen ions (H^+).

Aortic reflex A reflex that helps maintain normal systemic blood pressure; initiated by baroreceptors in the wall of the ascending aorta and arch of the aorta.

Apex (Ā-peks) The pointed end of a conical structure, such as the apex of the heart.

Aphasia (a-FĀ-zē-a) Loss of ability to express oneself properly through speech or loss of verbal comprehension.

Apnea (AP-nē-a) Temporary cessation of breathing.

Apneustic area (ap-NOO-stik) A part of the respiratory center in the pons that sends stimulatory nerve impulses to the inspiratory area that activate and prolong inhalation and inhibit exhalation.

Aponeurosis (ap′-ō-noo-RŌ-sis) A sheetlike tendon joining one muscle with another or with bone.

Apoptosis (ap′-ō-TŌ-sis *or* ap′-ōp-TŌ-sis) Programmed cell death; a normal type of cell death that removes unneeded cells during embryological development, regulates the number of cells in tissues, and eliminates many potentially dangerous cells such as cancer cells. During apoptosis, the DNA fragments, the nucleus condenses, mitochondria cease to function, and the cytoplasm shrinks, but the plasma membrane remains intact. Phagocytes engulf and digest the apoptotic cells, and an inflammatory response does not occur.

Aqueous humor (AK-wē-us HŪ-mer) The watery fluid, similar in composition to cerebrospinal fluid, that fills the anterior cavity of the eye.

Arachnoid mater (a-RAK-noyd MĀ-ter) The middle of the three meninges (coverings) of the brain and spinal cord. Also termed the **arachnoid.**

Arachnoid villus (VIL-us) Berrylike tuft of the arachnoid mater that protrudes into the superior sagittal sinus and through which cerebrospinal fluid is reabsorbed into the bloodstream.

Areola (a-RĒ-ō-la) Any tiny space in a tissue. The pigmented ring around the nipple of the breast.

Arm The part of the upper limb from the shoulder to the elbow.

Arrhythmia (a-RITH-mē-a) An irregular heart rhythm. Also called a **dysrhythmia.**

Arteriole (ar-TĒ-rē-ōl) A small, almost microscopic, artery that delivers blood to a capillary.

Arteriosclerosis (ar-tē-rē-ō-skle-RŌ-sis) Group of diseases characterized by thickening of the walls of arteries and loss of elasticity.

Artery (AR-ter-ē) A blood vessel that carries blood away from the heart.

Arthritis (ar-THRĪ-tis) Inflammation of a joint.

Arthrology (ar-THROL-ō-jē) The study or description of joints.

Arthroscopy (ar-THROS-co-pē) A procedure for examining the interior of a joint, usually the knee, by inserting an arthroscope into a small incision; used to determine extent of damage, remove torn cartilage, repair cruciate ligaments, and obtain samples for analysis.

Arthrosis (ar-THRŌ-sis) A joint or articulation.

Articular capsule (ar-TIK-ū-lar) Sleevelike structure around a synovial joint composed of a fibrous capsule and a synovial membrane.

Articular cartilage (KAR-ti-lij) Hyaline cartilage attached to articular bone surfaces.

Articular disc Fibrocartilage pad between articular surfaces of bones of some synovial joints. Also called a **meniscus** (men-IS-kus).

Articulation (ar-tik′-ū-LĀ-shun) A joint; a point of contact between bones, cartilage and bones, or teeth and bones.

Ascending colon (KŌ-lon) The part of the large intestine that passes superiorly from the cecum to the inferior border of the liver, where it bends at the right colic (hepatic) flexure to become the transverse colon.

Ascites (as-SĪ-tēz) Abnormal accumulation of serous fluid in the peritoneal cavity.

Association areas Large cortical regions on the lateral surfaces of the occipital, parietal, and temporal lobes and on the frontal lobes anterior to the motor areas connected by many motor and sensory axons to other parts of the cortex. The association areas are concerned with motor patterns, memory, concepts of word-hearing and word-seeing, reasoning, will, judgment, and personality traits.

Asthma (AZ-ma) Usually allergic reaction characterized by smooth muscle spasms in bronchi resulting in wheezing and difficult breathing. Also called **bronchial asthma.**

Astigmatism (a-STIG-ma-tizm) An irregularity of the lens or cornea of the eye causing the image to be out of focus and producing faulty vision.

Astrocyte (AS-trō-sīt) A neuroglial cell having a star shape that participates in brain development and the metabolism of neurotransmitters, helps form the blood–brain barrier, helps maintain the proper balance of K^+ for generation of nerve impulses, and provides a link between neurons and blood vessels.

Ataxia (a-TAK-sē-a) A lack of muscular coordination, lack of precision.

Atherosclerotic plaque (ath′-er-ō-skle-RO-tic PLAK) A lesion that results from accumulated cholesterol and smooth muscle fibers (cells) of the tunica media of an artery; may become obstructive.

Atom Unit of matter that makes up a chemical element; consists of a nucleus (containing positively charged protons and uncharged neutrons) and negatively charged electrons that orbit the nucleus.

Atomic mass (weight) Average mass of all stable atoms of an element, reflecting the relative proportion of atoms with different mass numbers.

Atomic number Number of protons in an atom.

Atrial fibrillation (Ā-trē-al fib-ri-LĀ-shun) Asynchronous contraction of cardiac muscle fibers in the atria that results in the cessation of atrial pumping.

Atrial natriuretic peptide (ANP) (na′-trē-ū-RET-ik) Peptide hormone, produced by the atria of the heart in response to stretching, that inhibits aldosterone production and thus lowers blood pressure; causes natriuresis, increased urinary excretion of sodium.

Atrioventricular (AV) bundle (ā′-trē-ō-ven-TRIK-ū-lar) The part of the conduction system of the heart that begins at the atrioventricular (AV) node, passes through the cardiac skeleton separating the atria and the ventricles, then extends a short distance down the interventricular septum before splitting into right and left bundle branches. Also called the **bundle of His** (HISS).

Atrioventricular (AV) node The part of the conduction system of the heart made up of a compact mass of conducting cells located in the septum between the two atria.

Atrioventricular (AV) valve A heart valve made up of membranous flaps or cusps that allows blood to flow in one direction only, from an atrium into a ventricle.

Atrium (Ā-trē-um) A superior chamber of the heart.

Atrophy (AT-rō-fē) Wasting away or decrease in size of a part, due to a failure, abnormality of nutrition, or lack of use.

Auditory ossicle (AW-di-tō-rē OS-si-kul) One of the three small bones of the middle ear called the **malleus, incus,** and **stapes.**

Auditory tube The tube that connects the middle ear with the nose and nasopharynx region of the throat. Also called the **eustachian tube** (ū-STĀ-shun *or* ū-STĀ-kē-an) or **pharyngotympanic tube.**

Auscultation (aws-kul-TĀ-shun) Examination by listening to sounds in the body.

Autoimmunity An immunological response against a person's own tissues.

Autolysis (aw-TOL-i-sis) Self-destruction of cells by their own lysosomal digestive enzymes after death or in a pathological process.

Autonomic ganglion (aw′-tō-NOM-ik GANG-lē-on) A cluster of cell bodies of sympathetic or parasympathetic neurons located outside the central nervous system.

Autonomic nervous system (ANS) Visceral sensory (afferent) and visceral motor (efferent) neurons. Autonomic motor neurons, both sympathetic and parasympathetic, conduct nerve impulses from the central nervous system to smooth muscle, cardiac muscle, and glands. So named because this part of the nervous system was thought to be self-governing or spontaneous.

Autophagy (aw-TOF-a-jē) Process by which worn-out organelles are digested within lysosomes.

Autopsy (AW-top-sē) The examination of the body after death.

Autosome (AW-tō-sōm) Any chromosome other than the X and Y chromosomes (sex chromosomes).

Axilla (ak-SIL-a) The small hollow beneath the arm where it joins the body at the shoulders. Also called the **armpit.**

Axon (AK-son) The usually single, long process of a nerve cell that propagates a nerve impulse toward the axon terminals.

Axon terminal Terminal branch of an axon where synaptic vesicles undergo exocytosis to release neurotransmitter molecules.

B

B cell A lymphocyte that can develop into a clone of antibody-producing plasma cells or memory cells when properly stimulated by a specific antigen.

Babinski sign (ba-BIN-skē) Extension of the great toe, with or without fanning of the other toes, in response to stimulation of the outer margin of the sole; normal up to 18 months of age and indicative of damage to descending motor pathways such as the corticospinal tracts after that.

Back The posterior part of the body; the dorsum.

Ball-and-socket joint A synovial joint in which the rounded surface of one bone moves within a cup-shaped depression or socket of another bone, as in the shoulder or hip joint. Also called a **spheroid joint** (SFĒ-royd).

Baroreceptor (bar′-ō-re-SEP-tor) Neuron capable of responding to changes in blood, air, or fluid pressure. Also called a **pressoreceptor.**

Basal ganglia (GANG-glē-a) Paired clusters of gray matter deep in each cerebral hemisphere including the globus pallidus, putamen, and caudate nucleus. Nearby structures that are functionally linked to the basal ganglia are the substantia nigra of the midbrain and the subthalamic nuclei of the diencephalon.

Basal metabolic rate (BMR) (BĀ-sal met′-a-BOL-ik) The rate of metabolism measured under standard or basal conditions (awake, at rest, fasting).

Base A nonacid or a proton acceptor, characterized by excess of hydroxide ions (OH^-) and a pH greater than 7. A ring-shaped, nitrogen-containing organic molecule that is one of the components of a nucleotide, namely, adenine, guanine, cytosine, thymine, and uracil; also known as a **nitrogenous base.**

Basement membrane Thin, extracellular layer between epithelium and connective tissue consisting of a basal lamina and a reticular lamina.

Basilar membrane (BĀS-i-lar) A membrane in the cochlea of the internal ear that separates the cochlear duct from the scala tympani and on which the spiral organ (organ of Corti) rests.

Basophil (BĀ-sō-fil) A type of white blood cell characterized by a pale nucleus and large granules that stain blue-purple with basic dyes.

Belly The abdomen. The gaster or prominent, fleshy part of a skeletal muscle.

Beta cell (BĀ-ta) A type of cell in the pancreatic islets (islets of Langerhans) in the pancreas that secretes the hormone insulin.

Bicuspid valve (bī-KUS-pid) Atrioventricular (AV) valve on the left side of the heart. Also called the **mitral valve.**

Bilateral (bī-LAT-er-al) Pertaining to two sides of the body.

Bile (BĪL) A secretion of the liver consisting of water, bile salts, bile pigments, cholesterol, lecithin, and several ions; it emulsifies lipids prior to their digestion.

Bilirubin (bil-ē-ROO-bin) An orange pigment that is one of the end products of hemoglobin breakdown in the hepatocytes and is excreted as a waste material in bile.

Blastocyst (BLAS-tō-sist) In the development of an embryo, a hollow ball of cells that consists of a blastocyst cavity (the internal cavity), trophoblast (outer cells), and inner cell mass.

Blastomere (BLAS-tō-mēr) One of the cells resulting from the cleavage of a fertilized ovum.

Blind spot Area in the retina at the end of the optic (II) nerve in which there are no photoreceptors.

Blood The fluid that circulates through the heart, arteries, capillaries, and veins and that constitutes the chief means of transport within the body.

Blood–brain barrier (BBB) A barrier consisting of specialized brain capillaries and astrocytes that prevents the passage of materials from the blood to the cerebrospinal fluid and brain.

Blood pressure (BP) Force exerted by blood against the walls of blood vessels due to contraction of the heart and influenced by the elasticity of the vessel walls; clinically, a measure of the pressure in arteries during ventricular systole and ventricular diastole. *See also* **mean arterial blood pressure.**

Blood–testis barrier (BTB) A barrier formed by Sertoli cells that prevents an immune response against antigens produced by spermatogenic cells by isolating the cells from the blood.

Body cavity A space within the body that contains various internal organs.

Body fluid Body water and its dissolved substances; constitutes about 60% of total body mass.

Bolus (BŌ-lus) A soft, rounded mass, usually food, that is swallowed.

Brachial plexus (BRĀ-kē-al PLEK-sus) A network of nerve axons of the ventral rami of spinal nerves C5, C6, C7, C8, and T1. The nerves that emerge from the brachial plexus supply the upper limb.

Bradycardia (brād′-i-KAR-dē-a) A slow resting heart or pulse rate (under 50 beats per minute).

Brain The part of the central nervous system contained within the cranial cavity.

Brain stem The portion of the brain immediately superior to the spinal cord, made up of the medulla oblongata, pons, and midbrain.

Brain waves Electrical signals that can be recorded from the skin of the head due to electrical activity of brain neurons.

Broca's area (BRŌ-kaz) Motor area of the brain in the frontal lobe that translates thoughts into speech. Also called the **motor speech area.**

Bronchi (BRONG-kē) Branches of the respiratory passageway including primary bronchi (the two divisions of the trachea), secondary or lobar bronchi (divisions of the primary bronchi that are distributed to the lobes of the lungs), and tertiary or segmental bronchi (divisions of the secondary bronchi that are distributed to bronchopulmonary segments of the lungs). *Singular* is **bronchus.**

Bronchial tree The trachea, bronchi, and their branching structures up to and including the terminal bronchioles.

Bronchiole (BRONG-kē-ōl) Branch of a tertiary bronchus further dividing into terminal bronchioles (distributed to lobules of the lungs), which divide into respiratory bronchioles (distributed to alveolar sacs).

Bronchitis (brong-KĪ-tis) Inflammation of the mucous membrane of the bronchial tree; characterized by hypertrophy and hyperplasia of seromucous glands and goblet cells that line the bronchi and which results in a productive cough.

Buccal (BUK-al) Pertaining to the cheek or mouth.

Buffer system (BUF-er) A pair of chemicals—one a weak acid and the other the salt of the weak acid, which functions as a weak base—that resists changes in pH.

Bulbourethral gland (bul′-bō-ū-RĒ-thral) One of a pair of glands located inferior to the prostate on either side of the urethra that secretes an alkaline fluid into the urethra. Also called a **Cowper's gland** (KOW-perz).

Bulimia (boo-LIM-ē-a *or* boo-LĒ-mē-a) A disorder characterized by overeating at least twice a week followed by purging by self-induced vomiting, strict dieting or fasting, vigorous exercise, or use of laxatives or diuretics. Also called **binge–purge syndrome.**

Bulk flow The movement of large numbers of ions, molecules, or particles in the same direction due to pressure differences (osmotic, hydrostatic, or air pressure).

Bulk-phase endocytosis (pi′-nō-sī-TŌ-sis) A process by which most body cells can ingest membrane-surrounded droplets of interstitial fluid.

Bundle branch One of the two branches of the atrioventricular (AV) bundle made up of specialized muscle fibers (cells) that transmit electrical impulses to the ventricles.

Bursa (BUR-sa) A sac or pouch of synovial fluid located at friction points, especially about joints.

Buttocks (BUT-oks) The two fleshy masses on the posterior aspect of the inferior trunk, formed by the gluteal muscles.

C

Calcaneal tendon (kal-KĀ-nē-al) The tendon of the soleus, gastrocnemius, and plantaris muscles at the back of the heel. Also called the **Achilles tendon** (a-KIL-ēz).

Calcification (kal-si-fi-KĀ-shun) Deposition of mineral salts, primarily hydroxyapatite, in a framework formed by collagen fibers in which the tissue hardens. Also called **mineralization** (min′-e-ral-i-ZĀ-shun).

Calcitonin (CT) (kal-si-TŌ-nin) A hormone produced by the parafollicular cells of the thyroid gland that can lower the amount of blood calcium and phosphates by inhibiting bone resorption (breakdown of bone extracellular matrix) and by accelerating uptake of calcium and phosphates into bone extracellular matrix.

Calculus (KAL-kū-lus) A stone, or insoluble mass of crystallized salts or other material, formed within the body, as in the gallbladder, kidney, or urinary bladder.

Callus (KAL-lus) A growth of new bone tissue in and around a fractured area, ultimately replaced by mature bone. An acquired, localized thickening.

Calorie (KAL-ō-rē) A unit of heat. A calorie (cal) is the standard unit and is the amount of heat needed to raise the temperature of 1 g of water from 14 °C to 15 °C. The **kilocalorie (kcal)** or **Calorie** (spelled with an uppercase C), used to express the caloric value of foods and to measure metabolic rate, is equal to 1000 cal.

Calyx (KĀL-iks) Any cuplike division of the kidney pelvis. *Plural is* **calyces** (KĀ-li-sēz).

Canal (ka-NAL) A narrow tube, channel, or passageway.

Canaliculus (kan′-a-LIK-ū-lus) A small channel or canal, as in bones, where they connect lacunae. *Plural is* **canaliculi** (kan′-a-LIK-ū-lī).

Cancellous (KAN-sel-us) Having a reticular or latticework structure, as in spongy bone tissue.

Capacitation (ka′-pas-i-TĀ-shun) The functional changes that sperm undergo in the female reproductive tract that allow them to fertilize a secondary oocyte.

Capillary (KAP-i-lar′-ē) A microscopic blood vessel located between an arteriole and venule through which materials are exchanged between blood and interstitial fluid.

Carbohydrate (kar′-bō-HĪ-drāt) An organic compound containing carbon, hydrogen, and oxygen in a particular amount and arrangement and composed of monosaccharide subunits; usually has the general formula $(CH_2O)_n$.

Carcinogen (kar-SIN-ō-jen) A chemical substance or radiation that causes cancer.

Cardiac arrest (KAR-dē-ak) Cessation of an effective heartbeat in which the heart is completely stopped or in ventricular fibrillation.

Cardiac cycle A complete heartbeat consisting of systole (contraction) and diastole (relaxation) of both atria plus systole and diastole of both ventricles.

Cardiac muscle Striated muscle fibers (cells) that form the wall of the heart; stimulated by an intrinsic conduction system and autonomic motor neurons.

Cardiac notch An angular notch in the anterior border of the left lung into which part of the heart fits.

Cardiac output (CO) The volume of blood pumped from one ventricle of the heart (usually measured from the left ventricle) in 1 min; normally about 5.2 liters/min in an adult at rest.

Cardiology (kar-dē-OL-ō-jē) The study of the heart and diseases associated with it.

Cardiovascular center (kar-dē-ō-VAS-kū-lar) Groups of neurons scattered within the medulla oblongata that regulate heart rate, force of contraction, and blood vessel diameter.

Carotene (KAR-o-tēn) Antioxidant precursor of vitamin A, which is needed for synthesis of photopigments; yellow-orange pigment present in the stratum corneum of the epidermis. Accounts for the yellowish coloration of skin. Also termed **beta-carotene.**

Carotid body (ka-ROT-id) Cluster of chemoreceptors on or near the carotid sinus that respond to changes in blood levels of oxygen, carbon dioxide, and hydrogen ions.

Carotid sinus A dilated region of the internal carotid artery just above the point where it branches from the common carotid artery; it contains baroreceptors that monitor blood pressure.

Carotid sinus reflex A reflex that helps maintain normal blood pressure in the brain. Nerve impulses propagate from the carotid sinus baroreceptors over sensory axons in the glossopharyngeal (IX) nerves to the cardiovascular center in the medulla oblongata.

Carpal bones The eight bones of the wrist. Also called **carpals.**

Carpus (KAR-pus) A collective term for the eight bones of the wrist.

Cartilage (KAR-ti-lij) A type of connective tissue consisting of chondrocytes in lacunae embedded in a dense network of collagen and elastic fibers and an extracellular matrix of chondroitin sulfate.

Cartilaginous joint (kar′-ti-LAJ-i-nus) A joint without a synovial (joint) cavity where the articulating bones are held tightly together by cartilage, allowing little or no movement.

Cast A small mass of hardened material formed within a cavity in the body and then discharged from the body; can originate in different areas and can be composed of various materials.

Catabolism (ka-TAB-ō-lizm) Chemical reactions that break down complex organic compounds into simple ones, with the net release of energy.

Cataract (KAT-a-rakt) Loss of transparency of the lens of the eye or its capsule or both.

Cation (KAT-ī-on) A positively charged ion. An example is a sodium ion (Na^+).

Cauda equina (KAW-da ē-KWĪ-na) A tail-like array of roots of spinal nerves at the inferior end of the spinal cord.

Cecum (SĒ-kum) A blind pouch at the proximal end of the large intestine that attaches to the ileum.

Cell The basic structural and functional unit of all organisms; the smallest structure capable of performing all the activities vital to life.

Cell cycle Growth and division of a single cell into two identical cells; consists of interphase and cell division.

Cell division Process by which a cell reproduces itself that consists of a nuclear division (mitosis) and a cytoplasmic division (cytokinesis); types include somatic and reproductive cell division.

Cell-mediated immunity That component of immunity in which specially sensitized T lymphocytes (T cells) attach to antigens to destroy them. Also called **cellular immunity.**

Cementum (se-MEN-tum) Calcified tissue covering the root of a tooth.

Center of ossification (os′-i-fi-KĀ-shun) An area in the cartilage model of a future bone where the cartilage cells hypertrophy and then secrete enzymes that result in the calcification of their matrix, resulting in the death of the cartilage cells, followed by the invasion of the area by osteoblasts that then lay down bone.

Central canal A microscopic tube running the length of the spinal cord in the gray commissure. A circular channel running longitudinally in the center of an osteon (haversian system) of mature compact bone, containing blood and lymphatic vessels and nerves. Also called a **haversian canal** (ha-VER-shan).

Central fovea (FŌ-vē-a) A depression in the center of the macula lutea of the retina, containing cones only and lacking blood vessels; the area of highest visual acuity (sharpness of vision).

Central nervous system (CNS) That portion of the nervous system that consists of the brain and spinal cord.

Centrioles (SEN-trē-ōlz) Paired, cylindrical structures of a centrosome, each consisting of a ring of microtubules and arranged at right angles to each other.

Centromere (SEN-trō-mēr) The constricted portion of a chromosome where the two chromatids are joined; serves as the point of attachment for the microtubules that pull chromatids during anaphase of cell division.

Centrosome (SEN-trō-sōm) A dense network of small protein fibers near the nucleus of a cell, containing a pair of centrioles and pericentriolar material.

Cephalic (se-FAL-ik) Pertaining to the head; superior in position.

Cerebellum (ser-e-BEL-um) The part of the brain lying posterior to the medulla oblongata and pons; governs balance and coordinates skilled movements.

Cerebral aqueduct (SER-ē-bral AK-we-dukt) A channel through the midbrain connecting the third and fourth ventricles and containing cerebrospinal fluid. Also termed the **aqueduct of Sylvius.**

Cerebral arterial circle A ring of arteries forming an anastomosis at the base of the brain between the internal carotid and basilar arteries and arteries supplying the cerebral cortex. Also called the **circle of Willis.**

Cerebral cortex The surface of the cerebral hemispheres, 2–4 mm thick, consisting of gray matter; arranged in six layers of neuronal cell bodies in most areas.

Cerebrospinal fluid (CSF) (se-rē′-brō-SPĪ-nal) A fluid produced by ependymal cells that cover choroid plexuses in the ventricles of the brain; the fluid circulates in the ventricles, the central canal, and the subarachnoid space around the brain and spinal cord.

Cerebrovascular accident (CVA) (se rē′-brō-VAS-kū-lar) Destruction of brain tissue (infarction) resulting from obstruction or rupture of blood vessels that supply the brain. Also called a **stroke** or **brain attack.**

Cerebrum (SER-e-brum *or* se-RĒ-brum) The two hemispheres of the forebrain (derived from the telencephalon), making up the largest part of the brain.

Cerumen (se-ROO-men) Waxlike secretion produced by ceruminous glands in the external auditory meatus (ear canal). Also termed **ear wax.**

Ceruminous gland (se-ROO-mi-nus) A modified sudoriferous (sweat) gland in the external auditory meatus that secretes cerumen (ear wax).

Cervical ganglion (SER-vi-kul GANG-glē-on) A cluster of cell bodies of postganglionic sympathetic neurons located in the neck, near the vertebral column.

Cervical plexus (PLEK-sus) A network formed by nerve axons from the ventral rami of the first four cervical nerves and receiving gray rami communicates from the superior cervical ganglion.

Cervix (SER-viks) Neck; any constricted portion of an organ, such as the inferior cylindrical part of the uterus.

Chemical bond Force of attraction in a molecule or compound that holds its atoms together. Examples include ionic and covalent bonds.

Chemical element Unit of matter that cannot be broken apart into a simpler substance by ordinary chemical reactions. Examples include hydrogen (H), carbon (C), and oxygen (O).

Chemical reaction The combination or separation of atoms in which chemical bonds are formed or broken and new products with different properties are produced.

Chemoreceptor (kē′-mō-rē-SEP-tor) Sensory receptor that detects the presence of a specific chemical.

Chemotaxis (kē-mō-TAK-sis) Attraction of phagocytes to microbes by a chemical stimulus.

Chiasm (KĪ-azm) A crossing; especially the crossing of axons in the optic (II) nerve as they extend toward the opposite optic tract.

Chief cell The secreting cell of a gastric gland that produces pepsinogen, the precursor of the enzyme pepsin, and the enzyme gastric lipase. Also called a **zymogenic cell** (zī′-mō-JEN-ik). Cell in the parathyroid glands that secretes parathyroid hormone (PTH). Also called a **principal cell.**

Chiropractic (kī-rō-PRAK-tik) A system of treating disease by using one's hands to manipulate body parts, mostly the vertebral column.

Cholecystectomy (kō′-lē-sis-TEK-tō-mē) Surgical removal of the gallbladder.

Cholecystitis (kō′-lē-sis-TĪ-tis) Inflammation of the gallbladder.

Cholesterol (kō-LES-te-rol) Classified as a lipid, the most abundant steroid in animal tissues; located in cell membranes and used for the synthesis of steroid hormones and bile salts.

Cholinergic neuron (kō′-lin-ER-jik) A neuron that liberates acetylcholine as its neurotransmitter.

Chondrocyte (KON-drō-sīt) Cell of mature cartilage.

Chondroitin sulfate (kon-DROY-tin) An amorphous extracellular matrix material found outside connective tissue cells.

Chordae tendineae (KOR-dē TEN-di-nē-ē) Tendonlike, fibrous cords that connect atrioventricular valves of the heart with papillary muscles.

Chorion (KŌ-rē-on) The most superficial fetal membrane that becomes the principal embryonic portion of the placenta; serves a protective and nutritive function.

Chorionic villi (kō-rē-ON-ik VIL-lī) Fingerlike projections of the chorion that grow into the endometrium and contain fetal blood vessels.

Choroid (KŌ-royd) One of the vascular coats of the eyeball.

Choroid plexus (PLEK-sus) A network of capillaries located in the roof of each of the four ventricles of the brain; ependymal cells around choroid plexuses produce cerebrospinal fluid.

Chromatid (KRŌ-ma-tid) One of a pair of identical connected nucleoprotein strands that are joined at the centromere and separate during cell division, each becoming a chromosome of one of the two resulting cells.

Chromatin (KRŌ-ma-tin) The threadlike mass of genetic material, consisting of DNA and histone proteins, that is present in the nucleus of a nondividing or interphase cell.

Chromatolysis (krō-ma-TOL-i-sis) The breakdown of Nissl bodies into finely granular masses in the cell body of a neuron whose axon has been damaged.

Chromosome (KRŌ-mō-sōm) One of the small, threadlike structures in the nucleus of a cell, normally 46 in a human diploid cell, that bears the genetic material; composed of DNA and proteins (histones) that form a delicate chromatin thread during interphase; becomes packaged into compact rodlike structures that are visible under the light microscope during cell division.

Chronic (KRON-ik) Long term or frequently recurring; applied to a disease that is not acute.

Chronic obstructive pulmonary disease (COPD) A disease, such as bronchitis or emphysema, in which there is some degree of obstruction of airways and consequent increase in airway resistance.

Chyle (KĪL) The milky-appearing fluid found in the lacteals of the small intestine after absorption of lipids in food.

Chylomicron (kī-lō-MĪ-kron) Protein-coated sphericalstructure that contains triglycerides, phospholipids, and cholesterol and is absorbed into the lacteal of a villus in the small intestine.

Chyme (KĪM) The semifluid mixture of partly digested food and digestive secretions found in the stomach and small intestine during digestion of a meal.

Ciliary body (SIL-ē-ar′-ē) One of the three parts of the vascular tunic of the eyeball, the others being the choroid and the iris; includes the ciliary muscle and the ciliary processes.

Cilium (SIL-ē-um) A hair or hairlike process projecting from a cell that may be used to move the entire cell or to move substances along the surface of the cell. *Plural is* **cilia.**

Circadian rhythm (ser-KĀ-dē-an) A cycle of active and nonactive periods in organisms determined by internal mechanisms and repeating about every 24 hours.

Circular folds Permanent, deep, transverse folds in the mucosa and submucosa of the small intestine that increase the surface area for absorption. Also called **plicae circulares** (PLĪ-kē SER-kū-lar-ēs).

Circulation time Time required for blood to pass from the right atrium, through pulmonary circulation, back to the left ventricle, through systemic circulation to the foot, and back again to the right atrium; normally about 1 min.

Circumduction (ser′-kum-DUK-shun) A movement at a synovial joint in which the distal end of a bone moves in a circle while the proximal end remains relatively stable.

Cirrhosis (si-RŌ-sis) A liver disorder in which the parenchymal cells are destroyed and replaced by connective tissue.

Cisterna chyli (sis-TER-na-KĪ-lē) The origin of the thoracic duct.

Cleavage The rapid mitotic divisions following the fertilization of a secondary oocyte, resulting in an increased number of progressively smaller cells, called blastomeres.

Climacteric (klī-mak-TER-ik) Cessation of the reproductive function in the female or decreased testicular activity in the male.

Climax The peak period or moments of greatest intensity during sexual excitement.

Clitoris (KLI-to-ris) An erectile organ of the female, located at the anterior junction of the labia minora, that is homologous to the male penis.

Clone (KLŌN) A population of identical cells.

Clot The end result of a series of biochemical reactions that changes liquid plasma into a gelatinous mass; specifically, the conversion of fibrinogen into a tangle of polymerized fibrin molecules.

Clot retraction (rē-TRAK-shun) The consolidation of a fibrin clot to pull a damaged tissue together.

Clotting Process by which a blood clot is formed. Also known as **coagulation** (cō-ag-ū-LĀ-shun).

Coccyx (KOK-six) The fused bones at the inferior end of the vertebral column.

Cochlea (KŌK-lē-a) A winding, cone-shaped tube forming a portion of the inner ear and containing the spiral organ (organ of Corti).

Cochlear duct The membranous cochlea consisting of a spirally arranged tube enclosed in the bony cochlea and lying along its outer wall. Also called the **scala media** (SCA-la MĒ-dē-a).

Coenzyme A nonprotein organic molecule that is associated with and activates an enzyme; many are derived from vitamins. An example is nicotinamide adenine dinucleotide (NAD), derived from the B vitamin niacin.

Coitus (KŌ-i-tus) Sexual intercourse.

Collagen (KOL-a-jen) A protein that is the main organic constituent of connective tissue.

Collateral circulation The alternate route taken by blood through an anastomosis.

Colliculus (ko-LIK-ū-lus) A small elevation.

Colloid (KOL-loyd) The material that accumulates in the center of thyroid follicles, consisting of thyroglobulin and stored thyroid hormones.

Colon The portion of the large intestine consisting of ascending, transverse, descending, and sigmoid portions.

Colostrum (kō-LOS-trum) A thin, cloudy fluid secreted by the mammary glands a few days prior to or after delivery before true milk is produced.

Column (KOL-um) Group of white matter tracts in the spinal cord.

Common bile duct A tube formed by the union of the common hepatic duct and the cystic duct that empties bile into the duodenum at the hepatopancreatic ampulla (ampulla of Vater).

Compact (dense) bone tissue Bone tissue that contains few spaces between osteons (haversian systems); forms the external portion of all bones and the bulk of the diaphysis (shaft) of long bones; is found immediately deep to the periosteum and external to spongy bone.

Complement (KOM-ple-ment) A group of at least 20 normally inactive proteins found in plasma that forms a component of innate and adaptive nonspecific resistance and immunity by bringing about cytolysis, inflammation, and opsonization.

Compound A substance that can be broken down into two or more other substances by chemical means.

Concha (KONG-ka) A scroll-like bone found in the skull. *Plural is* **conchae** (KONG-kē).

Concussion (kon-KUSH-un) Traumatic injury to the brain that produces no visible bruising but may result in abrupt, temporary loss of consciousness.

Conduction system A group of autorhythmic cardiac muscle fibers that generates and distributes electrical impulses to stimulate coordinated contraction of the heart chambers; includes the sinoatrial (SA) node, the atrioventricular (AV) node, the atrioventricular (AV) bundle, the right and left bundle branches, and the Purkinje fibers.

Conductivity (kon′-duk-TIV-i-tē) The ability of a cell to propagate (conduct) action potentials along its plasma membrane; characteristic of neurons and muscle fibers (cells).

Condyloid joint (KON-di-loyd) A synovial joint structured so that an oval-shaped condyle of one bone fits into an elliptical cavity of another bone, permitting side-to-side and back-and-forth movements, such as the joint at the wrist between the radius and carpals.

Cone (KŌN) The type of photoreceptor in the retina that is specialized for highly acute color vision in bright light.

Congenital (kon-JEN-i-tal) Present at the time of birth.

Conjunctiva (kon′-junk-TĪ-va) The delicate membrane covering the eyeball and lining the eyes.

Connective tissue One of the most abundant of the four basic tissue types in the body, performing the functions of binding and supporting; consists of relatively few cells in a generous extracellular matrix (the ground substance and fibers between the cells).

Consciousness (KON-shus-nes) A state of wakefulness in which an individual is fully alert, aware, and oriented, partly as a result of feedback between the cerebral cortex and reticular activating system.

Continuous conduction (kon-DUK-shun) Propagation of an action potential (nerve impulse) in a step-by-step depolarization of each adjacent area of an axon membrane.

Contraception (kon′-tra-SEP-shun) The prevention of fertilization or impregnation without destroying fertility.

Contractility (kon′-trak-TIL-i-tē) The ability of cells or parts of cells to actively generate force to undergo shortening for movements. Muscle fibers (cells) exhibit a high degree of contractility.

Contralateral (kon′-tra-LAT-er-al) On the opposite side; affecting the opposite side of the body.

Control center The component of a feedback system, such as the brain, that determines the point at which a controlled condition, such as body temperature, is maintained.

Conus medullaris (KŌ-nus med-ū-LAR-is) The tapered portion of the spinal cord inferior to the lumbar enlargement.

Convergence (con-VER-jens) A synaptic arrangement in which the synaptic end bulbs of several presynaptic neurons terminate on one postsynaptic neuron. The medial movement of the two eyeballs so that both are directed toward a near object being viewed in order to produce a single image.

Convulsion (con-VUL-shun) Violent, involuntary contractions or spasms of an entire group of muscles.

Cornea (KOR-nē-a) The nonvascular, transparent fibrous coat through which the iris of the eye can be seen.

Corona radiata The innermost layer of granulosa cells that is firmly attached to the zona pellucida around a secondary oocyte.

Coronary artery disease (CAD) A condition such as atherosclerosis that causes narrowing of coronary arteries so that blood flow to the heart is reduced. The result is **coronary heart disease**

(CHD), in which the heart muscle receives inadequate blood flow due to an interruption of its blood supply.

Coronary circulation The pathway followed by the blood from the ascending aorta through the blood vessels supplying the heart and returning to the right atrium. Also called **cardiac circulation.**

Coronary sinus (SĪ-nus) A wide venous channel on the posterior surface of the heart that collects the blood from the coronary circulation and returns it to the right atrium.

Corpus albicans (KOR-pus AL-bi-kanz) A white fibrous patch in the ovary that forms after the corpus luteum regresses.

Corpus callosum (kal-LŌ-sum) The great commissure of the brain between the cerebral hemispheres.

Corpus luteum (LOO-tē-um) A yellowish body in the ovary formed when a follicle has discharged its secondary oocyte; secretes estrogens, progesterone, relaxin, and inhibin.

Corpuscle of touch The sensory receptor for the sensation of touch; found in the dermal papillae, especially in palms and soles. Also called a **Meissner corpuscle** (MĪZ-ner).

Cortex (KOR-teks) An outer layer of an organ. The convoluted layer of gray matter covering each cerebral hemisphere.

Costal (KOS-tal) Pertaining to a rib.

Costal cartilage (KAR-ti-lij) Hyaline cartilage that attaches a rib to the sternum.

Cramp A spasmodic, usually painful contraction of a muscle.

Cranial cavity (KRĀ-nē-al) A body cavity formed by the cranial bones and containing the brain.

Cranial nerve One of 12 pairs of nerves that leave the brain; pass through foramina in the skull; and supply sensory and motor neurons to the head, neck, part of the trunk, and viscera of the thorax and abdomen. Each is designated by a Roman numeral and a name.

Cranium (KRĀ-nē-um) The skeleton of the skull that protects the brain and the organs of sight, hearing, and balance; includes the frontal, parietal, temporal, occipital, sphenoid, and ethmoid bones.

Creatine phosphate (KRĒ-a-tin FOS-fāt) Molecule in striated muscle fibers that contains high-energy phosphate bonds; used to generate ATP rapidly from ADP by transfer of a phosphate group. Also called **phosphocreatine** (fos′-fō-KRĒ-a-tin).

Crenation (krē-NĀ-shun) The shrinkage of red blood cells into knobbed, starry forms when they are placed in a hypertonic solution.

Crista (KRIS-ta) A crest or ridged structure. A small elevation in the ampulla of each semicircular duct that contains receptors for dynamic equilibrium.

Crossing-over The exchange of a portion of one chromatid with another during meiosis. It permits an exchange of genes among chromatids and is one factor that results in genetic variation of progeny.

Cryptorchidism (krip-TOR-ki-dizm) The condition of undescended testes.

Cupula (KUP-ū-la) A mass of gelatinous material covering the hair cells of a crista; a sensory receptor in the ampulla of a semicircular canal stimulated when the head moves.

Cushing's syndrome Condition caused by a hypersecretion of glucocorticoids characterized by spindly legs, “moon face,” “buffalo hump,” pendulous abdomen, flushed facial skin, poor wound healing, hyperglycemia, osteoporosis, hypertension, and increased susceptibility to disease.

Cutaneous (kū-TĀ-nē-us) Pertaining to the skin.

Cyanosis (sī-a-NŌ-sis) A blue or dark purple discoloration, most easily seen in nail beds and mucous membranes, that results from an increased concentration of deoxygenated (reduced) hemoglobin (more than 5 gm/dL).

Cyclic AMP (cyclic adenosine-3′,5′-monophosphate) Molecule formed from ATP by the action of the enzyme adenylate cyclase; serves as second messenger for some hormones and neurotransmitters.

Cyst (SIST) A sac with a distinct connective tissue wall, containing a fluid or other material.

Cystic duct (SIS-tik) The duct that carries bile from the gallbladder to the common bile duct.

Cystitis (sis-TĪ-tis) Inflammation of the urinary bladder.

Cytolysis (sī-TOL-i-sis)The rupture of living cells in which the contents leak out.

Cytokinesis (sī-tō-ki-NĒ-sis) Distribution of the cytoplasm into two separate cells during cell division; coordinated with nuclear division (mitosis).

Cytoplasm (SĪ-tō-plazm) Cytosol plus all organelles except the nucleus.

Cytoskeleton Complex internal structure of cytoplasm consisting of microfilaments, microtubules, and intermediate filaments.

Cytosol (SĪ-tō-sol) Fluid located within cells. Also called **intracellular fluid (ICF)** (in′-tra-SEL-ū-lar).

D

Deciduous (dē-SID-ū-us) Falling off or being shed seasonally or at a particular stage of development. In the body, referring to the first set of teeth.

Deep Away from the surface of the body or an organ.

Deep fascia (FASH-ē-a) A sheet of connective tissue wrapped around a muscle to hold it in place.

Defecation (def-e-KĀ-shun) The discharge of feces from the rectum.

Deglutition (dē-gloo-TISH-un) The act of swallowing.

Dehydration (dē-hī-DRĀ-shun) Excessive loss of water from the body or its parts.

Demineralization (de-min′-er-al-i-ZĀ-shun) Loss of calcium and phosphorus from bones.

Denaturation (de-nā-chur-Ā-shun) Disruption of the tertiary structure of a protein by heat, changes in pH, or other physical or chemical methods, in which the protein loses its physical properties and biological activity.

Dendrite (DEN-drīt) A neuronal process that carries electrical signals toward the cell body.

Dendritic cell (den-DRIT-ik) One type of antigen-presenting cell with long branchlike projections that commonly is present in mucosal linings such as the vagina, in the skin (Langerhans cells in the epidermis), and in lymph nodes.

Dental caries (KA-rēz) Gradual demineralization of the enamel and dentin of a tooth that may invade the pulp and alveolar bone. Also called **tooth decay.**

Dentin (DEN-tin) The bony tissues of a tooth enclosing the pulp cavity.

Dentition (den-TI-shun) The eruption of teeth. The number, shape, and arrangement of teeth.

Deoxyribonucleic acid (DNA) (dē-ok′-sē-rī′-bō-noo-KLĒ-ik) A nucleic acid constructed of nucleotides consisting of one of four bases (adenine, cytosine, guanine, or thymine), deoxyribose, and a phosphate group; encoded in the nucleotides is genetic information.

Depolarization (dē-pō-lar-i-ZĀ-shun) Areductionofvoltage across a plasma meembrane; expressed as a change toward less negative (more positive) voltages on the interior surface of the plasma membrane.

Depression (de-PRESH-un) Movement in which a part of the body moves inferiorly.

Dermal papilla (pa-PILL-a) Fingerlike projection of the papillary region of the dermis that may contain blood capillaries or corpuscles of touch (Meissner corpuscles).

Dermatology (der-ma-TOL-ō-jē) The medical specialty dealing with diseases of the skin.

Dermatome (DER-ma-tōm) The cutaneous area developed from one embryonic spinal cord segment and receiving most of its sensory innervation from one spinal nerve. An instrument for incising the skin or cutting thin transplants of skin.

Dermis (DER-mis) A layer of dense irregular connective tissue lying deep to the epidermis.

Descending colon (KŌ-lon) The part of the large intestine descending from the left colic (splenic) flexure to the level of the left iliac crest.

Detritus (de-TRĪ-tus) Particulate matter produced by or remaining after the wearing away or disintegration of a substance or tissue; scales, crusts, or loosened skin.

Detrusor muscle (de-TROO-ser) Smooth muscle that forms the wall of the urinary bladder.

Developmental biology The study of development from the fertilized egg to the adult form.

Diagnosis (dī-ag-NŌ-sis) Distinguishing one disease from another or determining the nature of a disease from signs and symptoms by inspection, palpation, laboratory tests, and other means.

Dialysis (dī-AL-i-sis) The removal of waste products from blood by diffusion through a selectively permeable membrane.

Diaphragm (DĪ-a-fram) Any partition that separates one area from another, especially the dome-shaped skeletal muscle between the thoracic and abdominal cavities. Also a dome-shaped device that is placed over the cervix, usually with a spermicide, to prevent conception.

Diaphysis (dī-AF-i-sis) The shaft of a long bone.

Diarrhea (dī-a-RĒ-a) Frequent defecation of liquid feces caused by increased motility of the intestines.

Diarthrosis (dī-ar-THRŌ-sis) A freely movable joint; types are gliding, hinge, pivot, condyloid, saddle, and ball-and-socket.

Diastole (dī-AS-tō-lē) In the cardiac cycle, the phase of relaxation or dilation of the heart muscle, especially of the ventricles.

Diastolic blood pressure (dī-as-TOL-ik) The force exerted by blood on arterial walls during ventricular relaxation; the lowest blood pressure measured in the large arteries, normally about 70 mmHg in a young adult.

Diencephalon (dī′-en-SEF-a-lon) A part of the brain consisting of the thalamus, hypothalamus, epithalamus, and subthalamus.

Diffusion (dif-Ū-zhun) A passive process in which there is a net or greater movement of molecules or ions from a region of high concentration to a region of low concentration until equilibrium is reached.

Digestion (dī-JES-chun) The mechanical and chemical breakdown of food to simple molecules that can be absorbed and used by body cells.

Dilate (DĪ-lāt)To expand or swell.

Diploid (DIP-loyd) Having the number of chromosomes characteristically found in the somatic cells of an organism; having two haploid sets of chromosomes, one each from the mother and father. Symbolized $2n$.

Direct motor pathways Collections of upper motor neurons with cell bodies in the motor cortex that project axons into the spinal cord, where they synapse with lower motor neurons or interneurons in the anterior horns. Also called the **pyramidal pathways.**

Disease Any change from a state of health.

Dislocation (dis-lō-KĀ-shun) Displacement of a bone from a joint with tearing of ligaments, tendons, and articular capsules. Also called **luxation** (luks-Ā-shun).

Dissect (di-SEKT) To separate tissues and parts of a cadaver or an organ for anatomical study.

Distal (DIS-tal) Farther from the attachment of a limb to the trunk; farther from the point of origin or attachment.

Diuretic (dī-ū-RET-ik) A chemical that increases urine volume by decreasing reabsorption of water, usually by inhibiting sodium reabsorption.

Diverticulum (dī-ver-TIK-ū-lum) A sac or pouch in the wall of a canal or organ, especially in the colon.

Dominant allele An allele that overrides the influence of an alternate allele on the homologous chromosome; the allele that is expressed.

Dorsiflexion (dor′-si-FLEK-shun) Bending the foot in the direction of the dorsum (upper surface).

Ductus arteriosus (DUK-tus ar-tē-rē-Ō-sus) A small vessel connecting the pulmonary trunk with the aorta; found only in the fetus.

Ductus (vas) deferens (DEF-er-ens) The duct that carries sperm from the epididymis to the ejaculatory duct. Also called the **seminal duct.**

Ductus epididymis (ep′-i-DID-i-mis) A tightly coiled tube inside the epididymis, distinguished into a head, body, and tail, in which sperm undergo maturation.

Ductus venosus (ve-NŌ-sus) A small vessel in the fetus that helps the circulation bypass the liver.

Duodenum (doo′-ō-DĒ-num *or* doo-OD-e-num) The first 25 cm (10 in.) of the small intestine, which connects the stomach and the ileum.

Dura mater (DOO-ra MĀ-ter) The outermost of the three meninges (coverings) of the brain and spinal cord.

Dynamic equilibrium (ē-kwi-LIB-rē-um) The maintenance of body position, mainly the head, in response to sudden movements such as rotation.

Dysfunction (dis-FUNK-shun) Absence of completely normal function.

Dysmenorrhea (dis′-men-ō-RĒ-a) Painful menstruation.

Dyspnea (DISP-nē-a) Shortness of breath.

E

Eardrum A thin, semitransparent partition of fibrous connective tissue between the external auditory meatus and the middle ear. Also called the **tympanic membrane.**

Ectoderm The primary germ layer that gives rise to the nervous system and the epidermis of skin and its derivatives.

Ectopic (ek-TOP-ik) Out of the normal location, as in ectopic pregnancy.

Edema (e-DĒ-ma) An abnormal accumulation of interstitial fluid.

Effector (e-FEK-tor) An organ of the body, either a muscle or a gland, that is innervated by somatic or autonomic motor neurons.

Efferent arteriole (EF-er-ent ar-TĒ-rē-ōl) A vessel of the renal vascular system that carries blood from a glomerulus to a peritubular capillary.

Efferent ducts (EF-er-ent) A series of coiled tubes that transport sperm from the rete testis to the epididymis.

Ejaculation (e-jak-ū-LĀ-shun) The reflex ejection or expulsion of semen from the penis.

Ejaculatory duct (e-JAK-ū-la-tō-rē) A tube that transports sperm from the ductus (vas) deferens to the prostatic urethra.

Elasticity (e-las-TIS-i-tē) The ability of a tissue to return to its original shape after contraction or extension.

Electrocardiogram (ECG or **EKG)** (e-lek′-trō-KAR-dē-ō-gram) A recording of the electrical changes that accompany the cardiac cycle that can be detected at the surface of the body; may be resting, stress, or ambulatory.

Electroencephalogram (EEG) (e-lek′-trō-en-SEF-a-lō-gram) A recording of the electrical activity of the brain from the scalp surface; used to diagnose certain diseases (such as epilepsy), furnish information regarding sleep and wakefulness, and confirm brain death.

Electrolyte (ē-LEK-trō-līt) Any compound that separates into ions when dissolved in water and that conducts electricity.

Electromyography (e-lek′-trō-mī-OG-ra-fē) Evaluation of the electrical activity of resting and contracting muscle to ascertain causes of muscular weakness, paralysis, involuntary twitching, and abnormal levels of muscle enzymes; also used as part of biofeedback studies.

Electron transport chain A sequence of electron carrier molecules on the inner mitochondrial membrane that undergo oxidation and reduction as they synthesize ATP.

Elevation (el-e-VĀ-shun) Movement in which a part of the body moves superiorly.

Embolism (EM-bō-lizm) Obstruction or closure of a vessel by an embolus.

Embolus (EM-bō-lus) A blood clot, bubble of air or fat from broken bones, mass of bacteria, or other debris or foreign material transported by the blood.

Embryo (EM-brē-ō) The young of any organism in an early stage of development; in humans, the developing organism from fertilization to the end of the eighth week of development.

Embryology (em′-brē-OL-ō-jē) The study of development from the fertilized egg to the end of the eighth week of development.

Emesis (EM-e-sis) Vomiting.

Emigration (em′-e-GRĀ-shun) Process whereby white blood cells (WBCs) leave the bloodstream by rolling along the endothelium, sticking to it, and squeezing between the endothelial cells. Adhesion molecules help WBCs stick to the endothelium. Also known as **migration** or **extravasation.**

Emission (ē-MISH-un) Propulsion of sperm into the urethra due to peristaltic contractions of the ducts of the testes, epididymides, and ductus (vas) deferens as a result of sympathetic stimulation.

Emmetropia (em′-e-TRŌ-pē-a) Normal vision in which light rays are focused exactly on the retina.

Emphysema (em′-fi′-SĒ-ma) A lung disorder in which alveolar walls disintegrate, producing abnormally large air spaces and loss of elasticity in the lungs; typically caused by exposure to cigarette smoke.

Emulsification (ē-mul′-si-fi-KĀ-shun) The dispersion of large lipid globules into smaller, uniformly distributed particles in the presence of bile.

Enamel (e-NAM-el) The hard, white substance covering the crown of a tooth.

Endocardium (en-dō-KAR-dē-um) The layer of the heart wall, composed of endothelium and smooth muscle, that lines the inside of the heart and covers the valves and tendons that hold the valves open.

Endochondral ossification (en′-dō-KON-dral os′-i-fi-KĀ-shun) The replacement of cartilage by bone. Also called **intracartilaginous ossification** (in′-tra-kar′-ti-LAJ-i-nus).

Endocrine gland (EN-dō-krin) A gland that secretes hormones into interstitial fluid and then the blood; a ductless gland.

Endocrinology (en′-dō-kri-NOL-ō-jē) The science concerned with the structure and functions of endocrine glands and the diagnosis and treatment of disorders of the endocrine system.

Endocytosis (en′-dō-sī-TŌ-sis) The uptake into a cell of large molecules and particles in which a segment of plasma membrane surrounds the substance, encloses it, and brings it in; includes phagocytosis, bulk phase endocytosis, and receptor-mediated endocytosis.

Endoderm (EN-dō-derm) A primary germ layer of the developing embryo; gives rise to the gastrointestinal tract, urinary bladder, urethra, and respiratory tract.

Endodontics (en′-dō-DON-tiks) The branch of dentistry concerned with the prevention, diagnosis, and treatment of diseases that affect the pulp, root, periodontal ligament, and alveolar bone.

Endogenous (en-DOJ-e-nus) Growing from or beginning within the organism.

Endolymph (EN-dō-limf′) The fluid within the membranous labyrinth of the internal ear.

Endometrium (en′-dō-MĒ-trē-um) The mucous membrane lining the uterus.

Endoplasmic reticulum (ER) (en′-do-PLAZ-mik re-TIK-ū-lum) A network of channels running through the cytoplasm of a cell that serves in intracellular transportation, support, storage, synthesis, and packaging of molecules. Portions of ER where ribosomes are attached to the outer surface are called **rough ER;** portions that have no ribosomes are called **smooth ER.**

Endorphin (en-DOR-fin) A neuropeptide in the central nervous system that acts as a painkiller.

Endosteum (en-DOS-tē-um) The membrane that lines the medullary (marrow) cavity of bones, consisting of osteogenic cells and scattered osteoclasts.

Endothelium (en′-dō-THĒ-lē-um) The layer of simple squamous epithelium that lines the cavities of the heart, blood vessels, and lymphatic vessels.

Energy The capacity to do work.

Enkephalin (en-KEF-a-lin) A peptide found in the central nervous system that acts as a painkiller.

Enteric nervous system (EN-ter-ik) The part of the nervous system that is embedded in the submucosa and muscularis of the gastrointestinal (GI) tract; governs motility and secretions of the GI tract.

Enzyme (EN-zīm) A substance that accelerates chemical reactions, usually a protein.

Eosinophil (ē′-ō-SIN-ō-fil) A type of white blood cell characterized by granules that stain red or pink with acid dyes.

Ependymal cells (e-PEN-de-mal) Neuroglial cells that cover choroid plexuses and produce cerebrospinal fluid (CSF); they also line the ventricles of the brain and probably assist in the circulation of CSF.

Epicardium (ep′-i-KAR-dē-um) The thin outer layer of the heart wall, composed of serous tissue and mesothelium. Also called the **visceral pericardium.**

Epidemiology (ep′-i-dē-mē-OL-ō-jē) Study of the occurrence and distribution of diseases and disorders in human populations.

Epidermis (ep-i-DERM-is) The superficial, thinner layer of skin, composed of keratinized stratified squamous epithelium.

Epididymis (ep′-i-DID-i-mis) A comma-shaped organ that lies along the posterior border of the testis and contains the ductus epididymis, in which sperm undergo maturation. *Plural is* **epididymides** (ep′-i-DID-i-mi-dēz).

Epidural space (ep′-i-DOO-ral) A space between the spinal dura mater and the vertebral canal, containing areolar connective tissue and a plexus of veins.

Epiglottis (ep′-i-GLOT-is) A large, leaf-shaped piece of cartilage lying on top of the larynx, attached to the thyroid cartilage and its unattached portion is free to move up and down to cover the glottis (vocal folds and rima glottidis) during swallowing.

Epinephrine (ep-ē-NEF-rin) Hormone secreted by the adrenal medulla that produces actions similar to those that result from sympathetic stimulation. Also called **adrenaline** (a-DREN-a-lin).

Epineurium (ep′-i-NOO-rē-um) The superficial connective tissue covering around an entire nerve.

Epiphyseal line (ep′-i-FIZ-ē-al) The remnant of the epiphyseal plate in the metaphysis of a long bone.

Epiphyseal plate (ep′-i-FIZ-ē-al) The hyaline cartilage plate in the metaphysis of a long bone; site of lengthwise growth of long bones.

Epiphysis (ē-PIF-i-sis) The end of a long bone, usually larger in diameter than the shaft (diaphysis).

Episiotomy (e-piz′-ē-OT-ō-mē) A cut made with surgical scissors to avoid tearing of the perineum at the end of the second stage of labor.

Epistaxis (ep′-i-STAK-sis) Loss of blood from the nose due to trauma, infection, allergy, neoplasm, and bleeding disorders. Also called **nosebleed.**

Epithelial tissue (ep′-i-THĒ-lē-al) The tissue that forms innermost and outermost surfaces of body structures and forms the secreting portion of glands.

Erectile dysfunction Failure to maintain an erection long enough for sexual intercourse. Also known as **impotence** (IM-pō-tens).

Erection (ē-REK-shun) The enlarged and stiff state of the penis or clitoris resulting from the engorgement of the spongy erectile tissue with blood.

Erythema (er′-i-THĒ-ma) Skin redness usually caused by dilation of the capillaries.

Erythrocyte (e-RITH-rō-sīt) A mature red blood cell.

Erythropoiesis (e-rith′-rō-poy-Ē-sis) The process by which red blood cells are formed.

Erythropoietin (e-rith′-rō-POY-e-tin) A hormone released by the kidneys that stimulates red blood cell production.

Esophagus (e-SOF-a-gus) The hollow muscular tube that connects the pharynx and the stomach.

Essential amino acids Those 10 amino acids that cannot be synthesized by the human body at an adequate rate to meet its needs and therefore must be obtained from the diet.

Estrogens (ES-tro-jenz) Feminizing sex hormones produced by the ovaries; govern development of oocytes, maintenance of female reproductive structures, and appearance of secondary sex characteristics; also affect fluid and electrolyte balance, and protein anabolism. Examples are β-estradiol, estrone, and estriol.

Etiology (ē′-tē-OL-ō-jē) The study of the causes of disease, including theories of the origin and organisms (if any) involved.

Eupnea (ŪP-nē-a) Normal quiet breathing.

Eversion (ē-VER-zhun) The movement of the sole laterally at the ankle joint or of an atrioventricular valve into an atrium during ventricular contraction.

Excitability (ek-sīt′-a-BIL-i-tē) The ability of muscle fibers to receive and respond to stimuli; the ability of neurons to respond to stimuli and generate nerve impulses.

Excrement (EKS-kre-ment) Material eliminated from the body as waste, especially fecal matter.

Excretion (eks-KRĒ-shun) The process of eliminating waste products from the body; also the products excreted.

Exhalation (eks-ha-LĀ-shun) Breathing out; expelling air from the lungs into the atmosphere. Also called **expiration.**

Exocrine gland (EK-sō-krin) A gland that secretes its products into ducts that carry the secretions into body cavities, into the lumen of an organ, or to the outer surface of the body.

Exocytosis (ex′-ō-sī-TŌ-sis) A process in which membrane-enclosed secretory vesicles form inside the cell, fuse with the plasma membrane, and release their contents into the interstitial fluid; achieves secretion of materials from a cell.

Exogenous (ex-SOJ-e-nus) Originating outside an organ or part.

Expiratory (eks-PĪ-ra-tō-rē) **reserve volume** The volume of air in excess of tidal volume that can be exhaled forcibly; about 1200 mL in males and 700 mL in females.

Extensibility (ek-sten′-si-BIL-i-tē) The ability of muscle tissue to stretch when it is pulled.

Extension (ek-STEN-shun) An increase in the angle between two bones; restoring a body part to its anatomical position after flexion.

External Located on or near the surface.

External auditory canal (AW-di-tōr-ē) or **meatus** (mē-Ā-tus) A curved tube in the temporal bone that leads to the middle ear.

External ear The outer ear, consisting of the pinna, external auditory canal, and tympanic membrane (eardrum).

External nares (NĀ-rez) The external nostrils, or the openings into the nasal cavity on the exterior of the body.

External respiration The exchange of respiratory gases between the lungs and blood. Also called **pulmonary respiration.**

Exteroceptor (eks′-ter-ō-SEP-tor) A sensory receptor adapted for the reception of stimuli from outside the body.

Extracellular fluid (ECF) Fluid outside body cells, such as interstitial fluid and plasma.

Extracellular matrix (MĀ-triks) The ground substance and fibers between cells in a connective tissue.

Extrinsic (ek-STRIN-sik) Of external origin.

Extrinsic pathway (of blood clotting) Sequence of reactions leading to blood clotting that is initiated by the release of tissue factor (TF), also known as thromboplastin, that leaks into the blood from damaged cells outside the blood vessels.

Exudate (EKS-oo-dāt) Escaping fluid or semifluid material that oozes from a space and that may contain serum, pus, and cellular debris.

Eyebrow The hairy ridge superior to the eye.

F

Face The anterior aspect of the head.

Facilitated diffusion (fa-SIL-i-tā-ted dif-Ū-zhun) Diffusion in which a substance not soluble by itself in lipids diffuses across a selectively permeable membrane with the help of a transporter protein.

Fascia (FASH-ē-a) A fibrous membrane covering, supporting, and separating muscles.

Fascicle (FAS-i-kul) A small bundle or cluster, especially of nerve or muscle fibers (cells). Also called a **fasciculus** (fa-SIK-ū-lus). *Plural is* **fasciculi** (fa-SIK-yū-lī).

Fasciculation (fa-sik′-ū-LĀ-shun) Abnormal, spontaneous twitch of all skeletal muscle fibers in one motor unit that is visible at the skin surface; not associated with movement of the affected muscle; present in progressive diseases of motor neurons, for example, poliomyelitis.

Fauces (FAW-sēz) The opening from the mouth into the pharynx.

Feces (FĒ-sēz) Material discharged from the rectum and made up of bacteria, excretions, and food residue. Also called **stool.**

Feedback system A sequence of events in which information about the status of a situation is continually reported (fed back) to a control center.

Female reproductive cycle General term for the ovarian and uterine cycles, the hormonal changes that accompany them, and cyclic changes in the breasts and cervix; includes changes in the endometrium of a nonpregnant female that prepares the lining of the uterus to receive a fertilized ovum.

Fertilization (fer′-ti-li-ZĀ-shun) Penetration of a secondary oocyte by a sperm cell, meiotic division of secondary oocyte to form an ovum, and subsequent union of the nuclei of the gametes.

Fetal circulation The cardiovascular system of the fetus, including the placenta and special blood vessels involved in the exchange of materials between fetus and mother.

Fetus (FĒ-tus) In humans, the developing organism *in utero* from the beginning of the ninth week to birth.

Fever An elevation in body temperature above the normal temperature of 37 °C (98.6 °F) due to a resetting of the hypothalamic thermostat.

Fibrillation (fi-bri-LĀ-shun) Abnormal, spontaneous twitch of a single skeletal muscle fiber (cell) that can be detected with electromyography but is not visible at the skin surface; not associated with movement of the affected muscle; present in certain disorders of motor neurons, for example, amyotrophic lateral sclerosis (ALS). With reference to cardiac muscle, *see* **Atrial fibrillation** and **Ventricular fibrillation.**

Fibrin (FĪ-brin) An insoluble protein that is essential to blood clotting; formed from fibrinogen by the action of thrombin.

Fibrinogen (fī-BRIN-ō-jen) A clotting factor in blood plasma that by the action of thrombin is converted to fibrin.

Fibrinolysis (fī-bri-NOL-i-sis) Dissolution of a blood clot by the action of a proteolytic enzyme, such as plasmin (fibrinolysin), that dissolves fibrin threads and inactivates fibrinogen and other blood-clotting factors.

Fibroblast (FĪ-brō-blast) A large, flat cell that secretes most of the extracellular matrix material of areolar and dense connective tissues.

Fibrous joint (FĪ-brus) A joint that allows little or no movement, such as a suture or a syndesmosis.

Fibrous tunic (TOO-nik) The superficial coat of the eyeball, made up of the posterior sclera and the anterior cornea.

Fight-or-flight response The effects produced upon stimulation of the sympathetic division of the autonomic nervous system.

Filtration (fil-TRĀ-shun) The flow of a liquid through a filter (or membrane that acts like a filter) due to a hydrostatic pressure; occurs in capillaries due to blood pressure.

Filtration membrane Site of blood filtration in nephrons of the kidneys, consisting of the endothelium and basement membrane of the glomerulus and the epithelium of the visceral layer of the glomerular (Bowman's) capsule.

Fissure (FISH-ur) A groove, fold, or slit that may be normal or abnormal.

Fixed macrophage (MAK-rō-fāj) Stationary phagocytic cell found in the liver, lungs, brain, spleen, lymph nodes, subcutaneous tissue, and red bone marrow. Also called a **histiocyte** (HIS-tē-ō-sīt).

Flaccid (FLAS-sid) Relaxed, flabby, or soft; lacking muscle tone.

Flagellum (fla-JEL-um) A hairlike, motile process on the extremity of a bacterium, protozoan, or sperm cell. *Plural is* **flagella** (fla-JEL-a).

Flatus (FLĀ-tus) Gas in the stomach or intestines, commonly used to denote expulsion of gas through the anus.

Flexion (FLEK-shun) Movement in which there is a decrease in the angle between two bones.

Flexor reflex A protective reflex in which flexor muscles are stimulated while extensor muscles are inhibited.

Follicle (FOL-i-kul) A small secretory sac or cavity; the group of cells that contains a developing oocyte in the ovaries.

Follicle-stimulating hormone (FSH) Hormone secreted by the anterior pituitary that initiates development of ova and stimulates the ovaries to secrete estrogens in females, and initiates sperm production in males.

Fontanel (fon′-ta-NEL) A space filled with mesenchyme where bone formation is not yet complete, especially between the cranial bones of an infant's skull.

Foot The terminal part of the lower limb, from the ankle to the toes.

Foramen (fō-RĀ-men) A passage or opening; a communication between two cavities of an organ, or a hole in a bone for passage of vessels or nerves. *Plural is* **foramina** (fō-RĀM-i-na).

Foramen ovale (fō-RĀ-men ō-VAL-ē) An opening in the fetal heart in the septum between the right and left atria. A hole in the greater wing of the sphenoid bone that transmits the mandibular branch of the trigeminal (V) nerve.

Forearm (FOR-arm) The part of the upper limb between the elbow and the wrist.

Fossa (FOS-a) A furrow or shallow depression.

Fourth ventricle (VEN-tri-kul) A cavity filled with cerebrospinal fluid within the brain lying between the cerebellum and the medulla oblongata and pons.

Fracture (FRAK-choor) Any break in a bone.

Frenulum (FREN-ū-lum) A small fold of mucous membrane that connects two parts and limits movement.

Frontal plane A plane at a right angle to a midsagittal plane that divides the body or organs into anterior and posterior portions. Also called a **coronal plane** (kō-RŌ-nal).

Functional residual capacity (re-ZID-ū-al) The sum of residual volume plus expiratory reserve volume; about 2400 mL in males and 1800 mL in females.

Fundus (FUN-dus) The part of a hollow organ farthest from the opening.

Fungiform papilla (FUN-ji-form pa-PIL-a) A mushroomlike elevation on the upper surface of the tongue appearing as a red dot; most contain taste buds.

G

Gallbladder A small pouch, located inferior to the liver, that stores bile and empties by means of the cystic duct.

Gallstone A solid mass, usually containing cholesterol, in the gallbladder or a bile-containing duct; formed anywhere between bile canaliculi in the liver and the hepatopancreatic ampulla (ampulla of Vater), where bile enters the duodenum. Also called a **biliary calculus.**

Gamete (GAM-ēt) A male or female reproductive cell; a sperm cell or secondary oocyte.

Ganglion (GANG-glē-on) Usually, a group of neuronal cell bodies lying outside the central nervous system (CNS). *Plural is* **ganglia** (GANG-glē-a).

Gastric glands (GAS-trik) Glands in the mucosa of the stomach composed of cells that empty their secretions into narrow channels called gastric pits. Types of cells are chief cells (secrete pepsinogen), parietal cells (secrete hydrochloric acid and intrinsic factor), surface mucous and mucous neck cells (secrete mucus), and G cells (secrete gastrin).

Gastroenterology (gas′-trō-en′-ter-OL-ō-jē) The medical specialty that deals with the structure, function, diagnosis, and treatment of diseases of the stomach and intestines.

Gastrointestinal (gas-trō-in-TES-ti-nal) **(GI) tract** A continuous tube extending from the mouth to the anus. Also called the **alimentary canal** (al′-i-MEN-tar-ē).

Gastrulation (gas′-troo-LĀ-shun) The migration of groups of cells from the epiblast that transform a bilaminar embryonic disc into a trilaminar embryonic disc that consists of the three primary germ layers; transformation of the blastula into the gastrula.

Gene (JĒN) Biological unit of heredity; a segment of DNA located in a definite position on a particular chromosome; a sequence of DNA that codes for a particular mRNA, rRNA, or tRNA.

Generator potential The graded depolarization that results in a change in the resting membrane potential in a receptor (specialized neuronal ending); may trigger a nerve action potential (nerve impulse) if depolarization reaches threshold.

Genetic engineering The manufacture and manipulationof genetic material.

Genetics The study of genes and heredity.

Genitalia (jen′-i-TĀ-lē-a) Reproductive organs.

Genome (JĒ-nōm) The complete set of genes of an organism.

Genotype (JĒ-nō-tīp) The genetic makeup of an individual; the combination of alleles present at one or more chromosomal locations, as distinguished from the appearance, or phenotype, that results from those alleles.

Geriatrics (jer′-ē-AT-riks) The branch of medicine devoted to the medical problems and care of elderly persons.

Gestation (jes-TĀ-shun) The period of development from fertilization to birth.

Gingivae (jin-JI-vē) Gums. They cover the alveolar processes of the mandible and maxilla and extend slightly into each socket.

Gland Specialized epithelial cell or cells that secrete substances; may be exocrine or endocrine.

Glans penis (GLANZ PĒ-nis) The slightly enlarged region at the distal end of the penis.

Glaucoma (glaw-KŌ-ma) An eye disorder in which there is increased intraocular pressure due to an excess of aqueous humor.

Gliding joint A synovial joint having articulating surfaces that are usually flat, permitting only side-to-side and back-and-forth movements, as between carpal bones, tarsal bones, and the scapula and clavicle. Also called an **arthrodial joint** (ar-THRŌ-dē-al).

Glomerular capsule (glō-MER-ū-lar) A double-walled globe at the proximal end of a nephron that encloses the glomerular capillaries. Also called **Bowman's capsule** (BŌ-manz).

Glomerular filtrate (glō-MER-ū-lar FIL-trāt) The fluid produced when blood is filtered by the filtration membrane in the glomeruli of the kidneys.

Glomerular filtration The first step in urine formation in which substances in blood pass through the filtration membrane and the filtrate enters the proximal convoluted tubule of a nephron.

Glomerular filtration rate (GFR) The total volume of fluid that enters all the glomerular (Bowman's) capsules of the kidneys in 1 min; about 100–125 mL/min.

Glomerulus (glō-MER-ū-lus) A rounded mass of nerves or blood vessels, especially the microscopic tuft of capillaries that is surrounded by the glomerular (Bowman's) capsule of each kidney tubule. *Plural* is **glomeruli.**

Glottis (GLOT-is) The vocal folds (true vocal cords) in the larynx plus the space between them (rima glottidis).

Glucagon (GLOO-ka-gon) A hormone produced by the alpha cells of the pancreatic islets (islets of Langerhans) that increases blood glucose level.

Glucocorticoids (gloo-kō-KOR-ti-koyds) Hormones secreted by the cortex of the adrenal gland, especially cortisol, that influence glucose metabolism.

Gluconeogenesis (gloo′-kō-nē-ō-JEN-e-sis) The synthesis of glucose from certain amino acids or lactic acid.

Glucose (GLOO-kōs) A six-carbon sugar, $C_6H_{12}O_6$, that is a major energy source for the production of ATP by body cells.

Glucosuria (gloo′-kō-SOO-rē-a) The presence of glucose in the urine; may be temporary or pathological.

Glycogen (GLĪ-kō-jen) A highly branched polymer of glucose containing thousands of subunits; functions as a compact store of glucose molecules in liver and muscle fibers (cells).

Glycogenesis (glī′-kō-JEN-e-sis) The chemical reactions by which many molecules of glucose are used to synthesize glycogen.

Glycogenolysis (glī-kō-je-NOL-i-sis) The breakdown of glycogen into glucose.

Glycolysis (glī-KOL-i-sis) Series of chemical reactions in the cytosol of a cell in which a molecule of glucose is split into two molecules of pyruvic acid with the net production of two ATPs.

Goblet cell A goblet-shaped unicellular gland that secretes mucus; present in epithelium of the airways and intestines.

Goiter (GOY-ter) An enlarged thyroid gland.

Golgi complex (GOL-jē) An organelle in the cytoplasm of cells consisting of three to twenty flattened sacs (cisternae), stacked on one another, with expanded areas at their ends; functions in processing, sorting, packaging, and delivering proteins and lipids to the plasma membrane, lysosomes, and secretory vesicles.

Gomphosis (gom-FŌ-sis) A fibrous joint in which a cone-shaped peg fits into a socket.

Gonad (GŌ-nad) A gland that produces gametes and hormones; the ovary in the female and the testis in the male.

Gonadotropic hormone Anterior pituitary hormone that affects the gonads.

Gray matter Area in the central nervous system and ganglia containing neuronal cell bodies, dendrites, unmyelinated axons, axon terminals, and neuroglia; Nissl bodies impart a gray color and there is little or no myelin in gray matter.

Greater omentum (ō-MEN-tum) A large fold in the serosa of the stomach that hangs down like an apron anterior to the intestines.

Greater vestibular glands (ves-TIB-ū-lar) A pair of glands on either side of the vaginal orifice that open by a duct into the space between the hymen and the labia minora. Also called **Bartholin's glands** (BAR-to-linz).

Groin (GROYN) The depression between the thigh and the trunk; the inguinal region.

Gross anatomy The branch of anatomy that deals with structures that can be studied without using a microscope. Also called **macroscopic anatomy.**

Growth An increase in size due to an increase in (1) the number of cells, (2) the size of existing cells as internal components increase in size, or (3) the size of intercellular substances.

Gustatory (GUS-ta-tō′-rē) Pertaining to taste.

Gynecology (gī′-ne-KOL-ō-jē) The branch of medicine dealing with the study and treatment of disorders of the female reproductive system.

Gynecomastia (gīn′-e-kō-MAS-tē-a) Excessive growth (benign) of the male mammary glands due to secretion of estrogens by an adrenal gland tumor (feminizing adenoma).

Gyrus (JĪ-rus) One of the folds of the cerebral cortex of the brain. *Plural* is **gyri** (JĪ-rī). Also called a **convolution.**

H

Hair A threadlike structure composed of dead, keratinized cells produced by hair follicles that develops in the dermis. Also called a **pilus** (PĪ-lus).

Hair follicle (FOL-li-kul) Structure, composed of epithelium and surrounding the root of a hair, from which hair develops.

Hair root plexus (PLEK-sus) A network of dendrites arranged around the root of a hair as free or naked nerve endings that are stimulated when a hair shaft is moved.

Hand The terminal portion of an upper limb, including the carpus, metacarpus, and phalanges.

Haploid (HAP-loyd) Having half the number of chromosomes characteristically found in the somatic cells of an organism; characteristic of mature gametes. Symbolized *n*.

Hard palate (PAL-at) The anterior portion of the roof of the mouth, formed by the maxillae and palatine bones and lined by mucous membrane.

Head The superior part of a human, cephalic to the neck. The superior or proximal part of a structure.

Heart A hollow muscular organ lying slightly to the left of the midline of the chest that pumps the blood through the cardiovascular system.

Heart block An arrhythmia (dysrhythmia) of the heart in which the atria and ventricles contract independently because of a blocking of electrical impulses through the heart at some point in the conduction system.

Heart murmur (MER-mer) An abnormal sound that consists of a flow noise that is heard before, between, or after the normal heart sounds, or that may mask normal heart sounds.

Heat exhaustion Condition characterized by cool, clammy skin, profuse perspiration, and fluid and electrolyte (especially sodium and chloride) loss that results in muscle cramps, dizziness, vomiting, and fainting. Also called **heat prostration.**

Heat stroke Condition produced when the body cannot easily lose heat and characterized by reduced perspiration and elevated body temperature. Also called **sunstroke.**

Hematocrit (Hct) (hē-MAT-ō-krit) The percentage of blood made up of red blood cells. Usually measured by centrifuging a blood sample in a graduated tube and then reading the volume of red blood cells and dividing it by the total volume of blood in the sample.

Hematology (hē′-ma-TOL-ō-jē) The study of blood.

Hemiplegia (hem-i-PLĒ-jē-a) Paralysis of the upper limb, trunk, and lower limb on one side of the body.

Hemoglobin (Hb) (hē′-mō-GLŌ-bin) A substance in red blood cells consisting of the protein globin and the iron-containing red pigment heme that transports most of the oxygen and some carbon dioxide in blood.

Hemolysis (hē-MOL-i-sis) The escape of hemoglobin from the interior of a red blood cell into the surrounding medium; results from disruption of the cell membrane by toxins or drugs, freezing or thawing, or hypotonic solutions.

Hemolytic disease of the newborn A hemolytic anemia of a newborn child that results from the destruction of the infant's erythrocytes (red blood cells) by antibodies produced by the mother; usually the antibodies are due to an Rh blood type incompatibility. Also called **erythroblastosis fetalis** (e-rith′-rō-blas-TŌ-sis fe-TAL-is).

Hemophilia (hē′-mō-FIL-ē-a) A hereditary blood disorder where there is a deficient production of certain factors involved in blood clotting, resulting in excessive bleeding into joints, deep tissues, and elsewhere.

Hemopoiesis (hē-mō-poy-Ē-sis) Blood cell production, which occurs in red bone marrow after birth. Also called **hematopoiesis** (hem′-a-tō-poy-Ē-sis).

Hemorrhage (HEM-or-rij) Bleeding; the escape of blood from blood vessels, especially when the loss is profuse.

Hemorrhoids (HEM-ō-royds) Dilated or varicosed blood vessels (usually veins) in the anal region. Also called **piles.**

Hemostasis (hē-MŌ-stā-sis) The stoppage of bleeding.

Heparin (HEP-a-rin) An anticoagulant given to slow the conversion of prothrombin to thrombin, thus reducing the risk of blood clot formation; found in basophils, mast cells, and various other tissues, especially the liver and lungs.

Hepatic (he-PAT-ik) Refers to the liver.

Hepatic duct A duct that receives bile from the bile capillaries. Small hepatic ducts merge to form the larger right and left hepatic ducts that unite to leave the liver as the common hepatic duct.

Hepatic portal circulation The flow of blood from the gastrointestinal organs to the liver before returning to the heart.

Hepatocyte (he-PAT-ō-cyte) A liver cell.

Hernia (HER-nē-a) The protrusion or projection of an organ or part of an organ through a membrane or cavity wall, usually the abdominal cavity.

Herniated disc (HER-nē-ā′-ted) A rupture of an intervertebral disc so that the nucleus pulposus protrudes into the vertebral cavity. Also called a **slipped disc.**

Heterozygous (he-ter-ō-ZĪ-gus) Possessing different alleles on homologous chromosomes for a particular hereditary trait.

Hiatus (hī-Ā-tus) An opening; a foramen.

Hinge joint A synovial joint in which a convex surface of one bone fits into a concave surface of another bone, such as the elbow, knee, ankle, and interphalangeal joints. Also called a **ginglymus joint** (JIN-gli-mus).

Hirsutism (HER-soot-izm) An excessive growth of hair in females and children, with a distribution similar to that in adult males, due to the conversion of vellus hairs into large terminal hairs in response to higher-than-normal levels of androgens.

Histamine (HISS-ta-mēn) Substance found in many cells, especially mast cells, basophils, and platelets, released when the cells are injured; results in vasodilation, increased permeability of blood vessels, and constriction of bronchioles.

Histology (hiss-TOL-ō-jē) Microscopic study of the structure of tissues.

Homeostasis (hō′-mē-ō-STĀ-sis) The condition in which the body's internal environment remains relatively constant, within physiological limits.

Hyperthermia (hī′-per-THERM-ē-a) An elevated body temperature.

Hypertonia (hī′-per-TŌ-nē-a) Increased muscle tone that is expressed as spasticity or rigidity.

Hypertonic (hī′-per-TON-ik) Solution that causes cells to shrink due to loss of water by osmosis.

Hypertrophy (hī-PER-trō-fē) An excessive enlargement or overgrowth of tissue without cell division.

Hyperventilation (hī′-per-ven-ti-LĀ-shun) A rate of respiration higher than that required to maintain a normal partial pressure of carbon dioxide in the blood.

Hypoglycemia (hī′-pō-glī-SĒ-mē-a) An abnormally low concentration of glucose in the blood; can result from excess insulin (injected or secreted).

Hypokalemia (hī′-pō-ka-LĒ-mē-a) Deficiency of potassium ions in the blood.

Hyponatremia (hī′-pō-na-TRĒ-mē-a) Deficiency of sodium ions in the blood.

Hypophyseal fossa (hī′-pō-FIZ-ē-al FOS-a) A depression on the superior surface of the sphenoid bone that houses the pituitary gland.

Hypophysis (hī-POF-i-sis) Pituitary gland.

Hyposecretion (hī′-pō-se-KRĒ-shun) Underactivity of glands resulting in diminished secretion.

Hypothalamus (hī′-pō-THAL-a-mus) A portion of the diencephalon, lying beneath the thalamus and forming the floor and part of the wall of the third ventricle.

Hypothermia (hī′-pō-THER-mē-a) Lowering of body temperature below 35 °C (95 °F); in surgical procedures, it refers to deliberate cooling of the body to slow down metabolism and reduce oxygen needs of tissues.

Hypotonia (hī′-pō-TŌ-nē-a) Decreased or lost muscle tone in which muscles appear flaccid.

Hypotonic (hī′-pō-TON-ik) Solution that causes cells to swell and perhaps rupture due to gain of water by osmosis.

Hypoventilation (hī-pō-ven-ti-LĀ-shun) A rate of respiration lower than that required to maintain a normal partial pressure of carbon dioxide in plasma.

Hypovolemic shock (hī-pō-vō-LĒ-mik) A type of shock characterized by decreased blood volume; may be caused by acute hemorrhage or excessive loss of other body fluids, for example, by vomiting, diarrhea, or excessive sweating.

Hypoxia (hī-POKS-ē-a) Lack of adequate oxygen at the tissue level.

Hysterectomy (hiss-te-REK-tō-mē) The surgical removal of the uterus.

Homologous chromosomes Two chromosomes that belong to a pair.

Homozygous (hō-mō-ZĪ-gus) Possessing the same alleles on homologous chromosomes for a particular hereditary trait.

Hormone (HOR-mōn) A secretion of endocrine cells that alters the physiological activity of target cells of the body.

Horn An area of gray matter (anterior, lateral, or posterior) in the spinal cord.

Human chorionic gonadotropin (hCG) (kō-rē-ON-ik gō-nad-ō-TRŌ-pin) A hormone produced by the developing placenta that maintains the corpus luteum.

Human growth hormone (hGH) Hormone secreted by the anterior pituitary that stimulates growth of body tissues, especially skeletal and muscular tissues. Also known as **somatotropin** and **somatotropic hormone (STH).**

Hyaluronic acid (hī′a-loo-RON-ik) A viscous, amorphous extracellular material that binds cells together, lubricates joints, and maintains the shape of the eyeballs.

Hymen (HĪ-men) A thin fold of vascularized mucous membrane at the vaginal orifice.

Hypercalcemia (hī′-per-kal-SĒ-mē-a) An excess of calcium in the blood.

Hypercapnia (hī′-per-KAP-nē-a) An abnormal increase in the amount of carbon dioxide in the blood.

Hyperextension (hī′-per-ek-STEN-shun) Continuation of extension beyond the anatomical position, as in bending the head backward.

Hyperglycemia (hī′-per-glī-SĒ-mē-a) An elevated blood glucose level.

Hyperkalemia (hī′-per-kā-LĒ-mē-a) An excess of potassium ions in the blood.

Hypermetropia (hī′-per-mē-TRŌ-pē-a) A condition in which visual images are focused behind the retina, with resulting defective vision of near objects; farsightedness.

Hyperplasia (hī′-per-PLĀ-zē-a) An abnormal increase in the number of normal cells in a tissue or organ, increasing its size.

Hyperpolarization (hī′-per-PŌL-a-ri-zā′-shun) Increase in the internal negativity across a cell membrane, thus increasing the voltage and moving it farther away from the threshold value.

Hypersecretion (hī′-per-se-KRĒ-shun) Overactivityof glands resulting in excessive secretion.

Hypersensitivity (hī′-per-sen-si-TI-vi-tē) Overreaction to an allergen that results in pathological changes in tissues. Also called **allergy**.

Hypertension (hī′-per-TEN-shun) High blood pressure.

I

Ileocecal sphincter (il-ē-ō-SĒ-kal) A fold of mucous membrane that guards the opening from the ileum into the large intestine. Also called the **ileocecal valve**.

Ileum (IL-ē-um) The terminal part of the small intestine.

Immunity (im-Ū-ni-tē) The state of being resistant to injury, particularly by poisons, foreign proteins, and invading pathogens.

Immunoglobulin (Ig) (im-ū-nō-GLOB-ū-lin) An antibody synthesized by plasma cells derived from B lymphocytes in response to the introduction of an antigen. Immunoglobulins are divided into five kinds (IgG, IgM, IgA, IgD, IgE).

Immunology (im′-ū-NOL-ō-jē) The study of the responses of the body when challenged by antigens.

Implantation (im-plan-TĀ-shun) The insertion of a tissue or a part into the body. The attachment of the blastocyst to the stratum basalis of the endometrium about 6 days after fertilization.

Incontinence (in-KON-ti-nens) Inability to retain urine, semen, or feces through loss of sphincter control.

Indirect motor pathways Motor tracts that convey information from the brain down the spinal cord for automatic movements, coordination of body movements with visual stimuli, skeletal muscle tone and posture, and balance. Also known as **extrapyramidal pathways.**

Infarction (in-FARK-shun) A localized area of necrotic tissue, produced by inadequate oxygenation of the tissue.

Infection (in-FEK-shun) Invasion and multiplication of microorganisms in body tissues, which may be inapparent or characterized by cellular injury.

Inferior (in-FĒR-ē-or) Away from the head or toward the lower part of a structure. Also called **caudad** (KAW-dad).

Inferior vena cava (IVC) (VĒ-na CĀ-va) Large vein that collects deoxygenated blood from parts of the body inferior to the heart and returns it to the right atrium.

Infertility Inability to conceive or to cause conception. Also called **sterility.**

Inflammation (in′-fla-MĀ-shun) Localized, protective response to tissue injury designed to destroy, dilute, or wall off the infecting agent or injured tissue; characterized by redness, pain, heat, swelling, and sometimes loss of function.

Inflation reflex Reflex that prevents overinflation of the lungs. Also called the **Hering–Breuer reflex.**

Ingestion (in-JES-chun) The taking in of food, liquids, or drugs, by mouth.

Inguinal (IN-gwi-nal) Pertaining to the groin.

Inguinal canal An oblique passageway in the anterior abdominal wall just superior and parallel to the medial half of the inguinal ligament that transmits the spermatic cord and ilioinguinal nerve in the male and round ligament of the uterus and ilioinguinal nerve in the female.

Inhalation (in-ha-LĀ-shun) The act of drawing air into the lungs. Also termed **inspiration.**

Inheritance The acquisition of body traits by transmission of genetic information from parents to offspring.

Inhibin A hormone secreted by the gonads that inhibits release of follicle-stimulating hormone (FSH) by the anterior pituitary.

Inhibiting hormone Hormone secreted by the hypothalamus that can suppress secretion of hormones by the anterior pituitary.

Inner cell mass A region of cells of a blastocyst that differentiates into the three primary germ layers—ectoderm, mesoderm, and endoderm—from which all tissues and organs develop.

Inorganic compound (in′-or-GAN-ik) Compound that usually lacks carbon, usually is small, and often contains ionic bonds. Examples include water and many acids, bases, and salts.

Insertion (in-SER-shun) The attachment of a muscle tendon to a movable bone or the end opposite the origin.

Inspiratory capacity (in-SPĪ-ra-tor-ē) Total inspiratory capacity of the lungs; the total of tidal volume plus inspiratory reserve volume; averages 3600 mL in males.

Inspiratory reserve volume (in-SPĪ-ra-tor-ē) Additional inspired air over and above tidal volume; averages 3100 mL.

Insulin (IN-soo-lin) A hormone produced by the beta cells of a pancreatic islet (islet of Langerhans) that decreases the blood glucose level.

Insulinlike growth factor (IGF) Small protein, produced by the liver and other tissues in response to stimulation by human growth hormone (hGH), that mediates most of the effects of human growth hormone. Previously called **somatomedin** (sō′-ma-tō-MĒ-din).

Integumentary (in-teg′-ū-MEN-tar-e) Relating to the skin.

Intercalated disc (in-TER-ka-lāt-ed) An irregular transverse thickening of sarcolemma that contains desmosomes, which hold cardiac muscle fibers (cells) together, and gap junctions, which aid in conduction of muscle action potentials from one fiber to the next.

Intercostal nerve (in′-ter-KOS-tal) A nerve supplying a muscle located between the ribs.

Interferons (IFNs) (in′-ter-FĒR-ons) Antiviral proteins produced by virus-infected host cells; induce uninfected host cells to synthesize proteins that inhibit viral replication and enhance phagocytic activity of macrophages; types include alpha interferon, beta interferon, and gamma interferon.

Internal Away from the surface of the body.

Internal capsule A large tract of projection fibers lateral to the thalamus that is the major connection between the cerebral cortex and the brain stem and spinal cord; contains axons of sensory neurons carrying auditory, visual, and somatic sensory signals to the cerebral cortex plus axons of motor neurons descending from the cerebral cortex to the thalamus, subthalamus, brain stem, and spinal cord.

Internal ear The inner ear or labyrinth, lying inside the temporal bone, containing the organs of hearing and balance.

Internal nares (NĀ-rez) The two openings posterior to the nasal cavities opening into the nasopharynx. Also called the **choanae** (kō-Ā-nē).

Internal respiration The exchange of respiratory gases between blood and body cells. Also called **tissue respiration.**

Interneurons (in′-ter-NOO-ronz) Neurons whose axons extend only for a short distance and contact nearby neurons in the brain, spinal cord, or a ganglion; they comprise the vast majority of neurons in the body.

Interoceptor (in′-ter-ō-SEP-tor) Sensory receptor located in blood vessels and viscera that provides information about the body's internal environment.

Interphase (IN-ter-fāz) The period of the cell cycle between cell divisions, consisting of the G_1-(gap or growth) phase, when the cell is engaged in growth, metabolism, and production of substances required for division; S-(synthesis) phase, during which chromosomes are replicated; and G_2-phase.

Interstitial fluid (in′-ter-STISH-al) The portion of extracellular fluid that fills the microscopic spaces between the cells of tissues; the internal environment of the body. Also called **intercellular** or **tissue fluid.**

Interstitial growth Growth from within, as in the growth of cartilage. Also called **endogenous growth** (en-DOJ-e-nus).

Intervertebral disc (in′-ter-VER-te-bral) A pad of fibrocartilage located between the bodies of two vertebrae.

Intestinal gland A gland that opens onto the surface of the intestinal mucosa and secretes digestive enzymes. Also called a **crypt of Lieberkühn** (LĒ-ber-kun).

Intracellular fluid (ICF) (in′-tra-SEL-ū-lar) Fluid located within cells. Also called **cytosol** (SĪ-tō-sol).

Intramembranous ossification (in′-tra-MEM-bra-nus os′-i′-fi-KĀ-shun) The method of bone formation in which the bone is formed directly within mesenchyme arranged in sheetlike layers that resemble membranes.

Intraocular pressure (IOP) (in′-tra-OK-ū-lar) Pressure in the eyeball, produced mainly by aqueous humor.

Intrapleural pressure Air pressure between the two pleurae of the lungs, usually subatmospheric. Also called **intrathoracic pressure.**

Intrinsic (in-TRIN-sik) Of internal origin.

Intrinsic pathway (of blood clotting) Sequence of reactions leading to blood clotting that is initiated by damage to blood vessel endothelium or platelets; activators of this pathway are contained within blood itself or are in direct contact with blood.

Intrinsic factor (IF) A glycoprotein, synthesized and secreted by the parietal cells of the gastric mucosa, that facilitates vitamin B_{12} absorption in the small intestine.

In utero (Ū-ter-ō) Within the uterus.

Invagination (in-vaj′-i-NĀ-shun) The pushing of the wall of a cavity into the cavity itself.

Inversion (in-VER-zhun) The movement of the sole medially at the ankle joint.

Ion (Ī-on) Any charged particle or group of particles; usually formed when a substance, such as a salt, dissolves and dissociates.

Ionization (ī′-on-i-ZĀ-shun) Separation of inorganic acids, bases, and salts into ions when dissolved in water. Also called **dissociation.**

Iris The colored portion of the vascular tunic of the eyeball seen through the cornea that contains circular and radial smooth muscle; the hole in the center of the iris is the pupil.

Irritable bowel syndrome (IBS) Disease of the entire gastrointestinal tract in which a person reacts to stress by developing symptoms (such as cramping and abdominal pain) associated with alternating patterns of diarrhea and constipation. Excessive amounts of mucus may appear in feces, and other symptoms include flatulence, nausea, and loss of appetite. Also known as **irritable colon** or **spastic colitis.**

Ischemia (is-KĒ-mē-a) A lack of sufficient blood to a body part due to obstruction or constriction of a blood vessel.

Isometric contraction A muscle contraction in which tension on the muscle increases, but there is only minimal muscle shortening so that no visible movement is produced.

Isotonic (i′-sō-TON-ik) Having equal tension or tone. A solution having the same concentration of impermeable solutes as cytosol.

Isotonic contraction Contraction in which the tension remains the same; occurs when a constant load is moved through the range of motions possible at a joint.

Isotopes (Ī-sō-tōps′) Chemical elements that have the same number of protons but different numbers of neutrons. Radioactive isotopes change into other elements with the emission of alpha or beta particles or gamma rays.

Isthmus (IS-mus) A narrow strip of tissue or narrow passage connecting two larger parts.

J

Jaundice (JAWN-dis) A condition characterized by yellowness of the skin, the white of the eyes, mucous membranes, and body fluids because of a buildup of bilirubin.

Jejunum (je-JOO-num) The middle part of the small intestine.

Joint kinesthetic receptor (kin′-es-THET-ik) A proprioceptive receptor located in a joint, stimulated by joint movement.

K

Keratin (KER-a-tin) An insoluble protein found in the hair, nails, and other keratinized tissues of the epidermis.

Keratinocyte (ke-RAT-in′-ō-sīt) The most numerous of the epidermal cells; produces keratin.

Ketone bodies (KĒ-tōn) Substances produced primarily during excessive triglyceride catabolism, such as acetone, acetoacetic acid, and β-hydroxybutyric acid.

Ketosis (kē-TŌ-sis) Abnormal condition marked by excessive production of ketone bodies.

Kidney (KID-nē) One of the paired reddish organs located in the lumbar region that regulates the composition, volume, and pressure of blood and produces urine.

Kidney stone A solid mass, usually consisting of calcium oxalate, uric acid, or calcium phosphate crystals, that may form in any portion of the urinary tract. Also called a **renal calculus.**

Kinesiology (ki-nē′-sē-OL-ō-jē) The study of the movement of body parts.

Kinesthesia (kin-es-THĒ-zē-a) The perception of the extent and direction of movement of body parts; this sense is possible due to nerve impulses generated by proprioceptors.

Korotkoff sounds (kō-ROT-kof) The various sounds that are heard while taking blood pressure.

Krebs cycle A series of biochemical reactions that occurs in the matrix of mitochondria in which electrons are transferred to coenzymes and carbon dioxide is formed. The electrons carried by the coenzymes then enter the electron transport chain, which generates a large quantity of ATP. Also called the **citric acid cycle** or **tricarboxylic acid (TCA) cycle.**

Kyphosis (kī-FŌ-sis) An exaggeration of the thoracic curve of the vertebral column, resulting in a "round-shouldered" appearance. Also called **hunchback.**

L

Labia majora (LĀ-bē-a ma-JŌ-ra) Two longitudinal folds of skin extending downward and backward from the mons pubis of the female.

Labia minora (min-OR-a) Two small folds of mucous membrane lying medial to the labia majora of the female.

Labial frenulum (LĀ-bē-al FREN-ū-lum) A medial fold of mucous membrane between the inner surface of the lip and the gums.

Labor The process of giving birth in which a fetus is expelled from the uterus through the vagina.

Lacrimal canal A duct, one on each eyelid, beginning at the punctum at the medial margin of an eyelid and conveying tears medially into the nasolacrimal sac.

Lacrimal gland Secretory cells, located at the superior anterolateral portion of each orbit, that secrete tears into excretory ducts that open onto the surface of the conjunctiva.

Lacrimal sac The superior expanded portion of the nasolacrimal duct that receives the tears from a lacrimal canal.

Lactation (lak-TĀ-shun) The secretion and ejection of milk by the mammary glands.

Lacteal (LAK-tē-al) One of many lymphatic vessels in villi of the intestines that absorb triglycerides and other lipids from digested food.

Lacuna (la-KOO-na) A small, hollow space, such as that found in bones in which the osteocytes lie. *Plural* is **lacunae** (la-KOO-nē).

Lamellae (la-MEL-ē) Concentric rings of hard, calcified extracellular matrix found in compact bone.

Lamellated corpuscle Oval-shaped pressure receptor located in the dermis or subcutaneous tissue and consisting of concentric layers of connective tissue wrapped around the dendrites of a sensory neuron. Also called a **pacinian corpuscle** (pa-SIN-ē-an).

Lamina propria (PRŌ-prē-a) The connective tissue layer of a mucosa.

Langerhans cell (LANG-er-hans) Epidermal dendritic cell that functions as an antigen-presenting cell (APC) during an immune response.

Lanugo (la-NOO-gō) Fine downy hairs that cover the fetus.

Large intestine The portion of the gastrointestinal tract extending from the ileum of the small intestine to the anus, divided structurally into the cecum, colon, rectum, and anal canal.

Laryngopharynx (la-rin′-gō-FAR-inks) The inferior portion of the pharynx, extending downward from the level of the hyoid bone that divides posteriorly into the esophagus and anteriorly into the larynx. Also called the **hypopharynx.**

Larynx (LAR-inks) The voice box, a short passageway that connects the pharynx with the trachea.

Lateral (LAT-er-al) Farther from the midline of the body or a structure.

Lateral ventricle (VEN-tri-kul) A cavity within a cerebral hemisphere that communicates with the lateral ventricle in the other cerebral hemisphere and with the third ventricle by way of the interventricular foramen.

Leg The part of the lower limb between the knee and the ankle.

Lens A transparent organ constructed of proteins (crystallins) lying posterior to the pupil and iris of the eyeball and anterior to the vitreous body.

Lesion (LĒ-zhun) Any localized, abnormal change in a body tissue.

Leukemia (loo-KĒ-mē-a) A malignant disease of the blood-forming tissues characterized by either uncontrolled production and accumulation of immature leukocytes in which many cells fail to reach maturity (acute) or an accumulation of mature leukocytes in the blood because they do not die at the end of their normal life span (chronic).

Leukocyte (LOO-kō-sīt) A white blood cell.

Leukocytosis (loo′-kō-sī-TŌ-sis) An increase in the number of white blood cells, above 10,000 per μL, characteristic of many infections and other disorders.

Leukopenia (loo-kō-PĒ-nē-a) A decrease in the number of white blood cells below 5000 cells per μL.

Leydig cell (LĪ-dig) A type of cell that secretes testosterone; located in the connective tissue between seminiferous tubules in a mature testis. Also known as **interstitial cell of Leydig.**

Libido (li-BĒ-dō) Sexual desire.

Ligament (LIG-a-ment) Dense regular connective tissue that attaches bone to bone.

Ligand (LĪ-gand) A chemical substance that binds to a specific receptor.

Limbic system A part of the forebrain, sometimes termed the visceral brain, concerned with various aspects of emotion and behavior; includes the limbic lobe, dentate gyrus, amygdala, septal nuclei, mammillary bodies, anterior thalamic nucleus, olfactory bulbs, and bundles of myelinated axons.

Lipase An enzyme that splits fatty acids from triglycerides and phospholipids.

Lipid (LIP-id) An organic compound composed of carbon, hydrogen, and oxygen that is usually insoluble in water, but soluble in alcohol, ether, and chloroform; examples include triglycerides (fats and oils), phospholipids, steroids, and eicosanoids.

Lipid bilayer Arrangement of phospholipid, glycolipid, and cholesterol molecules in two parallel layers in which the hydrophilic "heads" face outward and the hydrophobic "tails" face inward; found in cellular membranes.

Lipogenesis (li-pō-GEN-e-sis) The synthesis of triglycerides.

Lipolysis (lip-OL-i-sis) The splitting of fatty acids from a triglyceride or phospholipid.

Lipoprotein (lip′-ō-PRŌ-tēn) One of several types of particles containing lipids (cholesterol and triglycerides) and proteins that make it water soluble for transport in the blood; high levels of **low-density lipoproteins (LDLs)** are associated with increased risk of atherosclerosis, whereas high levels of **high-density lipoproteins (HDLs)** are associated with decreased risk of atherosclerosis.

Liver Large organ under the diaphragm that occupies most of the right hypochondriac region and part of the epigastric region. Functionally, it produces bile and synthesizes most plasma proteins; interconverts nutrients; detoxifies substances; stores glycogen, iron, and vitamins; carries on phagocytosis of worn-out blood cells and bacteria; and helps synthesize the active form of vitamin D.

Lordosis (lor-DŌ-sis) An exaggeration of the lumbar curve of the vertebral column. Also called **hollow back.**

Lower limb The appendage attached at the pelvic (hip) girdle, consisting of the thigh, knee, leg, ankle, foot, and toes. Also called **lower extremity.**

Lumbar (LUM-bar) Region of the back and side between the ribs and pelvis; loin.

Lumbar plexus (PLEK-sus) A network formed by the anterior (ventral) branches of spinal nerves L1 through L4.

Lumen (LOO-men) The space within an artery, vein, intestine, renal tubule, or other tubular structure.

Lungs Main organs of respiration that lie on either side of the heart in the thoracic cavity.

Lunula (LOO-noo-la) The moon-shaped white area at the base of a nail.

Luteinizing hormone (LH) (LOO-tē-in′-īz-ing) A hormone secreted by the anterior pituitary that stimulates ovulation, stimulates progesterone secretion by the corpus luteum, and readies the mammary glands for milk secretion in females; stimulates testosterone secretion by the testes in males.

Lymph (LIMF) Fluid confined in lymphatic vessels and flowing through the lymphatic system until it is returned to the blood.

Lymph node An oval or bean-shaped structure located along lymphatic vessels.

Lymphatic capillary (lim-FAT-ik) Closed-ended microscopic lymphatic vessel that begins in spaces between cells and converges with other lymphatic capillaries to form lymphatic vessels.

Lymphatic tissue A specialized form of reticular tissue that contains large numbers of lymphocytes.

Lymphatic vessel A large vessel that collects lymph from lymphatic capillaries and converges with other lymphatic vessels to form the thoracic and right lymphatic ducts.

Lymphocyte (LIM-fō-sīt) A type of white blood cell that helps carry out cell-mediated and antibody-mediated immune responses; found in blood and in lymphatic tissues.

Lysosome (LĪ-sō-sōm) An organelle in the cytoplasm of a cell, enclosed by a single membrane and containing powerful digestive enzymes.

Lysozyme (LĪ-sō-zīm) A bactericidal enzyme found in tears, saliva, perspiration, nasal secretions, and tissue fluids.

M

Macrophage (MAK-rō-fāj) Phagocytic cell derived from a monocyte; may be fixed or wandering.

Macula (MAK-ū-la) A discolored spot or a colored area. A small, thickened region on the wall of the utricle and saccule that contains receptors for static equilibrium.

Macula lutea (MAK-ū-la LOO-tē-a) The yellow spot in the center of the retina.

Major histocompatibility (MHC) antigens Surface proteins on white blood cells and other nucleated cells that are unique for each person (except for identical siblings); used to type tissues and help prevent rejection of transplanted tissues. Also known as **human leukocyte antigens (HLAs).**

Malignant (ma-LIG-nant) Referring to diseases that tend to become worse and cause death, especially the invasion and spreading of cancer.

Mammary gland (MAM-ar-ē) Modified sudoriferous (sweat) gland of females that produces milk for the nourishment of the young.

Marrow (MAR-ō) Soft, spongelike material in the cavities of bones. Red bone marrow produces blood cells; yellow bone marrow contains adipose tissue that stores triglycerides.

Mast cell A cell found in areolar connective tissue that releases histamine, a dilator of small blood vessels, during inflammation.

Mastication (mas′-ti-KĀ-shun) Chewing.

Matter Anything that occupies space and has mass.

Mature follicle A large, fluid-filled follicle containing a secondary oocyte and surrounding granulosa cells that secrete estrogens. Also called a **Graafian follicle** (GRAF-ē-an).

Meatus (mē-Ā-tus) A passage or opening, especially the external portion of a canal.

Mechanoreceptor (me-KAN-ō-rē-sep-tor) Sensory receptor that detects mechanical deformation of the receptor itself or adjacent cells; stimuli so detected include those related to touch, pressure, vibration, proprioception, hearing, equilibrium, and blood pressure.

Medial (MĒ-dē-al) Nearer the midline of the body or a structure.

Mediastinum (mē′-dē-as-TĪ-num) The broad, median partition between the pleurae of the lungs, that extends from the sternum to the vertebral column in the thoracic cavity.

Medulla (me-DUL-la) An inner layer of an organ, such as the medulla of the kidneys.

Medulla oblongata (me-DUL-la ob′-long-GA-ta) The most inferior part of the brain stem. Also termed the **medulla.**

Medullary cavity (MED-ū-lar′-ē) The space within the diaphysis of a bone that contains yellow bone marrow. Also called the **marrow cavity.**

Medullary rhythmicity area (rith-MIS-i-tē) The neurons of the respiratory center in the medulla oblongata that control the basic rhythm of respiration.

Meiosis (mē-Ō-sis) A type of cell division that occurs during production of gametes, involving two successive nuclear divisions that result in daughter cells with the haploid *(n)* number of chromosomes.

Melanin (MEL-a-nin) A dark black, brown, or yellow pigment found in some parts of the body such as the skin, hair, and pigmented layer of the retina.

Melanocyte (MEL-a-nō-sīt′) A pigmented cell, located between or beneath cells of the deepest layer of the epidermis, that synthesizes melanin.

Melanocyte-stimulating hormone (MSH) A hormone secreted by the anterior pituitary that stimulates the dispersion of melanin granules in melanocytes in amphibians; continued administration produces darkening of skin in humans.

Melatonin (mel-a-TŌN-in) A hormone secreted by the pineal gland that helps set the timing of the body's biological clock.

Membrane A thin, flexible sheet of tissue composed of an epithelial layer and an underlying connective tissue layer, as in an epithelial membrane, or of areolar connective tissue only, as in a synovial membrane.

Membranous labyrinth (mem-BRA-nus LAB-i-rinth) The part of the labyrinth of the internal ear that is located inside the bony labyrinth and separated from it by the perilymph; made up of the semicircular ducts, the saccule and utricle, and the cochlear duct.

Menarche (me-NAR-kē) The first menses (menstrual flow) and beginning of ovarian and uterine cycles.

Meninges (me-NIN-jēz) Three membranes covering the brain and spinal cord, called the dura mater, arachnoid mater, and pia mater. *Singular* is **meninx** (MEN-inks).

Menopause (MEN-ō-pawz) The termination of the menstrual cycles.

Menstruation (men′-stroo-Ā-shun) Periodic discharge of blood, tissue fluid, mucus, and epithelial cells that usually lasts for 5 days; caused by a sudden reduction in estrogens and progesterone. Also called the **menstrual phase** or **menses.**

Merkel cell (MER-kel) Type of cell in the epidermis of hairless skin that makes contact with a tactile (Merkel) disc, which functions in touch.

Mesenchyme (MEZ-en-kīm) An embryonic connective tissue from which all other connective tissues arise.

Mesentery (MEZ-en-ter′-ē) A fold of peritoneum attaching the small intestine to the posterior abdominal wall.

Mesoderm The middle primary germ layer that gives rise to connective tissues, blood and blood vessels, and muscles.

Metabolism (me-TAB-ō-lizm) All the biochemical reactions that occur within an organism, including the synthetic (anabolic) reactions and decomposition (catabolic) reactions.

Metacarpus (met′-a-KAR-pus) A collective term for the five bones that make up the palm.

Metaphase (MET-a-phāz) The second stage of mitosis, in which chromatid pairs line up on the metaphase plate of the cell.

Metaphysis (me-TAF-i-sis) Region of a long bone between the diaphysis and epiphysis that contains the epiphyseal plate in a growing bone.

Metastasis (me-TAS-ta-sis) The spread of cancer to surrounding tissues (local) or to other body sites (distant).

Metatarsus (met′-a-TAR-sus) A collective term for the five bones located in the foot between the tarsals and the phalanges.

Micelle (mī-SEL) A spherical aggregate of bile salts that dissolves fatty acids and monoglycerides so that they can be absorbed into small intestinal epithelial cells.

Microglia (mī-krō-GLĒ-a) Neuroglial cells that carry on phagocytosis.

Microvilli (mī′-krō-VIL-ē) Microscopic, fingerlike projections of the plasma membranes of cells that increase surface area for absorption, especially in the small intestine and proximal convoluted tubules of the kidneys.

Micturition (mik′-choo-RISH-un) The act of expelling urine from the urinary bladder. Also called **urination** (ū-ri-NĀ-shun).

Midbrain The part of the brain between the pons and the diencephalon. Also called the **mesencephalon** (mes′-en-SEF-a-lon).

Middle ear A small, epithelial-lined cavity hollowed out of the temporal bone, separated from the external ear by the eardrum and from the internal ear by a thin bony partition containing the oval and round windows; extending across the middle ear are the three auditory ossicles. Also called the **tympanic cavity** (tim-PAN-ik).

Midline An imaginary vertical line that divides the body into equal left and right sides.

Midsagittal plane A vertical plane through the midline of the body that divides the body or organs into *equal* right and left sides. Also called a **median plane.**

Milk ejection reflex Contraction of alveolar cells to force milk into ducts of mammary glands, stimulated by oxytocin (TO), which is released from the posterior pituitary in response to suckling action. Also called the **milk letdown reflex.**

Mineral Inorganic, homogeneous solid substance that may perform a function vital to life; examples include calcium and phosphorus.

Mineralocorticoids (min′-er-al-ō-KOR-ti-koyds) A group of hormones of the adrenal cortex that help regulate sodium and potassium balance.

Minute ventilation (MV) Total volume of air inhaled and exhaled per minute; about 6000 mL at rest.

Mitochondrion (mī′-tō-KON-drē-on) A double-membraned organelle that plays a central role in the production of ATP; known as the "powerhouse" of the cell.

Mitosis (mī-TŌ-sis) The orderly division of the nucleus of a cell that ensures that each new nucleus has the same number and kind of chromosomes as the original parent nucleus. The process includes the replication of chromosomes and the distribution of the two sets of chromosomes into two separate and equal nuclei.

Mitotic spindle Collective term for a football-shaped assembly of microtubules that is responsible for the movement of chromosomes during cell division.

Modality (mō-DAL-i-tē) Any of the specific sensory entities, such as vision, smell, taste, or touch.

Molecule (MOL-e-kūl) The chemical combination of two or more atoms covalently bonded together.

Monocyte (MON-ō-sīt′) The largest type of white blood cell, characterized by agranular cytoplasm.

Monounsaturated fat A fatty acid that contains one double covalent bond between its carbon atoms; it is not completely saturated with hydrogen atoms. Plentiful in triglycerides of olive and peanut oils.

Mons pubis (MONZ Pū-bis) The rounded, fatty prominence over the pubic symphysis, covered by coarse pubic hair.

Morula (MOR-ū-la) A solid sphere of cells produced by successive cleavages of a fertilized ovum about four days after fertilization.

Motor end plate Region of the sarcolemma of a muscle fiber (cell) that includes acetylcholine (ACh) receptors, which bind ACh released by synaptic end bulbs of somatic motor neurons.

Motor neurons (NOO-ronz) Neurons that conduct impulses from the brain toward the spinal cord or out of the brain and spinal cord into cranial or spinal nerves to effectors that may be either muscles or glands. Also called **efferent neurons.**

Motor unit A motor neuron together with the muscle fibers (cells) it stimulates.

Mucin (MŪ-sin) A protein found in mucus.

Mucous cell (MŪ-kus) A unicellular gland that secretes mucus. Two types are mucous neck cells and surface mucous cells in the stomach.

Mucous membrane A membrane that lines a body cavity that opens to the exterior. Also called the **mucosa** (mū-KŌ-sa).

Mucus The thick fluid secretion of goblet cells, mucous cells, mucous glands, and mucous membranes.

Muscle An organ composed of one of three types of muscular tissue (skeletal, cardiac, or smooth), specialized for contraction to produce voluntary or involuntary movement of parts of the body.

Muscle action potential A stimulating impulse that propagates along the sarcolemma and transverse tubules; in skeletal muscle, it is generated by acetylcholine, which increases the permeability of the sarcolemma to sodium ions (Na^+).

Muscle fatigue (fa-TĒG) Inability of a muscle to maintain its strength of contraction or tension; may be related to insufficient oxygen, depletion of glycogen, and/or lactic acid buildup.

Muscular tissue A tissue specialized to produce motion in response to muscle action potentials by its qualities of contractility, extensibility, elasticity, and excitability; types include skeletal, cardiac, and smooth.

Muscle tone A sustained, partial contraction of portions of a skeletal or smooth muscle in response to activation of stretch receptors or a baseline level of action potentials in the innervating motor neurons.

Muscular dystrophies (DIS-trō-fēz′) Inherited muscle-destroying diseases, characterized by degeneration of muscle fibers (cells), which causes progressive atrophy of the skeletal muscle.

Muscularis (MUS-kū-la′-ris) A muscular layer (coat or tunic) of an organ.

Muscularis mucosae (mū-KŌ-sē) A thin layer of smooth muscle fibers that underlie the lamina propria of the mucosa of the gastrointestinal tract.

Mutation (mū-TĀ-shun) Any change in the sequence of bases in a DNA molecule resulting in a permanent alteration in some inheritable trait.

Myasthenia gravis (mī-as-THĒ-nē-a) Weakness and fatigue of skeletal muscles caused by antibodies directed against acetylcholine receptors.

Myelin sheath (MĪ-e-lin) Multilayered lipid and protein covering, formed by Schwann cells and oligodendrocytes, around axons of many peripheral and central nervous system neurons.

Myocardial infarction (MI) (mī′-ō-KAR-dē-al in-FARK-shun) Gross necrosis of myocardial tissue due to interrupted blood supply. Also called a **heart attack.**

Myocardium (mī′-ō-KAR-dē-um) The middle layer of the heart wall, made up of cardiac muscle tissue, lying between the epicardium and the endocardium and constituting the bulk of the heart.

Myofibril (mī-ō-FĪ-bril) A threadlike structure, extending longitudinally through a muscle fiber (cell) consisting mainly of thick filaments (myosin) and thin filaments (actin, troponin, and tropomyosin).

Myoglobin (mī-ō-GLŌ-bin) The oxygen-binding, iron-containing protein present in the sarcoplasm of muscle fibers (cells); contributes the red color to muscle.

Myogram (MĪ-ō-gram) The record or tracing produced by a myograph, an apparatus that measures and records the force of muscular contractions.

Myology (mī-OL-ō-jē) The study of muscles.

Myometrium (mī′-ō-MĒ-trē-um) The smooth muscle layer of the uterus.

Myopathy (mī-OP-a-thē) Any abnormal condition or disease of muscle tissue.

Myopia (mī-Ō-pē-a) Defect in vision in which objects can be seen distinctly only when very close to the eyes; nearsightedness.

Myosin (MĪ-ō-sin) The contractile protein that makes up the thick filaments of muscle fibers.

N

Nail A hard plate, composed largely of keratin, that develops from the epidermis of the skin to form a protective covering on the dorsal surface of the distal phalanges of the fingers and toes.

Nail matrix (MĀ-triks) The part of the nail beneath the body and root from which the nail is produced.

Nasal cavity (NĀ-zal) A mucosa-lined cavity on either side of the nasal septum that opens onto the face at the external nares and into the nasopharynx at the internal nares.

Nasal septum (SEP-tum) A vertical partition composed of bone (perpendicular plate of ethmoid and vomer) and cartilage, covered with a mucous membrane, separating the nasal cavity into left and right sides.

Nasolacrimal duct (nā′-zō-LAK-ri-mal) A canal that transports the lacrimal secretion (tears) from the nasolacrimal sac into the nose.

Nasopharynx (nā′-zō-FAR-inks) The superior portion of the pharynx, lying posterior to the nose and extending inferiorly to the soft palate.

Neck The part of the body connecting the head and the trunk. A constricted portion of an organ such as the neck of the femur or uterus.

Necrosis (ne-KRŌ-sis) A pathological type of cell death that results from disease, injury, or lack of blood supply in which many adjacent cells swell, burst, and spill their contents into the interstitial fluid, triggering an inflammatory response.

Negative feedback The principle governing most control systems; a mechanism of response in which a stimulus initiates actions that reverse or reduce the stimulus.

Neonatal (nē-ō-NĀ-tal) Pertaining to the first four weeks after birth.

Neoplasm (NĒ-ō-plazm) A new growth that may be benign or malignant.

Nephron (NEF-ron) The functional unit of the kidney.

Nerve A cordlike bundle of neuronal axons and/or dendrites and associated connective tissue coursing together outside the central nervous system.

Nerve fiber General term for any process (axon or dendrite) projecting from the cell body of a neuron.

Nerve impulse A wave of depolarization and repolarization that self-propagates along the plasma membrane of a neuron; also called a **nerve action potential.**

Nervous tissue Tissue containing neurons that initiate and conduct nerve impulses to coordinate homeostasis, and neuroglia that provide support and nourishment to neurons.

Net filtration pressure (NFP) Net pressure that promotes fluid outflow at the arterial end of a capillary, and fluid inflow at the venous end of a capillary; net pressure that promotes glomerular filtration in the kidneys.

Neural plate A thickening of ectoderm, induced by the notochord, that forms early in the third week of development and represents the beginning of the development of the nervous system.

Neuralgia (noo-RAL-jē-a) Attacks of pain along the entire course or branch of a peripheral sensory nerve.

Neuritis (noo-RĪ-tis) Inflammation of one or more nerves.

Neuroglia (noo-RŌG-lē-a) Cells of the nervous system that perform various supportive functions. The neuroglia of the central nervous system are the astrocytes, oligodendrocytes, microglia, and ependymal cells; neuroglia of the peripheral nervous system include Schwann cells and satellite cells. Also called **glial cells** (GLĒ-al).

Neurolemma (noo-rō-LEM-ma) The peripheral, nucleated cytoplasmic layer of the Schwann cell. Also called **sheath of Schwann** (SCHVON).

Neurology (noo-ROL-ō-jē) The study of the normal functioning and disorders of the nervous system.

Neuromuscular junction (noo-rō-MUS-kū-lar) A synapse between the axon terminals of a somatic motor neuron and the sarcolemma of a muscle fiber (cell).

Neuron (NOO-ron) A nerve cell, consisting of a cell body, dendrites, and an axon.

Neurosecretory cell (noo-rō-SEC-re-tō-rē) A neuron that secretes a hypothalamic releasing hormone or inhibiting hormone into blood capillaries of the hypothalmus; a neuron that secretes oxytocin or antidiuretic hormone into blood capillaries of the posterior pituitary.

Neurotransmitter One of a variety of molecules within axon terminals that are released into the synaptic cleft in response to a nerve impulse, and that change the membrane potential of the postsynaptic neuron.

Neurulation (noor-oo-LĀ-shun) The process by which the neural plate, neural folds, and neural tube form.

Neutrophil (NOO-trō-fil) A type of white blood cell characterized by granules that stain pale lilac with a combination of acidic and basic dyes.

Nipple A pigmented, wrinkled projection on the surface of the breast that is the location of the openings of the lactiferous ducts for milk release.

Nociceptor (nō′-sē-SEP-tor) A free (naked) nerve ending that detects painful stimuli.

Node of Ranvier (ron-vē-Ā) A space, along a myelinated axon, between the individual Schwann cells that form the myelin sheath and the neurolemma. Also called a **neurofibral node.**

Norepinephrine (NE) (nor′-ep-ē-NEF-rin) A hormone secreted by the adrenal medulla that produces actions similar to those that result from sympathetic stimulation. Also called **noradrenaline** (nor-a-DREN-a-lin).

Notochord (NŌ-tō-cord) A flexible rod of mesodermal tissue that helps form part of the backbone and intervertebral discs.

Nuclear medicine The branch of medicine concerned with the use of radioisotopes in the diagnosis and therapy of disease.

Nucleic acid (noo-KLĒ-ic) An organic compound that is a long polymer of nucleotides, with each nucleotide containing a pentose sugar, a phosphate group, and one of four possible nitrogenous bases (adenine, cytosine, guanine, and thymine or uracil).

Nucleolus (noo-KLĒ-ō-lus) Spherical body within a cell nucleus composed of protein, DNA, and RNA that is the site of the assembly of small and large ribosomal subunits.

Nucleus (NOO-klē-us) A spherical or oval organelle of a cell that contains the hereditary factors of the cell, called genes. A cluster of unmyelinated nerve cell bodies in the central nervous system. The central part of an atom made up of protons and neutrons.

Nucleus pulposus (pul-PŌ-sus) A soft, pulpy, highly elastic substance in the center of an intervertebral disc; a remnant of the notochord.

Nutrient A chemical substance in food that provides energy, forms new body components, or assists in various body functions.

O

Obesity (ō-BĒS-i-tē) Body weight more than 20% above a desirable standard due to excessive accumulation of fat.

Oblique plane (ō-BLĒK) A plane that passes through the body or an organ at an angle between the transverse plane and either the midsagittal, parasagittal, or frontal plane.

Obstetrics (ob-STET-riks) The specialized branch of medicine that deals with pregnancy, labor, and the period of time immediately after delivery (about 6 weeks).

Olfactory (ōl-FAK-tō-rē) Pertaining to smell.

Olfactory bulb A mass of gray matter containing cell bodies of neurons that form synapses with neurons of the olfactory (I) nerve, lying inferior to the frontal lobe of the cerebrum on either side of the crista galli of the ethmoid bone.

Olfactory receptor A bipolar neuron with its cell body lying between supporting cells located in the mucous membrane lining the superior portion of each nasal cavity; transduces odors into neural signals.

Olfactory tract A bundle of axons that extends from the olfactory bulb posteriorly to olfactory regions of the cerebral cortex.

Oligodendrocyte (ol′-i-gō-DEN-drō-sīt) A neuroglial cell that supports neurons and produces a myelin sheath around axons of neurons of the central nervous system.

Oligospermia (ol′-i-gō-SPER-mē-a) A deficiency of sperm cells in the semen.

Oncogenes (ONG-kō-jēnz) Cancer-causing genes; they derive from normal genes, termed proto-oncogenes, that encode proteins involved in cell growth or cell regulation but have the ability to transform a normal cell into a cancerous cell when they are mutated or inappropriately activated. One example is *p53.*

Oncology (ong-KOL-ō-jē) The study of tumors.

Oogenesis (ō′-ō-JEN-e-sis) Formation and development of female gametes (oocytes).

Oophorectomy (ō′-of-ō-REK-tō-me) Surgical removal of the ovaries.

Ophthalmic (of-THAL-mik) Pertaining to the eye.

Ophthalmologist (of′-thal-MOL-ō-jist) A physician who specializes in the diagnosis and treatment of eye disorders using drugs, surgery, and corrective lenses.

Ophthalmology (of′-thal-MOL-ō-jē) The study of the structure, function, and diseases of the eye.

Opsin (OP-sin) The glycoprotein portion of a photopigment.

Opsonization (op-sō-ni-ZĀ-shun) The action of some antibodies that renders bacteria and other foreign cells more susceptible to phagocytosis.

Optic (OP-tik) Refers to the eye, vision, or properties of light.

Optic chiasm (KĪ-azm) A crossing point of the optic (II) nerves, anterior to the pituitary gland. Also called the **optic chiasma.**

Optic disc A small area of the retina containing openings through which the axons of the ganglion cells emerge as the optic nerve (cranial nerve II). Also called the **blind spot.**

Optician (op-TISH-an) A technician who fits, adjusts, and dispenses corrective lenses on prescription of an ophthalmologist or optometrist.

Optic tract A bundle of axons that carry nerve impulses from the retina of the eye between the optic chiasm and the thalamus.

Optometrist (op-TOM-e-trist) Specialist with a doctorate degree in optometry who is licensed to examine and test the eyes and treat visual defects by prescribing corrective lenses.

Orbit (OR-bit) The bony, pyramidal-shaped cavity of the skull that holds the eyeball.

Organ A structure composed of two or more different kinds of tissues with a specific function and usually a recognizable shape.

Organelle (or-gan-EL) A permanent structure within a cell with characteristic shape that is specialized to serve a specific function in cellular activities.

Organic compound (or-GAN-ik) Compound that always contains carbon in which the atoms are held together by covalent bonds. Examples include carbohydrates, lipids, proteins, and nucleic acids (DNA and RNA).

Organism (OR-ga-nizm) A total living form; one individual.

Orgasm (OR-gazm) Sensory and motor events involved in ejaculation for the male and involuntary contraction of the perineal muscles in the female at the climax of sexual intercourse.

Orifice (OR-i-fis) Any aperture or opening.

Origin (OR-i-jin) The attachment of a muscle tendon to a stationary bone or the end opposite the insertion.

Oropharynx (or′-ō-FAR-inks) The intermediate portion of the pharynx, lying posterior to the mouth and extending from the soft palate to the hyoid bone.

Orthopedics (or′-thō-PĒ-diks) The branch of medicine that deals with the preservation and restoration of the skeletal system, articulations, and associated structures.

Osmoreceptor (oz′-mō-re-SEP-tor) Receptor in the hypothalamus that is sensitive to changes in blood osmolarity and, in response to high osmolarity (low water concentration), stimulates synthesis and release of antidiuretic hormone (ADH).

Osmosis (os-MŌ-sis) The net movement of water molecules through a selectively permeable membrane from an area of higher water concentration to an area of lower water concentration until equilibrium is reached.

Osmotic pressure The pressure required to prevent the movement of pure water into a solution containing solutes when the solutions are separated by a selectively permeable membrane.

Osseous (OS-ē-us) Bony.

Ossicle (OS-si-kul) One of the small bones of the middle ear (malleus, incus, stapes).

Ossification (os′-i-fi-KĀ-shun) Formation of bone. Also called **osteogenesis.**

Osteoblast (OS-tē-ō-blast) Cell formed from an osteogenic cell that participates in bone formation by secreting some organic components and inorganic salts.

Osteoclast (OS-tē-ō-clast′) A large, multinuclear cell that resorbs (destroys) bone extracellular matrix.

Osteocyte (OS-tē-ō-sīt′) A mature bone cell that maintains the daily activities of bone tissue.

Osteogenic layer (os′-tē-ō-JEN-ik) The inner layer of the periosteum that contains cells responsible for forming new bone during growth and repair.

Osteology (os′-tē-OL-ō-jē) The study of bones.

Osteon (OS-tē-on) The basic unit of structure in adult compact bone, consisting of a central (haversian) canal with its concentrically arranged lamellae, lacunae, osteocytes, and canaliculi. Also called an **haversian system** (ha-VER-shan).

Osteoporosis (os′-tē-ō-pō-RO-sis) Age-related disorder characterized by decreased bone mass and increased susceptibility to fractures, often as a result of decreased levels of estrogens.

Otic (Ō-tik) Pertaining to the ear.

Otolith (Ō-tō-lith) A particle of calcium carbonate embedded in the otolithic membrane that functions in maintaining static equilibrium.

Otolithic membrane (ō-tō-LITH-ik) Thick, gelatinous, glycoprotein layer located directly over hair cells of the macula in the saccule and utricle of the internal ear.

Otorhinolaryngology (ō′-tō-rī-nō-lar′-in-GOL-ō-jē) The branch of medicine that deals with the diagnosis and treatment of diseases of the ears, nose, and throat.

Oval window A small, membrane-covered opening between the middle ear and inner ear into which the footplate of the stapes fits.

Ovarian cycle (ō-VAR-ē-an) A monthly series of events in the ovary associated with the maturation of a secondary oocyte.

Ovarian follicle (FOL-i-kul) A general name for oocytes (immature ova) in any stage of development, along with their surrounding epithelial cells.

Ovary (Ō-var-ē) Female gonad that produces oocytes and hormones (estrogens, progesterone, inhibin, and relaxin).

Ovulation (ov-ū-LĀ-shun) The rupture of a mature ovarian (graafian) follicle with discharge of a secondary oocyte into the pelvic cavity.

Ovum (Ō-vum) The female reproductive or germ cell; an egg cell; arises through completion of meiosis in a secondary oocyte after penetration by a sperm.

Oxidation (ok-si-DĀ-shun) The removal of electrons from a molecule or, less commonly, the addition of oxygen to a molecule that results in a decrease in the energy content of the molecule. The oxidation of glucose in the body is called **cellular respiration.**

Oxyhemoglobin ($Hb-O_2$) (ok′-sē-HĒ-mō-glō-bin) Hemoglobin combined with oxygen.

Oxytocin (OT) (ok′-sē-TŌ-sin) A hormone secreted by neurosecretory cells in the hypothalamus that stimulates contraction of smooth muscle in the pregnant uterus and myoepithelial cells around the ducts of mammary glands.

P

P wave The deflection wave of an electrocardiogram that signifies atrial depolarization.

Palate (PAL-at) The horizontal structure separating the oral and the nasal cavities; the roof of the mouth.

Palpate (PAL-pāt) To examine by touch; to feel.

Pancreas (PAN-krē-as) A soft, oblong organ lying along the greater curvature of the stomach and connected by a duct to the duodenum. It is both an exocrine gland (secreting pancreatic juice) and an endocrine gland (secreting insulin and glucagon).

Pancreatic duct (pan′-krē-AT-ik) A single large tube that unites with the common bile duct from the liver and gallbladder and drains pancreatic juice into the duodenum.

Pancreatic islet A cluster of endocrine gland cells in the pancreas that secretes insulin, glucagon, somatostatin, and pancreatic polypeptide. Also called an **islet of Langerhans** (LANG-er-hanz).

Papanicolaou test (pap′-a-NIK-ō-la-oo) A cytological staining test for the detection and diagnosis of premalignant and malignant conditions of the female genital tract. Cells scraped from the epithelium of the cervix of the uterus are examined microscopically. Also called a **Pap test** or **Pap smear.**

Papilla (pa-PIL-a) A small nipple-shaped projection or elevation.

Paralysis (pa-RAL-a-sis) Loss or impairment of motor function due to a lesion of nervous or muscular origin.

Paranasal sinus (par′-a-NĀ-zal SĪ-nus) A mucus-lined air cavity in a skull bone that communicates with the nasal cavity. Paranasal sinuses are located in the frontal, maxillary, ethmoid, and sphenoid bones.

Paraplegia (par-a-PLĒ-jē-a) Paralysis of both lower limbs.

Parasagittal plane (par-a-SAJ-i-tal) A vertical plane that does not pass through the midline and that divides the body or organs into *unequal* left and right portions.

Parasympathetic division (par′-a-sim-pa-THET-ik) One of the two subdivisions of the autonomic nervous system, having cell bodies of preganglionic neurons in nuclei in the brain stem and in the lateral gray horn of the sacral portion of the spinal cord; primarily concerned with activities that conserve and restore body energy.

Parathyroid gland (par′-a-THĪ-royd) One of usually four small endocrine glands embedded in the posterior surfaces of the lateral lobes of the thyroid gland.

Parathyroid hormone (PTH) A hormone secreted by the chief (principal) cells of the parathyroid glands that increases blood calcium level and decreases blood phosphate level.

Parenchyma (par-EN-ki-ma) The functional parts of any organ, as opposed to tissue that forms its stroma or framework.

Parietal (pa-RĪ-e-tal) Pertaining to or forming the outer wall of a body cavity.

Parietal cell A type of secretory cell in gastric glands that produces hydrochloric acid and intrinsic factor. Also called an **oxyntic cell.**

Parkinson disease (PD) Progressive degeneration of the basal ganglia and substantia nigra of the cerebrum resulting in decreased production of dopamine (DA) that leads to tremor, slowing of voluntary movements, and muscle weakness.

Parotid gland (pa-ROT-id) One of the paired salivary glands located inferior and anterior to the ears and connected to the oral cavity via a duct that opens into the inside of the cheek opposite the maxillary (upper) second molar tooth.

Parturition (par′-too-RISH-un) Act of giving birth to young; childbirth, delivery.

Patellar reflex (pa-TELL-ar) Extension of the leg by contraction of the quadriceps femoris muscle in response to tapping the patellar ligament. Also called the **knee jerk reflex.**

Patent ductus arteriosus Congenital anatomical heart defect in which the fetal connection between the aorta and pulmonary trunk remains open instead of closing completely after birth.

Pathogen (PATH-ō-jen) A disease-producing microbe.

Pathological anatomy (path′-ō-LOJ-i-kal) The study of structural changes caused by disease.

Pectoral (PEK-tō-ral) Pertaining to the chest or breast.

Pediatrician (pē′-dē-a-TRISH-un) A physician who specializes in the care and treatment of children.

Pedicel (PED-i-sel) Footlike structure, as on podocytes of a glomerulus.

Pelvic cavity (PEL-vik) Inferior portion of the abdominopelvic cavity that contains the urinary bladder, sigmoid colon, rectum, and internal female and male reproductive structures.

Pelvis The basinlike structure formed by the two hip bones, the sacrum, and the coccyx. The expanded, proximal portion of the ureter, lying within the kidney and into which the major calyces open.

Penis (PĒ-nis) The organ of urination and copulation in males; used to deposit semen into the female vagina.

Pepsin Protein-digesting enzyme secreted by chief cells of the stomach in the inactive form pepsinogen, which is converted to active pepsin by hydrochloric acid.

Peptic ulcer An ulcer that develops in areas of the gastrointestinal tract exposed to hydrochloric acid; classified as a gastric ulcer if in the lesser curvature of the stomach and as a duodenal ulcer if in the first part of the duodenum.

Percussion (per-KUSH-un) The act of striking (percussing) an underlying part of the body with short, sharp blows as an aid in diagnosing the part by the quality of the sound produced.

Perforating canal A minute passageway by means of which blood vessels and nerves from the periosteum penetrate into compact bone. Also called **volkmann's canal** (FŌLK-manz).

Pericardial cavity (per′-i-KAR-dē-al) Small potential space between the visceral and parietal layers of the serous pericardium that contains pericardial fluid.

Pericardium (per′-i-KAR-dē-um) A loose-fitting membrane that encloses the heart, consisting of a superficial fibrous layer and a deep serous layer.

Perichondrium (per′-i-KON-drē-um) The membrane that covers cartilage.

Perilymph (PER-i-limf) The fluid contained between the bony and membranous labyrinths of the inner ear.

Perineum (per′-i-NĒ-um) The pelvic floor; the space between the anus and the scrotum in the male and between the anus and the vulva in the female.

Periodontal disease (per-ē-ō-DON-tal) A collective term for conditions characterized by degeneration of gingivae, alveolar bone, periodontal ligament, and cementum.

Periodontal ligament The periosteum lining the alveoli (sockets) for the teeth in the alveolar processes of the mandible and maxillae.

Periosteum (per′-ē-OS-tē-um) The membrane that covers bone and consists of connective tissue, osteogenic cells, and osteoblasts; is essential for bone growth, repair, and nutrition.

Peripheral (pe-RIF-er-al) Located on the outer part or a surface of the body.

Peripheral nervous system (PNS) The part of the nervous system that lies outside the central nervous system, consisting of nerves and ganglia.

Peristalsis (per′-i-STAL-sis) Successive muscular contractions along the wall of a hollow muscular structure.

Peritoneum (per′-i-tō-NĒ-um) The largest serous membrane of the body that lines the abdominal cavity and covers the viscera within the cavity.

Peritonitis (per′-i-tō-NĪ-tis) Inflammation of the peritoneum.

Peroxisome (per-OK-si-sōm) Organelle similar in structure to a lysosome that contains enzymes that use molecular oxygen to oxidize various organic compounds; such reactions produce hydrogen peroxide; abundant in liver cells.

Perspiration Sweat; produced by sudoriferous (sweat) glands and containing water, salts, urea, uric acid, amino acids, ammonia, sugar, lactic acid, and ascorbic acid. Helps maintain body temperature and eliminate wastes.

pH A measure of the concentration of hydrogen ions (H^+) in a solution. The pH scale extends from 0 to 14, with a value of 7 expressing neutrality, values lower than 7 expressing increasing acidity, and values higher than 7 expressing increasing alkalinity.

Phagocytosis (fag′-ō-sī-TŌ-sis) The process by which phagocytes ingest particulate matter; the ingestion and destruction of microbes, cell debris, and other foreign matter.

Phalanx (FĀ-lanks) The bone of a finger or toe. *Plural* is **phalanges** (fa-LAN-jēz).

Pharmacology (far′-ma-KOL-ō-jē) The science that deals with the effects and uses of drugs in the treatment of disease.

Pharynx (FAR-inks) The throat; a tube that starts at the internal nares and runs partway down the neck, where it opens into the esophagus posteriorly and the larynx anteriorly.

Phenotype (FĒ-nō-tīp) The observable expression of genotype; physical characteristics of an organism determined by genetic makeup and influenced by interaction between genes and internal and external environmental factors.

Phlebitis (fle-BĪ-tis) Inflammation of a vein, usually in a lower limb.

Photopigment A substance that can absorb light and undergo structural changes that can lead to the development of a receptor potential. An example is rhodopsin. In the eye, also called **visual pigment.**

Photoreceptor Receptor that detects light shining on the retina of the eye.

Physiology (fiz′-ē-OL-ō-jē) Science that deals with the functions of an organism or its parts.

Pia mater (PĪ-a MĀ-ter *or* PĒ -a MĀ-ter) The innermost of the three meninges (coverings) of the brain and spinal cord.

Pineal gland (PĪN-ē-al) A cone-shaped gland located in the roof of the third ventricle that secretes melatonin.

Pinna (PIN-na) The projecting part of the external ear composed of elastic cartilage and covered by skin and shaped like the flared end of a trumpet. Also called the **auricle** (AW-ri-kul).

Pituitary gland (pi-TOO-i-tār-ē) A small endocrine gland occupying the hypophyseal fossa of the sphenoid bone and attached to the hypothalamus by the infundibulum. Also called the **hypophysis** (hī-POF-i-sis).

Pivot joint A synovial joint in which a rounded, pointed, or conical surface of one bone articulates with a ring formed partly by another bone and partly by a ligament, as in the joint between the atlas and axis and between the proximal ends of the radius and ulna.

Placenta (pla-SEN-ta) The special structure through which the exchange of materials between fetal and maternal circulations occurs. Also called the **afterbirth.**

Plantar flexion (PLAN-tar FLEK-shun) Bending the foot in the direction of the plantar surface (sole).

Plaque (PLAK) A layer of dense proteins on the inside of a plasma membrane in adherens junctions and desmosomes. A mass of bacterial cells, dextran (polysaccharide), and other debris that adheres to teeth (dental plaque). See also atherosclerotic plaque.

Plasma (PLAZ-ma) The extracellular fluid found in blood vessels; blood minus the formed elements.

Plasma cell Cell that develops from a B cell (lymphocyte) and produces antibodies.

Plasma (cell) membrane Outer, limiting membrane that separates the cell's internal parts from extracellular fluid or the external environment.

Platelet (PLĀT-let) A fragment of cytoplasm enclosed in a cell membrane and lacking a nucleus; found in the circulating blood; plays a role in hemostasis. Also called a **thrombocyte** (THROM-bō-sīt).

Platelet plug Aggregation of platelets at a site where a blood vessel is damaged that helps stop or slow blood loss.

Pleura (PLOOR-a) The serous membrane that covers the lungs and lines the walls of the chest and the diaphragm.

Pleural cavity Small potential space between the visceral and parietal pleurae.

Plexus (PLEK-sus) A network of nerves, veins, or lymphatic vessels.

Pluripotent stem cell Immature stem cell in red bone marrow that gives rise to precursors of all the different mature blood cells.

Pneumotaxic area (noo-mō-TAK-sik) A part of the respiratory center in the pons that continually sends inhibitory nerve impulses to the inspiratory area, limiting inhalation and facilitating exhalation.

Podiatry (pō-DĪ-a-trē) The diagnosis and treatment of foot disorders.

Polar body The smaller cell resulting from the unequal division of primary and secondary oocytes during meiosis. The polar body has no function and degenerates.

Polycythemia (pol′-ē-sī-THĒ-mē-a) Disorder characterized by an above-normal hematocrit (above 55%) in which hypertension, thrombosis, and hemorrhage can occur.

Polysaccharide (pol′-ē-SAK-a-rīd) A carbohydrate in which three or more monosaccharides are joined chemically.

Polyunsaturated fat A fatty acid that contains more than one double covalent bond between its carbon atoms; abundant in triglycerides of corn oil, safflower oil, and cottonseed oil.

Polyuria (pol′-ē-U-rē-a) An excessive production of urine.

Pons (PONZ) The part of the brain stem that forms a "bridge" between the medulla oblongata and the midbrain, anterior to the cerebellum.

Positive feedback A feedback mechanism in which the response enhances the original stimulus.

Postcentral gyrus A gyrus of the cerebral cortex located immediately posterior to the central sulcus; contains the primary somatosensory area.

Posterior (pos-TĒR-ē-or) Nearer to or at the back of the body. Equivalent to **dorsal** in bipeds.

Posterior column–medial lemniscus pathways Sensory pathways that carry information related to proprioception, fine touch, two-point discrimination, pressure, and vibration. First-order neurons project from the spinal cord to the ipsilateral (same side) medulla in the posterior columns. Second-order neurons project from the medulla to the contralateral (opposite side) thalamus in the medial lemniscus. Third-order neurons project from the thalamus to the somatosensory cortex (postcentral gyrus) on the same side.

Posterior pituitary Posterior lobe of the pituitary gland. Also called the **neurohypophysis** (noo-rō-hī-POF-i-sis).

Posterior root The structure composed of sensory axons lying between a spinal nerve and the dorsolateral aspect of the spinal cord. Also called the **dorsal (sensory) root.**

Posterior root ganglion (GANG-glē-on) A group of cell bodies of sensory neurons and their supporting cells located along the posterior root of a spinal nerve. Also called a **dorsal (sensory) root ganglion.**

Postganglionic neuron (pōst′-gang-lē-ON-ik NOO-ron) The second autonomic motor neuron in an autonomic pathway, having its cell body and dendrites located in an autonomic ganglion and its unmyelinated axon ending at cardiac muscle, smooth muscle, or a gland.

Postsynaptic neuron (pōst-sin-AP-tik) The nerve cell that is activated by the release of a neurotransmitter from another neuron and carries nerve impulses away from the synapse.

Precapillary sphincter (SFINGK-ter) A ring of smooth muscle fibers (cells) at the site of origin of true capillaries that regulate blood flow into true capillaries.

Precentral gyrus A gyrus of the cerebral cortex located immediately anterior to the central sulcus; contains the primary motor area.

Preganglionic neuron (prē′-gang-lē-ON-ik) The first autonomic motor neuron in an autonomic pathway, with its cell body and dendrites in the brain or spinal cord and its myelinated axon ending at an autonomic ganglion, where it synapses with a postganglionic neuron.

Pregnancy Sequence of events that normally includes fertilization, implantation, embryonic growth, and fetal growth and terminates in birth.

Premenstrual syndrome (PMS) Severe physical and emotional stress ocurring late in the postovulatory phase of the menstrual cycle and sometimes overlapping with menstruation.

Prepuce (PRĒ-poos) The loose-fitting skin covering the glans of the penis and clitoris. Also called the **foreskin.**

Presbyopia (prez-bē-Ō-pē-a) A loss of elasticity of the lens of the eye due to advancing age with resulting inability to focus clearly on near objects.

Presynaptic neuron (prē-sin-AP-tik) A neuron that propagates nerve impulses toward a synapse.

Prevertebral ganglion (prē-VER-te-bral GANG-lē-on) A cluster of cell bodies of postganglionic sympathetic neurons anterior to the spinal column and close to large abdominal arteries. Also called a **collateral ganglion.**

Primary germ layer One of three layers of embryonic tissue, called ectoderm, mesoderm, and endoderm, that give rise to all tissues and organs of the body.

Primary motor area A region of the cerebral cortex in the precentral gyrus of the frontal lobe of the cerebrum that controls specific muscles or groups of muscles.

Primary somatosensory area A region of the cerebral cortex posterior to the central sulcus in the postcentral gyrus of the parietal lobe of the cerebrum that localizes exactly the points of the body where somatic sensations originate.

Prime mover The muscle directly responsible for producing a desired motion. Also called an **agonist** (AG-ō-nist).

Primitive gut Embryonic structure formed from the dorsal part of the yolk sac that gives rise to most of the gastrointestinal tract.

Proctology (prok-TOL-ō-jē) The branch of medicine concerned with the rectum and its disorders.

Progeny (PROJ-e-nē) Offspring or descendants.

Progesterone (prō-JES-te-rōn) A female sex hormone produced by the ovaries that helps prepare the endometrium of the uterus for implantation of a fertilized ovum and the mammary glands for milk secretion.

Prognosis (prog-NŌ-sis) A forecast of the probable results of a disorder; the outlook for recovery.

Prolactin (PRL) (prō-LAK-tin) A hormone secreted by the anterior pituitary that initiates and maintains milk secretion by the mammary glands.

Prolapse (PRŌ-laps) A dropping or falling down of an organ, especially the uterus or rectum.

Proliferation (prō-lif′-er-Ā-shun) Rapid and repeated reproduction of new parts, especially cells.

Pronation (prō-NĀ-shun) A movement of the forearm in which the palm is turned posteriorly.

Prophase (PRŌ-fāz) The first stage of mitosis during which chromatid pairs are formed and aggregate around the metaphase plate of the cell.

Proprioception (prō-prē-ō-SEP-shun) The perception of the position of body parts, especially the limbs, independent of vision; this sense is possible due to nerve impulses generated by proprioceptors.

Proprioceptor (prō′-prē-ō-SEP-tor) A receptor located in muscles, tendons, joints, or the internal ear (muscle spindles, tendon organs, joint kinesthetic receptors, and hair cells of the vestibular apparatus) that provides information about body position and movements.

Prostaglandin (PG) (pros′-ta-GLAN-din) A membrane-associated lipid; released in small quantities and acts as a local hormone.

Prostate (PROS-tāt) A doughnut-shaped gland inferior to the urinary bladder that surrounds the superior portion of the male urethra and secretes a slightly acidic solution that contributes to sperm motility and viability.

Protein An organic compound consisting of carbon, hydrogen, oxygen, nitrogen, and sometimes sulfur and phosphorus; synthesized on ribosomes and made up of amino acids linked by peptide bonds.

Proteasome Tiny cellular organelle in the cytosol and nucleus containing proteases that destroy unneeded, damaged, or faulty proteins.

Prothrombin (prō-THROM-bin) An inactive blood-clotting factor synthesized by the liver, released into the blood, and converted to active thrombin in the process of blood clotting by the activated enzyme prothrombinase.

Protraction (prō-TRAK-shun) The movement of the mandible or clavicle forward on a plane parallel with the ground.

Proximal (PROK-si-mal) Nearer the attachment of a limb to the trunk; nearer to the point of origin or attachment.

Pseudopods (SOO-dō-pods) Temporary protrusions of the leading edge of a migrating cell; cellular projections that surround a particle undergoing phagocytosis.

Ptosis (TŌ-sis) Drooping, as of the eyelid or the kidney.

Puberty (PŪ-ber-tē) The time of life during which the secondary sex characteristics begin to appear and the capability for sexual reproduction is possible; usually occurs between the ages of 10 and 17.

Pubic symphysis A slightly movable cartilaginous joint between the anterior surfaces of the hip bones.

Puerperium (pū′-er-PER-ē-um) The period immediately after childbirth, usually 4–6 weeks.

Pulmonary (PUL-mo-ner′-ē) Concerning or affected by the lungs.

Pulmonary circulation The flow of deoxygenated blood from the right ventricle to the lungs and the return of oxygenated blood from the lungs to the left atrium.

Pulmonary edema (e-DĒ-ma) An abnormal accumulation of interstitial fluid in the tissue spaces and alveoli of the lungs due to increased pulmonary capillary permeability or increased pulmonary capillary pressure.

Pulmonary embolism (PE) (EM-bō-lizm) The presence of a blood clot or a foreign substance in a pulmonary arterial blood vessel that obstructs circulation to lung tissue.

Pulmonary ventilation The inflow (inhalation) and outflow (exhalation) of air between the atmosphere and the lungs. Also called **breathing**.

Pulp cavity A cavity within the crown and neck of a tooth, which is filled with pulp, a connective tissue containing blood vessels, nerves, and lymphatic vessels.

Pulse (PULS) The rhythmic expansion and elastic recoil of a systemic artery after each contraction of the left ventricle.

Pupil The hole in the center of the iris, the area through which light enters the posterior cavity of the eyeball.

Purkinje fiber (pur-KIN-jē) Muscle fiber (cell) in the ventricular tissue of the heart specialized for conducting an action potential to the myocardium; part of the conduction system of the heart.

Pus The liquid product of inflammation containing leukocytes or their remains and debris of dead cells.

Pyloric sphincter (pī-LOR-ik) A thickened ring of smooth muscle through which the pylorus of the stomach communicates with the duodenum. Also called the **pyloric valve.**

Pyramid (PIR-a-mid) A pointed or cone-shaped structure. One of two roughly triangular structures on the anterior aspect of the medulla oblongata composed of the largest motor tracts that run from the cerebral cortex to the spinal cord. A triangular structure in the renal medulla.

Pyramidal tracts (pathways) (pi-RAM-i-dal) *See* **Direct motor pathways**.

Q

QRS wave The deflection wave of an electrocardiogram that represents the onset of ventricular depolarization.

Quadriplegia (kwod′-ri-PLĒ-jē-a) Paralysis of four limbs: two upper and two lower.

R

Radiographic anatomy (rā′-dē-ō-GRAF-ic) Diagnostic branch of anatomy that includes the use of x rays.

Rapid eye movement (REM) sleep Stage of sleep in which dreaming occurs, lasting for 5 to 10 minutes several times during a sleep cycle; characterized by rapid movements of the eyes beneath the eyelids.

Receptor A specialized cell or a distal portion of a neuron that responds to a specific sensory modality, such as touch, pressure, cold, light, or sound, and converts it to an electrical signal (generator or receptor potential). A specific molecule or cluster of molecules that recognizes and binds a particular ligand.

Receptor-mediated endocytosis A highly selective process whereby cells take up specific ligands, which usually are large molecules or particles, by enveloping them within a sac of plasma membrane. Ligands are eventually broken down by enzymes in lysosomes.

Recessive allele An allele whose presence is masked in the presence of a dominant allele on the homologous chromosome.

Recombinant DNA Synthetic DNA, formed by joining a fragment of DNA from one source to a portion of DNA from another.

Recovery oxygen consumption Elevated oxygen use after exercise ends due to metabolic changes that start during exercise and continue after exercise. Previously called **oxygen debt.**

Recruitment (rē-KROOT-ment) The process of increasing the number of active motor units. Also called **motor unit summation.**

Rectum (REK-tum) The last 20 cm (8 in.) of the gastrointestinal tract, from the sigmoid colon to the anus.

Reduction The addition of electrons to a molecule or, less commonly, the removal of oxygen from a molecule that results in an increase in the energy content of the molecule.

Referred pain Pain that is felt at a site remote from the place of origin.

Reflex Fast response to a change (stimulus) in the internal or external environment that attempts to restore homeostasis.

Reflex arc The most basic conduction pathway through the nervous system, connecting a receptor and an effector and consisting of a receptor, a sensory neuron, an integrating center in the central nervous system, a motor neuron, and an effector.

Refraction (rē-FRAK-shun) The bending of light as it passes from one medium to another.

Refractory period (re-FRAK-to-rē) A time period during which an excitable cell (neuron or muscle fiber) cannot respond to a stimulus that is usually adequate to evoke an action potential.

Regional anatomy The division of anatomy dealing with a specific region of the body, such as the head, neck, chest, or abdomen.

Regurgitation (rē-gur′-ji-TĀ-shun) Return of solids or fluids to the mouth from the stomach; backward flow of blood through incompletely closed heart valves.

Relaxin (RLX) A female hormone produced by the ovaries and placenta that increases flexibility of the pubic symphysis and helps dilate the uterine cervix to ease delivery of a baby.

Releasing hormone Hormone secreted by the hypothalamus that can stimulate secretion of hormones of the anterior pituitary.

Remodeling Replacement of old bone by new bone tissue.

Renal (RE-nal) Pertaining to the kidneys.

Renal corpuscle (KOR-pus-l) A glomerular (Bowman's) capsule and its enclosed glomerulus.

Renal pelvis A cavity in the center of the kidney formed by the expanded, proximal portion of the ureter, lying within the kidney, and into which the major calyces open.

Renal pyramid A triangular structure in the renal medulla containing the straight segments of renal tubules and the vasa recta.

Renin (RĒ-nin) An enzyme released by the kidney into the plasma, where it converts angiotensinogen into angiotensin I.

Renin–angiotensin–aldosterone (RAA) pathway A mechanism for the control of blood pressure, initiated by the secretion of renin by the kidney in response to low blood pressure; renin catalyzes formation of angiotensin I, which is converted to angiotensin II by angiotensin-converting enzyme (ACE), and angiotensin II stimulates secretion of aldosterone.

Repolarization (rē-pō-lar-i-ZĀ-shun) Restoration of a resting membrane potential after depolarization.

Reproduction (rē-prō-DUK-shun) The formation of new cells for growth, repair, or replacement; the production of a new individual.

Reproductive cell division Type of cell division in which gametes (sperm and oocytes) are produced; consists of meiosis and cytokinesis.

Residual volume (re-ZID-ū-al) The volume of air still contained in the lungs after a maximal exhalation; about 1200 mL in males and 1100 mL in females.

Resistance (re-ZIS-tans) Hindrance (impedance) to blood flow as a result of higher viscosity, longer total blood vessel length, and smaller blood vessel radius. Ability to ward off disease. The hindrance encountered by electrical charges as they move from one point to another. The hindrance encountered by air as it moves through the respiratory passageways.

Respiration (res-pi-RĀ-shun) Overall exchange of gases between the atmosphere, blood, and body cells consisting of pulmonary ventilation, external respiration, and internal respiration.

Respiratory center Neurons in the pons and medulla oblongata of the brain stem that regulate the rate and depth of pulmonary ventilation.

Respiratory membrane Structure in the lungs consisting of the alveolar wall and its basement membrane and a capillary endothelium and its basement membrane through which the diffusion of respiratory gases occurs.

Resting membrane potential The voltage difference between the inside and outside of a cell membrane when the cell is not responding to a stimulus; in many neurons and muscle fibers it is -70 to -90 mV, with the inside of the cell negative relative to the outside.

Retention (rē-TEN-shun) A failure to void urine due to obstruction, nervous contraction of the urethra, or absence of sensation of desire to urinate.

Reticular activating system (RAS) (re-TIK-ū-lar) A portion of the reticular formation that has many ascending connections with the cerebral cortex; when this area of the brain stem is active, nerve impulses pass to the thalamus and widespread areas of the cerebral cortex, resulting in generalized alertness or arousal from sleep.

Reticular formation A network of small groups of neuronal cell bodies scattered among bundles of axons (mixed gray and white matter) beginning in the medulla oblongata and extending superiorly through the central part of the brain stem.

Reticulocyte (re-TIK-ū-lō-sīt) An immature red blood cell.

Reticulum (re-TIK-ū-lum) A network.

Retina (RET-i-na) The deep coat of the posterior portion of the eyeball consisting of nervous tissue (where the process of vision begins) and a pigmented layer of epithelial cells that contact the choroid.

Retinal (RE-ti-nal) A derivative of vitamin A that functions as the light-absorbing portion of the photopigment rhodopsin.

Retraction (rē-TRAK-shun) The movement of a protracted part of the body posteriorly on a plane parallel to the ground, as in pulling the lower jaw back in line with the upper jaw.

Retrograde degeneration (RE-trō-grād dē-jen-er-Ā-shun) Changes that occur in the proximal portion of a damaged axon only as far as the first node of Ranvier; similar to changes that occur during Wallerian degeneration.

Retroperitoneal (re′-trō-per-i-tō-NĒ-al) External to the peritoneal lining of the abdominal cavity.

Rh factor An inherited antigen on the surface of red blood cells in Rh^+ individuals; not present in Rh^- individuals.

Rhinology (rī-NOL-ō-jē) The study of the nose and its disorders.

Rhodopsin (rō-DOP-sin) The photopigment in rods of the retina, consisting of a glycoprotein called opsin and a derivative of vitamin A called retinal.

Ribonucleic acid (RNA) (rī-bō-noo-KLĒ-ik) A single-stranded nucleic acid made up of nucleotides, each consisting of a nitrogenous base (adenine, cytosine, guanine, or uracil), ribose, and a phosphate group; three types are messenger RNA (mRNA), transfer RNA (tRNA), and ribosomal RNA (rRNA), each of which has a specific role during protein synthesis.

Ribosome (RĪ-bō-sōm) An organelle in the cytoplasm of cells, composed of a small subunit and a large subunit that contain ribosomal RNA and ribosomal proteins; the site of protein synthesis.

Rigidity (ri-JID-i-tē) Hypertonia characterized by increased muscle tone, but reflexes are not affected.

Rigor mortis State of partial contraction of muscles after death due to lack of ATP; myosin heads (crossbridges) remain attached to actin, thus preventing relaxation.

Rod One of two types of photoreceptor in the retina of the eye; specialized for vision in dim light.

Root canal A narrow extension of the pulp cavity lying within the root of a tooth.

Rotation (rō-TĀ-shun) Moving a bone around its own axis, with no other movement.

Round window A small opening between the middle and internal ear, directly inferior to the oval window, covered by the secondary tympanic membrane.

Rugae (ROO-gē) Large folds in the mucosa of an empty hollow organ, such as the stomach and vagina.

S

Saccule (SAK-ūl) The inferior and smaller of the two chambers in the membranous labyrinth inside the vestibule of the internal ear containing a receptor organ for static equilibrium.

Sacral plexus (SĀ-kral PLEK-sus) A network formed by the ventral branches of spinal nerves L4 through S3.

Sacral promontory (PROM-on-tor′-ē) The superior surface of the body of the first sacral vertebra that projects anteriorly into the pelvic cavity; a line from the sacral promontory to the superior border of the pubic symphysis divides the abdominal and pelvic cavities.

Saddle joint A synovial joint in which the articular surface of one bone is saddle shaped and the articular surface of the other bone is shaped like the legs of the rider sitting in the saddle, as in the joint between the trapezium and the metacarpal of the thumb.

Sagittal plane (SAJ-i-tal) A plane that divides the body or organs into left and right portions. Such a plane may be **midsagittal (median),** in which the divisions are equal, or **parasagittal,** in which the divisions are unequal.

Saliva (sa-LĪ-va) A clear, alkaline, somewhat viscous secretion produced mostly by the three pairs of salivary glands; contains various salts, mucin, lysozyme, salivary amylase, and lingual lipase (produced by glands in the tongue).

Salivary amylase (SAL-i-ver-ē AM-i-lās) An enzyme in saliva that initiates the chemical breakdown of starch.

Salivary gland One of three pairs of glands that lie external to the mouth and pour their secretory product (saliva) into ducts that empty into the oral cavity; the parotid, submandibular, and sublingual glands.

Salt A substance that, when dissolved in water, ionizes into cations and anions, neither of which are hydrogen ions (H^+) nor hydroxide ions (OH^-).

Saltatory conduction (sal-ta-TŌ-rē) The propagation of an action potential (nerve impulse) along the exposed parts of a myelinated axon. The action potential appears at successive nodes of Ranvier and therefore seems to leap from node to node.

Sarcolemma (sar′-kō-LEM-ma) The cell membrane of a muscle fiber (cell), especially of a skeletal muscle fiber.

Sarcomere (SAR-kō-mēr) A contractile unit in a striated muscle fiber (cell) extending from one Z disc to the next Z disc.

Sarcoplasm (SAR-kō-plazm) The cytoplasm of a muscle fiber (cell).

Sarcoplasmic reticulum (sar′-kō-PLAZ-mik re-TIK-ū-lum) A network of saccules and tubes surrounding myofibrils of a muscle fiber (cell), comparable to endoplasmic reticulum; functions to reabsorb calcium ions during relaxation and to release them to cause contraction.

Saturated fat A fatty acid that contains only single bonds (no double bonds) between its carbon atoms; all carbon atoms are bonded to the maximum number of hydrogen atoms; prevalent in triglycerides of animal products such as meat, milk, milk products, and eggs.

Scala tympani (SKA-la TIM-pan-ē) The inferior spiral-shaped channel of the bony cochlea, filled with perilymph.

Scala vestibuli (ves-TIB-ū-lē) The superior spiral-shaped channel of the bony cochlea, filled with perilymph.

Schwann cell (SCHVON) A neuroglial cell of the peripheral nervous system that forms the myelin sheath and neurolemma around a nerve axon by wrapping around the axon in a jelly-roll fashion.

Sciatica (sī-AT-i-ka) Inflammation and pain along the sciatic nerve; felt along the posterior aspect of the thigh extending down the inside of the leg.

Sclera (SKLE-ra) The white coat of fibrous tissue that forms the superficial protective covering over the eyeball except in the most anterior portion; the posterior portion of the fibrous tunic.

Scleral venous sinus A circular venous sinus located at the junction of the sclera and the cornea through which aqueous humor drains from the anterior chamber of the eyeball into the blood. Also called the **canal of Schlemm** (SHLEM).

Sclerosis (skle-RŌ-sis) A hardening with loss of elasticity of tissues.

Scoliosis (skō′-lē-Ō-sis) An abnormal lateral curvature from the normal vertical line of the backbone.

Scrotum (SKRŌ-tum) A skin-covered pouch that contains the testes and their accessory structures.

Sebaceous gland (se-BĀ-shus) An exocrine gland in the dermis of the skin, almost always associated with a hair follicle, that secretes sebum. Also called an **oil gland.**

Sebum (SĒ-bum) Secretion of sebaceous (oil) glands.

Secondary response Accelerated, more intense cell-mediated or antibody-mediated immune response upon a subsequent exposure to an antigen after the primary response.

Secondary sex characteristic A characteristic of the male or female body that develops at puberty under the influence of sex hormones but is not directly involved in sexual reproduction; examples are the distribution of body hair, voice pitch, body shape, and muscle development.

Second messenger An intracellular mediator molecule that is produced in response to a first messenger (hormone or neurotransmitter) binding to its receptor in the plasma membrane of a target cell. Initiates a cascade of chemical reactions that produce characteristic effects for that particular target cell.

Secretion (se-KRĒ-shun) Production and release from a cell or a gland of a physiologically active substance.

Selective permeability (per′-mē-a-BIL-i-tē) The property of a membrane by which it permits the passage of certain substances but restricts the passage of others.

Semen (SĒ-men) A fluid discharged at ejaculation by a male that consists of a mixture of sperm and the secretions of the seminiferous tubules, seminal vesicles, prostate, and bulbourethral (Cowper's) glands.

Semicircular canals Three bony channels (anterior, posterior, lateral), filled with perilymph, in which lie the membranous semicircular canals filled with endolymph. They contain receptors for equilibrium.

Semicircular ducts The membranous semicircular canals filled with endolymph and floating in the perilymph of the bony semicircular canals; they contain cristae that are concerned with dynamic equilibrium.

Semilunar valve (sem′-ē-LOO-nar) A valve between the aorta or the pulmonary trunk and a ventricle of the heart.

Seminal vesicle (SEM-i-nal VES-i-kul) One of a pair of convoluted, pouchlike structures, lying posterior and inferior to the urinary bladder and anterior to the rectum, that secrete a component of semen into the ejaculatory ducts. Also termed a **seminal gland.**

Seminiferous tubule (sem′-i-NI-fer-us TOO-būl) A tightly coiled duct, located in the testis, where sperm are produced.

Sensation A state of awareness of external or internal conditions of the body.

Sensory neurons (NOO-ronz) Neurons that carry sensory information from cranial and spinal nerves into the brain and spinal cord or from a lower to a higher level in the spinal cord and brain. Also called **afferent neurons.**

Septal defect An opening in the septum (interatrial or interventricular) between the left and right sides of the heart.

Septum (SEP-tum) A wall dividing two cavities.

Serous membrane (SIR-us) A membrane that lines a body cavity that does not open to the exterior. The external layer of an organ formed by a serous membrane. The membrane that lines the pleural, pericardial, and peritoneal cavities. Also called a **serosa** (se-RŌ-sa).

Sertoli cell (ser-TŌ-lē) A supporting cell in the seminiferous tubules that secretes fluid for supplying nutrients to sperm and the hormone inhibin, removes excess cytoplasm from spermatogenic cells, and mediates the effects of FSH and testosterone on spermatogenesis. Also called a **sustentacular cell** (sus′-ten-TAK-ū-lar).

Serum Blood plasma minus its clotting proteins.

Sesamoid bones (SES-a-moyd) Small bones usually found in tendons.

Sex chromosomes The twenty-third pair of chromosomes, designated X and Y, which determine the genetic sex of an individual; in males, the pair is XY; in females, XX.

Sexual intercourse The insertion of the erect penis of a male into the vagina of a female. Also called **coitus** (KŌ-i-tus).

Shivering Involuntary contraction of skeletal muscles that generates heat.

Shock Failure of the cardiovascular system to deliver adequate amounts of oxygen and nutrients to meet the metabolic needs of the body due to inadequate cardiac output. It is characterized by hypotension; clammy, cool, and pale skin; sweating; reduced urine formation; altered mental state; acidosis; tachycardia; weak, rapid pulse; and thirst. Types include hypovolemic, cardiogenic, vascular, and obstructive.

Shoulder joint A synovial joint where the humerus articulates with the scapula.

Sigmoid colon (SIG-moyd KŌ-lon) The S-shaped part of the large intestine that begins at the level of the left iliac crest, projects medially, and terminates at the rectum at about the level of the third sacral vertebra.

Sign Any objective evidence of disease that can be observed or measured such as a lesion, swelling, or fever.

Sinoatrial (SA) node (si-nō-Ā-trē-al) A small mass of cardiac muscle fibers (cells) located in the right atrium inferior to the opening of the superior vena cava that spontaneously depolarize and generate a cardiac action potential about 100 times per minute. Also called the **pacemaker.**

Sinus (SĪ-nus) A hollow in a bone (paranasal sinus) or other tissue; a channel for blood (vascular sinus); any cavity having a narrow opening.

Sinusoid (SĪ-nū-soyd) A large, thin-walled, and leaky type of capillary, having large intercellular clefts that may allow proteins and blood cells to pass from a tissue into the bloodstream; present in the liver, spleen, anterior pituitary, parathyroid glands, and red bone marrow.

Skeletal muscle An organ specialized for contraction, composed of striated muscle fibers (cells), supported by connective tissue, attached to a bone by a tendon or an aponeurosis, and stimulated by somatic motor neurons.

Skin The external covering of the body that consists of a superficial, thinner epidermis (epithelial tissue) and a deep, thicker dermis (connective tissue) that is anchored to the subcutaneous layer.

Skull The skeleton of the head consisting of the cranial and facial bones.

Sleep A state of partial unconsciousness from which a person can be aroused; associated with a low level of activity in the reticular activating system.

Sliding-filament mechanism The explanation of how thick and thin filaments slide relative to one another during striated muscle contraction to decrease sarcomere length.

Small intestine A long tube of the gastrointestinal tract that begins at the pyloric sphincter of the stomach, coils through the central and inferior part of the abdominal cavity, and ends at the large intestine; divided into three segments: duodenum, jejunum, and ileum.

Smooth muscle A tissue specialized for contraction, composed of smooth muscle fibers (cells), located in the walls of hollow internal organs, and innervated by autonomic motor neurons.

Sodium-potassium pump An active transport pump located in the plasma membrane that transports sodium ions out of the cell and potassium ions into the cell at the expense of cellular ATP. It functions to keep the ionic concentrations of these ions at physiological levels.

Soft palate (PAL-at) The posterior portion of the roof of the mouth, extending from the palatine bones to the uvula. It is a muscular partition lined with mucous membrane.

Solution A homogeneous molecular or ionic dispersion of one or more substances (solutes) in a dissolving medium (solvent) that is usually liquid.

Somatic cell division (sō-MAT-ik) Type of cell division in which a single starting parent cell duplicates itself to produce two identical cells; consists of mitosis and cytokinesis.

Somatic nervous system (SNS) The portion of the peripheral nervous system consisting of somatic sensory (afferent) neurons and somatic motor (efferent) neurons.

Spasm (SPAZM) A sudden, involuntary contraction of large groups of muscles.

Spasticity (spas-TIS-i-tē) Hypertonia characterized by increased muscle tone, increased tendon reflexes, and pathological reflexes (Babinski sign).

Spermatic cord (sper-MAT-ik) A supporting structure of the male reproductive system, extending from a testis to the deep inguinal ring, that includes the ductus (vas) deferens, arteries, veins, lymphatic vessels, nerves, cremaster muscle, and connective tissue.

Spermatogenesis (sper′-ma-tō-JEN-e-sis) The formation and development of sperm in the seminiferous tubules of the testes.

Sperm cell A mature male gamete. Also termed a **spermatozoon** (sper′-ma-tō-ZŌ-on).

Spermiogenesis (sper′-mē-ō-JEN-e-sis) The maturation of spermatids into sperm.

Sphincter (SFINGK-ter) A circular muscle that constricts an opening.

Sphygmomanometer (sfig′-mō-ma-NOM-e-ter) An instrument for measuring arterial blood pressure.

Spinal cord (SPĪ-nal) A mass of nerve tissue located in the vertebral cavity from which 31 pairs of spinal nerves originate.

Spinal nerve One of the 31 pairs of nerves that originate on the spinal cord from posterior and anterior roots.

Spinal shock A period from several days to several weeks following transection of the spinal cord and characterized by the abolition of all reflex activity.

Spinothalamic tracts (spī-nō-tha-LAM-ik) Sensory (ascending) tracts that convey information up the spinal cord to the thalamus for sensations of pain, temperature, crude touch, and deep pressure.

Spiral organ The organ of hearing, consisting of supporting cells and hair cells that rest on the basilar membrane and extend into the endolymph of the cochlear duct. Also called the **organ of Corti** (KOR-tē).

Spirometer (spī-ROM-e-ter) An apparatus used to measure lung volumes and capacities.

Spleen (SPLĒN) Large mass of lymphatic tissue between the fundus of the stomach and the diaphragm that functions in formation of blood cells during early fetal development, phagocytosis of ruptured blood cells, and proliferation of B cells during immune responses.

Spongy (cancellous) bone tissue Bone tissue that consists of an irregular latticework of thin plates of bone called trabeculae; spaces between trabeculae of some bones are filled with red bone marrow; found inside short, flat, and irregular bones and in the epiphyses (ends) of long bones.

Sprain Forcible wrenching or twisting of a joint with partial rupture or other injury to its attachments without dislocation.

Squamous (SKWĀ-mus) Flat or scalelike.

Starvation (star-VĀ-shun) The loss of energy stores in the form of glycogen, triglycerides, and proteins due to inadequate intake of nutrients or inability to digest, absorb, or metabolize ingested nutrients.

Stasis (STĀ-sis) Stagnation or halt of normal flow of fluids, as blood or urine, or of the intestinal contents.

Static equilibrium (ē-kwi-LIB-rē-um) The maintenance of posture in response to changes in the orientation of the body, mainly the head, relative to the ground.

Stellate reticuloendothelial cell (STEL-āt re-tik′-ū-lō-en′-dō-THĒ-lē-al) Phagocytic cell bordering a sinusoid of the liver. Also called a **Kupffer cell** (KOOP-fer).

Stem cell Unspecialized cell that has the ability to divide for indefinite periods and give rise to specialized cells.

Stenosis (sten-Ō-sis) An abnormal narrowing or constriction of a duct or opening.

Sterile (STE-ril) Free from any living microorganisms. Unable to conceive or produce offspring.

Sterilization (ster′-i-li-ZĀ-shun) Elimination of all living microorganisms. Any procedure that renders an individual incapable of reproduction (for example, castration, vasectomy, hysterectomy, or oophorectomy).

Stimulus Any stress that changes a controlled condition; any change in the internal or external environment that excites a sensory receptor, a neuron, or a muscle fiber.

Stomach The J-shaped enlargement of the gastrointestinal tract directly inferior to the diaphragm in the epigastric, umbilical, and left hypochondriac regions of the abdomen, between the esophagus and small intestine.

Stratum (STRĀ-tum) A layer.

Stressor A stress that is extreme, unusual, or long-lasting and triggers the stress response.

Stress response Wide-ranging set of bodily changes, triggered by a stressor, that gears the body to meet an emergency. Also known as **general adaptation syndrome (GAS).**

Stretch receptor Receptor in the walls of blood vessels, airways, or organs that monitors the amount of stretching. Also termed a **baroreceptor.**

Stretch reflex A monosynaptic reflex triggered by sudden stretching of muscle spindles within a muscle that elicits contraction of that same muscle. Also called a **tendon jerk.**

Stroke volume The volume of blood ejected by either ventricle during one systole; about 70 mL in an adult at rest.

Subarachnoid space (sub′-a-RAK-noyd) A space between the arachnoid mater and the pia mater that surrounds the brain and spinal cord and through which cerebrospinal fluid circulates.

Subcutaneous (sub′-kū-TĀ-nē-us) Beneath the skin. Also called **hypodermic** (hi-pō-DER-mik).

Subcutaneous layer A continuous sheet of areolar connective tissue and adipose tissue between the dermis of the skin and the deep fascia of the muscles. Also called the **superficial fascia** (FASH-ē-a).

Subdural space (sub-DOO-ral) A space between the dura mater and the arachnoid mater of the brain and spinal cord that contains a small amount of fluid.

Sublingual gland (sub-LING-gwal) One of a pair of salivary glands situated in the floor of the mouth deep to the mucous membrane and to the side of the lingual frenulum, with a duct that opens into the floor of the mouth.

Submandibular gland (sub′-man-DIB-ū-lar) One of a pair of salivary glands found inferior to the base of the tongue deep to the mucous membrane in the posterior part of the floor of the mouth, posterior to the sublingual glands, with a duct situated to the side of the lingual frenulum. Also called the **submaxillary gland** (sub′-MAK-si-ler-ē).

Submucosa (sub-mū-KŌ-sa) A layer of connective tissue located deep to a mucous membrane, as in the gastrointestinal tract or the urinary bladder; the submucosa connects the mucosa to the muscularis layer.

Substrate A molecule upon which an enzyme acts.

Sudoriferous gland (soo′-dor-IF-er-us) An apocrine or eccrine exocrine gland in the dermis or subcutaneous layer that produces perspiration. Also called a **sweat gland.**

Sulcus (SUL-kus) A groove or depression between parts, especially between the convolutions of the brain. *Plural* is **sulci** (SUL-sī).

Summation (sum-MĀ-shun) The addition of the excitatory and inhibitory effects of many stimuli applied to a neuron. The increased strength of muscle contraction that results when stimuli follow one another in rapid succession.

Superficial (soo′-per-FISH-al) Located on or near the surface of the body or an organ.

Superficial fascia (FASH-ē-a) A continuous sheet of fibrous connective tissue between the dermis of the skin and the deep fascia of the muscles. Also called **subcutaneous layer** (sub′-kū-TĀ-nē-us).

Superior (soo-PĒR-ē-or) Toward the head or upper part of a structure.

Superior vena cava (VĒ-na CĀ-va) **(SVC)** Large vein that collects blood from parts of the body superior to the heart and returns it to the right atrium.

Supination (soo-pi-NĀ-shun) A movement of the forearm in which the palm is turned anteriorly.

Surface anatomy The study of the structures that can be identified from the outside of the body.

Surfactant (sur-FAK-tant) Complex mixture of phospholipids and lipoproteins, produced by type II alveolar (septal) cells in the lungs, that decreases surface tension.

Susceptibility (sus-sep′-ti-BIL-i-tē) Lack of resistance to the damaging effects of an agent such as a pathogen.

Suspensory ligament (sus-PEN-so-rē LIG-a-ment) A fold of peritoneum extending laterally from the surface of the ovary to the pelvic wall.

Sutural bone (SOO-cher-al) A small bone located within a suture between certain cranial bones.

Suture (SOO-cher) An immovable fibrous joint that joins skull bones.

Sympathetic division (sim′-pa-THET-ik) One of the two subdivisions of the autonomic nervous system, having cell bodies of preganglionic neurons in the lateral gray columns of the thoracic segment and the first two or three lumbar segments of the spinal cord; primarily concerned with processes involving the expenditure of energy.

Sympathetic trunk ganglion (GANG-glē-on) A cluster of cell bodies of sympathetic postganglionic neurons lateral to the vertebral column, close to the body of a vertebra. These ganglia extend inferiorly through the neck, thorax, and abdomen to the coccyx on both sides of the vertebral column and are connected to one another to form a chain on each side of the vertebral column. Also called **sympathetic chain** or **vertebral chain ganglia.**

Sympathomimetic (sim′-pa-thō-mi-MET-ik) Producing effects that mimic those brought about by the sympathetic division of the autonomic nervous system.

Symphysis (SIM-fi-sis) A line of union. A slightly movable fibrocartilaginous joint such as the pubic symphysis.

Symptom (SIMP-tum) A subjective change in body function not apparent to an observer, such as pain or nausea, that indicates the presence of a disease or disorder of the body.

Synapse (SYN-aps) The functional junction between two neurons or between a neuron and an effector, such as a muscle or gland; may be electrical or chemical.

Synaptic cleft (sin-AP-tik) The narrow gap at a chemical synapse that separates the axon terminal of one neuron from another neuron or muscle fiber (cell) and across which a neurotransmitter diffuses to affect the postsynaptic cell.

Synaptic end bulb Expanded distal end of an axon terminal that contains synaptic vesicles. Also called a **synaptic knob.**

Synaptic vesicle Membrane-enclosed sac in a synaptic end bulb that stores neurotransmitters.

Synarthrosis (sin′-ar-THRŌ-sis) An immovable joint such as a suture, gomphosis, and synchondrosis.

Synchondrosis (sin′-kon-DRŌ-sis) A cartilaginous joint in which the connecting material is hyaline cartilage.

Syndesmosis (sin′-dez-MŌ-sis) A slightly movable joint in which articulating bones are united by fibrous connective tissue.

Syndrome (SIN-drōm) A group of signs and symptoms that occur together in a pattern that is characteristic of a particular disease or abnormal condition.

Synergist (SIN-er-jist) A muscle that assists the prime mover by reducing undesired action or unnecessary movement.

Synostosis (sin′-os-TŌ-sis) A joint in which the dense fibrous connective tissue that unites bones at a suture has been replaced by bone, resulting in a complete fusion across the suture line.

Synovial cavity (si-NŌ-vē-al) The space between the articulating bones of a synovial joint, filled with synovial fluid. Also called a **joint cavity.**

Synovial fluid Secretion of synovial membranes that lubricates joints and nourishes articular cartilage.

Synovial joint A fully movable or diarthrotic joint in which a synovial (joint) cavity is present between the two articulating bones.

Synovial membrane The deeper of the two layers of the articular capsule of a synovial joint, composed of areolar connective tissue that secretes synovial fluid into the synovial (joint) cavity.

System An association of organs that have a common function.

Systemic (sis-TEM-ik) Affecting the whole body; generalized.

Systemic anatomy The anatomic study of particular systems of the body, such as the skeletal, muscular, nervous, cardiovascular, or urinary systems.

Systemic circulation The routes through which oxygenated blood flows from the left ventricle through the aorta to all the organs of the body and deoxygenated blood returns to the right atrium.

Systemic vascular resistance (SVR) All the vascular resistance offered by systemic blood vessels. Also called **total peripheral resistance.**

Systole (SIS-tō-lē) In the cardiac cycle, the phase of contraction of the heart muscle, especially of the ventricles.

Systolic blood pressure (sis-TOL-ik) The force exerted by blood on arterial walls during ventricular contraction; the highest pressure measured in the large arteries, about 110 mmHg under normal conditions for a young adult.

T

T cell A lymphocyte that becomes immunocompetent in the thymus and can differentiate into a helper T cell or a cytotoxic T cell, both of which function in cell-mediated immunity.

T wave The deflection wave of an electrocardiogram that represents ventricular repolarization.

Tachycardia (tak′-i-KAR-dē-a) An abnormally rapid resting heartbeat or pulse rate (over 100 beats per minute).

Tactile (TAK-tīl) Pertaining to the sense of touch.

Tactile disc Modified epidermal cell in the stratum basale of hairless skin that functions as a cutaneous receptor for discriminative touch. Also called a **Merkel disc** (MER-kel).

Target cell A cell whose activity is affected by a particular hormone.

Tarsal bones The seven bones of the ankle. Also called **tarsals.**

Tarsal gland Sebaceous (oil) gland that opens on the edge of each eyelid. Also called a **Meibomian gland** (mī-BŌ-mē-an).

Tarsal plate A thin, elongated sheet of connective tissue, one in each eyelid, giving the eyelid form and support. The aponeurosis of the levator palpebrae superioris is attached to the tarsal plate of the superior eyelid.

Tarsus (TAR-sus) A collective term for the seven bones of the ankle.

Tectorial membrane (tek-TŌ-rē-al) A gelatinous membrane projecting over and in contact with the hair cells of the spiral organ (organ of Corti) in the cochlear duct.

Teeth (TĒTH) Accessory structures of digestion, composed of calcified connective tissue and embedded in bony sockets of the mandible and maxilla, that cut, shred, crush, and grind food. Also called **dentes** (DEN-tēz).

Telophase (TEL-ō-fāz) The final stage of mitosis in which two nuclei become established.

Tendon (TEN-don) A white fibrous cord of dense regular connective tissue that attaches muscle to bone.

Tendon organ A proprioceptive receptor, sensitive to changes in muscle tension and force of contraction, found chiefly near the junctions of tendons and muscles. Also called a **Golgi tendon organ** (GOL-jē).

Tendon reflex A polysynaptic, ipsilateral reflex that protects tendons and their associated muscles from damage that might be brought about by excessive tension. The receptors involved are called tendon organs (Golgi tendon organs).

Teratogen (TER-a-tō-jen) Any agent or factor that causes physical defects in a developing embryo.

Testis (TES-tis) Male gonad that produces sperm and the hormones testosterone and inhibin. Also called a **testicle.**

Testosterone (tes-TOS-te-rōn) A male sex hormone (androgen) secreted by interstitial endocrinocytes (Leydig cells) of a mature testis; needed for development of sperm; together with a second androgen termed dihydrotestosterone (DHT), controls the growth and development of male reproductive organs, secondary sex characteristics, and body growth.

Tetany (TET-a-nē) Hyperexcitability of neurons and muscle fibers (cells) caused by hypocalcemia and characterized by intermittent or continuous tonic muscular contractions; may be due to hypoparathyroidism.

Thalamus (THAL-a-mus) A large, oval structure located bilaterally on either side of the third ventricle, consisting of two masses of gray matter organized into nuclei; main relay center for sensory impulses ascending to the cerebral cortex.

Thermoreceptor (THER-mō-rē-sep-tor) Sensory receptor that detects changes in temperature.

Thigh The portion of the lower limb between the hip and the knee.

Third ventricle (VEN-tri-kul) A slitlike cavity between the right and left halves of the thalamus and between the lateral ventricles of the brain.

Thirst center A cluster of neurons in the hypothalamus that is sensitive to the osmotic pressure of extracellular fluid and brings about the sensation of thirst.

Thoracic cavity (thō-RAS-ik) A cavity that contains two pleural cavities, the mediastinum, and the pericardial cavity.

Thoracic duct A lymphatic vessel that begins as a dilation called the cisterna chyli, receives lymph from the left side of the head, neck, and chest, the left arm, and the entire body below the ribs, and empties into the left subclavian vein. Also called the **left lymphatic duct** (lim-FAT-ik).

Thoracolumbar outflow (thō′-ra-kō-LUM-bar) The axons of sympathetic preganglionic neurons, which have their cell bodies in the lateral gray columns of the thoracic segments and first two or three lumbar segments of the spinal cord.

Thorax (THŌ-raks) The chest.

Threshold potential The membrane voltage that must be reached to trigger an action potential.

Threshold stimulus Any stimulus strong enough to initiate an action potential or activate a sensory receptor.

Thrombin (THROM-bin) The active enzyme formed from prothrombin that converts fibrinogen to fibrin during the formation of a blood clot.

Thrombolytic agent (throm-bō-LIT-ik) Chemical substance injected into the body to dissolve blood clots and restore circulation; mechanism of action is direct or indirect activation of plasminogen; examples include tissue plasminogen activator (t-PA), streptokinase, and urokinase.

Thrombosis (throm-BŌ-sis) The formation of a clot in an unbroken blood vessel, usually a vein.

Thrombus A stationary clot formed in an unbroken blood vessel, usually a vein.

Thymus (THĪ-mus) A bilobed organ, located in the superior mediastinum posterior to the sternum and between the lungs, in which T cells develop immunocompetence.

Thyroglobulin (TGB) (thī-rō-GLŌ-bū-lin) A large glycoprotein molecule produced by follicular cells of the thyroid gland in which some tyrosines are iodinated and coupled to form thyroid hormones.

Thyroid cartilage (THĪ-royd KAR-ti-lij) The largest single cartilage of the larynx, consisting of two fused plates that form the anterior wall of the larynx.

Thyroid follicle (FOL-i-kul) Spherical sac that forms the parenchyma of the thyroid gland and consists of follicular cells that produce thyroxine (T_4) and triiodothyronine (T_3).

Thyroid gland An endocrine gland with right and left lateral lobes on either side of the trachea connected by an isthmus; located anterior to the trachea just inferior to the cricoid cartilage; secretes thyroxine (T_4), triiodothyronine (T_3), and calcitonin.

Thyroid-stimulating hormone (TSH) A hormone secreted by the anterior pituitary that stimulates the synthesis and secretion of thyroxine (T_4) and triiodothyronine (T_3).

Thyroxine (T_4) (thī-ROK-sēn) A hormone secreted by the thyroid gland that regulates metabolism, growth and development, and the activity of the nervous system.

Tic Spasmodic, involuntary twitching of muscles that are normally under voluntary control.

Tidal volume The volume of air breathed in and out in any one breath; about 500 mL in quiet, resting conditions.

Tissue A group of similar cells and their intercellular substance joined together to perform a specific function.

Tissue factor (TF) A factor, or collection of factors, whose appearance initiates the blood clotting process. Also called **thromboplastin** (throm-bō-PLAS-tin).

Tissue plasminogen activator (t-PA) An enzyme that dissolves small blood clots by initiating a process that converts plasminogen to plasmin, which degrades the fibrin of a clot.

Tongue A large skeletal muscle covered by a mucous membrane located on the floor of the oral cavity.

Tonicity (tō-NIS-i-tē) A measure of the concentration of impermeable solute particles in a solution relative to cytosol. When cells are bathed in an isotonic solution, they neither shrink nor swell.

Tonsil (TON-sil) An aggregation of large lymphatic nodules embedded in the mucous membrane of the throat.

Torn cartilage A tearing of an articular disc (meniscus) in the knee.

Total lung capacity The sum of tidal volume, inspiratory reserve volume, expiratory reserve volume, and residual volume; about 6000 mL in males.

Trabecula (tra-BEK-ū-la) Irregular latticework of thin plates of spongy bone. Fibrous cord of connective tissue serving as supporting fiber by forming a septum extending into an organ from its wall or capsule. *Plural* is **trabeculae** (tra-BEK-ū-lē).

Trachea (TRĀ-kē-a) Tubular air passageway extending from the larynx to the fifth thoracic vertebra. Also called the **windpipe.**

Tract A bundle of nerve axons in the central nervous system.

Transcription (trans-KRIP-shun) The first step in the expression of genetic information in which a single strand of DNA serves as a template for the formation of an RNA molecule.

Translation (trans-LĀ-shun) The synthesis of a new protein on a ribosome as dictated by the sequence of codons in messenger RNA.

Transverse colon (trans-VERS KŌ-lon) The portion of the large intestine extending across the abdomen from the right colic (hepatic) flexure to the left colic (splenic) flexure.

Transverse fissure (FISH-er) The deep cleft that separates the cerebrum from the cerebellum.

Transverse plane A plane that divides the body or organs into superior and inferior portions. Also called a **horizontal plane.**

Transverse tubules (T tubules) (TOO-būls) Small, cylindrical invaginations of the sarcolemma of striated muscle fibers (cells) that conduct muscle action potentials toward the center of the muscle fiber.

Trauma (TRAW-ma) An injury, either a physical wound or psychic disorder, caused by an external agent or force, such as a physical blow or emotional shock; the agent or force that causes the injury.

Tremor (TREM-or) Rhythmic, involuntary, purposeless contraction of opposing muscle groups.

Tricuspid valve (trī-KUS-pid) Atrioventricular (AV) valve on the right side of the heart.

Triglyceride (trī-GLI-cer-īd) A lipid formed from one molecule of glycerol and three molecules of fatty acids that may be either solid (fats) or liquid (oils) at room temperature; the body's most highly concentrated source of chemical potential energy. Found mainly within adipocytes. Also called a **neutral fat** or a **triacylglycerol.**

Triiodothyronine (trī-ī-ō-dō-THĪ-rō-nēn) **(T_3)** A hormone produced by the thyroid gland that regulates metabolism, growth and development, and the activity of the nervous system.

Trophoblast (TRŌF-ō-blast) The superficial covering of cells of the blastocyst.

Tropic hormone (TRŌ-pik) A hormone whose target is another endocrine gland.

Trunk The part of the body to which the upper and lower limbs are attached.

Tubal ligation (lī-GĀ-shun) A sterilization procedure in which the uterine (fallopian) tubes are tied and cut.

Tubular reabsorption The movement of filtrate from renal tubules back into blood in response to the body's specific needs.

Tubular secretion The movement of substances in blood into renal tubular fluid in response to the body's specific needs.

Tumor suppressor gene A gene coding for a protein that normally inhibits cell division; loss or alteration of a tumor suppressor gene called *p53* is the most common genetic change in a wide variety of cancer cells.

Tunica externa (eks-TER-na) The superficial coat of an artery or vein, composed mostly of elastic and collagen fibers. Also called the **adventitia.**

Tunica interna (in-TER-na) The deep coat of an artery or vein, consisting of a lining of endothelium, basement membrane, and internal elastic lamina. Also called the **tunica intima** (IN-ti-ma).

Tunica media (MĒ-dē-a) The intermediate coat of an artery or vein, composed of smooth muscle and elastic fibers.

Twitch contraction Brief contraction of all muscle fibers (cells) in a motor unit triggered by a single action potential in its motor neuron.

Type II cutaneous mechanoreceptor A sensory receptor embedded deeply in the dermis and deeper tissues that detects stretching of skin. Also called a **Ruffini corpuscle.**

U

Umbilical cord The long, ropelike structure containing the umbilical arteries and vein that connect the fetus to the placenta.

Umbilicus (um-BIL-i-kus *or* um-bil-Ī-kus) A small scar on the abdomen that marks the former attachment of the umbilical cord to the fetus. Also called the **navel.**

Upper limb The appendage attached at the shoulder girdle, consisting of the arm, forearm, wrist, hand, and fingers. Also called **upper extremity.**

Uremia (ū-RĒ-mē-a) Accumulation of toxic levels of urea and other nitrogenous waste products in the blood, usually resulting from severe kidney malfunction.

Ureter (Ū-rē-ter) One of two tubes that connect the kidney with the urinary bladder.

Urethra (ū-RĒ-thra) The duct from the urinary bladder to the exterior of the body that conveys urine in females and urine and semen in males.

Urinary bladder (Ū-ri-ner-ē) A hollow, muscular organ situated in the pelvic cavity posterior to the pubic symphysis; receives urine via two ureters and stores urine until it is excreted through the urethra.

Urine The fluid produced by the kidneys that contains wastes and excess materials; excreted from the body through the urethra.

Urology (ū-ROL-ō-jē) The specialized branch of medicine that deals with the structure, function, and diseases of the male and female urinary systems and the male reproductive system.

Uterine tube (Ū-ter-in) Duct that transports ova from the ovary to the uterus. Also called the **fallopian tube** (fal-LŌ-pē-an) or **oviduct.**

Uterus (Ū-te-rus) The hollow, muscular organ in females that is the site of menstruation, implantation, development of the fetus, and labor. Also called the **womb.**

Utricle (Ū-tri-kul) The larger of the two divisions of the membranous labyrinth located inside the vestibule of the inner ear, containing a receptor organ for static equilibrium.

Uvula (Ū-vū-la) A soft, fleshy mass, especially the V-shaped pendant part, descending from the soft palate.

V

Vagina (va-JĪna) A muscular, tubular organ that leads from the uterus to the vestibule, situated between the urinary bladder and the rectum of the female.

Vallate papilla (VAL-āt pa-PIL-a) One of the circular projections that is arranged in an inverted V-shaped row at the back of the tongue; the largest of the elevations on the upper surface of the tongue that contains taste buds. Also called **circumvallate papilla.**

Valence (VĀ-lens) The combining capacity of an atom; the number of deficit or extra electrons in the outermost electron shell of an atom.

Varicose (VAR-i-kōs) Pertaining to an unnatural swelling, as in the case of a varicose vein.

Vasa recta (VĀ-sa REK-ta) Extensions of the efferent arteriole of a juxtamedullary nephron that run alongside the loop of the nephron (Henle) in the medullary region of the kidney.

Vasa vasorum (va-SŌ-rum) Blood vessels that supply nutrients to the larger arteries and veins.

Vascular (VAS-kū-lar) Pertaining to or containing many blood vessels.

Vascular spasm Contraction of the smooth muscle in the wall of a damaged blood vessel to prevent blood loss.

Vascular (venous) sinus A vein with a thin endothelial wall that lacks a tunica media and externa and is supported by surrounding tissue.

Vascular tunic (TOO-nik) The middle layer of the eyeball, composed of the choroid, ciliary body, and iris. Also called the **uvea** (Ū-ve-a).

Vasectomy (va-SEK-tō-mē) A means of sterilization of males in which a portion of each ductus (vas) deferens is removed.

Vasoconstriction (vāz-ō-kon-STRIK-shun) A decrease in the size of the lumen of a blood vessel caused by contraction of the smooth muscle in the wall of the vessel.

Vasodilation (vāz′-ō-DĪ-lā-shun) An increase in the size of the lumen of a blood vessel caused by relaxation of the smooth muscle in the wall of the vessel.

Vein A blood vessel that conveys blood from tissues back to the heart.

Vena cava (VĒ-na KĀ-va) One of two large veins that open into the right atrium, returning to the heart all of the deoxygenated blood from the systemic circulation except from the coronary circulation.

Ventral (VEN-tral) Pertaining to the anterior or front side of the body; opposite of dorsal.

Ventricle (VEN-tri-kul) A cavity in the brain filled with cerebrospinal fluid. An inferior chamber of the heart.

Ventricular fibrillation (ven-TRIK-ū-lar fib-ri-LĀ-shun) Asynchronous ventricular contractions; unless reversed by defibrillation, results in heart failure.

Venule (VEN-ūl) A small vein that collects blood from capillaries and delivers it to a vein.

Vermiform appendix (VER-mi-form a-PEN-diks) A twisted, coiled tube attached to the cecum.

Vertebral cavity (VER-te-bral) A space within the vertebral column formed by the vertebral foramina of all the vertebrae and containing the spinal cord.

Vertebral column The 26 vertebrae of an adult and 33 vertebrae of a child; encloses and protects the spinal cord and serves as a point of attachment for the ribs and back muscles. Also called the **backbone, spine,** or **spinal column.**

Vesicle (VES-i-kul) A small bladder or sac containing liquid.

Vestibular apparatus (ves-TIB-ū-lar) Collective term for the organs of equilibrium, which includes the saccule, utricle, and semicircular ducts.

Vestibular membrane The membrane that separates the cochlear duct from the scala vestibuli.

Vestibule (VES-ti-būl) A small space or cavity at the beginning of a canal, especially the inner ear, larynx, mouth, nose, and vagina.

Villus (VIL-lus) A projection of the intestinal mucosal cells containing connective tissue, blood vessels, and a lymphatic vessel; functions in the absorption of the end products of digestion. *Plural* is **villi** (VIL-ī).

Viscera (VIS-er-a) The organs inside the thoracic and abdominopelvic cavities. *Singular* is **viscus** (VIS-kus).

Visceral (VIS-er-al) Pertaining to the organs or to the covering of an organ.

Visceral effectors (e-FEK-torz) Organs of the thoracic and abdominopelvic cavities that respond to neural stimulation, including cardiac muscle, smooth muscle, and glands.

Vital capacity The sum of inspiratory reserve volume, tidal volume, and expiratory reserve volume; about 4800 mL in males.

Vital signs Signs necessary to life that include temperature (T), pulse (P), respiratory rate (RR), and blood pressure (BP).

Vitamin An organic molecule necessary in trace amounts that acts as a catalyst in normal metabolic processes in the body.

Vitreous body (VIT-rē-us) A soft, jellylike substance that fills the vitreous chamber of the eyeball, lying between the lens and the retina.

Vocal folds Pair of mucous membrane folds below the ventricular folds that function in voice production. Also called **true vocal cords.**

Voltage-gated channel An ion channel in a plasma membrane composed of integral proteins that functions like a gate to permit or restrict the movement of ions across the membrane in response to changes in the voltage.

Vulva (VUL-va) Collective designation for the external genitals of the female. Also called the **pudendum** (poo-DEN-dum).

W

Wallerian degeneration (wal-LE-rē-an) Degeneration of the portion of the axon and myelin sheath of a neuron distal to the site of injury.

Wandering macrophage (MAK-rō-fāj) Phagocytic cell that develops from a monocyte, leaves the blood, and migrates to infected tissues.

Wave summation (sum-MĀ-shun) The increased strength of muscle contraction that results when muscle action potentials occur one after another in rapid succession.

White matter Aggregations or bundles of myelinated and unmyelinated axons located in the brain and spinal cord.

X

Xiphoid (ZĪ-foyd) Sword-shaped. The inferior portion of the sternum is the **xiphoid process.**

Y

Yolk sac An extraembryonic membrane composed of the exocoelomic membrane and hypoblast. It transfers nutrients to the embryo, is a source of blood cells, contains primordial germ cells that migrate into the gonads to form primitive germ cells, forms part of the gut, and helps prevent desiccation of the embryo.

Z

Zona pellucida (pe-LOO-si-da) Clear glycoprotein layer between a secondary oocyte and the surrounding granulosa cells of the corona radiata.

Zygote (ZĪ-g-ot) The single cell resulting from the union of male and female gametes; the fertilized ovum.

CREDITS

ART CREDITS

Chapter 1
Figure 1.1: Kevin Somerville. 1.2–1.3: Jared Schneidman Design. 1.4: Molly Borman. 1.5: Kevin Somerville. 1.6: Molly Borman. 1.7–1.9: Imagineering. 1.10a–b: Kevin Somerville. 1.11: Kevin Somerville. Table 1.1a–k: Keith Kasnot.

Chapter 2
Figure 2.1: Imagineering. 2.2–2.8: Jared Schneidman Design. 2.9: Imagineering. 2.10: Jared Schneidman Design. 2.11: Imagineering. 2.12–2.14: Jared Schneidman Design. 2.15: Imagineering. 2.16: Jared Schneidman Design.

Chapter 3
Figure 3.1–3.2: Tomo Narashima. 3.4–3.6: Imagineering. 3.7: Jared Schneidman Design. 3.8–3.11: Imagineering. 3.12–3.17: Tomo Narashima. 3.18–3.21: Imagineering. Table 3.2: Tomo Narashima.

Chapter 4
Figure 4.1–4.2: Imagineering. Table 4.1a–k: Imagineering. Table 4.2a–i: Imagineering.

Chapter 5
Figure 5.1–5.4: Kevin Somerville.

Chapter 6
Figure 6.1: Imagineering. 6.2a,c: Kevin Somerville/Imagineering. 6.2b: Nadine Sokol/Imagineering. 6.3–6.4: Kevin Somerville. 6.5: Jared Schneidman Design. 6.6–6.7a–c: Imagineering. 6.8–6.12a–b: Imagineering. 6.13–6.20a–b: Imagineering. 6.21–6.22a–b: Imagineering. 6.23–6.27: Imagineering. Table 6.2a–b: Imagineering. Table 6.3a–b: Imagineering. Table 6.4: Imagineering.

Chapter 7
Figure 7.1–7.3: Imagineering. 7.09a–f: Imagineering. 7.10: Imagineering.

Chapter 8
Figure 8.1: Kevin Somerville. 8.2a,c: Kevin Somerville/Imagineering. 8.3–8.4: Imagineering. 8.5: Kevin Somerville. 8.6–8.7: Imagineering. 8.8–8.9: Jared Schneidman Design. 8.10–8.11: Imagineering. 8.12: Leonard Dank. 8.13a,b: Leonard Dank. 8.14–8.21a–c: Leonard Dank. 8.22–8.24a–d: Leonard Dank. Table 8.1a–c: Kevin Somerville.

Chapter 9
Figure 9.1: Kevin Somerville/Imagineering. 9.2: Jared Schneidman Design. 9.3: Kevin Somerville. 9.4–9.5: Imagineering. 9.6: Jared Schneidman Design. 9.7: Imagineering. Table 9.1: Kevin Somerville.

Chapter 10
Figure 10.1–10.2: Kevin Somerville. 10.3–10.4: Kevin Somerville/Imagineering. 10.5: Leonard Dank. 10.6–10.10: Kevin Somerville/Imagineering. 10.11: Kevin Somerville. 10.12: Kevin Somerville/Imagineering. 10.13–10.15: Kevin Somerville. Table 10.1a–f: Imagineering.

Chapter 11
Figure 11.1–11.3: Imagineering.

Chapter 12
Figure 12.1–12.2: Kevin Somerville. 12.3: Tomo Narashima. 12.4: Molly Borman. 12.5: Sharon Ellis. 12.6: Tomo Narashima/Imagineering. 12.7: Imagineering. 12.8: Steve Oh/Imagineering. 12.9–12.10: Imagineering. 12.11: Nadine Sokol. 12.12–12.13: Tomo Narashima/Imagineering. 12.14: Tomo Narashima. 12.15: Hilda Muinos/Imagineering. 12.16: Tomo Narashima/Imagineering. Table 12.2–12.3: Imagineering.

Chapter 13
Figure 13.1: Steve Oh/Imagineering. 13.2–13.3: Jared Schneidman Design. 13.4–13.5: Lynn O'Kelley/Imagineering. 13.6: Jared

Schneidman Design. 13.7: Molly Borman/Imagineering. 13.8: Jared Schneidman Design. 13.9: Molly Borman/Imagineering. 13.10: Imagineering. 13.11a–c: Molly Borman/Imagineering. 13.12: Jared Schneidman Design. 13.13: Molly Borman/Imagineering. 13.14: Imagineering.

Chapter 14

Figure 14.1–14.2a: Imagineering. 14.3–14.4: Jared Schneidman Design. 14.5–14.6: Imagineering. Table 14.2: Imagineering

Chapter 15

Figure 15.1: Kevin Somerville. 15.2: Kevin Somerville/Imagineering. 15.3a: Kevin Somerville. 15.3b–15.4: Kevin Somerville/Imagineering. 15.4: Kevin Somerville. 15.5a–b Hilda Muinos/Imagineering. 15.6: Kevin Somerville/Imagineering. 15.7–15.8: Imagineering. 15.9: Kevin Somerville.

Chapter 16

Figure 16.1: Kevin Somerville. 16.2a–b: Jared Schneidman Design. 16.3: Imagineering. 16.4: Kevin Somerville. 16.5–16.6: Imagineering. 16.7: Jared Schneidman Design. 16.8–16.16a: Kevin Somerville. 16.16b: Nadine Sokol. 16.17: Kevin Somerville.

Chapter 17

Figure 17.1: Molly Borman. 17.2: Sharon Ellis. 17.3: Nadine Sokol. 17.4–17.5: Molly Borman. 17.6: Jared Schneidman Design/Imagineering. 17.7: Jared Schneidman Design. 17.8–17.10: Imagineering. 17.11: Jared Schneidman Design

Chapter 18

Figure 18.1: Molly Borman. 18.2: Kevin Somerville/Imagineering. 18.3: Molly Borman/Imagineering. 18.4: Molly Borman. 18.5–18.6: Kevin Somerville/Imagineering. 18.7: Kevin Somerville. 18.8–18.13: Jared Schneidman Design.

Chapter 19

Figure 19.1: Steve Oh. 19.2: Kevin Somerville. 19.3: Steve Oh. 19.4: Nadine Sokol. 19.5: Steve Oh/Imagineering. 19.6: Nadine Sokol. 19.7: Steve Oh. 19.8–19.9: Kevin Somerville. 19.10: Steve Oh. 19.11–19.13: Kevin Somerville. 19.14: Jared Schneidman Design. 19.15: Molly Borman. 19.16: Kevin Somerville.

Chapter 20

Figure 20.1–20.5: Imagineering. 20.6: Jared Schneidman Design/Imagineering.

Chapter 21

Figure 21.1: Kevin Somerville. 21.2–21.3: Steve Oh/Imagineering. 21.4: Imagineering. 21.5: Nadine Sokol. 21.6: Kevin Somerville. 21.7: Imagineering. 21.8: Jared Schneidman Design. 21.9: Steve Oh/Imagineering.

Chapter 22

Figure 22.1–22.6: Jared Schneidman Design.

Chapter 23

Figure 23.1–23.2: Kevin Somerville/Imagineering. 23.3: Imagineering. 23.4: Kevin Somerville. 23.5: Imagineering. 23.6–23.7: Kevin Somerville/Imagineering. 23.38: Imagineering. 23.9: Kevin Somerville/Imagineering. 23.10: Kevin Somerville. 23.11: Kevin Somerville/Imagineering. 23.12–23.13: Imagineering.

Chapter 24

Figure 24.1–24.7: Kevin Somerville. 24.8: Imagineering. 24.10–24.12: Jared Schneidman Design. Table 24.1: Kevin Somerville.

PHOTOS

Chapter 1

Fig. 1.1: Rubberball Productions/Getty Images; Fig. 1.1: Rubberball Productions/Getty Images; Fig. 1.7a: Stephen A. Kieffer and E. Robert Heitzman, An Atlas of Cross-Sectional Anatomy. Harper & Row, New York, 1979.; Fig. 1.7b: Lester Bergman/The Bergman Collection; Fig. 1.7c: Martin Rotker; Opener and Focus on Wellness: 1: The Image Bank/Getty Images

Chapter 2

Opener and Focus on Wellness: Spencer Jones/Taxi/Getty Images

Chapter 3

Fig. 3.3: Andy Washnik; Fig. 3.8 left: David Phillips/Photo Researchers; Fig. 3.8 mid: David Phillips/Photo Researchers; Fig. 3.8 right: David Phillips/Photo Researchers; Fig. 3.21a: Courtesy Michael Ross, University of Florida; Fig. 3.21b left: Courtesy Michael Ross, University of Florida; Fig. 3.21b right: Courtesy Michael Ross, University of Florida; Fig. 3.21 c: Courtesy Michael Ross, University of Florida; Fig. 3.21d right: Courtesy Michael Ross, University of Florida; Fig. 3.21d left: Courtesy Michael Ross, University of Florida; Fig. 3.21e: Courtesy Michael Ross, University of Florida; Fig. 3.21f: Courtesy Michael Ross, University of Florida; Opener and Focus on Wellness: EyeWire, Inc./Getty Images

Chapter 4

Table: 4.1a top: Biophoto Associates/Photo Researchers, Inc.; Table: 4.1a middle: Courtesy Michael Ross, University of Florida; Table: 4.1B: Courtesy Michael Ross, University of Florida; Table: 4.1C: Courtesy Michael Ross, University of Florida; Table: 4.1D top: Courtesy Michael Ross, University of Florida; Table: 4.1E: Courtesy Michael Ross, University of Florida; Table: 4.1F: Biophoto Associates/Photo Researchers; Table: 4.1G: Courtesy Michael Ross, University of Florida; Table: 4.1H: Courtesy Michael Ross, University of Florida; Table: 4.1I: Courtesy Michael Ross, University of Florida; Table: 4.1J: Lester Bergman/The Bergman Collection; Table: 4.1K: Courtesy Michael Ross, University of Florida; Table: 4.2A: Courtesy Michael Ross, University of Florida; Table: 4.2B: Courtesy Michael Ross, University of Florida; Table: .2c: Courtesy Michael Ross, University of Florida; Table: 4.2d: Courtesy Andrew J. Kuntzman; Table: 4.2e: Ed Reschke; Table: 4.2f: Courtesy Michael Ross, University of Florida; Table: 4.2g: Courtesy Michael Ross, University of Florida; Table: 4.2h: Courtesy Michael Ross, University of Florida; Table: 4.2i: Courtesy

Michael Ross, University of Florida; Opener and Focus on Wellness: P. Motta/Photo Researchers, Inc.

Chapter 5
Fig. 5.2b: Courtesy Michael Ross, University of Florida; Opener and Focus on Wellness: Digital Vision; Fig. 5.5a: Alain Dex/Photo Researchers; Fig. 5.5b: Biophoto Associates/Photo Researchers

Chapter 6
Fig. 6.1b: Mark Nielsen; Fig. 6.2b: John Burbidge/Photo Researchers; Opener and Focus on Wellness: Digital Vision; Fig. 6.28a: P. Motta, Dept. of Anatomy, University Lapienza, Rome/Photo Researchers, Inc.; Fig. 6.28b: P. Motta, Dept. of Anatomy, University Lapienza, Rome/Photo Researchers, Inc.

Chapter 7
Fig. 7.4 a–f: John Wilson White; Fig. 7.5a–c: John Wilson White; Fig. 7.6a–b: John Wilson White; Fig. 7.7a–b: John Wilson White; Fig. 7.8a–h: John Wilson White; Opener and Focus on Wellness: PhotoDisc, Inc./Getty Images

Chapter 8
Opener and Focus on Wellness: PhotoDisc, Inc./Getty Images

Chapter 9
Opener and Focus on Wellness: Chris Bayley/Stone/Getty Images

Chapter 10
Fig. 10.6: Mark Nielsen; Opener and Focus on Wellness: James D'Addio/Corbis Stock Market

Chapter 11
Opener and Focus on Wellness: PhotoDisc, Inc./Getty Images

Chapter 12
Opener and Focus on Wellness: Zigy Kaluzny/Stone/Getty Images

Chapter 13
Fig. 13.7b: Courtesy Michael Ross, University of Florida; Fig. 13.11c: Courtesy Michael Ross, University of Florida; Fig. 13.13b: Courtesy Michael Ross, University of Florida; Fig. 13.15a: From New England Journal of Medicine, February 18, 1999, vol. 340, No. 7, page 524. Photo provided courtesy of Robert Gagel, Department of Internal Medicine, University of Texas M.D. Anderson Cancer Center, Houston Texas; Fig. 13.15b: The Bergman Collection/Project Masters, Inc.; Fig. 13.15c: Martin Rotker/Phototake; Fig. 13.15d: The Bergman Collection/Project Masters, Inc.; Fig. 15.e: Biophoto Associates/Photo Researchers, Inc.; Opener and Focus on Wellness: Donna Day/Stone/Getty Images

Chapter 14
Fig. 14.2b top left: Courtesy Michael Ross, University of Florida; Fig. 14.2b top middle: Courtesy Michael Ross, University of Florida; Fig. 14.2b top right: Courtesy Michael Ross, University of Florida; Fig. 14.2b center: Courtesy Michael Ross, University of Florida; Fig. 14.2b left bottom: Courtesy Michael Ross, University of Florida; Fig. 14.2b bottom right: Courtesy Michael Ross, University of Florida; Fig. 14.5: Dennis Kunkel/Phototake; Opener and Focus on Wellness: PhotoDisc, Inc./Getty Images

Chapter 15
Fig. 15.2c: Courtesy Michael Ross, University of Florida; Fig. 15.10a: Vu/Cabisco/Visuals Unlimited; Fig. 15.10b: W. Ober/Visuals Unlimited; Opener and Focus on Wellness: Charles Thatcher/Stone/Getty Images

Chapter 16
Fig. 16.2b: Courtesy Michael Ross, University of Florida; Opener and Focus on Wellness: Jim Cummins/Taxi/Getty Images

Chapter 17
Opener and Focus on Wellness: Stephen Simpson/Taxi/Getty Images

Chapter 18
Fig. 18.5b: Biophoto Associates/Photo Researchers, Inc.; Opener and Focus on Wellness: Steve Taylor/Stone/Getty Images

Chapter 19
Fig. 19.12b: Mark Nielsen; Opener and Focus on Wellness: Donna Day/Stone/Getty Images

Chapter 20
Opener and Focus on Wellness: 518

Chapter 21
Opener and Focus on Wellness: Jacqui Hurst/Corbis Images

Chapter 22
Opener and Focus on Wellness: Jim Cummins/Taxi/Getty Images

Chapter 23
Opener and Focus on Wellness: John Kelly/The Image Bank/Getty Images

Chapter 24
Fig. 24.8b: Siu, Biomedical Comm. /Custom Medical Stock Photo, Inc.; Fig. 24.9a: Photo provided courtesy of Kohei Shiote, Congenital Anomaly Research Center, Kyoto University, Graduate School of Medicine; Fig. 24.9b: Courtesy National Museum of Health and Medicine, Armed Forces Institute of Pathology; Fig. 24.9c: Courtesy National Museum of Health and Medicine, Armed Forces Institute of Pathology; Fig. 24.9d: Courtesy National Museum of Health and Medicine, Armed Forces Institute of Pathology; Fig. 24.9e: Courtesy National Museum of Health and Medicine, Armed Forces Institute of Pathology; Fig. 24.9f: ∏Lennart Nilsson, from A CHILD IS BORN; Fig. 24.9g: Photo provided courtesy of Kohei Shiote, Congenital Anomaly Research Center, Kyoto University, Graduate School of Medicine; Fig. 24.9h: Photo provided courtesy of Kohei Shiote, Congenital Anomaly Research Center, Kyoto University, Graduate School of Medicine; Opener and Focus on Wellness: PhotoDisc, Inc./Getty Images

INDEX

Note: A page number followed by the letter E indicates terms to be found in Exhibits and corresponding art, a page number followed by the letter *f* indicates terms to be found in figures and corresponding captions, and a page number followed by the letter T indicates terms to be found in Tables.

A

A band, 175, 176*f*, 177*f*
Abdominal aorta, 396E, 397*f*
Abdominal cavity, 15, 17
Abdominal muscles, 198E, 199*f*
Abdominal thrust maneuver, 468
Abdominopelvic cavity, 15*f*, 17
  regions and quadrants of, 16*f*, 17, 17*f*
Abducens nerve, 252*f*, 264T
Abduction, 161, 162*f*
Abductor pollicis longus, 209*f*
ABO blood group, 357–358
Abortion, 577
Abrasion, 108
Abscess, 427
Absorption
  of bile salts, 489–490
  cuboidal cells in, 75
  in GI tract, 474
  in large intestine, 493
  skin in, 105
  in small intestine, 490*f*, 499–491
  in stomach, 482–483
Absorptive cell, in small intestine, 487
Abstinence, periodic, for birth control, 575
Accessory digestive organs, 473, 473*f*
Accessory ligaments, 159
Accessory nerve, 252*f*, 264T
Accessory sex glands, 556, 557*f*, 562–563
Accommodation, 299, 299*f*
Acetabulum, 142*f*, 143*f*, 144
Acetylcholine (ACh)
  in neuromuscular junction, 178–179, 178*f*
  as neurotransmitter, 237
    in ANS, 277
Acetylcholinesterase (AChE)
  in neuromuscular junction, 179
  as neurotransmitter, in ANS, 277
Acetyl coenzyme A, 511
Acetyl group, 511
Acid, 30, 30*f*
Acid-base balance, 30–31, 550–552. *See also* pH balance
  aging and, 553
  buffer systems in, 550–551
  exhalation of CO_2 in, 551–552
  imbalances in, 552
  ions in, 548
  kidneys in, 552
  mechanisms of, 553T
  prolonged physical activity and, 551
Acidosis, 393, 512, 552
Acini, 483
Acquired immunodeficiency syndrome (AIDS). *See* HIV/AIDS
Acromegaly, 337, 337*f*
Acromion, 139, 139*f*
Acrosome, 560–561, 560*f*
Actin, 175, 177*f*
Action potential, 230–234
  generation of, 232–233
  of heart, 372, 374
  impulse conduction, 233–234, 234*f*
  ion channels as, 230, 232
  resting membrane potential as, 232, 232*f*
Active site, 37, 37*f*
Active transport, 47, 50–51, 53T
Activity-adjusted pacemaker, 373
Acute prostatitis, 579
Acute renal failure, 539
Adam's apple (thyroid cartilage), 447*f*, 448, 449*f*
Adaptation, in sensory receptors, 280
Adaptive immunity, 420, 428–437
  antibodies in, 429–430, 429*f*
  antibody-mediated, B cells in, 435–436, 435*f*, 436T
  antigens in, 429–430, 429*f*
  cell-mediated, T cells and, 432–434, 433*f*, 434*f*, 436T
  immune responses in, 429
  immunological memory in, 436–437
  processing and presenting antigens in, 430–432
  types of, 437T
Addison's disease, 338
Adductor longus, 212E, 213*f*, 215E
Adductor magnus, 212E, 213*f*, 215E
Adenoid, 425
Adenoma
  feminizing, 340
  virilizing, 340
Adenosine diphosphate (ADP), 38, 39, 40*f*
  in muscle contraction, 179–180, 179*f*, 182, 182*f*
Adenosine triphosphate (ATP), 29, 38, 40, 40*f*
  in glucose metabolism, 510, 510*f*
  in metabolic reactions, 507
  in muscle contraction, 179–180, 179*f*, 181*f*, 182–183, 182*f*
  as organic compound, 38, 40
  synthesis of, 510
Adhesions, 92
Adipocytes, 83, 83*f*, 84, 85T
Adipose tissue, 33, 84–85, 85T
  excess, 91

Adiposity, excess, 91
Adrenal cortex, 329, 330*f*, 331, 333
  hormones of, 329, 331, 333
Adrenal glands, 329–331, 330*f*, 333
  adrenal cortex as, 329, 330*f*, 331, 333
  adrenal medulla as, 333
  disorders of, 338
Adrenal medulla, 273*f*, 274, 329, 330*f*
  hormones of, 333
Adrenocorticotropic hormone (ACTH), 320, 323T, 331, 333
Aerobic cellular respiration, 183, 511
Aerobic exercise, 378
Afferent arteriole, 526, 527*f*, 528*f*
Afferent lymphatic vessel, 425
Afferent neuron, 227
Afterbirth, 412, 594
After-hyperpolarizing phase, 233
Age-related macular disease (AMD), 310
Age spots, 101
Agglutination, 435
Agglutinins, 358
Aggregated lymphatic follicles, 422*f*, 425
Aging
  acid-base balance and, 55
  cardiovascular system and, 412
  cells and, 65
  digestive system and, 495
  electrolyte balance and, 553
  endocrine system and, 336
  fluid balance and, 553
  homeostasis and, 9
  immune system and, 437
  integumentary system and, 106
  joints and, 165, 167
  metabolic rate and, 516
  muscular system and, 187–188
  nervous system and, 263
  reproductive system and, 577
  respiratory system and, 465
  skeletal system and, 147
  tissue and, 92
  urinary system and, 537
Agranular leukocytes, 353, 355T
Airways, irritation of, in respiration control, 465
Albinism, 101
Albumin, 37, 346
  in urine, 534T
Alcohol
  abuse of, 472
  diuresis and, 533
Aldosterone, 331
  in blood pressure/blood flow regulation, 393
  in nephron regulation, 533
  in water loss regulation, 546–547, 546*f*, 547T
Alkalosis, 552
Allantois, 592, 595*f*, 602*f*
Allele, 604
Allergen, 440
Allergic reactions, 440
Allograft, 441
All-or-none principle, 233
Alopecia, androgenic, 103
Alpha cell, 327, 328*f*
Alveolar cell, 451–452, 453*f*
Alveolar duct, 451, 452*f*
Alveolar fluid, 452, 453*f*
Alveolar macrophage, 452, 453*f*
Alveolar pressure, 455, 455*f*
Alveolar process, 127*f*, 129–130, 131
Alveolar sacs, 451
Alveolus(i), 131, 451–452, 452*f*, 453*f*, 569*f*, 570
Alzheimer disease, 266–267
Amenorrhea, 581
  in female athletes, 576
Amino-acid-based hormones, 317
Amino acids, 35–36, 35*f*
  absorption of, in small intestine, 489, 490*f*
  in lipid anabolism, 512, 513*f*
  in protein metabolism, 514
  in proteins, 35–36
Amino group, 35, 35*f*
Amniocentesis, 607
Amnion, 590, 591*f*, 595*f*
Amniotic cavity, 590, 591*f*
Amniotic fluid, 590
Ampulla, 302, 303*f*, 307*f*
Amyotrophic lateral sclerosis (ALS), 265–266
Anabolic steroid, 220
Anabolism, 28, 507
  glucose, 511–512, 511*f*
  lipid (lipogenesis), 512, 513*f*, 515T
  protein, 514, 515T
  stimulation of, testicular hormones in, 562
Anaerobic cellular respiration, 183, 511
Anal canal, 491*f*, 492
Analgesia, 267, 288
Anaphylactic shock, 440
Anaplasia, 67
Anastomoses, 372
Anatomical adjectives, 10, 11*f*
Anatomical neck of humerus, 140, 140*f*
Anatomical position, 9, 11*f*
Anatomical terms, 9–15
  body regions, 9–10, 11*f*
  directional, 10, 12, 12E, 13*f*
  planes and sections, 14–15, 14*f*
Anatomic dead space, 456
Anatomy, 2
Androgenic alopecia, 103
Androgens, 329, 333, 558
  hair and, 103
Andrology, 556
Anemia, 351, 360
  in premature newborns, 351
Anencephaly, 592
Anesthesia, 267
  local, 234
Aneurysm, 415
Angina pectoris, 380
Angiocardiography, 381
Angiogenesis, 66, 415
Angiotensin converting enzyme (ACE), 331
Angiotensin I and II, 331
Angiotensin II
  in blood pressure/blood flow regulation, 393
  in nephron regulation, 533
  in water loss regulation, 546–547, 546*f*, 547T
Angular movement of joint, 160–161, 161*f*, 162*f*
Anhydrases, 37
Anions, 26
  in body fluids, 548, 548*f*
Ankle, 145
Anorexia nervosa, 498
Anosmia, 310
Antagonist muscle, 189
Anterior cavity, 295*f*, 297, 299T
Anterior compartment of foot, 216E, 217E
Anterior corticospinal tract, 261*f*
Anterior cruciate ligament (ACL), 165, 166*f*
Anterior (directional term), 12E, 13*f*
Anterior (extensor) compartment, 214E, 215E
Anterior (flexor) compartment, 206E
  muscles of, 208E, 209*f*
Anterior lobe, 319. *See also* Pituitary gland
Anterior median fissure, 243, 245*f*
Anterior spinothalamic tract, 260, 260*f*
Anterior thoracic muscles, 202E, 203*f*
Anterior tibial arteries, 400E, 401*f*
Anterior tibial veins, 408E, 409*f*
Anterior (ventral) gray horns, 245, 245*f*
Anterior (ventral) root, 245*f*, 246
Anterior white column, 245*f*
Anti-A antibody, 358
Anti-B antibody, 358
Antibody(ies), 346
  in ABO blood group, 358, 358*f*
  in adaptive immunity, 429–430, 429*f*
  functions of, 435
Antibody-mediated immune response, 429
Anticoagulant drug, 357
Anticodon, 62
Antidiuretic hormone (ADH)
  actions of, 323T
  in blood pressure/blood flow regulation, 393

function of, 322–323
synthesis of, 321*f*
in water loss regulation, 547, 547T
in water reabsorption regulation, 533, 533*f*
Antigen(s)
in ABO blood group, 358*f*
in adaptive immunity, 429–430, 429*f*
immunity and, 428
processing and presenting of, 430–432
Antigen-binding site, 430
Antigen-presenting cell, 431–432, 432*f*, 436T
Antimicrobial proteins, 428T
protective functions of, 426
Antioxidant, 25, 64
Antioxidant vitamins, 505
Antypyretics, 519
Anuria, 530
Anus, 491*f*, 492
Anvil (incus), 302, 302*f*
Aorta, 394
Aortic body, 393
Aortic insufficiency, 370
Aortic stenosis, 370
Aortic valve, 368*f*, 369, 370*f*
Aortography, 415
Apex
of heart, 365, 365*f*
of lung, 450*f*, 451
Aphasia, 259
Aplastic anemia, 360
Apnea, 465
Apneustic area, 462
Apocrine sweat gland, 103
Apoptosis, 67
Appendicitis, 497
Appendicular skeleton, 124, 125T
Appendix, 491*f*, 492
Aqueous humor, 295*f*, 297
Arachnoid mater, 243, 243*f*, 250, 251*f*
Arachnoid villi, 250, 251*f*
Arches of foot, 146, 146*f*
Arch of aorta, 396E, 397*f*
Arcuate artery, 527*f*, 528*f*
Arcuate popliteal ligament, 165, 166*f*
Arcuate veins, 527*f*, 528*f*
Areola, 569*f*, 570
Areolar connective tissue, 84, 85T
Arm. *See* Upper limbs
Arrector pili, 102*f*, 103
Arrhythmia, 380
Arterial stick, 352T
Arteriole, 386, 388*f*
Artery, 386
Arthralgia, 169
Arthritis, 168
Arthrology, 156
Arthroplasty, 168
total hip, 151
Arthroscopy, 160
Articular capsule, 158–159, 159*f*, 166*f*
Articular cartilage, 115*f*, 116, 120*f*, 121, 158, 159*f*
Articular fat pad, 159, 166*f*
Articular process, 135
Articulations, 156–169. *See also* Joints
Artificial pacemaker, 373
Arytenoid cartilage, 448, 449*f*
Ascending aorta, 367*f*, 368*f*, 369, 396E, 397*f*
Ascending colon, 491*f*, 492
Ascending limb of loop of Henle, 526, 528*f*
Ascending tracts, 245
Ascorbic acid, 509T
Aspartate, as neurotransmitter, 237
Asphyxia, 468
Aspiration, 468
Association areas, of cerebral cortex, 258
Asthma, 467
Asthma attack, 451
Astigmatism, 300
Astrocytes, 231T
Ataxia, 255
Atherosclerosis, 379
blood clotting and, 357
prevention of, 410
Atherosclerotic plaques, 379–380, 379*f*
Athletes
female, menstrual irregularity and, 576
training of, 518
Athlete's foot, 108
Atlas, 135*f*, 136
Atomic number, 24, 24*f*
Atoms, 2, 23, 24*f*
atomic structure, 24*f*
ATPases, 37
ATP synthase, 38
Atria, 367, 367*f*, 368*f*, 369
Atrial fibrillation, 380
Atrial flutter, 380
Atrial natriuretic peptide (ANP)
in blood pressure/blood flow regulation, 393
in blood pressure regulation, 334T
in nephron function, 531
in nephron regulation, 533
in water loss regulation, 546–547, 546*f*, 547T
Atrial septal defect, 380
Atrial systole, 374, 375*f*
Atrioventricular (AV) block, 380
Atrioventricular (AV) bundle, 372
Atrioventricular (AV) node, 372
Atrioventricular (AV) valves, 369
Atrophy, 67
Auditory association area, 258, 259*f*
Auditory ossicles, 302
Auditory pathway, 305
Auditory tube, 301, 304*f*
Auricle, 301, 302*f*, 368*f*, 369
Auscultation, 9
Autograft, 100, 441
Autoimmune diseases, 93, 428
Autoimmune response, 65
Autologous preoperative transfusion, 361
Autologous skin transplantation, 100
Autolysis, 56
Autonomic dysreflexia, 280
Autonomic ganglion, 272, 273*f*
Autonomic motor neuron, 246, 272, 273*f*
Autonomic nervous system (ANS), 227–228, 271–280
disorders of, 280
functions of, 277–280, 279T
parasympathetic activities as, 278–280, 279T
sympathetic activities as, 277–278, 279T
in heart rate regulation, 376
hypothalamus and, 255
neurotransmitters in, 277
somatic nervous system compared with, 272–273, 272T, 273*f*
structure of, 273–277
Autonomic sensory neuron, 272
Autonomic (visceral) reflex, 247
Autophagy, 56
Autoregulation of capillary blood flow, 389
Autorhythmicity, 186
Autosomes, 605
Avascular tissue, epithelium, 73
Axial skeleton, 124, 125T
Axillary artery, 398E, 399*f*
Axillary nerve, 244*f*
Axillary vein, 406E, 407*f*
Axis, 135, 136
Axon, 177, 178*f*, 228, 229*f*
Axon collateral, 228, 229*f*
Axon hillock, 228, 229*f*
Axon terminal, 177–178, 178*f*, 228, 229*f*

B

Backbone. *See* Vertebral column
"Bad" cholesterol, 513
Balance, 305–306
Baldness, 103
Ball-and-socket joints, 164*f*, 165
Baroreceptors
in cardiovascular center, 392
in heart rate regulation, 377*f*, 378
reflex, 392
Barrier methods of birth control, 575
Basal cell carcinomas, 107
Basal cells, gustatory, 297
Basal ganglia, 256, 261
damage to, 258
Basal metabolic rate (BMR), 515
thyroid hormones and, 325

Basal state, 515
Basal stem cell, 290, 291*f*
Base, 30, 30*f*
  of heart, 365
  of lung, 450*f*, 451
  metacarpal, 141*f*, 143
  metatarsal, 145*f*
  phalangeal
    of foot, 146
    of hand, 143
Basement membrane, 73
Base triplet, 58, 60*f*
Basilar membrane, 303*f*, 304, 304*f*
Basilic vein, 406E, 407*f*
Basophils, 353, 355T
B cells, 353, 355T
  antibody-mediated immunity and, 435–436, 435*f*, 436T
  maturation of, 429
Behavioral patterns, hypothalamus and, 255
Bell's palsy, 194E
Belly of muscle, 188, 189*f*
Benign prostatic hyperplasia (BPH), 577
Benign tumor, 66
Beta cell, 327, 328*f*
Bicarbonate ions
  in acid-base balance, 550
  in carbon dioxide transport, 460*f*, 461
Biceps brachii, 205*f*, 206, 207*f*, 209*f*
Biceps femoris, 213*f*, 215E
Bicuspid valve, 368*f*, 369, 370*f*
Bifid spinous process, 135*f*, 136
Bifidus factor in breast milk, 603
Big toe, 145*f*
Bilaminar embryonic disc, 590, 591*f*
Bile, 484
Bile canaliculi, 484, 485*f*
Bile salts
  absorption of, in small intestine, 489–490
  functions of, 35
Bilirubin, 351
  excretion of, liver in, 486
  in liver function, 484
  in urine, 534T
Biliverdin, 351
Binge-purge syndrome, 498
Binocular vision, 300
Biopsy, 67
Biotin, 509T
Bipolar cell layer of retina, 296, 296*f*
Birth control methods, 574–576
Birth control skin patch, 575
Blackheads, 103
Bladder, urinary, 536, 536*f*
Blastocyst, 588, 589*f*
Blastocyst cavity, 588
Blastomeres, 588, 589*f*
Blasts, 116
Blind spot, 297
Blister, 108
Blood, 345–361. *See also* Platelet(s); Red blood cells; White blood cells
  blood tissue, 90
  components of, 346–354, 347*f*
  disorders of, 360
  flow of (*See* Blood flow)
  formed elements in, 346, 347*f*, 348–354
  functions of, 346
  hemostasis and, 354–357
    clotting in, 354–356
  pH of, regulation of, kidneys in, 524
  plasma, 346
  supply of (*See* Blood supply)
  transfusion of, 358–359
  viscosity of, 391
  volume of, regulation of, kidneys in, 524
Blood bank, 345, 361
Blood-brain barrier (BBB), 250
Blood cells
  formation of, 346, 348, 348*f*, 350*f*
  production of, in bone marrow, 114
  red, 349–351 (*See also* Red blood cells)
  white, 353
Blood cholesterol. *See* Cholesterol
Blood colloid osmotic pressure, 388
Blood flow. *See also* Blood pressure
  blood pressure and, 390–391
  lymphatic system and, 421, 423*f*
  regulation of, 391–394
    blood pressure and, 391–394
  through heart, 371–372, 371*f*
Blood glucose. *See* Glucose
Blood plasma. *See* Plasma
Blood poisoning, 361
Blood pressure. *See also* Blood flow
  blood flow and, 390–391
  elevated, 415
  homeostasis of, negative feedback system in, 8, 8*f*
  measurement of, 394
  regulation of, 391–394
    kidneys in, 524
Blood reservoir, 386
Blood samples for medical testing, 352T
Blood supply
  to brain, 250
  to kidney, 526, 527*f*, 528*f*
  to skeletal muscle tissue, 174*f*, 175
Blood tissue, 90
Blood vessels
  and circulation, 385–416
    blood flow (*See also* Blood flow)
    circulation checking, 394
    disorders of, 415
    structure and function of, 386–390, 387*f*
  clotting in, 357
  great, of heart, 369
  repair of, 356
Blood vessel walls, 544
Body
  sternum, 137, 138*f*
  of stomach, 480*f*, 481
  vertebral, 134, 134*f*
Body cavities, 15–17
Body (connecting) stalk, 592, 593*f*, 594*f*
Body fluid compartment, 544–547
Body fluids, 543
  aging and, 553
  electrolytes in, 548–550
    imbalances of, 549T
Body heat
  loss of, 516
  production of, 515–516
Body mass, fluids and, 544, 544*f*
Body regions, 9–10, 11*f*
Body temperature
  elevated, 427
  heat measurement and, 515
  homeostasis of, 515–516
  hypothalamus and, 255
  loss of, 516–517
  metabolic rate and, 515–518
  production of, 515–516
  regulation of, 516–518, 517*f*
    skin in, 104
  in respiration control, 465
  thermal sensation, 287
Boils, 103
Bolus, 478
Bone. *See also* Skeletal system
  deposition of, 121
  exercise and, 123–124, 124T
  formation of, 118–123
  growth of, 121
  metabolism of, 124T
  remodeling of, 121–122
  resorption of, 121
  structure of, 114–118, 115*f*
    microscopic, 116–118
  types of, 114
Bone marrow
  red, 114 (*See also* Red bone marrow)
    in long bones, 115*f*
  transplant of, 354
  yellow, 114
Bone scan, 118
Bone tissue, 89–90. *See also* Skeletal system
Bony labyrinth, 302, 303*f*
Botox, 179
Botulinum toxin (Botox), 179
  as anti-aging treatment, 106
Bowman's capsule, 526, 528*f*
Brachial artery, 398E, 399*f*

Brachialis, 205f, 206E, 207f, 209f
Brachial plexus, 244f, 246
Brachial vein, 406E, 407f
Brachiocephalic trunk, 396E, 397f, 398E, 399f
Brachioradialis, 206E, 209f
Bradycardia, 394
Bradykinesia, 266
Brain, 226, 226f, 248–263
  blood supply to, 250
  brain stem of, 248, 249f, 250, 252–254, 262T
  cerebellum of, 249f, 250, 251f, 255, 262T
  cerebrum of, 248, 249f, 250, 255–263, 262T (*See also* Cerebrum)
  diencephalon of, 248, 249f, 254–255, 262T
  major parts of, 248–250, 249f
  planes and sections of, 14–15, 14f
  principal parts of, functions of, 262T
  protective coverings of, 250
Brain attack, 266
Brain stem, 248, 249f, 250, 252–254, 262
Brain wave, 262
Breast cancer, 579–580
Breast-feeding, 603–604
Breast milk, 586
  benefits of, 603–604
Breathing. *See* Respiration; Ventilation
Breathing patterns, 457
Breech presentation, 602, 607
Broca's speech area, 258, 259f
  damage to, 259
Bronchi, 450f, 451
Bronchial arteries, 396E, 397f
Bronchial tree, 450f, 451
Bronchiole, 450f, 451
Bronchitis, 463, 467
Bronchoscopy, 468
Browlift, as anti-aging treatment, 106
Buccinator, 194E, 195f
Buffer systems, 30, 31T, 553T
Buffy coat, 346
Bulb, hair, 102, 102f
Bulbourethral gland, 557f, 562
Bulimia, 498
Bulk-phase endocytosis, 52, 53T
Bundle branches, 372
Bundle of His, 372
Bunion, 151
Burns, 107–108
Bursae, 160, 166f
Bursectomy, 169
Bursitis, 160

C

Caffeine
  diuresis and, 533
  health risks of, 257
Calcaneus, 145, 145f
Calcification, 116
  in intramembranous ossification, 119, 119f
Calcitonin (CT), 123, 325, 327f
Calcitriol, 326, 327f
  production of, kidneys in, 525
Calcium, 506T
  bone, 123f
  as electrolyte, 548f, 549
    imbalances of, 549T
  homeostasis of, bone in, 122–123
  in muscle contraction and relaxation, 179, 179f, 180, 181f
  regulation of, 326, 327f
  in synaptic transmission, 235
Calcium homeostasis, 123f
Callus, 100
Calorie (Cal), 515
calorie (cal), 515
Canaliculi, 116, 117f
Canal of Schlemm, 297
Cancer, 66
  of breast, 579–580
  of cervix, 580
  colorectal, 497
  of larynx, 449
  of lung, 467
  of ovaries, 580
  of pancreas, 484
  of prostate, 579
  of skin, 107
  smoking and, 463
  of testes, 579
Candidiasis, vulvovaginal, 580
Canker sore, 498
Capacitation, 587
Capillary, 386–389, 388f
  lymphatic, 421, 423f
Capillary blood pressure, 388
Capillary exchange, 387–389, 389f
Capillary loop, 100
Capitate, 142
Capitulum, 140, 140f, 141f
Capsular space, 530, 530f
Carbaminohemoglobin, 461
Carbohydrate, 31–32
  complex, 31, 32, 33f
  metabolism of, 510–512, 510f, 515T
    glucose anabolism in, 511–512, 511f
    glucose catabolism in, 510–511, 510f
    liver in, 485–486
    in small intestine, 488–489, 489T
  as neurotransmitters, 237
Carbohydrate loading, 512
Carbon dioxide
  exhalation of, 553T
    in acid-base balance, 551–552
  transport of, 460f, 461
Carbon dioxide exchange, 457–459, 458f
Carbonic acid-bicarbonate buffer system, 550, 553T
Carbonic anhydrase, 461
Carbon monoxide poisoning, 461
Carboxyl group, 35, 35f
Carboxypeptidase, 484, 489T
Carcinogenesis, 66
Carcinogen, 66
Cardiac accelerator nerve, 377, 377f
Cardiac arrest, 381
Cardiac catheterization, 381
Cardiac circulation, 372
Cardiac cycle, 374–375, 375f
Cardiac muscle tissue, 90, 173, 186
  autonomic nervous system and, 279T
Cardiac notch, 450f, 451
Cardiac output, 375–378. *See also* Heart
Cardiac rehabilitation, 381
Cardiac tamponade, 366
Cardia of stomach, 480f, 481
Cardiology, 364
Cardiomegaly, 381
Cardiopulmonary resuscitation (CPR), 381
Cardiovascular (CV) center
  in blood pressure regulation, 391–393, 392f
  in heart rate regulation, 250, 377, 377f
Cardiovascular system, 4T, 345–419. *See also* Blood; Blood vessels and circulation; Heart
  aging and, 412
  blood in, 345–361
  blood vessels and circulation in, 385–416
  contribution of, to homeostasis, 6, 414
  heart in, 364–381
  in pregnancy, 600
Caries, dental, 497
Carotene, 101
Carotid body, 392
Carotid foramen, 128, 128f
Carpals, 141–143, 141f
Carpal tunnel syndrome, 142, 208E
Carpus, 141
Carrier, 604
Cartilage, 88–89T, 89
  torn, 160
Cartilage model, 120f
  development and growth of, 121
Cartilaginous joints, 157, 158, 159f
Casts, in urine, 534T
Catabolism, 28, 507
  glucose, 510–511, 510f
  lipid (lipolysis), 512, 513f, 515T
  protein, 514, 515T
Catalase, in peroxisomes, 56
Catalysts, 37
  enzymes as, 37

Cataracts, 310
Catheterization, cardiac, 381
Cation, 26
Cauda equina, 243, 244*f*
Caudal anesthesia, 136–137
Caudate nucleus, 254*f*, 257
CCK cell, in small intestine, 487
Cecum, 491*f*, 492
Celiac ganglion, 274, 275*f*
Celiac trunk, 396E, 397*f*
Cells, 2, 44–47
  aging, 65
  connective tissue, 83
  cytoplasm of, 45, 45*f*, 52–57
  differentiation of, 6
  disorders of, 66
  generalized view of, 45
  nucleus of, 45, 57–58
  parts and functions of, 59T
  plasma membrane of, 45, 46–47, 46*f*
  protein synthesis in, 58, 60–62
  somatic cell division, 62–65, 63*f*
  transport of, 47–52
  transport of materials into and out of, 53T
Cell biology, 44
Cell body, 229*f*
Cell cycle, 62
Cell division, 44, 62
Cell identity marker, 46
Cell junction, 73
Cell-mediated immune response, 429
Cellular level of structural organization, 2, 3*f*
Cellular respiration, 510–511, 510*f*
Cellulose, 32
Cementum, 477–478, 477*f*
Center of ossification, 119, 120*f*
Central canal, 116, 117*f*, 243, 245*f*
Central chemoreceptor of medulla, 463
Central fovea, 295*f*, 297
Central nervous system (CNS), 226*f*, 227, 242–263
  brain in, 248–263 (*See also* Brain)
  disorders of, 265–267
  neuroglia in, 231T
  spinal cord in, 243–247, 244*f* (*See also* Spinal cord)
Central sulcus, 255, 256*f*
Central vein of liver, 484, 485*f*
Centriole, 54, 54*f*
Centromere, 64
Centrosome, 54, 54*f*, 59T
Cephalic phase, of digestion, 493–494
Cephalic vein, 406E, 407*f*
Cerebellar cortex, 255
Cerebellar hemisphere, 255
Cerebellar nuclei, 255
Cerebellar peduncle, 255
Cerebellum, 249*f*, 250, 251*f*, 252*f*, 255, 256*f*, 261, 262T
Cerebral cortex, 250, 254*f*, 255
  association areas of, 258
  functional areas of, 258–259, 259*f*
  motor areas of, 258
  sensory areas of, 258
Cerebral hemispheres, 255
Cerebral peduncle, 252*f*, 253, 253*f*
Cerebral white matter, 254*f*, 256
Cerebrospinal fluid (CSF), 250
Cerebrovascular accident (CVA), 266
Cerebrum, 248, 249*f*, 250, 251*f*, 252*f*, 254*f*, 255–263, 262
  basal ganglia of, 256*f*
  cerebral cortex (*See* Cerebral cortex)
  electroencephalogram and, 262
  hemispheric lateralization in, 261–262
  limbic system, 257, 258*f*
  memory and, 262
  somatic pathways of, 259–261, 260*f*
Cerumen, 104, 301, 302*f*
Ceruminous gland, 104, 301
Cervical curve, 133*f*, 134
Cervical enlargement, 243, 244*f*
Cervical plexus, 244*f*, 246
Cervical vertebrae, 133*f*, 135–136, 135*f*
Cervix, 564*f*, 567*f*
  cancer of, 580
  dysplasia of, 580
Cesarean section, 602
Chancre, 580
Channel, 46*f*
  ion, 230, 232
Charley horse, 220
Chemical barriers to pathogens, 425–426
Chemical bonds, 25–28
  covalent, 26, 28
    formation of, 27*f*
  hydrogen, 28
  ionic, 26, 26*f*
Chemical compounds, 29–40
  classes of, 29
  inorganic, 29–30
  organic, 31–40 (*See also* Organic compounds)
Chemical digestion, 474
  in small intestine, 488
  in stomach, 483
Chemical elements, 23, 23T
Chemical energy, 28
Chemical level of structural organization, 2, 3*f*
Chemical peel, as anti-aging treatment, 106
Chemical reactions, 28–29
  decomposition as, 28–29
  exchange as, 29
  reversible, 29
  synthesis as, 28
Chemical symbols, 23
Chemistry, 22–40
  chemical bonds in, 25–28
  elements and atoms in, 23–25, 23T
  ions, molecules, and compounds in, 25
  reactions in, 28–29
Chemoreceptor reflexes, in blood pressure regulation, 392–393
Chemoreceptor, 286
  in blood pressure regulation, 392–393
  in heart rate regulation, 378
  in regulation of respiration, 463–464, 464*f*
Chemotaxis, 426
  in inflammation, 427
Chemotherapy, 66
  hair and, 102
Chief cell
  in gastric glands, 481, 481*f*, 482*f*
  in parathyroid glands, 326, 327*f*
Chlamydia, 580
Chloride, 506T
  as electrolyte, 548*f*, 558–549
    imbalances of, 549T
Chloride shift, in carbon dioxide transport, 461
Cholecystectomy, 484
Cholecystitis, 498
Cholecystokinin (CCK), 334T, 487
  in digestion, 495, 495T
Cholesterol, 35, 35*f*, 513
  estrogens and, 570
Chondritis, 169
Chondrocyte, 89
Chordae tendineae, 368*f*, 369, 370*f*
Chorion, 591*f*, 592, 595*f*
Chorionic villi, 592–593, 594*f*, 595*f*
Chorionic villi sampling (CVS), 607
Choroid, 294, 295*f*, 299T
Choroid plexus, 250, 251*f*
Chromatid, 62
Chromatin, 58*f*
Chromium, 506T
Chromosome, 58, 63*f*
  homologous, 558, 604
  sex, 605
Chronic bronchitis, 467
Chronic fatigue syndrome (CFS), 441
Chronic obstructive pulmonary disease (COPD), 467
Chronic pain, management of, 289
Chronic prostatitis, 579
Chronic renal failure, 539
Chylomicrons, in lipid transport, 513
Chyme, 482
Chymotrypsin, 484, 489T
Cigarette smoking. *See* Smoking

Cilia, 45*f*, 54, 59T
  mucous membrane, 428T
  of mucous membrane in immunity, 426
  nasal, movement of, inhibition of, 448
Ciliary body, 294, 295*f*, 299T
Ciliated simple columnar epithelium, 75, 78T
Circadian rhythms, hypothalamus and, 255
Circular fold, in small intestine, 486*f*, 487
Circulation
  of blood (*See* Blood vessels and circulation)
  fetal, 412, 413*f*
Circulation time, 415
Circulatory route, 394, 395*f*, 396–412
Circumcision, 563
Circumduction, 161, 162*f*
Cirrhosis, 498
cis-fatty acid, 34
Cisterns, 56, 56*f*
Clasts, bone, 116
Claudication, 415
Clavicle, 139, 139*f*
Clawfoot, 151
Cleavage, 588, 589*f*, 590*f*
Cleavage furrow, 63*f*, 64–65
Cleft lip, 130
Cleft palate, 130
Clitoris, 564*f*, 568*f*, 569
Clone (T cell), 433
Cloning, 588
Closed (simple) fracture, 122
Clostridium botulinum, 179
Clot, 354
  retraction of, 356
Clotting, blood, 354–356. *See also* Blood
  in blood vessels, 357
Clotting factors, 355, 356*f*
Coagulation, blood, 354–356.
    See also Blood
Cobalt, 506T
Cocaine, 237
Coccygeal vertebrae, 133
Coccyx, 133*f*, 137, 137*f*
Cochlea, 302, 302*f*, 303*f*, 304, 304*f*
Cochlear duct, 303*f*, 304, 304*f*
Cochlear implant, 310
Codon, 58
Coenzyme, 37
  in metabolism, 507
Cofactor, 37
Cold receptor, 287
Cold sore, 108
Collagen, 84
Collagen fiber, 84
Collecting duct, 526, 528*f*
Colon, 491*f*, 492
  polyps of, 492
Color blindness, 300
  inheritance of, 606, 606*f*
Colorectal cancer, 497
Colostomy, 498
Colostrum, 603
Columnar cells, 74
Columnar epithelium
  pseudostratified, 75, 78T
  simple, 75, 78T
    ciliated, 78T
    nonciliated, 77T
  stratified, 80T, 82
Common bile duct, 483*f*, 484
Common cold, 467
Common hepatic artery, 396E, 397*f*
Common hepatic duct, 483*f*, 484
Common iliac artery, 396E, 397*f*,
    400E, 401*f*
Common integrative area, 258–259, 259*f*
Compact bone tissue, 116–118
Compartment, 206E
Compensation, 552
Complement system, 428T
  protective functions of, 426
Complete blood count (CBC), 352T, 353
Complete fracture, 122
Complex carbohydrate (polysaccharide), 31,
    32, 33*f*
Compound, 25
Compound fracture, 122
Concentration, 47
Concentration gradient, 47
Concentric lamellae, 116, 117*f*
Conceptus, 608
Condom, 575
Conducting zone of respiratory system, 447
Conduction
  body heat loss by, 516
  of nerve impulses, 233–234, 234*f*
Conduction deafness, 310
Conduction system of heart, 372–373, 373*f*
Condylar process, 126*f*, 131
Condyloid joint, 164*f*, 165
Cones, 296–297, 297*f*
  stimulation of, 309
Congenital adrenal hyperplasia (CAH), 333
Congenital defects, of heart, 380
Congenital hypothyroidism, 337–338
Congestive heart failure (CHF), 377
Conjoined twins, 587
Conjunctiva, 294, 295*f*
Connecting stalk, 592, 593*f*
Connective tissue(s), 73, 82–90, 85–89T
  blood, 90 (*See also* Blood)
  bone, 89–90
  cartilage, 88–89T, 89
  cells of, 83
  dense, 86–87, 86–87T, 89
  extracellular matrix of, 82–83, 83–84
  features of, 82–83
  fibers of, 84
  functions of, 82–83
  level of organization, 2
  loose, 84–86, 85–86T
  lymph, 90
  repair of, 92
  in skeletal muscle, 173–175, 174*f*
  types of, 84–90
Connective tissue sheath, 102, 102*f*
Consciousness, 254, 267
  states of, hypothalamus and, 255
Constipation, 493
Contact dermatitis, 108
Continuous conduction, 233, 234*f*
Contraceptive, 574
Contraction
  of urinary bladder, 537
  of heart, 374–375, 376
  isometric, 185
  isotonic, 185
  of skeletal muscle, 177–180
    physiology of, 179*f*, 181*f*,
      179–180
  twitch, 184
Contraction cycle, 179, 179*f*
Contraction period, in twitch
    contraction, 184
Control center of feedback system,
    7–8, 7*f*
Controlled condition, 7, 7*f*
Convection, body heat loss by, 516
Convergence, 300
Cooper's ligament, 569*f*, 570
Copper, 506T
Coracobrachialis, 204E, 205*f*
Coracoid process, 139, 139*f*
Corn, 108
Cornea, 294, 295*f*, 299
Coronal plane, 14
Coronal suture, skull, 126*f*, 127*f*, 129*f*
Corona radiata, 587, 587*f*
Coronary artery disease (CAD),
    379–380
Coronary circulation, 372
Coronary sinus, 368*f*, 369, 372, 394,
    402E, 403*f*
Coronoid fossa, 140, 140*f*
Coronoid process, 140, 141*f*
Corpora cavernosa penis, 557*f*, 563
Cor pulmonale, 381
Corpus albicans, 565, 565*f*, 571*f*
Corpus callosum, 254*f*, 255
Corpuscles of touch, 100, 287
Corpus luteum, 565, 565*f*, 571*f*
Corpus spongiosum penis, 557*f*, 563
Corticobulbar axon, 253*f*
Corticomedullary junction, 528*f*
Corticopontine axons, 253*f*

Corticospinal axon, 253*f*
Corticotropin, 320, 323T
Corticotropin-releasing hormone (CRH), 331
  in pregnancy, 600
  in resistance reaction, 335
Cortisol, 331, 333
  functions of, 35
Coryza, 467
Costal breathing, 457
Costal cartilage, 138, 138*f*
Costimulation, in T cell activation, 432, 433*f*
Coughing, 457T
Coumadin, 357
Covalent bond, 26, 28
  formation of, 27*f*
Covering and lining epithelium, 73–82, 76–81T. *See also* Epithelial tissue (epithelium)
Coxal bones, 143
Cramp, 219
Cranial bones, 125, 126*f*, 127–129, 127*f*, 129*f*, 130*f*
Cranial cavity, 15, 15*f*
Cranial meninges, 243, 250, 251*f*
Cranial nerves, 226, 226*f*, 252*f*, 263–265T
Cranial reflex, 247
Craniosacral division of autonomic nervous system, 274. *See also* Parasympathetic division
C-reactive proteins, coronary artery disease and, 379
Creatine phosphate, 180, 182
Creatine supplementation, 183
Crenation, 50, 50*f*
Cretinism, 338
Cribriform plate, 127*f*, 128–129, 129*f*, 130*f*
Cricoid cartilage, 448, 449*f*
Crista, 57, 57*f*, 305–306, 307*f*
Crista galli, 127*f*, 129, 129*f*, 130*f*
Crossbridge, 179*f*, 180
Crossing over, in meiosis, 558, 560*f*
Cross-sectional plane, 14
Crown, 477, 477*f*
Crying, 457T
Cryptorchidism, 558
Cuboid, 145, 145*f*
Cuboidal cell, 74
Cuboidal epithelium
  simple, 75, 77T
  stratified, 79T, 82
Cumulative trauma disorder (CTD), 167
Cuneiform bones, 145, 145*f*
Cupula, 306, 307*f*
Curettage, endocervical, 581
Cushing's syndrome, 337*f*, 338
Cuspid, 478
Cutaneous membrane, 98. *See also* Skin
Cutaneous sensations, 105
Cuticle, 104, 104*f*
Cyanocobalamin, 509T
Cyanosis, 101, 351, 361
Cyclic AMP (cAMP), 318, 318*f*
Cyclosporine, 433
Cyst(s)
  bone, 116
  ovarian, 565, 581
Cystic acne, 108
Cystic duct, 483*f*, 484
Cystic fibrosis, 468
Cytokinesis, 62, 63*f*, 64–65
Cytolysis, 426
Cytoplasm, 45, 45*f*, 52–57, 59T
Cytoplasmic division, 64–65
Cytoskeleton, 53, 54*f*, 59T
Cytosol (intracellular fluid), 45, 47, 52, 59T, 544
  electrolytes in, 548, 548*f*
Cytotoxic T cell, 433, 436T
Cytotrophoblast, 590, 591*f*

D

Dandruff, 100
DASH diet, 543
Deafness, 310
Deamination, 514
Deciduous teeth, 478
Decomposition reaction, 28–29
Decubitus ulcers, 108
Deep (directional term), 12E
Deep palmar arch, 398E, 399*f*
Deep veins
  of lower limbs, 408E, 409*f*
  of upper limbs, 406E, 407*f*
Deep venous thrombosis (DVT), 415
Defecation, 474
  protective function of, 426, 428T
Defecation reflex, 493
Dehydration, 322, 552–553
  prolonged physical activity and, 551
  sucrose, 32*f*
Dehydration synthesis, 31–32, 32*f*
Dehydrogenase, 37
Delayed hypersensitivity reactions, 440
Deltoid, 204E, 205*f*, 207*f*
Deltoid tuberosity, 140, 140*f*
Dementia, 267
Demineralization, bone, 123
Demyelination, 238
Denaturation, 37
Dendrites, 228, 229*f*
Denervation atrophy, 175
Dens, 135*f*, 136
Dense body, 186
Dense connective tissue, 86–87, 86–87T, 89
Dental caries, 497
Dentes, 477–478
Dentin, 477, 477*f*
Deoxygenated blood
  in heart vessels, 369
  in oxygen-carbon dioxide exchange, 458
Deoxyhemoglobin, 459
Deoxyribonuclease, 484
  in small intestine, 488, 489T
Deoxyribonucleic acid (DNA), 2, 38, 39*f*
  in protein synthesis, 58, 60–62
Depolarizing phase, 235, 236
  of action potential, 232
Depo-provera, 575
Depression, 225
Depression (joint movement), 162, 163*f*
Depressor labii inferioris, 195*f*
Dermabrasion, 106
Dermal filler, as anti-aging treatment, 106
Dermal papilla, 100
Dermatitis, contact, 108
Dermatology, 97
Dermis, 98*f*, 99, 100–101
Descending colon, 491*f*, 492
Descending limb of loop of Henle, 526, 528*f*
Descending tract, 245
Detached retina, 310
Detrusor muscle, 536, 536*f*
Developmental biology, 586
Deviated nasal septum, 131
Diabetes insipidus, 337, 533
Diabetes mellitus, 338
  insulin resistance in, 332
Dialysis, 539
Diaphragm, 15*f*, 17, 200E, 201*f*, 575
Diaphragmatic breathing, 457
Diaphysis, 114, 115*f*
Diarrhea, 493
  traveler's, 498
Diarthrosis, 157
Diastole, 374
Diastolic blood pressure, 394
Diencephalon, 248, 249*f*, 254–255, 262T
Diet
  blood circulation and, 359
  DASH, 543
  heart and, 364
Dietary fiber, 497
Differential white blood cell count, 352T, 353
Differentiation, 6
Diffusion, 47–49, 47*f*, 53T
Digestion, 472, 474

cephalic phase of, 493–494
gastric phase of, 494
intestinal phase of, 494–495
in large intestine, 493
in mouth, 478
in stomach, 482–483
Digestive enzymes, 489T
Digestive system, 5T, 472–498
aging and, 495
contribution of, to homeostasis, 496
disorders of, 497
esophagus in, 478–480, 479*t*
functions of, 473–474, 473*f*
gallbladder in, 483*f*, 484
gastrointestinal tract in, 473–476, 473*f*, 474*f* (*See also* Gastrointestinal (GI) tract)
large intestine in, 491–493, 491*f*, 492*f*
liver in, 483*f*, 484–486, 485*f*
mouth in, 476–478, 476*f*
organs of, 473–474, 473*f*
pancreas in, 483–484, 483*f*
pharynx in, 478–480, 479*f*
small intestine in, 486–491, 486*f*, 487*f*
stomach in, 480–483, 480*f*, 481*f*, 482*f*
Dihydrogen phosphate, 550
Dihydrotestosterone (DHT), 561
Dilation, stage of, 602
Dipeptide, 35*f*, 36
Diplegia, 265
Diploid cell, 558
Directional terms, 10, 12, 12E, 13*f*
Disaccharides, 31–32
Disease, 9
homeostasis and, 8–9
resistance to (*See* Lymphatic and immune system)
Dislocation, 169
of knee, 168
of radial head, 168
Disorder, 9
Distal convoluted tubule, 526, 528*f*
Distal (directional term), 12E, 13*f*
Disuse atrophy, 175
Diuresis, 533
Diuretic, 533
Diverticulitis, 497
Diverticulosis, 497
Dizygotic twins, 587
Dominant allele, 604
Dopamine (DA), as neurotransmitter, 237
Dorsal gray horns, 245
Dorsal root, 245*f*, 246
Dorsal root ganglion, 246
Dorsiflexion, 163, 163*f*
Double helix, 38, 39*f*
Down syndrome, 607
Drinking, hypothalamus and, 255
Drugs
administration of, transdermal, 105
processing of, liver in, 486
Dual innervation, 272
Duchenne muscular dystrophy (DMD), 219
Ducts, 556
in male reproductive system, 562
Ductus arteriosus, 369, 412, 413*f*
Ductus epididymis, 559*f*, 562
Ductus (vas) deferens, 557*f*, 559*f*, 562
Ductus venosus, 412, 413*f*
Duodenal gland, 487
Duodenum, 486, 486*f*
Dura mater, 243, 243*f*, 250, 251*f*
Dwarfism, 337
Dynamic equilibrium, 305–306
Dysautonomia, 280
Dysmenorrhea, 581
Dysplasia, 67
Dyspnea, 468
Dysrhythmia, 380
Dystocia, 602
Dysuria, 539

E

Ear, 301–304. *See also* Hearing
disorders of, 310
inner, 302, 303*f*, 304, 308T
middle, 301–302, 304*f*, 308T
outer, 301, 308T
Eardrum, 301, 302*f*, 304*f*
perforated, 301
Earwax, 104, 301
Eating
emotional, 494
hypothalamus and, 255
Eccrine sweat gland, 103
Eclampsia, 608
Ectoderm, embryo, 592, 593*f*
Ectopic pregnancy, 589
Edema, 421, 547
Effector(s), 227, 247, 248*f*
in feedback system, 7*f*, 8
Efferent arteriole, 526, 527*f*, 528*f*
Efferent lymphatic vessel, 425
Efferent neuron, 227
Ejaculation, 563
Ejaculatory duct, 557*f*, 562
Elastic arteries, 386
Elastic cartilage, 89, 89T
Elastic connective tissue, 87, 87T, 89
Elastic fiber, 83T, 84
Elasticity, 84, 89, 100
muscle, 190
Elastic recoil, 454
Elastin, 84
Elbow
little-league, 168
tennis, 168
Electrical excitability, 232
Electrocardiogram (ECG), 374, 374*f*
Electroencephalogram (EEG), 262
Electrolytes, 26, 545, 548–550
absorption of, 489
aging and, 553
concentrations of, in fluid compartments, 548, 548*f*
imbalances of, 549T, 550
Electromyography (EMG), 220
Electron, 24, 24*f*
Electron shell, 24*f*, 25
Electron transport chain, 511
Elevation (joint movement), 162, 163*f*
Embolus, 357
Embryo, 586
Embryology, 586
Embryonic and fetal development. *See also* Pregnancy
changes associated with, 596T, 598–599*f*
Down syndrome and, 607
embryonic period in, 587–596
1st week, 588–590, 589*f*, 590*f*, 596T
2nd week, 590–592, 591*f*, 596T
3rd week, 592–594, 593*f*, 596T
4th through 8th weeks, 594–596, 596T
fetal period of, 596–597, 596T, 598–599*f*
placenta and umbilical cord in, 593–594, 595*f*
Embryonic folding, 594
Embryonic period, 586
Embryo transfer, 607
Emergency contraception, 574–575
Emesis gravidarum, 608
Emigration, in inflammation, 427
Emission, 563
Emmotropic eye, 299, 299*f*
Emotional eating, 494
Emotions, hypothalamus and, 255
Emphysema, 463, 467
Emulsification, bile in, 484
Enamel, 477, 477*f*
Encapsulated nerve ending, 286
Encephalitis, 267
Encephalomyelitis, 267
Endocardium, 366*f*, 367
Endocervical curettage, 581
Endochondral ossification, 118, 120–121, 120*f*
Endocrine glands, 81T, 82, 316–317, 317*f*

Endocrine system, 4T, 315–340
  adrenal glands in, 329–331, 330*f*, 333
  aging, 336
  contribution of, to homeostasis, 339
  disorders of, 337–338, 337*f*
  glands of, 316–317, 316*f*, 317*f*
  hormones of, 317–318, 334T (*See also* Hormones)
  ovaries and testes in, 333–334
  pancreas in, 327–339, 328*f*
  parathyroid glands in, 325–327, 326*f*
  pineal gland in, 334
  pituitary gland, 319–323, 323T
  stress response, 335–336
  thyroid gland of, 323–325, 324*f*
Endocrinology, 317
Endocytosis, 52, 53T
Endoderm, embryo, 592, 593*f*
Endolymph, 302
Endometrial glands, 567
Endometriosis, 579
Endometrium, 567, 567*f*, 571*f*
Endomysium, 173, 174*f*
Endoneurium, 246, 246*f*
Endoplasmic reticulum, 55, 55*f*, 59T
Endorphins, as neurotransmitters, 237
Endosteum, 116
Endothelium, 75
  of artery, 386, 387*f*
End-stage renal failure, 539
Energy, 28
Enteric nervous system (ENS), 227
  of GI tract, 475
Enteric plexus, 226, 226*f*
Enterokinase, 484
Enuresis, 539
Enzymes, 36–37, 37*f*, 46
  digestive, summary of, 489T
  in metabolism, 507
  as organic compounds, 37–38
  pancreatic, 484
  properties of, 37–38
Enzyme-substrate complex, 37, 37*f*
Eosinophil, 353, 355T
Ependymal cell, 231T
Epiblast, 590, 591*f*
Epicardium, 365, 366, 366*f*
Epicranial aponeurosis, 195*f*
Epidemiology, 18
Epidermis, 98*f*, 99–100, 99–101, 99*f*, 428T
  as barrier in immunity, 425–426
Epididymis, 557*f*, 559*f*, 562
Epidural block, 136–137, 272
Epidural space, 243
Epiglottis, 447*f*, 448, 449*f*
Epilepsy, 237–238
Epileptic seizure, 237–238
Epimysium, 173, 174*f*
Epinephrine, 274, 333
  in blood pressure/blood flow regulation, 393
Epineurium, 246, 246*f*
Epiphyseal line, 116, 121
Epiphyseal plate, 116, 120*f*, 121
Epiphysis, 114, 115*f*
Episiotomy, 569
Epistaxis, 468
Epithelial membrane, 90
Epithelial tissue (epithelium), 73–82
  classification of
    by cell shape, 74
    by layer arrangement, 73–74
  covering and lining, 73–82, 76–81T
  disorders of, 93
  features of, 73
  functions of, 73
  glandular, 73, 81T, 82
  level of organization, 2
  pseudostratified columnar, 75, 78T
  repair of, 92
  simple, 75
  stratified, 75, 79–80T, 82
  transitional, 80T, 82
Epoetin alfa, 350
Epstein-Barr virus, 440
Equilibrium, 47, 47*f*, 308T
  pathways of, 306, 308
  physiology of, 305–306
Erectile dysfunction (ED), 563
Erection, 563
Erector spinae, 210E, 211*f*
Erythema, 101
Erythrocytes. *See* Red blood cells
Erythropoiesis, 350, 351
Erythropoietin (EPO), 334T, 351
  production of, kidneys in, 525
Escherichia coli, urinary tract infections from, 535
Esophageal stage, of swallowing, 479*f*, 480
Esophagus, 478–480, 479*f*
Essential amino acid, 514
Essential nutrients, 504
Estradiol, 35, 35*f*
Estrogens, 333
  in female reproductive cycle, 570, 571*f*, *572*, *573*f
  functions of, 570
  in labor and delivery, 601
Ethmoidal sinuses, 128, 131, 132*f*
Ethmoid bone, 126*f*, 128, 130*f*
Eupnea, 457
Eustachian (auditory) tube, 301
Evaporation, body heat loss by, 516
Eversion, 162, 163*f*
Exchange reaction, 29
Exchange vessel, 386
Excretion
  kidneys in, 525
  skin in, 105
Exercise
  benefits of, 385
  blood circulation and, 359
  bone metabolism and, 123–124, 124T
  heart and, 378
  joints and, 156
  metabolic rate and, 516
  mind-body, 278
  muscles and, 172, 186
  oxygen consumption after, 183
  pregnancy and, 601
  prolonged, fluid and electrolyte balance and, 551
  respiratory system and, 465
  stretching, 190
  sudden cardiac death and, 376
Exhalation, 454–455, 454*f*
  of carbon dioxide, 553T
    in acid-base balance, 551–552
  muscles of, 200E, 201*f*
Exhaustion, 335
Exocrine gland cells of stomach, 481
Exocrine glands, 82, 82T, 316
Exocytosis, 51, 52, 53T
Exophthalmos, 337*f*, 338
Expiratory area, 462*f*
Expiratory reserve volume, 457
Expulsion, stage of, 602
Extensibility of skin, 100–101
Extension, 160, 161*f*
Extensor carpi radialis longus, 208E, 209*f*
Extensor carpi ulnaris, 208E, 209*f*
Extensor digitorum, 208E, 209*f*
Extensor digitorum longus, 217E, 217*f*
Extensor retinaculum, 208E, 209*f*
External auditory canal, 301, 304*f*
External auditory meatus, 126*f*, 127
External carotid artery, 398E, 399*f*
External iliac artery, 396E, 397*f*, 400E, 401*f*
External intercostals, 200E, 201*f*, 203*f*
External jugular veins, 402E, 403*f*
External nares, 447, 447*f*
External oblique, 198E, 199*f*, 201*f*
External respiration, 446, 458–459, 458*f*
External root sheath, 102, 102*f*
External urethral orifice, 536*f*, 537, 557*f*, 562, 568*f*
External urethral sphincter, 536*f*, 537
Extracellular fluid, 47, 544, 544*f*
  electrolytes in, 548, 548*f*
Extracellular matrix, of connective tissue, 82–83, 83–84
Extraembryonic mesoderm, 591*f*, 592
Extrinsic muscle, 196E, 197*f*

Extrinsic pathway, 356, 356*f*
Eye, 293–300
accessory structures of, 293–294, 294*f*
disorders of, 310
image formation in, 297, 299–300
interior of, 295*f*, 297
layers of, 294–297, 295*f*
muscles of, 196E, 197*f*, 293–294
photoreceptor stimulation in, 300
visual pathway, 300–301, 301*f*
Eyebrows, 293
Eyelashes, 293
Eyelids, 293

F

Face, 9
bones of, 129–131, 131*f*
expressions of, muscles of, 194E, 195*f*
Facelift
as anti-aging treatment, 106
nonsurgical, radio frequency, as anti-aging treatment, 106
Facets, vertebral, 135
Facial nerve, 252*f*, 264T
Facial paralysis, 194E
Facilitated diffusion, 48–49, 48*f*, 49*f*, 53T
Fallopian tube, 566, 567*f*
False labor, 602
False pelvis, 142*f*, 143
False ribs, 138
False vocal cords, 447*f*, 448
Fascia lata, 212E
Fascicle, 173, 174*f*, 246, 246*f*
Fasciculation, 219
Fast glycolytic (FG) fiber, 184–185
Fast oxidative-glycolytic (FOG) fiber, 184
Fast pain, 288
Fat, transplantation of, as anti-aging treatment, 106
Fat cell. *See* Adipocyte
Fat-soluble vitamins, 505, 508T
Fatty acid, 22, 33, 33*f*
essential, 33–34
Fatty streak, 380
Feces, 474, 493
Feedback, positive, in labor contractions, 601
Feedback system
components of, 7–8, 7*f*
negative, 7*f*, 8, 8*f*
in acid-base balance, 551, 552, 552*f*
in blood pressure regulation, 392, 393*f*
body temperature regulation, 516–517, 517*f*
respiration control, 464, 464*f*
positive, 7*f*, 8
in water reabsorption regulation, 533, 533*f*
Feet. *See* Foot
Female athlete triad, 576
Female infertility, 606
Female reproductive cycle, 570–574, 571*f*
hormonal regulation of, 570–572, 571*f*, 573*f*
phases of, 572–573
Female reproductive system, 5T. *See also* Male reproductive system; Reproductive system
aging and, 577
disorders of, 579–581, 607
functions of, 564*f*
mammary glands in, 569–570, 569*f*
organs of, 564–570, 564*f*
ovaries in, 564–566, 564*f*
perineum of, 568
reproductive cycle in, 570–574, 571*f*
uterine (fallopian) tubes in, 566, 567*f*
uterus in, 566–567, 567*f*
vagina in, 567*f*, 568
vulva in, 568–569, 568*f*
Feminizing adenoma, 340
Femoral arteries, 400E, 401*f*
Femoral nerve, 244*f*
Femoral vein, 408E, 409*f*
Femur, 115*f*, 144, 144*f*
muscles of, 212E, 213*f*, 214–215E
Fertilization, 556, 587, 587*f*
in vitro, 607
Fertilization age, 608
Fetal alcohol syndrome (FAS), 608
Fetal circulation, 412, 413*f*
Fetal development. *See* Embryonic and fetal development
Fetal period, 586
Fetal ultrasonography, 608
Fetus, 586
surgery of, 608
Fever, 428T, 519
protective functions of, 427–428
Fever blister, 108
Fiber(s)
connective tissue, 84
dietary, 497
Fibrillation, 219
Fibrillin, 84
Fibrin, 354
Fibrinogen, 346, 356, 356*f*
coronary artery disease and, 379
Fibrinolysis, 357
Fibroblast, 83*f*, 83T
Fibrocartilage, 88T, 89
Fibrocystic disease, 570
Fibroid, 581
Fibromyalgia, 219
Fibronectin in breast milk, 603
Fibrosis, in tissue repair, 92
Fibrous capsule, 159, 159*f*
Fibrous joint, 157, 158*f*
Fibrous pericardium, 365, 366*f*
Fibrous tunic, 294, 299T
Fibula, 145, 145*f*
muscles of, 214–215E
Fibular collateral ligament, 165, 166*f*
Fibularis (peroneus) longus, 216*f*, 217E, 217*f*
Fibular notch, 145, 145*f*
Fight-or-flight response, 271, 277–278, 335
Filiform papillae, 292*f*, *293*
Filtration
in capillary function, 388
as nephron function, 528, 529T
Filtration membrane, 530
Fimbriae, 566
of uterine tube, 567*f*
Fine touch, 260
Fingers, muscles of, 208E, 209*f*
Fingerstick, 352T
First messenger, 318
First polar body, 566, 566*f*
Fissures, 255, 256*f*
of lung, 450*f*, 451
Fixator, 189
Fixed macrophages, protective functions of, 426
Flaccid, 180
Flaccid paralysis, 261
Flagella, 45*f*, 54, 59T
Flat bone, 114
Flatfoot, 146
Flatulence, 493
Flexibility
of bone, 116
of muscle, 190
Flexion, 160, 161*f*
Flexor carpi radialis, 208E, 209*f*
Flexor carpi ulnaris, 208E, 209*f*
Flexor digitorum longus, 216*f*, 217*f*
Flexor digitorum profundus, 208E
Flexor digitorum superficialis, 208E, 209*f*
Flexor retinaculum, 208E, 209*f*
Floating rib, 138
Fluent aphasia, 259
Fluid balance, 544
aging and, 553
water gain in, 545
water loss in, 545–547, 546*f*
water movement in, 547
Fluoride, 506T
Fluoxetine (Prozac), 236
Foam cell, 380
Folacin, 509T
Folate, 509T

Folic acid, 509T
Follicle
  hair, 102, 102*f*
  thyroid, 323, 324*f*
Follicle-stimulating hormone (FSH), 320, 323T
  in female reproductive cycle, 570, 571*f*, 572, 573*f*
  in male reproduction, 561, 561*f*
Follicular cell, 323, 324*f*
Follicular phase, 571*f*, 572
Fontanel, 132, 132T
Food
  ingestion of, metabolic rate and, 516
  neurotransmitters and, 236
Foot
  arches of, 146, 146*f*
  health of, 148
  muscles of, 216–217E, 216*f*, 217*f*
Foramen magnum, 126*f*, 128
Foramen ovale, 128, 128*f*, 369, 412, 413*f*
Forearm
  extensors of, 206E, 209*f*
  flexors of, 206E, 207*f*
  muscles of, 206E, 207*f*
Foreskin
  female, 569
  male, 557*f*, 563
Formed elements of blood, 346, 347*f*, 348–354
Fornix, 567*f*, 568
Fossa ovalis, 369, 412
Fourth ventricle, 250, 251*f*
Fracture
  clavicular, 139
  hip, 151
  rib, 138
  types of, 122
Frank-Starling law of heart, 376
Fraternal twins, 587
Freckles, 101
Free edge, 104, 104*f*
Free nerve ending, 100, 286
Free radicals, 25, 372, 505
Frequency of stimulation, 183, 184, 185*f*
Frontal belly, 194E, 195*f*
Frontal bone, 126*f*, 127, 127*f*
Frontal eye field area, 259, 259*f*
Frontal lobe, 255, 256*f*, 259*f*
Frontal plane, 14, 14*f*
Frontal sinus, 127, 131, 132*f*
Functional residual capacity, 457
Fundus
  of stomach, 480*f*, 481
  of uterus, 567, 567*f*
Fungiform papilla, 292*f*, 293
Fused tetanus, 184, 185*f*

G

Gallbladder, 483*f*, 484
Gallstones, 484
Gamete intrafallopian transfer (GIFT), 607
Gamete, 556
Gamma aminobutyric acid (GABA), as neurotransmitter, 237
Gamma globulin, 441
Ganglia, 226, 226*f*
Ganglion cell layer of retina, 296, 296*f*
Gap junctions, 186, 366
Gastric emptying, 482
Gastric gland, 481, 481*f*, 482*f*
Gastric juice, 428T, 481, 483
  protective functions of, 426
Gastric phase, of digestion, 494
Gastric pit, 481, 481*f*
Gastrin, 334T, 481
  in digestion, 494, 495T
Gastrocnemius, 216*f*, 217E, 217*f*
Gastroenterology, 472
Gastroesophageal reflux disease (GERD), 480
Gastrointestinal (GI) tract, 473–476, 473*f*, 474*f*. *See also* Digestive system
  mouth in, 476–478, 476*f*
  mucosa of, 474*f*, 475
  muscularis of, 474*f*, 475
  in pregnancy, 601
  serosa and peritoneum of, 474*f*, 475
  submucosa of, 474*f*, 475
Gastrulation, 592, 593*f*
Gated channel, 48, 48*f*, 232
G cells, in gastric glands, 481, 482*f*
Gender, inheritance of, 605, 606*f*
Generalized seizures, 238
General senses, 285
Genes, 38, 58
Genetic counseling, 604
Genetic recombination, in meiosis, 558, 560
Genital herpes, 581
Genome, 58
Genomics, 58
Genotype, 604
Geriatrics, 18, 65
Germinal epithelium, 564, 565*f*
Gerontology, 65
Gestational age, 608
Giantism, 337, 337*f*
Gigantism, 337
Gingivae, 477, 477*f*
Gland
  autonomic nervous system and, 279T
  endocrine, 81T, 82
  exocrine, 81T, 82
  hair, 102*f*
Glandular epithelium, 73, 82, 82T. *See also* Epithelial tissue (epithelium)
Glans penis, 557*f*, 563
Glaucoma, 310
Glenoid cavity, 139, 139*f*
Gliding of joint, 160
Glioma, 230
Globulin, 346
Globus pallidus, 254*f*, 257
Glomerular (Bowman's) capsule, 526, 528*f*
Glomerular filtrate, 526
Glomerular filtration, 528, 529T, 530–531, 530*f*
  net filtration pressure in, 530
  rate of, 531
Glomerulonephritis, 539
Glomerulus, 526, 527*f*, 528*f*
Glossopharyngeal nerve, 252*f*, 264T
Glucagon, 327, 327*f*, 329*f*
Glucocorticoid, 331, 333, 339
Gluconeogenesis, 512
Glucose
  anabolism of, 511–512, 511*f*
  catabolism of, 510–511, 510*f*
  homeostasis and, 7
  regulation of, 327, 329, 329*f*
    liver in, 485–486
  in urine, 534T
Glucose-dependent insulinotropic peptide (GIP), 334T, 487
Glucosuria, 531
Glutamate, as neurotransmitter, 237
Gluteal nerve, 244*f*
Gluteus maximus, 212E, 213*f*
Gluteus medius, 212E, 213*f*
Glycerol, 33, 33*f*
Glycine, as neurotransmitter, 237
Glycogen, 32, 33*f*, 511
Glycogen granule, 52
Glycolysis, 511
  in muscle contraction, 183
Glycoprotein, 46, 46*f*
Goblet cells, 75
  in small intestine, 487
Goiter, 337*f*, 338
Golgi complex, 56, 56*f*, 59T
Gomphosis, 157, 158*f*
Gonadal artery, 396E, 397*f*
Gonadotropin-releasing hormone (GnRH), 320
  in female reproductive cycle, 570, 571*f*, 572, 573*f*
  in male reproduction, 561, 561*f*
Gonad, 333, 556
Gonorrhea, 580
"Good" cholesterol, 513
Gouty arthritis, 168
Graafian follicle, 565, 565*f*, 571*f*, 572
Gracilis, 213*f*, 215E
Graft, 441
  skin, 100

Granular leukocytes, 353, 355T
Granulysin, in cytotoxic T cell action, 433, 434*f*
Granzymes, in cytotoxic T cell action, 433, 434*f*
Graves disease, 337*f*, 338
Gray matter, 228, 230
Greater omentum, 475–476, 475*f*
Greater pelvis, 142*f*, 143
Greater sciatic notch, 143*f*, 144
Greater trochanter, 144, 144*f*
Greater vestibular gland, 569
Great saphenous veins, 408E, 409*f*
Great toe, 145*f*
Groin, 10, 11*f*
  pulled, 212E
Ground substance, 83–84
Growth, 6
Growth hormone-inhibiting hormone (GHIH), 320
Growth hormone-releasing hormone (GHRH), 320
Guillain-Barre Syndrome (GBS), 238
Gustation, 291–293
Gustatory hair, 292*f*, 293
Gustatory pathway, 293
Gustatory receptor cell, 292*f*, 293
Gynecology, 556
Gynecomastia, 340
Gyri, 255, 256*f*

H

Hair, 101–103
  chemotherapy and, 102
  mucous membrane, 428T
  of mucous membrane, protective functions of, 426
Hair cells
  of crista, 306, 307*f*
  of maculae, 305, 306f
  of spiral organ, 303*f*, 304
Hair root plexuses, 287
Hallux (big toe), 146
Hamate, 142
Hammer (malleus), 302, 302*f*
Hamstrings, 214E, 215E
  strains of, 214E
Hand, muscles of, 208E, 209*f*
Haploid cell, 558
Hardness of bone, 116
Hard palate, 476, 476*f*
Haversian canal, 116, 117*f*
Haversian system, 116, 117
Head
  of femur, 144, 144*f*
  of fibula, 145, 145*f*
  of humerus, 140, 140*f*
  metacarpal, 141*f*, 143
  metatarsal, 145*f*
  phalangeal, of foot, 146
  of radius, 141
    dislocation of, 168
  of sperm, 560, 560*f*
  veins of, 404E, 405*f*
Head (body regions), 9, 11*f*
Health
  adiposity and, 91
  atherosclerosis and, 410
  breast milk and, 603
  caffeine and, 257
  diabetes and, 332
  exercise and metabolism in, 518
  feet, 148
  female athletes, 576
  food and emotion and, 494
  herbal supplements and, 36
  homeostasis and, 10
  lifestyle and blood circulation, 359
  neurotransmitters, food and, 236
  phytochemicals and, 64
  repetitive motion injuries and, 167
  skin care, 105
  smoking and, 463
  stress and, 431
  stress and exercise and, 278
  urinary tract infections, 535
Hearing, 301–305, 308T. *See also* Ear
  auditory pathway in, 305
  disorders of, 310
  loss of, 310
  physiology of, 304–305, 304*f*
  sound waves in, 304–305
Heart, 364–381
  blood supply to, 372
  cardiac cycle and, 374–375, 375*f*
  chambers of, 367–368*f*, 367–369
  conduction system of, 372–373, 373*f*
  disorders of, 379–380
  electrocardiogram and, 374, 374*f*
  exercise and, 378
  great vessels of, 369
  location and coverings of, 365–366
  structure of, 365–370, 367–368*f*
  valves of, 368*f*, 369–370, 370*f*, 375*f*
  wall and chambers of, 366–369
Heart attack, 380
Heartbeat, 374–375
  sound of, 375
Heart block, 380
Heart cell replacement, 367
Heart murmur, 375
Heart prominence, 594
Heart rate, 375
  regulation of, 377–378, 377*f*
Heart sound, 375
Heat
  body (*See* Body heat)
  in inflammation, 427
  loss of, 516–517
  measurement of, 515
Heat cramp, 519
Heatstroke, 519
Heimlich maneuver, 468
Helper T cell, 433, 436T
Hemangioma, 108
Hematocrit, 346, 352T
Hematology, 345
Hemiarthroplasty, 151
Hemiplegia, 265
Hemispheric lateralization, 261–262
Hemochromatosis, 361
Hemodialysis, 539
Hemoglobin, skin color and, 101
Hemolysis, 50, 50*f*
Hemolytic anemia, 364
Hemolytic disease of newborn (HDN), 360
Hemopoiesis, 114, 346, 348, 348*f*, 350*f*
Hemorrhage, 354
Hemorrhagic anemia, 360
Hemospermia, 563
Hemostasis, 354–357
  clotting in, 354–356
  control mechanisms in, 357
  vascular spasm in, 354
Heparin, 357
Hepatic portal circulation, 410–412, 411*f*
Hepatitis, 497
Hepatocyte, 484, 485*f*, 486
Herbal supplements, 36
Hernia, 198E
Herniated disc, 150
Herpes, 581
Herpes zoster, 265
Heterozygous trait, 604
Hiccupping, 457T
High altitude sickness, 459
High-density lipoprotein (HDL), 379
  in lipid metabolism, 513
Highly active antiretroviral therapy (HAART), 440
Hinge joint, 164–165, 164*f*
Hip, 142*f*, 143–144
  fracture of, 151
Hip bone, 142*f*, 143
Hip joint, 142*f*, 143–144, 144
Hirsutism, 103, 340
Histocompatibility, 429
Histology, 72
  of nervous tissue, 228–230
  of skeletal muscle tissue, 174*f*, 175, 176*f*
  of smooth muscle tissue, 186–187, 187*f*
  of testes, 558, 559*f*
HIV/AIDS, 345, 439–440

Hives, 108
Hodgkin disease, 440
Homeostasis, 1, 6–9
  aging and, 9
  of blood calcium level, 326, 327*f*
  body temperature, 515–516
  cardiovascular system in, 6, 414
  control of, feedback systems in, 7–8, 7*f*
  digestive system in, 496
  disease and, 8–9
  endocrine system in, 339
  health and, 10
  lymphatic system in, 438
  mineral, 114
  muscular system in, 218
  nervous system in, 309
  reproductive system in, 578
  respiratory system in, 466
  skeletal system in, 149
  urinary system in, 538
Homocysteine, coronary artery disease and, 379
Homolog, 604
Homologous chromosome, 558, 604
Homozygous trait, 604
Horizontal plane, 14
Hormones, 7, 82, 315, 334T. *See also under specific hormone*
  actions of, 317–318
    for lipid-soluble hormones, 317–318, 317*f*
    for water-soluble hormones, 318, 318*f*
  of adrenal glands, 222, 329, 331
  in birth control methods, 574–575
  in blood pressure regulation, 393–394
  in body temperature regulation, 516–517
  bone metabolism and, 124T
  chemistry of, 317
  in digestion, 495, 495t
  of endocrine cells, 334T
  female reproductive cycle and, 570–572, 571*f*, 572, 573*f*
  in heart rate regulation, 378
  in labor and delivery, 601
  metabolic rate and, 516
  in nephron function regulation, 533, 533*f*
  of ovaries, 333
  of pancreas, 327, 329
  of parathyroid glands, 326, 327*f*
  of pineal gland, 334
  of pituitary gland, 319–323, 323T
  in pregnancy, maternal changes and, 600
  processing of, liver in, 486
  production of
    in kidneys, 524
    kidneys in, 525
  secretion of, control of, 318
  stress response, 335–336
  of testes, 333
    function of, 561–562
  of thyroid gland, 323–325
  in water loss regulation, 546–547, 546*f*
Horner's syndrome, 274
Horn, 245, 245*f*
Human chorionic gonadotropin (hCG), 334T
  pregnancy, 600
Human development. *See* Embryonic and fetal development
Human Genome Project, 58
Human growth hormone (hGH), 320, 323T
  bone growth and, 122
  disorders of, 337
Human immunodeficiency virus (HIV). *See* HIV/AIDS
Human placental lactogen (hPL), 600
Humerus (arm bone), 139*f*, 140, 141*f*
  muscles of, 204E, 205*f*
Huntington's disease, 604–605
Hyaline cartilage, 88T, 89
Hyaluronic acid, 83–84
Hydrocephalus, 250
Hydrogen, renal excretion of, in acid-base balance, 551
Hydrogenation, *trans* fatty acids and, 34
Hydrogen bond, 28
Hydrolysis, 30
  of sucrose, 32*f*
Hydrophilic compound, 29
Hydrophobic compound, 29
Hymen, 568, 568*f*
Hyoid bone, 126*f*, 127*f*, 132–133
Hypercapnia, 393, 464
Hyperextension, 160–161, 161*f*
Hyperinsulinism, 338
Hypermetropia, 300
Hyperopia, 299*f*, 300
Hyperplasia, 67
Hyperpolarization, 236
Hypersecretion, 337
Hypertension, 415
Hyperthyroidism, 325
Hypertonia, 220
Hypertonic solution, 50, 50*f*
Hypertrophy, 67
Hyperventilation, 464
Hypoblast, 590, 591*f*
Hypocapnia, 464
Hypodermis, 99
Hypoglossal nerve, 252*f*, 264T
Hypoglycemia, 338
Hypokinesia, 266
Hypoparathyroidism, 338
Hypophyseal fossa, 127*f*, 128, 319, 319*f*
Hypophyseal portal vein, 320
Hypophysis, 319. *See* Pituitary gland
Hyposecretion, 337
Hyposmia, 291
Hypotension, 415
Hypothalamus, 249*f*, 254–255, 254*f*, 262T
  pituitary gland and, 319–323
Hypothermia, 518
Hypotonia, 220
Hypotonic solution, 50, 50*f*
Hypovolemia, 547
Hypoxia, 351, 380, 393, 464, 468
Hysterectomy, 567
H zone, 175, 176*f*, 177

I

I band, 175, 176*f*, 177*f*
Ibuprofen, 335
Identical twins, 587
Ileocecal sphincter, 486, 491*f*, 492
Ileum, 486, 486*f*
Iliac crest, 142*f*, 143, 143*f*
Iliacus, 210E, 212E, 213*f*
Iliocostalis group, 210E, 211*f*
Ilioinguinal nerve, 244*f*
Iliopsoas, 212E
Iliotibial tract, 212E, 213*f*
Ilium, 142*f*, 143
Image formation, 297, 299–300
Immune system, 428
Immunity, 420
  adaptive, 420, 428–437 (*See also* Adaptive immunity)
  aging and, 437
  breast milk and, 603
  cell-mediated, T cells and, 432–434, 433*f*, 434*f*
  innate, 420, 425–428
  naturally and artificially acquired, 437T
  types of, 437T
Immunoglobulins, 430
  classes of, 430T
Immunological memory, 436–437
Immunology, 428
Impetigo, 108
Impingement syndrome, 204E
Implantation, 588, 590*f*
Impulse. *See* Action potentials
Incisors, 478
Incompetence, 370
Incomplete dominance, 605
Incontinence, 537
Incus, 302, 302*f*, 304*f*
Induced polycythemia, 350
Infant. *See* Newborn
Infections, urinary tract, 535
Infectious mononucleosis, 440
Inferior articular facet, 136
Inferior articular process, 135

Inferior colliculus, 253
Inferior (directional term), 12E, 13*f*
Inferior extensor retinaculum, 216E, 217*f*
Inferior gluteal nerve, 244*f*
Inferior mesenteric artery, 396E, 397*f*
Inferior mesenteric ganglion, 274, 275*f*
Inferior nasal concha, 126*f*, 127*f*, 130*f*, 131
Inferior oblique, 196E, 197*f*
Inferior rectus, 196E, 197*f*
Inferior vena cava, 368*f*, 369, 394, 402E, 403*f*
Infertility, 606
Inflammation, 428T
  protective functions of, 426–427
Inflammatory bowel disease (IBD), 498
Inflation reflex, 465
Influenza, 467
Infraspinatus, 204E
Infundibulum, 319*f*, 320, 566, 567*f*
Ingestion, 473
  metabolic rate and, 516
Inguinal hernia, 198E
Inguinal ligament, 213*f*
Inhalation, 453–454, 454*f*
  muscles of, 200E, 201*f*
Inheritance, 612–616
  autosomes and sex chromosomes in, 605–606
  of gender, 605, 606*f*
  genotype and phenotype in, 604–605
  of male-pattern baldness, 103
  of phenylketonuria (PKU), 604, 605*f*
  sex-linked, 606
Inhibin, 333
  in female reproduction, 572
  in male reproduction, 561
Inhibiting hormones, 320
Innate defenses. *See* Nonspecific resistance
Innate immunity, 420, 425–428
Inner cell mass, 588, 589*f*
Inner ear, 302, 303*f*, 304, 308T
Inorganic compounds, 29–30
Input in feedback system, 7, 7*f*
Insertion, of muscle, 188, 189*f*
Insoluble fiber, 497
Inspection, 9
Inspiration. *See* Inhalation
Inspiratory area, 462
Inspiratory capacity, 456*f*, 457
Inspiratory reserve volume, 456, 456*f*
Insufficiency, 370
Insula, 254*f*, 256*f*
Insulin, 327, 329*f*
Insulinlike growth factor (IGF), 320
Insulin resistance, 332
Insulin shock, 338
Integral protein, 46, 46*f*
Integrating center, 247, 248*f*
Integrative functions of nervous system, 227
Integumentary system, 4T, 97–109. *See also* Skin
  aging of, 106
  contribution of, to homeostasis, 109
  disorders of, 107–108
Intercalated discs, 186, 366, 366*f*
Intercostal space, 138, 138*f*
Intercostal (thoracic) nerve, 244*f*, 247
Interferons, protective functions of, 426, 428T
Interleukin-2, in cell-mediated immunity, 432, 433*f*
Interleukin-10 in breast milk, 603
Interlobar artery, 527*f*
Interlobar vein, 527*f*
Interlobular artery, 527*f*, 528*f*
Interlobular vein, 527*f*, 528*f*
Intermediate filament, 53, 54*f*, 186
Internal capsule, 254*f*
Internal carotid artery, 398E, 399*f*
Internal iliac artery, 396E, 397*f*, 400E, 401*f*
Internal intercostals, 200E, 201*f*, 203*f*
Internal jugular vein, 402E, 403*f*
Internal nares, 447, 447*f*
Internal oblique, 198E, 199*f*
Internal respiration, 446, 458*f*, 459
Internal root sheath, 102, 102*f*
Internal urethral sphincter, 536*f*, 537
Interneurons, 227
Interphase, 62, 63*f*
Interstitial fluid, 6, 47, 316, 421, 544, 544*f*
Interventricular septum, 367, 368*f*, 369
Intervertebral disc, 133*f*, 134
Intervertebral foramen, 134, 134*f*, 246
Intestinal glands, 487, 487*f*, 492
Intestinal juice, 488
Intestinal phase, of digestion, 494–495
Intracellular fluid, 47, 52, 544, 544*f*
  electrolytes in, 548, 548*f*
Intracutaneous, 108
Intracytoplasmic sperm injection, 607
Intradermal, 108
Intramembranous ossification, 118–120, 119*f*
Intraocular pressure, 297
Intrauterine device, 575
Intravenous pyelogram (IVP), 539
Intravenous solution, isotonicity of, 50
Intrinsic factor, 350, 481
Intrinsic muscle, 196E, 197*f*
Intrinsic pathway, 356, 356*f*
In-utero surgery, 608
Inversion, 162, 163*f*
In vitro fertilization, 607
Involuntary muscles
  cardiac muscle tissue as, 173, 186
  smooth muscle tissue as, 173, 186
Iodide, 506T
Ion channel, 46, 48, 48*f*, 230, 232
Ionic bond, 26, 26*f*
Ion, 25, 26*f*
  absorption of, in small intestine, 489
  functions of, 548
  in heart rate regulation, 378
  regulation of, kidneys in, 524
Iris, 295, 295*f*, 299T
Iron, 506T
Iron-deficiency anemia, 360
Irregular bone, 114
Irregular connective tissue, dense, 87, 87T
Irritable bowel syndrome (IBS), 498
Ischium, 142*f*, 144
Islets of Langerhans, 327–329, 328*f*, 484. *See* Pancreatic islets
Isograft, 100
Isometric contraction, 185
Isoproterenol (Isuprel), 237
Isotonic contraction, 185
Isotonic solution, 50, 50*f*
Isuprel, 237
Itch and tickle, 287

J

Jaundice, 101, 361
Jejunum, 486, 486*f*
Joints, 156–169
  aging and, 165, 167
  cartilaginous, 157, 158, 159*f*
  classification of, 157
  disorders of, 168
  exercise and, 156
  fibrous, 157, 158*f*
  knee, 165, 166*f*
    dislocation of, 168
  repetitive motion injuries to, 167
  structure and function of, 157
  synovial, 157, 158–165
    movement types, 160–163
    types of, 164–165, 164*f*

K

K cell, in small intestine, 487
Keratin, 75, 99
Keratinization, 100
Keratinized stratified squamous epithelium, 75
Keratinocyte, 99, 99*f*
Keratosis, 108
Ketoacidosis, 338
Ketone bodies
  in lipid metabolism, 512
  in urine, 534T
Ketosis, 512

Kidneys
  blood supply to, 526, 527*f*, 528*f*
  external anatomy of, 525–526
  failure of, 539
  functions of, 524–525
  internal anatomy of, 526
  nephrons of, 526 (*See also* Nephrons)
  in pH balance, 524, 553T
Kidney stone, 523, 539
Kilocalorie (kcal), 515
Kinase, 37
Kinesiology, 156
Kinesthesia, 260, 289
Kinetic energy, 28
Knee joint, 165, 166*f*
  dislocation of, 168
  swollen, 168
Krebs cycle, 510*f*, 511
Kupffer cell, 484, 485, 485*f*
Kwashiorkor, 519
Kyphosis, 151

L

Labia majora, 568, 568*f*
Labia minora, 568, 568*f*
Labor and delivery, 601–602
Laceration, 108
Lacrimal apparatus, 294, 428T
  protective functions of, 426
Lacrimal bone, 126*f*, 131
Lacrimal canal, 294, 294*f*
Lacrimal duct, 294, 294*f*
Lacrimal fluid, 294
Lacrimal gland, 294, 294*f*
Lactase, 488, 489T
Lactation, 322, 570, 602–604
Lacteal, 487, 487*f*
Lactose intolerance, 38, 488
Lacunae, 89, 116, 117*f*, 592, 594*f*
Lacunar network, 591*f*, 592
Lambdoid suture, 126*f*, 127*f*, 128*f*, 129*f*
Lamellae, bone tissue, 91, 116, 117*f*
Lamellar granule, 99*f*, 100
Lamellated corpuscle, 99, 287
Laminae, 134, 134*f*, 135*f*
Langerhans cell, 99, 99*f*
Large intestine, 491–493, 491*f*, 492*f*
Laryngitis, 449
Laryngopharynx, 447*f*, 448
Larynx, 446*f*, 447*f*, 448, 449*f*
  cancer of, 449
Laser resurfacing, as anti-aging treatment, 106
LASIK surgery, 310
Latent period, in twitch contraction, 184
Lateral cerebral sulcus, 256*f*, 259*f*
Lateral compartment of foot, 216E, 217*f*
Lateral condyle
  of femur, 144, 144*f*
  of tibia, 145, 145*f*
Lateral corticospinal tract, 261
Lateral (directional term), 12E, 13*f*
Lateral gray horn, 245, 245*f*
Lateral malleolus, 145, 145*f*, 146*f*
Lateral meniscus, 165, 166*f*
Lateral rectus, 196E, 197*f*
Lateral spinothalamic tract, 260, 260*f*, 261*f*
Lateral ventricle, 250, 254*f*
Latissimus dorsi, 204E, 205*f*
Laughing, 457T
Laughter, 420
Leakage channel, 232
Left bundle branch, 372
Left common carotid artery, 396E, 397*f*, 398E, 399*f*
Left coronary artery, 367*f*, 372, 396E, 397*f*
Left hepatic duct, 483*f*, 484
Left primary bronchus, 450*f*, 451
Left pulmonary artery, 367*f*, 368*f*, 369, 395*f*, 410
Left subclavian artery, 396E, 397*f*, 398E, 399*f*
Leg bones, muscles of, 214–215E
Lens, 294–295, 295*f*, 299T
Lens placode, 594
Leptin, 334T
Lesser pelvis, 143*f*
Lethal gene, 608
Leukemia, 360
Leukocytes. *See* White blood cells
Leukocytosis, 353
Leukopenia, 353
Leukotrienes, 335
Levator palpebrae superioris, 194E
Levator scapulae, 202E, 203*f*
Leydig cells, 558, 559*f*, 561, 561*f*
Lidocaine, 234
Life processes, 6
Lifestyle, immune function and, 431
Lifting, improper, 210E
Ligament, 159
Ligamentum arteriosum, 367*f*, 368*f*, 369, 412
Ligamentum teres, 412
Ligamentum venosum, 412
Limb bud, 594
Limbic system, 257, 258*f*
  in respiration control, 464
Linea alba, 199*f*
Lingual frenulum, 466*f*, 467
Lingual tonsil, 425, 447*f*, 448
Lipectomy, suction, 85
Lipid, 32–35
  absorption of, in small intestine, 489–490
  metabolism of, 512–513, 513*f*, 515T
    anabolism (lipogenesis) in, 512, 513*f*, 515T
    catabolism (lipolysis) in, 512, 513*f*, 515T
    lipid transport in, 512–513
    liver in, 486
    in small intestine, 489, 489T
  organic compounds, 33–34
Lipid bilayer, 46, 46*f*
Lipid droplet, 52
Lipid-soluble hormones, action of, 317–318, 317*f*
Lipolysis, 512, 513*f*, 515T
Lipoprotein (a), coronary artery disease and, 379
Lipoprotein, 379, 512–513
Liposuction, 85
Lips, 476, 476*f*
Lithotripsy, for gallstones, 484
Little-league elbow, 168
Liver, 483*f*, 484–486, 485*f*
Liver spot, 101
Lobe
  of lung, 450*f*, 451
  of mammary glands, 570
Lobules
  of liver, 484, 485*f*
  of lungs, 451, 452*f*
  of testes, 558, 559*f*
Local anesthetic, 234
Local interneuron, 261
Long bone, 114, 115*f*
Longissimus group, 210E, 211*f*
Longitudinal arch, 146, 146*f*
Longitudinal fissure, 254*f*, 255, 256*f*
Loop of Henle, 526, 528*f*
Loose connective tissue, 84–86, 85–86T
Lordosis, 151
Lou Gehrig's disease, 265–266
Low-density lipoprotein (LDL), 379
  diabetes and, 332
  in lipid metabolism, 513
Lower esophageal sphincter (LES), 478
Lower jaw. *See* Mandible
Lower limb bud, 594
Lower limbs, 10, 11*f*, 144–146
  arteries of, 400E, 401*f*
  veins of, 408E, 409*f*
Lower motor neuron, 261, 261*f*
Lower respiratory system, 446
Lumbar curve, 133*f*, 134
Lumbar enlargement, 243, 244*f*
Lumbar plexus, 244*f*, 246
Lumbar puncture, 243, 244*f*
Lumbar vertebrae, 133, 136, 136*f*
Lumen
  of artery, 386, 387*f*
  size of, vascular resistance and, 391

Lumpectomy, 580
Lunate, 141*f*, 142
Lunelle, 575
Lungs, 446*f*, 450*f*, 451–452, 452*f*
cancer of, 467
function of, in pregnancy, 600–601
volume and capacity of, 456–457, 456*f*
Lunula, 104, 104*f*
Lupus, 93, 441
Luteal (postovulatory) phase, 571*f*, 572
Luteinizing hormone (LH), 320, 323T
in female reproductive cycle, 570, 571*f*, 572, 573*f*
in male reproduction, 561, 561*f*
Luxation, 169
Lymph, 47, 90, 421
Lymphadenopathy, 441
Lymphatic and immune system, 5T, 420–441
barriers in, physical and chemical, 425–426
components of, 421, 422*f*
contribution of, to homeostasis, 438
disorders of, 439–441
functions of, 421
immunity and (*See also* Immunity)
innate immunity and, 420, 425–428
internal defenses in, 426–428
lymph and interstitial fluid in, relationship between, 421
organs and tissues of, 424–425
lymphatic nodules as, 425
lymph nodes as, 423*f*, 424–425
spleen as, 425
thymus as, 424
vessels and circulation in, 421
Lymphatic capillary, 421, 423*f*
Lymphatic nodule, 424
Lymphatic tissue, 421
Lymphatic vessel, 421, 422*f*, 423*f*
Lymph node, 421, 422*f*, 423*f*, 424–425, 424*f*
Lymphocytes, 353, 355T
Lymphoid stem cell, 348, 348*f*
Lymphomas, 440
Lymphotoxin, in cytotoxic T cell action, 433, 434*f*
Lysosome, 56, 59T
Lysozyme, 294, 428T
protective functions of, 426

M

Macromolecule, 29
Macrophage
alveolar, 452, 453*f*
connective tissue, 83, 83*f*, 84
protective functions of, 426
wandering, 353
Maculae, 305, 306*f*
Macula lutae, 297
Macular degeneration, 310
Magnesium, 506T
Major caly, 525*f*, 526
Major histocompatibility (MHC) antigen, 353, 429
Male infertility, 607
Male-pattern baldness, 103
Male reproductive system, 5T, 557–563, 557*f*. *See also* Female reproductive system; Reproductive system
accessory sex glands in, 562–563
aging and, 577
disorders of, 579, 607
ducts of, 562
functions of, 557*f*
organs of, 557*f*
penis in, 557*f*, 563
scrotum in, 557–558, 557*f*
testes in, 557*f*, 558–562
Male sexual characteristics, testicular hormones in, 561–562
Malignancy, 66
Malignant melanoma, 107, 107*f*
Malignant tumor, 66
Malleus, 302, 302*f*, 304*f*
Malnutrition, 519
Malocclusion, 498
Maltase, 488, 489T
Mammary gland, 569–570, 569*f*
Mammogram, 579
Mammography, 579
Mandible, 126*f*, 127*f*, 131
muscles of, 195E, 195*f*
Mandibular fossa, 126*f*, 127, 128*f*
Manganese, 506T
Manubrium, 137, 138*f*
Marasmus, 519
Marfan syndrome, 84
Marrow, bone
red, 114 (*See also* Red bone marrow)
yellow, 114
Marrow cavity, 116
Mass, 23
Masseter, 195E, 195*f*
Mass number, 25
Mass peristalsis, 493
Mast cells, 83, 83*f*
Mastectomy, 580
Mastication, 195E, 478
Mastoid process, 126*f*, 127
Matrix, 57, 57*f*
of connective tissue (*See also* Connective tissue(s))
extracellular, of connective tissue, 82–83
of hair, 102, 102*f*
Matter, 23
Mature follicle, 565, 565*f*
Maxilla, 126*f*, 127*f*, 128*f*, 129–130, 130*f*
Maxillary sinus, 129, 131, 132*f*
Meatus, auditory, 126*f*
Mechanical digestion, 474
in small intestine, 488
Mechanical ventilation, 468
Mechanoreceptor, 281
Medial (adductor) compartment, 214E, 215E
Medial condyle
of femur, 144, 144*f*
of tibia, 145, 145*f*
Medial (directional term), 12E, 13*f*
Medial lemniscus, 253*f*
Medial malleolus, 145, 145*f*
Medial meniscus, 165, 166*f*
Medial rectus, 196E, 197*f*
Medial umbilical ligament, 412
Median antebrachial vein, 406E, 407*f*
Median nerve, 244*f*
Mediastinum, 15*f*, 16–17, 16*f*
Medical history, 9
Medulla oblongata, 249*f*, 250, 251*f*, 252, 252*f*, 262T
Medullary cavity, 115*f*, 116
Medullary rhythmicity area, 250, 462, 462*f*
damage to, 252
Megakaryoblast, 353
Meiosis, 62
in oogenesis, 565–566, 566*f*
in spermatogenesis, 558, 560, 560*f*
Meissner corpuscle, 100, 287
Melanin, 99, 101
in hair, 103
Melanocyte, 99, 99*f*
Melanocyte-stimulating hormone (MSSH), 321, 323T
Melatonin, 334
secretion of, 255
Membrane, 90–92
Membrane potential, 230
Membranous labyrinth, 302, 303*f*
Memory, 262
Memory B cell, 435, 436T
Memory T cell, 434, 436T
Menarch, 577
Ménière's disease, 310
Meninges, 243
cranial, 250, 251*f*
Meningitis, 267
Menisci, 159
Menopause, 577
Menorrhagia, 581
Menses, 572
Menstrual cycle, 570, 571*f*
in female athletes, 576
Menstrual phase, 172, 571*f*

Menstruation, 572
Mental foramen, 126*f*, 131
Merkel cell, 99*f*, 100
Merkel disc, 100, 287
Mesentery, 476
Mesothelium, 75, 91
Messenger RNA, 60
Metabolic rate, 515
Metabolic syndrome, 332
Metabolic water, 545
Metabolism, 6, 507, 510–518. *See also* Nutrition
  anabolism in, 507
  body heat and, 515–518 (*See also* Body temperature)
  of bone, 125T
  of carbohydrates, 510–512, 510*f*, 515T (*See also* Carbohydrate(s), metabolism of)
  catabolism in, 507
  disorders of, 519
  of lipids, 512–513, 513*f*, 515T (*See also* Lipid(s), metabolism of)
  in liver, 485–486
  of proteins, 514, 515T
  of skeletal muscle tissue, 180, 182–183
  thyroid hormones and, 325
Metacarpals, 141*f*, 143
Metacarpus, 143
Metafemale syndrome, 608
Metaphase, 63*f*, 64
Metaphase plate, 64
Metaphysis, 115*f*, 116
Metaplasia, 67
Metastasis, 66, 425
Metatarsals, 145*f*, 146
Metatarsus, 146
Miacalcin, 30, 325
Micelle, 489
Microbes, in urine, 534T
Microdermabrasion, as anti-aging treatment, 106
Microfilaments, 53, 54*f*
Microglia, 231T
Microtubule, 53, 54*f*
Microvilli, 45*f*, 53, 54*f*, 75
  of small intestine, 487–488, 487*f*
Micturition, 537
Micturition reflex, 537
Midbrain, 249*f*, 251*f*, 253, 262T
Middle ear, 301–302, 302*f*, 304*f*, 308T
Middle nasal concha, 126*f*, 128*f*, 129, 130*f*
Middle piece of sperm, 560*f*, 561
Midsagittal plane, 14, 14*f*
Mifepristone (RU 486), 317, 577
Milking, 390
Mineral homeostasis, 114
Mineralocorticoids, 329, 331
Minerals, 504–505, 506T
  bone metabolism and, 124T
  storage of, liver in, 486
  supplemental, 507
Minor caly, 525*f*, 526
Minoxidil (Rogaine), 103
Minute ventilation, 456
Mitochondria, 57, 57*f*, 59T
Mitosis, 62–64, 63*f*
Mitotic phase, 62–65, 63*f*
Mitotic spindle, 63*f*, 64
Mitral stenosis, 370
Mitral valve, 369
Mitral valve prolapse (MVP), 370
Mixed nerve, 246
Mixing waves in digestion, 482
Molar, 478
Mole, 101, 107*f*
Molecular formula, 25
Molecule, 2, 25, 25*f*
  structure of, 27*f*
Monoclonal antibody, 435–436
Monocytes, 353, 355T
Monohydrogen phosphate, 550
Monomers, 29
Mononucleosis, infectious, 440
Monoplegia, 265
Monosaccharides, 31
  absorption of, 488–489, 490*f*
Monounsaturated fats, 33, 33*f*
Monozygotic twins, 587
Mons pubis, 568, 568*f*
Mood, food and, 236
Morning sickness, 608
Morula, 588, 589*f*, 590*f*
Motility
  in GI tract, 474
  in small intestine, 488
Motor area, of cerebral cortex, 258
Motor (descending) tract, 245
Motor end plate, 178, 178*f*
Motor function, of nervous system, 227
Motor neuron, 177, 178*f*, 227, 247, 248*f*
Motor unit, 177
  recruitment, 184
Motrin, 335
Mouth, 476–478, 476*f*
  digestion in, 478
Movement, 6
Mucosa, 90–91
  of GI tract, 474*f*, 475
Mucous membrane, 90–91, 428T
  as barrier in immunity, 426
  of gastrointestinal tract, 474*f*, 475
Mucous neck cell, in gastric glands, 481, 482*f*, 483
Mucus
  goblet cells and, 75
  in immunity, 426, 428T
Multiple-allele inheritance, 605
Multiple sclerosis (MS), 237
Multipotent stem cell, 588
Multiunit smooth muscle tissue, 187
Muscle, strain of, 220
Muscle action potential, 177
Muscle fatigue, 183
Muscle fiber, 90, 173, 175, 176*f*
Muscle spindles, 247
Muscle tissue, 73, 90. *See also* Cardiac muscle tissue; Skeletal muscle tissue; Smooth muscle tissue
  aging and, 187–188
  levels of organization of, 2
  repair of, 92
  summary of, 188T
Muscle tone, 180
Muscular artery, 386
Muscular atrophy, 175
Muscular dystrophy, 219
Muscular hypertrophy, 175
Muscularis of GI tract, 474*f*, 475
Muscular system, 4T, 172–220
  cardiac muscle tissue in, 90, 173, 186
    autonomic nervous system, 279T
  contribution of, to homeostasis, 218
  disorders of, 219–220
  skeletal muscle tissue in, 90, 173–183 (*See also* Skeletal muscle tissue)
  smooth muscle tissue in, 90, 173, 186–187
    autonomic nervous system, 279T
  tissue functions in, 173
  tissue types in, 17
Musculocutaneous nerve, 244*f*
Mutation, 38, 66, 604
Myalgia, 220
Myasthenia gravis, 219
Mycobacterium tuberculosis, 440, 467
Myelinated axon, 228
Myelination, 228
Myelin sheath, 228, 229*f*, 237
Myeloid stem cell, 348, 348*f*
Myocardial infarction, 380
Myocardial ischemia, 380
Myocardium, 366, 366*f*
Myofibril, 175, 176*f*, 178*f*
Myoglobin, 175, 176*f*
  in muscle contraction, 183
Myogram, 184, 184*f*, 185*f*
Myology, 172
Myoma, 220
Myomalacia, 220
Myometrium, 567, 567*f*
Myopathy, 219

Myopia, 299*f*, 300
Myosin, 175, 177*f*
Myosin-binding site, 175, 177*f*
Myosin head, 175, 177*f*
Myosin tail, 175, 177*f*
Myositis, 220
Myotonia, 220
My Pyramid, 504, 505*f*
Myxedema, 338

N

Nail body, 104, 104*f*
Nail matrix, 104, 104*f*
Nails, 104
Nasal bones, 126*f*, 127
Nasal cavity, 446*f*, 447
Nasal concha, 447*f*, 448
Nasal septum, 131, 447
  deviated, 131
Nasolacrimal duct, 294, 294*f*
Nasopharynx, 447*f*, 448
Natriuresis, 546–547
Natural family planning, 575
Natural killer cell, 353, 355T, 428T
  protective functions of, 426
Nausea, 498
Navicular, 145, 145*f*
Neck
  of femur, 144, 144*f*
  of tooth, 477, 477*f*
  veins of, 404E, 405*f*
Neck (body region), 9, 11*f*
Necklift, as anti-aging treatment, 106
Necrosis, 67
Negative feedback systems. *See* Feedback systems
Neonatal period, 586
Neoplasm, 66
Nephrology, 524
Nephron, 526
  damaged, 530
  functions of, 528–534, 529*f*
    glomerular filtration as, 528, 529T, 530–531, 530*f*
    tubular reabsorption as, 528–529, 529T, 531–534, 532*f*, 533*f*
    tubular secretion as, 529, 529T, 531–534, 532*f*
  hormonal regulation of, 533
  number of, 526
Nerve, 226. *See also specific nerve, e.g.* Facial nerve, Optic nerve
Nerve block, 267
Nerve endings, free, 100
Nerve impulse, 7, 7*f*
Nerve supply of skeletal muscle tissue, 174*f*, 175, 176*f*
Nervous system, 4T, 226–228. *See also* Sensation(s)
  aging and, 263
  autonomic, 271–280 (*See also* Autonomic nervous system (ANS))
  central, 226*f*, 227, 242–263
  contribution of, to homeostasis, 309
  disorders of, 237–238
  functions of, 226–227
  metabolic rate and, 515
  organization of, 227–228
  peripheral, 226*f*
  somatic, 227
    autonomic nervous system compared with, 272–273, 272T, 273*f*
Nervous tissue, 73, 90, 225–238
  action potentials, 230–234
  histology of, 228–230
  level of organization of, 2
  repair of, 92
  structures of, 226, 226*f*
  synaptic transmission in, 234–237
Net diffusion, 47, 47*f*
Net filtration pressure, 530
Neural fold, 592
Neuralgia, 267
Neural groove, 592
Neural layer of retina, 296
Neural plate, 592
Neural tube, 592
  defects in, 592
Neuritis, 267
Neuroglia, 90, 228, 230, 231T
Neurology, 225
Neuromuscular disease, 219
Neuromuscular junction (NMJ), 178, 178*f*
Neuron, 90, 228, 229*f*
  of autonomic nervous system, 277
Neuropathy, 238
Neuropeptides, 237
Neurosecretory cells, 319*f*, 320
Neurosyphilis, 580
Neurotransmitter receptors, 235, 235*f*
Neurotransmitters, 237
  in autonomic nervous system, 277
  exocytosis and, 52
  food and, 236
  in synaptic vesicles, 178
Neurulation, 592
Neutron, 24, 24*f*
Neutrophils, 353, 355T
Nevus, 101, 107*f*
Newborn
  hemolytic disease of, 360
  premature, 602
    anemia in, 351
Niacin, 509T
Nicotinamide, 509T
Night blindness, 300
Nipple, 569–570, 569*f*
Nitric oxide (NO)
  as hormone, 317
  as neurotransmitter, 237
Nitrogenous base, 38, 39*f*
Nociceptor, 286, 287
Nocturia, 539
Nocturnal enuresis, 539
Nodes of Ranvier, 228, 229*f*, 231T
Nonciliated simple columnar epithelium, 75, 77T
Nonessential amino acid, 514
Nonfluent aphasia, 259
Non-Hodgkin's lymphoma, 440
Nonkeratinized stratified squamous epithelium, 75
Nonmelanoma skin cancer, 107
Nonpolar covalent bond, 27*f*, *28*
Nonspecific immunity, 420, 425–428
Nonsteroidal anti-inflammatory drug (NSAID), 219, 335
Nonstriated (smooth) muscle, 173, 186–187
Norepinephrine, 274, 333
  in blood pressure/blood flow regulation, 393
  metabolic rate and, 516
  as neurotransmitter, 237
    in ANS, 277
Normal saline solution, 50
Norplant, 575
Nose, 446*f*, 447–448, 447*f*
Nosebleed, 468
Nostrils, 447, 447*f*
Notochord, 592
Novocaine, 234
Nuclear envelope, 45*f*, 57, 58*f*
Nuclear pore, 57, 58*f*
Nucleases in small intestine, 488, 489T
Nucleic acid, 38
Nucleolus, 57, 58*f*
Nucleotides, 38, 38*f*
Nucleus(i), 24, 24*f*, 45, 45*f*, 57–58, 58*f*, 59T
  in nervous tissue, 230
Nulliparity, 580
Nutrient, 504–507
Nutrition. *See also* Metabolism
  guidelines for, 504
  minerals in, 504–505, 506T
  nutrients in, 504–507
  vitamins in, 505, 507, 508–509T
Nystagmus, 310

O

Obesity, 519
Oblique plane, 14, 14*f*
Oblique popliteal ligament, 165, 166*f*
Obstetrics, 586

Obturator foramen, 142*f*, 143*f*, 144
Obturator nerve, 244*f*
Occipital belly, 194E, 195*f*
Occipital bone, 126*f*, 127*f*, 128
Occipital condyles, 128
Occipital lobe, 255, 256*f*, 259*f*
Occipitofrontalis, 194E, 195*f*
Occlusion, 416
Octet rule, 25
Oculomotor nerve, 252*f*, 253*f*, 263T
Oculomotor nucleus, 253*f*
Odorants, 290, 291*f*
Office hypertension, 416
Oil glands, 102*f*, 103
Olecranon, 140, 141*f*
Olecranon fossa, 140, 140*f*
Oleic acid, 33*f*
Olfaction, 290–291
Olfactory bulb, 252*f*, 290, 291*f*
Olfactory epithelium, 290, 291*f*, 447*f*, 448
Olfactory foramina, 129, 129*f*, 130*f*
Olfactory gland, 290, 291*f*
Olfactory hair, 290, 291*f*
Olfactory nerve, 252*f*, 263T, 290, 291*f*
Olfactory pathway, 290–291
Olfactory receptor, 290, 291*f*
Olfactory tract, 252*f*, 290
Oligodendrocyte, 231T
Oliguria, 530
Omega-3 fatty acids, 34
Omega-6 fatty acids, 34
Omentum, 475–476 , 475*f*
Oncogenes, 66
Oncogenic virus, 66
Oncology, 66
One-way information transfer, 236
Oocyte, 565, 566, 566*f*
Oogenesis, 565–566, 566*f*
Oogonia, 565
Oophorectomy, 581
Open (compound) fracture, 122
Ophthalmology, 290
Ophthalmoscope, 295
Opportunistic infection, 439
Opsonization, 426
Optic chiasm, 300, 301*f*
Optic disc, 297
Optic foramen, 126*f*, 128
Optic nerve, 252*f*, 263T, 300, 301*f*
Optic tract, 252*f*, 300–301, 301*f*
Oral cavity. *See* Mouth
Oral contraceptive, 574
Oral rehydration therapy, 547
Orbicularis oculi, 194E, 195*f*
Orbicularis oris, 194E, 195*f*
Orbits, 126*f*, 127, 130*f*
Organelle, 2, 45, 45*f*, 52–57, 59T
Organic compound, 29, 31–40
  adenosine triphosphate (ATP) as, 38, 40
  carbohydrates as, 31–32
  enzymes as, 37–38
  lipids as, 33–34
  nucleic acids as, 38
  proteins as, 35–37
Organism, 3
Organismal level of structural organization, 3, 3*f*
Organ level of structural organization, 2, 3*f*
Organ of Corti, 303*f*, 304, 304*f*
Organ, 2
Organ transplant rejection, 433
Origin of muscle, 188, 189*f*
Oropharynx, 447*f*, 448
Orthodontics, 122
Orthopedics, 172
Orthostatic hypotension, 416
Osmoreceptor, 286, 322–323
Osmosis, 49–50, 49*f*, 53T
  body water, 545
Osmotic pressure, 49–50, 547
Osseous tissue, 89–90, 116. *See also* Bone; Bone tissue
Ossification, 118–123, 120*f*
  endochondral, 118, 120–121, 120*f*
  intramembranous, 118–120, 119*f*
Osteoarthritis, 151, 168
Osteoblast, 116, 117*f*, 122
Osteoclast, 116, 122, 177*f*
Osteocyte, 116, 117*f*
Osteogenic cell, 116, 117*f*
Osteogenic sarcoma, 151
Osteology, 113
Osteomalacia, 122, 150
Osteomyelitis, 151
Osteons, 116, 117*f*
Osteopenia, 151
Osteoporosis, 122, 150, 150*f*
  female athletes, 576
Otalgia, 310
Otic placode, 594
Otitis media, 310
Otoacoustic emissions, 305
Otolithic membrane, 305, 306*f*
Otolith, 305, 306*f*
Otorhinolaryngology, 290, 446
Outer ear, 301, 302*f*, 308T
Output in feedback system, 7*f*, 8
Oval window, 302, 302*f*, 303*f*, 304*f*
Ovarian artery, 396E, 397*f*
Ovarian cortex, 564–565, 565*f*
Ovarian cycle, 565–566, 566*f*, 570, 571*f*
  lactation and, 602–604
Ovarian follicle, 565, 565*f*, 571*f*
Ovarian medulla, 565, 565*f*
Ovary, 564–566, 564*f*
  cancer of, 580
  cyst of, 565, 581
  hormones of, 333
  menstruation and, 572
  in preovulatory phase, 572
Overuse injuries, 167
Ovulation, 565*f*, 566, 571*f*, 572
Ovum, 566, 566*f*
Oxidase, 37
Oxidative damage, 64
Oxygen
  muscle contraction and, 183
  transport of, 459–461, 460*f*
Oxygenated blood
  in heart vessels, 369
  in oxygen-carbon dioxide exchange, 458–459
Oxygen debt, 183
Oxyhemoglobin, 459
Oxytocin, 321, 321*f*, 322, 323T
  in labor and delivery, 601–602
  in lactation, 602–603

P

Pacemaker
  activity-adjusted, 373
  artificial, 373
  heart, 372, 373
Pacinian corpuscle, 99, 287
Pain
  chronic, management of, 289
  fast, 288
  in inflammation, 427
  referred, 288, 288*f*
  in respiration control, 465
  sensations of, 287–288
  slow, 288
Palatine bones, 127*f*, 128*f*, 131
Palatine tonsil, 422*f*, 425, 447*f*, 448, 476*f*, 477
Pallor, 101
Palm, 143
Palmaris longus, 208E, 209*f*
Palmitic acid, 33, 33*f*
Palpation, 9
Palpitation, 381
Pancreas, 327–329, 328*f*, 483–484, 483*f*
  cancer of, 484
Pancreatic amylase, 484, 488, 489T
Pancreatic duct, 483, 483*f*
Pancreatic islet, 327–329, 484
  disorders of, 338
Pancreatic juice, 483, 484
Pancreatic lipase, 484, 488, 489T
Pantothenic acid, 509T
Papanicolaou (PAP) test, 82, 581

Papilla
of hair, 102, 102*f*
of tongue, 292–293, 292*f*, 477
Papillary muscles, 368*f*, 369, 370*f*
Pap smear, 82, 581
Parafollicular cells, 323, 324*f*
Paralysis, 261
facial, 194T
Paranasal sinus, 131, 132*f*
Paraplegia, 265
Parasagittal plane, 14, 14*f*
Parasympathetic division of autonomic nervous system, 228, 272
activities of, 278–280
organization of, 274, 276*f*, 277
Parathyroid glands, 326*f*, 335–327
disorders of, 338
Parathyroid hormone (PTH), 123*f*, 326, 327*f*
in aging, 336
in calcium homeostasis, 122–123
Parenchyma in tissue repair, 92
Parietal bones, 126*f*, 127, 127*f*
Parietal cell, in gastric glands, 481, 482*f*
Parietal layer, 365
serous membranes, 91
Parietal lobe, 255, 256*f*, 259*f*
Parietal peritoneum, of GI tract, 475
Parietal pleura, 450*f*, 451
Parkinson disease, 258, 266
Parotid glands, 473*f*, 477
Paroxysmal tachycardia, 381
Partial fracture, 122
Partial pressure, 457, 458*f*
Partial seizures, 238
Parturition. *See* Labor and delivery
Passive transport, 47–50
Passive transport processes, 53T
Patella, 144, 144*f*
Patellar ligament, 165, 166*f*, 214E
Patellar reflex, 247, 248*f*
absence of, 247
Patellar surface, 144
Patellofemoral stress syndrome, 144
Patent ductus arteriosus, 385
Pathogen, 420
Pathologist, 72
Pathology, 18
Pectineus, 212E, 213*f*, 215E
Pectoral girdle, 139, 139*f*, 142*f*
muscles of, 202E, 203*f*
Pectoralis major, 204E, 205*f*
Pectoralis minor, 201*f*, 202E, 203*f*
Pedicle, 134, 134*f*, 135*f*
Peirpheral edema, 377
Pelvic axis, 142*f*, 143
Pelvic brim, 142*f*, 143
Pelvic cavity, 15*f*, 17
Pelvic girdle, 142*f*, 143–144
Pelvic inflammatory disease (PID), 581
Pelvic inlet, 143
Pelvic outlet, 142*f*, 143
Pelvimetry, 143
Pelvis, 142*f*, 143
arteries of, 400E, 401*f*
sex differences in, 146, 147T
Penis, 557*f*, 563
Pepsin, 481, 483, 489T
Pepsinogen, 481, 483
Peptic ulcer disease (PUD), 497
Peptidase, 488, 489T
Peptide, 36
Peptide bond, 35, 35*f*
Perception, 227
sensory areas of cerebral cortex in, 258
sensory receptors in, 280
Perceptions. *See* Sensation(s)
Percussion, 9
Perforated eardrum, 301
Perforating canal, 117*f*
Perforin, in cytotoxic T cell action, 433, 434*f*
Pericardial cavity, 15*f*, 16*f*, 17, 365, 366*f*
Pericardial fluid, 365
Pericarditis, 366
Pericardium, 91, 365, 366*f*
Pericentriolar materials, 54, 55*f*
Perichondrium, 89, 120*f*, 121
Perilymph, 302, 304*f*
Perimysium, 173, 174*f*
Perineurium, 246, 246*f*
Periodontal disease, 497
Periodontal ligament, 477, 477*f*
Periosteum, 115*f*, 116, 120*f*, 121
development of, in intramembranous ossification, 119*f*, 120
Peripheral chemoreceptor, 463
Peripheral nervous system (PNS), 226*f*, 227
neuroglia in, 231T
Peripheral protein, 46, 46*f*
Peristalsis
in large intestine, 493
in swallowing, 479*f*, 480
Peritoneum, 91
of GI tract, 475
Peritonitis, 476
Peritubular capillary, 526, 527*f*, 528*f*
Permanent teeth, 478
Permeability increase, in inflammation, 427
Pernicious anemia, 360
Peroxisomes, 56, 59T
Perpendicular plate, 126*f*, 127*f*, 128, 130*f*
Perspiration, protective functions of, 424, 428T
Peyer's patches, 422*f*, 425
Phagocyte, 428T
protective functions of, 426
Phagocytosis, 51, 52*f*, 53T, 426
antibodies and, 435
in spleen, 425
white blood cells in, 353
Phalanges
of foot, 145*f*, 146
of hand, 143
Phantom limb sensation, 287
Pharmacology, 18
Pharyngeal stage, of swallowing, 478, 479*f*
Pharyngeal tonsil, 425, 447*f*, 448
Pharynx, 446*f*, 448, 478–480, 479*f*
pH balance
in acid-base balance, 550–552
concept of, 30–31
kidneys in, 524
Phenotype, 604
Phenylketonuria (PKU), 514
inheritance of, 604, 605*f*
Pheochromocytomas, 338
Phlebitis, 416
Phlebotomist, 361
pH of blood, regulation of, kidneys in, 524
Phosphate buffer system, 550–551, 553T
Phospholipid, 34–35, 34*f*
Phosphorus, 506T
Photodamage, 107
Photopigment, 300
Photoreceptor layer of retina, 296, 296*f*
Photoreceptors, 286, 296–297
stimulation of, 300
Photosensitivity, 107
Phrenic nerve, 244*f*
pH scale, 30, 31*f*
Physical barriers to pathogens, 425–426
Physical examination, 9
Physiology, 2
Phytochemical, 44
health and, 64
Pia mater, 243, 243*f*, 250, 251*f*
Pigmented layer of retina, 296, 296*f*
Pigment of skin, 101
Pili, 101. *See also* Hair
arrector, 102*f*, 103
Pimple, 103
Pineal gland, 249*f*, 255, 262T, 334
Pinkeye, 310
Pinocytosis, 52
Piriformis, 212E, 213*f*
Pisiform, 141*f*, 142
Pitocin, 322. *See* Oxytocin
Pituitary dwarfism, 337
Pituitary gland, 249*f*, 252*f*, 319–323
disorders of, 337
hormones of (*See* Pituitary hormones)
hypothalamus and, 255

Pituitary hormones, 323T
  anterior, 320–321, 323T
    adrenocorticotropic hormone (ACTH) as, 320, 323T
    follicle-stimulating hormone (FSH) as, 320, 323T
    human growth hormone (hGH) and growth factors as, 320, 323T
    insulinlike growth factors (IGFs) as, 320
    luteinizing hormone (LH) as, 320, 323T
    melanocyte-stimulating hormone (MSH) as, 321, 323T
    prolactin (PRL) as, 320, 323T
    thyroid-stimulating hormone (TSH) as, 320, 323T
  posterior, 321–323, 323T
    antidiuretic hormone (ADH) as, 312*f*, 321–323, 322*f*, 323T
    oxytocin as, 321, 321*f*, 322, 323T
Pivot joint, 164*f*, 165
Placenta, 412, 413*f*, 593–594, 595*f*
  hormones produced by, 600
Placental stage, 602
Placenta previa, 594
Planar joint, 164, 164*f*
Planes and sections, 14–15, 14*f*
  of brain, 14*f*
Plantar flexion, 163, 163*f*
Plasma, 47, 346, 348*f*
  blood, 90
  body fluid and, 544
Plasma cells, 83, 83*f*, 435, 436T
Plasma membrane, 45, 45*f*, 46–47, 46*f*, 59T, 544
  transport across, 47–52
    active, 50–51
    diffusion in, 47–49
    osmosis in, 49–50
    passive, 47–50
Plasmin, 357
Plasminogen, 357
Platelet plug formation, in hemostasis, 354
Platelets, 90, 353, 355T
Platysma, 194E, 195*f*
Pleura, 91
Pleural cavity, 16*f*, 17, 17*f*, 450*f*, 451
Pleural membrane, 450*f*, 451
Pleurisy, 468
Pleuritis, 468
Pluripotent stem cell, 348, 348*f*, 588
Pneumonia, 467
Pneumonitis, 467
Pneumotaxic area, 462
Podocytes, nephron, 530
Poison ivy toxin, 440
Polar covalent bond, 27*f*, 28
Polarized cell, 230
Poliomyelitis, 266
Polycystic kidney disease, 539
Polycythemia, 361
  induced, 350
Polydipsia, in diabetes, 338
Polypeptide, 36
Polyphagia, in diabetes, 338
Polyps, of colon, 492
Polyribosome, 58*f*
  in protein synthesis, 62
Polysaccharide, 32, 33*f*
Polyunsaturated fat, 33
Polyuria, 531
  in diabetes, 338
Pons, 249*f*, 251*f*, 252*f*, 253, 262T
Popliteal artery, 400E, 401*f*
Popliteal fossa, 214E
Popliteal vein, 408E, 409*f*
Portal vein, 410, 411*f*
Port-wine stain, 108
Postcentral gyrus, 256, 256*f*
Posterior column-medial lemniscus pathway, 259
Posterior compartment of foot, 216E
Posterior cruciate ligament (PCL), 165, 166*f*
Posterior (directional term), 12E
Posterior (dorsal) gray horn, 245, 245*f*
Posterior (dorsal) root, 246
Posterior (dorsal) root ganglion, 246
Posterior (extensor) compartment, 206E
  muscles of, 208E, 209*f*
Posterior (flexor) compartment, 214E, 215E
Posterior intercostal arteries, 396E, 397*f*
Posterior lobe, 319. *See also* Pituitary gland
Posterior median sulcus, 243, 245*f*
Posterior tibial artery, 400E, 401*f*
Posterior tibial vein, 408E, 409*f*
Postganglionic neuron, 272, 273*f*, 275*f*
Postovulatory phase, 571*f*, 572–573
Post-polio syndrome, 266
Postsynaptic neuron, 234, 235*f*
Post term baby, 602
Posttraumatic stress disorder (PTSD), 336
Postural hypotension, 416
Potassium, 506T
  as electrolyte, 548*f*, 549
    imbalances of, 549T
Potential energy, 28
Power stoke, 179*f*, 180
Precapillary sphincter, 387, 388*f*
Precentral gyrus, 255, 256*f*
Preeclampsia, 608
Preganglionic neuron, 272, 273*f*, 275*f*
Pregnancy, 586, 600–601
  early, tests for, 600
  ectopic, 589
  exercise and, 601
  maternal changes in, 600–601
  pelvic measurements in, 143
Premature newborn, 602
  anemia in, 351
Premenstrual syndrome (PMS), 579
Premolar, 478
Premotor area, 259, 259*f*
Prenatal development, 586
  male, testicular hormones in, 561
Preovulatory phase, 571*f*, 572
Prepuce, 557*f*, 563, 569
Presbyopia, 300
Pressure, 287
Pressure ulcer, 108
Presynaptic neuron, 234, 235*f*
Prevertebral ganglia, 274, 275*f*
Primary auditory area, 258, 259*f*
Primary germ layer, 592
Primary gustatory area, 258, 259*f*, 293
Primary motor area, 258, 259*f*
Primary olfactory area, 258
Primary oocyte, 566, 566*f*
Primary ossification center, 120*f*, 121
Primary response, in immunological memory, 436
Primary somatosensory area, 260*f*, 285
  of cerebral cortex, 258, 259*f*
Primary spermatocyte, 558, 559*f*, 560*f*
Primary visual area, 258, 259*f*
Prime mover, 189
Primitive ectoderm (epiblast), 590, 591*f*
Primitive endoderm (hypoblast), 590, 591*f*
Procaine (Novocaine), 234
Processes, vertebral, 134, 134*f*, 135
Proctology, 472
Products, 37
Progeny, 67
Progeria, 65
Progesterone, 333
  in female reproductive cycle, 571*f*, 572, 573*f*
Prolactin-inhibiting hormone (PIH), 320
Prolactin (PRL), 320, 323T, 602
Prolactin-releasing hormone (PRH), 320
Promotor, 60, 64
Pronation
  of forearm, 163, 163*f*
Pronator teres, 206E, 209*f*
Propagation of nerve impulses, 233–234, 234*f*
Prophase, 62, 64. *See also* Mitosis
Proprioception, 260
Proprioceptive sensation, 289–290
Proprioceptors, 289
  in blood pressure regulation, 392
  in respiration control, 465
Prostaglandin, 335
  in labor and delivery, 601

Prostate, 557*f*, 562–563
cancer of, 579
disorders of, 579
hyperplasia of, benign, 577
Prostatitis, 579
Protease inhibitor, 439–440
Proteases, 37
Proteasomes, 57, 59T
Protein, 35–37
metabolism of, 514, 515T
anabolism in, 514, 515T
catabolism in, 514, 515T
in liver, 486
in small intestine, 489, 489T
as organic compounds, 35–37
plasma membrane of, 46–47
synthesis of, 58, 60–62
estrogens and, 570
Protein buffer system, 550, 553T
Protein concentration of body fluid, 548, 548*f*
Proteomics, 67
Prothrombin, 356, 356*f*
Prothrombinase, 356, 356*f*
Protons, 23, 23T, 24*f*
Proto-oncogene, 66
Protraction, 162, 163*f*
Provitamin, 505
Proximal convoluted tubule, 526, 528*f*, 530*f*
Proximal (directional term), 12E, 13*f*
Prozac, 236
Pruritis, 108
Pseudopod, 51, 52*f*
Pseudostratified columnar epithelium, 75, 78T
Pseudostratified epithelium, 74
Psoas major, 210E, 213E
Psoriasis, 108
Psychoneuroimmunology, 431
Puberty, 577
hair and, 103
Pubic symphysis, 142*f*, 143, 557*f*
Pubis, 142*f*, 144
Pudendal nerve, 244*f*
Pudendum, 569
Puerperal fever, 608
Puerperium, 602
Pulled groin, 212E
Pulmonary circulation, 395*f*, 410
Pulmonary edema, 377, 467
Pulmonary embolism, 357
Pulmonary gas exchange, 458–459, 458*f*
Pulmonary trunk, 367*f*, 368*f*, 369, 395*f*, 410
Pulmonary valve, 368*f*, 369, 370*f*
Pulmonary vein, 367*f*, 368*f*, 369, 395*f*, 410
Pulmonary ventilation, 446, 453–457
pressure changes during, 455–456, 455*f*
Pulmonologist, 446
Pulp, 477*f*, 478
Pulp cavity, 477*f*, 478
Pulse, 394
Pump, 50–51
Punnett square, 604
Pupil, 295, 295*f*, 300*f*
constriction of, 300
Pupillary light reflex, absence of, 247
Purkinje fiber, 372
Pus in inflammation, 427
Putamen, 254*f*
P wave, 374, 374*f*
Pyelogram, intravenous, 539
Pyloric sphincter, 480*f*, 481
Pylorus, 480*f*, 481
Pyridoxine, 509T
Pyrogen, 519

Q

QRS complex wave, 374, 374*f*
Quadratus lumborum, 201*f*, 210E
Quadriceps femoris, 215E
Quadriceps tendon, 214E
Quadriplegia, 265

R

Radial artery, 398E, 399*f*
Radial fossa, 140, 140*f*
Radial nerve, 244*f*
Radial notch, 140, 141*f*
Radial tuberosity, 141, 141*f*
Radial vein, 406E, 407*f*
Radiation, body heat loss by, 516
Radiation therapy, 66
Radical mastectomy, 58
Radio frequency nonsurgical facelift, as anti-aging treatment, 106
Radius, 140–141, 140*f*, 141*f*
head of, dislocation of, 168
muscles of, 206E, 207*f*
Rales, 468
Range of motion (ROM), 190
Raynaud disease, 280
Reabsorption, in capillary function, 389, 389*f*
Receptor, 46
in feedback system, 7, 7*f*
hormone, 317
Recessive allele, 604
Recovery oxygen uptake, 183
Rectal exam, digital, 579
Rectum, 491*f*, 492
Rectus abdominis, 198E, 199*f*, 201*f*, 203*f*, 210E
Rectus femoris, 213*f*, 215E
Red blood cells, 90, 349–351, 355T
formation of, 350*f*, 351
life cycle of, 350–351, 350*f*
structure of, 349–350
in urine, 534T
Red bone marrow, 114
in blood cell production, 348
in long bones, 115*f*
in lymphatic system, 424
T cell and B cell maturation in, 429
Red fiber, 184
Redness in inflammation, 427
Red nucleus, 253, 253*f*
Red pulp, 425
Referred pain, 288, 288*f*
Reflex, 247
Reflex arc, 247
Refraction, of light, 297, 299, 299*f*
Refractory period, 233
Regeneration, 230
Regeneration tube, 230
Regular connective tissue, dense, 86–87, 86T
Regulation
of blood flow, 391–394
of respiration, 461–465
of water and solute loss, 545–547, 546*f*
Rehabilitation, cardiac, 381
Relaxation
of bladder, 537
of skeletal muscle, 180, 181*f*
Relaxation period
in cardiac cycle, 374, 375*f*
in twitch contraction, 184
Relaxin, 333
in female reproductive cycle, 572
in labor and delivery, 601
in pregnancy, 600
Releasing hormone, 320
Remodeling of bone, 121–122
Renal arteries, 396E, 397*f*, 525*f*, 526, 527*f*
Renal calculi, 539
Renal capsule, 525*f*, 526, 527*f*, 528*f*
Renal column, 525*f*, 526
Renal compensation, 552
Renal corpuscle, 528*f*, 5226
Renal cortex, 525*f*, 526, 527*f*, 528*f*
Renal failure, 539
Renal hilum, 525–526, 525*f*
Renal medulla, 525*f*, 526, 527*f*, 528*f*
Renal papilla, 528*f*
Renal pelvis, 525*f*, 526
Renal pyramids, 525*f*, 526, 527*f*
Renal tubule, 526, 528*f*
Renal vein, 525*f*, 526, 527*f*
Renin, 331
Renin-angiotensin-aldosterone pathway, 331, 331*f*
in blood pressure/blood flow regulation, 393
Reperfusion, 372

Repetitive motion injuries, 167
Repolarizing phase, of action potential, 232
Reproduction, 6
Reproductive cell division, 62. *See* Meiosis
Reproductive system, 5T, 556–581
  abortion, 577
  aging and, 577
  birth control methods and, 574–576
  contribution of, to homeostasis, 578
  disorders of, 579–581, 607
  female, 564–574 (*See* Female reproductive system)
  male, 557–563 (*See* Male reproductive system)
Residual body, 51
Residual volume, 457
Resistance, 420
Resistance reaction, 335
Resolution, 297
Resorption, bone, 116
Respiration
  cellular, 510–511, 510*f*
  muscles of, 200E, 201*f*
Respiratory bronchiole, 451
Respiratory center, 462–465
Respiratory compensation, 552
Respiratory distress syndrome (RDS), 468
Respiratory failure, 468
Respiratory membrane, 452, 453*f*
Respiratory pump, 390, 421
Respiratory system, 5T, 445–468
  aging and, 465
  breathing patterns in, 457
  contribution of, to homeostasis, 466
  disorders of, 467
  exercise and, 465
  functions of, 446*f*
  inhalation and exhalation in, 453–455
  lung volumes and capacities in, 456–457, 456*f*
  muscles of, 200E, 201*f*, 453–455, 454*f*
  organs of, 446–452, 446*f*
    bronchi as, 450*f*, 451
    larynx as, 447*f*, 448, 449*f*
    lungs as, 446*f*, 450*f*, 451–452, 452*f*
    nose as, 446*f*, 447–448, 447*f*
    pharynx as, 448
    trachea as, 446*f*, 447*f*, 450–451, 450*f*
  oxygen and carbon dioxide exchange in, 457–459, 458*f*
  oxygen and carbon dioxide transport in, 459–461, 460*f*
  pressure changes, 455–456, 455*f*
  respiration control in, 461–465
  respiratory movements in, modified, 457T
  ventilation in, 446, 453–457
Respiratory zone (respiratory system), 447
Response in feedback system, 7*f*, 8
Responsiveness, 6
Resting membrane potential, 232, 232*f*
Reticular activating system (RAS), 254
Reticular connective tissue, 86, 86T
Reticular fiber, 84, 425
Reticular formation, 253–254, 253*f*
Reticulocyte, 351
Reticulocyte count, 352T
Retina, 295*f*, 296, 296*f*, 299T
  detached, 310
Retinacula, 208E
Retinoblastoma, 310
Retinoic acids in breast milk, 603
Retraction, 162, 163*f*
Reverse transcriptase inhibitors, 439
Reversible reaction, 29
Reye's syndrome, 267
Rh blood group, 358
Rheumatic fever, 381
Rheumatism, 168
Rheumatoid arthritis (RA), 168
Rhinitis, 468
Rhinoplasty, 447
Rhinovirus, 467
Rhodopsin, 300
Rhomboid major, 202E, 203*f*
Rhythm method of birth control, 575
Rib(s), 138
  fractures of, 138
Riboflavin, 509T
Ribonuclease, 484
  in small intestine, 488, 489T
Ribonucleic acid (RNA), 38
  in protein synthesis, 58, 60–62
Ribosomal RNA, 60
Ribosome, 54–55, 55*f*, 57*f*, 59T
Rickets, 122, 150
Right bundle branch, 372
Right common carotid artery, 396E, 397*f*, 398E, 399*f*
Right coronary artery, 367*f*, 368*f*, 372, 396E, 397*f*
Right hepatic duct, 483*f*, 484
Right lymphatic duct, 421, 422*f*
Right primary bronchus, 450*f*, 451
Right pulmonary artery, 367*f*, 368*f*, 369, 410
Right subclavian artery, 396E, 397*f*, 398E, 399*f*
Rigor mortis, 180
RNA polymerase, 60
Rod, 296, 296*f*
  stimulation of, 300
Rogaine (Minoxidil), 103
Root
  axon, 246
  of hair, 102, 102*f*
  of penis, 563
  of tooth, 477, 477*f*
Root canal, 477*f*, 478
Root canal therapy, 478
Rotation, 161–162, 163*f*
Rotator cuff, 204E
  injury to, 168
Rough ER, 58*f*, 55, 55*f*
Round ligament, 412
Round window, 303*f*, 304, 304*f*
Ruffini corpuscle, 287
Rugae, of gastric mucosa, 480*f*, 481
RU 486 (mifepristone), 577
"Runner's knee," 144
Running injuries, 219–220

S

Saccule, 302, 303*f*, 306*f*
Sacral canal, 136
Sacral curve, 133*f*, 134
Sacral foramina, 136, 137*f*
Sacral hiatus, 136, 137*f*
Sacral plexus, 244*f*, 246
Sacral promontory, 136, 137*f*
Sacrum, 133, 136, 137*f*
Saddle joint, 164*f*, 165
Safe sex, 556
Sagittal plane, 14
Sagittal suture, 126*f*
Saliva, 428T, 477
  protective functions of, 426
Salivary amylase, 477, 478, 489T
Salivary glands, 473*f*, 477
Salivation, 477
Salpingectomy, 581
Salt, 30, 30*f*
Saltatory conduction, 233, 234*f*
Sarcolemma, 175, 176*f*, 178*f*
Sarcoma, osteogenic, 151
Sarcomere, 175, 176*f*, 177*f*
  muscle contraction, 179*f*, 180
Sarcoplasm, 175, 176*f*
Sarcoplasmic reticulum, 175, 176*f*
Sartorius, 213*f*, 215E
Satellite cell, 231T
Saturated fat, 33, 33*f*
Scala tympani, 303*f*, 304, 340*f*
Scala vestibuli, 303*f*, 304, 304*f*
Scaphoid, 141*f*, 142
Scapula, 139, 139*f*
Scapular muscles, 204E, 205*f*
Scar tissue, 92
S cell, in small intestine, 487
Schwann cells, 229*f*, 231T
Sciatica, 267
Sciatic nerve, 244*f*
Sclera, 294, 295*f*, 299T
Scleral venous sinus, 297

Scleroses, 237
Scoliosis, 151
Scotoma, 310
Scrotum, 557–558, 557*f*
Seasonal affective disorder (SAD), 333
Sebaceous gland, 102*f*, 103
Sebum, 103, 428T
protective functions of, 426
Secondary bronchi, 450*f*, 451
Secondary oocyte, 566, 566*f*
Secondary ossification center, 120*f*, 121
Secondary response, in immunological memory, 437
Secondary spermatocyte, 559*f*, 560, 560*f*
Second messenger, 318
Second polar body, 566, 566*f*
Secretin
in digestion, 495, 495T
intestinal, 487
pancreas, 334T
Secretion, 74, 75
by GI tract, 473
Secretory immunoglobulin A (IgA) in breast milk, 603
Secretory vesicles, exocytosis and, 52
Sections, 14–15, 15*f*
Segmental artery, 527*f*
Segmentations, 488
Selective permeability, 46
Selective serotonin reuptake inhibitor (SSRI), 236
Selenium, 506T
Self-tolerance, 428
Semen, 563
Semicircular canal, 302, 302*f*, 303*f*
Semicircular duct, 302, 303*f*
in equilibrium, 305–306, 307*f*
Semilunar valve, 369
Semimembranosus, 213*f*, 215E
Seminal vesicles, 557*f*, 562
Seminiferous tubule, 558, 559*f*
Semitendinosus, 213*f*, 215E
Sensation(s), 285
characteristics of, 285
gustation as, 291–293
hearing as (*See also* Ear; Equilibrium; Hearing)
olfaction as, 290–291
pain, 287–288
proprioceptive, 289–290
sensory receptor classification and, 285T
somatic (*See* Somatic senses)
tactile, 287
thermal, 287
thirst as, 255, 545, 546*f*
vision as, 293–300 (*See also* Eye; Vision)
Sensorineural deafness, 310
Sensory area, of cerebral cortex, 258
Sensory (ascending) tract, 245
Sensory function of nervous system, 227
Sensory neuron, 227, 247, 248*f*
Sensory receptor, 226, 226*f*, 247, 248*f*
adaptation in, 285
classification of, 285T
sensation and, 285
types of, 280–281
Separated shoulder, 168
Septicemia, 361
Serosa, of GI tract, 474*f*, 475
Serotonin, as neurotransmitter, 236, 237
Serous fluid, 91
Serous membranes, 91
Serous pericardium, 365, 366*f*
Serratus anterior, 202E, 203*f*, 205*f*
Sertoli cell, 558, 559*f*, 561, 561*f*
Serum, 354
Sex chromosome, 605
Sex-determining region of Y chromosomes (SRY), 605
Sex differences
in osteoporosis, 150
in skeletal system, 146, 147T
Sex-linked inheritance, 606
Sexual characteristics, male, testicular hormones in, 561–562
Sexual function, male, development of, testicular hormones in, 562
Sexually transmitted disease, 580–581
Shaft, of hair, 102, 102*f*
Shingles, 265
Shin splints, 145, 216E
Shivering, 517
Shock, 415
anaphylactic, 440
Short bone, 114
Shoulder, separated, 168
Shoulder blade, 139
Shoulder girdle, 139, 139*f*
muscles of, 202E, 203*f*
Sickle cell disease (SCD), 36, 360, 605
Side chain, 35, 35*f*
Sighing, 457T
Sigmoid colon, 491*f*, 492
Signs, 9
Sildenafil (Viagra), 563
Silent myocardial ischemia, 380
Simple diffusion, 48, 48*f*, 53T
Simple epithelium, 74, 76–78T
Simple sugar, 31–32
Single-unit muscle tissue, 187
Sinoatrial (SA) node, 372
Sinusitis, 131
Sinusoid, of liver, 484, 485*f*
Sjögren's syndrome, 93
Skeletal muscle. *See also specific muscle*
of GI tract, 475
group action of, 189
naming of, 191T
principal, 189–217
superficial, 192–193*f*
Skeletal muscle pump, 390. 390*f*, 421
Skeletal muscle tissue, 90, 173–183
connective tissue components of, 173–175, 174*f*
contraction and relaxation of, 177–180
energy for, 180, 182–183
muscle tone and, 180
neuromuscular junction in, 177–179, 178*f*
physiology of, 179–180
sliding-filament mechanism in, 177, 177*f*
fiber type, 184–185
histology of, 175, 176*f*
metabolism, 180, 182–183
nerve and blood supply of, 174*f*, 175, 176*f*
stretching of, 190
tension of, control of, 183–185
Skeletal system, 4T, 113–151. *See also* Bone
aging and, 149
contribution of, to homeostasis, 149
disorders of, 150–151
divisions of, 124–125, 125*f*
functions of, 114, 115*f*
hyoid bone in, 126*f*, 127*f*
lower limbs in, 144–146
pectoral girdle in, 139, 139*f*
pelvic girdle in, 142*f*, 143–144
sex differences in, 146, 147T
skull in, 125, 126*f*, 127 (*See also* Skull)
thorax in, 137–138, 137*f*, 138*f*
upper limbs in, 140–143
vertebral column in, 133–137 (*See also* Vertebral column)
Skin, 97–101
accessory structures of, 101–104
glands as, 103–104
hair as, 101–103
aging of, 106
anti-aging treatments for, 106
in body temperature regulation, 104
cancer of, 107
care of, 105
color of, 101
contribution of, to homeostasis, 109
cutaneous sensations in, 105
dermis of, 100–101
disorders of, 107–108 (*See also* Psoriasis)
epidermis of, 99–100, 99*f*
functions of, 104–106
as physical barrier in immunity, 425–426
in pregnancy, 600

Skin (*Continued*)
structure of, 98–99, 98*f*
sun damage to, 107
transplantation of, 100
Skin graft, 100
Skin patch, birth control, 575
Skull, 9
cranial bones, 125, 126*f*, 127–129, 127*f*, 128*f*, 129*f*, 130*f*
facial bones, 126*f*, 127*f*, 129–131, 132*f*
fontanels, 132, 132T
hyoid bone and, 126*f*, 127*f*
sutures of, 126*f*, 127*f*, 128*f*, 129*f*, 131
Sleep, 254
Sliding-filament mechanism, 177, 177*f*
Slipped disc, 150
Slow oxidative (SO) fiber, 184
Slow pain, 288
"SLUDD" responses, 279
Small intestine, 486–491, 486*f*, 487*f*
absorption in, 488–491, 490*f*
digestion in, 488
enzymes in, 489T
structure of, 486–488, 486*f*, 487*f*
Small saphenous vein, 408E, 409*f*
Smegma, 581
Smell, sense of, 290–291
Smoking
blood and, 359
cilia and, 448
health and, 445, 463
Smooth ER, 56, 56*f*
Smooth muscle, of GI tract, 475
Smooth muscle tissue, 90, 173, 186–187
autonomic nervous system, 279T
Smooth muscle tone, 187
Sneezing, 457T
Sobbing, 457T
Sodium, 506T
as electrolyte, 548, 548*f*, 549T
urinary salt and, 545–546
Sodium-potassium pump, 50–51, 51*f*
Soft palate, 476, 476*f*
Solar keratosis, 108
Soleus, 216*f*, 217E, 217*f*
Soluble fiber, 497
Solute, 29, 47
Solution, 29
acidic, 30
basic (alkaline), 30
hypertonic, 50, 50*f*
hypotonic, 50, 50*f*
isotonic, 50, 50*f*
normal saline, 50
Solvent, 29, 47
Somatic cell division, 62–65, 63*f*
Somatic motor neurons, 246
Somatic motor pathways, 260–261, 261*f*
Somatic nervous system (SNS), 227
autonomic nervous system compared with, 272–273, 272T, 273*f*
Somatic reflex, 247
Somatic sense, 284–290
disorders of, 310
pain sensation as, 287, 288
proprioceptive sensations as, 289–290
sensation and, 285
tactile sensations as, 287
itch and tickle as, 287
pressure and vibration as, 287
touch as, 287
thermal sensation as, 287
Somatic sensory pathway, 259, 260*f*
Somatomedin, 320
Somatosensory association area, 258, 259*f*
Spasm, 219
Spastic paralysis, 261
Special movements of joints, 162–163, 163*f*
Special senses, 285, 290–308
disorders of, 310
gustation as, 291–293
hearing as, 301–305 (*See also* Ear; Equilibrium; Hearing)
olfaction as, 290–291
vision as, 293–300 (*See also* Eye; Vision)
Specific immunity, 420, 428–437. *See also* Adaptive immunity
Specific resistance. *See* Immunity
Sperm, 559*f*, 560–561, 560*f*
Spermatic cord, 562
Spermatids, 559*f*, 560
Spermatogenesis, 558, 560
Spermatogenic cell, 558, 561*f*
Spermatogonia, 558, 559*f*
Spermatozoon, 558, 559*f*
Sperm cell, 558, 559*f*
Spermicide, 575
Spermiogenesis, 560
Sphenoidal sinus, 128, 130*f*, 131, 132*f*
Sphenoid bone, 126*f*, 127*f*, 128, 128*f*, 129*f*
Sphincter, 173
Sphygmomanometer, 394
Spina bifida, 150–151
Spinal cavity, 15
Spinal column. *See* Vertebral column
Spinal cord, 243–247, 244*f*
anatomy of, gross, 243, 244*f*
central nervous system, 226, 226*f*
functions of, 247
injury to, 265
structure of, 243–245, 244*f*
internal, 245, 245*f*
Spinalis group, 210E, 211*f*
Spinal meninges, 243, 243*f*
Spinal nerve, 226, 226*f*, 244*f*, 245*f*, 246–247, 246*f*
coverings of, 246, 246*f*
distribution of, 246–247
intercostal, 247
plexuses of, 246–247
Spinal reflex, 247
Spinal segments, of spinal cord, 243
Spinal tap, 243
Spine. *See* Vertebral column
scapula, 139, 139*f*
Spinothalamic pathway, 260
Spinous process (spine), 134, 134*f*, 135, 135*f*
Spiral organ, 303*f*, 304, 304*f*
Spirogram, 456, 456*f*
Spirometer, 456
Spleen, 422*f*, 424
Splenectomy, 425
Splenic artery, 396E, 397*f*
Splenic tissue, 425
Splenomegaly, 441
Spongy bone tissue, 117*f*, 118
Sports drinks, 551
Sprain, 84, 168
Spurs, bone, 122
Squamous cell carcinomas, 107
Squamous cell, 74
Squamous epithelium
simple, 75, 76T
stratified, 75, 79T
Squamous suture, 126*f*, 127*f*, 129*f*
Stapes, 302, 302*f*
Starch, 32
Static equilibrium, 305
Stearic acid, 33, 33*f*
Stem cell
in blood cell formation, 348, 348*f*
in epidermis, 100
for research, 588, 590*f*
in tissue repair, 92
Stenosis, 370, 380
Stercobilin, 351
Sterility, 607
Sterilization, 574
Sternocleidomastoid, 195*f*, 210E
Sternum, 137, 138*f*
Steroid hormones, 317, 329
Steroid, 35, 35*f*
anabolic, 220
Stimulation frequency, muscle, 183, 184, 185*f*
Stimulus, 7*f*, 232, 285
in feedback system, 7
Stirrup (stapes), 302, 302*f*
Stomach, 480–483, 480*f*, 481*f*, 482*f*
absorption in, 482–483
cells of, 481–482, 482*f*
digestion in, 482–483
structure of, 480*f*, 481–482

Strabismus, 196E, 310
Strain, 168
  hamstring, 214E
  muscle, 220
Stratified epithelium, 74, 79–80T, 82
Stratum basale, 99*f*, 100
Stratum corneum, 99*f*, 100
Stratum lucidum, 99*f*, 100
Stratum spinosum, 99*f*, 100
Streptokinase, 357
Stress
  blood circulation and, 359
  immune function and, 431
Stress incontinence, 537
Stressor, 335
Stress response, 335–336
Stretching exercises, 190
Stretch receptors, in respiration control, 465
Striae, 101
Striation
  in cardiac muscle tissue, 173, 186
  in skeletal muscle tissue, 173, 174*f*
Stroke, 266
Stroke volume, 375–377
Stroma, 84
  in tissue repair, 92
Structural organization, levels of, 3*f*
  cellular, 2, 3*f*
  chemical, 2, 3*f*
  organ, 2, 3*f*
  organismal, 3, 3*f*
  system, 2–3, 3*f*
  tissue, 2, 3*f*
Styloid process, 126*f*, 127, 127*f*
  of radius, 141, 141*f*
  of ulna, 140, 141*f*
Subarachnoid space, 243, 243*f*, 251*f*
Subclavian vein, 406E, 407*f*
Subcutaneous layer, 99
  of connective tissue, 84
Subcutaneous sensation, 105
Sublingual glands, 473*f*, 477
Subluxation, 169
Submandibular glands, 473*f*, 476*f*, 477
Submucosa, of GI tract, 474*f*, 475
Subscapularis, 204E, 205*f*
Substantia nigra, 253, 253*f*
Substrate, 37
Sucrase, 488, 489T
Sucrose, 32*f*
Suction lipectomy, 85
Sudden cardiac death, 381
  exercise and, 376
Sudoriferous glands, 103–104
Sulci, 255, 256*f*
Sulfur, 506T
Sunburn, 107
Sunscreen, 105
Sunstroke, 519
Superficial (directional term), 12E
Superficial palmar arch, 398E, 399*f*
Superficial vein
  of lower limbs, 408E, 409*f*
  of upper limbs, 406E, 407*f*
Superior articular facet, 136, 137*f*
Superior articular process, 134*f*, 135, 137*f*
Superior colliculi, 253, 253*f*
Superior (directional term), 12E, 13*f*
Superior extensor retinaculum, 216E, 217*f*
Superior gluteal nerve, 244*f*
Superior mesenteric artery, 396E, 397*f*
Superior mesenteric ganglion, 274, 275*f*
Superior nasal concha, 129, 130*f*
Superior oblique, 196E, 197*f*
Superior phrenic artery, 396E, 397*f*
Superior rectus, 196E, 197*f*
Superior sagittal sinus, 250, 251*f*
Superior vena cava, 367*f*, 368*f*, 369, 394, 402E, 403*f*
Superoxide, 25
Supination of forearm, 163, 163*f*
Supinator, 206E
Supporting cells
  of crista, 306, 307*f*
  gustatory, 292*f*, 293
  of maculae, 305, 306*f*
  of nose, 290, 291*f*
  of spiral organ, 303*f*, 304
Supporting structures of reproductive system, 556
Suprarenal artery, 396E, 397*f*
Supraspinatus, 204E, 205*f*
Surface mucous cell, in gastric mucosa, 481, 482*f*
Surfactant, 452
Surfactant-secreting cell, 452, 453*f*
Suspensory ligaments of breast, 569*f*, 570
Sutures of skull, 126*f*, 127*f*, 128*f*, 129*f*, 131, 157, 158*f*
Swallowing, 478–480, 479*f*
Sweat, protective functions of, 426
Sweat gland, 102*f*, 103–104
Swelling in inflammation, 427
Swollen knee, 168
Sympathetic division of autonomic nervous system, 228, 272
  activities of, 277–278
  in body temperature regulation, 516–517
  nephron function, 531
  organization of, 274, 275*f*
Sympathetic trunk ganglia, 274, 275*f*
Symphysis, 158, 159*f*
Symptoms, 9
Sympto-thermal method of birth control, 575
Synapse, 228
Synaptic cleft, 178, 178*f*, 235, 235*f*
Synaptic end bulb, 178, 178*f*, 228, 229*f*, 235*f*
Synaptic transmission, 234–237
  neurotransmitters in, 237
Synaptic vesicle, 178, 178*f*, 228
Synarthrosis, 157
Synchondrosis, 158, 159*f*
Syncope, 416
Syncytiotrophoblast, 590, 591*f*
Syndesmosis, 157, 158*f*
Synergist muscle, 189
Synovial cavity, 158, 159*f*
Synovial fluid, 91–92, 159, 159*f*
Synovial joints, 157, 158–165
  structure of, 158–160, 159*f*
  types of, 160–163, 164–165, 164*f*
Synovial membrane, 91–92, 159, 159*f*
Synovitis, 169
Synthesis reactions, 28
  sucrose, 32*f*
Synthetic oxytocin, 322
Syphilis, 580
Systemic circulation, 394, 395*f*
Systemic gas exchange, 458*f*, 459
Systemic lupus erythematosus (SLE), 93, 441
System level of organization, 2–3, 4–5T. *See also individual systems*
  cardiovascular, 4T
  digestive system, 5T
  endocrine system, 4T
  integumentary system, 4T
  lymphatic system, 5T
  muscular system, 4T
  nervous system, 4T
  reproductive systems, 5T
  respiratory system, 5T
  skeletal system, 4T
  structural, 2, 3*f*
  urinary system, 5T
Systole, 374
Systolic blood pressure, 394

T

Tachycardia, 394
Tachypnea, 468
Tactile disc, 100
Tactile sensation, 105, 287
Tail, 594
  of sperm, 560*f*, 561
Talus, 145, 145*f*
Target cell, 317
Tarsals, 145, 145*f*
Tarsus, 145

Tastants, 293
Taste, sense of, 291–293
Taste aversion, 293
Taste bud, 292–293, 292*f*
Taste pore, 292*f*, 293
Tay-Sachs disease, 56
T cells, 353, 355T
  in cell-mediated immunity, 432–434, 433*f*, 434*f*
  maturation of, 429
Tears, 294
Tectorial membrane, 303*f*, 304, 304*f*
Teeth, 477–478, 477*f*
Telomere, 65
Telophase, 63*f*, 64
Temperature. *See* Body temperature
Temporal bones, 126*f*, 127–128
Temporalis, 195E, 195*f*
Temporal lobe, 255, 256*f*, 259*f*
Temporomandibular joint (TMJ), 127, 131
Temporomandibular joint (TMJ) syndrome, 131
Tendinitis, 188–189
Tendon, 173, 174*f*, 175
Tennis elbow, 168, 188
Tenosynovitis, 188–189
Tensile strength of bone, 116
Tensor fasciae latae, 212E, 213*f*
Teratogen, 608
Teres major, 204E, 205*f*, 207*f*
Teres minor, 204E
Terminal bronchiole, 450*f*, 451
Terminal ganglia, 276*f*, 277
Terminator, in transcription of DNA, 60, 60*f*
Tertiary bronchus, 450*f*, 451
Testes, 557*f*, 558–562
  anatomy of, 557*f*, 558, 559*f*
  cancer of, 579
  hormonal control of, 561–562, 561*f*
  hormones of, 333
Testicular artery, 396E, 397*f*
Testosterone, 333, 558, 561, 561*f*
  functions of, 35
Tetany, 338
Tetralogy of Fallot, 380
Thalamus, 249*f*, 254, 254*f*, 260*f*, 262T
Thalassemia, 360
Therapeutic cloning, 588
Thermal sensations, 287
Thermoreceptor, 286
Thiamine, 508T
Thick filaments, 175, 176*f*, 177*f*
Thick skin, 100
Thigh bone, muscles of, 212E, 213*f*, 214–215E
Thin filament, 175, 176*f*, 177*f*
Thin skin, 100
Third ventricle, 250, 251*f*, 254*f*
Thirst center, 255, 545, 546*f*, 547T
Thoracic aorta, 396E, 397*f*
Thoracic cage, 137
Thoracic cavity, 15, 15*f*, 16*f*, 17
Thoracic curve, 133*f*, 134
Thoracic duct, 421, 422*f*
Thoracic muscles, 202E, 203*f*
Thoracic nerve, 244*f*
Thoracic vertebrae, 133, 136
Thoracolumbar division of autonomic nervous system, 274, 275*f*. *See also* Sympathetic division
Thorax, 137–138, 137*f*, 138*f*
Threshold, 232
Thrombin, 356, 356*f*
Thrombocytes. *See* Platelets
Thrombocytopenia, 361
Thrombolytic agent, 357
Thrombophlebitis, 416
Thrombosis, 355, 357
Thrombus, 357
Thymosin, 334T
Thymus, 422*f*, 424
Thyroid cartilage, 447*f*, 448, 449*f*
Thyroid crisis (storm), 340
Thyroid gland, 323–325, 324*f*
  in body temperature regulation, 516–517
  disorders of, 337–338
  follicles in, 323, 324*f*
  hormones of, 317, 323–325
    calcitonin as, 325
    metabolic rate and, 516
Thyroid-stimulating hormone (TSH), 320, 323T, 325
  in body temperature regulation, 516–517
Thyrotropin-releasing hormone (TRH), 320, 325
  in body temperature regulation, 516–517, 517*f*
  in resistance reaction, 335
Thyroxine (T_4), 323
Tibia, 145, 145*f*
  muscles of, 214–215E
Tibial collateral ligament, 165, 166*f*
  rupture of, 168
Tibialis anterior, 217E, 217*f*
Tibialis posterior, 216*f*, 217E
Tibial tuberosity, 145, 145*f*
Tic, 219
Tidal volume, 456, 456*f*
Tinnitus, 310
Tissue(s), 2, 72–93. *See also* Muscular system; Skeletal muscle tissue
  aging, 92
  bone, 89–90
  cardiac muscle, 186 (*See also* Cardiac muscle tissue)
  connective, 73 (*See also* Connective tissue(s))
    repair of, 92
  disorders of, 93
  epithelial, 73–82 (*See also* Epithelial tissue (epithelium))
  membranes, 90–92
  muscular, 73, 90, 173, 188T
    repair of, 92
  nervous, 73, 90 (*See also* Nervous tissue)
    repair of, 92
  osseous, 89–90
  repair of, 92
  tissue engineering, 90
  transplantation of, 93
Tissue engineering, 90
Tissue factor (TF), 356, 356*f*
Tissue level of structural organization, 2, 3*f*
Tissue plasminogen activator (tPA), 357
Tissue regeneration, 92
Tissue rejection, 93
TMJ syndrome, 131
Tocopherols, 508T
Tongue, 476*f*, 477
Tonsillectomy, 441
Tonsils, 425
Topical, 108
Total hip arthroplasty, 151
Total lung capacity, 456*f*, 457
Totipotent stem cell, 588
Touch, 287
Trabeculae, 117*f*, 118, 119*f*
  formation of, in intramembranous ossification, 119, 119*f*
Trace elements, 23
Trachea, 446*f*, 447*f*, 450–451, 450*f*
Tracheostomy, 451
Tracheotomy, 451
Trachoma, 310
Tract, 245
  in CNS white matter, 230
Transcription, protein synthesis, 60*f*
Transcription in protein synthesis, 58, 60
Transdermal drug administration, 105
Trans fatty acids, 33
Transferrin, protective functions of, 426, 428T
Transfer RNA, 60
Transfer vesicle, 56*f*
Transfusion, blood, 358–359
Transient ischemic attack (TIA), 266
Transitional cells, 74
Transitional epithelium, 80T, 82
  in ureter and urinary bladder, 536
Translation in protein synthesis, 60–62, 61*f*
Transplantation
  bone marrow, 354
  skin, autologous, 100

Transporters, 46
Transverse abdomonis, 198E, 199*f*
Transverse arch, 146, 146*f*
Transverse colon, 491*f*, 492
Transverse fissure, 256*f*
Transverse plane, 14, 14*f*
Transverse process, 134*f*, 135, 135*f*
Transverse tubule, 175, 176*f*
Trapezium, 141*f*, 142
Trapezius, 202E, 203*f*
Trapezoid, 142
Traveler's diarrhea, 498
Tremor, 219, 266
Triceps brachii, 206E, 207*f*, 209*f*
Tricuspid valve, 368*f*, 369, 370*f*
Trigeminal nerve, 252*f*, 264, 264T
Trigger finger, 188
Triglyceride, 32–33, 33*f*
  metabolism of, 512–513
  storage of, in yellow bone marrow, 114
Triiodothyronine (T_3), 323
Tripeptide, 36
Triquetrum, 141*f*, 142
Trochlea, 140, 140*f*, 141*f*
Trochlear nerve, 252*f*, 264T
Trochlear notch, 140, 141*f*
Trophoblast, 588
Tropic hormone, 320
Tropins, 320
Tropomyosin, 175, 177*f*
  in muscle relaxation, 180, 181*f*
Troponin, 175, 177*f*
  in muscle relaxation, 180, 181*f*
True labor, 601–602
True pelvis, 142*f*, 143
True rib, 138
True vocal cord, 447*f*, 448
Trunk (body region), 9, 11*f*
Trypsin, 484, 489T
Tryptophan, as neurotransmitter, 236
T tubule, 175, 176*f*
Tubal ligation, 574
Tuberculosis, 467
Tubular fluid, 531
Tubular reabsorption, as nephron function, 528–529, 529T, 531–534, 532*f*, 533*f*
Tubular secretion, as nephron function, 529, 529T, 531–534, 532*f*
Tubulin, 54
Tumor, 66
Tumor marker, 67
T wave, 374, 374*f*
Twins, 587
Twitch contraction, 184
Tympanic membrane (eardrum), 301
Type I and II cutaneous mechanoreceptors, 287

U

Ulcer
  peptic, 497
  pressure, 108
Ulna, 140–141, 140*f*, 141*f*
  muscles of, 206E, 207f
Ulnar artery, 398E, 399*f*
Ulnar nerve, 244*f*
Ulnar veins, 406E, 407*f*
Ultrasonography, 579
Ultraviolet (UVA) rays, 107
Umbilical artery, 412, 413*f*
Umbilical cord, 412, 413*f*, 594, 595*f*
Umbilical vein, 412, 413*f*
Umbilicus, 594
Unfused tetanus, 184, 185*f*
Universal donor, 359
Universal recipient, 359
Unmyelinated axon, 228
Upper esophageal sphincter (UES), 478
Upper limb bud, 594
Upper limbs, 9–10, 11*f*, 140–143
  muscles of, 204E, 205*f*
  veins of, 406E, 407*f*
Upper motor neuron, 261, 261*f*
Upper respiratory system, 446
Ureteral opening, 536, 536*f*
Ureters, 535–536, 536*f*
Urethra, 536–537, 536*f*
  male, 557*f*, 562
Urinalysis, 533
Urinary bladder, 536, 536*f*
Urinary retention, 539
Urinary system, 5T, 523–539. *See also* Kidneys
  aging and, 537
  components of, 524*f*
  contribution of, to homeostasis, 538
  disorders of, 539
  kidneys in, 524–534 (*See also* Kidneys)
  nephrons of, 526 (*See also* Nephrons)
  organs of, 524*f*
  urine and, 524 (*See also* Urine)
Urinary tract infection (UTI), 535
Urination, 537
Urine, 525
  abnormal constituents of, 534T
  components of, 533–534
  physical characteristics or, 534T
  protective functions of, 426, 428T
  transportation, storage, and elimination of, 535–537
Urobilin, 351
Urobilinogen, 351
  in urine, 534T
Urologist, 524
Urology, 524, 556
Uterine cavity, 567, 567*f*
Uterine cycle, 570, 571*f*
Uterine tube, 566, 567*f*
Uterus, 566–567, 567*f*
  menstruation and, 572
  in postovulatory phase, 573
  in preovulatory phase, 572
Utricle, 302, 303*f*, 306*f*
Uvula, 476, 476*f*

V

Vaccination, immunological memory and, 437
Vagina, 564*f*, 567*f*, 568
Vaginal orifice, 568, 568*f*, 569
Vaginal pouch, 575
Vaginal ring, 575
Vaginal secretions, 428T
  protective functions of, 426
Vaginitis, 580
Vagus nerve, 252*f*, 264T, 377, 377*f*
Valence shell, 25
Vallate papillae, 292, 292*f*
Valves
  of heart, 368*f*, 369–370, 370*f*
  of veins, 389
Valvular stenosis, 380
Variable regions, of antibody, 429–430, 429*f*
Varicose veins, 389
Vascular resistance, 391
Vascular spasm, in hemostasis, 354
Vascular tunic, 299
Vas deferens, 557*f*, 559*f*, 562
Vasectomy, 574
Vasoconstriction, 386
Vasodilation, 386
  in inflammation, 427
Vasomotor tone, 393
Vasopressin, 322, 323T. *See* Antidiuretic hormone (ADH)
  in blood pressure/blood flow regulation, 393
Vastus intermedius, 213*f*, 215E
Vastus lateralis, 213*f*, 215E
Vastus medialis, 213*f*, 215E
Veins, 386, 389–390
  of head and neck, 404E, 405*f*
  of lower limbs, 408E, 409*f*
  of systemic circulation, 402E, 403*f*
  of upper limbs, 406E, 407*f*
  varicose, 389
Venipuncture, 352T
Venous return, 390
Venous sinus, 425
Ventilation, 446
  mechanical, 468
  muscles of, 200E, 201*f*
  pulmonary, 446, 453–457 (*See also* Pulmonary ventilation)

Ventral gray horn, 245
Ventral root, 245*f*, 246
Ventricle
 of brain, 250
 of heart, 367, 367*f*, 368*f*, 369
Ventricular fibrillation, 380
Ventricular septal defect, 380
Ventricular systole, 374–375, 375*f*
Venules, 386, 388*f*, 389–390
Vertebrae
 cervical, 133, 135–136, 135*f*
 coccygeal, 133
 lumbar, 133, 136, 136*f*
 sacral, 133, 136
 thoracic, 133, 136
Vertebral arch, 134, 134*f*
Vertebral artery, 398E, 399*f*
Vertebral cavity, 15, 15*f*, 134
Vertebral column, 133–137
 curves of, 133*f*, 134
 muscles of, 210E, 211*f*
 regions of, 133–134, 133*f*
 vertebrae of, 133–134 (*See also* Vertebrae)
Vertebral curves, normal, 133*f*, 134
Vertebral foramen, 134, 134*f*, 135*f*
Vertebral veins, 402E, 403*f*
Vertebra prominens, 136
Vertigo, 310
Very-low-density lipoprotein (VLDL)
 diabetes and, 332
 in lipid metabolism and, 513
Vesicle, 47
 of Golgi complex, 56, 56*f*
 lysosomes as, 56
 secretory, exocytosis and, 52
Vesicle transport, 51–52, 53T
Vessels
 blood (*See* Blood vessels and circulation; Heart)
 lymphatic, 421, 422*f*, 423*f*
Vestibular apparatus, 305
Vestibular branch, 305
Vestibular membrane, 303*f*, 304, 304*f*
Vestibule
 of ear, 302
 vulvar, 569
Vestibulocochlear nerve, 252*f*, 264T
Viagra, 563
Vibration, 287
Vibratory sensation, 260
Villi of small intestine, 487, 487*f*
Virilism, 333
Virilizing adenoma, 340
Viscera, 17
Visceral layer, 365, 366*f*
 serous membranes, 91
Visceral muscle tissue, 187
Visceral peritoneum, of GI tract, 475
Visceral pleura, 450*f*, 451, 452*f*
Visceral reflex, 247
Visceral senses, 285
Viscosupplementation, 168
Vision, 293–300. *See also* Eye
 binocular, 300
 disorder of, 310
 image formation and, 297, 299–300
Visual acuity, 297
Visual association area, 258, 259*f*
Visual pathway, 300–301
Vital capacity, 456*f*, 457
Vitamin(s), 505, 507, 508–509T
 absorption of, in small intestine, 491
 bone metabolism and, 124T
 storage of, liver in, 486
 supplemental, 507
Vitamin A, 508T
Vitamin B_1, 508T
Vitamin B_2, 509T
Vitamin B_6, 509T
Vitamin B_{12}, 509T
Vitamin C, 509T
Vitamin D, 508T
 activation of, liver in, 486
 functions of, 35
 skin and, 106
Vitamin E, 508T
Vitamin K, 508T
Vitreous body, 295*f*, 297
Vitreous chamber, 295*f*, 297, 299T
Vocal cords, 447*f*, 448
Voice production, 448–449
Volkmann's canal, 118
Voltage-gated channel, 232, 235, 235*f*
Voluntary stage of swallowing, 478, 479*f*
Vomer, 126*f*, 127*f*, 128*f*, 131
Vomiting, 482
 protective function of, 426, 428T
Vulvovaginal candidiasis, 580

W

Wander macrophage
 protective functions of, 426
 in white blood cells, 353
Warfarin (Coumadin), 357
Warm receptor, 287
Wart, 108
Waste, excretion of, kidneys in, 525
Water
 absorption of, 489
 balance of, 545–547, 547T
 functions of, 29–30
 gain and loss of, 545–547, 547T
 movement of, between fluid compartments, 547
Water intoxication, 547
Water-soluble hormones, action of, 318, 318*f*
Water-soluble vitamins, 505, 508–509T
Wave summation, 184, 185*f*
Weight loss, 503
Werner syndrome, 65
Wernicke's area, 258, 259*f*
Wheeze, 468
Whiplash injury, 151
White blood cells, 90, 353, 355T
 in urine, 534T
White coat hypertension, 416
White columns, 245, 245*f*
White fiber, 184–185
White fibrous capsule, 558, 559*f*
White matter, 228, 230, 255
White pulp, 425
Withdrawal reflex, 247
Womb, 567*f*
Wrist, 141–143, 141*f*
 muscles of, 208E, 209*f*

X

Xenograft, 441
Xenotransplantation, 93
Xiphoid process, 137, 138*f*

Y

Yawning, 457T
Yellow bone marrow, 114
Yolk sac, 591, 591*f*, 593*f*, 595*f*

Z

Z disc, 175, 176*f*, 177*f*
Zinc, 506T
Zona pellucida, 587, 587*f*
Zonular fiber, 295, 295*f*
Zygomatic arch, 126*f*, 127, 128*f*
Zygomatic bones, 126*f*, 128*f*, 131
Zygomaticus major, 194E, 195*f*
Zygote, 566, 566*f*, 587, 588, 589*f*
Zyprexa, 237

PRONUNCIATION KEY

The most strongly accented syllable appears in capital letters, for example, bilateral (bī-LAT-er-al) and diagnosis (dī′-ag-NŌ-sis).

If there is a secondary accent, it is noted by a prime (′), for example, constitution (kon′-sti-TOO-shun) and physiology (fiz′-ē-OL-ō-jē). Any additional secondary accents are also noted by a prime, for example, decarboxylation (dē′-kar-bok′-si-LĀ-shun).

Vowels marked with a line above the letter are pronounced with the long sound, as in the following common words:

ā as in *māke*　　*ē* as in *bē*
ī as in *īvy*　　*ō* as in *pōle*

Vowels not so marked are pronounced with the short sound, as in the following words:

a as in *above*　　*e* as in *bet*
i as in *sip*　　*o* as in *not*
u as in *bud*

Other phonetic symbols are used to indicate the following sounds:

oo as in *sue*
yoo as in *cute*
oy as in *oil*

Many medical terms are "compound" words; that is, they are made up of one or more word roots or combining forms of word roots with prefixes or suffixes. For example, *leukocyte* (white blood cell) is a combination of *leuko*, the combining form for the word root meaning "white," and *cyt*, the word root meaning "cell." Learning the medical meanings of the fundamental word roots will enable you to analyze many long, complicated terms.

The following list includes the most commonly used combining forms, word roots, prefixes, and suffixes used in medical terms and an example for each.

COMBINING FORMS, WORD ROOTS, PREFIXES, AND SUFFIXES

Many terms used on anatomy and physiology are compound words; that is, they are made up of word roots and one or more prefixes or suffixes. For example, *leukocyte* is formed from the word roots *leuko-*, meaning "white," and *-cyte*, meaning "cell." Thus, a leukocyte is a white blood cell. The following list includes some of the most commonly used combining forms, word roots, prefixes, and suffixes used in the study of anatomy and physiology. Each entry includes a usage example. Learning the meanings of these fundamental word parts will help you to remember the terms that, at first glance, may seem long or complicated.

Combining Forms and Word Roots

Acr-, Acro- extremity; Acromegaly
Aden- gland; Adenoma
Angi- vessel; Angiocardiography
Arthr-, Arthro- joint; Arthritis, arthroscopy
Aut-, Auto- self; Autolysis
Audit- hearing; Auditory canal
Bio- life, living; Biopsy
Blast- germ, bud; Blastomere
Brachi- arm; Brachial plexus
Bucc- cheek; Buccal
Capit- head; Decapitate
Carcin- cancer; Carcinogen
Cardi-, Cardia-, Cardio- heart; Cardiogram
Cephal- head; Hydrocephalus
Cerebro- brain; Cerebrospinal fluid
Chondr-, Chond- cartilage; Chondrocyte
Cost- rib; Intercostal
Crani- skull; Cranial cavity
Derma-, Dermat- skin; Dermatology
Dur- hard; Dura mater
Enter, Entero- intestine; Gastroenterology
Erythr-, Erythro- red; Erythrocyte
Gastr-, Gastro- stomach; Gastrointestinal tract
Gloss-, Glosso- tongue; Hypoglossal
Glyco- sugar; Glycogen
Gon- seed; Gonad
Gyn- female, woman; Gynecology
Hem-, Hemato- blood; Hematoma
Hepar-, Hepat- liver; Hepatic duct
Hist-, Histo- tissue; Histology